z	.00	.01	.02	.03	.04	.05	.06	.07	.08	.09
0.0	.5000	.5040	.5080	.5120	.5160	.5199	.5239	.5279	.5319	.5359
0.1	.5398	.5438	.5478	.5517	.5557	.5596	.5636	.5675	.5714	.5753
0.2	.5793	.5832	.5871	.5910	.5948	.5987	.6026	.6064	.6103	.6141
0.3	.6179	.6217	.6255	.6293	.6331	.6368	.6406	.6443	.6480	.6517
0.4	.6554	.6591	.6628	.6664	.6700	.6736	.6772	.6808	.6844	.6879
0.5	.6915	.6950	.6985	.7019	.7054	.7088	.7123	.7157	.7190	.7224
0.6	.7257	.7291	.7324	.7357	.7389	.7422	.7454	.7486	.7517	.7549
0.7	.7580	.7611	.7642	.7673	.7704	.7734	.7764	.7794	.7823	.7852
0.8	.7881	.7910	.7939	.7967	.7995	.8023	.8051	.8078	.8106	.8133
0.9	.8159	.8186	.8212	.8238	.8264	.8289	.8315	.8340	.8365	.8389
1.0	.8413	.8438	.8461	.8485	.8508	.8531	.8554	.8577	.8599	.8621
1.1	.8643	.8665	.8686	.8708	.8729	.8749	.8770	.8790	.8810	.8830
1.2	.8849	.8869	.8888	.8907	.8925	.8944	.8962	.8980	.8997	.9015
1.3	.9032	.9049	.9066	.9082	.9099	.9115	.9131	.9147	.9162	.9177
1.4	.9192	.9207	.9222	.9236	.9251	.9265	.9278	.9292	.9306	.9319
1.5	.9332	.9345	.9357	.9370	.9382	.9394	.9406	.9418	.9429	.9441
1.6	.9452	.9463	.9474	.9484	.9495	.9505	.9515	.9525	.9535	.9545
1.7	.9554	.9564	.9573	.9582	.9591	.9599	.9608	.9616	.9625	.9633
1.8	.9641	.9649	.9656	.9664	.9671	.9678	.9686	.9693	.9699	.9706
1.9	.9713	.9719	.9726	.9732	.9738	.9744	.9750	.9756	.9761	.9767
2.0	.9772	.9778	.9783	.9788	.9793	.9798	.9803	.9808	.9812	.9817
2.1	.9821	.9826	.9830	.9834	.9838	.9842	.9846	.9850	.9854	.9857
2.2	.9861	.9864	.9868	.9871	.9875	.9878	.9881	.9884	.9887	.9890
2.3	.9893	.9896	.9898	.9901	.9904	.9906	.9909	.9911	.9913	.9916
2.4	.9918	.9920	.9922	.9925	.9927	.9929	.9931	.9932	.9934	.9936
2.5	.9938	.9940	.9941	.9943	.9945	.9946	.9948	.9949	.9951	.9952
2.6	.9953	.9955	.9956	.9957	.9959	.9960	.9961	.9962	.9963	.9964
2.7	.9965	.9966	.9967	.9968	.9969	.9970	.9971	.9972	.9973	.9974
2.8	.9974	.9975	.9976	.9977	.9977	.9978	.9979	.9979	.9980	.9981
2.9	.9981	.9982	.9982	.9983	.9984	.9984	.9985	.9985	.9986	.9986
3.0	.9987	.9987	.9987	.9988	.9988	.9989	.9989	.9989	.9990	.9990
3.1	.9990	.9991	.9991	.9991	.9992	.9992	.9992	.9992	.9993	.9993
3.2	.9993	.9993	.9994	.9994	.9994	.9994	.9994	.9995	.9995	.9995
3.3	.9995	.9995	.9995	.9996	.9996	.9996	.9996	.9996	.9996	.9997
3.4	.9997	.9997	.9997	.9997	.9997	.9997	.9997	.9997	.9997	.9998

APPLIED MATHEMATICS
for the Management, Life, and Social Sciences

LAWRENCE E. SPENCE
Illinois State University

CHARLES VANDEN EYNDEN
Illinois State University

DANIEL GALLIN

SCOTT, FORESMAN/LITTLE, BROWN HIGHER EDUCATION
A Division of Scott, Foresman and Company
Glenview, Illinois London, England

TO THE STUDENT

If you want further help with this course, you may want to obtain a copy of the *Student's Solutions Manual* that accompanies this textbook. This manual provides detailed step-by-step solutions to the odd-numbered exercises in the textbook and can help you study and understand the course material. Your college bookstore either has this manual or can order it for you.

Library of Congress Cataloging-in-Publication Data

Spence, Lawrence E.
Applied mathematics for the management, life, and social sciences / Lawrence E. Spence, Charles Vanden Eynden, Daniel Gallin. p. cm.
ISBN 0-673-18837-X
1. Mathematics. I. Vanden Eynden. Charles, II. Gallin, Daniel. III. Title.
QA39.2.S6818 1990 89-29500
510—dc20 CIP

Portions of this book have been adapted from *Finite Mathematics* by Daniel Gallin.

ISBN 0-673-18837-X

1 2 3 4 5 6 - RRW - 95 94 93 92 91 90

PREFACE

Applied Mathematics for the Management, Life, and Social Sciences is intended for use in a one-year sequence of courses that emphasizes mathematical applications and models in the management, life, and social sciences. Such courses have become commonplace at most American colleges and universities during the past fifteen years, a development that reflects the increased use of quantitative concepts and techniques in these disciplines.

The mathematical topics presented here are those that have proven to have the broadest usage in the disciplines for which this book is intended. These come primarily from two branches of mathematics, finite mathematics and calculus. The book is organized into two parts along these lines.

Flexibility

Applied Mathematics contains ample material for two four-semester-hour courses. Consequently, instructors will have considerable flexibility in designing a course to meet the needs of their students.

Prerequisites

Although it is assumed that the reader has completed the equivalent of one and a half years of high school algebra, essential algebraic concepts are reviewed when necessary throughout the text. Additional topics from algebra are contained in an optional appendix that can be used by instructors who desire more complete coverage. The need for review, however, has been balanced against the risk of boredom. We prefer to avoid the discussion of algebra for its own sake in favor of treating mathematics that is more readily applicable to the disciplines for which the book is written.

APPROACH

The mathematics presented in this book is developed intuitively, yet without loss of mathematical precision. Above all we have tried to provide students with conceptual understanding rather than a collection of mechanical procedures. In our treatment of calculus, for example, we use graphs of functions extensively to reinforce the analytic concepts under discussion. In addition, we include a wide variety of applications which will motivate students to learn mathematics as well as to see its power and utility in other disciplines.

Algebra Review

As mentioned above, the review of algebra is dispersed throughout the text as needed, rather than concentrated at the beginning. We have chosen to disperse the material for two reasons. First, a book that begins with a long review of algebra is not exciting to students; we prefer to introduce useful and interesting mathematics as soon as possible. Second, a review that occurs far in advance of the need for a topic will probably be wasted, in that some students will have forgotten it in the interim. Since most of the need for algebra occurs in connection with calculus rather than finite mathematics, we have chosen to defer many of the algebraic concepts until Chapter 8. This chapter presents the essential ideas about functions that are required for the remainder of the book. For students with strong algebra backgrounds, Chapter 8 can be covered quickly or even omitted.

Exponential and Logarithmic Functions

We have also chosen to disperse the presentation of exponential and logarithmic functions throughout Chapters 8–15, rather than to confine it to a single chapter. In our experience, students who are exposed to these functions throughout the calculus develop a much deeper understanding of them than when the functions are discussed only in one brief chapter. Moreover, the early introduction of the exponential and logarithmic functions allows for more varied examples and applications than could be given using only rational functions. Our desire to introduce exponential and logarithmic functions as soon as possible has led us to present the chain rule prior to the rules for differentiating products and quotients. This enables us to use the exponential and logarithmic functions as examples when discussing the product and quotient rules.

Calculators

We assume that students will have access to a calculator when working exercises, especially those involving natural logarithms and values of the exponential function e^x. Nevertheless, most exercises are designed so that the necessary computations can be performed by hand, and tables giving the values of $\ln x$ and e^x are included. Only in Chapter 4 is a calculator essential; the financial calculations there require a calculator capable of evaluating powers of a number.

KEY FEATURES

Applications

Special emphasis has been placed on applications to business, economics, and the life and social sciences in special sections called ''Mathematics in Action.'' These sections present realistic uses of mathematics rather than contrived or artificial examples. At the discretion of the instructor, they may be omitted, however, as no other parts of the text depend on them. Other sections, marked "optional" in the table of contents, contain mathematics that can be skipped without loss of continuity. The extent to which the application sections and the optional sections are covered will depend on the length of the course and the nature of the students being taught.

Examples and Exercises

Applied Mathematics contains over 400 numbered examples that are carefully chosen to demonstrate both the mathematical techniques under consideration and their usefulness in solving real-world problems. A large number of exercises are included in the book. The exercises within each section range from easy to difficult, and many contain realistic applications of the mathematical ideas presented.

Practice Problems

Nearly 300 practice problems occur throughout the book. These are designed to test a student's understanding of the mathematical content being developed by providing an immediate opportunity to work a problem involving the concepts and techniques being discussed. As a further help to the student, both answers (at the end of each section) and *complete solutions* (at the back of the book) to practice problems are provided.

Chapter Reviews

Each chapter concludes with a comprehensive chapter review that lists the new terms and important formulas introduced in that chapter. These reviews also contain exercises that are intended as a practice chapter test for students. Although these exercise sets may sometimes require more than an hour to complete, they cover all the principal ideas in the chapter and should enable students to test their understanding prior to an examination.

Format

The format of this book is intended to help students recognize important results and techniques. Terminology being defined appears in boldface. Boxes are used to highlight theorems and procedures for emphasis and ease of reference. A second color is also used functionally in the artwork and in the text (for example, to distinguish the graph of a function from a tangent line or to annotate calculations).

COURSE ORGANIZATION

Applied Mathematics can be subdivided into two major parts: Chapters 1–7 present topics from finite mathematics, and Chapters 8–15 contain material from calculus. Each part contains enough material for use in a four-credit-hour semester course. The algebra appendix can be used with either part or omitted altogether if students have an adequate background in algebra.

The content from finite mathematics can be subdivided into four areas: linear equations and matrices (Chapters 1–2); linear programming (Chapter 3); the mathematics of finance (Chapter 4); and probability and statistics (Chapters 5–7). The calculus chapters can also be subdivided into four major subjects: functions of one variable (Chapter 8); differential calculus (Chapters 9–11); integral calculus (Chapters 12–13); and multivariable calculus (Chapter 15). In Chapter 14, concepts from integral calculus are applied to probability. The diagram below shows the logical dependence of the chapters.

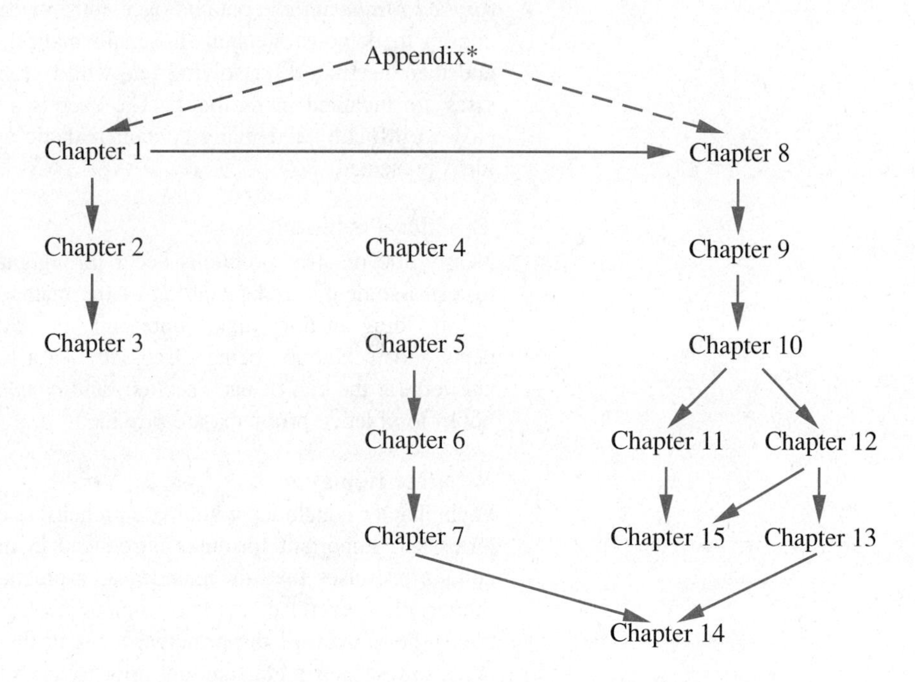

*The appendix is optional

SUPPLEMENTS

Manuals

The following resources were prepared by John I. Hill, Vanessa F. Miller, Linda Ann Spence, and Joan Vanden Eynden of Illinois State University.

The **Instructor's Guide and Solutions Manual** gives a lengthy set of test questions for each chapter, organized by section, plus answers to all the questions. It also provides complete solutions to all of the even-numbered text exercises.

The **Instructor's Answer Manual** gives the answers to every text exercise, collected in one convenient location.

The **Student's Solutions Manual,** available for purchase by students, provides detailed, worked-out solutions to all of the odd-numbered text exercises, plus a chapter test for each chapter. Answers to the chapter test questions are given at the back of the book.

For the calculus portions of the book, some students may find the **Short Calculus Workbook,** prepared by Walter Turner, of Western Michigan University, helpful. Worked-out examples are presented so that students may follow the solutions step-by-step. These examples are followed by problems progressing in level of difficulty; only part of the solution is presented, and the student is asked to complete the rest. The *Workbook* supplements the problems presented in the original textbook and provides many additional worked examples.

A brief supplement on *Logic* is also available to adopters by writing to the publisher.

Visuals

A set of **overhead transparencies** showing charts, figures, and portions of examples is available and can be used to accompany lectures. The transparencies are especially useful in large classroom situations.

Computerized Testing

The **Scott, Foresman/Little, Brown Test Generator for Mathematics** enables instructors to select questions by section or chapter or to use a ready-made test for each chapter. Instructors may generate tests in multiple-choice (IBM^R/ Macintosh^R versions) or open-response formats, scramble the order of questions, and produce multiple versions of each test (up to 9 with Apple II^R and up to 25 with IBM^R and Macintosh^R). The system features a preview option that allows instructors to view questions before printing, to regenerate variables, and to replace or skip questions.

Software

Computer Applications for Finite Mathematics and Calculus by Donald R. Coscia is a softbound textbook packaged with two diskettes (in Apple II and IBM-PC versions) with programs and exercises keyed to the text. The programs allow students to solve meaningful problems without the difficulties of extensive computation. This book bridges the gap between the text and the computer by providing additional explanations and exercises for solution using a microcomputer.

Spreadsheet workbooks will be available for the finite mathematics and calculus portions of this textbook. **Matrices, Statistics, and the Mathematical Functions of Finance: An Introduction to the Electronic Spreadsheet** by Samuel W. Spero is a workbook for students which comes packaged with the LOTUS-compatible electronic spreadsheet, VP-Planner Plus for the IBM-PC and compatibles.

The many applications of Finite Mathematics to business and industry are bound up with computing. To learn matrices, statistics, and the mathematical functions of finance effectively students should have available to them a computational tool. This workbook not only provides such a computational tool, but it also introduces the student to the electronic spreadsheet which is the tool most often used in business and industry for these applications.

The workbook is closely linked to the material in the text by Spence and Vanden Eynden and can be profitably used from Chapter 1 through Chapter 8.

Graphing Functions: An Introduction to the Electronic Spreadsheet by Samuel W. Spero is a workbook for students that will be accompanied by the LOTUS-compatible electronic spreadsheet, VP-Planner Plus for the IBM-PC and compatibles.

Because one cannot learn calculus without graphing functions, it is important that a graphing tool be made available to students. The *Graphing Functions* workbook not only instructs the student in the use of a very powerful, computer-based graphing tool, it also introduces him or her to the electronic spreadsheet. The electronic spreadsheet is the preferred computational tool in business and industry.

The workbook is closely linked to the material in the text and can be profitably used from Chapter 8 on.

RELATED BOOKS IN THE SERIES

This book is the parent text including finite mathematics and elementary calculus. The other books in the Spence/Vanden Eynden series are:

- *Finite Mathematics*
- *Calculus with Applications to the Management, Life, and Social Sciences*

ACKNOWLEDGMENTS

This book is based in part on *Finite Mathematics* by Daniel Gallin and *Finite Mathematics and Calculus* by Lawrence E. Spence.

We are grateful for the enthusiasm and advice of Pam Carlson, Leslie Borns, George Duda, and Barbara Schneider of Scott, Foresman/Little, Brown Higher Education Division as this project was developed. We also appreciate the fine work of Carol Leon during the production process.

While preparing this text, we have benefited greatly from the advice of many reviewers. We are especially indebted to those who provided thoughtful comments about the early drafts:

James Angelos *Central Michigan University*
Paul Britt *Louisiana State University*
James Daly *University of Colorado at Colorado Springs*
Frank C. Denney *Chabot College*
Alan Gorfin *Western New England College*

Joseph Harkin *SUNY-Brockport*
Richard W. Marshall *Eastern Michigan University*
Caroline Woods *Marquette University*

We would also like to thank the following market reviewers for their insights:

Paul Britt *Louisiana State University*
Charles Clever *South Dakota State University*
John T. Gresser *Bowling Green State University*
John Haverhals *Bradley University*
Charlotte Lewis *University of New Orleans*
Thomas J. Miles *Central Michigan University*
Harald M. Ness *University of Wisconsin, Fond du Lac*

We hope to have created a text that instructors can teach from and students can learn from with both success and enjoyment. We welcome and appreciate your comments.

Lawrence E. Spence
Charles Vanden Eynden

CONTENTS

INDEX OF APPLICATIONS

Management Science and Economics

Life and Physical Science

Behavioral and Social Science

General Interest

1

LINEAR RELATIONSHIPS

The simplest possible relationship between two quantities is the linear one; yet this is often a good approximation to reality. In this chapter we investigate linear relationships and how they arise.

1.1 Graphing in the Euclidean Plane

When Sarah Marchant's husband died early in 1985 she used some of the insurance money to buy IBM stock, hoping that it would increase in value. After several years she began to wonder how her investment was doing. She was able to find most of the reinvestment reports she had received. These listed the selling price of the stock at various times as follows.

Date	*Price*
3–31–87	150 1/8
9–30–85	123 7/8
3–29–85	127
6–28–85	123 3/4
9–30–86	134 1/2
12–31–85	155 1/2
6–30–86	146 1/2

These seemed to her to be a jumble of numbers and dates until she arranged the reports in chronological order, then plotted the stock prices against time as in Figure 1.1. From this picture she could see that the general trend of the price was upward for the past two years, although there was a downturn in most of 1986. Even so, the stock was worth approximately 20 percent more than when she bought it. When she called her broker for advice, she already had an overview of the stock's performance since her purchase.

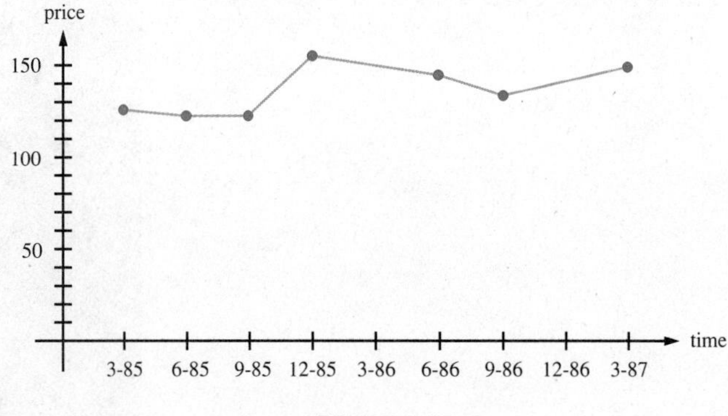

FIGURE 1.1

THE REAL NUMBER LINE

In this example time is measured horizontally and stock price vertically along **real number lines**. To introduce a coordinate system on a given straight line: (i) designate a particular point O on the line as the **origin** of the system; (ii) select one of the two directions on the line to be the **positive** direction; and (iii) choose a **unit length** on the line for measuring distances. With such a coordinate system established, each real number x can be represented as a unique point on the line. To plot the value x, we start at the origin and move x units along the line, taking into account the sign (positive or negative) of x. As an illustration, several values are plotted as points on the number line of Figure 1.2. The positive direction is chosen (by convention) to the right, as indicated by the arrow. In this way every real number corresponds to exactly one point, and conversely every point corresponds to exactly one number, called its **coordinate**.

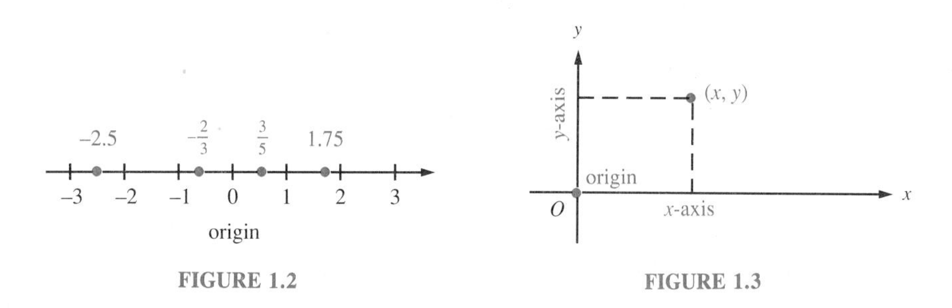

FIGURE 1.2

FIGURE 1.3

RECTANGULAR COORDINATES IN THE PLANE

To establish a coordinate system in the plane we choose two perpendicular lines, or **coordinate axes.** By convention these axes are horizontal and vertical, and are called the ***x*-axis** and ***y*-axis** respectively. We introduce a coordinate system on each axis in such a way that the lines share a common **origin** O at their point of intersection. (See Figure 1.3.) The positive direction is usually taken to the right on the x-axis and upward on the y-axis. The unit length may or may not be the same on the two axes.

The location of a point P is now specified by an **ordered pair** (x, y) of real numbers. To plot the ordered pair (x, y), we start at the origin and move x units horizontally and then y units vertically, taking into account the signs of x and y. Figure 1.4 shows an xy-coordinate system in the plane, with several points plotted for illustration.

For example, the ordered pair $(3, 2)$ corresponds to point A in the figure, obtained by starting at the origin and moving 3 units to the right, then 2 units upward. In the same way, every ordered pair (x, y) determines a unique point

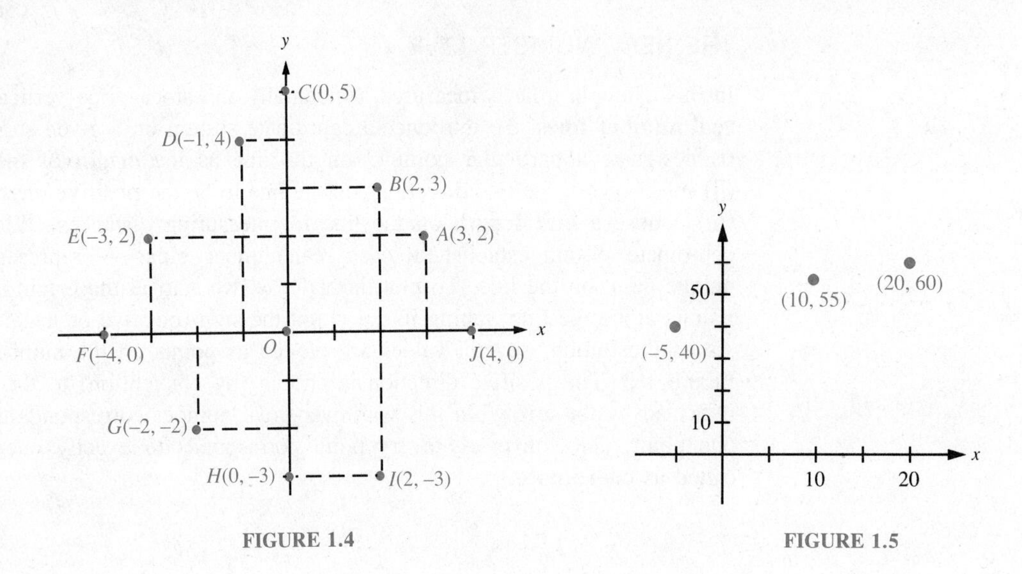

FIGURE 1.4

FIGURE 1.5

P in the plane, and conversely. The numbers x and y are called the **coordinates** of the point. For example, point D in Figure 1.4 has x-coordinate -1 and y-coordinate 4; we would therefore describe D as "the point $(-1, 4)$." Notice that points on the x-axis, such as $F(-4, 0)$ and $J(4, 0)$, have y-coordinate zero, while points on the y-axis, such as $C(0, 5)$ and $H(0, -3)$, have x-coordinate zero. The origin O has coordinates $(0, 0)$.

For example, a scientist interested in how temperature affects the distance a football can be kicked finds that a kicking machine propels the ball 60 feet when the outside temperature is 20° Celsius, 55 feet at 10° C, and 40 feet at $-5°$ C. Letting x be the temperature and y the distance leads to the points $(20, 60)$, $(10, 55)$, and $(-5, 40)$ plotted in Figure 1.5. Notice that different scales are used on the two axes.

GRAPHING MATHEMATICAL STATEMENTS

A mathematical statement involving x and y may be true for certain pairs x, y of numbers and false for others. For example the statement $x + y = 7$ is true for the pair $(x, y) = (3, 4)$, and also for the pair $(-2, 9)$, but false for the pair $(4, 5)$. The **graph** of a statement is the set of all points (x, y) in the plane such that the statement is true for the given x and y. For example, the graph of the statement $x > 0$ and $y > 0$ is the set of all points to the right of the y-axis and above the x-axis, as shown in Figure 1.6(a). Likewise, the graph of the statement $x = 3$ and $1 \le y \le 4$ is the vertical line segment shown in Figure 1.6(b).

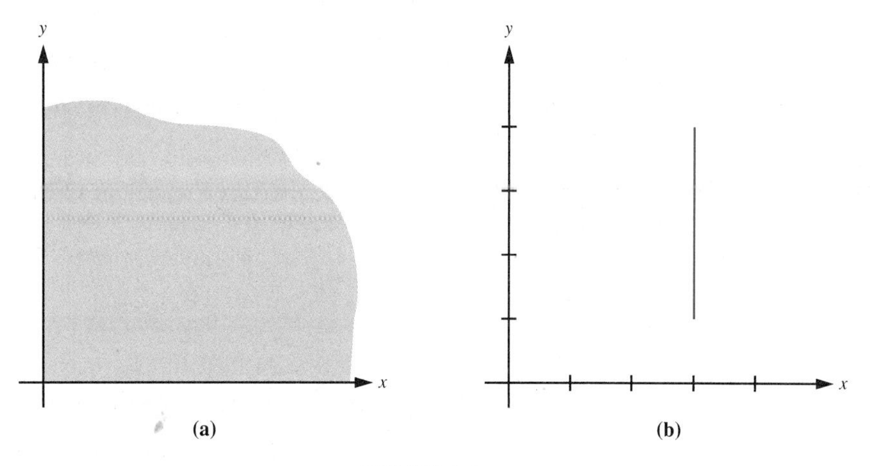

(a) (b)

FIGURE 1.6

To get an idea of the graph of the statement $x + y = 7$ we will make a table of various pairs for which it is true, using the fact that for any value of x, the number y satisfies the equation if and only if $y = 7 - x$.

x	1	2	3	4	5	6	7	8	9	0	−1	−2
y	6	5	4	3	2	1	0	−1	−2	7	8	9

The corresponding points are plotted in Figure 1.7.

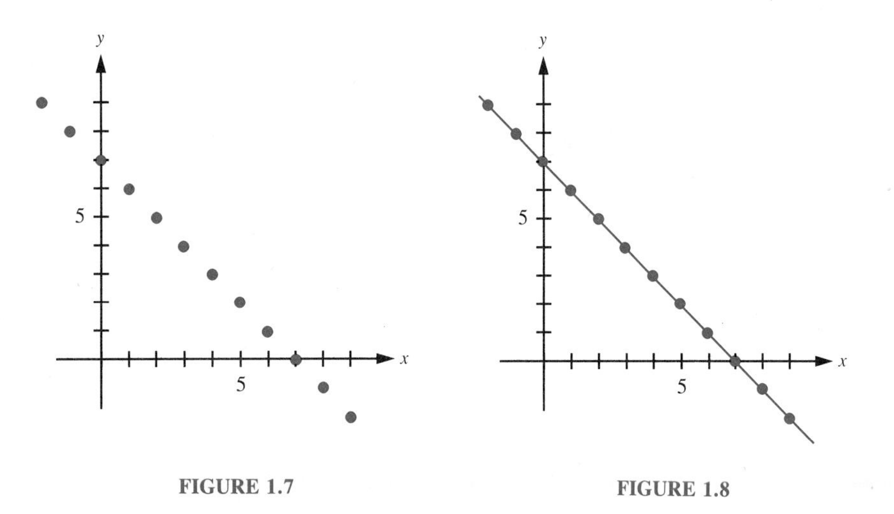

FIGURE 1.7 FIGURE 1.8

Of course there are many other points on the graph of $x + y = 7$, infinitely many, in fact. As one might guess, they are the set of all points on the line shown in Figure 1.8.

It can be shown that if A, B, and C are real numbers, with A and B not both 0, then the graph of the equation

$$Ax + By = C \tag{1.1}$$

is a straight line. Conversely, any straight line is the graph of an equation of this form. For this reason equations of the form (1.1) are called **linear equations**. For example, the equation $x + y = 7$ is a linear equation with $A = 1$, $B = 1$, and $C = 7$, and $3y = -5/6$ is a linear equation with $A = 0$, $B = 3$, and $C = -5/6$.

GRAPHING BY INTERCEPTS

The line determined by a given equation can be graphed very easily by finding its **intercepts,** the points where it crosses the coordinate axes. The procedure is described below.

Graphing by Intercepts

To graph a linear equation $Ax + By = C$:

1. Set $y = 0$ and solve for x. This determines the **x-intercept** of the line.
2. Set $x = 0$ and solve for y. This determines **y-intercept** of the line.
3. The graph is the straight line through these points.

EXAMPLE 1.1 Graph the equation $2x + 3y = 6$.

Solution We set $y = 0$ in the equation and solve for x.

$$2x + 3(0) = 6$$
$$2x = 6$$
$$x = 3$$

The point (3, 0) thus lies on the graph of $2x + 3y = 6$, and it also lies on the x-axis (since its y-coordinate is zero). We have therefore found the x-intercept of the line, the point where it crosses the x-axis. To find the y-intercept of the line, we set $x = 0$ and solve for y.

$$2(0) + 3y = 6$$
$$3y = 6$$
$$y = 2$$

The line therefore crosses the y-axis at (0, 2). Since two points determine a single straight line, we have only to plot the intercepts and draw the line through them to obtain the graph in Figure 1.9. ■

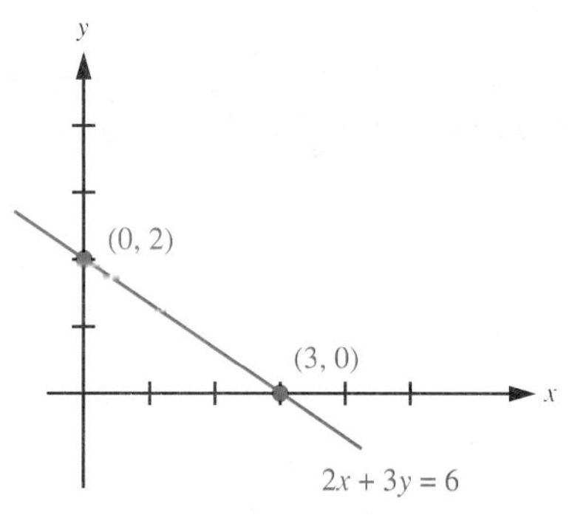

FIGURE 1.9

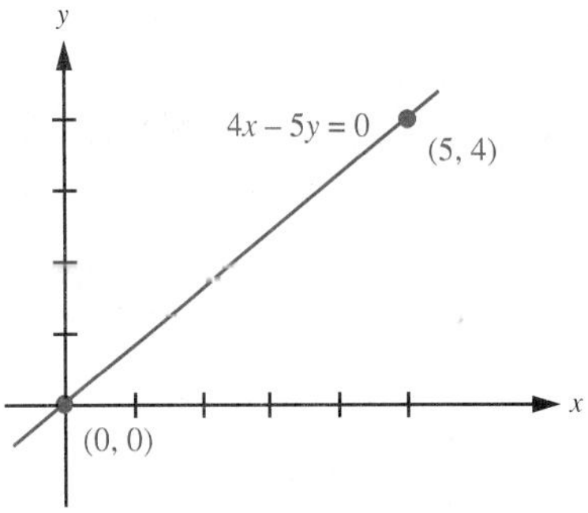

FIGURE 1.10

Practice Problem 1 Find the x-intercept and y-intercept, and graph the equation $3x - 4y = 6$.

If one or two of the constants A, B, C in the linear equation $Ax + By = C$ are zero, then the method of graphing by intercepts must be modified slightly.

EXAMPLE 1.2

Graph the equation $4x - 5y = 0$.

Solution Here we have $Ax + By = C$, where $A = 4$, $B = -5$, and $C = 0$. When we try to find the intercepts we encounter a slight problem: $y = 0$ yields $x = 0$, and vice-versa. This is because *the line goes through the origin* and thus has only one intercept, namely the point (0, 0). To graph the equation we need only find one other point on the line. This can be accomplished by arbitrarily assigning a value to x in the equation, for instance, the value $x = 5$.

$$\begin{aligned} 4(5) - 5y &= 0 \\ 20 - 5y &= 0 \\ 5y &= 20 \\ y &= 4 \end{aligned}$$

We now have a second point (5, 4) on the line. Joining (0, 0) and (5, 4) we obtain the graph in Figure 1.10. ■

EXAMPLE 1.3

Graph the equation $3x = 4$.

Solution If we write this equation in the form $3x + 0y = 4$, we see that it is of the form $Ax + By = C$, with $A = 3$, $B = 0$, and $C = 4$. Therefore, it must have a straight line as its graph. Setting $y = 0$ we find that $x = 4/3$, so the x-intercept of the line is $(4/3, 0)$. But setting $x = 0$ leads to an absurdity,

$0 = 4$. This tells us that the equation $3x = 4$ *cannot be satisfied* by any point $(0, y)$ on the y-axis. In other words, the line has no y-intercept. From this observation we conclude that the line must be parallel to the y-axis, that is, vertical. (See Figure 1.11(a).) ■

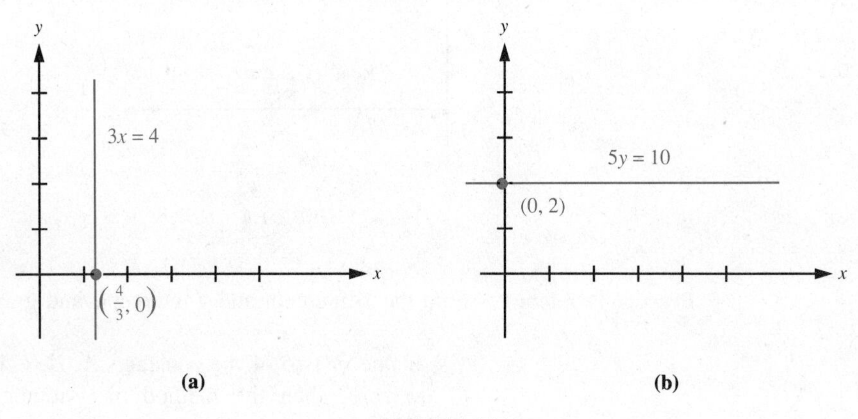

FIGURE 1.11

In a way similar to Example 1.3 we can see that the equation $5y = 10$ describes a line that has y-intercept $(0, 2)$ but no x-intercept. Thus it is parallel to the x-axis, that is, horizontal. (See Figure 1.11(b).)

Notice that the equation $3x = 4$ can be written equivalently as $x = 4/3$, while the equation $5y = 10$ is equivalent to $y = 2$. In general,

> The equation $x = k$ (k constant) represents a *vertical* line that crosses the x-axis at $(k, 0)$.
>
> The equation $y = k$ (k constant) represents a *horizontal* line that crosses the y-axis at $(0, k)$.

EXERCISES 1.1

In Exercises 1–6, plot the listed points in the xy-plane. Use appropriate units on the axes.

1. $(2, 5)$, $(-1, 3)$, and $(0, -4)$

2. $(-3, -2)$, $(3, 1)$, and $(2, 0)$

3. $(25, 30)$, $(-5, -15)$, and $(20, 0)$

4. $(.05, 0.1)$, $(.03, -.08)$, and $(.11, -.02)$

5. $(350, -400)$, $(0, 800)$, and $(100, 0)$

6. $(4000, 1000)$, $(2000, 0)$, and $(0, -1000)$

In Exercises 7–10, give the ordered pair corresponding to the indicated point in the following figures.

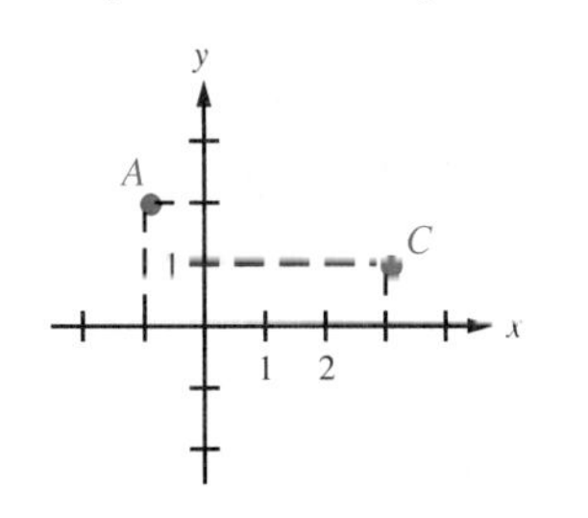

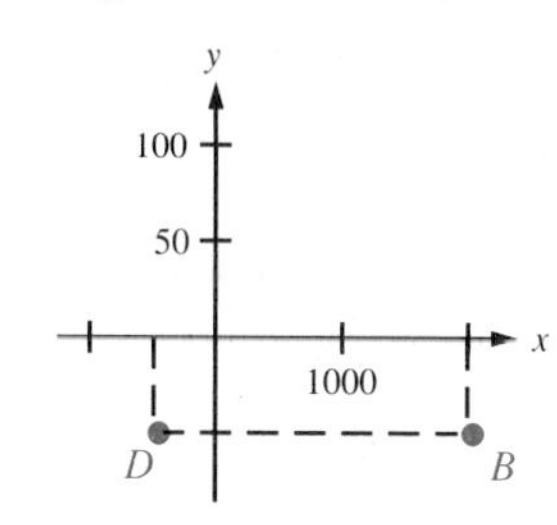

7. Point A in the first graph

8. Point B in the second graph

9. Point D in the second graph

10. Point C in the first graph

11. By going through his old income tax returns, a farmer found he had made \$35,000 in 1984, \$17,000 in 1985, and \$29,000 in 1987, and had lost \$7,000 in 1986. Represent this information graphically.

12. A textbook author finds she received royalty checks for \$350.75 in January 1986, \$202.20 in January 1985, \$157.50 in July 1986, and \$13.25 in January 1987. Represent this information graphically.

In Exercises 13–18, draw the graph of the given statement.

13. $x \geq 0$ and $y \geq 0$

14. $x \geq 3$ and $y = 2$

15. $1 \leq y$ and $y \leq 3$

16. $0 \leq x$, $x \leq 3$, and $y \geq -1$

17. $2 \leq y$, $y \leq 3$, $-1 \leq x$, and $x \leq 2$

18. $xy = 0$

In Exercises 19–22, write the given equation in the form $Ax + By = C$.

19. $y = 5x + 7$

20. $5x = x + y + 7$

21. $2x + 3y + 4 = x - y$

22. $x - 4y = \dfrac{x}{2} + \dfrac{y}{3} + 4$

In Exercises 23–36, find the x-intercept and y-intercept of the given equation. Then graph the equation.

23. $4x + 6y = 24$

24. $6x + 4y = 24$

25. $4x - 6y = 24$

26. $4x + 6y = -24$

27. $3x - 5y = 10$

28. $4y - 3x = 9$

29. $2x + 3y + 12 = 0$

30. $3x = 5y + 30$

31. $x + 4y + 4 = 3x - y + 24$

32. $3y - 2x = 5x - 4y + 14$

33. $5x + 3y = 450$

34. $4x - 7y = 140$

35. $4x + 5y = 1$

36. $20y - 25x = 2$

Answer to Practice Problem **1.** The x-intercept is $(2, 0)$, the y-intercept is $(0, -3/2)$, and the graph is shown below.

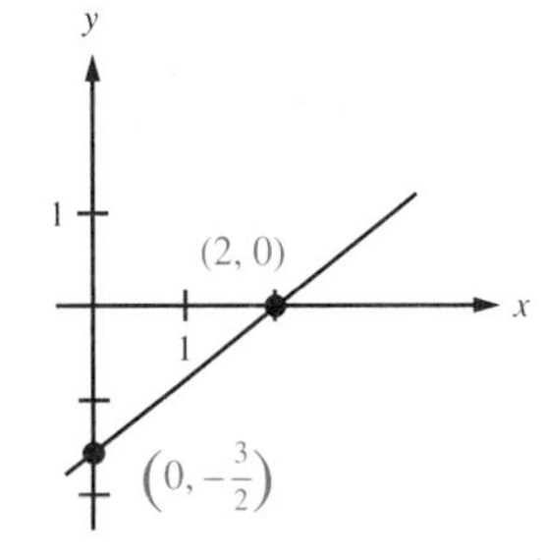

1.2 Linear Functions

THE SLOPE-INTERCEPT FORM

If the quantities x and y satisfy the linear equation $Ax + By = C$ with $B \neq 0$, then we can solve for y, getting an equation of the form

$$y = mx + b,$$

where m and b are constants. In this case we say that y is a **linear function** of x. We usually think of y as depending on x, and call y the **dependent** variable and x the **independent** variable.

EXAMPLE 1.4 Write the equation $2y + 6x = 5$ in the form $y = mx + b$.

Solution We have the equations:

$$\begin{aligned} 2y + 6x &= 5 \\ 2y &= -6x + 5 \\ y &= -3x + \frac{5}{2}. \end{aligned}$$

The last equation is in the form $y = mx + b$ with $m = -3$ and $b = 5/2$. ■

Notice that if we set $x = 0$ in the equation $y = mx + b$, we get $y = b$. Thus the y-intercept of the graph of this equation is the point $(0, b)$. The constant m also has a geometric interpretation. Consider a pair (x_1, y_1) satisfying the equation $y = mx + b$, so that $y_1 = mx_1 + b$. How does y change if we increase x by 1, to $x_1 + 1$? Then

$$y = m(x_1 + 1) + b = mx_1 + m + b = (mx_1 + b) + m = y_1 + m.$$

We see that increasing x by 1 makes y change by the amount m. This will represent an increase or a decrease, depending on whether m is positive or negative. If $m = 0$, then y does not change at all. Figure 1.12 illustrates the three cases when m is positive, negative, or zero.

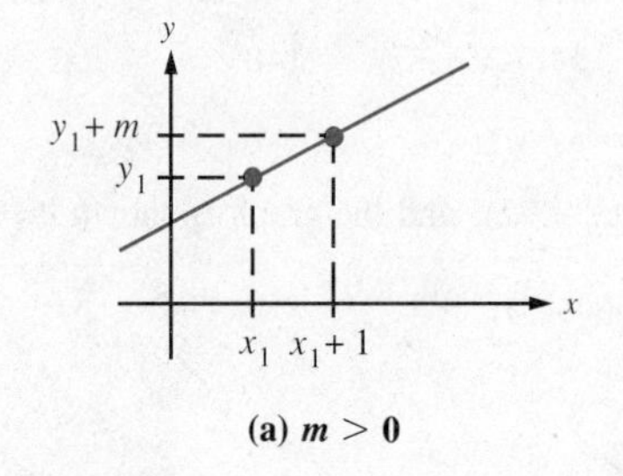

(a) $m > 0$

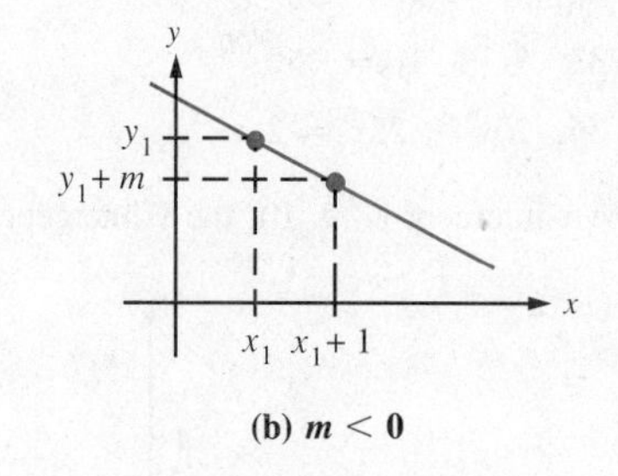

(b) $m < 0$

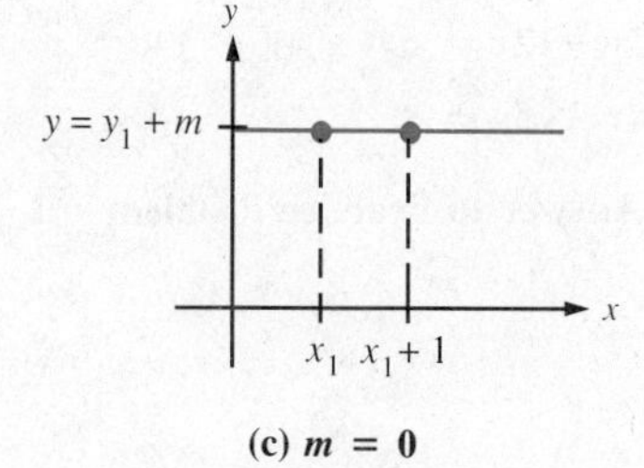

(c) $m = 0$

FIGURE 1.12

The number m is called the **slope** of the line, and the equation

$$y = mx + b$$

is said to be in **slope-intercept form.**

In general, if (x_1, y_1) and (x_2, y_2) are any two points on the line, then

$$y_2 = mx_2 + b \qquad \text{and} \qquad y_1 = mx_1 + b.$$

Subtracting these equations gives

$$y_2 - y_1 = mx_2 + b - (mx_1 + b) = m(x_2 - x_1),$$

or

Slope Formula

$$m = \frac{y_2 - y_1}{x_2 - x_1} \qquad \text{if } x_2 \neq x_1.$$

The slope of a line is a measure of how fast the line rises (or, if $m < 0$, falls) as we move to the right. Some examples are shown in Figure 1.13. The slope of a vertical line is undefined, since the slope formula would entail division by zero.

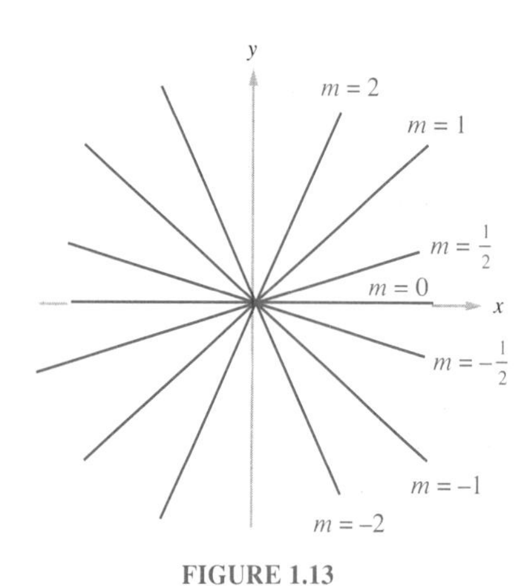

FIGURE 1.13

EXAMPLE 1.5 Find the slope of the line through the points $(2, -5)$ and $(-3, 4)$.

Solution By the slope formula with $(x_1, y_1) = (2, -5)$ and $(x_2, y_2) = (-3, 4)$, we have

$$m = \frac{4 - (-5)}{-3 - 2} = \frac{9}{-5} = -\frac{9}{5}. \quad ■$$

EXAMPLE 1.6 Find the slope and y-intercept of the line with equation $2x + 3y = 6$.

Solution Solving for y gives

$$3y = -2x + 6$$
$$y = -\frac{2}{3}x + 2.$$

Since this equation is in the form $y = mx + b$, the slope is $-2/3$ and the y-intercept is $(0, 2)$. ■

THE POINT-SLOPE FORM

Suppose we know the slope m of a line and a point (x_1, y_1) on the line. Then if (x, y) is any other point on the line, from the slope formula we must have

$$m = \frac{y - y_1}{x - x_1},$$

or

Point-Slope Form

$$y - y_1 = m(x - x_1).$$

Practice Problem 2 What is the slope-intercept form of the equation of the line with slope -2 through the point $(5, 3)$?

EXAMPLE 1.7 A manufacturer finds that the cost y of making x light bulbs is a linear function of x. If making 2000 light bulbs costs $900, and if each additional bulb costs $.20 to make, express y as a linear function of x. How much would it cost to make 4000 light bulbs?

Solution We are given that x and y satisfy a linear equation. Since increasing x by 1 increases y by $.20, we have $m = .20$. Also, if $x = 2000$, then $y = 900$, so the point $(x_1, y_1) = (2000, 900)$ is on the corresponding line. Then the point-slope form of the equation is

$$y - 900 = .20(x - 2000).$$

Thus

$$y - 900 = .20x - 400$$
$$y = .20x + 500,$$

which is the required equation. If $x = 4000$, then

$$y = .20(4000) + 500 = 800 + 500 = 1300,$$

so 4000 lightbulbs would cost $1300. ■

In Example 1.7 the cost of making one more bulb, namely $.20, is called the **marginal cost.** Notice that if the total cost y of making x items is a linear function of x, then the marginal cost is the slope of the corresponding line.

If we are given two points on a line, we can find the equation of the line by first computing its slope, and then using either of the points in the point-slope formula.

EXAMPLE 1.8 Find the equation of the line through the points $(-1, 3)$ and $(4, 2)$. Where does this line cross the coordinate axes?

Solution Let $(x_1, y_1) = (-1, 3)$ and $(x_2, y_2) = (4, 2)$. Then the slope of the line is

$$m = \frac{y_2 - y_1}{x_2 - x_1} = \frac{2 - 3}{4 - (-1)} = -\frac{1}{5}.$$

From the point-slope form its equation is

$$\begin{aligned} y - 3 &= -\frac{1}{5}[x - (-1)] \\ &= -\frac{1}{5}(x + 1) \\ 5(y - 3) &= -(x + 1) \\ 5y - 15 &= -x - 1 \\ 5y + x &= 14. \end{aligned}$$

Setting $y = 0$ gives $x = 14$, and setting $x = 0$ gives $y = 14/5$. Thus the intercepts are $(14, 0)$ and $(0, 14/5)$, as shown in Figure 1.14. ■

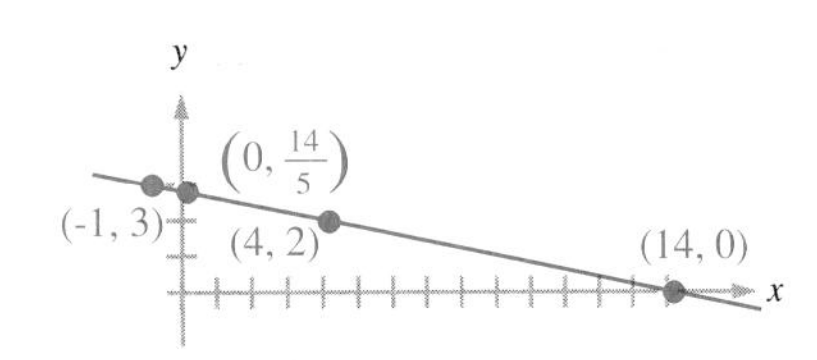

FIGURE 1.14

Practice Problem 3 Find the slope-intercept form of the equation of the line through the points $(1, 3)$ and $(2, -5)$.

EXAMPLE 1.9 It is known that the Fahrenheit temperature y of an object is a linear function of its Celsius temperature x. Because the Celsius scale sets 0 degrees as the freezing point of water and 100 degrees as its boiling point, 0 degrees Celsius corresponds to 32 degrees Fahrenheit and 100 degrees Celsius corresponds to 212 degrees Fahrenheit. What Fahrenheit temperature corresponds to 50 degrees Celsius? What is the Celsius temperature on a hot day when it is 100 degrees Fahrenheit?

Solution The Fahrenheit temperature is a linear function of the Celsius temperature, and the graph of the corresponding equation is a straight line containing the points (0, 32) and (100, 212). Thus its slope is

$$m = \frac{212 - 32}{100 - 0} = \frac{180}{100} = \frac{9}{5}.$$

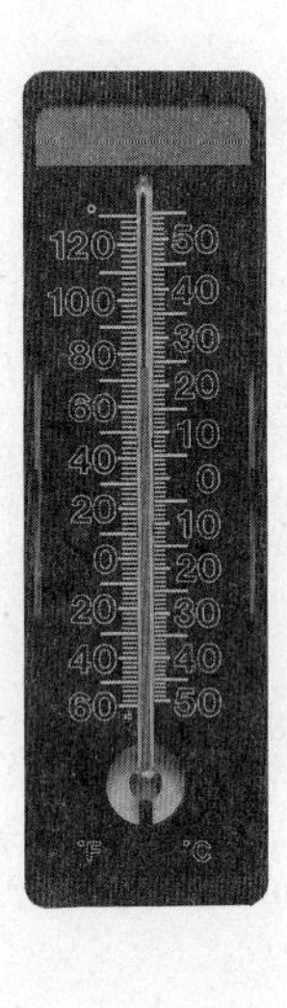

Using the point $(x_1, y_1) = (0, 32)$ in the point-slope form we get the equation

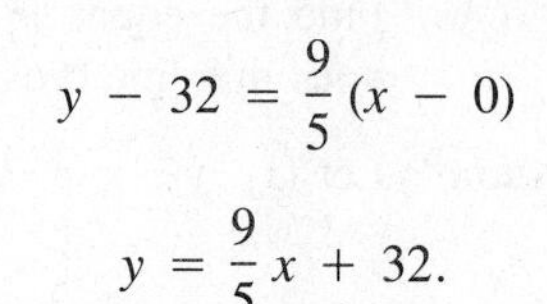

$$y - 32 = \frac{9}{5}(x - 0)$$

or

$$y = \frac{9}{5}x + 32.$$

Corresponding to the Celsius temperature $x = 50$ we have

$$\begin{aligned} y &= \frac{9}{5}(50) + 32 \\ &= 90 + 32 = 122. \end{aligned}$$

Thus, 50 degrees Celsius corresponds to 122 degrees Fahrenheit. Conversely, if the Fahrenheit temperature is $y = 100$, then we have

$$100 = \frac{9}{5}x + 32$$

$$\frac{9}{5}x = 100 - 32 = 68$$

$$x = \frac{68}{\left(\frac{9}{5}\right)} = 68\left(\frac{5}{9}\right) = \frac{340}{9} = 37\frac{7}{9}.$$

We see that 100 degrees Fahrenheit corresponds to almost 38 degrees Celsius. ■

EXAMPLE 1.10 A middle-eastern country exports a constant amount of oil each year, so its oil reserves have been decreasing linearly with time. Industry experts estimate that the country's reserves were 350 million barrels in 1975 and 252 million barrels in 1982. If this trend continues, when will the oil run out?

Solution Letting R denote the remaining oil reserves, in millions of barrels, and t denote the time (in years) since 1975, we have a linear function $R = mt + b$. (Notice that here we are using the more suggestive variable names, t for time and R for reserves, in place of x and y.) The given conditions say that $R = 350$ when $t = 0$ (in 1975), and $R = 252$ when $t = 7$ (in 1982). Thus, the points (0, 350) and (7, 252) lie on the corresponding line, as illustrated in Figure 1.15. By the slope formula the slope of this line is

$$m = \frac{252 - 350}{7 - 0} = \frac{-98}{7} = -14.$$

Since the line crosses the R-axis at $b = 350$, the slope-intercept equation is

$$R = -14t + 350.$$

Note that $m = -14$ represents the rate of change of the function, that is, the amount of oil pumped out each year. According to the above equation, the reserves will reach zero when

$$0 = -14t + 350.$$

Solving this equation, we obtain $14t = 350$, or $t = 25$. In other words, if the linear trend continues, the oil will run out in the year 2000. (See Figure 1.15.) ■

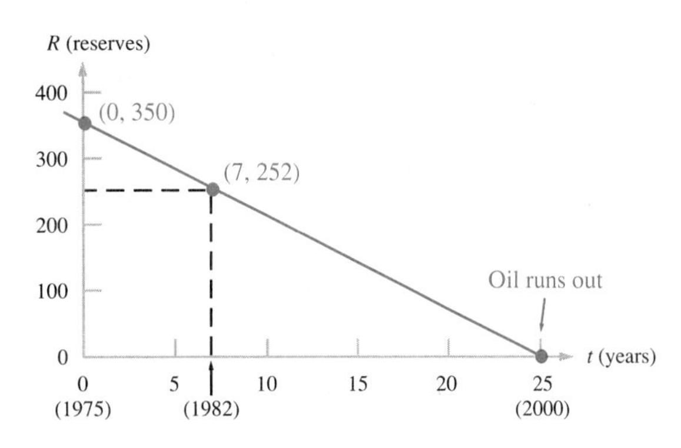

FIGURE 1.15

EXERCISES 1.2

In Exercises 1–6 change the equation given to slope-intercept form.

1. $3x + 4y = 18$

2. $5x - 6y = 15$

3. $x = 3y + 5$

4. $5x = 6y$

5. $4y + 7 = 0$

6. $2x + 3y = -4x + 9$

In Exercises 7–14 give the slope and y-intercept, if they exist, of the line with the given equation.

7. $7x + 8y = 42$

8. $-3x - 6y = 8$

9. $3y + 5 = 0$

10. $\frac{1}{2}x + \frac{1}{3}y = 5$

11. $2y + 3 = 4x$

12. $-3x + 6 = 0$

13. $3(x + 1) = 4$

14. $4x - 2y = 5 + x$

In Exercises 15–20 give the slope of the line through the two points, if possible.

15. $(2, 3)$ and $(4, 7)$

16. $(5, -7)$ and $(1, -9)$

17. $(-2, 5)$ and $(6, -1)$

18. $(-3, 6)$ and $(2, 6)$

19. $(5, 4)$ and $(5, -2)$

20. $(1/2, 2)$ and $(5/2, 4/3)$

In Exercises 21–28 give the equation of the line described, in slope-intercept form if possible.

21. slope 4, y-intercept $(0, -3)$

22. slope -3, y-intercept $(0, 5)$

23. slope -3, through $(1, -4)$

24. slope 2, through $(-2, 5)$

25. x-intercept $(2, 0)$, y-intercept $(0, -2)$

26. x-intercept $(-4, 0)$, slope 6

27. slope undefined, through $(7, 5)$

28. horizontal, through $(2, -3)$

In Exercises 29–34 give the equation of the line through the two points, in slope-intercept form if possible.

29. $(3, 4)$ and $(5, 2)$

30. $(6, 1)$ and $(3, 7)$

31. $(-1, 3)$ and $(2, -5)$

32. $(3, 4)$ and $(3, -2)$

33. $(4, 2)$ and $(-1, 2)$

34. $(3, 1/2)$ and $(-1, 5/2)$

35. A motion picture company finds that the number of people who will see one of its pictures the first week it is released is a linear function of the amount of money it spends on pre-release television advertising. If it spends \$500,000 on such advertising, 200,000 tickets will be sold the first week, and if it spends \$700,000, 250,000 tickets will be sold.

(a) How many tickets will be sold if \$1,000,000 is spent?
(b) How much advertising is needed to sell 1,000,000 tickets the first week?

The number of people who will see this movie the first week is a linear function of the money spent on prerelease advertising.

36. A psychologist is testing the hypothesis that the number of mistakes per hour a subject will make copying nonsense words is a linear function of the room temperature. The subject makes 13 mistakes per hour at 70° F and 17 mistakes per hour at 80° F. If the hypothesis is correct, how many mistakes per hour can be expected at 95° F?

37. An automobile manufacturer finds that the number of complaints it gets per month is a linear function of the number of inspections given to cars before they are shipped. If there are 40 inspections, there are 60 complaints per month; and if there are 30 inspections, there are 85 complaints per month. How many complaints per month would there be if there were 45 inspections?

38. It is found that the rate at which baby rats gain weight is a linear function of the amount of food they are fed each day. If fed 30 grams per day, they gain 2 ounces per week; and if fed 36 grams per day, they gain 2.5 ounces per week.

(a) How much would a rat gain per week if fed 40 grams per day?
(b) What daily ration would produce a gain of 2.8 ounces per week?

39. A machine purchased for \$10,000 is depreciated linearly over a 15-year period, at the end of which it has a scrap value of \$2500. In other words, the value of the machine decreases linearly from \$10,000 to \$2500 over fifteen years.

(a) Let V be the value of the machine t years after its purchase. Express V as a linear function of t.
(b) What will the machine be worth after three years? Five years?
(c) How long will it take for the value to reach \$4000?

40. The level of oxidant pollution in a certain area has been decreasing linearly since 1980, when an intensive pollution control program began. At that time the oxidant level in the air was measured at .15 parts per million (ppm); by 1990 the figure was .09 ppm.

(a) Let P be the oxidant level (in ppm) and let t be the time (in years) since 1980. Express P as a linear function of t.
(b) What will the oxidant level be in 1995? In 1997?
(c) Air with an oxidant level of .03 ppm or less is considered clean. If the present linear trend continues, when will this be achieved?

Answers to Practice Problems

2. $y = -2x + 13$

3. $y = -8x + 11$

1.3 Intersecting Lines

The geometric problem of finding the point where two given lines intersect corresponds to the algebraic problem of solving two linear equations in two unknowns. In this section we review the methods of substitution and elimination, the generalizations of which will play an important role in Chapter 2.

SYSTEMS OF EQUATIONS

We begin with the fact, familiar from Euclidean geometry, that two nonparallel lines intersect at a point. If we know the equations of the lines, we should be able to find the coordinates of the point.

EXAMPLE 1.11 Find the point where the lines $2x - 3y = 9$ and $x + y = 2$ intersect.

Discussion Since the point of intersection lies on *both* lines, its coordinates (x, y) must satisfy both equations. We must therefore solve the **system of two linear equations in two unknowns**

$$\begin{aligned} 2x - 3y &= 9 \\ x + y &= 2. \end{aligned}$$

Two different methods of solution are illustrated.

Substitution Method

To solve a system of two linear equations in two unknowns:

1. Use one of the equations to solve for one variable in terms of the other.
2. Substitute this quantity into the remaining equation and solve for the other variable.

In our example we can solve for y in terms of x using the second equation.

$$\begin{aligned} x + y &= 2 \\ y &= 2 - x \end{aligned} \qquad (1.2)$$

Substituting $2 - x$ for y in the first equation of the system, we obtain

$$\begin{aligned} 2x - 3(2 - x) &= 9 \\ 2x - 6 + 3x &= 9 \\ 5x &= 15 \\ x &= 3. \end{aligned}$$

Finally, letting $x = 3$ in equation (1.2), we obtain y.

$$y = 2 - x = 2 - 3 = -1$$

The point of intersection is therefore $(3, -1)$. We can verify this solution by direct substitution in the original equations.

Check $2x - 3y = 2(3) - 3(-1) = 6 + 3 = 9$

$x + y = 3 + (-1) = 2$

The situation is illustrated in Figure 1.16.

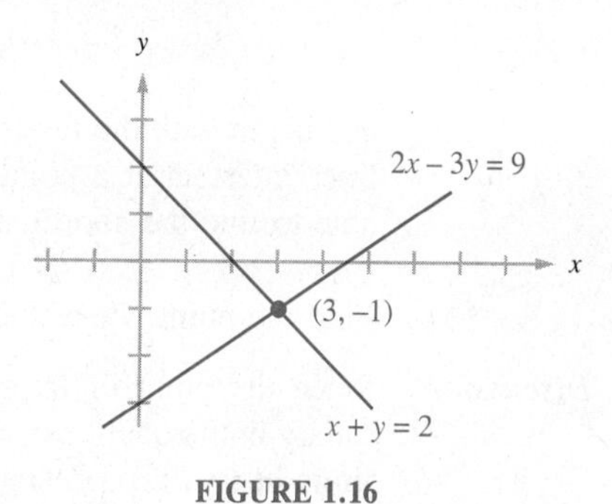

FIGURE 1.16

The second method of solution is called **elimination.**

Elimination Method

To solve a system of two linear equations in two unknowns:

1. Multiply one or both equations by appropriate constants so that one of the variables has coefficient c in one equation and $-c$ in the other.
2. Add the equations together to eliminate this variable.

To apply this method to the system of Example 1.11 we begin by writing down the given equations.

$$\begin{aligned} 2x - 3y &= 9 \\ x + y &= 2 \end{aligned}$$

Multiplying the second equation by 3, we obtain an **equivalent system,** that is, a system having exactly the same solutions, if any.

$$\begin{aligned} 2x - 3y &= 9 \\ 3x + 3y &= 6 \end{aligned}$$

If we now add these equations, the term involving y is eliminated.

$$\begin{aligned} 5x &= 15 \\ x &= 3 \end{aligned}$$

Finally, we substitute $x = 3$ into either of the original equations—say, the second—and solve for y.

$$\begin{aligned} 3 + y &= 2 \\ y &= -1 \end{aligned}$$

We thus arrive at the solution $(3, -1)$, as before. ■

EXAMPLE 1.12 Find the point where the lines $3x + 7y = 27$ and $5x + 4y = 22$ intersect.

Solution We must solve the following system of equations:

$$\begin{aligned} 3x + 7y &= 27 \\ 5x + 4y &= 22. \end{aligned}$$

If we multiply the first equation by 5 and the second by -3, we obtain an equivalent system.

$$\begin{aligned} 15x + 35y &= 135 \\ -15x - 12y &= -66 \end{aligned}$$

When we add these equations the term involving x is eliminated.

$$\begin{aligned} 23y &= 69 \\ y &= 3 \end{aligned}$$

We can now find x by substituting $y = 3$ into the first equation.

$$\begin{aligned} 3x + 7(3) &= 27 \\ 3x + 21 &= 27 \\ 3x &= 6 \\ x &= 2 \end{aligned}$$

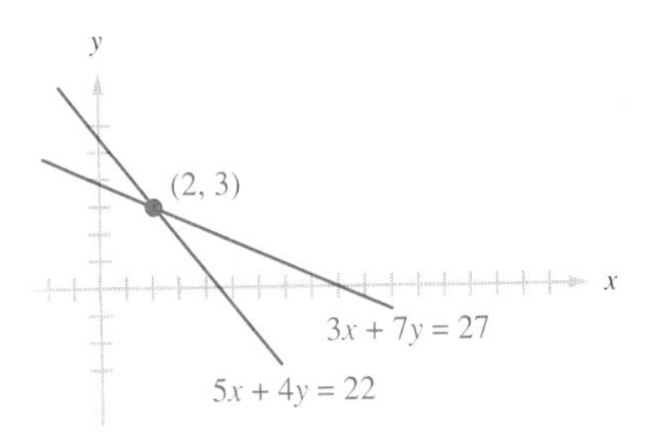

FIGURE 1.17

The lines intersect at the point $(2, 3)$, as shown in Figure 1.17. ■

Practice Problem 4 Find the point where the lines $x + 3y = -5$ and $3x - 2y = 7$ intersect.

EXAMPLE 1.13 Find the point where the lines $2x - 4y = 7$ and $-3x + 6y = 8$ intersect.

Solution We apply the elimination method to the system

$$2x - 4y = 7$$
$$-3x + 6y = 8.$$

Multiplying the first equation by 3 and the second by 2, we transform the system into

$$6x - 12y = 21$$
$$-6x + 12y = 16.$$

When we add these equations the terms involving both x and y drop out, leaving us with an absurdity:

$$0 = 37.$$

We say that the given equations form an **inconsistent system,** that is, they have no common solution (x, y). This means that the lines they represent are parallel, as in Figure 1.18. ■

Note that both the lines in Example 1.13 have slope $m = 1/2$. It is easy to show that *two nonvertical lines are parallel if and only if they have the same slope*.

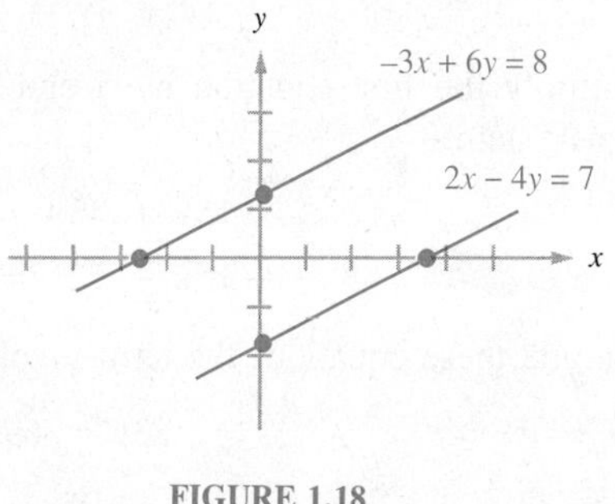

FIGURE 1.18

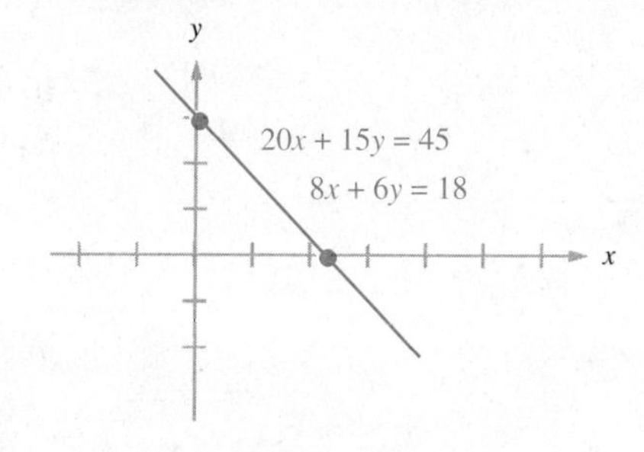

FIGURE 1.19

EXAMPLE 1.14 Find the point where the lines $8x + 6y = 18$ and $20x + 15y = 45$ intersect.

Solution Here we must solve the following system:

$$8x + 6y = 18$$
$$20x + 15y = 45.$$

We multiply the first equation by 5 and the second by -2.

$$40x + 30y = 90$$
$$-40x - 30y = -90$$

When we add these together we obtain

$$0 = 0,$$

which is true but not very informative. The problem here is that the given system is a **dependent system:** the first equation implies the second, as we see by multiplying it by 5/2. Both equations thus have the straight line graph shown in Figure 1.19, and every point on this line is a common solution to the given equations. ■

These examples illustrate all possible cases that can arise in solving two linear equations in two unknowns. The results may be summarized as follows.

Every system

$$a_1x + b_1y = c_1, \qquad a_1, b_1 \text{ not both } 0,$$
$$a_2x + b_2y = c_2, \qquad a_2, b_2 \text{ not both } 0,$$

of linear equations leads to exactly one of the following three cases:

Systems with a unique solution: The system has only one solution, and the corresponding lines intersect in one point.

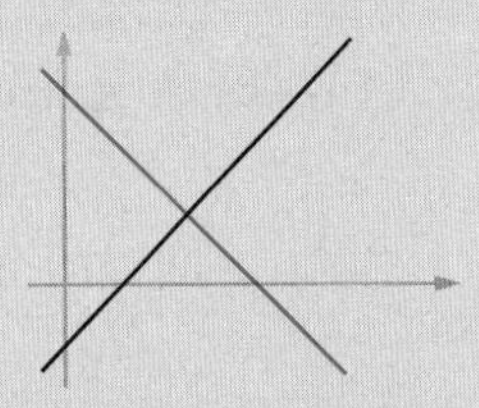

Inconsistent systems: The system has no solution, and the corresponding lines are parallel.

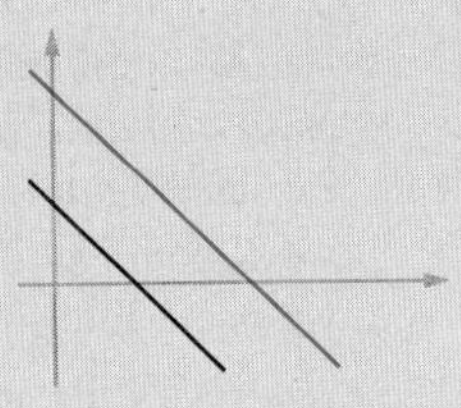

Dependent systems: The system has infinitely many solutions, and the corresponding lines coincide.

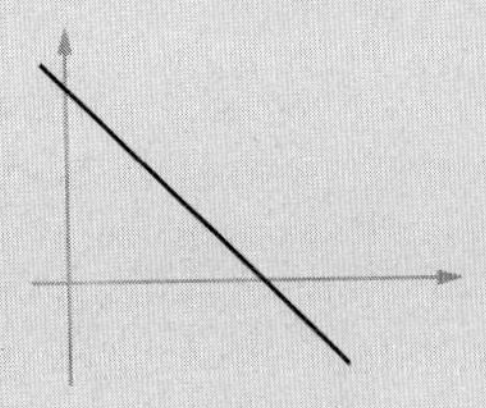

In the next chapter we shall solve systems involving more than two equations in more than two unknowns, using a general technique based on the elimination method. We shall see, however, that even in this more general setting, the only possible cases are the three listed in the box on page 21.

APPLICATIONS

Linear equations play a central role in many practical problems. This is illustrated by the following example.

EXAMPLE 1.15 A furniture manufacturer makes two styles of chair, ''Antique'' and ''Baroque,'' for sale to retail outlets. Each Antique chair requires two work-hours of construction and three work-hours of finishing, while each Baroque chair requires three work-hours of construction and five work-hours of finishing. The company has a total of 60 work-hours of construction labor and 95 work-hours of finishing labor available each day. Is it possible to use up exactly the available amount of labor?

Solution We begin by introducing two unknowns. Let

$x =$ the number of Antique chairs produced daily,
$y =$ the number of Baroque chairs produced daily.

Next, we summarize the given data in the form of a table, indicating the total available amounts of labor in the last column.

		Style		
		Antique	*Baroque*	*Available*
	Number of units	x	y	
Labor	*Construction*	2 work-hours	3 work-hours	60 work-hours
per unit	*Finishing*	3 work-hours	5 work-hours	95 work-hours

We now make the following simple but crucial observation: *If one Antique chair requires 2 work-hours of construction, then x Antique chairs will require $2x$ work-hours of construction*. Similarly, y Baroque chairs will require $3y$ work-hours of construction. Hence, in making x Antique and y Baroque chairs we use a total of $2x + 3y$ work-hours of construction labor. By the same kind of reasoning, we arrive at the expression $3x + 5y$ for the total amount of finishing labor used. The condition that exactly the available amount of labor be used up therefore translates into the following pair of equations:

$$\begin{aligned} 2x + 3y &= 60 \\ 3x + 5y &= 95. \end{aligned}$$

We will solve this system by the elimination method. Multiply the first equation by 5 and the second by -3.

$$\begin{aligned} 10x + 15y &= 300 \\ -9x - 15y &= -285 \end{aligned}$$

Add these to eliminate the term involving y.

$$x = 15$$

Finally, substitute $x = 15$ back into the first equation.

$$\begin{aligned} 2(15) + 3y &= 60 \\ 30 + 3y &= 60 \\ 3y &= 30 \\ y &= 10 \end{aligned}$$

We now have our answer: the company can use up all the available labor by producing 15 Antique chairs and 10 Baroque chairs each day. ■

Question What if either x or y had come out negative in Example 1.15?

Answer If we had obtained negative values for x or y we would have had to conclude that *the problem has no solution,* since the physical interpretation of x and y makes such values meaningless. For example, if the manufacturer had only 80 work-hours of finishing labor available instead of 95, the corresponding equations would be

$$\begin{aligned} 2x + 3y &= 60 \\ 3x + 5y &= 80, \end{aligned}$$

which have the solution $x = 60$, $y = -20$. In this case it would not be possible to use up exactly the available supply of labor.

Practice Problem 5 A bag of apples weighs 6 pounds and costs $4, while a bag of pears weighs 4 pounds and costs $3. A man bought 112 pounds of fruit costing $79. How many bags of apples and pears did he buy?

EXAMPLE 1.16

A dietitian has available two types of flour: Type 1 with 20% protein and Type 2 with 24% protein. How many ounces of each type must she mix together to get 16 ounces of flour that contains 21% protein?

Solution Suppose she mixes x ounces of Type 1 flour and y ounces of Type 2. We summarize our data in the table below.

	Flour		
	Type 1	*Type 2*	*Desired*
Ounces	x	y	16
Protein proportion	.20	.24	.21

Then we must have

$$\begin{aligned} x + \quad y &= 16 \\ .20x + .24y &= .21(16) \end{aligned}$$

Multiplying the second equation by 5 yields

$$x + 1.2y = 16.8.$$

We subtract the first equation to get

$$\begin{aligned} .2y &= .8 \\ y &= \frac{.8}{.2} = 4. \end{aligned}$$

Then $x = 16 - y = 12$. She should use 12 ounces of Type 1 flour and 4 ounces of Type 2. ■

EXERCISES 1.3

In Exercises 1–20, find the point (if any) where the lines intersect. Identify the cases when there are no solutions or infinitely many.

1. $\begin{aligned} 3x \qquad &= 6 \\ 2x + 3y &= -2 \end{aligned}$

2. $\begin{aligned} 2x + y &= 7 \\ -2y &= 8 \end{aligned}$

3. $\begin{aligned} .5y \qquad &= 5 \\ 2x - 4y &= 6 \end{aligned}$

4. $\begin{aligned} 3x - 2y &= 4 \\ \tfrac{1}{3}x &= \tfrac{2}{3} \end{aligned}$

5. $\begin{aligned} 3x + 2y &= 8 \\ x - \quad y &= 1 \end{aligned}$

6. $\begin{aligned} x + 2y &= 7 \\ 3x + y &= 6 \end{aligned}$

7. $\begin{aligned} 6x + 5y &= 21 \\ 3x - 4y &= 30 \end{aligned}$

8. $\begin{aligned} 2x - 5y &= 14 \\ 5x - 2y &= 14 \end{aligned}$

9. $\begin{aligned} 7x + 5y &= 41 \\ 4x - 3y &= 0 \end{aligned}$

10. $\begin{aligned} 5x + 4y &= 0 \\ 4x + 5y &= 9 \end{aligned}$

11. $\begin{aligned} .9x + .8y &= 72 \\ .1x + .2y &= 10 \end{aligned}$

12. $\begin{aligned} 1.6x - .8y &= 8 \\ -1.2x + .6y &= 0 \end{aligned}$

13. $\begin{aligned} x + y &= 8 \\ x - y &= 3 \end{aligned}$

14. $\begin{aligned} 2x - 3y &= 6 \\ 3x - 5y &= 15 \end{aligned}$

15. $\begin{aligned} -\tfrac{2}{3}x + \tfrac{1}{2}y &= 15 \\ \tfrac{4}{5}x - \tfrac{3}{5}y &= 12 \end{aligned}$

16. $\begin{aligned} \tfrac{2}{3}x + \quad y &= 8 \\ x + \tfrac{1}{2}y &= 8 \end{aligned}$

17. $\begin{aligned} 2x + 3y &= 75 \\ 4y &= 20 \end{aligned}$

18. $\begin{aligned} -x + \tfrac{2}{5}y &= -2 \\ \tfrac{5}{2}x - \quad y &= 5 \end{aligned}$

19. $\begin{aligned} x - 3y &= 14 \\ 4y &= 16 - 3x \end{aligned}$

20. $\begin{aligned} 3x &= 4y + 1 \\ x &= y + 6 \end{aligned}$

21. Each ounce of Food A contains 2 gm of carbohydrate and 2 gm of protein, while each ounce of Food B contains 4 gm of carbohydrate and 2 gm of protein. By combining these foods, is it possible to obtain exactly 60 gm of carbohydrate and 50 gm of protein?

22. An oil company operates two eastern refineries. The Albany facility produces 100 barrels of high-grade oil and 300 barrels of medium-grade oil per day. The Boston facility produces 200 barrels of high-grade and 100 barrels of medium-grade per day. By operating each facility for a certain number of days, can the company produce exactly 12,000 barrels of high-grade and 24,000 barrels of medium-grade oil?

23. The House of Coffee sells two different blends of Brazilian and Colombian beans. Each pound of Blend A contains 1/4 lb of Brazilian and 3/4 lb of Colombian. Each pound of Blend B contains 2/3 lb of Brazilian and 1/3 lb of Colombian. The store would like to use up its entire supply of Brazilian and Colombian beans by producing a certain amount of each blend. Is this possible

    (a) if the store has 1200 lb of Brazilian and 1500 lb of Colombian?
    (b) if the store has 3000 lb of Brazilian and 1400 lb of Colombian?

24. World Travel Service has agreed to transport a group of 1800 football fans from Los Angeles to Chicago for one of the season's big games. The agency can charter two different types of aircraft for the trip. Type A has 20 first-class seats and 80 economy seats. Type B has 40 first-class seats and 60 economy seats. By using a certain number of planes of each type, the agency hopes to accommodate the group's needs exactly, with no seats left empty on any plane.

    (a) Is this possible if 600 fans want first-class seats and 1200 want economy seats?
    (b) Is this possible if 650 fans want first-class seats and 1150 want economy seats?

25. A chemist has available two strengths of acid solution. Solution A is 10% acid (by volume), while Solution B is 20% acid. How many liters of each solution should the chemist mix together to obtain a total of 100 liters of 12% acid solution?

26. A marine biologist uses two different fish nutrients. Nutrient 1 contains 7 units of protein per gm, while Nutrient 2 contains 3 units of protein per gm. How many grams of each should be mixed together, if the biologist needs 300 gm of nutrient containing 4 units of protein per gm?

27. Customs officials suspect a man of smuggling jade rings and bracelets into the U.S., where he sells them for $150 and $250, respectively. Their investigation shows that his latest shipment sold here for a total of $50,000, and officials in Taiwan, where the items cost 40% and 50% less, respectively, have proof that he paid $28,000 there for the merchandise. How many rings and how many bracelets did the shipment contain?

28. During a certain year, Laura had a profit of $5600 from two investments, the first earning 5% and the second 12%. Had she put all her money into the second investment, her profit would have been $8400. How much did she put into each investment?

29. A woman holds 45 shares of stock A and 24 shares of stock B. On the morning of a certain day the value of her holdings was $4800. During the day the price of stock A doubled, while the price of stock B fell by 50%. At day's end her holdings were worth $7800. What was the initial price of each stock?

30. A traveling salesperson represents two companies. The first company pays a weekly salary of $200 plus a 6% commission on gross sales, and the second pays a 10% commission on gross sales, but no salary. One week the salesperson earned a total of $440 on $2800 of gross sales. How much did he sell for each company?

Answers to Practice Problems **4.** $(1, -2)$

5. He bought 10 bags of apples and 13 bags of pears.

1.4 Break-Even Analysis

In any production process there are both **fixed costs** and **variable costs.** Fixed costs include such items as rent, equipment, depreciation, and interest, which remain the same no matter how much output is produced. Variable costs include such items as material and labor, which depend on the amount of output. The **total cost** of an operation is the sum of the fixed and variable costs, which depends on the number of units produced. The **revenue,** or sales income, also depends on the amount produced. The number of items to be produced so that cost exactly equals revenue is called the **break-even point.**

EXAMPLE 1.17 The Zeno Corporation has a division that makes color TV sets. The fixed costs of the operation are $15,000 per day, and the variable cost is $125 for each unit produced. The sets sell for $325 each. How many sets must be produced and sold daily for the operation to break even?

Solution If the firm produces x units per day, the total production cost C (in dollars) will be

$$C = 125x + 15{,}000,$$

and the total revenue R (assuming all the units are sold) will be

$$R = 325x.$$

Note that both cost and revenue are linear functions of x. If they are graphed in a common coordinate system, with y representing both cost and revenue, the **cost** and **revenue lines** in Figure 1.20 are obtained.

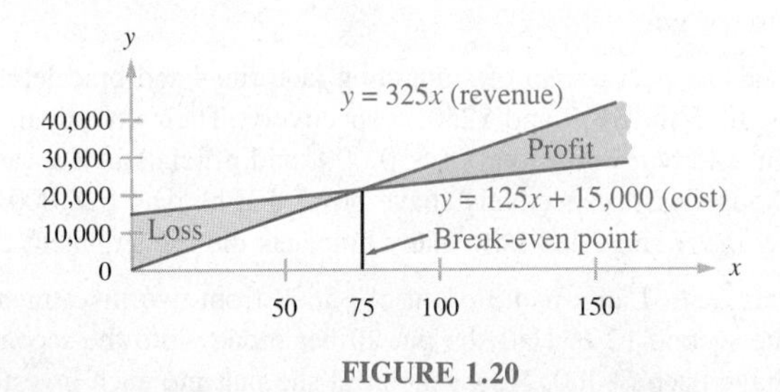

FIGURE 1.20

The break-even point is the value of x for which cost equals revenue.

$$C = R$$
$$125x + 15{,}000 = 325x$$
$$200x = 15{,}000$$
$$x = 75$$

This is the x-coordinate of the point where the cost and revenue lines intersect. For $x = 75$ we have $C = R = \$24{,}375$, so the operation shows neither a profit nor a loss. For $x < 75$ the cost line lies above the revenue line ($C > R$), so the operation is losing money. For $x > 75$ the cost line lies below the revenue line ($C < R$), so the operation shows a net profit. ■

In general, **profit** equals the difference between revenue and cost.

Profit

$$P = R - C$$

If revenue and cost are linear functions of x, then so is profit. In the example above,

$$\begin{aligned} P &= R - C \\ &= 325x - (125x + 15{,}000) \\ &= 200x - 15{,}000. \end{aligned}$$

The graph of this function is the **profit line** shown in Figure 1.21. At the break-even point $x = 75$, the profit line crosses the x-axis. For $x < 75$ the profit line lies below the x-axis, indicating a *negative* profit, that is, a net loss. For $x > 75$ the line lies above the x-axis, indicating a *positive* profit, or a net gain.

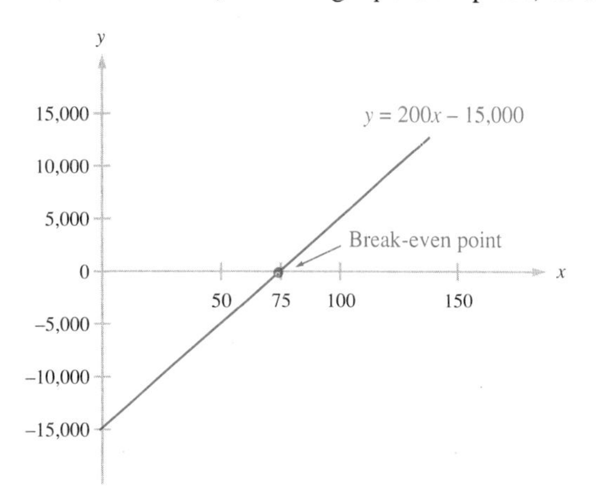

FIGURE 1.21

Using the profit function, we can also answer some other questions of interest.

EXAMPLE 1.17 Continued

How many sets must be produced and sold daily:

(a) to make a profit of $17,000?
(b) to make a profit of $125 on each unit?
(c) to make a profit of 20% on sales?

Solution In each case we begin with the profit function $P = 200x - 15{,}000$.

(a) Setting the profit equal to 17,000, we solve for x.

$$\begin{aligned} P &= 17{,}000 \\ 200x - 15{,}000 &= 17{,}000 \\ 200x &= 32{,}000 \\ x &= 160 \end{aligned}$$

Thus, producing and selling 160 units will yield a profit of \$17,000.

(b) The profit per unit is P/x. Hence, we must solve the equation

$$\begin{aligned} \frac{P}{x} &= 125. \\ P &= 125x \\ 200x - 15{,}000 &= 125x \\ 75x &= 15{,}000 \\ x &= 200 \end{aligned}$$

Thus, producing and selling 200 units will yield a profit of \$125 per unit.

(c) The condition that profit is 20% of sales translates into the equation

$$\begin{aligned} P &= .20R. \\ 200x - 15{,}000 &= .20(325x) \\ 200x - 15{,}000 &= 65x \\ 135x &= 15{,}000 \\ x &\approx 111.1 \end{aligned}$$

(The symbol $\approx$ indicates that the answer has been rounded off.) In practical terms, this means that the company must product at least 112 units per day to achieve a profit margin of at least 20%. ■

Practice Problem 6 A lemonade stand sells lemonade for 50 cents per glass. The ingredients in a glass of lemonade cost 10 cents, and the fixed costs of running the stand are \$44 per day.

(a) How many glasses of lemonade must be made and sold in a day to break even?

(b) Write the profit P in dollars as a linear function of the number x of glasses sold.

(c) How many glasses must be sold in a day to make \$100 profit?

COMPARISON OF PROFIT FUNCTIONS

In some applications we are presented with several alternative methods of production. By comparing their associated profit functions we can determine the conditions under which one method is preferable to another.

EXAMPLE 1.18 The firm of Ziferstein & Son manufactures a dental floss dispenser that sells for \$1.55. Under the present production method the operation has fixed costs of \$200 a day and a variable cost of 75¢ per unit. The junior owner wants to lease some new equipment that will save labor and bring the variable cost down to 35¢ per unit. The senior owner is opposed to the new plan, since payments for the equipment will increase the daily fixed costs by \$250. The junior owner argues that the savings in variable costs will more than offset the increase in fixed costs. Is he right?

Discussion Let x be the number of units produced per day. The cost functions for the two production methods are given by

$$C_1 = .75x + 200 \quad \text{(old method)},$$
$$C_2 = .35x + 450 \quad \text{(new method)},$$

while the revenue function is the same for both.

$$R = 1.55x$$

We calculate the profit function for each method.

$$P_1 = R - C_1 = .80x - 200 \quad \text{(old method)}$$
$$P_2 = R - C_2 = 1.20x - 450 \quad \text{(new method)}$$

The two profit lines are shown in Figure 1.22.

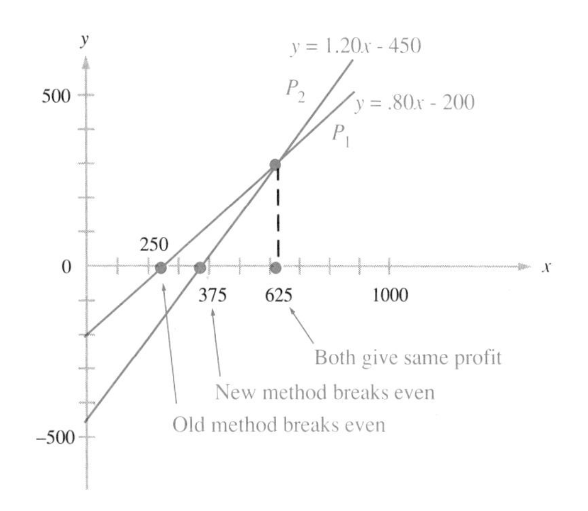

FIGURE 1.22

To find the break-even point under the old method, we set $P_1 = 0$.

$$.80x - 200 = 0$$
$$.80x = 200$$
$$x = 250$$

This is the point where the line giving P_1 crosses the x-axis in Figure 1.22. Similarly, the new method will break even when $P_2 = 0$.

$$1.20x - 450 = 0$$
$$1.20x = 450$$
$$x = 375$$

For values of x in the range $250 < x \leq 375$, therefore, only the old method will show a profit. We see from Figure 1.22 that the old method is more profitable (that is, $P_1 > P_2$) up to the point where the two profit lines cross, but beyond that point the new method is more profitable ($P_2 > P_1$). To find this intersection point we set $P_1 = P_2$ and solve for x.

$$.80x - 200 = 1.20x - 450$$
$$.40x = 250$$
$$x = 625$$

When $x = 625$ both methods yield the same profit: $P = \$300$. When $x < 625$ we have $P_1 > P_2$, and when $x > 625$ we have $P_2 > P_1$. Our answer to the problem, therefore, must be conditional: if it is possible to produce more than 625 units per day under the new method, and they can all be sold at the price stated, then the new method is preferable; otherwise the old method is preferable. ■

EXAMPLE 1.19 Sylvia plans to run a private shuttle service between the Astor Hotel and the local airport. Limousine rental, insurance, and license costs total \$3600 a year. Each round-trip will average \$14 in revenues. Gasoline and maintenance costs average \$2 per run. In addition, the city council intends to charge her either a flat franchise fee of \$1200 a year or else a franchise tax of 50% of revenues received. Which plan will be more profitable for Sylvia?

Solution Suppose she makes x round-trip runs per year. Under the franchise fee (Plan 1), the fixed yearly costs will be $\$3600 + \$1200 = \$4800$, and the variable costs will be $2x$ dollars. The total annual cost is therefore given by

$$C_1 = 2x + 4800,$$

while annual revenue is given by

$$R = 14x.$$

The profit function under Plan 1 is therefore

$$P_1 = R - C_1 = 12x - 4800.$$

Under the franchise tax (Plan 2), the fixed cost is reduced to \$3600, but $.50R = .50(14x) = 7x$ dollars must be added to the variable cost, for a total annual cost of

$$C_2 = 9x + 3600.$$

Since the revenue function remains the same, the profit function under Plan 2 is given by

$$P_2 = R - C_2 = 5x - 3600.$$

The profit lines are shown in Figure 1.23.

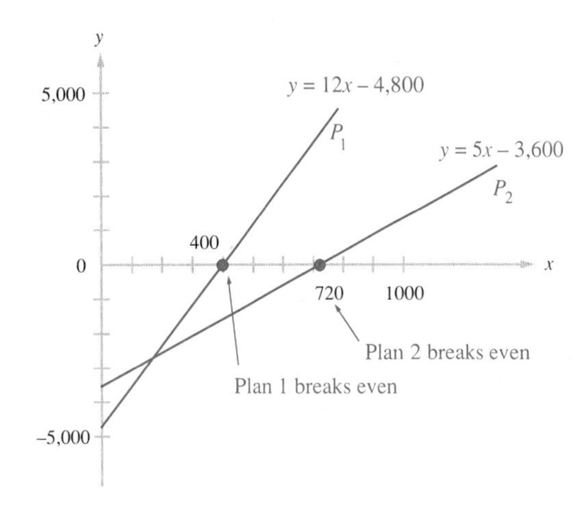

FIGURE 1.23

Under Plan 1, the operation will break even when $P_1 = 0$.

$$12x - 4800 = 0$$
$$12x = 4800$$
$$x = 400$$

Thus, Plan 1 breaks even when $x = 400$. Similarly, Plan 2 breaks even when $P_2 = 0$, or $x = 720$. From the graph, however, we see that Plan 2 never becomes more profitable than Plan 1, since $P_1 > P_2$ for all values of x beyond the first break-even point. Thus Sylvia should favor the franchise fee instead of the franchise tax; and under this plan she will have to make 400 runs per year, or about 34 per month, to break even. ■

EXAMPLE 1.20 A bank offers its customers three different checking plans. Under Plan 1, you pay a flat monthly service charge of \$4.00, regardless of how many checks you write. Under Plan 2 you pay \$1.75 a month, plus 5¢ for each check written. Under Plan 3 you pay 12¢ per check, with no fixed service charge. Under what conditions will Plan 2 be the least expensive of the three plans?

Solution Let x be the number of checks written per month. The monthly cost functions for the three plans (in dollars) are as follows:

$$C_1 = 4$$
$$C_2 = .05x + 1.75$$
$$C_3 = .12x.$$

These functions are graphed in Figure 1.24. The cost lines for Plans 2 and 3 cross when $C_2 = C_3$, that is, when

$$.05x + 1.75 = .12x$$
$$.07x = 1.75$$
$$x = 25.$$

Similarly, the cost lines for Plans 1 and 2 cross when $C_1 = C_2$:

$$4 = .05x + 1.75$$
$$.05x = 2.25$$
$$x = 45.$$

From Figure 1.24 we see that the cost line for Plan 2 lies below the other two lines over the interval $25 \le x \le 45$. Hence, Plan 2 is the least expensive when the customer writes between 25 and 45 checks per month. ■

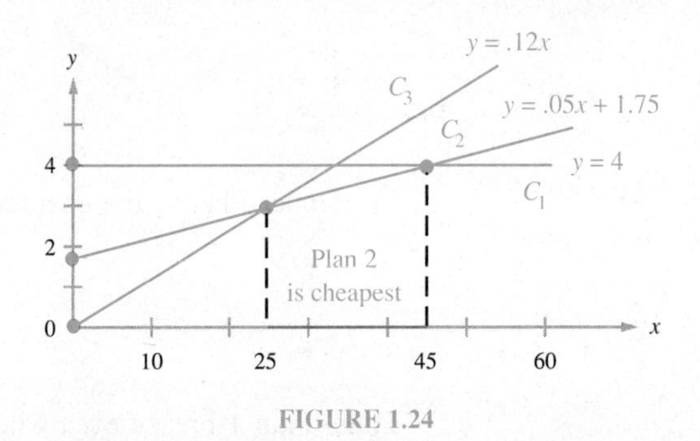

FIGURE 1.24

Remark It is important in these problems to choose the right scale on each axis so your graph will show all the relevant intersection points. If your first attempt is unsuccessful, try a larger unit of measurement. As in the examples above, you may want to use different units of measurement on the x- and y-axes.

Practice Problem 7 A wig manufacturer has fixed costs of $21,000 per week and variable costs of $25 per wig. The wigs wholesale for $60. How many wigs must the company sell per week to break even?

EXERCISES 1.4

1. A manufacturing operation has cost and revenue functions

$$C = 1.25x + 8400$$
$$R = 2x,$$

where x is the number of units of output. How many units must be produced in order for the operation to break even? Draw a graph of the profit function.

2. A power company has fixed costs of $35,000 a day, with a variable cost of 2¢ for each kilowatt-hour (kwh) of electricity produced. The company charges its customers 4¢ per kwh. How many kwh must be sold daily in order for the company to break even? Draw a graph of the profit function.

Customers pay the power company 2¢ per kilowatt-hour of electricity.

3. An electronics firm has a plant that produces a single type of silicon chip, which the firm sells for $1.40. If the plant's fixed costs are $1200 a day, and the chips cost 60¢ each to produce, how many must be made daily in order for the plant to break even? Draw a graph of the profit function.

4. An agricultural corporation finds that its asparagus processing division has fixed costs of $14,000 a year, with a variable cost of 15¢ for each pound processed. If the asparagus is sold to a distributor at 35¢/lb, how much must be processed per year in order to break even? Draw a graph of the profit function.

5. In Exercise 1, how many units must be produced
(a) to make a profit of $6000?
(b) to make a profit of 25¢ per unit?
(c) to make a profit of 25% on sales?

6. In Exercise 2, how many kwh must be sold daily
(a) to make a profit of $25,000 a day?
(b) to make a profit of 1¢ per kwh?
(c) to make a profit of 15% on sales?

7. In Exercise 3, how many silicon chips must be produced daily
(a) to make a profit of $3000 a day?
(b) to make a profit of 50¢ on each chip?
(c) to make a profit of 40% on sales?

8. In Exercise 4, how many pounds of asparagus must be processed per year
(a) to make an annual profit of $100,000?
(b) to make a profit of 15¢/lb?
(c) to make a profit of 30% on sales?

9. A manufacturer's profit functions using two different production methods are given by

$$P_1 = .75x - 8400 \qquad \text{(method 1)}$$
$$P_2 = .90x - 10{,}800 \qquad \text{(method 2)},$$

where x is the number of units of output.
(a) How many units must be produced in order to break even under method 1? Under method 2?
(b) Draw a graph of the two profit functions. Which method is better for the manufacturer?

10. An ecology group is planning a fund-raising dinner to be held in one of the two local hotels. The group must pay \$1200 for the speaker and \$200 for promotion. In addition, the Astor Hotel wants \$400 plus \$10 a plate to provide the meal, while the Belmont Hotel is asking \$800 plus \$7 a plate. Admission to the event is set at \$15 a plate.

(a) How many people must attend for the event to break even, if it is held at the Astor? At the Belmont?

(b) Draw a graph of the two profit functions involved. Which hotel's offer is better?

11. An auto plant has fixed costs of \$60,000 a day and a variable cost of \$3000 for each car produced. The cars are sold to the distributor for \$4250. A new computerized system could cut the variable cost by 20%, but installing the system would double the present fixed costs.

(a) How many cars must be produced daily to break even, under the present operation? Under the computerized system?

(b) Draw a graph of the two profit functions involved. Under what conditions would it pay to install the new system?

12. A corporation's annual operating costs and revenue (in dollars) are given by

$$C = 5x + 90{,}000$$
$$R = 20x,$$

where x is the number of units of output per year. In addition to its operating costs the company must pay corporate tax. Under Tax Plan 1, the company pays a tax T equal to 15% of revenues received. Under Tax Plan 2, the tax T is equal to 40% of gross profit $(R - C)$. The company's *net* profit P, after taxes, is given by

$$P = R - C - T.$$

(a) Express the net profits, P_1 and P_2, under the two plans as linear functions of x.

(b) Compare these functions by graphing.

(c) What is the company's break-even point under Plan 1? Under Plan 2?

(d) For what values of x will Plan 1 be better for the company than Plan 2?

13. A private school is planning to produce a desk calendar to be mailed out to the community. Printer 1 charges \$1000 for design and layout, plus \$2.80 a copy. Printer 2 charges \$4000 for design and layout, plus \$1.80 a copy. Which offer is better, in terms of the number of copies the school needs? Illustrate with a graph.

14. Three different lawyers are willing to handle a certain lawsuit. Lawyer 1 charges \$1000 plus 20% of the judgment to be awarded. Lawyer 2 charges \$500 plus 40% of the award. Lawyer 3 charges 60% of the award, with no fixed fee. Let x be the dollar amount awarded to the plaintiff, and let P_1, P_2, and P_3 be his net gain (award minus legal costs) if he accepts lawyer 1, 2, or 3, respectively.

(a) Express P_1, P_2, and P_3 as linear functions of x.

(b) Compare these functions by graphing.

(c) What is the plaintiff's break-even point with each of the three lawyers?

(d) Assuming that the suit is won, for what values of x will each lawyer's offer be the best?

15. Taxi rates are not regulated in Baker County, and they vary among the three competing companies. Red Cab charges a fixed fee of 65¢ plus a variable fee of 20¢ a mile. White Cab charges 85¢ plus 15¢ a mile, and Blue Cab charges \$1.15 plus 12¢ a mile. A sales representative who uses cabs frequently would like a simple rule to determine which cab is least expensive, given the number x of miles she has to travel. Solve her problem, and illustrate with a graph.

16. A woman is comparing three hospitalization insurance plans. Plan 1 pays all hospital charges for an annual premium of $750. Plan 2 pays 90% of the charges for a premium of $500. Plan 3 pays 60% of the charges for a premium of $200. Let x be the hospital charges the woman might incur during the year. Express her net loss (premium plus uncovered charges) as a linear function of x, under each of the three plans. For what values of x will each plan be best? Illustrate with a graph.

17. Three recording companies are bidding on production rights for a demo featuring a country singer. The first company has offered a flat payment of $1 million. The second has offered $100,000 plus 20% of gross receipts. The third has offered $250,000 plus 10% of gross receipts. Which offer is best, in terms of the expected gross receipts x? Illustrate with a graph.

18. A famous diplomat is trying to decide between two publishers interested in his memoirs. Publisher 1 offers royalties of 90¢ per copy after the first 2000 copies sold in a given year, with no royalties paid on the first 2000 copies. Publisher 2 offers 50¢ per copy after the first 1000 copies sold. Which offer is better, in terms of the number x of copies he expects to be sold per year? Illustrate with a graph.

Answers to Practice Problems **6.** (a) 110 (b) $P = .4x - 44$ (c) 360
7. 600

MATHEMATICS IN ACTION

1.5 Market Equilibrium

SUPPLY AND DEMAND

As the price of an item goes up, more producers are willing to supply the item, and the amount produced, or **supply** of the item, rises also. For example, when the price of corn increases, so does the supply, because many farmers will grow corn rather than some other crop to take advantage of the better price. Conversely, if the price of corn drops, then farmers turn to other crops and the supply goes down.

The amount of a product consumed, or **demand** for the product, depends on the price of the product in the opposite manner. An increase in price tends to restrict demand. For example, if the price of corn goes up, users of corn products may try to find less expensive substitutes. Specifically, soft drink manufacturers use sugars derived from not only corn, but also sugar cane, beets, and other crops. They will use more or less corn syrup according to whether its price is low or high compared to other sugar sources.

Often supply and demand are approximately linear functions of price, and the techniques of the previous sections may be used to analyze how the market forces determine price.

EXAMPLE 1.21 A company manufacturing kitchen graters estimates that if the price of a grater is x (in dollars), then the monthly demand will be D units, where

$$D = 1200 - 150x.$$

The number of graters the company is willing to make also depends on the price. If the price is low, it will turn to more profitable items. The company is willing to supply S units per month at the price x, where

$$S = 300x - 600.$$

Notice that both supply and demand are linear functions of the price x. The corresponding lines are graphed in Figure 1.25.

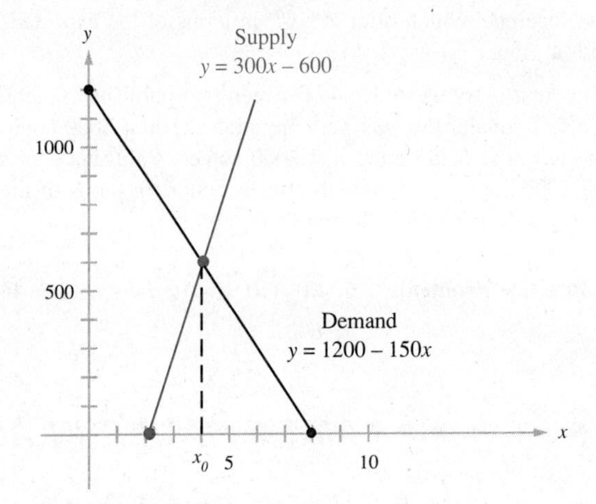

FIGURE 1.25

Discussion The point where the two lines cross has a special significance. To its right supply exceeds demand, and the company will have to lower its price to sell what it is producing. To the left of the intersection point demand exceeds supply, and the company will raise its price to take advantage of the situation. Only at the point where the two lines intersect does the price tend to remain the same, and the value of x at this point is called the **equilibrium price,** denoted x_0 in Figure 1.25. At this price we have $S = D$, so that

$$300x_0 - 600 = 1200 - 150x_0$$
$$450x_0 = 1800$$
$$x_0 = 4.$$

As long as neither the supply nor demand function changes, the price should remain a steady \$4 per grater.

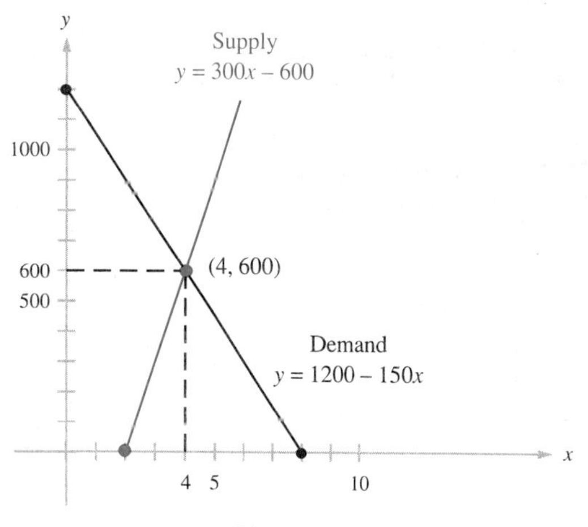

FIGURE 1.26

At the equilibrium price the supply and demand have a common value, called the **equilibrium supply,** or **equilibrium demand.** This can be found by substituting x_0 into either the supply or demand function. In our example at $x_0 = 4$ we have

$$S = 300(4) - 600 = 600.$$

Thus the equilibrium price and equilibrium demand are both 600 units per month. The geometric meaning of these values is shown in Figure 1.26. ■

EXAMPLE 1.22 A factory manufactures cigars. At a price of x dollars per box it is willing to supply S boxes per week, where

$$S = 120x - 240,$$

and the demand is D boxes per week, where

$$D = -75x + 1125.$$

What is the equilibrium price? How many boxes will be manufactured and sold at this price, and what will the total revenue to the company be?

Solution To find the equilibrium price we set $S = D$.

$$\begin{aligned} 120x - 240 &= -75x + 1125 \\ 195x &= 1365 \\ x &= \frac{1365}{195} = 7 \end{aligned}$$

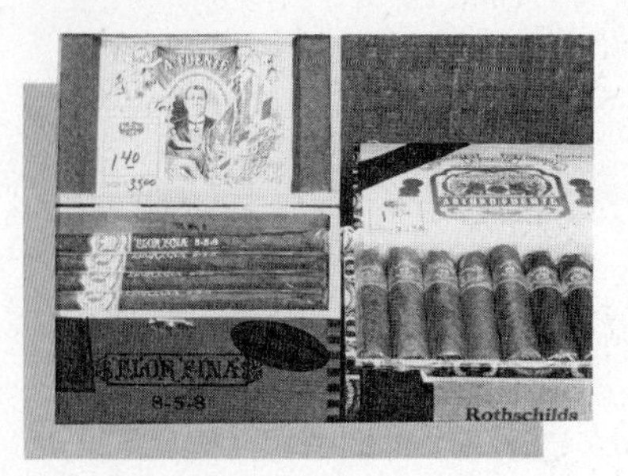

We see that the equilibrium price is \$7 per box. At this price the equilibrium supply (which equals the equilibrium demand) is

$$120(7) - 240 = 600.$$

Thus 600 boxes would be sold each week, for a total revenue of $600(7) =$ \$4200 per week. The supply and demand functions and equilibrium price are indicated in Figure 1.27. ■

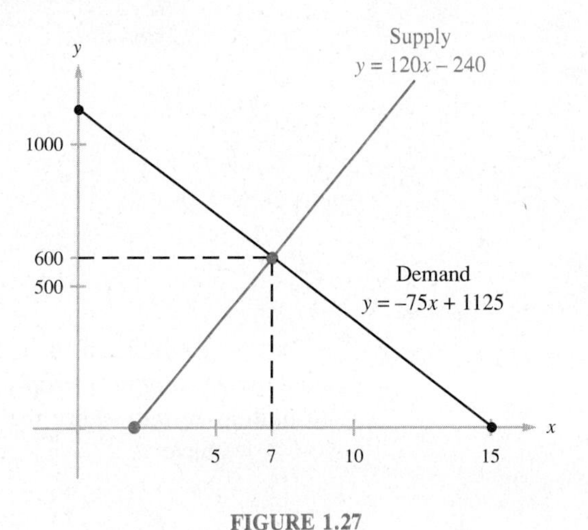

FIGURE 1.27

Practice Problem 8 The supply of a certain computer chip is $S = 10{,}000x - 7000$ at a price of x dollars, and the demand is $D = -8000x + 20{,}000$.

(a) What is the equilibrium price?

(b) How many chips will be supplied at this price?

THE EFFECT OF TAXATION

Suppose that the city containing the cigar factory in Example 1.22 decides to tax the boxes of cigars. It imposes a 20% **sales tax** on each box. Since the factory is selling 600 boxes per week at \$7 a box, the city officials expect to take in $.20(7) =$ \$1.40 tax on each box, for a total of $600(1.40) =$ \$840 per week. They have not taken into account the laws of supply and demand, however.

For simplicity, let x denote the *before-tax* price of each box. Since this is still the price the factory gets for each box, the supply function is the same:

$$S = 120x - 240.$$

The demand, on the other hand, depends on the price the consumer actually pays for each box, which includes the 20% tax. Thus the new demand function is given by

$$D_1 = -75(x + .20x) + 1125$$
$$D_1 = -90x + 1125.$$

We find the equilibrium price by setting $S = D_1$.

$$120x - 240 = -90x + 1125$$
$$210x = 1365$$
$$x = \frac{1365}{210} = 6.50$$

Because the tax has reduced demand, the factory will be able to get only $6.50 for a box of cigars, and the 20% tax will net the city only .20(6.50) = $1.30 per box instead of the expected $1.40. Notice that the equilibrium supply at $6.50 will be

$$120(6.50) - 240 = 540$$

boxes per week, so that the weekly tax revenue will be 540($1.30) = $702 per week, rather than the expected $840.

Frustrated that the tax revenue on boxes of cigars is not as great as expected, the city changes from the 20% tax to a flat tax of $1.40 per box. Such a flat tax, not depending on the price of the item sold, is called a **stamp tax.** An example of such a tax is the state gasoline tax, which is a fixed amount per gallon, independent of the selling price of gasoline.

Again letting x be the before-tax price of a box of cigars, we still have the supply function

$$S = 120x - 240.$$

Now, however, a consumer is paying $x + 1.40$ per box of cigars, so the demand is

$$D_2 = -75(x + 1.40) + 1125,$$

or

$$D_2 = -75x + 1020.$$

Setting $S = D_2$ gives

$$120x - 240 = -75x + 1020$$
$$195x = 1260$$
$$x = \frac{1260}{195} \approx 6.46.$$

Now the factory gets about \$6.46 per box, and will produce

$$120\left(\frac{1260}{195}\right) - 240 \approx 535$$

boxes per week. Since production is reduced, the city will still not get the \$840 per week in taxes it originally expected. The three demand functions D, D_1, and D_2 are shown in Figure 1.28.

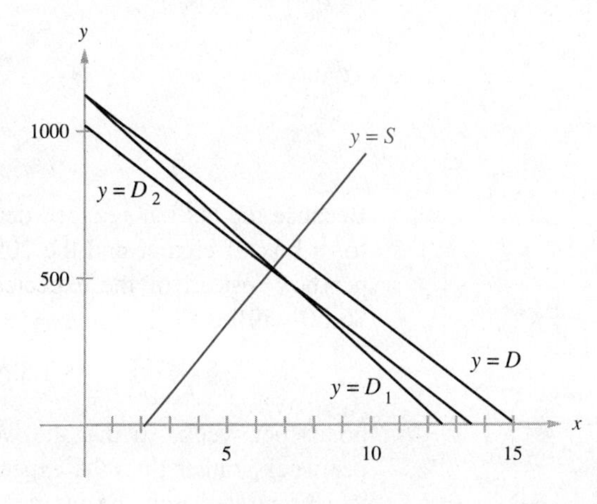

FIGURE 1.28

Practice Problem 9 Example 1.21 considered the production of graters, for which there was a demand

$$D = 1200 - 150x,$$

where x is the price in dollars. Find the demand functions D_1, when a 10% sales tax is imposed, and D_2, when a 20¢ stamp tax is imposed.

EXERCISES 1.5

In Exercises 1–8 find the equilibrium price and demand (supply) for the following demand and supply functions D(x) and S(x), where x is the price in dollars.

1. $D(x) = 4500 - 500x, \quad S(x) = 250x$

2. $D(x) = 7000 - 200x, \quad S(x) = 300x$

3. $D(x) = 1200 - 60x, \quad S(x) = 100x$

4. $D(x) = 9900 - 400x, \quad S(x) = 800x$

5. $D(x) = 1500 - 100x, \quad S(x) = 400x - 2000, \quad \text{for } x \geq 6$

6. $D(x) = 15{,}300 - 450x, \quad S(x) = 300x - 2700, \quad \text{for } x \geq 12$

7. $D(x) = 5040 - 280x, \quad S(x) = 220x - 1660, \quad \text{for } x \geq 8$

8. $D(x) = 19{,}000 - 7600x, \quad S(x) = 2400x - 1800, \quad \text{for } x \geq 1$

In Exercises 9–14 determine how the equilibrium price and demand (supply) change in the given exercise if the government imposes a sales tax as indicated. Find the total tax collected.

9. 10% tax in Exercise 1

10. 20% tax in Exercise 2

11. 25% tax in Exercise 5

12. 15% tax in Exercise 6

13. 30% tax in Exercise 7

14. 8% tax in Exercise 8

In Exercises 15–20 determine how the equilibrium price and demand (supply) change in the given exercise if the government imposes a stamp tax as indicated. Find the total tax collected.

15. \$0.75 in Exercise 1

16. \$1 in Exercise 2

17. \$0.50 in Exercise 5

18. \$2 in Exercise 6

19. \$0.90 in Exercise 7

20. \$0.20 in Exercise 8

21. A manufacturer has found that the demand for one of his products is 1600 units when the price per unit is \$8 and is 1000 units when the price per unit is \$10. If the manufacturer is willing to supply $S(x) = 200x - 1200$ units at any price x exceeding \$7 per unit, determine the equilibrium price and supply if demand is a linear function of x.

22. The manager of a popcorn stand is willing to supply $100x - 2000$ bags of popcorn per week at a price of x cents per bag. She sells 2000 bags per week if the price is 75¢ and 1400 bags per week if the price is 90¢. If the demand is a linear function of price, find the equilibrium price and supply.

Raising the price of popcorn from 75¢ to 90¢ decreases the weekly sales from 2000 bags to 1400 bags.

23. A part-time barber is willing to give 578 haircuts per year at \$8 each, but can sell only 364 at that price. He is willing to give 155 haircuts at \$5 each, and can sell 583 at that price. If the supply and demand are both linear functions of price, find the equilibrium price and supply.

24. A manufacturer is willing to supply 800 units of a product at a price per unit of \$6 and 1200 units at a price per unit of \$7. The manufacturer has found that the demand for the product is 7100 units at a price of \$6 per unit and 6800 units at a price of \$7 per unit. If the supply and demand functions are both linear, determine the equilibrium price and demand.

25. Suppose in Example 1.22 the city pays the cigar factory a subsidy of \$1 per box of cigars it makes and sells. What is the new equilibrium price and demand?

26. In Example 1.22 what flat tax per box would give the city its \$840 per week revenue?

Answers to Practice Problems

8. (a) \$1.50 (b) 8000

9. $D_1 = 1200 - 165x$

$D_2 = 1170 - 150x$

CHAPTER 1 REVIEW

IMPORTANT TERMS

- **coordinate axes** *(1.1)*
- **graph**
- **linear equation**
- **ordered pair**
- **origin**
- **real number line**
- **x-axis, y-axis**
- **x-intercept, y-intercept**
- **dependent variable** *(1.2)*
- **independent variable**
- **linear function**
- **marginal cost**
- **point-slope form**
- **slope**
- **slope-intercept form**
- **dependent system** *(1.3)*
- **elimination method**
- **equivalent system**
- **inconsistent system**
- **substitution method**
- **system of equations**
- **break-even point** *(1.4)*
- **cost and revenue lines**
- **fixed, variable, and total cost**
- **profit, profit line**
- **revenue**
- **demand** *(1.5)*
- **equilibrium price**
- **equilibrium supply, demand**
- **sales tax**
- **stamp tax**
- **supply**

IMPORTANT FORMULAS

Slope-intercept form

The equation of the line with slope m and y-intercept $(0, b)$ is $y = mx + b$.

Slope of a line through two points

The slope of the line through the points (x_1, y_1) and (x_2, y_2) is

$$m = \frac{y_2 - y_1}{x_2 - x_1}, \quad \text{if } x_2 \neq x_1.$$

Point-slope form

The equation of the line through the point (x_1, y_1) with slope m is $y - y_1 = m(x - x_1)$.

Profit

If P = profit, R = revenue, and C = cost, then $P = R - C$.

REVIEW EXERCISES

In Exercises 1–6 graph the given statement.

1. $x \geq 0$ and $y \leq 0$

2. $x \geq 1$ and $y \leq 3$

3. $3x + 5y = 600$

4. $x = -2$

5. $2x + 5y = 0$

6. $x + y = 3$ and $x \geq 1$

7. Give the slope, x-intercept, and y-intercept of $3x + 1 = 3 - 2y$.

8. Give the equation of the line through $(2, -3)$ and $(1, 5)$ in slope-intercept form.

9. The number of ice cream cones sold in a day at the Dairy Queen is a linear function of the maximum temperature on that day. If the maximum temperature is 80° F, 55 cones are sold; and if the maximum temperature is 100° F, 95 cones are sold. How many cones are sold when the maximum temperature is 65° F?

10. Solve the system

$$\begin{aligned} 2x + y &= 2 \\ 6x + 4y &= 3. \end{aligned}$$

11. Solve the system

$$\begin{aligned} 2a - b &= 0 \\ 12a + 4b &= 2. \end{aligned}$$

12. In planning a birthday party for 11 children it is decided that each child should have 9 balloons and 4 noisemakers. A Fun-Pac contains 7 balloons and 3 noisemakers, while an Economy Box contains 15 balloons and 8 noisemakers. How many Fun-Pacs and Economy Boxes should be bought to supply exactly the right number of balloons and noisemakers?

13. An investor has $10,000 to invest and wants an 8% return. Investment A pays an 11% return, while investment B, which is safer, pays a 6% return. How should the $10,000 be divided between investments A and B?

14. A company produces toasters which it sells for $25. There are fixed costs of $30,000 each month, and variable costs of $13 per toaster.

(a) Plot the cost and revenue lines.
(b) How many toasters must be made and sold to break even?

How many toasters must be produced and sold monthly

(c) to make a profit of $12,000?
(d) to make a profit of $4 on each toaster?
(e) to make a profit of 24% on sales?

15. A singing telegram business has fixed office expenses of $400 per week, and each telegram delivered costs an average of $7.50 for labor and transportation. It charges $20 for a singing telegram.

(a) Graph the cost and revenue lines.
(b) How many telegrams must it deliver per week to break even?

How many telegrams must it deliver per week

(c) to make a profit of $175?
(d) to make a profit of $2.50 on each telegram?
(e) to make a profit of 22.5% on sales?

16. A photographer is willing to supply $280x - 500$ cheap passport photographs per year at a price of x dollars each. The public will buy $1000 - 120x$ such photographs at x dollars. What is the equilibrium price and equilibrium demand?

17. Work the previous exercise if there is a 10% tax on passport pictures.

18. Work Exercise 16 if there is a stamp tax of 50¢ per passport picture.

2

SYSTEMS OF LINEAR EQUATIONS AND MATRICES

In this chapter we will generalize the methods of Chapter 1 to solve systems of more than two linear equations. It turns out that many applications entail a large number of variables subject to linear constraints. Analyzing such a system can be simplified by organizing and manipulating the numbers involved in a table, called a *matrix* or *tableau*.

2.1 Gauss-Jordan Elimination

In Chapter 1 we presented two different methods for solving a system of two linear equations in two unknowns, substitution and elimination. While the substitution technique can be applied to larger systems, it becomes increasingly cumbersome. The elimination idea, however, can be generalized to give a systematic procedure known as the *Gauss-Jordan elimination* method. In this section this method is used to solve a system of n linear equations in n unknowns.

LINEAR SYSTEMS AND ELEMENTARY OPERATIONS

An equation of the form

$$a_1x_1 + a_2x_2 + \dots + a_nx_n = c,$$

where $a_1, a_2, \dots, a_n$, and c are constants and at least one of the coefficients a_i is nonzero, is called a **linear equation** in the unknowns $x_1, x_2, \dots, x_n$. A system of m linear equations in n unknowns is called an ***m* by *n* linear system.** For instance, the systems

$$\begin{aligned} 2x + 3y &= 9 \\ 3x - 5y &= 7 \end{aligned} \qquad \begin{aligned} x + 2y - z &= 4 \\ 3x - y + 5z &= 6 \\ 2x + y - 3z &= 1 \end{aligned} \qquad \begin{aligned} 5x + y + z &= 8 \\ -x + 2y + 2z &= 5 \end{aligned}$$

are 2 by 2, 3 by 3, and 2 by 3, respectively.

When a system has the same number of equations as unknowns (when $m = n$), it will usually have a unique solution $(x_1, x_2, \dots, x_n)$. For the case $n = 2$ this is reflected in the fact that two lines usually intersect in a unique point. We now take up the general problem of finding the solution of an n by n linear system when it exists.

One approach to solving a system of linear equations is to transform it into a simpler algebraically equivalent system, that is, a simpler system having exactly the same solutions. The types of transformations we shall need, called **elementary operations,** are the following:

Type 1 An equation can be multiplied by a nonzero number.

Type 2 A multiple of one equation can be added to another.

Type 3 Two equations can be interchanged.

EXAMPLE 2.1 Solve the following 3 by 3 system:

$$\begin{aligned} 2x + 4y + 2z &= 6 \\ -2x + y + 3z &= 4 \\ 3x + 2y + z &= 5. \end{aligned} \tag{2.1}$$

Discussion The first task is to eliminate the unknown x from all but the first equation in the system. We multiply the first equation by 1/2 in order to change the coefficient of x to $+1$, so that this variable can be eliminated from the other equations. This Type 1 operation transforms (2.1) into the following system:

$$\begin{aligned} x + 2y + z &= 3 \\ -2x + y + 3z &= 4 \\ 3x + 2y + z &= 5. \end{aligned} \tag{2.2}$$

Note that the x-term in the first equation now has coefficient $+1$. Next, we add 2 times the (new) first equation to the second, in order to eliminate the x-term in the second equation. The scratch work for this Type 2 operation follows.

$$\begin{array}{rrcl} & -2x + y + 3z &=& 4 \quad \text{(second equation)} \\ (+) & \underline{2x + 4y + 2z} &=& \underline{6} \quad \text{(2 times first equation)} \\ & 5y + 5z &=& 10 \quad \text{(sum)} \end{array}$$

This transforms (2.2) into the equivalent system (2.3).

$$\begin{aligned} x + 2y + z &= 3 \\ 5y + 5z &= 10 \\ 3x + 2y + z &= 5 \end{aligned} \tag{2.3}$$

Remark Notice that it is the second equation, not the first, which is changed by this operation. We merely *use* the first equation to eliminate x from the second equation.

Working now with the system (2.3), we add -3 times the first equation to the third to eliminate the x-term in the third equation, yielding the system:

$$\begin{aligned} x + 2y + z &= 3 \\ 5y + 5z &= 10 \\ -4y - 2z &= -4. \end{aligned} \tag{2.4}$$

We have achieved our first objective: the unknown x has now been isolated so that it occurs in one and only one equation of the system, and has coefficient $+1$ in that equation. Our next goal is to isolate y in one of the remaining equations; we will use the second. To begin, we simplify the y-coefficient by multiplying the second equation by 1/5.

$$\begin{aligned} x + 2y + z &= 3 \\ y + z &= 2 \\ -4y - 2z &= -4 \end{aligned} \tag{2.5}$$

Next, we use the second equation to eliminate the y-terms in the other two equations. Specifically, we add -2 times the second equation to the first and add 4 times the second equation to the third. These Type 2 operations transform (2.5) into the following system in which both x and y have been isolated.

$$\begin{aligned} x - z &= -1 \\ y + z &= 2 \\ 2z &= 4 \end{aligned} \tag{2.6}$$

To isolate z, multiply the third equation by 1/2 to make the z-coefficient 1.

$$\begin{aligned} x - z &= -1 \\ y + z &= 2 \\ z &= 2 \end{aligned} \tag{2.7}$$

Then add (1 times) the third equation to the first and add -1 times the third equation to the second in (2.7). The result is the system

$$\begin{aligned} x &= 1 \\ y &= 0 \\ z &= 2, \end{aligned} \tag{2.8}$$

which announces its own solution, the ordered triple (1, 0, 2). Since all the systems (2.1) through (2.8) are algebraically equivalent, we conclude that (1, 0, 2) is the unique solution to the original system. This can be checked by direct substitution in (2.1):

$$\begin{aligned} 2(1) + 4(0) + 2(2) &= 6 \\ -2(1) + (0) + 3(2) &= 4 \\ 3(1) + 2(0) + (2) &= 5. \end{aligned}$$ ■

TABLEAU FORM

The elimination procedure can be streamlined by writing the given system in **tableau form,** as follows:

Linear system	***Tableau form***
$\begin{aligned} 2x + 4y + 2z &= 6 \\ -2x + y + 3z &= 4 \\ 3x + 2y + z &= 5 \end{aligned}$	$\left[\begin{array}{rrr\|r} 2 & 4 & 2 & 6 \\ -2 & 1 & 3 & 4 \\ 3 & 2 & 1 & 5 \end{array}\right]$

This is merely a shorthand device which enables us to avoid rewriting the variables x, y, z and the equals sign after each transformation. The coefficients of the variables are listed to the left of the vertical line. The first column corresponds to x, the second to y, and the third to z. The three rows of the tableau correspond to the three equations of the system.

Our earlier operations on equations correspond to **elementary row operations,** defined as follows:

Elementary Row Operations

Type 1 A row can be multiplied by a nonzero number.
Type 2 A multiple of one row can be added to another.
Type 3 Two rows can be interchanged.

EXAMPLE 2.1 Revisited

Solve the linear system in Example 2.1 using the tableau form.

Solution The sequence of transformations (2.1)-(2.8) in Example 2.1 can be carried out in tableau form, as shown below. (In the annotation to the right R_1 stands for the old first row, etc.) The first tableau represents the original system (2.1).

$$\left[\begin{array}{rrr|r} 2 & 4 & 2 & 6 \\ -2 & 1 & 3 & 4 \\ 3 & 2 & 1 & 5 \end{array}\right]$$

Multiply the first row by 1/2; this is indicated by the notation $\frac{1}{2}R_1$ to the right of this row.

$$\left[\begin{array}{rrr|r} 1 & 2 & 1 & 3 \\ -2 & 1 & 3 & 4 \\ 3 & 2 & 1 & 5 \end{array}\right] \qquad (\tfrac{1}{2}R_1)$$

Add 2 times the first row to the second row. (We indicate this with the notation $R_2 + 2R_1$ to the right of row 2. Note that the row that is changed is written first.)

$$\left[\begin{array}{rrr|r} 1 & 2 & 1 & 3 \\ 0 & 5 & 5 & 10 \\ 3 & 2 & 1 & 5 \end{array}\right] \qquad (R_2 + 2R_1)$$

Add -3 times the first row to the third row.

$$\left[\begin{array}{rrr|r} 1 & 2 & 1 & 3 \\ 0 & 5 & 5 & 10 \\ 0 & -4 & -2 & -4 \end{array}\right] \qquad (R_3 + (-3)R_1)$$

Multiply the second row by 1/5.

$$\left[\begin{array}{rrr|r} 1 & 2 & 1 & 3 \\ 0 & 1 & 1 & 2 \\ 0 & -4 & -2 & -4 \end{array}\right] \qquad (\tfrac{1}{5}R_2)$$

Add -2 times the second row to the first row, and 4 times the second row to the third row.

$$\left[\begin{array}{ccc|c} 1 & 0 & -1 & -1 \\ 0 & 1 & 1 & 2 \\ 0 & 0 & 2 & 4 \end{array}\right] \quad \begin{array}{l} (R_1 + (-2)R_2) \\ \\ (R_3 + 4R_2) \end{array}$$

Multiply the third row by 1/2.

$$\left[\begin{array}{ccc|c} 1 & 0 & -1 & -1 \\ 0 & 1 & 1 & 2 \\ 0 & 0 & 1 & 2 \end{array}\right] \quad (\tfrac{1}{2}R_3)$$

Add the third row to the first row, and add -1 times the third row to the second row.

$$\left[\begin{array}{ccc|c} 1 & 0 & 0 & 1 \\ 0 & 1 & 0 & 0 \\ 0 & 0 & 1 & 2 \end{array}\right] \quad \begin{array}{l} (R_1 + R_3) \\ (R_2 + (-1)R_3) \\ \end{array}$$

This final tableau corresponds to the linear system (2.8). The solution is therefore $x = 1$, $y = 0$, $z = 2$. ■

THE PIVOT TRANSFORMATION

When a linear system is written in tableau form, the process of isolating its unknowns can be carried out as follows: In each column but the last, we look for a nonzero entry a. Using row operations, we change this entry to $+1$ and change the other entries in the column to 0. This combination of row operations is called a **pivot transformation** on the entry a. If we call the row and column of a the **pivot row** and **pivot column,** respectively, we can describe this transformation as follows:

Pivot Transformation on a Nonzero Entry *a*

Step 1 Multiply the pivot row by $1/a$.

Step 2 For every other nonzero entry b in the pivot column, add $-b$ times the (new) pivot row to the row in which it occurs.

The pivot transformation is the heart of the Gauss-Jordan elimination method. Before going further, we illustrate it with an example.

EXAMPLE 2.2 Perform a pivot transformation on the entry -2 (circled) in the following tableau:

$$\left[\begin{array}{ccc|c} 3 & 1 & -1 & 3 \\ \textcircled{-2} & 4 & 2 & 10 \\ 4 & 1 & 0 & 6 \end{array}\right].$$

Solution Since the pivot entry is in the second row and first column, we designate row 2 as the pivot row and column 1 as the pivot column.

Step 1 The pivot entry is not equal to 1. Hence, we multiply the second row by $-1/2$.

$$\left[\begin{array}{ccc|c} 3 & 1 & -1 & 3 \\ \textcircled{1} & -2 & -1 & -5 \\ 4 & 1 & 0 & 6 \end{array}\right] \qquad (-\tfrac{1}{2}R_2)$$

Notice that this changes the pivot entry to $+1$.

Step 2 There are two other nonzero entries in the pivot column, namely 3 (first row) and 4 (third row). Hence, we add -3 times the second row to the first row, and add -4 times the second row to the third row.

$$\left[\begin{array}{ccc|c} 0 & 7 & 2 & 18 \\ 1 & -2 & -1 & -5 \\ 0 & 9 & 4 & 26 \end{array}\right] \qquad \begin{array}{l} (R_1 + (-3)R_2) \\ \\ (R_3 + (-4)R_2) \end{array}$$

This completes the pivot. Note that the pivot entry is now $+1$ and all other entries in the pivot column are zero. This means that the first unknown is now isolated (with coefficient 1) in the second equation of the system. ■

Practice Problem 1 Perform the pivot transformation using the entry circled below.

$$\left[\begin{array}{ccc|c} 1 & 2 & 5 & 0 \\ 0 & \textcircled{3} & -6 & 9 \\ 0 & -1 & -1 & 2 \end{array}\right]$$

Gauss-Jordan Elimination When using Gauss-Jordan elimination to solve an n by n linear system with a unique solution, we perform a sequence of pivots until all the unknowns have been isolated. The procedure is as follows:

Gauss-Jordan Elimination

To solve an n by n system:

1. Write the system in tableau form.
2. Perform a sequence of pivots on nonzero entries, using a different row and column for each pivot.
3. If the system has a unique solution, this process will terminate after at most n pivots. The system is then solved.

Note that the pivot entries can be chosen arbitrarily, subject only to the condition that we use a new row and new column each time. Whenever possible it is desirable to use 1 as a pivot entry, since Step 1 of the pivot transformation then becomes unnecessary.

EXAMPLE 2.3 Solve the linear system

$$\begin{aligned} 2y + 3z &= 8 \\ x + 2y + z &= 6 \\ 3x + 2y &= 8. \end{aligned}$$

Solution This 3 by 3 system has the following tableau form:

$$\left[\begin{array}{ccc|c} 0 & 2 & 3 & 8 \\ \textcircled{1} & 2 & 1 & 6 \\ 3 & 2 & 0 & 8 \end{array}\right].$$

For the first pivot, we choose the entry 1 (circled) in the first column. Step 1, transforming the pivot entry to 1, is unnecessary. We proceed to Step 2, transforming the other nonzero entries in the pivot column to zero. Add -3 times the second row to the third row.

$$\left[\begin{array}{ccc|c} 0 & \textcircled{2} & 3 & 8 \\ 1 & 2 & 1 & 6 \\ 0 & -4 & -3 & -10 \end{array}\right] \qquad (\mathbf{R}_3 + (-3)\mathbf{R}_2)$$

For the second pivot we must use an entry in a different row and column. For the entry 2 (circled), Step 1 of the pivot is to multiply the first row by 1/2.

$$\left[\begin{array}{ccc|c} 0 & \textcircled{1} & \frac{3}{2} & 4 \\ 1 & 2 & 1 & 6 \\ 0 & -4 & -3 & -10 \end{array}\right] \qquad (\tfrac{1}{2}\mathbf{R}_1)$$

Step 2 needs two operations. Add -2 times the first row to the second, and add 4 times the first row to the third.

$$\left[\begin{array}{ccc|c} 0 & 1 & \frac{3}{2} & 4 \\ 1 & 0 & -2 & -2 \\ 0 & 0 & \textcircled{3} & 6 \end{array}\right] \qquad \begin{array}{l} (\mathbf{R}_2 + (-2)\mathbf{R}_1) \\ (\mathbf{R}_3 + 4\mathbf{R}_1) \end{array}$$

For the last pivot we *must* use the circled entry 3. (Why?) For Step 1, multiply the third row by 1/3.

$$\left[\begin{array}{ccc|c} 0 & 1 & \frac{3}{2} & 4 \\ 1 & 0 & -2 & -2 \\ 0 & 0 & \textcircled{1} & 2 \end{array}\right] \qquad (\tfrac{1}{3}\mathbf{R}_3)$$

For Step 2, add $-3/2$ times the third row to the first, and add 2 times the third row to the second.

$$\left[\begin{array}{ccc|c} 0 & 1 & 0 & 1 \\ 1 & 0 & 0 & 2 \\ 0 & 0 & 1 & 2 \end{array}\right] \qquad \begin{array}{l} (\mathbf{R}_1 + (-\frac{3}{2})\,\mathbf{R}_3) \\ (\mathbf{R}_2 + 2\,\mathbf{R}_3) \end{array}$$

The system is now solved. However, in order to make interpreting the solution easier we *interchange the first and second rows,* so the pivot entries lie along the diagonal from top-left to bottom-right.

$$\left[\begin{array}{ccc|c} 1 & 0 & 0 & 2 \\ 0 & 1 & 0 & 1 \\ 0 & 0 & 1 & 2 \end{array}\right] \quad \begin{array}{l} (\mathbf{R_2}) \\ (\mathbf{R_1}) \\ \\ \end{array}$$

The tableau corresponds to the linear system

$$\begin{aligned} x \quad &= 2 \\ y \quad &= 1 \\ z &= 2. \end{aligned}$$

The reader should check this result by direct substitution in the original equations. ■

Remark The Gauss-Jordan elimination procedure—and the pivot transformation on which it is based—will be needed for several later topics. For this reason *you should practice the technique until you have thoroughly mastered it.* Here are some suggestions.

1. Annotations are the key to success. Before performing a row operation, *write down its annotation.*
2. Don't try to combine Steps 1 and 2 of the pivot transformation. As in the examples above, *first* do Step 1, *then* do all Step 2 operations.
3. Be especially wary of mistakes involving signs (+ or −) and fractions. Bear in mind that one careless error throws the entire procedure off.
4. Always check your final answer by substitution in the original equations.

Practice Problem 2 Solve the linear system

$$\begin{aligned} 3x \qquad + z &= 0 \\ x + y + z &= 4 \\ 3y - z &= 3. \end{aligned}$$

EXAMPLE 2.4 A piggy bank contains 90 coins, each of which is a penny, nickel, or dime. If the total value of the coins is \$3.30 and the value of the nickels equals the value of the other two coins, how many coins are there of each type?

Solution Let there be x pennies, y nickels, and z dimes. Then since there are 90 coins with total value \$3.30, we have

$$\begin{aligned} x + y + z &= 90 \\ x + 5y + 10z &= 330. \end{aligned}$$

The 90 coins shown here are worth $3.30.

Because the value of the nickels equals that of the other two coins, we have $5y = x + 10z$, or

$$-x + 5y - 10z = 0.$$

Thus we have the tableau

$$\left[\begin{array}{rrr|r} \textcircled{1} & 1 & 1 & 90 \\ 1 & 5 & 10 & 330 \\ -1 & 5 & -10 & 0 \end{array}\right].$$

The computations that follow are self-explanatory.

$$\left[\begin{array}{rrr|r} 1 & 1 & 1 & 90 \\ 0 & \textcircled{4} & 9 & 240 \\ 0 & 6 & -9 & 90 \end{array}\right] \quad \begin{array}{l} \\ (R_2 + (-1)R_1) \\ (R_3 + R_1) \end{array}$$

$$\left[\begin{array}{rrr|r} 1 & 1 & 1 & 90 \\ 0 & \textcircled{1} & \frac{9}{4} & 60 \\ 0 & 6 & -9 & 90 \end{array}\right] \quad (\tfrac{1}{4}R_2)$$

$$\left[\begin{array}{rrr|r} 1 & 0 & -\frac{5}{4} & 30 \\ 0 & 1 & \frac{9}{4} & 60 \\ 0 & 0 & -\frac{45}{2} & -270 \end{array}\right] \quad \begin{array}{l} (R_1 + (-1)R_2) \\ \\ (R_3 + (-6)R_2) \end{array}$$

$$\left[\begin{array}{rrr|r} 1 & 0 & -\frac{5}{4} & 30 \\ 0 & 1 & \frac{9}{4} & 60 \\ 0 & 0 & \textcircled{1} & 12 \end{array}\right] \quad \begin{array}{l} \\ \\ (-\tfrac{2}{45}R_3) \end{array}$$

$$\left[\begin{array}{rrr|r} 1 & 0 & 0 & 45 \\ 0 & 1 & 0 & 33 \\ 0 & 0 & 1 & 12 \end{array}\right] \quad \begin{array}{l} (R_1 + (\tfrac{5}{4})R_3) \\ (R_2 + (-\tfrac{9}{4})R_3) \\ \\ \end{array}$$

Since the last tableau corresponds to the system

$$\begin{aligned} x \qquad &= 45 \\ y \quad &= 33 \\ z &= 12, \end{aligned}$$

we conclude that 45 pennies, 33 nickels, and 12 dimes are in the bank. This checks, since

$$\begin{aligned} 45 + 33 + 12 &= 90 \\ 45 + 5\cdot 33 + 10\cdot 12 &= 330 \\ -45 + 5\cdot 33 - 10\cdot 12 &= 0. \end{aligned}$$ ■

Practice Problem 3 A toaster weighs 6 pounds and costs $20, a blender weighs 5 pounds and costs $15, and a mixer weighs 3 pounds and costs $15. If a shipment of 12 of these small appliances weighs 57 pounds and costs $205, how many of each appliance does it contain?

EXERCISES 2.1

Exercises 1–12 refer to the following three tableaus:

$$A = \left[\begin{array}{cc|c} 2 & 3 & 12 \\ 4 & -1 & 3 \end{array}\right] \qquad B = \left[\begin{array}{ccc|c} 1 & -2 & \frac{1}{2} & 4 \\ 0 & \frac{1}{2} & -1 & -5 \\ 2 & 0 & -\frac{2}{3} & 6 \end{array}\right] \qquad C = \left[\begin{array}{cccc|c} 2 & -1 & 1 & 0 & 6 \\ 0 & \frac{1}{2} & \frac{5}{4} & -1 & 2 \\ \frac{3}{4} & -1 & 1 & 2 & 2 \\ -1 & 0 & 1 & 1 & -5 \end{array}\right]$$

In Exercises 1–6 perform the indicated row operation on the given tableau.

1. $\frac{1}{4}R_2$, on R_2 of A

2. $R_2 + \left(-\frac{1}{2}R_1\right)$, on R_2 of A

3. $R_3 + (-2)R_1$, on R_3 of B

4. $R_1 + (-1)R_3$, on R_1 of B

5. $2R_2$, on R_2 of C

6. $R_4 + 4R_2$, on R_4 of C

In Exercises 7–12 perform a pivot on the indicated entry in the given tableau.

7. pivot on -1 in row 2, column 2 of A

8. pivot on 3 in row 1, column 2 of A

9. pivot on 1/2 in row 1, column 3 of B

10. pivot on -2 in row 1, column 2 of B

11. pivot on 2 in row 3, column 4 of C

12. pivot on 3/4 in row 3, column 1 of C

In Exercises 13–22 solve the linear system by Gauss-Jordan elimination. Annotate each row operation and indicate pivot entries with a circle.

13. $\begin{aligned} x + 2y &= 18 \\ 5x + 4y &= 60 \end{aligned}$

14. $\begin{aligned} 3x + y &= 7 \\ -2x + 4y &= 7 \end{aligned}$

15. $\begin{aligned} .5x + y + .5z &= 4 \\ x + 2y - .5z &= 2 \\ -.5x + z &= 3 \end{aligned}$

16. $\begin{aligned} x + y &= 1 \\ x + z &= 2 \\ y + z &= 3 \end{aligned}$

17. $\begin{aligned} x + 2y - z &= 2 \\ -x + y + 2z &= 1 \\ x + y - z &= 1 \end{aligned}$

18. $\begin{aligned} x + 2y + 3z &= 3 \\ 3x + y + 2z &= 7 \\ 2x + 3y + z &= 2 \end{aligned}$

19. $\begin{aligned} x + y + z &= 6 \\ x - y + z &= 0 \\ x + y - z &= 2 \end{aligned}$

20. $\begin{aligned} x + 2y + z &= 1 \\ 2x + 5y + z &= 2 \\ y + 2z &= 3 \end{aligned}$

21. $\begin{aligned} x + y + z &= 0 \\ x + y + w &= 1 \\ x + z + w &= 2 \\ y + z + w &= 3 \end{aligned}$

22. $\begin{aligned} -x + y + z + w &= 20 \\ x - y + z + w &= 40 \\ x + y - z + w &= 60 \\ x + y + z - w &= 80 \end{aligned}$

23. Three herbicides are available that contain the chemicals A, B, and C. One gallon of the first herbicide contains 1 unit of chemical A, 1 unit of chemical C, and none of chemical B; one gallon of the second herbicide contains none of chemical A, 2 units of chemical B, and 1 unit of chemical C; and one gallon of the third herbicide contains 1 unit of chemical A, 1 unit of chemical B, and none of chemical C. To control a certain weed, a farmer wants to apply 11 units of chemical A, 12 units of chemical B, and 14 units of chemical C per acre. How many gallons of each herbicide should be used per acre?

24. An investor asked her investment counselor to invest \$18,000 for her in a mutual fund, in stocks, and in a speculative land development. She asked only that the counselor invest 5 times as much in the mutual fund and stocks combined as in land. One year later she learned that her \$18,000 investment had increased \$580 in value, with the mutual fund yielding a 5% return, the stocks yielding a 3% return, and the land losing 1%. How much money had been invested in each way?

25. A caterer is to supply slices of apple pie, at \$1 each, strawberry shortcakes, at \$2 each, and baked Alaska, at \$3 each. There must be a total of 70 slices of pie and baked Alaska, and 40 more shortcakes than pie slices. The total cost of the two more expensive desserts is to be \$240. How can this be done?

26. A 1-pound bag of grass seed costs \$2, a 2-pound bag costs \$3, and a 4-pound bag costs \$4. If 35 bags cost a total of \$103 and weigh 73 pounds, how many of each type are included?

27. In a certain city there are two high schools. Eastern High School has 1400 white students and 1200 black students, and Western High School has 1600 white students and 800 black students. In order to achieve a racial balance in the two schools some of the students attending Eastern are going to be bused to Western, and some of those now attending Western will be bused to Eastern. The city's school board has suggested the following guidelines:

(i) The total number of students to be bused should be 20% of the total population of the two schools.

(ii) The total number of white students who are bused should equal the total number of black students who are bused.

(iii) The proportion of white to black students who are bused from Western to Eastern should equal the present proportion of white students to black students at Western.

(iv) After the busing plan is implemented the proportion of black students to white students should be the same at both schools.

Determine the number of black students and the number of white students now attending each school who should be bused to the other school.

28. A 200-acre farm is used to grow corn and soybeans. For the coming year the owner estimates yields of 100 bushels per acre for corn and 50 bushels per acre for soybeans and selling prices of \$2.30 per bushel for corn and \$5.00 per bushel for soybeans. The owner wishes to use all of his 550 bags of available fertilizer on the land where the corn and soybeans are planted and intends to apply 4 bags of fertilizer per acre of corn and 3 bags of fertilizer per acre of soybeans. In addition, some land may be left fallow, in which case a federal price support program will pay \$85 per acre to the owner. To receive an equitable return on his investment for land and equipment while remaining in a desirable tax bracket, the owner would like a gross return of \$42,000 for his farm. How many acres should be planted with corn, how many acres should be planted with soybeans, and how many acres should lie fallow to meet these conditions?

29. If a tableau T is transformed into T' by a Type 1 row operation replacing R_i with aR_i ($a \neq 0$), what row operation will transform T' back into T?

30. If a tableau T is transformed into T' by a Type 2 row operation replacing R_i with $R_i + aR_j$, what row operation will transform T' back into T?

31. Show that the effect of a Type 3 operation interchanging R_i and R_j can be achieved by a sequence of four operations of Types 1 and 2.

Answers to Practice Problems

1. $\left[\begin{array}{ccc|c} 1 & 0 & 9 & -6 \\ 0 & 1 & -2 & 3 \\ 0 & 0 & -3 & 5 \end{array}\right]$

2. $x = -1, y = 2, z = 3$

3. 5 toasters, 3 blenders, and 4 mixers

2.2 Systems without Unique Solutions

It often happens that a system of linear equations fails to have a unique solution. For an n by n system this is somewhat exceptional, but when the system contains fewer equations than unknowns it is always the case. In this section we consider the general m by n linear system, and show that in all cases—unique solution or not—the Gauss-Jordan elimination procedure tells us everything we might want to know about the system.

INCONSISTENT SYSTEMS

Beneath a healthy-looking system of linear equations there may lurk several types of pathology, as we see in the following examples.

EXAMPLE 2.5 Solve the following system:

$$\begin{aligned} x - y + z &= 1 \\ x + 2y - z &= 4 \\ 3x + 3y - z &= 6. \end{aligned}$$

Solution We write the system in tableau form and proceed by Gauss-Jordan elimination (pivot entries are circled).

$$\left[\begin{array}{ccc|c} \textcircled{1} & -1 & 1 & 1 \\ 1 & 2 & -1 & 4 \\ 3 & 3 & -1 & 6 \end{array}\right]$$

We add -1 times the first row to the second row, and -3 times the first row to the third row.

$$\left[\begin{array}{ccc|c} 1 & -1 & 1 & 1 \\ 0 & \textcircled{3} & -2 & 3 \\ 0 & 6 & -4 & 3 \end{array}\right] \quad \begin{array}{l} \\ (R_2 + (-1)R_1) \\ (R_3 + (-3)R_1) \end{array}$$

Now we multiply the second row by 1/3.

$$\left[\begin{array}{ccc|c} 1 & -1 & 1 & 1 \\ 0 & \textcircled{1} & -\frac{2}{3} & 1 \\ 0 & 6 & -4 & 3 \end{array}\right] \quad (\tfrac{1}{3}R_2)$$

We add the second row to the first row, and add -6 times the second row to the third row.

$$\left[\begin{array}{ccc|c} 1 & 0 & \frac{1}{3} & 2 \\ 0 & 1 & -\frac{2}{3} & 1 \\ 0 & 0 & 0 & -3 \end{array}\right] \qquad \begin{array}{l} (\mathbf{R}_1 + \mathbf{R}_2) \\ \\ (\mathbf{R}_3 + (-6)\mathbf{R}_2) \end{array}$$

For the final pivot, in the third column, we are now forced to use an entry in the third row, since the first and second rows have already been used as pivot rows. But *zero cannot be used as a pivot entry,* so there is no way to complete the Gauss-Jordan procedure.

If we translate the final tableau back into the language of linear equations, we can see where the problem lies.

$$\begin{aligned} (1)x + (0)y + (\tfrac{1}{3})z &= 2 \\ (0)x + (1)y + (-\tfrac{2}{3})z &= 1 \\ (0)x + (0)y + (0)z &= -3 \end{aligned}$$

The third equation in this system says $0 = -3$, which is false no matter what values x, y, and z represent. Since this system is equivalent to the one we started with, we conclude that the original system has no solution (x, y, z). ■

We call a linear system an **inconsistent system** when, as in the last example, it has no solution. The Gauss-Jordan procedure always signals such a system by producing a tableau row of the form $0\ 0\ .\ .\ .\ 0 \mid a$, where $a \neq 0$.

Practice Problem 4 Solve the system

$$\begin{aligned} x + 2y - z &= 3 \\ 2x \qquad + z &= 2 \\ 4x + 4y - z &= 6. \end{aligned}$$

A GEOMETRICAL INTERPRETATION OF A 3 BY 3 SYSTEM

We saw in Chapter 1 that equations involving two variables x and y could be graphed using a coordinate system based on two perpendicular axes. In the same way a 3-dimensional coordinate system can be established, based on three mutually perpendicular axes, labeled x, y, and z. Although we will not go into the details of such a coordinate system now, it can be used to give a geometrical meaning to a system of three equations in three unknowns. In such a coordinate system each point corresponds to a triple (x, y, z), as shown in Figure 2.1(a).

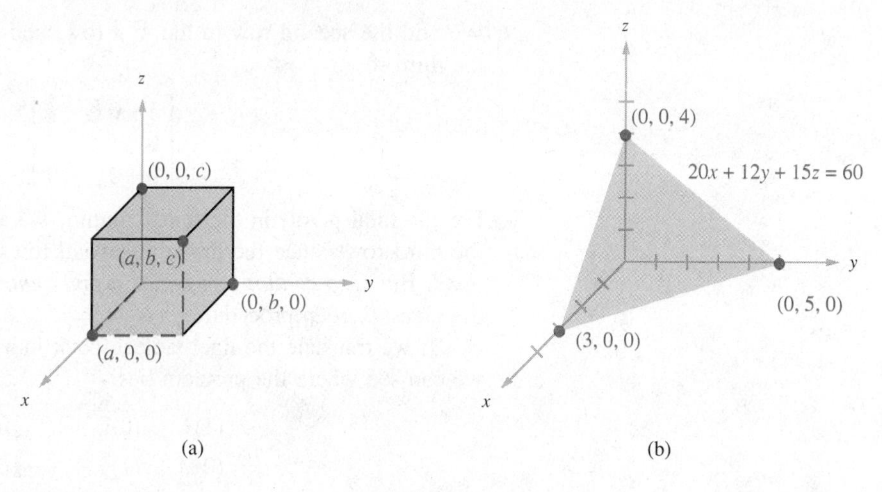

FIGURE 2.1

It turns out that just as the graph of a linear equation in x and y is always a line, the graph of a linear equation in x, y, and z is always a plane. An example is shown in Figure 2.1(b), in which a portion of the plane with equation $20x + 12y + 15z = 60$ is depicted. Notice that each of the points (3, 0, 0), (0, 5, 0), and (0, 0, 4) on the x-axis, y-axis, and z-axis, respectively, satisfies this equation.

Ordinarily two planes intersect in a line, and a third plane will intersect this line in a single point, as in Figure 2.2. This corresponds to the fact that three linear equations in three unknowns usually have a unique solution. It is possible, however, for three planes to have no point in common, even though any two of them intersect in a line. Such a situation is illustrated in Figure 2.3, and corresponds to an inconsistent 3 by 3 system of linear equations.

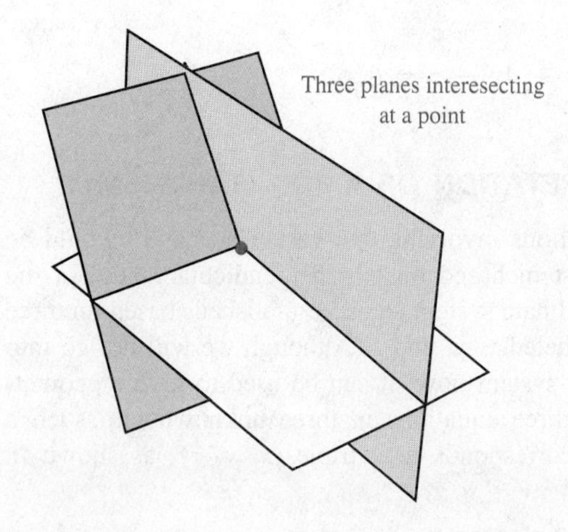

FIGURE 2.2 3 by 3 System with a Unique Solution

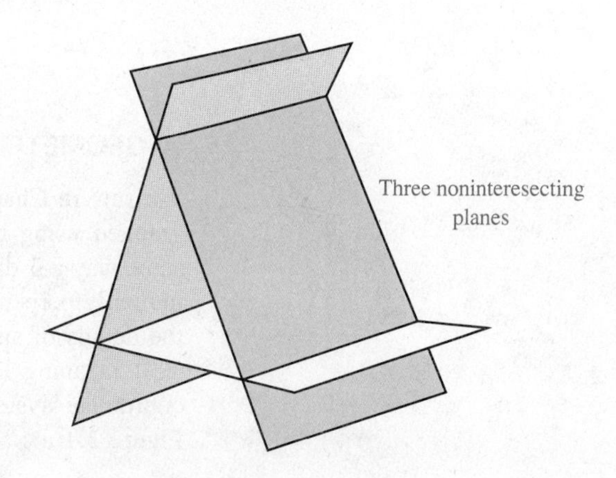

FIGURE 2.3 Inconsistent System

DEPENDENT SYSTEMS

The other type of exceptional behavior for an n by n system is *dependency,* which we encounter in the following example.

EXAMPLE 2.6 Solve the system

$$\begin{aligned} -x + 3y - 2z &= 1 \\ 2x - 5y + 2z &= 0 \\ 2x - y - 6z &= 8. \end{aligned}$$

Solution Again, we proceed by Gauss-Jordan elimination.

$$\left[\begin{array}{ccc|c} \textcircled{-1} & 3 & -2 & 1 \\ 2 & -5 & 2 & 0 \\ 2 & -1 & -6 & 8 \end{array}\right]$$

We multiply the first row by -1.

$$\left[\begin{array}{ccc|c} \textcircled{1} & -3 & 2 & -1 \\ 2 & -5 & 2 & 0 \\ 2 & -1 & -6 & 8 \end{array}\right] \qquad (-\mathbf{R}_1)$$

Now we add -2 times the first row to the second row, and -2 times the first row to the third row.

$$\left[\begin{array}{ccc|c} 1 & -3 & 2 & -1 \\ 0 & \textcircled{1} & -2 & 2 \\ 0 & 5 & -10 & 10 \end{array}\right] \qquad \begin{array}{l} \\ (\mathbf{R}_2 + (-2)\mathbf{R}_1) \\ (\mathbf{R}_3 + (-2)\mathbf{R}_1) \end{array}$$

We add 3 times the second row to the first row, and -5 times the second row to the third row.

$$\left[\begin{array}{ccc|c} 1 & 0 & -4 & 5 \\ 0 & 1 & -2 & 2 \\ 0 & 0 & 0 & 0 \end{array}\right] \qquad \begin{array}{l} (\mathbf{R}_1 + 3\mathbf{R}_2) \\ \\ (\mathbf{R}_3 + (-5)\mathbf{R}_2) \end{array}$$

Once again the method breaks down. We cannot use zero as a pivot in the third column, and the other two entries are in rows that have already been used for pivots. This time the third row represents the equation

$$(0)x + (0)y + (0)z = 0,$$

which is true for *every* triple (x, y, z). Since it tells us nothing, we can ignore this row in the final tableau. Its appearance indicates that the system is a **dependent system;** that is, one of the equations is implied by the others. (In our case, it happens that the third equation is equal to 8 times the first plus 5 times the second.)

Ignoring the third row in the final tableau, we conclude that the original system is equivalent to the system

$$\begin{aligned} x \quad - 4z &= 5 \\ y - 2z &= 2 \end{aligned}$$

of two equations in three unknowns. We can solve these equations for x and y in terms of z, to obtain the general solution:

$$\begin{aligned} x &= 5 + 4z \\ y &= 2 + 2z \\ z &= \text{any real number.} \end{aligned}$$

Note that z can be assigned any value we please in the general solution to yield a specific solution of the original system. For instance

$z = 0$ gives $x = 5$, $y = 2$, and thus the solution $(5, 2, 0)$;
$z = 2$ gives $x = 13$, $y = 6$, and thus the solution $(13, 6, 2)$;
$z = -1$ gives $x = 1$, $y = 0$, and thus the solution $(1, 0, -1)$.

The original system therefore has *infinitely many solutions* (x, y, z), one for each value of z. We represent the solution we have found by writing

$$(x, y, z) = (5 + 4z, 2 + 2z, z),$$

where it is understood that z can be any real number.

Notice that we can check our solution by plugging it back into the original equations, just as we check a unique solution. Here the original system is

$$\begin{aligned} -x + 3y - 2z &= 1 \\ 2x - 5y + 2z &= 0 \\ 2x - \quad y - 6z &= 8. \end{aligned}$$

Substituting $5 + 4z$ for x and $2 + 2z$ for y yields

$$\begin{aligned} -(5 + 4z) + 3(2 + 2z) - 2z &= 1 \\ 2(5 + 4z) - 5(2 + 2z) + 2z &= 0 \\ 2(5 + 4z) - \quad (2 + 2z) - 6z &= 8, \end{aligned}$$

or

$$\begin{aligned} -5 - 4z + \ 6 + \ 6z - 2z &= 1 \\ 10 + 8z - 10 - 10z + 2z &= 0 \\ 10 + 8z - \ 2 - \ 2z - 6z &= 8; \end{aligned}$$

and each of these equations is true for all values of z. ■

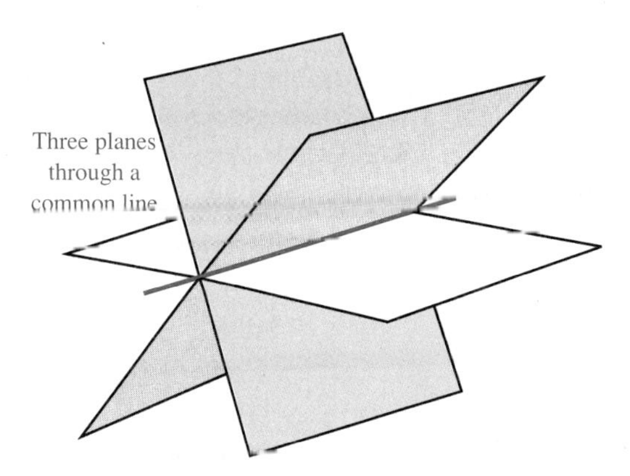

FIGURE 2.4 3 by 3 Dependent System

The geometric meaning of the last example is illustrated in Figure 2.4, which shows three planes intersecting along a common line. Any point on this line lies on all three planes and therefore satisfies all three equations.

BASIC AND FREE VARIABLES

The preceding examples, together with those in the previous section, illustrate the three possibilities that can arise in solving a system of m equations in n unknowns.

Every m by n linear system leads to *exactly one* of the following three cases.

System with a unique solution: The system has exactly one solution, and can be solved for all the unknowns.

Inconsistent system: The system has no solutions.

Dependent system: The system has infinitely many solutions, which can be expressed by writing some of the unknowns in terms of the others.

In Example 2.6 we saw that the dependent system

$$\begin{aligned} -x + 3y - 2z &= 1 \\ 2x - 5y + 2z &= 0 \\ 2x - y - 6z &= 8 \end{aligned}$$

has the general solution

$$(x, y, z) = (5 + 4z, 2 + 2z, z), \tag{2.9}$$

where z is any real number. Since z is unrestricted in (2.9), we call z a **free variable** in this form of the solution. The variables x and y are called **basic variables** when the solution is written this way. That x and y turned out to be basic variables and z a free variable depended on various choices we made while applying the Gauss-Jordan procedure. For example it can be shown that

$$(x, y, z) = \left(1 + 2y, y, -1 + \frac{1}{2}y\right) \tag{2.10}$$

represents exactly the same set of ordered triples as (2.9). In particular,

$$\begin{aligned} y &= 2 \text{ gives } x = 5, z = 0, \text{ and thus the solution } (5, 2, 0); \\ y &= 6 \text{ gives } x = 13, z = 2, \text{ and thus the solution } (13, 6, 2); \\ y &= 0 \text{ gives } x = 1, z = -1, \text{ and thus the solution } (1, 0, -1); \end{aligned}$$

and these are the same three particular solutions we computed in Example 2.6. When the solution is expressed as in (2.10), y is the free variable and x and z are basic variables.

Practice Problem 5 Solve the system

$$\begin{aligned} x + 2y + z &= 3 \\ y - z &= -2. \end{aligned}$$

In our examples and solutions to exercises, we will always choose our pivot columns as far to the left as possible so that solutions can be easily compared. For example, given the system

$$\begin{aligned} x + y + z &= 3 \\ x + y + w &= 4, \end{aligned}$$

with tableau

$$\left[\begin{array}{cccc|c} 1 & 1 & 1 & 0 & 3 \\ 1 & 1 & 0 & 1 & 4 \end{array}\right],$$

we first pivot on the 1 in the first row and column, which produces

$$\left[\begin{array}{cccc|c} 1 & 1 & 1 & 0 & 3 \\ 0 & 0 & -1 & 1 & 1 \end{array}\right]. \qquad (\mathbf{R_2 + (-1)R_1})$$

Since we have already pivoted using an element in the first row, we cannot pivot on the 1 in the first row and second column. Instead we pivot on the -1 in row 2 and column 3.

$$\left[\begin{array}{cccc|c} 1 & 1 & 1 & 0 & 3 \\ 0 & 0 & 1 & -1 & -1 \end{array}\right] \qquad (\mathbf{-R_2})$$

$$\left[\begin{array}{cccc|c} 1 & 1 & 0 & 1 & 4 \\ 0 & 0 & 1 & -1 & -1 \end{array}\right] \qquad (\mathbf{R_1 + (-1)R_2})$$

Since we pivoted in columns 1 and 3 we take x and z as the basic variables, using the equivalent system

$$\begin{aligned} x + y + \phantom{z - {}} w &= 4 \\ z - w &= -1. \end{aligned}$$

Thus $x = -y - w + 4$, $z = w - 1$, and we have the general solution

$$(x, y, z, w) = (-y - w + 4, y, w - 1, w),$$

with y and w free variables.

Although pivoting from left to right allows us to compare solutions, it may not be the quickest path to an answer. For example in the system

$$\begin{aligned} x + y + z \phantom{{} + w} &= 3 \\ x + y + \phantom{z + {}} w &= 4 \end{aligned}$$

just treated, we can see immediately that $z = 3 - x - y$ and $w = 4 - x - y$, giving the general solution

$$(x, y, z, w) = (x, y, 3 - x - y, 4 - x - y),$$

with x and y free variables.

SUMMARY OF GAUSS-JORDAN ELIMINATION

Gauss-Jordan Elimination

To solve an m by n linear system:

1. Write the system in tableau form.
2. Perform a sequence of pivots on nonzero entries, using a different row and column for each pivot. Eventually this process will terminate.
3. Look for rows of the form $0\ 0\ \ldots\ 0 \mid a$, where $a \neq 0$. If any such row appears, *the system has no solution*.
4. Look for rows of the form $0\ 0\ \ldots\ 0 \mid 0$. Ignore all such rows in the final tableau.
5. If the left side of the tableau now has the same number of rows and columns, *the system has a unique solution,* which can be read from the final tableau.
6. If the left side of the tableau has fewer rows than columns, *the system has infinitely many solutions*. Use the final tableau to solve for the basic variables in terms of the free variables.

EXAMPLE 2.7 Solve the following 3 by 5 system:

$$\begin{aligned} x - 2y + z \qquad\qquad &= 1 \\ y - 2z + u \qquad &= 2 \\ z - 2u + v &= 3. \end{aligned}$$

Solution In tableau form, the system becomes

$$\begin{array}{c} \begin{array}{ccccc} x & y & z & u & v \end{array} \\ \left[\begin{array}{ccccc|c} 1 & -2 & 1 & 0 & 0 & 1 \\ 0 & 1 & -2 & 1 & 0 & 2 \\ 0 & 0 & 1 & -2 & 1 & 3 \end{array}\right]. \end{array}$$

Since we already have a 1 and all zeros in the x-column, we proceed to the y-column, using the circled 1 as pivot. We add twice the second row to the first row.

$$\left[\begin{array}{ccccc|c} 1 & 0 & -3 & 2 & 0 & 5 \\ 0 & 1 & -2 & 1 & 0 & 2 \\ 0 & 0 & \textcircled{1} & -2 & 1 & 3 \end{array}\right] \qquad (\mathbf{R_1 + 2R_2})$$

Now we add 3 times the third row to the first row, and add 2 times the third row to the second row.

$$\left[\begin{array}{ccccc|c} 1 & 0 & 0 & -4 & 3 & 14 \\ 0 & 1 & 0 & -3 & 2 & 8 \\ 0 & 0 & 1 & -2 & 1 & 3 \end{array}\right] \qquad \begin{array}{l} (\mathbf{R_1 + 3R_3}) \\ (\mathbf{R_2 + 2R_3}) \end{array}$$

This tableau corresponds to the equations

$$\begin{aligned} x \qquad - 4u + 3v &= 14 \\ y \quad - 3u + 2v &= 8 \\ z - 2u + v &= 3, \end{aligned}$$

or

$$\begin{aligned} x &= 4u - 3v + 14 \\ y &= 3u - 2v + 8 \\ z &= 2u - v + 3. \end{aligned}$$

We see that the general solution is

$$(x, y, z, u, v) = (4u - 3v + 14, 3u - 2v + 8, 2u - v + 3, u, v),$$

where u and v are free variables. ■

EXAMPLE 2.8 Chocolate kisses come in 2-pound bags costing \$3, jelly beans come in pound boxes costing \$2, and mints come in 3-pound cans costing \$5. Thirty-two pounds of candy costing \$53 are needed for a school party. Express the number of bags of kisses and boxes of jelly beans in terms of the number of cans of mints bought.

Solution Let x bags of kisses, y boxes of jelly beans, and z cans of mints be bought. Then

$$\begin{aligned} 2x + y + 3z &= 32 \\ 3x + 2y + 5z &= 53. \end{aligned}$$

The corresponding tableau is

$$\left[\begin{array}{ccc|c} \textcircled{2} & 1 & 3 & 32 \\ 3 & 2 & 5 & 53 \end{array}\right].$$

First we pivot on the first row and column.

$$\left[\begin{array}{ccc|c} \textcircled{1} & \frac{1}{2} & \frac{3}{2} & 16 \\ 3 & 2 & 5 & 53 \end{array}\right] \qquad (\tfrac{1}{2}R_1)$$

$$\left[\begin{array}{ccc|c} 1 & \frac{1}{2} & \frac{3}{2} & 16 \\ 0 & \textcircled{\tfrac{1}{2}} & \frac{1}{2} & 5 \end{array}\right] \qquad (R_2 + (-3)R_1)$$

Next we pivot on the second row and column.

$$\left[\begin{array}{ccc|c} 1 & \frac{1}{2} & \frac{3}{2} & 16 \\ 0 & \textcircled{1} & 1 & 10 \end{array}\right] \qquad (2R_2)$$

$$\left[\begin{array}{ccc|c} 1 & 0 & 1 & 11 \\ 0 & 1 & 1 & 10 \end{array}\right] \qquad (R_1 + (-\tfrac{1}{2})R_2)$$

The corresponding equations are

$$\begin{aligned} x \quad + z &= 11 \\ y + z &= 10, \end{aligned}$$

or $x = 11 - z$, $y = 10 - z$. These give the number of bags of kisses and boxes of jelly beans in terms of z, the number of cans of mints. Note that at most 10 cans of mints may be bought, since otherwise y is negative. ■

Practice Problem 6 Wrapping paper comes in three sizes. The standard roll is 9 square feet, the large is 12 square feet, and the jumbo is 18 square feet. A woman needs 30 rolls of paper with an area of 357 square feet. Express the number of standard and large rolls in terms of the number of jumbo rolls she must buy.

EXERCISES 2.2

In Exercises 1–16 solve the given linear system using the Gauss-Jordan procedure. At each stage choose pivots in the left-most column possible.

1. $$\begin{aligned} 2x + y - 3z &= 2 \\ x + 2y + 3z &= 4 \end{aligned}$$

2. $$\begin{aligned} 2x + 4y - 2z &= 6 \\ 3x + y + 2z &= -1 \end{aligned}$$

3. $$\begin{aligned} x + 2y - z &= 3 \\ 2x + 4y + 3z &= 1 \end{aligned}$$

4. $$\begin{aligned} 3x - y + 6z &= 1 \\ x + y + 2z &= 3 \end{aligned}$$

5. $$\begin{aligned} w \quad + 3y + 2z &= 1 \\ 4w + 2x + 2y - 2z &= 2 \end{aligned}$$

6. $$\begin{aligned} 2p - 3q \quad + s &= 6 \\ -6p + 9q + 2r + 5s &= -2 \end{aligned}$$

7. $$\begin{aligned} u + 2v \phantom{{}+3w} &= 5 \\ 2u \phantom{{}+2v} + 3w &= 11 \\ v - 2w &= -4 \end{aligned}$$

8. $$\begin{aligned} x - 2y + z &= 2 \\ -2x + 3y \phantom{{}+z} &= -1 \\ x - 3y + 3z &= 5 \end{aligned}$$

9. $$\begin{aligned} 2r + 3s - 4t &= 1 \\ -r + 2s + t &= -2 \\ r + 12s - 5t &= -4 \end{aligned}$$

10. $$\begin{aligned} r + 2s + 5t &= 2 \\ r + 4s - t &= -3 \\ r - 2s + 17t &= 10 \end{aligned}$$

11. $$\begin{aligned} a + 5b - c &= 4 \\ -2a + 2b \phantom{{}-c} &= -3 \\ -2a + 26b - 4c &= 5 \end{aligned}$$

12. $$\begin{aligned} 2X + Y - Z &= 4 \\ -X + 2Y + Z &= -4 \\ X + 3Y - 3Z &= -3 \end{aligned}$$

13. $$\begin{aligned} y + 2z + u + 3v &= 3 \\ x + 2y - z + 2v &= 1 \end{aligned}$$

14. $$\begin{aligned} -x + y + 2z + 3u + v &= 2 \\ 3x - 3y - 6z - 2u + 4v &= 1 \end{aligned}$$

15. $$\begin{aligned} x - 2y + 3z &= 1 \\ 2x + y - z &= -2 \\ 5x \phantom{{}+y} + z &= -3 \\ 4x - 3y + 5z &= 0 \end{aligned}$$

16. $$\begin{aligned} 2x - y + 2z &= -1 \\ -3x + 3y - z &= 2 \\ x + y + 3z &= 0 \\ -x + 5y - 7z &= 3 \end{aligned}$$

17. Use Gauss-Jordan elimination to find the general solution to the system

$$\begin{aligned} x + 2y + 3z &= 2 \\ 3x + 2y + z &= 10 \end{aligned}$$

(a) with x and y basic, z free

(b) with x and z basic, y free

(c) with y and z basic, x free.

18. Use Gauss-Jordan elimination to find the general solution to the system

$$\begin{aligned} x + 2y + 3z + 4w &= 5 \\ x + 3y + 5z + 7w &= 9 \end{aligned}$$

(a) with x and y basic, z and w free

(b) with x and z basic, y and w free

(c) with z and w basic, x and y free.

19. An airline uses a 200-passenger plane on a certain flight. A first-class ticket costs \$360, a coach ticket \$180, and a super-saver ticket \$90. It is desired to take in \$27,000 in fares on the flight. Suppose x first-class tickets, y coach tickets, and z super-saver tickets are sold, and w seats are empty. Express x and y in terms of z and w.

20. A college student has 48 hours available to work at her summer job at a discount store. She gets \$3.50 per hour ordinarily, but \$5.25 per hour on Sunday. One week she worked x hours at ordinary pay, y hours on Sunday, and did not work z of the 48 hours, earning \$140. Express x and y in terms of z.

21. A merchant wants to mix peanuts costing \$0.60 per pound, cashews costing \$1.50 per pound, and almonds costing \$2.00 per pound, to make 40 pounds of mixed nuts costing \$1.28 per pound. Express the number of pounds of peanuts and cashews he must use in terms of the number of pounds of almonds.

22. At his next party, a host intends to serve a punch made from three types of mixed fruit juices. Mix A contains 0% apple juice, Mix B contains 20% apple juice, and Mix C contains 40% apple juice. He wants to make 200 fluid ounces of punch containing 15.2% apple juice. Express the number of ounces of Mixes A and B he must use in terms of the number of ounces of Mix C.

Answers to Practice Problems

4. The system is inconsistent, and has no solution.
5. $(x, y, z) = (7 - 3z, z - 2, z)$, where z is any real number.
6. If she buys z jumbo rolls, then she must buy $1 + 2z$ standard rolls and $29 - 3z$ large rolls.

2.3 Applications of Systems of Linear Equations

This section gives some examples of problems that are solved using systems of linear equations. More applications of linear systems will be presented in later sections, and in the next chapter.

PRODUCT-MIX PROBLEMS

There is a large class of applications referred to as **product-mix** problems, since they involve finding a suitable combination of products or available resources in order to satisfy certain given conditions. Some 2 by 2 problems of this type were treated in Chapter 1.

EXAMPLE 2.9

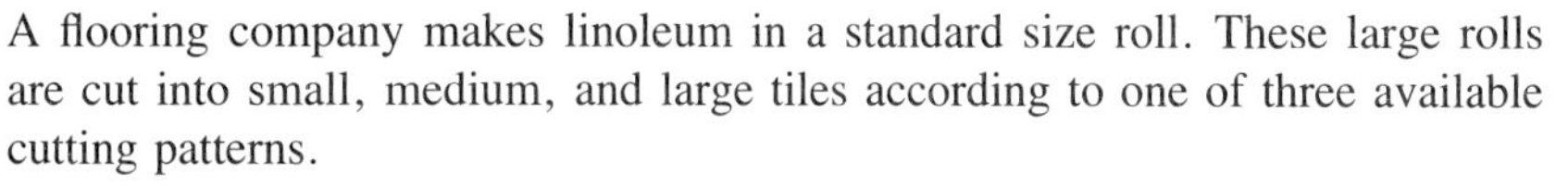
A flooring company makes linoleum in a standard size roll. These large rolls are cut into small, medium, and large tiles according to one of three available cutting patterns.

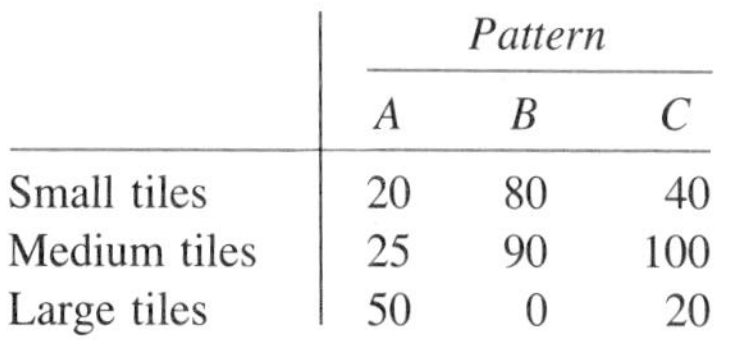

	Pattern		
	A	*B*	*C*
Small tiles	20	80	40
Medium tiles	25	90	100
Large tiles	50	0	20

For example, if a roll is cut using pattern A, then 20 small, 25 medium, and 50 large tiles are produced. How many rolls should be cut according to each pattern, if the company wants to fill an order for 2000 small tiles, 3400 medium tiles, and 1400 large tiles, without wasting any linoleum?

Solution The three unknowns in the problem are:

x = number of rolls cut according to Pattern A
y = number of rolls cut according to Pattern B
z = number of rolls cut according to Pattern C.

The condition that we fill the order without any waste—that is, without any tiles left over—translates into a 3 by 3 system of linear equations.

$$\begin{aligned} 20x + 80y + 40z &= 2000 \text{ (small tiles)} \\ 25x + 90y + 100z &= 3400 \text{ (medium tiles)} \\ 50x + 20z &= 1400 \text{ (large tiles)} \end{aligned}$$

For example, the first equation comes from the fact that 2000 small tiles are needed, while each roll cut to Pattern A yields 20 small tiles, each roll cut to Pattern B yields 80 small tiles, and each roll cut to Pattern C yields 40 small tiles.

We write the system in tableau form and apply Gauss-Jordan elimination. The annotations will explain the row operations done at each step.

$$\left[\begin{array}{ccc|c} \textcircled{20} & 80 & 40 & 2000 \\ 25 & 90 & 100 & 3400 \\ 50 & 0 & 20 & 1400 \end{array}\right]$$

$$\left[\begin{array}{ccc|c} \textcircled{1} & 4 & 2 & 100 \\ 25 & 90 & 100 & 3400 \\ 50 & 0 & 20 & 1400 \end{array}\right] \qquad (\tfrac{1}{20}R_1)$$

$$\left[\begin{array}{ccc|c} 1 & 4 & 2 & 100 \\ 0 & \textcircled{-10} & 50 & 900 \\ 0 & -200 & -80 & -3600 \end{array}\right] \qquad \begin{array}{l} (R_2 + (-25)R_1) \\ (R_3 + (-50)R_1) \end{array}$$

$$\left[\begin{array}{ccc|c} 1 & 4 & 2 & 100 \\ 0 & \textcircled{1} & -5 & -90 \\ 0 & -200 & -80 & -3600 \end{array}\right] \qquad (-\tfrac{1}{10}R_2)$$

$$\left[\begin{array}{ccc|c} 1 & 0 & 22 & 460 \\ 0 & 1 & -5 & -90 \\ 0 & 0 & \textcircled{-1080} & -21{,}600 \end{array}\right] \qquad \begin{array}{l} (R_1 + (-4)R_2) \\ \\ (R_3 + 200R_2) \end{array}$$

$$\left[\begin{array}{ccc|c} 1 & 0 & 22 & 460 \\ 0 & 1 & -5 & -90 \\ 0 & 0 & \textcircled{1} & 20 \end{array}\right] \qquad (-\tfrac{1}{1080}R_3)$$

$$\left[\begin{array}{ccc|c} 1 & 0 & 0 & 20 \\ 0 & 1 & 0 & 10 \\ 0 & 0 & 1 & 20 \end{array}\right] \qquad \begin{array}{l} (R_1 + (-22)R_3) \\ (R_2 + 5R_3) \end{array}$$

The solution is therefore (20, 10, 20), that is, the company should cut 20 rolls according to Pattern A, 10 according to Pattern B, and 20 according to Pattern C. ■

Practice Problem 7 A slice of bread contains 70 calories and 2 grams of protein, while an egg contains 80 calories and 6 grams of protein. At a certain orphanage the egg and toast breakfasts are to provide 1090 calories and 46 grams of protein per week for each child.

(a) If x slices of bread and y eggs are provided per week, express the above information as a 2 by 2 linear system.

(b) Solve the linear system.

A MODEL FOR ALLOCATING SERVICE CHARGES*

Manufacturing firms typically are composed of two types of departments having entirely different functions. Service departments such as data processing and shipping provide services to other departments within the firm, whereas production departments produce commodities for sale outside the firm. Before the commodities made by the production departments can be priced for sale, the firm must know the total cost for each production department. These costs consist not only of direct costs such as salaries and material costs but also of indirect costs such as charges for services provided by the service departments. Standard accounting procedures require that the direct costs of the service departments be charged to the production departments in accordance with the amount of service used. The allocation of these charges is complicated by the fact that each service department contributes to the expenses of the other service departments, as well as to its own expenses.

EXAMPLE 2.10 An appliance manufacturer has three service departments (maintenance, shipping, and data processing), and two production departments (small appliances and large appliances). The five departments each consume a proportion of the output of the three service departments as given in Table 2.1.

	Provider of Service		
User of services	*Maintenance*	*Shipping*	*Data processing*
Maintenance	0.3	0.3	0.2
Shipping	0.3	0.3	0.2
Data processing	0.1	0.1	0.4
Small appliances	0.1	0.1	0.1
Large appliances	0.2	0.2	0.1

TABLE 2.1

For example, the maintenance department expends 30% of its effort on itself, 30% on the shipping department, 10% on each of data processing and small appliances, and 20% on large appliances. Notice that each column in Table 2.1 adds up to 1, since this sum accounts for the total effort of the corresponding department. The direct costs per week for the five departments are $d_1 = \$1000$, $d_2 = \$800$, $d_3 = \$1200$, $d_4 = \$2000$, and $d_5 = \$3000$, for maintenance, shipping, data processing, small appliances, and large appliances, respectively. These include the salaries, materials, etc. expended directly by each department per week. How should the direct cost

$$d_1 + d_2 + d_3 = \$3000$$

of the service departments be allocated between the two production departments?

*For a more detailed explanation of the problems of allocating service charges see Don O. Koehler, "The Cost Accounting Problem," *Mathematics Magazine*, Vol. 53, No. 1 (1980) 3–12.

Solution Let x_4 and x_5 stand for the total costs of the small and large appliance departments, respectively, including both direct and indirect costs. Since all costs are to be allocated to the two production departments, we should have $x_4 + x_5 = d_1 + d_2 + d_3 + d_4 + d_5 = \$8000$. On the other hand, from the last two rows of Table 2.1 we should have

$$\begin{aligned} x_4 &= d_4 + 0.1x_1 + 0.1x_2 + 0.1x_3 \\ x_5 &= d_5 + 0.2x_1 + 0.2x_2 + 0.1x_3, \end{aligned} \tag{2.11}$$

where x_1, x_2, and x_3 are defined in a similar way from the first three rows of Table 2.1:

$$\begin{aligned} x_1 &= d_1 + 0.3x_1 + 0.3x_2 + 0.2x_3 \\ x_2 &= d_2 + 0.3x_1 + 0.3x_2 + 0.2x_3 \\ x_3 &= d_3 + 0.1x_1 + 0.1x_2 + 0.4x_3. \end{aligned} \tag{2.12}$$

The meaning of x_1, x_2, and x_3 is somewhat fictitious, since if $x_4 + x_5$ is the total direct cost of all five departments, some of these costs must be counted again in x_1, x_2, and x_3. In fact, if we add up the equations of (2.11) and (2.12), we get

$$x_1 + x_2 + x_3 + x_4 + x_5 = d_1 + d_2 + d_3 + d_4 + d_5 + x_1 + x_2 + x_3,$$

from which we see that

$$x_4 + x_5 = d_1 + d_2 + d_3 + d_4 + d_5,$$

as desired.

Since neither x_4 nor x_5 appears in the three equations of (2.12), we start by solving this 3 by 3 system. Plugging in the values of d_1, d_2, and d_3 and arranging the variables on the left and constants on the right yields

$$\begin{aligned} 0.7x_1 - 0.3x_2 - 0.2x_3 &= 1000 \\ -0.3x_1 + 0.7x_2 - 0.2x_3 &= 800 \\ -0.1x_1 - 0.1x_2 + 0.6x_3 &= 1200, \end{aligned}$$

and the corresponding tableau

$$\left[\begin{array}{rrr|r} 0.7 & -0.3 & -0.2 & 1000 \\ -0.3 & 0.7 & -0.2 & 800 \\ -0.1 & -0.1 & 0.6 & 1200 \end{array}\right].$$

The reader should confirm that Gauss-Jordan elimination leads to the unique solution $x_1 = 4000$, $x_2 = 3800$, and $x_3 = 3300$. Substituting these values and the given $d_4 = 2000$ and $d_5 = 3000$ into (2.11) gives $x_4 = 3110$ and $x_5 = 4890$. Thus of the total of \$8000 of direct costs per week, \$3110 should be allocated to small appliances and \$4890 to large appliances. ■

Practice Problem 8 A retailer has two service departments, data processing and maintenance, and two production departments, direct sales and mail order. The direct costs and distribution of outputs of the service departments to each department are shown below.

User of Services	*Direct cost*	*Provider of Service*	
		Data processing	*Maintenance*
Data processing	\$400	.2	.1
Maintenance	\$210	.2	.2
Direct sales	\$500	.2	.4
Mail order	\$600	.4	.3

(a) Write a system of equations for this problem corresponding to (2.12).

(b) How should the total costs of \$1710 be allocated to direct sales and mail order?

TRAFFIC FLOW THROUGH A ROAD NETWORK

Network analysis has been used for many years in physics and engineering to study the flow of currents through electric circuits. More recently, this subject has proved useful in other fields as well. The next application involves a technique for analyzing traffic flow through a road network.

EXAMPLE 2.11 A one-way drive circles a reservoir, with two one-way streets coming into it and two leaving it, as shown in Figure 2.5. During rush hour 350 cars enter the drive per hour from North Avenue. Likewise 250, 200, and 300 cars per

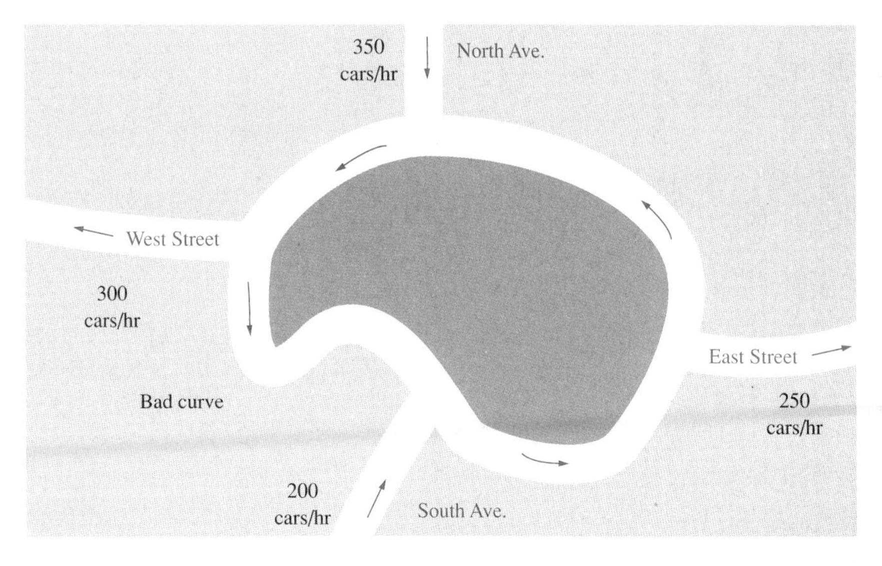

FIGURE 2.5

hour use East Street, South Avenue, and West Street, respectively. There is a bad curve between West Street and South Avenue that should be straightened out, but the construction should not impede rush hour traffic. How many cars per hour must be able to take the portion of the drive between West Street and South Avenue in order to maintain this traffic flow?

Solution Let

w = number of cars per hour from North Avenue to West Street

x = number of cars per hour from East Street to North Avenue

y = number of cars per hour from South Avenue to East Street

z = number of cars per hour from West Street to South Avenue,

and let A, B, C, and D be the intersections shown in Figure 2.6.

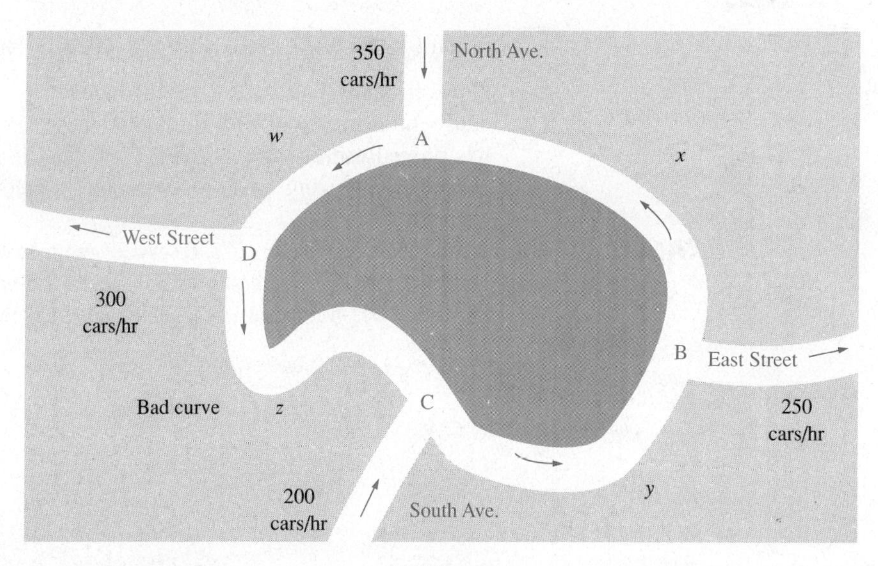

FIGURE 2.6

Now at intersection A there are x cars entering from the east and 350 cars entering from the north, while w cars leave the intersection to the west. Since the total number of cars entering an intersection must equal the number leaving it, we have

$$x + 350 = w.$$

Applying a similar analysis to intersections B, C, and D leads to the system

$$\begin{aligned} x + 350 &= w \quad \text{(intersection A)} \\ y &= x + 250 \quad \text{(intersection B)} \\ z + 200 &= y \quad \text{(intersection C)} \\ w &= z + 300 \quad \text{(intersection D).} \end{aligned}$$

If we arrange the variables on the left and constants on the right, we get

$$\begin{aligned} -w + x \qquad\qquad &= -350 \\ -x + y \qquad &= 250 \\ -y + z &= -200 \\ w \qquad\qquad - z &= 300, \end{aligned}$$

which translates into the following tableau.

$$\left[\begin{array}{rrrr|r} -1 & 1 & 0 & 0 & -350 \\ 0 & -1 & 1 & 0 & 250 \\ 0 & 0 & -1 & 1 & -200 \\ 1 & 0 & 0 & -1 & 300 \end{array}\right]$$

The reader should confirm that Gauss-Jordan elimination transforms the tableau into the following.

$$\left[\begin{array}{rrrr|r} 1 & 0 & 0 & -1 & 300 \\ 0 & 1 & 0 & -1 & -50 \\ 0 & 0 & 1 & -1 & 200 \\ 0 & 0 & 0 & 0 & 0 \end{array}\right]$$

Thus taking z as a free variable gives

$$\begin{aligned} w &= z + 300 \\ x &= z - 50 \\ y &= z + 200. \end{aligned} \tag{2.13}$$

Of course w, x, y, and z must all be nonnegative since they represent a number of cars per hour. Since $x = z - 50$, this means z must be at least 50. Note that z corresponds to the traffic past the proposed construction, which must be handled so that at least 50 cars per hour can pass the construction site. Indeed, taking $z = 50$ gives $(w, x, y, z) = (350, 0, 250, 50)$, which satisfies the conditions of the problem. Since $z \geq 50$ we can draw additional information from (2.13), namely that $w \geq 350$ (which was obvious from the start, since at least 350 cars per hour enter intersection A), and that $y \geq 250$ (which was not obvious before our analysis). ■

Practice Problem 9 Figure 2.7 shows the number of vehicles per hour using certain portions of four one-way streets during rush hour. The city would like to repave Low Street between Vine and Elm. Determine the minimum number of vehicles per hour that must use this portion of Low Street in order to avoid congestion.

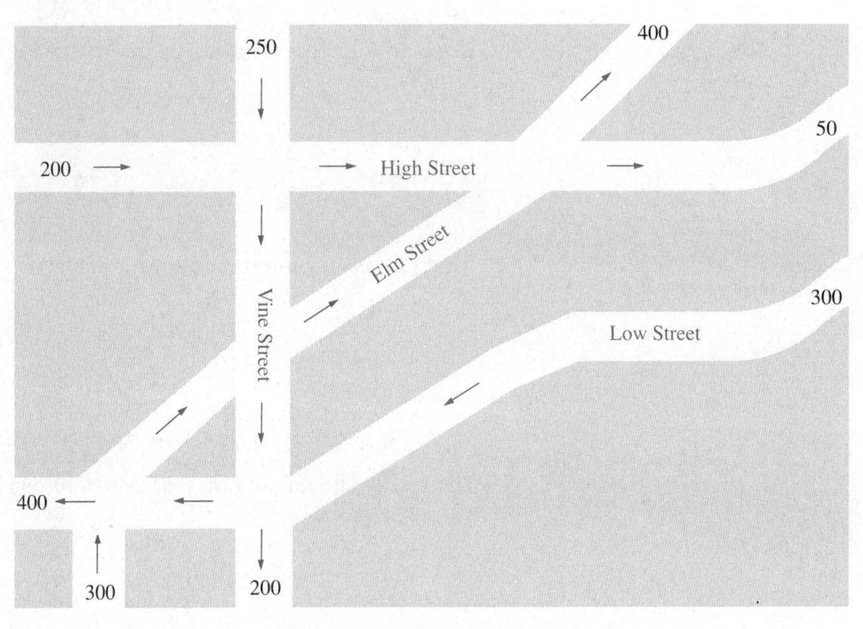

FIGURE 2.7

EXERCISES 2.3

1. The Allegheny Mining Company operates three mines which produce high-, medium-, and low-grade ore. The daily yield of each mine is given in the table below.

	Ore yield per day		
Mine	1	2	3
High-grade	2 tons	1 ton	0 tons
Medium-grade	2 tons	3 tons	1 ton
Low-grade	1 ton	2 tons	2 tons

The company must fill an order for a total of 40 tons of high-grade, 75 tons of medium-grade, and 50 tons of low-grade ore. How many days should each mine be operated to fill this order exactly?

2. The Consolidated Steel Corporation has been ordered to reduce the three main pollutants—particulates, sulfur oxide, and hydrocarbons—emitted at its Pittsburgh plant, by 5400, 6000, and 9600 units per month, respectively. Three filtering methods are avail-

able. The monthly reduction in emissions, per dollar spent, for each of the filtering methods is given below.

Method	*Monthly reduction per dollar spent* A	B	C
Particulates	9 units	10 units	10 units
Sulfur oxide	15 units	10 units	5 units
Hydrocarbons	15 units	20 units	15 units

How much must the company allocate per month to each method, in order to meet the reduction order exactly?

3. An agricultural supply house makes three types of fertilizer. Type A contains 10% nitrogen and 2% phosphates by weight. Type B contains 5% nitrogen and 1% phosphates, and Type C contains 8% nitrogen and 4% phosphates. The company has received an order for 600 pounds of fertilizer which is to contain 9% nitrogen and 2% phosphates. Can the order be filled by mixing together the three available types? If so, how?

4. Each ounce of Food A contains 3 gm of carbohydrate, 2 gm of protein, and 4 gm of fat, while an ounce of Food B contains 3 gm of carbohydrate, 4 gm of protein, and 2 gm of fat. By combining these foods, is it possible to obtain exactly 75 gm of carbohydrate, 60 gm of protein, and 60 gm of fat?

5. Four grades of crude oil are to be blended in an oil storage tank. The cost and sulfur content of each grade are given in the table below.

Grade	A	B	C	D
Cost per barrel	$20	$25	$40	$50
Sulfur per barrel	125 gm	100 gm	75 gm	25 gm

The tank initially contains 12,000 barrels of crude which costs $15 per barrel and has a sulfur content of 225 gm per barrel. Suppose that x, y, z, and w barrels of the four respective grades are added to the tank to produce a blend which costs $25 per barrel and contains 150 gm per barrel of sulfur.

(a) What equations must x, y, z, and w satisfy?
(b) Solve these equations for x and y in terms of z and w.
(c) What values of z and w will yield a feasible solution?

6. A metallurgist has melted down three bronze alloys, whose composition by weight is given in the table below.

Alloy	A	B	C
Copper	95%	90%	80%
Tin	4%	2%	10%
Zinc	1%	8%	10%

A new alloy is to be formed by adding x, y, and z tons of the three respective alloys to a smelting kiln that already contains 100 tons of Alloy A.

(a) Is it possible to obtain a mixture whose composition is 90% copper, 5% tin, and 5% zinc?
(b) Is the solution unique?

7. A manufacturing firm that produces desk calculators and typewriters has three service departments—data processing, maintenance, and payroll. The table below gives the direct cost of each department and the proportion of services used by each department. Determine the total costs for each department.

User of Services	*Direct cost*	*Provider of Service*		
		Data processing	*Maintenance*	*Payroll*
Data processing	$3000	0.3	0.1	0.3
Maintenance	$4000	0.2	0.4	0.2
Payroll	$1500	0.1	0.2	0.1
Calculators	$6000	0.2	0.2	0.2
Typewriters	$7000	0.2	0.1	0.2

8. A firm manufactures laboratory equipment and culture media, which it supplies to hospitals. Its three service departments (data processing, research and development (R&D), and electronics) provide services to the five departments as indicated in the table below. Using the direct costs in this table, determine the total costs for each department.

User of Services	*Direct cost*	*Provider of Service*		
		Data processing	*R&D*	*Electronics*
Data processing	$4,000	0.4	0.0	0.2
R&D	$10,000	0.1	0.8	0.1
Electronics	$3,000	0.2	0.0	0.1
Equipment	$12,000	0.1	0.1	0.5
Media	$15,000	0.2	0.1	0.1

9. A men's clothing factory manufactures trousers, jackets, and suits. These three production departments use services from the accounts receivable, data processing, and maintenance departments as indicated in the table below. Determine the total costs for each department.

User of Services	*Direct cost*	*Provider of Service*		
		Accounts receivable	*Data processing*	*Maintenance*
Accounts receivable	$22,000	0.1	0.2	0.1
Data processing	$40,000	0.2	0.1	0.3
Maintenance	$2,000	0.1	0.2	0.1
Trousers	$20,000	0.2	0.2	0.1
Jackets	$20,000	0.2	0.1	0.1
Suits	$30,000	0.2	0.2	0.3

10. The diagram at the top of the next page shows the number of vehicles per hour using certain portions of five one-way streets during rush hour. The city would liked to repave East Street between North Avenue and South Avenue. Determine the minimum number of vehicles per hour that must use this portion of East Street in order to avoid congestion.

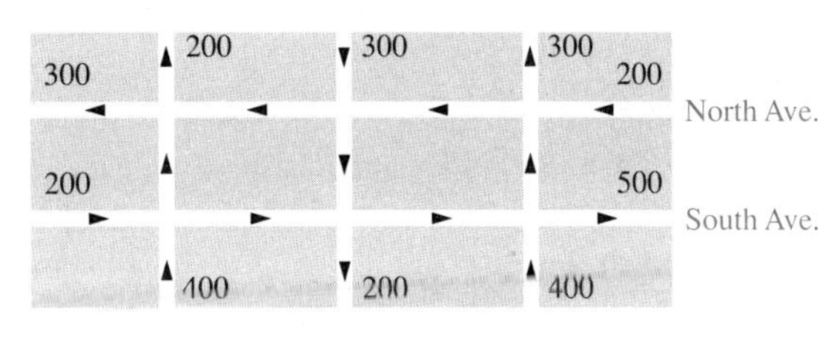

11. The diagram below shows the number of vehicles per hour using certain portions of five one-way streets during rush hour. Determine the minimum number of vehicles per hour that must use Third Avenue between Washington Street and Adams Street to avoid congestion.

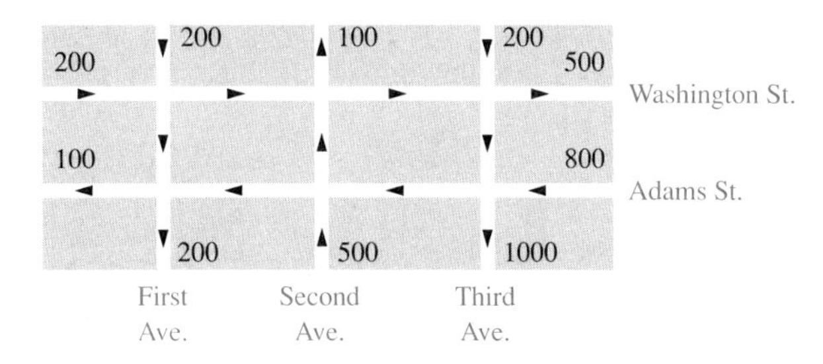

12. The diagram below shows the number of vehicles per hour using certain portions of five one-way streets during rush hour. Determine the minimum number of vehicles per hour that must use Pine Street between Jefferson Avenue and Monroe Avenue in order to avoid congestion.

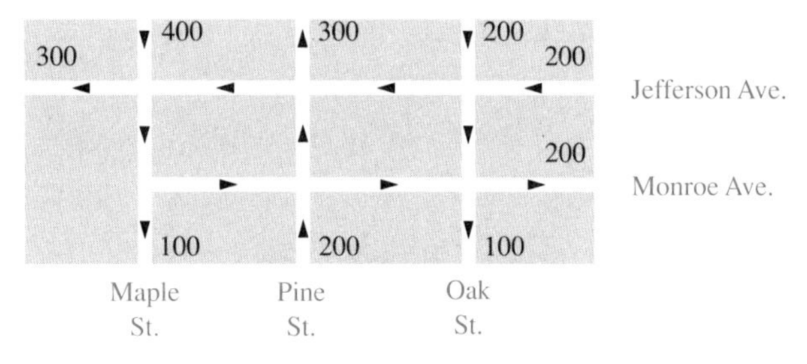

Answers to Practice Problems

7. (a)
$$\begin{aligned} 70x + 80y &= 1090 \\ 2x + 6y &= 46 \end{aligned}$$

(b) 11 slices of bread and 4 eggs

8. (a)
$$\begin{aligned} x_1 &= 400 + .2x_1 + .1x_2 \\ x_2 &= 210 + .2x_1 + .2x_2 \end{aligned}$$

(b) \$770 to Direct Sales and \$940 to Mail Order

9. 100

2.4 Matrices

The first use of matrices is generally attributed to the English mathematician Arthur Cayley in 1858. The widespread use of matrices is much more recent, however. In fact, the development of high-speed computers capable of doing matrix calculations is the primary reason for the present use of matrices in business and the social sciences.

In the remainder of this chapter we introduce matrices and the fundamental matrix operations and show their connection to the solution of linear systems.

MATRICES AND VECTORS

A **matrix** (plural: matrices) is simply a rectangular array of numbers, enclosed in brackets. (Some authors use parentheses.) The numbers are called the **entries** of the matrix. The following are three examples.

$$\begin{bmatrix} 3 & 5 & -2 \\ 4 & 0 & 0 \\ 0 & 22 & \frac{1}{2} \end{bmatrix} \qquad \begin{bmatrix} 3.4 & 3.3 & 5.6 & 1.2 \\ 9.1 & 0.0 & -1.3 & 0.4 \end{bmatrix} \qquad \begin{bmatrix} 76 \\ 44 \\ 9 \\ 0 \\ 3 \end{bmatrix}$$

The **size** of a matrix is determined by the number of rows and columns it has. A matrix with m rows and n columns is said to be an $\boldsymbol{m \times n}$ **matrix** (read "m by n"). For example, the three matrices above are 3×3, 2×4, and 5×1 matrices. We have already dealt with matrices under a different name. Example 2.10 in the previous section involved the linear system

$$\begin{aligned} 0.7x_1 - 0.3x_2 - 0.2x_3 &= 1000 \\ -0.3x_1 + 0.7x_2 - 0.2x_3 &= 800 \\ -0.1x_1 - 0.1x_2 + 0.6x_3 &= 1200, \end{aligned}$$

which we wrote in tableau form as follows.

$$\left[\begin{array}{rrr|r} 0.7 & -0.3 & -0.2 & 1000 \\ -0.3 & 0.7 & -0.2 & 800 \\ -0.1 & -0.1 & 0.6 & 1200 \end{array}\right]$$

This can just as well be regarded as a matrix, and in fact we call this the **augmented matrix** of the system of equations. Vertical or horizontal lines are often used to segregate certain entries of a matrix, as the vertical line in the last matrix separates the coefficients in our linear system from the constants. They have no effect on matrix definitions or operations.

Two matrices are considered to be **equal** when they have the same size and all their corresponding entries are equal. No two of the following matrices are equal.

$$\begin{bmatrix} 2 & 3 \\ 1 & 4 \\ 0 & 2 \end{bmatrix} \qquad \begin{bmatrix} 2 & 1 & 0 \\ 3 & 4 & 2 \end{bmatrix} \qquad \begin{bmatrix} 2 & 1 & 0 \\ 3 & 5 & 2 \end{bmatrix}$$

The first two are not equal because they have different sizes; the first is 3×2 and the second is 2×3. The second and third matrices are not equal because the entries in the second row and second column are not the same. In general the entry in row i and column j of a matrix is called the ***i,j* entry** of the matrix. For example the 3,2 entry of the first matrix above is 2, and its 1,2 entry is 3.

Often matrices are denoted by a single letter. For example we could define

$$A = \begin{bmatrix} 2 & 1 & 0 \\ 3 & 5 & 2 \end{bmatrix}.$$

We will usually use capital letters to denote matrices and small letters for their entries. Sometimes a system of double subscripts is used to name the entries in a matrix. We might write

$$A = \begin{bmatrix} 2 & 1 & 0 \\ 3 & 5 & 2 \end{bmatrix} = \begin{bmatrix} a_{11} & a_{12} & a_{13} \\ a_{21} & a_{22} & a_{23} \end{bmatrix},$$

where $a_{11} = 2$, $a_{12} = 1$, etc. Notice that in the notation a_{ij} the first subscript i is the row number, and the second subscript j is the column number. We may indicate this situation symbolically by writing $A = [a_{ij}]$; this means that A is a matrix whose i,j entry is denoted by a_{ij}.

EXAMPLE 2.12 Suppose we have

$$B = [b_{ij}] = \begin{bmatrix} 3 & 5 & 6 & -1 \\ 4 & -3 & 0 & 7 \end{bmatrix} = \begin{bmatrix} 5 - x & 5 & 6 & -1 \\ 4 & -3 & 0 & 7 \end{bmatrix}.$$

(a) What is the size of B?
(b) What is the 2,3 entry of B? What is b_{14}?
(c) What is x?

Solution (a) Since B has 2 rows and 4 columns, B is a 2×4 matrix.
(b) The 2,3 entry of B is 0, and $b_{14} = -1$.
(c) Since for two matrices to be equal their corresponding entries must be the same, we have $3 = 5 - x$. Thus $x = 2$. ■

Matrices with a single row or column are given special names. A matrix with a single row is called a **row vector,** and a matrix with a single column is called a **column vector.** For example, the 1×5 matrix C below is a row vector, and the 3×1 matrix D is a column vector.

$$C = [3 \quad 4 \quad 11 \quad -2 \quad 6] \qquad D = \begin{bmatrix} 81 \\ 43 \\ 30 \end{bmatrix}.$$

MATRIX OPERATIONS

We compute the **sum** of two matrices of the same size by adding their corresponding entries. For example if

$$A = \begin{bmatrix} 1 & -1 & 3 \\ 4 & 5 & -2 \end{bmatrix} \quad \text{and} \quad B = \begin{bmatrix} 3 & 4 & -2 \\ 3 & 0 & -1 \end{bmatrix},$$

then

$$A + B = \begin{bmatrix} 1+3 & -1+4 & 3+(-2) \\ 4+3 & 5+0 & -2+(-1) \end{bmatrix} = \begin{bmatrix} 4 & 3 & 1 \\ 7 & 5 & -3 \end{bmatrix}.$$

We can indicate the definition of matrix addition symbolically by

$$[a_{ij}] + [b_{ij}] = [a_{ij} + b_{ij}].$$

Practice Problem 10 Compute

$$\begin{bmatrix} 3 & 0 \\ -1 & 5 \\ -2 & 1 \end{bmatrix} + \begin{bmatrix} 2 & 2 \\ 6 & 4 \\ 5 & -3 \end{bmatrix}.$$

EXAMPLE 2.13 The matrices P_0 and P_1 below give the number of millions of tons of various pollutants emitted into the air of the United States by various sources during the years 1970 and 1971, respectively.

	Transportation	Stationary fuel combustion	Industrial processes	Solid waste disposal	Miscel-laneous
Carbon monoxide	82.3	1.1	11.8	5.5	6.6
Sulfur oxides	0.7	27.0	6.4	0.1	0.1
Hydrocarbons	14.7	1.6	2.9	1.4	11.5
Particulates	1.2	8.3	15.7	1.1	1.2
Nitrogen oxides	9.3	10.1	0.6	0.3	0.1

$= P_0$

	Transportation	Stationary fuel combustion	Industrial processes	Solid waste disposal	Miscel-laneous
Carbon monoxide	77.5	1.0	11.4	3.8	4.9
Sulfur oxides	1.0	26.3	5.1	0.1	0.1
Hydrocarbons	14.7	0.3	5.6	1.0	4.7
Particulates	1.0	6.5	13.5	0.7	4.9
Nitrogen oxides	11.2	10.2	0.2	0.2	0.2

$= P_1$

Compute $P_0 + P_1$, and interpret the meaning of its entries.

Solution By adding the corresponding entries, we get

$$P_0 + P_1 =$$

	Transportation	Stationary fuel combustion	Industrial processes	Solid waste disposal	Miscel-laneous
Carbon monoxide	159.8	2.1	23.2	9.3	11.5
Sulfur oxides	1.7	53.3	11.5	0.2	0.2
Hydrocarbons	29.4	1.9	8.5	2.4	16.2
Particulates	2.2	14.8	29.2	1.8	6.1
Nitrogen oxides	20.5	20.3	0.8	0.5	0.3

The entries in this matrix represent the total number of millions of tons of pollutants emitted by different sources in the years 1970 and 1971 combined. ■

The **difference** between two matrices of the same size is formed by subtracting their corresponding entries. Thus

$$\begin{bmatrix} 2 & 3 & -1 \\ 5 & 3 & 9 \end{bmatrix} - \begin{bmatrix} 1 & 8 & -6 \\ 7 & -2 & 0 \end{bmatrix} = \begin{bmatrix} 2-1 & 3-8 & -1-(-6) \\ 5-7 & 3-(-2) & 9-0 \end{bmatrix}$$

$$= \begin{bmatrix} 1 & -5 & 5 \\ -2 & 5 & 9 \end{bmatrix}.$$

When we talk about matrices, the word **scalar** is used to denote a real number, as opposed to a matrix. If c is a scalar and A is a matrix, we define the **product** cA of c and A to be the matrix formed by multiplying each entry of A by c. For example

$$3\begin{bmatrix} 2 & 3 & -1 \\ 5 & 3 & 9 \end{bmatrix} = \begin{bmatrix} 3(2) & 3(3) & 3(-1) \\ 3(5) & 3(3) & 3(9) \end{bmatrix} = \begin{bmatrix} 6 & 9 & -3 \\ 15 & 9 & 27 \end{bmatrix}.$$

Matrix subtraction and multiplication of a matrix by a scalar can be defined symbolically by

$$[a_{ij}] - [b_{ij}] = [a_{ij} - b_{ij}] \quad \text{and} \quad c[a_{ij}] = [ca_{ij}].$$

It is easily checked that

$$A - B = A + (-1)B.$$

Notice that matrices A and B can only be added or subtracted when they are of the same size.

Practice Problem 11 Compute $2A - 3B$, where

$$A = \begin{bmatrix} 1 & 3 & -1 \\ 0 & 2 & 1 \end{bmatrix} \quad \text{and} \quad B = \begin{bmatrix} -2 & 0 & 1 \\ 3 & -3 & 5 \end{bmatrix}.$$

EXAMPLE 2.14 (a) Use the matrices P_0 and P_1 of Example 2.13 to compute a matrix giving the decrease from 1970 to 1971 in the number of millions of tons of each pollutant from each source listed.

(b) Suppose the goal for 1990 is to have each source decrease the amount of each pollutant by 40% from its 1970 level. Compute a matrix giving the goal for each pollutant and source in millions of tons.

Solution (a) The decrease from 1970 to 1971 for each source and pollutant will be given by the entries in the matrix

$$P_0 - P_1 =$$

	Transportation	Stationary fuel combustion	Industrial processes	Solid waste disposal	Miscellaneous
Carbon monoxide	4.8	0.1	0.4	1.7	1.7
Sulfur oxides	−0.3	0.7	1.3	0.0	0.0
Hydrocarbons	0.0	1.3	−2.7	0.4	6.8
Particulates	0.2	1.8	2.2	0.4	−3.7
Nitrogen oxides	−1.9	−0.1	0.4	0.1	−0.1

Of course negative entries indicate an *increase* between 1970 and 1971.

(b) If the amount of each pollutant from each source is to decrease by 40%, then this amount will be 60%, or 0.6 times its 1970 level. Thus we compute the matrix

$$0.6P_0 =$$

	Transportation	Stationary fuel combustion	Industrial processes	Solid waste disposal	Miscellaneous
Carbon monoxide	49.38	0.66	7.08	3.30	3.96
Sulfur oxides	0.42	16.20	3.84	0.06	0.06
Hydrocarbons	8.82	0.96	1.74	0.84	6.90
Particulates	0.72	4.98	9.42	0.66	0.72
Nitrogen oxides	5.58	6.06	0.36	0.18	0.06

We could also have computed $P_0 - 0.4P_0$, which would have given the same result. ■

MATRIX ALGEBRA

In the last example we claimed that the matrices $0.6P_0$ and $P_0 - 0.4P_0$ were the same. We might try to justify this with the calculation

$$\begin{aligned} P_0 - 0.4P_0 &= 1P_0 - 0.4P_0 \\ &= (1 - 0.4)P_0 \\ &= 0.6P_0. \end{aligned} \tag{2.14}$$

If P_0 were a real number these steps would certainly be valid, but since P_0 is a matrix we need to be careful. Actually, matrices obey many algebraic laws similar to those for real numbers. Some of these laws involving matrix addition and multiplication by a scalar are given below. In these rules whenever matrices are added they are assumed to be of the same size. The symbol O stands for a **zero matrix** of the appropriate size, that is, a matrix in which all the entries are 0.

Laws for Matrix Addition and Multiplication by a Scalar

M1. $A + B = B + A$

M2. $A + (B + C) = (A + B) + C$

M3. $A + O = A$

M4. $k(A + B) = kA + kB$

M5. $(c + d)A = cA + dA$

M6. $c(dA) = (cd)A$

M7. $1A = A$

The reader should check that the first two equations of (2.14) are justified by M7 and M5. Other familiar-looking rules, such as $c(A - B) = cA - cB$ and $0A = O$, may be derived from M1 through M7. In the next section we will explain how to multiply matrices of appropriate sizes. We will find that matrix multiplication satisfies some, but not all, of the real number multiplication properties.

EXAMPLE 2.15 Find a 2×2 matrix X satisfying the equation

$$2(B + 3X) = 3B - 4(A - X),$$

where

$$A = \begin{bmatrix} 1 & -5 \\ 3 & 0 \end{bmatrix} \quad \text{and} \quad B = \begin{bmatrix} 8 & -6 \\ 4 & 2 \end{bmatrix}.$$

Solution First, we solve the equation for the unknown X.

$$\begin{aligned} 2(B + 3X) &= 3B - 4(A - X) \\ 2B + 6X &= 3B - 4A + 4X \\ 2X &= B - 4A \\ X &= \frac{1}{2}(B - 4A) = \frac{1}{2}B - 2A \\ X &= \frac{1}{2}\begin{bmatrix} 8 & -6 \\ 4 & 2 \end{bmatrix} - 2\begin{bmatrix} 1 & -5 \\ 3 & 0 \end{bmatrix} \\ &= \begin{bmatrix} 4 & -3 \\ 2 & 1 \end{bmatrix} - \begin{bmatrix} 2 & -10 \\ 6 & 0 \end{bmatrix} = \begin{bmatrix} 2 & 7 \\ -4 & 1 \end{bmatrix} \end{aligned}$$

We remark (without proof) that each of the steps used in solving the equation—such as multiplying out and transposing—can be justified on the basis of the laws M1–M7. ■

EXERCISES 2.4

Exercises 1–16 refer to the matrices

$$A = \begin{bmatrix} 3 & -1 & 0 & 4 \\ -3 & 5 & 2 & 1 \end{bmatrix}, \quad B = \begin{bmatrix} 6 & 3 & 11 & 8 \\ 2 & -1 & 0 & 3 \end{bmatrix},$$

$$C = \begin{bmatrix} 2 & 4 \\ 6 & 7 \\ 3 & -2 \end{bmatrix}, \quad D = \begin{bmatrix} 4 & 9 \\ -1 & 0 \\ 2 & 3 \end{bmatrix}.$$

In Exercises 1–12 compute the expression, if it is defined.

1. $A + B$

2. $C + D$

3. $A + C$

4. $B + D$

5. $B - A$

6. $D - C$

7. $3B$

8. $-2C$

9. $2A + B$

10. $D + 3A$

11. $2B + 4A$

12. $5D - 2C$

In Exercises 13–16 solve for the matrix X.

13. $X + 2A = B$

14. $3B - X = A$

15. $3A + 2X = 4B$

16. $2(X - 2A) = 3B$

In Exercises 17–22 write the augmented matrix of the given linear system.

17. $\begin{aligned} x + 3y + z &= 5 \\ x + 7y - z &= 0 \end{aligned}$

18. $\begin{aligned} x - y + 2z &= 6 \\ 2x + 2y - 3z &= 8 \\ 3x + 2y + z &= 5 \end{aligned}$

19. $\begin{aligned} x + 3y \quad &= 9 \\ x + \quad 7z &= 12 \end{aligned}$

20. $\begin{aligned} x + y &= 1 \\ y + z &= 3 \end{aligned}$

21. $\begin{aligned} x_1 + 2x_3 &= -3 \\ x_2 - x_4 &= 5 \\ x_1 + 4x_3 - x_2 &= 7 \end{aligned}$

22. $\begin{aligned} a + b &= c \\ 2a - b + c &= 4 \\ 4a + 4c &= b \end{aligned}$

23. Write a row vector with 5 entries, all 3's.

24. Write a column vector with 4 entries, all 2's.

25. Write a 3×2 matrix with all entries equal to 5.

26. Write a 2×4 matrix with all entries equal to 3.

The matrices E and F below give the average monthly cost in dollars of housing and food for a family of 4 in 1986 and 1987 in the cities of Elkville and Foxtown, respectively. All questions in Exercises 27–32 refer to monthly costs.

$$E = \begin{bmatrix} 300 & 420 \\ 335 & 440 \end{bmatrix} \begin{matrix} 1986 \\ 1987 \end{matrix} \qquad F = \begin{bmatrix} 275 & 350 \\ 300 & 370 \end{bmatrix} \begin{matrix} 1986 \\ 1987 \end{matrix}$$

(columns: housing, food)

27. What is the average housing cost for a family of four in Foxtown in 1987?

28. What is the average food cost for a family of four in Elkville in 1986?

29. Express in terms of E and F a matrix giving the average excess cost for housing and food for a family living in Elkville instead of Foxtown in 1986 and 1987.

30. Express in terms of E and F the average cost for three families for housing and food in Elkville in 1986 and 1987.

31. Express in terms of E and F the average housing and food cost per family for 2 families to live in Foxtown and 3 families in Elkville in 1986 and 1987.

32. Express in terms of E and F the housing and food cost for a family in Foxtown for 1986 and 1987 if the state welfare department pays 20% of their housing and food bills.

In Elkville, housing cost a family of four $300 per month in 1986 and $335 in 1987.

Answers to Practice Problems

10. $\begin{bmatrix} 5 & 2 \\ 5 & 9 \\ 3 & -2 \end{bmatrix}$

11. $\begin{bmatrix} 8 & 6 & -5 \\ -9 & 13 & -13 \end{bmatrix}$

2.5 Matrix Multiplication

In this section we will define the product of two matrices. The definition may seem unnatural and unnecessarily complicated at first, but it turns out to be very useful. We will start by defining the product AB when A is a $1 \times n$ matrix (row vector) and B is an $n \times 1$ matrix (column vector). The **product** AB is defined to be a 1×1 matrix with a single entry, formed by multiplying the corresponding entries of A and B and adding the resulting products. Symbolically,

$$[a_1 \quad a_2 \; . \; . \; . \; a_n]\begin{bmatrix} b_1 \\ b_2 \\ \cdot \\ \cdot \\ \cdot \\ b_n \end{bmatrix} = [a_1b_1 + a_2b_2 + \; . \; . \; . \; + a_nb_n].$$

For example

$$[2 \quad -3 \quad 4 \quad 1]\begin{bmatrix} 3 \\ 3 \\ 0 \\ 4 \end{bmatrix} = [2\cdot 3 + (-3)3 + 4\cdot 0 + 1\cdot 4]$$

$$= [6 - 9 + 0 + 4] = [1].$$

Sums of products of this sort arise in many situations. For example, the Sporting Chance, a retail sporting goods store, finds that its downtown store has 20 metal, 12 graphite, and 48 wood tennis rackets in stock. We could summarize this as a row matrix, namely

$$R = \begin{matrix} \text{metal} & \text{graphite} & \text{wood} \\ [\;20 & 12 & 48\;]. \end{matrix}$$

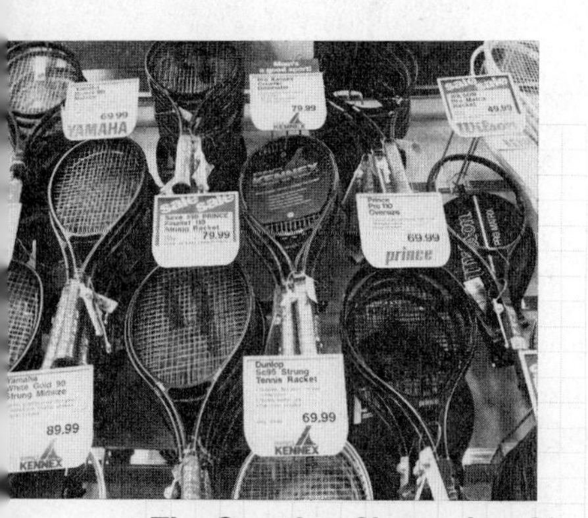

The Sporting Chance has 20 metal, 12 graphite, and 48 wood rackets.

The metal rackets retail for \$60, the graphite for \$110, and the wood rackets for \$40. We put these prices in a column vector:

$$C = \begin{matrix} \text{metal} \\ \text{graphite} \\ \text{wood} \end{matrix}\begin{bmatrix} 60 \\ 110 \\ 40 \end{bmatrix}.$$

Notice now that the total retail value of all the rackets is

$$20(\$60) + 12(\$110) + 48(\$40) = \$1200 + \$1320 + \$1920 = \$4440.$$

We can get the same answer by computing RC.

$$RC = [20 \quad 12 \quad 48]\begin{bmatrix} 60 \\ 110 \\ 40 \end{bmatrix}$$

$$= [20\cdot 60 + 12\cdot 110 + 48\cdot 40] = [4440]$$

EXAMPLE 2.16 Compute AB, where

$$A = [-1 \quad 2 \quad 2 \quad 0] \qquad \text{and} \qquad B = \begin{bmatrix} 4 \\ 3 \\ 2 \\ 5 \end{bmatrix}.$$

Solution The answer is the 1×1 matrix $[-1 \cdot 4 + 2 \cdot 3 + 2 \cdot 2 + 0 \cdot 5] = [6]$. ■

We will now show how to multiply matrices that are not necessarily row or column vectors. We will give several examples before giving the formal definition of matrix multiplication.

The Sporting Chance actually has three stores in all. Besides the downtown store there is also a suburban store and a store in the mall. The numbers of metal, graphite, and wood tennis rackets at each store are listed in the following matrix.

	metal	graphite	wood
downtown	20	12	48
suburban	16	9	30
mall	25	18	36

To compute the total retail value of the rackets at each store, we need to do three calculations like the one above. We organize them as follows.

$$\begin{bmatrix} 20 & 12 & 48 \\ \hdashline 16 & 9 & 30 \\ \hdashline 25 & 18 & 36 \end{bmatrix} \begin{bmatrix} 60 \\ 110 \\ 40 \end{bmatrix} = \begin{bmatrix} 20 \cdot 60 + 12 \cdot 110 + 48 \cdot 40 \\ 16 \cdot 60 + 9 \cdot 110 + 30 \cdot 40 \\ 25 \cdot 60 + 18 \cdot 110 + 36 \cdot 40 \end{bmatrix} = \begin{bmatrix} 4440 \\ 3150 \\ 4920 \end{bmatrix}$$

We have introduced some horizontal lines in the matrix on the left to indicate how the rows are used one at a time, like row vectors. Each row on the left contributes to one entry in the 3×1 answer, which indicates that the rackets have a retail value of \$4400 at the downtown store, \$3150 at the suburban store, and \$4920 at the mall store. This example shows how to compute a product AB, where B is a column vector with as many entries as the matrix A has columns.

EXAMPLE 2.17 Compute AB, where

$$A = \begin{bmatrix} 3 & 0 \\ 2 & 3 \\ -5 & 2 \\ 4 & -1 \end{bmatrix} \qquad \text{and} \qquad B = \begin{bmatrix} 5 \\ -2 \end{bmatrix}.$$

Solution Using the rows of A one at a time, we get

$$AB = \begin{bmatrix} 3 & 0 \\ 2 & 3 \\ -5 & 2 \\ 4 & -1 \end{bmatrix} \begin{bmatrix} 5 \\ -2 \end{bmatrix}$$

$$= \begin{bmatrix} 3 \cdot 5 + 0(-2) \\ 2 \cdot 5 + 3(-2) \\ (-5)5 + 2(-2) \\ 4 \cdot 5 + (-1)(-2) \end{bmatrix} = \begin{bmatrix} 15 \\ 4 \\ -29 \\ 22 \end{bmatrix}. \quad \blacksquare$$

Practice Problem 12 Compute AB, where

$$A = \begin{bmatrix} 1 & 3 & 2 \\ 0 & -1 & 1 \end{bmatrix} \quad \text{and} \quad B = \begin{bmatrix} -1 \\ 3 \\ 0 \end{bmatrix}.$$

Now we return to our example of the sporting goods stores. For insurance purposes the Sporting Chance needs to know not only the retail, but also the total *wholesale* value of the tennis rackets at each store. The wholesale prices of metal, graphite, and wood rackets are \$35, \$60, and \$25, respectively. These and the retail prices can be summarized in a matrix as follows.

$$\begin{array}{l} \\ \text{metal} \\ \text{graphite} \\ \text{wood} \end{array} \begin{array}{c} \begin{array}{cc} \text{retail} & \text{wholesale} \end{array} \\ \begin{bmatrix} 60 & 35 \\ 110 & 60 \\ 40 & 25 \end{bmatrix} \end{array}$$

Now we calculate both the retail and wholesale values of the rackets at each store by using both columns of the last matrix. The calculation goes as follows.

$$\begin{array}{l} \text{downtown} \\ \text{suburban} \\ \text{mall} \end{array} \overset{\begin{array}{ccc} \text{metal} & \text{graphite} & \text{wood} \end{array}}{\begin{bmatrix} 20 & 12 & 48 \\ 16 & 9 & 30 \\ 25 & 18 & 36 \end{bmatrix}} \overset{\begin{array}{cc} \text{retail} & \text{wholesale} \end{array}}{\begin{bmatrix} 60 & 35 \\ 110 & 60 \\ 40 & 25 \end{bmatrix}} \begin{array}{l} \text{metal} \\ \text{graphite} \\ \text{wood} \end{array}$$

$$= \begin{bmatrix} 20 \cdot 60 + 12 \cdot 110 + 48 \cdot 40 & 20 \cdot 35 + 12 \cdot 60 + 48 \cdot 25 \\ 16 \cdot 60 + 9 \cdot 110 + 30 \cdot 40 & 16 \cdot 35 + 9 \cdot 60 + 30 \cdot 25 \\ 25 \cdot 60 + 18 \cdot 110 + 36 \cdot 40 & 25 \cdot 35 + 18 \cdot 60 + 36 \cdot 25 \end{bmatrix}$$

$$= \begin{array}{l} \text{downtown} \\ \text{suburban} \\ \text{mall} \end{array} \overset{\begin{array}{cc} \text{retail} & \text{wholesale} \end{array}}{\begin{bmatrix} 4440 & 2620 \\ 3150 & 1850 \\ 4920 & 2855 \end{bmatrix}}$$

THE GENERAL DEFINITION OF MATRIX MULTIPLICATION

The entries of the **product** AB of the matrices A and B are computed by multiplying the rows of A, regarded as row vectors, by the columns of B, regarded as column vectors. More precisely, the i,j entry of AB is the product of the ith row of A and the jth column of B. This is indicated graphically in Figure 2.8.

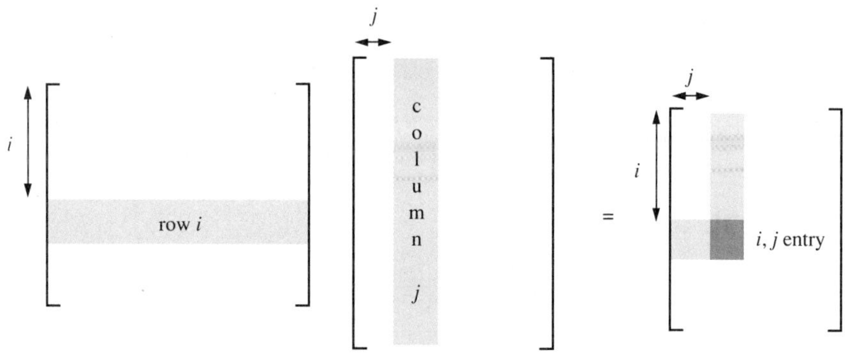

FIGURE 2.8

Notice that in order to multiply a row by a column, they must have the same number of entries. This means that in order for the product AB to be defined, the number of columns of A must equal the number of rows of B.

Suppose A is an $m \times n$ matrix and B is an $r \times s$ matrix.

1. The product AB is defined if and only if $n = r$.
2. If AB is defined, then AB is an $m \times s$ matrix.

EXAMPLE 2.18 Compute any of the products AB, AC, BA, BC, CA, and CB that are defined, where

$$A = \begin{bmatrix} 3 & 0 \\ 4 & -2 \\ 1 & 6 \end{bmatrix}, \quad B = \begin{bmatrix} -3 & 8 \\ 5 & -1 \end{bmatrix}, \quad \text{and} \quad C = \begin{bmatrix} 0 & 2 & 9 \\ 4 & -5 & 7 \end{bmatrix}.$$

Solution Notice that A, B, and C are 3×2, 2×2, and 2×3 matrices, respectively. Since in order for an $m \times n$ matrix times an $r \times s$ matrix to be defined we need that $n = r$, only AB, AC, BC, and CA are defined. We will show the computation of AB and AC.

$$\begin{aligned} AB &= \begin{bmatrix} 3 & 0 \\ 4 & -2 \\ 1 & 6 \end{bmatrix} \begin{bmatrix} -3 & 8 \\ 5 & -1 \end{bmatrix} \\ &= \begin{bmatrix} 3(-3) + 0(5) & 3(8) + 0(-1) \\ 4(-3) + (-2)5 & 4(8) + (-2)(-1) \\ 1(-3) + 6(5) & 1(8) + 6(-1) \end{bmatrix} \end{aligned}$$

$$= \begin{bmatrix} -9 & 24 \\ -22 & 34 \\ 27 & 2 \end{bmatrix}$$

$$AC = \begin{bmatrix} 3 & 0 \\ \hdashline 4 & -2 \\ \hdashline 1 & 6 \end{bmatrix} \left[\begin{array}{c:c:c} 0 & 2 & 9 \\ 4 & -5 & 7 \end{array}\right]$$

$$= \begin{bmatrix} 3(0) + 0(4) & 3(2) + 0(-5) & 3(9) + 0(7) \\ 4(0) + (-2)4 & 4(2) + (-2)(-5) & 4(9) + (-2)7 \\ 1(0) + 6(4) & 1(2) + 6(-5) & 1(9) + 6(7) \end{bmatrix}$$

$$= \begin{bmatrix} 0 & 6 & 27 \\ -8 & 18 & 22 \\ 24 & -28 & 51 \end{bmatrix}$$

The reader should check that

$$BC = \begin{bmatrix} 32 & -46 & 29 \\ -4 & 15 & 38 \end{bmatrix} \quad \text{and} \quad CA = \begin{bmatrix} 17 & 50 \\ -1 & 52 \end{bmatrix}. \quad ■$$

Practice Problem 13 Compute AB, where

$$A = \begin{bmatrix} 1 & 2 \\ 3 & -1 \end{bmatrix} \quad \text{and} \quad B = \begin{bmatrix} 1 & 0 & -1 \\ 2 & 3 & 0 \end{bmatrix}.$$

PROPERTIES OF MATRIX MULTIPLICATION

We have seen that, given matrices A and B, the product AB may or may not be defined. And if AB is defined, BA may not be; in fact this was the case in Example 2.18. Even when two matrices can be multiplied in either order, the results are usually not the same. We saw in Example 2.18 that the products AC and CA were both defined, but not equal. In fact AC was a 3×3 matrix, while CA was a 2×2 matrix. According to law M1 of the previous section, whenever A and B have the same size, then $A + B = B + A$. As we have seen, however, the corresponding property for matrix multiplication need not be true, even when both products are defined. Nevertheless, matrix multiplication does obey many familiar rules.

Laws of Matrix Multiplication

M8. $A(BC) = (AB)C$
M9. $A(B + C) = AB + AC$
M10. $(A + B)C = AC + BC$
M11. $c(AB) = (cA)B = A(cB)$

It is assumed in the above laws that A, B, and C are matrices, c is a scalar, and the matrix sums and products indicated are all defined.

EXERCISES 2.5

Exercises 1–24 refer to the following matrices.

$$A = [3 \quad -2 \quad 2] \qquad B = [-1 \quad 0 \quad 5] \qquad C = \begin{bmatrix} 4 \\ 3 \\ 0 \end{bmatrix} \qquad D = \begin{bmatrix} 3 \\ -2 \\ 1 \end{bmatrix}$$

$$E = \begin{bmatrix} 2 & -1 & 5 \\ 3 & 1 & 0 \\ 4 & 0 & 3 \end{bmatrix} \qquad F = \begin{bmatrix} 2 & 0 & -4 \\ 2 & 9 & 3 \\ 3 & 1 & 0 \end{bmatrix} \qquad G = \begin{bmatrix} 4 & 0 \\ 0 & -2 \\ -3 & 6 \end{bmatrix} \qquad H = \begin{bmatrix} 0 & 3 & -3 \\ 5 & 2 & 8 \end{bmatrix}$$

In Exercises 1–24 compute each indicated matrix product that is defined.

1. AB **2.** BC **3.** AD **4.** BD **5.** EC **6.** ED
7. FC **8.** FD **9.** HC **10.** HD **11.** EF **12.** FE
13. EG **14.** GE **15.** EH **16.** HE **17.** GH **18.** HG
19. BE **20.** BF **21.** CA **22.** CB **23.** $(EF)G$ **24.** $E(FG)$

Exercises 25–27 refer to the following three matrices.

$$A = \begin{array}{c} \text{cake} \quad \text{bread} \\ \begin{bmatrix} 2 & .5 \\ 2 & 3 \\ 1 & 2 \end{bmatrix} \end{array} \begin{array}{l} \text{cups sugar} \\ \text{cups flour} \\ \text{number eggs} \end{array} \qquad B = \begin{bmatrix} 13 \\ 17 \end{bmatrix} \begin{array}{l} \text{number cakes} \\ \text{number loaves} \end{array} \qquad C = \begin{array}{c} \text{cup sugar} \quad \text{cup flour} \quad \text{one egg} \\ \begin{bmatrix} .10 & .08 & .06 \\ .09 & .08 & .07 \end{bmatrix} \end{array} \begin{array}{l} \text{Chicago cost} \\ \text{Detroit cost} \end{array}$$

Matrix A gives the amount of sugar, flour and eggs needed to make a cake or a loaf of bread. For example a loaf of bread uses 2 eggs. Matrix B indicates that 13 cakes and 17 loaves of bread are needed. Matrix C gives the cost of a unit of each ingredient, both in Chicago and Detroit. In Exercises 25–27 indicate symbolically a matrix product presenting the information listed, then compute the product.

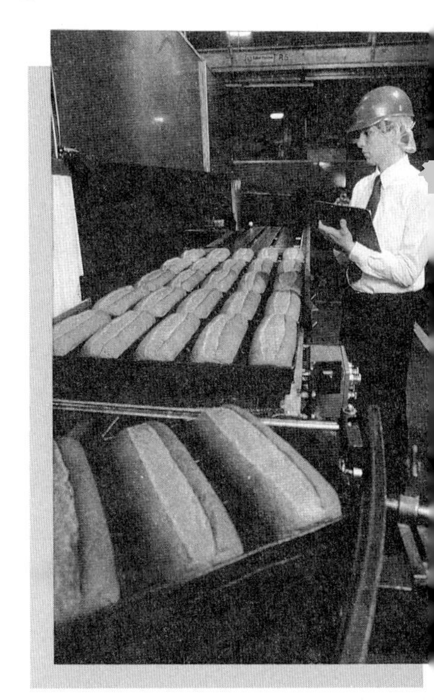

25. The amounts of sugar, flour, and eggs to make 13 cakes and 17 loaves of bread.

26. The cost of ingredients for making a cake and for making a loaf of bread in Chicago and Detroit.

27. The ingredient cost of making 13 cakes and 17 loaves of bread, both for Chicago and Detroit.

Exercises 28–31 refer to the following four matrices.

$$W = \begin{array}{c} \text{bluegrass} \quad \text{rye} \\ [.7 \qquad .4] \end{array} \text{ germination} \qquad X = \begin{array}{c} \text{mix A} \quad \text{mix B} \\ \begin{bmatrix} .75 & .60 \\ .25 & .40 \end{bmatrix} \end{array} \begin{array}{l} \text{bluegrass} \\ \text{rye} \end{array}$$

$$Y = \begin{bmatrix} 30 \\ 50 \end{bmatrix} \begin{array}{l} \text{lbs mix A} \\ \text{lbs mix B} \end{array} \qquad Z = \begin{array}{c} \text{mix A} \quad \text{mix B} \\ \begin{bmatrix} 3 & 2 \\ 4.5 & 3 \end{bmatrix} \end{array} \begin{array}{l} \text{wholesale} \\ \text{retail} \end{array}$$

Matrix W gives the proportion of bluegrass and rye seed that will germinate. Matrix X indicates that mix A contains 75% bluegrass and 25% rye, while mix B contains 60% bluegrass and 40% rye. Matrix Y indicates the number of pounds of the two mixes to be used in a city park. Matrix Z gives the wholesale and retail cost in dollars of a pound of each mix. In Exercises 28–31 indicate symbolically a matrix product presenting the information listed, then compute the product.

28. The number of pounds of bluegrass and rye to be used in the park.

29. The wholesale and retail cost of the seed used in the park.

30. The proportion of mix A and mix B that will germinate.

31. The number of pounds of seed used in the park that will germinate.

Answers to Practice Problems **12.** $\begin{bmatrix} 8 \\ -3 \end{bmatrix}$ **13.** $\begin{bmatrix} 5 & 6 & -1 \\ 1 & -3 & -3 \end{bmatrix}$

2.6 Inverse Matrices

In this section we will explore more of the algebra of matrix multiplication and show its connection with the solution of linear systems.

SQUARE MATRICES

The product of two matrices of different sizes can have a different size than either of the original two. For example, the product of a 3×2 matrix and a 2×4 matrix is a 3×4 matrix. If we restrict ourselves to matrices with the same number of rows as columns, that is, to **square matrices,** then the situation is much simpler. The product of an $m \times m$ matrix and an $n \times n$ matrix is only defined when $m = n$, and then the result is another matrix of the same size. For square matrices of a given size, one matrix is special. The $n \times n$ **identity matrix,** I_n, is the matrix with 1's down the top-left to bottom-right diagonal and 0's everywhere else. For example,

$$I_2 = \begin{bmatrix} 1 & 0 \\ 0 & 1 \end{bmatrix} \quad \text{and} \quad I_4 = \begin{bmatrix} 1 & 0 & 0 & 0 \\ 0 & 1 & 0 & 0 \\ 0 & 0 & 1 & 0 \\ 0 & 0 & 0 & 1 \end{bmatrix}.$$

The special property of the identity matrix is the following law.

M12. If A is an $n \times n$ matrix, then $I_nA = AI_n = A$.

We illustrate this in the 2×2 case:

$$\begin{bmatrix} 1 & 0 \\ 0 & 1 \end{bmatrix} \begin{bmatrix} a & b \\ c & d \end{bmatrix} = \begin{bmatrix} 1 \cdot a + 0 \cdot c & 1 \cdot b + 0 \cdot d \\ 0 \cdot a + 1 \cdot c & 0 \cdot b + 1 \cdot d \end{bmatrix} = \begin{bmatrix} a & b \\ c & d \end{bmatrix}.$$

The reader should check that the result of this computation is the same if the matrix factors on the left are reversed. Sometimes we denote an identity matrix as simply I, instead of I_n.

LINEAR SYSTEMS IN MATRIX FORM

Linear systems can be expressed using matrices. We illustrate this with the following example involving a system of three equations in three unknowns.

EXAMPLE 2.19 Consider the 3 by 3 system

$$\begin{aligned} x - 2y + 3z &= 1 \\ 2x - y + 4z &= 25 \\ -x + 3y - 4z &= 5. \end{aligned} \tag{2.15}$$

If we let

$$A = \begin{bmatrix} 1 & -2 & 3 \\ 2 & -1 & 4 \\ -1 & 3 & -4 \end{bmatrix}, \quad X = \begin{bmatrix} x \\ y \\ z \end{bmatrix}, \quad \text{and} \quad B = \begin{bmatrix} 1 \\ 25 \\ 5 \end{bmatrix},$$

then the system (2.15) is equivalent to the single matrix equation

$$AX = B \tag{2.16}$$

because (2.16) says that

$$AX = \begin{bmatrix} 1 & -2 & 3 \\ 2 & -1 & 4 \\ -1 & 3 & -4 \end{bmatrix} \begin{bmatrix} x \\ y \\ z \end{bmatrix} = \begin{bmatrix} x - 2y + 3z \\ 2x - y + 4z \\ -x + 3y - 4z \end{bmatrix} = \begin{bmatrix} 1 \\ 25 \\ 5 \end{bmatrix} = B,$$

and these two 3×1 vectors can be equal only if all three corresponding entries are equal; that is, only if equations (2.15) hold. ■

Equation (2.16) is called the **matrix form of the linear system.** We call A the **matrix of coefficients,** X the **vector of unknowns,** and B the **vector of right-hand constants.** Note that X and B are both written as *column* vectors.

Any m by n linear system can be expressed in matrix form as

$$AX = B,$$

where

$$\begin{aligned} A &= \text{matrix of coefficients } (m \times n) \\ X &= \text{vector of unknowns } (n \times 1) \\ B &= \text{vector of right-hand constants } (m \times 1). \end{aligned}$$

We saw in Section 2.1 that an $n \times n$ linear system normally has a unique solution. In other words, when A is square the matrix equation $AX = B$ usually has one and only one solution vector X. Before we discuss solving this equation by matrix methods, let us review briefly the Gauss-Jordan elimination procedure as applied to the linear system of Example 2.19.

The first step is to write the system in tableau form.

$$\left[\begin{array}{rrr|r} 1 & -2 & 3 & 1 \\ 2 & -1 & 4 & 25 \\ -1 & 3 & -4 & 5 \end{array}\right]$$

This is just the matrix $[A|B]$, the augmented matrix of the system. It consists of the matrix A of coefficients, "augmented" by the vector B of right-hand constants.

When an n by n linear system has a unique solution, the Gauss-Jordan procedure amounts to applying elementary row operations to $[A|B]$, until finally the identity matrix I appears on the left side of the line. In our example the calculation goes as follows.

$$\left[\begin{array}{rrr|r} \textcircled{1} & -2 & 3 & 1 \\ 2 & -1 & 4 & 25 \\ -1 & 3 & -4 & 5 \end{array}\right]$$

$$\left[\begin{array}{rrr|r} 1 & -2 & 3 & 1 \\ 0 & 3 & -2 & 23 \\ 0 & \textcircled{1} & -1 & 6 \end{array}\right] \quad \begin{array}{l} \\ (R_2 + (-2)R_1) \\ (R_3 + R_1) \end{array}$$

$$\left[\begin{array}{rrr|r} 1 & 0 & 1 & 13 \\ 0 & 0 & \textcircled{1} & 5 \\ 0 & 1 & -1 & 6 \end{array}\right] \quad \begin{array}{l} (R_1 + 2R_3) \\ (R_2 + (-3)R_3) \\ \\ \end{array}$$

$$\left[\begin{array}{rrr|r} 1 & 0 & 0 & 8 \\ 0 & 0 & 1 & 5 \\ 0 & 1 & 0 & 11 \end{array}\right] \quad \begin{array}{l} (R_1 + (-1)R_2) \\ \\ (R_3 + R_2) \end{array}$$

$$\left[\begin{array}{rrr|r} 1 & 0 & 0 & 8 \\ 0 & 1 & 0 & 11 \\ 0 & 0 & 1 & 5 \end{array}\right] \quad \begin{array}{l} \\ (R_3) \\ (R_2) \end{array}$$

The solution is therefore $x = 8$, $y = 11$, $z = 5$, or, in vector form,

$$X = \begin{bmatrix} x \\ y \\ z \end{bmatrix} = \begin{bmatrix} 8 \\ 11 \\ 5 \end{bmatrix}.$$

Note that the final tableau has the form $[I \mid C]$, where I is the identity matrix and $X = C$ is the unique solution. More generally, we have the following result.

Theorem 2.1

Consider the equation

$$AX = B,$$

where A is a square matrix. If $[A \mid B]$ can be transformed into $[I \mid C]$ by elementary row operations, then $AX = B$ has the unique solution $X = C$.

In fact, it is not difficult to show that Theorem 2.1 holds not only when X and B are column vectors, but more generally whenever A is $n \times n$ and X and B are both $n \times k$ matrices. The next example illustrates this for $n = k = 2$.

EXAMPLE 2.20 Solve the matrix equation

$$\begin{bmatrix} 1 & 2 \\ 3 & 2 \end{bmatrix} \begin{bmatrix} x & y \\ z & w \end{bmatrix} = \begin{bmatrix} 7 & 4 \\ 9 & 8 \end{bmatrix}.$$

Solution This has the form $AX = B$, where X is an unknown 2×2 matrix. Accordingly, we form the augmented matrix $[A \mid B]$ and try to reach $[I \mid C]$ by elementary row operations.

$$[A \mid B]$$

$$\left[\begin{array}{cc|cc} \textcircled{1} & 2 & 7 & 4 \\ 3 & 2 & 9 & 8 \end{array}\right]$$

$$\left[\begin{array}{cc|cc} 1 & 2 & 7 & 4 \\ 0 & \textcircled{-4} & -12 & -4 \end{array}\right] \quad (R_2 + (-3)R_1)$$

$$\left[\begin{array}{cc|cc} 1 & 2 & 7 & 4 \\ 0 & \textcircled{1} & 3 & 1 \end{array}\right] \quad (-\tfrac{1}{4}R_2)$$

$$\left[\begin{array}{cc|cc} 1 & 0 & 1 & 2 \\ 0 & 1 & 3 & 1 \end{array}\right] \quad (R_1 + (-2)R_2)$$

$$[I \mid C]$$

Since the identity matrix appears on the left, the solution appears on the right.

$$X = \begin{bmatrix} x & y \\ z & w \end{bmatrix} = \begin{bmatrix} 1 & 2 \\ 3 & 1 \end{bmatrix}$$

Check

$$AX = \begin{bmatrix} 1 & 2 \\ 3 & 2 \end{bmatrix} \begin{bmatrix} 1 & 2 \\ 3 & 1 \end{bmatrix} = \begin{bmatrix} 7 & 4 \\ 9 & 8 \end{bmatrix} = B. \quad ■$$

MATRIX INVERSION

Let us now consider the matrix equation $AX = B$ from the point of view of matrix algebra. Since matrices behave somewhat like numbers, we should be able to "divide" both sides of the equation by the matrix A, to obtain the solution as $X =$ "B/A." We must proceed carefully, though, because matrix division is not defined in general, and even matrix multiplication does not obey all the rules of ordinary algebra, as we have seen.

The first step is to define the "reciprocal" of a square matrix A. Since the identity matrix I plays the role of the number 1 in matrix algebra, we define the **inverse** of A, written A^{-1}, to be a square matrix satisfying the conditions

$$AA^{-1} = I \quad \text{and} \quad A^{-1}A = I.$$

If such a matrix exists, we say that A is **invertible.** As we shall see, there are noninvertible square matrices. However, it is an important fact that *the inverse of a square matrix is unique when it exists.* That is, no matrix can have two different inverses.

Since the inverse of A is a solution of the matrix equation $AX = I$, we can find it by the method of Theorem 2.1. This procedure is called **matrix inversion.**

Matrix Inversion

To find the inverse of an $n \times n$ matrix A:

1. Form the augmented matrix $[A \mid I]$, where I is the $n \times n$ identity matrix.
2. Apply elementary row operations until I appears to the left of the line. The inverse A^{-1} then appears to the right of the line.

To illustrate the method, let us apply it to the coefficient matrix A in Example 2.19.

EXAMPLE 2.19 Revisited Invert the matrix

$$A = \begin{bmatrix} 1 & -2 & 3 \\ 2 & -1 & 4 \\ -1 & 3 & -4 \end{bmatrix}.$$

Solution We form the augmented matrix $[A \mid I]$ and try to reach $[I \mid A^{-1}]$ by elementary row operations.

$$[A \mid I]$$

$$\left[\begin{array}{rrr|rrr} \textcircled{1} & -2 & 3 & 1 & 0 & 0 \\ 2 & -1 & 4 & 0 & 1 & 0 \\ -1 & 3 & -4 & 0 & 0 & 1 \end{array}\right]$$

Note that the row operations are exactly the same as before.

$$\left[\begin{array}{rrr|rrr} 1 & -2 & 3 & 1 & 0 & 0 \\ 0 & 3 & -2 & -2 & 1 & 0 \\ 0 & \textcircled{1} & -1 & 1 & 0 & 1 \end{array}\right] \quad \begin{array}{l} \\ (R_2 + (-2)R_1) \\ (R_3 + R_1) \end{array}$$

$$\left[\begin{array}{rrr|rrr} 1 & 0 & 1 & 3 & 0 & 2 \\ 0 & 0 & \textcircled{1} & -5 & 1 & -3 \\ 0 & 1 & -1 & 1 & 0 & 1 \end{array}\right] \quad \begin{array}{l} (R_1 + 2R_3) \\ (R_2 + (-3)R_3) \\ \end{array}$$

$$\left[\begin{array}{rrr|rrr} 1 & 0 & 0 & 8 & -1 & 5 \\ 0 & 0 & 1 & -5 & 1 & -3 \\ 0 & 1 & 0 & -4 & 1 & -2 \end{array}\right] \quad \begin{array}{l} (R_1 + (-1)R_2) \\ \\ (R_3 + R_2) \end{array}$$

$$\left[\begin{array}{rrr|rrr} 1 & 0 & 0 & 8 & -1 & 5 \\ 0 & 1 & 0 & -4 & 1 & -2 \\ 0 & 0 & 1 & -5 & 1 & -3 \end{array}\right] \quad \begin{array}{l} \\ (R_3) \\ (R_2) \end{array}$$

$$[I \mid A^{-1}]$$

The identity matrix now appears on the left side of the line, and so the inverse appears on the right.

$$A^{-1} = \begin{bmatrix} 8 & -1 & 5 \\ -4 & 1 & -2 \\ -5 & 1 & -3 \end{bmatrix}$$

Check

$$AA^{-1} = \begin{bmatrix} 1 & -2 & 3 \\ 2 & -1 & 4 \\ -1 & 3 & -4 \end{bmatrix} \begin{bmatrix} 8 & -1 & 5 \\ -4 & 1 & -2 \\ -5 & 1 & -3 \end{bmatrix} = \begin{bmatrix} 1 & 0 & 0 \\ 0 & 1 & 0 \\ 0 & 0 & 1 \end{bmatrix} = I \quad \blacksquare$$

It can be shown that for a square matrix X, the equation $AX = I$ implies $XA = I$, so that only the "one-sided" check of the inverse is necessary.

Practice Problem 14 Find A^{-1}, where

$$A = \begin{bmatrix} 0 & 1 & 3 \\ 0 & 1 & 2 \\ 1 & 0 & -1 \end{bmatrix}.$$

MATRIX SOLUTION OF $AX = B$

If A is invertible and we know A^{-1}, then the equation $AX = B$ can be solved very simply. A little matrix algebra confirms that a solution is $X = A^{-1}B$. For then

$$\begin{aligned} AX &= A(A^{-1}B) & \\ &= (AA^{-1})B & \text{(by M8)} \\ &= IB & \text{(definition of } A^{-1}\text{)} \\ &= B. & \text{(by M12)} \end{aligned}$$

For the system $AX = B$ of Example 2.19, this gives

$$X = A^{-1}B = \begin{bmatrix} 8 & -1 & 5 \\ -4 & 1 & -2 \\ -5 & 1 & -3 \end{bmatrix} \begin{bmatrix} 1 \\ 25 \\ 5 \end{bmatrix} = \begin{bmatrix} 8 \\ 11 \\ 5 \end{bmatrix},$$

which agrees with the result we obtained by Gauss-Jordan elimination. Note that only a single matrix multiplication is required to solve $AX = B$, once the inverse of A has been calculated.

Matrix Solution of $AX = B$

If A is an invertible matrix, then the equation $AX = B$ has the unique solution $X = A^{-1}B$.

Practice Problem 15 Solve

$$\begin{bmatrix} 0 & 1 & 3 \\ 0 & 1 & 2 \\ 1 & 0 & -1 \end{bmatrix} X = \begin{bmatrix} 3 \\ -1 \\ 5 \end{bmatrix}.$$

(Note that the inverse of the matrix on the left was computed in Practice Problem 14.)

EXAMPLE 2.21 The Allegheny Mining Company produces two grades of ore. Grade 1 contains 25% copper and 50% iron by weight, while Grade 2 contains 30% copper and 20% iron. How many tons of each grade should be sent to the refinery to fill the following orders exactly, without wasting any copper or iron?

(a) 1000 tons of copper and 1000 tons of iron
(b) 1000 tons of copper and 1500 tons of iron
(c) 2000 tons of copper and 1000 tons of iron

Solution If the company sends x tons of Grade A ore and y tons of Grade B ore, the amount of copper and iron in the shipment is given by

$$\begin{aligned} \text{number of tons of copper} &= .25x + .30y \\ \text{number of tons of iron} &= .50x + .20y. \end{aligned}$$

To fill an order for b_1 tons of copper and b_2 tons of iron, x and y must satisfy the equations

$$\begin{aligned} .25x + .30y &= b_1 \\ .50x + .20y &= b_2, \end{aligned}$$

or, in matrix form,

$$AX = \begin{bmatrix} .25 & .30 \\ .50 & .20 \end{bmatrix} \begin{bmatrix} x \\ y \end{bmatrix} = \begin{bmatrix} b_1 \\ b_2 \end{bmatrix} = B.$$

Notice that *the matrix A stays the same for all three orders; only the vector B changes*. We will use our method to invert the matrix A.

$$[A \mid I]$$

$$\left[\begin{array}{cc|cc} \textcircled{.25} & .30 & 1 & 0 \\ .50 & .20 & 0 & 1 \end{array}\right]$$

$$\left[\begin{array}{cc|cc} \textcircled{1} & 1.2 & 4 & 0 \\ .50 & .20 & 0 & 1 \end{array}\right] \qquad (4R_1)$$

$$\left[\begin{array}{cc|cc} 1 & 1.2 & 4 & 0 \\ 0 & \textcircled{-.4} & -2 & 1 \end{array}\right] \qquad (R_2 + (-.5)R_1)$$

$$\left[\begin{array}{cc|cc} 1 & 1.2 & 4 & 0 \\ 0 & \textcircled{1} & 5 & -2.5 \end{array}\right] \qquad (-2.5R_2)$$

$$\left[\begin{array}{cc|cc} 1 & 0 & -2 & 3 \\ 0 & 1 & 5 & -2.5 \end{array}\right] \qquad (R_1 + (-1.2)R_2)$$

$$[I \mid A^{-1}]$$

Since the identity matrix is on the left, the inverse is on the right.

$$A^{-1} = \begin{bmatrix} -2 & 3 \\ 5 & -2.5 \end{bmatrix}$$

Substituting the appropriate vector B in the equation $X = A^{-1}B$, we now obtain each of the three solutions.

$$\text{(a)} \quad \begin{bmatrix} x \\ y \end{bmatrix} = X = A^{-1}B = \begin{bmatrix} -2 & 3 \\ 5 & -2.5 \end{bmatrix} \begin{bmatrix} 1000 \\ 1000 \end{bmatrix} = \begin{bmatrix} 1000 \\ 2500 \end{bmatrix}$$

$$\text{(b)} \quad \begin{bmatrix} x \\ y \end{bmatrix} = X = A^{-1}B = \begin{bmatrix} -2 & 3 \\ 5 & -2.5 \end{bmatrix} \begin{bmatrix} 1000 \\ 1500 \end{bmatrix} = \begin{bmatrix} 2500 \\ 1250 \end{bmatrix}$$

$$\text{(c)} \quad \begin{bmatrix} x \\ y \end{bmatrix} = X = A^{-1}B = \begin{bmatrix} -2 & 3 \\ 5 & -2.5 \end{bmatrix} \begin{bmatrix} 2000 \\ 1000 \end{bmatrix} = \begin{bmatrix} -1000 \\ 7500 \end{bmatrix}$$

Note that the solution in (c) calls for a negative amount of Grade 1 ore, which means that this order cannot be filled exactly. That is, there is no way to provide exactly 2000 tons of copper and 1000 tons of iron. ■

NONINVERTIBLE MATRICES

The method of matrix inversion can only be used to solve n by n linear systems, because only a square matrix can have an inverse. But we may even encounter problems in the n by n case if the coefficient matrix A is noninvertible. When we try to invert such a matrix, our procedure will eventually produce *a row in which all the entries to the left of the line are zero*. This tells us that the matrix has no inverse.

EXAMPLE 2.22 Invert the matrix

$$A = \begin{bmatrix} 1 & -1 & 0 \\ 1 & 0 & -1 \\ 0 & 1 & -1 \end{bmatrix}.$$

Solution We form $[A \mid I]$ as usual and try to obtain an identity matrix I on the left side.

$$\left[\begin{array}{ccc|ccc} \textcircled{1} & -1 & 0 & 1 & 0 & 0 \\ 1 & 0 & -1 & 0 & 1 & 0 \\ 0 & 1 & -1 & 0 & 0 & 1 \end{array}\right]$$

$$\left[\begin{array}{ccc|ccc} 1 & -1 & 0 & 1 & 0 & 0 \\ 0 & \textcircled{1} & -1 & -1 & 1 & 0 \\ 0 & 1 & -1 & 0 & 0 & 1 \end{array}\right] \quad (R_2 + (-1)R_1)$$

$$\left[\begin{array}{ccc|ccc} 1 & 0 & -1 & 0 & 1 & 0 \\ 0 & 1 & -1 & -1 & 1 & 0 \\ 0 & 0 & 0 & 1 & -1 & 1 \end{array}\right] \quad \begin{array}{l} (R_1 + R_2) \\ \\ (R_3 + (-1)R_2) \end{array}$$

We cannot continue, because the third row of the left-hand matrix consists entirely of zeros. The matrix A is therefore noninvertible. ■

Practice Problem 16 Find the inverses of the following matrices, if possible.

(a) $\begin{bmatrix} 2 & 1 & 0 \\ 0 & 1 & -1 \\ 1 & 0 & 1 \end{bmatrix}$ (b) $\begin{bmatrix} 2 & 1 & 0 \\ 1 & 2 & 1 \\ 4 & 5 & 2 \end{bmatrix}$

EXERCISES 2.6

Exercises 1 and 2 refer to the following matrices.

$$A = \begin{bmatrix} 2 & 0 & -2 \\ 2 & 1 & 0 \\ 3 & 2 & 4 \end{bmatrix} \qquad B = \begin{bmatrix} 4 & 1 & -3 & 0 & 5 \\ 0 & 3 & 2 & -1 & 0 \\ 3 & 0 & 0 & 1 & 2 \\ 2 & 0 & 1 & -1 & 3 \\ 3 & 3 & 4 & -3 & 0 \end{bmatrix}$$

1. Write out I_3, and check that $AI_3 = A$ and $I_3A = A$.

2. Write out I_5, and check that $BI_5 = B$ and $I_5B = B$.

In Exercises 3–8 a linear system is given that can be expressed as AX = B in matrix form. Tell what A, X, and B are.

3. $\begin{aligned} x + 2y + z &= 3 \\ 2x + 2z &= 5 \\ 3y - z &= 0 \end{aligned}$

4. $\begin{aligned} 3x + 2y + z &= 1 \\ x - z &= 4 \\ 5x &= 9 \end{aligned}$

5. $\begin{aligned} r + s + t &= 5 \\ s + t &= 8 \end{aligned}$

6. $\begin{aligned} u + v &= 6 \\ u - v &= -2 \\ 3u + v &= 10 \end{aligned}$

7. $\begin{aligned} x_1 + x_3 &= 5 \\ x_1 + x_2 &= x_3 \\ x_1 + x_3 &= 0 \end{aligned}$

8. $\begin{aligned} a + 2b + 3c + d &= 4 \\ a + c + 3 &= b \\ a + e &= 5 \end{aligned}$

In Exercises 9–12 rewrite AX = B as a linear system of equations, where A, X, and B are as given.

9. $A = \begin{bmatrix} 2 & -2 \\ 3 & 5 \end{bmatrix}, X = \begin{bmatrix} x \\ y \end{bmatrix}, B = \begin{bmatrix} 0 \\ 2 \end{bmatrix}$

10. $A = \begin{bmatrix} 1 & 0 & 2 \\ 2 & 1 & 0 \end{bmatrix}, X = \begin{bmatrix} x \\ y \\ z \end{bmatrix}, B = \begin{bmatrix} 5 \\ -1 \end{bmatrix}$

11. $A = \begin{bmatrix} 1 & 2 \\ 2 & 1 \\ 0 & 3 \end{bmatrix}, X = \begin{bmatrix} u \\ v \end{bmatrix}, B = \begin{bmatrix} 0 \\ 2 \\ 5 \end{bmatrix}$

12. $A = \begin{bmatrix} 0 & 1 & 2 \\ 3 & 0 & -1 \\ -1 & 2 & 0 \end{bmatrix}, X = \begin{bmatrix} r \\ s \\ t \end{bmatrix}, B = \begin{bmatrix} 9 \\ 1 \\ 0 \end{bmatrix}$

In Exercises 13–22 find the inverse of the given matrix, if possible.

13. $\begin{bmatrix} 1 & 3 \\ 2 & 5 \end{bmatrix}$

14. $\begin{bmatrix} 2 & 7 \\ 1 & 4 \end{bmatrix}$

15. $\begin{bmatrix} 3 & 2 & 0 \\ 1 & 1 & 0 \\ 3 & 2 & 1 \end{bmatrix}$

16. $\begin{bmatrix} 1 & 1 & 1 \\ 0 & 1 & -2 \\ 1 & 1 & 2 \end{bmatrix}$

17. $\begin{bmatrix} 1 & -1 & 0 \\ 3 & -1 & 3 \\ 1 & -1 & 1 \end{bmatrix}$

18. $\begin{bmatrix} 1 & 2 & 0 \\ 2 & 1 & 1 \\ -1 & -5 & 1 \end{bmatrix}$

19. $\begin{bmatrix} 3 & 5 & 0 \\ 0 & 2 & -1 \\ 3 & 1 & 2 \end{bmatrix}$

20. $\begin{bmatrix} 3 & 2 & 1 \\ 2 & 2 & 1 \\ 2 & 1 & 1 \end{bmatrix}$

21. $\begin{bmatrix} 1 & 1 & 0 & 0 \\ 0 & 1 & 0 & 0 \\ 0 & 0 & 1 & 1 \\ 1 & 1 & 0 & 1 \end{bmatrix}$

22. $\begin{bmatrix} 1 & -1 & 0 & 0 \\ 0 & 1 & 0 & 0 \\ 3 & 0 & 1 & 0 \\ 2 & 0 & 0 & 1 \end{bmatrix}$

Use the given products to solve the linear systems in Exercises 23–30, using the matrix solution to AX = B.

$$\begin{bmatrix} 1 & 0 & 1 \\ -1 & 1 & -1 \\ -2 & 0 & -1 \end{bmatrix} \begin{bmatrix} -1 & 0 & -1 \\ 1 & 1 & 0 \\ 2 & 0 & 1 \end{bmatrix} = I_3 \qquad \begin{bmatrix} 4 & 3 & -3 \\ -1 & 0 & 1 \\ -1 & -1 & 1 \end{bmatrix} \begin{bmatrix} 1 & 0 & 3 \\ 0 & 1 & -1 \\ 1 & 1 & 3 \end{bmatrix} = I_3$$

23. $\begin{aligned} x + z &= 3 \\ -x + y - z &= 0 \\ -2x - z &= 5 \end{aligned}$

24. $\begin{aligned} x + z &= -1 \\ -x + y - z &= 2 \\ -2x - z &= -2 \end{aligned}$

25. $\begin{aligned} 4x + 3y - 3z &= 2 \\ -x + z &= 5 \\ -x - y + z &= -6 \end{aligned}$

26. $\begin{aligned} 4x + 3y - 3z &= 8 \\ -x \qquad\quad + z &= 2 \\ -x - y + z &= 0 \end{aligned}$

27. $\begin{aligned} -u \qquad\quad - w &= 3 \\ u + v \qquad\quad &= 2 \\ 2u \qquad\quad + w &= 1 \end{aligned}$

28. $\begin{aligned} a \qquad\quad + 3c &= 0 \\ b - c &= 5 \\ a + b + 3c &= -2 \end{aligned}$

29. $\begin{aligned} x_1 \qquad\quad + 3x_3 &= 1 \\ x_2 - x_3 &= 2 \\ x_1 + x_2 + 3x_3 &= -2 \end{aligned}$

30. $\begin{aligned} -r \qquad\quad - t &= 2 \\ r + s \qquad\quad &= 1 \\ 2r \qquad\quad + t &= -3 \end{aligned}$

Answers to Practice Problems

14. $\begin{bmatrix} 1 & -1 & 1 \\ -2 & 3 & 0 \\ 1 & -1 & 0 \end{bmatrix}$

15. $X = \begin{bmatrix} 9 \\ -9 \\ 4 \end{bmatrix}$

16. (a) $\begin{bmatrix} 1 & -1 & -1 \\ -1 & 2 & 2 \\ -1 & 1 & 2 \end{bmatrix}$ (b) Doesn't exist

MATHEMATICS IN ACTION

2.7 The Leontief Input-Output Model

The technique of matrix inversion gives us a powerful new tool for the solution of n by n linear systems. The method is particularly useful when we have a number of large systems to solve, all of which share the same coefficient matrix. In this section we encounter an important application of this type.

THE ANALYSIS OF NATIONAL ECONOMIES

The economies of modern industrial nations are extremely complex. When we break such systems down into their components, we find hundreds or even thousands of different industries, each supplying the others with goods and services needed in the production process. The coal industry, for example, supplies the steel industry with one of its main raw materials, but at the same time steel—in the form of machinery, vehicles, and so on—is needed to mine and transport coal. It is this constant flow of goods between industries that characterizes an interacting economy, and at the same time constitutes one of the main obstacles to effective economic planning. If we set arbitrary production goals for each industry we may wind up with shortages; and a price increase in one key industry may send a ripple through the entire economy.

Is it possible to describe mathematically the interaction of an economic system? This question occurred to Wassily Leontief, a Russian-born economist, while he was a student in Berlin in the 1920s. Using matrix algebra, Leontief developed a very elegant model for the working of an economy, known as *input-output theory*. After emigrating to the U.S. in 1931 he concentrated on the task of putting his theoretical ideas into practice. By the late 1940s, with the help of the newly developed computer, Leontief was able to distill the parameters he needed from a mountain of government statistics. His model proved to be remarkably accurate in predicting the needs of the postwar American economy, and today input-output theory is used to forecast the production needs of large corporations, as well as advanced and developing nations. The theory is also used to set prices for various commodities, to predict the effects of price changes, and to analyze the impact of production on the environment. In recognition of the widespread application of his ideas, Leontief was awarded the Nobel Prize for Economics in 1973.

INPUT-OUTPUT THEORY

We begin by dividing an economy into a certain number of **producing sectors** $P_1, P_2, \ldots, P_n$, each of which is thought of as producing a single good, commodity, or service. In his famous study of the 1947 U.S. economy, Leontief identified 500 industries, which he combined into 42 producing sectors such as ''textile mill products,'' ''petroleum and coal products,'' ''electrical machinery,'' and so on.

Once the producing sectors are determined, we must decide on an appropriate unit of measurement for the output of each sector. These units may be either *physical* or *monetary;* for example, one unit of steel might represent 10,000 tons (physical) or $1 million (monetary). The advantage of monetary units is that the output of different sectors can be directly compared, but physical units are sometimes essential, as in consideration of pricing. For our present discussion it does not matter which type of unit we adopt, so we shall use the term *unit* without specifying whether it is physical or monetary.

Central to Leontief's theory is a certain set of ratios. If the sectors $P_1, \ldots, P_n$ produce respective goods $G_1, \ldots, G_n$, then the **input-output ratio** a_{ij} specifies the amount of G_i needed to produce one unit of G_j. The $n \times n$ matrix $A = [a_{ij}]$ whose entries are the input-output ratios is called the **input-output matrix** (or **technology matrix**) for the economy. The meaning of the input-output matrix is illustrated in the next example.

EXAMPLE 2.23 Consider a highly simplified economy with three producing sectors: agriculture, manufacturing, and energy. Suppose the input-output matrix is as follows.

$$\begin{array}{c} \text{To output one unit of} \\ \begin{array}{ccc} \text{agr} & \text{mfg} & \text{engy} \end{array} \end{array}$$

$$A = \begin{bmatrix} .2 & .1 & .3 \\ .4 & .2 & .1 \\ .3 & .5 & .5 \end{bmatrix} \begin{array}{l} \text{agr} \\ \text{mfg} \\ \text{engy} \end{array} \quad \begin{array}{l} \text{Input} \\ \text{units} \\ \text{needed} \end{array}$$

The entries in each *column* indicate how many input units are needed from each of the three sectors in order to produce one unit of output. For example, the second column (manufacturing) shows that we need .1, .2, and .5 units of input from the agriculture, manufacturing, and energy sectors, respectively, to produce one unit of manufactured goods. If the units are monetary, this means that for each $1 worth of manufactured goods produced, the manufacturing sector consumes 10¢ worth of agricultural products, 20¢ worth of (other) manufactured goods, and 50¢ worth of energy. ■

In practice, these input-output ratios could be determined by looking at how much each sector buys from the others. For instance, if the manufacturing sector bought $45 billion worth of energy in a given year and produced $90 billion worth of goods, then each $1 worth of manufactured goods required 45/90 = .5 = 50¢ worth of energy. This is essentially how Leontief calculated his input-output ratios.

It is a fundamental assumption of the input-output model that *these ratios remain the same regardless of the levels of production*. Thus, if one unit of manufactured output requires .5 units of energy, then 1000 units of output should require 500 units of energy, and more generally x units of output should require $.5x$ units of energy. This assumption must be carefully investigated in real applications. For example, for large values of x there may be a more energy-efficient manufacturing process available. In practice, however, the assumption has been found to be reasonable, provided that production levels stay within certain limits. Thus, barring any sudden technological changes—such as the discovery of a new, cheap energy source—the input-output matrix will be a fairly stable feature of the economy.

GROSS PRODUCTION AND SURPLUS

Returning to Example 2.23, let us represent the total output of the three sectors by x_1, x_2, x_3.

$$x_1 = \text{total units of } G_1 \text{ (agriculture) produced}$$

$$x_2 = \text{total units of } G_2 \text{ (manufactured goods) produced}$$

$$x_3 = \text{total units of } G_3 \text{ (energy) produced}$$

Then the flow of goods in the economy can be represented graphically by means of the diagram in Figure 2.9.

In addition to the three sectors and their gross outputs, the diagram shows the number of units of output sent from each sector to the others. For example, since we are producing x_1 units of agricultural products, the input-output matrix shows that the agriculture sector will need as input $.2x_1$ units from itself, $.4x_1$ units from manufacturing, and $.3x_1$ units from energy, as indicated in Figure 2.9. The other arcs are labeled similarly, corresponding to the input-output matrix A.

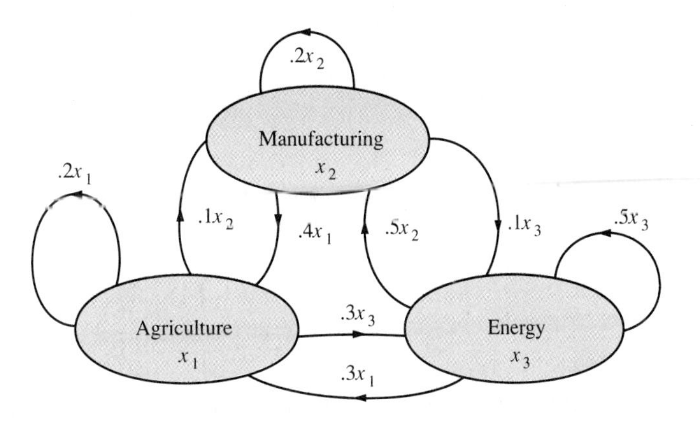

FIGURE 2.9

Looking at the three outgoing arcs from the agriculture sector, we see that the amount of agricultural output *used up internally by the three sectors* (the number of units needed for the production process itself) is given by the following equation:

$$u_1 = .2x_1 + .1x_2 + .3x_3.$$

There are similar equations for the other two sectors:

$$u_2 = .4x_1 + .2x_2 + .1x_3$$
$$u_3 = .3x_1 + .5x_2 + .5x_3.$$

In matrix terms these three equations say that

$$\begin{bmatrix} u_1 \\ u_2 \\ u_3 \end{bmatrix} = \begin{bmatrix} .2 & .1 & .3 \\ .4 & .2 & .1 \\ .3 & .5 & .5 \end{bmatrix} \begin{bmatrix} x_1 \\ x_2 \\ x_3 \end{bmatrix}$$

or

$$U = AX, \tag{2.17}$$

where

$$U = \begin{bmatrix} u_1 \\ u_2 \\ u_3 \end{bmatrix} = \textbf{internal consumption vector,}$$

$$X = \begin{bmatrix} x_1 \\ x_2 \\ x_3 \end{bmatrix} = \textbf{gross production vector,} \quad \text{and}$$

$$A = \begin{bmatrix} .2 & .1 & .3 \\ .4 & .2 & .1 \\ .3 & .5 & .5 \end{bmatrix} = \textbf{input-output matrix.}$$

If we are given the gross production X, we can calculate $U = AX$ and then find the surplus, or net production, by subtracting.

$$S = X - U = \textbf{surplus or net production vector} \tag{2.18}$$

The entries of S represent the amount of output available for consumer demand or export.

EXAMPLE 2.23 Continued Suppose the gross production of the economy is 800 units of agricultural products, 900 units of manufactured goods, and 1500 units of energy. What will the net production be?

Solution The gross production vector is given by

$$X = \begin{bmatrix} x_1 \\ x_2 \\ x_3 \end{bmatrix} = \begin{bmatrix} 800 \\ 900 \\ 1500 \end{bmatrix}.$$

We substitute this into formula (2.17).

$$U = AX = \begin{bmatrix} .2 & .1 & .3 \\ .4 & .2 & .1 \\ .3 & .5 & .5 \end{bmatrix} \begin{bmatrix} 800 \\ 900 \\ 1500 \end{bmatrix} = \begin{bmatrix} 700 \\ 650 \\ 1440 \end{bmatrix}$$

Hence, by (2.18),

$$\text{net production} = S = X - U = \begin{bmatrix} 800 \\ 900 \\ 1500 \end{bmatrix} - \begin{bmatrix} 700 \\ 650 \\ 1440 \end{bmatrix} = \begin{bmatrix} 100 \\ 250 \\ 60 \end{bmatrix}.$$

In other words, at the given gross production the economy will be able to keep itself running and have 100 units of agriculture, 250 units of manufactured goods, and 60 units of energy left over. ■

Practice Problem 17 Suppose in Example 2.23 the gross production is 1000 units of agricultural products, 1200 units of manufactured goods, and 2000 units of energy. What are the internal consumption vector U and the surplus production vector S?

MEETING A GIVEN DEMAND

We can also turn the problem around. Suppose we want the economy to produce just enough to meet a given **demand vector:**

$$D = \begin{bmatrix} d_1 \\ d_2 \\ d_3 \end{bmatrix} \begin{matrix} \text{units of } G_1 \text{ (agriculture)} \\ \text{units of } G_2 \text{ (manufactured goods)} \\ \text{units of } G_3 \text{ (energy).} \end{matrix}$$

At what gross production X must the economy operate?

To answer this, we first observe that by (2.17) and (2.18)

$$\text{surplus} = X - U = X - AX.$$

Setting the surplus (net production) exactly equal to the demand, we obtain the matrix equation

$$\begin{aligned} D = S &= X - AX \\ &= IX - AX \qquad \textbf{(by M12)} \\ &= (I - A)X \qquad \textbf{(by M10).} \end{aligned}$$

If we multiply both sides on the left by $(I - A)^{-1}$, we arrive at the solution

$$X = (I - A)^{-1}D, \tag{2.19}$$

provided, of course, that the matrix $I - A$ is invertible.

EXAMPLE 2.23 Continued What must the gross production of the economy be to meet a demand for 100 units of agriculture, 200 units of manufactured goods, and 150 units of energy?

Solution The demand vector is given by

$$D = \begin{bmatrix} 100 \\ 200 \\ 150 \end{bmatrix}.$$

To find the gross production vector X using formula (2.19), we first compute the matrix $I - A$.

$$I - A = \begin{bmatrix} 1 & 0 & 0 \\ 0 & 1 & 0 \\ 0 & 0 & 1 \end{bmatrix} - \begin{bmatrix} .2 & .1 & .3 \\ .4 & .2 & .1 \\ .3 & .5 & .5 \end{bmatrix} = \begin{bmatrix} .8 & -.1 & -.3 \\ -.4 & .8 & -.1 \\ -.3 & -.5 & .5 \end{bmatrix}$$

Although the decimals make the computation tedious, this matrix can be inverted by the method of Section 2.6 to obtain

$$(I - A)^{-1} = \begin{bmatrix} 2.80 & 1.60 & 2.0 \\ 1.84 & 2.48 & 1.6 \\ 3.52 & 3.44 & 4.8 \end{bmatrix}.$$

Formula (2.19) now gives

$$X = (I - A)^{-1}D = \begin{bmatrix} 2.80 & 1.60 & 2.0 \\ 1.84 & 2.48 & 1.6 \\ 3.52 & 3.44 & 4.8 \end{bmatrix} \begin{bmatrix} 100 \\ 200 \\ 150 \end{bmatrix} = \begin{bmatrix} 900 \\ 920 \\ 1760 \end{bmatrix}.$$

That is, the three sectors must produce 900, 920, and 1760 units of output, respectively, in order to meet the specified demand. ■

Practice Problem 18 In Example 2.23 what gross production is needed for a surplus of 200 units of agricultural products, 100 units of manufactured goods, and 300 units of energy?

The derivation of equation (2.19) can be generalized to any n-sector economy as follows.

Theorem 2.2

Suppose an economy has input-output matrix A such that $I - A$ is invertible. Then in order to meet a given demand vector D, the economy must operate at gross production X, where

$$X = (I - A)^{-1}D.$$

EXAMPLE 2.24

The economy of a Pacific island can be broken down into goods and services. To produce \$1 worth of goods requires \$.20 worth of goods and \$.20 worth of services. To produce \$1 worth of services requires \$.70 worth of goods and \$.20 worth of services. How many dollars worth of goods and services must be produced to have an export surplus of \$40,000 worth of goods and \$8000 worth of services?

Solution First we set up the input-output matrix A.

To produce \$1 worth of
goods services

$$A = \begin{bmatrix} .2 & .7 \\ .2 & .2 \end{bmatrix} \begin{matrix} \text{goods} \\ \text{services} \end{matrix} \quad \begin{matrix} \text{Input} \\ \text{value} \\ \text{needed} \end{matrix}$$

Thus we have

$$I - A = \begin{bmatrix} 1 & 0 \\ 0 & 1 \end{bmatrix} - \begin{bmatrix} .2 & .7 \\ .2 & .2 \end{bmatrix} = \begin{bmatrix} .8 & -.7 \\ -.2 & .8 \end{bmatrix}.$$

Using the method of the previous section we compute

$$(I - A)^{-1} = \begin{bmatrix} 1.6 & 1.4 \\ .4 & 1.6 \end{bmatrix}.$$

Now the demand vector is given by

$$D = \begin{bmatrix} 40{,}000 \\ 8{,}000 \end{bmatrix},$$

so by Theorem 2.2 we can compute the gross production vector

$$X = (I - A)^{-1}D = \begin{bmatrix} 1.6 & 1.4 \\ .4 & 1.6 \end{bmatrix} \begin{bmatrix} 40{,}000 \\ 8{,}000 \end{bmatrix} = \begin{bmatrix} 75{,}200 \\ 28{,}800 \end{bmatrix}.$$

Thus \$75,200 worth of goods and \$28,800 worth of services must be produced to supply the desired surplus. This can be checked by computing the internal consumption vector

$$U = AX = \begin{bmatrix} .2 & .7 \\ .2 & .2 \end{bmatrix} \begin{bmatrix} 75{,}200 \\ 28{,}800 \end{bmatrix} = \begin{bmatrix} 35{,}200 \\ 20{,}800 \end{bmatrix}$$

Then the surplus is

$$S = X - U = \begin{bmatrix} 75{,}200 \\ 28{,}800 \end{bmatrix} - \begin{bmatrix} 35{,}200 \\ 20{,}800 \end{bmatrix} = \begin{bmatrix} 40{,}000 \\ 8{,}000 \end{bmatrix},$$

as was desired. ■

EXERCISES 2.7

Exercises 1–4 refer to an interacting economy with four producing sectors: metals, nonmetals, energy, and services, with the following input-output matrix, where the units are dollars.

To produce \$1 worth of

metals	nonmetals	energy	services		
.32	.04	.09	.02	metals	
.05	.22	.41	.06	nonmetals	Input
.14	.03	.12	.02	energy	needed
.07	.11	.08	.05	services	

1. On what sector is the energy sector most dependent? Least dependent?
2. Which sector is most dependent on energy? Least dependent on energy?
3. If the nonmetals sector produces \$7.5 million worth of output, how much input must it consume from each of the four sectors?
4. If the services sector produces \$5.8 million worth of output, how much input must it consume from each of the four sectors?

Exercises 5–8 refer to an interacting economy with five producing sectors: agriculture, textiles, chemicals, machinery, and transportation, with the following input-output matrix, where the units are dollars.

To produce \$1 worth of

agriculture	textiles	chemicals	machinery	transportation		
.25	.21	.12	0	0	agriculture	
0	.13	0	0	0	textiles	Input
.02	.09	.35	.03	0	chemicals	needed
.05	.03	.01	.08	.14	machinery	
.07	.10	.02	.04	.10	transportation	

5. On what sector is the textiles sector most dependent? Least dependent?
6. Which sector is most dependent on machinery? Least dependent on machinery?

7. If the agriculture sector produces \$20 billion worth of output, how much input must it consume from each of the five sectors?

8. If the machinery sector produces \$12 billion worth of output, how much input must it consume from each of the five sectors?

9. An economy has two interacting sectors, manufacturing and energy. It takes .3 units of manufacturing and .4 units of energy to produce one unit of manufactured goods, and it takes .2 units of manufacturing and .1 units of energy to produce one unit of energy. What is the input-output matrix for this economy?

10. An economy has two interacting sectors, agriculture and chemicals. In a typical year, the agriculture sector spends a total of \$30 billion on its own goods and \$6 billion on chemicals and produces \$60 billion worth of output, while the chemicals sector spends \$20 billion on its own products and \$10 billion on agricultural goods and produces \$50 billion worth of output. Using monetary units, what is the input-output matrix for this economy?

11. A two-sector economy has the following input-output matrix:

$$A = \begin{bmatrix} .02 & .08 \\ .12 & .03 \end{bmatrix}.$$

The gross production of the economy is 1250 units for the first sector and 1800 units for the second sector.

(a) Find the gross production vector X.

(b) Compute the internal consumption vector U.

(c) Compute the surplus, or net production, S.

(d) Make a diagram like Figure 2.9, showing the flow of goods in the economy.

12. A three-sector economy has the following input-output matrix:

$$A = \begin{bmatrix} 0 & .11 & .05 \\ .15 & .02 & .08 \\ .09 & .14 & .01 \end{bmatrix}.$$

The gross production of the economy is $x_1 = 1500$, $x_2 = 1400$, $x_3 = 1700$.

(a) Find the gross production vector X.

(b) Compute the internal consumption vector U.

(c) Compute the surplus, or net production, S.

(d) Make a diagram like Figure 2.9, showing the flow of goods in the economy.

13. In a two-sector economy each dollar of goods requires \$.20 of goods and \$.20 of services to produce, and each dollar of services requires \$.30 of goods and \$.30 of services to produce. What must the gross production of the economy be to meet an external demand for

(a) \$5000 of goods and \$6000 of services?

(b) \$20,000 of goods and \$10,000 of services?

14. In a two-sector economy each megawatt of electricity needs .5 barrels of oil to produce, and each barrel of oil needs .6 megawatts of electricity and .3 barrels of oil to produce. What must the gross production of the economy be to meet an external demand for

(a) 2000 megawatts of electricity and 2000 barrels of oil?

(b) 0 megawatts and 4000 barrels of oil?

15. A certain simple society produces housing, clothing, and food. Production of 1 unit of housing or 1 unit of clothing requires 0.2 units of housing, 0.2 units of clothing, and 0.1 units of food. Production of 1 unit of food requires 0.4 units of housing, 0.4 units of clothing, and 0.2 units of food. What gross production is necessary to meet an external demand for 10 units of housing, 10 units of clothing, and 20 units of food?

16. Suppose that a small country produces three forms of energy—electricity, natural gas, and oil. Production of 1 unit of electricity requires 0.4 units of electricity, 0.2 units of natural gas, and 0.1 units of oil. Production of 1 unit of natural gas requires 0.1 units of electricity, 0.1 units of natural gas, and 0.05 units of oil. Finally, production of 1 unit of oil requires 0.4 units of natural gas and 0.2 units of oil. What gross production is required to meet an external demand for 400 units of electricity, 200 units of natural gas, and 300 units of oil?

17. A small country produces three forms of energy—electricity, oil, and coal. Production of 1 unit of electricity requires 0.2 units of electricity, 0.2 units of oil, and 0.1 units of coal. Production of 1 unit of oil requires 0.4 units of electricity, 0.4 units of oil, and 0.2 units of coal. Finally, production of 1 unit of coal requires 0.1 units of electricity, 0.1 units of oil, and 0.3 units of coal. What gross production is necessary to meet an external demand for 100 units of electricity, 200 units of oil, and 100 units of coal?

18. Suppose that a simple economy produces transportation, food, and oil. Production of 1 unit of transportation or 1 unit of food requires 0.1 units of transportation, 0.25 units of food, and 0.1 units of oil. Production of 1 unit of oil requires 0.3 units of transportation, 0.25 units of food, and 0.3 units of oil. What gross production is necessary to meet an external demand for 40 units of transportation, 60 units of food, and 80 units of oil?

Answers to Practice Problems

17. $$U = \begin{bmatrix} 920 \\ 840 \\ 1900 \end{bmatrix}, \quad S = \begin{bmatrix} 80 \\ 360 \\ 100 \end{bmatrix}$$

18. 1320 units of agricultural products, 1096 units of manufactured goods, and 2488 units of energy

CHAPTER 2 REVIEW

IMPORTANT TERMS

- elementary operations (2.1)
- elementary row operations
- Gauss-Jordan elimination
- linear equation
- *m* by *n* linear system
- pivot row and column
- pivot transformation
- tableau form
- basic and free variables (2.2)
- dependent system
- inconsistent system
- augmented matrix of a linear system (2.4)
- *i, j* entry
- product of a matrix by a scalar
- row and column vectors
- sum and difference of matrices
- zero matrix
- product of matrices (2.5)
- identity matrix (2.6)
- inverse matrix
- invertible matrix
- matrix of a linear system
- square matrix
- demand vector (2.7)
- input-output matrix
- input-output ratio
- internal consumption vector
- gross production vector
- surplus or net production vector

IMPORTANT FORMULAS

Laws of Matrix Algebra

M1. $A + B = B + A$
M2. $A + (B + C) = (A + B) + C$
M3. $A + O = A$
M4. $k(A + B) = kA + kB$
M5. $(c + d)A = cA + dA$
M6. $c(dA) = (cd)A$
M7. $1A = A$
M8. $A(BC) = (AB)C$
M9. $A(B + C) = AB + AC$
M10. $(A + B)C = AC + BC$
M11. $c(AB) = (cA)B = A(cB)$
M12. $IA = AI = A$

Matrix Solution to a Linear System

If A is an invertible matrix, then $AX = B$ has the unique solution $X = A^{-1}B$.

Internal Consumption and Surplus

If A is the input-output matrix, X is the gross production vector, U is the internal consumption vector, and S is the surplus vector, then $U = AX$ and $S = X - U$.

Gross Production

The gross production needed to satisfy a demand D is $X = (I - A)^{-1}D$, assuming that the matrix inverse exists.

REVIEW EXERCISES

In Exercises 1–3 solve the given linear system using Gauss-Jordan elimination.

1. $$\begin{aligned} 3x + y + z &= 2 \\ x - y + z &= 0 \\ -x + 4y - 2z &= 3 \end{aligned}$$

2. $$\begin{aligned} 2x + y - z + w &= 3 \\ x - 2y + w &= 0 \\ -x + 7y - z - 2w &= 3 \end{aligned}$$

3. $$\begin{aligned} a + 2b - 3c + d &= 6 \\ 2a - b + c - d &= -1 \\ 4a + 3b - 5c + d &= 2 \end{aligned}$$

4. The chemicals in a fertilizer are denoted by three numbers, giving the percentage by weight of nitrogen, phosphorus, and potassium in the fertilizer. For example, 5-10-5 fertilizer is 5% nitrogen, 10% phosphorus, and 5% potassium. A gardener has available three fertilizers: 5-15-10, 20-5-5, and 10-10-10. How many pounds of each type should she use to get a mixture containing 55 pounds of nitrogen, 70 pounds of phosphorus, and 55 pounds of potassium?

Exercises 5–12 refer to the following matrices.

$$A = \begin{bmatrix} 0 & 1 & 1 \\ 2 & 0 & 1 \\ -3 & 1 & 0 \end{bmatrix} \quad B = \begin{bmatrix} 2 & 1 & 0 \\ 1 & 1 & -1 \\ -1 & -2 & 3 \end{bmatrix} \quad C = \begin{bmatrix} 2 & 1 & 0 \\ 3 & -1 & 5 \end{bmatrix} \quad D = \begin{bmatrix} -2 \\ 4 \\ 3 \end{bmatrix}$$

In Exercises 5–12 compute the indicated matrix, if possible.

5. $A - 2B$ **6.** AB **7.** BD **8.** AC

9. $A + I_3$ **10.** CI_3 **11.** A^{-1} **12.** B^{-1}

13. Each of the linear systems in Exercises 1–3 can be expressed as $AX = B$ in matrix form. Tell what A, X, and B are in each case.

14. Find the inverse of the coefficient matrix and use it to solve the system

$$\begin{aligned} x \qquad + z &= b_1 \\ y + z &= b_2 \\ x + 2y + z &= b_3 \end{aligned}$$

for (a) $b_1 = 3$, $b_2 = 5$, $b_3 = -2$, and (b) $b_1 = 350$, $b_2 = 400$, $b_3 = 100$.

15. A simple economy is based on goods and services. To produce a unit of goods requires .7 units of goods and .1 units of services, while to produce a unit of services requires .4 units of goods and .7 units of services.

(a) What is the surplus production if the gross production is 1000 units of goods and 600 units of services?

(b) What gross production is needed to produce a surplus of 500 units of goods and 800 units of services?

3

LINEAR PROGRAMMING

Probably the most widely applied mathematical theory in modern business and industry is one that did not even exist fifty years ago: the theory of *linear programming*. Today it is estimated that fully 25 percent of scientific computer use is devoted to linear programming, and the range of its applications seems almost unlimited.

In this chapter we study linear programming and present some basic applications of the theory. We start geometrically, then reexamine the subject from an algebraic viewpoint, explaining the *simplex algorithm*, a computational technique for solving linear programming problems efficiently and methodically.

3.1 Optimization with Linear Constraints

Many practical problems involve **optimization:** how can a certain quantity—profit or cost, for instance—be maximized or minimized, subject to given conditions? Questions of this type arise in every phase of industrial production. For example, materials must be obtained at minimum cost, subject to fixed production quotas; limited resources must be allocated in the most profitable way; finished goods must be transported as economically as possible, given fixed supplies and demands in different locations.

During World War II many optimization problems arose in connection with military logistics, and teams of research scientists were formed to investigate possible methods of solution. They noticed that many of their problems shared the same features, once they were formulated in mathematical terms. The name ''linear program'' was used to describe any abstract problem of maximization or minimization subject to linear constraints, and a new mathematical theory was born. By the 1950s, linear programming was being applied far beyond its original setting, to problems in business, industry, agriculture, and the natural and social sciences.

EXAMPLE 3.1 Albert's Smoke Shop has in stock 480 ounces of Virginia and 720 ounces of Latakia tobacco leaf. Albert sells two popular house blends in a standard 12-ounce can. Each can of Blend A (Albert's Mixture) contains 8 ounces of Virginia and 4 ounces of Latakia, and sells at \$16. Each can of Blend B (Balkan Intrigue) contains 3 ounces of Virginia and 9 ounces of Latakia, and sells at \$12. How many cans of each blend should Albert make in order to maximize his total sales revenue?

Formulation We assume that Albert can sell both blends at the stated prices, so that his problem is not limited demand, but rather his limited supply of tobacco leaf. The unknowns in the problem are

$$x = \text{number of cans of Blend A to be produced}$$
$$y = \text{number of cans of Blend B to be produced.}$$

We summarize the information we have in the following table.

	Blend A	*Blend B*	*Available*
Virginia per can	8 oz	3 oz	480 oz
Latakia per can	4 oz	9 oz	720 oz
Price per can	\$16	\$12	
Number of cans	x	y	

Albert's revenue R from x cans of Blend A at \$16 and y cans of Blend B at \$12 is

$$16x + 12y = R,$$

and this is the quantity he wants to maximize. The total number of ounces of Virginia needed is $8x + 3y$, and so we must have

$$8x + 3y \leq 480. \tag{3.1}$$

Likewise $4x + 9y$ ounces of Latakia will be used, so

$$4x + 9y \leq 720. \tag{3.2}$$

An implicit restriction on the variables is that neither can be negative, since x and y represent the number of cans of Blends A and B. Thus

$$x \geq 0, \tag{3.3}$$

$$y \geq 0. \tag{3.4}$$

We have reduced our problem to a purely mathematical formulation, namely:

Find x and y that maximize

$$16x + 12y = R,$$

subject to

$$\begin{aligned} 8x + 3y &\leq 480 && (3.1)\\ 4x + 9y &\leq 720 && (3.2)\\ x &\geq 0 && (3.3)\\ y &\geq 0. && (3.4) \end{aligned}$$

We will give a geometric method of solving this problem in Section 3.2. ■

The mathematical problem above is known as a **linear program.** The quantity $16x + 12y = R$ that is to be optimized (in this case maximized) is called the **objective function,** and the inequalities (3.1) through (3.4) are called the **constraints** of the program. A point (x, y) is said to be a **feasible point** in case it satisfies all the constraints. For example, the point (40, 50) is feasible in the above program because

$$\begin{aligned} 8(40) + 3(50) &= 470 \le 480 \\ 4(40) + 9(50) &= 610 \le 720 \\ 40 &\ge 0, \\ 50 &\ge 0. \end{aligned}$$

Thus Albert *could* make 40 cans of Blend A and 50 cans of Blend B, even though he would have tobacco left over, and so not maximize his revenue. On the other hand (50, 40) is *not* feasible, since for this point (3.1) becomes

$$8(50) + 3(40) = 520 \le 480,$$

which is false. A point (x, y) that optimizes the objective function is called an **optimal point.**

Practice Problem 1 (a) Tell which of the following points are feasible in the program of Example 3.1: (50, 30), (30, 50), (45, 40). (b) Among the feasible points listed, which gives the best value of the objective function?

In Example 3.1 it could be argued that another implicit restriction on x and y is that they be whole numbers, since Albert only sells full, 12-ounce cans of Blends A and B. In general we will ignore such restrictions when setting up linear programming problems since they make solving the problems much more difficult. Usually reasonable answers can be found by treating the variables as real numbers, then rounding to integers after the best real solution is found.

EXAMPLE 3.2

Nutritionists indicate that an adequate daily diet should provide at least 75 gm of carbohydrate, 60 gm of protein, and 60 gm of fat. An ounce of Food A contains 3 gm of carbohydrate, 2 gm of protein, and 4 gm of fat, and costs 10¢, while an ounce of Food B contains 3 gm of carbohydrate, 4 gm of protein, and 2 gm of fat, and costs 15¢. How many ounces of each food should be combined per day to meet the nutritional requirements at lowest cost?

Formulation The unknowns are

$$\begin{aligned} x &= \text{number of ounces of Food A per day,} \\ y &= \text{number of ounces of Food B per day.} \end{aligned}$$

The information is summarized in the table below.

	Food A	*Food B*	*Needed*
Carbohydrate/oz	3 gm	3 gm	75 gm
Protein/oz	2 gm	4 gm	60 gm
Fat/oz	4 gm	2 gm	60 gm
Cost/oz	10¢	15¢	
Number of ounces	x	y	

The objective function is the total cost

$$10x + 15y = C \text{ (in cents)},$$

which we want to *minimize* in this example. The number of grams of carbohydrates in x ounces of Food A and y ounces of Food B is $3x + 3y$, which must be at least 75 to meet the daily requirement. Thus we have the constraint

$$3x + 3y \geq 75.$$

Similarly, the protein and fat requirements lead to the following constraints:

$$2x + 4y \geq 60,$$
$$4x + 2y \geq 60.$$

Adding the obvious requirement that neither x nor y can be negative, we arrive at the following linear program:

$$\begin{aligned} \text{Minimize} \quad & 10x + 15y = C \\ \text{subject to} \quad & 3x + 3y \geq 75 \\ & 2x + 4y \geq 60 \\ & 4x + 2y \geq 60 \\ & x \geq 0 \\ & y \geq 0. \end{aligned}$$ ■

Practice Problem 2 A truck can carry at most 4000 pounds and 2000 cubic feet of cargo. A TV set weighs 40 pounds and occupies 3 cubic feet. A VCR weighs 25 pounds and occupies 2 cubic feet. The trucker gets \$5 for each TV set and \$4 for each VCR carried, and wants to load the truck so as to maximize revenue. Set up the corresponding linear program.

Remark Notice that the first order of business when setting up a linear program is to *state clearly what the variables represent*. Without this step anything that follows is meaningless. In Example 3.1 statements such as

$$x = \text{Blend A}, \qquad y = \text{Blend B}$$

would be inadequate. Do x and y count cans, ounces, or what? *Failure to identify completely and explicitly what the variables represent leads to confusion for both the person setting up a linear program and anyone trying to understand his or her work.*

Making a table as in the previous examples can be helpful in determining the constraints and objective function of a linear program.

LINEAR INEQUALITIES

Although a linear program may involve any number of variables, if there are precisely two, as in our examples so far, the constraints of the program can be given a geometric interpretation. They have the form of **linear inequalities,** that is, inequalities of the form

$$ax + by \geq c$$

or

$$ax + by \leq c,$$

where a and b are not both 0. Of course the graph of the corresponding *equality,*

$$ax + by = c,$$

is a straight line, as we saw in Chapter 1. It turns out that

> The graph of a linear inequality
>
> $$ax + by \geq c \qquad \text{or} \qquad ax + by \leq c,$$
>
> where a and b are not both 0, consists of all points in the plane on one side of the line with equation $ax + by = c$, along with the points on the line itself.

Consider, for example, the linear inequality

$$8x + 3y \leq 480, \tag{3.1}$$

which is one of the constraints in the linear program of Example 3.1. The line that is the graph of $8x + 3y = 480$ is graphed by finding its x- and y-intercepts, as shown in Figure 3.1(a).

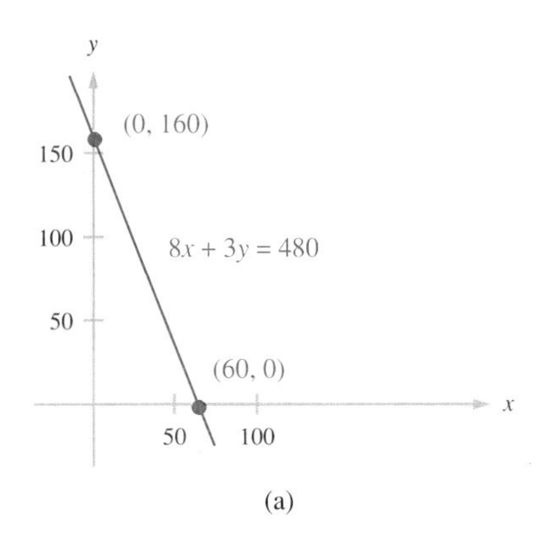

(a)

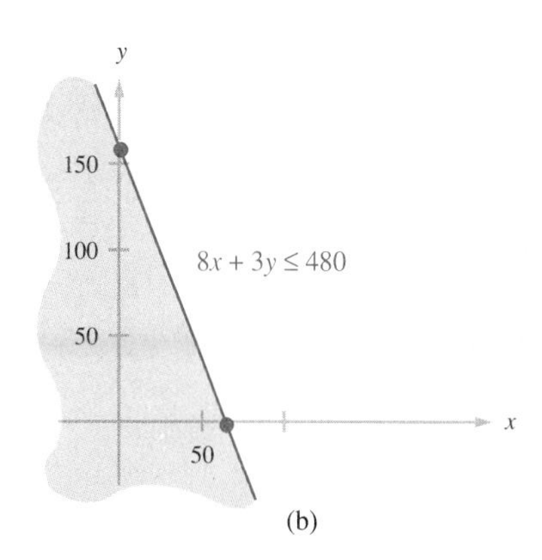

(b)

FIGURE 3.1

Figure 3.1(b) indicates with shading the set of points (x, y) satisfying the inequality

$$8x + 3y \leq 480. \tag{3.1}$$

Once one knows that the graph of a linear inequality consists of all points on one side of a line, graphing the inequality entails only graphing the line and deciding which side to shade. The latter decision can be made by testing in the inequality a single point P, not lying on the line. If the point's x- and y-coordinates satisfy the inequality, then the side of the line containing P is shaded; but if they do not, then the side of the line not containing P is shaded. For example, the origin (0, 0) may be tested in the inequality (3.1). Since

$$8(0) + 3(0) \leq 480$$

is a true statement, the side of the line containing (0, 0), that is, below and to its left, is the graph of the inequality. Such a set of points on one side of a line is called a **half-plane.**

EXAMPLE 3.3 Graph the inequalities (a) $3x - 4y \geq 12$ and (b) $4x + 5y \geq 0$.

Solution (a) We start graphing the line with equation $3x - 4y = 12$ by finding its intercepts, (4, 0) and (0, −3). Testing (0, 0) gives us

$$3(0) - 4(0) \geq 12,$$

which is false. Thus we shade the side of the line *not* containing the origin, as shown in Figure 3.2(a).

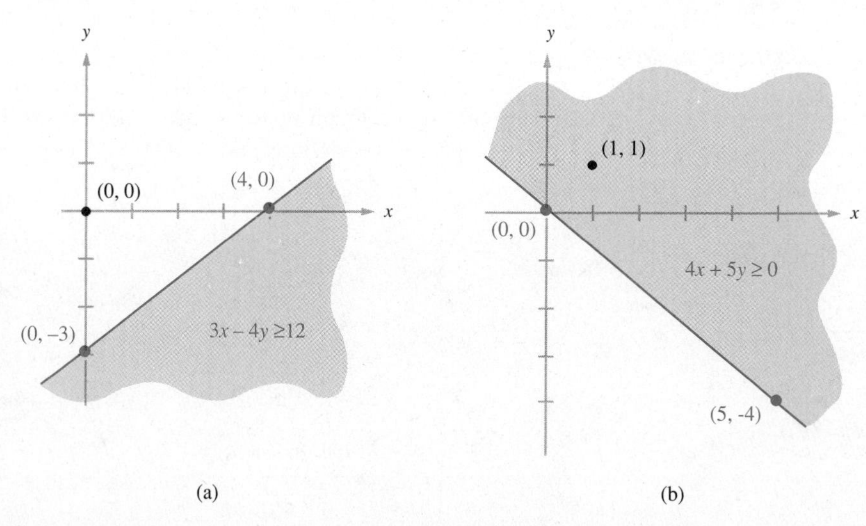

FIGURE 3.2

(b) Again we start by graphing the corresponding line, in this case with equation $4x + 5y = 0$. In this case, however, the point (0, 0) is *on* this line. Thus we choose another point to test, say (1, 1). Since

$$4(1) + 5(1) > 0$$

is true, we shade the side of the line containing (1, 1) as shown in Figure 3.2(b). ■

EXAMPLE 3.3 Continued Graph the set of points satisfying *both* $3x - 4y \geq 12$ and $4x + 5y \geq 0$.

Solution Using Figure 3.2, we now graph *both* inequalities on the same set of axes in Figure 3.3(a). To save time in graphing, instead of shading each inequality, we use arrows to point from each line to the half-plane satisfying each inequality as in Figure 3.3(b). The set of points where the two half-planes overlap is shaded in Figure 3.3(c). Notice that only the portions of the lines $3x - 4y = 12$ and $4x + 5y = 0$ adjacent to the hatched region belong to the graph. ■

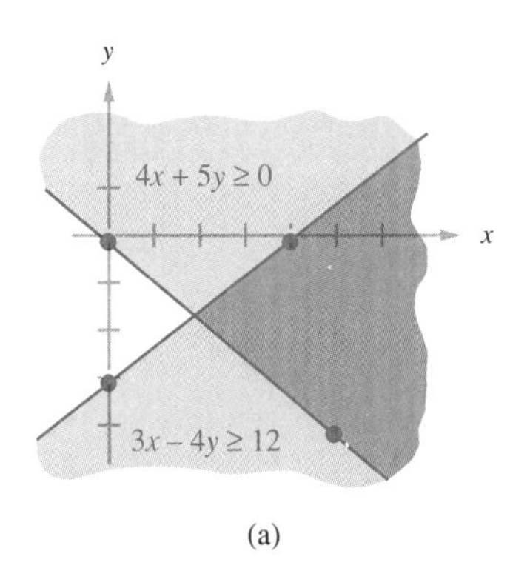

(a)

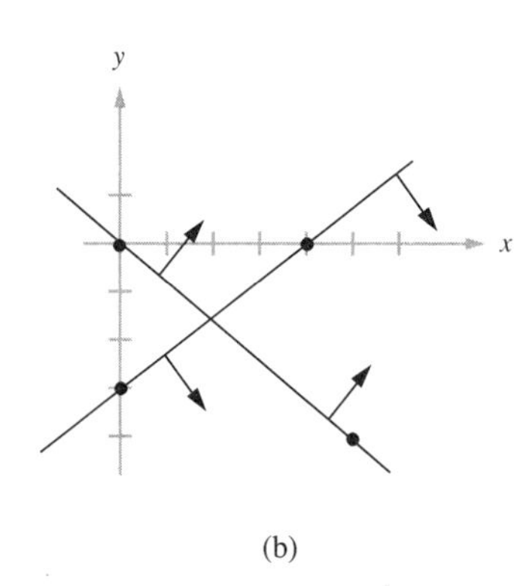

(b)

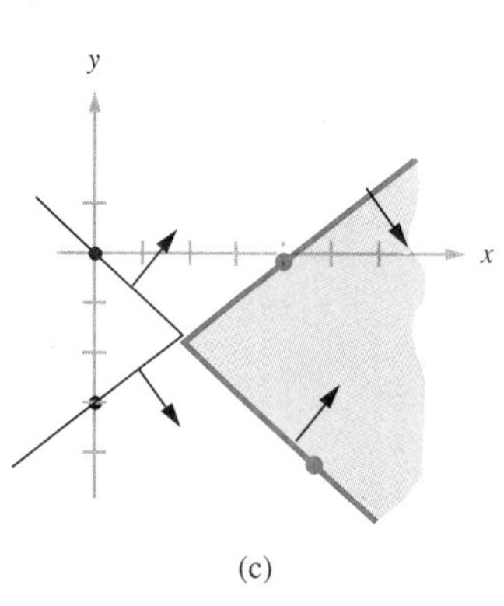

(c)

FIGURE 3.3

We see that to graph a system of linear inequalities we may use the following technique.

Graphing a System of Linear Inequalities

1. Graph the equation corresponding to each inequality.
2. Use a test point to decide which side of each line contains points satisfying the corresponding inequality, and indicate this side with an arrow.
3. Shade the region on the correct side of all the lines.

Practice Problem 3 Graph the set of points satisfying the inequalities $2x + 3y \leq 12$ and $x \geq 2$.

Practice Problem 4 Graph the region of feasible points for the linear program of Example 3.2; that is, graph the system of linear inequalities forming its constraints. These constraints are

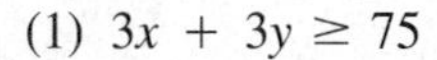

$$
\begin{aligned}
&(1)\quad 3x + 3y \geq 75\\
&(2)\quad 2x + 4y \geq 60\\
&(3)\quad 4x + 2y \geq 60\\
&(4)\quad x \geq 0\\
&(5)\quad y \geq 0.
\end{aligned}
$$

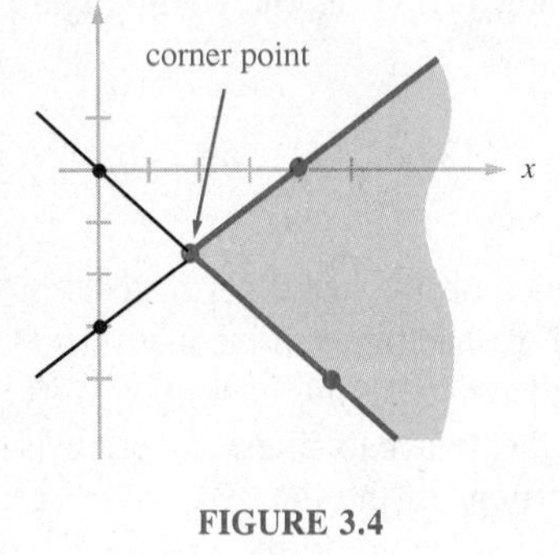

FIGURE 3.4

Regions formed by overlapping half-planes may have certain **corner points,** that is, points in all the half-planes and on more than one of the lines defining them. In Example 3.3 there is just one corner point, namely the point where the lines $3x - 4y = 12$ and $4x + 5y = 0$ intersect. (See Figure 3.4.) This point can be found by solving the system

$$
\begin{aligned}
3x - 4y &= 12\\
4x + 5y &= 0
\end{aligned}
$$

as in Chapters 1 and 2. We find the corner point to be $(60/31, -48/31)$.

EXAMPLE 3.4

Graph the set of points satisfying

$$
\begin{aligned}
&(1)\quad 3x + 3y \leq 15\\
&(2)\quad x + 2y \geq 2\\
&(3)\quad y \geq x,
\end{aligned}
$$

and find all corner points.

Solution The region is graphed in Figure 3.5.

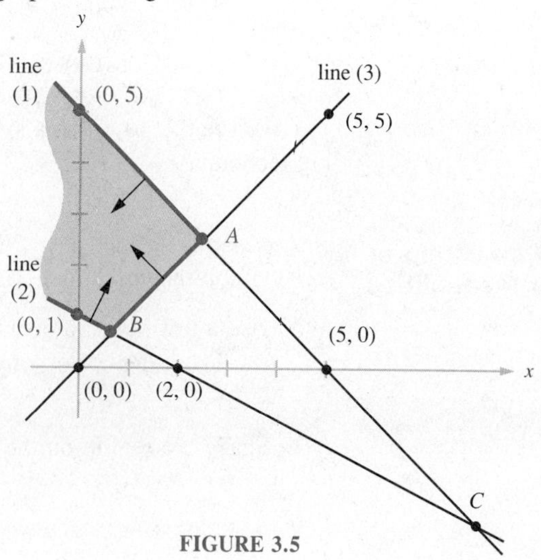

FIGURE 3.5

Corner point A is found to be (2.5, 2.5) by solving the system

$$\begin{aligned} &(1)\quad 3x + 3y = 15, \\ &(3)\qquad\qquad y = x. \end{aligned}$$

Likewise corner point B is found to be (2/3, 2/3) by solving the system

$$\begin{aligned} &(2)\quad x + 2y = 2, \\ &(3)\qquad\quad\; y = x. \end{aligned}$$

Notice that the point labeled C is *not* a corner point, since it does not satisfy all the inequalities. ■

Practice Problem 5 How many corner points does the region graphed in Practice Problem 4 have?

EXERCISES 3.1

In Exercises 1 and 2 set up a linear program corresponding to the problem.

1. Karl's Koffeehaus sells two coffee blends. Each pound of the Haus blend contains 8 ounces of Brazilian and 8 ounces of Colombian coffee, while each pound of the Special blend contains 4 ounces of Brazilian and 12 ounces of Colombian. Karl has 1000 ounces of Brazilian and 1500 ounces of Colombian coffee in stock. He makes a profit of $1 on each pound of the Haus blend and $1.25 on each pound of the Special blend. He would like to maximize his profit by making up x pounds of the Haus blend and y pounds of the Special blend. This information is summarized in the following table.

	Haus	*Special*	*Available*
Brazilian per pound	8 oz	4 oz	1000 oz
Colombian per pound	8 oz	12 oz	1500 oz
Profit per pound	$1	$1.25	
Number of pounds	x	y	

2. A farmer grows beets for both their roots and green tops. Each acre of Big Red beet costs $400 to plant and yields 2000 pounds of roots and 1500 pounds of tops. Each acre of Cannonball beet costs $600 to plant and yields 2500 pounds of roots and 1200 pounds of tops. The farmer has a contract to supply 100,000 pounds of roots and 75,000 pounds of tops. He plants x acres of Big Red and y acres of Cannonball, and would like to minimize his total cost. This information is summarized in the following table.

	Big Red	*Cannonball*	*Needed*
Roots per acre	2000 lbs	2500 lbs	100,000 lbs
Tops per acre	1500 lbs	1200 lbs	75,000 lbs
Cost per acre	$400	$600	
Number of acres	x	y	

In Exercises 3 and 4 make a table summarizing the information given. Then set up the corresponding linear program.

3. A teacher is buying fried chicken for a class picnic. A Big Barrel of chicken contains 10 pieces of white meat and 10 pieces of dark meat, and costs \$9. A Supersack contains 7 pieces of white meat and 4 pieces of dark meat, and costs \$6. The class demands at least 100 pieces of white meat and 80 pieces of dark meat. The teacher would like to provide this at the least possible cost, buying x Big Barrels and y Supersacks.

4. Patty's Party Service gives birthday parties for \$40 and Halloween parties for \$50. A birthday party requires 20 balloons and 15 noisemakers, while a Halloween party requires 25 balloons and 20 noisemakers. Only 200 balloons and 150 noisemakers are available. Patty would like to schedule x birthday parties and y Halloween parties so as to take in as much money as possible.

In Exercises 5–12 set up a linear program corresponding to each problem. Explain what the variables represent in Exercises 9–12.

5. A steel mill makes two grades of stainless steel, both of which are sold in 100-pound bars. The standard-grade bar contains 90% steel and 10% chromium by weight, and sells for \$110. The high-grade bar contains 80% steel and 20% chromium by weight, and sells for \$120. If the company has a supply of 72,000 pounds of steel and 10,000 pounds of chromium, how many bars of each grade should it produce to maximize total revenue? Suppose it produces x standard-grade bars and y high-grade bars.

6. A farmer has a 2000-acre plot of land on which she intends to plant some combination of alfalfa and beets. Alfalfa requires \$20 capital and 25 inches of water for each acre planted, while beets require \$30 capital and 75 inches of water for each acre planted. The net profit per acre is \$30 for alfalfa and \$40 for beets. If the farmer has a total of \$45,000 in capital and a total water allotment of 90,000 acre-inches, how many acres of each crop should she plant to maximize total profit? (*Note:* An acre-inch is the amount of water needed to irrigate an acre of land to a depth of one inch.) Suppose she grows x acres of alfalfa and y acres of beets.

7. The House of Coffee sells two different mixtures of Mocha, Java, and Santos beans. Each pound of Mixture A contains $\frac{1}{4}$ pound Mocha, $\frac{1}{2}$ pound Java, and $\frac{1}{4}$ pound Santos and sells for \$3.50. Each pound of Mixture B contains $\frac{3}{5}$ pound Mocha, $\frac{1}{5}$ pound Java, and $\frac{1}{5}$ pound Santos and sells for \$4.00. If the store has a supply of 300 pounds of Mocha, 250 pounds of Java, and 150 pounds of Santos beans, how many pounds of each mixture should be made in order to maximize total revenue? Suppose x pounds of Mixture A and y pounds of Mixture B are made.

8. World Travel Service has agreed to transport a group of 1800 football fans from Los Angeles to Chicago for one of the season's big games. The agency can charter two different types of aircraft for the trip. Type A has 20 first-class seats and 80 economy seats and costs \$3000 to operate for one trip. Type B has 40 first-class seats and 60 economy seats and costs \$4000 to operate. If 600 of the passengers want to go first-class while the remaining 1200 have ordered economy seats, how many planes of each type should be assigned to the charter to minimize total operating cost? Suppose x Type A and y Type B planes are used.

9. A gem dealer buys rough-cut amethysts and beryls of uniform size, which she finishes for sale to retail jewelers. Each amethyst requires one hour on a grinder, one hour on a sander, and two hours on a polisher and yields a net profit of \$75. Each beryl requires two hours on a grinder, five hours on a sander, and two hours on a polisher and yields a profit of \$100. If the grinder, sander, and polisher are each available for 40 hours a week, how many amethysts and beryls should be finished each week to maximize total profit?

10. A cannery has decided to replace its old machines with newer equipment. The manager has been allocated \$300,000 of capital for the new machines. He has 36 available operators and space for, at most, 28 new units. Of the two models under consideration, Unit A costs \$10,000, requires one operator, and produces 600 cans per hour, while Unit B costs \$15,000, requires two operators, and produces 1000 cans per hour. How many units of each type should be purchased in order to maximize can production?

11. A printing house orders newsprint in bulk rolls. The rolls supplied by Company A cost \$100; each roll gives a two-day supply of paper and occupies 35 square feet of storage space. The rolls supplied by Company B cost \$300; each roll gives a five-day supply of paper and occupies 25 square feet of storage space. The printing house has a total of 10,500 square feet of storage space and needs enough paper for 1000 days. How can this be achieved at lowest cost?

12. The Heine-Borel Corporation is trying to decide between two comprehensive insurance policies covering fire, theft, and liability. The Acme Insurance Company offers a policy that provides \$10,000, \$5000, and \$20,000 of fire, theft, and liability coverage, respectively, for each unit bought, at a premium of \$500 per unit. The Baltimore Insurance Company offers a policy that provides \$10,000 each of fire, theft, and liability coverage for each unit bought, at a premium of \$450 per unit. The corporation needs coverage of at least \$400,000 each for fire and theft and \$1,000,000 for liability. How many units of each policy should be purchased to minimize cost?

In Exercises 13–18 tell which of the points gives the best value of the objective function, among those points given that are feasible.

13. Points: $(3, 5)$, $(4, 4)$, $(5, 3)$, $(4, 5)$, $(3, 6)$

Program: Maximize $5x + 6y$ subject to

$$\begin{aligned} 3x + 4y &\le 30 \\ 5x + y &\le 35 \\ x \ge 0,\ y &\ge 0. \end{aligned}$$

14. Points: $(10, 20)$, $(15, 12)$, $(13, 17)$, $(20, 0)$, $(4, 25)$

Program: Maximize $7x + 3y$ subject to

$$\begin{aligned} 3x + 4y &\ge 100 \\ x + 2y &\ge 30 \\ x - y &\ge -5 \\ x \ge 0,\ y &\ge 0. \end{aligned}$$

15. Points: $(2, -5)$, $(5, 1)$, $(7, 0)$, $(3, 3)$, $(1, 4)$

Program: Minimize $3x + 5y$ subject to

$$\begin{aligned} x + y &\ge 5 \\ 3x + 2y &\ge 12 \\ x \ge 0,\ y &\ge 0. \end{aligned}$$

16. Points: $(1.1, 3.2)$, $(2.0, 4.3)$, $(1.5, -1.5)$, $(0.9, 3.3)$, $(4.2, 1.0)$

Program: Minimize $10x + 12y$ subject to

$$\begin{aligned} 2x + 4y &\ge 12 \\ 3x - 2y &\ge 0 \\ 5x + y &\le 5 \\ x \ge 0,\ y &\ge 0. \end{aligned}$$

17. Points: (1, 2, 3), (2, 3, 1), (2, 2, 2), (3, 0, 2), (1, 4, 0)

$$\begin{array}{lll} \text{Program:} & \text{Maximize} & 2x + 4y + 3z \\ & \text{subject to} & x + y + 2z \le 6 \\ & & 2x + z \le 7 \\ & & x \ge 0,\ y \ge 0,\ z \ge 0. \end{array}$$

18. Points: (11, 0, 20), (30, 0, 0), (12, 10, 10), (0, 13, 18), (5, 15, 9)

$$\begin{array}{lll} \text{Program:} & \text{Minimize} & x + 3y + 2z \\ & \text{subject to} & 2x + y + 2z \ge 45 \\ & & x + 2y + 2z \ge 40 \\ & & x \ge 0,\ y \ge 0,\ z \ge 0. \end{array}$$

In Exercises 19–28 graph the given inequality.

19. $5x + 7y \ge 35$ **20.** $8x + 5y \ge 20$ **21.** $3x + 8y \le 12$ **22.** $10x + 7y \le 35$

23. $4x - 6y \le 15$ **24.** $6x - 5y \ge -15$ **25.** $2x - 5y \ge 30$ **26.** $-8x + 6y \le 36$

27. $3x + 5y \le 0$ **28.** $12x - 7y \ge 0$

In Exercises 29–36 graph the system of inequalities and identify the coordinates of all corner points of the region graphed.

29. $4x + 5y \le 10,\ x \ge 0$

30. $2x + 3y \ge 6,\ y \ge 0$

31. $3x + 2y \le 12,\ x + y \ge 2,\ x \ge 0,\ y \ge 0$

32. $x + 2y \ge 1,\ x + 3y \le 9$

33. $4x + 7y \le 14,\ x \ge 0,\ y \ge 0$

34. $7x + 3y \ge 21,\ x \ge 0,\ y \ge 0$

35. $3x + 4y \le 12,\ x - y \ge 3,\ x \ge 0,\ y \ge 0$

36. $x + y \ge 2,\ 2x + y \le 10,\ x \ge 0,\ y \ge 0$

Answers to Practice Problems

1. (a) The points (30, 50) and (45, 40) are feasible. (b) The point (45, 40) is better.

2. If x = the number of TV sets carried and y = the number of VCRs carried,

then the linear program is:

$$\begin{array}{ll} \text{Maximize} & 5x + 4y = R \\ \text{subject to} & 40x + 25y \le 4000 \\ & 3x + 2y \le 2000 \\ & x \ge 0 \\ & y \ge 0. \end{array}$$

3.

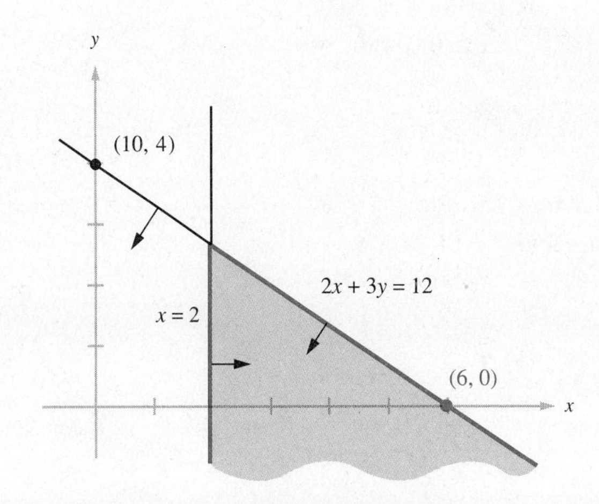

4.

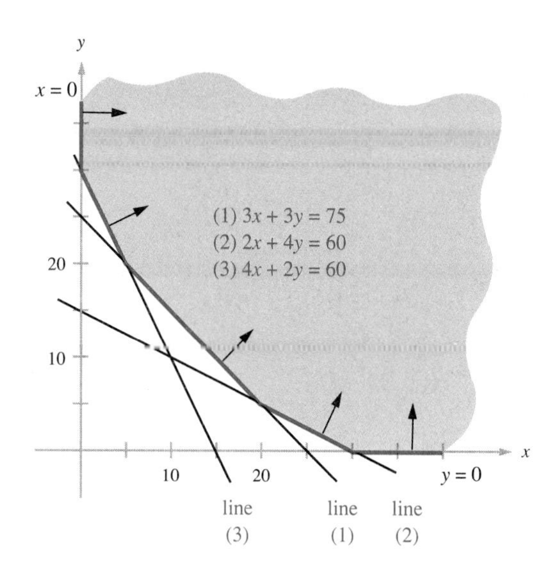

5. 4

3.2 The Graphical Method

In this section we will actually solve some linear programs. Our method involves graphing the set of points satisfying the constraints of the linear program. This set is called the **feasible region** of the program. Because our method is graphical we are limited to problems where there are two variables.

EXAMPLE 3.1 Revisited (a) Graph the feasible region for Example 3.1, which involved Albert's Smoke Shop.

(b) Can Albert choose x and y so as to have a revenue of \$1000? Of \$1500?

Solution (a) Recall that the constraints in Example 3.1 are

$$\begin{aligned} 8x + 3y &\le 480 \\ 4x + 9y &\le 720 \\ x &\ge 0 \\ y &\ge 0. \end{aligned}$$

Using the methods of the last section we graph the feasible region as in Figure 3.6 on the next page.

(b) To determine whether Albert can achieve revenues of \$1000 and \$1500, we recall that his revenue is given by

$$16x + 12y = R.$$

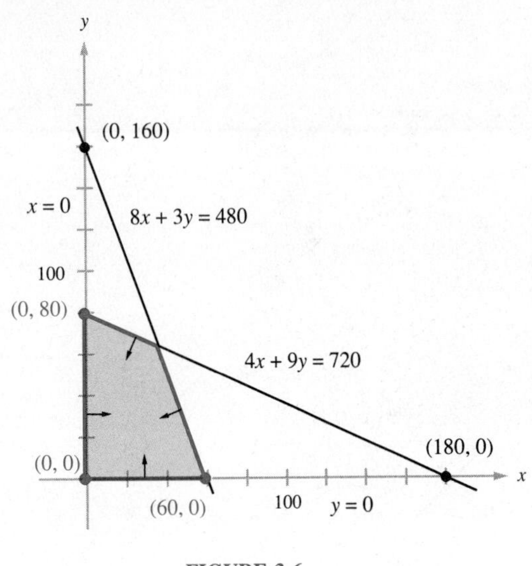

FIGURE 3.6

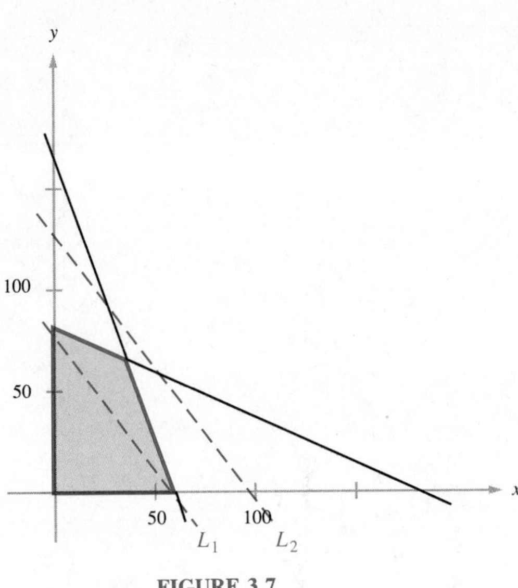

FIGURE 3.7

Thus the set of points giving a revenue of \$1000 is exactly the set of points on the line L_1: $16x + 12y = 1000$. Likewise those points giving a revenue of \$1500 lie on the line L_2: $16x + 12y = 1500$. Including these lines in our diagram yields Figure 3.7. Since there are feasible points on the line L_1, a revenue of \$1000 is possible for Albert. The line L_2 does not touch the feasible region, however, so it is not possible for him to take in \$1500. ■

OPTIMIZING THE OBJECTIVE FUNCTION

An examination of Figure 3.7 shows us how Albert may maximize his revenue. Clearly he can do better than \$1000, since there are points (x, y) in the feasible region having both x and y bigger than points on L_1, and increasing both x and y must increase $16x + 12y = R$. The lines

$$L_1\colon 16x + 12y = 1000$$

$$L_2\colon 16x + 12y = 1500$$

appear to be parallel. Indeed, it is not hard to show that all lines with equations of the form

$$16x + 12y = R,$$

where R is some constant, are parallel to each other. These are called the **isolines** of the objective function. Increasing R simply moves the line upward and to the right, as Figure 3.8 shows.

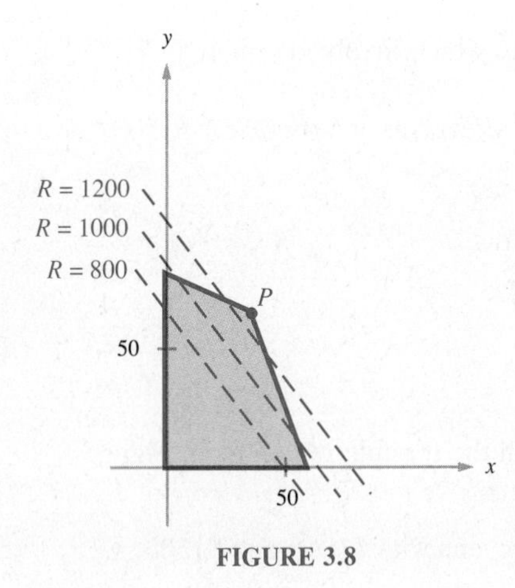

FIGURE 3.8

To achieve maximal revenue we need to find a line parallel to those shown that is as far to the top and right of the feasible region as possible while still touching it. Clearly in this case such a line will just touch the corner point P, and so the x- and y-coordinates of P will produce the most revenue for Albert. The point P is easily found, since it is on the lines with equations

$$8x + 3y = 480$$
$$4x + 9y = 720.$$

Solving these equations simultaneously gives the unique solution $x = 36$, $y = 64$. Thus Albert should make up 36 cans of Blend A and 64 cans of Blend B to maximize his revenue. This will bring in

$$R = 16(36) + 12(64) = \$1344.$$

THE GRAPHICAL METHOD

We summarize the graphical method of linear programming as follows.

1. Identify what variables x and y are sought.
2. Write down a linear program consisting of an objective function to be optimized and a set of linear constraints.
3. Graph the feasible region.
4. Set the objective function equal to two different arbitrarily chosen constants, and graph the corresponding isolines.
5. Imagine one of the isolines in step 4 moved parallel to itself so as to optimize the objective function at a point in the feasible region. *If this optimal point exists, it can always be taken to be a corner point of the feasible region.*
6. If necessary, solve simultaneously the equations of the lines intersecting at the optimal point to find the values of x and y there.

EXAMPLE 3.5 A university alumni association is planning a trip to a bowl game. Two types of buses are available. Type A can carry 40 people and 1000 pounds of baggage, and costs \$2000. Type B can carry 50 people and 750 pounds of baggage, and costs \$2400. A total of 800 people are going on the trip, and they have 18,000 pounds of baggage. How many of each type bus should be chartered to minimize the cost of the trip?

Solution Let

$$x = \text{the number of Type A buses}$$
$$y = \text{the number of Type B buses.}$$

We summarize the given information in a table.

	Type A	*Type B*	*Needed*
People carried	40	50	800
Baggage carried	1000	750	18,000
Cost	\$2000	\$2400	
Number	x	y	

Our linear program is:

$$\begin{aligned} \text{Minimize} \quad & 2000x + 2400y = C \\ \text{subject to} \quad & 40x + 50y \geq 800 \\ & 1000x + 750y \geq 18,000 \\ & x \geq 0 \\ & y \geq 0. \end{aligned}$$

The feasible region is graphed in Figure 3.9.

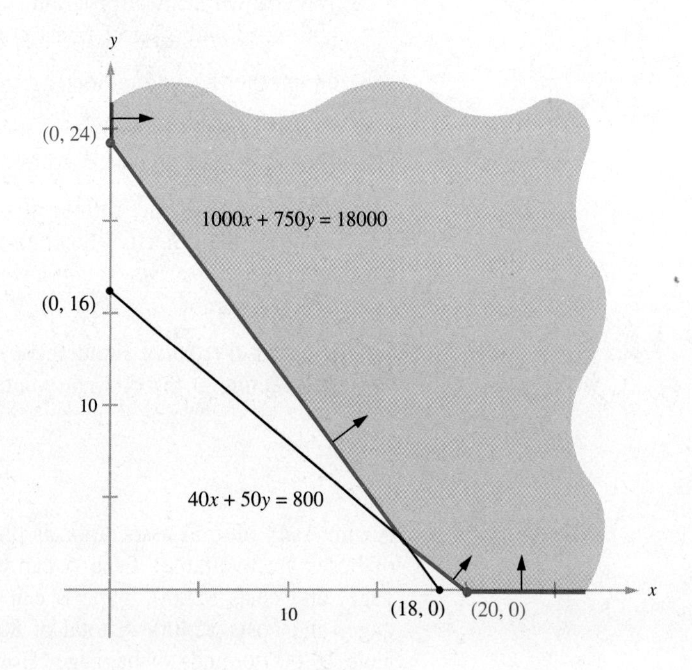

FIGURE 3.9

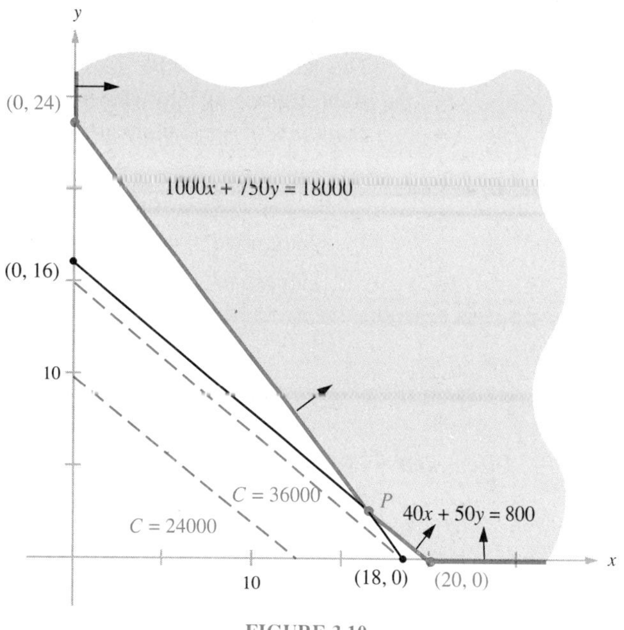

FIGURE 3.10

Now we graph two isolines $2000x + 2400y = C$, where we take C to be 24,000 and 36,000. (These values were chosen to make the intercepts of the corresponding isolines easy to compute.) The result is shown in Figure 3.10. Since we wish to minimize the cost C, we imagine these lines moved parallel to themselves as far to the lower left as possible while still touching the feasible region. The last point touched will be the corner point P where the two lines $40x + 50y = 800$ and $1000x + 750y = 18{,}000$ intersect. This is easily determined to be $(x, y) = (15, 4)$ by the methods of the previous chapter. Thus the alumni association should charter 15 Type A buses and 4 Type B buses. The cost will then be

$$C = 2000(15) + 2400(4) = \$39{,}600. \quad ■$$

Practice Problem 6 Use the graphical method to solve the following linear program:

$$\begin{aligned} \text{Maximize} \quad & 2x + y = R \\ \text{subject to} \quad & 4x + 3y \le 12 \\ & x + 3y \le 6 \\ & x \ge 0 \\ & y \ge 0. \end{aligned}$$

SPECIAL CASES

The advantage of the graphical method of solving a linear program is the insight it gives us into what such problems are really about. We now give several examples of special situations that can arise.

EXAMPLE 3.6 Solve the linear program:

$$\begin{aligned} &\text{Maximize} & 6x + 3y &= R \\ &\text{subject to} & x + 2y &\geq 8 \\ && 2x + y &\geq 10 \\ && x &\geq 0 \\ && y &\geq 0. \end{aligned}$$

Discussion Figure 3.11 shows the feasible region, along with isolines corresponding to $R = 12$ and $R = 24$. Since we are trying to maximize R, we want to find an isoline as far to the top and right as possible that still touches the feasible region. But this region is unbounded in this direction, and we can find lines

$$6x + 3y = R$$

with R as big as we please containing feasible points. Thus, there is no maximum value of R, and the linear program has no solution. ■

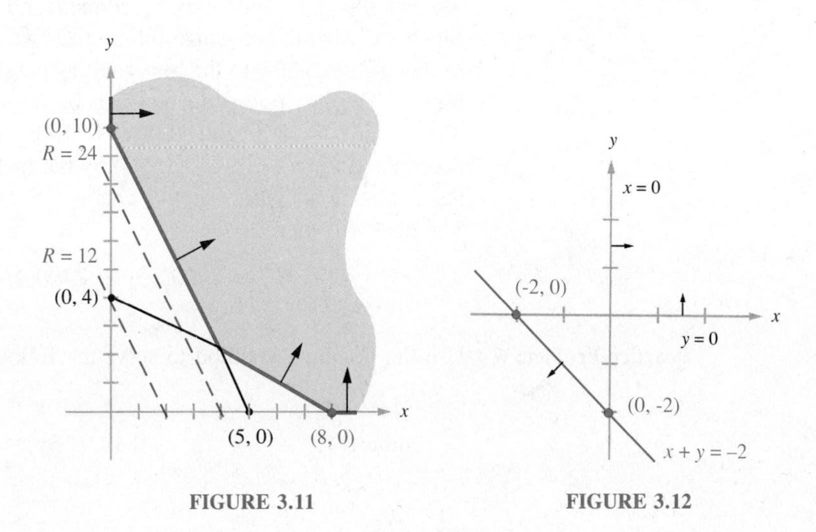

FIGURE 3.11

FIGURE 3.12

EXAMPLE 3.7 Solve the linear program:

$$\begin{aligned} \text{Maximize} \quad & 3x + 7y = R \\ \text{subject to} \quad & x + y \leq -2 \\ & x \geq 0 \\ & y \geq 0. \end{aligned}$$

Discussion Figure 3.12 shows the graphs of the lines corresponding to our constraints, with arrows indicating the half-planes defined. Clearly no points satisfy all the inequalities. The feasible region is empty, and there is no solution to the linear program. ■

EXAMPLE 3.8 Solve the linear program:

$$\begin{aligned} \text{Maximize} \quad & 7x + 9y = R \\ \text{subject to} \quad & 10x + 13y \leq 130 \\ & x + 2y \leq 18 \\ & x \geq 0 \\ & y \geq 0. \end{aligned}$$

Solution In Figure 3.13 we have graphed the feasible region, along with the isolines corresponding to $R = 45$ and $R = 63$.

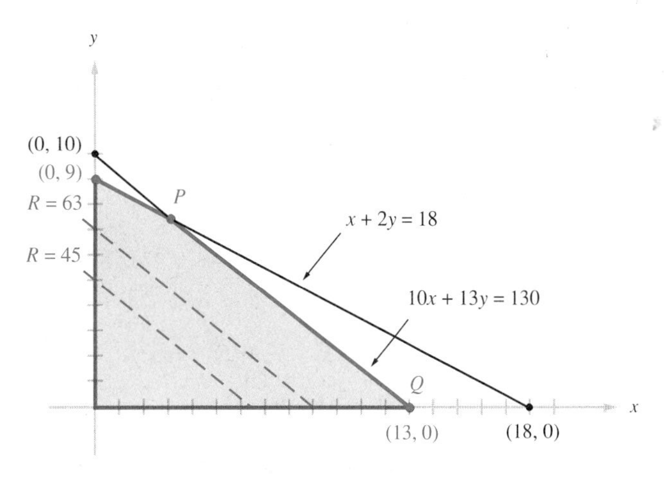

FIGURE 3.13

When we move an isoline to the top and right it is not clear whether the point P or the point Q will be the last feasible point touched. Our graph is not sufficiently precise to tell whether P or Q is optimal. Thus we resort to computing the objective function at both P and Q. We find P by solving the system

$$\begin{aligned} 10x + 13y &= 130 \\ x + 2y &= 18 \end{aligned}$$

getting $P = (26/7, 50/7)$. Likewise Q is the solution of the system

$$\begin{aligned} 10x + 13y &= 130 \\ y &= 0, \end{aligned}$$

which is easily seen to be (13, 0). Now we test these points in the objective function.

(x, y)	$7x + 9y = R$
$P = (26/7, 50/7)$	$7(26/7) + 9(50/7) = 90\frac{2}{7}$
$Q = (13, 0)$	$7(13) + 9(0) = 91 \longleftarrow$ maximum

Since we want to maximize R, the optimal point is (13, 0). ■

EXAMPLE 3.9 Solve the linear program:

$$\begin{aligned} \text{Maximize} \quad & 6x + 3y = R \\ \text{subject to} \quad & x + 2y \le 8 \\ & 2x + y \le 10 \\ & x \ge 0 \\ & y \ge 0. \end{aligned}$$

Solution The feasible region and the lines $R = 12$ and $R = 24$ are graphed in Figure 3.14.

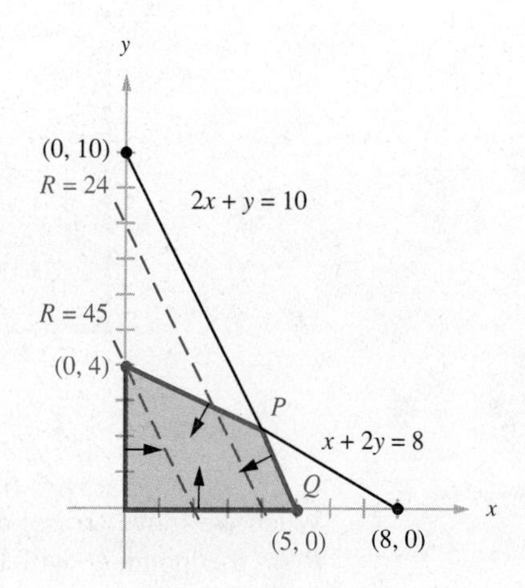

FIGURE 3.14

It is not clear from our graph whether point P or point Q will be the last point in the feasible region touched by an isoline moving to the top and right. We will decide by testing both points in the objective function $6x + 3y = R$. We can find P by solving the system

$$\begin{aligned} x + 2y &= 8 \\ 2x + y &= 10, \end{aligned}$$

getting $P = (4, 2)$. Likewise Q satisfies the system

$$\begin{aligned} 2x + y &= 10 \\ y &= 0, \end{aligned}$$

and so $Q = (5, 0)$. We compute R at both these points.

(x, y)	$6x + 3y = R$
$P = (4, 2)$	$6(4) + 3(2) = 30$
$Q = (5, 0)$	$6(5) + 3(0) = 30$

Since we get $R = 30$ at both P and Q, the isolines must be parallel to the line through P and Q. Any of the infinitely many points on the segment from P to Q gives the same maximal value of R. Note that two of the optimal points are corner points of the feasible region, namely P and Q. Even though there are many optimal points, there is only one maximal value of R, namely 30. ■

LIMITATIONS OF THE GRAPHICAL METHOD

The main limitation of the graphical method is that it can only be used when there are two variables to be determined. (Although a geometric interpretation can be given to three-variable problems in which the feasible region is a solid bounded by planes, graphing in three dimensions is too difficult for this to be a practical method for solving three-variable problems.) Even in two-variable problems graphing may not be a reasonable method. For example, when there are many constraints the feasible region may be quite complicated, or if three or more lines intersect at points close to each other, it may be hard to tell which intersections are corners of the feasible region. Furthermore imagining an isoline moved parallel to itself involves an "eyeball" estimate that may not be accurate enough to produce the correct answer, as we saw in our last two examples. In the next section we will introduce an algebraic method of solving linear programming problems that applies no matter how many variables there are.

EXERCISES 3.2

Use the graphical method to solve Exercises 1–14. Give the optimal point and corresponding value of the objective function.

1. Maximize $3x + 2y = R$
subject to $4x + y \leq 100$
$x + 4y \leq 100$
$x \geq 0, y \geq 0.$

2. Maximize $4x + 5y = R$
subject to $5x + 3y \leq 28$
$3x + 4y \leq 30$
$x \geq 0, y \geq 0.$

3. Minimize $6x + 4y = C$
subject to $5x + 3y \geq 15$
$2x + 5y \geq 10$
$x \geq 0, y \geq 0.$

4. Minimize $x + 2y = C$
subject to $x + y \geq 30$
$2x + y \geq 40$
$x \geq 0, y \geq 0.$

5. Maximize $x + 2y = R$
subject to $x + y \leq 50$
$x - y \geq -30$
$x \leq 40$
$x \geq 0, y \geq 0.$

6. Minimize $x - y = C$
subject to $4x + y \geq 4$
$x + 2y \leq 4$
$x \geq 0, y \geq 0.$

7. Exercise 5, Section 3.1

8. Exercise 6, Section 3.1

9. Exercise 7, Section 3.1

10. Exercise 8, Section 3.1

11. Exercise 9, Section 3.1

12. Exercise 10, Section 3.1

13. Exercise 11, Section 3.1

14. Exercise 12, Section 3.1

15. For an experiment on animal behavior a psychologist must train rats and mice to run two mazes. The amount of time each rat and each mouse is given in the mazes is shown below.

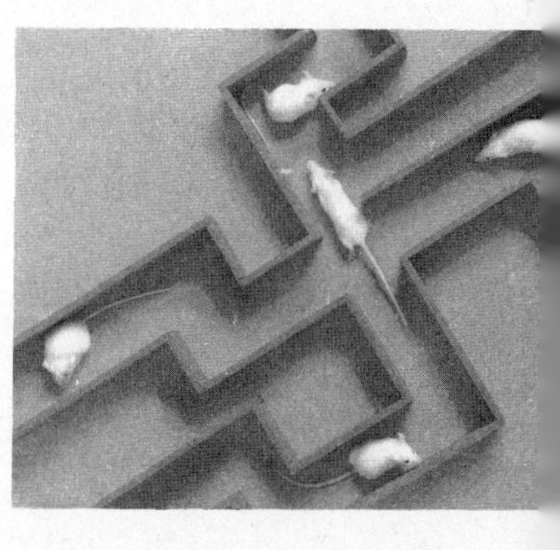

	Time per rat	*Time per mouse*	*Available*
Maze A	12 min	8 min	240 min
Maze B	10 min	15 min	300 min

Under these conditions what is the maximum number of animals that can be used in this experiment, and what combination of rats and mice will produce this maximum?

16. A manufacturer makes two sizes of glass vases. The taller ones, being more sturdy than the shorter ones, are less expensive to ship but more expensive to produce. If the production and shipping costs for each type of vase are as given in the table below, and if the manufacturer earns a $4 profit on the taller vases and a $3 profit on the shorter vases, what combination of vases will yield the maximum profit? What is this profit?

	Taller vase	*Shorter vase*	*Available*
Production cost	$10	$4	$6000
Shipping cost	$ 1	$2	$ 800

17. Using two foods, a hospital dietician must prepare a special meal that contains required amounts of vitamin A and calcium. The amount of vitamin A and calcium supplied by one ounce of each food and the required amounts are shown in the table below. (*Note:* i.u. = international units)

	Units per ounce		
	Food 1	*Food 2*	*Needed*
Vitamin A	800 i.u.	300 i.u.	6000 i.u.
Calcium	50 mg	150 mg	900 mg

If each ounce of Food 1 contains 200 calories and each ounce of Food 2 contains 150 calories, what combination of the foods will supply the required amounts of vitamin A and calcium with the fewest calories? What is the minimum number of calories?

18. A homeowner wants to mix two lawn fertilizers, Nourish and Greenup, to obtain a mixture of nitrogen and phosphoric acid for use on her lawn. The amounts of each chemical provided by the two fertilizers are given in the following table.

	Pounds per bag		
	Nourish	*Greenup*	*Needed*
Nitrogen	2.25	1.5	36 lbs
Phosphoric acid	0.25	0.5	7 lbs

If each bag of Nourish costs \$5 and each bag of Greenup costs \$6, what combination of the two fertilizers will provide the necessary amounts of nitrogen and phosphoric acid at the least cost?

19. An oil company operates two eastern refineries. The Albany facility produces 100 barrels of high-grade oil and 300 barrels of medium-grade oil per day. The Boston facility produces 200 barrels of high-grade and 100 barrels of medium-grade per day. The company has determined that the combined total output of high-grade oil should fall somewhere between 12,000 and 18,000 barrels (inclusive), and the combined output of medium-grade should fall between 24,000 and 30,000 barrels. If the Albany and Boston refineries cost \$2000 and \$4200 per day to operate, respectively, how many days should each be run to meet the requirement at lowest cost?

20. A chemical firm has been trying to reduce sulfur oxide emissions from its two Chicago plants. Plant I produces 10 tons of sulfuric acid, 6 tons of ammonia, and 7.5 tons of ammonium sulfate per day, with 100 pounds of sulfur oxide emitted. Plant II produces no sulfuric acid, 12 tons of ammonia, and 5 tons of ammonium sulfate per day, with 72 pounds of sulfur oxide emitted. How many days should each plant operate in order to produce at least 600 tons of each of the three chemicals with a minimum total emission of sulfur oxide?

Answer to Practice Problem **6.** The maximum is 6 when $x = 3$ and $y = 0$.

3.3 The Simplex Method

In this section we will introduce an algebraic method, called the **simplex algorithm,** for solving linear programs. This method was developed by the American mathematician George B. Dantzig in the 1940s. It relies heavily on the theory of linear systems explained in Chapter 2, and in particular on the pivot transformation.

We first apply the simplex method to a linear program with two variables, so that our work will have a geometric interpretation. Let us consider the problem:

$$\begin{array}{ll} \text{Maximize} & 20x + 30y = R \\ \text{subject to} & x + 2y \leq 14 \\ & 3x + 2y \leq 18 \\ & x \geq 0, \quad y \geq 0. \end{array}$$

The feasible region for this program is shown in Figure 3.15, along with the isolines corresponding to $R = 60$ and $R = 90$.

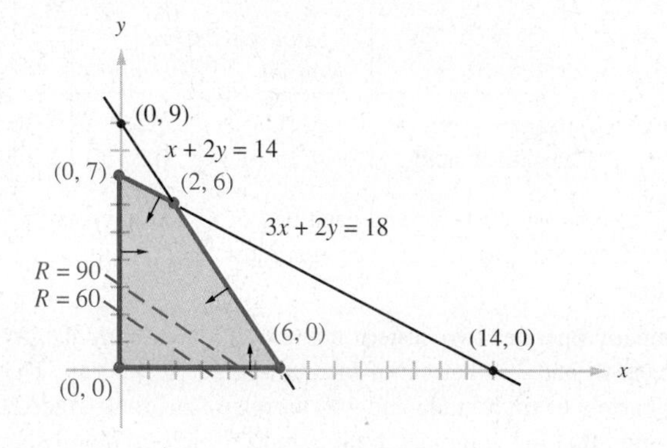

FIGURE 3.15

The feasible region is bounded by the lines $x + 2y = 14$, $3x + 2y = 18$, $x = 0$, and $y = 0$, corresponding to the constraints of the program. We know that the solution to the program will be one of the corners of this region, and each such corner is the intersection of a pair of these lines. We have marked the coordinates of all intersections of two of these lines in Figure 3.15. Four of these intersections are corners of the feasible region, but two, namely (0, 9) and (14, 0), are not. Since R is to be maximized, it appears that (2, 6) is the optimal point, although it is somewhat of a close call between (2, 6) and (0, 7).

Now we will show how the simplex method arrives at this solution algebraically. First we will replace the constraints

$$x + 2y \leq 14$$
$$3x + 2y \leq 18$$

with equations by introducing what are called **slack variables.** If $x + 2y$ is less than or equal to 14, then it falls short of 14 by some nonnegative quantity s; that is

$$x + 2y + s = 14, \qquad s \geq 0.$$

Likewise $3x + 2y \leq 18$ means that

$$3x + 2y + t = 18, \qquad t \geq 0.$$

Thus our constraints can be replaced by a set of equations and inequalities involving two new slack variables s and t:

$$\begin{aligned} x + 2y + s &= 14 \\ 3x + 2y + t &= 18 \\ x \geq 0,\ y \geq 0,\ s \geq 0,\ t &\geq 0. \end{aligned}$$

If we neglect the requirement that x, y, s, and t all be nonnegative, we have a system of two equations in four unknowns,

$$\begin{aligned} x + 2y + s &= 14 \\ 3x + 2y + t &= 18, \end{aligned} \tag{3.5}$$

which will have infinitely many solutions. Notice that setting any one of the unknowns equal to zero restricts the point (x, y) to lie on one of the lines bounding the feasible region. For setting x or y equal to zero means that (x, y) is on the line $x = 0$ or $y = 0$, while setting s equal to zero makes the first equation $x + 2y = 14$, and setting t equal to zero makes the second equation $3x + 2y = 18$. Of course the feasible region is bounded by the lines $x = 0$, $y = 0$, $x + 2y = 14$, and $3x + 2y = 18$.

Setting *two* of the variables equal to zero thus means that (x, y) must be on two of the bounding lines; that is, (x, y) must be one of the six intersection points labeled in Figure 3.15. These include the corner points of the feasible region, at least one of which optimizes our objective function. Thus we need a method for finding which variables must be set equal to zero in order to specify an optimal point.

In Chapter 2 we saw how in the system of two equations in four unknowns (3.5) we could choose any two of the variables as *free* variables and solve for the other two variables (called *basic* variables) in terms of them. As the system stands, s and t are isolated, so we could easily solve for them:

$$\begin{aligned} s &= 14 - x - 2y \\ t &= 18 - 3x - 2y. \end{aligned}$$

Here we are thinking of x and y as free and s and t as basic variables. Setting the free variables x and y both equal to 0 gives the point $(x, y) = (0, 0)$ in Figure 3.15. (Of course our graph does not show s and t.)

By performing a pivot transformation we could make a different variable basic. Recall from Section 2.1 that the tableau form of system (3.5) is

$$\begin{array}{c} \\ s \\ t \end{array} \begin{array}{c} \begin{array}{cccc} x & y & s & t \end{array} \\ \left[\begin{array}{cccc|c} \textcircled{1} & 2 & \boxed{1} & 0 & 14 \\ 3 & 2 & 0 & \boxed{1} & 18 \end{array}\right] \end{array}$$

The columns corresponding to the basic variables s and t consist of a single 1 (outlined with a square), with the other entries 0. Setting both the free variables x and y equal to 0 makes $s = 14$ and $t = 18$. These values can be read from the rightmost column, with 14 in the row where there is a 1 in the s column, and 18 in the row where there is a 1 in the t column. We indicate this by marking each row on the left with the corresponding basic variable. These **row labels** indicate which basic variables take on the values in the rightmost column when all free variables equal 0.

We will pivot on the circled entry in row 1 and column 1. Since column 1 corresponds to the variable x, after pivoting x will become the basic variable corresponding to row 1.

$$\begin{array}{c} \\ x \\ t \end{array} \begin{array}{c} \begin{array}{cccccc} x & y & s & t & & \end{array} \\ \left[\begin{array}{cccc|c} \boxed{1} & 2 & 1 & 0 & 14 \\ 0 & -4 & -3 & \boxed{1} & -24 \end{array}\right] \end{array} \qquad (\mathbf{R}_2 + (-3)\mathbf{R}_1)$$

Now x and t are the basic variables, and we can solve for them in terms of y and s:

$$\begin{aligned} x &= 14 - 2y - s \\ t &= -24 + 4y + 3s. \end{aligned}$$

Setting the new free variables y and s equal to zero gives $x = 14$ and $t = -24$, as can be read from the rightmost column. This corresponds to the point $(x, y) = (14, 0)$ in Figure 3.15, again the intersection of two lines defining the feasible region.

Of course the point (14, 0) is not even a corner of the feasible region, so our pivot does not seem to have done us much good. *The idea of the simplex method is to perform pivot operations in such an order that when all the pivots have been done, the solution to the linear program is obtained by setting the free variables equal to 0.*

THE SIMPLEX METHOD DEMONSTRATED

Now we will show how we can find the solution to our linear program by a sequence of pivots. To keep things simple *we will delay explaining how to choose the pivot entries until Section 3.4.* Before starting we will add another row to our tableau, a row corresponding to the equation

$$20x + 30y = R$$

defining the objective function. Thus our linear system is

$$\begin{aligned} x + 2y + s \quad\quad &= 14 \\ 3x + 2y \quad\quad + t &= 18 \\ 20x + 30y \quad\quad\quad &= R, \end{aligned}$$

and our tableau is

$$\begin{array}{c} \\ s \\ t \\ \\ \end{array}\left[\begin{array}{cccc|c} x & y & s & t & \\ 1 & \textcircled{2} & \boxed{1} & 0 & 14 \\ 3 & 2 & 0 & \boxed{1} & 18 \\ \hline 20 & 30 & 0 & 0 & R \end{array}\right].$$

Here the horizontal line sets off the last row because the corresponding equation involving the objective function is special. *We will never pivot on an element in the bottom row.* We start by pivoting on the circled 2 in row 1 and column 2. Row 1 had corresponded to the basic variable s, but after the pivot s will be replaced by y (the variable of column 2) as a basic variable.

$$\left[\begin{array}{cccc|c} .5 & \textcircled{1} & .5 & 0 & 7 \\ 3 & 2 & 0 & 1 & 18 \\ \hline 20 & 30 & 0 & 0 & R \end{array}\right] \qquad (.5\mathbf{R}_1)$$

$$\begin{array}{c} \\ y \\ t \\ \\ \end{array}\left[\begin{array}{cccc|c} x & y & s & t & \\ .5 & \boxed{1} & .5 & 0 & 7 \\ \textcircled{2} & 0 & -1 & \boxed{1} & 4 \\ \hline 5 & 0 & -15 & 0 & R - 210 \end{array}\right] \qquad \begin{array}{l} \\ \\ (\mathbf{R}_2 + (-2)\mathbf{R}_1) \\ (\mathbf{R}_3 + (-30)\mathbf{R}_1) \end{array}$$

Now y and t are basic variables and x and s are free. Setting x and s equal to 0 gives $x = 0$ (of course), and $y = 7$, since 7 is the element in the rightmost column opposite the 1 in the y-column. Thus we have moved to the point $(x, y) = (0, 7)$ in Figure 3.16, another corner of the feasible region.

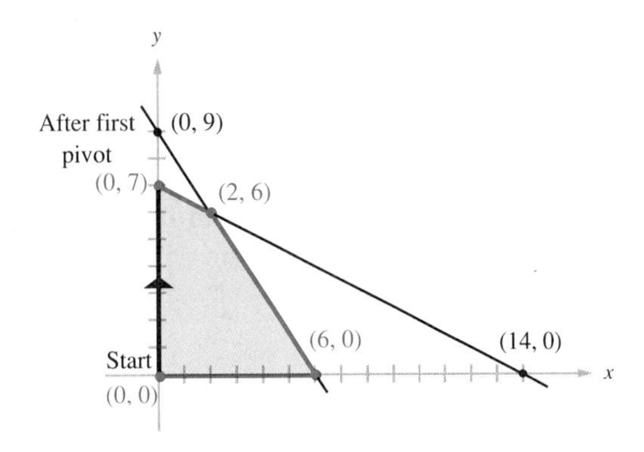

FIGURE 3.16

We continue by pivoting on the entry in row 2 and column 1. Note that column 1 corresponds to x, which will become the new basic variable for row 2.

$$\left[\begin{array}{cccc|c} .5 & 1 & .5 & 0 & 7 \\ \textcircled{1} & 0 & -.5 & .5 & 2 \\ \hline 5 & 0 & -15 & 0 & R - 210 \end{array}\right] \quad (.5\text{R}_2)$$

$$\begin{array}{c} \\ y \\ x \\ \\ \end{array} \begin{array}{c} \begin{array}{cccc} x & y & s & t \end{array} \\ \left[\begin{array}{cccc|c} 0 & \boxed{1} & .75 & -.25 & 6 \\ \boxed{1} & 0 & -.5 & .5 & 2 \\ \hline 0 & 0 & -12.5 & -2.5 & R - 220 \end{array}\right] \end{array} \quad \begin{array}{l} (\text{R}_1 + (-.5)\text{R}_2) \\ \\ (\text{R}_3 + (-5)\text{R}_2) \end{array}$$

Now x and y are the basic variables and s and t are free. Setting $s = t = 0$ gives $y = 6$ and $x = 2$, as we can read from the rightmost column. Thus we have reached the corner $(x, y) = (2, 6)$. Our progress from corner to corner of the feasible region is indicated in Figure 3.17.

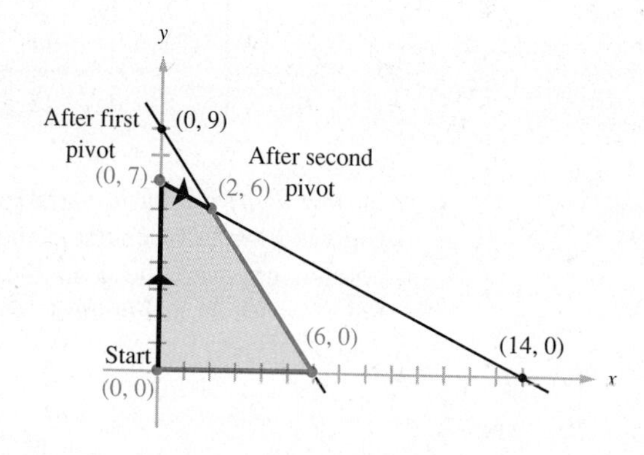

FIGURE 3.17

It turns out that now we can determine that (2, 6) is the solution to our linear program without reference to the graph. Consider the bottom row of our last tableau. It translates into the equation

$$-12.5s - 2.5t = R - 220,$$

or

$$220 - 12.5s - 2.5t = R. \tag{3.6}$$

Recall that the point $(x, y) = (2, 6)$ corresponds to setting $s = t = 0$. Also recall that we must have $s \geq 0$ and $t \geq 0$. The only way we could change s and t would be to make one or both of them larger. *But that would decrease R*, and R is the quantity we are trying to maximize.

We see that the solution to our linear program must occur when $s = t = 0$, and so $(x, y) = (2, 6)$. Moreover we see from (3.6) that the maximum value of R is 220. Notice that this is precisely the number subtracted from R in the bottom right corner of our final tableau. In fact the rightmost column of this tableau, when properly interpreted, gives not only the maximal value of R but also the x and y coordinates achieving this value.

EXAMPLE 3.10 Solve the linear program:

$$\begin{array}{ll} \text{Maximize} & 7x + 4y = R \\ \text{subject to} & x + y \le 160 \\ & 3x + 2y \le 360 \\ & x \ge 0,\ y \ge 0, \end{array}$$

by introducing slack variables and pivoting on the entry in row 2 and column 1 of the resulting tableau.

Solution Introducing slack variables s and t leads to the system

$$\begin{aligned} x + y + s \quad &= 160 \\ 3x + 2y \quad + t &= 360 \\ 7x + 4y \quad &= R \\ x \ge 0,\ y \ge 0,\ s \ge 0,&\ t \ge 0, \end{aligned}$$

with tableau

$$\begin{array}{c} \\ s \\ t \\ \\ \end{array} \begin{array}{c} \begin{array}{cccc|c} x & y & s & t & \end{array} \\ \left[\begin{array}{cccc|c} 1 & 1 & \boxed{1} & 0 & 160 \\ \textcircled{3} & 2 & 0 & \boxed{1} & 360 \\ \hline 7 & 4 & 0 & 0 & R \end{array}\right]. \end{array}$$

We proceed to pivot on the circled entry. This pivot is in the column headed by x, which will replace t as the basic variable for the row of the pivot.

$$\left[\begin{array}{cccc|c} 1 & 1 & 1 & 0 & 160 \\ \textcircled{1} & \frac{2}{3} & 0 & \frac{1}{3} & 120 \\ \hline 7 & 4 & 0 & 0 & R \end{array}\right] \qquad (\tfrac{1}{3}R_2)$$

$$\begin{array}{c} \\ s \\ x \\ \\ \end{array} \begin{array}{c} \begin{array}{cccc|c} x & y & s & t & \end{array} \\ \left[\begin{array}{cccc|c} 0 & \frac{1}{3} & \boxed{1} & -\frac{1}{3} & 40 \\ \boxed{1} & \frac{2}{3} & 0 & \frac{1}{3} & 120 \\ \hline 0 & -\frac{2}{3} & 0 & -\frac{7}{3} & R - 840 \end{array}\right] \end{array} \qquad \begin{array}{l} (R_1 + (-1)R_2) \\ \\ (R_3 + (-7)R_2) \end{array}$$

Here the last row corresponds to the equation

$$840 - \frac{2}{3}y - \frac{7}{3}t = R.$$

Since neither y nor t can be negative, we see that R is maximal when $y = t = 0$. From the second row of our tableau, we see that $x = 120$ here. Thus the maximal value of R is 840 at the point $(x, y) = (120, 0)$. ■

Practice Problem 7 Solve the linear program:

$$\begin{array}{ll} \text{Maximize} & 4x + 3y = R \\ \text{subject to} & x + 2y \le 320 \\ & 5x + 4y \le 1000 \\ & x \ge 0,\ y \ge 0 \end{array}$$

by introducing slack variables and pivoting on the entry in row 2 and column 1 of the corresponding tableau.

EXAMPLE 3.11 Solve the linear program:

$$\begin{array}{ll} \text{Maximize} & 6x + 4y = R \\ \text{subject to} & x + y \le 100 \\ & 2x + y \le 180 \\ & x + 3y \le 240 \\ & x \ge 0,\ y \ge 0, \end{array}$$

by pivoting first on the entry in row 2 and column 1, and then on the entry in row 1 and column 2.

Solution We change the first three constraints to equations by introducing three slack variables s, t, and u. This gives the system

$$\begin{aligned} x + y + s \qquad\qquad &= 100 \\ 2x + y \qquad + t \qquad &= 180 \\ x + 3y \qquad\qquad + u &= 240 \\ 6x + 4y \qquad\qquad\quad &= R \end{aligned}$$

$$x \ge 0,\ y \ge 0,\ s \ge 0,\ t \ge 0,\ u \ge 0,$$

and the tableau

$$\begin{array}{c} \\ s \\ t \\ u \\ \\ \end{array} \begin{array}{c} \begin{array}{ccccc} x & y & s & t & u \end{array} \\ \left[\begin{array}{ccccc|c} 1 & 1 & 1 & 0 & 0 & 100 \\ \textcircled{2} & 1 & 0 & 1 & 0 & 180 \\ 1 & 3 & 0 & 0 & 1 & 240 \\ \hline 6 & 4 & 0 & 0 & 0 & R \end{array}\right], \end{array}$$

where the indicated first pivot has been circled. It is in the column of x, which will replace t as the basic variable of row 2. The computation proceeds as follows.

$$\left[\begin{array}{ccccc|c} 1 & 1 & 1 & 0 & 0 & 100 \\ \textcircled{1} & .5 & 0 & .5 & 0 & 90 \\ 1 & 3 & 0 & 0 & 1 & 240 \\ \hline 6 & 4 & 0 & 0 & 0 & R \end{array}\right] \qquad (.5R_2)$$

$$\begin{array}{c} \\ s \\ x \\ u \\ \\ \end{array} \begin{array}{c} \begin{array}{ccccc} x & y & s & t & u \end{array} \\ \left[\begin{array}{ccccc|c} 0 & \textcircled{.5} & 1 & -.5 & 0 & 10 \\ 1 & .5 & 0 & .5 & 0 & 90 \\ 0 & 2.5 & 0 & -.5 & 1 & 150 \\ \hline 0 & 1 & 0 & -3 & 0 & R - 540 \end{array}\right] \end{array} \qquad \begin{array}{l} (R_1 + (-1)R_2) \\ \\ (R_3 + (-1)R_2) \\ (R_4 + (-6)R_2) \end{array}$$

Now we pivot on the .5 in row 1 and column 2, which will cause y to replace s as the basic variable for row 1.

$$\left[\begin{array}{ccccc|c} 0 & \textcircled{1} & 2 & -1 & 0 & 20 \\ 1 & .5 & 0 & .5 & 0 & 90 \\ 0 & 2.5 & 0 & -.5 & 1 & 150 \\ \hline 0 & 1 & 0 & -3 & 0 & R - 540 \end{array}\right] \qquad (2R_1)$$

$$\begin{array}{c} \\ y \\ x \\ u \\ \\ \end{array} \begin{array}{c} \begin{array}{ccccc} x & y & s & t & u \end{array} \\ \left[\begin{array}{ccccc|c} 0 & \boxed{1} & 2 & -1 & 0 & 20 \\ \boxed{1} & 0 & -1 & 1 & 0 & 80 \\ 0 & 0 & -5 & 2 & \boxed{1} & 100 \\ \hline 0 & 0 & -2 & -2 & 0 & R - 560 \end{array}\right] \end{array} \qquad \begin{array}{l} \\ (R_2 + (-.5)R_1) \\ (R_3 + (-2.5)R_1) \\ (R_4 + (-1)R_1) \end{array}$$

The last row corresponds to the equation

$$560 - 2s - 2t = R,$$

and so R takes on the maximum value 560 when $s = t = 0$. Also from the first two rows of the tableau we have $(x, y) = (80, 20)$ as the optimal point. ■

EXAMPLE 3.12 Solve the linear program:

$$\begin{array}{ll} \text{Maximize} & x + y + z = R \\ \text{subject to} & x + 2y + z \le 4 \\ & 2x + y + 3z \le 6 \\ & x \ge 0,\ y \ge 0,\ z \ge 0, \end{array}$$

by pivoting first on the entry in row 2, column 1, and then on the entry in row 1, column 2.

Solution Notice that this is a three-variable problem, so a graphical solution is not practical. We introduce nonnegative slack variables s and t such that

$$x + 2y + z + s = 4 \quad \text{and} \quad 2x + y + 3z + t = 6,$$

leading to the tableau

$$\begin{array}{c} \\ s \\ t \\ \\ \end{array} \begin{array}{c} \begin{array}{ccccc} x & y & z & s & t \end{array} \\ \left[\begin{array}{ccccc|c} 1 & 2 & 1 & \boxed{1} & 0 & 4 \\ \textcircled{2} & 1 & 3 & 0 & \boxed{1} & 6 \\ \hline 1 & 1 & 1 & 0 & 0 & R \end{array}\right]. \end{array}$$

When we pivot on the circled entry, x will replace t as the basic variable for row 2.

$$\left[\begin{array}{ccccc|c} 1 & 2 & 1 & 1 & 0 & 4 \\ \textcircled{1} & \frac{1}{2} & \frac{3}{2} & 0 & \frac{1}{2} & 3 \\ \hline 1 & 1 & 1 & 0 & 0 & R \end{array}\right] \qquad (\tfrac{1}{2}\mathbf{R}_2)$$

$$\begin{array}{c} \\ s \\ x \\ \\ \end{array} \begin{array}{c} \begin{array}{ccccc} x & y & z & s & t \end{array} \\ \left[\begin{array}{ccccc|c} 0 & \textcircled{$\frac{3}{2}$} & -\frac{1}{2} & \boxed{1} & -\frac{1}{2} & 1 \\ \boxed{1} & \frac{1}{2} & \frac{3}{2} & 0 & \frac{1}{2} & 3 \\ \hline 0 & \frac{1}{2} & -\frac{1}{2} & 0 & -\frac{1}{2} & R - 3 \end{array}\right] \end{array} \qquad \begin{array}{l} (\mathbf{R}_1 + (-1)\mathbf{R}_2) \\ \\ (\mathbf{R}_3 + (-1)\mathbf{R}_2) \end{array}$$

Now we pivot on the 3/2 in row 1 and column 2, causing y to replace s as the basic variable for row 1.

$$\left[\begin{array}{ccccc|c} 0 & \textcircled{1} & -\frac{1}{3} & \frac{2}{3} & -\frac{1}{3} & \frac{2}{3} \\ 1 & \frac{1}{2} & \frac{3}{2} & 0 & \frac{1}{2} & 3 \\ \hline 0 & \frac{1}{2} & -\frac{1}{2} & 0 & -\frac{1}{2} & R - 3 \end{array}\right] \qquad (\tfrac{2}{3}\mathbf{R}_1)$$

$$\begin{array}{c} \\ y \\ x \\ \\ \end{array} \begin{array}{c} \begin{array}{ccccc} x & y & z & s & t \end{array} \\ \left[\begin{array}{ccccc|c} 0 & \boxed{1} & -\frac{1}{3} & \frac{2}{3} & -\frac{1}{3} & \frac{2}{3} \\ \boxed{1} & 0 & \frac{5}{3} & -\frac{1}{3} & \frac{2}{3} & \frac{8}{3} \\ \hline 0 & 0 & -\frac{1}{3} & -\frac{1}{3} & -\frac{1}{3} & R - \frac{10}{3} \end{array}\right] \end{array} \qquad \begin{array}{l} (\mathbf{R}_2 + (-\frac{1}{2})\mathbf{R}_1) \\ (\mathbf{R}_3 + (-\frac{1}{2})\mathbf{R}_1) \end{array}$$

The last row of this tableau implies the equation

$$\frac{10}{3} - \frac{1}{3}z - \frac{1}{3}s - \frac{1}{3}t = R,$$

from which we see that R is maximal when $z = s = t = 0$. The first two rows of the tableau then give $y = 2/3$ and $x = 8/3$. Thus R achieves the maximum value 10/3 when $(x, y, z) = (8/3, 2/3, 0)$. ■

EXERCISES 3.3

In Exercises 1–8 set up the simplex tableaus for the given linear program.

1. Maximize $6x + 4y = R$
subject to
$$x + 3y \le 200$$
$$2x + 5y \le 500$$
$$x \ge 0,\ y \ge 0.$$

2. Maximize $3x + 5y = R$
subject to
$$4x + 3y \le 200$$
$$x + y \le 60$$
$$x \ge 0,\ y \ge 0.$$

3. Maximize $3x + 8y = R$
subject to
$$2x + 4y \le 1200$$
$$x + 3y \le 1000$$
$$x \ge 0,\ y \ge 0.$$

4. Maximize $4x + 3y = R$
subject to
$$2x + 3y \le 200$$
$$2x + y \le 300$$
$$x \ge 0,\ y \ge 0.$$

5. Maximize $5x + 12y = R$
subject to
$$3x + 4y \le 300$$
$$x + 2y \le 120$$
$$x \le 30$$
$$x \ge 0,\ y \ge 0.$$

6. Maximize $6x + 5y = R$
subject to
$$x + 4y \le 240$$
$$2x + 3y \le 200$$
$$3x + 5y \le 360$$
$$x \ge 0,\ y \ge 0.$$

7. Maximize $5x + 4y + 3z = R$
subject to
$$2x + y + z \le 30$$
$$y + 3z \le 40$$
$$x \ge 0,\ y \ge 0,\ z \ge 0.$$

8. Maximize $4x + 2y + 3z = R$
subject to
$$0.1x + 0.25y \le 40$$
$$0.2x + 0.3y + 0.4z \le 100$$
$$x \ge 0,\ y \ge 0,\ z \ge 0.$$

In Exercises 9–12 pivot on the indicated entry of the tableau below, changing the appropriate row label. Perform each pivot on the original *tableau shown. If the free variables are set equal to zero after the pivot, what is the point (x, y)?*

$$\begin{array}{c} \\ s \\ t \\ \\ \end{array}\left[\begin{array}{cccc|c} x & y & s & t & \\ 3 & 4 & 1 & 0 & 7 \\ 1 & 2 & 0 & 1 & 10 \\ \hline 3 & 2 & 0 & 0 & R \end{array}\right]$$

9. The entry 3 in row 1 and column 1.

10. The entry 4 in row 1 and column 2.

11. The entry 1 in row 2 and column 1.

12. The entry 2 in row 2 and column 2.

In Exercises 13–16 pivot on the indicated entry of the tableau below, changing the appropriate row label. Perform each pivot on the original *tableau shown. If the free variables are set equal to zero after the pivot, what is the point (x, y)?*

$$\begin{array}{c} \\ x \\ t \\ \\ \end{array}\left[\begin{array}{cccc|c} x & y & s & t & \\ 1 & \frac{2}{3} & \frac{1}{3} & 0 & \frac{8}{3} \\ 0 & \frac{8}{3} & -\frac{2}{3} & 1 & \frac{20}{3} \\ \hline 1 & 0 & -1 & 0 & R - 8 \end{array}\right]$$

13. The entry 2/3 in row 1 and column 2.

14. The entry 1/3 in row 1 and column 3.

15. The entry 8/3 in row 2 and column 2.

16. The entry $-2/3$ in row 2 and column 3.

In Exercises 17–24 solve the given linear program by pivoting on the indicated entries. Give the maximum value of R and the optimal point that produces it.

17. The entry in row 1 and column 1 of the tableau for Exercise 1.

18. The entry in row 2 and column 2 of the tableau for Exercise 2.

19. The entry in row 1 and column 2 of the tableau for Exercise 3.

20. The entry in row 1 and column 1 of the tableau for Exercise 4.

21. The entry in row 2 and column 2 of the tableau for Exercise 5.

22. The entry in row 2 and column 1 of the tableau for Exercise 6.

23. The entry in row 1 and column 1, then the entry in row 1 and column 2, of the tableau for Exercise 7.

24. The entry in row 1 and column 1, then the entry in row 2 and column 3, of the tableau for Exercise 8.

In Exercises 25 and 26 set up the simplex tableau and solve the linear program by pivoting on a single entry. Find the entry to pivot on by trial and error.

25. Maximize $4.5x + 3.5y = R$
subject to
$$6x + 5y \le 600$$
$$3x + 4y \le 240$$
$$x \ge 0,\ y \ge 0.$$

26. Maximize $2.5x + 0.9y = R$
subject to
$$5x + 2y \le 3000$$
$$0.2x + 0.4y \le 160$$
$$x \ge 0,\ y \ge 0.$$

Answer to Practice Problem **7.** The final tableau is

$$\begin{array}{c} \\ s \\ x \\ \\ \end{array}\left[\begin{array}{cccc|c} x & y & s & t & \\ 0 & 1.2 & \boxed{1} & -0.2 & 120 \\ \boxed{1} & 0.8 & 0 & 0.2 & 200 \\ \hline 0 & -0.2 & 0 & -0.8 & R - 800 \end{array}\right],$$

and we conclude from it that R has the maximum value of 800 at $(x, y) = (200, 0)$.

3.4 Maximization Problems

In this section we will answer two questions that arise in applying the simplex method, namely,

(1) How do we choose the pivot entry of the simplex tableau?
(2) How do we know when we are finished, so that we can find the optimal point by setting the free variables equal to zero?

To illustrate the answers to these questions, we will return to the example of Albert's Smoke Shop (Example 3.1 from Sections 3.1 and 3.2), which led to the linear program

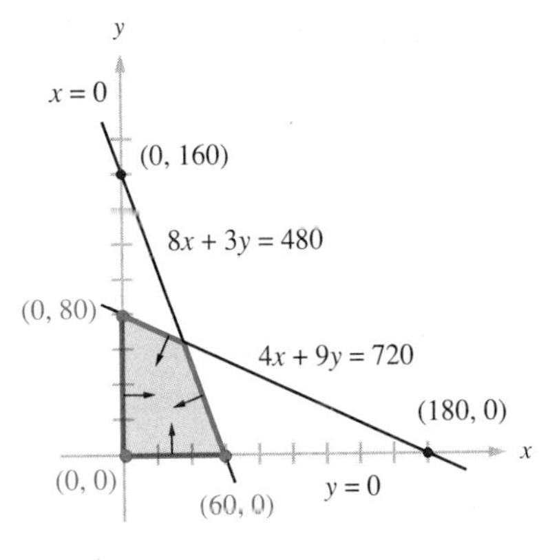

FIGURE 3.18

$$
\begin{array}{llr}
\text{Maximize} & 16x + 12y = R & \\
\text{subject to} & 8x + 3y \le 480 & (3.1)\\
& 4x + 9y \le 720 & (3.2)\\
& x \ge 0 & (3.3)\\
& y \ge 0. & (3.4)
\end{array}
$$

The feasible region is shown in Figure 3.18.

We will start by treating only **standard maximum programs,** that is, programs where

(a) The objective function is to be maximized.
(b) The variables are constrained to be nonnegative (as in (3.3) and (3.4)).
(c) All other constraints are of the $\le$ type (as (3.1) and (3.2) are in this problem).
(d) The right-hand constants of the constraints are all nonnegative (480 and 720 in this problem).

We introduce slack variables s and t to transform the program to:

$$
\begin{array}{ll}
\text{Maximize} & 16x + 12y = R \\
\text{subject to} & 8x + 3y + s \qquad = 480 \\
& 4x + 9y \qquad + t = 720 \\
& x \ge 0,\ y \ge 0,\ s \ge 0,\ t \ge 0,
\end{array}
$$

with the tableau

$$
\text{Tableau 1}\qquad
\begin{array}{c}
\\ s \\ t \\ \\
\end{array}
\begin{array}{cccc|c}
x & y & s & t & \\
\textcircled{8} & 3 & \boxed{1} & 0 & 480 \\
4 & 9 & 0 & \boxed{1} & 720 \\
\hline
16 & 12 & 0 & 0 & R
\end{array}.
$$

Note that the original row labels are simply the slack variables. It turns out that we can solve the program by pivoting first on the entry in row 1 and column 1, and then on the entry in row 2 and column 2. We will exhibit only the tableaus after each of these pivots, leaving it for the reader to check the arithmetic details. We pivot on the 8 in row 1 and column 1 (causing x to replace s as the label of row 1) to get

$$
\text{Tableau 2}\qquad
\begin{array}{c}
\\ x \\ t \\ \\
\end{array}
\begin{array}{cccc|c}
x & y & s & t & \\
\boxed{1} & \frac{3}{8} & \frac{1}{8} & 0 & 60 \\
0 & \textcircled{$\frac{15}{2}$} & -\frac{1}{2} & \boxed{1} & 480 \\
\hline
0 & 6 & -2 & 0 & R - 960
\end{array}.
$$

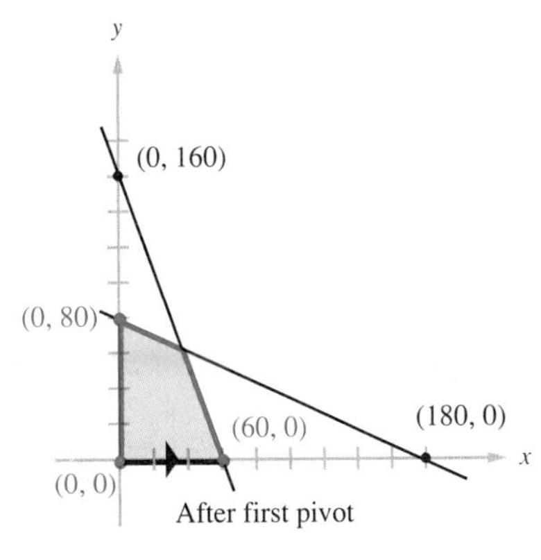

FIGURE 3.19

Notice that setting the free variables y and s equal to 0 gives $x = 60$ and $t = 480$, so that we are now at the point $(x, y) = (60, 0)$ shown in Figure 3.19.

Now we pivot on the entry 15/2 in row 2 and column 2 (causing y to replace t as the label of row 2), to get

$$\text{Tableau 3} \quad \begin{array}{c} \\ x \\ y \\ \\ \\ \end{array} \begin{array}{c} \begin{array}{cccc} x & y & s & t \end{array} \\ \left[\begin{array}{cccc|c} \boxed{1} & 0 & \frac{3}{20} & -\frac{1}{20} & 36 \\ 0 & \boxed{1} & -\frac{1}{15} & \frac{2}{15} & 64 \\ \hline 0 & 0 & -\frac{8}{5} & -\frac{4}{5} & R - 1344 \end{array}\right] \end{array}.$$

We know we are done at this point, because the last row of Tableau 3 is equivalent to the equation

$$R = 1344 - \frac{8}{5}s - \frac{4}{5}t,$$

so setting s and t both equal to 0 is the best we can do. This gives the maximal value of 1344 when $(x, y) = (36, 64)$. (See Figure 3.20.)

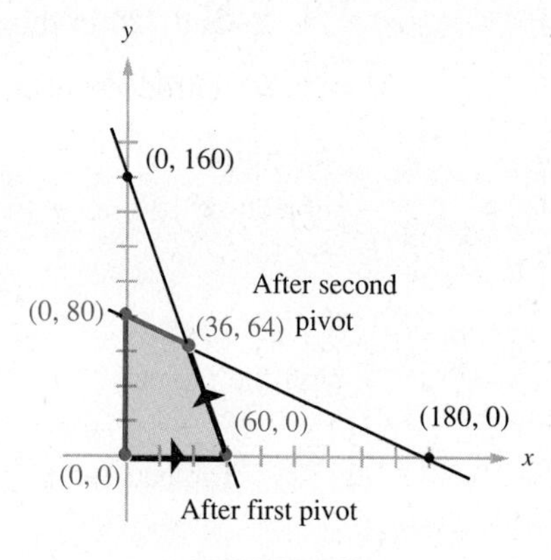

FIGURE 3.20

Why not stop at Tableau 2? Translating its last row into an equation provides the answer. An equivalent equation is

$$R = 960 + 6y - 2s. \tag{3.7}$$

It appears that making the free variable y positive instead of 0 would increase R. The positive coefficient 6 in (3.7) corresponds to the 6 in the last row of the Tableau 2. The entries in the bottom row of the simplex tableau to the left of the vertical line are called the **indicators** of the tableau. This brings us to our first rule.

The simplex algorithm is completed when no indicator is positive.

Since we want to get rid of all positive indicators, it seems reasonable to pivot on an entry in a column with a positive indicator, so as to change it to 0. In Tableau 1 there are two positive indicators, 16 and 12. Which of their columns should we pivot in? Notice that the last row of Tableau 1 corresponds to the equation

$$16x + 12y = R.$$

Increasing x by 1 increases R by 16, but increasing y by 1 only increases R by 12. Since we are trying to maximize R, we adopt the following rule to choose the column to pivot in.

Always pivot on a column with a positive indicator. If more than one column has a positive indicator, choose the column with the indicator that is largest. If there is a tie among columns, choose any one of them.

Applying this rule to Tableau 1, we choose our first pivot in column 1, and we mark this column with a star.

Now we must decide whether to pivot on the 8 or the 4. The effect of each of these pivots is easier to see if we multiply the first row by $\frac{1}{8}$ and the second row by $\frac{1}{4}$.

Tableau 1a

$$\left[\begin{array}{cccc|c} x & y & s & t & \\ \textcircled{1} & \frac{3}{8} & \frac{1}{8} & 0 & \frac{480}{8} = 60 \\ \textcircled{1} & \frac{9}{4} & 0 & \frac{1}{4} & \frac{720}{4} = 180 \\ \hline 16^* & 12 & 0 & 0 & R \end{array}\right] \; ?$$

If the pivot entry is in the first row, we would complete the pivot transformation by subtracting this row from the second. This would make the entries to the right of the vertical line in these rows 60 and 120. Likewise to pivot using the 1 in the second row, we would subtract the second row from the first, making the entries to the right of the vertical line -120 and 180. But *we do not want negative entries to the right of the vertical line,* for then setting the free variables equal to 0 would give a negative basic variable, contradicting the original conditions of the program. This is what happened at our first attempt at a pivot in Section 3.3, when we moved to a point that was not in the feasible region.

Therefore in Tableau 1 we pivot on the 8 in the first row instead of the 4 because 480/8 is less than 720/4. This leads us to our third rule.

Once the pivot column has been decided, divide each positive entry in that column above the horizontal line into the corresponding entry in the rightmost column. Pivot on the entry that gives the smallest positive quotient.

For example, knowing that in Tableau 1 we want to pivot on an entry in column 1, we divide the entries in that column into the corresponding rightmost entries, writing the results to the right of the display.

$$\text{Tableau 1} \quad \begin{array}{c} \\ \\ \\ \\ \end{array}\left[\begin{array}{cccc|c} x & y & s & t & \\ 8 & 3 & 1 & 0 & 480 \\ 4 & 9 & 0 & 1 & 720 \\ \hline 16^* & 12 & 0 & 0 & R \end{array}\right] \qquad \begin{array}{l} 480/8 = 60^* \\ 720/4 = 180 \end{array}$$

Since $60 < 180$ our pivot will be chosen from row 1. Thus we pivot on the 8 in the first row and column, producing, as we have already seen, the following tableau.

$$\text{Tableau 2} \quad \left[\begin{array}{cccc|c} x & y & s & t & \\ 1 & \frac{3}{8} & \frac{1}{8} & 0 & 60 \\ 0 & \frac{15}{2} & -\frac{1}{2} & 1 & 480 \\ \hline 0 & 6^* & -2 & 0 & R - 960 \end{array}\right] \qquad \begin{array}{l} 60/(3/8) = 160 \\ 480/(15/2) = 64^* \end{array}$$

Since the only positive indicator is the 6 in column 2, our next pivot will be chosen from this column. We divide the entries in the second column into the entries 60 and 480 in the rightmost column, getting the quotients 160 and 64. Since $64 < 160$ we mark the second row with a star, and pivot on the 15/2 in row 2 and column 2. This leads to Tableau 3 and the solution of the linear program, as we have already seen.

EXAMPLE 3.11 Revisited Show how one would choose the pivots suggested in Example 3.11 of the previous section.

Solution The initial tableau in Example 3.11 was

$$\begin{array}{c} \\ \\ s \\ t \\ u \\ \text{Indicators} \end{array} \left[\begin{array}{ccccc|c} \multicolumn{2}{c}{\overbrace{\quad\quad}^{\text{Free}}} & \multicolumn{3}{c}{\overbrace{\quad\quad\quad}^{\text{Basic}}} & \\ x & y & s & t & u & \\ 1 & 1 & 1 & 0 & 0 & 100 \\ 2 & 1 & 0 & 1 & 0 & 180 \\ 1 & 3 & 0 & 0 & 1 & 240 \\ \hline 6^* & 4 & 0 & 0 & 0 & R \end{array}\right]. \qquad \begin{array}{l} 100/1 = 100 \\ 180/2 = 90^* \\ 240/1 = 240 \end{array}$$

Since 6 is the largest indicator, we will choose a pivot in the column 1. Dividing the other entries of column 1 into the corresponding entries in the rightmost column gives the quotients 100, 90, and 240. Since 90 is the smallest of these, we pivot on the 2 in row 2 and column 1. As we saw in Section 3.3, this yields the tableau

$$
\begin{array}{c} \\ s \\ x \\ u \\ \\ \end{array}
\begin{array}{c} x \quad\; y \quad\; s \quad\;\; t \quad u \\
\left[\begin{array}{ccccc|c} 0 & .5 & 1 & -.5 & 0 & 10 \\ 1 & .5 & 0 & .5 & 0 & 90 \\ 0 & 2.5 & 0 & -.5 & 1 & 150 \\ \hline 0 & 1^* & 0 & -3 & 0 & R - 540 \end{array}\right]. \end{array}
\qquad
\begin{array}{l} 10/.5 = 20^* \\ 90/.5 = 180 \\ 150/2.5 = 60 \end{array}
$$

Now 1 is the only positive indicator, so we will pivot on an entry of column 2. Dividing the other entries of this column into 10, 90, and 150 gives 20, 180, and 60, and the smallest of these is 20. Thus we pivot on the entry .5 in row 1 and column 2. As we saw in Section 3.3, this produces the tableau

$$
\begin{array}{c} \\ y \\ x \\ u \\ \\ \end{array}
\begin{array}{c} x \quad y \quad\; s \quad\;\; t \quad u \\
\left[\begin{array}{ccccc|c} 0 & 1 & 2 & -1 & 0 & 20 \\ 1 & 0 & -1 & 1 & 0 & 80 \\ 0 & 0 & -5 & 2 & 1 & 100 \\ \hline 0 & 0 & -2 & -2 & 0 & R - 560 \end{array}\right]. \end{array}
$$

Since this has no positive indicators, we are done. We read from the tableau the solution $(x, y) = (80, 20)$, which yields the maximum $R = 560$. ■

Practice Problem 8 What pivot should be chosen for the following tableau?

$$
\left[\begin{array}{ccccc|c} 2 & 6 & 5 & 1 & 0 & 34 \\ 3 & 8 & 2 & 0 & 1 & 16 \\ 4 & 12 & 8 & 0 & 0 & 70 \\ \hline 3 & 5 & 7 & 0 & 0 & R \end{array}\right]
$$

Practice Problem 9 Which are the basic variables and which are the free variables in the following tableau? What value of R results from setting all free variables equal to 0? What value of (x, y, z) produces this R?

$$
\begin{array}{c} \\ z \\ s \\ x \\ \\ \end{array}
\begin{array}{c} x \quad\;\; y \quad z \quad s \quad\;\; t \\
\left[\begin{array}{ccccc|c} 0 & 3 & 1 & 0 & 6 & 4 \\ 0 & -2 & 0 & 1 & 0 & 6 \\ 1 & 5 & 0 & 0 & -2 & 10 \\ \hline 0 & -3 & 0 & 0 & -5 & R - 22 \end{array}\right] \end{array}
$$

We now summarize the method we have developed.

The Simplex Method for Standard Maximum Programs

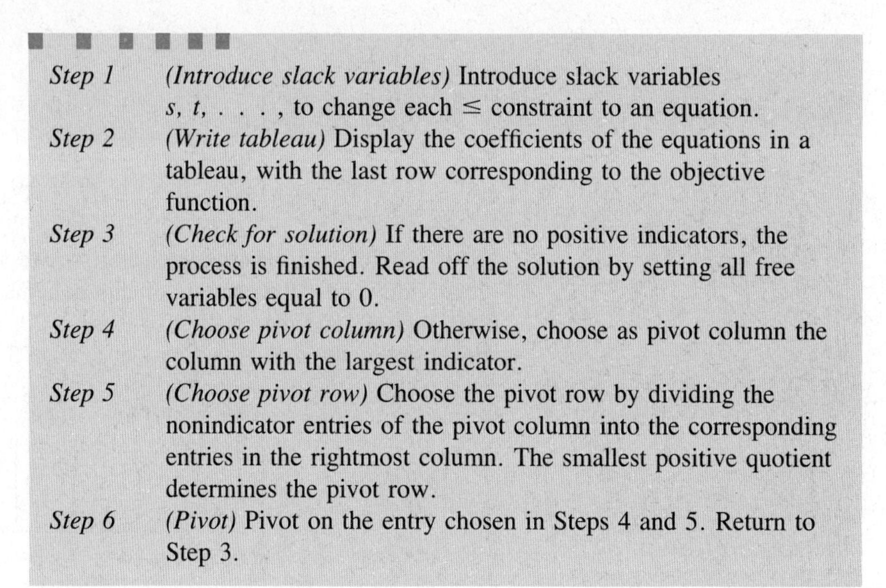

Step 1 *(Introduce slack variables)* Introduce slack variables $s, t, \ldots$, to change each $\le$ constraint to an equation.

Step 2 *(Write tableau)* Display the coefficients of the equations in a tableau, with the last row corresponding to the objective function.

Step 3 *(Check for solution)* If there are no positive indicators, the process is finished. Read off the solution by setting all free variables equal to 0.

Step 4 *(Choose pivot column)* Otherwise, choose as pivot column the column with the largest indicator.

Step 5 *(Choose pivot row)* Choose the pivot row by dividing the nonindicator entries of the pivot column into the corresponding entries in the rightmost column. The smallest positive quotient determines the pivot row.

Step 6 *(Pivot)* Pivot on the entry chosen in Steps 4 and 5. Return to Step 3.

It should be noted that if in Step 5 there is a tie for the smallest positive quotient, complications may occur. We will not treat this possibility. The next example illustrates the simplex method from beginning to end.

EXAMPLE 3.13 A small furniture finishing factory produces desks, buffets, and cocktail tables. Each desk requires 0.3 hours of sanding, 0.1 hours of staining, and 0.2 hours of varnishing. Each buffet requires 0.2 hours of sanding, 0.4 hours of staining, and 0.4 hours of varnishing. In addition, each cocktail table requires 0.1 hours of sanding, 0.1 hours of staining, and 0.2 hours of varnishing. During a particular week the factory will have at most 44 hours of time available for sanding, at most 40 hours of time available for staining, and at most 48 hours available for varnishing. If the factory earns profits of \$10 on each desk, \$14 on each buffet, and \$7 on each cocktail table, how should production be scheduled in order to obtain a maximum profit?

Solution Let

$$x = \text{the number of desks to be made}$$
$$y = \text{the number of buffets to be made}$$
$$z = \text{the number of cocktail tables to be made.}$$

Then we have the following standard maximum linear program:

$$\begin{array}{ll} \text{Maximize} & 10x + 14y + 7z = P \\ \text{subject to} & 0.3x + 0.2y + 0.1z \le 44 \\ & 0.1x + 0.4y + 0.1z \le 40 \\ & 0.2x + 0.4y + 0.2z \le 48 \\ & x \ge 0,\ y \ge 0,\ z \ge 0. \end{array}$$

Introducing nonnegative slack variables s, t, and u, we can rewrite the problem in the form

$$\begin{aligned} \text{Maximize} \quad & 10x + 14y + 7z &&= P \\ \text{subject to} \quad & 0.3x + 0.2y + 0.1z + s &&= 44 \\ & 0.1x + 0.4y + 0.1z + t &&= 40 \\ & 0.2x + 0.4y + 0.2z + u &&= 48 \\ & x \geq 0,\ y \geq 0,\ z \geq 0,\ s \geq 0,\ t \geq 0,\ u \geq 0. \end{aligned}$$

Thus our simplex tableau is

$$\begin{array}{c} \\ s \\ t \\ u \\ \\ \end{array} \begin{array}{c} \begin{array}{cccccc} x & y & z & s & t & u \end{array} \\ \left[\begin{array}{cccccc|c} 0.3 & 0.2 & 0.1 & 1 & 0 & 0 & 44 \\ 0.1 & \bigcirc\!\!\!\!\!\!0.4 & 0.1 & 0 & 1 & 0 & 40 \\ 0.2 & 0.4 & 0.2 & 0 & 0 & 1 & 48 \\ \hline 10 & 14^* & 7 & 0 & 0 & 0 & P \end{array}\right] \end{array} \qquad \begin{array}{l} 44/0.2 = 220 \\ 40/0.4 = 100^* \\ 48/0.4 = 120 \end{array}$$

Since 14 is the largest indicator, we will pivot in column 2. The smallest of the corresponding quotients is the 100 in row 2, so we will pivot at the circled entry 0.4. This will change the label of row 2 from t to y.

$$\left[\begin{array}{cccccc|c} 0.3 & 0.2 & 0.1 & 1 & 0 & 0 & 44 \\ 0.25 & \textcircled{1} & 0.25 & 0 & 2.5 & 0 & 100 \\ 0.2 & 0.4 & 0.2 & 0 & 0 & 1 & 48 \\ \hline 10 & 14 & 7 & 0 & 0 & 0 & P \end{array}\right] \qquad ((1/0.4)R_2)$$

$$\begin{array}{c} \\ s \\ y \\ u \\ \\ \end{array} \begin{array}{c} \begin{array}{cccccc} x & y & z & s & t & u \end{array} \\ \left[\begin{array}{cccccc|c} 0.25 & 0 & 0.05 & 1 & -0.5 & 0 & 24 \\ 0.25 & 1 & 0.25 & 0 & 2.5 & 0 & 100 \\ 0.1 & 0 & 0.1 & 0 & -1 & 1 & 8 \\ \hline 6.5 & 0 & 3.5 & 0 & -35 & 0 & P - 1400 \end{array}\right] \end{array} \qquad \begin{array}{l} (R_1 + (-0.2)R_2) \\ \\ (R_3 + (-0.4)R_2) \\ (R_4 + (-14)R_2) \end{array}$$

We repeat this tableau, indicating how we choose to pivot next on the entry in column 1 and row 3.

$$\left[\begin{array}{cccccc|c} 0.25 & 0 & 0.05 & 1 & -0.5 & 0 & 24 \\ 0.25 & 1 & 0.25 & 0 & 2.5 & 0 & 100 \\ 0.1 & 0 & 0.1 & 0 & -1 & 1 & 8 \\ \hline 6.5^* & 0 & 3.5 & 0 & -35 & 0 & P - 1400 \end{array}\right] \qquad \begin{array}{l} 24/0.25 = 96 \\ 100/0.25 = 400 \\ 8/0.1 = 80^* \end{array}$$

The actual pivot transformation proceeds as follows.

$$\left[\begin{array}{cccccc|c} 0.25 & 0 & 0.05 & 1 & -0.5 & 0 & 24 \\ 0.25 & 1 & 0.25 & 0 & 2.5 & 0 & 100 \\ \textcircled{1} & 0 & 1 & 0 & -10 & 10 & 80 \\ \hline 6.5 & 0 & 3.5 & 0 & -35 & 0 & P - 1400 \end{array}\right] \qquad (10R_3)$$

$$
\begin{array}{c|cccccc|c|}
 & x & y & z & s & t & u & \\
\hline
s & 0 & 0 & -0.2 & 1 & 2 & -2.5 & 4 \\
y & 0 & 1 & 0 & 0 & 5 & -2.5 & 80 \\
x & 1 & 0 & 1 & 0 & -10 & 10 & 80 \\
\hline
 & 0 & 0 & -3 & 0 & 30 & -65 & P - 1920
\end{array}
\qquad
\begin{array}{l}
(\mathbf{R_1 + (-0.25)R_3}) \\
(\mathbf{R_2 + (-0.25)R_3}) \\
\\
(\mathbf{R_4 + (-6.5)R_3})
\end{array}
$$

We repeat this tableau, showing how we choose the next pivot to be the entry 2 in column 5 and row 1.

$$
\left[\begin{array}{cccccc|c}
0 & 0 & -0.2 & 1 & \textcircled{2} & -2.5 & 4 \\
0 & 1 & 0 & 0 & 5 & -2.5 & 80 \\
1 & 0 & 1 & 0 & -10 & 10 & 80 \\
\hline
0 & 0 & -3 & 0 & 30^* & -65 & P - 1920
\end{array}\right]
\qquad
\begin{array}{l}
4/2 = 2^* \\
80/5 = 16 \\
\text{(negative quotient)}
\end{array}
$$

The pivot transformation proceeds as follows.

$$
\left[\begin{array}{cccccc|c}
0 & 0 & -0.1 & 0.5 & \textcircled{1} & -1.25 & 2 \\
0 & 1 & 0 & 0 & 5 & -2.5 & 80 \\
1 & 0 & 1 & 0 & -10 & 10 & 80 \\
\hline
0 & 0 & -3 & 0 & 30 & -65 & P - 1920
\end{array}\right]
\qquad
(\mathbf{0.5R_1})
$$

$$
\begin{array}{c|cccccc|c|}
 & x & y & z & s & t & u & \\
\hline
t & 0 & 0 & -0.1 & 0.5 & 1 & -1.25 & 2 \\
y & 0 & 1 & 0.5 & -2.5 & 0 & 3.75 & 70 \\
x & 1 & 0 & 0 & 5 & 0 & -2.5 & 100 \\
\hline
 & 0 & 0 & 0 & -15 & 0 & -27.5 & P - 1980
\end{array}
\qquad
\begin{array}{l}
\\
(\mathbf{R_2 + (-5)R_1}) \\
(\mathbf{R_3 + 10R_1}) \\
(\mathbf{R_4 + (-30)R_1})
\end{array}
$$

Since there are no positive indicators, we are done. From the last row we see that the maximum value of P is 1980, which occurs when the free variables z, s, and u are all 0. From the upper part of the tableau we see this makes $t = 2$, $y = 70$, and $x = 100$. The optimal point is $(x, y, z) = (100, 70, 0)$. The factory should make 100 desks, 70 buffets, and no cocktail tables and will then realize a profit of $1980. ■

EXERCISES 3.4

In Exercises 1–10 state the row and column, if any, where a pivot should be made in the given tableau.

1.
$$
\left[\begin{array}{cccc|c}
2 & 4 & 1 & 0 & 7 \\
6 & 3 & 0 & 1 & 20 \\
\hline
5 & 3 & 0 & 0 & R
\end{array}\right]
$$

2.
$$
\left[\begin{array}{cccc|c}
6 & 2 & 1 & 0 & 12 \\
2 & 3 & 0 & 1 & 15 \\
\hline
6 & 9 & 0 & 0 & R
\end{array}\right]
$$

3. $\left[\begin{array}{cccc|c} 5 & 1 & 2 & 0 & 45 \\ 2 & 0 & 9 & 1 & 21 \\ \hline 1 & -2 & .5 & 0 & R-32 \end{array}\right]$

4. $\left[\begin{array}{cccc|c} 12 & 0 & 7 & 1 & 33 \\ 6 & 1 & 4 & 0 & 18 \\ \hline 5 & 0 & 11 & -1 & R-22 \end{array}\right]$

5. $\left[\begin{array}{ccccc|c} 3.3 & 4 & 1 & 0 & 0 & 12 \\ 7 & 2 & 0 & 1 & 0 & 25 \\ 2 & 5 & 0 & 0 & 1 & 7 \\ \hline 3 & -6 & 0 & 0 & 0 & R \end{array}\right]$

6. $\left[\begin{array}{ccccc|c} 2 & 3 & 1 & 0 & 0 & 11 \\ 3 & -1 & 0 & 1 & 0 & 16 \\ 5 & 4 & 0 & 0 & 1 & 27 \\ \hline .3 & .2 & 0 & 0 & 0 & R \end{array}\right]$

7. $\left[\begin{array}{ccccc|c} \frac{2}{3} & \frac{1}{5} & 3 & 1 & 0 & 13 \\ 4 & \frac{2}{3} & -2 & 0 & 1 & 44 \\ \hline -3 & 1 & \frac{1}{2} & 0 & 0 & R \end{array}\right]$

8. $\left[\begin{array}{cccc|c} 3 & \frac{3}{2} & 1 & 0 & 12 \\ -2 & 4 & 0 & 1 & 9 \\ \hline 4 & \frac{7}{2} & 0 & 0 & R \end{array}\right]$

9. $\left[\begin{array}{cccc|c} 3 & 0 & 4 & 1 & 13 \\ 5 & 1 & 7 & 0 & 9 \\ \hline -2 & -1 & 0 & 0 & R-6 \end{array}\right]$

10. $\left[\begin{array}{cccc|c} 3 & 5 & 0 & 1 & 33 \\ 0 & 4 & 1 & 0 & 21 \\ \hline 4 & 3 & 0 & 0 & R \end{array}\right]$

In Exercises 11–14 final tableaus are given after the simplex method has been applied. Tell what the maximal value of R is, and what values of x and y or x, y, and z produce it.

11. $\begin{array}{c} \\ t \\ y \\ \\ \end{array} \left[\begin{array}{cccc|c} x & y & s & t & \\ 4 & 0 & 3 & 1 & 5 \\ 5 & 1 & -2 & 0 & 7 \\ \hline -1 & 0 & -3 & 0 & R-13 \end{array}\right]$

12. $\begin{array}{c} \\ y \\ x \\ \\ \end{array} \left[\begin{array}{cccc|c} x & y & s & t & \\ 0 & 1 & 5 & -2 & 6 \\ 1 & 0 & 0 & 4 & 3 \\ \hline 0 & 0 & -3 & -2 & R-7 \end{array}\right]$

13. $\begin{array}{c} \\ z \\ x \\ \\ \end{array} \left[\begin{array}{ccccc|c} x & y & z & s & t & \\ 0 & 3 & 1 & 2 & 1 & 5 \\ 1 & -2 & 0 & 4 & 6 & 7 \\ \hline 0 & -1 & 0 & -3 & -3 & R-22 \end{array}\right]$

14. $\begin{array}{c} \\ t \\ y \\ \\ \end{array} \left[\begin{array}{ccccc|c} x & y & z & s & t & \\ 0 & 0 & 3 & 1 & 1 & 11 \\ 6 & 1 & -2 & 4 & 0 & 3 \\ \hline -1 & 0 & -4 & -1 & 0 & R-2 \end{array}\right]$

In Exercises 15–18 solve the linear program by the simplex method and also by the graphical method. Indicate on your graph the edge-path described by the simplex pivots.

15. Maximize $5x + 2y = R$

subject to
$$\begin{aligned} 2x + y &\leq 30 \\ x + 2y &\leq 30 \\ x \geq 0,\ y &\geq 0. \end{aligned}$$

16. Maximize $10x + 15y = R$

subject to
$$\begin{aligned} x + 2y &\leq 4 \\ x + y &\leq 3 \\ x \geq 0,\ y &\geq 0. \end{aligned}$$

17. Maximize $10x + 5y = R$

subject to
$$\begin{aligned} x + y &\leq 5 \\ x &\leq 3 \\ y &\leq 4 \\ x \geq 0,\ y &\geq 0. \end{aligned}$$

18. Maximize $2x - y = R$

subject to
$$\begin{aligned} x + y &\leq 10 \\ x - y &\leq 4 \\ x &\leq 6 \\ x \geq 0,\ y &\geq 0. \end{aligned}$$

In Exercises 19–26 solve the standard maximum program by the simplex method.

19. Maximize $3x - 2y = R$

subject to
$$\begin{aligned} x &\le 20 \\ y &\le 20 \\ x - y &\le 10 \\ -x + y &\le 10 \\ x \ge 0,\ y &\ge 0. \end{aligned}$$

20. Maximize $3x + 4y = R$

subject to
$$\begin{aligned} 2x + 3y &\le 180 \\ x - y &\le 15 \\ y &\le 50 \\ x &\le 30 \\ x \ge 0,\ y &\ge 0. \end{aligned}$$

21. Maximize $2x + y + z = R$

subject to
$$\begin{aligned} x + 2y + 3z &\le 80 \\ x + y + z &\le 60 \\ x \ge 0,\ y \ge 0,\ z &\ge 0. \end{aligned}$$

22. Maximize $2x + 3y + 4z = R$

subject to
$$\begin{aligned} 2x + y + z &\le 40 \\ x + y + 2z &\le 50 \\ x \ge 0,\ y \ge 0,\ z &\ge 0. \end{aligned}$$

23. Maximize $2x + y - 3z = R$

subject to
$$\begin{aligned} x + y + z &\le 4 \\ x - y + z &\le 2 \\ -x + y + z &\le 1 \\ x \ge 0,\ y \ge 0,\ z &\ge 0. \end{aligned}$$

24. Maximize $5x - 4y + z = R$

subject to
$$\begin{aligned} x - y + z &\le 20 \\ x + y &\le 10 \\ y + z &\le 15 \\ x \ge 0,\ y \ge 0,\ z &\ge 0. \end{aligned}$$

25. Maximize $x + 2y - z + w = R$

subject to
$$\begin{aligned} 2x + y + 3z + w &\le 10 \\ x + z + 2w &\le 2 \\ y + 2z + w &\le 4 \\ x \ge 0,\ y \ge 0,\ z \ge 0,\ w &\ge 0. \end{aligned}$$

26. Maximize $x + 2y + 4z + w = R$

subject to
$$\begin{aligned} x - y + z - w &\le 1 \\ x - z &\le 1 \\ y + w &\le 1 \\ x \ge 0,\ y \ge 0,\ z \ge 0,\ w &\ge 0. \end{aligned}$$

27. A farmer has 1000 acres of land on which he intends to plant asparagus, brussels sprouts, and cauliflower. These three crops require, respectively, 15, 25, and 20 inches of water per acre to grow to maturity, and the farmer has been allotted a total of 20,000 acre-inches of water for the season. (An acre-inch is the amount of water needed to irrigate an acre of land to a depth of one inch.) The respective profits on the three crops are \$200, \$250, and \$220 an acre. How much of each crop should be planted to maximize total profit?

28. The Zeno Corporation makes three models of stereo receivers. There are three stages in the manufacture of each unit, as shown in the following table:

	Work-hours per unit		
Model	A	B	C
Assembly	0.2 wh	0.3 wh	0.1 wh
Wiring	0.2 wh	0.2 wh	0.1 wh
Testing	0.1 wh	0.2 wh	0.0 wh

The respective profits on the three models are \$50, \$75, and \$30 per unit. The company has available a total of 40 work-hours of labor per day for assembly, 28 for wiring, and 16 for testing. How many receivers of each type should be made to maximize total profit?

By looking at the terminal tableau one can determine whether or not there is more than one optimal solution as follows: If every column headed by a free variable has a negative *indicator, then the optimal solution is unique. But if some such column has indicator* zero, *there is more than one optimal solution. In fact another solution can be found by pivoting in this column. Exercises 29 and 30 illustrate this fact.*

29. Show that the solution $(x, y, z) = (100, 70, 0)$ found in Example 3.13 is not unique by pivoting in column 3 of the final tableau. What is the new solution and value of the objective function?

30. Solve Example 3.9 of Section 3.2 by the simplex method. Then find another optimal point by pivoting in a column of the final tableau having indicator 0.

Answers to Practice Problems

8. The 5 in row 1 and column 3.

9. The variables x, z, and s are basic and y and t are free. Setting $y = t = 0$ gives $R = 22$. This corresponds to $(x, y, z) = (10, 0, 4)$.

3.5 Minimization Problems and Duality

Many linear programs that arise in practice involve minimizing the objective function, with constraints of the $\geq$ variety. In Section 3.2 we solved such a problem graphically in Example 3.5. The problem involved hiring buses for an alumni association excursion, and led to the linear program

$$\begin{array}{lrcl} \text{Minimize} & 2000x + 2400y & = & C \\ \text{subject to} & 40x + 50y & \geq & 800 \\ & 1000x + 750y & \geq & 18{,}000 \\ & x \geq 0, \; y \geq 0. & & \end{array}$$

We could summarize the information in this problem in a **compact tableau** as follows.

$$\begin{array}{cc} x & y \\ \end{array}$$
$$\left[\begin{array}{cc|l} 40 & 50 & (\geq) \quad 800 \\ 1000 & 750 & (\geq) \; 18{,}000 \\ \hline 2000 & 2400 & (\text{minimize}) \; C \end{array}\right] \tag{3.8}$$

Here the conditions $x \geq 0$ and $y \geq 0$ are understood. It turns out that corresponding to each minimization program with $\geq$ constraints, there is a maximization program with $\leq$ constraints, called the **dual program.** The numbers involved in the dual program are found by interchanging rows and columns in the compact tableau of the original program. For example the tableau just presented translates into

$$\begin{array}{cc} X & Y \\ \end{array}$$
$$\left[\begin{array}{cc|r} 40 & 1000 & (\leq) \; 2000 \\ 50 & 750 & (\leq) \; 2400 \\ \hline 800 & 18{,}000 & (\text{maximize}) \; R \end{array}\right]. \tag{3.9}$$

Notice how the first row of this tableau comes from the first column of (3.8), the second row comes from the second column of (3.8), etc. Tableau (3.9) corresponds to the linear program

$$\begin{array}{ll} \text{Maximize} & 800X + 18{,}000Y = R \\ \text{subject to} & 40X + 1000Y \leq 2000 \\ & 50X + 750Y \leq 2400 \\ & X \geq 0,\ Y \geq 0. \end{array}$$

Of course this is a standard maximum program, solvable with the simplex method. It is a surprising fact that *applying the simplex method to this dual program leads also to a solution of the original minimization problem.*

Let us demonstrate the last statement. To solve the dual problem we introduce nonnegative slack variables s and t satisfying

$$\begin{aligned} 40X + 1000Y + s &= 2000 \\ 50X + 750Y + t &= 2400, \end{aligned}$$

which leads to the tableau

$$\begin{array}{c} \\ s \\ t \\ \\ \end{array} \begin{array}{c} \begin{array}{cccc|c} X & Y & s & t & \\ \end{array} \\ \left[\begin{array}{cccc|c} 40 & \textcircled{1000} & 1 & 0 & 2000 \\ 50 & 750 & 0 & 1 & 2400 \\ \hline 800 & 18000^* & 0 & 0 & R \end{array}\right] \end{array} \qquad \begin{array}{l} 2000/1000 = 2^* \\ 2400/750 = 3.2 \end{array}$$

In the usual way we choose to pivot on the entry 1000 in row 1 and column 2 as follows.

$$\begin{array}{c} X \quad Y \quad s \quad t \\ \left[\begin{array}{cccc|c} .04 & \textcircled{1} & .001 & 0 & 2 \\ 50 & 750 & 0 & 1 & 2400 \\ \hline 800 & 18000 & 0 & 0 & R \end{array}\right] \end{array} \qquad (.001\mathbf{R}_1)$$

$$\begin{array}{c} \\ Y \\ t \\ \\ \end{array} \begin{array}{c} X \quad Y \quad s \quad t \\ \left[\begin{array}{cccc|c} .04 & 1 & .001 & 0 & 2 \\ \textcircled{20} & 0 & -.75 & 1 & 900 \\ \hline 80^* & 0 & -18 & 0 & R - 36000 \end{array}\right] \end{array} \qquad \begin{array}{ll} & 2/.04 = 50 \\ (\mathbf{R}_2 + (-750)\mathbf{R}_1) & 900/20 = 45^* \\ (\mathbf{R}_3 + (-18000)\mathbf{R}_1) & \end{array}$$

Our next pivot will be on the entry 20 in row 2 and column 1, producing the following tableaus.

$$\begin{array}{c} X \quad Y \quad s \quad t \\ \left[\begin{array}{cccc|c} .04 & 1 & .001 & 0 & 2 \\ \textcircled{1} & 0 & -.0375 & .05 & 45 \\ \hline 80 & 0 & -18 & 0 & R - 36000 \end{array}\right] \end{array} \qquad (.05\mathbf{R}_2)$$

$$\begin{array}{c} \\ Y \\ X \\ \\ \end{array} \left[\begin{array}{cccc|c} X & Y & s & t & \\ 0 & 1 & .0025 & -.002 & .2 \\ 1 & 0 & -.0375 & .05 & 45 \\ \hline 0 & 0 & -15 & -4 & R - 39600 \end{array}\right] \quad \begin{array}{l} \\ (R_1 + (-.04)R_2) \\ \\ R_3 + (-80)R_2 \end{array}$$

Since there are no positive indicators, we are done. From the last row of our tableau the maximum value of R is 39600. But *39600 is also the minimal value of C in our original problem*. The optimal point (x, y) from the original minimization problem can also be read from the last tableau. It is $(x, y) = (15, 4)$. The values of x and y can be found by simply taking the negatives of the indicators (in color) corresponding to the slack variables s and t in our final tableau. (In general the optimal values of the original variables will be the negatives of the indicators corresponding to the slack variables of the dual problem after the simplex method has been applied.) The reader should check that in Example 3.5 we did find graphically that the minimal value of C was 39600 at $(x, y) = (15, 4)$. Notice that row labels are not necessary when solving the dual program because the solution is read from the bottom row of the final tableau.

THE DUALITY THEOREM

The method just illustrated is justified by a theorem first proved by the famous Hungarian-born American mathematician John von Neumann.

The Duality Theorem

Consider a linear program with $\geq$ constraints where the objective function C is to be minimized. Such a program has an optimal solution if and only if the dual program, which has $\leq$ constraints and an objective function R to be maximized, has an optimal solution. In this case the maximal value of R is the minimal value of C. Furthermore, if the dual program is solved with the simplex method, then the negatives of the indicators corresponding to the slack variables in the final tableau give the optimal values of the variables in the original problem.

EXAMPLE 3.2 Revisited

In Example 3.2 in Section 3.1 we considered a problem of combining two foods. There we translated the problem into the linear program:

$$\begin{array}{ll} \text{Minimize} & 10x + 15y = C \\ \text{subject to} & 3x + 3y \geq 75 \\ & 2x + 4y \geq 60 \\ & 4x + 2y \geq 60 \\ & x \geq 0,\ y \geq 0. \end{array}$$

Let us solve this program by solving its dual.

Our minimizing program has the compact tableau

$$\begin{array}{c} \begin{array}{cc} \ x & \ y \end{array} \\ \left[\begin{array}{cc|c} 3 & 3 & (\geq)\ 75 \\ 2 & 4 & (\geq)\ 60 \\ 4 & 2 & (\geq)\ 60 \\ \hline 10 & 15 & (\text{minimize})\ C \end{array}\right]. \end{array}$$

Interchanging rows and columns gives us

$$\begin{array}{c} \begin{array}{ccc} X & Y & Z \end{array} \\ \left[\begin{array}{ccc|c} 3 & 2 & 4 & (\leq)\ 10 \\ 3 & 4 & 2 & (\leq)\ 15 \\ \hline 75 & 60 & 60 & (\text{maximize})\ R \end{array}\right], \end{array}$$

the first two rows of which yield the system

$$\begin{aligned} 3X + 2Y + 3Z + s \qquad &= 10 \\ 3X + 4Y + 2Z \qquad + t &= 15 \end{aligned}$$

when nonnegative slack variables s and t are introduced. Thus we have the simplex tableau

$$\begin{array}{c} \begin{array}{ccccc} X & Y & Z & s & t \end{array} \\ \left[\begin{array}{ccccc|c} \textcircled{3} & 2 & 4 & 1 & 0 & 10 \\ 3 & 4 & 2 & 0 & 1 & 15 \\ \hline 75^* & 60 & 60 & 0 & 0 & R \end{array}\right]. \end{array} \qquad \begin{array}{l} 10/3^* \\ 15/3 \\ \\ \end{array}$$

By the methods of the previous section we see that we should pivot on the entry 3 in the first row and column.

$$\left[\begin{array}{ccccc|c} \textcircled{1} & \frac{2}{3} & \frac{4}{3} & \frac{1}{3} & 0 & \frac{10}{3} \\ 3 & 4 & 2 & 0 & 1 & 15 \\ \hline 75 & 60 & 60 & 0 & 0 & R \end{array}\right] \qquad (\tfrac{1}{3}\mathbf{R}_1)$$

$$\left[\begin{array}{ccccc|c} 1 & \frac{2}{3} & \frac{4}{3} & \frac{1}{3} & 0 & \frac{10}{3} \\ 0 & \textcircled{2} & -2 & -1 & 1 & 5 \\ \hline 0 & 10^* & -40 & -25 & 0 & R - 250 \end{array}\right] \qquad \begin{array}{l} \\ (\mathbf{R}_2 + (-3)\mathbf{R}_1) \\ (\mathbf{R}_3 + (-75)\mathbf{R}_1) \end{array} \qquad \begin{array}{l} (10/3)/(2/3) = 5 \\ 5/2^* \\ \\ \end{array}$$

The next pivot should be on the entry 2 in row 2 and column 2.

$$\left[\begin{array}{ccccc|c} 1 & \frac{2}{3} & \frac{4}{3} & \frac{1}{3} & 0 & \frac{10}{3} \\ 0 & \textcircled{1} & -1 & -\frac{1}{2} & \frac{1}{2} & \frac{5}{2} \\ \hline 0 & 10^* & -40 & -25 & 0 & R - 250 \end{array}\right] \qquad (\tfrac{1}{2}\mathbf{R}_2)$$

$$\begin{array}{c} \begin{array}{ccccc} X & Y & Z & s & t \end{array} \\ \left[\begin{array}{ccccc|c} 1 & 0 & 2 & \frac{2}{3} & -\frac{1}{3} & \frac{5}{3} \\ 0 & 1 & -1 & -\frac{1}{2} & \frac{1}{2} & \frac{5}{2} \\ \hline 0 & 0 & -30 & \mathbf{-20} & \mathbf{-5} & R - 275 \end{array}\right] \end{array} \qquad \begin{array}{l} \\ \\ (\mathbf{R}_1 + (-\tfrac{2}{3})\ \mathbf{R}_2) \\ (\mathbf{R}_3 + (-10)\mathbf{R}_2) \end{array}$$

Since there are no positive indicators, we are done. The maximum value of R, and thus the minimum value of C in the original problem, is 275. Furthermore, from the indicators corresponding to the slack variables s and t we see that C achieves its minimum at $(x, y) = (20, 5)$. In terms of the original problem this means that 20 ounces of Food A and 5 ounces of Food B should be used. See Figure 3.21 for a graphical interpretation of this problem. ■

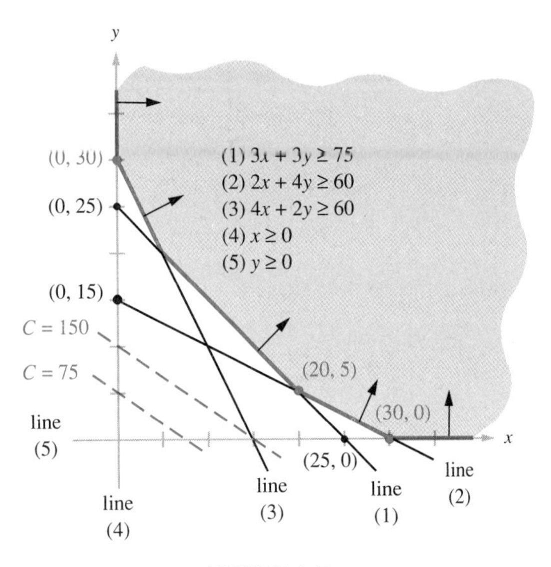

FIGURE 3.21

Practice Problem 10 What is the dual of the following linear program?

$$\begin{array}{ll} \text{Minimize} & 2x + 5y + 3z = C \\ \text{subject to} & x + 2y + 4z \geq 12 \\ & 3x + y + 2z \geq 15 \\ & x \geq 0,\ y \geq 0,\ z \geq 0. \end{array}$$

EXAMPLE 3.14 During the ski season, World Travel Service offers a one-month vacation in Aspen, including round-trip fare and accommodations at the Aspen Lodge. The numbers of rooms needed at the lodge for each of the months January through March are as follows:

Month	*Number of rooms needed*
January	25
February	12
March	20

The Lodge leases its rooms to World Travel at a discount rate of \$800 for two consecutive months or \$1000 for three consecutive months. What leasing plan will be least expensive?

Solution The Travel Service must decide how many rooms to lease for two months and how many for three months, and also which periods the various leases should cover. The three unknowns are illustrated in Figure 3.22.

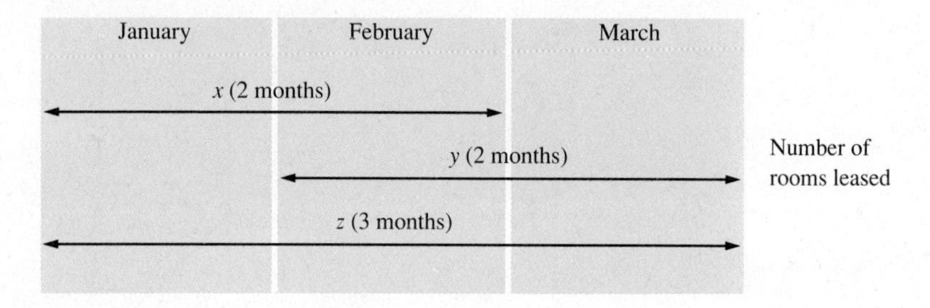

FIGURE 3.22

For example, x represents the number of rooms leased for two months beginning January 1. The total cost in dollars is given by

$$C = 800x + 800y + 1000z.$$

From Figure 3.22 we see that the total number of rooms available for January will be $x + z$. Thus, the condition that 25 rooms are needed for this month translates into the constraint

$$x + z \geq 25.$$

The requirements for February and March lead in the same way to the conditions

$$\begin{aligned} x + y + z &\geq 12 \\ y + z &\geq 20. \end{aligned}$$

The problem can thus be formulated as the three-variable linear program:

$$\begin{array}{lrcrcrcl} \text{Minimize} & 800x & + & 800y & + & 1000z & = & C \\ \text{subject to} & x & & & + & z & \geq & 25 \\ & x & + & y & + & z & \geq & 12 \\ & & & y & + & z & \geq & 20 \\ & \multicolumn{7}{l}{x \geq 0,\ y \geq 0,\ z \geq 0,} \end{array}$$

with compact tableau

$$\begin{array}{ccc|c} x & y & z & \\ \hline 1 & 0 & 1 & (\geq)\ 25 \\ 1 & 1 & 1 & (\geq)\ 12 \\ 0 & 1 & 1 & (\geq)\ 20 \\ \hline 800 & 800 & 1000 & (\text{minimize})\ C \end{array}.$$

The corresponding tableau for the dual program is

$$\begin{array}{ccc|c} X & Y & Z & \\ \hline 1 & 1 & 0 & (\leq)\ 800 \\ 0 & 1 & 1 & (\leq)\ 800 \\ 1 & 1 & 1 & (\leq)\ 1000 \\ \hline 25 & 12 & 20 & \text{(maximize) } R \end{array}.$$

We introduce nonnegative slack variables s, t, and u such that

$$\begin{aligned} X + Y \quad\quad + s \quad\quad\quad &= 800 \\ Y + Z \quad\quad + t \quad &= 800 \\ X + Y + Z \quad\quad\quad + u &= 1000, \end{aligned}$$

giving the simplex tableau

$$\left[\begin{array}{cccccc|c} X & Y & Z & s & t & u & \\ \textcircled{1} & 1 & 0 & 1 & 0 & 0 & 800 \\ 0 & 1 & 1 & 0 & 1 & 0 & 800 \\ 1 & 1 & 1 & 0 & 0 & 1 & 1000 \\ \hline 25^* & 12 & 20 & 0 & 0 & 0 & R \end{array}\right]. \qquad \begin{array}{l} 800/1 = 800^* \\ \\ 1000/1 = 1000 \end{array}$$

Pivoting on the entry 1 in row 1 and column 1 produces

$$\left[\begin{array}{cccccc|c} 1 & 1 & 0 & 1 & 0 & 0 & 800 \\ 0 & 1 & 1 & 0 & 1 & 0 & 800 \\ 0 & 0 & \textcircled{1} & -1 & 0 & 1 & 200 \\ \hline 0 & -13 & 20 & -25 & 0 & 0 & R - 20{,}000 \end{array}\right]. \qquad \begin{array}{l} \\ \\ (R_3 + (-1)R_1) \\ (R_4 + (-25)R_1) \end{array}$$

It is easily seen that the next pivot should be on the entry in row 3 and column 3, producing the tableau

$$\left[\begin{array}{cccccc|c} X & Y & Z & s & t & u & \\ 1 & 1 & 0 & 1 & 0 & 0 & 800 \\ 0 & 1 & 0 & 1 & 1 & -1 & 600 \\ 0 & 0 & 1 & -1 & 0 & 1 & 200 \\ \hline 0 & -13 & 0 & -5 & 0 & -20 & R - 24{,}000 \end{array}\right] \qquad \begin{array}{l} \\ (R_2 + (-1)R_3) \\ \\ (R_4 + (-20)R_3). \end{array}$$

Since there are no more positive indicators, we are done. In the dual problem the maximum of R is 24,000, and so this is also the minimum value of C in the original problem. Furthermore, by taking the negatives of the indicators corresponding to the slack variables s, t, and u we see that C achieves its minimum when $(x, y, z) = (5, 0, 20)$. The Travel Service should rent 5 rooms for January and February, and 20 rooms for the 3-month period, for a total cost of $24,000. ■

INTERPRETING THE DUAL VARIABLES

The variables in the dual of a linear program can be given a meaning. We consider again Example 3.5 of Section 3.2, which leads to the program:

$$\begin{array}{lrl} \text{Minimize} & 2000x + 2400y &= C \\ \text{subject to} & 40x + 50y &\geq 800 \\ & 1000x + 750y &\geq 18{,}000 \\ & x \geq 0,\ y \geq 0. & \end{array}$$

Here x and y represent the number of Type A and Type B buses hired by an alumni association. These cost \$2000 and \$2400 per day, respectively, and the cost $2000x + 2400y$ is to be minimized. The first constraint comes about because the buses carry 40 and 50 people, respectively, and 800 people are going on the trip. Likewise the buses carry 1000 and 750 pounds of baggage, respectively, and 18,000 pounds must be carried.

The dual program is:

$$\begin{array}{lrl} \text{Maximize} & 800X + 18{,}000Y &= R \\ \text{subject to} & 40X + 1000Y &\leq 2000 \\ & 50X + 750Y &\leq 2400 \\ & X \geq 0,\ Y \geq 0, & \end{array}$$

as we saw at the beginning of this section. The variables X and Y can be interpreted as follows. Suppose a rival transportation company is being formed to compete with the company that charters buses. The new company plans to charge a certain amount X for each person transported, and an amount Y for each pound of baggage carried. Recall that the bus company rents a Type A bus, carrying 40 people and 1000 pounds of baggage, for \$2000. The new company will charge $40X + 1000Y$ to carry the same 40 people and 1000 pounds of baggage. Since it wants to be competitive with the first bus company, X and Y should be chosen such that

$$40X + 1000Y \leq 2000, \tag{3.10}$$

for if we had $40X + 1000Y > 2000$, then customers would hire the Type A bus whenever it could be used to capacity. A similar analysis with respect to the Type B bus, which carries 50 people and 750 pounds of baggage for \$2400, leads to the constraint

$$50X + 750Y \leq 2400. \tag{3.11}$$

Of course (3.10) and (3.11) are exactly the constraints of the dual program.

Finally, the new company would like to maximize its revenue. Of course this will depend on how many passengers and pounds of baggage it carries. It is decided that the ratio of 800 people to 18,000 pounds of baggage of the alumni trip is about the usual proportion. Thus it is decided to maximize the revenue that would be generated for carrying 800 people and 18,000 pounds of baggage, which is

$$800X + 18{,}000Y = R.$$

This is exactly the objective function for the dual program.

At the beginning of this section we found the following final tableau in the simplex solution to this dual program.

$$\begin{array}{c} \\ Y \\ X \\ \\ \end{array}\begin{array}{c} \begin{array}{cccc} X & Y & s & t \end{array} \\ \left[\begin{array}{cccc|c} 0 & 1 & .0025 & .002 & .2 \\ 1 & 0 & -0.375 & .05 & 45 \\ \hline 0 & 0 & -15 & -4 & R - 39600 \end{array}\right] \end{array}$$

Thus the revenue is maximized when $(X, Y) = (45, .2)$. This means the new company should charge \$45 per passenger and 20¢ per pound of baggage. By doing so for the alumni trip it would take in \$39,600, the same as the minimal cost of renting buses found for the original problem. This is perhaps not surprising, since the new company's prices are set to be as high as possible without exceeding those of the first bus company.

LIMITATIONS OF THE DUAL PROGRAM METHOD

In order for the method of passing to the dual to be successful, the dual program must be a standard maximum program, since this was the only type treated in Section 3.4. In particular, the right-hand constants of the new constraints must all be nonnegative, since this is one of the conditions for a standard maximum program. This means that the coefficients of the objective function of the original minimization problem must all be nonnegative. In the next section we treat programs that cannot be transformed into standard maximum programs.

EXERCISES 3.5

In Exercises 1–6 write the dual of the given program.

1. Minimize $5x + 4y = C$
subject to
$x - y \geq -5$
$x + 2y \geq 20$
$x + y \geq 15$
$x \geq 0,\ y \geq 0.$

2. Minimize $x + y = C$
subject to
$2x + y \geq 24$
$x + 3y \geq 27$
$x \geq 0,\ y \geq 0.$

3. Minimize $6x + 2y = C$
subject to
$x + 4y \geq 40$
$4x + y \geq 40$
$x + y \geq 25$
$x \geq 0,\ y \geq 0.$

4. Minimize $5x + 4y = C$
subject to
$x + y \geq 20$
$-x + y \geq -10$
$x - y \geq -10$
$x \geq 0,\ y \geq 0.$

5. Minimize $10x + 20y = C$

subject to
$$\begin{aligned} y &\geq 5 \\ -x + y &\geq 0 \\ 5x + 2y &\geq 20 \\ x \geq 0,\ y &\geq 0. \end{aligned}$$

6. Minimize $2x + 4y + z = C$

subject to
$$\begin{aligned} x + 2y + 3z &\geq 80 \\ x + y + z &\geq 60 \\ x \geq 0,\ y \geq 0,\ z &\geq 0. \end{aligned}$$

In Exercises 7–12 solve the program in the listed exercise by applying the simplex algorithm to the dual program.

7. Exercise 1

8. Exercise 2

9. Exercise 3

10. Exercise 4

11. Exercise 5

12. Exercise 6

In Exercises 13–16 solve the program by applying the simplex algorithm to the dual program.

13. Minimize $5x + 4y + 3z = C$

subject to
$$\begin{aligned} 2x + y + z &\geq 40 \\ x + y + 2z &\geq 50 \\ x \geq 0,\ y \geq 0,\ z &\geq 0. \end{aligned}$$

14. Minimize $x + 4y + 3z = C$

subject to
$$\begin{aligned} x + y + z &\geq 20 \\ y + z &\geq 15 \\ z \geq 10,\ x \geq 0,\ y \geq 0,\ z &\geq 0. \end{aligned}$$

15. Minimize $2x + y + z + 2w = C$

subject to
$$\begin{aligned} z + w &\geq -50 \\ x + z &\geq 60 \\ y + w &\geq 40 \\ x \geq 0,\ y \geq 0,\ z \geq 0,\ w &\geq 0. \end{aligned}$$

16. Minimize $3x + y + 2z = C$

subject to
$$\begin{aligned} x + y &\geq -10 \\ y + z &\geq 15 \\ -x - z &\geq -10 \\ x \geq 0,\ y \geq 0,\ z &\geq 0. \end{aligned}$$

In Exercises 17–22 formulate the given problem as a linear program and solve by applying the simplex algorithm to the dual program.

17. The heating plant for a large government building must produce at least 300 million Btu's of heat per hour. The plant is equipped to burn any combination of fuel oil and coal, but it must observe government pollution limits of at most 300 pounds of sulfur dioxide and 125 pounds of nitrogen dioxide emitted per hour. The cost, heat yield, and pollution content per ton for each of the fuels are summarized in the following table:

Fuel	*Oil*	*Coal*
Heat (million Btu/ton)	40	20
SO_2 (lb/ton)	20	30
NO_2 (lb/ton)	5	20
Cost ($/ton)	$290	$65

How many tons of each fuel should the plant burn per hour in order to meet all the requirements at minimum cost?

18. Nutritionists indicate that an adequate daily diet should provide at least 75 grams of carbohydrate, 60 grams of protein, and 60 grams of fat. Two foods are available; their nutritional yields and costs, per ounce, are as follows:

Food	A	B
Carb/oz	3 gm	3 gm
Protein/oz	2 gm	4 gm
Fat/oz	4 gm	2 gm
Cost/oz	10¢	15¢

How many ounces of each food should be combined per day to meet the nutritional requirements at lowest cost?

19. World Travel Service has agreed to transport a group of 1800 football fans from Los Angeles to Chicago for one of the season's big games. The agency can charter three different types of aircraft. Their operating costs for the trip, and their seating capacities, are as follows.

	Number of Seats per Plane		
Type	A	B	C
First-class	20	40	50
Economy	80	60	50
Cost	$3000	$4000	$5000

If 600 of the passengers want to go first-class while the remaining 1200 have ordered economy seats, how many planes of each type should be assigned to the charter to minimize the total operating cost?

20. The Allegheny Mining Company operates three mines which produce both high-grade and low-grade ore. The daily yield of each mine, and the daily operating cost, are given in the table below.

	Ore Yield per Day		
Mine	1	2	3
High-grade	2 tons	1 ton	0 tons
Low-grade	1 ton	2 tons	2 tons
Cost per day	$300	$200	$100

The company must fill an order for a total of 40 tons of high-grade ore and 50 tons of low-grade ore. How many days should each mine be operated in order to fill the order at lowest total cost?

21. A hospital dietician must prepare a special diet from two foods. The first food provides 1 unit of iron, 1 unit of magnesium, and 20 calories at a cost of 3¢ per gram. The second food provides 1 unit of iron, 4 units of magnesium, and 40 calories at a cost of 4¢ per gram. If the diet must provide at least 100 units of iron, 160 units of magnesium, and 2600 calories, what combination of the two foods will satisfy these conditions at the least cost? What is the least cost?

22. A jeweler makes rings, stickpins, and pendants from semiprecious stones. For each ring, 1 hour is required to cut the stones and 3 hours are needed to polish them; for each stickpin, 2 hours are required to cut the stones and 1 hour is needed to polish them; and for each pendant, 2 hours are required to cut the stones and 5 hours are required to polish them. During a particular week the jeweler wants to devote at least 9 hours to cutting and at least 30 hours to polishing. If it costs the jeweler $34 to cut and polish the stones for each ring, $28 to cut and polish the stones for each stickpin, and $60 to cut and polish the stones for each pendant, how many rings, stickpins, and pendants should be made in order to minimize the jeweler's costs? What is the minimum cost?

Answer to Practice Problem **10.** Maximize $12X + 15Y = R$

subject to

$$X + 3Y \le 2$$
$$2X + Y \le 5$$
$$4X + 2Y \le 3$$
$$X \ge 0,\ Y \ge 0$$

3.6 Artificial Variables

So far we have only applied the simplex method to standard maximum linear programs, that is, programs with the four properties below.

(a) The objective function is to be maximized.
(b) All variables are to be nonnegative.
(c) The other constraints are of the $\le$ variety.
(d) The righthand constants in the $\le$ constraints are all nonnegative.

Restriction (a) is not hard to get around, since minimizing an objective function C is the same as maximizing the function $-C$. Thus any linear program can be written so as to involve maximizing the objective function. Likewise restriction (b) is automatic in almost all practical applications of linear programming; and we will not consider any programs where it does not hold.

Restrictions (c) and (d) are harder to deal with; and there are problems for which one or the other of them fails no matter how we set up the program. The algebraic fact that

$$A \ge B \quad \text{if and only if} \quad -A \le -B \tag{3.12}$$

enables us to switch between $\ge$ and $\le$ constraints, but the constant on the right may become negative in the process. For example, the $\ge$ constraint

$$3x + 5y \ge 7$$

is equivalent to the $\le$ constraint

$$-3x - 5y \le -7,$$

but a program with such a constraint would not be a standard maximum program because of restriction (d).

The reader may wonder why restriction (d) is necessary. The answer is that when the simplex method is applied to a standard maximum program, we start at the point where all the original variables are 0 and move by a sequence of pivots to an optimal corner of the feasible region. That is, the slack variables start out as basic and the original variables free. Setting the (free) original variables equal to 0 makes each slack variable equal to the corresponding constant in the rightmost column of the simplex tableau. But if one of these constants is negative, then so is the corresponding slack variable, contradicting the assumption made about the slack variables when they are introduced. What this means is that we are starting at a point that is not in the feasible region. We are in "left field", so to speak, and though pivoting may take us from vertex to vertex, there is no guarantee we will ever get to the feasible region.

In this section we will describe a method that can be used to solve linear programming problems that cannot be put into standard maximum form. The method will involve two phases. In **phase one** we will introduce not only slack variables, but also new "artificial variables." These artificial variables and a new objective function will be used to move to a corner of the feasible region. Then the artificial variables will be discarded, and in **phase two** we will use the simplex method and our original objective function to complete the solution of the problem.

We will assume that we have a linear program satisfying (a), (b), and (d), but not necessarily (c). That is:

1. If the problem is to minimize C, change it to the problem of maximizing $-C$.
2. If the right-hand constant of any constraint is negative, change it to a positive constant by multiplying both sides of the constraint by -1 and reversing the inequality sign.

Practice Problem 11 Change the following linear program to one satisfying (a), (b), and (d) at the beginning of this section.

$$\begin{array}{ll} \text{Minimize} & -3x + y = C \\ \text{subject to} & -x + y \geq -2 \\ & -2x - 3y \geq -6 \\ & x + y \geq 1 \\ & x \geq 0,\ y \geq 0. \end{array}$$

HOW PHASE ONE WORKS

We will demonstrate phase one with the program.

$$\begin{array}{ll} \text{Maximize} & 3x - y = R \\ \text{Subject to} & x - y \leq 2 \\ & 2x + 3y \leq 6 \\ & x + y \geq 1 \\ & x \geq 0,\ y \geq 0. \end{array}$$

The first two constraints, which are $\leq$ inequalities, we will treat as usual, introducing nonnegative slack variables s and t such that

$$\begin{aligned} x - y + s \quad &= 2 \\ 2x + 3y \quad + t &= 6. \end{aligned}$$

Since according to the third constraint $x + y$ may exceed 1, it would be natural to introduce a nonnegative slack variable u such that

$$x + y - u = 1.$$

The problem with this is that setting the free variables x and y both equal to 0 gives $-u = 1$, or $u = -1$, contrary to the condition that u should be nonnegative. Thus at the same time we introduce u we will also introduce an **artificial variable** $a \geq 0$, such that

$$x + y - u + a = 1.$$

Thus we can have $x = y = u = 0$, so long as $a = 1$.

Now we have the system

$$\begin{aligned} x - y + s \quad\quad\quad &= 2 \\ 2x + 3y \quad + t \quad\quad &= 6 \\ x + y \quad\quad - u + a &= 1, \end{aligned}$$

with all these variables nonnegative, and we wish to maximize

$$3x - y = R.$$

This gives the tableau

$$\begin{array}{c} \\ s \\ t \\ a \\ \\ \end{array} \begin{array}{c} \begin{array}{cccccc} x & y & s & t & u & a \end{array} \\ \left[\begin{array}{cccccc|c} 1 & -1 & 1 & 0 & 0 & 0 & 2 \\ 2 & 3 & 0 & 1 & 0 & 0 & 6 \\ 1 & 1 & 0 & 0 & -1 & 1 & 1 \\ \hline 3 & -1 & 0 & 0 & 0 & 0 & R \end{array}\right]. \end{array}$$

Notice that row 3 has the label a rather than u, since the column headed by a contains a 1 in row 3 and zeros elsewhere. Our first order of business will be to get rid of the artificial variable a. As we have seen, setting x, y, and u equal to 0 makes $a = 1$. We would like to make $a = 0$, so that we could drop this variable. Thus we want to minimize a, or, since the simplex method is set up for maximization problems,

$$\text{maximize} \quad -a = Q. \tag{3.13}$$

Thus we have a new objective function that we will deal with before we try to maximize R. Maximizing Q will have the effect of moving to a corner of the original feasible region; that is, it will complete phase one of our method.

In order to maximize Q we will add a row corresponding to (3.13) to our tableau.

$$\begin{array}{c} \\ s \\ t \\ a \\ \\ \\ \end{array} \begin{array}{c} \begin{array}{cccccc} x & y & s & t & u & a \end{array} \\ \left[\begin{array}{cccccc|c} 1 & -1 & 1 & 0 & 0 & 0 & 2 \\ 2 & 3 & 0 & 1 & 0 & 0 & 6 \\ 1 & 1 & 0 & 0 & -1 & 1 & 1 \\ \hline 3 & -1 & 0 & 0 & 0 & 0 & R \\ 0 & 0 & 0 & 0 & 0 & -1 & Q \end{array}\right] \end{array}$$

We have come far enough that we will summarize the steps necessary in phase one of our method. We have already illustrated steps 1 and 2 listed below.

Phase One of the Method of Artificial Variables

It is assumed that we have a linear program that is a maximization problem, and that the right-hand constants of all constraints are nonnegative.

Step 1 Make each constraint into an equation as follows. On the left side of each $\leq$ constraint introduce a slack variable with coefficient $+1$. On the left side of each $\geq$ constraint introduce a slack variable with coefficient -1, and also an artificial variable with coefficient $+1$.

Step 2 Set up the simplex tableau for the equations from step 1, along with the objective function. Add a row at the bottom with a -1 under each artificial variable, a 0 under every other variable, and the letter Q in the bottom right corner.

Step 3 To the last row of the tableau add each row above the horizontal line that corresponds to an equation with an artificial variable.

Step 4 Proceed with the simplex method as usual to get rid of all positive indicators in the bottom row of the tableau.

After phase one is completed we will throw away the last row of our tableau, along with all the columns corresponding to artificial variables, and finish the simplex algorithm in the usual way. This will be phase two.

Continuing with our example, we now apply step 3 to the tableau by adding row 3 to the last row.

$$\begin{array}{c} \\ s \\ t \\ a \\ \\ \\ \end{array} \begin{array}{c} \begin{array}{cccccc} x & y & s & t & u & a \end{array} \\ \left[\begin{array}{cccccc|c} 1 & -1 & 1 & 0 & 0 & 0 & 2 \\ 2 & 3 & 0 & 1 & 0 & 0 & 6 \\ \textcircled{1} & 1 & 0 & 0 & -1 & 1 & 1 \\ \hline 3 & -1 & 0 & 0 & 0 & 0 & R \\ 1^* & 1 & 0 & 0 & -1 & 0 & Q+1 \end{array}\right] \end{array} \begin{array}{l} \\ 2/1 = 2 \\ 6/2 = 3 \\ 1/1 = 1^* \\ \\ (R_5 + R_3) \end{array}$$

Notice that at this stage x, y, and u are free variables. Setting them all equal to 0 gives $s = 2$, $t = 6$, and $a = 1$. Although

$$(x, y, s, t, u, a) = (0, 0, 2, 6, 0, 1)$$

is feasible with respect to our new system, $(x, y) = (0, 0)$ is not a feasible point with respect to our original maximization problem. The feasible region of this problem is illustrated in Figure 3.23.

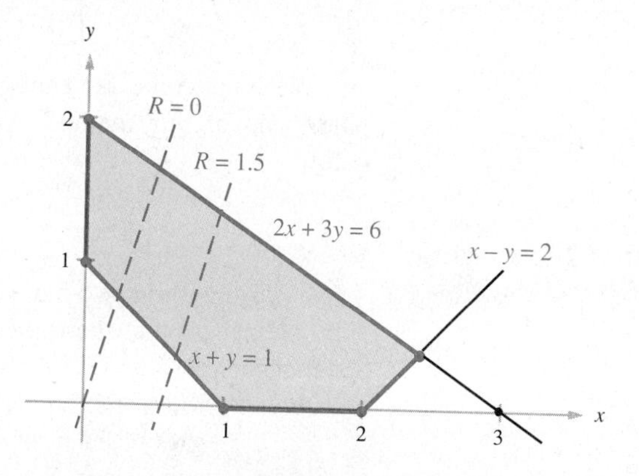

FIGURE 3.23

We proceed with step 4 to get rid of all positive indicators in the bottom row of our tableau. We arbitrarily choose to pivot in the first column instead of the second, and pick as the pivot the entry in row 3 and column 1 in the usual way. Pivoting produces the following tableau.

	x	y	s	t	u	a		
s	0	−2	1	0	1	−1	1	$(R_1 + (-1)R_3)$
t	0	1	0	1	2	−2	4	$(R_2 + (-2)R_3)$
x	1	1	0	0	−1	1	1	
	0	−4	0	0	3	−3	$R - 3$	$(R_4 + (-3)R_3)$
	0	0	0	0	0	−1	Q	$(R_5 + (-1)R_3)$

At this point there are no positive indicators in the bottom row, so we have finished phase one. To start phase two, we eliminate the last row and the column corresponding to the artificial variable a.

	x	y	s	t	u		
s	0	−2	1	0	(1)	1	1/1 = 1*
t	0	1	0	1	2	4	4/2 = 2
x	1	1	0	0	−1	1	(negative)
	0	−4	0	0	3*	$R - 3$	

Notice that now y and u are free variables and x, s, and t are basic. Setting y and u equal to 0 gives $x = 1$, $s = 1$, and $t = 4$. Furthermore, the point $(x, y) = (1, 0)$ is a corner of our original feasible region. (See Figure 3.23.)

Pivoting on the next indicated pivot in row 1 and column 5 produces the following tableau:

$$\begin{array}{c} \\ u \\ t \\ x \\ \\ \end{array}
\begin{array}{c}
\begin{array}{ccccc|c} x & y & s & t & u & \end{array} \\
\left[\begin{array}{ccccc|c}
0 & -2 & 1 & 0 & 1 & 1 \\
0 & \textcircled{5} & -2 & 1 & 0 & 2 \\
1 & -1 & 1 & 0 & 0 & 2 \\
\hline
0 & 2^* & -3 & 0 & 0 & R - 6
\end{array}\right]
\end{array}
\begin{array}{ll}
 & \text{(negative)} \\
(R_2 + (-2)R_1) & 2/5^* \\
(R_3 + R_1) & \text{(negative)} \\
 & \\
(R_4 + (-3)R_1) &
\end{array}$$

Notice that now setting the free variables y and s equal to 0 gives $x = 2$. Thus we are at the point $(x, y) = (2, 0)$ in the feasible region.

Our last pivot is on the entry 5 in row 2 and column 2, and results in the following tableaus.

$$\left[\begin{array}{ccccc|c}
0 & -2 & 1 & 0 & 1 & 1 \\
0 & \textcircled{1} & -.4 & .2 & 0 & .4 \\
1 & -1 & 1 & 0 & 0 & 2 \\
\hline
0 & 2 & -3 & 0 & 0 & R - 6
\end{array}\right] \quad (.2R_2)$$

$$\begin{array}{c} \\ u \\ y \\ x \\ \\ \end{array}
\begin{array}{c}
\begin{array}{ccccc|c} x & y & s & t & u & \end{array} \\
\left[\begin{array}{ccccc|c}
0 & 0 & .2 & .4 & 1 & 1.8 \\
0 & 1 & -.4 & .2 & 0 & 0.4 \\
1 & 0 & .6 & .2 & 0 & 2.4 \\
\hline
0 & 0 & -2.2 & -.4 & 0 & R - 6.8
\end{array}\right]
\end{array}
\begin{array}{l}
 \\
(R_1 + 2R_2) \\
(R_3 + R_2) \\
 \\
(R_4 + (-2)R_2)
\end{array}$$

There are no more positive indicators, and we are done. The maximum value of R is 6.8 and occurs when $(x, y) = (2.4, 0.4)$. The geometrical progress of our two-phase method is shown in Figure 3.24.

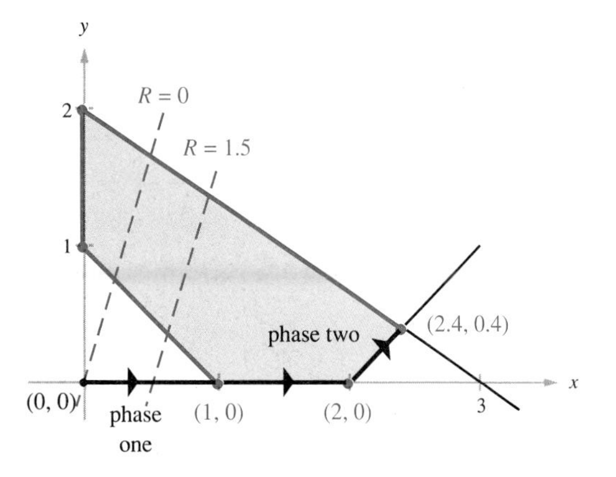

FIGURE 3.24

EXAMPLE 3.15 A shoe manufacturer makes two types of tennis shoes, the Court Ace and Great Get. The manufacturer has a contract with a large retailer to provide at least 5000 pairs of shoes, of which at least 2000 must be the Court Ace model. Only 18,000 minutes of machine time are available on the machines that make the tennis shoes, and a pair of the Court Ace model takes 2 minutes of machine time, while a pair of the Great Get model takes 3 minutes of machine time. The profit is \$3 for each pair of Court Ace shoes, and \$4 for each pair of Great Get shoes. Use the simplex method to determine how many pairs of each model should be made in order to maximize the manufacturer's profit.

Solution Let us suppose

$$x = \text{the number of pairs of Court Ace shoes made}$$
$$y = \text{the number of pairs of Great Get shoes made.}$$

Then the conditions of the problem lead us to the linear program:

$$\begin{array}{ll} \text{Maximize} & 3x + 4y = P \\ \text{subject to} & x + y \geq 5000 \\ & x \geq 2000 \\ & 2x + 3y \leq 18000 \\ & x \geq 0,\ y \geq 0. \end{array}$$

Since the constraints are of both the $\geq$ and $\leq$ variety, we will use the method of artificial variables. We introduce nonnegative slack variables s, t, and u and artificial variables a and b such that

$$\begin{aligned} x + y - s \qquad\qquad + a &= 5000 \\ x \qquad - t \qquad + b &= 2000 \\ 2x + 3y \qquad + u \qquad &= 18000. \end{aligned}$$

Performing step 2 of phase one gives us the tableau

$$\begin{array}{c} \\ a \\ b \\ u \\ \\ \\ \end{array} \begin{array}{c} \begin{array}{ccccccc} x & y & s & t & u & a & b \end{array} \\ \left[\begin{array}{ccccccc|c} 1 & 1 & -1 & 0 & 0 & 1 & 0 & 5000 \\ 1 & 0 & 0 & -1 & 0 & 0 & 1 & 2000 \\ 2 & 3 & 0 & 0 & 1 & 0 & 0 & 18000 \\ \hline 3 & 4 & 0 & 0 & 0 & 0 & 0 & P \\ 0 & 0 & 0 & 0 & 0 & -1 & -1 & Q \end{array}\right] \end{array}.$$

Now we perform step 3 by adding rows 1 and 2 to the last row.

$$\begin{array}{c} \\ a \\ b \\ u \\ \\ \\ \end{array} \begin{array}{c} \begin{array}{ccccccc} x & y & s & t & u & a & b \end{array} \\ \left[\begin{array}{ccccccc|c} 1 & 1 & -1 & 0 & 0 & 1 & 0 & 5000 \\ \textcircled{1} & 0 & 0 & -1 & 0 & 0 & 1 & 2000 \\ 2 & 3 & 0 & 0 & 1 & 0 & 0 & 18000 \\ \hline 3 & 4 & 0 & 0 & 0 & 0 & 0 & P \\ 2^* & 1 & -1 & -1 & 0 & 0 & 0 & Q + 7000 \end{array}\right] \end{array} \quad \begin{array}{l} 5000/1 = 5000 \\ 2000/1 = 2000^* \\ 18000/2 = 9000 \\ \\ \\ \end{array}$$

Here x and y are free variables so we are at the point $(x, y) = (0, 0)$. Our first pivot will be on the entry in row 2 and column 1. We will show the result of each pivot, but not the details of the calculations. Pivoting produces the following tableau.

$$\begin{array}{c} \\ a \\ x \\ u \\ \\ \\ \end{array}\begin{array}{c} \begin{array}{ccccccc|c} x & y & s & t & u & a & b & \end{array} \\ \left[\begin{array}{ccccccc|c} 0 & \textcircled{1} & -1 & 1 & 0 & 1 & -1 & 3000 \\ 1 & 0 & 0 & -1 & 0 & 0 & 1 & 2000 \\ 0 & 3 & 0 & 2 & 1 & 0 & -2 & 14000 \\ \hline 0 & 4 & 0 & 3 & 0 & 0 & -3 & P - 6000 \\ 0 & 1^* & -1 & 1 & 0 & 0 & -2 & Q + 3000 \end{array}\right] \end{array} \quad \begin{array}{l} 3000/1 = 3000^* \\ \text{(undefined)} \\ 14000/3 \approx 4667 \\ \\ \\ \end{array}$$

Notice that now setting the free variables equal to 0 gives $(x, y) = (2000, 0)$, which is not feasible because $x + y < 5000$. Thus we continue with phase one by pivoting on the entry in row 1 and column 2 to obtain the next tableau.

$$\begin{array}{c} \\ y \\ x \\ u \\ \\ \\ \end{array}\begin{array}{c} \begin{array}{ccccccc|c} x & y & s & t & u & a & b & \end{array} \\ \left[\begin{array}{ccccccc|c} 0 & 1 & -1 & 1 & 0 & 1 & -1 & 3000 \\ 1 & 0 & 0 & -1 & 0 & 0 & 1 & 2000 \\ 0 & 0 & 3 & -1 & 1 & -3 & 1 & 5000 \\ \hline 0 & 0 & 4 & -1 & 0 & -4 & 1 & P - 18000 \\ 0 & 0 & 0 & 0 & 0 & -1 & -1 & Q \end{array}\right] \end{array}$$

Since there are no more positive entries in the last row, phase one is completed. We erase the last row and the columns corresponding to the artificial variables a and b.

$$\begin{array}{c} \\ y \\ x \\ u \\ \\ \end{array}\begin{array}{c} \begin{array}{ccccc|c} x & y & s & t & u & \end{array} \\ \left[\begin{array}{ccccc|c} 0 & 1 & -1 & 1 & 0 & 3000 \\ 1 & 0 & 0 & -1 & 0 & 2000 \\ 0 & 0 & \textcircled{3} & -1 & 1 & 5000 \\ \hline 0 & 0 & 4^* & -1 & 0 & P - 18000 \end{array}\right] \end{array} \quad \begin{array}{l} \text{(negative)} \\ \text{(undefined)} \\ 5000/3^* \\ \\ \end{array}$$

Notice that now we are at the feasible point $(x, y) = (2000, 3000)$. The next pivot will be on the entry in row 3 and column 3. The result is the following tableau.

$$\begin{array}{c} \\ y \\ x \\ s \\ \\ \end{array}\begin{array}{c} \begin{array}{ccccc|c} x & y & s & t & u & \end{array} \\ \left[\begin{array}{ccccc|c} 0 & 1 & 0 & \textcircled{\tfrac{2}{3}} & \tfrac{1}{3} & 14000/3 \\ 1 & 0 & 0 & -1 & 0 & 2000 \\ 0 & 0 & 1 & -\tfrac{1}{3} & \tfrac{1}{3} & 5000/3 \\ \hline 0 & 0 & 0 & \tfrac{1}{3}^* & -\tfrac{4}{3} & P - 74000/3 \end{array}\right] \end{array} \quad \begin{array}{l} (14000/3)/(2/3) = 7000^* \\ \text{(negative)} \\ \text{(negative)} \\ \\ \end{array}$$

Now we are at the point $(x, y) = (2000, 14000/3)$, and we pivot on the entry in row 1 and column 4.

$$\begin{array}{c} \\ t \\ x \\ s \\ \\ \end{array} \begin{array}{c} \begin{array}{ccccc|c} x & y & s & t & u & \end{array} \\ \left[\begin{array}{ccccc|c} 0 & \frac{3}{2} & 0 & 1 & \frac{1}{2} & 7000 \\ 1 & \frac{3}{2} & 0 & 0 & \frac{1}{2} & 9000 \\ 0 & \frac{1}{2} & 1 & 0 & \frac{1}{2} & 4000 \\ \hline 0 & -\frac{1}{2} & 0 & 0 & -\frac{3}{2} & P - 27000 \end{array}\right] \end{array}$$

Finally there are no positive indicators, and we are done. We see that the maximum profit P is \$27,000, achieved when $(x, y) = (9000, 0)$. That is, 9000 pairs of Court Ace shoes should be made, and no Great Get shoes. Figure 3.25 shows geometrically how our method reaches the solution, pivoting twice in phase one and twice in phase two. ■

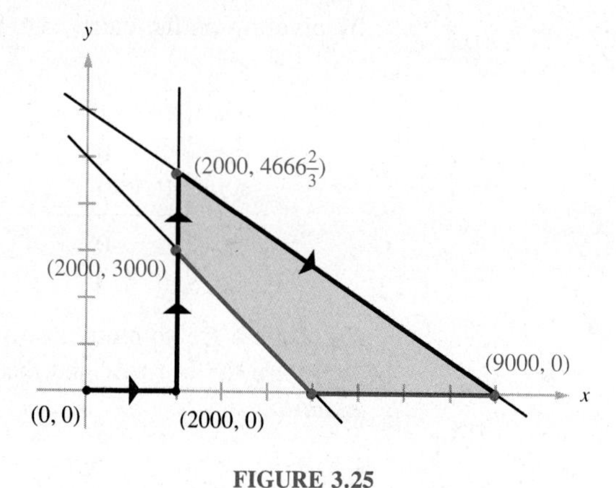

FIGURE 3.25

Practice Problem 12 Set up the initial phase one tableau for the following linear program:

$$\begin{aligned} \text{Maximize } & x + 2y - 3z = R \\ \text{subject to } & x + z \le 30 \\ & 2x + y + z \ge 10 \\ & x + 3y + 2z \ge 15 \\ & x \ge 0,\ y \ge 0,\ z \ge 0. \end{aligned}$$

What should be done in step 3?

LINEAR PROGRAMS WITH NO SOLUTION

The purpose of phase one of the method of artificial variables is to reach a point of the feasible region. If this region is empty, as in Example 3.7 of Section 3.2, then obviously phase one cannot be completed. We would rewrite the program of that example as

$$\begin{array}{ll} \text{Maximize} & 3x + 7y = R \\ \text{subject to} & -x - y \geq 2 \\ & x \geq 0,\ y \geq 0, \end{array}$$

then introduce slack and artificial variables $s \geq 0$ and $a \geq 0$ such that

$$-x - y - s + a = 2.$$

The corresponding tableau is as follows.

$$\begin{array}{c} \\ a \\ \\ \\ \end{array} \left[\begin{array}{cccc|c} x & y & s & a & \\ -1 & -1 & -1 & 1 & 2 \\ \hline 3 & 7 & 0 & 0 & R \\ 0 & 0 & 0 & -1 & Q \end{array}\right]$$

Step 3 of phase one gives the following tableau.

$$\begin{array}{c} \\ a \\ \\ \\ \end{array} \left[\begin{array}{cccc|c} x & y & s & a & \\ -1 & -1 & -1 & 1 & 2 \\ \hline 3 & 7 & 0 & 0 & R \\ -1 & -1 & -1 & 0 & Q + 2 \end{array}\right]$$

The last row corresponds to the equation

$$Q = -2 - x - y - s,$$

and since we are trying to maximize Q the best we can do is set $x = y = s = 0$. This makes $a = 2$. Since we cannot make the artificial variable a equal to 0, phase one cannot be completed.

In cases when the feasible region is nonempty, but the objective function cannot be optimized because the feasible region is unbounded, the simplex algorithm will not break down until phase two. Consider, for instance, Example 3.6 of Section 3.2, which is the program

$$\begin{array}{ll} \text{Maximize} & 6x + 3y = R \\ \text{subject to} & x + 2y \geq 8 \\ & 2x + y \geq 10 \\ & x \geq 0,\ y \geq 0. \end{array}$$

Introducing nonnegative variables s, t, a, and b such that

$$\begin{aligned} x + 2y - s \quad + a \quad &= 8 \\ 2x + y \quad - t \quad + b &= 10, \end{aligned}$$

leads to the following tableau.

$$\begin{array}{c} \\ a \\ b \\ \\ \\ \end{array}\begin{array}{c} \begin{array}{cccccc} x & y & s & t & a & b \end{array} \\ \left[\begin{array}{cccccc|c} 1 & 2 & -1 & 0 & 1 & 0 & 8 \\ 2 & 1 & 0 & -1 & 0 & 1 & 10 \\ \hline 6 & 3 & 0 & 0 & 0 & 0 & R \\ 0 & 0 & 0 & 0 & -1 & -1 & Q \end{array}\right] \end{array}$$

Then step 3 of phase one gives

$$\begin{array}{c} \\ a \\ b \\ \\ \\ \end{array}\begin{array}{c} \begin{array}{cccccc} x & y & s & t & a & b \end{array} \\ \left[\begin{array}{cccccc|c} 1 & 2 & -1 & 0 & 1 & 0 & 8 \\ \textcircled{2} & 1 & 0 & -1 & 0 & 1 & 10 \\ \hline 6 & 3 & 0 & 0 & 0 & 0 & R \\ 3^* & 3 & -1 & -1 & 0 & 0 & Q+18 \end{array}\right]. \end{array} \qquad \begin{array}{l} 8/1 = 8 \\ 10/2 = 5^* \end{array}$$

First we pivot on the entry in row 2 and column 1. The details are omitted.

$$\begin{array}{c} \\ a \\ x \\ \\ \\ \end{array}\begin{array}{c} \begin{array}{cccccc} x & y & s & t & a & b \end{array} \\ \left[\begin{array}{cccccc|c} 0 & \textcircled{\frac{3}{2}} & -1 & \frac{1}{2} & 1 & -\frac{1}{2} & 3 \\ 1 & \frac{1}{2} & 0 & -\frac{1}{2} & 0 & \frac{1}{2} & 5 \\ \hline 0 & 0 & 0 & 3 & 0 & -3 & R-30 \\ 0 & \frac{3}{2}^* & -1 & \frac{1}{2} & 0 & -\frac{3}{2} & Q+3 \end{array}\right] \end{array} \qquad \begin{array}{l} 3/(3/2) = 2^* \\ 5/(1/2) = 10 \end{array}$$

The next pivot entry is in row 1 and column 2.

$$\begin{array}{c} \\ y \\ x \\ \\ \\ \end{array}\begin{array}{c} \begin{array}{cccccc} x & y & s & t & a & b \end{array} \\ \left[\begin{array}{cccccc|c} 0 & 1 & -\frac{2}{3} & \frac{1}{3} & \frac{2}{3} & -\frac{1}{3} & 2 \\ 1 & 0 & \frac{1}{3} & -\frac{2}{3} & -\frac{1}{3} & \frac{2}{3} & 4 \\ \hline 0 & 0 & 0 & 3 & 0 & -3 & R-30 \\ 0 & 0 & 0 & 0 & -1 & -1 & Q \end{array}\right] \end{array}$$

Since there are no positive indicators in the bottom row, phase one is finished. Notice that we are at the point $(x, y) = (4, 2)$, a corner of the feasible region. (See Figure 3.11.)

Dropping the last row and the columns corresponding to artificial variables produces the tableau

$$\begin{array}{c} \\ y \\ x \\ \\ \end{array}\begin{array}{c} \begin{array}{cccc} x & y & s & t \end{array} \\ \left[\begin{array}{cccc|c} 0 & 1 & -\frac{2}{3} & \textcircled{\frac{1}{3}} & 2 \\ 1 & 0 & \frac{1}{3} & -\frac{2}{3} & 4 \\ \hline 0 & 0 & 0 & 3^* & R-30 \end{array}\right], \end{array} \qquad \begin{array}{l} 2/(1/3) = 6^* \\ \text{(negative)} \end{array}$$

and we proceed to pivot on row 1 and column 4.

$$\begin{array}{c} \\ t \\ x \\ \\ \end{array} \begin{array}{c} \begin{array}{cccc} x & y & s & t \end{array} \\ \left[\begin{array}{cccc|c} 0 & 3 & -2 & 1 & 6 \\ 1 & 2 & -1 & 0 & 8 \\ \hline 0 & -9 & 6^* & 0 & R-48 \end{array}\right] \end{array} \begin{array}{l} \\ \text{(negative)} \\ \text{(negative)} \\ \\ \end{array}$$

Now a pivot in column 3 is indicated, but impossible because all the corresponding quotients are negative. Thus phase two breaks down, and the program has no solution.

Although we have illustrated the artificial variable algorithm with two-variable problems so that we could show pictorially how the method works, its main usefulness is with many-variable problems for which no graphical method is practical. The simplex algorithm proceeds by well-determined rules, and so may be programmed on a computer, where it is common to solve linear programming problems with hundreds of variables and constraints in seconds.

EQUALITY CONSTRAINTS

The artificial variable method may also be applied when some of the constraints are equations, rather than inequalities. All that need be done is to *introduce an artificial variable for each equality constraint*. (No slack variable is needed for such a constraint.) The following example illustrates the method.

EXAMPLE 3.16 Ralph plans to spend at most \$1.40 for a 30-day supply of vitamin C pills. Healthcaps cost 4¢ each and provide 300 units of the vitamin. Likewise No-colds cost 3¢ and provide 200 units, and Orangos cost 5¢ and provide 450 units. How can he get the most vitamin C for his money?

Solution Suppose Ralph buys x Healthcaps, y No-colds, and z Orangos. Then we have the linear program:

$$\begin{array}{ll} \text{Maximize} & 300x + 200y + 450z = R \\ \text{subject to} & 4x + 3y + 5z \le 140 \\ & x + y + z = 30 \\ & x \ge 0,\ y \ge 0,\ z \ge 0. \end{array}$$

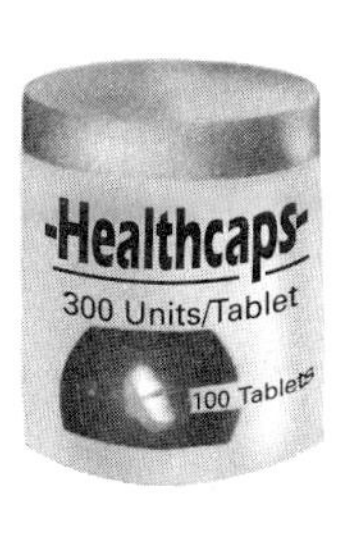

\$4.00

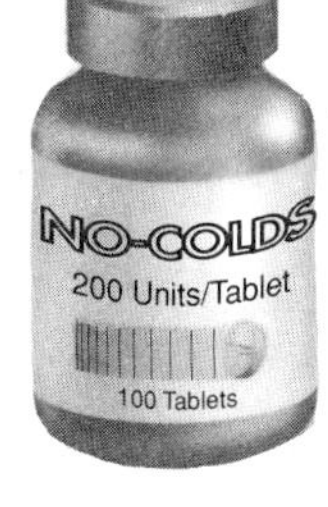

\$3.00

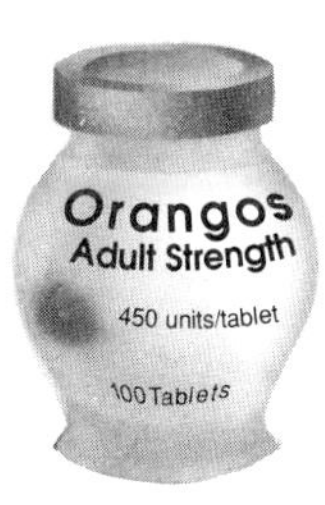

\$5.00

As usual we replace the inequality constraint with the equation

$$4x + 3y + 5z + s = 140, \qquad s \geq 0.$$

We introduce the artificial variable a to produce the equation

$$x + y + z + a = 30.$$

The reason for this is that the given equation $x + y + z = 30$ is not satisfied when we set the originally free variables x, y, and z equal to 0. The new equation is satisfied, however, if $x = y = z = 0$ and $a = 30$.

From this point we proceed as in the previous examples in this section. Our tableau is

$$\begin{array}{c} \\ \\ \\ \\ \\ \end{array}\begin{array}{ccccc|c} x & y & z & s & a & \\ \hline 4 & 3 & 5 & 1 & 0 & 140 \\ 1 & 1 & 1 & 0 & 1 & 30 \\ \hline 300 & 200 & 450 & 0 & 0 & R \\ 0 & 0 & 0 & 0 & -1 & Q \end{array}.$$

Step 3 produces the tableau

$$\begin{array}{c|ccccc|c|l} & x & y & z & s & a & & \\ \hline s & 4 & 3 & 5 & 1 & 0 & 140 & 140/4 = 35 \\ a & \textcircled{1} & 1 & 1 & 0 & 1 & 30 & 30/1 = 30^* \\ \hline & 300 & 200 & 450 & 0 & 0 & R & \\ & 1^* & 1 & 1 & 0 & 0 & Q + 30 & \end{array}.$$

We choose to pivot first in column 1, and thus in row 2. The result is as follows.

$$\begin{array}{c|ccccc|c} & x & y & z & s & a & \\ \hline s & 0 & -1 & 1 & 1 & -4 & 20 \\ x & 1 & 1 & 1 & 0 & 1 & 30 \\ \hline & 0 & -100 & 150 & 0 & -300 & R - 9000 \\ & 0 & 0 & 0 & 0 & -1 & Q \end{array}$$

Now phase one is finished, and we remove the last row and the column headed by a.

$$\begin{array}{c|cccc|c|l} & x & y & z & s & & \\ \hline s & 0 & -1 & \textcircled{1} & 1 & 20 & 20/1 = 20^* \\ x & 1 & 1 & 1 & 0 & 30 & 30/1 = 30 \\ \hline & 0 & -100 & 150^* & 0 & R - 9000 & \end{array}$$

Pivoting on the element in row 1 and column 3 produces the following tableau.

$$\begin{array}{c|cccc|c|l} & x & y & z & s & & \\ \hline z & 0 & -1 & 1 & 1 & 20 & \text{negative} \\ x & 1 & \textcircled{2} & 0 & -1 & 10 & 10/2 = 5^* \\ \hline & 0 & 50^* & 0 & -150 & R - 12000 & \end{array}$$

The last pivot is on the element in row 2 and column 2. We omit the intermediate step.

$$\begin{array}{c} \\ z \\ y \\ \\ \end{array} \begin{bmatrix} x & y & z & s & \\ \frac{1}{2} & 0 & 1 & \frac{1}{2} & 25 \\ \frac{1}{2} & 1 & 0 & -\frac{1}{2} & 5 \\ \hline -25 & 0 & 0 & -125 & R - 12250 \end{bmatrix}$$

The maximum value of R is 12250, and occurs when $(x, y, z) = (0, 5, 25)$. Ralph should buy 5 No-colds and 25 Orangos. ■

EXERCISES 3.6

In Exercises 1–8 introduce slack and artificial variables as appropriate for the two-phase method, and give the resulting tableau.

1. Maximize $3x + 5y = R$ subject to $x + y \le 20$, $3x + y \ge 36$, $x \ge 0, y \ge 0$.

2. Maximize $3x + 4y = R$ subject to $x + y \le 40$, $2x + y \ge 50$, $x \ge 0, y \ge 0$.

3. Maximize $3x + y = R$ subject to $x + 2y \le 200$, $2x + y \ge 250$, $x \ge 0, y \ge 0$.

4. Maximize $3x + 2y = R$ subject to $2x + y \le 75$, $x + y \ge 60$, $x \ge 0, y \ge 0$.

5. Minimize $3x + 6y = C$ subject to $2x + y \le 75$, $x + y \ge 60$, $x \ge 0, y \ge 0$.

6. Minimize $4x + y = C$ subject to $x + 2y \le 200$, $2x + y \ge 250$, $x \ge 0, y \ge 0$.

7. Minimize $6x + 4y = C$ subject to $x + 2y \le 150$, $2x + 3y \ge 240$, $x \ge 0, y \ge 0$.

8. Minimize $5x + 3y = C$ subject to $4x + y \le 400$, $2x + y \ge 300$, $x \ge 0, y \ge 0$.

In Exercises 9 and 10 a tableau for the two-phase method is presented. Apply the method, finding the maximum value of R and the optimal values of x and y.

9.
$$\begin{bmatrix} x & y & s & t & a & \\ 1 & 1 & 1 & 0 & 0 & 10 \\ 5 & 3 & 0 & -1 & 1 & 15 \\ \hline -2 & -1 & 0 & 0 & 0 & R \\ 0 & 0 & 0 & 0 & -1 & Q \end{bmatrix}$$

10.
$$\begin{bmatrix} x & y & s & t & a & b & \\ 1 & 2 & -1 & 0 & 1 & 0 & 2 \\ 2 & 1 & 0 & -1 & 0 & 1 & 2 \\ \hline -3 & -2 & 0 & 0 & 0 & 0 & R \\ 0 & 0 & 0 & 0 & -1 & -1 & Q \end{bmatrix}$$

In Exercises 11–18 use the two-phase method to solve the program of the indicated exercise.

11. Exercise 1
12. Exercise 2
13. Exercise 3
14. Exercise 4
15. Exercise 5
16. Exercise 6
17. Exercise 7
18. Exercise 8

Solve the programs of Exercises 19–24 using the two-phase method.

19. Maximize $3x + 2y = R$
subject to
$$4x + y \leq 400$$
$$2x + y \geq 300$$
$$x \geq 0,\ y \geq 0.$$

20. Maximize $5x + 2y = R$
subject to
$$x + 2y \leq 150$$
$$2x + 3y \geq 240$$
$$x \geq 0,\ y \geq 0.$$

21. Minimize $2x + 5y = C$
subject to
$$x + 3y \leq 180$$
$$3x + 5y \geq 480$$
$$x \geq 0,\ y \geq 0.$$

22. Minimize $10x + 2y = C$
subject to
$$5x + 3y \leq 600$$
$$3x + 2y \geq 390$$
$$x \geq 0,\ y \geq 0.$$

23. Minimize $2x + 3y + 5z = C$
subject to
$$x - y - z \leq 30$$
$$x - y + z \geq 40$$
$$x \geq 0,\ y \geq 0,\ z \geq 0.$$

24. Minimize $2x + 6y + 3z = C$
subject to
$$x + y + z \geq 120$$
$$-2x + 2y + z \geq 108$$
$$x \geq 0,\ y \geq 0,\ z \geq 0.$$

25. A manufacturer would like to produce a total of at least 30 ceramic tiles, which are available in three styles. The first style requires 5 minutes of shaping and no painting time; the second style requires 2 minutes of shaping and 3 minutes of painting; and the third style requires 2 minutes of shaping and 5 minutes of painting. Profits on the three styles are \$3, \$4, and \$6 per tile, respectively. If at most 160 minutes are available for shaping and at most 125 minutes are available for painting, how many tiles of each style should be made for a maximum profit? What is the maximum profit?

26. A certain diet specifies a minimum daily requirement of 50 units of food supplement R and 60 units of supplement S. These supplements are obtained most economically by eating three foods. One gram of the first food provides 2 units of R and 5 units of S, one gram of the second food provides 1 unit of R and 3 units of S, and one gram of the third food provides 1 unit of R and 6 units of S. If a gram of each food costs 4¢, 5¢, and 3¢, respectively, what amounts of each food should be eaten in order to satisfy the minimum daily requirements at the least cost? What is the least daily cost?

27. The owner of a nut shop has 50 pounds of a mixture which is 40% cashews and 50% peanuts by weight. To this amount she adds certain quantities of the mixtures described below.

Mixture	A	B
Cashews	50%	80%
Peanuts	5%	10%
Cost per pound	\$4	\$6

The final blend is to contain at least 60% cashews and at most 20% peanuts. How can this be achieved at lowest cost?

28. A metallurgist wants to make a new bronze alloy by combining certain amounts of the three alloys described below (percentages by weight).

Alloy	A	B	C
Copper	95%	90%	75%
Zinc	1%	8%	5%
Cost per ton	\$500	\$450	\$200

There must be exactly 10 tons of the new alloy, with at least 8 tons of copper and at most ½ ton of zinc.

In Exercises 29–32 solve the linear programs using the two-phase method.

29. Maximize $x + y + z = R$
subject to
$$2x + y + 3z \le 6$$
$$x - y = 1$$
$$x \ge 0,\ y \ge 0,\ z \ge 0.$$

30. Maximize $2x + y - z = R$
subject to
$$x + y + z = 12$$
$$-x + y \ge 1$$
$$x \ge 0,\ y \ge 0,\ z \ge 0.$$

31. Minimize $x + y + 2z = C$
subject to
$$x + z \ge 3$$
$$x - y + z = 4$$
$$x \ge 0,\ y \ge 0,\ z \ge 0.$$

32. Minimize $3x + z = C$
subject to
$$x + y = 10$$
$$x + y + z \ge 3$$
$$x \ge 0,\ y \ge 0,\ z \ge 0.$$

Answers to Practice Problems

11. Maximize $3x - y = R$
subject to
$$x - y \le 2$$
$$2x + 3y \le 6$$
$$x + y \ge 1$$
$$x \ge 0,\ y \ge 0.$$

12.

	x	y	z	s	t	u	a	b	
s	1	0	1	1	0	0	0	0	30
a	2	1	1	0	−1	0	1	0	10
b	1	3	2	0	0	−1	0	1	15
	1	2	−3	0	0	0	0	0	R
	0	0	0	0	0	0	−1	−1	Q

Rows 2 and 3 should be added to the bottom row.

CHAPTER 3 REVIEW

IMPORTANT TERMS

- **constraint** *(3.1)*
- **corner point**
- **feasible point**
- **half-plane**
- **linear inequality**
- **linear program**
- **objective function**
- **optimal point**
- **feasible region** *(3.2)*
- **isoline**
- **row label** *(3.3)*
- **simplex algorithm**
- **slack variable**
- **indicator** *(3.4)*
- **standard maximum program**
- **compact tableau** *(3.5)*
- **dual program**
- **artificial variable** *(3.6)*
- **two-phase method**

REVIEW EXERCISES

In Exercises 1 and 2 introduce variables and set up the problem as a linear program.

1. To make a stuffed bear requires 15 minutes of sewing, 6 minutes of painting, and 12 minutes of finishing, while a stuffed dog requires 18 minutes of sewing, 5 minutes of painting, and 10 minutes of finishing. Only 300 minutes of sewing, 120 minutes of painting, and 240 minutes of finishing are available. If the profit is \$3 on each bear and \$3.20 on each dog, how many of each should be made to maximize the profit?

2. One cup of cooked snap beans contains 30 calories, 63 mg calcium, and 2 grams protein; one cup of canned red kidney beans contains 230 calories, 74 mg calcium, and 15 grams protein; and one cup of baked beans with pork and molasses contains 385 calories, 161 mg calcium, and 16 grams of protein. A mixture of the three beans is to be made containing at least 350 mg calcium, at least 50 grams protein, and as few calories as possible. How should this be done?

In Exercises 3 and 4 graph the given system of linear inequalities, and find the coordinates of all corner points of the region.

3.
$$\begin{aligned} 2x + 3y &\geq 6 \\ 2x - 3y &\leq 2 \\ y &\leq 3 \end{aligned}$$

4.
$$\begin{aligned} x + y &\geq 10 \\ 2x - y &\geq 20 \\ x &\geq 20 \end{aligned}$$

In Exercises 5–8 solve the given linear program graphically, if possible. Otherwise tell why the program has no solution.

5. Minimize $4x + 5y = C$
subject to
$$\begin{aligned} 3x + 5y &\geq 15 \\ 12x + 5y &\geq 24 \\ x \geq 0,\ y &\geq 0. \end{aligned}$$

6. Minimize $-20x + 10y = C$
subject to
$$\begin{aligned} 2x + 5y &\geq 10 \\ -x + y &\geq 10 \\ x \geq 0,\ y &\geq 0. \end{aligned}$$

7. Maximize $3x + 4y = R$
subject to
$$\begin{aligned} x + 2y &\leq 10 \\ 3x + y &\leq 15 \\ x + y &\geq 8 \\ x \geq 0,\ y &\geq 0 \end{aligned}$$

8. Maximize $3x + 2y = R$
subject to
$$\begin{aligned} 7x + 8y &\leq 28 \\ y &\leq 2 \\ x - y &\leq 2 \\ x \geq 0,\ y &\geq 0. \end{aligned}$$

Solve the programs in Exercises 9 and 10 using the simplex algorithm.

9. Maximize $6x + 5y + z = R$
subject to
$$\begin{aligned} x + 2z &\leq 10 \\ 2x + y + z &\leq 12 \\ x \geq 0,\ y \geq 0,\ z &\geq 0. \end{aligned}$$

10. Maximize $x + 2y + 2z = R$
subject to
$$\begin{aligned} x + y + z &\leq 6 \\ x + 2y - z &\leq 8 \\ x \geq 0,\ y \geq 0,\ z &\geq 0. \end{aligned}$$

Solve the programs in Exercises 11 and 12 by applying the simplex algorithm to the dual program.

11. Minimize $3x + 2y + z = C$
subject to
$$\begin{aligned} x + 3y &\geq 12 \\ x + 2y + 2z &\geq 18 \\ x \geq 0,\ y \geq 0,\ z &\geq 0. \end{aligned}$$

12. Minimize $x + 3y + z = C$
subject to
$$\begin{aligned} x + y + z &\geq 50 \\ 2y + 4z &\geq 100 \\ x \geq 0,\ y \geq 0,\ z &\geq 0. \end{aligned}$$

Solve Exercises 13 and 14 by the two-phase method.

13. Maximize $3x + 2y = R$
subject to $5x + 3y \leq 600$
$3x + 2y \geq 390$
$x \geq 0, y \geq 0.$

14. Maximize $4x + 6y = R$
subject to $x + 3y \leq 180$
$3x + 5y \geq 480$
$x \geq 0, y \geq 0.$

15. A mining company operates three mines which produce titanium ore. The ore from Mine A is 5% titanium by weight and requires 4 work-hours per ton to mine. The ore from Mine B is 3% titanium and requires 3 work-hours per ton, while the ore from Mine C is 2% titanium and requires 1 work-hour per ton. The company has 800 work-hours of mining labor available per day, and its refinery can handle up to 500 tons of ore per day. How much ore should be taken daily from each mine in order to maximize the total amount of titanium produced?

16. An amusement park has "A" rides, which cost \$1, and "B" rides, which cost \$2. An economy book containing 10 A-ride tickets and 5 B-ride tickets can be purchased for \$18 (a 10% discount), and a super economy book containing 25 A-ride tickets and 10 B-ride tickets can be purchased for \$36 (a 20% discount). A teacher is bringing a class of 20 students to the park and each has been promised 5 A-rides and 3 B-rides. How many books and individual tickets of each type should be purchased in order to minimize the cost of the rides?

4

MATHEMATICS OF FINANCE

Anyone who has a savings account or makes installment purchases has contact with the world of finance. In recent years high interest rates have made knowledge of finance essential for every consumer. In this chapter we will study basic financial topics such as simple and compound interest, installment loans, and annuities. The mathematical foundation for this study involves the concept of a *difference equation*. As we will see, difference equations can be used to analyze not only financial matters but any periodic process.

Because of the nature of the calculations that are required in this chapter, we assume that the reader has access to a calculator with the capability of computing powers of any positive number. (The keys that perform these types of calculations are usually denoted y^x or x^y.)

4.1 Simple Interest

Interest is a fee paid for the use of someone else's money. Interest is paid not only by individuals, who borrow money by making installment purchases, but also by savings institutions and governmental units, which borrow money by offering savings accounts and issuing bonds. In this section we will develop the basic equations involving simple interest, which underlies almost all financial calculations. We begin by studying the special type of equation that occurs in all of the financial calculations that we will consider.

ARITHMETIC AND GEOMETRIC PROGRESSIONS

An infinite list of numbers

$$x_0, x_1, x_2, \ldots$$

is called a **sequence.** The individual numbers in the sequence are called its **terms.** The first term in the sequence, x_0, is called the **initial value** of the sequence. In this chapter we will be concerned with sequences that are defined by an equation of the form

$$x_{n+1} = ax_n + b, \tag{4.1}$$

where a and b are constants. An equation of this type is called a **first-order linear difference equation.** Given the initial value x_0, we can use (4.1) to calculate all subsequent terms in the sequence as demonstrated in Example 4.1.

EXAMPLE 4.1 If a sequence defined by the difference equation $x_{n+1} = 2x_n - 3$ has an initial value of 7, find the next five terms.

Solution The first term in the sequence is the initial value x_0, which is 7. The next term x_1 is found by taking $n = 0$ in the defining equation $x_{n+1} = 2x_n - 3$; this yields

$$x_1 = 2x_0 - 3 = 2(7) - 3 = 11.$$

Similarly, by taking n to be 1, 2, 3, and 4 in the defining equation, we obtain

$$\begin{aligned} x_2 &= 2x_1 - 3 = 2(11) - 3 = 19, \\ x_3 &= 2x_2 - 3 = 2(19) - 3 = 35, \\ x_4 &= 2x_3 - 3 = 2(35) - 3 = 67, \\ x_5 &= 2x_4 - 3 = 2(67) - 3 = 131. \end{aligned}$$

Hence the first six terms of the sequence are 7, 11, 19, 35, 67, and 131. ■

Practice Problem 1 If the sequence defined by the difference equation $x_{n+1} = -3x_n + 10$ has an initial value of 4, find the next five terms.

Although it is possible to find any term of (4.1) as in Example 4.1, the calculations will be quite tedious if n is large. Therefore it is desirable to have an explicit formula giving the value of x_n in terms of n. This will enable us to compute x_n directly without finding all the preceding terms in the sequence. For instance, we will see in Section 4.3 that the formula $x_n = 4 \cdot 2^n + 3$ gives the terms of the sequence in Example 4.1, as the reader can check for $n = 0, 1, 2, 3, 4, 5$. Thus we can compute $x_5 = 4 \cdot 2^5 + 3 = 4(32) + 3 = 131$ and $x_9 = 4 \cdot 2^9 + 3 = 4(512) + 3 = 2051$ directly.

In this section we will develop formulas for sequences defined by two special cases of (4.1). The first case that we will consider is the equation

$$x_{n+1} = x_n + b,$$

which occurs when $a = 1$ in (4.1). Each term of a sequence governed by this equation differs from its predecessor by a fixed amount b; such a sequence is known as an **arithmetic progression.** Given the initial value x_0, the equation $x_{n+1} = x_n + b$ generates the successive terms

$$\begin{aligned} x_1 &= x_0 + b, \\ x_2 &= x_1 + b = (x_0 + b) + b = x_0 + 2b, \\ x_3 &= x_2 + b = (x_0 + 2b) + b = x_0 + 3b, \\ x_4 &= x_3 + b = (x_0 + 3b) + b = x_0 + 4b, \\ &\vdots \end{aligned}$$

Thus for the equation $x_{n+1} = x_n + b$, the formula relating x_n to n is

$$x_n = x_0 + nb.$$

Theorem 4.1

The general term of the arithmetic progression defined by the difference equation $x_{n+1} = x_n + b$ and having initial value x_0 is $x_n = x_0 + nb$.

EXAMPLE 4.2 Find a formula for the general term of the sequence defined by the difference equation $x_{n+1} = x_n + 5$ and having initial term -8.

Solution The first few terms in this sequence are

$$0, \quad 3, 2, 7, 12, 17, \ldots .$$

This is an arithmetic progression because successive terms differ by 5. By Theorem 4.1 the general term of this progression is

$$x_n = -8 + 5n.$$

Therefore, for instance, we see that $x_{10} = 42$, $x_{50} = 242$, and $x_{100} = 492$. ■

Practice Problem 2 Find a formula for the general term of the sequence defined by the difference equation $x_{n+1} = x_n - 3$ and having initial term 36.

The second special case of (4.1) that we will consider is the equation

$$x_{n+1} = ax_n,$$

which occurs when $b = 0$ in (4.1). Each term of a sequence governed by this equation is obtained by multiplying its predecessor by the fixed amount a; such a sequence is called a **geometric progression.** Given the initial value x_0, the equation $x_{n+1} = ax_n$ generates the successive terms

$$\begin{aligned} x_1 &= ax_0, \\ x_2 &= ax_1 = a(ax_0) = a^2x_0, \\ x_3 &= ax_2 = a(a^2x_0) = a^3x_0, \\ x_4 &= ax_3 = a(a^3x_0) = a^4x_0, \\ &\ \ \vdots \end{aligned}$$

Hence for the equation $x_{n+1} = ax_n$, the formula relating x_n to n is

$$x_n = a^nx_0.$$

Theorem 4.2 ■ ■ ■ ■ ■ ■

The general term of the geometric progression defined by the difference equation $x_{n+1} = ax_n$ and having initial value x_0 is $x_n = a^nx_0$.

EXAMPLE 4.3 Find a formula for the general term of the sequence defined by the difference equation $x_{n+1} = 3x_n$ and having initial term 7.

Solution The first few terms in this sequence are

$$7, 21, 63, 189, 567, \ldots$$

This is a geometric progression, because each term is 3 times its predecessor.

Theorem 4.2 shows that the general term of this sequence is

$$x_n = 3^n \cdot 7.$$

The terms in this sequence grow very quickly. For instance, x_{10} is almost half a million, x_{15} is more than 100 million, and x_{50} exceeds the number of grains of sand on all the beaches in the world! ■

Practice Problem 3 Find a formula for the general term of the sequence defined by the difference equation $x_{n+1} = 2x_n$ and having initial term 5.

Arithmetic and geometric progressions arise quite naturally in the mathematics of finance.

INTEREST

When money is invested in a savings account or loaned to a commercial borrower, the initial amount is called the **principal.** The fee charged for the use of the money is called **interest.** The amount of interest to be charged is usually stated as a percentage called the **interest rate.** Normally interest rates are stated as annual (yearly) rates, and we will follow common practice and assume that *all interest rates are annual rates unless it is explicitly stated otherwise*. The sum of the principal and the interest is called the **balance** of the account or loan.

Of the many ways that interest can be charged, the most basic is simple interest. **Simple interest** is charged as a percentage of the principal, per unit time. The amount of simple interest owed is given by the formula below.

Simple Interest Charge

$$I = Prt \tag{4.2}$$

where I denotes the interest charge,
P denotes the principal,
r denotes the annual interest rate, and
t denotes the time in years for which the money is borrowed.

EXAMPLE 4.4 Suppose that an individual borrows $5000 at 10% simple interest. What amount of interest is due when the loan is repaid at the end of three years?

Solution In this situation time is measured in years and the annual interest rate is 10%. So taking $P = \$5000$, $r = 0.10$, and $t = 3$ in the simple interest formula (4.2), we see that the amount of interest to be paid after three years is

$$I = Prt = \$5000(0.10)(3) = \$1500.$$

Thus the borrower must pay a total of \$6500 at the end of three years, \$5000 to repay the principal and an interest charge of \$1500. ■

EXAMPLE 4.5 Suppose that an individual borrows \$4000 at 12% simple interest. What amount of interest is due when the loan is repaid at the end of six months?

Solution In this case, time is measured in months and the annual simple interest rate is 12%. Since formula (4.2) requires that t be measured in years, we must convert 6 months to $6/12 = 1/2$ year. Thus we take $P = \$4000$, $r = 0.12$, and $t = 1/2$ in formula (4.2). Therefore

$$I = Prt = \$4000(0.12)\frac{1}{2} = \$240,$$

and hence the interest charge will be \$240 at the end of six months. ■

Practice Problem 4 Suppose that \$8000 is borrowed at 16% simple interest. What amount of interest is due when the loan is repaid at the end of 30 months?

More generally, suppose that a principal P earns simple interest at a rate of r per year, and let A_t denote the balance after t years. Now the balance after $t + 1$ years equals the balance after t years plus the interest charge for one year. Since the interest for one year is Pr by (4.2), we obtain the equation

$$A_{t+1} = A_t + Pr.$$

This is a difference equation of the form (4.1) with $a = 1$ and $b = Pr$. Since its initial value is $A_0 = P$, Theorem 4.1 gives the following formula:

$$A_t = A_0 + t(rP) = P + trP = P(1 + rt).$$

Simple Interest Amount

$$A = P(1 + rt) \qquad (4.3)$$

where A denotes the amount after t years,
P denotes the principal,
r denotes the annual interest rate, and
t denotes the time in years for which the money is borrowed.

Thus for the loan in Example 4.4 with $P = \$5000$, $r = 0.10$, and $n = 3$, formula (4.3) yields

$$A = \$5000[1 + 0.10(3)] = \$5000(1.30) = \$6500.$$

Consequently the amount to be repaid after three years (principal plus interest) is \$6500, which agrees with our earlier calculation.

EXAMPLE 4.6 If a manufacturer borrows \$250,000 for 9 months at 15% simple interest, how much will be owed when the loan is due?

Solution Since time is measured in months, we must convert it to 9/12 = 3/4 years as in Example 4.5. Thus we take $P = \$250{,}000$, $r = 0.15$, and $t = 3/4$ in (4.3).

$$\begin{aligned} A &= P(1 + rt) \\ &= \$250{,}000\left[1 + (0.15)\frac{3}{4}\right] \\ &= \$250{,}000(1.1125) = \$278{,}125. \end{aligned}$$

Hence the manufacturer must repay \$278,125 after 9 months. Of this, \$250,000 is the principal and the remaining \$28,125 is interest. ■

Practice Problem 5 Suppose that \$10,000 is borrowed at 15% simple interest. How much will be owed when the loan is due at the end of 18 months? How much of this amount is interest?

In some situations we must solve formula (4.3) for r, the simple interest rate.

EXAMPLE 4.7 A bond that is redeemable for \$12,100 in 18 months was sold for \$10,000. What annual simple interest rate does this bond pay?

Solution In this problem we must solve (4.3) for r knowing that $A = \$12{,}100$, $P = \$10{,}000$, and $t = 18/12 = 3/2$.

$$\begin{aligned} A &= P(1 + rt) \\ \$12{,}100 &= \$10{,}000\left[1 + r\left(\frac{3}{2}\right)\right] \\ 1.21 &= 1 + r\left(\frac{3}{2}\right) \\ 0.21 &= r\left(\frac{3}{2}\right) \\ 0.14 &= r \end{aligned}$$

Thus the bond is paying 14% simple interest per year. ■

Practice Problem 6 A bond that is redeemable for \$22,650 in 15 months was sold for \$20,000. What annual simple interest rate does this bond pay?

EXERCISES 4.1

In Exercises 1–8 generate the first six terms of the sequence defined by each difference equation. Then determine whether the sequence is an arithmetic progression, a geometric progression, or neither.

1. $x_{n+1} = 5x_n - 1, \quad x_0 = 3$

2. $x_{n+1} = \frac{1}{2}x_n, \quad x_0 = 1$

3. $x_{n+1} = x_n + \frac{3}{2}, \quad x_0 = 0$

4. $x_{n+1} = -0.2x_n + 40, \quad x_0 = 5$

5. $x_{n+1} = -1.8x_n, \quad x_0 = 625$

6. $x_{n+1} = 2x_n - 7, \quad x_0 = 6$

7. $x_{n+1} = 1.1x_n + 100, \quad x_0 = 5000$

8. $x_{n+1} = x_n - 3.2, \quad x_0 = 24.8$

In Exercises 9–16 find a formula for the general term of the sequence satisfying each difference equation. Then use your formula to compute x_{10} *and* $x_{[illegible]}$

9. $x_{n+1} = x_n + 6, \quad x_0 = 2$

10. $x_{n+1} = 2x_n, \quad x_0 = 3$

11. $x_{n+1} = x_n - \frac{5}{2}, \quad x_0 = 35$

12. $x_{n+1} = -\frac{1}{3}x_n, \quad x_0 = 1$

13. $x_{n+1} = 1.5x_n, \quad x_0 = 1$

14. $x_{n+1} = x_n + 12, \ x_0 = -32$

15. $x_{n+1} = 0.9x_n, \quad x_0 = 100$

16. $x_{n+1} = x_n - 0.18, \ x_0 = 5$

In Exercises 17–20 compute the amount of interest earned if a principal P is invested at simple interest for t years at an annual interest rate r.

17. $P = \$5000, \quad t = 1\frac{1}{2}, \quad r = 7\%$

18. $P = \$1,200, \quad t = 3, \quad r = 8\%$

19. $P = \$40,000, \quad t = 6, \quad r = 10\%$

20. $P = \$2000, \quad t = \frac{1}{2}, \quad r = 9\%$

In Exercises 21–24 compute the amount necessary to repay a loan of P dollars borrowed at simple interest for t years at an annual interest rate r.

21. $P = \$900, \quad t = 4, \quad r = 15\%$

22. $P = \$2400, \quad t = 8, \quad r = 16\%$

23. $P = \$3000, \quad t = 5, \quad r = 8\%$

24. $P = \$5000, \quad t = 2, \quad r = 9\%$

25. An individual borrows \$6800 for 18 months at 14% simple interest. How much must be repaid when this loan is due? How much of this amount is interest?

26. A corporation borrows \$2 million for 3½ years at 12% simple interest. How much must be repaid when this loan is due? How much of this amount is interest?

27. A lender offers \$750 on the condition that the borrower repay \$900 in 12 months. What simple interest rate is involved in this transaction?

28. A buyer pays \$4000 for a bond that is redeemable at its face value of \$5000 in another 15 months. What simple interest rate does this bond pay?

29. An item costing \$50 was charged at a department store. Three months later the store issued a bill for \$52. What annual rate of simple interest was the store charging?

30. If \$3000 is borrowed at 12% simple interest, how much will be due in 1½ years?

31. A bank loaned \$500 and collected \$620 in payment 18 months later. What annual rate of simple interest did it charge?

32. If money is invested at 18% simple interest, how long will it take for the money to double?

Answers to Practice Problems

1. -2, 16, -38, 124, and -362

2. $x_n = 36 - 3n$

3. $x_n = 2^n \cdot 5$

4. \$3200

5. Of the \$12,250 that is owed, \$2250 is interest.

6. 10.6%

4.2 Compound Interest

Most transactions involving interest charges do not use simple interest, which is a percentage of the principal. Instead, the interest is computed as a percentage of the *balance* of the loan. In this situation, interest is charged not only on the original amount, but also on any previous interest charges. In this case the interest is said to have been **compounded,** and this type of interest is called **compound interest.**

EXAMPLE 4.8 Suppose that an individual borrows \$5000 at 10% interest compounded annually. What will the total debt be at the end of three years?

Solution During the first year the interest charge is 10% of the principal. Using the formula for a simple interest amount, we see that at the end of the first year the borrower owes

$$P(1 + rt) = \$5000[1 + 0.10(1)] = \$5000(1.10) = \$5500.$$

The lender now treats this amount as the new principal during the second year. Thus the balance of the loan at the end of the second year is

$$P(1 + rt) = \$5500[1 + 0.10(1)] = \$5500(1.10) = \$6050.$$

Using \$6050 as the principal for the third year, we see that the amount due after three years is

$$P(1 + rt) = \$6050[1 + 0.10(1)] = \$6050(1.10) = \$6655.$$

Note that this amount is considerably greater than the \$6500 that was due in Example 4.4 when simple interest was used. In fact an interest rate of 10% compounded annually over a three-year period is equivalent to a simple interest rate of more than 11%. ■

In order to derive a general formula for the balance of a loan when interest is compounded, we will assume that a principal P earns interest at the rate of i per period. If A_n denotes the balance after n periods, then the interest charge for the next period will be iA_n. Hence the balance after $n + 1$ interest periods is

$$A_{n+1} = A_n + iA_n = (1 + i)A_n.$$

This is a difference equation of the form (4.1) with $b = 0$, and so the sequence of balances is a geometric progression with initial value $A_0 = P$. By Theorem 4.2 we see that $A_n = (1 + i)^n A_0 = P(1 + i)^n$. Thus we have the following general formula.

Compound Interest Amount

$$A = P(1 + i)^n \tag{4.4}$$

where A denotes the amount after n periods,
P denotes the principal,
i denotes the interest rate per period, and
n denotes the number of interest periods for which the money is borrowed.

Applying this formula to the loan in Example 4.8 with $P = \$5000$, $i = 0.10$, and $n = 3$, we find that the amount due after three years is

$$\begin{aligned} A &= P(1 + i)^n = \$5000(1 + 0.10)^3 \\ &= \$5000(1.10)^3 = \$5000(1.331) = \$6655. \end{aligned}$$

Of course, this is the amount we obtained in Example 4.8 when we performed the calculation the long way.

The compound interest rate on savings accounts and installment loans is usually stated as an annual rate. If interest is compounded more than once a year, however, then the rate at which the balance grows is somewhat greater.

EXAMPLE 4.9 Compare the balance when \$1000 is deposited for five years at 9% interest compounded (a) annually, (b) quarterly, and (c) monthly.

Solution (a) If interest is compounded annually, then the interest period is one year and the interest rate is 9% per period. Thus after five years the balance will be

$$\begin{aligned} A &= P(1 + i)^5 = \$1000(1 + 0.09)^5 \\ &= \$1000(1.09)^5 \approx \$1000(1.53862) \\ &= \$1538.62. \end{aligned}$$

(b) With quarterly compounding the interest period is three months rather than one year. Thus one-fourth of the interest is paid every three months rather than paying the entire amount once a year. So because there are four quarters per year, the interest rate per period is 9%/4 = 2.25%, and the principal is earning interest for 5(4) = 20 quarters. Hence by (4.4) the balance after 20 quarters is

$$\begin{aligned} A &= P(1 + i)^{20} = \$1000(1 + 0.0225)^{20} \\ &= \$1000(1.0225)^{20} \approx \$1000(1.56051) = \$1560.51. \end{aligned}$$

(c) If interest is compounded monthly, then the interest rate per month is 9%/12 = 0.75% and the principal will earn interest for 5(12) = 60 months. Hence the balance after 60 months is

$$\begin{aligned} A &= P(1 + i)^{60} = \$1000(1 + 0.0075)^{60} \\ &= \$1000(1.0075)^{60} \approx \$1000(1.56568) = \$1565.68. \end{aligned}$$ ■

Practice Problem 7 If $6000 is invested in a certificate of deposit at 8% interest compounded quarterly and the interest is left to accumulate, how much will the certificate be worth after five years?

Notice that in Example 4.9 the balance grows more quickly when interest is compounded more frequently, but with a diminishing gain. That is, the largest balance occurs when interest is compounded most frequently. However, the difference between yearly compounding and quarterly compounding is about $22, whereas the difference between quarterly compounding and monthly compounding is only about $5. If interest were compounded daily (365 times per year), the balance would be $1568.23 after five years, which is less than $3 more than the balance obtained when interest is compounded monthly. Thus, unless very large sums of money are involved or the length of the loan is long, the difference between monthly compounding and daily compounding is insignificant.

EXAMPLE 4.10 At what annual rate must inflation continue so that in twelve years the purchasing power of a dollar will be half as much as it is now?

Solution To say that in twelve years the purchasing power of a dollar will be half as much as it is now means that goods costing $1 today will cost $2 in twelve years. Since the inflation rate i is in effect for each of the twelve years, the effects of inflation are "compounded" throughout the period, and we assume that the effects are compounded annually. Thus we must solve equation (4.4) for i if $A = \$2$, $P = \$1$, and $n = 12$.

$$A = P(1 + i)^n$$
$$\$2 = \$1(1 + i)^{12}$$
$$2 = (1 + i)^{12}$$
$$2^{1/12} = 1 + i$$
$$i = 2^{1/12} - 1 \approx 0.0595$$

Hence an inflation rate of approximately 5.95% will cut the purchasing power of a dollar by half in twelve years. ■

Fresh Anxiety About Inflation Deflates Rallies

MONDAY'S MARKETS

By DOUGLAS R. SEASE
Staff Reporter of THE WALL STREET JOURNAL

Renewed inflation worries pooped Wall Street's party.

After pushing stock and bond prices sharply higher Friday because a rip-roaring rise in wholesale prices has moderated, investors began the new week hunkered down for an expected dose of bad news about climbing consumer prices.

NOMINAL AND EFFECTIVE RATES

In Example 4.9 we saw that the same interest rate will yield different amounts depending on the frequency with which the interest is compounded. The stated rate of interest is called the **nominal rate.** Thus in Example 4.9 the nominal rate of interest is 9%, the stated annual rate. In order to compare the yields for different frequencies of compounding, it is useful to convert the nominal rate into an equivalent simple interest rate, called the **effective rate** (or **annual yield**). The effective rate can be computed by finding the percentage gain for an arbitrary principal, as in the example on the next page.

EXAMPLE 4.11 What is the effective rate of interest corresponding to a nominal rate of 8% compounded quarterly?

Solution To answer this question, we see how much of an increase an arbitrary principal experiences in one year. Suppose we choose a principal of \$100. By (4.4) we see that the balance after one year is

$$A = P(1 + i)^4 = \$100(1.02)^4 \approx \$108.24.$$

Thus the \$100 principal grew to \$108.24 in one year, an increase of 8.24%. So the effective rate of interest for a nominal rate of 8% compounded quarterly is 8.24% per year. This means that each dollar invested at 8% compounded quarterly actually earns 8.24% in interest during a year. ■

Practice Problem 8 Approximate the effective rate of interest corresponding to a nominal rate of 6% compounded monthly.

PRESENT VALUE

In our examples using simple and compound interest we have known the principal and been interested in finding the value of that principal at some future time. Other applications require knowing what principal P must be deposited now at a specified interest rate in order to accumulate a specific amount of money A at some future time. In this situation P is called the **present value** of A. To find present values, we use formulas (4.3) and (4.4) to solve for P.

Present Value of a Future Amount

simple interest $$P = \frac{A}{1 + rt} \tag{4.5}$$

compound interest $$P = \frac{A}{(1 + i)^n} \tag{4.6}$$

where P denotes the present value,
A denotes the future amount,
r denotes the annual interest rate,
i denotes the interest rate per period, and
n denotes the number of interest periods for which the money is borrowed.

EXAMPLE 4.12 How much money must be invested now in order to accumulate \$12,000 in 15 years at the following interest rates:

(a) 18% simple interest?

(b) 10% interest compounded quarterly?

Solution (a) We must find the present value P of \$12,000 in 15 years at 18% simple interest. Using (4.5), we obtain

$$\begin{aligned} P &= \frac{A}{1 + rt} \\ &= \frac{\$12{,}000}{1 + 15(0.18)} \\ &= \frac{\$12{,}000}{3.7} \approx \$3243.24. \end{aligned}$$

(b) With quarterly compounding, the interest rate per period is $i = 0.10/4 = 0.025$ and the number of interest periods is $n = 15(4) = 60$. Thus by (4.6) we see that the present value of \$12,000 in 15 years at 10% interest compounded quarterly is

$$\begin{aligned} P &= \frac{A}{(1 + i)^n} \\ &= \frac{\$12{,}000}{(1.025)^{60}} \\ &\approx \$2727.40. \end{aligned}$$

Note that even though the interest rate in (b) is significantly lower than in (a), less money must be deposited in (b). This demonstrates again how much more quickly an amount grows when interest is compounded. ■

Practice Problem 9 Approximately how much money must be invested now in order to accumulate \$8000 in 6 years at the following interest rates:
(a) 15% simple interest? (b) 10% interest compounded quarterly?

EXERCISES 4.2

In Exercises 1–4 compute the amount in m years if a principal P is invested at a nominal annual interest rate of r compounded quarterly.

1. $P = \$6000, \quad m = 5, \quad r = 4\%$

2. $P = \$800, \quad m = 2, \quad r = 6\%$

3. $P = \$300, \quad m = 1\frac{1}{2}, \quad r = 5\%$

4. $P = \$400, \quad m = 2\frac{1}{2}, \quad r = 8\%$

In Exercises 5–8 compute the present value of an amount A invested for m years in an account paying interest compounded semiannually at a nominal annual rate r.

5. $A = \$1000, \quad m = 3, \quad r = 5\%$

6. $A = \$500, \quad m = 8, \quad r = 4\%$

7. $A = \$3000, \quad m = 1, \quad r = 6\%$

8. $A = \$2000, \quad m = 2, \quad r = 8\%$

In Exercises 9–12 compute the effective rate of interest for a nominal annual rate r under the given conditions.

9. $r = 6\%$ compounded monthly

10. $r = 8\%$ compounded quarterly

11. $r = 7\%$ compounded daily

12. $r = 9\%$ compounded monthly

13. A consumer makes a $250 purchase with a credit card that charges 18% interest compounded monthly. If no payments are made for three months, how much will be owed at that time?

14. A savings and loan pays 9% interest compounded quarterly on its 2½-year certificate of deposit. If $3200 is invested in such a certificate and interest is left to accumulate, how much will be available when the certificate matures?

15. Compare the balance when $600 is deposited for three years at 7.2% interest compounded (a) annually, (b) quarterly, (c) monthly, and (d) daily.

16. Which return is better for the investor: 15.6% compounded semiannually or 14.6% compounded daily?

17. In nine years a company expects to replace a piece of machinery that will cost $2.6 million. If money can be invested at 10% interest compounded quarterly, how much must be invested now so that the necessary amount will be available in nine years?

18. If money can be invested at 6% interest compounded quarterly, how much should be deposited in order to have $10,000 for college expenses in eight years?

19. Suppose that an executive deposits $20,000 in an account paying 8% interest compounded quarterly. How much money will be available at retirement in 25 years?

20. If you have $500 to invest for 2 years, which option is better: 6% interest compounded semiannually or 6¼% simple interest?

21. How much must be borrowed now at 18% simple interest so that the amount owed in six months is $100?

22. What annual simple interest rate is equivalent to a rate of 8.4% compounded monthly?

23. If money can be invested at 7% interest compounded quarterly, how much should be paid for a note that will be worth $5000 in six years?

24. An insurance company allows a policyholder the option of collecting $1000 now or $1125 in two years. If money can be invested at 6% interest compounded quarterly, which option is better for the policyholder?

25. A company with current sales of $3 billion is predicting sales of $4.1 billion in three years. What annual rate of increase in sales is this company expecting?

26. At what annual rate must inflation continue so that a dollar will lose half its value in ten years?

27. If food prices increase 4% per year, how much will food cost in 1995 that costs $100 in 1990?

28. A commodities speculator wants cash. He is willing to borrow money at the rate of 14.4% compounded monthly, and will repay the loan with the proceeds from the sale of property. If the property is expected to be sold in 15 months for $180,000, how much can the speculator afford to borrow?

29. A municipal bond selling for $8000 can be redeemed in 6 years for $10,000. If interest is compounded quarterly, what nominal annual interest rate is this bond paying?

30. In 1975 a town had a population of 20,000, and by 1985 it had grown to 30,000. Assuming that the town's annual percentage of growth remains constant throughout the period 1975–1995, estimate the town's population in 1995.

31. The consumer price index was 118.3 in 1970 (using 1965 as the base year for which the consumer price index is set at 100.0). Assuming that the annual rate of inflation was constant throughout this period, determine this annual inflation rate.

32. The consumer price index was 118.3 in 1970 and 159.1 in 1975 (using 1965 as the base year for which the consumer price index is set at 100.0). Assuming that the annual rate of inflation was constant throughout the period 1970–1975, determine this annual inflation rate.

33. Show that the value of an account in which a principal P is invested for n years at an annual rate of r compounded m times per year is

$$A = P\left(1 + \frac{r}{m}\right)^{mn}$$

34. Find the effective rate of interest corresponding to a nominal annual rate r compounded m times per year.

Answers to Practice Problems

7. \$8915.68

8. 6.17%

9. (a) \$4210.53 (b) \$4423.00

4.3 Annuities and Installment Loans

Loans for the purchase of a home or an automobile are usually repaid in equal monthly installments. Likewise most life insurance companies offer whole life polices that are structured so that the policyholder makes fixed payments to the company for 20 or 30 years, after which time the sum of all the payments is available to the policyholder. Such a set of equal payments at regular intervals of time is called an **annuity.** In this section we will return to the general first-order linear difference equation (4.1) to analyze annuities.

THE GENERAL FIRST-ORDER LINEAR DIFFERENCE EQUATION

In Section 4.1 we saw how to solve the first-order linear difference equation

$$x_{n+1} = ax_n + b \tag{4.1}$$

in the special cases in which $a = 1$ or $b = 0$. If we are given such an equation with $a \neq 1$, we can reduce it to the case $b = 0$ by an appropriate substitution. Specifically, for $n = 0, 1, 2, \ldots$ we let

$$y_n = x_n - c,$$

where c is a constant that we will determine later. Since $y_n + c = x_n$ and $y_{n+1} + c = x_{n+1}$, substitution into (4.1) produces

$$y_{n+1} + c = a(y_n + c) + b,$$

or equivalently,

$$y_{n+1} + c = ay_n + (ac + b).$$

If we now choose c so that $c = ac + b$, then the preceding equation will have the same constant term on both sides. Now if $c = ac + b$, then:

$$c - ac = b$$
$$c(1 - a) = b$$
$$c = \frac{b}{1 - a}.$$

Thus if we choose

$$c = \frac{b}{1 - a},$$

the equation $y_{n+1} + c = ay_n + (ac + b)$ simplifies to

$$y_{n+1} = ay_n.$$

By Theorem 4.2 the general term of a sequence defined by this equation has the form

$$y_n = a^n y_0.$$

Substituting $y_n = x_n - c$ and $y_0 = x_0 - c$, we obtain

$$x_n - c = a^n(x_0 - c),$$

or

$$x_n = a^n(x_0 - c) + c.$$

Thus we have obtained the following result.

Theorem 4.3

If $a \neq 1$, then the general term of the sequence satisfying the difference equation

$$x_{n+1} = ax_n + b \tag{4.1}$$

and having initial value x_0 is

$$x_n = a^n(x_0 - c) + c, \tag{4.7}$$

where $c = \dfrac{b}{1 - a}$.

EXAMPLE 4.13 Find a formula for the general term of the sequence with initial value 7 that satisfies the difference equation $x_{n+1} = 2x_n - 3$.

Solution Since the given equation is of the form $x_{n+1} = ax_n + b$ with $a = 2$, we can apply Theorem 4.3. In this case we have

$$c = \frac{b}{1-a} = \frac{-3}{1-2} = 3.$$

Thus the desired formula is

$$\begin{aligned} x_n &= a^n(x_0 - c) + c \\ &= 2^n(7 - 3) + 3 \\ &= 4 \cdot 2^n + 3. \end{aligned}$$

We can see from this formula that the terms of the sequence x_n increase without bound as n increases. ■

Practice Problem 10 Find a formula for the general term of the sequence with initial value 8 that satisfies the difference equation $x_{n+1} = 3x_n - 4$.

EXAMPLE 4.14 Suppose that a lumber company owns a stand of 8000 redwood trees. Each year the company plans to harvest 10% of these trees and plant 650 new ones.

(a) How many trees will there be in the stand after 10 years? After 20 years?
(b) How many trees will there be in the long run?

Solution (a) Let x_n denote the number of trees in the stand after n years. During year $n + 1$, 10% of the trees existing in year n will be harvested; this number is $0.10x_n$. Since 650 additional trees will be planted during year $n + 1$, the number of trees after $n + 1$ years will be described by the equation

$$x_{n+1} = x_n - 0.10x_n + 650$$

or

$$x_{n+1} = 0.90x_n + 650.$$

Since this is a difference equation of the form $x_{n+1} = ax_n + b$ with $a = 0.90$ and $b = 650$, we can use Theorem 4.3 to find a formula for x_n in terms of n. Now

$$c = \frac{b}{1-a} = \frac{650}{1-0.90} = 6500.$$

Hence by Theorem 4.3

$$\begin{aligned} x_n &= a^n(x_0 - c) + c \\ &= (0.9)^n(8000 - 6500) + 6500 \\ &= 1500(0.9)^n + 6500. \end{aligned}$$

Therefore after 10 years the number of trees will be

$$x_{10} = 1500(0.9)^{10} + 6500 \approx 7023,$$

and after 20 years the number of trees will be

$$x_{20} = 1500(0.9)^{20} + 6500 \approx 6682.$$

(b) As n increases, the quantity $(0.9)^n$ will decrease steadily to zero. Hence the formula

$$x_n = 1500(0.9)^n + 6500$$

implies that the number of trees in the stand will approach 6500. (Note that as the number of trees in the stand approaches 6500, the number of trees being harvested each year becomes equal to the number of new trees being planted.) ■

THE FUTURE VALUE OF AN ANNUITY

Suppose that equal payments are made at regular intervals into an annuity to accumulate money. It is possible that the interval between payments may be different from the interest period. For instance, we may deposit \$100 *per month* into a savings account paying *quarterly* interest. However, for installment loans the two time periods are always the same. Thus for simplicity *we will consider only annuities in which the regular payments are made at the end of each interest period.* We would like to be able to calculate the amount of money in the annuity at any time. Since we are determining the total sum of money in the annuity at some future time, this sum is called the **future value of the annuity.**

To analyze this situation, we let p be the regular periodic payment, i be the interest rate per period, and F_n be the amount of money accumulated in the account after n payments. Now the accumulated amount after $n + 1$ payments equals the amount accumulated after n payments plus interest plus the regular periodic payment. Hence we have the following equation.

$$\begin{aligned} F_{n+1} &= F_n + iF_n + p \\ &= (1 + i)F_n + p \end{aligned}$$

This is a difference equation of the form $x_{n+1} = ax_n + b$ with $a = 1 + i$ and $b = p$. Also,

$$c = \frac{b}{1 - a} = -\frac{p}{i}.$$

Thus by Theorem 4.3 we have

$$F_n = (1 + i)^n\left(F_0 + \frac{p}{i}\right) - \frac{p}{i}.$$

In the important special case in which the initial value of the fund is $F_0 = 0$, we obtain the following result.

Future Value of an Annuity

$$F = p\left[\frac{(1 + i)^n - 1}{i}\right] \tag{4.8}$$

where F denotes the future value after n payments,
p denotes the amount of the periodic payment,
i denotes the interest rate per period, and
n denotes the number of payments.

The expression in brackets in formula (4.8) occurs often in financial analysis and is denoted by the symbol $s_{\overline{n}|i}$, which is read "s angle n at i." Thus $s_{\overline{n}|i}$ denotes the future value of an annuity consisting of n payments of $1 with an interest rate of i per period. Note that with this notation (4.8) can be written as

$$F = ps_{\overline{n}|i}.$$

EXAMPLE 4.15 An individual makes quarterly payments of $100 into a retirement fund which earns 9% interest, compounded quarterly. How much will the fund contain after 30 years?

Solution Applying (4.8) with $p = \$100$, $i = 0.09/4 = 0.0225$, and $n = 30(4) = 120$, we obtain

$$\begin{aligned} F &= p\left[\frac{(1 + i)^n - 1}{i}\right] \\ &= \$100\left[\frac{(1.0225)^{120} - 1}{0.0225}\right] \\ &\approx \$59{,}737.89. \end{aligned}$$

Thus the fund will contain approximately $59,737.89 after 30 years. ■

Practice Problem 11 In order to obtain money for a down payment on their first house, a newly married couple has decided to deposit $150 per month into a savings account paying 6% interest compounded monthly. How much money will this account contain after five years?

THE PRESENT VALUE OF AN ANNUITY

In Section 4.2 we learned how to find the value of the principal P that, if invested now, will accumulate to a particular amount A at some future time. (Recall that the amount P is called the *present value* of A.) Likewise it is often necessary to find the amount of money that must be deposited now in order to provide enough money to make a series of future periodic payments. For example, instant winners of the Illinois State Lottery receive $1 million in 20 yearly payments of $50,000. Lottery officials must determine how much money must be invested in order for these payments to be made. More specifically, the amount of money P necessary to make n periodic payments p from an account earning interest at the rate of i per period before exhausting the account is called the **present value of an annuity.** By (4.6) we have

$$P = \frac{A}{(1 + i)^n}.$$

So taking $A = F$ and substituting from (4.8) gives

$$P = \frac{p\left[\dfrac{(1 + i)^n - 1}{i}\right]}{(1 + i)^n} = p\left[\frac{1 - \dfrac{1}{(1 + i)^n}}{i}\right] = p\left[\frac{1 - (1 + i)^{-n}}{i}\right].$$

We have obtained the following formula.

Present Value of an Annuity

$$P = p\left[\frac{1 - (1 + i)^{-n}}{i}\right] \tag{4.9}$$

where P denotes the present value of the annuity,
p denotes the amount of the periodic payment,
i denotes the interest rate per period, and
n denotes the number of payments.

Again the expression in brackets in formula (4.9) has a special symbol because of the frequency with which it occurs in financial analysis; it is denoted $a_{\overline{n}|i}$, which is read "a angle n at i." Thus $a_{\overline{n}|i}$ denotes the present value of an annuity consisting of n payments of $1 with an interest rate of i per period. Note that with this notation (4.9) can be written as

$$P = pa_{\overline{n}|i}.$$

EXAMPLE 4.16 A teenager's grandparents have decided to give her $5000 per year for the next four years to pay for her college education. How much must they deposit in an account paying 8% interest compounded annually in order to begin making these $5000 payments next year?

Solution This question asks for the present value of 4 yearly payments of \$5000. Using (4.9) with $p = \$5000$, $i = 0.08$, and $n = 4$, we see that

$$P = p\left[\frac{1 - (1 + i)^{-n}}{i}\right]$$

$$= \$5000\left[\frac{1 - (1.08)^{-4}}{0.08}\right]$$

$$\approx \$16{,}560.63.$$

Therefore, by depositing this amount, the grandparents will be able to withdraw \$5000 each year for the next 4 years and will leave no money in the account after the last withdrawal. ■

EXAMPLE 4.17 If money can be invested at the rate of 6% compounded quarterly, which is worth more: a gift of \$32,000 now, or an installment gift of \$1000 every three months for ten years?

Solution The installment gift is an annuity with $4(10) = 40$ payments of \$1000. By (4.9) its present value is

$$P = p\left[\frac{1 - (1 + i)^{-n}}{i}\right]$$

$$= \$1000\left[\frac{1 - (1.015)^{-40}}{0.015}\right]$$

$$\approx \$29{,}915.85.$$

Since the present value of the installment is approximately \$29,915.85, the outright gift of \$32,000 is worth more. ■

EXAMPLE 4.18 The Gilmore family is considering the purchase of their first home. A conventional 30-year mortgage is available at a rate of 10.8% compounded monthly. After considering their financial status, the loan officer at their bank has advised the Gilmores that they can afford a down payment of \$10,000 and mortgage payments of \$1000 per month. What is the most expensive house they can afford?

Solution The most expensive house that the Gilmores can afford will cost \$10,000 (their down payment) plus the present value of their mortgage payments. Using (4.9), we see that this present value is

$$P = p\left[\frac{1 - (1 + i)^{-n}}{i}\right]$$

$$= \$1000\left[\frac{1 - (1.009)^{-360}}{0.009}\right]$$

$$\approx \$106{,}696.04.$$

Hence the most expensive house that the Gilmores can afford costs approximately \$116,700. ■

Practice Problem 12 Instant winners of the Illinois State Lottery receive \$1 million in 20 yearly payments of \$50,000. If money can be invested at 6% interest compounded annually, how much must be deposited in order for these payments to be made?

EXERCISES 4.3

In Exercises 1–8 find a formula for the general term of the given difference equation with initial value x_0. Then evaluate x_{10}.

1. $x_{n+1} = 3x_n + 4, \quad x_0 = 5$

2. $x_{n+1} = \frac{1}{2}x_n + 1, \quad x_0 = 34$

3. $x_{n+1} = -0.6x_n + 24, \quad x_0 = 10$

4. $x_{n+1} = 2x_n - 3, \quad x_0 = 8$

5. $x_{n+1} = x_n - 5, \quad x_0 = 6$

6. $x_{n+1} = -\frac{2}{3}x_n + 5, \quad x_0 = 4$

7. $x_{n+1} = 1.01x_n + 25, \quad x_0 = 100$

8. $x_{n+1} = 1.2x_n - 2.4, \quad x_0 = 12$

In Exercises 9–14 compute the future value of an annuity after n payments if p is the amount of the periodic payment and i is the interest rate per period.

9. $n = 12, \quad p = \$200, \quad i = 2\%$

10. $n = 18, \quad p = \$500, \quad i = 3\%$

11. $n = 36, \quad p = \$100, \quad i = 0.5\%$

12. $n = 25, \quad p = \$1000, \quad i = 7\%$

13. $n = 24, \quad p = \$600, \quad i = 1.2\%$

14. $n = 30, \quad p = \$400, \quad i = 9\%$

In Exercises 15–20 compute the present value of an annuity after n payments if p is the amount of the periodic payment and i is the interest rate per period.

15. $n = 48, \quad p = \$200, \quad i = 0.5\%$

16. $n = 10, \quad p = \$12{,}000, \quad i = 7\%$

17. $n = 20, \quad p = \$3000, \quad i = 2\%$

18. $n = 24, \quad p = \$500, \quad i = 1.1\%$

19. $n = 18, \quad p = \$25, \quad i = 1.5\%$

20. $n = 16, \quad p = \$80, \quad i = 0.8\%$

21. Suppose that \$500 is deposited initially in an account that earns 8% interest compounded quarterly, and at the end of each quarter an additional \$50 is deposited into the account. Let B_n denote the balance of the account after n quarters.
(a) Write a difference equation that expresses B_{n+1} in terms of B_n.
(b) Find a formula for the general term of the difference equation in part (a).
(c) How much will this account be worth at the end of five years?

22. A university has raised \$1.5 million for a scholarship endowment fund. This money is invested at 10% interest compounded annually, and at the end of each year \$120,000 is withdrawn from the fund for scholarships. Let B_n denote the balance in the fund after n years.

(a) Write a difference equation that expresses B_{n+1} in terms of B_n.
(b) Find a formula for the general term of the difference equation in part (a).
(c) How much will the fund contain at the end of ten years?

23. The Small Business Administration offers business loans at 9% interest compounded monthly for 7 years. If the owner of a restaurant can afford monthly payments of \$800, what is the maximum amount the owner can borrow?

24. If monthly deposits of \$75 are made into an account paying 9% interest compounded monthly, how much will the account contain after ten years?

25. A life insurance policy calls for quarterly payments of \$110 to be made for 25 years. When the last payment is made, the insured may elect to receive as a lump sum payment an amount equal to all the payments plus 6% interest compounded quarterly. How much will this lump sum payment be?

26. A medical institute is attempting to raise funds to support a \$25,000 research fellowship, to be awarded every year for the next twenty years. If the institute can invest its money at 11% interest compounded annually, how much money must be raised to support this project?

27. In his will, a rich uncle stipulated that his niece should receive \$10,000 per year for 25 years, with the first payment made one year after his death. If money can be invested at 9% interest compounded annually, what is the equivalent cash value of this inheritance?

28. If monthly deposits of \$25 are made into a Christmas Club account paying 6% interest compounded monthly, how much money will be available after the twelfth deposit?

29. Beginning on his thirtieth birthday, an accountant deposited \$2000 per year into an individual retirement account (IRA). If this money earns 8.5% interest compounded annually, how much will the account be worth on his sixtieth birthday?

30. A publishing house sweepstakes winner will receive \$1000 per month for the next ten years. How much must the company deposit at 9% interest compounded monthly in order to be able to meet these payments if the first payment will be made next month?

31. A newspaper ad states that a new car can be purchased for a down payment of \$2000.00 and monthly payments of \$189.23. If interest is charged at the rate of 3.9% compounded monthly for 60 months, how much does the car cost?

32. The owner of oriental rugs worth \$7500 pays \$85 per quarter for insurance. Show that if these premiums were invested instead in an account paying 10% interest compounded quarterly, then after twelve years the account would contain enough money to cover the loss of the rugs.

33. A company must decide whether to replace a piece of machinery or to keep it for another three years. The cost of leasing replacement machinery for three years will be \$2800, and there will be no additional expenses during this period because the machinery will be serviced by the lessor. On the other hand, if the machinery is not replaced, the company can expect to make semiannual repairs costing \$500 each. If money can be deposited at 6% interest compounded semiannually, which option is cheaper for the company?

34. The owner of an insurance policy has the option of collecting $25,000 now or receiving yearly payments of $5000 for the next six years. If money can be invested at 6% interest compounded annually, which option is worth more?

35. Karen has $3000 in a savings account paying 5% interest compounded quarterly. Beginning with the next interest period, she intends to make quarterly deposits of $600 into this account. How much money will be in this account after four years?

36. An insurance company would like to establish a permanent scholarship for an actuarial science major. If money can be invested at an annual rate of 10%, how much money must be deposited in order to provide $2000 to each year's recipient?

37. Theorem 4.3 gives the general term of a sequence satisfying the difference equation $x_{n+1} = ax_n + b$ when $a \neq 1$. What is the general term of such a sequence if $a = 1$?

38. Suppose the sequence $x_0, x_1, x_2, \ldots$ satisfies the difference equation $x_{n+1} = ax_n + b$, where $a \neq 1$. Let $c = b/(1 - a)$.

(a) Show that if $x_0 = c$, then x_n is a constant.
(b) Show that if $|a| < 1$, then x_n approaches c as n increases.

39. As a birthday present for her grandson's twelfth birthday, a wealthy widow has decided to establish a trust fund to pay for his college education. She would like the trust to provide $8000 payable on each of his 18th, 19th, 20th, and 21st birthdays. If the trust's money can be invested at the rate of 8% compounded annually, how much money must she place in the trust on his twelfth birthday in order for it to provide the four $8000 payments?

40. Evaluate $1/a_{\overline{n}|i} - 1/s_{\overline{n}|i}$.

Answers to Practice Problems

10. $x_n = 6 \cdot 3^n + 2$

11. $10,465.50

12. $573,496.06

4.4 Sinking Funds and Amortization

In Section 4.3 we learned how to compute the future and present value of an annuity using formulas (4.8) and (4.9). These formulas can also be used when we know the total amount of money for the annuity and must determine the size of the periodic payment.

SINKING FUNDS

Businesses and governments are frequently faced with the problem of accumulating capital to meet future expenses. Consider, for instance, a local government that issues bonds maturing in 20 years to fund a $10 million courthouse and jail. The revenue from the sale of the bonds is immediately available to pay for the construction project, and the government will have 20 years to save enough money to pay off the bonds when they mature. A fund of money that is being accumulated to meet some future expense (such as the redeeming of the bonds) is called a **sinking fund.** Often sinking funds are established in the form of annuities. For instance, the local government in our example may decide that it would like to make 20 equal deposits into a sinking fund, one per year, to accumulate enough capital to redeem its bonds. It needs to know the amount of the yearly deposit that will raise the needed $10 million after the last payment is made.

This type of calculation requires solving the future value of an annuity formula (4.8) for the periodic payment p. Performing this calculation produces the following formula.

Sinking Fund Payment

$$p = \frac{iF}{(1+i)^n - 1} \tag{4.10}$$

where p denotes the amount of the periodic payment,
F denotes the future amount to be accumulated,
i denotes the interest rate per period, and
n denotes the number of payments.

This formula can also be written as

$$p = \frac{F}{s_{\overline{n}|i}}.$$

EXAMPLE 4.19 A local government would like to raise $10 million in 20 years by making equal yearly deposits into a sinking fund. If interest is earned at the rate of 8% compounded annually, how much must the yearly deposit be in order to have the needed $10 million after the last payment is made?

Solution The size of the yearly payment can be found by using the sinking fund payment formula with $F = \$10{,}000{,}000$, $i = 0.08$, and $n = 20$.

$$\begin{aligned} p &= \frac{iF}{(1+i)^n - 1} \\ &= \frac{0.08(\$10{,}000{,}000)}{(1.08)^{20} - 1} \\ &\approx \$218{,}522.09. \end{aligned}$$

Hence yearly payments of $218,522.09 will grow to $10 million in 20 years. ■

EXAMPLE 4.20 Under a payroll savings plan, an employee can set aside a part of her monthly salary. If these savings earn 8.4% interest compounded monthly, how much must be set aside per month in order to accumulate $12,000 in two years?

Solution Using the sinking fund payment formula again, we see that the monthly amount to be set aside is

$$p = \frac{iF}{(1 + i)^n - 1} = \frac{0.007(\$12{,}000)}{(1.007)^{24} - 1} \approx \$460.92.$$

Practice Problem 13 An airline estimates that its present fleet of airplanes will have to be replaced in eight years at a cost of $120 million. In order to have this money available, the airline intends to make quarterly payments into an account earning 9% interest compounded quarterly. What should the quarterly payments be in order to pay for the replacement of the planes at the end of eight years?

$120 million needed in 8 years, 9% interest rate, compounded quarterly

AMORTIZATION

Home mortgages and many other loans are repaid in installments. That is, the borrower makes regular payments of a fixed amount, part of which repays the amount borrowed and the remainder of which is interest. This method of repaying a loan is called **amortization.** As an illustration of amortization, suppose that a consumer borrows $4500 for a home improvement loan at an interest rate of 12% compounded monthly. The loan is to be repaid by twelve monthly payments of $400. The following table, called an **amortization schedule,** shows the interest charge and balance of the loan for each month that the money is borrowed.

Month	*Old Balance*	*Interest*	*Payment*	*New Balance*
1	$4500.00	$45.00	$400	$4145.00
2	$4145.00	$41.45	$400	$3786.45
3	$3786.45	$37.86	$400	$3424.31
4	$3424.31	$34.24	$400	$3058.55
5	$3058.55	$30.59	$400	$2689.14
6	$2689.14	$26.89	$400	$2316.03
7	$2316.03	$23.16	$400	$1939.19
8	$1939.19	$19.39	$400	$1558.58
9	$1558.58	$15.59	$400	$1174.17
10	$1174.17	$11.74	$400	$785.91
11	$785.91	$7.86	$400	$393.77
12	$393.77	$3.94	$397.71	$0.00

Each new balance is computed by adding the current interest charge (1% of the old balance) to the old balance and subtracting the current payment. (The monthly payments are always \$400 except for the final payment, which is slightly less.) Note that in later months the interest charge decreases, so that a larger portion of each monthly payment is used to repay the amount borrowed. The total interest paid on this loan is the sum of the twelve monthly interest charges, which is \$297.71. An easier way to compute the total interest paid is by means of the relation

$$\text{total interest paid} = \text{sum of payments} - \text{amount borrowed}.$$

In this case the sum of the payments is

$$11(\$400.00) + \$397.71 = \$4797.71;$$

so the total interest paid is

$$\$4797.71 - \$4500.00 = \$297.71.$$

How was the size of the monthly payment determined so that it repays not only the amount borrowed but also the accumulated interest? Since the twelve monthly payments constitute an annuity which has a present value of \$4500 (the amount borrowed), the size of the payment can be found by applying the present value of an annuity formula (4.9). Specifically, we must solve (4.9) for the amount of the periodic payment p. Thus we obtain the following formula.

Amortization

$$p = \frac{iA}{1 - (1 + i)^{-n}} \tag{4.11}$$

where p denotes the amount of the periodic payment,
A denotes the amount borrowed,
i denotes the interest rate per period, and
n denotes the number of payments.

This formula can also be written as

$$p = \frac{A}{a_{\overline{n}|i}}.$$

Often students experience difficulty in determining whether the sinking fund payment formula or the amortization formula should be used. To distinguish between these two, note whether the total amount of money in the problem is a future amount or a present amount. *If it is a future amount, then the sinking fund payment formula should be used, but if it is a present amount, then the amortization formula should be used.*

EXAMPLE 4.21 A home buyer is purchasing a $120,000 house. The down payment will be $20,000, and the remainder will be financed by a 30-year mortgage at a rate of 9.6% interest compounded monthly. What will the monthly payment be?

Solution The buyer will be borrowing $120,000 − $20,000 = $100,000. Using the amortization formula with $A = \$100,000$, $i = 0.008$, and $n = 30(12) = 360$, we see that

$$\begin{aligned} p &= \frac{iA}{1 - (1 + i)^{-n}} \\ &= \frac{0.008(\$100,000)}{1 - (1.008)^{-360}} \\ &\approx \$848.16. \end{aligned}$$

Hence the monthly payment will be approximately $848.16. ■

EXAMPLE 4.22 Because of special incentives in 1987 to attract buyers, it was possible to finance a car at 3.9% interest compounded monthly. If the buyer of a car costing $13,300 made a $1000 down payment and financed the rest at this rate over 48 months, how much is the monthly payment? What is the total amount of interest that the buyer will pay?

Solution By using the amortization formula with $A = \$12,300$, $i = 0.00325$, and $n = 48$, we see that the monthly payment is

$$\begin{aligned} P &= \frac{iA}{1 - (1 + i)^{-n}} \\ &= \frac{0.00325(\$12,300)}{1 - (1.00325)^{-48}} \\ &\approx \$277.17. \end{aligned}$$

Therefore

$$\begin{aligned} \text{total interest paid} &= \text{sum of payments} - \text{amount borrowed} \\ &= 48(\$277.17) - \$12,300 \\ &= \$1004.16. \end{aligned}$$ ■

EXAMPLE 4.21 Revisited In recent years home buyers are increasingly choosing 15-year mortgages instead of the more traditional 25-year and 30-year mortgages. For the home buyer in Example 4.21, compare the monthly payments and the total amounts of interest paid if a 15-year mortgage is chosen instead of a 30-year mortgage.

Solution For a 15-year mortgage, the amortization formula gives

$$\begin{aligned} p &= \frac{iA}{1 - (1 + i)^{-n}} \\ &= \frac{0.008(\$100{,}000)}{1 - (1.008)^{-180}} \\ &\approx \$1050.27. \end{aligned}$$

Hence the monthly payment for a 15-year mortgage will be approximately \$1050.27 instead of the \$848.16 monthly payment for a 30-year mortgage.

The total interest charge for the 30-year mortgage is

$$\begin{aligned} \text{total interest paid} &= \text{sum of payments} - \text{amount borrowed} \\ &= 360(\$848.16) - \$100{,}000 \\ &= \$205{,}337.60. \end{aligned}$$

On the other hand, for the 15-year mortgage the total interest charge is

$$\begin{aligned} \text{total interest paid} &= \text{sum of payments} - \text{amount borrowed} \\ &= 180(\$1050.27) - \$100{,}000 \\ &= \$89{,}048.60. \end{aligned}$$

Thus \$116,289 in additional interest is paid with the 30-year mortgage. It is easy to see why the 15-year mortgages are becoming popular! ■

Practice Problem 14 A Pontiac Fiero is advertised with a sale price of \$9946. A prospective buyer anticipates that his present car will bring \$1500 as a trade-in. If the Fiero can be financed for 60 months at 9.9% interest compounded monthly, how large will the prospective buyer's monthly payment be?

EXERCISES 4.4

In Exercises 1–6 compute the value of the sinking fund payment needed to accumulate the amount F in n payments if interest is paid at the rate of i per period.

1. $F = \$5000,\ n = 16,\ i = 1.5\%$

2. $F = \$10{,}000,\ n = 10,\ i = 3\%$

3. $F = \$2000,\ n = 8,\ i = 2\%$

4. $F = \$8000,\ n = 36,\ i = 0.5\%$

5. $F = \$180{,}000,\ n = 24,\ i = 0.8\%$

6. $F = \$16{,}000,\ n = 60,\ i = 0.7\%$

In Exercises 7–12 compute the value of the payment necessary to amortize the amount A in n payments if the interest rate is i per period.

7. $A = \$6000,\ n = 36,\ i = 1\%$

8. $A = \$2000,\ n = 8,\ i = 2\%$

9. $A = \$10{,}000,\ n = 60,\ i = 0.75\%$

10. $A = \$500,\ n = 12,\ i = 1.5\%$

11. $A = \$80{,}000,\ n = 180,\ i = 1.1\%$

12. $A = \$90{,}000,\ n = 300,\ i = 0.9\%$

13. An executive intends to retire in twenty years. At the time of her retirement she would like to have a retirement account of $1 million. In order to accumulate this amount, she intends to make yearly deposits into an account paying 8.25% interest compounded annually. How large should her yearly deposits be?

14. Work question 13 if the executive intends to retire in 25 years.

15. A home buyer borrows $90,000 for 25 years at 12% interest compounded monthly.
(a) How large will the monthly payment be?
(b) How much interest will be paid during the course of this mortgage?

16. Work question 15 if the mortgage lasts for 15 years instead of 25 years.

17. The Small Business Administration offers business loans at 14% interest compounded monthly. If the owner of a hardware store wants to borrow $400,000 for five years, how large will the monthly payments be to repay the loan?

18. What amount of money should be deposited every six months into an account paying 8% interest compounded semiannually in order to accumulate $5000 in three years?

19. A university has issued bonds worth $12.5 million in order to pay for the construction of a new arena. In order to have this money available, it will make annual deposits into an account paying 7.5% interest compounded annually. How large must its payments be in order to redeem the bonds at the end of 20 years?

20. A couple puts 20% down on furniture costing $1800. The balance is to be financed in equal payments over 24 months. If interest is charged at the rate of 18% compounded monthly, what will the monthly payment be?

21. In order to purchase a new car, Ms. Hawkins borrowed $9800 and agreed to repay the loan with 48 equal monthly payments. If interest is charged at the rate of 10.8% compounded monthly, how much will her payments be?

22. A vacationer traveling to Hawaii paid for an $800 airline ticket with a credit card charging monthly interest of 1.5%. If the vacationer decides to repay the loan in six equal monthly payments, how large should each payment be?

23. In ten years a municipality must retire an airport bond issue in the amount of $1,250,000. To meet this obligation it intends to make semiannual payments into an account paying 7% interest compounded semiannually. How large should the payments be in order to retire the debt?

24. A homeowner has decided to set aside a certain amount each month to meet his annual $1300 real estate tax bill. If money can be invested at 6% compounded monthly, how much should be set aside each month?

25. Mr. and Mrs. Washington are interested in building their dream house, which will cost $160,000. They intend to use the $40,000 equity in their present house as a down payment on the new one and will finance the rest with a 30-year mortgage. If the present interest rate is 11.4% compounded monthly, how large will their monthly payments be on the new house?

26. To prepare for their daughter's college education, a couple has decided to deposit a fixed amount each year on her birthday into an account earning interest at the rate of 8% compounded quarterly. If these deposits start on her first birthday and continue until their child turns eighteen, how much should be deposited in order to accumulate $20,000?

27. In three years a small company would like to purchase a minicomputer costing \$80,000. To raise this money, the company has decided to make quarterly payments into an account paying 8% interest compounded quarterly. How much should each payment be in order to raise the amount needed?

28. Mr. Gonzalez intends to buy a Buick Skyhawk for \$10,219. His present car is worth \$2200 as a trade-in, and he intends to borrow the rest at 10.2% interest compounded monthly. If he repays this loan with sixty equal monthly payments, how much interest will he pay?

29. Alice charged stereo equipment costing \$1000 to her credit card, which charges interest at the rate of 1.5% per month. Prepare an amortization schedule that will repay this loan in 6 months.

30. A \$6000 automobile loan was taken at an interest rate of 3.9% compounded monthly. Create an amortization schedule to repay this loan in 12 months.

31. An executive intends to retire in twenty years. During his retirement, he would like to receive a monthly income of \$5000 for fifteen years. In order to make this possible, he intends to deposit equal amounts each month until his retirement into an account paying 7.2% interest compounded monthly. How large should his monthly deposit be? (*Hint:* First determine the lump sum of money he must have at the time of his retirement in order to receive \$5000 per month for fifteen years.)

32. A government-subsidized student loan program provides a college student \$2000 per year for four years. One year after receiving the last \$2000 check, the student must begin to repay the loan by making the first of five equal yearly payments. If interest is charged at the rate of 5% compounded annually, how large must the monthly payments be to repay this loan? What total interest will the student pay under this program?

33. On his twelfth birthday, a boy received \$5000 from his grandmother to be used for his college education. The money will be paid in four equal installments on his 18th, 19th, 20th and 21st birthdays. If the money is invested at the rate of 8% compounded annually, how large will each installment be?

34. A professional athlete estimates that his playing career will last another five years. For each of these five years he has decided to put some of his salary into a retirement fund earning 10% interest compounded annually. When he retires, he wants to be able to draw \$30,000 per year from this fund for 35 years. How much should each of his five payments be?

35. Suppose that a loan of L dollars is to be repaid by a sequence of n payments of p dollars each with an interest rate of i per period.

(a) Write a difference equation describing the balance due after $k + 1$ payments.

(b) Find a formula for the balance due after the kth payment.

Answers to Practice Problems **13.** \$2,600,897.92

14. \$179.04

CHAPTER 4 REVIEW

IMPORTANT TERMS

- **arithmetic progression** *(4.1)*
- **balance of a loan**
- **difference equation**
- **geometric progression**
- **initial value of a sequence**
- **interest**
- **interest rate**
- **principal**
- **sequence**
- **simple interest**
- **terms of a sequence**
- **compound interest** *(4.2)*
- **effective rate of interest**
- **nominal interest rate**
- **present value of an amount**
- **annuity** *(4.3)*
- **future value of an annuity**
- **present value of an annuity**
- **amortization** *(4.4)*
- **amortization schedule**
- **sinking fund**

IMPORTANT FORMULAS

Simple Interest Charge

$$I = Prt \qquad (4.2)$$

where I denotes the interest charge,
P denotes the principal,
r denotes the annual interest rate, and
t denotes the time in years for which the money is borrowed.

Simple Interest Amount

$$A = P(1 + rt) \qquad (4.3)$$

where A denotes the amount after t years,
P denotes the principal,
r denotes the annual interest rate, and
t denotes the time in years for which the money is borrowed.

Compound Interest Amount

$$A = P(1 + i)^n \qquad (4.4)$$

where A denotes the amount after n periods,
P denotes the principal,
i denotes the interest rate per period, and
n denotes the number of interest periods for which the money is borrowed.

Present Value of a Future Amount

for simple interest $$P = \frac{A}{1 + rt} \qquad (4.5)$$

for compound interest $$P = \frac{A}{1 + i^n} \qquad (4.6)$$

where P denotes the present value,
A denotes the future amount,
r denotes the annual interest rate,
i denotes the interest rate per period, and
n denotes the number of interest periods for which the money is borrowed.

Future Value of an Annuity

$$F = p\left[\frac{(1 + i)^n - 1}{i}\right] \tag{4.8}$$

where F denotes the future value after n payments,
p denotes the amount of the periodic payment,
i denotes the interest rate per period, and
n denotes the number of payments.

Present Value of an Annuity

$$P = p\left[\frac{1 - (1 + i)^{-n}}{i}\right] \tag{4.9}$$

where P denotes the present value of the annuity,
p denotes the amount of the periodic payment,
i denotes the interest rate per period, and
n denotes the number of payments.

Sinking Fund Payment

$$p = \frac{iF}{(1 + i)^n - 1} \tag{4.10}$$

where p denotes the amount of the periodic payment,
F denotes the future amount to be accumulated,
i denotes the interest rate per period, and
n denotes the number of payments.

Amortization

$$p = \frac{iA}{1 - (1 + i)^{-n}} \tag{4.11}$$

where p denotes the amount of the periodic payment,
A denotes the amount borrowed,
i denotes the interest rate per period, and
n denotes the number of payments.

REVIEW EXERCISES

In Exercises 1–8 find a formula for the general term of the given difference equation with initial value x_0. Then evaluate x_8.

1. $x_{n+1} = x_n + 2, \quad x_0 = -1$

2. $x_{n+1} = 0.8x_n, \quad x_0 = 125$

3. $x_{n+1} = 4x_n + 9, \quad x_0 = 0$

4. $x_{n+1} = \frac{1}{2}x_n + 1, \quad x_0 = 256.$

5. $x_{n+1} = -2x_n, \quad x_0 = 3$

6. $x_{n+1} = 1.02x_n - 50, \quad x_0 = 1000$

7. $x_{n+1} = -\frac{2}{3}x_n + 5/3, \quad x_0 = 8$

8. $x_{n+1} = x_n + 0.75, \quad x_0 = 0.5$

9. A company borrowed \$75,000 for six months at 16% simple interest.

(a) How much must be repaid when the loan is due?
(b) How much of the amount in part (a) is interest?

10. Suppose that $6810 is required to repay a loan of $6000. If the money is borrowed for nine months, what simple interest rate is being charged?

11. Suppose that $2500 is deposited for two years in an account paying interest at the nominal annual rate of 9%. Compute the balance of the account if interest is compounded (a) annually, (b) quarterly, and (c) monthly.

12. What is the effective rate of interest corresponding to 15% interest compounded monthly?

13. How much invested now will grow to $30,000 in seven years at 12% interest compounded quarterly?

14. Suppose that you hold a note that promises to pay $15,000 in three years. If money can be invested at 7.2% interest compounded monthly, what is the present value of this note?

15. Valerie has just received an inheritance of $25,000 from her aunt. She intends to deposit the money in an account that earns 8.4% interest compounded monthly. At the end of each month she will withdraw $700 to cover her living expenses at college.

(a) Write a difference equation involving B_n, the balance of the account after n months.
(b) Find a formula for the general term of the difference equation in part (a).
(c) How much money will the account contain when Valerie graduates in three years?

16. A buyer makes a $1500 down payment on a car costing $7200. The balance will be repaid in equal monthly installments over three years. If the interest rate is 10.2% compounded monthly, what is the monthly payment?

17. A savings and loan offers 30-year mortgages at 11.1% interest compounded monthly. If a home buyer can afford a down payment of $20,000 and monthly payments of $625, what is the most expensive house he can buy?

18. A manufacturer estimates that his plant's present machinery will need to be replaced in ten years at a cost of $750,000. In order to have this amount of money available, he has decided to make quarterly deposits into an account earning 8% interest compounded quarterly. How much must be deposited each quarter?

19. A power company spends $1.5 million per month for fuel. If it can invest money at 7.5% interest compounded monthly, how much money does the company need now in order to pay its fuel costs for the next twelve months?

20. Mr. and Mrs. Cohen intend to purchase a lot costing $33,800 in a new subdivision. They have arranged to borrow this amount at 10.5% interest compounded monthly. If they repay this loan with equal monthly payments over an eight-year period, how much interest will they pay on this loan?

21. Mr. Snedeker opened an individual retirement account on his thirtieth birthday and intends to continue depositing $2000 on every birthday until he becomes 65. If his IRA earns 8.5% interest compounded annually, how much will Mr. Snedeker have available after his last deposit?

22. After Mr. Snedeker in question 21 retires, what amount can he withdraw from his IRA each year if he wants to remove all the money from this account in twenty equal withdrawals?

23. Mr. Armstrong pays $180 per month on his car loan, which is amortized over 48 months at 9.6% interest compounded monthly. How much interest will Mr. Armstrong pay during the course of this loan?

24. In question 23, what will the unpaid balance of Mr. Armstrong's car loan be after three years?

5

SETS AND COUNTING TECHNIQUES

Counting is one of the most basic of all mathematical skills, one which most persons learn before beginning their formal schooling. This type of counting involves the enumeration of a specific collection of displayed objects. Other counting problems are more difficult because we must determine the number of objects of a certain type without having them exhibited before us. In this chapter we will develop some sophisticated methods for performing the latter type of counting. These techniques will be quite useful in our study of probability (Chapter 6). We will begin by introducing the concept of a set, which will also play an important role in the next chapter.

5.1 Sets and Venn Diagrams

We will regard a **set** as a collection of objects called its **elements** or **members.** If an object x is an element of a set S, we write $x \in S$; otherwise we write $x \notin S$. A set can be specified by listing its elements between braces. For example, the set S having as its elements the letters a, e, i, o, and u can be written as

$$S = \{a, e, i, o, u\}.$$

Thus $a \in S$ and $i \in S$ but $c \notin S$.

Another way to describe a set is by characterizing its elements in terms of some property that they possess. For example, the set

$$A = \{1, 3, 5, 7, 9\}$$

consists of all the odd positive integers less than 10. This set can be written as

$$A = \{x | x \text{ is an odd positive integer less than } 10\},$$

which is read "the set of all elements x such that x is an odd positive integer less than 10." Here the symbol x denotes a typical element of the set and can be replaced by any other letter without changing the set being described.

We call sets A and B **equal** when A and B contain precisely the same elements. If A and B are equal, we write $A = B$; otherwise we write $A \neq B$.

EXAMPLE 5.1 The set

$$B = \{n | n \text{ is an even positive integer less than } 9\}$$

consists of the four elements 2, 4, 6, and 8. Thus B can also be written as

$$B = \{2, 4, 6, 8\}. \blacksquare$$

EXAMPLE 5.2 The set

$$\{\text{Alaska, California, Hawaii, Oregon, Washington}\}$$

is equal to the set

$$\{s|s \text{ is a state bordering the Pacific Ocean}\}. \quad ■$$

EXAMPLE 5.3 The set

$$\{y|y \text{ is a positive integer less than or equal to } 100\}$$

can be written

$$\{1, 2, 3, \ldots, 100\},$$

where the three dots indicate that the pattern established by the elements 1, 2, and 3 continues through the element 100. ■

A set which contains all the objects under consideration is called a **universal set** and will usually be denoted U. Since there can be many possible universal sets for a particular problem, we must always specify the particular one that we have in mind. The set that contains no elements is called the **empty set** and is denoted $\varnothing$.

EXAMPLE 5.4 If the universal set U is the set of all real numbers, determine the elements that belong to the following sets:

(a) $\{x|x^2 + x - 2 = 0\}$ (b) $\{x|x^2 - 2x + 1 \geq 0\}$ (c) $\{x|x^2 + 1 = 0\}$.

Solution (a) The given set consists of all real numbers x such that $x^2 + x - 2 = 0$. Since $x^2 + x - 2 = (x - 1)(x + 2)$, we see that this equation has solutions $x = 1$ and $x = -2$. Hence

$$\{x|x^2 + x - 2 = 0\} = \{1, -2\}.$$

(b) Because $x^2 - 2x + 1 = (x - 1)^2$ and the square of every real number is nonnegative, *every* real number satisfies the condition $x^2 - 2x + 1 \geq 0$. Thus

$$\{x|x^2 - 2x + 1 \geq 0\} = U.$$

(c) If $x^2 + 1 = 0$, then $x^2 = -1$. But since the square of every real number is nonnegative, no real number satisfies the equation $x^2 + 1 = 0$. Therefore

$$\{x|x^2 + 1 = 0\} = \varnothing. \quad ■$$

Practice Problem 1 If the universal set U is the set of all real numbers, determine the elements that belong to the following sets:

(a) $\{x|x^2 - 2x + 1 < 0\}$ (b) $\{x|x^2 = 1\}$ (c) $\{x|(x - 2)^2 \geq 0\}$.

If every element of set A is contained in set B, then we say that A is a **subset** of B or that A is **included** in B. In this case we write $A \subseteq B$. For instance, if

$$A = \{2, 3, 5\} \qquad \text{and} \qquad B = \{1, 2, 3, 4, 5\},$$

then A is a subset of B because each element of A is also an element of B. But B is not a subset of A since there are elements of B that are not elements of A (namely, 1 and 4).

EXAMPLE 5.5 List all the subsets of the set $S = \{a, b, c\}$.

Solution Since S contains three elements, a subset of S must contain 0, 1, 2, or 3 elements. Thus there are eight subsets of S:

$$\varnothing, \{a\}, \{b\}, \{c\}, \{a, b\}, \{a, c\}, \{b, c\}, \{a, b, c\}. \quad ■$$

Notice that in Example 5.5 the empty set $\varnothing$ and S itself are two of the subsets of the given set S. In fact, for any set A we have $\varnothing \subseteq A$ and $A \subseteq A$.

SET OPERATIONS

Two sets A and B can be combined in various ways to form new sets. The **union** of A and B, denoted $A \cup B$, is the set consisting of all the elements that belong to either A *or* B or both. The **intersection** of A and B, denoted $A \cap B$, is the set consisting of all the elements that belong to both A *and* B. And the **complement** of A is the set denoted $\overline{A}$ that consists of all the elements in the universal set that do *not* belong to A. Symbolically, we have

union	$A \cup B = \{x \mid x \in \mathrm{A} \text{ or } x \in \mathrm{B}\}$,	
intersection	$A \cap B = \{x \mid x \in \mathrm{A} \text{ and } x \in \mathrm{B}\}$,	and
complement	$\overline{A} = \{x \mid x \in \mathrm{A}\}$.	

If A and B have no elements in common, that is, if $A \cap B = \varnothing$, then A and B are called **disjoint.**

For three or more sets the union and intersection are defined analogously. Thus $A \cup B \cup C$ is the set consisting of the elements that are in at least one of the sets A, B, or C, and $A \cap B \cap C \cap D$ is the set consisting of the elements that are in all four of the sets A, B, C, and D.

EXAMPLE 5.6 Let $A = \{1, 2, 3, 4\}$, $B = \{2, 4, 6\}$, and $C = \{1, 3, 5, 7\}$. If the universal set is

$$U = \{1, 2, 3, 4, 5, 6, 7, 8, 9, 10\},$$

find $A \cup B$, $A \cap B$, $\overline{A}$, $A \cup C$, $B \cap C$, $\overline{(A \cap C)}$, and $A \cup B \cup C$.

Solution We see from the preceding definitions that

$$\begin{aligned} A \cup B &= \{1, 2, 3, 4, 6\}, \\ A \cap B &= \{2, 4\}, \\ \overline{A} &= \{5, 6, 7, 8, 9, 10\}, \\ A \cup C &= \{1, 2, 3, 4, 5, 7\} \\ B \cap C &= \varnothing, \\ \overline{(A \cap C)} &= \overline{\{1, 3\}} = \{2, 4, 5, 6, 7, 8, 9, 10\}, \text{ and} \\ A \cup B \cup C &= \{1, 2, 3, 4, 5, 6, 7\}. \end{aligned}$$

Since $B \cap C = \varnothing$, the sets B and C are disjoint. ■

Practice Problem 2 Let $A = \{1, 2, 8\}$, $B = \{2, 3, 4, 6\}$, and $C = \{3, 4, 6, 7\}$. If the universal set is

$$U = \{1, 2, 3, 4, 5, 6, 7, 8\},$$

find $A \cup B$, $A \cap B$, $\overline{B}$, $A \cap C$, $B \cup C$, $\overline{(B \cup C)}$, and $A \cup B \cup C$.

The three set operations defined above can be illustrated by use of **Venn diagrams,** which are named for the English logician John Venn (1834–1923). In a Venn diagram the universal set is depicted by a rectangle, and the sets under consideration are represented by regions within this rectangle. Figure 5.1(a) shows two overlapping circles that represent subsets A and B of the

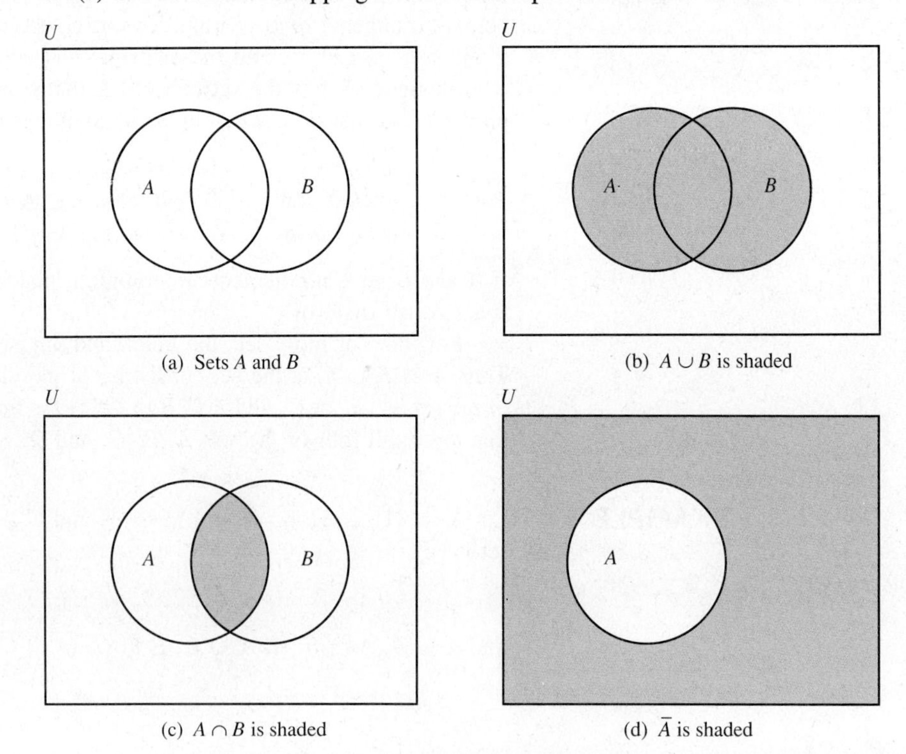

(a) Sets A and B

(b) $A \cup B$ is shaded

(c) $A \cap B$ is shaded

(d) $\overline{A}$ is shaded

FIGURE 5.1

universal set U. The union, intersection, and complement operations can be illustrated by shading the appropriate regions as in Figures 5.1(b), (c), and (d).

Venn diagrams are helpful in visualizing the logical combinations of several properties. Consider, for instance, the sets A of all adults and M of all males if the universal set is the set of all living people. The sets A and M divide U into four disjoint regions, which are numbered I, II, III, and IV in Figure 5.2.

Each of the four numbered regions can be described as an intersection of A or $\overline{A}$ with M or $\overline{M}$:

Region I, which is $A \cap \overline{M}$, contains all women (that is, adult nonmales).

Region II, which is $A \cap M$, contains all men (that is, adult males).

Region III, which is $\overline{A} \cap M$, contains all boys (that is, nonadult males).

Region IV, which is $\overline{A} \cap \overline{M}$, contains all girls (that is, nonadult nonmales).

If we start with three sets instead of only two, the corresponding Venn diagram will divide the universal set into eight disjoint regions as in the example below.

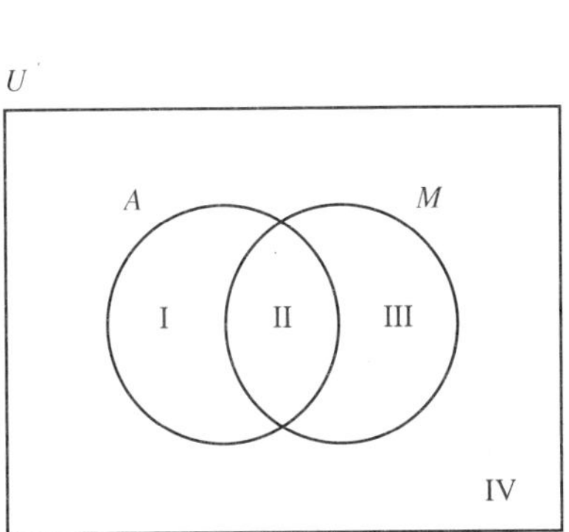

FIGURE 5.2

FIGURE 5.3

EXAMPLE 5.7 A nutritionist is conducting a survey of college athletes to determine which athletes regularly take vitamins A, B, and C. Let U denote the set of all athletes surveyed, and let A, B, and C denote the sets of athletes who regularly take vitamins A, B, and C, respectively. Use a Venn diagram to depict the sets of athletes who regularly take

(a) at least one of the three vitamins and
(b) exactly one of the three vitamins.

Solution The three sets A, B, and C are pictured as circular regions in Figure 5.3. Note that these three sets divide U into eight disjoint subsets, numbered I through VIII. Each of these subsets is an intersection of A or $\overline{A}$ with B or $\overline{B}$ and C or

$\overline{C}$. Region I, for instance, depicts the subset $A \cap \overline{B} \cap \overline{C}$ and represents the athletes who regularly take vitamin A but not B or C. Also region IV depicts the subset $\overline{A} \cap B \cap C$, which consists of the athletes who regularly take vitamins B and C but not A. Likewise region VIII depicts the subset $\overline{A} \cap \overline{B} \cap \overline{C}$, which consists of the athletes who do not take any of the three vitamins regularly.

(a) The athletes who regularly take at least one of vitamins A, B, or C are those who belong to one of the subsets numbered I through VII. These are just the elements of $A \cup B \cup C$, which is shaded in Figure 5.4.

(b) The athletes who regularly take exactly one of the vitamins are found in subsets I, III, and V. Thus the desired set is the union of these three subsets, which is

$$(A \cap \overline{B} \cap \overline{C}) \cup (\overline{A} \cap B \cap \overline{C}) \cup (\overline{A} \cap \overline{B} \cap C).$$

This set is shaded in Figure 5.5. ■

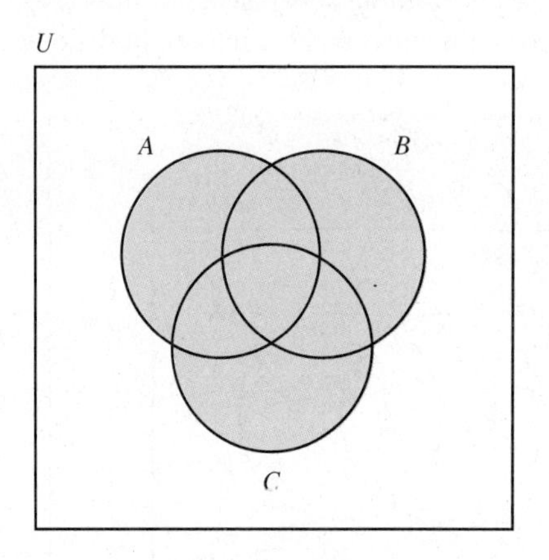

FIGURE 5.4

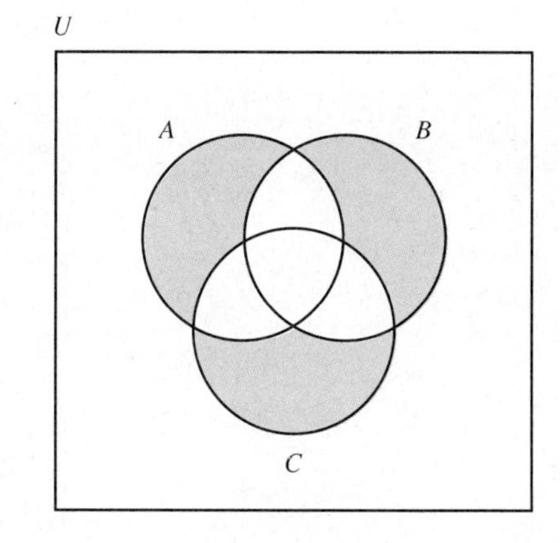

FIGURE 5.5

Practice Problem 3 (a) Describe regions III and VI in Figure 5.3 as intersections of A or $\overline{A}$ with B or $\overline{B}$ and C or $\overline{C}$. (b) Use a Venn diagram to depict the sets of athletes in Example 5.7 who regularly take exactly two of the three vitamins.

EXAMPLE 5.8 Human blood is classified according to the presence or absence of three antigens called A, B, and Rh. As we have seen, these three sets of antigens will divide human blood into eight types. These types are called AB+, AB−, A+, A−, B+, B−, O+, and O−. In this classification the letters A and B denote the presence of the corresponding antigens, and O denotes the absence of both the A and B antigens. The presence or absence of the Rh antigen is denoted by + or −, respectively. Thus blood of type AB− contains antigens

A and B but not Rh, blood of type A− contains only the A antigen, and blood of type O+ contains only the Rh antigen. The Venn diagram in Figure 5.6 displays the eight blood types. ■

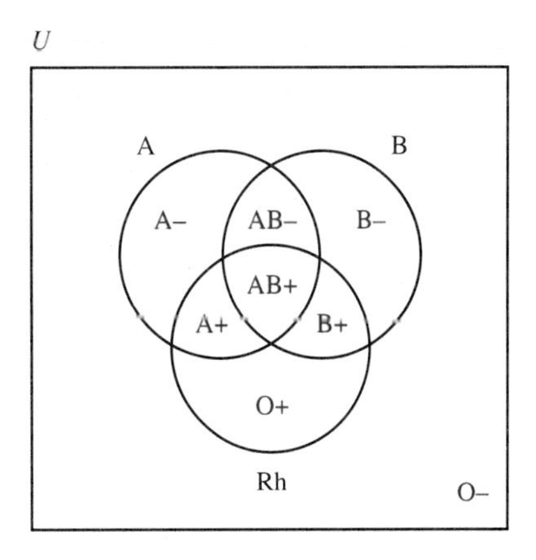

FIGURE 5.6

EXERCISES 5.1

In Exercises 1–10, list the elements in each of the following subsets of the universal set $U = \{1, 2, 3, \ldots\}$ of all positive integers.

1. $\{x \in U | 2 < x \leq 8\}$
2. $\{x \in U | x^2 - 5x + 6 = 0\}$
3. $\{x \in U | x^2 - 11x + 30 = 0\}$
4. $\{x \in U | (x - 2)(x - 9)(x - 17) = 0\}$
5. $\{x \in U | 2x + 8 > 5x - 1 \text{ or } 3x + 5 = 20\}$
6. $\{x \in U | \text{both } 24/x \text{ and } 60/x \text{ are integers}\}$
7. $\{x \in U | (x + 10)/x \text{ is an integer}\}$
8. $\{x \in U | x = 5n + 3 \text{ for some integer } n \leq 4\}$
9. $\{x \in U | x \text{ is odd and } x^2 \text{ is even}\}$
10. $\{x \in U | x \leq 5 \text{ or } 2x + 3 = 17\}$

Let $A = \{u, w, x, y\}$, $B = \{u, v, y\}$, and $C = \{v, x, z\}$ be subsets of the universal set $U = \{u, v, w, x, y, z\}$. In Exercises 11–20, list the elements in each of the following sets.

11. $A \cap B$
12. $A \cup C$
13. $A \cap B \cap C$
14. $\overline{B}$
15. $\overline{A} \cup \overline{C}$
16. $\overline{(A \cup C)}$
17. $\overline{(A \cup B)}$
18. $(A \cap \overline{B}) \cup (\overline{A} \cap B)$
19. $A \cup (B \cap C)$
20. $A \cap \overline{B} \cap C$

Let $A = \{1, 2, 3, 4\}$, $B = \{5, 6, 7, 8\}$, *and* $C = \{1, 2, 7, 8\}$ *be subsets of the universal set* $U = \{1, 2, 3, \ldots, 9, 10\}$. *In Exercises 21–32, determine whether each statement is true or false.*

21. $\varnothing \subseteq B$

22. $\overline{C} \subseteq \overline{C}$ 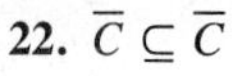

23. $2 \in A \cap (B \cup C)$

24. $7 \notin (A \cup B) \cap \overline{C}$

25. $A \cap C \subseteq \overline{B}$

26. $\varnothing \in B \cup C$

27. $A \cup B \cup C = U$

28. $A \cap B \cap C = \varnothing$

29. $7 \subseteq B \cap C$

30. $C \neq \{2, 8, 7, 1\}$

31. A and B are disjoint.

32. $\overline{A}$ and $\overline{B}$ are disjoint.

33. List all the subsets of $S = \{a, b, c, d\}$ containing
(a) exactly one element.
(b) exactly two elements.
(c) exactly three elements.

34. List all the subsets of $S = \{1, 2, 3, 4, 5\}$ containing
(a) no elements.
(b) exactly three elements.
(c) exactly five elements.

35. In History 152, a midsemester examination and a final examination were given. Let M and F denote the sets of students who passed the midsemester and final examinations, respectively. Use set operations to describe the following sets of students, and illustrate each set with a Venn diagram.
(a) Those who passed the midsemester examination but failed the final exam.
(b) Those who passed both examinations.
(c) Those who passed exactly one of the examinations.
(d) Those who failed both examinations.

36. As part of its monthly unemployment figures, the census bureau compiles information about the employment of married couples. Let H denote the set of couples in which the husband is employed and W denote the set of couples in which the wife is employed. Use set operations to describe the following sets, and illustrate each with a Venn diagram.
(a) Couples in which both spouses are employed.
(b) Couples in which only the wife is employed.
(c) Couples in which neither spouse is employed.
(d) Couples in which exactly one spouse is employed.

37. A marketing consultant classifies persons by sex, marital status, and employment status. Let F, S, and E denote the sets of females, single persons, and employed persons, respectively. Use set operations to describe the following sets, and illustrate each one with a Venn diagram.
(a) Working wives.
(b) Unemployed bachelors.
(c) Single women.
(d) Persons who are single or employed.

38. Let U denote the set of students at Illinois State University, and let M, B, and E denote the subsets of U consisting of the students who are taking a course in mathematics, business, and economics, respectively. Use set operations to describe the following sets, and illustrate each one with a Venn diagram.

(a) Students taking a business course or an economics course.

(b) Students taking a mathematics course and a business course.

(c) Students taking an economics course but not taking a mathematics course.

(d) Students not taking any courses in mathematics, business, or economics.

(e) Students taking a course in exactly one of these three subjects.

(f) Students taking a course in exactly two of these three subjects.

Answers to Practice Problems

1. (a) $\varnothing$ (b) $\{-1, 1\}$ (c) U

2. $A \cup B = \{1, 2, 3, 4, 6, 8\}$, $A \cap B = \{2\}$, $\overline{B} = \{1, 5, 7, 8\}$, $A \cap C = \varnothing$, $B \cup C = \{2, 3, 4, 6, 7\}$, $\overline{(B \cup C)} = \{1, 5, 8\}$, and $A \cup B \cup C = \{1, 2, 3, 4, 6, 7, 8\}$

3. (a) Regions III and VI denote the sets $\overline{A} \cap B \cap \overline{C}$ and $A \cap \overline{B} \cap C$, respectively.

(b) Regions II, IV, and VI in Figure 5.3 contain the athletes who regularly take exactly two of the three vitamins. Thus the desired Venn diagram is shown below.

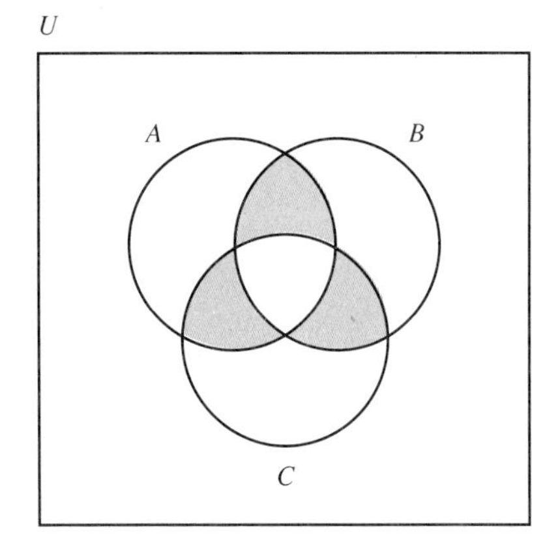

5.2 Counting with Venn Diagrams and Tree Diagrams

We saw in Section 5.1 that Venn diagrams are helpful in visualizing different combinations of properties in which we are interested. We begin this section by discussing the use of Venn diagrams to solve certain counting problems.

We will denote the number of elements in a finite set S by $|S|$. Thus if

$$A = \{1, 2, 3, 4\}, \qquad B = \{3, 4, 6\},$$

and the universal set is

$$U = \{1, 2, 3, 4, 5, 6, 7, 8\},$$

then $|A| = 4$ and $|B| = 3$. Moreover, since

$$\overline{A} = \{5, 6, 7, 8\} \quad \text{and} \quad \overline{B} = \{1, 2, 5, 7, 8\},$$

we see that $|\overline{A}| = 4$ and $|\overline{B}| = 5$. Note that

$$|\overline{A}| = |U| - |A| \quad \text{and} \quad |\overline{B}| = |U| - |B|.$$

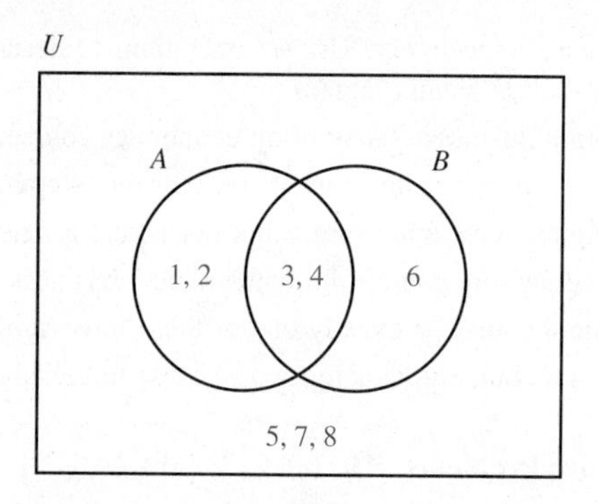

FIGURE 5.7

Now consider the problem of determining the number of elements in the union of two sets. In the example above, $A \cup B = \{1, 2, 3, 4, 6\}$. Note that $|A \cup B| \neq |A| + |B|$. The Venn diagram in Figure 5.7 shows why these numbers are different: the elements 3 and 4, which belong to both A and B, are counted only once by $|A \cup B|$ but twice by $|A| + |B|$. Hence $|A| + |B|$ counts each element of $A \cap B$ twice, and so $|A \cup B| = |A| + |B| - |A \cap B|$. We call this the **principle of inclusion-exclusion.**

This discussion illustrates the following general relationships.

Theorem 5.1

Let A and B be any subsets of a finite universal set U. Then

(a) $|\overline{A}| = |U| - |A|$

(b) $|A \cup B| = |A| + |B| - |A \cap B|$. *(principle of inclusion-exclusion)*

EXAMPLE 5.9 In a certain group of 50 people, 16 have blond hair and 20 have blue eyes. If exactly 9 of these persons have both blond hair and blue eyes, determine

(a) the number of people without blue eyes,
(b) the number of people with blond hair or blue eyes, and
(c) the number of people with neither blond hair nor blue eyes.

Solution Let U denote the set of all 50 people, and let H and E denote the subsets of U consisting of the people having blond hair and blue eyes, respectively.

(a) The set of people without blue eyes is $\overline{E}$, and so the number of people without blue eyes is $|\overline{E}|$. By Theorem 5.1(a) this number is

$$|\overline{E}| = |U| - |E| = 50 - 20 = 30.$$

(b) The set of people with blond hair or blue eyes is $H \cup E$. By the principle of inclusion-exclusion, Theorem 5.1(b), this number is

$$|H \cup E| = |H| + |E| - |H \cap E| = 16 + 20 - 9 = 27.$$

(c) Since the set of people with neither blond hair nor blue eyes is $\overline{H \cup E}$, we can use part (b) and Theorem 5.1(a) to find the desired number. Thus there are

$$|\overline{H \cup E}| = |U| - |H \cup E| = 50 - 27 = 23$$

persons in this group with neither blond hair nor blue eyes.

The Venn diagram in Figure 5.8 shows the number of elements in the various subsets determined by H and E. ■

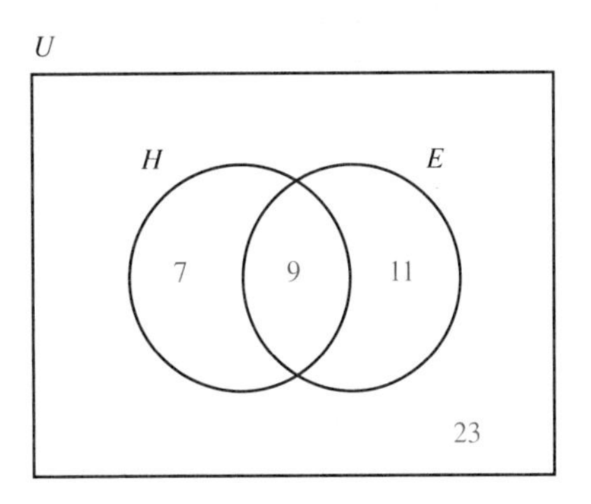

FIGURE 5.8

Practice Problem 4 In a group of 100 executives of small companies, 36 read *Forbes* and 28 read *Business Week*. If 7 of these executives read both *Forbes* and *Business Week*, how many of the executives

(a) do not read *Forbes?*
(b) read *Forbes* or *Business Week?*
(c) read neither *Forbes* nor *Business Week?*

The preceding ideas can be used to analyze data obtained from surveys. To illustrate the technique, let us consider the following data from a survey of 350 athletes.

90 regularly take vitamin A.
88 regularly take vitamin B.
97 regularly take vitamin C.
53 regularly take vitamins A and B.
55 regularly take vitamins A and C.
57 regularly take vitamins B and C.
32 regularly take all three vitamins.

How many of these athletes

(a) take only vitamin A regularly?
(b) take only one vitamin regularly?
(c) take none of the three vitamins regularly?

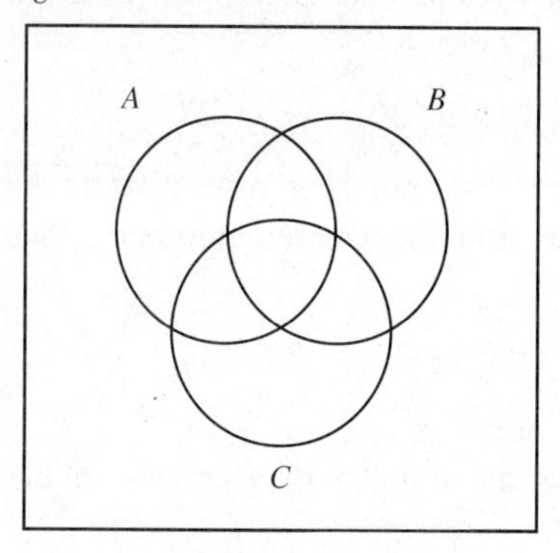

FIGURE 5.9

To answer these questions, we use the Venn diagram from Example 5.7, which is shown in Figure 5.9. We start at the center of the Venn diagram and work outward. Since 32 athletes take all three vitamins, we place this number in the region representing $A \cap B \cap C$, as shown in Figure 5.10(a). We are told that 53 athletes regularly take both vitamins A and B, and 32 of these also take vitamin C. Thus there must be $53 - 32 = 21$ who take A and B but not C. Similarly, $55 - 32 = 23$ take A and C but not B, and $57 - 32 = 25$ take B and C but not A. We place these numbers in the corresponding regions of the Venn diagram, as shown in Figure 5.10(b).

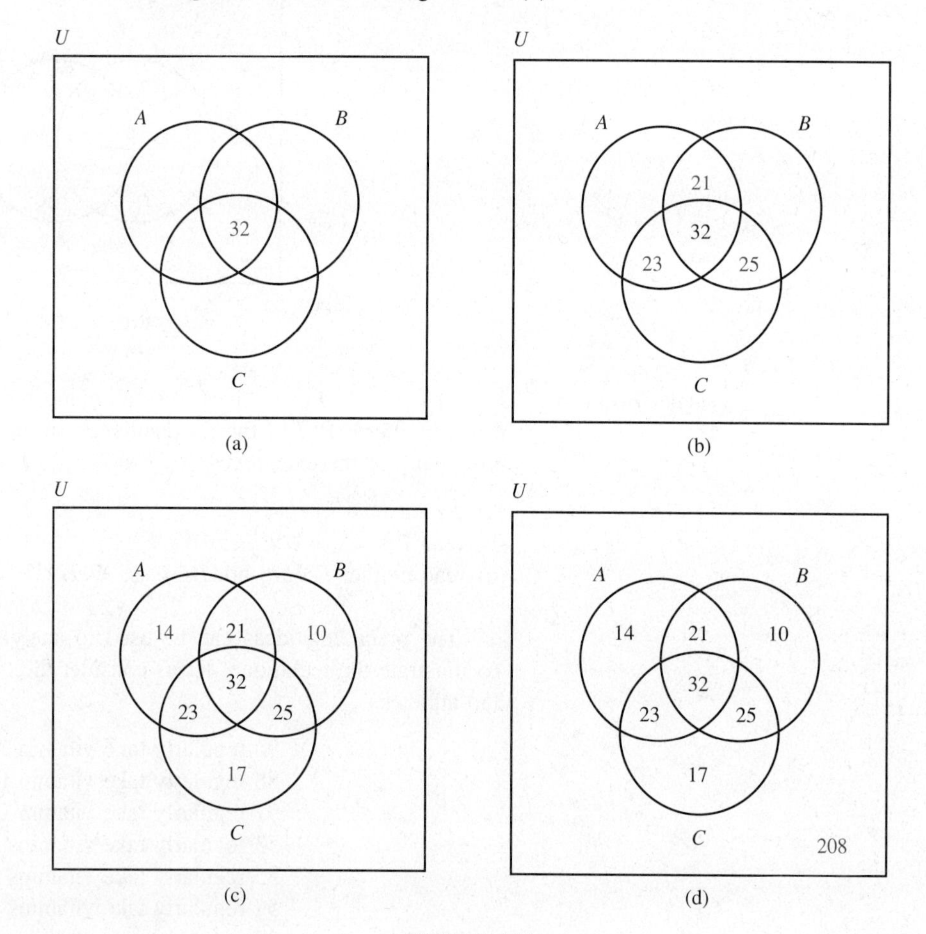

FIGURE 5.10

We see in Figure 5.10(b) that 21 + 32 + 23 = 76 of the athletes who take vitamin A regularly have been accounted for. Since there are 90 athletes in all who take vitamin A, there must be 90 − 76 = 14 who take vitamin A but not B or C. Similarly 88 − 78 = 10 athletes take only vitamin B, and 97 − 80 = 17 take only vitamin C. We now place these numbers in the corresponding regions of the Venn diagram, as shown in Figure 5.10(c).

In Figure 5.10(c) we see that $A \cup B \cup C$ contains a total of

$$14 + 21 + 10 + 23 + 32 + 25 + 17 = 142$$

elements, leaving 350 − 142 = 208 athletes who take none of the three vitamins regularly. The completed Venn diagram is shown in Figure 5.10(d). Thus we see that:

(a) 14 of these athletes take only vitamin A regularly.
(b) 14 + 10 + 17 = 41 of these athletes take only one vitamin regularly.
(c) 208 of these athletes take none of the vitamins regularly.

EXAMPLE 5.10 A firm purchased three mailing lists from a consultant dealing in such lists. The price is 10¢ per *distinct* name. The first list contains 1500 names, the second 3300, and the third 2800. A computer check shows that the first and second lists contain 382 names in common, the first and third contain 417 names in common, the second and third contain 741 names in common, and 219 names occur on all three lists. How much should the firm pay for these lists?

Solution Let U denote the set of all the names on the three lists. By proceeding as in Figure 5.10, we obtain the Venn diagram in Figure 5.11.

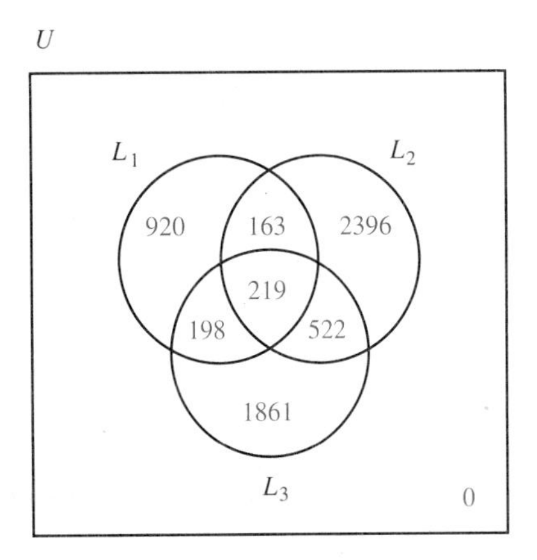

FIGURE 5.11

If we add all of the numbers in this diagram, we see that there are 6279 distinct names on the three lists. Thus the firm should pay $627.90 for the lists. ■

Practice Problem 5 Consider the following data from a survey of students at a certain college.

244 were taking a business course.
208 were taking a mathematics course.
152 were taking an economics course.
72 were taking courses in business and mathematics.
46 were taking courses in business and economics.
60 were taking courses in mathematics and economics.
24 were taking courses in business, mathematics, and economics.
150 were not taking a course in business, mathematics, or economics.

Determine:

(a) the number of students taking a business course who were not taking a course in either mathematics or economics,
(b) the number of students taking a course in exactly two of the three disciplines,
(c) the number of students who were surveyed.

TREE DIAGRAMS

Many counting problems involve a sequence of choices. Consider, for instance, the problem of counting the number of teams consisting of one man and one woman that can be formed from among three men and two women. A pictorial device called a **tree diagram** is helpful in keeping track of the different ways in which a sequence of choices can be made. The tree diagram in Figure 5.12 shows the 6 possible teams that can be formed by choosing one man and then one woman.

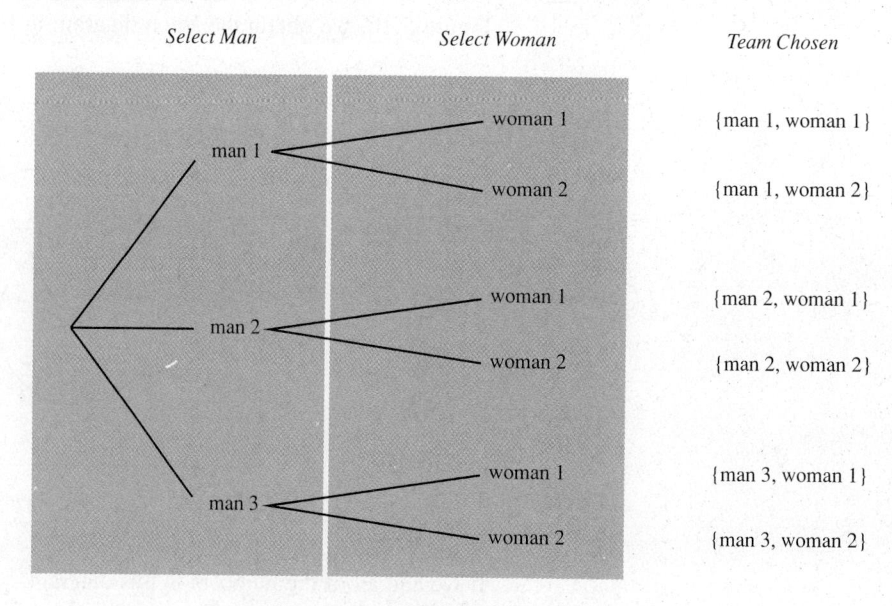

FIGURE 5.12

EXAMPLE 5.11 A new subcompact car is available in red, blue, or gray. The buyer also has a choice of automatic or manual transmission and bench or bucket seats. In how many different ways can these three options be ordered?

Solution There are three choices of color. For each choice of color there are two choices of transmission followed by two choices of seats. Thus in ordering these three options a buyer follows one of the twelve paths through the tree diagram in Figure 5.13. For example, the buyer who selects a blue car with automatic transmission and bucket seats follows the path shown in color. ■

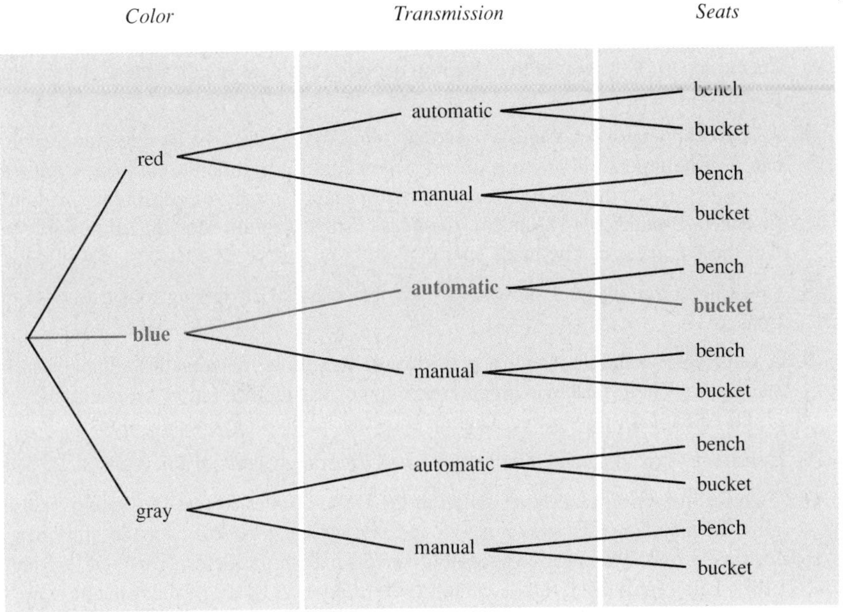

FIGURE 5.13

Practice Problem 6 In Example 5.5 we listed all the subsets of the set $S = \{a, b, c\}$. Draw a tree diagram to consider systematically all the possibilities.

EXERCISES 5.2

1. Construct a tree diagram showing the possible ways to order a sweater that is available in three colors (red, green, or blue) and three sizes (small, medium, or large).
2. A die is a cube with faces numbered 1 through 6. Construct a tree diagram showing the possible results when a single die is rolled and then a single coin is flipped.
3. Construct a tree diagram showing the possible results when a coin is flipped three times in succession.

4. Baron, in reference [1], hypothesizes that a subject cannot attend to more than one task at a time. An experiment is presented in which there are 3 possible states of attention (focused on the first task, the second task, or neither task), 2 possible responses to the first task (correct or incorrect), and 2 possible responses to the second task (correct or incorrect). Construct a tree diagram showing the possible cases for this experiment.

5. At a certain ice cream shop, a sundae can be ordered with one, two, or three scoops of ice cream, a choice of hot fudge or marshmallow, and a choice of whipped cream or nuts. Construct a tree diagram showing the possible ways of ordering a sundae.

6. Broadhurst, in reference [2], describes an experiment with male albino rats. The experiment involves 4 levels of motivation (air deprivation for 0, 2, 4 or 8 seconds), 3 levels of difficulty (easy, moderate, and difficult), and two levels of emotionality (emotional or unemotional). Construct a tree diagram showing the possible classifications for this experiment.

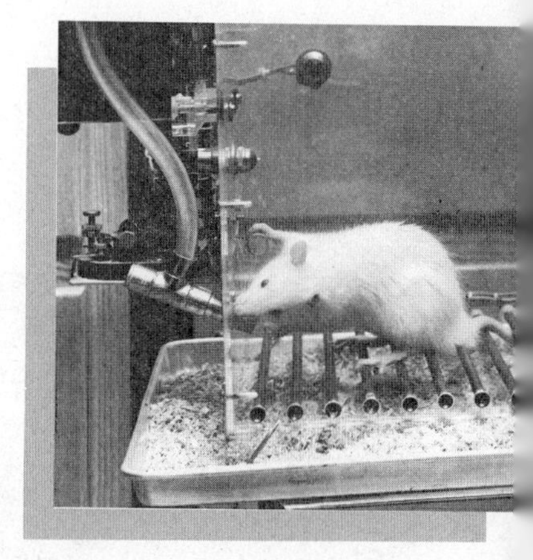

7. A soft drink company would like to determine the demand for its new mineral water. It can test the product in any one of 5 different regions (northeast, southeast, midwest, southwest, or northwest) and can bottle the product in one of 3 different size bottles (6-ounce, 8-ounce, or 12-ounce). Construct a tree diagram showing all possible ways that these choices can be made.

8. Construct a tree diagram showing all possible ways of answering four true-false questions.

9. A researcher examining student performance in a finite mathematics course has classified students by class (freshman, sophomore, junior, or senior) and by course grade (A, B, C, D, or F). Construct a tree diagram showing all possible classifications.

10. Construct a tree diagram showing all possible arrangements of the digits 1, 2, 3, and 4.

11. Two welfare-reform measures failed in the United States Senate last year by close margins. In analyzing the votes it was found that 45 senators voted for the first bill and 47 for the second. Moreover, 39 senators voted for both measures. This year a compromise bill is being introduced that is certain to be supported by all of the senators who voted for either of the two bills last year. If this bill needs 51 votes to pass, will it succeed?

12. A medical research team is comparing two different diagnostic tests for diabetes. When both tests were given to a group of known diabetics, test A indicated diabetes in 94% and test B indicated diabetes in 91%. Moreover, both tests indicated diabetes in 87% of the group. In what percentage did neither test indicate diabetes?

13. In a recent mayoral election the winner received 7431 votes. After analyzing the results, it was found that the winner received 4957 votes from liberals and 3893 votes from women. If the winner received 2315 votes from liberal women, how many votes did the winner get from men who are not liberal?

14. In a survey of 100 moviegoers it was found that 44 liked films directed by Bergman, 37 liked films directed by Fellini, and 23 liked both men's films. How many of these moviegoers liked neither man's films?

15. The following data were obtained from 25 fast-food restaurants in a certain city:

12 served hamburgers.
10 served roast beef sandwiches.
9 served pizza.

4 served hamburgers and roast beef sandwiches.
2 served hamburgers and pizza.
3 served roast beef sandwiches and pizza.
1 served all three of these foods.

(a) How many of these restaurants served hamburgers but neither roast beef sandwiches nor pizza?
(b) How many of these restaurants served pizza but neither hamburgers nor roast beef sandwiches?

16. The following information was obtained from persons at a shopping mall:

563 were adults.
414 were female.
310 had come alone.
296 were women (adult females).
213 adults had come alone.
124 females had come alone.
92 women had come alone.
54 were neither adult nor female and had not come alone.

(a) How many men (adult males) had come alone?
(b) How many children had come alone?
(c) How many people were questioned?

17. A congressman sent a questionnaire to his constituents, and his staff prepared the following summary of the replies for release to the local media.

3688 replies were received.
2471 wanted welfare reform.
2952 wanted a balanced budget.
1936 wanted gun control.
1997 wanted welfare reform and a balanced budget.
1713 wanted welfare reform and gun control.
1504 wanted a balanced budget and gun control.
1376 wanted welfare reform, a balanced budget, and gun control.

In preparing an article about the questionnaire, the local newspaper needs the additional information below. Supply this information.
(a) How many of the respondents favored none of the three issues?
(b) How many of the respondents favored only one of the three issues?

18. On May 7, 1970, Miami University in Oxford, Ohio was closed for ten days because of student protests following the invasion of Cambodia by the United States. An attitude questionnaire was prepared to sample student views regarding the closing of the university, and the results were reported in reference [4]. Among the statements to which the students responded were the three below.

(1) Miami University will have to change drastically its educational policies in the future.
(2) Trouble will never end at Miami until the whole university administrative structure is overturned.
(3) Violence seems to be the only way to get the administration to listen to us.

The following data were reported from the 434 responses.

193 persons thought the first statement was probably true.
82 persons thought the second statement was probably true.
112 persons thought the third statement was probably true.
62 persons thought the first two statements were probably true.
69 persons thought the first and third statements were probably true.
36 persons thought the last two statements were probably true.
30 persons thought all three statements were probably true.

How many respondents indicated that none of the three statements were probably true?

19. Among a random sample of 400 blacks, the following distribution of antigens can be expected:

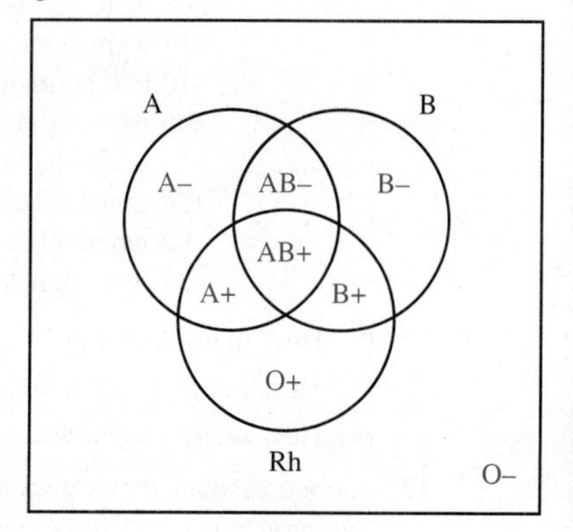

140 will have antigen A.
100 will have antigen B.
340 will have antigen Rh.
20 will have both antigens A and B.
119 will have both antigens A and Rh.
85 will have both antigens B and Rh.
17 will have all three antigens.

Using the figure at the right, determine the number of persons of each blood type.

20. Among a random sample of 2000 whites, the following distribution of antigens can be expected:

860 will have antigen A.
340 will have antigen B.
1700 will have antigen Rh.
100 will have both antigens A and B.
731 will have both antigens A and Rh.
289 will have both antigens B and Rh.
85 will have all three antigens.

Using the figure above, determine the number of persons of each blood type.

21. An insurance company claimed to have 900 new policyholders, of which

796 bought automobile insurance.
402 bought life insurance.
667 bought home insurance.
347 bought automobile and life insurance.
580 bought automobile and home insurance.
291 bought life and home insurance.
263 bought automobile, life, and home insurance.

Explain why the state insurance regulatory office ordered an audit of the company's records.

22. A congresswoman hired an opinion research firm to survey voter attitudes in her district. Among the results in the report were these: 80% of the voters favored gun control, 53% favored nuclear power, and 69% favored tighter pollution limits. Both gun control and nuclear power were favored by 21% of those surveyed, both gun control and tighter pollution limits were favored by 46%, both nuclear power and tighter pollution limits were favored by 34%, and 9% favored all three. After studying the report, the congresswoman accused the research firm of incompetence, claiming that these survey results are impossible. Why did she reach this conclusion?

23. Determine the set S if $|S| = 0$.

24. What can be said about sets A and B if $|A \cup B| = |A| + |B|$?

25. If $|U| = 100$, $|A| = 36$, $|B| = 61$, and $|A \cap B| = 23$, determine each of the following.
(a) $|\overline{A}|$ (b) $|\overline{B}|$ (c) $|A \cup B|$

26. If $|U| = 160$, $|A| = 87$, $|B| = 52$, and $|A \cap B| = 39$, determine each of the following.
(a) $|\overline{A}|$ (b) $|\overline{B}|$ (c) $|A \cup B|$

27. In a sociological study, 500 married couples were asked if they were satisfied with their marriages. In 210 cases both spouses were satisfied, while in 65 cases neither spouse was satisfied. Twice as many husbands as wives responded that they were satisfied with their marriages. How many husbands and how many wives were satisfied with their marriages? (*Hint:* Put unknowns into the regions of a Venn diagram determined by the sets of husbands and satisfied persons.)

28. During a broadcast of a baseball game, the announcer wanted to know the number of games last year in which the team's ace pitcher had lost a game that he failed to complete. The team's yearbook contained the following statistics about this pitcher.

Games started: 36; Wins: 17; Losses: 8;
Complete games: 14; Complete games won: 9.

Determine the answer to the announcer's question. (*Hint:* A complete game must result in either a win or a loss.)

Answers to Practice Problems

4. (a) 64 (b) 57 (c) 43

5. (a) 150 (b) 106 (c) 600

6. One possible tree diagram is shown here. The last column contains the eight possible subsets: $\varnothing$, $\{a\}$, $\{b\}$, $\{c\}$, $\{a, b\}$, $\{a, c\}$, $\{b, c\}$, $\{a, b, c\}$.

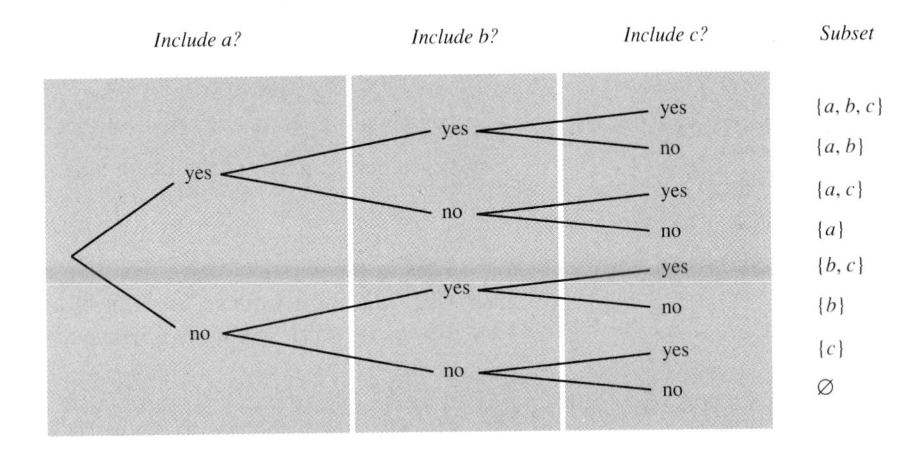

5.3 The Multiplication and Addition Principles

In this section we will discuss two basic principles of counting. These principles form the foundation for all of the counting problems that we will consider. Moreover, as we will see in Chapter 6, they lead to two important rules for computing probabilities.

THE MULTIPLICATION PRINCIPLE

We have seen that many counting problems involve a sequence of choices, each of which can be made in several ways. For example, the following question, which we considered in Section 5.2, is of this type.

> In order to resolve a labor dispute, a team of federal mediators is to be selected from 3 men and 2 women. If the team must consist of exactly one man and one woman, how many different teams can be chosen?

Since the number of men and women involved is small, it is easy to answer this question by listing all of the possibilities. As the tree diagram in Figure 5.12 shows, there are 6 possible teams that can be formed by selecting one man and one woman; these are

{man 1, woman 1}, {man 2, woman 1}, {man 3, woman 1},
{man 1, woman 2}, {man 2, woman 2}, {man 3, woman 2}.

However, there is a simple rule called the multiplication principle which can be applied in such cases to count the number of possibilities without actually enumerating them.

Theorem 5.2 (the multiplication principle)

Suppose that there are k operations to be performed in sequence, and the first operation can occur in n_1 ways, and for each of these the second operation can occur in n_2 ways, and for each of these the third operation can occur in n_3 ways, and so forth. Then there are

$$n_1 \cdot n_2 \cdot \cdots \cdot n_k$$

ways to perform all k of the operations in sequence.

Thus in the question above there are two choices to be made: choosing a man and choosing a woman. Since there are 3 ways in which to choose a man and 2 ways in which to choose a woman, there are

$$3 \cdot 2 = 6$$

ways in which both choices can be made. These correspond to the six possible teams of mediators.

EXAMPLE 5.12 A woman has 6 different pairs of slacks, 8 different blouses, 5 different pairs of shoes, and 3 different purses. How many outfits consisting of one pair of slacks, one blouse, one pair of shoes, and one purse can she create?

Solution In this situation there are four separate choices to be made: selecting a pair of slacks, selecting a blouse, selecting a pair of shoes, and selecting a purse. Since there are 6 ways to select slacks, 8 ways to select a blouse, 5 ways to select shoes, and 3 ways to select a purse, there are

$$6 \cdot 8 \cdot 5 \cdot 3 = 720$$

different ways in which all four choices can be made. Each of these choices corresponds to an outfit that can be created. ■

6 pairs of slacks
8 blouses
5 pairs of shoes,
and 3 purses

EXAMPLE 5.13 An automobile license plate consists of three letters (A–Z) followed by three digits (0–9). How many different license plates are possible?

Solution In constructing a license plate, six separate choices must be made: select three letters and three digits. Since there are 26 ways in which to choose each letter and 10 ways to choose each digit, the multiplication principle shows that the number of possible ways of making all six choices is

$$26 \cdot 26 \cdot 26 \cdot 10 \cdot 10 \cdot 10 = 17{,}576{,}000.$$

Thus there are 17,576,000 possible license plates consisting of three letters followed by three digits. ■

Practice Problem 7 An examiner wishes to create a test consisting of six questions. She has 8 variations of the first question, 7 variations of the second question, 10 variations of the third question, 5 variations of the fourth question, 6 variations of the fifth question, and 8 variations of the sixth question. How many different tests can be created by selecting one of the variations of each question?

EXAMPLE 5.14 At Wendy's Old Fashioned Hamburger Restaurants a hamburger can be ordered with any combination of 8 toppings—cheese, ketchup, lettuce, mayonnaise, mustard, onion, pickles, and tomatoes. How many different ways are there to order a hamburger and toppings?

Solution Since there are eight possible toppings, there are eight separate choices to be made. Moreover, there are two ways to order each topping, with or without. We will assume that the order in which the toppings are applied is irrelevant. Then the multiplication principle shows that the number of different ways to make all eight choices is

$$2 \cdot 2 \cdot 2 \cdot 2 \cdot 2 \cdot 2 \cdot 2 \cdot 2 = 2^8 = 256.$$

Consequently there are 256 different ways to order a hamburger and toppings at Wendy's. ■

Practice Problem 8 A quiz is to be composed of ten questions to be answered true or false. How many different ways are there to answer all ten questions?

Notice that in order to use the multiplication principle, the number of ways to perform an operation must not depend on how the previous operations are performed. This restriction may complicate the solution of a problem, as in the example below.

EXAMPLE 5.15 Suppose that we are to use the digits 1–7 without repetition to make five-digit numbers.

(a) How many different five-digit integers can be made?
(b) How many of the numbers in part (a) begin with 6?
(c) How many of the numbers in part (a) contain both 1 and 2?

Solution (a) We will construct five-digit numbers by choosing a value for each of the five digits in the number. This amounts to filling each of the blanks below

$$\underset{1}{\underline{\quad}}\ \underset{2}{\underline{\quad}}\ \underset{3}{\underline{\quad}}\ \underset{4}{\underline{\quad}}\ \underset{5}{\underline{\quad}}$$

with one of the digits 1–7. Now there are 7 ways in which the first digit can be selected because any of the digits 1–7 can be used. There are only 6 ways to choose the second digit, however, because the first digit cannot be repeated. Similar reasoning shows that there are 5 ways to choose the third digit, 4 ways to choose the fourth digit, and 3 ways to choose the fifth digit. Hence the multiplication principle shows that the number of possible ways of making all five choices is

$$7 \cdot 6 \cdot 5 \cdot 4 \cdot 3 = 2520.$$

So this is the number of five-digit numbers that can be formed from the digits 1–7 without repetition.

(b) To answer question (b), we proceed as above except that there is only one way to choose the first digit (since it must be 6). Therefore of the 2520 integers in part (a),

$$1 \cdot 6 \cdot 5 \cdot 4 \cdot 3 = 360$$

begin with 6.

(c) The method used in part (a) cannot be used to count the number of five-digit numbers containing both 1 and 2. The reason this method fails is that the number of choices for the fourth and fifth digits will depend on earlier choices. For example, if the first three digits are 231, then there are 4 ways in which the fourth digit can be chosen (namely, 4, 5, 6, or 7); but if the first three digits are 567, then there are only two ways in which the fourth digit can be chosen (namely, 1 or 2).

Consequently we must look for another approach. Since the digits 1 and 2 must be used, we begin by deciding where to put them. We can then fill the three remaining positions with any of the digits 3–7. Thus we will proceed as follows.

Choose a position for the 1 (in 5 possible ways).
Choose a position for the 2 (in 4 possible ways).
Fill the first unfilled position (using one of 5 possible digits).
Fill the second unfilled position (using one of 4 possible digits).
Fill the third unfilled position (using one of 3 possible digits).

Thus by the multiplication principle there are

$$5 \cdot 4 \cdot 5 \cdot 4 \cdot 3 = 1200$$

ways of making all five choices. Hence 1200 of the integers in part (a) contain both of the digits 1 and 2. ■

As the last example shows, use of the multiplication principle requires careful attention. Although the principle is easy to understand, its application may require a bit of ingenuity. In order to use the multiplication principle correctly, you must clearly understand the objects being counted and create a systematic procedure for generating all of them. Be certain to specify precisely what is being chosen at each step, and only when you are convinced that your procedure is sound should you start multiplying numbers.

Practice Problem 9 How many ways are there to seat four married couples in a row of eight seats if

(a) each husband is to sit beside his wife? (*Hint:* When the first seat is filled, who can sit in the second seat?)
(b) each husband is to sit beside his wife and the first and last seats are to be filled by men?

THE ADDITION PRINCIPLE

Consider the following question.

> After much pleading by Alison, her parents have agreed to permit her to have a pet. However, the type of pet she chooses must be approved by her mother or father. Her mother, a cat fancier, will permit an Angora, Manx, or Siamese. Her father, a dog lover, will permit a cocker spaniel, collie, Dalmatian, Kerry blue terrier, or schnauzer. If Alison must choose her pet from one of these types of animals, how many different pets can she choose?

The answer is obviously 8, for Alison can choose among 3 different types of cats and 5 different types of dogs.

This question can be formulated as a question about sets by letting

$$C = \{\text{Angora, Manx, Siamese}\}$$

and

$$D = \{\text{cocker spaniel, collie, Dalmatian, Kerry blue terrier, schnauzer}\}$$

be the sets of pets acceptable to her mother and father, respectively. Then the number of different pets that Alison can choose is $|C \cup D|$, the number of elements in $C \cup D$. This number can be determined by using the principle of inclusion-exclusion [Theorem 5.1(b)]. For in this case $C \cap D = \emptyset$, and so

$$|C \cup D| = |C| + |D| - |C \cap D| = 3 + 5 - 0 = 8.$$

Notice that when the sets A and B are disjoint, the principle of inclusion-exclusion reduces to $|A \cup B| = |A| + |B|$. In fact, an analogous statement is true for a collection of k sets if all pairs of the sets are disjoint:

$$|A_1 \cup A_2 \cup \cdots \cup A_k| = |A_1| + |A_2| + \cdots + |A_k|.$$

This observation can be formulated in the following way.

Theorem 5.3 (the addition principle)

Suppose that there are k sets, each pair of which is disjoint. (That is, no two of the sets have an element in common.) If the first set contains n_1 elements, the second n_2 elements, and so forth, then the union of the sets contains

$$n_1 + n_2 + \cdots + n_k$$

elements.

EXAMPLE 5.16 The snack bar at a movie theatre sells 5 different sizes of popcorn, 12 different candy bars, and 4 different beverages. In how many different ways can one snack be selected?

Solution There are three sets to consider: the set of popcorn snacks, the set of candy bars, and the set of beverages. Clearly each pair of these sets is disjoint. Thus since they contain 5, 12, and 4 elements, respectively, the number of elements in their union is

$$5 + 12 + 4 = 21.$$

This is the number of different ways exactly one snack can be chosen. ■

Practice Problem 10 A college student needs to choose one more course to complete next semester's schedule. She is considering 4 business courses, 7 physical education courses, and 3 economics courses. How many different courses can she select?

The addition principle is often used together with the multiplication principle. Recall that when using the multiplication principle, the number of ways to perform an operation must not depend on how previous operations are performed. If this condition is not met, it may be possible to subdivide the problem into cases where the multiplication principle can be used. If no two cases produce the same result, then they can be combined using the addition principle. The following examples illustrate this technique.

EXAMPLE 5.17 In some states a license plate may be of three different types: three letters followed by three digits, two letters followed by four digits, or no letters and six digits. How many such license plates are there?

Solution Although each license plate contains six characters, the number of choices for the first three characters depends on the type of license plate being constructed. For example, the first character on a plate containing three letters and three digits can be any of the 26 letters A–Z, but the first character on a plate containing no letters and six digits must be one of the ten digits 0–9. However, the number of ways to choose each character is the same for all plates of the same type.

Consequently we consider three cases: plates containing three letters followed by three digits, plates containing two letters followed by four digits, and plates containing no letters and six digits. By Example 5.13 there are

$$26 \cdot 26 \cdot 26 \cdot 10 \cdot 10 \cdot 10 = 17{,}576{,}000$$

plates containing three letters and three digits. Similar reasoning shows that there are

$$26 \cdot 26 \cdot 10 \cdot 10 \cdot 10 \cdot 10 = 6{,}760{,}000$$

plates containing two letters and four digits and

$$10 \cdot 10 \cdot 10 \cdot 10 \cdot 10 \cdot 10 = 1{,}000{,}000$$

plates containing no letters and six digits. Since no plate can be of two different types, the addition principle shows that there are

$$17{,}576{,}000 + 6{,}760{,}000 + 1{,}000{,}000 = 25{,}336{,}000$$

plates containing three letters followed by three digits, two letters followed by four digits, or no letters and six digits. ■

EXAMPLE 5.18 How many integers between 1500 and 8000 (inclusive) contain no repeated digits?

Solution Notice that the number of ways in which the second (hundreds) digit can be chosen depends on the choice of the first (thousands) digit. For example, if the first digit is 6, then there are 9 possible choices for the second digit (any digit except 6). On the other hand, if the first digit is 1, then the second digit must be 5–9; so there are only 5 possible choices for the second digit. Thus we will consider two cases according to whether the first digit is 1 or not.

If the first digit is 1, then the second digit must be 5–9, and the third and fourth digits can be any unused value. Thus by the multiplication principle there are

$$1 \cdot 5 \cdot 8 \cdot 7 = 280$$

numbers between 1500 and 8000 that begin with 1 and have no repeated digits. On the other hand, if the first digit is not 1, then it must be 2, 3, 4, 5, 6, or 7. In this case the second, third, and fourth digits can be any unused value. So there are

$$6 \cdot 9 \cdot 8 \cdot 7 = 3024$$

numbers between 1500 and 8000 that do not begin with 1 and have no repeated digits. Clearly no integer is of both types because an integer cannot begin with 1 and also not begin with 1. Thus by the addition principle there are

$$280 + 3024 = 3304$$

integers between 1500 and 8000 that have no repeated digits. ■

Practice Problem 11 In Bogart's restaurant the entrees include prime rib, filet mignon, ribeye steak, scallops, and a fish-of-the-day. Each dinner is served with salad and a vegetable. Customers may choose from 4 salad dressings and 5 vegetables, except that the seafood dishes are served with wild rice instead of the choice of vegetable. In how many different ways can a dinner be ordered?

EXERCISES 5.3

1. A men's clothing store has a sale on selected suits and blazers. If there are 30 different suits and 40 different blazers on sale, in how many different ways can a customer purchase exactly one item that is on sale?
2. A restaurant offers a choice of three green vegetables or a potato prepared in one of 5 ways. In how many different ways can one vegetable be ordered?
3. There are 16 nominees for the outstanding teacher award from the College of Arts and Sciences, 4 from the College of Business, and 3 from the College of Education. In how many different ways can the outstanding teacher award be given?
4. An actuarial science student has job offers from 4 insurance companies with headquarters in Chicago, 2 companies with headquarters in Bloomington, and 3 companies with headquarters in Boston. How many different choices of company are available to her?
5. A particular automobile manufacturer makes 4 subcompact models, 6 compact models, and 5 intermediate-sized models. In how many different ways can a car be ordered from this manufacturer?
6. Susan wants to buy one birthday present for her boyfriend. She is considering 5 different record albums, 3 different sweaters, and 6 different shirts. From these possibilities, how many different presents can she select?
7. A high school basketball player has been offered scholarships at 8 eastern universities, 5 southern universities, 6 midwestern universities, and 2 western universities. How many choices of scholarship are possible?
8. A student needing a physical education course can take 6 different sections of tennis, 3 sections of bowling, 4 sections of golf, and 8 sections of swimming. In how many different ways can a section of physical education be selected?
9. At a motel restaurant breakfast consists of a choice of cereal, scrambled eggs, fried eggs, or soft-boiled eggs; a choice of toast or muffins; and a choice of orange, tomato, or apple juice. In how many different ways can breakfast be ordered?
10. From among sixteen finalists in a contest the judges must select a winner and a runner-up. How many possible outcomes are there?
11. A tourist can take one of three different routes from her hotel to the museum and one of five routes from the museum to the park. In how many ways can she go from her hotel to the park with a stop at the museum?
12. An organization with 25 members is electing a president, secretary, and treasurer (all different). Assuming there are no ties, how many outcomes can the election have?
13. A consumer rating service is testing twelve new cars and will rate the three best in order of preference. How many different ratings are possible?
14. In how many different sequences can fifteen numbered billiard balls be hit into the pockets of a pool table?
15. In how many different ways can each of six court cases be assigned to one of four judges if a judge may hear more than one case?

16. Five rooms are to be painted one of twelve different colors. In how many different ways can the rooms be painted if

(a) different rooms can be painted the same color?

(b) each room must be a different color?

17. In a baseball league with six teams, each team is scheduled to play every other team four times, twice on each team's home field. How many games will be played in this league?

18. In a chess tournament with eight players, each player is scheduled to play two games against every other player, once as white and once as black. How many different games will be played in this tournament?

19. In how many different ways can a student answer each question on a twelve-question examination if each question is

(a) to be answered true or false?

(b) to be answered with one of five possible multiple-choice responses?

20. A die is a cube with faces numbered 1 through 6.

(a) How many different sequences of numbers are possible if one die is rolled n times?

(b) How many different results are possible if n differently colored dice are rolled at the same time?

21. At Avanti's a pizza can be ordered in three different sizes with any combination of seven toppings: green pepper, ham, hamburger, mushrooms, onions, pepperoni, and sausage. How many different pizzas can be ordered?

22. The following items can be ordered as optional equipment on a certain model of new car: air conditioning, automatic transmission, bucket seats, power steering, AM/FM radio, and rear window defroster. How many different ways are there to order the optional equipment on this car?

23. (a) A local telephone number is a sequence of seven digits (0–9) with the restriction that the first and second digits cannot be 0 or 1. How many local telephone numbers are possible?

(b) An area code is a three-digit number that cannot begin with 0 or 1 and must have 0 or 1 as its middle digit. How many telephone numbers consisting of a local number and an area code are possible?

24. In the United States, radio station call letters consist of a sequence of three or four letters beginning with the letter K or W.

(a) How many call letters are possible?

(b) How many of the call letters contain no repeated letters?

(c) How many of the call letters end in Z and contain no repeated letters?

25. How many five-digit numbers consisting of only even digits or only odd digits can be formed from the digits 2, 3, 4, 5, 6, 7, and 8?

26. How many four-digit numbers (integers between 999 and 10,000) either start with a 3 or do not contain a 3?

27. In how many different orders can three married couples be seated in a row of six seats under the following conditions?

(a) Anyone may sit in any seat.

(b) The first and last seats must be filled by men.

(c) Each husband must sit beside his wife.

(d) All members of the same sex are seated in adjacent seats.

(e) Men and women are seated alternately.

28. Three women and two men are to be presented awards. In how many different orders can the awards be presented under the following conditions?

(a) The awards can be presented in any order.

(b) The awards are presented to the women before the men.

(c) The first award is presented to a woman, and subsequent awards are given alternately to a man and then a woman.

(d) The first and last awards are made to women.

(e) The first and last awards are made to men.

29. Two different awards are to be given to business students. The 14 nominees include 5 accounting majors, 3 marketing majors, and 6 business administration majors.

(a) In how many different ways can the awards be given if it is possible for the same student to win both awards?

(b) In how many different ways can the awards be presented if they must be given to two different students?

(c) In how many different ways can the awards be given to students with different majors?

30. A committee is to be formed from among 7 physicians, 10 engineers, and 5 lawyers. How many committees consisting of two members from different professions are possible?

31. A restaurant menu lists six appetizers, eight entrees, and four desserts.

(a) How many complete meals consisting of one appetizer, one entree, and one dessert can be ordered?

(b) How many meals consisting of one appetizer and one entree can be ordered?

(c) How many meals consisting of one appetizer and one entree and either no dessert or one dessert can be ordered?

32. How many subsets are there of a set with n elements?

33. Consider all the five-digit numbers that do not begin with 0.

(a) How many such numbers are there?

(b) How many of these numbers contain only odd digits?

(c) How many of them contain five different odd digits?

(d) How many of these numbers begin and end with an even digit?

(e) How many of them contain the digit 5 at least once? (*Hint:* Determine how many do not contain the digit 5.)

34. The letters in the word COMPUTERS are to be used no more than once each to form sequences of four letters.

(a) How many different sequences can be formed?

(b) How many of these sequences contain no vowels?

(c) How many of these sequences begin with a consonant and end with a vowel?

(d) How many of these sequences contain the letter M? (*Hint:* Determine the number of sequences that do not contain M.)

Answers to Practice Problems

7. 134,400 **8.** 1024

9. (a) 384 (b) 96 **10.** 14

11. 68

5.4 Permutations and Combinations

Two types of counting problems occur so frequently that they deserve special attention. These problems are:

(1) How many different arrangements (ordered lists) of r objects can be formed from a set of n distinct objects?
(2) How many different selections (unordered lists) of r objects can be made from a set of n distinct objects?

In this section we will develop formulas for solving these two types of counting problems.

PERMUTATIONS

To illustrate the ideas involved, let us consider a newly formed club having five members. At the first meeting it is decided to elect a slate of officers consisting of a president, vice-president, and treasurer (all different). How many slates are possible?

A slate of officers can be formed by making three successive choices: choosing a president, choosing a vice-president, and choosing a treasurer. The multiplication principle shows that the number of ways in which all three choices can be made is

$$5 \cdot 4 \cdot 3 = 60.$$

Thus there are 60 possible slates of officers. If we represent the five members of the club by the letters A, B, C, D, and E, then each slate of officers is represented by an arrangement of three of these letters. The 60 possible slates are listed below.

ABC	ABD	ABE	ACD	ACE	ADE	BCD	BCE	BDE	CDE
ACB	ADB	AEB	ADC	AEC	AED	BDC	BEC	BED	CED
BAC	BAD	BAE	CAD	CAE	DAE	CBD	CBE	DBE	DCE
BCA	BDA	BEA	CDA	CEA	DEA	CDB	CEB	DEB	DEC
CAB	DAB	EAB	DAC	EAC	EAD	DBC	EBC	EBD	ECD
CBA	DBA	EBA	DCA	ECA	EDA	DCB	ECB	EDB	EDC

Notice that the slate ABC (in which A is president, B is vice-president, and C is treasurer) is different from the slate BCA (in which B is president, C is vice-president, and A is treasurer). This is true even though the same three individuals are involved in each slate. Since the order of choice determines the office to be held, we are interested not only in *which* individuals are chosen but also in the *specific order* in which they are chosen.

Ordered arrangements of this type arise often in applications and are called **permutations.** We will denote the number of permutations of r objects selected from a set of n distinct objects by $P(n, r)$. Since the list above contains all permutations of the letters A, B, C, D, and E taken three at a time, we see that

$$P(5, 3) = 5 \cdot 4 \cdot 3 = 60.$$

More generally, suppose that there are n distinct objects and we are given an integer r such that $1 \leq r \leq n$. To form a permutation of r of these n objects, we must fill in each of the blanks below with a different object.

$$\underset{1}{\_\_\_}\ \underset{2}{\_\_\_}\ \underset{3}{\_\_\_}\ \cdots\ \underset{r-1}{\_\_\_}\ \underset{r}{\_\_\_}$$

The first object can be chosen in n ways, the second in $n - 1$ ways, the third in $n - 2$ ways, and so forth. When we reach the last blank, we will have used $r - 1$ of the n objects, leaving $n - (r - 1) = n - r + 1$ objects from which to fill the last blank. Thus by the multiplication principle the number of permutations of n objects taken r at a time is

$$n(n - 1)(n - 2) \cdots (n - r + 1).$$

Notice that this "descending product" starts with n and contains exactly r factors.

Theorem 5.4

For any positive integers r and n such that $1 \leq r \leq n$,

$$P(n, r) = \underbrace{n(n - 1)(n - 2) \cdots (n - r + 1).}_{r \text{ factors}}$$

Thus, for instance, we find by Theorem 5.4 that

$$P(10, 1) = 10,$$
$$P(10, 2) = 10 \cdot 9 = 90,$$
$$P(10, 3) = 10 \cdot 9 \cdot 8 = 720, \quad \text{and}$$
$$P(10, 4) = 10 \cdot 9 \cdot 8 \cdot 7 = 5040.$$

EXAMPLE 5.19 A pianist participating in a Chopin competition has decided to perform five of the fourteen Chopin waltzes. How many different programs are possible consisting of five waltzes played in a certain order?

Solution The pianist must select five different waltzes from fourteen and arrange them in a specific order. By Theorem 5.4 the number of such ordered arrangements (permutations) is

$$P(14, 5) = \underbrace{14 \cdot 13 \cdot 12 \cdot 11 \cdot 10}_{5 \text{ factors}} = 240{,}240. \blacksquare$$

EXAMPLE 5.20 In how many ways can six people be seated in a row of six chairs?

Solution A seating arrangement is merely a permutation of the people to be seated. Hence by Theorem 5.4 there are

$$P(6, 6) = 6 \cdot 5 \cdot 4 \cdot 3 \cdot 2 \cdot 1 = 720$$

possible seating arrangements. ■

EXAMPLE 5.21 How many three-digit integers can be formed using the digits 1–7 without repetition?

Solution A three-digit integer using the digits 1–7 without repetition is simply an ordered arrangement of three of the digits 1–7. Thus by Theorem 5.4 there are

$$P(7, 3) = 7 \cdot 6 \cdot 5 = 210$$

such integers. ■

Practice Problem 12 Nine athletes are entered in the conference high jump competition. In how many different ways can the gold, silver, and bronze medals be awarded?

FACTORIAL NOTATION

Computing the number of permutations of n objects taken r at a time involves products of consecutive integers. There is a convenient notation for representing such products.

If n is a positive integer, the symbol $n!$ (read ***n* factorial**) denotes the product of the positive integers less than or equal to n. That is,

$$n! = n(n - 1)(n - 2) \cdots 3 \cdot 2 \cdot 1.$$

In addition, we define

$$0! = 1.$$

The first few values of $n!$ are

$$\begin{aligned} 0! &= 1, \\ 1! &= 1, \\ 2! &= 2 \cdot 1 = 2, \\ 3! &= 3 \cdot 2 \cdot 1 = 6, \\ 4! &= 4 \cdot 3 \cdot 2 \cdot 1 = 24, \text{ and} \\ 5! &= 5 \cdot 4 \cdot 3 \cdot 2 \cdot 1 = 120. \end{aligned}$$

For larger integers, $n!$ can be computed using the relation

$$n! = n \cdot (n - 1)!,$$

which follows immediately from the definition of the factorial symbol. Thus

$$6! = 6(5!) = 6(120) = 720,$$
$$7! = 7(6!) = 7(720) = 5040,$$
$$8! = 8(7!) = 8(5040) = 40{,}320,$$

and so forth. Notice that the value of $n!$ grows very quickly; in fact, $10!$ exceeds 3 million and $13!$ exceeds 6 billion.

The number of permutations of n objects taken r at a time can be expressed in a very compact form using factorial notation. For by Theorem 5.4

$$P(n, r) = n(n-1)(n-2)\cdots(n-r+1)$$
$$= \frac{n(n-1)(n-2)\cdots(n-r+1)(n-r)(n-r-1)\cdots 2\cdot 1}{(n-r)(n-r-1)\cdots 2\cdot 1}$$

Thus we have the following formula.

$$P(n, r) = \frac{n!}{(n-r)!}$$

Notice that $P(n, n) = n!$. In addition, if $r = 0$, the expression

$$\frac{n!}{(n-r)!}$$

has a value of 1. Accordingly we define

$$P(n, 0) = 1.$$

Although the formula

$$P(n, r) = \frac{n!}{(n-r)!}$$

is important theoretically and provides a convenient way of writing $P(n, r)$, it should not normally be used for computations because it requires more calculations than the expression in Theorem 5.4:

$$P(n, r) = n(n-1)(n-2)\cdots(n-r+1).$$

COMBINATIONS

Now we will consider the second type of counting problem mentioned above. Suppose that the club with five members A, B, C, D, and E decides to appoint an executive committee consisting of three members having equal status instead of electing a president, a vice-president, and a treasurer. How many different executive committees can be formed?

If the three members of the executive committee are chosen in order, one after the other, the result will be one of the 60 permutations obtained before. (This list is reproduced below.)

ABC	ABD	ABE	**ACD**	ACE	ADE	BCD	BCE	BDE	CDE
ACB	ADB	AEB	**ADC**	AEC	AED	BDC	BEC	BED	CED
BAC	BAD	BAE	**CAD**	CAE	DAE	CBD	CBE	DBE	DCE
BCA	BDA	BEA	**CDA**	CEA	DEA	CDB	CEB	DEB	DEC
CAB	DAB	EAB	**DAC**	EAC	EAD	DBC	EBC	EBD	ECD
CBA	DBA	EBA	**DCA**	ECA	EDA	DCB	ECB	EDB	EDC

Now, however, since the members of the executive committee all have the same status, we are only interested in knowing which three individuals are chosen, not the specific order of their selection. Hence all of the permutations in any column above represent exactly the same executive committee. For instance, the permutations ACD, ADC, CAD, CDA, DAC, and DCA (shown in color above) all represent the committee with A, C, and D as its members.

We call such an unordered list of objects a **combination.** Thus a combination of n objects taken r at a time is simply a set containing r of the given n objects. The number of combinations of r objects selected from a set of n distinct objects will be denoted* by $C(n, r)$.

From the list above we see that each combination corresponds to

$$P(3, 3) = 3!$$

permutations, namely those in the same column of the list. Hence

$$C(5, 3) = \frac{P(5, 3)}{3!} = \frac{5 \cdot 4 \cdot 3}{3 \cdot 2 \cdot 1} = 10.$$

Thus there are 10 combinations of A, B, C, D, and E taken 3 at a time.

In general, each permutation of n objects taken r at a time can be obtained by first choosing r of the n objects and then arranging them in order. Since there are $C(n, r)$ ways to choose r objects from among n distinct objects and $P(r, r)$ ways to arrange these r objects in order, the multiplication principle shows that

$$P(n, r) = C(n, r) \cdot P(r, r).$$

Therefore

$$C(n, r) = \frac{P(n, r)}{P(r, r)} = \frac{P(n, r)}{r!}.$$

*Another common notation for $C(n, r)$ is $\binom{n}{r}$.

Theorem 5.5

For any positive integers r and n such that $0 \leq r \leq n$,

$$C(n, r) = \frac{P(n, r)}{r!}.$$

Thus, for instance, we find by Theorem 5.5 that

$$C(10, 0) = \frac{P(10, 0)}{0!} = \frac{1}{1} = 1,$$

$$C(10, 1) = \frac{P(10, 1)}{1!} = \frac{10}{1} = 10,$$

$$C(10, 2) = \frac{P(10, 2)}{2!} = \frac{10 \cdot 9}{2 \cdot 1} = 45,$$

$$C(10, 3) = \frac{P(10, 3)}{3!} = \frac{10 \cdot 9 \cdot 8}{3 \cdot 2 \cdot 1} = 120, \text{ and}$$

$$C(10, 4) = \frac{P(10, 4)}{4!} = \frac{10 \cdot 9 \cdot 8 \cdot 7}{4 \cdot 3 \cdot 2 \cdot 1} = 210.$$

Since

$$P(n, r) = \frac{n!}{(n - r)!},$$

we can also write the formula for $C(n, r)$ as shown below.

$$C(n, r) = \frac{n!}{r!(n - r)!}$$

But as was true for the corresponding permutation formula, this formula should not normally be used for computations.

EXAMPLE 5.22 How many ways are there to select a subcommittee of five members from among a committee of 12?

Solution The number of committees is just the number of combinations of 12 members taken 5 at a time. By Theorem 5.5 this number is

$$C(12, 5) = \frac{P(12, 5)}{5!} = \frac{12 \cdot 11 \cdot 10 \cdot 9 \cdot 8}{5 \cdot 4 \cdot 3 \cdot 2 \cdot 1} = 792. \quad \blacksquare$$

EXAMPLE 5.23 How many ways are there to select 4 novels from a list of 16 novels to be read for a literature class?

Solution Since we are interested in knowing only which novels are selected (and not in their order of selection), this is a problem involving combinations. By Theorem 5.5 the answer is

$$C(16, 4) = \frac{P(16, 4)}{4!} = \frac{16 \cdot 15 \cdot 14 \cdot 13}{4 \cdot 3 \cdot 2 \cdot 1} = 1820. \blacksquare$$

Practice Problem 13 Suppose that eight people raise their glasses in a toast. If every person clinks glasses exactly once with everyone else, how many clinks will be heard?

In practice one of the most difficult aspects of solving a counting problem is to determine whether to use permutations or combinations. Remember that although both involve choosing objects from a set of distinct objects, *permutations are used when the order of selection matters, and combinations are used when the order of selection does not matter.*

EXAMPLE 5.24 An investor is going to invest \$16,000 in four stocks from a list of twelve prepared by his broker. How many different investments are possible if

(a) \$4000 is to be invested in each stock?
(b) \$6000 is to be invested in one stock, \$5000 in another, \$3000 in a third, and \$2000 in the fourth?
(c) \$5000 is to be invested in each of two stocks and \$3000 is to be invested in each of two others?

Solution (a) Since the same amount of money is to be invested in each stock, the order in which the four stocks are selected is unimportant. Thus part (a) is a problem involving *combinations,* and so there are

$$C(12, 4) = \frac{P(12, 4)}{4!} = \frac{12 \cdot 11 \cdot 10 \cdot 9}{4 \cdot 3 \cdot 2 \cdot 1} = 495$$

different investments that can be made.

(b) In this case we will choose four stocks in sequence. In the first stock chosen we will invest \$6000, in the second we will invest \$5000, in the third we will invest \$3000, and in the fourth we will invest \$2000. Thus the order of selection matters, and so part (b) is a problem involving *permutations*. Hence there are

$$P(12, 4) = 12 \cdot 11 \cdot 10 \cdot 9 = 11{,}880$$

different investments that can be made.

(c) In this case we will divide the problem into two parts: first we will choose the two stocks in which \$5000 will be invested, and then we will choose the two stocks in which \$3000 will be invested. As in part (a) there are

$$C(12, 2) = \frac{P(12, 2)}{2!} = \frac{12 \cdot 11}{2 \cdot 1} = 66$$

ways to select the two stocks in which \$5000 will be invested. Now we must select two of the remaining ten stocks in which to invest \$3000 each. This selection can be made in

$$C(10, 2) = \frac{P(10, 2)}{2!} = \frac{10 \cdot 9}{2 \cdot 1} = 45$$

different ways. Thus by the multiplication principle there are

$$66 \cdot 45 = 2970$$

different investments that can be made. ■

Often, as in part (c) of Example 5.24, it is necessary to use permutations or combinations together with the addition or multiplication principles. The following examples are of this type.

EXAMPLE 5.25 A committee of four is to be chosen from among five women and six men. How many different committees are possible that contain at least three women?

Solution A committee containing at least three women must consist of either three women and one man or four women and no men. Thus we will consider two cases. The number of committees containing three women and one man can be computed using the multiplication principle because we can construct such a committee by first choosing the women and then choosing the man. Since there are $C(5, 3)$ ways to select three women from among five and $C(6, 1)$ ways to select one man from among six, there are

$$C(5, 3) \cdot C(6, 1) = 10 \cdot 6 = 60$$

committees containing three women and one man. Similar reasoning shows that there are

$$C(5, 4) \cdot C(6, 0) = 5 \cdot 1 = 5$$

committees containing four women and no men. Hence by the addition principle there are

$$60 + 5 = 65$$

possible committees containing at least three women. ■

EXAMPLE 5.26 An election is being held to fill two faculty seats and three student seats on the University Curriculum Committee. From among the five candidates for the faculty seat, the person receiving the most votes will serve a three-year term, and the runner-up will serve a two-year term. From among the seven student candidates, the three winners will each serve one-year terms. How many different election results are possible?

Solution The possible election results can be obtained by first choosing the faculty winners and then choosing the student winners. Because the order of finish is important for the faculty seats, the number of possible election results for the faculty seats involves *permutations*. Thus there are

$$P(5, 2) = 5 \cdot 4 = 20$$

possible results in the faculty election. On the other hand, the student winners all have the same length of term; so the selection of the student winners involves *combinations*. Hence there are

$$C(7, 3) = \frac{P(7, 3)}{3!} = \frac{7 \cdot 6 \cdot 5}{3 \cdot 2 \cdot 1} = 35$$

possible results in the student election. It now follows from the multiplication principle that there are

$$20 \cdot 35 = 700$$

possible election results. ■

Practice Problem 14 How many different lists of five letters contain three different consonants and two different vowels? (Regard "y" as a consonant.)

EXERCISES 5.4

Evaluate each of the numbers in Exercises 1–16.

1. $P(7, 5)$ **2.** $P(6, 6)$ **3.** $P(16, 3)$ **4.** $P(15, 5)$

5. $4!$ **6.** $7!$ **7.** $8!$ **8.** $10!$

9. $C(9, 6)$ **10.** $C(15, 5)$ **11.** $C(8, 4)$ **12.** $C(6, 3)$

13. $P(n, 0)$ **14.** $P(n, n)$ **15.** $C(n, n)$ **16.** $C(n, 0)$

17. (a) How many permutations are there of the letters a, b, c, d, and e taken two at a time? List them all.

(b) How many combinations are there of the letters a, b, c, d, and e taken two at a time? List them all.

18. (a) How many permutations are there of the numbers 1, 2, 3, and 4 taken three at a time? List them all.

(b) How many combinations are there of the numbers 1, 2, 3, and 4 taken three at a time? List them all.

19. Twenty applicants for an executive position are to be interviewed to narrow the list of candidates to the top five. How many possible results are there if

(a) the top five are ranked in order of preference?

(b) thc top five are unranked?

20. A newly formed consumer action group has thirty members. In how many ways can the group elect

(a) a president, vice-president, secretary, and treasurer (all different)?

(b) an executive committee consisting of five members?

21. Six speakers are scheduled to address a convention. In how many different orders can the speakers appear?

22. In how many different ways can the letters of the word CREAM be arranged?

23. A nautical signal consisting of five flags strung vertically on a rope is to be constructed. How many different signals can be made if there are nine flags of different colors?

24. Five chairs are placed in a row.

(a) How many different ways are there of seating five persons in these chairs?

(b) If there are eight people available, in how many different ways can the five seats be filled?

25. For marketing purposes a company has divided the United States into eight regions. It wishes to test a new product in three of these regions. How many different ways are there to select these three regions?

26. In how many different ways can the Supreme Court render a 5-to-4 decision?

27. In the state senate, thirteen of the twenty members come from urban districts, and the others come from rural districts. A committee of six is to be appointed to study educational reform.

(a) How many committees can be formed?

(b) If the committee is to be equally divided between urban and rural members, how many committees are possible?

28. A grievance committee consisting of five members is to be selected from among nine women and seven men.

(a) How many different committees can be formed?

(b) How many of these committees contain three women and two men?

29. Twelve of the houses on a particular street receive cable television and eight do not. A researcher studying the amount of time that people watch television intends to visit four of the houses that receive cable television and three that do not. How many such selections of houses are there?

30. A candy store classifies its candies as creams (ten types), nuts (seven types), and chews (eight types). A customer has ordered an assortment to consist of six types of creams, four types of nuts, and five types of chews. How many such assortments are possible?

31. A five-member committee is to be selected from among four representatives of management and five representatives of labor. In how many different ways can the committee be formed under the following circumstances?

(a) Anyone is eligible to serve on the committee.

(b) The committee must consist of three representatives of management and two representatives of labor.

(c) The committee must consist of two representatives of management and three representatives of labor.

(d) The committee must contain at least three representatives of management.

32. A woman must choose two pairs of shoes and either two skirts and three blouses or two dresses to pack for a business trip. If she has five pairs of shoes, four skirts, eight blouses, and six dresses from which to choose, in how many different ways can she pack her suitcase?

33. Of the 2170 stocks on the New York Stock Exchange, 1438 advanced, 522 declined, and 210 were unchanged during the last week of May, 1987. In how many ways can this happen? [Write your answer in terms of $C(n, r)$.]

34. In a literature class of twelve graduate students, the instructor will choose three students to analyze *Howards End,* four students to analyze *Room with a View,* and five students to analyze *A Passage to India*. In how many different ways can the students be chosen?

35. A Chinese restaurant offers a ''family dinner'' consisting of any four items on its menu. Its advertising claims that over 300 different family dinners are available. If this claim is true, what is the smallest number of items that can be listed on its menu?

36. The figure below shows a portion of the map of a city. How many different routes are there from City Hall to the Post Office without moving south or west?

37. Suppose that a coin is flipped n times and the resulting sequence of heads and tails is recorded.

(a) How many sequences are possible?

(b) If $0 \leq k \leq n$, how many of the sequences in part (a) contain exactly k heads?

(c) A sequence is an ordered arrangement. Why is it not possible to use permutations to solve part (b)?

38. Use Theorem 5.5 to show that $C(n, 0) = 1$, $C(n, n) = 1$, and $C(n, r) = C(n, n - r)$. Justify these results in terms of the meaning of combinations.

39. Use Theorem 5.5 to show that $r \cdot C(n, r) = n \cdot C(n-1, r-1)$ for all integers r and n such that $1 \leq r \leq n$.

40. Show that $C(n, r) \leq C(n, r+1)$ for all integers r and n such that $0 < r < n/2$.

41. Three Americans, four Russians, and two Chinese delegates are attending a conference on international trade.

(a) In how many different ways can these nine delegates line up for a group photograph with all persons of the same nationality standing together?

(b) How many of the ways in part (a) do not have a Russian standing beside a Chinese?

42. A book store wants to create a window display containing a row of two cookbooks, five novels, and three biographies. The books are to be chosen from among five cookbooks, eight novels, and seven biographies released in the last month.

(a) How many such displays are possible?

(b) How many such displays are possible if, in addition, all books of the same category must be side-by-side?

43. How many ways are there to seat n people around a *circular* table, taking into account only the position of each person with respect to the others?

44. (a) How many distinct numbers can be formed using the digits 1, 2, 3, 4, 5, 9, 9, 9, 9, 9, 9 if no two 9's can be adjacent?

(b) How many distinct numbers can be formed using the digits 1, 2, 3, 4, 5, 6, 7, 8, 9, 9, 9, 9, 9, 9 if no two 9's can be adjacent?

Answers to Practice Problems

12. 504

13. 28

14. 1,596,000

5.5 Pascal's Triangle and the Binomial Theorem

The numbers $C(n, r)$ defined in Section 5.4 arise naturally not only in counting problems but also in algebra, where they are referred to as **binomial coefficients.** The binomial coefficients occur when expressions of the form $(x + y)^n$ are multiplied out. In this section we will discuss the evaluation of such expressions.

Let us begin by multiplying out $(x + y)^n$ for several values of n to see if we can find a pattern.

$$\begin{aligned}
(x+y)^0 &= 1\\
(x+y)^1 &= x + y\\
(x+y)^2 &= x^2 + 2xy + y^2\\
(x+y)^3 &= x^3 + 3x^2y + 3xy^2 + y^3\\
(x+y)^4 &= x^4 + 4x^3y + 6x^2y^2 + 4xy^3 + y^4\\
(x+y)^5 &= x^5 + 5x^4y + 10x^3y^2 + 10x^2y^3 + 5xy^4 + y^5
\end{aligned}$$

Note that in each case the expansion of $(x + y)^n$ contains $n + 1$ terms of the form

$$cx^{n-r}y^r$$

for $r = 0, 1, 2, \ldots, n$, where the coefficient c is some positive integer that depends on the values of n and r.

PASCAL'S TRIANGLE

We will begin by concentrating on the coefficients. The array of coefficients obtained by expanding the expressions $(x + y)^n$ for all nonnegative integers n is called **Pascal's triangle** after the French mathematician and philosopher Blaise Pascal (1623–62), who discovered many of the triangle's properties. Thus the beginning of Pascal's triangle is shown below.

(row 0)	1
(row 1)	1 1
(row 2)	1 2 1
(row 3)	1 3 3 1
(row 4)	1 4 6 4 1
(row 5)	1 5 10 10 5 1

Notice that each row begins and ends with 1. Moreover, every other entry in a row is the sum of the two closest entries in the row above. Thus, for example, the third entry in row 5 is the sum of the second and third entries in row 4:

$$10 = 4 + 6.$$

This property enables us to generate the rest of the triangle very easily. For instance, row 6 is obtained by adding consecutive entries in row 5 and then putting a 1 at both ends.

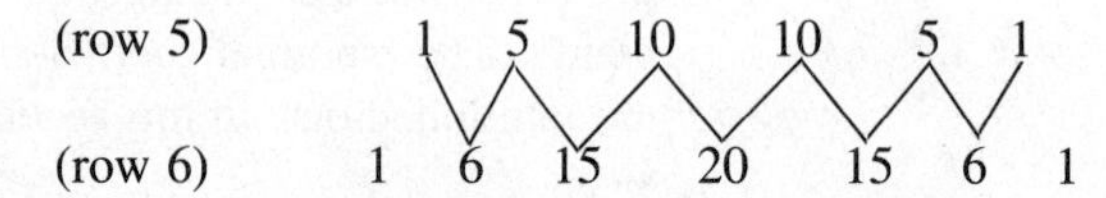

EXAMPLE 5.27 Find row 7 of Pascal's triangle.

Solution The row begins and ends with 1. The other entries are obtained by adding adjacent entries in row 6. The result is shown below.

$$1 \quad 7 \quad 21 \quad 35 \quad 35 \quad 21 \quad 7 \quad 1 \quad \blacksquare$$

Practice Problem 15 Find row 8 of Pascal's triangle.

Surprisingly, the numbers in Pascal's triangle turn out to be the numbers $C(n, r)$ from Section 5.4. For example, the numbers in row 4 of Pascal's triangle are

$$C(4, 0) = 1 \quad C(4, 1) = 4 \quad C(4, 2) = 6 \quad C(4, 3) = 4 \quad C(4, 4) = 1.$$

It is not difficult to see why the coefficient of the term $x^{n-r}y^r$ in the expansion of $(x + y)^n$ is $C(n, r)$. Since

$$(x + y)^n = \underbrace{(x + y)(x + y) \cdots (x + y)}_{n \text{ factors}},$$

a term of the form $x^{n-r}y^r$ results whenever we choose the y term from exactly r of the factors $x + y$. And by Theorem 5.5 the number of ways to choose exactly r terms of y from among the n factors of $x + y$ is $C(n, r)$.

THE BINOMIAL THEOREM

Having determined the coefficients that occur in the expansion of $(x + y)^n$, it is now easy to evaluate this expression.

Theorem 5.6 (the binomial theorem)

The coefficient of $x^{n-r}y^r$ in the expansion of $(x + y)^n$ is $C(n, r)$. Thus

$$(x + y)^n = C(n, 0)x^n + C(n, 1)x^{n-1}y^1 + \cdots + C(n, n)y^n.$$

Observe that in the expansion of $(x + y)^n$ the powers of x decrease from n to 0 while the powers of y increase from 0 to n. Moreover, the sum of the x and y exponents in each term is n. The coefficients of the terms are just the numbers from row n of Pascal's triangle.

EXAMPLE 5.28 Expand $(x + y)^5$ in powers of x and y.

Solution Applying the binomial theorem with $n = 5$, we obtain

$$(x + y)^5 = C(5, 0)x^5 + C(5, 1)x^4y^1 + C(5, 2)x^3y^2 + C(5, 3)x^2y^3 + C(5, 4)x^1y^4 + C(5, 5)y^5.$$

Thus, using row 5 of Pascal's triangle, we have

$$(x + y)^5 = x^5 + 5x^4y + 10x^3y^2 + 10x^2y^3 + 5xy^4 + y^5. \quad ■$$

EXAMPLE 5.29 Expand $(2a - b)^4$ in powers of x and y.

Solution Taking $n = 4$ in the binomial theorem gives

$$\begin{aligned}(x + y)^4 &= C(4, 0)x^4 + C(4, 1)x^3y^1 + C(4, 2)x^2y^2 + C(4, 3)x^1y^3 + C(4, 4)y^4 \\ &= x^4 + 4x^3y + 6x^2y^2 + 4xy^3 + y^4.\end{aligned}$$

If we now substitute $x = 2a$ and $y = -b$ in the expansion above, we obtain

$$\begin{aligned}(2a - b)^4 &= (2a)^4 + 4(2a)^3(-b) + 6(2a)^2(-b)^2 + 4(2a)(-b)^3 + (-b)^4 \\ &= 16a^4 - 32a^3b + 24a^2b^2 - 8ab^3 + b^4. \quad ■\end{aligned}$$

Practice Problem 16 Expand $(x - 2y)^5$ in powers of x and y.

In Chapter 6 we will use the binomial theorem to find a particular term in the expansion of $(x + y)^n$, as in the example below.

EXAMPLE 5.30 Find the coefficient of the term involving $x^{16}y^4$ in the expansion of $(x - 3y)^{20}$.

Solution The binomial theorem states that the term involving $x^{16}y^4$ in the expansion of $(x - 3y)^{20}$ is

$$C(20, 4)x^{16}(-3y)^4.$$

Since the value of only one binomial coefficient is needed, we will use Theorem 5.5 (rather than Pascal's triangle) to evaluate $C(20, 4)$. In this way we find that

$$C(20, 4) = \frac{P(20, 4)}{4!} = \frac{20 \cdot 19 \cdot 18 \cdot 17}{4 \cdot 3 \cdot 2 \cdot 1} = 4845.$$

Thus the term involving $x^{16}y^4$ in the expansion of $(x - 3y)^{20}$ is

$$C(20, 4)x^{16}(-3y)^4 = 4845x^{16}(81y^4) = 392{,}445x^{16}y^4,$$

and so the desired coefficient is 392,445. ■

Practice Problem 17 Find the coefficient of the term involving p^5q^{11} in the expansion of $(2p - q)^{16}$.

EXERCISES 5.5

1. Compute row 9 of Pascal's triangle.

2. Compute row 10 of Pascal's triangle.

In Exercises 3–14, use the binomial theorem to expand the given expression.

3. $(x + y)^6$

4. $(x + y)^7$

5. $(x - y)^7$

6. $(x - y)^6$

7. $(a + 4b)^4$

8. $(3p + q)^5$

9. $(3u - v)^5$

10. $(r - 2s)^6$

11. $(2x - \frac{1}{2})^6$

12. $(x - 1)^9$

13. $(1 - x)^8$

14. $(-2 + y)^8$

In Exercises 15–26, find the indicated coefficient.

15. the coefficient of x^{12} in $(x + 1)^{20}$

16. the coefficient of x^{18} in $(x + 1)^{28}$

17. the coefficient of x^{17} in $(x - 1)^{30}$

18. the coefficient of x^{15} in $(x - 1)^{24}$

19. the coefficient of $x^{14}y^{8}$ in $(x + y)^{22}$

20. the coefficient of $x^{6}y^{8}$ in $(x + y)^{14}$

21. the coefficient of $x^{19}y^{13}$ in $(x - y)^{32}$

22. the coefficient of $x^{10}y^{11}$ in $(x - y)^{21}$

23. the coefficient of $x^{12}y^{6}$ in $(x + 2y)^{18}$

24. the coefficient of $x^{4}y^{15}$ in $(2x + y)^{19}$

25. the coefficient of $x^{10}y^{5}$ in $(x - \frac{1}{2}y)^{15}$

26. the coefficient of $x^{3}y^{17}$ in $(\frac{1}{3}x - y)^{20}$

27. From among a group of eight disgruntled workers, a grievance committee of at least one and at most three persons is to be formed. How many different committees can be formed?

28. The university's alumni service award can be given to at most four people per year. This year there are seven nominees. In how many different ways can the recipients of the award be chosen?

29. Show that $C(n, 0) + C(n, 1) + C(n, 2) + \cdots + C(n, n) = 2^n$ for all positive integers n by using the binomial theorem.

30. Show that $C(n, 0) - C(n, 1) + C(n, 2) - \cdots + (-1)^n C(n, n) = 0$ for all positive integers n by using the binomial theorem.

31. How does the equation in Exercise 29 relate to the number of subsets of a set with n elements?

32. Let S be a set containing n elements. Show that the number of subsets of S containing an even number of elements equals the number of subsets of S containing an odd number of elements. (*Hint:* Use Exercise 30.)

33. Use Theorem 5.5 to show that $C(n + 1, r) = C(n, r - 1) + C(n, r)$.

Answers to Practice Problems

15. 1 8 28 56 70 56 28 8 1

16. $(x - 2y)^5 = x^5 - 10x^4y + 40x^3y^2 - 80x^2y^3 + 80xy^4 - 32y^5$

17. $-139{,}776$

CHAPTER 5 REVIEW

IMPORTANT TERMS

- **complement of a set ($\overline{A}$)** *(5.1)*
- **disjoint sets**
- **element of a set**
- **empty set ($\varnothing$)**
- **equality of sets ($A = B$)**
- **intersection of sets ($A \cap B$)**
- **set**
- **subset ($A \subseteq B$)**
- **union of sets ($A \cup B$)**
- **universal set**
- **Venn diagram**
- **number of elements in a set ($|S|$)** *(5.2)*
- **tree diagram**
- **combination** *(5.4)*
- **n factorial ($n!$)**
- **permutation**
- **binomial coefficients** *(5.5)*
- **Pascal's triangle**

IMPORTANT FORMULAS

Theorem 5.1

Let A and B be any subsets of a finite universal set U. Then
(a) $|\overline{A}| = |U| - |A|$
(b) $|A \cup B| = |A| + |B| - |A \cap B|$. *(principle of inclusion-exclusion)*

Theorem 5.2 (the multiplication principle)

Suppose that there are k operations to be performed in sequence, and the first operation can be performed in n_1 ways, and for each of these the second operation can occur in n_2 ways, and for each of these the third operation can occur in n_3 ways, and so forth. Then there are

$$n_1 \cdot n_2 \cdot \cdots \cdot n_k$$

ways to perform all k of the operations in sequence.

Theorem 5.3 (the addition principle)

Suppose that there are k sets, each pair of which is disjoint. (That is, no two of the sets have an element in common.) If the first set contains n_1 elements, the second n_2 elements, and so forth, then the union of the sets contains

$$n_1 + n_2 + \cdots + n_k$$

elements.

Factorial notation

For any positive integer n,

$$0! = 1$$
$$n! = n(n-1)(n-2) \cdots 3 \cdot 2 \cdot 1.$$

Theorem 5.4 (number of permutations of n objects taken r at a time)

For any positive integers r and n such that $0 \le r \le n$,

$$P(n, r) = n(n-1)(n-2) \cdots (n-r+1) = \frac{n!}{(n-r)!}.$$

Theorem 5.5 (number of combinations of n objects taken r at a time)

For any positive integers r and n such that $0 \le r \le n$,

$$C(n, r) = \frac{P(n, r)}{P(r, r)} = \frac{P(n, r)}{r!}.$$

Theorem 5.6 (the binomial theorem)

The coefficient of $x^{n-r}y^r$ in the expansion of $(x+y)^n$ is $C(n, r)$. Thus

$$(x+y)^n = C(n, 0)x^n + C(n, 1)x^{n-1}y^1 + C(n, 2)x^{n-2}y^2 + \cdots + C(n, n)y^n.$$

REVIEW EXERCISES

1. Let $A = \{2, 5, 6, 8\}$ and $B = \{1, 2, 3, 4, 5, 6\}$ be subsets of the universal set $U = \{1, 2, 3, 4, 5, 6, 7, 8, 9, 10\}$. Compute each of the following sets.

(a) $A \cup B$ (b) $A \cap B$ (c) $\overline{A}$ (d) $A \cup \overline{B}$

2. Let $A = \{1, 2, 3\}$, $B = \{2, 3, 4, 5\}$, and $C = \{4, 5, 6, 7, 8\}$ be subsets of the universal set $U = \{1, 2, 3, 4, 5, 6, 7, 8, 9, 10\}$. Determine if the following statements are true or false.

(a) $8 \in \overline{A} \cup B$ (b) $5 \not\in A \cup \overline{(B \cap C)}$
(c) $A \cap B = \varnothing$ (d) $C \subseteq \overline{A}$
(e) $A \cup B \cup C = U$ (f) $A \subseteq \{x \in U | x < 5\}$
(g) $\{3\} \not\subseteq B$ (h) $|C| = |\overline{C}|$

3. When a radio station surveyed 250 listeners of its evening programming, it found that 83 were female, 96 were teenagers, and 67 were both (that is, teenage females). How many of those surveyed were nonteenage males?

4. The following information was obtained about the houses in a particular subdivision.

195 have central air conditioning.
92 have a finished basement.
201 have a two-car garage.
72 have central air conditioning and a finished basement.
151 have central air conditioning and a two-car garage.
66 have a finished basement and a two-car garage.
54 have central air conditioning, a finished basement, and a two-car garage.
7 did not have central air conditioning, a finished basement, or a two-car garage.

(a) How many houses in this subdivision have a two-car garage but neither central air conditioning nor a finished basement?

(b) How many houses in this subdivision have a finished basement and a two-car garage but not central air conditioning?

(c) How many houses are in this subdivision?

5. A hospital classifies its patients according to sex and blood type. Draw a tree diagram showing all of the possible classifications. (See Example 5.8 for the possible blood types.)

6. Evaluate each of the following numbers.

(a) 5! (b) 9! (c) $P(11, 0)$ (d) $C(9, 8)$
(e) $P(8, 4)$ (f) $C(7, 3)$ (g) $P(7, 7)$ (h) $C(12, 0)$

7. A police artist makes a composite sketch of a crime suspect's face by asking witnesses to select one each of ten noses, twelve pairs of eyes, eight mouths, and seven hairstyles. How many different faces are possible?

8. In reference [3] Kotler discusses the division of markets into segments based on geographic, demographic, and psychographic variables. He lists 5 geographic variables, 11 demographic variables, and 8 psychographic variables. How many different divisions are possible using exactly one variable of each type?

9. How many different rearrangements of the letters in SPRING are there?

10. An investor intends to buy 100 shares of each of four stocks selected from a list of twenty. How many different selections are possible?

11. First, second, and third prizes are to be awarded in a pie-baking contest. If there are 12 entries, in how many different ways can the prizes be given?

12. The mathematics department must assign one course to each of nine part-time instructors. The courses to be assigned consist of four sections of college algebra, three sections of finite math, and two sections of calculus. How many different teaching assignments are possible? (Consider two teaching assignments the same if each teacher teaches the same course.)

13. Six liberal candidates and five conservative candidates are running for four seats on a city council. Assume that no two candidates receive the same number of votes.
(a) How many different orders of finish are possible?
(b) How many different sets of winners are possible?
(c) In how many of the sets in part (b) are four liberals elected?
(d) In how many of the sets in part (b) are two liberals and two conservatives elected?
(e) In how many of the sets in part (b) are at least two liberals elected?

14. Write rows 0, 1, 2, 3, 4, 5, and 6 of Pascal's triangle.

15. Expand $(3r + s)^6$ in powers of r and s.

16. Determine the coefficient of the term $x^{10}y^5$ in the expansion of $(x - 2y)^{15}$.

REFERENCES

1. Baron, Jonathan, "Division of Attention in Successiveness Discrimination," in *Attention and Performance IV,* Sylvan Kornblum, ed. New York: Academic Press, 1973, pp. 703–711.
2. Broadhurst, P. L., "Emotionality and the Yerkes-Dodson Law," *Journal of Experimental Psychology,* vol. 54(1957), pp. 345–352.
3. Kotler, Philip, *Marketing Management,* 3rd ed. Englewood Cliffs, N.J.: Prentice-Hall, 1976.
4. Rudestam, Kjell Erik and Bruce John Morrison, "Student Attitudes Regarding the Temporary Closing of a Major University," *American Psychologist,* vol. 26(1971), pp. 519–525.

6

PROBABILITY

Many historians regard an exchange of letters in 1654 between the great French mathematicians Blaise Pascal and Pierre de Fermat as the origin of probability theory. Their correspondence was prompted by two gambling problems posed to Pascal by the Chevalier de Méré, a noted dilettante in the French court. From this origin around the gambling tables of seventeenth century Europe, the theory of probability has gradually progressed to a position of prominence in modern science. Many of the greatest mathematicians in history, including Pascal, Bernoulli, Euler, Laplace, and Gauss, contributed to its early development. Yet the theory was rarely applied outside of gambling until the nineteenth century, when it began to be recognized as a powerful tool. Today probabilistic methods permeate the physical, biological, behavioral, and management sciences. In this chapter we will examine the basic principles of this theory and explore some of its applications.

6.1 Computing Probabilities by Counting

Uncertainties are part of life. When we flip a coin or take an airplane flight or start a new business, to some extent we must simply hope for the best. These actions are examples of *random processes,* in the sense that the outcome in each case cannot be predicted with absolute certainty. Probability can be defined as the mathematical theory of random processes, in other words, as the "science of uncertainty." It is paradoxical that mathematics, the most exact of the sciences, has something to say about the very nature of uncertainty. In this section we will begin to investigate this curious relationship.

THE CLASSICAL DEFINITION OF PROBABILITY

Among the many ways to think about probability, the oldest and most widely used is the *relative frequency* interpretation: The probability of an event represents the proportion of the time that the event tends to occur over the long run. Hence the probability of an event will be a number between 0 and 1 inclusive that measures the event's likelihood of happening. Thus, for instance, when a symmetric coin is flipped, we say that the probability of heads appearing is 1/2. This means that in a large number of flips the coin will tend to land with heads showing half of the time.

In many situations it is intuitively plausible that all possible outcomes are equally likely. In this case it is possible to calculate probabilities theoretically.

EXAMPLE 6.1 What is the probability that a 5 will appear in one roll of a uniform die?

Solution When a uniform die is rolled, each of its six faces is equally likely to appear. Thus the probability that a 5 appears is 1/6. (See Figure 6.1.) ■

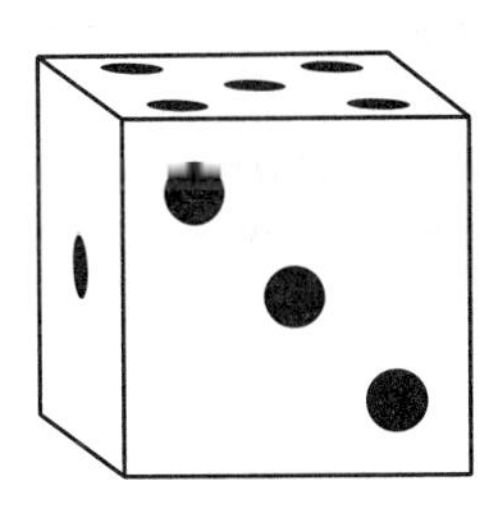

FIGURE 6.1

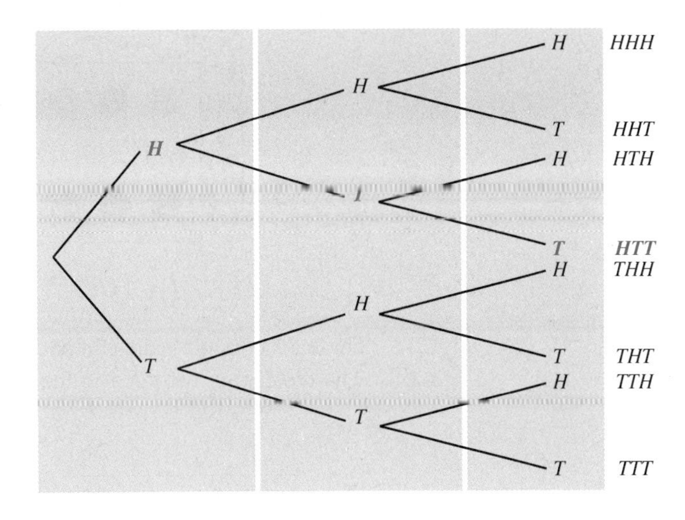

FIGURE 6.2

EXAMPLE 6.2 If a symmetric coin is flipped three times, what is the probability that it will land heads on the first flip and tails on the second and third flips?

Solution The possible sequences of heads and tails can be found by a tree diagram (see Figure 6.2, where H denotes heads and T denotes tails). Since the coin is symmetric, it is just as likely to land heads as tails on any flip. Thus each of the paths in this tree diagram is equally likely. Since there are eight sequences of heads and tails, the probability that the sequence H, T, T occurs is 1/8. ■

Practice Problem 1 The spinner in Figure 6.3 is to be spun twice.

(a) Construct a tree diagram showing the possible outcomes.
(b) Determine the probability that both spins are 2's.

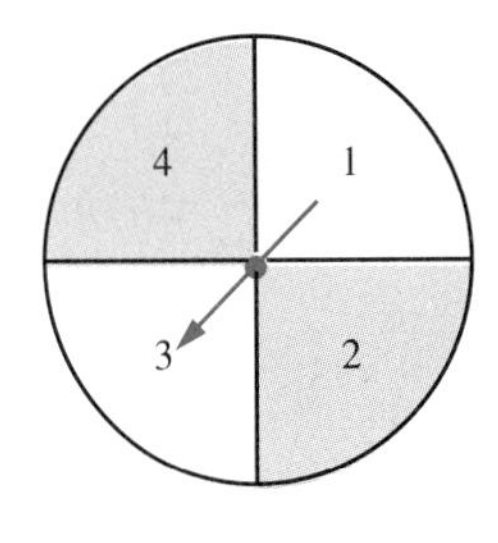

FIGURE 6.3

EXAMPLE 6.3 A pair of uniform dice, one white and one black, are rolled. What is the probability that 1 appears on each die?

Solution There are 6 values possible on the white die, and for each of these there are 6 possible values for the black die. (See Figure 6.4.) Hence the multiplication principle shows that there are $6 \cdot 6 = 36$ possible pairs of numbers that can result when these dice are rolled.

FIGURE 6.4

These 36 possible pairs can be found using a tree diagram similar to Figure 6.2. The results are recorded in the table below.

		Black					
		1	2	3	4	5	6
	1	(1, 1)	(1, 2)	(1, 3)	(1, 4)	(1, 5)	(1, 6)
	2	(2, 1)	(2, 2)	(2, 3)	(2, 4)	(2, 5)	(2, 6)
	3	(3, 1)	(3, 2)	(3, 3)	(3, 4)	(3, 5)	(3, 6)
White	4	(4, 1)	(4, 2)	(4, 3)	(4, 4)	(4, 5)	(4, 6)
	5	(5, 1)	(5, 2)	(5, 3)	(5, 4)	(5, 5)	(5, 6)
	6	(6, 1)	(6, 2)	(6, 3)	(6, 4)	(6, 5)	(6, 6)

Since the dice are uniform, we assume that each of these 36 pairs of numbers is equally likely to occur. Thus the probability is 1/36 that 1 appears on each die. ■

SAMPLE SPACES AND EVENTS

We see from the table in Example 6.3 that, when rolling a pair of uniform dice, the only way to obtain a sum of 2 is for a 1 to appear on each die. Hence the probability of obtaining a sum of 2 is 1/36. Notice that there are only 11 possible sums that can occur if a pair of dice are rolled (namely, 2, 3, 4, . . ., 12). Thus if we had considered only the *sums* that can occur (instead of considering the actual pairs of numbers), we might have incorrectly concluded that the probability of obtaining a sum of 2 is 1/11. To avoid this type of mistake, we must be certain to identify all of the possible results in the situation of interest, and we must be sure that these results are all equally likely to occur.

We call a situation having various possible results an **experiment.** Thus we consider all of the following to be experiments: flipping a coin, rolling a pair of dice, observing the frequency of automobile accidents in a particular age group of drivers, or counting the number of customers in line at a supermarket.

By a **sample space** for an experiment we mean a set containing possible results of the experiment, which are called **outcomes.** *The outcomes must be* *chosen to have the property that when the experiment is performed, one and only one of the outcomes occurs.*

What we consider as outcomes of an experiment will depend on the type of situations that are of interest. For example, a particular marketing experiment may require observing the customers in a particular store. But, depending on the situation, we may wish to observe the number of customers, the interval between arrival of customers, or the yearly income of the customers.

It is important to understand that there may be different ways of choosing the outcomes (and hence the sample space) of an experiment. For example, in the experiment of rolling a single die, one way of choosing the outcomes is to select the possible numbers that can result. For this choice the sample space is

$$\{1, 2, 3, 4, 5, 6\}.$$

Notice that when a die is rolled, *exactly one* of these outcomes must occur. On the other hand, we may be interested in knowing only if an even or odd number is rolled; in this case the sample space is

$$\{\text{even}, \text{odd}\}.$$

Again notice that *one and only one* of these outcomes must occur when a die is rolled. However, neither

$$A = \{1, \text{even}, 5\} \qquad \text{nor} \qquad B = \{1, 2, 4, 6, \text{odd}\}$$

is a sample space for this experiment: The outcome 3 is not contained in set A, and two outcomes in set B (namely, 1 and odd) occur when a 1 is rolled.

EXAMPLE 6.4 Construct a sample space for the experiment of flipping a coin, assuming that the coin will not land on edge.

Solution In this case there are two outcomes, namely heads and tails. Thus a sample space for this experiment is

$$\{\text{heads}, \text{tails}\}. \quad \blacksquare$$

EXAMPLE 6.5 Construct a sample space for the experiment of flipping a coin three times and observing the sequence of heads and tails that appears.

Solution In this experiment there are eight possible outcomes, shown in the tree diagram of Figure 6.2. The corresponding sample space is

$$\{HHH, HHT, HTH, HTT, THH, THT, TTH, TTT\}. \quad \blacksquare$$

EXAMPLE 6.6 An insurance company is going to promote two of its executives to manage regional offices. Those under consideration are Mr. Adams, Mrs. Bethea, Ms. Choi, Mr. Dunn, and Mr. Eovaldi. Construct a sample space for this experiment.

Solution Here we must select two individuals from among five for promotion. The number of possible outcomes is therefore $C(5, 2) = 10$. A sample space for this experiment is

$$\{AB, AC, AD, AE, BC, BD, BE, CD, CE, DE\},$$

where we have indicated by their initials the pairs of persons to be promoted. ■

Practice Problem 2 A distributor of business forms has a toll-free phone number on which its out-of-state customers can place orders. The company can handle as many as 15 calls per hour on this line. Construct a sample space for the experiment of observing the number of calls on this toll-free line on a Monday morning between 9 and 10 A.M.

Often we are interested in knowing the probability that one of several outcomes happens. For instance, in rolling a die we may want to know the probability that the die comes up greater than 3. In this case we are interested in finding the probability that one of the outcomes

$$4, \quad 5, \quad \text{or} \quad 6$$

occurs. We call any such set of outcomes an **event.** In other words, *an event is simply a subset of the sample space of an experiment*. Listed below are several other events and the corresponding sets of outcomes for the experiment of rolling a die.

Description of the event	*Set of outcomes*
1. The die comes up odd.	{1, 3, 5}
2. The die comes up less than 5.	{1, 2, 3, 4}
3. The die comes up 2.	{2}
4. The die comes up 7.	∅
5. The die comes up less than 8.	{1, 2, 3, 4, 5, 6}

We say that an event **occurs** if one of the outcomes in the event happens. Thus the event "the die comes up odd" occurs if the die comes up 1, 3, or 5, and it does not occur otherwise. Note that the fourth event above can never occur. We call the empty subset ∅ the **impossible** event. On the other hand, the fifth event above must occur, and so we call the sample space itself the **certain** event.

EXAMPLE 6.6 Revisited For the experiment described in Example 6.6, list the subsets of outcomes corresponding to each of the following events.

(a) Two women are promoted.
(b) Two men are promoted.
(c) Ms. Choi is promoted.
(d) All the men are promoted.

Solution Refer to the sample space in Example 6.6.

(a) Of the five candidates for promotion, only two are women, namely Mrs. Bethea and Ms. Choi. Thus there is only one outcome in which two women are promoted, namely, BC. Hence the desired subset of outcomes is $\{BC\}$.
(b) Of the candidates for promotion, Mr. Adams, Mr. Dunn, and Mr. Fovaldi are the men. Therefore the outcomes in which two men are promoted are the possible pairs of these three candidates; so the desired subset of outcomes is $\{AD, AE, DE\}$.
(c) An outcome in which Ms. Choi is promoted is a pair containing the letter C. Thus the desired set of outcomes is $\{AC, BC, CD, CE\}$.
(d) Since three of the candidates for promotion are men and only two persons are to be promoted, this subset of outcomes is $\varnothing$. Thus promoting all the men is the impossible event. ■

Practice Problem 3 Recall the toll-free telephone number in Practice Problem 2. List the sets of outcomes described by the following events.

(a) Exactly ten calls are received.
(b) At least twelve calls are received.
(c) Fewer than five calls are received.
(d) No calls are received.
(e) More than sixteen calls are received.

THE PROBABILITY OF AN EVENT

We are interested in computing the probability of an event. When the sample space consists of equally likely outcomes, this computation involves counting. For example, in the experiment of rolling a symmetric die, we can easily calculate the probability of the event "the die comes up less than 5." For as we have already seen, this event corresponds to the set of outcomes

$$E = \{1, 2, 3, 4\}.$$

Since the die is symmetric, we assume that each of the six outcomes is equally likely. Hence in the long run we expect that one of the outcomes in E will occur on 4/6 of the rolls. We indicate that the probability of E is 4/6 by writing

$$P(E) = \frac{4}{6} = \frac{2}{3}.$$

This calculation is based on the following fundamental formula.

The Counting Rule

Let S be a finite sample space consisting of equally-likely outcomes. Then for any event $E \subseteq S$, the probability of E is given by

$$P(E) = \frac{|E|}{|S|} = \frac{\text{number of outcomes in } E}{\text{number of outcomes in } S}.$$

In order to use the counting rule it is crucial that the sample space consist of equally likely outcomes. The early probabilists usually took this fact for granted on the assumption that two outcomes should be considered equally likely if there is no reason why one will occur rather than the other. Today mathematicians are more cautious. For example, hospital records indicate that the probability of a random child being born male is about .514, not .5 as one might suspect. Even gambling devices may have small but significant biases.* We will use phrases such as "fair die" and "chosen at random" to signify that the possible outcomes are equally likely.

EXAMPLE 6.7 What is the probability that exactly two heads appear in three flips of a fair coin?

Solution Here the experiment consists of flipping a coin three times and observing the sequence of heads and tails that results. Recall that we saw in Example 6.5 that the sample space for this experiment is

$$\{HHH, HHT, HTH, HTT, THH, THT, TTH, TTT\}.$$

Since the coin is fair, this sample space consists of equally likely outcomes. Let E be the set of outcomes for the event "exactly two heads appear." Then

$$E = \{HHT, HTH, THH\}.$$

Hence by the counting rule

$$P(\text{exactly two heads appear}) = P(E) = \frac{|E|}{|S|} = \frac{3}{8}.$$

Therefore the probability of obtaining exactly two heads in three flips of a fair coin is 3/8. ■

EXAMPLE 6.8 What is the probability of rolling a sum of 7 with a pair of fair dice?

Solution Recall from Example 6.3 that when rolling a pair of dice, the possible outcomes are as follows.

		Second Die					
		1	2	3	4	5	6
	1	(1, 1)	(1, 2)	(1, 3)	(1, 4)	(1, 5)	**(1, 6)**
	2	(2, 1)	(2, 2)	(2, 3)	(2, 4)	**(2, 5)**	(2, 6)
First	3	(3, 1)	(3, 2)	(3, 3)	**(3, 4)**	(3, 5)	(3, 6)
Die	4	(4, 1)	(4, 2)	**(4, 3)**	(4, 4)	(4, 5)	(4, 6)
	5	(5, 1)	**(5, 2)**	(5, 3)	(5, 4)	(5, 5)	(5, 6)
	6	**(6, 1)**	(6, 2)	(6, 3)	(6, 4)	(6, 5)	(6, 6)

---

*In 1947, two students from the California Institute of Technology, A. R. Hibbs and R. Walford, surveyed the roulette wheels in casinos. They found that one out of every four had a bias sufficient to eliminate the casino's expected profits from a player who bet correctly.

Let S denote the sample space of these 36 outcomes. Since the dice are fair, the outcomes in S are equally likely. Now the event "a sum of 7 is rolled" corresponds to the set of outcomes

$$E = \{(6, 1), (5, 2), (4, 3), (3, 4), (2, 5), (1, 6)\}.$$

Therefore by the counting rule

$$P(\text{sum is } 7) = P(E) = \frac{|E|}{|S|} = \frac{6}{36} = \frac{1}{6}. \quad ■$$

Practice Problem 4 A mouse is placed into the starting room of the maze pictured in Figure 6.5. If the mouse chooses a doorway at random, what is the probability that it will next enter room C?

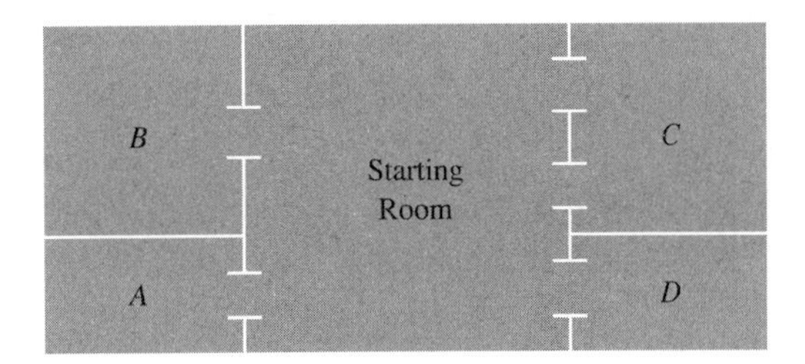

FIGURE 6.5

APPLICATIONS OF COUNTING TECHNIQUES

Let us consider the following problem. A firm has six accountants, three of whom are men (Mr. Allen, Mr. Babcock, and Mr. Casey) and three of whom are women (Ms. Dunbar, Mrs. Edmonds, and Ms. Fish). Two of these accountants are to be chosen at random to attend a financial seminar. What is the probability that the two persons selected are both men?

To solve this problem, we can proceed as before by enumerating the sample space S. In this case S consists of all possible pairs of the accountants, and so

$$S = \{AB, AC, AD, AE, AF, BC, BD, BE, BF, CD, CE, CF, DE, DF, EF\},$$

where we have identified each pair by their initials. Let M denote the event "two men are chosen." Then

$$M = \{AB, AC, BC\}.$$

Thus

$$P(M) = \frac{|M|}{|S|} = \frac{3}{15} = \frac{1}{5}.$$

There is another way to attack this problem, however, that does not require listing all the outcomes. Since S consists of all possible pairs chosen from among six accountants,

$$|S| = C(6, 2) = \frac{P(6, 2)}{2!} = \frac{6 \cdot 5}{2 \cdot 1} = 15.$$

Likewise M consists of all possible pairs chosen from among the three men (Allen, Babcock, and Casey); so

$$|M| = C(3, 2) = \frac{P(3, 2)}{2!} = \frac{3 \cdot 2}{2 \cdot 1} = 3.$$

Hence

$$P(M) = \frac{|M|}{|S|} = \frac{3}{15} = \frac{1}{5},$$

as above.

EXAMPLE 6.7 Revisited What is the probability that exactly two heads appear in three flips of a fair coin?

Solution Rather than enumerate the sample space S as in Example 6.7, we will solve the problem using techniques from Chapter 5. Since there are two possible results when a coin is flipped (heads or tails), the multiplication principle gives the number of outcomes when the coin is flipped three times to be

$$2 \cdot 2 \cdot 2 = 8.$$

Thus $|S| = 8$. To count the event E = "exactly two heads appear," we consider an outcome in E as a listing of two heads and one tail in some order. But then the number of elements in E is the number of ways to choose places for two heads from among three possible positions. Hence

$$|E| = C(3, 2) = 3.$$

Therefore

$$P(E) = \frac{|E|}{|S|} = \frac{3}{8}. \quad ■$$

Practice Problem 5 Use techniques from Chapter 5 to compute the probability of obtaining two different numbers on a single roll of a pair of fair dice.

When using the counting rule, the size of the sample space may make listing all of the outcomes impossible. In such cases we are forced to use the counting techniques discussed in Chapter 5.

EXAMPLE 6.9 A producer of computer software copies its programs onto diskettes to be sold to distributors. To check that the copying is being performed correctly, it randomly tests one fourth of its diskettes. What is the probability that there are 12 faulty diskettes among a lot of 40, and yet none of the faulty diskettes is chosen to be tested?

Solution Since the producer checks one fourth of all the diskettes, 10 diskettes from the lot of 40 will be tested. Thus for this experiment we choose as the sample space S the set consisting of all possible samples of 10 diskettes chosen from among the 40 produced. Then

$$|S| = C(40, 10).$$

Note that since the diskettes to be tested are chosen at random, each possible selection of 10 diskettes is equally likely. The event E of interest is the one in which no faulty diskettes are chosen for testing. Any outcome in E is obtained by selecting the 10 diskettes for testing from among the $40 - 12 = 28$ nondefective diskettes. Hence

$$|E| = C(28, 10).$$

Thus by the counting rule

$$\begin{aligned} P(\text{no faulty diskettes are chosen}) &= P(E) = \frac{|E|}{|S|} \\ &= \frac{C(28, 10)}{C(40, 10)} \\ &= \frac{13{,}123{,}110}{847{,}660{,}528} \approx .015, \end{aligned}$$

and so the probability that all the faulty diskettes go untested is about .015. ■

EXAMPLE 6.10 The semifinalists in a state lottery consist of 3 men and 6 women. If 4 of these persons are to be randomly selected to win \$1,000,000, what is the probability that 2 men and 2 women are selected?

Solution The sample space S consists of all possible selections of 4 persons from among the 9 semifinalists; so

$$|S| = C(9, 4) = 126.$$

The event E of interest consists of those selections containing 2 men and 2 women. Such a selection can be formed by choosing 2 of the 3 men and 2 of the 6 women. Thus the multiplication principle shows that

$$|E| = C(3, 2) \cdot C(6, 2) = 3 \cdot 15 = 45.$$

Hence by the counting rule

$$P(E) = \frac{|E|}{|S|} = \frac{45}{126} = \frac{5}{14}. \quad ■$$

Notice that in both Examples 6.9 and 6.10 we are sampling from a population of two types of objects. In Example 6.9 the diskettes are either faulty or nondefective, and we are interested in choosing 0 faulty diskettes and 10 nondefective diskettes. Similarly in Example 6.10 the people are either men or women, and we must select 2 men and 2 women. This type of sampling problem is quite common.

Practice Problem 6 From a committee consisting of 12 representatives of management and 10 representatives of labor, a subcommittee of 5 is to be selected. If members of the subcommittee are chosen at random, what is the probability that 3 representatives of management and 2 representatives of labor are chosen?

EXAMPLE 6.11 Forty persons are attending a seminar discussing financial investments. From among those attending, three *different* persons will be selected at random to receive door prizes.

(a) What is the probability that a particular person at the seminar wins the first door prize?

(b) What is the probability that a particular person at the seminar wins the third door prize?

Solution (a) The probability that someone wins the first door prize can be computed directly from the counting rule:

$$P(\text{someone wins first prize}) = \frac{\text{number of winners}}{\text{number in attendance}} = \frac{1}{40}.$$

(b) In order to compute the probability that someone wins the third door prize, let us regard the winners of the prizes as a permutation of three of the persons attending the seminar. Thus, for instance, the permutation

(Ms. McCauley, Mr. Jeffreys, Mrs. Earls)

will signify that the first prize was won by Ms. McCauley, the second by Mr. Jeffreys, and the third by Mrs. Earls. Let the sample space S consist of all such permutations. Because the prize winners are selected at random, each permutation in S is equally likely. Note that

$$|S| = P(40, 3) = 40 \cdot 39 \cdot 38.$$

In order for Mr. Smith to win the third door prize, his name must appear in the third position in the list of winners. For example,

(Mrs. Clark, Ms. Kennedy, Mr. Smith)

is one configuration in which Mr. Smith receives the third prize. Let E denote the set of all the permutations in S in which Mr. Smith's name occurs in the third position. Since nobody can win more than one prize, each such permutation can be constructed as follows.

1. Choose the name of a first prize winner (other than Mr. Smith).
2. Select the name of a second prize winner (other than Mr. Smith and the first prize winner).
3. Choose Mr. Smith's name.

Hence the multiplication principle shows that

$$|E| = 39 \cdot 38 \cdot 1,$$

and so by the counting rule

$$\begin{aligned} P(\text{Mr. Smith wins third prize}) = P(E) &= \frac{|E|}{|S|} \\ &= \frac{39 \cdot 38 \cdot 1}{40 \cdot 39 \cdot 38} \\ &= \frac{1}{40}. \end{aligned}$$

Thus we see that the probability of winning the third prize is the same as the probability of winning the first! With the benefit of hindsight, we see that this result is intuitively obvious: We have just as much chance of winning one prize as another. ■

Practice Problem 7 Six applicants for a particular job, four men and two women, are to be interviewed in a random order. What is the probability that the four men are interviewed before either woman?

EXERCISES 6.1

Describe a sample space for the experiments in Exercises 1–8.

1. One fair coin is flipped.

2. A two-headed coin is flipped.

3. Randomly selected persons are asked the day of the week on which they were born.

4. One marble is selected from an urn containing red, yellow, blue, green, and purple marbles.

5. A randomly selected voter is asked whether she voted for the Republican or Democratic candidate for governor.

6. A die is rolled, and then a coin is flipped.

7. The number of persons riding the subway in New York is counted on a particular day.

8. A coin is flipped until tails appears.

9. Fifteen automobile engines are to be tested and the number of defective units recorded. Describe a sample space for this experiment, and represent each of the following events as a subset of the sample space.

(a) None of the engines is defective.

(b) At most two of the engines are defective.

(c) More than half of the engines are defective.

10. Represent each of the following events as a subset of the sample space in Example 6.6.

(a) Mr. Adams is promoted.

(b) Mrs. Bethea and Mr. Dunn are promoted.

(c) Mr. Eovaldi is not promoted.

11. A coin is to be flipped twice. Describe a sample space for this experiment, and represent each of the following events as a subset of the sample space.

(a) The coin lands heads on the second flip.

(b) The coin comes up heads at least once.

(c) The coin comes up heads twice as many times as it comes up tails.

12. During practice, a basketball player is going to shoot three free throws. Describe a sample space for this experiment, and represent each of the following events as a subset of the sample space.

(a) The player misses the third shot.

(b) The player makes exactly two shots.

(c) The player makes some shot and misses the next.

13. If a fair die is rolled, what is the probability that it will come up

(a) odd? (b) less than 3?

(c) both odd and less than 3? (d) both odd and greater than 5?

14. American roulette wheels have 38 equal compartments numbered 00, 0, 1, . . ., 36. A ball is spun and comes to rest in one of these compartments. What is the probability that the number obtained is

(a) the number 29? (b) in the "second dozen" (13–24)?

(c) positive and even? (d) an odd number in the second dozen?

15. Assuming that all birthdays are equally likely, what is the probability that a randomly chosen person will have a birthday

(a) in March?

(b) between January 1 and February 15, inclusive?

16. A two-digit number from 00 to 99 is to be randomly generated by a computer. What is the probability that it will be

(a) less than 47? (b) positive and divisible by 3?

17. A large assortment of glass beads of uniform shape and size are mixed together. In this mixture, 34% of the beads are clear and blue, 12% are opaque and blue, 10% are clear and green, 18% are opaque and green, and 26% are clear and beige. If a bead is chosen from this mixture in the dark, what is the probability that it will be

(a) opaque and green? (b) blue?

(c) clear? (d) either blue or beige?

18. There were 250,000 tickets sold in a particular lottery. One ticket will win the first prize of $100,000; ten tickets will win second prizes of $10,000; fifty tickets will win third prizes of $2000; and one hundred tickets will win fourth prizes of $1000. If the winning tickets are selected at random, what is the probability that a random ticket will win

(a) $100,000? (b) $1000?

(c) $2000 or more? (d) nothing?

19. The registered voters in Baker County are 65% white, 25% black, and 10% Hispanic. Of the white voters, 60% are male, compared to 48% of the black voters and 50% of the Hispanic voters. Registered voters are selected at random for jury duty. What is the probability that someone selected for jury duty will be

(a) black? (b) a nonwhite? (c) an Hispanic male? (d) a female?

20. From a standard deck of 52 playing cards, one card is selected at random. Determine the probability of obtaining

(a) the ace of spades (b) any red card (c) any diamond (d) any king.

In Exercises 21–30 use Example 6.3 to determine the probability of the indicated event when two fair dice are rolled.

21. The sum is 5.

22. Both dice show the same number.

23. The white die comes up 6.

24. The sum exceeds 15.

25. Each die is greater than 2.

26. The sum is at least 8.

27. The white die is 3 and the sum is 10.

28. The numbers on both dice are even.

29. The number on the white die exceeds the number on the black die.

30. The number on one die is twice the number on the other die.

31. To win the Illinois State Lottery's daily game, a player must correctly pick a three-digit number, i.e., 000, 001, . . ., 999. What is the probability of winning the daily game?

32. A student guesses the answers to three true-or-false questions. What is the probability that

(a) all three answers are correct?

(b) exactly one answer is correct?

33. If a fair coin is flipped four times, find the probability that

(a) at least three heads appear

(b) exactly two heads appear

(c) no two heads occur on successive flips

(d) tails does not appear on either of the last two flips.

34. The integers 1–7 are written once each on separate slips of paper, placed in a box, and thoroughly mixed. If two slips are chosen simultaneously, find the probability that

(a) the number 4 is chosen (b) the number 3 is not chosen

(c) both 6 and 7 are chosen (d) some even number is chosen

(e) two odd numbers are chosen (f) the sum of the numbers chosen is 15.

Exercises 35–44 require the use of counting techniques from Chapter 5.

35. To play the Illinois State Lottery, a player must pick six numbers which match six numbers chosen at random from the integers 1–44. What is the probability of winning this lottery on a single play?

36. In an 8-horse race, a bettor bet the trifecta, which requires that the first three horses be identified in order of their finish. What is the probability of winning the trifecta by randomly selecting three numbers?

37. A supervisor randomly selected the personnel files of four employees. What is the probability that the files were selected in alphabetical order?

38. In a taste test, three crackers were spread with butter and three with margarine. Someone was asked to identify the three crackers spread with butter. What is the probability of doing so by guessing?

39. If three men and three women are seated at random in a row of six seats, what is the probability that no two people of the same gender are in adjacent seats?

40. To win the second prize in the Illinois State Lottery, a player must pick six numbers and correctly match exactly five of six numbers chosen at random from the integers 1–44. What is the probability of winning the second prize in this lottery on a single play?

41. A committee of four is to be chosen at random from a group of six men and six women. What is the probability that the committee will contain

(a) no men? (b) two men and two women?

42. An urn contains 4 red marbles, 6 blue marbles, and 5 green marbles. If two of the marbles are selected at random from the urn, what is the probability that

(a) both are blue? (b) neither is green?
(c) one is red and one is blue? (d) the two are different colors?

43. In a lot of twenty microcomputer diskettes, exactly four are defective. If the diskettes are packaged in two boxes of ten, what is the probability that

(a) all of the defective diskettes are packed in a particular box?
(b) 3 defective diskettes are packed in the same box?
(c) 2 defective diskettes are packed in each box?

44. A player is dealt five cards from a standard deck of 52 playing cards. What is the probability that the cards form

(a) a flush (all cards of the same suit)?
(b) a full house (3 cards of one denomination and 2 of another)?
(c) a pair (2 cards of one denomination and 1 each of three other denominations)?
(d) a straight (5 cards of consecutive denominations, where an ace is the highest denomination)?

Answers to Practice Problems

1. (a) The tree diagram is shown at the right.
(b) $\frac{1}{16}$

2. $\{0, 1, 2, \ldots, 15\}$

3. (a) $\{10\}$ (b) $\{12, 13, 14, 15\}$
(c) $\{0, 1, 2, 3, 4\}$ (d) $\{0\}$ (e) $\varnothing$

4. $\frac{2}{5}$ **5.** $\frac{30}{36} = \frac{5}{6}$

6. $\frac{9900}{26{,}334} = \frac{50}{133}$ **7.** $\frac{48}{720} = \frac{1}{15}$

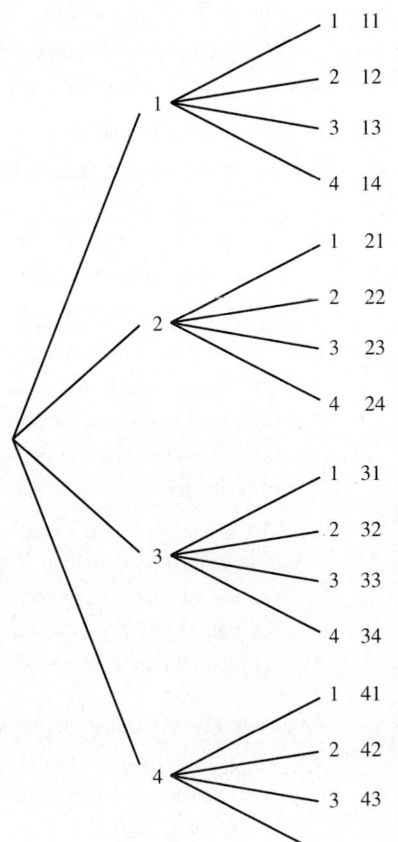

6.2 Assigning Probabilities

In Section 6.1 we learned how to calculate the probability of an event using a sample space consisting of equally likely outcomes. In this section we will extend the computation of probabilities to any finite sample space, whether consisting of equally likely outcomes or not.

PROBABILITY DISTRIBUTIONS

We saw in Section 6.1 that when a fair die is rolled, the theoretical probability of each outcome is 1/6. It is useful to record this information in the table below.

Outcome	*Probability*
1	1/6
2	1/6
3	1/6
4	1/6
5	1/6
6	1/6

This same type of table can be used even if the outcomes of an experiment are not equally likely. For example, in Example 6.3 we considered the experiment of rolling a pair of fair dice. Proceeding as in Example 6.8, we can compute the probability that the sum of the two dice is any of the possible values from 2 through 12. The results of these computations are shown below.

Outcome	*Probability*
2	1/36
3	2/36
4	3/36
5	4/36
6	5/36
7	6/36
8	5/36
9	4/36
10	3/36
11	2/36
12	1/36

Notice that each of the probabilities is a number between 0 and 1 and that the sum of all the probabilities is 1.

Generalizing from this example, we define a **probability distribution** for an experiment to be a sample space $\{s_1, s_2, \ldots, s_n\}$ for the experiment together with a set of numbers $p_1, p_2, \ldots, p_n$ that satisfy the following two conditions:

1. Each number is between 0 and 1; that is, $0 \leq p_i \leq 1$ for each i.
2. The sum of the numbers is 1; that is, $p_1 + p_2 + \ldots + p_n = 1$.

The number p_i is the probability of outcome s_i.

The first condition in the definition of a probability distribution requires that the probability of each outcome be a number between 0 and 1 (inclusive). The second condition says that the sum of the probabilities of the outcomes must be 1; this reflects that when the experiment is performed, one of the outcomes $s_1, s_2, \ldots, s_n$ must occur. Notice that nothing is said about *how* the numbers p_i are to be obtained; from the mathematical point of view they are regarded as being given. In practice, it may be difficult to determine an appropriate probability distribution for a particular experiment.

EXAMPLE 6.12 A die was rolled 500 times, and the outcome of each trial was recorded with the results shown in the frequency table below. Use this data to obtain a probability distribution for the experiment of rolling this die.

Outcome	*Frequency*	
1	79	
2	83	
3	85	
4	74	
5	91	
6	88	
	500	Total

Solution If we divide each frequency by the number of rolls (500), we obtain the relative frequency table below.

Outcome	*Relative Frequency*	
1	.158	
2	.166	
3	.170	
4	.148	
5	.182	
6	.176	
	1.000	Total

We call these relative frequencies the *empirical* probabilities of the six outcomes. For instance, a 4 appeared on 74 of the 500 rolls, and so we take the probability of rolling a 4 to be

$$\frac{74}{500} = .148.$$

Since each of the relative frequencies is a number between 0 and 1, and since the sum of all the relative frequencies is 1, the preceding table is a probability distribution for the experiment of rolling the die. ■

Two questions arise from Example 6.12. Recall that according to the relative frequency interpretation, the probability of an event represents the proportion of the time that the event tends to occur over the long run. How do we know in Example 6.12 that 500 rolls are enough to define "the long run"? We don't. In general, the frequency distribution constructed by this process is only as good as the empirical data on which it is based. Obviously, we would have more confidence in the distribution in Example 6.12 if the die had been rolled 5000 or 10,000 times rather than 500. But no matter how often it is rolled, we can only hope that our data accurately portray the general behavior of the die.

Second, is the probability distribution constructed in Example 6.12 more accurate than the theoretical distribution in which the probability of each outcome is 1/6? Not necessarily. Since the distribution we constructed is only an approximation to the true behavior of the die, we cannot be certain that the theoretical distribution is not a better description of this behavior. In fact, the probabilities in the distribution we constructed are close enough to 1/6 that it is reasonable to use 1/6 as the probability of each outcome. Which probability distribution to use in a particular application is a decision that must be made based on experience, intuition, or personal taste. However, in this book we will use the theoretical distribution whenever appropriate.

Once a probability distribution has been established for an experiment, we can find the probability of an event E by simply adding the probabilities of the outcomes for which E occurs. For example, if E is the event "the die comes up even," then $E = \{2, 4, 6\}$. Thus for the probability distribution in Example 6.12 we find that

$$\begin{aligned} P(E) &= P(2) + P(4) + P(6) \\ &= .166 + .148 + .176 = .490. \end{aligned}$$

Likewise if F is the event "the die comes up less than 5," then

$$\begin{aligned} P(F) &= P(1) + P(2) + P(3) + P(4) \\ &= .158 + .166 + .170 + .148 \\ &= .642. \end{aligned}$$

More generally, suppose that a probability distribution has been established for an experiment with sample space S. If $E = \varnothing$, then we define $P(E) = 0$. In all other cases we define $P(E)$, the probability of event E, to be the sum of the probabilities of the outcomes for which E occurs. Whereas in Section 6.1 we were able to compute probabilities only if we had a sample space consisting of equally likely outcomes, this definition enables us to compute the probabilities of events in any sample space for which we have a probability distribution.

EXAMPLE 6.13 A mail-order electronics supplier has recorded the number of toll-free telephone calls received between noon and 1 P.M. on 250 successive weekdays. These data are recorded below.

Number of calls	*Frequency*
0	28
1	42
2	59
3	47
4	28
5	19
6	12
7	7
8	5
9	2
10	1
	250 Total

(a) Use these data to create a probability distribution.
(b) Compute the probability that the supplier received fewer than five calls on a weekday between noon and 1 P.M.

Solution (a) First we divide each frequency by the number of observations (250) to obtain the following probability distribution.

Number of calls	*Probability*
0	.112
1	.168
2	.236
3	.188
4	.112
5	.076
6	.048
7	.028
8	.020
9	.008
10	.004
	1.000 Total

(b) Using this distribution, we see that

$$\begin{aligned} P(\text{fewer than 5 calls}) &= P(0) + P(1) + P(2) + P(3) + P(4) \\ &= .112 + .168 + .236 + .188 + .112 \\ &= .816. \end{aligned}$$

Therefore 81.6% of the time the supplier received fewer than five calls on weekdays between noon and 1 P.M. ■

Practice Problem 8 A bank tabulated its cashiers' transactions for the week as follows.

Type of Transaction	*Frequency*	
Cashed a check	1592	
Made a deposit	1144	
Made a withdrawal	936	
Paid a bill	256	
Requested change	44	
Bought a money order	28	
	4000	Total

(a) Use these data to create a probability distribution.

(b) Compute the probability that a transaction was either a deposit or withdrawal.

ODDS

Probabilities are often expressed in the language of "odds." For instance, we might say that the odds are 3 to 2 in favor of an event E. In terms of relative frequencies, this means that over the long run E will occur 3 times for every 2 times that it does not occur. In other words, in the long run E will occur 3 times out of 5. Therefore

$$P(E) = \frac{3}{5}.$$

More generally, to say that the **odds in favor of E are m to n** means that

$$P(E) = \frac{m}{m + n}.$$

Conversely, if we write $P(E)$ in this form, then the odds in favor of E are m to n. When the odds in favor of E are m to n, we also say that the **odds against E are n to m.**

Converting odds into probabilities requires nothing more than simple arithmetic. The example below demonstrates the method.

EXAMPLE 6.14 Find the probability of an event E in each of the following cases.

(a) The odds in favor of E are 4 to 1.
(b) The odds in favor of E are 5 to 3.
(c) The odds against E are 13 to 7.

Solution (a) Since the odds in favor of E are 4 to 1,

$$P(E) = \frac{4}{4 + 1} = \frac{4}{5} = .8.$$

(b) In this case the odds in favor of E are 5 to 3; so

$$P(E) = \frac{5}{5 + 3} = \frac{5}{8} = .625.$$

(c) Here the odds are 13 to 7 *against* E. This means that the odds *in favor of* E are 7 to 13. Consequently

$$P(E) = \frac{7}{7 + 13} = \frac{7}{20} = .35. \quad ■$$

Practice Problem 9 Find the probability of an event E if

(a) the odds in favor of E are 7 to 4
(b) the odds against E are 9 to 5.

It is also easy to convert probabilities into odds; for if

$$P(E) = \frac{a}{b},$$

then

$$P(E) = \frac{a}{a + (b - a)}.$$

Hence if $P(E) = a/b$, then the odds in favor of E are a to $b - a$.

EXAMPLE 6.15 (a) If $P(E) = 11/16$, find the odds in favor of E.
(b) If $P(E) = 2/7$, find the odds against E.
(c) If $P(E) = .76$, find the odds against E.

Solution (a) Since

$$P(E) = \frac{11}{16},$$

the odds in favor of E are 11 to $16 - 11$, that is, 11 to 5.
(b) Because

$$P(E) = \frac{2}{7},$$

the odds in favor of E are 2 to $7 - 2$, that is, 2 to 5. Thus the odds against E are 5 to 2.
(c) Here we must first write the given probability as a fraction, which we will reduce to lowest terms. In this way we obtain

$$P(E) = .76 = \frac{76}{100} = \frac{19}{25},$$

and so the odds in favor of E are 19 to $25 - 19$, that is, 19 to 6. Therefore the odds against E are 6 to 19. ■

Practice Problem 10 (a) If $P(E) = .75$, find the odds in favor of E.
(b) If $P(E) = 2/9$, find the odds against E.

EXERCISES 6.2

In Exercises 1–8 determine if the given values form a probability distribution for an experiment with sample space $S = \{a, b, c, d\}$.

1.

Outcome	*Probability*
a	.1
b	.2
c	.3
d	.4

2.

Outcome	*Probability*
a	1/5
b	1/4
c	1/3
d	1/2

3.

Outcome	*Probability*
a	.27
b	.53
c	−.22
d	.42

4.

Outcome	*Probability*
a	3/8
b	1/4
c	5/16
d	1/16

5.

Outcome	*Probability*
a	.06
b	.14
c	.39
d	.41

6.

Outcome	*Probability*
a	.6
b	−.1
c	.2
d	.3

7.

Outcome	*Probability*
a	.31
b	.17
c	.22
d	.25

8.

Outcome	*Probability*
a	1/4
b	1/4
c	1/4
d	1/4

9. Determine the probability of the given event using the probability distribution below.

Outcome	*Probability*
a	1/2
b	1/4
c	1/8
d	1/16
e	1/16

(a) $\{b, c\}$
(b) $\{a, d, e\}$
(c) $\{a, b, c, d, e\}$
(d) $\{b, c, d, e\}$

10. Determine the probability of the given event using the probability distribution below.

Outcome	*Probability*
a	1/5
b	1/5
c	1/5
d	1/5
e	1/5

(a) $\{b, c, e\}$
(b) $\{a, d\}$
(c) $\{b\}$
(d) $\{a, c, d, e\}$

11. Determine the probability of the given event using the probability distribution below.

Outcome	*Probability*
a	.15
b	.20
c	.25
d	.30
e	.10

(a) $\varnothing$
(b) $\{a, e\}$
(c) $\{c, d, e\}$
(d) $\{a, b, c, d\}$

12. Determine the probability of the given event using the probability distribution below.

Outcome	*Probability*
a	.4
b	.1
c	.2
d	.2
e	.1

(a) $\varnothing$
(b) $\{a, d\}$
(c) $\{b, c, e\}$
(d) $\{a, b, c, d, e\}$

13. Of the last 568 bids submitted by a construction company, 93 were accepted. What is the empirical probability that the next bid submitted by this company will be accepted?

14. A basketball player has made 27 free throws out of 35 attempts. What is the empirical probability that she makes her next free throw?

15. A Labor Department statistician interviewed 120 unemployed persons in an inner-city area to determine how many years of high school each had completed. Their responses are shown below.

Number of years	*Frequency*	
0	6	
1	24	
2	39	
3	30	
4	21	
	120	Total

(a) Use this frequency table to construct a probability distribution for the number of years of high school completed.
(b) What is the probability that someone completed four years of high school?
(c) What is the probability that someone completed no more than one year of high school?
(d) What is the probability that someone completed at least three years of high school?

16. An insurance company checked the records of 80 random motorists to determine how many moving violations each had been cited for during the past five years. The results are shown below.

Violations	*Frequency*	
0	34	
1	28	
2	12	
3	4	
4	2	
	80	Total

(a) Use this frequency table to construct a probability distribution for the number of moving violations.

(b) What is the probability that someone was guilty of three violations?

(c) What is the probability that someone was guilty of at least two violations?

(d) What is the probability that someone was guilty of no more than one violation?

17. The speeds of 100 cars selected at random from I-74 were distributed as shown below.

Speed s (mph)	*Cars*	
$45 \leq s < 55$	15	
$55 \leq s < 65$	31	
$65 \leq s < 75$	35	
$75 \leq s < 85$	17	
$85 \leq s < 95$	2	
	100	Total

(a) Use this frequency table to construct a probability distribution for the speed of the cars.

(b) What is the probability that a car was traveling from 55 to 65 mph?

(c) What is the probability that a car was below the 65 mph speed limit?

(d) What is the probability that a car was traveling at least 75 mph?

18. A test of the tensile strength (breaking point under stress) of a sample of 80 cables produced the frequency distribution shown below.

Tensile strength t (lbs)	*Number of cables*	
$1400 \leq t < 1500$	11	
$1500 \leq t < 1600$	29	
$1600 \leq t < 1700$	26	
$1700 \leq t < 1800$	9	
$1800 \leq t < 1900$	5	
	80	Total

(a) Use this frequency table to construct a probability distribution for the tensile strength of the cables.

(b) What is the probability that a cable had a tensile strength between 1400 and 1500 pounds?

(c) What is the probability that a cable had a tensile strength of at least 1700 pounds?

(d) What is the probability that a cable had a tensile strength of at most 1600 pounds?

19. A baker finds that the number of chocolate cakes requested on a given day has the following distribution.

Requests	*Frequency*	
0	33	
1	60	
2	96	
3	57	
4	36	
5	15	
6	3	
	300	Total

(a) Use this frequency table to construct a probability distribution for the number of cakes requested.

(b) What is the probability of selling exactly three cakes?

(c) What is the probability of selling at most 2 cakes?

(d) What is the probability of selling no more than 1 cake?

(e) What is the maximum number of cakes the baker can prepare in order to be at least 50% certain of selling them all?

20. A stockbroker observes that the delivery time for her local mailings has the distribution below.

Delivery time (working days)	*Number of items*	
1	109	
2	159	
3	136	
4	65	
5	31	
	500	Total

(a) Use this frequency table to construct a probability distribution for the delivery time of local mailings.

(b) What is the probability that an item mailed on Monday will arrive on Wednesday?

(c) What is the probability that an item mailed on Monday will arrive on Wednesday, Thursday, or Friday?

(d) What is the probability that an item mailed on Monday will arrive by Thursday at the latest?

(e) By what day must an item be mailed in order to be at least 80% certain that it will arrive before the weekend?

In Exercises 21–26 complete the missing entries in the table.

	Probability	*Odds in favor*	*Odds against*
21.	____	3 to 1	____
22.	.85	____	____
23.	____	____	18 to 7
24.	____	____	12 to 13
25.	4/9	____	____
26.	____	9 to 11	____

In Exercises 27–32 determine the indicated odds when a single die is rolled.

27. the odds in favor of rolling a 5 or 6

28. the odds against rolling a 5 or 6

29. the odds against rolling a number greater than 1

30. the odds in favor of rolling a number less than or equal to 4

31. the odds in favor of rolling a multiple of 3

32. the odds against rolling an even number

33. A lawyer has informed her client that, in her judgment, a stiff sentence is twice as likely as a light sentence and that some sentence (either stiff or light) is twice as likely as receiving probation. What is the probability of receiving probation?

34. A die is weighted in such a way that 1, 2, and 3 are equally likely to appear; 4, 5, and 6 are equally likely to appear; and 1 is 50% more likely to appear than 6. What is the probability of rolling an even number with this die?

35. An observer watching cars at a certain intersection has noticed that they go straight 50% more often than they turn left and 40% more often than they turn right. What is the probability that the next car approaching this intersection will turn left?

36. Two East German swimmers and three Americans are competing in the women's 200 meter freestyle event. One of the East Germans and two of the Americans are given exactly the same chance of winning. The other East German is twice as likely to win as the third American but only half as likely as her East German teammate. What is the probability that an East German will win the event?

Answers to Practice Problems

8. (a) The probability distribution is given below.

Type of Transaction	*Probability*	
Cashed a check	.398	
Made a deposit	.286	
Made a withdrawal	.234	
Paid a bill	.064	
Requested change	.011	
Bought a money order	.007	
	1.000	Total

(b) .520

9. (a) $\frac{7}{11}$ (b) $\frac{5}{14}$

10. (a) 3 to 1 (b) 7 to 2

6.3 Basic Laws of Probability

When events are logically related to one another, it is reasonable to expect their probabilities to be mathematically related. In this section we will develop two fundamental implications of this type.

OPERATIONS ON EVENTS

Since events are sets, we can combine them using the set operations of union, intersection, and complement that were introduced in Section 5.1. From the Venn diagram in Figure 6.6(a) we see that $A \cup B$ contains those outcomes in

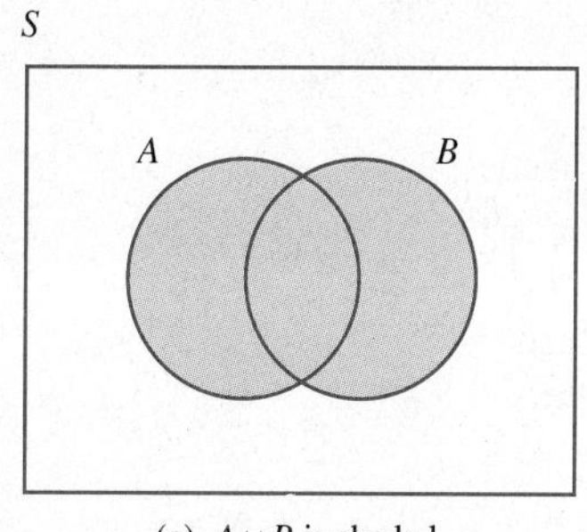

(a) $A \cup B$ is shaded

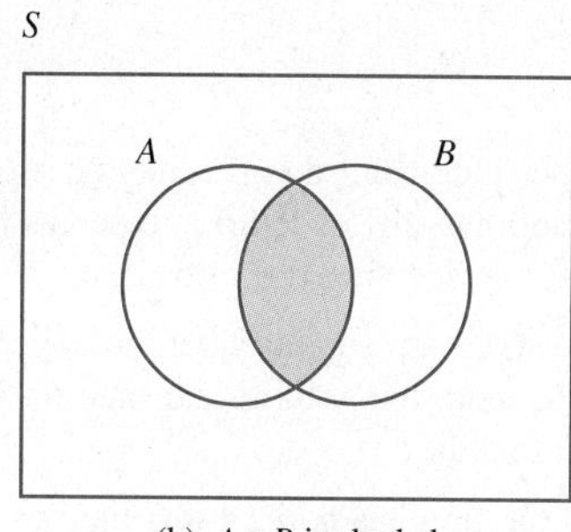

(b) $A \cap B$ is shaded

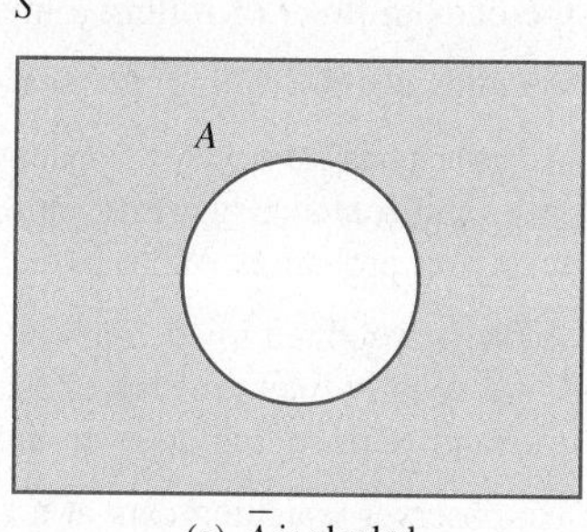

(c) $\overline{A}$ is shaded

FIGURE 6.6

either A or B or both. Likewise from Figures 6.6(b) and (c) we see that $A \cap B$ contains those outcomes in both A and B, and $\overline{A}$ contains those outcomes not in A. Hence

$A \cup B$ represents the event "*either* A occurs *or* B occurs (or both)," that is, the event "*at least one* of A or B occurs"

$A \cap B$ represents the event "*both* A *and* B occur"

$\overline{A}$ represents the event "A does *not* occur."

Since these set operations will enter into many of our later formulas, the meaning of each of these symbols should be thoroughly understood.

EXAMPLE 6.16 A fair coin is to be flipped three times. Consider the following events:

$$A = \text{"the first flip is heads" and}$$
$$B = \text{"the third flip is tails."}$$

Express each of the events $A \cup B$, $A \cap B$, $\overline{A}$, and $\overline{B}$ in words, and calculate its probability.

Solution The meaning of the events $A \cup B$, $A \cap B$, $\overline{A}$, and $\overline{B}$ is as follows.

$A \cup B$ is "the first flip is heads *or* the third flip is tails."

$A \cap B$ is "the first flip is heads *and* the third flip is tails."

$\overline{A}$ is "the first flip is *not* heads," that is, "the first flip is tails."

$\overline{B}$ is "the third flip is *not* tails," that is, "the third flip is heads."

Recall from Example 6.5 that for this experiment a sample space of equally likely outcomes is

$$\{HHH, HHT, HTH, HTT, THH, THT, TTH, TTT\}.$$

Hence

$$A \cup B = \{HHH, HHT, HTH, HTT, THT, TTT\},$$

and so the counting rule gives

$$P(A \cup B) = \frac{6}{8} = \frac{3}{4}.$$

Similarly we find that

$$A \cap B = \{HHT, HTT\}, \qquad \overline{A} = \{THH, THT, TTH, TTT\},$$

and

$$\overline{B} = \{HHH, HTH, THH, TTH\}.$$

Therefore

$$P(A \cap B) = \frac{2}{8} = \frac{1}{4}, \qquad P(\overline{A}) = \frac{4}{8} = \frac{1}{2}, \qquad \text{and} \qquad P(\overline{B}) = \frac{4}{8} = \frac{1}{2}. \quad ■$$

EXAMPLE 6.17 In the experiment of rolling a pair of fair dice, consider the following events:

A = "the sum rolled is 8" and

B = "each die comes up greater than 2."

Calculate the probability of each of the events $A \cup B$, $A \cap B$, $\overline{A}$, and $\overline{B}$.

Solution The meaning of the events $A \cup B$, $A \cap B$, $\overline{A}$, and $\overline{B}$ is given below.

$A \cup B$	is "the sum rolled is 8 or each die comes up greater than 2."
$A \cap B$	is "the sum rolled is 8 and each die comes up greater than 2."
$\overline{A}$	is "the sum rolled is not 8."
$\overline{B}$	is "1 or 2 comes up on at least one die."

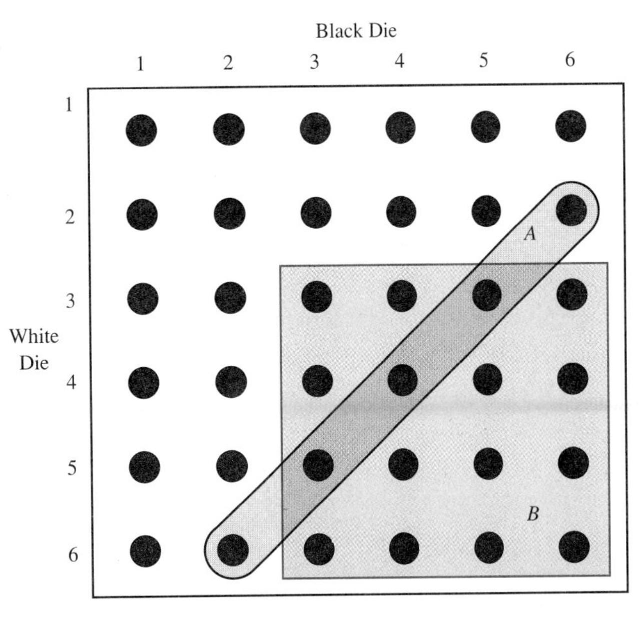

FIGURE 6.7

Recall from Example 6.3 that the possible outcomes are as shown in Figure 6.7. By counting the appropriate dots in this figure, we see that

$$P(A \cup B) = \frac{18}{36} = \frac{1}{2}, \qquad P(A \cap B) = \frac{3}{36} = \frac{1}{12},$$

$$P(\overline{A}) = \frac{31}{36}, \qquad P(\overline{B}) = \frac{20}{36} = \frac{5}{9}. \quad ■$$

Practice Problem 11 In the experiment of rolling a pair of fair dice, consider the following events:

A = "the sum rolled is 6" and

B = "at least one die comes up greater than 4."

Calculate the probability of each of the events A, B, $A \cup B$, $A \cap B$, $\overline{A}$, and $\overline{B}$.

THE COMPLEMENT RULE

In Example 6.17 notice that $P(A) = 5/36$ and $P(\overline{A}) = 31/36$. It is not difficult to see that the sum of the probabilities of an event and its complement must always be 1. For let A be any event in an experiment having sample space S, and let

$$A = \{r_1, r_2, \ldots, r_m\} \quad \text{and} \quad \overline{A} = \{s_1, s_2, \ldots, s_n\}$$

be the sets of distinct outcomes corresponding to A and $\overline{A}$. By definition of the complement of a set, we see that $r_1, r_2, \ldots, r_m, s_1, s_2, \ldots, s_n$ must be distinct and comprise all of the sample space S. But in any probability distribution the sum of the probabilities of all the outcomes must be 1, and so we have

$$P(A) + P(\overline{A}) = P(r_1) + \cdots + P(r_m) + P(s_1) + \cdots + P(s_n) = 1.$$

We have obtained the following result.

Theorem 6.1 (the complement rule)

For any event A, we have

$$P(A) + P(\overline{A}) = 1.$$

EXAMPLE 6.18 The American roulette wheel has 38 compartments numbered 0, 00, 1, 2, . . ., 36. The numbers 0 and 00 are green, and the remaining 36 numbers are equally divided between red and black. Assuming that each of the 38 numbers is equally likely to appear, what is the probability that someone betting on a red number will lose?

Solution Since the player wins whenever a red number appears, we have

$$P(\text{player wins}) = \frac{18}{38} = \frac{9}{19}.$$

Thus by the complement rule

$$P(\text{player wins}) + P(\text{player loses}) = 1.$$

Hence

$$\begin{aligned} P(\text{player loses}) &= 1 - P(\text{player wins}) \\ &= 1 - \frac{9}{19} \\ &= \frac{10}{19}. \end{aligned}$$ ■

When computing the probability of an event containing a large number of outcomes, the complement rule can often be used to simplify the calculation.

EXAMPLE 6.13 Revisited In Example 6.13 we constructed a probability distribution for the number of toll-free telephone calls received on a weekday between noon and 1 P.M. by a mail-order electronics supplier. What is the probability that this supplier receives at least one call during this time?

Solution The probability distribution from Example 6.13 is reproduced below.

Number of calls	*Probability*
0	.112
1	.168
2	.236
3	.188
4	.112
5	.076
6	.048
7	.028
8	.020
9	.008
10	.004
	1.000 Total

The event of interest is A = "at least one call is received." Now the set of outcomes corresponding to A is

$$\{1, 2, \ldots, 10\}.$$

Proceeding as in Section 6.2, we see that

$$\begin{aligned} P(A) &= P(1) + P(2) + \cdots + P(10) \\ &= .168 + .236 + \cdots + .004. \end{aligned}$$

There is an easier method of obtaining this answer that uses the complement rule. Note that the complement of A = "at least one call is received" is $\overline{A}$ = "no calls are received." Hence

$$P(A) = 1 - P(\overline{A}) = 1 - .112 = .888. \blacksquare$$

Practice Problem 12 For the probability distribution above, what is the probability that (a) at most 8 calls are received? (b) no more than 9 calls are received?

If the probability of an event seems hard to calculate directly, a good strategy is to consider use of the complement rule.

EXAMPLE 6.19 A subcommittee of four is to be randomly chosen from among the 27 members of the county board. If the board contains 17 men and 10 women, what is the probability that the subcommittee will contain at least one woman?

Solution Here the event of interest is A = "the subcommittee contains at least one woman." Since the number of subcommittees that contain at least one woman is difficult to compute directly, we consider the complementary event $\overline{A}$ = "the subcommittee contains only men." Now

$$P(\overline{A}) = \frac{C(17, 4)}{C(27, 4)} = \frac{2380}{17{,}550},$$

and so

$$P(A) = 1 - \frac{2380}{17{,}500} = \frac{15{,}170}{17{,}550}. \blacksquare$$

Practice Problem 13 A fair die is rolled five times. What is the probability that an even number appears on at least one roll?

THE UNION RULE

The complement rule is analogous to Theorem 5.1(a) in Section 5.2. There is also an analog to Theorem 5.1(b), the principle of inclusion-exclusion.

Theorem 6.2 (the union rule)

For any events A and B,

$$P(A \cup B) = P(A) + P(B) - P(A \cap B).$$

Recall that in Example 6.16 we considered the events

A = "the first flip is heads" and

B = "the third flip is tails."

For these events we found that $P(A \cap B) = 1/4$. Since the probability is 1/2 that a fair coin will land heads or tails on any flip, we see that $P(A) = 1/2$ and $P(B) = 1/2$. Thus by the union rule, we have

$$P(A \cup B) = P(A) + P(B) - P(A \cap B)$$
$$= \frac{1}{2} + \frac{1}{2} - \frac{1}{4} = \frac{3}{4},$$

as we found earlier by use of the counting rule.

EXAMPLE 6.20 A certain construction project requires carpenters and masons. Construction workers say that the probability of a strike by the carpenters is .4, the probability of a strike by the masons is .5, and the probability of a strike by both groups is .2. What is the probability that the project will be delayed by some form of strike?

The probability of a strike by the carpenters is .4. The probability of a strike by the masons is .5.

Solution Let C and M denote the events "the carpenters strike" and "the masons strike," respectively. We are told that

$$P(C) = .4, \quad P(M) = .5, \quad \text{and} \quad P(C \cap M) = .2.$$

The probability of a delay in the project due to a strike is $P(C \cup M)$, which we can compute by the union rule. Since

$$P(C \cup M) = P(C) + P(M) - P(C \cap M) = .4 + .5 - .2 = .7,$$

there is a .7 probability of delay because of a strike. ■

EXAMPLE 6.21 In a survey conducted by the fire department it was found that 62% of all houses had a smoke detector and 35% had a fire extinguisher. Moreover, 24% had both a smoke detector and a fire extinguisher. What percentage of houses do not have either a smoke detector or a fire extinguisher?

Solution Recall that we can regard percentages as probabilities. Therefore if S and F are the events "has a smoke detector" and "has a fire extinguisher," respectively, then $P(S) = .62$, $P(F) = .35$, and $P(S \cap F) = .24$. The event $S \cup F$ is "has a smoke detector or a fire extinguisher," and so we must compute $P(\overline{S \cup F})$. By the union rule

$$P(S \cup F) = P(S) + P(F) - P(S \cap F) = .62 + .35 - .24 = .73.$$

Hence by the complement rule

$$P(\overline{S \cup F}) = 1 - P(S \cup F) = 1 - .73 = .27.$$

Thus 27% of the homes have neither a smoke detector nor a fire extinguisher. ■

Practice Problem 14 A factory requires both aluminum and steel. There is a probability of .06 that it will be short of aluminum, a probability of .05 that it will be short of steel, and a probability of .01 that it will be short of both metals. What is the probability that the factory will have an adequate supply of both metals?

EXAMPLE 6.22 An oil company has undertaken exploratory drilling in northern Canada. Preliminary tests indicate that there is a 90% chance of finding oil and an 85% chance of finding natural gas. What can be said about the probability of finding both oil and natural gas?

Solution By the union rule

$$P(\text{oil or gas}) = P(\text{oil}) + P(\text{gas}) - P(\text{oil and gas}).$$

Since a probability cannot exceed 1, we see that

$$1 \geq P(\text{oil}) + P(\text{gas}) - P(\text{oil and gas}).$$

Hence

$$P(\text{oil and gas}) \geq P(\text{oil}) + P(\text{gas}) - 1 \geq .90 + .85 - 1 = .75.$$

Moreover, since finding both oil and gas is no easier than finding gas, we have

$$P(\text{oil and gas}) \leq P(\text{gas}) = .85.$$

Therefore the probability of finding both oil and gas is between .75 and .85. (We cannot determine this probability exactly without more information.) ■

MUTUALLY EXCLUSIVE EVENTS

An important special case of the union rule arises if A and B are events that cannot happen simultaneously. In this case A and B have no common outcomes, and so $A \cap B = \emptyset$. We call such events *mutually exclusive*. (See Figure 6.8.)

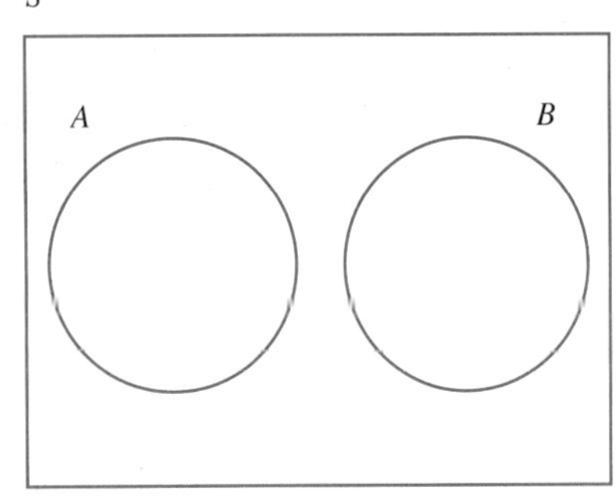

A and *B* are mutually exclusive

FIGURE 6.8

If A and B are mutually exclusive, then $P(A \cap B) = 0$. Hence the union rule reduces to the following simpler form.

If A and B are mutually exclusive events, then

$$P(A \cup B) = P(A) + P(B).$$

More generally, if all pairs of the events $A_1, A_2, \ldots, A_k$ are mutually exclusive, then we have the following analog of the addition principle (Theorem 5.2).

Theorem 6.3 (union rule for mutually exclusive events)

If all pairs of the events $A_1, A_2, \ldots, A_k$ are mutually exclusive, then

$$P(A_1 \cup A_2 \cup \cdots \cup A_k) = P(A_1) + P(A_2) + \cdots + P(A_k).$$

EXAMPLE 6.23 In a particular supermarket, the probabilities of having various numbers of checkout lines open are given below.

Precise Number of Open Checkout Lines	*Probability*
1	.12
2	.18
3	.21
4	.23
5 or more	.26

What is the probability that at most 3 checkout lines are open?

Solution Let A_1 denote the event ''exactly 1 checkout line is open,'' and let A_2 and A_3 be defined similarly. Now each pair of A_1, A_2, and A_3 is mutually exclusive because the supermarket cannot simultaneously have two different numbers of checkout lines open. Therefore by the union rule for mutually exclusive events

$$\begin{aligned} &P(\text{at most 3 are open}) \\ &\quad = P(\text{exactly 1 is open}) + P(\text{exactly 2 are open}) + P(\text{exactly 3 are open}) \\ &\quad = .12 + .18 + .21 = .51. \end{aligned}$$

Thus 51% of the time there are at most 3 checkout lines open. ■

EXAMPLE 6.24 From among a group of twelve persons with a certain disease, five are to be randomly selected to receive treatment with a new drug. If eight of the twelve persons are female, what is the probability that a majority of those selected will be female?

Solution In order that a majority of those selected be female, at least three of the five persons chosen must be women. The events ''exactly three women are chosen,'' ''exactly four women are chosen,'' and ''exactly five women are chosen'' are mutually exclusive. Thus

$$\begin{aligned} &P(\text{a majority is female}) \\ &\quad = P(\text{exactly 3 women}) + P(\text{exactly 4 women}) + P(\text{exactly 5 women}). \end{aligned}$$

Now a committee containing exactly three women can be formed by choosing three women (from among eight) and two men (from among four). Hence

$$P(\text{exactly 3 women}) = \frac{C(8, 3) \cdot C(4, 2)}{C(12, 5)} = \frac{56 \cdot 6}{792} = \frac{336}{792}.$$

Likewise

$$P(\text{exactly 4 women}) = \frac{C(8, 4) \cdot C(4, 1)}{C(12, 5)} = \frac{70 \cdot 4}{792} = \frac{280}{792}$$

and

$$P(\text{exactly 5 women}) = \frac{C(8, 5) \cdot C(4, 0)}{C(12, 5)} = \frac{56 \cdot 1}{792} = \frac{56}{792}.$$

It follows that

$$\begin{aligned} &P(\text{a majority is female}) \\ &\quad = P(\text{exactly 3 women}) + P(\text{exactly 4 women}) + P(\text{exactly 5 women}) \\ &\quad = \frac{336}{792} + \frac{280}{792} + \frac{56}{792} \\ &\quad = \frac{672}{792} = \frac{28}{33}. \end{aligned}$$

■

EXAMPLE 6.25 In the fourth race at Pimlico the newspaper handicapper has given respective odds of 5 to 3 against, 3 to 1 against, and 7 to 1 against the race being won by the number 4, 2, and 6 horses. What is the probability that one of these three horses wins?

Solution The probabilities that the number 4, 2, and 6 horses will win are 3/8, 1/4, and 1/8, respectively. Since only one horse can win the race, the events "number 4 horse wins," "number 2 horse wins," and "number 6 horse wins" are mutually exclusive. Hence by the union rule for mutually exclusive events the probability that one of these horses will win is

$$\begin{aligned} P(\text{number 4, 2, or 6 wins}) &= P(4 \text{ wins}) + P(6 \text{ wins}) + P(2 \text{ wins}) \\ &= \frac{3}{8} + \frac{1}{4} + \frac{1}{8} \\ &= \frac{3}{4}. \quad \blacksquare \end{aligned}$$

Practice Problem 15 A certain subdivision was developed by the Ark Corporation, a joint venture of the Armstrong Company, Rave Brothers Construction, and Kaisner Construction. In this subdivision 35% of the houses were built by Armstrong, 20% by Rave Brothers, 19% by Kaisner, 11% by Robert Rist, and 15% by others. What percentage of the houses were built by a member of the Ark Corporation?

EXERCISES 6.3

1. In the experiment of observing the U.S. economy for one year, consider the events

$$A = \text{"inflation increases"}$$
$$B = \text{"unemployment increases."}$$

Use set operations to symbolize each of the following events in terms of A and B, and illustrate each with a Venn diagram.

(a) Inflation does not increase.

(b) Both inflation and unemployment increase.

(c) Inflation increases but not unemployment.

(d) Neither inflation nor unemployment increases.

2. In the experiment of marketing three new products, consider the events

$$A = \text{"Product 1 is successful"}$$
$$B = \text{"Product 2 is successful"}$$
$$C = \text{"Product 3 is successful."}$$

Use set operations to symbolize each of the following events in terms of A, B, and C, and illustrate each with a Venn diagram.

(a) All three products are successful.

(b) None of the three products is successful.

(c) Only the second product is successful.

(d) At least two of the three products are successful.

3. In the experiment of selecting a member from a committee consisting of students, faculty, and administrators, let E be the event "a student is chosen" and F be the event "a female is chosen." Describe the following events in words.

(a) $\overline{E}$ (b) $\overline{F}$ (c) $E \cap F$ (d) $E \cup F$

4. In the experiment of flipping a coin twice, let E be the event "the same side of the coin shows on both flips" and let F be the event "the second flip is tails." Describe the following events in words.

(a) $\overline{E}$ (b) $\overline{F}$ (c) $E \cap F$ (d) $E \cup F$

5. In the experiment of generating a random three-digit number from 000 to 999, consider the events

A = "the number is less than 400"

B = "the number is even"

C = "the number is at least 200."

Calculate the probabilities of the following events.

(a) $\overline{A}$ (b) $A \cap C$ (c) $\overline{(A \cup B)}$ (d) $A \cap B \cap C$

6. In the experiment of randomly drawing a card from a standard deck of playing cards, consider the events

A = "the card is red"

B = "the card is higher than a ten"

C = "the card is a spade."

Calculate the probabilities of the following events if an ace is the highest card.

(a) $\overline{B}$ (b) $A \cap C$ (c) $\overline{(A \cap B)}$ (d) $(A \cup C) \cap B$

7. If the probability that A occurs is .53, the probability that B occurs is .72, and the probability that both occur is .48, what is the probability that

(a) A does not occur? (b) B does not occur?

(c) A or B occurs? (d) neither A nor B occurs?

8. If the probability that A occurs is .25, the probability that B occurs is .45, and the probability that both occur is .05, what is the probability that

(a) A does not occur? (b) B does not occur?

(c) A or B occurs? (d) neither A nor B occurs?

9. Suppose that A, B, and C are mutually exclusive events with respective probabilities .23, .42, and .31. What is the probability that

(a) A does not occur?

(b) at least one of the three events occurs?

(c) none of the three events occur?

(d) more than one of the three events occurs?

10. Suppose that A, B, and C are mutually exclusive events with respective probabilities 1/4, 1/5, and 1/2. What is the probability that

(a) B does not occur?

(b) at least one of the three events occurs?

(c) none of the three events occur?

(d) more than one of the three events occurs?

In Exercises 11–18 determine which pairs of events are mutually exclusive in the experiment of rolling a pair of dice.

11. "the sum is 2, 5, 8, or 11" and "the sum is 9, 10, 11, or 12"

12. "the sum is 7, 8, 9, or 10" and "the sum is 2, 3, 11, or 12"

13. "one die comes up 4" and "one die comes up 2"

14. "one die comes up 2" and "the sum of the dice is 10"

15. "the same number appears on both dice" and "the sum of the dice is 7"

16. "one die comes up 3 or less" and "an even number appears on one die"

17. "the sum of the dice is 2 or 3" and "1 does not appear on either die"

18. "the sum of the dice is a multiple of 3" and "odd numbers appear on both dice"

19. If a fair coin is flipped three times, what is the probability that it will come up heads at least once?

20. If three fair dice are rolled, what is the probability that at least one 6 will appear?

21* If five letters (not necessarily distinct) are typed at random, what is the probability that at least one will be a vowel?

22* Four women checked their hats at a restaurant. If the hats are returned at random to these women, what is the probability that someone receives the wrong hat?

23* A shipment of twenty refrigerators contains five defective units. If three units are chosen from the shipment at random, what is the probability that at least one defective unit will be among them?

24* If two cards are chosen at random from a standard deck of playing cards, what is the probability that at least one is a jack, queen, or king?

25. In a certain town the probability of reading *The Times* is .51 and the probability of reading *The Herald* is .43. If the probability of reading both papers is .16, what is the probability of reading *The Times* or *The Herald?*

26. In a certain subdivision 67% of the houses have central air conditioning, 32% have a fireplace, and 13% have both central air conditioning and a fireplace. What percentage of these houses have central air conditioning or a fireplace?

27. In a large sample of families it was found that 86% of the husbands and 63% of the wives worked outside the home, while in 54% of the cases both worked outside the home. What is the probability that in a random family from this sample

(a) at least one spouse works outside the home?

(b) neither spouse works outside the home?

---

*This exercise requires the use of counting techniques from Chapter 5.

28. During a certain year 53% of the cars sold in a large city were made in America and 63% were economy cars. If 21% of the cars sold in this city were American-made economy cars, what percentage of the cars sold were neither American nor economy cars?

29. According to a computer dating service, the odds are 3 to 1 that Norman will like Sylvia, 7 to 5 that she will like him, and 5 to 7 that they will both like each other. What are the odds that at least one will like the other?

30. Two mechanical systems must both function properly in order for a space probe to land successfully on Venus. According to NASA officials, the probabilities that the systems will function properly are .80 and .75, and the probability that both systems will fail is .05. What are the chances for a successful landing?

31. In the third race, the local handicapper has set the odds against Affirmation at 3 to 1 and the odds against Brobdingnag at 4 to 1. According to this handicapper, what is the probability that one of these two horses will win the race?

32. Three New York critics are screening a new French film. The film's producer rates the chances of favorable reviews from Armstrong, Babcock, and Craft to be 15%, 20%, and 25%, respectively. If no two of these reviewers ever like the same film, what is the probability that all three reviews are unfavorable?

33. An observant commuter noticed that the 5:30 train is usually between one and ten minutes late. Keeping records over the course of several months, she obtained the following probability distribution.

Minutes late	*Probability*
0	.03
1	.07
2	.12
3	.15
4	.17
5	.16
6	.13
7	.09
8	.05
9	.02
10	.01

What is the probability that on a random day the train will be

(a) at least five minutes late?

(b) at least two minutes late?

(c) at most seven minutes late?

34. In a particular barbershop, the probability distribution for the number of customers waiting for a haircut is as shown below.

Customers	*Probability*
0	.5
1	.2
2	.2
3	.1
4 or more	0

(a) What is the probability that someone is waiting?

(b) What is the probability that at least 2 are waiting?

(c) What is the probability that no more than two are waiting?

35.* If five people are chosen at random from among four men and six women, what is the probability that a majority of those selected will be women?

36.* If two marbles are chosen at random from an urn containing eight red marbles and four blue marbles, what is the probability that both will be the same color?

37.* A freezer contains three orange and five cherry Popsicles. If four Popsicles are chosen at random from the freezer, what is the probability that

(a) at least two will be cherry?

(b) no more than one will be orange?

38.* A city council consists of four Democrats and six Republicans. If a delegation of three is selected at random, what is the probability that

(a) none are Republicans?

(b) a majority are Republicans?

39. Prove the union rule for a sample space of equally likely outcomes.

40. Prove the union rule for three events: $P(A \cup B \cup C) =$
$P(A) + P(B) + P(C) - P(A \cap B) - P(A \cap C) - P(B \cap C) + P(A \cap B \cap C)$.

Answers to Practice Problems

11. $P(A) = \frac{5}{36}$, $P(B) = \frac{5}{9}$, $P(A \cup B) = \frac{23}{36}$, $P(A \cap B) = \frac{1}{18}$, $P(\overline{A}) = \frac{31}{36}$, $P(\overline{B}) = \frac{4}{9}$

12. (a) .988 (b) .996

13. $\frac{31}{32}$

14. .90

15. 74%

6.4 Conditional Probability, the Intersection Rule, and Independent Events

Our estimate of the probability of an event E reflects the amount of information we have about the event. If we obtain additional information, it may change our opinion of the likelihood that the event occurs. For instance, the probability of snow on a given day in Illinois is higher in January than in July. Thus knowledge of the time of year will cause us to adjust upward or downward our estimate of the probability of snow on a given day in Illinois. In this section we will discuss probabilities when such additional information is available.

---

*This exercise requires the use of counting techniques from Chapter 5.

REDUCED SAMPLE SPACES

Let S be a sample space consisting of equally likely outcomes, and let A be a particular event in S. (See Figure 6.9.) Suppose that we are interested in knowing whether A occurred when the experiment was performed. If we have not yet been told the outcome of the experiment, there are two possible situations.

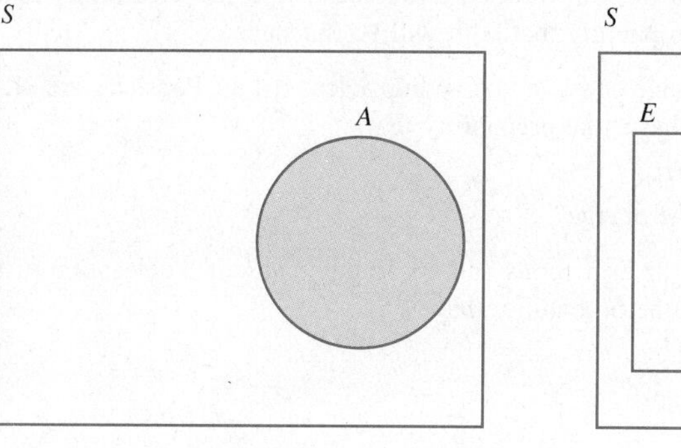

FIGURE 6.9

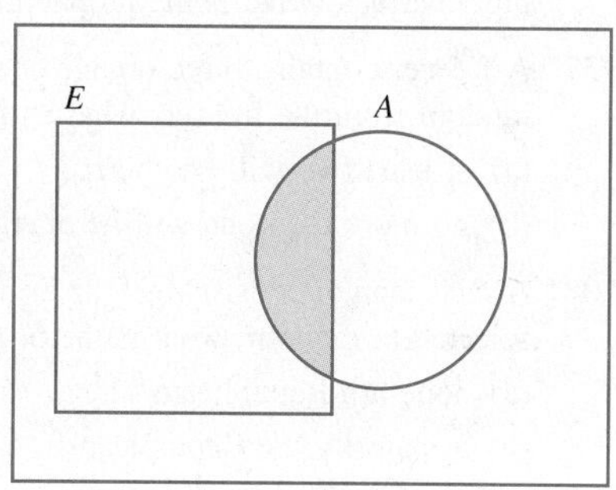

FIGURE 6.10

In the first case, we have no information whatsoever about the result of the experiment. In this case we are in the same position as before the experiment was performed; namely, as far as we know, all the outcomes in S are equally likely. Thus by the counting rule, the probability that A occurred is

$$P(A) = \frac{|A|}{|S|} = \frac{\text{number of outcomes in } A}{\text{number of outcomes in } S}.$$

In the second case, we are given information that some event E occurred, as illustrated in Figure 6.10. In this case the only possible outcomes for the experiment are those in E. Thus in effect E becomes a reduced sample space for the experiment. The outcomes in E for which A occurs are those in the intersection $A \cap E$, which is shaded in Figure 6.10. Applying the counting rule to the new sample space E, we see that the probability that A occurs given that E occurred is

$$\frac{\text{number of outcomes in } A \text{ and } E}{\text{number of outcomes in } E}.$$

We call the probability that A occurs given that E occurred the *conditional probability of A given E* and denote it by $P(A|E)$.

Counting Rule for Conditional Probabilities

If A and E are events in a finite sample space consisting of equally likely outcomes, then

$$P(A|E) = \frac{|A \cap E|}{|E|} = \frac{\text{number of outcomes in } A \text{ and } E}{\text{number of outcomes in } E}.$$

Recall that in part (b) of Example 6.11 we considered the following question.

> From among the forty persons attending a seminar discussing financial investments, three different persons will be selected at random to receive door prizes. What is the probability that a particular person at the seminar wins the third door prize?

As we have seen, the answer to this question is 1/40, not 1/38 as some people believe. This question illustrates the first case above. For let us consider the situation at the time when the third door prize is to be awarded. At this time two prizes have already been given, but *we have no information about the winners of the first two prizes*. Thus as far as we know, all forty persons in attendance are equally likely to win the third prize.

On the other hand, if we learn that Mr. Smith did not win either of the first two prizes, then the question becomes an illustration of the second case. With this new information, we can reduce the sample space to the 38 persons who did not win the first or second prize. And in this reduced sample space, we see that the probability of Mr. Smith's receiving the third door prize is 1/38. Thus the information we have about an experiment plays a crucial role in our calculation of probabilities.

EXAMPLE 6.26 In a pharmaceutical research study, half of a group of 500 patients suffering from chronic headaches were given a new painkilling drug, and the other half were given a placebo. The results are shown in the table below.

	Felt better	*Felt no better*	*Total*
Given the drug	183	67	250
Given placebo	117	133	250
Total	300	200	500

(a) What is the probability that a random patient felt better?
(b) What is the probability that a random patient felt better, given that he or she received the drug?
(c) What is the probability that a random patient received the placebo, given that he or she felt no better?

Solution (a) Of the 500 patients, 300 felt better. Hence the probability that a random patient felt better is given by the counting rule to be

$$P(\text{felt better}) = \frac{300}{500} = .600.$$

Notice that this is *not* a conditional probability.
(b) In question (b), we are told that the patient received the drug. Hence we restrict our attention to the 250 patients who received the drug. Of these, 183 felt better. Hence by the counting rule for conditional probabilities we see that

$$P(\text{felt better}|\text{given drug}) = \frac{\text{number who felt better given drug}}{\text{number given drug}}$$

$$= \frac{183}{250} = .732.$$

(c) In this case we know that the patient felt no better, and so we restrict our attention to the 200 patients who felt no better. Of this group, 133 received the placebo. Therefore

$$P(\text{placebo}|\text{felt no better}) = \frac{\text{number given placebo who felt no better}}{\text{number who felt no better}}$$

$$= \frac{133}{200} = .665.$$

That is, 66.5% of those who felt no better were given the placebo. ■

Problem Problem 16 In Example 6.26, what is the conditional probability that a patient felt no better given that the patient received a placebo?

EXAMPLE 6.27 Two fair dice are rolled. What is the probability that the sum is 7 if we are told that each die came up less than 5?

Solution Figure 6.11 shows the sample space S for the experiment of rolling two dice along with the events

$$A = \text{``the sum rolled is 7'' and}$$

$$E = \text{``each die came up less than 5.''}$$

Since we are told that each die came up less than 5, there are only 16 possible outcomes for the experiment instead of the usual 36. Of these 16, there are only 2 outcomes for which A occurs, namely the pairs (3, 4) and (4, 3). Thus by the counting rule for conditional probabilities, we find that

$$P(A|E) = \frac{|A \cap E|}{|E|} = \frac{2}{16} = \frac{1}{8}.$$

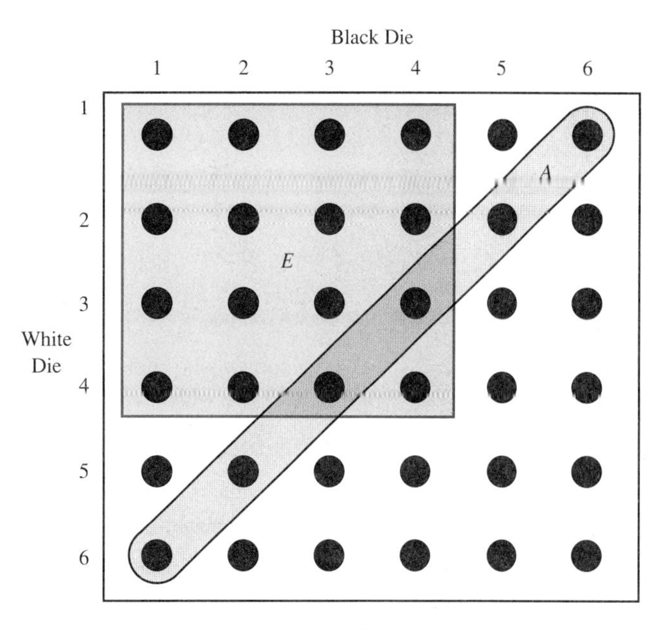

FIGURE 6.11

Observe that without the information that E occurred, we would have computed the probability of A to be

$$P(A) = \frac{|A|}{|S|} = \frac{6}{36} = \frac{1}{6}.$$

Hence knowing that E occurred decreases the probability that A occurred. ■

Practice Problem 17 Use the tree diagram in Figure 6.2 on page 273 to determine the probability that, in three flips of a fair coin, the first flip was heads given that two of the three flips were heads.

Recall that the various counting formulas are valid only for sample spaces consisting of equally likely outcomes. Since conditional probabilities may arise in experiments where the outcomes are not equally likely, we must define them in a more general context. Fortunately, a simple modification of the counting rule for conditional probabilities will yield this more general formula. For if we divide the numerator and denominator by $|S|$, the number of elements in the sample space, we obtain

$$\frac{\text{number of outcomes in } A \text{ and } E}{\text{number of outcomes in } E} = \frac{\dfrac{|A \cap E|}{|S|}}{\dfrac{|E|}{|S|}} = \frac{P(A \cap E)}{P(E)}.$$

Thus we define the **conditional probability of A given E** by

$$\frac{P(A \cap E)}{P(E)}$$

provided that $P(E) \neq 0$, and denote it, as before, by $P(A|E)$.

EXAMPLE 6.28 In a large survey of its customers, a nationwide retail store found that 20% made major purchases, 48% had a charge account, and 12% fell into both categories. What is the probability that

(a) a customer who makes a major purchase has a charge account?
(b) a customer with a charge account has made a major purchase?

Solution Assuming that the survey was representative of all of the store's customers, we see that

$$P(\text{major purchase}) = .20,\ P(\text{charge account}) = .48, \text{ and } P(\text{both}) = .12.$$

(a) In the first question we want to know the conditional probability that a random customer has a charge account *given the fact that the customer makes a major purchase*. Using the definition of conditional probability above, we find

$$P(\text{charge account}|\text{major purchase}) = \frac{P(\text{charge account and major purchase})}{P(\text{major purchase})} = \frac{.12}{.20} = .6.$$

In other words, among the customers who make major purchases, 60% have charge accounts as compared to 48% of all the store's customers.
(b) Here we want to determine the conditional probability that a random customer has made a major purchase *given the fact that the customer has a charge account*. As above, we obtain

$$P(\text{major purchase}|\text{charge account}) = \frac{P(\text{major purchase and charge account})}{P(\text{charge account})} = \frac{.12}{.48} = .25.$$

Thus 25% of the customers with charge accounts make a major purchase, as compared to 20% of all the store's customers. ■

Note that conditional probabilities can be disguised in ordinary discourse. In Example 6.28, for instance, the phrases "a customer who makes a major purchase" and "a customer with charge account" indirectly give information that limits the type of customers under consideration. For this reason, conditional probabilities are required in this problem.

Practice Problem 18 An insurance agent has found that, among her clients, 60% have purchased automobile insurance from her, 40% have purchased life insurance from her, and 15% have purchased both types of insurance from her. What is the probability that someone who has purchased life insurance from her has also purchased automobile insurance from her?

THE INTERSECTION RULE

In Section 6.3 we presented formulas for the probability of the complement and union of events. To obtain a similar formula for the probability of the intersection of events, we need only multiply the conditional probability formula

$$P(B|A) = \frac{P(B \cap A)}{P(A)}$$

by $P(A)$.

Theorem 6.4 (the intersection rule)

For any events A and B,

$$P(A \cap B) = P(A) \cdot P(B|A).$$

Thus the probability that two events will *both* occur is equal to the probability that the first occurs times the conditional probability that the second occurs given that the first occurred.

EXAMPLE 6.29 Ms. Caldwell, an insurance agent, often telephones persons at random in hopes of arranging appointments to discuss insurance. She has found that 8% of those who are called agree to an appointment, and 30% of those who agree to an appointment actually buy insurance from her. What is the probability that someone called at random will agree to an appointment and buy insurance?

Solution Let A and B be the following events:

A = "agrees to an appointment" and B = "buys insurance."

We must compute $P(A \cap B)$, the probability that someone will agree to an appointment and buy insurance. By the intersection rule

$$P(A \cap B) = P(A) \cdot P(B|A) = .08(.30) = .024.$$

Thus 2.4% of those who are called agree to an appointment and buy insurance. ■

Practice Problem 19 Engines on an automobile assembly line are inspected at two different points during the production process. The first inspection catches 70% of the defective units, and the second inspection catches 80% of the defective units that pass the first inspection. What is the probability that a defective engine will pass both inspections?

EXAMPLE 6.30 A firm has six accountants, three men and three women. If two of the accountants are randomly selected to attend an accounting seminar, what is the probability that the two who are chosen are both men?

Solution Instead of regarding the two persons as being chosen simultaneously, we will suppose that two persons are chosen in succession. (Clearly this will not affect the probability of obtaining two men.) Let

A = "a man is chosen first" and B = "a man is chosen second."

Then $P(A) = 3/6$ by the counting rule. Moreover, after the first man is selected there are 5 accountants remaining, 2 of whom are men. Thus $P(B|A) = 2/5$. Thus by the intersection rule

$$\begin{aligned} P(\text{two men are chosen}) &= P(A \cap B) \\ &= P(A) \cdot P(B|A) \\ &= \frac{3}{6} \cdot \frac{2}{5} = \frac{1}{5}. \end{aligned}$$

Compare this solution to that discussed in Section 6.1 after Example 6.8. ■

Practice Problem 20 In a box of ten transistors, three are defective. What is the probability that if two transistors are chosen at random from this box, both will be defective?

INDEPENDENT EVENTS

In the experiment of rolling two fair dice, one white and one black, consider the events

A = "the white die comes up 2, 3, or 4" and

B = "the black die comes up greater than 4."

It is intuitively obvious that what happens to the white die does not affect the result of the black die, and vice versa. Thus we expect that the occurrence of B will not change the likelihood that A occurs; that is, the probabilities $P(A)$ and $P(A|B)$ should be equal. Similarly we expect that $P(B) = P(B|A)$. To verify these statements, we can refer to the sample space S in Figure 6.12.

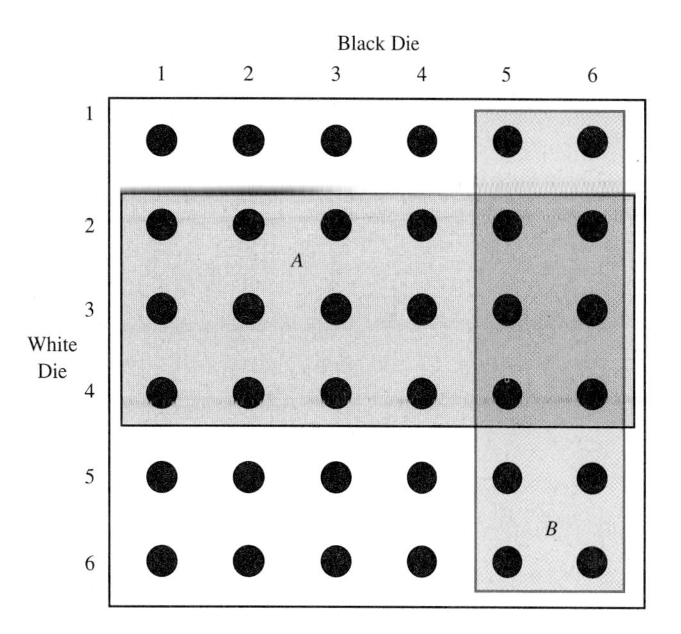

FIGURE 6.12

From this sample space we see that

$$P(A) = \frac{|A|}{|S|} = \frac{18}{36} = \frac{1}{2},$$

$$P(B) = \frac{|B|}{|S|} = \frac{12}{36} = \frac{1}{3},$$

$$P(A|B) = \frac{|A \cap B|}{|B|} = \frac{6}{12} = \frac{1}{2}, \text{ and}$$

$$P(B|A) = \frac{|B \cap A|}{|A|} = \frac{6}{18} = \frac{1}{3}.$$

Hence

$$P(A) = P(A|B) \qquad \text{and} \qquad P(B) = P(B|A),$$

as we anticipated.

Events A and B are called **independent** if both

$$P(A) = P(A|B) \qquad \text{and} \qquad P(B) = P(B|A).$$

We also call A and B independent if either of these conditional probabilities is undefined; that is, if either $P(A) = 0$ or $P(B) = 0$. Thus the preceding calculations have shown that the events ''the white die comes up 2, 3, or 4'' and ''the black die comes up greater than 4'' are independent.

The example above illustrates the distinction between *independent* events and *mutually exclusive* events. As we have seen, the events "the white die comes up 2, 3, 4" and "the black die comes up greater than 4" are independent, since the occurrence of one does not affect the probability of the other. However, these events are not mutually exclusive because it is possible for both events to occur at the same time. (For instance, the white die can come up 2 and the black die 5.) It can be shown that two independent events A and B can never be mutually exclusive unless $P(A) = 0$ or $P(B) = 0$. (See Exercise 43.)

When A and B are independent events, $P(B|A) = P(B)$. Hence the intersection rule reduces to the following simpler form.

If A and B are independent events, then $P(A \cap B) = P(A) \cdot P(B)$.

More generally, a collection of events $A_1, A_2, \ldots, A_k$ is independent if the probability of each of the events is unaffected by information concerning the others. In this case the intersection rule can be generalized as follows.

Theorem 6.5 (the intersection rule for independent events)

For any independent events $A_1, A_2, \ldots, A_k$,

$$P(A_1 \cap A_2 \cap \cdots \cap A_k) = P(A_1) \cdot P(A_2) \cdot \cdots \cdot P(A_k).$$

In applying this rule, remember that in order for the intersection of $A_1, A_2, \ldots, A_k$ to occur, *all* of the events must occur. Thus the intersection rule for independent events says that the probability that a collection of independent events all occur is equal to the product of their separate probabilities.

EXAMPLE 6.31 A computer system consists of the computer, a monitor, and a disk drive. Based on past experience, 2% of the computers, 1% of the monitors, and 5% of the disk drives can be expected to fail during the warranty period. If these components are independently produced, what percentage of these computer systems do not fail during the warranty period?

Solution Let C, M, and D be defined as follows:

C = "computer fails," M = "monitor fails," and D = "disk drive fails."

The given information is that

$$P(C) = .02, \quad P(M) = .01, \quad \text{and} \quad P(D) = .05.$$

We are interested in determining the probability that a system does not fail, that is, that none of the components fail. Therefore we must consider the complementary events $\overline{C}$, $\overline{M}$, and $\overline{D}$, for which the complement rule gives

$$P(\overline{C}) = .98, \qquad P(\overline{M}) = .99, \qquad \text{and} \qquad P(\overline{D}) = .95.$$

Since the components of the system are independently produced, we assume that $\overline{C}$, $\overline{M}$, and $\overline{D}$ are independent events. Hence

$$\begin{aligned} P(\text{system does not fail}) &= P(\text{no component fails}) \\ &= P(\overline{C} \cap \overline{M} \cap \overline{D}) \\ &= P(\overline{C}) \cdot P(\overline{M}) \cdot P(\overline{D}) \\ &= .98(.99)(.95) \\ &= .92169. \end{aligned}$$

Thus over 92% of the computer systems do not fail during the warranty period. ■

Practice Problem 21 A stereo system consists of a turntable, an amplifier, and two speakers. Suppose that 5% of the turntables, 2% of the amplifiers, and 3% of the pairs of speakers require servicing during the first year. If the turntables, amplifiers, and speakers are manufactured independently, what percentage of these systems do not require servicing during the first year?

The topics discussed in Sections 6.1–6.4 are essential to the understanding of probability. Almost all questions involving probability require use of one or more of the concepts and techniques from these sections. By comparison, the specialized techniques that we will consider in Sections 6.5 and 6.6 are used much less often. Thus the reader should place special importance on understanding the material presented up to this point.

EXERCISES 6.4

1. Suppose that the probability that A occurs is .53, the probability that B occurs is .72, and the probability that both occur is .48. Compute each of the following.

(a) $P(A|B)$ (b) $P(B|A)$

2. Suppose that the probability that A occurs is 1/3, the probability that B occurs is 1/4, and the probability that both occur is 1/8. Compute each of the following.

(a) $P(A|B)$ (b) $P(B|A)$

3. Two fair dice are rolled. What is the probability that the sum rolled is at least eight, given

(a) no information at all?

(b) that at least one die comes up five?

(c) that the same number appears on both dice?

4. A card is drawn at random from a standard deck of playing cards. What is the probability that the card is higher than a seven, given

(a) no information at all?

(b) that the card is not an ace?

(c) that the card is a club?

(d) that the card is a jack, queen, or king?

5. A fair coin is flipped three times. What is the probability that it comes up heads at least once, given

(a) no information at all?

(b) that the same side appears on all three flips?

(c) that it comes up heads at most once?

(d) that the second flip was tails?

6* Two letters, not necessarily distinct, were typed at random. What is the probability that both are vowels, given

(a) no information at all?

(b) that the two letters are different?

(c) that at least one is a vowel?

7. The following data from reference [2] shows the number of voters in Elmira, New York, who voted for the Republican candidate (Thomas E. Dewey) in the 1948 presidential election.

	Socioeconomic Status			
	High	*Middle*	*Low*	*Total*
Men	66	48	74	188
Women	60	80	90	230
Total	126	128	164	418

Suppose that one of these voters is selected at random.

(a) What is the probability that the voter is a man?

(b) What is the probability that the voter has high socioeconomic status?

(c) What is the probability that the voter is a woman given that the voter has low socioeconomic status?

(d) What is the probability that the voter has middle socioeconomic status given that the voter is a man?

8. In the Framingham study (see reference [6]) a group of 828 men in their thirties were classified according to their serum cholesterol level (in milligrams per deciliter). These cases were followed up many years later to see how many developed coronary disease. The results are shown at the top of page 323.

*This exercise requires the use of counting techniques from Chapter 5.

Cholesterol Level	*Coronary Disease*	*No Coronary Disease*	*Total*
Under 200	22	293	315
200–219	12	116	128
220–239	29	140	169
240–259	19	78	97
Over 259	35	84	119
Total	117	711	828

Assuming that this data is representative of men in their thirties, what is the probability that

(a) a man in his thirties will develop coronary disease in later life?

(b) a man in his thirties with a cholesterol level between 220 and 239 will develop coronary disease in later life?

(c) a man who develops coronary disease had a cholesterol level of at least 240 when he was in his thirties?

(d) a man without coronary disease had a cholesterol level below 200 when he was in his thirties?

The following data from reference [2] *shows the number of voters in Elmira, New York, categorized by religion, who voted for each 1948 presidential candidate.*

	Presidential Candidate Voted For		
Religion	*Democrat*	*Republican*	*Total*
Catholic	124	67	191
Protestant	90	345	435
Total	214	412	626

In Exercises 9–18 let C, T, D, and R denote the events of being Catholic, being Protestant, voting Democratic, and voting Republican, respectively.

9. Compute $P(C)$.

10. Compute $P(R)$.

11. Compute $P(C|D)$.

12. Compute $P(R|T)$.

13. Compute $P(R|C)$.

14. Compute $P(D|T)$.

15. Are C and D independent?

16. Are T and R independent?

17. Are C and R independent?

18. Are T and D independent?

19. Among army volunteers it was found that 17% fail the physical exam, 26% fail the written exam, and 8% fail both. What is the probability that a volunteer who fails the physical exam will also fail the written exam?

20. The percentage of voters favoring the incumbent went from 46% on August 1 to 52% on September 1, although only 38% favored him in both polls. What is the probability that a random voter who supported him on August 1 still supported him on September 1?

21. U.S. population life tables indicate that the probability that a white female will survive from birth to age 60 is .864 and the probability that she will survive from birth to age 70 is .733. What is the probability that a 60-year-old white female will survive to age 70?

22. Oddsmakers say that the odds against Brobdingnag's winning the Kentucky Derby are 3 to 2 and against winning the Triple Crown (the Kentucky Derby and two other races) are 9 to 1. If these odds are correct, what is the probability that if Brobdingnag wins the Kentucky Derby, he will go on to win the Triple Crown?

23. In a certain inner-city high school, 4000 students were asked (i) whether they were regular alcohol users, (ii) whether they were regular marijuana users, and (iii) whether they had ever tried hard drugs. Their responses are shown in the Venn diagram at the right, where U is the set of all students, A is the set of regular alcohol users, M is the set of regular marijuana users, and H is the set of students who have tried hard drugs. A marijuana advocate claimed that these data disproved the claim that marijuana use leads to hard drugs because an even higher percentage of those trying hard drugs were alcohol users (98.2%) than marijuana users (97.4%). Use the data to show that, in fact, marijuana users are more than twice as likely as alcohol users to try hard drugs.

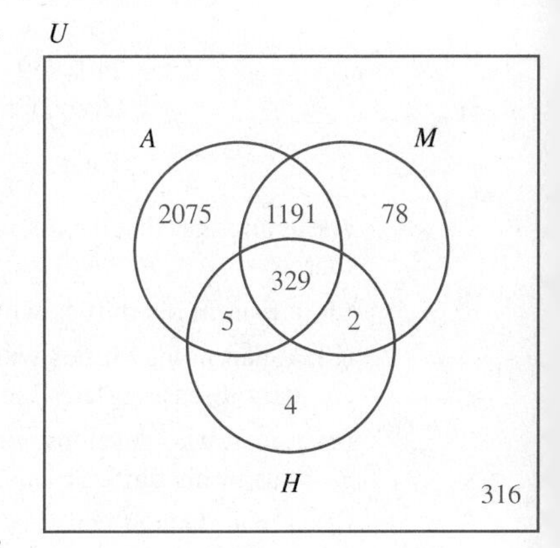

24. An argument similar to the following occurred during a murder trial.
Prosecutor: Ladies and gentlemen of the jury, you have heard the witnesses testify that the murderer drove a blue GM sedan with a license plate beginning with Z. Out of 735,000 cars registered in this state, there are only five which fit this description and could have been at the scene of the crime. Moreover, Snavely—the defendant—owns one of those five cars! The odds are therefore overwhelming that Snavely is the murderer.
As Snavely's attorney, how would you defend your client against this "mathematical" attack?

25. A plumber's welds are inspected twice for defects. The first inspection catches 80% of the defective welds, and of those that pass the first inspection, 90% are caught on the second inspection. What is the probability that a defective weld will pass both inspections?

26. A mathematics student would like to know the probability of not giving a wrong answer in class. She will avoid giving a wrong answer if the teacher does not call on her or if she answers correctly when questioned. She estimates that the probability of being questioned by the teacher is .2 and that the probability of answering correctly if questioned is .6. Assuming that she will not be questioned more than once per class, what is the probability that she will not give an incorrect answer during tomorrow's class?

27. Sixty percent of the graduates majoring in accounting at a certain university pass the CPA exam on the first try. Of those who fail on the first try, half pass on the second try; and of those who fail twice, 40% pass on the third attempt. What is the probability that a random graduate majoring in accounting at this university will pass the CPA exam within three tries?

28. The Motor Vehicle Department has found that the probabilities of passing the driver's license test on the first, second, and third tries are .80, .75, and .50, respectively. What are the probabilities of passing the test within two tries and within three tries?

29. A lawyer has told his client that there is a 30% chance of acquittal in the forthcoming trial. If convicted, the lawyer believes that there is a 60% chance of being acquitted when the verdict is appealed to a higher court. What is the probability that the client will eventually be acquitted?

30. A basketball player makes his first free throw 70% of the time. When this happens, he makes the second shot 90% of the time. But if he misses the first shot, then he makes the second shot only 80% of the time. If the player shoots two free throws, what is the probability that he will make

(a) both of them? (b) at least one of them?

31. From an audience of 15 boys and 10 girls two children are to be selected at random to assist a magician. What is the probability that

(a) two girls are chosen?

(b) a boy is chosen first and a girl is chosen second?

(c) one boy and one girl are chosen?

(d) a girl is chosen second?

32. A jeweler accidentally dropped two diamonds onto a tray containing three zircons of identical size and shape. In order to find the diamonds he must examine the five gems one by one until both diamonds are found. What is the probability that he will examine

(a) no more than two gems? (b) no more than three gems?

33. From an urn containing 10 red, 10 white, and 10 blue marbles, three are to be selected in sequence without replacement. What is the probability that they will be chosen in the order red, white, and blue?

34. From an urn containing 8 red, 5 white, and 6 yellow marbles, two are to be selected in sequence without replacement. What is the probability that two marbles of the same color will be chosen?

35. If $P(E) = .6$ and $P(F) = .2$, compute $P(E \cap F)$ and $P(E \cup F)$ if

(a) E and F are independent (b) E and F are mutually exclusive.

36. Let A and B be events for which $P(A) = .85$ and $P(B) = .90$.

(a) What is the smallest possible value for $P(A \cap B)$?

(b) Is it possible for A and B to be mutually exclusive?

(c) If A and B are independent, what is $P(A \cap B)$?

37. The probability that a female over 60 years of age will develop breast cancer is .04 and the probability that she will develop diabetes is .02. Assuming that these events are independent, what is the probability that a female over 60 will develop

(a) at least one of the diseases? (b) both of the diseases?

38. Three teams are making independent attempts to climb Mount Everest. If their chances of success are 40%, 45%, and 50%, respectively, what is the probability that at least one team will succeed?

39. During construction of a new office building, a contractor estimates that there is an 8% chance of a shortage of materials, a 10% chance of a strike, and a 20% chance of delays due to bad weather. If these events are independent, what is the probability that at least one of these problems occurs?

40. In blacks the probabilities of having blood types A and B are .30 and .20, respectively. Moreover, the probability that a black has the Rh antigen is .85. Use the fact that the presence or absence of the Rh antigen is independent of the presence or absence of the A and B antigens to compute the probability that a randomly selected black will have the following blood types. (Refer to Example 5.8 in Section 5.1 for a description of human blood types.)

(a) A+ (b) A− (c) B+ (d) B−

41. In their past games, chess grandmaster Lipner has defeated grandmaster Traub 40% of the time, lost 10% of the time, and drawn 50% of the time. If they play three games, what is the probability that

(a) all three games are drawn?

(b) Lipner will win one game and draw two?

42. An oil company is currently drilling in three different countries, where geologists estimate the chances of finding oil to be 30%, 50%, and 80%, respectively. What is the probability that

(a) oil will be found at all three sites?

(b) oil is found at exactly two sites?

43. Prove that if A and B are both independent and mutually exclusive, then $P(A) = 0$ or $P(B) = 0$.

44. Prove the intersection rule for three events A, B, and C:

$$P(A \cap B \cap C) = P(A) \cdot P(B|A) \cdot P(C|A \cap B).$$

45. Prove that if A and B are events such that $P(B) \neq 0$, then

$$P(\overline{A}|B) = 1 - P(A|B).$$

46. Prove that if events A and B are independent, then so are events A and $\overline{B}$.

Answers to Practice Problems

16. $\frac{133}{250} = .532$

17. $\frac{2}{3}$

18. .375

19. .06

20. $\frac{6}{90} = \frac{1}{15}$

21. approximately 90.3%

6.5 Bayes's Formula

Much of science is based on *inductive* rather than deductive inference, that is, instead of reasoning from hypotheses to conclusions we work backwards, assigning probabilities to various hypotheses on the basis of known conclusions. For example, a doctor making a diagnosis observes the patient's known symptoms and then assigns probabilities to various diseases that might produce those symptoms. A fundamental tool in the theory of inductive inference is Bayes's formula, which has widespread applications in statistics.

STOCHASTIC DIAGRAMS

Many experiments are performed in stages, each of which may have several possible outcomes. Such a multi-stage experiment is called a **stochastic process,** from the Greek word *stochos* meaning "guess." Such processes arise in many applications. We will begin by considering an example.

A regional producer of soft drinks fills and caps cola bottles on one of three machines in its plant in Decatur, Georgia. Twenty percent of the total output goes through machine 1, which has a 0.1% defective rate; that is, one out of every 1000 bottles that are processed on this machine is improperly filled or capped. Machine 2 handles 30% of the output with a 0.6% defective rate, and machine 3 processes the rest of the bottles with a 0.4% defective rate. What is the probability that a bottle of this cola is defective?

This problem describes a stochastic process. First, each bottle is sent to one of the three machines, and then it is filled and capped, possibly defectively. Let M_1, M_2, M_3, and D denote the events

M_1 = "bottle was processed by machine 1,"

M_2 = "bottle was processed by machine 2,"

M_3 = "bottle was processed by machine 3," and

D = "bottle is defective."

Since 20% of the bottles are processed by machine 1, we have $P(M_1) = .20$. Moreover, since 0.1% of the bottles processed by machine 1 are defective, $P(D|M_1) = .001$. Similarly we have

$$P(M_2) = .30,\ P(D|M_2) = .006,\ P(M_3) = .50,\ \text{and}\ P(D|M_3) = .004.$$

This process can be illustrated by using a tree diagram as in Figure 6.13. Because $\overline{D}$ = "bottle is satisfactory," the complement rule can be used to compute $P(\overline{D}|M_1)$, $P(\overline{D}|M_2)$, and $P(\overline{D}|M_3)$.

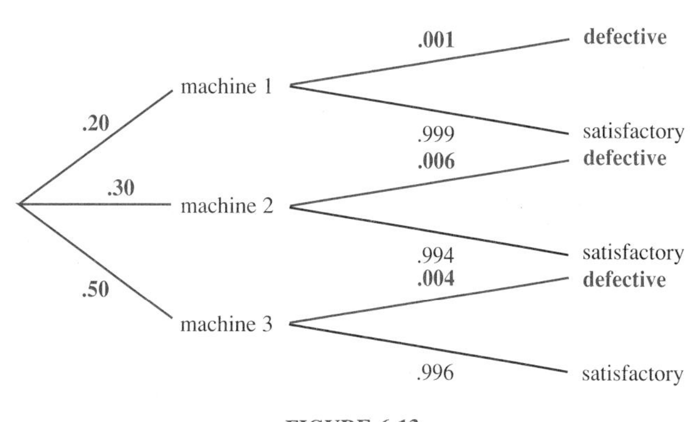

FIGURE 6.13

Notice that the three leftmost branches in this diagram are labeled with the probabilities that a bottle is processed by the corresponding machine, and the six rightmost branches are labeled with the *conditional probabilities* that a defective or satisfactory bottle is produced by the corresponding machine. A tree diagram of this sort, in which each branch is labeled with the appropriate probability or conditional probability, is called a **stochastic diagram.**

The event "bottle is defective" can be broken into three mutually exclusive cases corresponding to the three machines used to fill and cap the bottles. Each of these cases is represented by a different branch in the diagram that ends in "defective." Now the intersection rule states that

$$P(M_1 \cap D) = P(M_1) \cdot P(D|M_1),$$

and so $P(M_1 \cap D)$, the probability that a bottle is defective and was produced by machine 1, can be found by multiplying the probabilities along the corresponding branch of the tree. Thus

$$\begin{aligned} P(D) &= P(M_1 \cap D) + P(M_2 \cap D) + P(M_3 \cap D) \\ &= P(M_1) \cdot P(D|M_1) + P(M_2) \cdot P(D|M_2) + P(M_3) \cdot P(D|M_3) \end{aligned}$$

is the sum of all the branch probabilities that end in "defective." Hence

$$\begin{aligned} P(D) &= (.20)(.001) + (.30)(.006) + (.50)(.004) \\ &= .0002 + .0018 + .0020 = .0040. \end{aligned}$$

Therefore there is a .4% chance that a bottle will be defective.

It turns out that the probability of an event can always be computed in this manner if we have a set of events like M_1, M_2, and M_3. Accordingly we say that a set of events $A_1, A_2, \ldots, A_k$ is a **partition** of a sample space S if *exactly one* of the events occurs whenever the experiment is performed. In set-theoretic terms this means that the intersection of each pair of the events is $\varnothing$ and the union of all k of the events is S. (See Figure 6.14.)

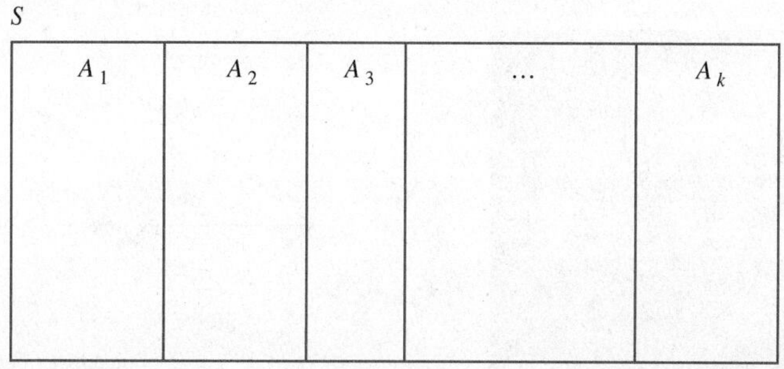

S

A_1	A_2	A_3	...	A_k

FIGURE 6.14

The effect of a partition on an arbitrary event E is to divide it into mutually exclusive subsets

$$A_1 \cap E, \quad A_2 \cap E, \quad \ldots, \quad A_k \cap E$$

as in Figure 6.15. Since

$$E = (A_1 \cap E) \cup (A_2 \cap E) \cup \cdots \cup (A_k \cap E),$$

we have

$$P(E) = P(A_1 \cap E) + P(A_2 \cap E) + \cdots + P(A_k \cap E)$$

by the union rule for mutually exclusive events. This equation is sometimes called the **law of complete probability.**

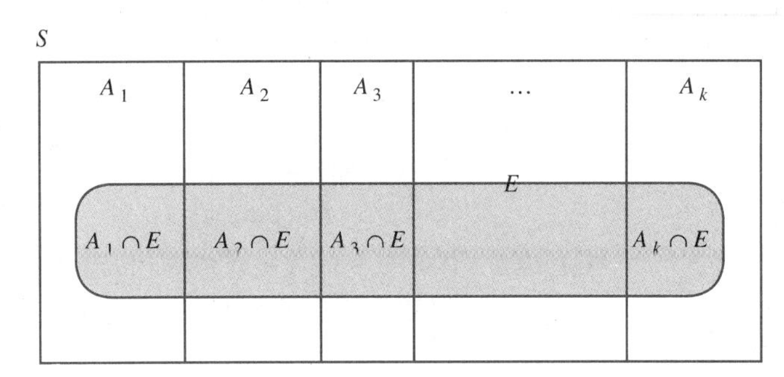

FIGURE 6.15

BAYES'S FORMULA

Given a partition $A_1, A_2, \ldots, A_k$ of a sample space, we can think of each event A_i as a "hypothesis" that has a certain probability of being true. Once an event E has occurred, we can ask how this new information alters the probability of A_i. In other words, what is the probability of A_i given that E has occurred? In the example above, for instance, we can ask the probability that a defectively capped bottle was processed on machine i. Of course, what we are seeking is the conditional probability

$$P(A_i|E) = \frac{P(A_i \cap E)}{P(E)}.$$

If we apply the intersection rule and the law of complete probability to the right side of this equation, we obtain our main result.

Theorem 6.6 (Bayes's formula*)

Let $A_1, A_2, \ldots, A_k$ be a partition of a sample space S. Then for any event $E \subseteq S$ with $P(E) \neq 0$

$$P(A_i|E) = \frac{P(A_i)P(E|A_i)}{P(A_1)P(E|A_1) + P(A_2)P(E|A_2) + \cdots + P(A_k)P(E|A_k)}.$$

*Thomas Bayes (1702–1761) was an English clergyman and mathematician. This theorem appeared in a famous paper entitled "An Essay Toward Solving a Problem in the Doctrine of Chances," which was published in 1763.

The significance of Bayes's formula is this: If we know the probability of E under each "hypothesis" A_i, then we can compute the conditional probability of each A_i given E. In other words, we can compute $P(A_i|E)$ in terms of each $P(E|A_i)$. Unfortunately, the notation required to express Bayes's Formula is quite complicated. However, for small values of n we can avoid the notation by using a stochastic diagram. The following example demonstrates this technique.

EXAMPLE 6.32 In the example above, a purchaser of the cola obtained a bottle that tasted flat because of improper sealing. What is the probability that the bottle was processed on machine 1?

Solution Here we are given that a bottle is defective, and we want to know the probability that it was processed on machine 1. Thus we must compute the conditional probability

$$P(\text{machine 1}|\text{defective}).$$

By definition of this conditional probability we have

$$P(\text{machine 1}|\text{defective}) = \frac{P(\text{machine 1 and defective})}{P(\text{defective})}.$$

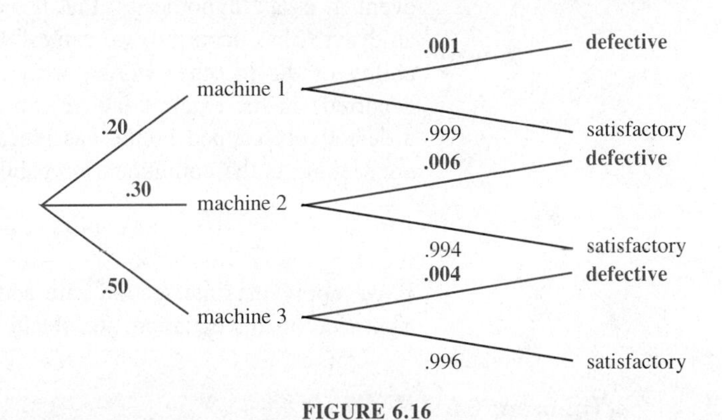

FIGURE 6.16

The right side of this equation can be easily evaluated using the stochastic diagram in Figure 6.16. Indeed, from the diagram we see that

$$P(\text{machine 1 and defective}) = (.20)(.001) = .0002,$$

and as we saw before

$$P(\text{defective}) = (.20)(.001) + (.30)(.006) + (.50)(.004) = .0040.$$

Hence

$$P(\text{machine 1}|\text{defective}) = \frac{.0002}{.0040} = \frac{2}{40} = .05.$$

Similarly we see that

$$P(\text{machine } 2|\text{defective}) = \frac{.0018}{.0040} = \frac{18}{40} = .45.$$

and

$$P(\text{machine } 3|\text{defective}) = \frac{.0020}{.0040} = \frac{20}{40} = .50.$$

Thus if we know that a bottle is defective, the probability that it was processed by machine 1 *decreases* from .20 to .05, the probability that it was processed by machine 2 *increases* from .30 to .45, and the probability that it was processed by machine 3 remains .50. In Bayesian terms, the probability of the events M_1 and M_2 has been altered by the knowledge that D occurred. ■

Notice that in Example 6.32 we are able to compute

$$P(\text{machine } 1|\text{defective})$$

by computing the probability along the branch containing "machine 1" and "defective" and dividing by the sum of all the branch probabilities that end in "defective." This is exactly what the complicated notation in Bayes's formula means in terms of our stochastic diagram.

EXAMPLE 6.33 A factory employs workers on three shifts. The first shift employs 40% of the workers, the second shift 35%, and the third shift 25%. The absentee rate for workers on the first shift is 1.5%, but increases to 4% on the second shift and 12% on the third. What percentage of the absentees are assigned to the first shift? To the second shift? To the third shift?

Solution In this problem we must compute

$$P(\text{first shift}|\text{absent}), \quad P(\text{second shift}|\text{absent}), \quad \text{and} \quad P(\text{third shift}|\text{absent}).$$

We are told the conditional probabilities of a worker's being absent given that he or she works on a particular shift. Thus since the three shifts partition the sample space of all workers, Bayes's Theorem will enable us to compute the desired probabilities. As above we will rely on a stochastic diagram, which is shown in Figure 6.17.

To compute $P(\text{first shift}|\text{absent})$, we divide the branch probability through "first shift" and "absent" by the sum of all the branch probabilities that end in "absent." In this way we find that

$$\begin{aligned} P(\text{first shift}|\text{absent}) &= \frac{.40(.015)}{.40(.015) + .35(.04) + .25(.12)} \\ &= \frac{.006}{.006 + .014 + .030} = \frac{.006}{.050} \\ &= \frac{6}{50} = .12. \end{aligned}$$

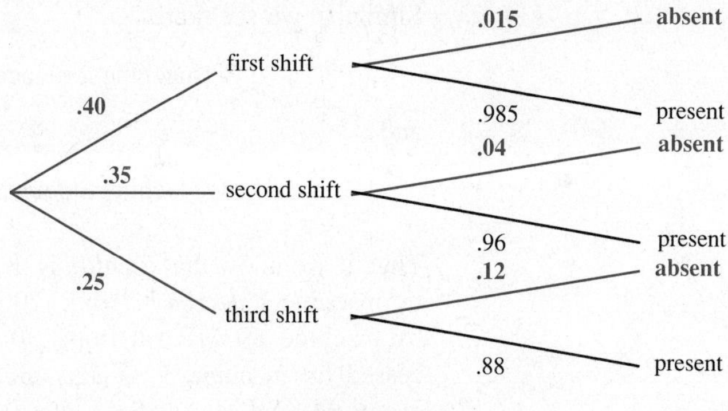

FIGURE 6.17

Note that by the law of complete probability the denominator of our first fraction,

$$.40(.015) + .35(.04) + .25(.12) = .050,$$

is the probability of a worker's being absent. This same denominator occurs in our other calculations

$$\begin{aligned} P(\text{second shift}|\text{absent}) &= \frac{.35(.04)}{.050} = \frac{.014}{.050} \\ &= \frac{14}{50} = .28 \end{aligned}$$

and

$$\begin{aligned} P(\text{third shift}|\text{absent}) &= \frac{.25(.12)}{.050} = \frac{.030}{.050} \\ &= \frac{30}{50} = .60. \end{aligned}$$

Thus 12% of the absentees work on the first shift, 28% on the second, and 60% on the third. ■

Practice Problem 22 In a particular county 50% of the registered voters are Democrats, 30% are Republicans, and 20% are independents. During a recent election, 40% of the Democrats voted, 60% of the Republicans voted, and 70% of the independents voted. What is the probability that someone who voted is a Democrat? A Republican? An independent?

RELIABILITY OF TESTS

Bayes's formula has important applications to the statistical evaluation of tests, particularly in medicine. In one typical situation we want to know how accurately a particular diagnostic test predicts a certain condition, that is, to what extent a positive test result can be believed.

EXAMPLE 6.34 Approximately 8% of all black Americans are carriers of sickle cell anemia, a hereditary form of anemia. A new test to detect this trait was administered to a large group of subjects known to be carriers and to another group known to be noncarriers. The test correctly identified 97% of the known carriers and 95% of the known noncarriers.* If this new test is randomly administered to a black American and the result is positive (indicating the presence of sickle cell anemia), what is the probability that the person really is a carrier?†

Solution We take as our sample space the set of all black Americans and consider the partition formed by the events "person is a carrier" and "person is a noncarrier." The probability that a randomly chosen black American is a carrier is equal to .08, the percentage of carriers in the sample space. Therefore by the complement rule the probability that he or she is a noncarrier is .92. Since the test is positive for 97% of the known carriers, the conditional probability that the test result is positive given that someone is a carrier is .97. (And therefore the conditional probability of a negative test result is .03.) Likewise the conditional probability of a *negative* test result given that someone is a noncarrier is .95, and so the conditional probability of a positive test result is .05 in this case. In this manner we obtain the stochastic diagram in Figure 6.18.

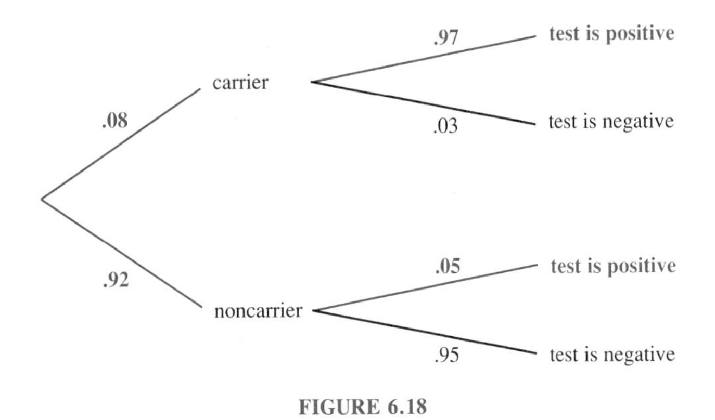

FIGURE 6.18

*In the terminology of diagnostic testing this test is said to have 97% *sensitivity* and 95% *specificity*.

†This probability is called the *predictive value* of the test.

What we want to know is *the probability that someone is a carrier given that he or she has a positive test result*. In other words, we want to compute

$$P(\text{carrier}|\text{test is positive}).$$

Since two branches lead to the event "test is positive," we see that

$$\begin{aligned} P(\text{carrier}|\text{test is positive}) &= \frac{.08(.97)}{.08(.97) + .92(.05)} \\ &= \frac{.0776}{.0776 + .0460} \\ &= \frac{.0776}{.1236} = \frac{776}{1236} \approx .628. \end{aligned}$$

Therefore a positive test result is correct only 62.8% of the time; in over one-third of the cases it is a false alarm! ■

The answer to Example 6.34 is surprising. In general, *the rarer a condition is among the population, the less reliable a diagnostic test will be*. In other words, as the percentage of actual carriers decreases, the percentage of inaccurate test results increases. As this example shows, our intuition can be very misleading in problems of statistical inference.

EXERCISES 6.5

In Exercises 1–8 find the indicated probability using the stochastic diagram shown below.

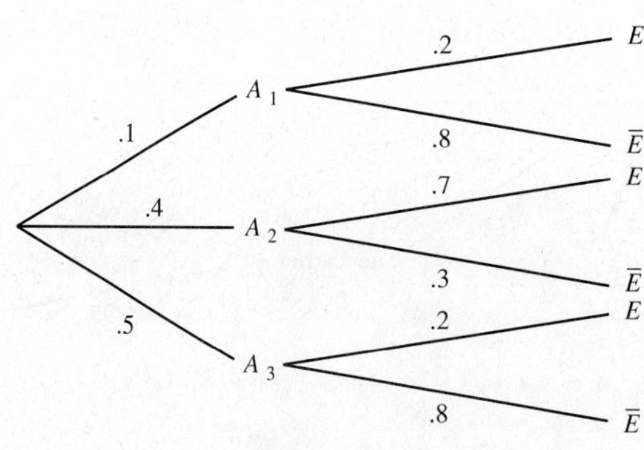

1. $P(E)$
2. $P(\overline{E})$
3. $P(A_1|E)$
4. $P(A_1|\overline{E})$
5. $P(A_2|E)$
6. $P(A_2|\overline{E})$
7. $P(A_3|E)$
8. $P(A_3|\overline{E})$

Suppose that A_1 and A_2 partition the sample space of an experiment. Find $P(E)$, $P(A_1|E)$, and $P(A_2|E)$ under the conditions in Exercises 9–12.

9. $P(A_1) = .7$, $P(E|A_1) = .4$, and $P(E|A_2) = .1$

10. $P(A_2) = .4$, $P(E|A_1) = .2$, and $P(E|A_2) = .7$

11. $P(A_2) = .8$, $P(E|A_1) = .3$, and $P(E|A_2) = .4$

12. $P(A_1) = .25$, $P(E|A_1) = .32$, and $P(E|A_2) = .12$

Suppose that A_1, A_2, and A_3 partition the sample space of an experiment. Find $P(E)$, $P(A_1|E)$, $P(A_2|E)$, and $P(A_3|E)$ under the conditions in Exercises 13–16.

13. $P(A_1) = .3$, $P(A_2) = .6$, $P(E|A_1) = .7$, $P(E|A_2) = .2$, and $P(E|A_3) = .4$

14. $P(A_1) = .5$, $P(A_3) = .1$, $P(E|A_1) = .4$, $P(E|A_2) = .6$, and $P(E|A_3) = .2$

15. $P(A_2) = .2$, $P(A_3) = .6$, $P(E|A_1) = .35$, $P(E|A_2) = .45$, and $P(E|A_3) = .15$

16. $P(A_1) = .4$, $P(A_2) = .3$, $P(E|A_1) = .5$, $P(E|A_2) = .2$, and $P(E|A_3) = .3$

17. From past experience an oil exploration firm has found that only 10% of the sites it drills yield oil. The firm is considering use of a new prediction method based on infrared analysis. When tested on a large number of randomly chosen drilling sites, the method correctly predicted oil in 82% of the sites where oil was subsequently found, and it correctly predicted no oil in 74% of the sites that subsequently proved to be dry. If the new method predicts oil at a particular site, what is the probability that oil will actually be found there?

18. A teacher found that only 60% of her class completed a particular examination. Of those who completed the examination, 90% passed, but only 75% of the others passed. What percentage of students passed this examination?

19. An automobile insurance company classifies its policyholders as either good risks (class G) or bad risks (class B). Among the company's current policyholders 95% are in class G and 5% are in class B. For drivers in class G the probability of having an accident in a given year is .01, and for drivers in class B the probability of having an accident in a given year is .06. If a policyholder reports that he or she has just had an accident, what is the probability that he or she belongs to class B?

20. An airline has hired a psychologist to determine whether prospective pilots will be able to react well in emergencies. The airline currently employs a large number of pilots, of whom 85% are known to react well. Each of the current pilots was interviewed anonymously and rated as either "satisfactory" or "unsatisfactory." Among the pilots known to react well, 92% were rated "satisfactory," compared to only 12% of the pilots known to react poorly. If a random applicant has a "satisfactory" rating, what is the probability that he or she will react well in emergencies?

21. A television consumer reporter finds that 20% of the complaints she investigates are satisfactorily resolved on her first inquiry to the company in question. The remaining 80% are presented on television. Of these, 70% are satisfactorily resolved after the complaint is aired and 30% remain unresolved. Of the complaints that are ultimately resolved, what percentage are aired on television?

22. A mail order business has an 18% response rate when it uses a mailing list purchased from a market research firm as compared to a 9% response rate for random mailings. If 60% of the items it mails are sent in random mailings, what percentage of the company's responses come from random mailings?

23. An automobile dealer sells Buicks, Oldsmobiles, and Pontiacs. His sales records show that 45% of his new car sales are Buicks, 30% are Oldsmobiles, and 25% are Pontiacs. Of those who buy Buicks, 60% want an AM-FM radio; for purchasers of Oldsmobiles and Pontiacs this percentage drops to 40% and 20%, respectively.

(a) What is the probability that someone buying a new car from this dealer will want an AM-FM radio?

(b) What is the probability that a car with an AM-FM radio bought from this dealer is a Buick? An Oldsmobile? A Pontiac?

24. In a particular county 40% of the registered voters are Democrats, 50% are Republicans, and 10% are independents. During a recent election, 50% of the Democrats voted, 60% of the Republicans voted, and 80% of the independents voted.

(a) What percentage of the eligible voters cast ballots in this election?

(b) What is the probability that someone who voted is a Democrat? A Republican? An independent?

25. An insurance company finds that 25% of its automobile insurance policies cover drivers under 25 years of age, 15% cover drivers 25–34, 40% cover drivers 35–60, and 20% cover drivers over 60. The probability of an accident during the next year for each age group is as follows: .020 for those under 25, .008 for those aged 25–34, .005 for those aged 35–60, and .012 for those over 60.

(a) What proportion of this company's policyholders will have an accident during the next year?

(b) What proportion of the accidents involving this company's policyholders involve drivers of each age group?

26. A physician knows that a patient's peculiar symptoms must be caused by one of four diseases. Research shows that the patient's symptoms occur in 80% of those with the first disease, in 90% of those with the second disease, in 95% of those with the third disease, and in 70% of those with the fourth disease. If these diseases are known to occur in 0.2%, 0.4%, 0.1%, and 0.3% of the population, respectively, what is the most probable cause of the patient's symptoms?

27. Three identical boxes each contain two coins. In the first box both coins are gold, in the second box one is gold and one is silver, and in the third box both coins are silver. One of the boxes is chosen at random, and a coin is randomly selected from this box. If this coin is gold, what is the probability that the other coin in this box is gold?

28. A box contains three coins, two ordinary coins and one two-headed coin. A coin is chosen at random from the box and flipped twice. If it comes up heads both times, what is the probability that the coin is two-headed?

29. Half the students in a certain class read the textbook after each lecture. The percentage of students receiving an A was twice as high among those who read the text after each lecture as among those who did not. If a student in this class got an A, what is the probability that he or she read the text after each lecture? (*Hint:* Let p be the probability that a nonreader got an A.)

30. A sales representative finds that 45% of the buyers she sees are male. Moreover, her success rate with male buyers is three times that with female buyers. If a meeting with a buyer results in a sale, what is the probability that the buyer was male? (*Hint:* Let p be the sales representative's success rate with female buyers.)

31. A do-it-yourself pregnancy test known as Ova II was evaluated in 1976 by the Center for Disease Control in Atlanta. The results, which were reported in reference [1], showed that when the test was administered to pregnant women, it was positive 50% of the time; and when the test was administered to women who were not pregnant, it was negative 41% of the time. Show that according to these findings a positive test result actually *lowers* the probability that a random woman is pregnant! (*Hint:* Let p be the probability of pregnancy.)

32. One percent of the inhabitants of an island suffer from a congenital disease having symptoms that appear only in adults. The disease is associated with the presence of the A and B antigens in the blood of those with the disease. Specifically, among those with the disease 60% have antigen A, 70% have antigen B, and 50% have both; whereas the corresponding rates of occurrence are 12%, 15%, and 9% for the nondiseased population. What is the probability that a child has the disease if his or her blood contains

(a) both antigens? (b) antigen A but not B?

Answer to Practice Problem **22.** $\frac{5}{13}, \frac{9}{26}$, and $\frac{7}{26}$

6.6 Bernoulli Trials

It often happens that an experiment is performed not once but repeatedly, and we are interested in the frequency of occurrence of a particular event E. In this section we will derive a useful formula for the probability that E will occur any specified number of times.

REPEATED INDEPENDENT TRIALS

The relative frequency interpretation of probability, which we encountered in Section 6.1, is based on the notion of an experiment as a *repeatable action* that has various possible outcomes. According to this interpretation, the probability of an event E predicts the proportion of the time E will occur over the long run.

Suppose, for instance, that two fair dice are rolled, and on each roll we are interested only in whether or not a sum of seven is obtained. Since

$$P(\text{sum} = 7) = \frac{6}{36} = \frac{1}{6},$$

over many trials we expect this event to occur one-sixth of the time. Thus out of 3000 rolls, we would expect to obtain a sum of seven approximately 500 times. But what if we obtained no sevens at all? Or what if we obtained a seven on every roll? Such occurrences are unlikely, of course, but they could happen. In general, if an event E is neither certain nor impossible, that is, if $0 < P(E) < 1$, then when the experiment is repeated n times, the number of occurrences of E might be any integer k from 0 to n. Our question now becomes: For a given value of k, what is the probability that E occurs exactly k times out of n repetitions of the experiment?

BERNOULLI PROCESSES

Since we are interested only in whether or not E occurs on each trial, we can simplify the problem by calling an occurrence of E a "success" and an occurrence of $\overline{E}$ a "failure." We call a sequence of repeated trials of an experiment a **Bernoulli process*** if the following three conditions are met:

1. Each trial has only two possible outcomes, success (denoted by S) and failure (denoted by F).
2. The trials are independent.
3. The probability of success is the same for each trial.

The terms "success" and "failure" should not be taken literally. As used here, a trial is a success if the particular outcome of interest occurs. Thus if we are counting the occurrence of adverse side effects produced by a new drug, the occurrence of adverse side effects is called a success, and their absence is called a failure. It is customary to denote the probability of success by p and the probability of failure by q. Note that since success and failure are complementary events, it follows from the complement rule that $p + q = 1$.

For example, suppose that we roll a pair of fair dice three times. If the sum is seven, the outcome is considered a success, otherwise it is considered a failure. The three rolls represent $n = 3$ Bernoulli trials, where for each trial we have

$$p = P(S) = P(\text{sum} = 7) = \frac{6}{36} = \frac{1}{6}$$

$$q = P(F) = 1 - P(S) = 1 - \frac{1}{6} = \frac{5}{6}.$$

This process can be represented by the stochastic diagram in Figure 6.19.

There are $2^3 = 8$ possible outcomes represented by the various paths through the stochastic diagram, but they are not all equally likely. For example, the probability of obtaining three sevens is

$$P(SSS) = p \cdot p \cdot p = p^3 = \left(\frac{1}{6}\right)^3 = \frac{1}{216},$$

whereas the probability of rolling three nonsevens is

$$P(FFF) = q \cdot q \cdot q = q^3 = \left(\frac{5}{6}\right)^3 = \frac{125}{216}.$$

---

*The Swiss mathematician Jacob Bernoulli (1654–1705) was among the first to study problems involving repeated trials. His work the *Ars Conjectandi*, published in 1713, was a landmark in probability theory.

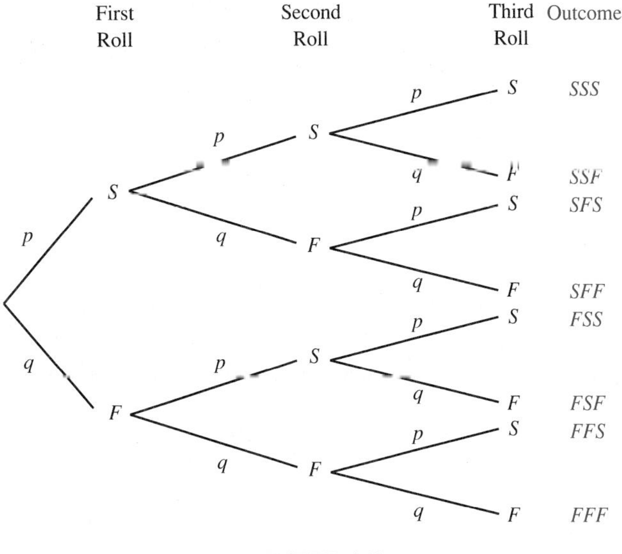

FIGURE 6.19

To find the probability of rolling *exactly one* seven, we add the probabilities of the three favorable branches:

$$
\begin{aligned}
P(\text{exactly one } 7) &= P(SFF) + P(FSF) + P(FFS) \\
&= pqq + qpq + qqp \\
&= pq^2 + pq^2 + pq^2 \\
&= 3pq^2 = \frac{75}{216} = \frac{25}{72}.
\end{aligned}
$$

Similarly, the probability of rolling exactly two sevens is

$$
\begin{aligned}
P(\text{exactly two 7s}) &= P(SSF) + P(SFS) + P(FSS) \\
&= ppq + pqp + qpp \\
&= p^2q + p^2q + p^2q \\
&= 3p^2q = \frac{15}{216} = \frac{5}{72}.
\end{aligned}
$$

Hence for $k = 0, 1, 2, 3$ we have computed the probability of obtaining exactly k sevens in $n = 3$ trials.

Now let us determine the probability of obtaining exactly two sevens in five rolls of a pair of fair dice. If we extend the tree diagram in Figure 6.19 by adding another two levels of branching, we will obtain a sample space for the experiment of rolling a pair of dice five times. Of the $2^5 = 32$ outcomes, there are 10 containing exactly two successes, namely, those listed below.

SSFFF	*SFSFF*	*SFFSF*	*SFFFS*	*FSSFF*
FSFSF	*FSFFS*	*FFSSF*	*FFSFS*	*FFFSS*

There are 10 such outcomes because the two positions that the successes occupy can be chosen in

$$C(5, 2) = \frac{P(5, 2)}{2!} = \frac{5 \cdot 4}{2 \cdot 1} = 10$$

different ways. Moreover, *all ten of these favorable outcomes have the same probability,* namely p^2q^3. For instance, by independence we have

$$P(SFFSF) = P(S) \cdot P(F) \cdot P(F) \cdot P(S) \cdot P(F) = p \cdot q \cdot q \cdot p \cdot q = p^2q^3,$$
$$P(FSFFS) = P(F) \cdot P(S) \cdot P(F) \cdot P(F) \cdot P(S) = q \cdot p \cdot q \cdot q \cdot p = p^2q^3,$$

and similarly for the other favorable outcomes. Therefore

$$\begin{aligned} P(\text{exactly 2 sevens in 5 rolls}) &= P(SSFFF) + \cdots + P(FFFSS) \\ &= p^2q^3 + \cdots + p^2q^3 \\ &= 10p^2q^3 = 10\left(\frac{1}{6}\right)^2\left(\frac{5}{6}\right)^3 \\ &= \frac{1250}{7776} = \frac{625}{3888} \approx .161. \end{aligned}$$

Consequently, about 16.1% of the time we expect to obtain exactly two sevens in five rolls of a pair of fair dice.

By generalizing the reasoning above, we obtain the following formula for the probability of obtaining exactly k successes in a sequence of n Bernoulli trials.

Theorem 6.7 (Bernoulli's formula)

Let p be the probability of success and $q = 1 - p$ the probability of failure on a single trial of a Bernoulli process. Then the probability of obtaining exactly k successes in a sequence of n Bernoulli trials, where $k = 0, 1, 2, \ldots, n$, is given by

$$P(\text{exactly } k \text{ successes in } n \text{ trials}) = C(n, k)p^kq^{n-k}.$$

Note that the expression $C(n, k)p^kq^{n-k}$ in Bernoulli's formula is a term in the binomial theorem expansion of $(p + q)^n$. (See Section 5.5.) The exponents k and $n - k$ that appear in this term are the number of successes and number of failures, respectively.

Bernoulli's formula agrees with our earlier calculation of the probabilities of obtaining k sevens in three rolls of a pair of fair dice; that is,

$$P(0 \text{ sevens}) = C(3, 0)p^0q^3 = q^3 = \left(\frac{5}{6}\right)^3 = \frac{125}{216},$$

$$P(\text{exactly 1 seven}) = C(3, 1)p^1q^2 = 3pq^2 = 3\left(\frac{1}{6}\right)\left(\frac{5}{6}\right)^2 = \frac{75}{216}$$

$$P(\text{exactly 2 sevens}) = C(3, 2)p^2q^1 = 3p^2q = 3\left(\frac{1}{6}\right)^2\left(\frac{5}{6}\right) = \frac{15}{216}$$

$$P(3 \text{ sevens}) = C(3, 3)p^3q^0 = p^3 = \left(\frac{1}{6}\right)^3 = \frac{1}{216}.$$

APPLICATIONS

Bernoulli's formula can be used in a wide variety of situations when an experiment is repeated under identical conditions.

EXAMPLE 6.35 What is the probability that out of ten flips, a fair coin will come up heads

(a) exactly five times?
(b) either four, five, or six times?

Solution (a) If we define "success" to mean that heads comes up when the coin is flipped, then $p = P(S) = \frac{1}{2}$ and $q = P(F) = \frac{1}{2}$. Hence by Bernoulli's formula

$$\begin{aligned} P(\text{exactly 5 heads in 10 flips}) &= P(\text{exactly 5 successes in 10 trials}) \\ &= C(10, 5)p^5q^5 = 252\left(\frac{1}{2}\right)^5\left(\frac{1}{2}\right)^5 \\ &= \frac{252}{1024} \approx .246. \end{aligned}$$

Thus half of the 10 flips are heads less than 25% of the time.

(b) As in part (a) we find that

$$\begin{aligned} P(\text{exactly 4 heads in 10 flips}) &= C(10, 4)p^4q^6 = 210\left(\frac{1}{2}\right)^4\left(\frac{1}{2}\right)^6 \\ &= \frac{210}{1024} \end{aligned}$$

and

$$\begin{aligned} P(\text{exactly 6 heads in 10 flips}) &= C(10, 6)p^6q^4 = 210\left(\frac{1}{2}\right)^6\left(\frac{1}{2}\right)^4 \\ &= \frac{210}{1024}. \end{aligned}$$

Since obtaining exactly 4, 5, or 6 heads are mutually exclusive events, the union rule for mutually exclusive events gives

$$\begin{aligned} P(\text{either 4, 5, or 6 heads}) &= P(\text{4 heads}) + P(\text{5 heads}) + P(\text{6 heads}) \\ &= \frac{210}{1024} + \frac{252}{1024} + \frac{210}{1024} \\ &= \frac{672}{1024} = .65625. \end{aligned}$$

Hence in 10 flips of a fair coin we obtain 4, 5, or 6 heads over 65% of the time. ■

EXAMPLE 6.36 Sixty percent of the voters in a large metropolitan area favor development of a proposed mass transit system. If seven voters are sampled at random, what is the probability that a majority of them will favor this system?

Solution Each time a voter is questioned we will declare a success if the voter favors the proposed system and a failure otherwise. Since the size of the metropolitan area is very large compared to the size of the sample, the probability of success on each trial remains approximately the same, namely, $p = .60$. Therefore the probability that a majority of the seven voters favor the project is given by

$$\begin{aligned} &P(\text{at least 4 successes in 7 trials}) \\ &\quad = P(\text{4 successes}) + P(\text{5 successes}) + P(\text{6 successes}) + P(\text{7 successes}) \\ &\quad = C(7, 4)p^4q^3 + C(7, 5)p^5q^2 + C(7, 6)p^6q^1 + C(7, 7)p^7q^0 \\ &\quad = 35(.6)^4(.4)^3 + 21(.6)^5(.4)^2 + 7(.6)^6(.4) + (.6)^7 \\ &\quad \approx .290 + .261 + .131 + .028 = .710. \end{aligned}$$

Note that there is an approximately 29% chance that a majority of those sampled will *not* favor development of the proposed mass transit system, and thus this small sample can give a misleading picture of the population at large. This example demonstrates the dangers of a prediction based on too small a sample. ■

Practice Problem 23 A 1987 poll by Louis Harris found that in any given year only 2% of American marriages end in divorce. If this trend continues, what is the probability that there will be exactly 1 divorce among 10 unacquainted married couples during the next year?

EXAMPLE 6.37 A manufacturer of light bulbs finds that 2% of the bulbs produced are defective. If the bulbs are packaged in boxes of four, what percentage of the boxes contains at least one defective bulb?

Solution Although it is possible to compute the desired probability directly as in the previous example, in this case we can shorten the calculation by using the complement rule. In this way we see that

$$\begin{aligned} P(\text{at least 1 defective}) &= 1 - P(0 \text{ defectives}) \\ &= 1 - C(4, 0)(.02)^0(.98)^4 \\ &= 1 - (.98)^4 \\ &\approx 1 - .922 = .078. \end{aligned}$$

Thus about 7.8% of the boxes will contain a defective bulb. ■

Practice Problem 24 An insurance salesman finds that 10% of her random telephone calls lead to a sale. What is the probability that at least one sale will result from placing calls to 8 persons chosen at random?

We will conclude this section with an application in which Bernoulli's formula is used to determine the number of trials.

EXAMPLE 6.38 From prior experience a manufacturer knows that the probability of making a successful bid on a government contract is .2. What is the smallest number of contracts that must be bid on in order for the manufacturer to have a 90% chance of receiving at least one?

Solution We will regard each submission of a bid as a Bernoulli trial in which the probability of success (receiving the contract) is .2 and the probability of failure is .8. If n bids are submitted, then the probability of receiving at least one contract is

$$\begin{aligned} P(\text{at least 1 contract received}) &= 1 - P(0 \text{ contracts received}) \\ &= 1 - C(n, 0)(.2)^0(.8)^n \\ &= 1 - (.8)^n. \end{aligned}$$

Since we want the manufacturer to have a 90% chance of receiving at least one contract, we must determine the smallest integer n for which

$$1 - (.8)^n \geq .90,$$

that is, the smallest integer n for which

$$.10 \geq (.8)^n.$$

By trial and error (or the use of logarithms), we see that $n = 11$ is the smallest number of contracts that must be bid on so as to assure a 90% chance of receiving one. ■

EXERCISES 6.6

1. Consider a sequence of $n = 6$ Bernoulli trials, where the probabilities of success and failure on each trial are $p = .7$ and $q = .3$, respectively. Determine the probability of obtaining

(a) exactly 3 successes (b) 6 successes

(c) at least 5 successes (d) at least one failure.

2. Consider a sequence of $n = 10$ Bernoulli trials, where the probabilities of success and failure on each trial are $p = 1/3$ and $q = 2/3$, respectively. Determine the probability of obtaining

(a) no successes (b) exactly 2 successes

(c) at least one success (d) at most two successes.

3. A fair die is rolled three times. What is the probability that the number 4 comes up

(a) no times? (b) once? (c) twice? (d) three times?

4. A coin is weighted in such a way that the probability of heads is .2 on each flip. If this coin is flipped four times, what is the probability of obtaining

(a) four tails? (b) three heads and one tail?

(c) two heads and two tails? (d) one head and three tails?

5. American roulette wheels have 38 equal compartments numbered 00, 0, 1, . . ., 36. The numbers 00 and 0 are green, and the numbers 1–36 are half red and half black. If a player bets four times in a row that red will come up, what is the probability that

(a) the player wins all four bets? (b) the player wins at least three times?

6. An archer has a 75% chance of hitting the target on each shot. If he takes five shots, what is the probability that he will hit the target on

(a) exactly four of them? (b) at least four of them?

7. The recovery rate for a certain disease is 75%; that is, on the average, 75% of those with the disease recover. If twelve patients have the disease, what is the probability that exactly nine of the twelve recover?

8. In a certain county, 40% of the registered voters are women. Citizens are selected at random for jury duty from among the registered voters. If twenty persons are called for jury duty, what is the probability that fewer than four are women?

9. Of the items produced on a certain machine, 5% are defective. If ten items are chosen at random from this machine's output, what is the probability that at most one item is defective?

10. A multiple-choice examination consists of ten questions, each having five possible answers, only one of which is correct. If a student randomly guesses the answer to each question, what is the probability that the student answers no more than two questions correctly?

11. A husband and wife are both carriers of a recessive gene for sickle-cell anemia. Although they are not affected by the disease, the probability is 1/4 that each of their children will have the disease. If they have four children, what is the probability that all are free of the disease?

12. Physicist Freeman Dyson, describing British bombing missions during World War II, writes in reference [4]:
The normal tour of duty for a crew in a regular squadron was thirty missions. The loss-rate during the middle years of the war [that is, the probability of being shot down during a mission] averaged four percent. This meant that a crewman had three chances in ten of completing a normal tour of duty.
Justify this conclusion.

13. A district attorney has the names of ten persons who witnessed an accident. She knows that in general only 20% of all witnesses can be convinced to testify. If she needs the testimony of at least two witnesses to get a conviction, what are her chances of obtaining a conviction?

14. The famous English diarist Samuel Pepys posed the following question to Isaac Newton in 1693 (see reference [8]). "*A* has six dice in a box, with which he is to fling [at least] one six. *B* has in another box twelve dice, with which he is to fling [at least] two sixes. *C* has in another box eighteen dice, with which he is to fling [at least] three sixes. Question: Whether *B* and *C* have not as easy a taske as *A*, at even luck?" Newton replied that "an easy computation" showed *A* to have the advantage. Justify Newton's conclusion.

15. A workshop has seven lathes and ten full-time employees who work independently of one another. If each worker uses a lathe 60% of the time, what is the probability that at any given moment there will not be enough lathes available for all who need them?

16. A restaurant has found that 20% of those with reservations fail to show up. If it books twelve reservations for ten available tables, what is the probability that everyone who shows up can be seated?

17. A sales representative has a 10% chance of making a sale at any given house. How many houses must be visited in order that the probability of making at least one sale is .50 or higher?

18. What is the smallest number of fair coins that must be flipped in order that the probability of obtaining at least one head is .99 or higher?

19. One of the two gambling problems posed to Pascal by the Chevalier de Méré that led to the development of probability theory was to compare the following probabilities:

(a) the probability that at least one six appears if a fair die is rolled four times

(b) the probability that at least one double-six appears if two fair dice are rolled 24 times.

Calculate these two probabilities.

20. In a certain city the weather bureau takes temperature readings in the center of the city and at the airport thirty miles away. From past experience it is known that the downtown readings are higher 60% of the time. What is the probability that, in ten independent readings, the downtown temperature is higher at least eight times?

21. On Main Street there are four traffic signals that operate independently. Each signal remains green for two minutes and red for one minute. What is the probability that a car traveling on Main Street will pass these four signals without encountering a red light?

22. Two evenly matched teams, *A* and *B*, are playing a best-of-seven series. What is the probability that team *A* will win the series in exactly seven games? (*Hint:* Team *A* must win exactly three of the first six games.)

23. A pharmaceutical company claims that its new drug is 90% effective in curing a certain disease. It is known that half of those with this disease recover without treatment. To test the manufacturer's claim, the drug has been administered to ten patients with the disease, and the claim will be accepted if eight or more recover.

(a) What is the probability that the drug will fail the test if the claim is true?

(b) What is the probability that the drug will pass the test even if it has no effect on the disease?

24. In a test of extrasensory perception, an experimenter concentrates on a card chosen at random from a standard deck of playing cards, and a subject attempts to guess its suit (spades, hearts, diamonds, or clubs). If a subject guesses the suit correctly at least four times out of five, he or she is considered to be clairvoyant.

(a) What is the probability that someone who guesses suits at random will be considered clairvoyant?

(b) What is the probability that out of 100 subjects, each of whom guesses suits at random, at least one will be considered clairvoyant?

Answers to Practice Problems **23.** approximately .167

24. approximately .570

MATHEMATICS IN ACTION

6.7 Markov Chains

In the preceding section we discussed Bernoulli trials, in which an experiment consists of a sequence of repeated trials and the outcome of each trial is unaffected by the outcome of any other. In this section we will encounter a repeated process called a *Markov chain** in which the outcome of one trial affects the next.

THE TRANSITION MATRIX

Consider two basketball players, Al and Bob, who are practicing free throws. Al is an 80% free-throw shooter and Bob a 70% shooter. Each player shoots until he misses, at which time the other begins to shoot.

This is an example of a simple system which is in one of two possible states, according to which person is shooting.

State 1: Al will shoot the next shot.

State 2: Bob will shoot the next shot.

*The Russian mathematician A. A. Markov (1856–1922) developed much of the modern theory of stochastic processes.

Since Al is an 80% shooter, there is a .8 probability that he will make any free throw and a .2 probability that he will miss. This means that if the system is in state 1, there is a .8 probability that it will remain in state 1 on the next shot and a .2 probability that it will change to state 2. Similarly, since Bob is a 70% shooter, when in state 2 there is a .3 probability that the system will change to state 1 and a .7 probability that it will remain in state 2. Thus we can calculate the probability of the next state of the system if we know its current state. This system is a simple example of a Markov chain.

Formally, a **Markov chain** is a process consisting of a sequence of trials, each of which results in the process being in one of a finite number of **states.** Moreover, if a given trial results in the process being in state i, there must be a fixed probability that the next trial will result in the process being in state j. This probability is denoted p_{ij} and is called a **transition probability.** Notice that instead of saying "the outcome of the nth trial is i," it is customary to say that "the system is in state i at time n." With this terminology, the transition probability p_{ij} represents the conditional probability that a system in state i moves to state j in one trial (time period). Of course, the time period between trials will vary according to the context of the problem.

Thus the free-throw shooting example is a Markov chain with two states in which each shot is a trial of the experiment. In this case the transition probabilities are $p_{11} = .8$, $p_{12} = .2$, $p_{21} = .3$, and $p_{22} = .7$. All of the necessary information for this Markov chain is conveyed in Figure 6.20, which is called a **transition diagram.** For example, the arrow pointing from state 1 to state 2 indicates that the probability of going from state 1 to state 2 in one trial is .2, that is, $p_{12} = .2$. If some transition probability p_{ij} is 0, then for simplicity we will omit the arrow from state i to state j in the transition diagram.

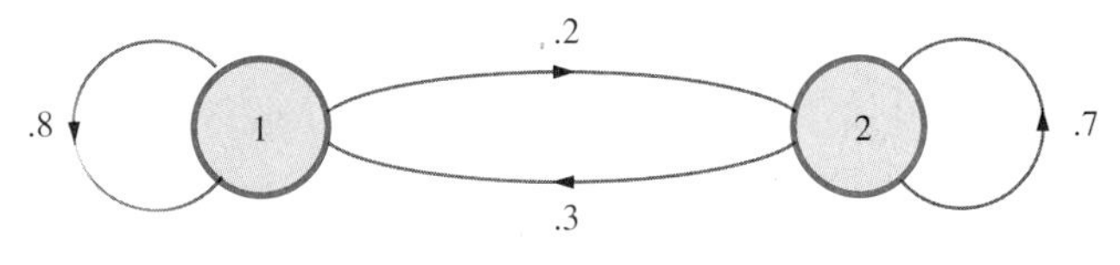

FIGURE 6.20

Another useful way to convey the same information is in a matrix. For a Markov chain with s states, the $s \times s$ matrix $[p_{ij}]$ is called the **transition matrix** of the Markov chain. Thus, for instance, the 2×2 matrix

$$\begin{array}{lc} & \text{Next state} \\ & \begin{array}{cc} 1 & 2 \end{array} \\ \begin{array}{l} \text{Current} \quad 1 \\ \text{state} \qquad 2 \end{array} & \begin{bmatrix} .8 & .2 \\ .3 & .7 \end{bmatrix} \end{array}$$

is the transition matrix for a Markov chain representing the free-throw shooting experiment. Notice that a transition matrix will always consist of nonnegative entries such that the sum of the entries in each *row* is 1. (As the transition matrix above demonstrates, the sum of the entries in each column need not be 1.) The transition matrix plays an important role in the theory of Markov chains because it enables us to use matrix methods to analyze a stochastic process. Before pursuing this idea, however, let us look at another example.

EXAMPLE 6.39 Most of the drivers in a certain community are insured by Allstate or State Farm. Recent data showed that of those who were insured by Allstate in 1987, 70% continued with Allstate in 1988, but 20% switched to State Farm, and 10% switched to another company. Similarly, of those who were insured by State Farm in 1987, 80% continued with State Farm in 1988, 10% switched to Allstate, and 10% switched to another company. Likewise, of those who were insured by another company in 1987, 30% switched to Allstate in 1988, 10% switched to State Farm, and 60% continued with another company. Form a transition diagram and a transition matrix for a Markov chain describing this process.

Solution Since each driver in the community can be classified according to the type of insurance carried, we can represent this situation as a Markov chain with 3 states:

State 1: insured by Allstate

State 2: insured by State Farm

State 3: insured by another company.

Here the time period between trials is one year because we are given information about the changes from 1987 to 1988. The transition diagram for this process is shown in Figure 6.21.

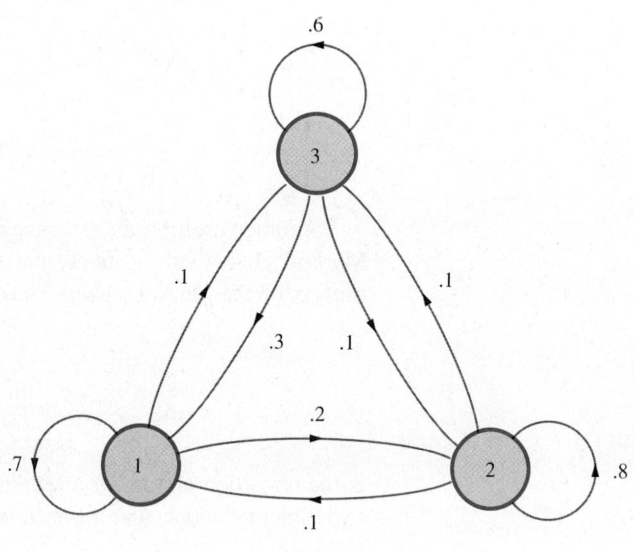

FIGURE 6.21

As the transition diagram shows, the probability of remaining in state 1 during one time period is .7, and the probability of moving from state 1 to state 2 is .2 and from state 1 to state 3 is .1. Since these probabilities are the entries in the first row of the transition matrix, we see that this row is

$$[.7 \quad .2 \quad .1].$$

Likewise in one year the probability of moving from state 2 to state 1 is .1, the probability of remaining in state 2 is .8, and the probability of moving from state 2 to state 3 is .1. Hence the second row of the transition matrix is

$$[.1 \quad .8 \quad .1].$$

Similarly the third row of the transition matrix is

$$[.3 \quad .1 \quad .6].$$

Thus the transition matrix for this Markov chain is

$$\begin{array}{cc} & \text{Next state} \\ & \begin{array}{ccc} 1 & 2 & 3 \end{array} \\ \text{Current state} \begin{array}{c} 1 \\ 2 \\ 3 \end{array} & \begin{bmatrix} .7 & .2 & .1 \\ .1 & .8 & .1 \\ .3 & .1 & .6 \end{bmatrix} \end{array}$$

Observe again that the entries in each row of the transition matrix are nonnegative numbers that sum to 1. ■

Practice Problem 25 In reference [7], a sociological study of class mobility, Prais found that children of upper-class parents were evenly divided between the upper and middle classes. Among children of middle-class parents, 10% moved to the upper class, 70% remained in the middle class, and 20% moved to the lower class; and children of lower-class parents were evenly divided between the middle and lower classes. Represent this process as a Markov chain and form the transition matrix.

STATE VECTORS

Our main interest in Markov chains centers on the question: *What is the probability that the system will be in a particular state at a given time?* We will denote by $x_i^{(n)}$ the probability that the system is in state i at time n, that is, after n trials. For a Markov chain with s states, the row vector

$$X^{(n)} = [x_1^{(n)} \quad x_2^{(n)} \quad \cdots \quad x_s^{(n)}]$$

is called the **state vector** at time n. Since its entries represent the probabilities of being in each of the respective states 1, 2, . . ., s at time n, their sum will always be 1. Note that the superscripts in this state vector are *not* exponents but merely indicate the time to which the state vector corresponds.

For example, suppose that Al and Bob flip a fair coin to determine who will shoot the first free throw. Then each player has a .5 probability of taking the first shot, and so at time 0 (after no shots) the state vector is

$$X^{(0)} = \begin{matrix} 1 & 2 \\ [.5 & .5] \end{matrix}.$$

We call this the **initial state vector** for the Markov chain. To find the corresponding probabilities after the first shot, we represent the first two stages of the process (namely, the coin toss followed by the first shot) by the stochastic diagram in Figure 6.22.

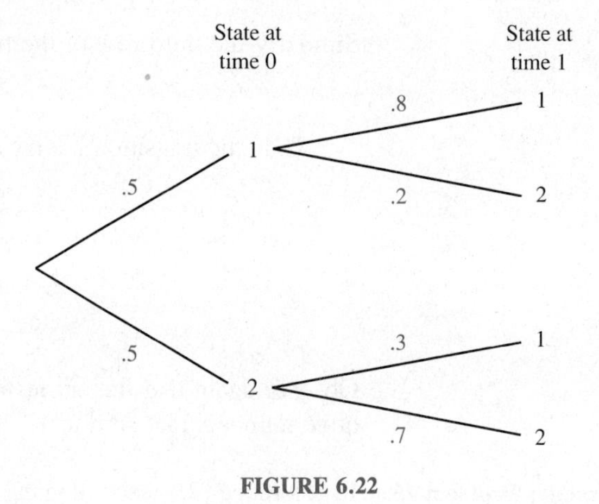

FIGURE 6.22

The probabilities .8, .2, .3, and .7 in this diagram are the entries of the transition matrix

$$P = \begin{matrix} & \begin{matrix} 1 & 2 \end{matrix} \\ \begin{matrix} 1 \\ 2 \end{matrix} & \begin{bmatrix} .8 & .2 \\ .3 & .7 \end{bmatrix} \end{matrix}.$$

Since two branches lead to state 1, we see as in Section 6.5 that the probability that the system will be in state 1 at time 1 is

$$.5(.8) + .5(.3) = .40 + .15 = .55.$$

Likewise the probability that the system will be in state 2 at time 1 is

$$.5(.2) + .5(.7) = .10 + .35 = .45.$$

Therefore the state vector at time 1 is

$$X^{(1)} = [.55 \quad .45].$$

Hence the probability is .55 that Al will take the second shot, and the probability is .45 that Bob will take the second shot.

The calculation performed above can be performed very simply by use of matrix multiplication:

$$[.5 \quad .5] \begin{bmatrix} .8 & .2 \\ .3 & .7 \end{bmatrix} = [.55 \quad .45].$$

state vector at time 0 — transition matrix — state vector at time 1

We can represent this calculation in symbols as

$$X^{(0)}P = X^{(1)}.$$

More generally, we have the following result.

Theorem 6.8

Let P be the transition matrix for a Markov chain. If the state vector at a given time is X, then the state vector one time period later is XP. In symbols,

$$X^{(n)}P = X^{(n+1)}.$$

We can use Theorem 6.8 to compute subsequent state vectors in our free-throw shooting example. For instance

$$[.55 \quad .45] \begin{bmatrix} .8 & .2 \\ .3 & .7 \end{bmatrix} = [.575 \quad .425]$$

state vector at time 1 — transition matrix — state vector at time 2

$$[.575 \quad .425] \begin{bmatrix} .8 & .2 \\ .3 & .7 \end{bmatrix} = [.5875 \quad .4125]$$

state vector at time 2 — transition matrix — state vector at time 3

$$[.5875 \quad .4125] \begin{bmatrix} .8 & .2 \\ .3 & .7 \end{bmatrix} = [.59375 \quad .40625].$$

state vector at time 3 — transition matrix — state vector at time 4

From these calculations we see that there is a .59375 probability that Al will take the fifth shot. We will see shortly that the state vectors in this example approach the "equilibrium" vector

$$X = [.6 \quad .4]$$

as the process continues. Over the long run, therefore, Al will take approximately 60% of the shots and Bob will take about 40%.

EXAMPLE 6.39 Revisited Suppose we are told that in 1987, 40% of the drivers in the community were insured by Allstate, 30% by State Farm, and 30% by another company. Assuming that the 1987–1988 trend continues, what percentage of the drivers will be insured by Allstate and State Farm in 1990?

Solution The distribution for 1987 tells us that the initial state vector is

$$X^{(0)} = [.40 \quad .30 \quad .30].$$

Then the distribution one year later (in 1988) will be given by $X^{(1)} = X^{(0)}P$, where P is the transition matrix for the Markov chain.

$$X^{(1)} = X^{(0)}P = [.40 \quad .30 \quad .30]\begin{bmatrix} .7 & .2 & .1 \\ .1 & .8 & .1 \\ .3 & .1 & .6 \end{bmatrix} = [.40 \quad .35 \quad .25]$$

The distribution for 1989 is given by $X^{(2)} = X^{(1)}P$:

$$X^{(2)} = X^{(1)}P = [.40 \quad .35 \quad .25]\begin{bmatrix} .7 & .2 & .1 \\ .1 & .8 & .1 \\ .3 & .1 & .6 \end{bmatrix} = [.390 \quad .385 \quad .225].$$

Finally the distribution for 1990 is given by $X^{(3)} = X^{(2)}P$:

$$\begin{aligned} X^{(3)} = X^{(2)}P &= [.390 \quad .385 \quad .225]\begin{bmatrix} .7 & .2 & .1 \\ .1 & .8 & .1 \\ .3 & .1 & .6 \end{bmatrix} \\ &= [.3790 \quad .4085 \quad .2125]. \end{aligned}$$

Hence in 1990, 37.9% of the drivers will be insured by Allstate and 40.85% by State Farm. ■

Practice Problem 26 In Practice Problem 25 suppose that the social distribution is currently 20% upper class, 40% middle class, and 40% lower class. Assuming that the birthrate for all three classes is the same, what will be the distribution in two generations?

EQUILIBRIUM VECTORS

Some Markov chains tend to stabilize over the long run in the sense that their state vectors approach an equilibrium vector as the process continues. For such Markov chains we can predict the proportion of the time over the long run that the system will be in each state.

Let us consider again the free-throw shooting example presented above. We have seen that if we know the initial state vector $X^{(0)}$, we can determine the state vector at any later time by computing

$$X^{(0)}P = X^{(1)},$$
$$X^{(1)}P = X^{(2)},$$
$$X^{(2)}P = X^{(3)},$$

and so forth. The table below shows the results of such calculations, rounded off to four decimal places. These calculations are performed for three different ways of beginning the process, according to whether Al shoots the first shot, Bob shoots the first shot, or they flip a fair coin to decide who shoots first.

Time	*Al starts*	*Bob starts*	*Coin flip*
0	[1 0]	[0 1]	[.5 .5]
1	[.8 .2]	[.3 .7]	[.55 .45]
2	[.70 .30]	[.45 .55]	[.575 .425]
3	[.650 .350]	[.525 .475]	[.5875 .4125]
4	[.6250 .3750]	[.5625 .4375]	[.5938 .4062]
5	[.6125 .3875]	[.5813 .4187]	[.5969 .4031]
6	[.6063 .3937]	[.5906 .4094]	[.5984 .4016]
7	[.6031 .3969]	[.5953 .4047]	[.5992 .4008]
8	[.6016 .3984]	[.5977 .4023]	[.5996 .4004]
9	[.6008 .3992]	[.5988 .4012]	[.5998 .4002]
10	[.6004 .3996]	[.5994 .4006]	[.5999 .4001]
11	[.6002 .3998]	[.5997 .4003]	[.6000 .4000]
12	[.6001 .3999]	[.5999 .4001]	[.6000 .4000]
13	[.6000 .4000]	[.5999 .4001]	[.6000 .4000]
14	[.6000 .4000]	[.6000 .4000]	[.6000 .4000]
15	[.6000 .4000]	[.6000 .4000]	[.6000 .4000]

From this table we see that in all three cases, as the process continues the state vectors approach

$$X = [.6 \quad .4].$$

If the state vectors reach this vector, they will not change. For if the state vector at a given time is X, then the state vector one step later is

$$\begin{aligned} XP &= [.6 \quad .4]\begin{bmatrix} .8 & .2 \\ .3 & .7 \end{bmatrix} \\ &= [.6(.8) + .4(.3) \quad .6(.2) + .4(.7)] \\ &= [.48 + .12 \quad .12 + .28] \\ &= [.6 \quad .4] \\ &= X. \end{aligned}$$

It turns out that *as the process continues, the probability that Al is shooting tends to stabilize at .6, no matter how the process began.* It follows that over the long run, approximately 60% of the shots will be taken by Al and about 40% by Bob.

A vector X is called an **equilibrium vector** for a Markov chain if

$$XP = X,$$

where P is the transition matrix for the Markov chain. Thus the preceding calculation shows that $[.6 \quad .4]$ is an equilibrium vector for the free-throw shooting Markov chain.

Two important questions now arise.

1. Do the state vectors of every Markov chain approach an equilibrium vector?
2. If an equilibrium vector exists, how do we find it?

It is not hard to show that the answer to the first question is negative. (See Exercises 33–34.) There are Markov chains in which the state vectors fail to stabilize no matter how long the process goes on, and there are others in which the state vectors stabilize at different equilibrium vectors depending on the initial state of the system. However, the next theorem, which we will not prove, shows that there is a large class of Markov chains that have a unique equilibrium vector.

Theorem 6.9

Let P be the transition matrix for a Markov chain, and suppose that some power of P contains only *positive* entries. Then the Markov chain has a unique equilibrium vector X, and regardless of the initial state of the system, the state vectors for the Markov chain will approach X as the process continues.

Note that in order to use Theorem 6.9 we must check that at least one of the matrices

$$P, P^2, P^3, P^4, \ldots$$

contains only positive entries, in other words, that one of these matrices contains no zero entries. It can be shown that if this does happen for a Markov chain with s states, then it must happen within the first $(s - 1)^2 + 1$ powers of P. Thus, for instance, to determine if a three-state Markov chain satisfies Theorem 6.9, we need only check at most the first $(3 - 1)^2 + 1 = 5$ powers of its transition matrix. The transition matrix P for the free-throw shooting example clearly satisfies this condition, because

$$P = \begin{bmatrix} .8 & .2 \\ .3 & .7 \end{bmatrix}$$

itself contains no zero entries.

We now turn to the second question posed above. To find an equilibrium vector if one exists, we need only solve the matrix equation $XP = X$. It will prove to be somewhat simpler to solve this equation in the following equivalent form:

$$\begin{aligned} O &= X - XP \\ &= XI - XP \\ &= X(I - P), \end{aligned}$$

where I is an identity matrix of the same size as P.

Consider, for instance, the transition matrix P for the free-throw shooting example. Since this Markov chain has two states, its equilibrium vector X will have two entries. Suppose that

$$X = [x_1 \quad x_2].$$

Now

$$I - P = I_2 - P = \begin{bmatrix} 1 & 0 \\ 0 & 1 \end{bmatrix} - \begin{bmatrix} .8 & .2 \\ .3 & .7 \end{bmatrix} = \begin{bmatrix} .2 & -.2 \\ -.3 & .3 \end{bmatrix},$$

and so the matrix equation $X(I - P) = O$ becomes

$$[x_1 \quad x_2] \begin{bmatrix} .2 & -.2 \\ -.3 & .3 \end{bmatrix} = [0 \quad 0].$$

Multiplying out the left side, we obtain

$$[.2x_1 - .3x_2 \quad -.2x_1 + .3x_2] = [0 \quad 0],$$

which is equivalent to the following system of linear equations.

$$\begin{aligned} .2x_1 - .3x_2 &= 0 \\ -.2x_1 + .3x_2 &= 0 \end{aligned}$$

We have not yet taken into account that the entries of X must sum to 1. This condition requires that

$$x_1 + x_2 = 1.$$

The resulting system of 3 equations in 2 unknowns

$$\begin{aligned} .2x_1 - .3x_2 &= 0 \\ -.2x_1 + .3x_2 &= 0 \\ x_1 + x_2 &= 1 \end{aligned}$$

is obviously redundant because the second equation is merely the first equation multiplied by -1; so we may discard the second equation and work with the system

$$\begin{aligned} .2x_1 - .3x_2 &= 0 \\ x_1 + x_2 &= 1 \end{aligned}$$

instead. This system can be solved either by substitution (see Section 1.3) or by the Gauss-Jordan method (see Sections 2.1–2.2). Applying the substitution method, for example, we first obtain

$$x_2 = 1 - x_1.$$

Now substitute this expression for x_2 in the first equation and simplify.

$$\begin{aligned} .2x_1 - .3(1 - x_1) &= 0 \\ .2x_1 - .3 + .3x_1 &= 0 \\ .5x_1 &= .3 \\ x_1 &= \frac{.3}{.5} = .6 \\ x_2 = 1 - x_1 &= 1 - .6 = .4 \end{aligned}$$

Therefore the equilibrium vector is

$$X = [.6 \quad .4],$$

which agrees with our earlier observation that the state vectors in the free-throw shooting example approach this vector.

EXAMPLE 6.40 For the Markov chain in Example 6.39 on page 348, determine the percentage of drivers in the long run who will be insured by each company.

Solution Recall that the states in this Markov chain are: insured by Allstate, insured by State Farm, and insured by another company, respectively. The transition matrix is

$$P = \begin{bmatrix} .7 & .2 & .1 \\ .1 & .8 & .1 \\ .3 & .1 & .6 \end{bmatrix}.$$

Since P contains no zero entries, Theorem 6.9 assures us that the state vectors of this Markov chain approach a unique equilibrium vector

$$X = [x_1 \quad x_2 \quad x_3].$$

The entries of X give the percentage of drivers in each state over the long run.

To compute X, we must solve the matrix equation $X(I - P) = O$, that is,

$$[x_1 \quad x_2 \quad x_3] \begin{bmatrix} .3 & -.2 & -.1 \\ -.1 & .2 & -.1 \\ -.3 & -.1 & .4 \end{bmatrix} = [0 \quad 0 \quad 0].$$

Multiplying out the left side and including the condition that the sum of the unknowns is 1, we obtain the system of equations at the top of the next page.

$$\begin{aligned} .3x_1 - .1x_2 - .3x_3 &= 0 \\ -.2x_1 + .2x_2 - .1x_3 &= 0 \\ -.1x_1 - .1x_2 + .4x_3 &= 0 \\ x_1 + x_2 + x_3 &= 1 \end{aligned}$$

Although it is not immediately apparent, this system is redundant. (The first equation equals -1 times the sum of the second and third equations.) Solving the system yields the solution

$$x_1 = .35, \qquad x_2 = .45, \qquad x_3 = .20.$$

Thus over the long run 35% of the drivers in this community will be insured by Allstate, 45% by State Farm, and 20% by another company. ■

Practice Problem 27 For the Markov chain in Practice Problem 25, determine the percentage of persons in each class over the long run.

Knowledge of the equilibrium vector for a Markov chain can be useful for long-term planning. In reference [3], for instance, Bourne studied changes in land usage in Toronto, Canada, from 1952 to 1962. Land was classified in terms of ten possible uses, resulting in a Markov chain with ten states. When the equilibrium vector for this process was determined, it was found that if the 1952–1962 trend continued, then 19% of the city's land would be used for parking and 9.1% would be vacant. Knowledge of these percentages can help the city government to take corrective action to prevent erosion of the city's tax base. A similar study in New Zealand (see reference [5]) showed that about 60% of the country's industry would eventually be located in the Auckland area, supporting the charge of the opposition party leader that the entire south island should be regarded as a depressed area suitable for government subsidies and other development incentives.

EXERCISES 6.7

In Exercises 1–6 draw a transition diagram for the Markov chain with the given transition matrix.

1. $\begin{bmatrix} .5 & .5 \\ .6 & .4 \end{bmatrix}$

2. $\begin{bmatrix} \frac{2}{3} & \frac{1}{3} \\ \frac{1}{4} & \frac{3}{4} \end{bmatrix}$

3. $\begin{bmatrix} .2 & .8 \\ 0 & 1 \end{bmatrix}$

4. $\begin{bmatrix} 0 & .6 & .4 \\ .7 & 0 & .3 \\ .4 & .4 & .2 \end{bmatrix}$

5. $\begin{bmatrix} 1 & 0 & 0 \\ \frac{1}{3} & \frac{1}{3} & \frac{1}{3} \\ 1 & 0 & 0 \end{bmatrix}$

6. $\begin{bmatrix} 0 & .5 & .5 \\ 0 & 0 & 1 \\ 1 & 0 & 0 \end{bmatrix}$

In Exercises 7–10 write the transition matrix for the Markov chain with the given transition diagram.

7.

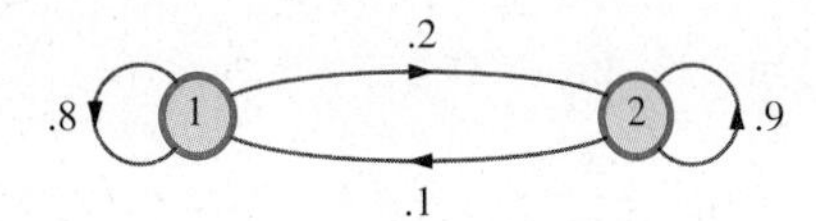

8.

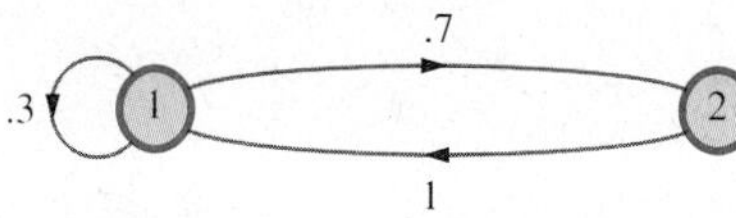

9.

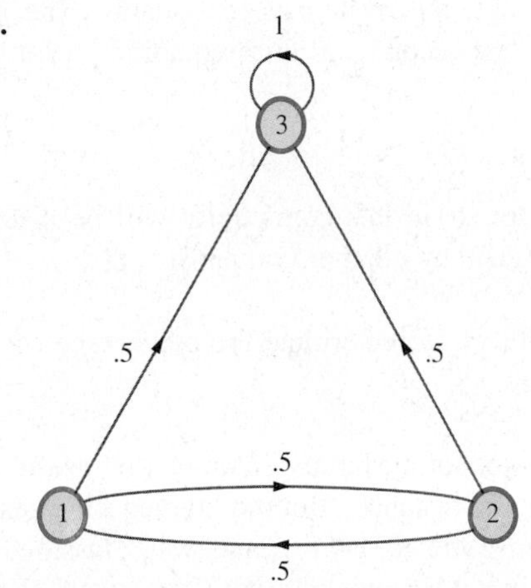

10.

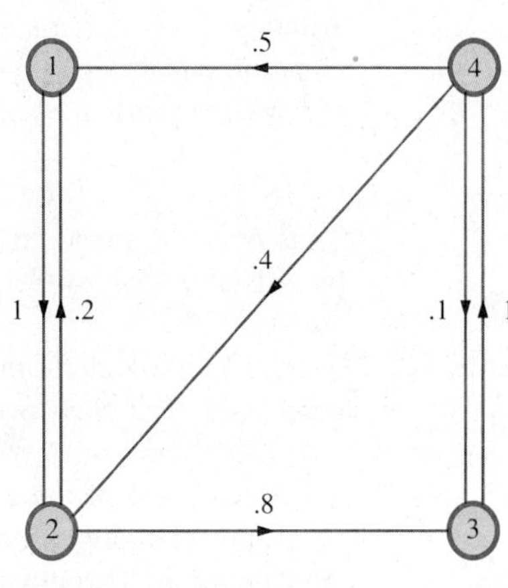

11. A two-state Markov chain has transition matrix

$$\begin{bmatrix} .6 & .4 \\ .1 & .9 \end{bmatrix}.$$

Find the state vector at time 3 (that is, after three steps) if the system is initially
(a) in state 1 (b) equally likely to be in state 1 or 2.

12. A three-state Markov chain has transition matrix

$$\begin{bmatrix} 0 & 1 & 0 \\ 0 & 0 & 1 \\ \frac{1}{3} & 0 & \frac{2}{3} \end{bmatrix}.$$

Find the state vector at time 3 (that is, after three steps) if the system is initially
(a) in state 2 (b) equally likely to be in state 1, 2, or 3.

13. An actor has found from experience that his performance on a given night affects his performance the following night. A good performance is followed by another good performance 90% of the time, but a bad performance is followed by a good performance only 70% of the time. If the actor performs well on Monday, what is the probability that he will perform well the following Friday?

14. An author has found that typing her manuscript over and over to eliminate errors yields diminishing returns. If a particular word contains an error, there is a 48% chance it will be corrected on the copy; however, if a word is correct on the original, there is a 2% chance that it will be retyped incorrectly on the copy. If a word is correct initially, what is the probability that it will still be correct after the manuscript has been retyped three times?

15. The sales manager for a metropolitan newspaper has observed an alarming trend: 15% of the subscribers at the beginning of any given month fail to renew their subscriptions the following month, whereas only 5% of all nonsubscribers become subscribers during the same period. If this trend continues, what will happen to the newspaper's current market share of 60% in another three months?

16. There are some recent signs that Booneville's high unemployment rate may be ending. A three-month study revealed that 45% of those unemployed on January 1 were employed on April 1, whereas only 5% of those employed on January 1 were unemployed on April 1. If this trend continues, what will happen to the town's current 20% unemployment rate in the next six months? In the next nine months?

17. Forty percent of an island's population lives in the cities, with the rest dwelling in rural areas. Suppose that, in any given year, 10% of the rural population moves to the cities and 5% of the city dwellers move to rural areas. How long will it be before a majority of the island's population lives in the cities?

18. A sample of American car owners were asked whether their current car was American-made or foreign-made, and which type of car they would buy next. Four out of five owners of foreign cars said that they would buy them again the next time, and two out of five owners of American cars said that they would switch to a foreign car the next time. If foreign cars currently have a 40% share of the market and car owners on the average buy a new car every five years, how long will it be before foreign cars capture 60% of the American market?

19. A mouse is to be placed in a box with three connecting compartments, as shown in the diagram at the right. When the mouse changes rooms, it selects at random one of the doors of the compartment that it is occupying and passes through that door into another compartment. If the mouse is placed in compartment 1 initially, what is the probability that it will still be there after changing rooms four times?

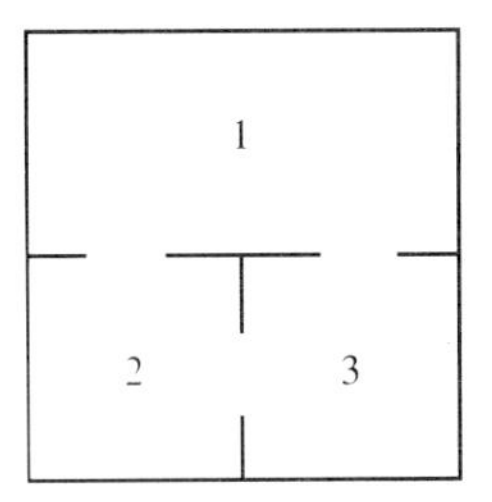

20. A political scientist observed that among children of registered Democrats, 70% become Democrats, 10% become Republicans, and 20% register as independents. For children of Republicans the distribution is 10% Democrats, 60% Republicans, and 30% independents, and for children of independents it is 10% Democrats, 10% Republicans, and 80% independents. If the current voter registration is 40% Democratic, 40% Republican, and 20% independent and if the birthrate for all three groups is the same, what will the distribution be two generations from now?

21. Suppose that 60% of the children of agricultural workers become agricultural workers, 30% become blue-collar workers, and 10% become white-collar workers; 20% of the children of blue-collar workers become agricultural workers, 50% become blue-collar workers, and 30% become white-collar workers; and 10% of the children of white-collar workers become agricultural workers, 30% become blue-collar workers, and 60% become white-collar workers. If the current generation consists of 20% agricultural workers, 50% blue-collar workers, and 30% white-collar workers and if the birthrate for all three groups is the same, what will the distribution be two generations from now?

22. In 1985 a survey of workers in a metropolitan area showed that 20% rode the subway to work, 30% took a bus, and 50% came in automobiles. A follow-up survey two years later showed that 60% of the subway riders in 1985 continued to ride the subway, but 10% rode a bus and 30% used automobiles in 1987. Among the 1985 bus riders, 90%

continued to ride the bus and 10% took the subway in 1987. Finally 70% of the 1985 automobile users continued to come by car, but 10% rode the subway and 20% took a bus in 1987. If this trend continues, what percentage of the workers will use each form of transportation in 1993?

23. A 1980 survey showed that 40% of the cars owned by Americans were large cars, 50% were mid-size cars, and 10% were small cars. A follow-up survey in 1985 showed that of those who owned large cars in 1980, 80% still owned large cars, but that 10% owned mid-size cars and 10% owned small cars. Of those who owned mid-size cars in 1980, 30% owned large cars, 60% owned mid-size cars, and 10% small cars in 1985. Of those with small cars in 1980, 10% owned mid-size cars and 90% owned small cars in 1985. If this trend continues, what percentage of the cars owned by Americans will be of each size in the year 2000?

24. In 1930, 10% of the land on a certain island was classified as urban, 60% was used for agriculture, and 30% was used for other purposes. Two decades later, 80% of the urban land remained urban, 10% was used for agriculture, and 10% was used for other purposes. Of the land used for agriculture in 1930, 80% continued to be used for agriculture, 10% had become urban, and 10% was used for other purposes in 1950. Of the land used for other purposes in 1930, 20% had become urban and 20% was used for agriculture in 1950. If this trend continues, what percentage of the island's land will be used in each way in the year 2010?

In Exercises 25–32 determine if some power of the given transition matrix contains only positive entries. If so, find its unique equilibrium vector.

25. $\begin{bmatrix} .2 & .8 \\ 1 & 0 \end{bmatrix}$

26. $\begin{bmatrix} 1 & 0 \\ .6 & .4 \end{bmatrix}$

27. $\begin{bmatrix} .7 & .3 \\ 0 & 1 \end{bmatrix}$

28. $\begin{bmatrix} 0 & 1 \\ .5 & .5 \end{bmatrix}$

29. $\begin{bmatrix} .4 & .6 & 0 \\ .5 & .5 & 0 \\ .3 & .3 & .4 \end{bmatrix}$

30. $\begin{bmatrix} .9 & 0 & .1 \\ .2 & .7 & .1 \\ .1 & .3 & .6 \end{bmatrix}$

31. $\begin{bmatrix} .6 & .2 & .2 \\ .1 & .8 & .1 \\ 0 & 1 & 0 \end{bmatrix}$

32. $\begin{bmatrix} 0 & 0 & 1 \\ .3 & .4 & .3 \\ .5 & .5 & 0 \end{bmatrix}$

33. Consider the two-state Markov chain with the transition diagram at the right.

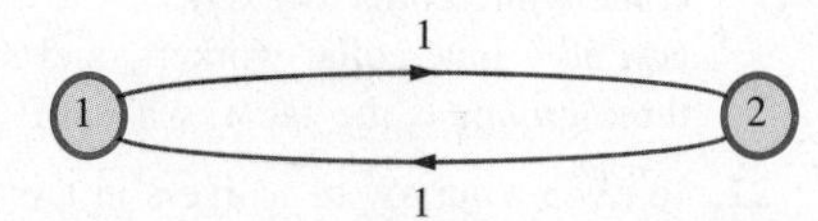

(a) If this system is initially in state 1, find the state vectors at times 0, 1, 2, and 3.

(b) Repeat part (a) if the system is initially in state 2.

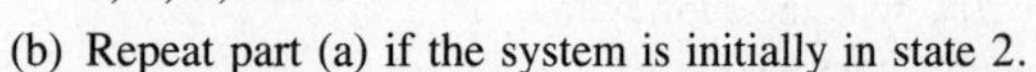

(c) Does this Markov chain have an equilibrium vector?

(d) Do the state vectors in parts (a) and (b) approach the equilibrium vector in part (c)?

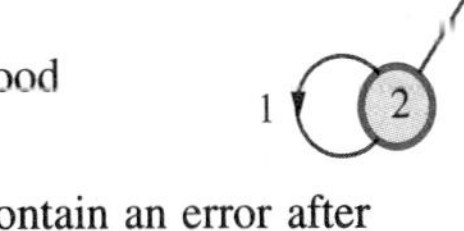

34. Consider the three-state Markov chain with the transition diagram at the right.

(a) If this system is initially in state 1, find the state vectors at times 0, 1, 2, and 3.

(b) Repeat part (a) if the system is initially in state 2. In state 3.

(c) Does this Markov chain have a unique equilibrium vector?

35. In Exercise 13 what proportion of the actor's performances will be good over the long run?

36. In Exercise 14 what proportion of the words in the manuscript will contain an error after many retypings?

37. In Exercise 15 what share of the market will the newspaper eventually have?

38. In Exercise 16 what unemployment rate will Booneville eventually have?

39. In Exercise 17 what proportion of the population will eventually live in the cities?

40. In Exercise 18 what share of the market will foreign cars eventually have?

41. In Exercise 19 what is the probability that the mouse will be in compartment 1 after many room changes?

42. In Exercise 20 what will the voter distribution be after many generations?

43. In Exercise 21 what will the distribution of workers be after many generations?

44. In Exercise 22 what percentage of the workers will eventually use each form of transportation?

45. In Exercise 23 what percentage of American cars will eventually be large? Mid-size? Small?

46. In Exercise 24 what proportion of the land on the island will eventually be used in each way?

Answers to Practice Problems

25. In this process there are three states.

State 1: belonging to the upper class
State 2: belonging to the middle class
State 3: belonging to the lower class

The time period between trials is a generation, and the transition matrix is

$$\begin{array}{cc} & \text{Next state} \\ & \begin{array}{ccc} 1 & 2 & 3 \end{array} \\ \text{Current state} \begin{array}{c} 1 \\ 2 \\ 3 \end{array} & \begin{bmatrix} .5 & .5 & 0 \\ .1 & .7 & .2 \\ 0 & .5 & .5 \end{bmatrix}. \end{array}$$

26. In two generations 12.8% of the people will be in the upper class, 61.6% in the middle class, and 25.6% in the lower class.

27. Over the long run 12.5% of the people will be in the upper class, 62.5% in the middle class, and 25% in the lower class.

CHAPTER 6 REVIEW

IMPORTANT TERMS

- **certain event** *(6.1)*
- **event**
- **experiment**
- **impossible event**
- **outcomes**
- **probability**
- **relative frequency**
- **sample space**
- **odds against** ***E*** *(6.2)*
- **odds in favor of** ***E***
- **probability distribution**
- **mutually exclusive events** *(6.3)*
- **conditional probability** ($P(A|E)$) *(6.4)*
- **independent events**
- **Bayes's formula** *(6.5)*
- **law of complete probability**
- **partition of a sample space**
- **stochastic diagram**
- **Bernoulli process** *(6.6)*
- **equilibrium vector** *(6.7)*
- **Markov chain**
- **state vector**
- **transition matrix**

IMPORTANT FORMULAS

The Counting Rule

Let S be a finite sample space consisting of equally-likely outcomes. Then for any event $E \subseteq S$, $P(E) = \dfrac{|E|}{|S|}$.

The Complement Rule

For any event A, $P(A) + P(\overline{A}) = 1$.

The Union Rule

For any events A and B, $P(A \cup B) = P(A) + P(B) - P(A \cap B)$.

The Union Rule for Mutually Exclusive Events

If each pair of the events $A_1, A_2, \ldots, A_k$ is mutually exclusive, then

$$P(A_1 \cup A_2 \cup \cdots \cup A_k) = P(A_1) + P(A_2) + \cdots + P(A_k).$$

The Counting Rule for Conditional Probabilities

If A and E are events in a finite sample space consisting of equally-likely outcomes, then

$$P(A|E) = \frac{|A \cap E|}{|E|}.$$

The Intersection Rule

For any events A and B, $P(A \cap B) = P(A) \cdot P(B|A)$.

The Intersection Rule for Independent Events

If $A_1, A_2, \ldots, A_k$ are independent events, then

$$P(A_1 \cap A_2 \cap \cdots \cap A_k) = P(A_1) \cdot P(A_2) \cdot \cdots \cdot P(A_k).$$

Bernoulli's Formula

If p and q are the probability of success and failure, respectively, on a single trial of a Bernoulli process, then $P(\text{exactly } k \text{ successes in } n \text{ trials}) = C(n, k)\, p^k q^{n-k}$.

REVIEW EXERCISES

1. The frequency table below shows the distribution of letter grades awarded over the last few semesters in Professor Gorki's Russian literature course.

Grade	*Frequency*	
A	36	
B	63	
C	129	
D	45	
F	27	
	300	Total

Based on this empirical data, what is the probability that a random student in this course will earn

(a) an A? (b) at least a B? (c) at most a C? (d) a passing grade?

2. If the odds in favor of an event are 18 to 7, what is its probability?

3. If an event E has probability .4, what are the odds against E?

4. If a person is chosen at random, what is the probability that he or she was born on a Tuesday?

5. A large TV station has narrowed its search for a new programming director to three candidates: Anderson, Bertolucci, and Chang. Bertolucci has been told that her chances of being chosen are twice as good as Anderson's but only half as good as Chang's. What is the probability that Bertolucci will get the job?

6. If two fair dice are rolled, what is the probability that the sum will be less than five?

7. If six cards are dealt at random from a standard 52-card deck, what is the probability that four are red and two are black?

8. Adriana and Barbara are classmates. If their class of twenty students is randomly divided into two groups of ten, what is the probability that they will be placed in the same group?

9. Suppose the probability that A occurs is .2, the probability that B occurs is .5, and the probability that both occur is .1. What is the probability that

(a) A or B occurs?

(b) neither A nor B occurs?

(c) B occurs given that A has occurred?

10. In a certain community 40% of the population reads the morning newspaper, 32% reads the afternoon newspaper, and 18% reads both.

(a) What percentage of the community reads at least one of the papers?

(b) If someone reads the morning paper, what is the probability that he or she also reads the afternoon paper?

11. Let A, B, and C be mutually exclusive events with probabilities .07, .23, and .12, respectively. What is the probability that none of these three events occurs?

12. A clothes dryer contains twelve identical gray socks and eight identical brown socks. If two socks are removed from the dryer at random, what is the probability of obtaining a matching pair?

13. Of the students entering a certain medical school, 36% will fail out during the first year and 48% will complete the full four-year program. If an entering student survives the first year, what is the probability that he or she will complete the full program?

14. An encyclopedia salesman has found that he has a 20% chance of getting his foot in the door when he calls at a given house. If he does get his foot in the door, then he has a 50% chance of making a sale; but if he does not get his foot in the door, then he has only a 10% chance of making a sale. What is the probability that he will make a sale at a randomly chosen house?

15. Two swimmers are independently attempting to swim from Cuba to Florida. If their chances of success are 20% and 30%, what is the probability that at least one of them succeeds?

16. A certain disease is known to occur in 5% of the population. Among those with the disease, 80% carry a blood antigen called the X-factor, whereas only 10% of those without the disease carry the X-factor. If a random person is found to carry the X-factor, what is the probability that he or she has the disease?

17. The voters in a certain congressional district are 60% white, 15% black, and 25% Hispanic. According to local polls, the Democratic candidate for Congress is supported by 55% of the white voters, 80% of the black voters, and 60% of the Hispanic voters.

(a) According to this poll, what percentage of the district's voters supports this candidate?

(b) What percentage of this candidate's support comes from the black community?

18. Assuming that a newborn kitten is equally likely to be male or female, what is the probability that in a litter of four, there are

(a) two males and two females?

(b) at least one kitten of each sex?

19. Each year 2% of the individual tax returns in a certain state are subject to a random audit. What is the probability that a given taxpayer

(a) will not undergo this audit during the next ten years?

(b) will undergo this audit at most once in the next ten years?

20. A television network is airing eight new situation comedies this season. If past experience shows that only one of every five new situation comedies is successful, what is the probability that

(a) at least one of the new sitcoms is successful?

(b) at least two of the new sitcoms are successful?

21. Suppose that physicians make up 5% of the current population, that 52% of the children of physicians become physicians, and that only 2% of the children of nonphysicians become physicians. Assume also that the birthrate for physicians and nonphysicians is the same.

(a) Write the transition matrix and initial state vector for this Markov chain.

(b) What percentage of the next generation will be physicians?

(c) What proportion of the population will eventually be physicians?

22. A small town has three supermarkets. During the first week in June, 40% of the residents shopped at Kroger, 50% shopped at Dominick's, and 10% shopped at Jewel. One week later, 70% of those who had shopped at Kroger the preceding week shopped at Kroger again, but 20% switched to Dominick's and 10% switched to Jewel. Of those who had shopped at Dominick's, 80% continued to shop at Dominick's, but 20% switched to Jewel. Of those who had shopped at Jewel, 60% continued to shop there, but 20% switched to Kroger and 20% to Dominick's.

(a) Write the transition matrix and initial state vector for this Markov chain.

(b) What percentage of the town shopped at each supermarket during the fourth week in June?

(c) If this trend continues, what percentage of the townspeople will eventually shop at each supermarket?

REFERENCES

1. Baker, L. D., et al., "Evaluation of a 'do-it-yourself' pregnancy test," *American Journal of Public Health,* vol. 66 (1976), p. 166.
2. Berelson, Bernard R., Paul F. Lazarsfeld, and William N. McPhee, *Voting*. Chicago: University of Chicago Press, 1954.
3. Bourne, Larry S., "Physical Adjustment Process and Land Use Succession: A Conceptual Review and Central City Example," *Economic Geography,* vol. 47 (1971), pp. 1–15.
4. Dyson, Freeman, "Reflections: Disturbing the Universe–I," *The New Yorker,* August 6, 1979, p. 46.
5. Hampton, P. "Regional Economic Development in New Zealand," *Journal of Regional Science,* vol. 8 (1968), pp. 41–51.
6. Kannel, William B., M.D., "Recent Findings of the Framingham Study," *Resident & Staff Physician,* January 1978, p. 57.
7. Prais, S. J., "Measuring Social Mobility," *Journal of the Royal Statistical Society,* vol. 118 (1955), pp. 56–66.
8. Schell, E. D., "Samuel Pepys, Isaac Newton and Probability," *The American Statistician,* October 1960, pp. 27–30.

7

STATISTICS

Statistics is the branch of mathematics concerned with classifying and interpreting numerical data. Because of its many important applications, this subject has become a specialized discipline in its own right. The branch of statistics that deals with collecting, organizing, and summarizing data is called *descriptive statistics,* and the branch that is concerned with drawing conclusions and making predictions is called *inferential statistics*. In this chapter we present a brief survey of some statistical techniques that are related to the discussion of probability in Chapter 6.

7.1 Descriptive Statistics

This section is concerned with the representation of data, which can be classified as either *qualitative* or *quantitative* according to the nature of the categories into which the measurements fall. For example, classifying "countries that produce steel" involves nonnumerical categories and hence is qualitative. On the other hand, classifying "age groups in the United States" involves numerical categories and is therefore quantitative.

*Largest steel producers** *(millions of metric tons)*	
U.S.S.R.	147.2
Japan	99.5
U.S.	67.7
China	37.1
West Germany	35.9

TABLE 7.1

U.S. population by age† *(millions)*	
0– 20	69.2
20– 40	69.6
40– 60	45.0
60– 80	29.2
80–100	5.0

TABLE 7.2

A common means for depicting qualitative data like those in Table 7.1 is with a **bar graph** as shown in Figure 7.1. Here the quantity of steel produced by a country is denoted by the length of a horizontal bar. For quantitative data like those in Table 7.2, a somewhat different type of graph is used. (See Figure 7.2.) Here the bars are constructed vertically, and the height of each bar represents the number of persons in the corresponding age bracket. This number is displayed either over the bar (as in Figure 7.2) or within the bar. This type of graph is called a **histogram.** Notice that unlike a bar graph, a histogram has no spaces between its bars.

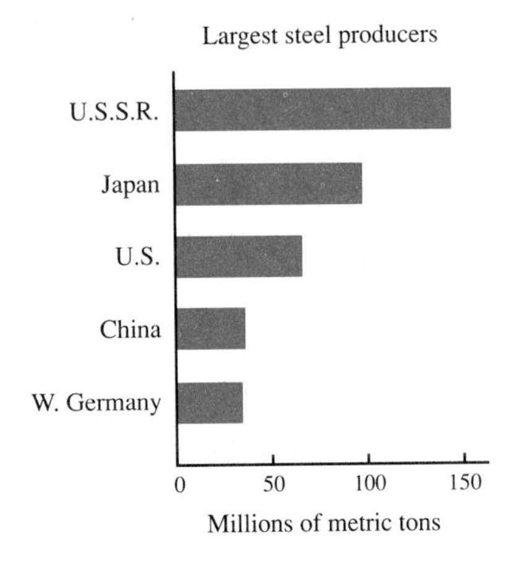

FIGURE 7.1

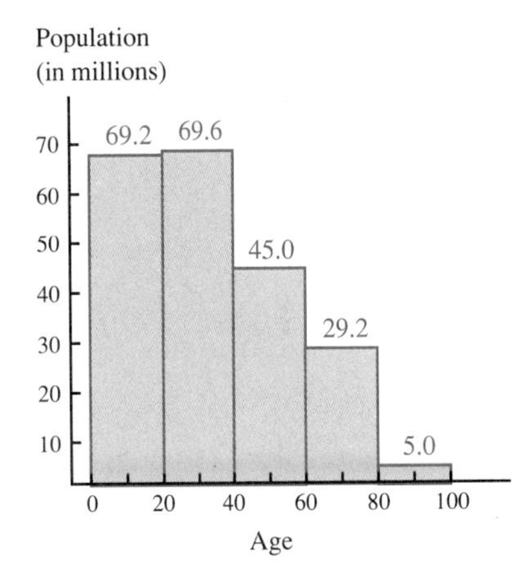

FIGURE 7.2

*1982 figures
†1980 figures

In working with quantitative data, the numerical categories (such as 0–20, 20–40, 40–60, 60–80, and 80–100 in Table 7.2) are called **class intervals.** It may sometimes happen that a data value falls on the boundary of two different class intervals. For instance, this occurs in Table 7.2 with someone 20 years old. *In this case we will follow the convention of assigning a data value to the higher of the class intervals into which it falls;* so a 20-year-old would be counted in the class interval 20–40 rather than the interval 0–20. This means that the class interval 20–40 in Table 7.2 includes all the ages greater than or equal to 20 and less than 40, and a similar interpretation holds for the other intervals in this table.

The data in Table 7.3 illustrate two common difficulties that occur with quantitative data. Notice that the fifth and sixth class intervals cover unequal income ranges ($5000 and $10,000, respectively). Moreover the last interval ("$50,000 or more") is open-ended, that is, no upper limit is specified. A histogram is normally used only for quantitative data such as in Table 7.2 in which *there are no open-ended intervals and each class interval covers an equal range of values.*

*Income of U.S. families**	
Below $5000	4.8%
$5000–$10,000	8.5%
$10,000–$15,000	10.2%
$15,000–$20,000	10.5%
$20,000–$25,000	10.3%
$25,000–$35,000	18.6%
$35,000–$50,000	18.8%
$50,000 or more	18.3%

TABLE 7.3

MEASURES OF CENTRAL TENDENCY

In describing a set of data the first task is to characterize its center or "typical" value. Numbers that measure the center of a set of data are called **measures of central tendency.**

The most widely used measure of central tendency is the mean (or average). The **mean** of a list of n numbers equals their sum divided by n. Thus the mean can be computed by the following formula.

Mean

The mean $\bar{x}$ of $x_1, x_2, \ldots, x_n$ is

$$\bar{x} = \frac{x_1 + x_2 + \cdots + x_n}{n}. \tag{7.1}$$

*1985 figures

By using the **summation sign** Σ, which is read "sum of", we can express this formula in the more compact form

$$\bar{x} = \frac{\Sigma x}{n}.$$

EXAMPLE 7.1 A student received grades of 84, 94, 87, and 79 on four hourly examinations. Find her mean (average) score.

Solution The student's mean score is

$$\bar{x} = \frac{\Sigma x}{n} = \frac{84 + 94 + 87 + 79}{4} = 86. \quad ■$$

Practice Problem 1 Find the mean of 26, 38, 17, 32, and 41.

EXAMPLE 7.2 An automobile assembly plant recorded the following numbers of absentees on fifty consecutive working days.

17	32	45	35	38	22	40	24	48	53
29	47	57	26	33	42	46	36	37	16
41	35	49	54	39	22	20	26	47	43
21	44	29	35	52	32	26	31	38	47
27	41	12	23	32	37	55	35	48	46

Find the mean number of absentees from the plant on these fifty days.

Solution Using the fifty values listed above, we find after much calculation that

$$\bar{x} = \frac{\Sigma x}{n} = \frac{17 + 32 + 45 + \cdots + 46}{50}$$

$$= \frac{1810}{50} = 36.2.$$

This is the mean of the fifty observed values. In Section 7.2 we will discuss a method for approximating this number that requires much less calculation. ■

Practice Problem 2 The following data show the number of tornadoes occurring in the United States during the years 1960–1984.

618	683	658	461	713
898	570	912	660	604
649	888	741	1102	947
920	835	852	788	852
866	783	1046	931	907

Determine the mean number of tornadoes reported in the United States during these years.

Another common measure of central tendency is the **median,** which is the middle value when the data are arranged in increasing order (or the average of the two middle values if there are an even number of data values). For example, to find the median of the seven numbers

$$10, 2, 8, 7, 4, 5, 3,$$

we must arrange them in increasing order

$$2, 3, 4, 5, 7, 8, 10.$$

The middle value, 5, is the median.

If there is another data value, say 9, then the ordered list is

$$2, 3, 4, 5, 7, 8, 9, 10.$$

In this case there are an even number of values, and so instead of a single middle value there are two, 5 and 7. Here the median is the average of these values

$$\frac{5 + 7}{2} = 6.$$

EXAMPLE 7.3 Find the median number of absentees from the automobile assembly line using the data presented in Example 7.2.

Solution If we arrange the raw data in increasing order (a tedious job), we obtain the list below.

12	16	17	20	21	22	22	23	24	26
26	26	27	29	29	31	32	32	32	33
35	35	35	35	36	37	37	38	38	39
40	41	41	42	43	44	45	46	46	47
47	47	48	48	49	52	53	54	55	57

The middle values are 36 and 37. Hence the median is

$$\frac{36 + 37}{2} = 36.5. \quad ■$$

Practice Problem 3 Determine the median number of tornadoes reported in the United States during the years 1960–1984 using the data given in Practice Problem 2.

When there are a large number of data (as in Example 7.3), the median can be more difficult to calculate than the mean because the data must be ordered. Unlike the mean, however, the median has the advantage of being unaffected by a few atypical large or small values. (See Exercises 15–16.)

MEASURES OF DISPERSION

When Mr. Smith learned that he was being promoted to a position as regional manager in Austin, Texas, he contacted two real estate agencies in order to sell his present house before moving. The first agency claimed that the last six houses comparable to Mr. Smith's that were listed with it had sold in an average of 60 days. The second agency said that the last six houses comparable to Mr. Smith's that were listed with it had sold in an average of 67 days. Which agency should Mr. Smith choose in order to sell his house within 90 days?

With no further information, Mr. Smith should choose the first agency because comparable houses listed recently with the first agency sold one week sooner on the average. In this case, however, Mr. Smith was able to request the data on which the agencies' claims were based. These data are shown below.

Number of days before sale for the first agency	*Number of days before sale for the second agency*
42	48
99	75
20	83
104	68
65	57
30	71

Looking at these data, Mr. Smith noticed that all six of the houses listed with the second agency had sold within 90 days. On the other hand, only four of the six houses listed with the first agency had sold within 90 days. On this basis Mr. Smith listed his house with the second agency.

In this example, Mr. Smith is concerned more with selling his house within 90 days than in selling it quickly. Thus for this situation the *distribution* of the data is more important than its center. In comparing data, we must be able to describe not only the center of the data but also its distribution. For reasons that will become clear later, it is most useful to measure the distribution of the data around the mean. One common such measure is the variance.

Variance

The **variance*** of $x_1, x_2, \cdots, x_n$ is

$$s^2 = \frac{(x_1 - \bar{x})^2 + (x_2 - \bar{x})^2 + \cdots + (x_n - \bar{x})^2}{n}. \tag{7.2}$$

*When the data are a sample of measurements from a larger population, the denominator in (7.2) is often replaced by $n - 1$ for technical reasons. For simplicity we will use only formula (7.2) in this book.

Note that formula (7.2) can be written using the summation sign as

$$s^2 = \frac{\Sigma(x - \bar{x})^2}{n}.$$

Thus the variance is the average of the squares of the distances between the data values and the mean. (The squaring is done to make all the terms in the summation positive. Without squaring, it would always turn out that

$$\frac{\Sigma(x - \bar{x})}{n} = 0$$

because $\bar{x}$ is, by definition, the average of the x values.) We see from the variance formula that when the values of x are close to the mean, the value of s^2 will be small; and when the values of x are far from the mean, the value of s^2 will be large.

Using the agencies' selling times above, we see that the mean selling time for the first agency is

$$\frac{42 + 99 + 20 + 104 + 65 + 30}{6} = \frac{360}{6} = 60$$

days, as it claimed. Hence the variance of these numbers is

$$\frac{(42 - 60)^2 + (99 - 60)^2 + (20 - 60)^2 + (104 - 60)^2 + (65 - 60)^2 + (30 - 60)^2}{6}$$

$$= \frac{324 + 1521 + 1600 + 1936 + 25 + 900}{6} = \frac{6306}{6} = 1051.$$

Because the variance for the first agency is large, its selling times are widely dispersed from the mean. On the other hand, the mean selling time for the second agency is

$$\frac{48 + 75 + 83 + 68 + 57 + 71}{6} = \frac{402}{6} = 67,$$

and so the variance of these numbers is

$$\frac{(48 - 67)^2 + (75 - 67)^2 + (83 - 67)^2 + (68 - 67)^2 + (57 - 67)^2 + (71 - 67)^2}{6}$$

$$= \frac{361 + 64 + 256 + 1 + 100 + 16}{6} = \frac{798}{6} = 133.$$

Because the variance for the second agency is small, its selling times are clustered near the mean.

Since the agencies' data are measured in days, the variances above are measured in square days, a meaningless unit. In order to measure the distribution of the selling times in more meaningful units, we take the square root of the variance. The square root of the variance is called the **standard deviation** and is denoted by s. Thus the standard deviations of the selling times for the two agencies are

$$\sqrt{1051} \approx 32.4 \text{ days} \qquad \text{and} \qquad \sqrt{133} \approx 11.5 \text{ days},$$

respectively. These numbers show that the selling times for the first agency are considerably more dispersed than those for the second agency.

EXAMPLE 7.4 The following lists contain estimates of the weight of a rock (in pounds). List (a) contains the estimates given by 10 sixth-graders, and list (b) contains the estimates given by 10 ninth-graders.

(a) 1, 1, 1, 2, 2, 4, 4, 5, 5, 5
(b) 2, 2, 2, 3, 3, 3, 3, 4, 4, 4

Find the mean and standard deviation of each set of estimates.

Solution Using formula (7.1), we see that the mean of list (a) is

$$\bar{x} = \frac{\Sigma x}{n} = \frac{1 + 1 + 1 + 2 + 2 + 4 + 4 + 5 + 5 + 5}{10} = \frac{30}{10} = 3$$

pounds, and the mean of list (b) is

$$\bar{x} = \frac{\Sigma x}{n} = \frac{2 + 2 + 2 + 3 + 3 + 3 + 3 + 4 + 4 + 4}{10} = \frac{30}{10} = 3$$

pounds. Thus each list has the same mean.

By formula (7.2) the variance of the values in data set (a) is

$$\begin{aligned} s^2 &= \frac{\Sigma(x - \bar{x})^2}{n} \\ &= \frac{(1 - 3)^2 + (1 - 3)^2 + \cdots + (5 - 3)^2 + (5 - 3)^2}{10} \\ &= \frac{4 + 4 + 4 + 1 + 1 + 1 + 1 + 4 + 4 + 4}{10} = \frac{28}{10} = 2.8. \end{aligned}$$

Therefore the standard deviation of the values in data set (a) is

$$s = \sqrt{2.8} \approx 1.67$$

pounds. On the other hand, the variance of the values in data set (b) is

$$\begin{aligned} s^2 &= \frac{\Sigma(x - \bar{x})^2}{n} \\ &= \frac{(2 - 3)^2 + (2 - 3)^2 + \cdots + (4 - 3)^2 + (4 - 3)^2}{10} \\ &= \frac{1 + 1 + 1 + 0 + 0 + 0 + 0 + 1 + 1 + 1}{10} = \frac{6}{10} = 0.6. \end{aligned}$$

Therefore the standard deviation of the estimates in data set (b) is

$$s = \sqrt{0.6} \approx 0.77$$

pounds. Thus we see that data set (a) has a greater standard deviation than (b). Hence the estimates of the weight of the rock made by the sixth-graders are more variable than those of the ninth-graders, even though the mean estimate of both groups of students is the same. ■

It is instructive to write each set of 10 estimates in Example 7.4 in table form.

(a)

Estimate	*Frequency*
1	3
2	2
3	0
4	2
5	3

(b)

Estimate	*Frequency*
1	0
2	3
3	4
4	3
5	0

These distributions are depicted in the histograms of Figure 7.3. Note that in each case the "center" of the distribution is 3, the mean value. The data in set (a), however, are more widely dispersed than those in set (b). This reflects the fact that a ninth-grader is a more accurate estimator than a sixth-grader.

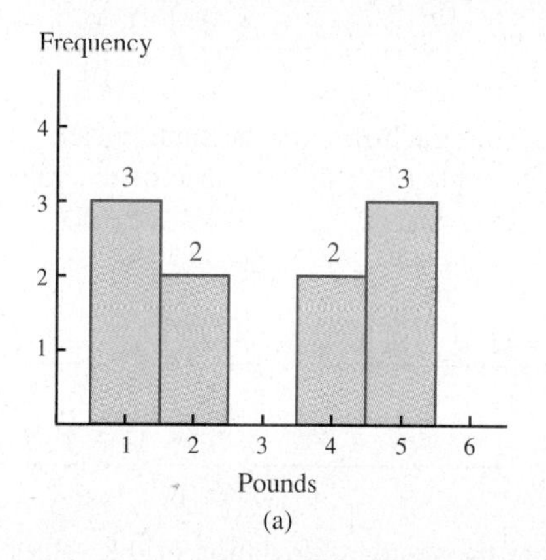

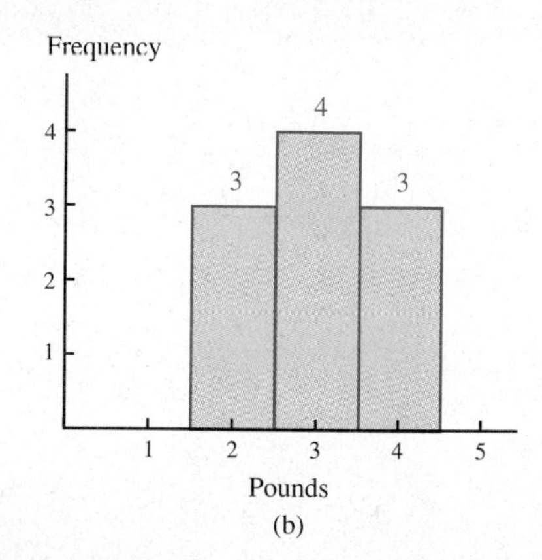

FIGURE 7.3

Practice Problem 4 Determine the standard deviation of the examination scores (84, 94, 87, and 79) for the student in Example 7.1.

For mathematical purposes the mean is the most useful measure of central tendency, and the standard deviation is the most useful measure of dispersion. One reason for their usefulness is that if each entry in a set of data is multiplied by the same factor c, then both the mean and standard deviation are also multiplied by c. As a result, when the data are measurements, a change in the unit of measurement gives the corresponding change in the mean and standard deviation. For example, if the unit of measurement in Example 7.4 is changed from pounds to ounces, then all the data values are multiplied by 16. For these new values, the mean and standard deviation will be 16 times greater than they are when the measurements are given in pounds. We will see applications of the mean and standard deviation throughout this chapter.

EXERCISES 7.1

1. Represent the data in the following table by a bar graph.

Advertising expenditures (in millions)	
1970	$11.9
1975	$15.3
1980	$19.6
1985	$28.2

2. Represent the data in the following table by a bar graph.

Health expenses per capita, 1977	
Hospital care	$297
Physicians	$146
Nursing homes	$57
Drugs	$57
Other	$89

3. A test of the tensile strength (breaking point under stress) of a sample of 80 cables produced the distribution shown below. Represent these data by a histogram.

Tensile strength (lbs)	*Number of cables*
1400–1500	11
1500–1600	29
1600–1700	26
1700–1800	9
1800–1900	5

4. The speeds of 100 cars selected at random from I–74 were distributed as shown below. Represent these data by a histogram.

Speed (mph)	*Cars*
40–50	15
50–60	31
60–70	35
70–80	17
80–90	2

For the data in Exercises 5–12 find: (a) the mean, (b) the median, and (c) the standard deviation.

5. 3, 7, 9, 4, 7

6. 3, 5, 7, 3, 7

7. 26, 38, 16, 32

8. 12, 9, 18, 21

9. 4, 10, 9, 4, 6, 9

10. 8, 7, 2, 6, 2, 11

11. 2, 6, 2, 7, 8, 6, 2, 7

12. 4, 12, 7, 9, 6, 5, 8, 13

13. The rates (in dollars per hour) charged by ten moving companies for local moves are as follows:

52, 57, 53, 59, 48, 50, 56, 52, 54, and 49.

Find: (a) the mean, (b) the median, and (c) the standard deviation of these data.

14. Twelve randomly selected owners of General Motors cars were asked how many years they had been driving their present automobiles. Their responses were as follows:

2, 8, 5, 11, 7, 6, 1, 14, 4, 5, 3, and 6.

Find: (a) the mean, (b) the median, and (c) the standard deviation of these data.

15. The salaries of ten randomly chosen major league pitchers are shown below.

\$185,000	\$200,000	\$340,000	\$405,000	\$1,500,000
\$360,000	\$315,000	\$210,000	\$170,000	\$205,000

Find the mean and median for these salaries. Which value gives a better idea of a representative salary for a major league pitcher?

16. In ten consecutive trips to Las Vegas, a gambler had the following net gains:

\$1500	\$1225	\$1375	\$2250	−\$4725
\$750	−\$8750	\$3150	−\$2725	\$2150

Find the mean and median for these winnings. Which of these measures gives a better idea of the long-term earnings that the gambler can expect?

17. Prove that if the mean and standard deviation of $x_1, x_2, \ldots, x_n$ are $\bar{x}$ and s, respectively, then the mean and standard deviation of $cx_1, cx_2, \ldots, cx_n$ are $c\bar{x}$ and cs, respectively.

18. Prove that the variance of $x_1, x_2, \ldots, x_n$ is equal to

$$\frac{x_1^2 + x_2^2 + \cdots + x_n^2}{n} - \bar{x}^2.$$

This formula simplifies the computation of the variance and standard deviation.

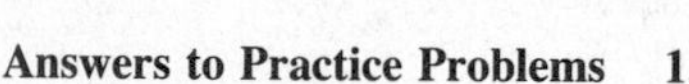

Answers to Practice Problems

1. 30.8
2. 795.36
3. 835
4. $s \approx 5.43$

7.2 Grouped Data

Once we have obtained a set of data, we need to organize it into a form that allows us to comprehend it easily. If the set is small, it is easy to deal with the raw data themselves. A large data set, however, is usually difficult to understand unless it is organized in some systematic way. Consider, for instance, the numbers of absentees from an automobile assembly plant on fifty consecutive working days that were introduced in Example 7.2. (These data are repeated below.)

17	32	45	35	38	22	40	24	48	53
29	47	57	26	33	42	46	36	37	16
41	35	49	54	39	22	20	26	47	43
21	44	29	35	52	32	26	31	38	47
27	41	12	23	32	37	55	35	48	46

In this form, it is very difficult to get a "feel" for the data. In order to understand these numbers better, it is helpful to classify the data in intervals of a uniform width. In this case the number of absentees varies from a low of 12 to a high of 57, for a range of 46 values. Thus all of the data can be accommodated in five class intervals containing ten values each. Of course, there are many different sets of class intervals that can be used to contain the data. Using a small number of wide intervals makes the data more manageable, but having a larger number of narrow intervals is more precise. The general practice is to use at least five and at most fifteen class intervals when grouping data.

Choosing the endpoints of the intervals often requires some thought. Although the intervals 10–20, 20–30, 30–40, 40–50, and 50–60 accommodate all the values in our data set, we encounter a problem with these intervals because values such as 20 or 40 could be placed in two different intervals. Whenever possible, we would like to choose our intervals so that the upper endpoint of one interval equals the lower endpoint of the next interval and each data value falls in exactly one interval. Unfortunately some types of data can (at least theoretically) assume all values in some interval of real numbers. This happens when the data represent ages, as in Table 7.2. In this instance, no matter how the endpoints are chosen, there can be data values that fall in two different intervals. This is not the case for the number of automobile plant absentees, however, and by using the intervals 10.5–20.5, 20.5–30.5, 30.5–40.5, 40.5–50.5, and 50.5–60.5, we can assure that each data value lies in exactly one interval.

In summary, whenever possible we choose the intervals into which the data will be grouped according to the following guidelines.

Guidelines for Choosing the Intervals when Grouping Data

1. The intervals should be of equal width.
2. The number of intervals should be between 5 and 15.
3. The upper endpoint of one interval should equal the lower endpoint of the next interval.
4. Every data value should lie in exactly one interval.

Tallying the number of data values that fall in each of the class intervals produces the **frequency distribution** shown in Table 7.4. Since each data value belongs to exactly one interval, the class frequencies add up to 50, the total number of observations. Note the similarity between a frequency distribution for a set of data and a probability distribution, which we encountered in Section 6.2.

Number of absentees	*Tally*	*Frequency*
10.5–20.5	||||	4
20.5–30.5	~~||||~~ ~~||||~~ |	11
30.5–40.5	~~||||~~ ~~||||~~ ~~||||~~ |	16
40.5–50.5	~~||||~~ ~~||||~~ ||||	14
50.5–60.5	~~||||~~	5

TABLE 7.4

The frequency distribution in Table 7.4 can be represented by a histogram, as in Figure 7.4. This histogram gives us a much better feel for the data than we had by looking at the raw data. For instance, we can see at a glance that the number of absentees is usually between 20.5 and 50.5; in fact, in our 50-

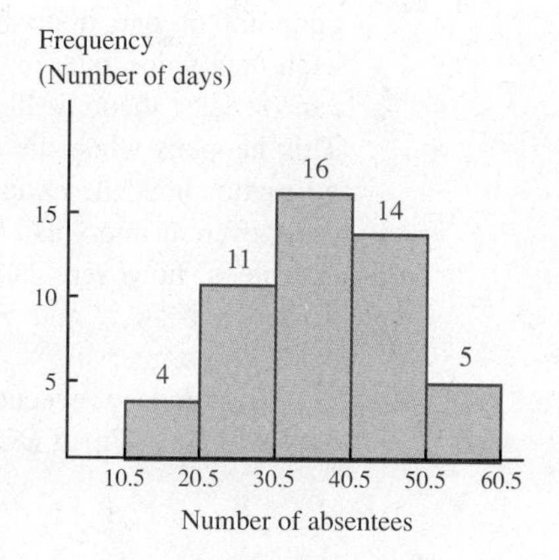

FIGURE 7.4

day sample the number of absentees fell in this range 11 + 16 + 14 = 41 times (82% of the time). The price that we pay for the convenience of grouped data is a loss of information. For example, if we want to know how often the number of absentees was between 25 and 45, our frequency distribution is of no use; we must go back to the original raw data.

Practice Problem 5 The data below show the number of tornadoes occurring in the United States during the years 1960–1984.

618	683	658	461	713
898	570	912	660	604
649	888	741	1102	947
920	835	852	788	852
866	783	1046	931	907

(a) Choose an appropriate set of intervals into which to group these data.

(b) Prepare a frequency distribution and histogram using the class intervals you chose in part (a).

THE MEAN AND STANDARD DEVIATION OF GROUPED DATA

We saw above that by grouping data into a frequency distribution, we could make it easier to understand. In order for a frequency distribution to prove really useful, however, we must be able to use it to estimate the mean and standard deviation without needing to return to the raw data. A simple approximation makes this possible: *For grouped data the values in each class interval are approximated by the midpoint of the interval.* Thus the mean can be estimated as follows.

Mean of Grouped Data

If data values are grouped into class intervals having midpoints $m_1, m_2, \ldots, m_k$ and respective frequencies $f_1, f_2, \ldots, f_k$, then

$$\bar{x} \approx \frac{m_1 f_1 + m_2 f_2 + \cdots + m_k f_k}{n} = \frac{\Sigma\, mf}{n}, \tag{7.3}$$

where $n = \Sigma f$ is the total number of data values.

EXAMPLE 7.5 Find the mean number of absentees from the automobile assembly plant using the frequency distribution in Table 7.4.

Solution The frequency distribution is reproduced below, showing the midpoint of each class interval. Note that the midpoint is simply the mean of the upper and lower endpoints of the class interval; for instance, the midpoint of the interval 10.5–20.5 is

$$\frac{10.5 + 20.5}{2} = 15.5.$$

Interval	*Midpoint* m	*Frequency* f	mf	
10.5–20.5	15.5	4	62.0	
20.5–30.5	25.5	11	280.5	
30.5–40.5	35.5	16	568.0	
40.5–50.5	45.5	14	637.0	
50.5–60.5	55.5	5	277.5	
		50	1825.0	Totals

The last column gives the product mf of the midpoint of each class interval with the corresponding class frequency. The sum of these products is shown at the bottom of the column. Applying formula (7.3), we find that the mean of the grouped data is

$$\bar{x} = \frac{\Sigma\, mf}{n} = \frac{1825}{50} = 36.5.$$

This value agrees fairly well with the true mean of 36.2 computed in Example 7.2. The discrepancy in the two numbers is due to the fact that information is lost when the data are grouped and all the values in each interval are replaced by the midpoint. ■

Practice Problem 6 Use the frequency distribution in Table 7.5 to estimate the mean number of tornadoes reported in the United States during the years 1953–1977.

Number of Tornadoes	*Frequency*	
399.5– 499.5	1	
499.5– 599.5	1	
599.5– 699.5	6	
699.5– 799.5	4	
799.5– 899.5	6	
899.5– 999.5	5	
999.5–1099.5	1	
1099.5–1199.5	1	
	25	Total

TABLE 7.5

For grouped data the midpoints $m_1, m_2, \ldots, m_k$ of the class intervals are also used to approximate the actual data in computing the variance and standard deviation. Thus the variance of grouped data can be approximated as follows.

Variance of Grouped Data

If data values are grouped into class intervals having midpoints m_1, m_2, . . . , m_k and respective frequencies $f_1, f_2, \ldots, f_k$, then

$$s^2 \approx \frac{(m_1 - \bar{x})^2 f_1 + (m_2 - \bar{x})^2 f_2 + \cdots + (m_k - \bar{x})^2 f_k}{n} \tag{7.4}$$
$$= \frac{\Sigma(m - \bar{x})^2 f}{n},$$

where $n = \Sigma f$ is the total number of data values.

EXAMPLE 7.6 Find the standard deviation of the number of absentees from the automobile assembly plant using the grouped data presented in Example 7.5.

Solution It can be shown by a very tedious calculation using formula (7.2) that the actual variance of the fifty observations is 121.84.

Calculating the variance of the grouped data is much simpler. In Example 7.5 we found that the mean of the grouped data is $\bar{x} = 36.5$. The table below shows the frequency distribution from Table 7.4 along with the midpoint of each class interval, the deviation from the mean $m - \bar{x}$, the squared deviation $(m - \bar{x})^2$, and the product $(m - \bar{x})^2 f$.

Interval	*Midpoint* m	*Frequency* f	$m - \bar{x}$	$(m - \bar{x})^2$	$(m - \bar{x})^2 f$	
10.5–20.5	15.5	4	−21	441	1764	
20.5–30.5	25.5	11	−11	121	1331	
30.5–40.5	35.5	16	−1	1	16	
40.5–50.5	45.5	14	9	81	1134	
50.5–60.5	55.5	5	19	361	1805	
		50			6050	Totals

Formula (7.4) estimates the variance as

$$s^2 \approx \frac{\Sigma(m - \bar{x})^2 f}{n} = \frac{6050}{50} = 121,$$

and so the standard deviation is

$$s \approx \sqrt{121} = 11.$$

Note that the grouped data have approximately the same variance and standard deviation as the ungrouped data. ■

EXAMPLE 7.7 Use the data in Table 7.2 on page 367 to approximate the mean and standard deviation for the age of someone in the U.S. population.

Solution The frequency distribution in Table 7.2 is reproduced below along with the midpoint of each class interval and the product of each interval's midpoint and frequency.

Interval	*Midpoint* m	*Frequency* f	mf	
0– 20	10	69.2×10^6	692×10^6	
20– 40	30	69.6×10^6	2088×10^6	
40– 60	50	45.0×10^6	2250×10^6	
60– 80	70	29.2×10^6	2044×10^6	
80–100	90	5.0×10^6	450×10^6	
		218.0×10^6	7524×10^6	Totals

Thus

$$\bar{x} \approx \frac{\Sigma mf}{n} = \frac{7524 \times 10^6}{218 \times 10^6} \approx 34.5$$

is approximately the average age of an individual in the U.S. population. To find the standard deviation, we extend the table above to include columns for $m - \bar{x}$, $(m - \bar{x})^2$, and $(m - \bar{x})^2 f$.

Interval	*Midpoint* m	*Frequency* f	$m - \bar{x}$	$(m - \bar{x})^2$	$(m - \bar{x})^2 f$	
0– 20	10	69.2×10^6	-24.5	600.75	$41{,}571.90 \times 10^6$	
20– 40	30	69.6×10^6	-4.5	20.25	$1{,}409.40 \times 10^6$	
40– 60	50	45.0×10^6	15.5	240.25	$10{,}811.25 \times 10^6$	
60– 80	70	29.2×10^6	35.5	1260.25	$36{,}799.30 \times 10^6$	
80–100	90	5.0×10^6	55.5	3080.25	$15{,}401.25 \times 10^6$	
		218.0×10^6			$105{,}993.10 \times 10^6$	Totals

By formula (7.4) we find the variance to be

$$s^2 \approx \frac{\Sigma(m - \bar{x})^2 f}{n} = \frac{105{,}993.10 \times 10^6}{218.0 \times 10^6} \approx 486,$$

and so the standard deviation is

$$s \approx \sqrt{486} \approx 22. \quad \blacksquare$$

Practice Problem 7 Determine the standard deviation of the number of tornadoes reported in the United States during the years 1960–1984 using the grouped data in Table 7.5. The mean of this distribution, as computed in Practice Problem 6, is 797.5.

EXERCISES 7.2

1. An insurance company checked the records of 80 motorists to determine how many moving violations each had been cited for during the past five years. The results are shown below.

Violations	*Frequency*	
0	34	
1	28	
2	12	
3	4	
4	2	
	80	Total

Use the given frequency distribution to find the mean and standard deviation of the number of moving violations.

2. A Labor Department statistician interviewed 120 unemployed persons in an inner-city area to determine how many years of high school each had completed. Their responses are shown below.

Number of Years	*Frequency*	
0	6	
1	24	
2	39	
3	30	
4	21	
	120	Total

Use the given frequency distribution to find the mean and standard deviation of the number of years of high school completed.

3. A test of the tensile strength (breaking point under stress) of a sample of 80 cables produced the frequency distribution shown below.

Tensile strength (lbs)	*Number of cables*	
1400–1500	11	
1500–1600	29	
1600–1700	26	
1700–1800	9	
1800–1900	5	
	80	Total

Use the given frequency distribution to find the mean and standard deviation of the tensile strength of the cables.

4. The speeds of 100 cars selected from I–74 were distributed as shown below.

Speed (mph)	*Cars*	
40–50	15	
50–60	31	
60–70	35	
70–80	17	
80–90	2	
	100	Total

Use this frequency distribution to find the mean and standard deviation of the speed of the cars.

5. The frequency distributions below show the weekly sales records for two realty companies during the past year.

Company A		*Company B*	
Number of houses sold	*Frequency*	*Number of houses sold*	*Frequency*
0	4	0	10
1	25	1	11
2	17	2	12
3	5	3	9
4	1	4	8
5	0	5	2

(a) Find the mean and standard deviation of each company's weekly sales.

(b) Which company had the better sales record? Which company was more consistent?

6. The life spans of 100 light bulbs of two brands were compared with the following results.

Brand S		*Brand W*	
Life span hours	*Frequency*	*Life span hours*	*Frequency*
500– 600	12	500– 600	5
600– 700	23	600– 700	30
700– 800	26	700– 800	47
800– 900	21	800– 900	16
900–1000	18	900–1000	2

(a) Find the mean and standard deviation of each brand's bulb life.

(b) Which brand of bulbs lasts longer, on the average? Which brand is more dependable?

In Exercises 7–14, (a) prepare a frequency distribution for the given data; (b) construct a histogram depicting the frequency distribution in part (a); and (c) approximate the mean and standard deviation of the given data.

7. The daily low temperature readings during the month of November (in degrees Fahrenheit) are shown below for Memphis, Tennessee.

41	43	39	34	38	42
44	47	48	53	58	54
51	59	53	46	47	41
36	31	36	43	43	47
49	48	49	54	57	54

8. The systolic blood pressures (in millimeters of mercury) are shown below for a group of 40 young males.

141	98	132	135	122	117	131	116
139	118	121	128	127	113	138	115
127	142	108	147	117	118	127	138
116	131	119	122	121	94	129	118
119	129	108	123	131	124	131	102

9. In reference [1] Anderson reports the IQ scores of 26 children prior to beginning nursery school. These data are shown below.

108	107	93	128	133	126	125	117	113
115	111	122	100	100	139	108	117	127
119	150	117	108	95	111	127	86	

10. The number of deaths caused by tornadoes in the United States during each year from 1953–1984 is shown below.

515	36	126	83	191	66	58	47
51	28	31	73	296	99	114	131
66	72	156	27	87	361	60	44
43	53	84	28	24	64	34	122

11. The ages of the 39 U.S. presidents at the time of their first inauguration are shown below.

57	57	49	52	50	51	51	55
61	61	64	56	47	56	50	61
57	54	50	46	55	55	62	52
57	68	48	54	54	51	43	69
58	51	65	49	42	54	55	

12. The ages at death of deceased presidents of the U.S. are shown below.

67	80	53	56	56	72	63
90	78	65	66	71	67	88
83	79	74	63	67	57	78
85	68	64	70	58	60	46
73	71	77	49	60	90	64

13. The number of home runs hit by the leading home run hitter in the American League during each season from 1937–1986 is shown below.

46	58	35	41	37	36	34	22	24	44
32	39	43	37	33	32	43	32	37	52
42	42	42	40	61	48	45	49	32	49
44	44	49	44	33	37	32	32	36	32
39	46	45	41	22	39	39	43	40	40

14. The number of points scored by the leading scorer in the National Football League during each season from 1936–1985 is shown below.

73	69	58	68	57	95	138	117	85	110
100	102	110	102	128	102	94	114	114	96
99	77	108	94	176	147	137	113	155	132
119	117	145	129	125	117	128	130	94	138
109	99	118	115	129	121	84	161	131	144

15. We noted in Section 7.1 that if all the raw data are multiplied by the same factor, then the mean and standard deviation are also multiplied by that factor. Show that the same is true when the mean and standard deviation are approximated using the grouped data formulas (7.3) and (7.4).

16. Show that if in formulas (7.3) and (7.4) all the frequencies are multiplied by the same factor, then the mean and variance (and hence the standard deviation) are unchanged.

Answers to Practice Problems

5. (a) There are many acceptable choices of intervals; one is 399.5–499.5, 499.5–599.5, 599.5–699.5, 699.5–799.5, 799.5–899.5, 899.5–999.5, 999.5–1099.5, and 1099.5–1199.5.

(b) The frequency distribution is given here on the left; it depends on the intervals chosen. The histogram using the intervals in part (a) is shown on the right.

Number of tornadoes	*Frequency*
399.5– 499.5	1
499.5– 599.5	1
599.5– 699.5	6
699.5– 799.5	4
799.5– 899.5	6
899.5– 999.5	5
999.5–1099.5	1
1099.5–1199.5	1
	25 Total

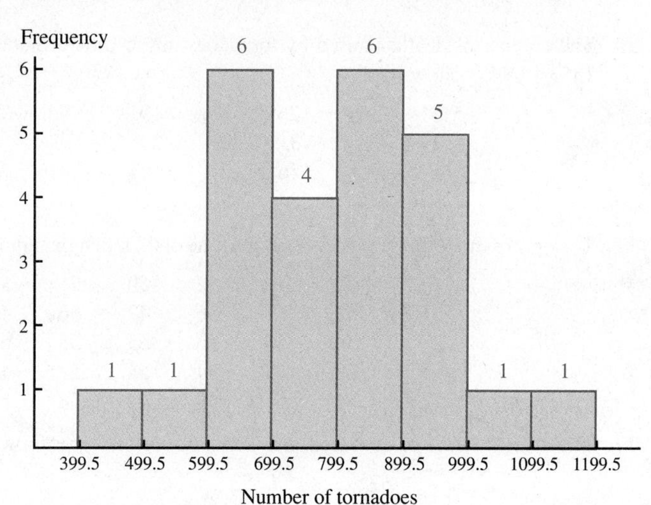

6. 797.5

7. $\sqrt{25696} \approx 160$

7.3 Random Variables: Expected Value and Variance

Quantitative data always represent measurements of some quantity over a certain population, such as the ages of individuals in the United States, the number of absentees at an auto plant on certain days, or the number of hours a light bulb will last before burning out. Note that the value of the quantity being measured depends on which member of the population is chosen. We call a numerical quantity whose value depends on the outcome of an experiment a **random variable.**

Consider, for instance, an automobile towing service that has recorded the number of service calls it received each day over a 60-day period. The results are shown in Table 7.6 and depicted in Figure 7.5.

Number of service calls	*Frequency*	
0	6	
1	15	
2	18	
3	12	
4	6	
5	3	
	60	Total

TABLE 7.6

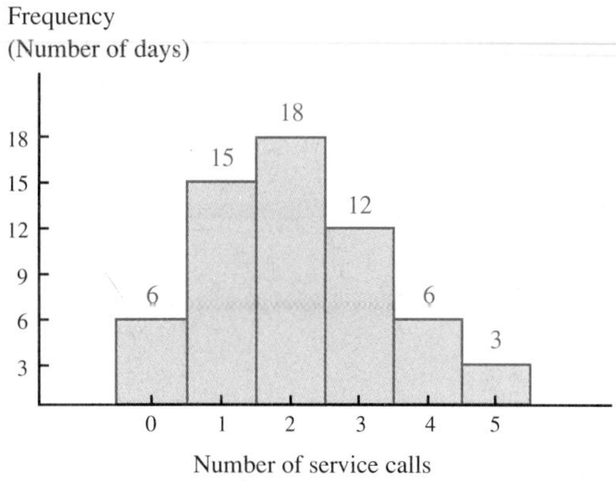

FIGURE 7.5

We can regard Table 7.6 as a frequency distribution for the random variable that counts the number of daily service calls. Recall that if each frequency is divided by the total number of observations (60), we obtain the corresponding relative frequency or empirical probability. This leads to the probability distribution shown in Table 7.7.

Number of service calls	*Probability*	
0	.10	
1	.25	
2	.30	
3	.20	
4	.10	
5	.05	
	1.00	Total

TABLE 7.7

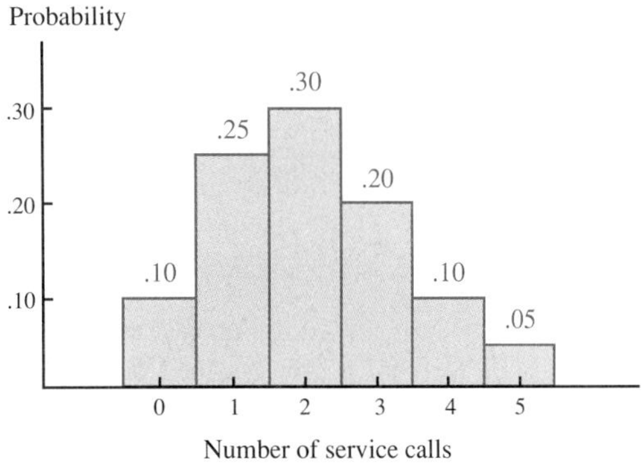

FIGURE 7.6

This distribution can be represented by a probability histogram as in Figure 7.6, where each value is the midpoint of a "class interval" extending one-half unit to the left and right. The height of each bar represents the probability of the corresponding value (the frequency divided by 60). Thus the probability histogram in Figure 7.6 has the same shape as the frequency histogram in Figure 7.5, the only difference being the units of measurement on the vertical axis.

EXAMPLE 7.8 Find the mean and the standard deviation of the number of daily service calls for the towing service described in Table 7.6.

Solution From Table 7.6 we can calculate the products mf of the number of calls times the corresponding frequencies.

Number of calls m	*Frequency f*	*mf*	$m - \bar{x}$	$(m - \bar{x})^2$	$(m - \bar{x})^2 f$	
0	6	0	−2.1	4.41	26.46	
1	15	15	−1.1	1.21	18.15	
2	18	36	−0.1	.01	0.18	
3	12	36	0.9	.81	9.72	
4	6	24	1.9	3.61	21.66	
5	3	15	2.9	8.41	25.23	
	60	126			101.40	Totals

Adding the products mf and dividing by the number of observations gives

$$\bar{x} = \frac{\Sigma mf}{n} = \frac{126}{60} = 2.10.$$

This is the mean of the number of daily service calls.

Having computed the mean, we can complete columns 4, 5, and 6 in the table above. Dividing the sum of the numbers in the last column by the number of observations gives the variance

$$s^2 = \frac{\Sigma(m - \bar{x})^2 f}{n} = \frac{101.40}{60} = 1.69.$$

Therefore the standard deviation of the number of service calls is

$$s = \sqrt{1.69} = 1.3. \quad \blacksquare$$

THE MEAN AND VARIANCE OF RANDOM VARIABLES

In a similar fashion we can construct frequency and probability distributions for any grouped data. In general we take the values of the random variable to be the midpoints $m_1, m_2, \ldots, m_k$ of the class intervals. The frequency and probability distributions for the random variable are constructed as in Figures 7.5 and 7.6 with the bars centered over the class intervals. If the data fall into the intervals with respective frequencies $f_1, f_2, \ldots, f_k$, then the heights of the bars in the frequency distribution represent $f_1, f_2, \ldots, f_k$ and the heights of the bars in the probability distribution represent the corresponding probabilities $p_i = f_i/n$, where $n = \Sigma f$ is the total number of observations.

Using formulas (7.3) and (7.4), we can estimate the mean and variance of the frequency distribution.

$$\bar{x} \approx \frac{\Sigma mf}{n} = \frac{m_1 f_1 + m_2 f_2 + \cdots + m_k f_k}{n}$$
$$= \frac{m_1 f_1}{n} + \frac{m_2 f_2}{n} + \cdots + \frac{m_k f_k}{n}$$
$$= m_1 p_1 + m_2 p_2 + \cdots + m_k p_k$$

$$s^2 \approx \frac{\Sigma(m - \bar{x})^2 f}{n}$$
$$= \frac{(m_1 - \bar{x})^2 f_1 + (m_2 - \bar{x})^2 f_2 + \cdots + (m_k - \bar{x})^2 f_k}{n}$$
$$= \frac{(m_1 - \bar{x})^2 f_1}{n} + \frac{(m_2 - \bar{x})^2 f_2}{n} + \cdots + \frac{(m_k - \bar{x})^2 f_k}{n}$$
$$= (m_1 - \bar{x})^2 p_1 + (m_2 - \bar{x})^2 p_2 + \cdots + (m_k - \bar{x})^2 p_k$$

These calculations express the mean and the variance of the frequency distribution in terms of the probabilities $p_i = f_i/n$ instead of the frequencies f_i. The reader should check that if we compute

$$m_1 p_1 + m_2 p_2 + \cdots + m_k p_k \quad \text{and}$$
$$(m_1 - \bar{x})^2 p_1 + (m_2 - \bar{x})^2 p_2 + \cdots + (m_k - \bar{x})^2 p_k$$

for the data in Table 7.7, then we get exactly the same values for $\bar{x}$ and s^2 that were obtained in Example 7.8.

In view of the above we can generalize the mean and variance to apply to any random variable as follows.

Mean and Variance of a Random Variable

For a random variable taking distinct values $x_1, x_2, \ldots, x_k$ with respective probabilities $p_1, p_2, \ldots, p_k$, the **mean** (or **expected value**) is defined by

$$\mu = \Sigma xp = x_1 p_1 + x_2 p_2 + \cdots + x_k p_k, \tag{7.5}$$

and the **variance** is defined by

$$\sigma^2 = \Sigma(x - \mu)^2 p \tag{7.6}$$
$$= (x_1 - \mu)^2 p_1 + (x_2 - \mu)^2 p_2 + \cdots + (x_k - \mu)^2 p_k.$$

As before, the square root of the variance is called the **standard deviation** and is denoted by σ.*

Recall that for grouped data we take the values of the random variable to be the midpoints $m_1, m_2, \ldots, m_k$ of the class intervals. Thus the calculation above shows that the expected average of the random variable over the long run is

$$m_1p_1 + m_2p_2 + \cdots + m_kp_k.$$

Hence the term "expected value" is also used in place of "mean."

EXAMPLE 7.9 Determine the mean and standard deviation for the number of heads appearing when three fair coins are flipped.

Solution The number of heads appearing when three fair coins are flipped is a random variable that takes the values 0, 1, 2, and 3. From Example 6.5 in Section 6.1 we see that the probabilities of these outcomes are 1/8, 3/8, 3/8, and 1/8, respectively. Hence the probability distribution is as shown in Figure 7.7. As in Example 7.8 we can perform the necessary calculations to determine the mean and variance in a table.

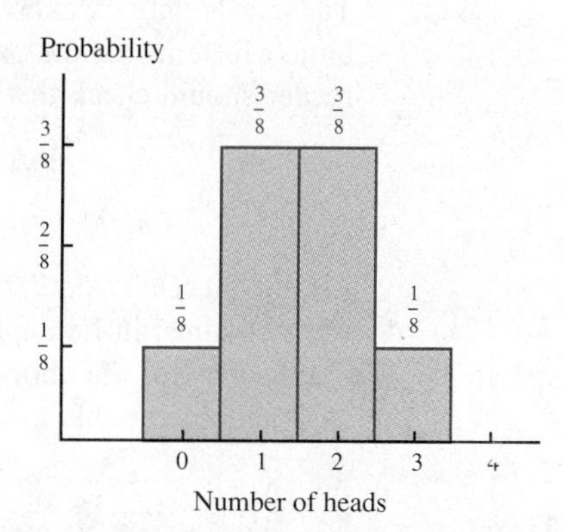

FIGURE 7.7

Number of heads x	Probability p	xp	$x - \mu$	$(x - \mu)^2$	$(x - \mu)^2p$	
0	1/8	0	−3/2	9/4	9/32	
1	3/8	3/8	−1/2	1/4	3/32	
2	3/8	6/8	1/2	1/4	3/32	
3	1/8	3/8	3/2	9/4	9/32	
	1.00	12/8			24/32	Totals

*The symbols μ and σ (the lower case Greek letters mu and sigma) are used to denote the mean and standard deviation of a random variable because these are *theoretical* quantities. We will continue to use the symbols $\bar{x}$ and s to denote the mean and standard deviation of *empirical* data.

The sum of the products xp is the mean, which is

$$\mu = 0\left(\frac{1}{8}\right) + 1\left(\frac{3}{8}\right) + 2\left(\frac{3}{8}\right) + 3\left(\frac{1}{8}\right) = \frac{12}{8} = 1.5.$$

Thus the expected value of the number of heads is 1.5; that is, over the long run we expect the average number of heads appearing to be 1.5. This value agrees with our intuition because we expect heads to appear on half of the flips of a fair coin.

Having computed the expected value, we can complete columns 4, 5, and 6 in the table above. As before, the sum of the numbers in the last column, 24/32 = 0.75, is the variance; so the standard deviation is

$$\sigma = \sqrt{0.75} \approx 0.87. \quad ■$$

EXAMPLE 7.10 Find the expected value and standard deviation for the sum obtained when two fair dice are rolled.

Solution The sum obtained when two fair dice are rolled is a random variable that takes the values 2, 3, 4, 5, 6, 7, 8, 9, 10, 11, and 12. Recall from Section 6.2 that it has the probability distribution shown on the left. This distribution is depicted in Figure 7.8.

Sum	*Probability*
2	1/36
3	2/36
4	3/36
5	4/36
6	5/36
7	6/36
8	5/36
9	4/36
10	3/36
11	2/36
12	1/36

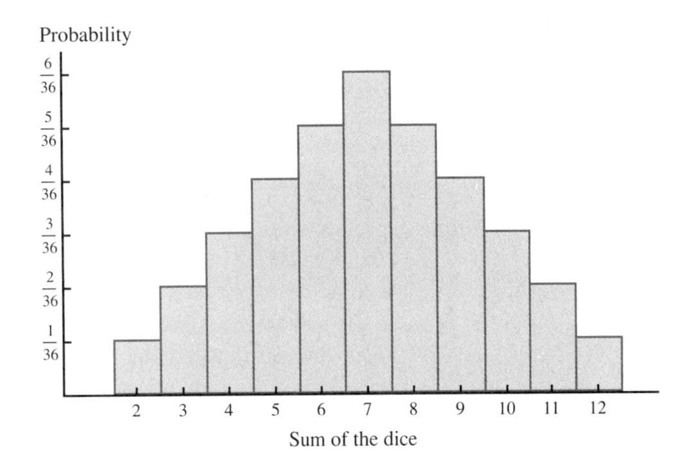

FIGURE 7.8

Thus by formula (7.5) the mean is

$$\mu = \Sigma xp = 2\left(\frac{1}{36}\right) + 3\left(\frac{2}{36}\right) + 4\left(\frac{3}{36}\right) + \cdots + 12\left(\frac{1}{36}\right)$$

$$= \frac{252}{36} = 7.$$

Hence the expected sum is 7. Using formula (7.6), we see that the variance is

$$\sigma^2 = (2-7)^2\left(\frac{1}{36}\right) + (3-7)^2\left(\frac{2}{36}\right) + \cdots + (12-7)^2\left(\frac{1}{36}\right)$$
$$= \frac{210}{36}.$$

Therefore the standard deviation is

$$\sigma = \sqrt{\frac{210}{36}} \approx 2.42. \blacksquare$$

Practice Problem 8 Draw a probability histogram for the number of spots appearing when a fair die is rolled; then determine its mean and standard deviation.

The expected value of a random variable arises naturally in applications involving decision making. We will explore some of these applications in the next section. To close this section, we present several more examples in which the expected value of a random variable occurs.

EXAMPLE 7.11 Kent found a ticket for a raffle in which 1000 tickets were sold. One ticket selected at random will win the first prize of $250, ten tickets will win second prizes of $25 each, and twenty tickets will win third prizes of $10 each. Unfortunately the holder of a winning ticket must be present in order to win, and Kent is about to leave town. Greg has offered to buy Kent's ticket at a fair price. What is a fair price for the ticket?

Solution Greg should pay as much for the ticket as it is expected to win; that is, the "true value" of the ticket is its expected winnings. The random variable that counts the expected winnings of the ticket can take four possible values: $250 if the first prize is won, $25 if a second prize is won, $10 if a third prize is won, and $0 if the ticket does not win any of the prizes. The probability distribution of the random variable is easily seen to be as follows.

Amount won x	*Probability p*	*xp*	
$250	.001	$.250	
$25	.010	$.250	
$10	.020	$.200	
$0	.969	$.000	
	1.00	$.700	Totals

Thus the expected winnings for each ticket are μ = $0.70, and so Greg should pay Kent 70¢ for the ticket. ■

EXAMPLE 7.12 In the raffle described in Example 7.11, the selling price of each ticket was $1. What are the expected net winnings for someone who buys a single ticket?

Solution The random variable that counts the *net* amount won takes four possible values. If the first prize is received, the net amount won is

$$\$250 - \$1 = \$249,$$

the prize money received minus the cost of the ticket. Likewise the net amount won is $24 if a second prize is received, $9 if a third prize is received, and −$1 if no money is received. Thus the probability distribution for this random variable is as shown in the first two columns of the table below.

Amount won x	*Probability p*	*xp*	
$249	.001	.249	
$24	.010	.240	
$9	.020	.180	
−$1	.969	−.969	
	1.00	−.300	Totals

Thus the expected winnings for each ticket are $\mu = -\$0.30$; in other words, each ticket is expected to *lose* 30¢. Hence the organization holding the raffle will gain an average of 30¢ on each of the 1000 tickets sold, for a net gain of $300. Note that this gain is the difference between the revenue derived from the sale of the tickets, $1000, and the prize money paid out,

$$\$250 + 10(\$25) + 20(\$10) = \$700. \quad \blacksquare$$

EXAMPLE 7.13 An insurance company charges $100 for a life insurance policy that pays $30,000 if a woman 25 years of age dies during the coming year. Mortality tables indicate the probability that a woman of age 25 will survive one more year is .998. What is the company's expected gain on such a policy?

Solution If a woman with this type of policy survives the year, then the company has a net gain of $100, the premium of the policy. If she dies, however, the company keeps the premium but pays out benefits of $30,000, for a net gain of

$$\$100 - \$30,000 = -\$29,900.$$

Thus the random variable that measures the company's gain takes on two values, $100 and −$29,900. The probability distribution for this random variable is as follows.

Gain	*Probability*
$100	.998
−$29,900	.002

Thus the company's expected net gain is

$$\begin{aligned}\mu &= (\$100)(.998) + (-\$29{,}900)(.002) \\ &= \$99.80 - \$59.80 = \$40.\end{aligned}$$

Hence over the long run the company will gain $40 on each of these policies that it sells. ■

Practice Problem 9 Barring complications, a contractor anticipates a $50,000 profit on a certain construction job. Severe winter weather, however, may raise his costs $20,000 more than expected (and lower his profits accordingly). If the probability of a severe winter is .4, what is the contractor's expected profit for this job?

EXERCISES 7.3

1. An insurance company checked the records of 80 motorists chosen at random to determine how many moving violations each had been cited for during the past five years. The results are shown below.

Violations	*Frequency*	
0	34	
1	28	
2	12	
3	4	
4	2	
	80	Total

(a) Construct the probability distribution for the number of moving violations.
(b) Find the mean and standard deviation for the number of moving violations using the probability distribution from part (a).

2. A Labor Department statistician interviewed 120 unemployed persons in an inner-city area to determine how many years of high school each had completed. Their responses are shown below.

Number of years	*Frequency*	
0	6	
1	24	
2	39	
3	30	
4	21	
	120	Total

(a) Construct the probability distribution for the number of years of high school completed.
(b) Find the mean and standard deviation for the number of years of high school completed using the probability distribution from part (a).

3. A test of the tensile strength (breaking point under stress) of a sample of 80 cables produced the frequency distribution shown at the top of the next page.

Tensile strength (lbs)	*Number of cables*	
1400–1500	11	
1500–1600	29	
1600–1700	26	
1700–1800	9	
1800–1900	5	
	80	Total

(a) Construct the probability distribution for the tensile strength of the cables.
(b) Find the mean and standard deviation for the tensile strength of the cables using the probability distribution from part (a).

4. The speeds of 100 cars selected at random from I–74 were distributed as shown below.

Speed (mph)	*Cars*	
40–50	15	
50–60	31	
60–70	35	
70–80	17	
80–90	2	
	100	Total

(a) Construct the probability distribution for the speed of the cars.
(b) Find the mean and standard deviation for the speed of the cars using the probability distribution from part (a).

5. A stockbroker observes that the delivery time for her local mailings is a random variable with the distribution below.

Delivery time (working days)	*Probability*
1	.218
2	.318
3	.272
4	.130
5	.062

Find the mean and standard deviation for the delivery time of the stockbroker's mailings.

6. A baker finds that the number of chocolate cakes requested on a given day has the following distribution.

Requests	*Probability*
0	.11
1	.20
2	.32
3	.19
4	.12
5	.05
6	.01

Find the mean and standard deviation for the number of chocolate cakes requested on a given day.

7. In a particular lottery, 161 winning tickets will be randomly selected from among the 250,000 tickets sold. One ticket will win \$10,000, ten tickets will win \$1,000, 50 tickets will win \$200, and 100 tickets will win \$100.

   (a) What are the expected winnings for each ticket?
   (b) If each ticket cost \$1, what is the expected *net* gain for someone who bought a single ticket?

8. A raffle offers a first prize of \$2000, 2 second prizes of \$500, and 10 third prizes of \$100. There were 10,000 tickets sold.

   (a) What are the expected winnings for each ticket?
   (b) If each ticket cost \$1, what is the expected *net* gain for someone who bought a single ticket?

9. A bank teller's cash drawer contains 100 one-dollar bills, 50 five-dollar bills, 50 ten-dollar bills, and 25 twenty-dollar bills. What is the expected value of a bill chosen at random from the drawer?

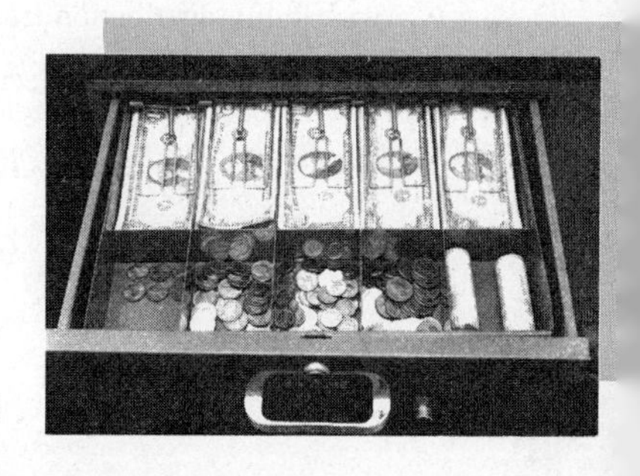

10. American roulette wheels have 38 slots numbered 00, 0, 1, 2, . . . , 36, all of which are equally likely to occur. A person who bets \$1 on a given number wins \$35 and receives his bet back if that number occurs and loses his bet otherwise. What is the expected value of such a bet?

11. An oil company has found that the cost of drilling for oil on a new site is \$5 million and that 10% of the sites that it drills yield oil. When oil is found, a site averages \$75 million in revenues. What is the expected net profit from a new site?

12. An investor has bought a tract of land for \$50,000. There is a 20% chance that an airport will be built nearby, increasing the value of the property to \$250,000. If the airport authority decides not to locate there, however, the value of the land will drop to \$35,000. What is the investor's expected net gain?

13. A business venture is expected to earn a \$250,000 profit if successful and to lose \$100,000 if unsuccessful. If the probability is .4 that this venture is successful, what is its expected value?

14. A farmer expects 110 bushels per acre of corn under normal weather conditions, 125 bushels per acre under excellent conditions, and 75 bushels per acre under poor conditions. If the probabilities of normal weather, excellent weather, and poor weather are .6, .1, and .3, respectively, what is the farmer's expected yield per acre?

15. An insurance company is issuing a new homeowner's policy. Based on their experience with similar policies, the company expects that 1.2% of their policyholders will file a claim each year, and the average amount paid on each claim will be \$12,000. How much should the company charge for such a policy in order to earn a \$50 profit from each policy sold?

16. A government agency has loaned \$25,000 to start a new business. The loan is to be repaid in one year at 10% interest, but there is a 15% chance that the borrower will default, resulting in a \$25,000 loss for the agency. What is the agency's expected net gain from such a loan?

17. A basketball player makes her first free throw 80% of the time. If she makes her first shot, her accuracy on the second shot increases to 90%. However if she misses her first shot, her accuracy on the second shot decreases to 70%. What is the expected number of free throws made when she shoots two shots?

18. An urn contains three blue and seven green marbles. Al will choose two marbles at random from the urn. If they are the same color, Bob will pay him \$7; but if they are different colors, Al will pay Bob \$8. What are Al's expected winnings in this game?

19. An urn contains two red marbles and three white marbles. Marbles are chosen at random from the urn, one after another, until a red marble is selected.
(a) Find the probability distribution for the number of marbles chosen.
(b) Find the mean and standard deviation of the number of marbles chosen.

20. Doug has agreed to pay Jennifer for the right to flip a fair coin until either one head or three tails occur. If heads appears on the first, second, or third flip, Jennifer will pay Doug \$1, \$2, or \$4, respectively; but if three tails occur, Doug will win nothing. How much should Doug pay to play this game?

21.* Three prize winners are to be randomly chosen from among six women and two men. What is the expected number of men selected?

22.* A bargaining committee is to be appointed from among six union members and three nonunion workers. If the committee is to consist of four of these persons chosen at random, what is the expected number of union members on the committee?

23.* A university committee of four faculty members and three students intends to form a three-member subcommittee. If the members of the subcommittee are chosen at random, what is the expected number of faculty on the subcommittee?

24.* A planning committee of three is to be selected at random from among the members of the city council. If there are nine Republicans and six Democrats on the city council, what is the expected number of Democrats on the subcommittee?

25. If a random variable takes values $x_1, x_2, \ldots, x_k$ with respective probabilities $p_1, p_2, \ldots, p_k$, prove that its variance equals

$$\Sigma x^2 p - (\Sigma xp)^2.$$

26. Let μ and σ^2 denote the mean and variance of a random variable X. Show that for any constant k, the mean and variance of kX are $k\mu$ and $k^2\sigma^2$, respectively.

27. Let μ and σ^2 denote the mean and variance of a random variable X. Show that for any constant k, the mean and variance of $X + k$ are $\mu + k$ and σ^2, respectively.

Answers to Practice Problems **8.** The probability histogram is shown below; $\mu = 3.5$ and $\sigma = \sqrt{\frac{70}{24}} \approx 1.71$.

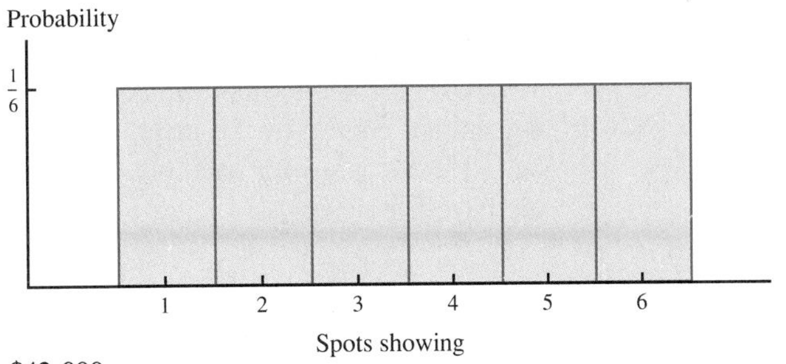

9. \$42,000

*This exercise requires use of the counting techniques in Chapter 5.

MATHEMATICS IN ACTION

7.4 Decision Theory

In earlier chapters we have seen how mathematical tools such as linear programming and matrix algebra can help us make optimal decisions in a wide range of practical problems. In this section we will use the techniques of probability theory to compare the different options available to a decision-maker.

DECISIONS UNDER UNCERTAINTY

Making a decision amounts to choosing among various possible courses of action. When the outcome of each action can be predicted exactly, we say that we are operating under *conditions of certainty*. In such a case the best decision is simply the one that produces the most desirable outcome, for example, the greatest profit or the least cost.

More often than not, however, we must operate under *conditions of uncertainty,* where the outcome of each action depends in part on factors beyond our control. To model such a situation, we assume that the decision-maker has available a number of possible actions and that one of various possible random events called **states of nature** will occur independently of the action taken. For each state of nature, the consequences of each action can be summarized in a **payoff table,** the entries of which show the decision-maker's net gain.

Consider, for example, an inventor who has patented a new, fuel-efficient carburetor. She would like to produce the device herself, but this is a costly gamble; for if the device is eventually adopted by a major automaker, she will make a profit of $2,000,000, but otherwise she will lose $500,000. A company has offered to finance the project for 70% of any profits; in this case the inventor will either earn $600,000 or break even. Another company has offered to buy the inventor's patent outright for $100,000. What is the inventor's best course of action?

In this situation the inventor has three options available:

1. produce the device herself,
2. retain a 30% interest in the device, or
3. sell the patent.

Her net gain depends on whether or not the device is adopted by a major automaker. Note that the decision as to whether or not the device is adopted is beyond the control of the inventor; these are the states of nature for this problem. The net gains for each option under each state of nature are the entries of the payoff table below.

	State of Nature	
Action	*Adopted*	*Not adopted*
Produce device	$2,000,000	−$500,000
Retain 30%	$600,000	$0
Sell patent	$100,000	$100,000

If the inventor knew that her device would be adopted, she would obviously do best to produce it herself to earn a payoff of $2,000,000. If, on the other hand, she knew that the device would not be adopted, her best course of action would be to sell the patent to earn a payoff of $100,000. Unfortunately her decision must be made under conditions of uncertainty because she does not know whether or not the device will be adopted, that is, she does not know the true state of nature. In this type of situation there are several different criteria for selecting the "best" course of action.

1. The **maximin criterion** chooses the action that *maxi*mizes the *mini*mum possible payoff.

In our example the first action has a minimum payoff of −$500,000, the second action has a minimum payoff of $0, and the third action has a minimum payoff of $100,000. Thus the third action (selling the patent) maximizes the minimum possible payoff, and so this is the action that the inventor will choose if she uses the maximin criterion. This strategy reflects a conservative or pessimistic attitude because it chooses the action that involves the least risk, that is, the action that offers the greatest payoff assuming that the worst happens.

2. The **maximax criterion** chooses the action that *maxi*mizes the *max*imum possible payoff.

In contrast to the maximin criterion, the maximax criterion is an adventurous or optimistic strategy; it chooses the action that yields the greatest payoff assuming that the best will happen. In our example the maximum possible payoffs for the three actions are $2,000,000, $600,000, and $100,000. Thus if the inventor uses the maximax criterion, she will choose the first action (producing the device herself) because this option offers the greatest opportunity for gain.

Since the state of nature is determined by external factors, a rational decision strategy should take into account all available information. If we can estimate the probability of each state of nature, then we can compare the options using the notion of expected value. This leads to a third criterion for selecting the "best" course of action.

3. **Bayes's criterion** chooses the action that maximizes the *expected* payoff.

Suppose that the inventor in our example estimates that there is a 25% chance of her device being adopted by a major automaker. Then the expected payoffs for her three actions are as follows.

$$
\begin{aligned}
E_1 &= \$2{,}000{,}000(.25) + (-\$500{,}000)(.75) &&= \$125{,}000 \\
E_2 &= \$600{,}000(.25) + \$0(.75) &&= \$150{,}000 \\
E_3 &= \$100{,}000(.25) + (\$100{,}000)(.75) &&= \$100{,}000
\end{aligned}
$$

Thus if the inventor uses Bayes's criterion, she will select the second option (retaining a 30% interest in her device) because this offers the greatest expected payoff of $150,000.

Notice that in this example the three decision criteria (maximin, maximax, and Bayes's) lead to three different decisions. Thus there is no single answer as to which course of action is "best." We will return to this matter shortly, but first let us look at another example.

EXAMPLE 7.14 A Houston-based firm keeps a number of trained crews on call to put out oil fires anywhere in the world. The number of calls it receives has been observed to vary between 0 and 5 per week according to the following distribution.

Number of calls	*Probability*
0	.1
1	.1
2	.2
3	.3
4	.2
5	.1

The firm charges $50,000 for its services. Each crew can handle one call per week and is paid a weekly salary of $10,000 whether it is used or not. Use Bayes's criterion to determine the optimal number of crews to keep on call.

Solution The firm's net profit depends both on the action it takes (deciding how many crews to keep on call) and the state of nature (the number of calls received). For instance, if the firm has 4 crews on call and 3 calls are received, its profit is

$$\begin{aligned} \text{Profit} &= \text{Revenue} - \text{Cost} \\ &= 3(\$50{,}000) - 4(\$10{,}000) \\ &= \$150{,}000 - \$40{,}000 \\ &= \$110{,}000. \end{aligned}$$

And if there are 4 crews on call and 4 calls are received, the firm's profit is

$$\begin{aligned} \text{Profit} &= \text{Revenue} - \text{Cost} \\ &= 4(\$50{,}000) - 4(\$10{,}000) \\ &= \$200{,}000 - \$40{,}000 \\ &= \$160{,}000. \end{aligned}$$

In this case, the firm's profit is $160,000 if 4 or 5 calls are received, because if there are only 4 crews available, only 4 calls can be accepted. Similar calculations show that the payoff table for this firm is as shown below, where the net profit is given in thousands of dollars per week.

Action: Number of crews	*State of Nature: Number of calls per week*					
	0	*1*	*2*	*3*	*4*	*5*
1	−10	40	40	40	40	40
2	−20	30	80	80	80	80
3	−30	20	70	120	120	120
4	−40	10	60	110	160	160
5	−50	0	50	100	150	200

Using this table and the given probability distribution for the number of calls per week, we can compute the expected profit for each action. For instance, with 4 crews on call the expected profit is

$$-40(.1) + 10(.1) + 60(.2) + 110(.3) + 160(.2) + 160(.1) = 90.$$

Hence over the long run the firm can expect an average weekly profit of \$90,000 by keeping 4 crews on call. Similar calculations produce the expected profits for the other courses of action; these are summarized in the table below.

Number of crews	*Expected profit (in thousands)*
1	\$35
2	\$65
3	\$85
4	\$90
5	\$85

Since the maximum expected profit is \$90,000 when there are 4 crews on call, 4 is the optimal number of crews according to Bayes's criterion. This is a somewhat surprising result because the expected number of calls per week is only 2.7, as the reader should check.

Observe that if we use the maximin criterion instead of Bayes's criterion, then the optimal number of crews is 1 since this gives the maximum (−\$10,000) of the minimum profits for each action. And if the maximax criterion is used, we would keep 5 crews on call since that gives the maximum (\$200,000) of the maximum possible profits. ■

Practice Problem 10 A farmer delivers eggs each Friday to his wholesale and retail customers in the city. From experience he knows that he will sell no fewer than 2 truckloads and no more than 5 truckloads each week; in fact, the probabilities that exactly 2, 3, 4, and 5 truckloads of eggs can be sold in any given week are .2, .4, .3, and .1, respectively. For each truckload that is sold, the farmer receives a profit of \$150, but each truckload that is not sold spoils with a resulting loss of \$50.

(a) Prepare a payoff table for this problem.

Continuing from part (a), determine the optimal number of truckloads of eggs that the farmer should send to the city each week using

(b) the maximin criterion,
(c) the maximax criterion, and
(d) Bayes's criterion.

What criterion should a rational decision-maker use in a particular situation? When relatively small amounts are at stake, Bayes's criterion makes the most sense because we know that it gives the greatest average payoff over the long run. In Example 7.14, for instance, the fire-fighting firm is not especially concerned with its profit for any single week but rather with the long-term profit if it repeatedly follows a particular course of action. However, in the case of the inventor discussed earlier in this section, it can be argued that the situation is different; here the decision-maker must decide how best to exploit her new invention, a nonrepeatable decision in which there is much at stake. If the inventor needs $80,000 immediately to pay for surgery, she may decide to take the conservative course (selling the patent) in order to be assured of $100,000 no matter what happens. On the other hand, if she is on the verge of bankruptcy and needs $800,000 to survive, she may well decide that there is no alternative but to hope for the largest possible payoff by producing the device herself.

These are not irrational decisions; they merely reflect the different utility of money in different situations. That is, *the utility that a person assigns to a possible gain or loss may not equal its monetary value*. For example, except for the very wealthy, most people agree that "the second $1000 gained is worth less than the first"; in other words, a gain of $2000 is not quite twice as valuable as a gain of $1000. Conversely, a *loss* of $2000 is *more* than twice as painful as a loss of $1000. Thus, given a choice between receiving an outright gift of $10,000 or receiving $20,000 if a fair coin lands heads and nothing otherwise, most people would choose the first option, even though the expected value is the same in each case. Likewise most people prefer a certain loss of $10 to a .001 probability of losing $10,000; indeed, if this were not so, the insurance industry would cease to exist. The average person who occasionally buys a lottery ticket or visits a casino probably feels that the most likely outcome, a small net loss, is more than compensated for by the possibility, however remote, of big winnings. Similarly the individual who buys an insurance policy is not interested in the average return on his investment, but rather in being protected against the worse possible case. In both of these illustrations the person's expected utility may be positive even though the expected net gain is negative.

There have been various attempts to measure subjective utilities, that is, how much a particular gain or loss is "really worth" to a given individual.

The interested reader should see reference [3], in which both the practical and theoretical aspects of utility theory are discussed.

In the case of a private or public corporation with large assets, it is obviously in the best long-term interest of the institution to use Bayes's criterion in making investment decisions. Although some investments will fail and some will succeed, the *overall* net profit will be maximized. Some interesting studies have shown, however, that corporate managers tend to show excessive risk aversion in their decisions, preferring safe courses of action to those promising the greatest expected gain. (See, for instance, reference [4].) The reason for this phenomenon lies in the psychology of the individual, who worries—often justifiably—that his or her performance will be evaluated on the basis of the success or failure of a few key decisions rather than on their inherent rationality. This may explain why the more bureaucratic such institutions become, the more conservative their decisions tend to be, with the result that millions of dollars in potential profits are lost!

DECISION TREES

Often a decision-maker must play a cat-and-mouse game against nature: The decision-maker makes a move and then nature responds, creating a new situation requiring a decision. Such sequential decision problems can be quite involved, but with the aid of a diagram called a **decision tree** they can be systematically analyzed.

Let us consider the case of an aircraft company that is bidding on a contract worth \$50 million. One of the company's directors wants to build a \$10 million prototype immediately because this will give the company a 50% chance of winning the contract at the preliminary stage of the bidding. Another director wants to submit the bid without the prototype, which will give the company a 20% chance of success at the preliminary stage. If this first-stage bid is rejected, the company will still have the option of building a prototype on a rush basis for \$15 million, which will restore a 50% chance of success in the final stage of the bidding. What decision strategy maximizes the company's expected net profit?

The decision tree for this problem is shown in Figure 7.9. The points where branching occurs are called **nodes.** Those points where a decision must be made (nodes A and D) are called **choice-nodes** and are denoted by squares, and those points where nature takes one of several states (B, C, and E) are called **chance-nodes** and are denoted by circles. This tree branches from left to right, each branch finally terminating in an **end-node** (heavy dot).

The initial choice-node A represents the decision whether or not to build the prototype immediately. The decision to build it leads to chance-node B, which branches according to whether the bid is accepted (with probability .5) or rejected (also with probability .5). If the bid is accepted, the company's payoff (in millions of dollars) will be $50 - 10 = 40$, the value of the contract

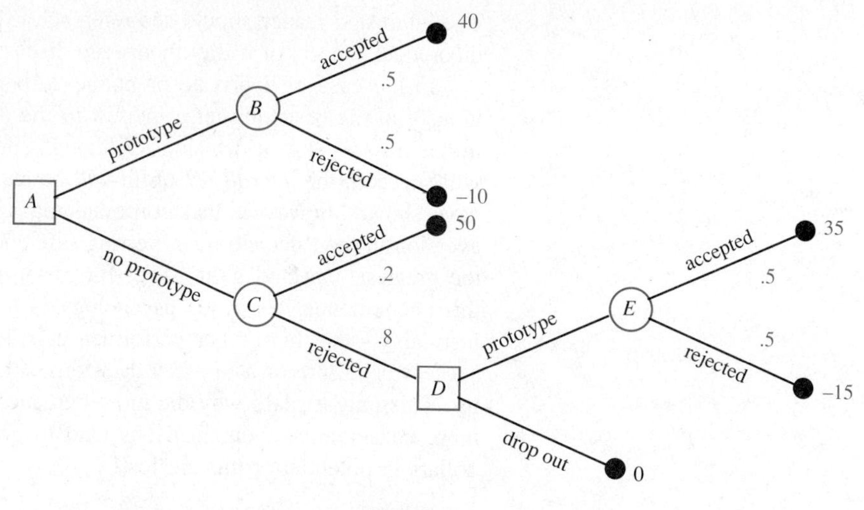

FIGURE 7.9

minus the cost of the prototype. If the bid is rejected, then the company will lose the cost of the prototype, for a payoff of -10.

On the other hand, if the prototype is not built immediately, we arrive at chance-node C, which branches according to whether the bid is accepted (with probability .2) or rejected (with probability .8). If the bid is accepted, the payoff is 50; if not, we must decide at choice-node D whether to build a rush prototype or drop out of the bidding and take a payoff of 0. If the rush prototype is built (chance-node E), the bid will either be accepted at the final stage (with a probability of .5 and a payoff of $50 - 15 = 35$) or rejected (with a probability of .5 and a payoff of -15).

Thus Figure 7.9 represents the entire sequential process described in our example. Note that each end-node is labeled with a numerical payoff denoting the company's net profit if that branch of the tree is followed. In order to compare the different strategies available, we work through the tree from right-to-left, assigning a number to each node according to the rules below. The number assigned to a node represents the maximum expected payoff at that point of the decision process.

Rules for Assigning Numbers to the Nodes in a Decision Tree

1. The number assigned to an end-node is the corresponding payoff.
2. The number assigned to a chance-node is the *expected value* of the numbers assigned to the nodes to which it branches.
3. The number assigned to a choice-node is the *maximum* of the numbers assigned to the nodes to which it branches.

In our example, this computation proceeds as follows. (Refer to Figure 7.9. You may wish to label each node with its number as it is assigned.) First, we assign the corresponding payoffs to all of the end-nodes; these numbers are already indicated in Figure 7.9. Second, we assign a number to node E. Since this is a chance-node, we use rule 2 above to assign to node E the expected value of the two end-nodes to which it branches, which is

$$35(.5) + (-15)(.5) = 10.$$

This number represents the expected payoff at this point of the decision problem.

Third, we assign a number to choice-node D in accordance with rule 3. Since node E (building the rush prototype) has been assigned the number 10 and dropping out of the bidding has been assigned the number 0, we assign to D the number 10. Note that at node D we must choose between building the rush prototype and dropping out of the bidding. Since node E (building the rush prototype) has an expected profit of 10 and dropping out of the bidding has an expected profit of 0, we obviously will choose to build the prototype, resulting in an expected profit of 10.

Fourth, we use rule 2 to assign node C the expected value of the numbers at the nodes to which it branches, which is

$$50(.2) + 10(.8) = 18.$$

Fifth, we use rule 2 to assign node B the number

$$40(.5) + (-10)(.5) = 15.$$

Finally, we assign choice-node A the larger of the numbers 15 (at node B) and 18 (at node C) in accordance with rule 3. Thus node A is assigned the number 18. This number represents the company's maximum expected profit (in millions of dollars), which it will realize by initially submitting the bid without the prototype and building a rush prototype later if necessary.

The sort of analysis used in the example above is called **Bayesian analysis** because it is based on Bayes's criterion. It provides the most rational decision method for an institution that wishes to maximize its long-term profits when facing many decisions of this type.

EXAMPLE 7.15 The county agricultural extension service has advised a farmer to apply a pesticide costing \$20 per acre to his soybean fields. Depending on the weather, this pesticide might have no effect on the soybean yield per acre, a 10% increase in yield, or a 15% increase in yield. The probability of each effect is shown below.

	Increased Yield		
Weather	*0%*	*10%*	*15%*
Dry	.1	.4	.5
Normal	.2	.6	.2
Wet	.3	.5	.2

Based on experience, the farmer can expect the following probabilities, yields, and soybean prices for the different weather conditions.

Weather	*Probability*	*Expected yield (bushels per acre)*	*Selling price per bushel*
Dry	.3	48	$7.00
Normal	.6	55	$6.20
Wet	.1	60	$5.80

Use a decision tree to decide whether or not the farmer should apply the pesticide.

Solution The farmer has only one decision to make—whether to use the pesticide or not. Thus the decision tree for this problem has only one choice-node, which is node A. (See Figure 7.10.) If the farmer does not use the pesticide, we arrive at the chance-node F. Here there are three branches according to the type of weather, each leading to an end-node. Things are more complicated, however, if the farmer elects to apply the pesticide. Using the pesticide leads to chance-node B, where there are again three branches according to the possible weather conditions. These branches lead to chance-nodes C, D, and E, each of which branches according to the three possible effects from using the pesticide (0% increase in yield, 10% increase in yield, and 15% increase in yield). These branches all terminate at end-nodes.

Computing the payoffs at the end-nodes requires care. If the pesticide is not applied, then the payoff is simply the expected yield for each weather condition multiplied by the corresponding price per bushel. Thus under dry conditions the payoff will be

$$48(7.00) = 336.00$$

dollars per acre. Likewise the payoff under normal conditions is

$$55(6.20) = 341.00$$

dollars per acre, and the payoff under wet conditions is

$$60(5.80) = 348.00$$

dollars per acre.

When the pesticide is used, we must perform a similar calculation in which the yield per acre is adjusted by the appropriate factor to account for the increase in yield. For example, if there is dry weather and the pesticide increases yields by 10%, then the yield per acre is

$$1.10(48) = 52.8$$

bushels per acre. Hence the return per acre in this case will be

$$52.8(7.00) = 369.60.$$

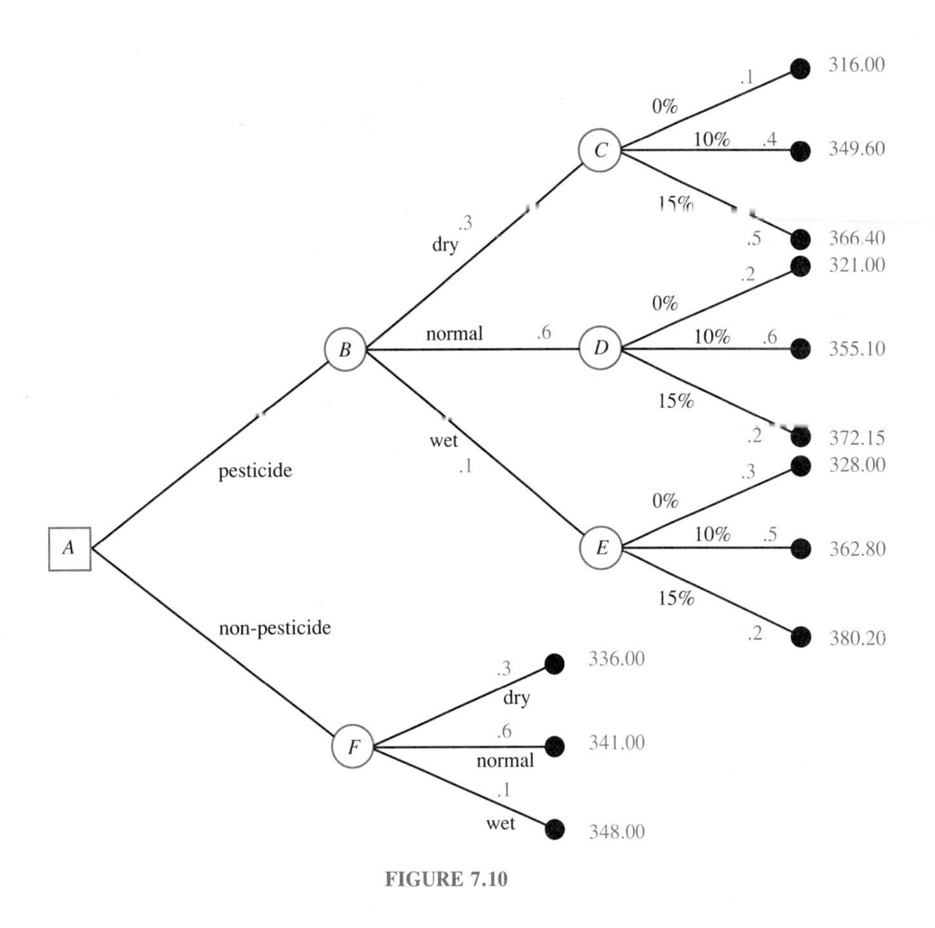

FIGURE 7.10

Finally, to obtain the payoff in dry weather when the pesticide increases yields by 10%, we must subtract the pesticide's cost per acre ($20.00) from the return per acre:

$$369.60 - 20.00 = 349.60.$$

Check that the payoffs for the other end-nodes are as shown in Figure 7.10.

We now assign numbers to the nodes using the rules on page 404. To chance-nodes C, D, E, and F, we assign the expected values of the numbers assigned to the end-nodes to which they branch; these expected values are 354.64, 351.69, 355.84, and 340.20, respectively. To chance-node B, we assign the expected value of the numbers assigned to nodes C, D, and E:

$$.3(354.64) + .6(351.69) + .1(355.84) = 352.99.$$

Finally, we assign to choice-node A the maximum of the numbers assigned to nodes B and F, which is $352.99. Thus the farmer's maximum expected gain is $352.99 per acre, and it is obtained by using the recommended pesticide. ■

Practice Problem 11 A contractor will receive \$200,000 in profits for completing a particular construction project. If the project is completed a month early, he will earn an additional \$40,000 bonus. However, the only way in which the project can be completed early is to begin in the fall instead of waiting until spring, and if the project is started in the fall, it faces an 80% risk of bad weather. If there is good fall weather, the project will surely be completed early, but bad weather will cost the contractor an additional \$20,000 in expenses as well as delaying the project. In this event, the project cannot be finished early unless the contractor chooses to pay his workers an additional \$10,000 for overtime, which will give him a 40% chance of finishing early.

(a) Represent this decision process by a decision tree.
(b) Determine the maximum expected payoff at each node.
(c) What strategy should the contractor use to maximize his expected profit?

EXERCISES 7.4

1. Consider the payoff table below.

	State of Nature			
Action	s_1	s_2	s_3	s_4
a_1	20	25	−10	0
a_2	30	15	30	−20
a_3	40	0	50	0

If the probabilities of the four states of nature are .3, .4, .2, and .1, respectively, determine the best course of action using (a) the maximin criterion; (b) the maximax criterion; and (c) Bayes's criterion.

2. Consider the payoff table below.

	State of Nature		
Action	s_1	s_2	s_3
a_1	10	15	25
a_2	−20	30	0
a_3	25	10	−10
a_4	−10	15	45

If the probabilities of the three states of nature are .2, .7, and .1, respectively, determine the best course of action using (a) the maximin criterion; (b) the maximax criterion; and (c) Bayes's criterion.

3. The owner of a newsstand gets a few requests each day for out-of-town papers. One of them, the Baltimore *Sun,* costs the dealer 40¢ and sells for 90¢ per copy. Unsold copies cannot be returned and therefore represent a loss for the dealer. The number of daily requests for the *Sun* has been observed to vary as follows.

Number requested	*Probability*
0	.1
1	.2
2	.3
3	.3
4	.1

How many copies should the dealer stock to maximize his long-term profit?

4. A fruit grower delivers peaches to a farmers market each Friday. From experience he knows that he can sell between 1 and 4 truckloads each week, as follows.

Number of truckloads	*Probability*
1	.4
2	.3
3	.2
4	.1

For each truckload of peaches that is sold, the grower will realize a profit of $200, but because of spoilage he will lose $80 for each truckload that is not sold. Under these conditions, how many truckloads should the grower send to the farmers market each week in order to maximize his long-term profit?

5. A fish market observes that the daily demand for fresh salmon varies between 5 and 9 pounds. In fact, the probabilities of selling exactly 5, 6, 7, 8, and 9 pounds of salmon per day are .1, .2, .4, .2, and .1, respectively. The salmon, which sells for $6 per pound, must be ordered the day before at a cost of $3 per pound. The dealer can sell day-old salmon to a local manufacturer of cat food for $2 per pound, and so unsold fish result in a $1 per pound loss to the market. How many pounds of salmon should be ordered each day in order to maximize the market's long-term profit?

6. A baker can sell as many as 3 birthday cakes per day besides those specially ordered. From experience he knows that the probabilities of selling 0, 1, 2, and 3 cakes are .1, .4, .3, and .2, respectively. If each cake that is sold results in a $3 profit and each cake that is not sold results in an $0.80 loss, how many cakes should the baker make in order to maximize his long-term profit?

7. A carnival operator is worried about the possibility of weekend rain. If he cancels the carnival, he will lose $2000 in promotional expenses. On the other hand, if he sets up his equipment, he will make $12,000 if it doesn't rain but will lose $8000 if it does. Assume that he applies Bayes's criterion to decide the best course of action.

(a) What should he do if the chance of rain is 65%?

(b) How great must the chance of rain be in order for the operator to decide to cancel the carnival?

8. An insurance company sells single-flight airplane accident policies in major airports. A $10,000 policy costs $10, and a $20,000 policy costs $15. Let p be the probability of an airplane accident.

(a) If $p = .0003$, which of the two policies is better for the company?

(b) What must the value of p be in order for the company's expected gain to be the same for both policies?

In Exercises 9–10, find the value of each node in the decision tree and determine the expected payoff if the decision-maker follows an optimal strategy.

9. **10.**

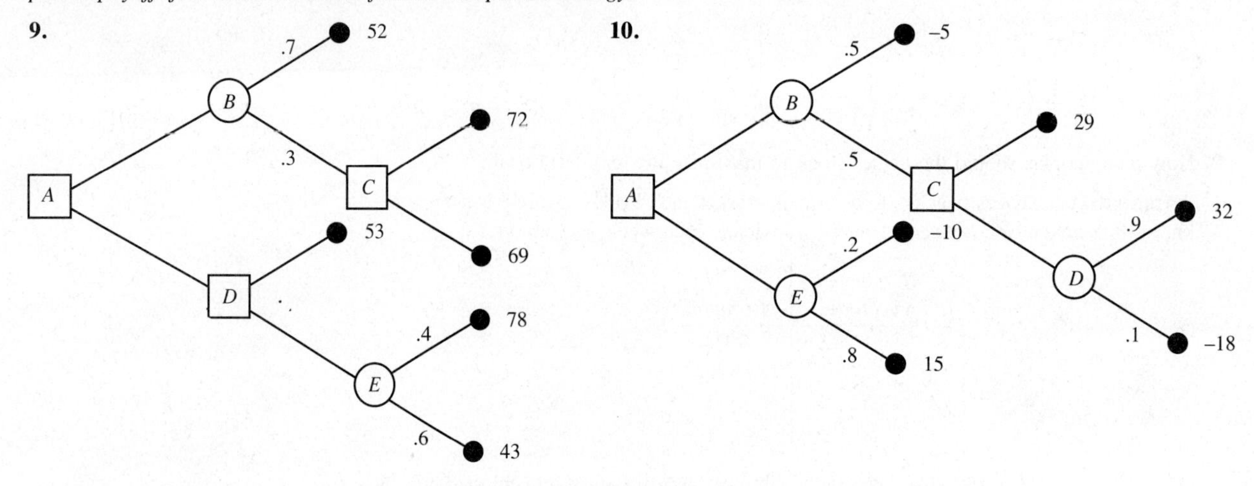

11. A special interest group is trying to decide whether or not to hire a lobbyist at a cost of \$250,000 to help pass a key piece of legislation. Hiring the lobbyist increases the bill's chance of passage from 20% to 30%, and passage of the bill is worth \$3 million in revenue to the group. What is the group's best course of action, and what is its expected net gain?

12. The director of the San Francisco Opera must decide whether or not to hire a famous tenor for the new season. If the tenor is not hired, the director expects a \$2 million profit for the company. On the other hand, if the tenor is hired, the company's profit should increase by \$800,000. In this case, however, there is a 50% chance that the company's star soprano will become temperamental and threaten to quit. If so, the director will either have to give her a \$500,000 increase in salary (which reduces profits by this amount) or fire her, in which case the net profits will be only \$1,900,000. What decision strategy maximizes the company's expected net profit?

13. Elaine often has to park downtown for an hour. She can pay 85¢ at a parking lot or look for a space on the street, which she finds 60% of the time. If she finds a space she can either put 25¢ in the meter or take a 5% risk of getting a \$10 ticket. If she can't find a space on the street, she must return to the parking lot, having wasted 50¢ worth of gasoline. What decision strategy minimizes her expected loss?

14. A homeowner is asking \$100,000 for his house, which he must sell within two weeks. A buyer has made what she claims is her final offer of \$80,000. The seller can either accept this offer or make a counteroffer of \$90,000. If he makes the counteroffer, the chances are even that it will be accepted. If it is not, there is an 80% chance that the buyer will lose interest in the house and a 20% chance that the buyer will propose a compromise offer of \$85,000, which the seller is prepared to accept. If the buyer loses interest in the house, however, the seller will have to accept a \$75,000 offer from another party. What should the seller do to maximize his expected gain from the sale?

15. Mrs. Martin is considering whether or not to file a civil suit against a local roofing company. If she files the suit, there is a 50% chance that she will be offered \$2000 to settle her claim out of court. If no offer is made, or if the offer is made and refused, the case will go to trial, in which case Mrs. Martin has a 70% chance of winning her full $4000 claim. If she loses in court, however, she will have to pay $3000 in court costs. What decision strategy should Mrs. Martin adopt in order to maximize her expected gain?

16. An oil company is trying to decide whether to do shallow or deep drilling at a new site. Shallow drilling costs $5 million and has a 12% chance of finding oil, and deep drilling costs $15 million and has a 23% chance of finding oil. If shallow drilling is unsuccessful, the company can either abandon the project or undertake deep drilling at an extra cost of $15 million, but the chance of finding oil in this case is only 12.5%. If oil is found at a site, the company will realize $90 million in revenues. What decision strategy maximizes the company's expected net profit?

Consider the payoff table below, where the states of nature have respective probabilities of .5, .3, and .2.

Action	*State of Nature* s_1	s_2	s_3	*Expected payoff*
a_1	$4000	$7000	−$1000	$3900
a_2	$5000	$4000	$3000	$4300
a_3	$6000	−$2000	$4000	$3200

According to Bayes's criterion, the best action is a_2 with an expected payoff of $4300. But suppose that the decision-maker knew in advance which state of nature would occur and acted accordingly to maximize the payoff. The expected gain in this case would be

$$\$6000(.5) + \$7000(.3) + \$4000(.2) = \$5900$$

because the maximum payoffs for the three states of nature are $6000, $7000, and $4000, respectively. The difference

$$\$5900 - \$4300 = \$1600$$

is called the ***value of perfect information.*** *This represents the maximum amount that the decision-maker should pay for information such as market forecasts or opinion polls that can be used to help choose an action.*

17. Find the value of perfect information in Exercise 1.

18. Find the value of perfect information in Exercise 2.

19. Find the value of perfect information in Exercise 3.

20. Find the value of perfect information in Exercise 7(a).

Answers to Practice Problems **10.** (a) The payoff table is shown below.

Action: Number of truckloads sent	*State of Nature: Demand for eggs (number of truckloads)* 2	3	4	5
2	$300	$300	$300	$300
3	$250	$450	$450	$450
4	$200	$400	$600	$600
5	$150	$350	$550	$750

(b) 2 (c) 5 (d) 4

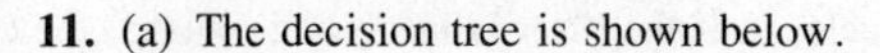

11. (a) The decision tree is shown below.

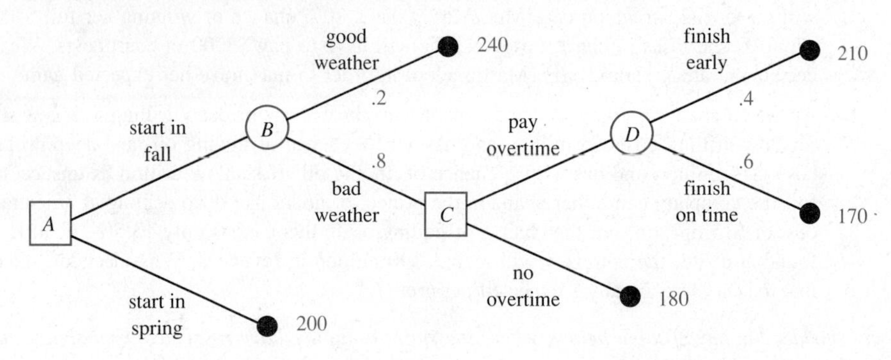

(b) The maximum expected payoffs are 186 at nodes D and C, 196.80 at node B, and 200 at node A.

(c) The contractor should begin construction in the spring.

7.5 Probability Distributions

In Section 7.3 we obtained the following distribution for the random variable that counts the number of daily service calls by an automobile towing service.

Number of service calls	*Probability*	
0	.10	
1	.25	
2	.30	
3	.20	
4	.10	
5	.05	
	1.00	Total

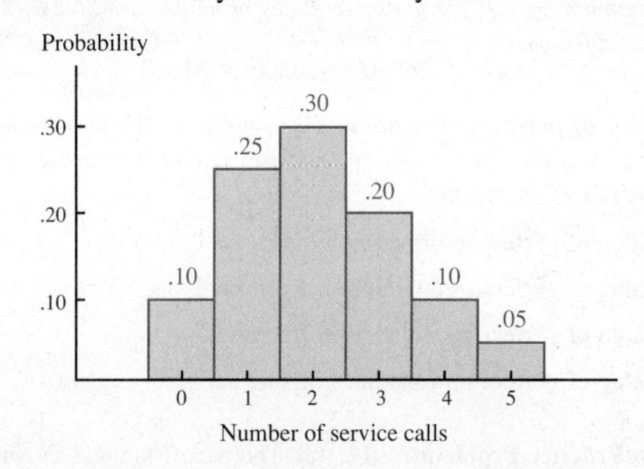

FIGURE 7.11

Recall that we represented this distribution by the histogram in Figure 7.11, where each value of the random variable is the midpoint of a class interval extending one-half unit to the left and right. The height of each bar represents the probability that the random variable equals the midpoint of the class interval corresponding to that bar. This histogram is called the **probability distribution** for the random variable; it is the graph of a function called the **probability density function.**

We can use the probability distribution in Figure 7.11 to find the probability that the towing service receives a given number of service calls on a particular day. For example, the probability that the number of service calls is between 1 and 3, inclusive, is

$$\begin{aligned} P(\text{between 1 and 3 calls}) &= P(\text{exactly 1, 2, or 3 calls}) \\ &= P(\text{exactly 1 call}) + P(\text{exactly 2 calls}) + P(\text{exactly 3 calls}) \\ &= .25 + .30 + .20 \\ &= .75. \end{aligned}$$

This calculation is depicted in Figure 7.12, where the part of the probability distribution corresponding to the values 1, 2, and 3 is shaded. Since each bar of the histogram has width 1, the total area of the shaded portion is

$$.25(1) + .30(1) + .20(1) = .75,$$

which is the probability in question. Notice that the total area represented by the probability distribution is 1.

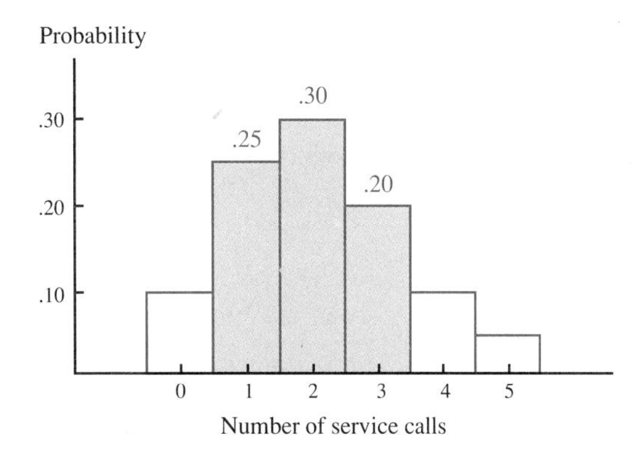

FIGURE 7.12

This example illustrates our first theorem.

Theorem 7.1

Consider a random variable that takes consecutive integer values.* The probability that the variable assumes a value between a and b (inclusive) is equal to the area under its probability distribution between $a - \frac{1}{2}$ and $b + \frac{1}{2}$.

---

*If the values of the variable are not consecutive integers, then Theorem 7.1 must be modified to account for the fact that the bars in its probability distribution do not have width 1. We will not consider this situation.

EXAMPLE 7.16 Use Figure 7.11 to find the probability that on a particular day the towing service receives

(a) between 1 and 4 service calls (inclusive);
(b) at least 3 service calls.

Solution (a) By Theorem 7.1 we see that the probability of receiving between 1 and 4 service calls in Figure 7.11 is

$$P(\text{between 1 and 4 calls}) = \text{area between 0.5 and 4.5} = .85.$$

The corresponding region of the probability distribution is shown in Figure 7.13(a).

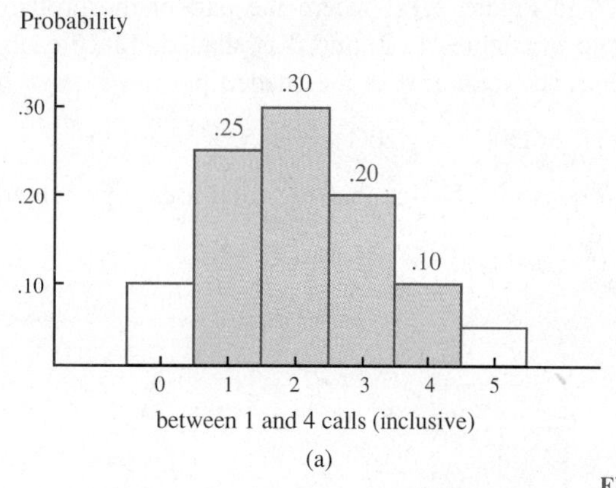

(a)

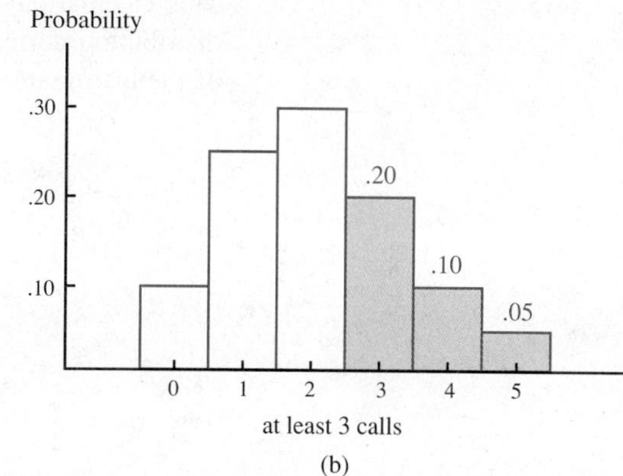

(b)

FIGURE 7.13

(b) The probability of receiving at least 3 requests is the same as the probability of receiving between 3 and 5 requests, inclusive. Thus we see that

$$P(\text{at least 3 calls}) = \text{area between 2.5 and 5.5} = \text{area to the right of 2.5} = .35.$$

See Figure 7.13(b). ■

Practice Problem 12 Use Figure 7.11 to find the probability that on a particular day the towing service receives

(a) between 3 and 4 service calls, inclusive;
(b) at most 1 service call.

THE BINOMIAL DISTRIBUTION

Recall that in Section 6.6 we defined a *Bernoulli process* to be a sequence of repeated trials of an experiment in which

1. each trial has only two possible outcomes, called *success* and *failure* (and denoted S and F, respectively);
2. the trials are independent; and
3. the probability of success is the same for each trial.

The principal result concerning Bernoulli processes is Theorem 6.7, which we reformulate below in the language of statistics.

Binomial Distribution

The number of successes in a sequence of n Bernoulli trials is a random variable that assumes the values 0, 1, 2, . . ., n. Its probability density function is given by Bernoulli's formula:

$$P(\text{exactly } k \text{ successes in } n \text{ trials}) = C(n, k)p^k q^{n-k} \ (k = 0, 1, \ldots, n),$$

where p and q denote the probabilities of success and failure, respectively, on each trial.

EXAMPLE 7.17 Find the probability density function and probability distribution for the number of heads obtained when a fair coin is flipped ten times.

Solution We must determine the distribution of the random variable that counts the number of successes in 10 trials, where success means that the coin lands heads. Since the probabilities of success and failure are $p = \frac{1}{2}$ and $q = \frac{1}{2}$ for a fair coin, Bernoulli's formula gives the following probability density function.

Number of heads	*Probability*
0	1/1024
1	10/1024
2	45/1024
3	120/1024
4	210/1024
5	252/1024
6	210/1024
7	120/1024
8	45/1024
9	10/1024
10	1/1024
	1 Total

For instance, the probability of obtaining exactly 3 heads is computed as follows:

$$\begin{aligned} P(\text{exactly 3 successes in 10 trials}) &= C(10, 3)p^3q^7 \\ &= 120\left(\frac{1}{2}\right)^3\left(\frac{1}{2}\right)^7 \\ &= \frac{120}{1024}. \end{aligned}$$

(See also Example 6.35 for the probability of obtaining exactly 5 heads.)

The probability distribution for this density function is shown in Figure 7.14. As we see from this figure, the *most likely* number of heads is 5, which has probability $252/1024 \approx .246$. Using Theorem 7.1 we find that the probability of obtaining between 4 and 6 heads is

$$P(4, 5, \text{ or } 6 \text{ heads}) = \text{area between } 3.5 \text{ and } 6.5$$
$$= \frac{672}{1024} = .65625.$$

Likewise the probability of obtaining between 3 and 7 heads is

$$P(3, 4, 5, 6, \text{ or } 7 \text{ heads}) = \text{area between } 2.5 \text{ and } 7.5$$
$$= \frac{912}{1024} = .890625. \quad ■$$

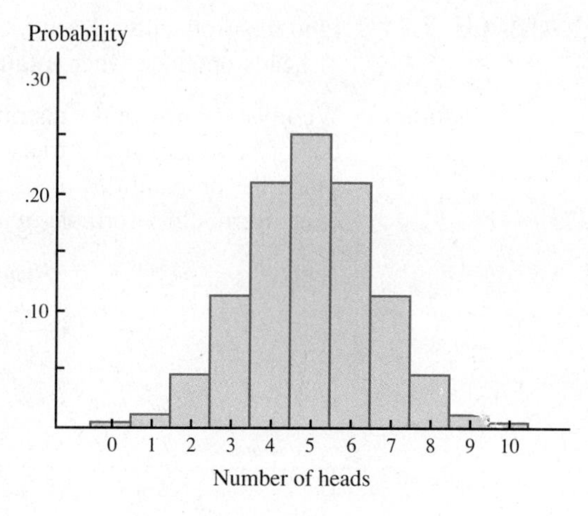

FIGURE 7.14

Practice Problem 13 (a) Find the probability density function for the number of sixes rolled when a fair die is tossed five times.
(b) Determine the probability of obtaining more than one six in five tosses.

Using Formulas (7.5) and (7.6), we can compute the mean and standard deviation for the probability density function in Example 7.17. Fortunately, however, there is a general result about binomial distributions that makes these calculations unnecessary.

Theorem 7.2

Consider the random variable that counts the number of successes in a sequence of n Bernoulli trials in which the probabilities of success and failure on each trial are p and q, respectively. Then the mean of this variable is

$$\mu = np,$$

and its standard deviation is

$$\sigma = \sqrt{npq}.$$

The formula for the mean in Theorem 7.2 makes sense intuitively: If the probability of success on each trial is p, then in n trials we should expect np successes.

EXAMPLE 7.18 Find the mean and standard deviation for the number of heads obtained when a fair coin is flipped three times.

Solution We can solve this problem by applying Theorem 7.2 with $n = 3$ and $p = q = ½$. The expected number of heads (the mean) is

$$\mu = np = 3(½) = 1.5,$$

and the standard deviation is

$$\sigma = \sqrt{npq} = \sqrt{3\left(\frac{1}{2}\right)\left(\frac{1}{2}\right)} = \sqrt{0.75} \approx 0.87.$$

Note that these are the same values obtained in Example 7.9 when we computed the mean and standard deviation directly from the probability density function. ■

EXAMPLE 7.19 Find the mean and standard deviation for the number of sixes obtained when a fair die is rolled 60 times.

Solution We can solve this problem by applying Theorem 7.2 with $n = 60$, $p = 1/6$, and $q = 5/6$. The expected number of sixes (the mean) is

$$\mu = np = 60\left(\frac{1}{6}\right) = 10$$

and the standard deviation is

$$\sigma = \sqrt{npq} = \sqrt{60\left(\frac{1}{6}\right)\left(\frac{5}{6}\right)} = \sqrt{\frac{300}{36}} \approx 2.89.$$

Note, however, that although the expected number of sixes is 10, the probability of obtaining *exactly* 10 sixes is only

$$C(60, 10)\left(\frac{1}{6}\right)^{10}\left(\frac{5}{6}\right)^{50} \approx .137.$$

Thus the probability that the mean actually occurs can be quite small. ■

Practice Problem 14 From past experience a construction company has found that its probability of making a successful bid on a contract is .2. Assuming that this probability continues, find the mean and standard deviation for the number of contracts obtained if it submits 30 bids per year.

Binomial distributions arise in a wide variety of applications. The next example illustrates their use in testing the effectiveness of drugs.

EXAMPLE 7.20 Suppose that 40% of those having a certain disease recover without treatment. Eight persons with this disease are given an experimental drug, and five of them recover. What is the probability that such a result occurs merely by chance?

Solution To answer this question, we will assume that the experimental drug has no effect and determine the probability that out of eight persons with the disease, five or more recover. Accordingly we will regard this situation as a Bernoulli process with $n = 8$ trials in which the probability of success (recovery) on each trial is $p = .4$. Since the number of successes is binomially distributed, the probability density function for the number of successes is as follows.

Number of successes	*Probability*
0	.017
1	.090
2	.209
3	.279
4	.232
5	.124
6	.041
7	.008
8	.001
Total	1.000

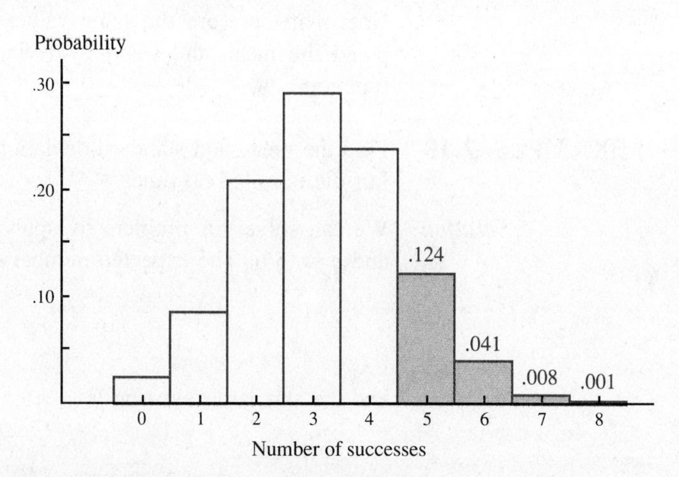

FIGURE 7.15

For instance, the probability that exactly three of the eight persons recover is

$$\begin{aligned} P(3 \text{ successes in 8 trials}) &= C(8,\, 3)p^3q^5 \\ &= 56(.4)^3(.6)^5 \\ &\approx .279. \end{aligned}$$

From the probability histogram in Figure 7.15, we see that

$$\begin{aligned} P(5 \text{ or more successes}) &= \text{area to the right of } 4.5 \\ &= .124 + .041 + .008 + .001 \\ &\approx .174. \end{aligned}$$

Thus the probability is .174 that five or more patients would have recovered even if they had not been given the experimental drug. Because this probability is high, few statisticians would regard this success rate as evidence of the drug's effectiveness. On the other hand, the probability that *six* or more of the eight persons recover is only

$$P(6 \text{ or more successes}) = \text{area to the right of } 5.5 = .05,$$

which many statisticians would accept as evidence of the drug's effectiveness. ■

Practice Problem 15 A graphologist (handwriting expert) claimed to be able to distinguish normal persons from psychotics on the basis of handwriting. A test was devised (see reference [2]) in which the graphologist was presented ten pairs of persons, one normal and one psychotic. In these ten pairs, the graphologist correctly identified the psychotic person six times.

(a) Find the probability of identifying the psychotic person in exactly six of the pairs by chance alone.
(b) Find the probability of identifying the psychotic person in exactly seven of the pairs by chance alone.
(c) Determine the probability of identifying the psychotic person in six or more pairs by chance alone.
(d) Would you accept the graphologist's claim?

When the number of trials is large, the binomial distribution becomes unwieldy. In such cases we can approximate the binomial probability distribution by a smooth curve called the *normal distribution*. We will discuss this curve in the next section.

EXERCISES 7.5

1. The probability distribution for a random variable taking the values 0, 1, 2, 3, and 4 is shown below.

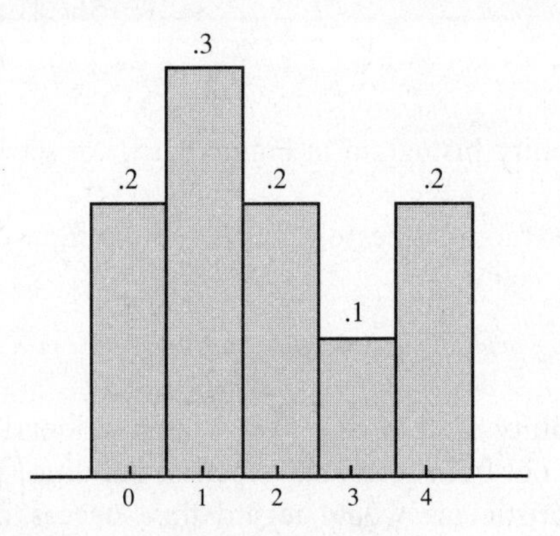

Find the probability that the variable assumes a value

(a) between 1 and 3 (inclusive) (b) at least 2.

(c) less than 3.

2. The probability distribution for a random variable taking the values 0, 1, 2, . . ., 10 is shown below.

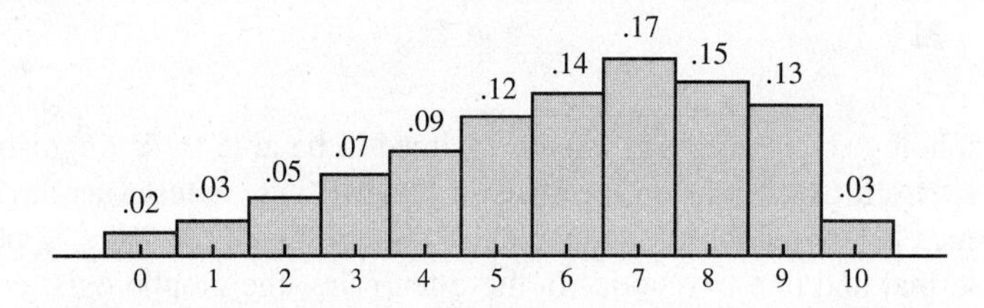

Find the probability that the variable assumes a value

(a) between 4 and 8 (inclusive) (b) greater than 5

(c) no more than 6.

3. An insurance company checked the records of 80 random motorists to determine how many moving violations each had been cited for during the past five years. The results are shown below.

Violations	*Frequency*	
0	34	
1	28	
2	12	
3	4	
4	2	
	80	Total

(a) Construct the probability density function and the corresponding probability distribution for this empirical data.

(b) Use Theorem 7.1 to determine the probability that someone had at most 2 violations.

(c) Use Theorem 7.1 to determine the probability that someone had between 1 and 3 violations (inclusive).

(d) Use Theorem 7.1 to determine the probability that someone had at least one violation.

4. A Labor Department statistician interviewed 120 unemployed persons in an inner-city area to determine how many years of high school each had completed. Their responses are shown below.

Number of years	*Frequency*	
0	6	
1	24	
2	39	
3	30	
4	21	
	120	Total

(a) Find the probability density function and the corresponding probability distribution for the number of years of high school completed.

(b) Use Theorem 7.1 to determine the probability that someone completed at most one year of high school.

(c) Use Theorem 7.1 to determine the probability that someone completed at least three years of high school.

(d) Use Theorem 7.1 to determine the probability that someone completed between one and three years of high school (inclusive).

5. A stockbroker observes that the delivery time for her local mailings is a random variable with the distribution below.

Delivery time (working days)	*Probability*
1	.218
2	.318
3	.272
4	.130
5	.062

(a) Construct the probability distribution for this data.

(b) Use Theorem 7.1 to determine the probability that an item mailed on Monday will arrive on Wednesday, Thursday, or Friday.

(c) Use Theorem 7.1 to determine the probability that an item mailed on Monday will arrive by Thursday at the latest.

6. A baker finds that the number of chocolate cakes requested on a given day has the following distribution.

Requests	*Probability*
0	.11
1	.20
2	.32
3	.19
4	.12
5	.05
6	.01

(a) Construct the probability distribution for this data.
(b) Use Theorem 7.1 to determine the probability that between 3 and 5 chocolate cakes will be requested on a given day.
(c) Use Theorem 7.1 to determine the probability that at most 2 chocolate cakes will be requested on a given day.
(d) Use Theorem 7.1 to determine the probability that at least 4 chocolate cakes will be requested on a given day.

In Exercises 7–10 determine the probability density function for the random variable that counts the number of successes in a sequence of n Bernoulli trials in which the probability of success on each trial is p.

7. $n = 3$ and $p = .4$

8. $n = 4$ and $p = .8$

9. $n = 5$ and $p = .25$

10. $n = 6$ and $p = .5$

In Exercises 11–18 find the mean and standard deviation of the indicated binomial distribution.

11. Two fair dice are rolled 90 times, and the number of times that a sum of seven occurs is counted.

12. An 80% shooter takes 50 shots at a target, and the number of hits is recorded.

13. A mail solicitation is sent to 10,000 individuals, each of whom has a 9% chance of responding, and the number of responses is counted.

14. A machine produces 10,000 items with a 0.4% defective rate, and the number of defective items is counted.

15. An airline has given 350 reservations for a flight on which there is a 10% chance that a person holding a reservation will fail to show up, and the number of no-shows is counted.

16. On a true-false test consisting of 100 questions, the number of correct answers obtained by random guessing is counted.

17. Of the registered voters in a certain district, 75% favor a tax cut. In a random survey of 200 of these voters, the number of persons favoring the cut is counted.

18. In American roulette, 18 of the 38 possible numbers are red. A player bets on red 95 times in a row, and the number of times he wins is counted.

19. A pharmaceutical house claims to have developed a drug that increases the chance of recovery from a certain disease from 60% to 80%. Suppose that the drug is given to 50 patients with the disease, and the number who recover is counted.

(a) Find the mean and standard deviation of this random variable if the manufacturer's claim is true.

(b) Find the mean and standard deviation of this random variable if the drug has no effect on the disease.

20. Suppose that there are 120 infants in a pediatric ward, each needing a nurse's attention 15% of the time. Find the mean and standard deviation of the number needing attention.

Answers to Practice Problems

12. (a) .30 (b) .35

13. (a)

Number of sixes	*Probability*
0	3125/7776
1	3125/7776
2	1250/7776
3	250/7776
4	25/7776
5	1/7776

(b) $\frac{1526}{7776}$

14. $\mu = 6$ and $\sigma \approx 2.19$

15. (a) $\frac{210}{1024}$

(b) $\frac{120}{1024}$

(c) $\frac{210}{1024} + \frac{120}{1024} + \frac{45}{1024} + \frac{10}{1024} + \frac{1}{1024} = \frac{386}{1024} \approx .377$

(d) no

7.6 The Normal Distribution

So far we have considered only random variables that take finitely many values. Probability distributions can also be defined for **continuous** random variables, those with values measured on a continuous scale, such as the set of all nonnegative real numbers. In this section we will discuss the most important case, random variables that are normally distributed.

CONTINUOUS DISTRIBUTIONS

When we measure the length of a javelin throw, the weight of a newborn infant, or the amount of cholesterol in a person's blood, we are dealing with continuous random variables. Although we can approximate such measurements by rounding them off to a finite number of decimal places, the fact remains that in theory their possible values cover an interval of real numbers.

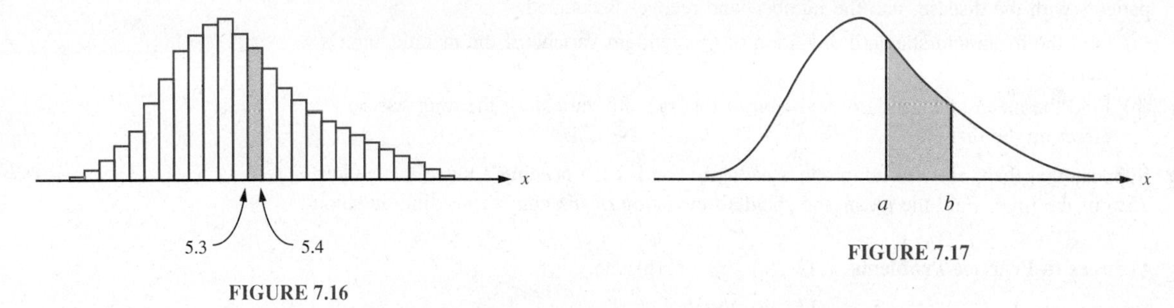

FIGURE 7.16

FIGURE 7.17

To obtain the probability distribution for a continuous random variable, we imagine dividing the interval of possible values into many small subintervals as in Figure 7.16. Over each subinterval (such as 5.3 to 5.4 in the figure) we construct a bar with *area* equal to the probability that the variable assumes a value in that subinterval. Using smaller and smaller subintervals, we find that the resulting histogram approaches a smooth curve, which is shown in Figure 7.17. If we think of the curve as a histogram with "infinitely thin" bars, we see that it has the following basic property.

> The area under the curve between any two values a and b represents the probability that the random variable assumes a value in this interval.

Note that the *total* area under a probability distribution is always 1. Furthermore, if $a = b$ in Figure 7.17, the area between them is zero. Hence *the probability that a continuous random variable takes any particular value is zero*. This is not as unreasonable as it may sound at first. For instance, it is highly unlikely—impossible for all practical purposes—that the length of the next throw of a javelin will be *exactly* 71.6325 meters. In any case this example points out an important difference between finite and continuous random variables: In the former we consider the probability of a given *value,* but for the latter we consider the probability of a given *interval*.

By using concepts from calculus (see Chapter 14), it is possible to define the mean and standard deviation of a continuous random variable. These numbers have the same significance as in the finite case; that is, the mean represents the "center" of the distribution and the standard deviation measures its dispersion around the mean.

THE NORMAL DISTRIBUTION

In his study of random experimental errors, the French mathematician Abraham De Moivre (1667–1754) discovered that repeated measurements of a physical quantity tend to be distributed according to a precise mathematical pattern called a **normal distribution.** Figure 7.18 shows a typical normal distribution with mean μ and standard deviation σ. There is a unique normal curve* for each value of μ and σ, and it can be shown that every normal curve has the following properties.

Properties of a Normal Curve

1. The curve is bell-shaped with its highest point at $x = \mu$.
2. The curve is symmetric with respect to the vertical line $x = \mu$.
3. The curve extends infinitely far in both directions and comes arbitrarily close to the horizontal axis.
4. The total area under the curve is 1.

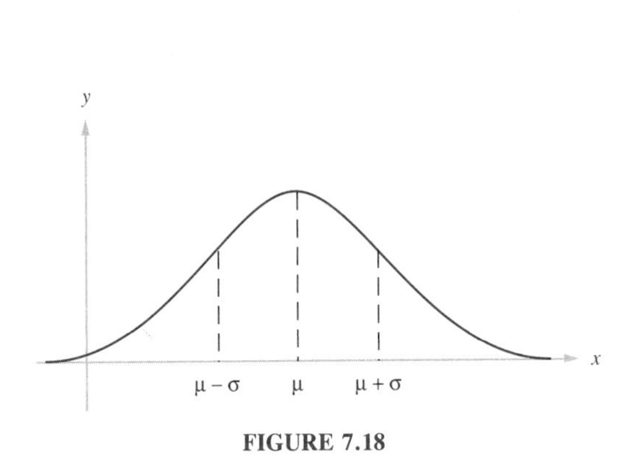

FIGURE 7.18

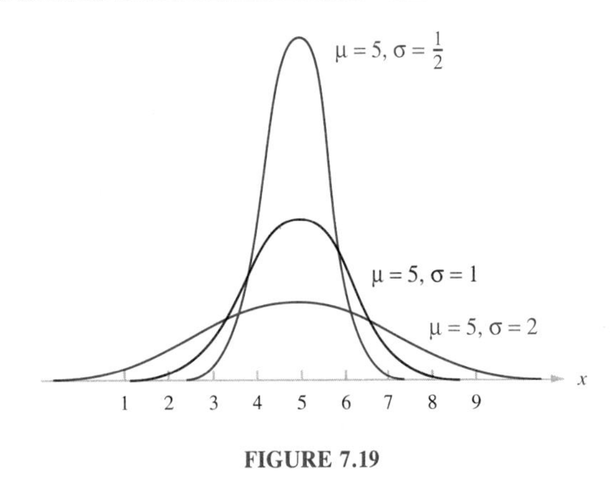

FIGURE 7.19

At the points $x = \mu \pm \sigma$ located at a distance σ on either side of the mean, the normal curve twists or *inflects*. That is, the curve is concave down (like an inverted bowl) between $\mu - \sigma$ and $\mu + \sigma$ and concave up outside this interval.

The mean μ determines the position of the normal curve, and the standard deviation σ determines its shape. Figure 7.19 shows three normal curves, all having the mean $\mu = 5$. Note that a small value of σ gives a tall, narrow curve, indicating that the values are concentrated near the mean. On the other hand, a large value of σ gives a lower, flatter curve, indicating a greater dispersion of the values.

---

*The equation of the normal curve with mean μ and standard deviation σ is

$$y = \left(\frac{1}{\sigma\sqrt{2\pi}}\right)e^{-(y-\mu)^2/2\sigma^2}, \text{ where } e \approx 2.71828 \text{ and } \pi \approx 3.14159.$$

STANDARD UNITS

In any normal distribution approximately 68.26% of the area under the curve lies within one standard deviation of the mean, and approximately 95.44% of the area under the curve lies within two standard deviations of the mean. (See Figure 7.20.) As a result, it is useful to measure distances along the x-axis in terms of standard deviations from the mean. We can convert any x-value into these **standard units** (or **z-values**) by the formula

$$z = \frac{x - \mu}{\sigma}. \tag{7.7}$$

This is merely a change of scale as illustrated in Figure 7.20. That is, the z-value measures the number of standard deviations between the corresponding x-value and the mean. Thus 68.26% of the area under any normal curve lies between $z = -1$ and $z = 1$, and 95.44% of the area lies between $z = -2$ and $z = 2$. More generally, *the area under a normal curve between any two z-values is a constant, independent of the mean and standard deviation of the distribution.*

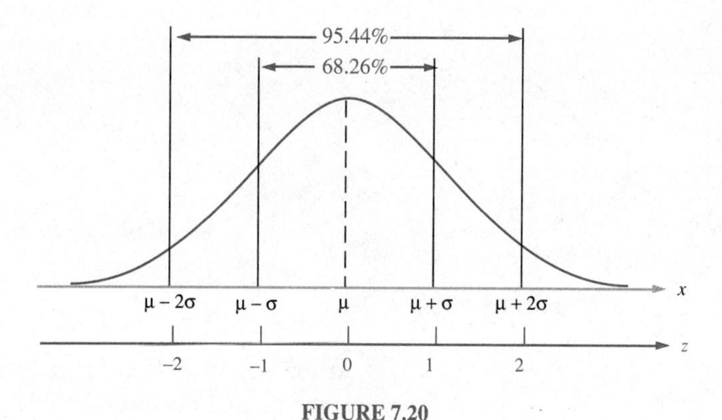

FIGURE 7.20

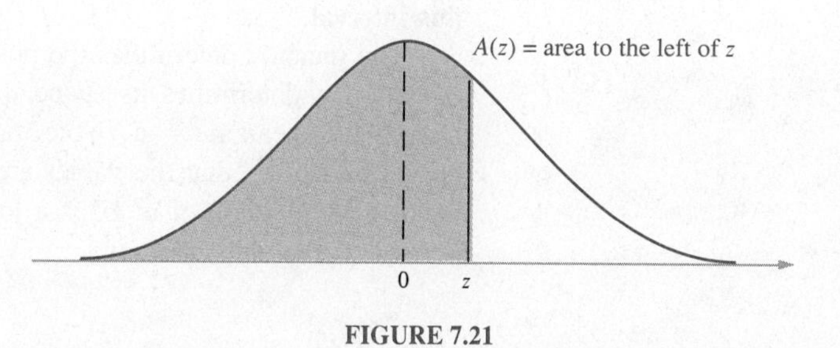

FIGURE 7.21

The table inside the front cover gives the area $A(z)$ under a normal curve to the left of z. (See Figure 7.21.) With this table we can find the area under any normal curve between two given z-values.

EXAMPLE 7.21 A normal distribution has mean 20 and standard deviation 4. Find the area under this curve between 17 and 22.

Solution The region in question is shaded in Figure 7.22. To compute its area, we must first convert the given values to standard units. By formula (7.7) $x = 17$ corresponds to

$$z = \frac{17 - 20}{4} = -.75,$$

and $x = 22$ corresponds to

$$z = \frac{22 - 20}{4} = .50.$$

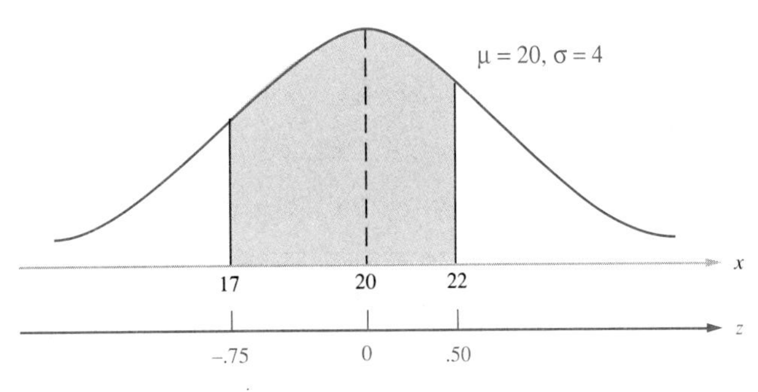

FIGURE 7.22

Thus $x = 17$ lies .75 standard deviations below the mean and $x = 22$ lies .50 standard deviations above the mean. From the table inside the front cover we find that

$$A(-.75) = .2266 \quad \text{and} \quad A(.50) = .6915.$$

Hence the area between $x = 17$ and $x = 22$ is given by

$$\begin{aligned} A(.50) - A(-.75) &= .6915 - .2266 \\ &= .4649. \end{aligned}$$

Therefore 46.49% of the total area under the curve lies between the values $x = 17$ and $x = 22$. ■

EXAMPLE 7.22 Find the area under the normal curve with mean 30 and standard deviation 6 that lies

(a) to the right of 36;
(b) between 21 and 24.

Solution (a) The region in question is shaded in Figure 7.23(a). In standard units $x = 36$ becomes

$$z = \frac{36 - 30}{6} = 1.00.$$

That is, $x = 36$ lics one standard deviation above the mean. From the table inside the front cover we find that the area to the *left* of $x = 36$ is $A(1.00) = .8413$. Since the total area is 1, the area to the *right* of $x = 36$ is

$$\begin{aligned} 1 - A(1.00) &= 1 - 0.8413 \\ &= 0.1587. \end{aligned}$$

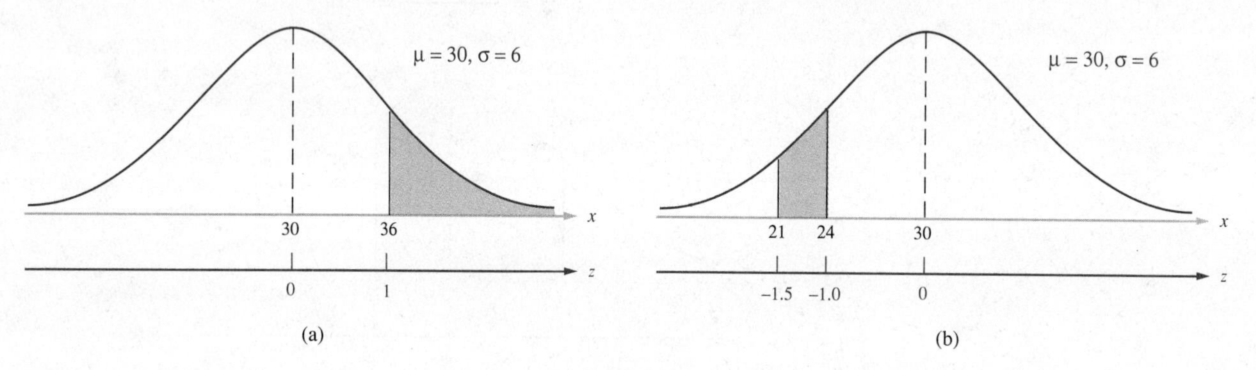

FIGURE 7.23

(b) The desired area is shaded in Figure 7.23(b). As in part (a) we find that $x = 21$ and $x = 24$ convert to $z = -1.50$ and $z = -1.00$, respectively. Thus the desired area is

$$\begin{aligned} A(-1.00) - A(-1.50) &= 0.1587 - 0.0668 \\ &= .0919. \end{aligned}$$

Hence 9.19% of the total area lies between $x = 21$ and $x = 24$. ■

Practice Problem 16 Find the area under the normal curve with mean 54 and standard deviation 8 that lies

(a) to the right of 48;
(b) between 68 and 72.

APPLICATIONS

Many of the random variables that arise in statistical applications turn out to be *normally distributed,* that is, distributed as in a normal distribution. In such cases we can calculate probabilities by finding areas under a normal curve.

EXAMPLE 7.23 The annual rainfall in a certain city is normally distributed with mean 48.5 inches and standard deviation 6.2 inches. Find the probability that the rainfall in a given year will be less than 42 inches.

Solution The amount of rainfall per year (in inches) has the normal distribution shown in Figure 7.24. In standard units, $x = 42$ becomes $z = -1.05$. Therefore from the table inside the front cover

$$
\begin{aligned}
P(\text{rainfall} < 42) &= \text{area to the left of } 42 \\
&= A(-1.05) \\
&= 0.1469.
\end{aligned}
$$

Hence the probability that the rainfall in a given year will be less than 42 inches is .1469. ■

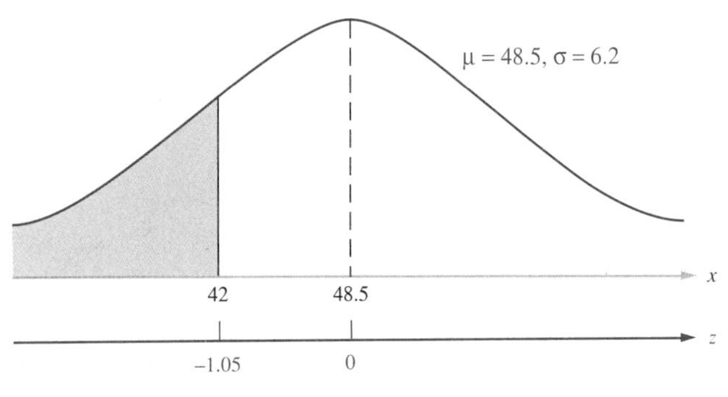

FIGURE 7.24

EXAMPLE 7.24 The mean systolic blood pressure of adult males is normally distributed with a mean of 138 (mm of mercury) and a standard deviation of 9.7. What percentage of adult males have blood pressure between 145 and 150?

Solution The systolic blood pressure of a randomly chosen adult male has the distribution shown in Figure 7.25.

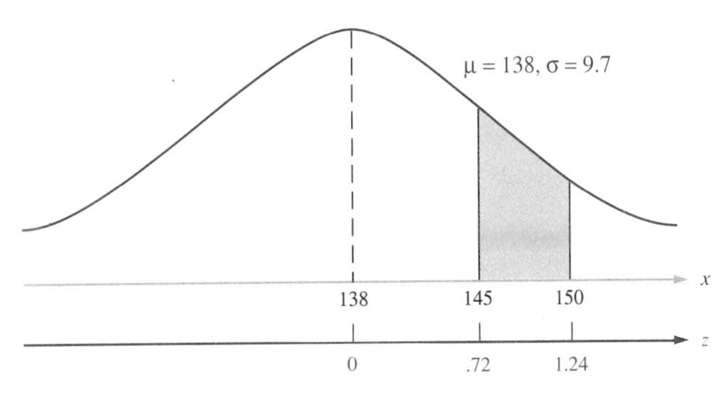

FIGURE 7.25

In standard units $x = 145$ and $x = 150$ become $z = 0.72$ and $z = 1.24$, respectively. Hence

$$\begin{aligned} P(145 \leq \text{systolic pressure} \leq 150) &= A(1.24) - A(.72) \\ &= .8925 - .7642 \\ &= .1283. \end{aligned}$$

Thus approximately 12.83% of all adult males have blood pressure in the given range. ■

THE NORMAL APPROXIMATION TO A BINOMIAL DISTRIBUTION

One of the most important applications of normal curves is based on the fact that they give good approximations to binomial distributions. For instance, the probability distribution for the number of heads obtained when a fair coin is tossed ten times is a binomial distribution with mean

$$\mu = np = 10(1/2) = 5$$

and standard deviation

$$\sigma = \sqrt{npq} = \sqrt{10\left(\frac{1}{2}\right)\left(\frac{1}{2}\right)} \approx 1.58.$$

As we see in Figure 7.26, the binomial histogram for the number of heads is approximated very well by the normal curve with mean 5 and standard deviation 1.58. Consequently we can approximate probabilities for this random variable by computing areas under this normal curve.

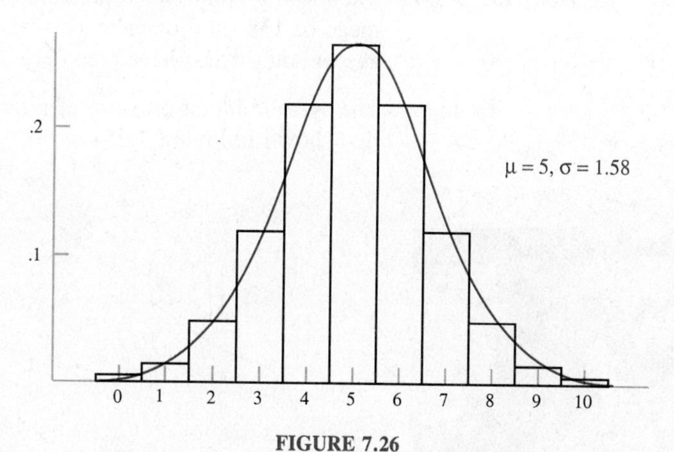

FIGURE 7.26

EXAMPLE 7.25 Use the normal curve with mean 5 and standard deviation 1.58 to approximate the probability that in ten flips of a fair coin

(a) exactly five heads appear;
(b) four, five, or six heads appear.

Solution (a) Consider the random variable that counts the number of heads obtained when a fair coin is flipped ten times. Recall that by Theorem 7.1 the probability of obtaining exactly five heads is equal to the area under the probability distribution for this random variable between 4.5 and 5.5. This area is approximately equal to the area under the normal curve in Figure 7.26 between 4.5 and 5.5. Converting these values to standard units, we obtain

$$z = \frac{4.5 - 5}{1.58} \approx -0.32 \qquad \text{for} \qquad x = 4.5$$

and

$$z = \frac{5.5 - 5}{1.58} \approx 0.32 \qquad \text{for} \qquad x = 5.5.$$

Hence by the table inside the front cover

$$\begin{aligned} P(\text{exactly 5 heads}) &\approx A(0.32) - A(-0.32) \\ &= .6255 - .3745 \\ &= .2510. \end{aligned}$$

The true probability was found to be $252/1024 \approx .246$ in Example 7.17.
(b) The probability of obtaining four, five, or six heads equals the area under the probability distribution between 3.5 and 6.5. Since $x = 3.5$ and $x = 6.5$ convert to -0.95 and 0.95 standard units, respectively, the normal approximation gives

$$\begin{aligned} P(\text{exactly 4, 5, or 6 heads}) &\approx A(0.95) - A(-0.95) \\ &= .8289 - .1711 \\ &= .6578. \end{aligned}$$

This value is quite close to the true value of .65625, which we calculated in Example 7.17. ■

As was true in Example 7.25, a normal curve generally approximates a binomial distribution more accurately when it is used over a larger interval of values. In addition, the approximation improves as the number of trials increases in the binomial distribution. A precise statement of this result, known as the De Moivre-Laplace Limit Theorem, is beyond the scope of this text; however, we can summarize it roughly as follows.

> Consider the random variable that counts the number of successes in n Bernoulli trials, and let p and q denote the probabilities of success and failure, respectively, on each trial. If n is large and neither p nor q is close to zero, then the probability distribution for this random variable can be approximated by the normal curve with mean $\mu = np$ and standard deviation $\sigma = \sqrt{npq}$.

As a rule of thumb, both np and nq should be greater than 5 in order to use a normal curve to approximate a binomial distribution.

EXAMPLE 7.26 Suppose that 40% of those having a certain disease recover without treatment. Assume that 20 individuals with this disease are given an experimental drug, and 13 of them recover. What is the probability that such a result occurs merely by chance?

Solution We proceed as in Example 7.20 by assuming that the drug had no effect. Consider the random variable that counts the number of successes (recoveries) among the 20 patients. Then the patients represent $n = 20$ Bernoulli trials with the probability of success being $p = .4$ on each trial. Moreover this random variable is *binomially distributed,* (that is, its probability distribution is a binomial distribution) with mean

$$\mu = np = 20(.4) = 8$$

and standard deviation

$$\sigma = \sqrt{npq} = \sqrt{20(.4)(.6)} \approx 2.19.$$

Since both $np = 8$ and $nq = 12$ exceed 5, we can approximate the distribution of this random variable by a normal curve with mean 8 and standard deviation 2.19. (See Figure 7.27.) The probability of 13 or more successes in

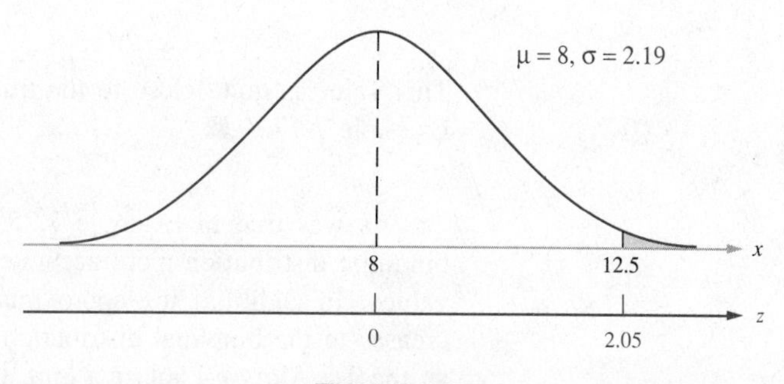

FIGURE 7.27

20 trials is approximately equal to the area under the normal curve to the right of 12.5. Because 12.5 converts to 2.05 standard units, the normal approximation gives

$$\begin{aligned} P(\text{at least 13 successes}) &\approx 1 - A(2.05) \\ &= 1 - .9798 \\ &= .0202. \end{aligned}$$

Thus there is about a 2% probability that the observed number of recoveries could have occurred by chance alone. ■

EXAMPLE 7.27 Actuarial tables indicate an accidental death rate in the United States of 56 persons per 100,000. If an insurance company covers 500,000 individuals for accidental death benefits, what is the probability that the number of claims in a given year will be between 260 and 300 (inclusive)?

Solution The number of claims by the policyholders is binomially distributed with $n = 500{,}000$, $p = .00056$, and $q = .99944$. The expected number of claims is

$$\mu = np = 280$$

with a standard deviation of

$$\sigma = \sqrt{npq} \approx 16.73.$$

Since both np and nq exceed 5, we can approximate the binomial distribution by a normal curve. (See Figure 7.28.) The probability that the number of claims will be between 260 and 300 is represented by the area under the normal curve between 259.5 and 300.5. Hence

$$\begin{aligned} P(260 \leq \text{number of claims} \leq 300) &\approx A(1.23) - A(-1.23) \\ &= .8907 - .1093 \\ &= .7814. \end{aligned}$$

Thus the number of claims will fall in the given range about 78% of the time. ■

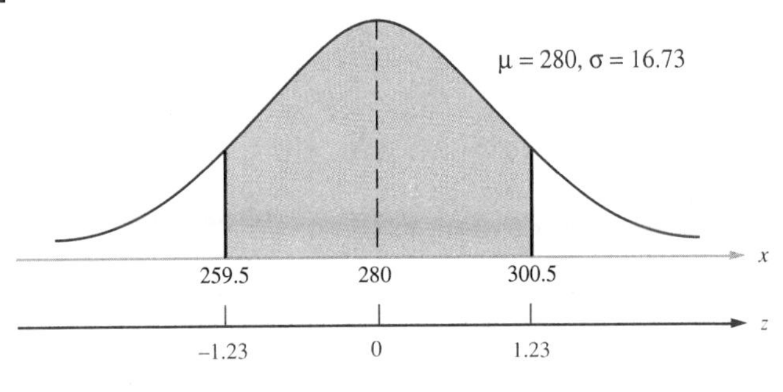

FIGURE 7.28

Practice Problem 17 A particular machine fills and caps cola bottles with a defective rate of 0.5%. That is, five out of every 1000 bottles processed on this machine are filled or capped improperly. What is the probability that more than 20 bottles are defective in a production run of 3000 bottles?

EXERCISES 7.6

In Exercises 1–8 find the area under a normal curve over the given z-interval.

1. to the left of $z = .58$

2. to the right of $z = -.02$

3. between $z = -1.25$ and $z = 1.34$

4. to the left of $z = -.27$

5. to the right of $z = 1.83$

6. between $z = .47$ and $z = 1.63$

7. between $z = -2.09$ and $z = -0.43$

8. between $z = -1.36$ and $z = 0.84$

In Exercises 9–12 find the value of z (standard units) for which the shaded region of a normal curve has the indicated area.

9.

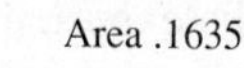

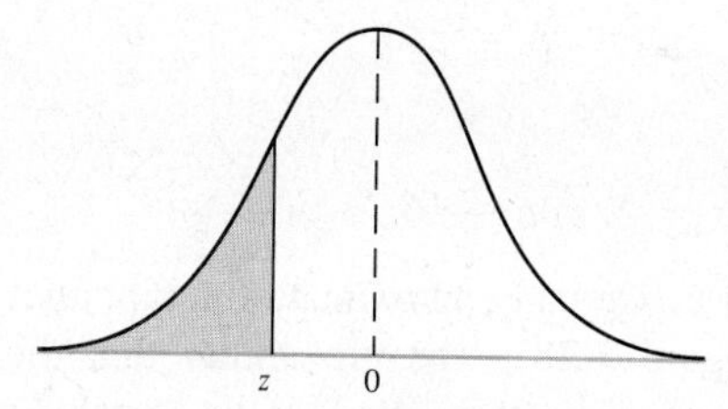

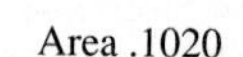

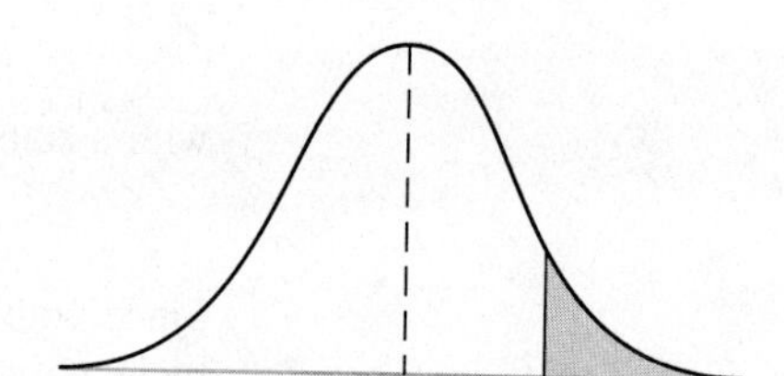

11.

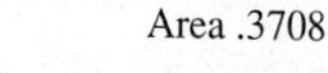

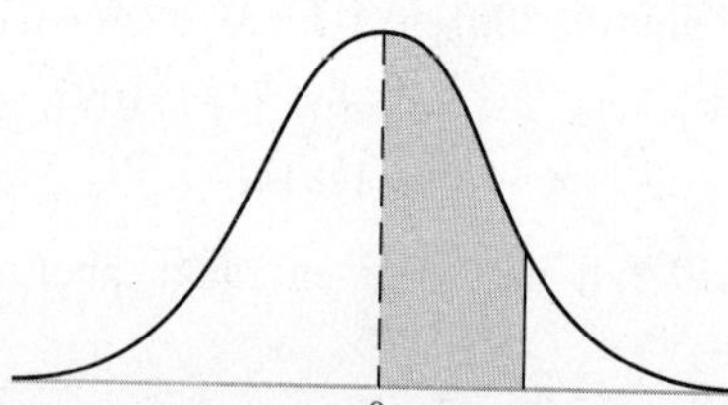

12.

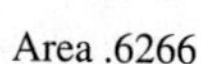

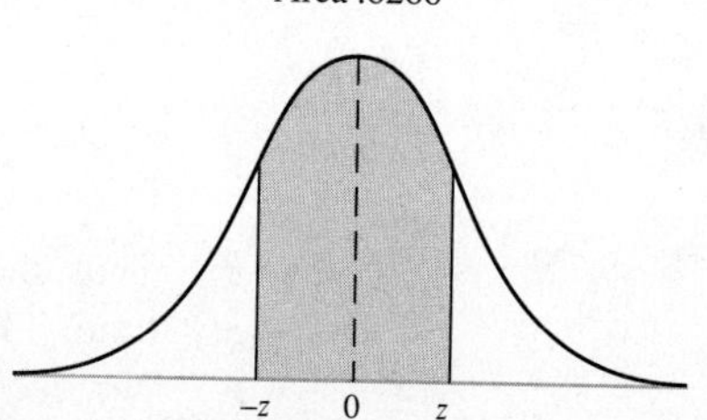

In Exercises 13–20 convert the given values of x into standard units for a normal curve with mean μ and standard deviation σ.

13. $x = 450$, $\mu = 600$, $\sigma = 100$

14. $x = 62$, $\mu = 50$, $\sigma = 12$

15. $x = 89$, $\mu = 80$, $\sigma = 5$

16. $x = 63$, $\mu = 90$, $\sigma = 12$

17. $x = 4.36$, $\mu = 3.5$, $\sigma = 1$

18. $x = 3.24$, $\mu = 6$, $\sigma = 2.4$

19. $x = 13.82$, $\mu = 17.85$, $\sigma = 1.55$

20. $x = 8.5$, $\mu = 6.1$, $\sigma = 1.5$

In Exercises 21–30 find the area under a normal curve with mean μ and standard deviation σ that lies in the indicated interval.

21. to the left of 99 if $\mu = 120$ and $\sigma = 12$

22. between 25 and 32 if $\mu = 25$ and $\sigma = 10$

23. between 38 and 50 if $\mu = 50$ and $\sigma = 12$

24. to the left of 8.77 if $\mu = 8.75$ and $\sigma = .05$

25. to the right of 54 if $\mu = 75$ and $\sigma = 15$

26. to the right of 90 if $\mu = 72$ and $\sigma = 24$

27. between 44 and 54 if $\mu = 48$ and $\sigma = 5$

28. between 28 and 36 if $\mu = 40$ and $\sigma = 10$

29. between 158 and 168 if $\mu = 150$ and $\sigma = 20$

30. between 50 and 60 if $\mu = 64$ and $\sigma = 8$

31. Scores on the 1960 revision of the Stanford-Binet intelligence test are normally distributed with a mean of 100 and a standard deviation of 16. What percentage of the population has IQ scores in the superior range (120–140)?

32. A slalom skier finds that her times on a certain course are normally distributed with mean 255 seconds and standard deviation 15 seconds. What is the probability that her time will be under 4 minutes on a given run?

mean of 255 seconds standard deviation of 15 seconds

33. A certain brand of 100-watt light bulbs has a mean life of 750 hours and a standard deviation of 50 hours. Assuming that the lives of these bulbs are normally distributed, what proportion of these bulbs burn at least 820 hours?

34. Before seeing the doctor, the patients in a certain doctor's office wait an average of 21 minutes with a standard deviation of 7 minutes. Assuming that the waiting times are normally distributed, what percentage of the patients wait between 15 minutes and 30 minutes?

35. A certain machine part must have a diameter between 11.99 and 12.01 cm. If the diameters of these parts are normally distributed with a mean of 12 cm and a standard deviation of 0.004 cm, what percentage of the parts have diameters in the acceptable range?

36. The life of a particular brand of automobile battery is normally distributed with a mean of 39 months and a standard deviation of 6 months. If these batteries are sold with a 36-month warranty, what proportion of them fail before the warranty expires?

In Exercises 37–44 use a normal curve to approximate the indicated binomial distribution. Refer to Exercises 11–19 in Section 7.5 for the mean and standard deviation.

37. If two fair dice are rolled 90 times, what is the probability that a sum of seven occurs between 10 and 20 times (inclusive)?

38. What is the probability that an 80% shooter hits a target between 35 and 45 times (inclusive) out of 50 shots?

39. Suppose that a mail solicitation is sent to 10,000 individuals, each of whom has a 9% chance of responding. What is the probability that at least 850 individuals respond?

40. A machine produces 0.4% defective items. What is the probability that in a day's production of 10,000 items, at most 50 are defective?

41. An airline has found that there is a 10% chance that a person holding a reservation will fail to show up. If it gives 350 reservations for a flight with 320 seats, what is the probability that everyone who shows up can be seated?

42. A true-false test consists of 100 questions, each worth one point. What is the probability of scoring 60 or more on this test by guessing at random?

43. In a certain district, 75% of the registered voters favor a tax cut. If 200 voters are surveyed at random, what is the probability that between 70% and 80% (inclusive) will favor the cut?

44. A pharmaceutical house claims to have developed a drug that increases the chance of recovery from a certain disease from 60% to 80%. Suppose that the drug is given to 50 patients with the disease, and 35 of them recover.

(a) What is the probability of a result this *low* if the maker's claim is true?

(b) What is the probability of a result this *high* if the drug has no effect?

45. Scholastic Aptitude Test scores are normally distributed with a mean of 500 and a standard deviation of 100.

(a) What percentage of scores are above 550?

(b) Approximately what score puts someone in the top 10%?

46. The weekly demand for poultry at a certain market is normally distributed with a mean of 2700 pounds and a standard deviation of 240 pounds.

(a) If the market stocks 2500 pounds per week, what proportion of the time will all the poultry be sold?

(b) Approximately how much poultry should the market stock per week in order to be 95% certain of selling it all?

47. The heights of adult males are normally distributed with mean 69 inches and standard deviation 2.5 inches. Approximately what height range contains the middle 50% of adult males?

48. The mean annual salary of beginning engineers is $28,500 with a standard deviation of $2300. Assuming that the salaries are normally distributed, approximately what salary range contains the middle 75% of these engineers?

Answers to Practice Problems **16.** (a) .7734 (b) .0279

17. .1210

CHAPTER 7 REVIEW

IMPORTANT TERMS

- **bar graph** *(7.1)*
- **class intervals**
- **histogram**
- **mean**
- **median**
- **standard deviation**
- **summation sign**
- **variance**
- **frequency distribution** *(7.2)*
- **expected value** *(7.3)*
- **random variable**
- **Bayes's criterion** *(7.4)*
- **decision tree**
- **maximax criterion**
- **maximin criterion**
- **payoff table**
- **states of nature**
- **binomial distribution** *(7.5)*
- **probability density function**
- **probability distribution**
- **continuous random variable** *(7.6)*
- **normal distribution**
- **standard units**

IMPORTANT FORMULAS

Mean of ungrouped data

The mean of $x_1, x_2, \ldots, x_n$ is

$$\bar{x} = \frac{x_1 + x_2 + \cdots + x_n}{n} = \frac{\Sigma x}{n}. \tag{7.1}$$

Variance of ungrouped data

The variance of $x_1, x_2, \ldots, x_n$ is

$$s^2 = \frac{(x_1 - \bar{x})^2 + (x_2 - \bar{x})^2 + \cdots + (x_n - \bar{x})^2}{n} = \frac{\Sigma(x - \bar{x})^2}{n}. \tag{7.2}$$

Mean of grouped data

If the class intervals have midpoints $m_1, m_2, \ldots, m_k$ and respective frequencies $f_1, f_2, \ldots, f_k$, then

$$\bar{x} \approx \frac{m_1 f_1 + m_2 f_2 + \cdots + m_k f_k}{n} = \frac{\Sigma mf}{n}, \tag{7.3}$$

where $n = \Sigma f$ is the total number of data values.

Variance of grouped data

If the class intervals have midpoints $m_1, m_2, \ldots, m_k$ and respective frequencies $f_1, f_2, \ldots, f_k$, then

$$s^2 \approx \frac{(m_1 - \bar{x})^2 f_1 + (m_2 - \bar{x})^2 f_2 + \cdots + (m_k - \bar{x})^2 f_k}{n} = \frac{\Sigma(x - \bar{x})^2 f}{n}, \tag{7.4}$$

where $n = \Sigma f$ is the total number of data values.

Mean and variance of a random variable

The mean of a random variable taking distinct values $x_1, x_2, \ldots, x_k$ with respective probabilities $p_1, p_2, \ldots, p_k$ is

$$\mu = \Sigma xp = x_1p_1 + x_2p_2 + \cdots + x_kp_k, \tag{7.5}$$

and the variance is

$$\sigma^2 = \Sigma(x - \mu)^2p = (x_1 - \mu)^2p_1 + (x_2 - \mu)^2p_2 + \cdots + (x_k - \mu)^2p_k. \tag{7.6}$$

Binomial distribution

The distribution of the number of successes in a sequence of n Bernoulli trials is a random variable that assumes the values $0, 1, 2, \ldots, n$. Its distribution is given by Bernoulli's formula:

$$P(\text{exactly } k \text{ successes in } n \text{ trials}) = C(n, k)p^kq^{n-k} \qquad (k = 0, 1, 2, \ldots, n),$$

where p and q denote the probabilities of success and failure, respectively, on each trial.

Mean and standard deviation of a binomial distribution

In a binomial distribution of n trials in which the probability of success and failure on each trial are p and q, respectively, the mean is

$$\mu = np$$

and the standard deviation is

$$\sigma = \sqrt{npq}.$$

Standard Units

In a normal distribution with mean μ and standard deviation σ, the number of standard units that a value x lies above or below the mean is

$$z = \frac{x - \mu}{\sigma}.$$

REVIEW EXERCISES

1. Five independent gem experts were asked to estimate the value of a diamond ring. Their appraisals were as follows: \$5000, \$7000, \$6500, \$8000, and \$7500. Find the mean, median, and standard deviation for these ungrouped data.
2. The percentage price increases in a certain year for ten selected consumer goods are as follows: 7%, 12%, 8%, 10%, 11%, 9%, 13%, 126%, 14%, and 10%. Find the mean and median for these ungrouped data. Which of these two measures better reflects a "typical" increase in consumer prices for the year?
3. The typing speeds (in words per minute) of 80 secretaries are distributed as follows.

Speed	*Secretaries*	
40–50	8	
50–60	12	
60–70	31	
70–80	22	
80–90	7	
	80	Total

(a) Represent these data by means of a histogram.

(b) Find the mean and standard deviation for these grouped data.

4. A class of 25 high school seniors had the following scores on a scholastic aptitude test.

672	419	523	614	531
724	367	431	602	612
569	472	763	525	643
675	798	664	372	599
518	349	617	558	593

(a) Write these data in grouped form.

(b) Represent the grouped data with a histogram.

(c) Find the mean and standard deviation for the grouped data in part (a).

5. A stereo equipment salesman found that the number of tape decks sold in a given day had the following distribution.

Units sold	*Probability*
0	.19
1	.35
2	.24
3	.12
4	.09
5	.01

Find the mean and standard deviation for the number of units sold per day.

6. An urn contains 10 red marbles, 20 white marbles, and 30 blue marbles. A player chooses a single marble at random from the urn. If it is red, he wins $4; if it is white, he wins $1; and if it is blue, he loses $2. Find the mean and standard deviation for the player's net gain.

7. Find the mean and standard deviation for the number of successes in 15 Bernoulli trials if the probability of success on each trial is 1/3.

8. Thirty percent of the registered voters in a certain area are senior citizens. A pool of 50 voters is chosen at random for jury duty. Find the mean and standard deviation for the number of senior citizens chosen.

9. A botanical research team found that the heights of a certain species of tree at a state park are normally distributed with a mean of 8.44 meters and a standard deviation of 2.4 meters. What percentage of these trees are more than 10 meters tall?

10. The speeds of cars on a two-lane highway were found to be normally distributed with a mean of 58.6 mph and a standard deviation of 3.0 mph. What percentage of cars are observing the 55 mph speed limit on this highway?

11. Estimate the probability that when a fair coin is flipped 100 times, heads will occur between 45 and 55 times (inclusive).

12. Of those having a certain disease, 25% recover without treatment. Suppose that 20 individuals with this disease are given an experimental drug, and 10 of them recover. What is the probability that a recovery rate this high is due merely to chance?

13. An investment analyst anticipates that during the next year the value of a certain stock will decrease by 10% with probability .1, remain unchanged with probability .2, increase by 10% with probability .4, and increase by 20% with probability .3. According to this analyst, what is the expected gain for this stock during the next year?

14. A company that sells solar heating units has found that on any given business day there is a 70% chance of selling no units, a 15% chance of selling one unit, a 10% chance of selling two units, and a 5% chance of selling three units. What is the company's average daily profit from these units if it earns a profit of $1000 on each solar heating unit that it sells?

15. If it rains during the next two weeks, a farmer will make $100,000 profit on his corn crop, but otherwise he will lose $80,000. The chance of rain during the period is estimated at 50%, but it can be increased to 60% by cloud-seeding. If the cloud-seeding costs $15,000, is it worthwhile in terms of the farmer's expected net gain?

16. A textbook publisher estimates that next year's demand for a new calculus text (to the nearest 5000) will be as follows.

Number ordered	*Probability*
5000	.2
10,000	.5
15,000	.2
20,000	.1

The publisher must decide whether to print 5, 10, 15, or 20 thousand copies in the first printing. Each printed copy costs the publisher $12, and the book sells for $25. In addition, each unsold copy costs an additional $4 in inventory expenses. How large should the first printing be to maximize the publisher's expected net profit?

17. Each morning a baker makes a certain number of apple pies to be sold during the day. The number of daily requests for apple pies has been observed to vary as follows.

Number requested	*Probability*
0	.2
1	.4
2	.3
3	.1

If each pie sold yields a $2 profit and each pie not sold results in a $1 loss, how many pies should the baker prepare in order to maximize his long-term profit?

18. A contestant on a quiz show has already won $4200. He can either keep this money, or risk it by trying to answer a jackpot question. If he answers the jackpot question correctly, he will win another $5000, but if he answers incorrectly, he will lose the $4200. In addition, by answering the jackpot question correctly, he becomes eligible for a superjackpot question. As before, he may choose to take the money already won or risk it by trying to answer the superjackpot question. The contestant will win another $20,000 if he answers the superjackpot question correctly, but he will lose all his previous winnings if he answers incorrectly. If the contestant feels that the probability of his answering the jackpot question correctly is .5 and the probability of his answering the superjackpot question correctly is .25, what decision strategy maximizes the contestant's expected net gain?

REFERENCES

1. Anderson, L. Dewey, ''A Longitudinal Study of the Effects of Nursery-School Training on Successive Intelligence-Test Ratings,'' in *Intelligence: Its Nature and Nurture,* Guy M. Whipple, ed. Bloomington, IL: National Society for the Study of Education, 1940.
2. Pascal, Gerald R. and Barbara Suttell, ''Testing the Claims of a Graphologist,'' *Journal of Personality,* vol. 16 (1947), pp. 192–197.
3. Raiffa, Howard, *Decision Analysis.* Reading, MA: Addison-Wesley, 1968.
4. Swalm, Ralph O., ''Utility Theory—Insights into Risk Taking,'' *Harvard Business Review,* November-December 1966, pp. 123–126.

8

FUNCTIONS

In Section 1.2 we introduced the concept of a linear function and noted its importance in describing relationships between variables. In this chapter we will introduce the principal types of functions that will be encountered in our study of calculus in Chapters 9–15. The reader should carefully review Sections 1.1 and 1.2, which form the basis for this discussion.

8.1 Functions and Their Graphs

Functions describe the relationships among variables. We have already worked with functions several times in this book. For example, in Section 1.2 we introduced the idea of a linear function, and in Section 1.4 we used linear cost and revenue functions in our discussion of break-even analysis. Then in Section 1.5 we considered functions that relate the supply and demand for a product to its selling price. Later, in Section 3.1, we encountered the objective function in our discussion of linear programming.

As used in this book, a **function** is a rule that associates to a given input *one and only one* output. We normally denote functions by lower or upper case italic letters such as f, g, P, or R. In Chapter 15 we will discuss functions for which the input is an ordered pair (x, y), an ordered triple (x, y, z), or, more generally, an ordered n-tuple $(x_1, x_2, \ldots, x_n)$. Until then we will restrict our attention to functions for which the input is a single real number.

If f is a function, the unique output assigned by f to the input x is called **the value of f at x** and is denoted by $f(x)$. The symbol $f(x)$ is read "f of x." In mathematics we are often concerned with the relationship between two variables defined by an equation of the form

$$y = f(x),$$

where f is a function. In this situation we refer to x as the **independent variable** and y as the **dependent variable.**

The rule that assigns to every Celsius temperature x the corresponding Fahrenheit temperature $F(x)$ is

$$F(x) = \frac{9}{5}x + 32.$$

(See Example 1.9 in Section 1.2 for details.) This rule provides us with a function in which the independent variable is the Celsius temperature and the dependent variable is the corresponding Fahrenheit temperature. Note that this function gives a rule for converting Celsius temperatures into Fahrenheit temperatures: To obtain the Fahrenheit temperature, multiply the Celsius temperature by 9/5 and add 32.

As another example of a function, consider the balance of an account into which a principal of \$1000 is deposited at 6% interest compounded annually. By formula (4.4) in Section 4.2, we see that the balance of this account after n years is given by

$$B(n) = \$1000(1.06)^n.$$

This formula expresses the account balance $B(n)$ as a function of the number of years n that the money is in the account; in other words, the value of B at n is the balance of this account after n years. In this case the time that the money is left in the account is the independent variable, and the value of the account is the dependent variable.

When a function f is defined by an equation such as

$$f(x) = x^2 - 3x + 5,$$

we can determine the output assigned to any input by substituting the input for each occurrence of x in the equation. Thus the output assigned to $x = -1$ is

$$f(-1) = (-1)^2 - 3(-1) + 5 = 1 + 3 + 5 = 9,$$

and the output assigned to 2 is

$$f(2) = (2)^2 - 3(2) + 5 = 4 - 6 + 5 = 3.$$

Practice Problem 1 Let $g(x) = x^3 - 2x^2 + x - 4$. Evaluate each of the following.
(a) $g(-2)$ (b) $g(3)$

Sometimes a function is defined piecewise; that is, more than one formula is involved in the definition of the function. Consider, for example, a student who earns $4 per hour but is paid time-and-a-half for any work beyond 40 hours in a given week. Let x denote the number of hours worked per week and $P(x)$ denote the student's weekly pay. If the student works up to 40 hours in a given week, then his pay will be $4x$ dollars (the hourly wage times the number of hours worked). On the other hand, if the student works more than 40 hours in one week, he will be paid at the rate of $4 per hour for the first 40 hours and $6 per hour for each hour worked beyond 40. Thus his pay is $4(40) = 160$ dollars for the first 40 hours and $6(x - 40)$ dollars for the hours beyond 40, for a total of

$$160 + 6(x - 40) = 6x - 80$$

dollars.

This function involves two separate formulas for computing $P(x)$ that depend on the value of x. If $0 \leq x \leq 40$, then $P(x) = 4x$; whereas if $40 < x$, then $P(x) = 6x - 80$. Hence we write

$$P(x) = \begin{cases} 4x & \text{if } 0 \leq x \leq 40 \\ 6x - 80 & \text{if } 40 < x. \end{cases}$$

To evaluate such a function, we must choose the formula corresponding to the particular value of x in which we are interested. For example, if the student works 30 hours in a given week, we apply the formula $P(x) = 4x$ to determine his pay, which is

$$P(30) = 4(30) = 120$$

dollars. But if the student works 50 hours in a given week, we must apply the formula $P(x) = 6x - 80$; in this case his pay is

$$P(50) = 6(50) - 80 = 300 - 80 = 220$$

dollars.

EXAMPLE 8.1 For first-class mail weighing up to 5 ounces, the U.S. Postal Service charges 25¢ for the first ounce and 20¢ for each additional ounce or fraction thereof. If x denotes the weight of a letter in ounces and $C(x)$ is the cost in cents for sending it via first-class mail, find a formula expressing $C(x)$ in terms of x and evaluate $C(2.5)$.

Solution The required formula has five parts. If $0 < x \leq 1$, then $C(x) = 25$. If $1 < x \leq 2$, then $C(x) = 25 + 20 = 45$ (25¢ for the first ounce and 20¢ for the additional weight). Continuing in this manner, we see that

$$C(x) = \begin{cases} 25 & \text{if } 0 < x \leq 1 \\ 45 & \text{if } 1 < x \leq 2 \\ 65 & \text{if } 2 < x \leq 3 \\ 85 & \text{if } 3 < x \leq 4 \\ 105 & \text{if } 4 < x \leq 5. \end{cases}$$

To evaluate $C(x)$ for a particular value of x, we simply apply the appropriate part of the formula. Thus $C(2.5) = 65$, and so the cost of mailing a letter weighing 2½ ounces is 65¢. ■

Practice Problem 2 A mutual fund assesses sales charges of 8% on investments of less than \$10,000, 6% on amounts of at least \$10,000 but less than \$25,000, and 5% on amounts of \$25,000 or more.

(a) Describe $S(x)$, the amount of sales charge for an investment of x dollars, where $x \geq 0$.

(b) Evaluate $S(8000)$, $S(12{,}000)$, and $S(30{,}000)$.

INTERVAL NOTATION

When working with functions, we will frequently be concerned with special types of sets of real numbers. In Example 8.1, for instance, the function C was defined piecewise on the intervals $0 < x \leq 1$, $1 < x \leq 2$, $2 < x \leq 3$, $3 < x \leq 4$, and $4 < x \leq 5$. It will prove useful to have special notations for the types of sets that occur most often. These notations are indicated in the table at the top of page 446, in which a and b denote fixed real numbers such that $a < b$.

Inequality Notation	*Interval Notation*	*Graph*
$a < x$	(a, ∞)	a
$a < x < b$	(a, b)	a b
$x < b$	$(-\infty, b)$	b
$a \leq x$	$[a, \infty)$	a
$a \leq x \leq b$	$[a, b]$	a b
$x \leq b$	$(-\infty, b]$	b
$a < x \leq b$	$(a, b]$	a b
$a \leq x < b$	$[a, b)$	a b

In this table, notice that:

1. A strict inequality ($<$) is denoted by a *parenthesis* in the interval notation. An *open circle* is used in the graph to denote that the corresponding endpoint of the interval is *not* part of the graph.
2. A less than or equal to inequality ($\leq$) is denoted by a *bracket* in the interval notation. A *filled circle* is used in the graph to denote that the corresponding endpoint of the interval is part of the graph.

Thus the set of numbers satisfying the conditions $3 < x \leq 5$ is denoted in interval notation as $(3, 5]$, and the set of numbers satisfying $2 < x < 9$ is denoted $(2, 9)$. Note that the symbols ∞ and $-\infty$, which are read "infinity" and "negative infinity," respectively, do not represent real numbers; instead these symbols are used to denote certain unbounded sets of real numbers.

Practice Problem 3 Express the following sets of numbers in interval notation, and graph the corresponding sets on a number line.

(a) $-3 \leq x < 8$ (b) $-5 \leq x \leq 0$ (c) $-4 \leq x$ (d) $x < 6$

THE DOMAIN OF A FUNCTION

In working with a function it is necessary to know the acceptable values of the independent variable. The set of all such values is called the **domain** of the function. Sometimes this set is described explicitly; for instance, the function C in Example 8.1 is defined for $0 < x \leq 5$, and the function S in Practice Problem 2 is defined for $x \geq 0$.

Most often, however, the domain is not explicitly given. In this case we will regard the domain of the function to be the set of all values of the independent variable for which the rule defining the function makes sense. For example, the function g defined by

$$g(x) = \frac{1}{x}$$

makes sense for all real numbers except $x = 0$. (Division by 0 is prohibited.) Hence if the domain of g is not described explicitly, we will regard the domain of g as consisting of all real numbers except 0. Likewise functions such as

$$h(x) = \sqrt{x} \qquad \text{or} \qquad H(x) = \sqrt[4]{x}$$

are defined only when $x \geq 0$ because we cannot extract even roots of negative numbers. Hence $[0, \infty)$ is the domain of both h and H.

EXAMPLE 8.2 Determine the domain of the function defined by

$$f(x) = \frac{\sqrt{x + 1}}{x - 2}.$$

Solution Since division by 0 is prohibited, we must have

$$x - 2 \neq 0,$$

that is, $x \neq 2$. Furthermore, the numerator $\sqrt{x + 1}$ is defined only when

$$x + 1 \geq 0,$$

that is, when $x \geq -1$. Hence the domain of f consists of all numbers in the interval $[-1, \infty)$ except $x = 2$. Thus the domain of f can be written in interval notation as

$$[-1, 2) \cup (2, \infty). \quad \blacksquare$$

Practice Problem 4 Find the domain of the function

$$g(x) = \frac{3}{\sqrt{4 - x}}.$$

GRAPHING FUNCTIONS

For many purposes it is useful to have a geometric representation of a function. Such a representation allows us to determine where a function is increasing, where it assumes a maximum or minimum value, and so forth. By the **graph** of a function f we mean the set of all ordered pairs (x, y) that satisfy the equation $y = f(x)$. By plotting enough of these ordered pairs, we can sketch the graph of the function. In fact, this is the method for graphing functions that is used by microcomputer software. In Chapter 11 we will use sophisticated techniques from calculus to sketch the graph of a function accurately by determining its key features. For the present, however, we will be able to obtain a reasonable approximation to the graph of a simple function by plotting enough points.

EXAMPLE 8.3 Graph the function $f(x) = 4 - x^2$.

Solution We must determine ordered pairs that satisfy the equation

$$y = 4 - x^2.$$

Since this equation is unfamiliar, it provides us no information about the graph of f. Consequently our only method for graphing this function is to plot enough ordered pairs that we obtain a good idea about the shape of the graph.

We obtain ordered pairs to plot by evaluating y at convenient values of x. The table below lists several ordered pairs satisfying the equation $y = 4 - x^2$.

x	-4	-3	-2	-1	0	1	2	3	4
y	-12	-5	0	3	4	3	0	-5	-12

By plotting these points and joining them with a smooth curve, we obtain the graph in Figure 8.1. ■

Practice Problem 5 Graph the function $g(x) = x^2 - 2x - 5$.

EXAMPLE 8.4 Graph the function

$$P(x) = \begin{cases} 4x & \text{if } 0 \leq x \leq 40 \\ 6x - 80 & \text{if } 40 < x \end{cases}$$

discussed earlier in this section.

Solution To graph P, we must plot ordered pairs (x, y) satisfying the equation $y = P(x)$. Thus we will evaluate P at convenient values of x. Note that since P is defined piecewise on the intervals $[0, 40]$ and $(40, \infty)$, we must choose several values of x lying in each of these intervals.

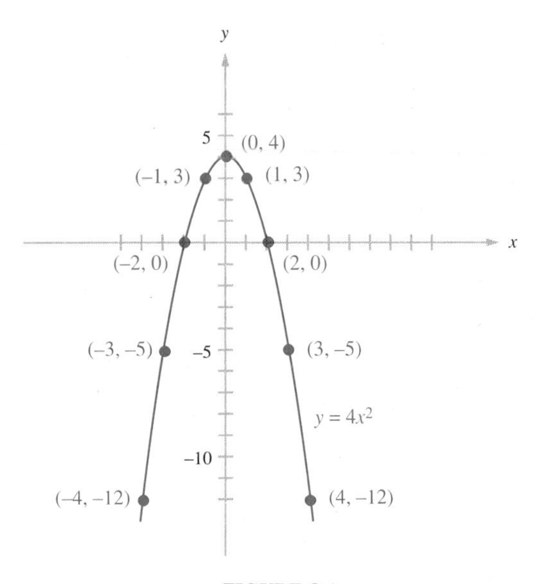

FIGURE 8.1

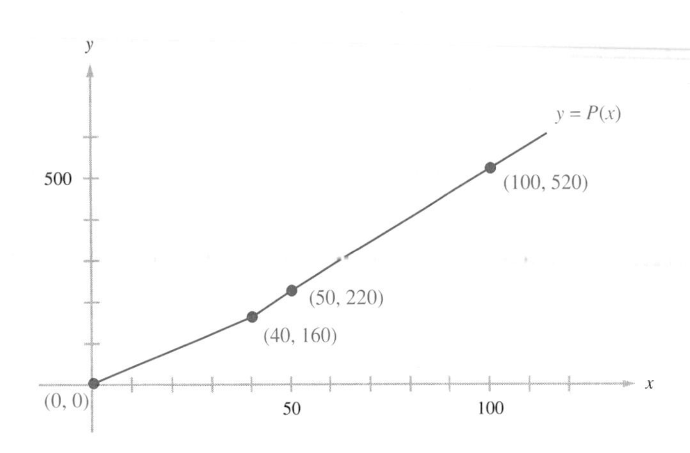

FIGURE 8.2

First consider the graph of P on $[0, 40]$. On this interval $P(x) = 4x$, and hence we must graph

$$y = 4x$$

for $0 \le x \le 40$. From Chapter 1 we recognize this equation as a linear equation, and so its graph is a straight line. Thus we can accurately graph P on the interval $[0, 40]$ by plotting two distinct points and drawing the line through them. Although any two points having x-coordinates in this interval will suffice, let us choose the endpoints $x = 0$ and $x = 40$. For $x = 0$ we have

$$y = 4x = 4(0) = 0,$$

and for $x = 40$ we have

$$y = 4x = 4(40) = 160.$$

Hence the graph of P on the interval $[0, 40]$ is the line segment joining $(0, 0)$ to $(40, 160)$. See Figure 8.2.

On the interval $(40, \infty)$, $P(x) = 6x - 80$. Since this is another linear function, we see that the graph of P can be sketched accurately by plotting any two points with x-coordinates greater than 40. Let us arbitrarily choose $x = 50$ and $x = 100$. For $x = 50$ we have

$$y = P(50) = 6(50) - 80 = 220,$$

and for $x = 100$ we have

$$y = P(100) = 6(100) - 80 = 520.$$

Hence the graph of P on the interval $(40, \infty)$ is a portion of the line through $(50, 220)$ and $(100, 520)$. See Figure 8.2 again. ■

Practice Problem 6 Graph the function C discussed in Example 8.1:

$$C(x) = \begin{cases} 25 & \text{if } 0 < x \leq 1 \\ 45 & \text{if } 1 < x \leq 2 \\ 65 & \text{if } 2 < x \leq 3 \\ 85 & \text{if } 3 < x \leq 4 \\ 105 & \text{if } 4 < x \leq 5. \end{cases}$$

EXAMPLE 8.5

Graph the absolute value function $f(x) = |x|$.

Solution We must graph the equation $y = |x|$. Evaluating y for various values of x, we obtain the table below.

x	−3	−2	−1	0	1	2	3
y	3	2	1	0	1	2	3

Plotting the pairs (x, y) and connecting the points with a smooth curve, we obtain the graph in Figure 8.3. Notice that the graph consists of two linear pieces joined at the origin. This description of the graph is easily seen from the definition of absolute value:

$$|x| = \begin{cases} x & \text{if } x \geq 0 \\ -x & \text{if } x < 0. \end{cases} \quad ■$$

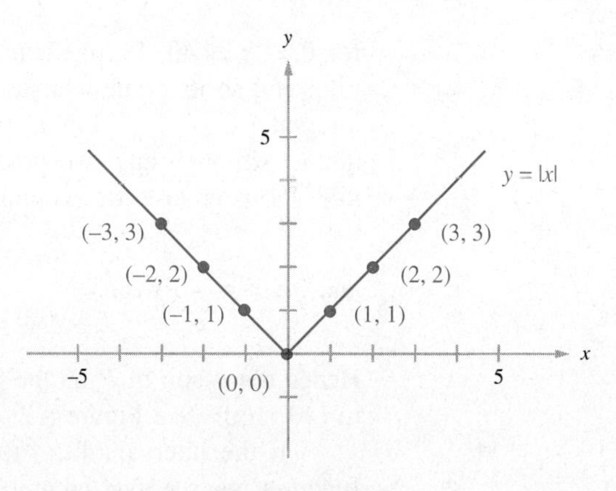

FIGURE 8.3

Practice Problem 7 Graph the function $F(x) = |x - 2| + 1$.

If y is a function of x, then each value of x in the domain of the function produces one and only one value of y. Thus there can never be two ordered pairs on the graph of a function of x with the same first coordinate and different second coordinates. This observation can be restated in geometric terms as follows.

Vertical Line Test

A curve in the xy-plane is the graph of a function of x if and only if every vertical line intersects the curve at no more than one point.

Observe that the vertical line test is satisfied by the graphs in Figures 8.1–8.3. On the other hand, the circle in Figure 8.4 does not satisfy the vertical line test because the vertical line $x = 4$ intersects the circle twice. Hence the circle is not the graph of a function.

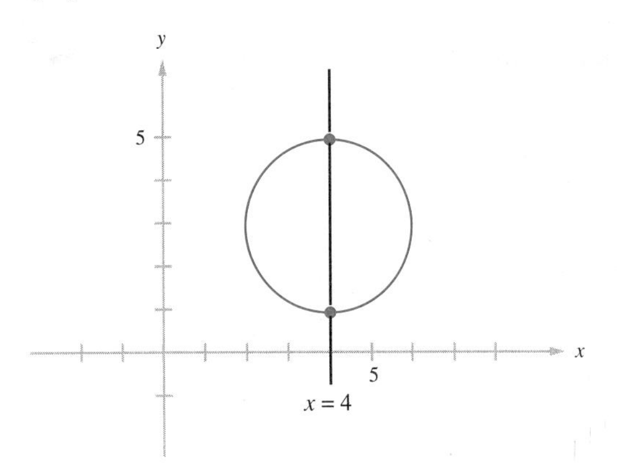

FIGURE 8.4

In Example 8.4 we knew from the equation $y = 4x$ that the graph of P would be a straight line on the interval [0, 40]. As a result we were able to graph P accurately on the interval by plotting only two ordered pairs. In the rest of this chapter we will discuss other functions with characteristic shapes that are defined by special types of equations. By recognizing these special equations, it is possible to graph these types of functions by plotting only a few points, just as it is easy to graph a linear function by plotting only two points.

EXERCISES 8.1

In Exercises 1–10 determine which curves can be the graphs of a function of x.

1.

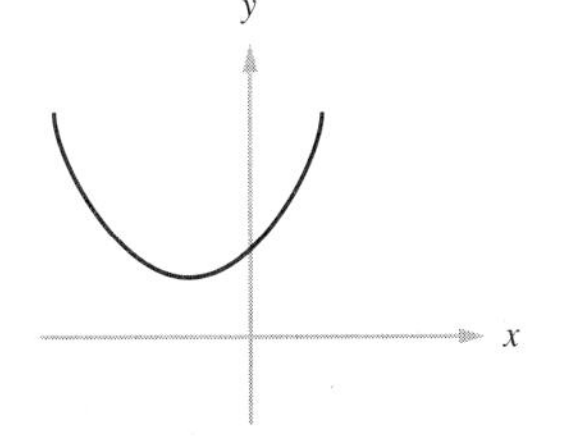

2.

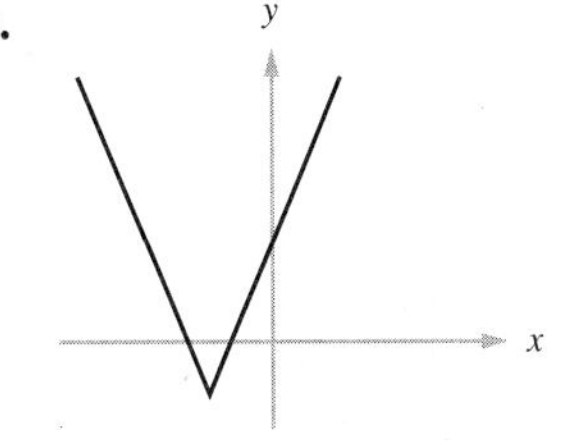

3.

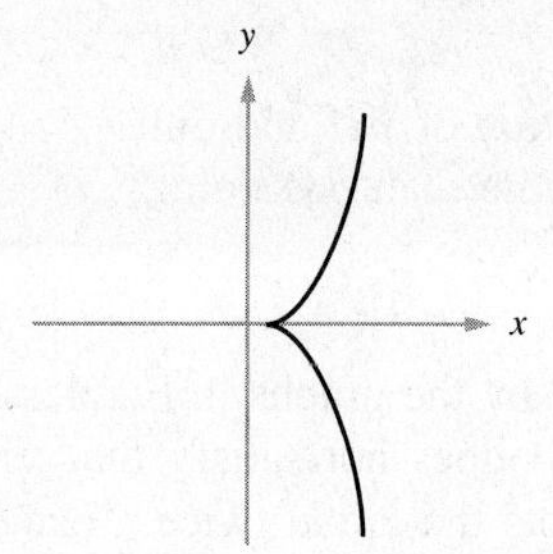

4.

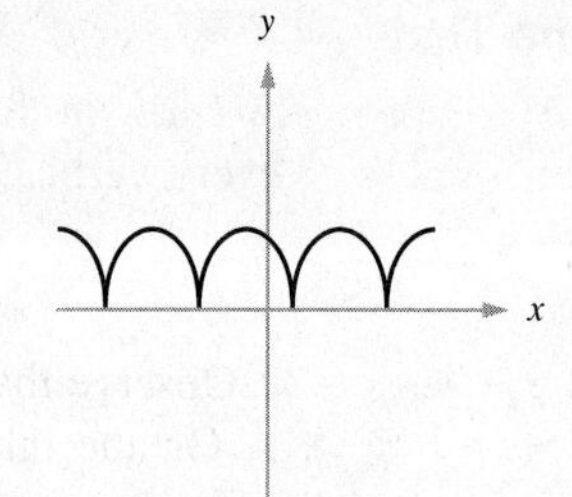

5.

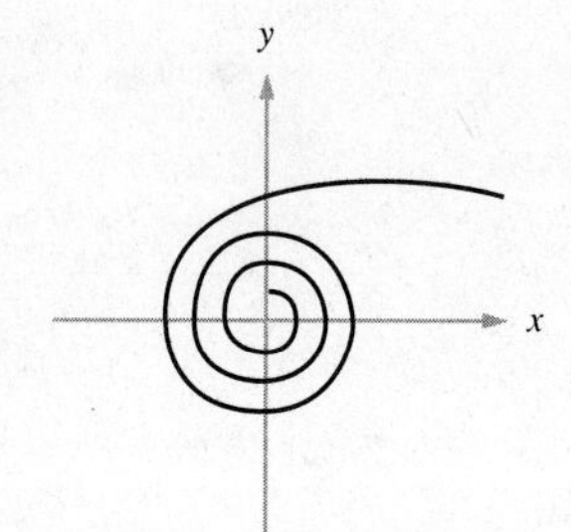

6.

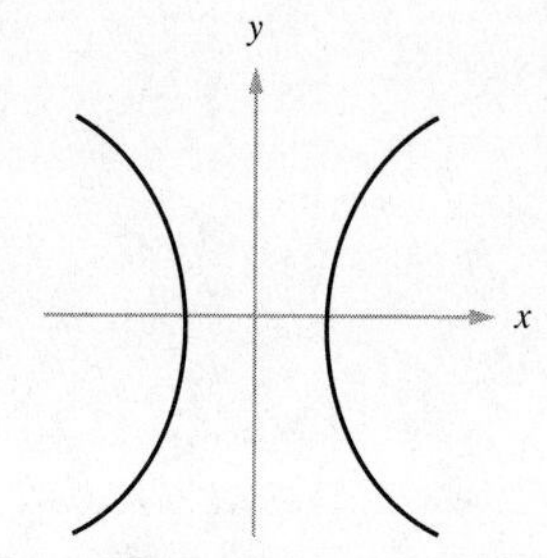

7.

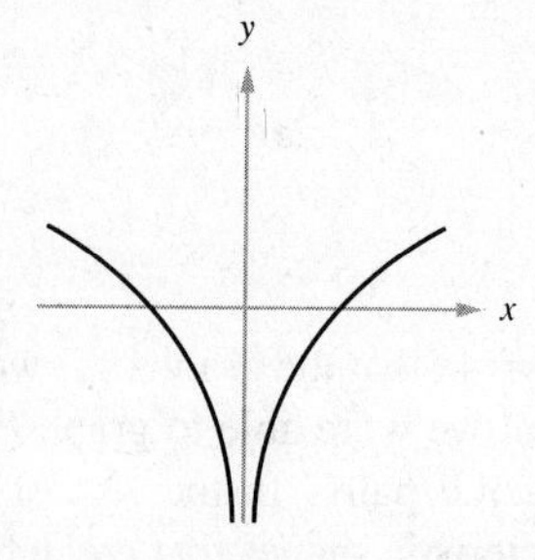

8.

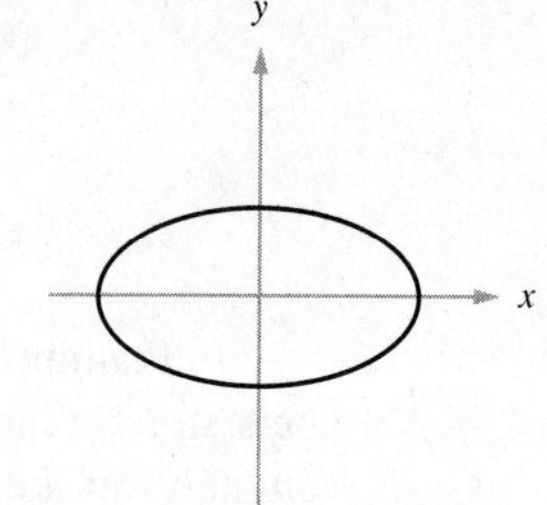

9.

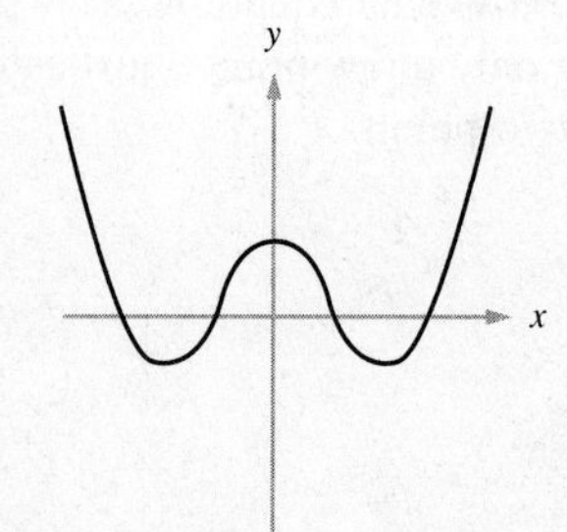

10.

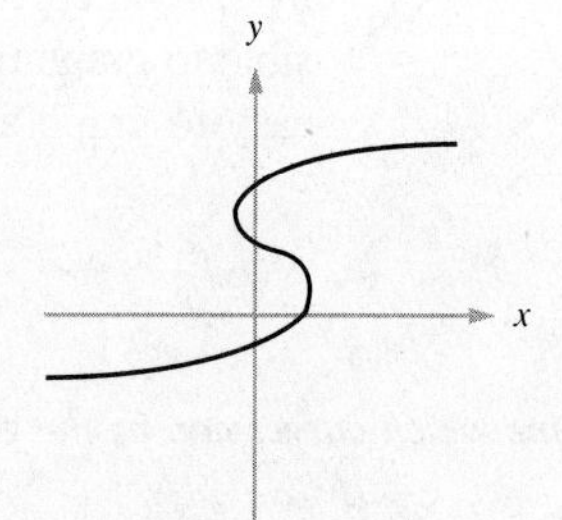

In Exercises 11–26 use interval notation to describe the set of numbers satisfying the given inequality, and graph each set on a number line.

11. $3 \leq x$

12. $x < 17$

13. $-4 < x < 8$

14. $-3 < x \leq 9$

15. $x \leq 13$

16. $-2 \leq x \leq 4$

17. $-1 \leq x < 5$

18. $16 < x$

19. $2 < x \leq 15$ **20.** $11 \leq x$ **21.** $1 \leq x \leq 10$ **22.** $x \leq -8$

23. $-6 < x$ **24.** $-6 < x < -1$ **25.** $x < -5$ **26.** $5 \leq x < 9$

In Exercises 27–36 compute f(a) for the given function f and number a.

27. $f(t) = \dfrac{t}{t^2 + 4}$ and $a = 6$

28. $f(x) = |3x - 1|$ and $a = 5$

29. $f(x) = 3^x$ and $a = -2$

30. $f(t) = 2t^3 - t^2 + 1$ and $a = -1$

31. $f(u) = |2 + 3u|$ and $a = -5$

32. $f(s) = \sqrt{2s - 3}$ and $a = 14$

33. $f(x) = \begin{cases} 2x - 5 & \text{if } x < 0 \\ x^2 + 3 & \text{if } x \geq 0 \end{cases}$ and $a = 4$

34. $f(x) = \begin{cases} -x^2 + 2x & \text{if } x \leq 1 \\ 4x - 3 & \text{if } x > 1 \end{cases}$ and $a = -2$

35. $f(t) = \begin{cases} 2 + t^3 & \text{if } t \leq -2 \\ 3t + 8 & \text{if } t > -2 \end{cases}$ and $a = -2$

36. $f(t) = \begin{cases} 2t & \text{if } t \leq 5 \\ \dfrac{1}{t} & \text{if } t > 5 \end{cases}$ and $a = -2$

In Exercises 37–46 determine the domain of the given function. Write your answers in interval notation.

37. $f(x) = x^3 - 4x^2 + 5x - 6$

38. $F(x) = \dfrac{2}{x - 2}$

39. $g(x) = \dfrac{2x}{x^2 - 9}$

40. $G(x) = 3x^4 - 5x^2 + 7$

41. $h(x) = \sqrt{6 - 3x}$

42. $H(x) = \dfrac{x - 3}{x^2 - 2x - 35}$

43. $F(x) = \dfrac{\sqrt[3]{x + 4}}{x}$

44. $f(x) = \dfrac{\sqrt{2x - 8}}{x - 9}$

45. $G(x) = \dfrac{\sqrt{4x + 12}}{x^2 - 5x}$

46. $g(x) = \dfrac{x^2 - 1}{\sqrt[3]{x - 5}}$

Graph the functions in Exercises 47–60.

47. $f(x) = 10 - 2x$

48. $g(x) = 3x - 6$

49. $F(x) = 1 - x^2$

50. $G(x) = |2x + 4|$

51. $g(x) = 5 - |x + 1|$

52. $h(x) = \sqrt{x - 2}$

53. $G(x) = \sqrt{4 - x}$

54. $H(x) = x^2 - x^3$

55. $h(x) = \begin{cases} 2x + 6 & \text{if } x < -1 \\ 3 - x & \text{if } x \geq -1 \end{cases}$

56. $f(x) = \begin{cases} 3 - 2x & \text{if } x \leq 1 \\ 4x - 3 & \text{if } x > 1 \end{cases}$

57. $H(x) = \begin{cases} -2x + 4 & \text{if } x < 0 \\ \dfrac{x}{2} - 2 & \text{if } x \geq 0 \end{cases}$

58. $F(x) = \begin{cases} x - 3 & \text{if } x < -1 \\ 5 - 2x & \text{if } x \geq -1 \end{cases}$

59. $f(x) = \begin{cases} -x & \text{if } x \leq 0 \\ x^2 & \text{if } x > 0 \end{cases}$

60. $g(x) = \begin{cases} -\dfrac{x}{2} + 1 & \text{if } x < 2 \\ 4 - x^2 & \text{if } x \geq 2 \end{cases}$

Answers to Practice Problems

1. (a) $g(-2) = -22$ (b) $g(3) = 8$

2. (a) The required function is

$$S(x) = \begin{cases} 0.08x & \text{if } 0 \leq x < 10{,}000 \\ 0.06x & \text{if } 10{,}000 \leq x < 25{,}000 \\ 0.05x & \text{if } 25{,}000 \leq x. \end{cases}$$

(b) $S(8000) = 640$, $S(12{,}000) = 720$, and $S(30{,}000) = 1500$

3. (a) $[-3, 8)$

(b) $[-5, 0]$

(c) $[-4, \infty)$

(d) $(-\infty, 6)$

4. The domain is $(-\infty, 4)$.

5.

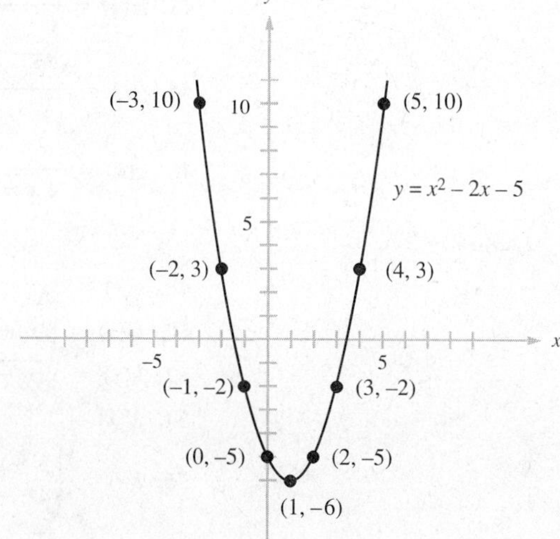

6.

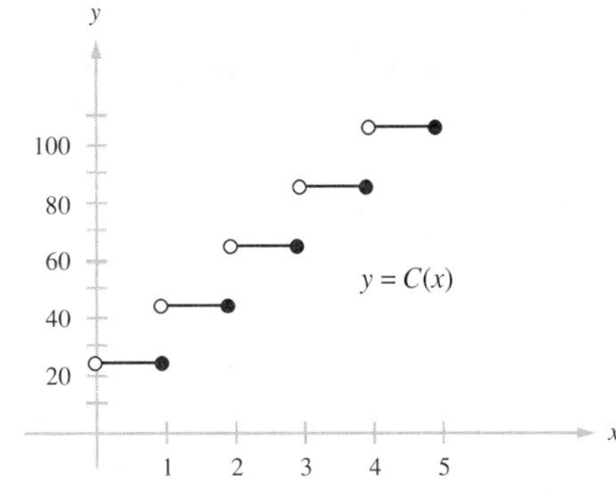

7.

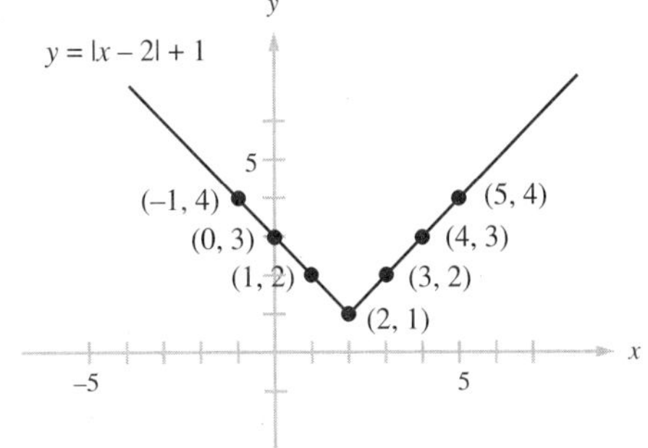

8.2 Polynomial and Rational Functions

The usefulness of linear functions in applications has been illustrated many times in this book. Linear functions are a special case of an important class of functions that will now be discussed.

By a **polynomial function** we mean a function expressible in the form

$$f(x) = a_n x^n + a_{n-1} x^{n-1} + \cdots + a_1 x + a_0,$$

where n is a nonnegative integer and the coefficients $a_n, a_{n-1}, \ldots, a_1, a_0$ are constants. Note that a polynomial function is defined for all real numbers.

If in the representation

$$f(x) = a_n x^n + a_{n-1} x^{n-1} + \cdots + a_1 x + a_0$$

all the coefficients $a_n, a_{n-1}, \cdots, a_1, a_0$ are zero, then f has the form $f(x) = 0$ and is called the **zero polynomial.** Otherwise the largest n for which $a_n \neq 0$ is called the **degree** of f. Thus

$f(x) = 5 - 2x + 7x^3$ is a polynomial function of degree 3,

$g(x) = 4x^5 - \sqrt{2}x^2$ is a polynomial function of degree 5, and

$h(x) = -8$ is a polynomial function of degree 0.

Practice Problem 8 Which of the following are polynomial functions? Give the degrees of those that are, and explain why the others are not.

(a) $f(x) = -3 + 6x$ (b) $g(x) = 5x^{-2} + 2x - 7$

(c) $h(x) = 3x^4 - 6x^3 + 5\sqrt{x} - 8$ (d) $F(x) = 3x^3 + 8x^7 - 2x^4$

(e) $G(x) = \dfrac{12}{x^4} - \dfrac{6}{x^2} + 15$

If the degree of a polynomial function

$$f(x) = a_n x^n + a_{n-1}x^{n-1} + \cdots + a_1 x + a_0$$

is 0, then f is a constant function

$$f(x) = a_0.$$

Moreover, if the degree of f is 1, then f is a linear function

$$f(x) = a_1 x + a_0.$$

Thus polynomial functions of degrees 0 and 1 have already been discussed in Chapter 1.

QUADRATIC FUNCTIONS

A polynomial function of degree 2,

$$f(x) = a_2 x^2 + a_1 x + a_0, \qquad \text{where} \quad a_2 \neq 0,$$

is called a **quadratic function.** The graph of a quadratic function has a characteristic shape called a **parabola,** which can be seen in Figure 8.5. Whether the parabola opens upward or downward is determined by the coefficient of the x^2 term.

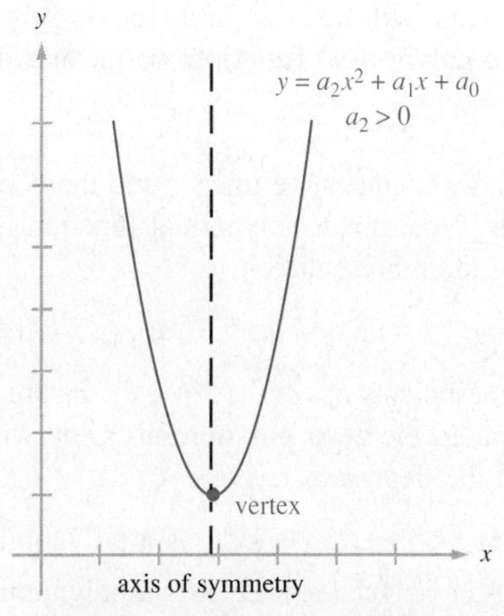

(a) An upward-opening parabola

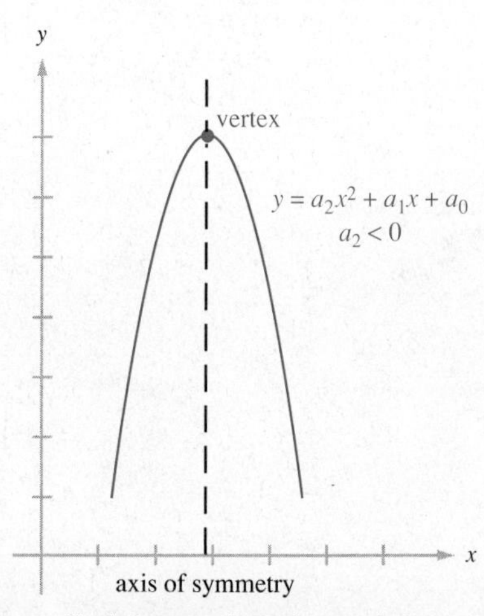

(b) A downward-opening parabola

FIGURE 8.5

If $a_2 > 0$, the parabola opens upward.

If $a_2 < 0$, the parabola opens downward.

For parabolas defined by an equation of the form

$$y = a_2x^2 + a_1x + a_0,$$

the lowest point on an upward-opening parabola or the highest point on a downward-opening parabola is called the **vertex.** Note that either parabola is symmetric with respect to a vertical line passing through the vertex. Determining the vertex of a parabola (and therefore its axis of symmetry) is a great help in graphing the parabola, and we will learn how to do this in Chapter 11. Until then, when graphing a parabola we must plot enough points on the graph to determine where the parabola changes direction.

EXAMPLE 8.6 Graph the function $g(x) = x^2 + 4x - 3$.

Solution Since g is a quadratic function, its graph is a parabola. Moreover, the coefficient of the x^2 term is positive, and so the parabola opens upward. To graph g accurately, we must compute some ordered pairs satisfying the equation $y = g(x)$.

x	-5	-4	-3	-2	-1	0	1
y	2	-3	-6	-7	-6	-3	2

From the graph of g, which is shown in Figure 8.6, it appears that the vertex of the parabola is the point $(-2, -7)$. In Chapter 11 we will develop a method for obtaining the vertex of a parabola without graphing. ■

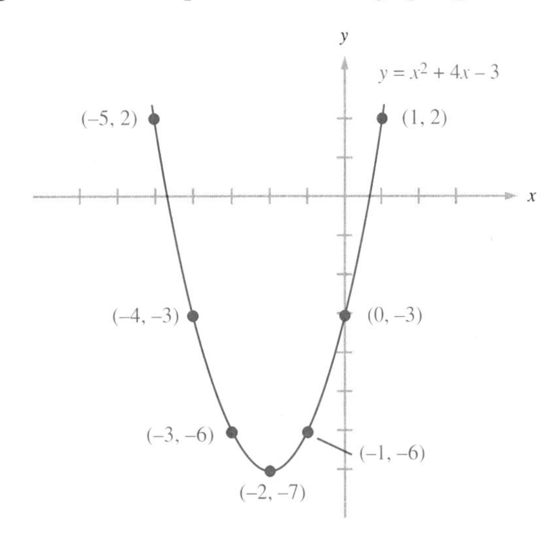

FIGURE 8.6

EXAMPLE 8.7 Graph the function $h(x) = 5 + 2x - x^2$.

Solution Since this is a quadratic function in which the coefficient of the x^2 term is negative, its graph is a downward-opening parabola. To obtain some ordered pairs on the graph, we evaluate $y = h(x)$ at the integers between -2 and 4.

x	-2	-1	0	1	2	3	4
y	-3	2	5	6	5	2	-3

The graph of h is shown in Figure 8.7. ■

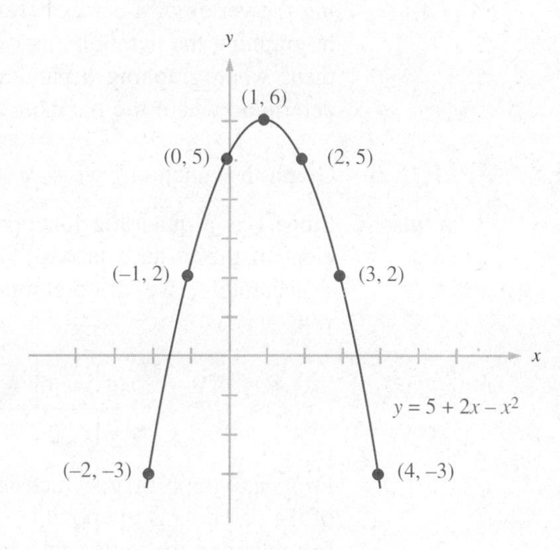

FIGURE 8.7

Practice Problem 9 Graph the function $f(x) = 3 + 4x - 2x^2$.

Quadratic functions arise in connection with certain common economic situations. Suppose, for example, that the number of units x of a product or service that can be sold is a linear function of its price p, say $x = mp + b$. (Since sales decrease as the price increases, m will be negative.) Then the revenue derived from the sale of x units at price p is

$$R = px = p(mp + b) = mp^2 + bp,$$

which is a quadratic function of p.

EXAMPLE 8.8 Suppose that the number of units of a particular model of television set that can be sold per month is

$$x = 16{,}000 - 40p,$$

where p is the price of the set in dollars. If the total cost of producing x sets is

$$C = 500{,}000 + 100x,$$

for which selling prices will the manufacturer earn a profit?

At a price of p dollars, the monthly demand is $x = 16{,}000 - 40p$. The cost of producing x sets is $C = 500{,}000 + 100x$.

Solution To determine the relationship between the total cost and the selling price of the television sets, we must substitute $x = 16{,}000 - 40p$ into the formula for C to obtain

$$\begin{aligned} C &= 500{,}000 + 100x \\ &= 500{,}000 + 100(16{,}000 - 40p) \\ &= 2{,}100{,}000 - 4000p. \end{aligned}$$

Furthermore, the total revenue received from the sale of x units at price p is

$$\begin{aligned} R &= px \\ &= p(16{,}000 - 40p) \\ &= 16{,}000p - 40p^2. \end{aligned}$$

The cost and revenue functions are shown in Figure 8.8. Since the cost function is a linear function of p, its graph is a line. Likewise because the revenue function is quadratic, its graph is a parabola.

As we can see, production cost exceeds revenue if the selling price per unit is too low ($p < p_1$) or too high ($p > p_2$). However, if $p_1 < p < p_2$, then the revenue exceeds the total cost, and there will be a profit. Thus in order to

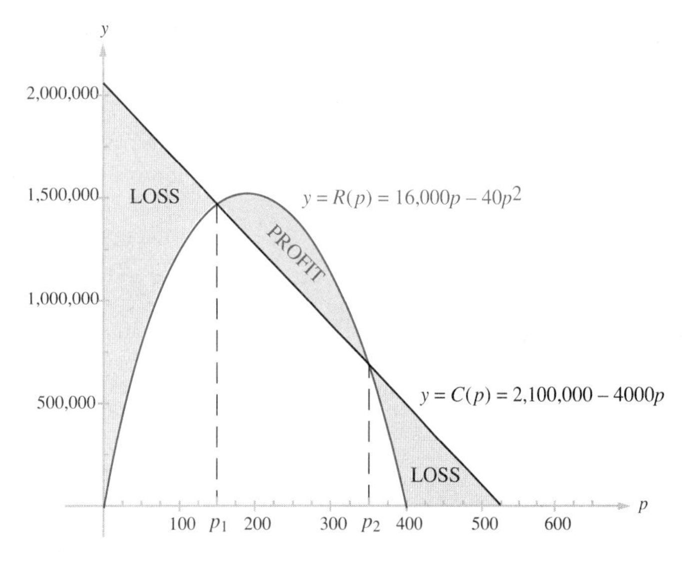

FIGURE 8.8

earn a profit the television sets must be priced so that $p_1 < p < p_2$. Note in Figure 8.8 that p_1 and p_2 are the break-even prices at which revenue equals cost. This observation enables us to find p_1 and p_2.

$$\begin{aligned} R &= C \\ 16{,}000p - 40p^2 &= 2{,}100{,}000 - 4000p \\ 0 &= 40p^2 - 20{,}000p + 2{,}100{,}000 \\ 0 &= p^2 - 500p + 52{,}500 \end{aligned}$$

By using the quadratic formula, we see that $p = 150$ or $p = 350$. (Hence $p_1 = 150$ and $p_2 = 350$ in Figure 8.8.) Therefore the manufacturer will earn a profit if the selling price of a television set is between \$150 and \$350. ■

POLYNOMIAL FUNCTIONS OF HIGHER DEGREE

Polynomial functions of degree three or higher are more difficult to graph than linear or quadratic polynomials because their graphs do not have a characteristic shape. In Chapter 11 we will discuss sophisticated techniques that enable us to graph any polynomial function accurately. For the present we will make a rough sketch of the graph of a polynomial of degree greater than two by plotting enough points to show its behavior.

EXAMPLE 8.9 Graph the function $f(x) = x^3$.

Solution To obtain some ordered pairs on the graph, we evaluate $f(x)$ for the values shown in the table below.

x	-2	-1.5	-1	-0.5	0	0.5	1	1.5	2
y	-8	-3.375	-1	-0.125	0	0.125	1	3.375	8

Plotting these ordered pairs produces the graph in Figure 8.9. ■

Practice Problem 10 Graph $f(x) = -x^3 + 3x^2$.

RATIONAL FUNCTIONS

A function is called a **rational function** if it is expressible in the form

$$\frac{p}{q},$$

where p and q are polynomial functions and q is not the zero polynomial. Thus

$$\frac{x+6}{x-3}, \quad \frac{x-4}{x^2-x}, \quad x^3 + 7x, \quad \frac{x^2-9}{x^3+1}, \quad \text{and} \quad \frac{5}{(x-2)^2}$$

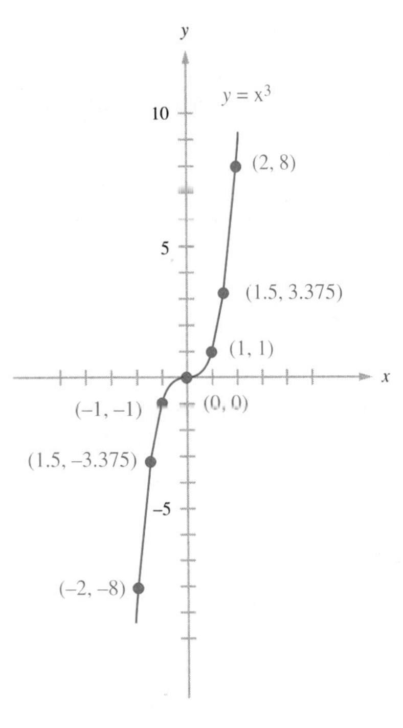

FIGURE 8.9

are rational functions. Note that every polynomial function is a rational function since q can be taken as the constant polynomial 1.

Unlike polynomial functions, a rational function need not be defined for all real numbers. It is easy to see, however, that the domain of a rational function consists of all real numbers for which its denominator is nonzero. Thus, for instance, the domain of $f(x) = 5/(x - 2)^2$ is the set of all real numbers except $x = 2$, that is,

$$(-\infty, 2) \cup (2, \infty).$$

Likewise the domain of $g(x) = (x^2 - 9)/(x^3 + 1)$ is the set of all real numbers except -1, which can be written as

$$(-\infty, -1) \cup (-1, \infty).$$

Practice Problem 11 Determine the domain of the rational functions below.

(a) $h(x) = \dfrac{x + 6}{x - 3}$ (b) $H(x) = \dfrac{x - 4}{x(x - 1)}$

Because a rational function need not be defined for all real numbers, the graph of a rational function may consist of several separate pieces. Consider, for instance, the rational function

$$f(x) = \frac{1}{x + 2},$$

which is not defined if $x = -2$. Evaluating f at convenient values of x produces the following table.

x	-5	-4	-3	-2.5	-2.25	-2.1	-1.9	-1.75	-1.5	-1	0	1	2
y	$-\frac{1}{3}$	$-\frac{1}{2}$	-1	-2	-4	-10	10	4	2	1	$\frac{1}{2}$	$\frac{1}{3}$	$\frac{1}{4}$

Plotting these points and connecting those corresponding to $x < -2$ and those corresponding to $x > -2$ yields the graph in Figure 8.10. Notice that we do not connect the two ordered pairs $(-2.5, -2)$ and $(-1.5, 2)$ because there must be a break in the graph at $x = -2$. This graph consists of two separate pieces, one for $x < -2$ and another for $x > -2$.

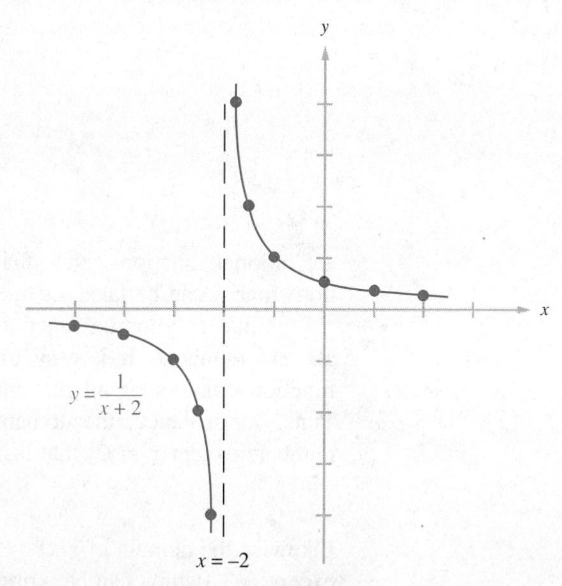

FIGURE 8.10

In Figure 8.10, we see that the values of $f(x) = 1/(x + 2)$ get very large in the positive direction as x approaches -2 from the right, and the values of $f(x)$ get very large in the negative direction as x approaches -2 from the left. When the values of a function get very large in either the positive or negative direction as x approaches a number k from one side or the other, then we say that the line $x = k$ is a **vertical asymptote** of the graph. Thus in Figure 8.10, $x = -2$ is a vertical asymptote of the graph of f. For the graph of a rational function, vertical asymptotes can occur only at values of x for which the function is undefined. However, there need not be a vertical asymptote at every point where a rational function is undefined. (See Exercise 47.)

Determining all the vertical asymptotes of a rational function is necessary if we are to graph the function accurately. Note also that when graphing a rational function, *it is essential to plot points on both sides of each value at which the function is undefined in order to determine the behavior of the entire function.*

It is also possible that the graph of a rational function approaches a horizontal line as we move far to the right or far to the left along the x-axis. In such a case we call the horizontal line a **horizontal asymptote** of the graph. Since the graph of $f(x) = 1/(x + 2)$ in Figure 8.10 gets close to the horizontal line $y = 0$ (the x-axis) as we move far to the right or far to the left, $y = 0$ is a horizontal asymptote of the graph of f.

Note that a polynomial function never has vertical asymptotes, and a polynomial function of degree 1 or more never has horizontal asymptotes.

EXAMPLE 8.10 Graph the function $f(x) = \dfrac{x - 2}{x - 4}$.

Solution The given function is undefined if $x = 4$, and so we will plot points on either side of this value. Evaluating f at convenient values around $x = 4$ produces the following table.

x	0	1	2	3	3.5	3.75	3.9	4.1	4.25	4.5	5	6	7	8
y	$\frac{1}{2}$	$\frac{1}{3}$	0	-1	-3	-7	-17	21	9	5	3	2	$\frac{5}{3}$	$\frac{3}{2}$

The graph is shown in Figure 8.11. From the graph it appears that $x = 4$ is a vertical asymptote and $y = 1$ is a horizontal asymptote. ■

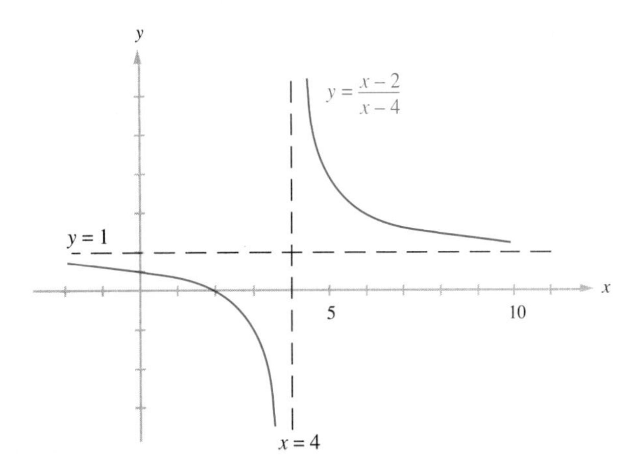

FIGURE 8.11

EXAMPLE 8.11 Graph the function $g(x) = \dfrac{x^2 - 3x + 1}{x - 3}$.

Solution Notice that $g(x)$ is not defined for $x = 3$. Thus we must evaluate g at values less than 3 and greater than 3. The table below shows such a set of values.

x	-2	-1	0	1	2	2.5	2.75	2.9	3.1	3.25	3.5	4	5	6	7	8
y	-2.2	-1.25	$-\frac{1}{3}$	$\frac{1}{2}$	1	$\frac{1}{2}$	-1.25	-7.1	13.1	7.25	5.5	5	5.5	$\frac{19}{3}$	$\frac{29}{7}$	8.2

Plotting all these points produces Figure 8.12. Notice that the line $x = 3$ is a vertical asymptote of the graph of g. In this case the graph of g does not get close to any horizontal line as we move far to the left or right along the x-axis; consequently the graph has no horizontal asymptotes. ■

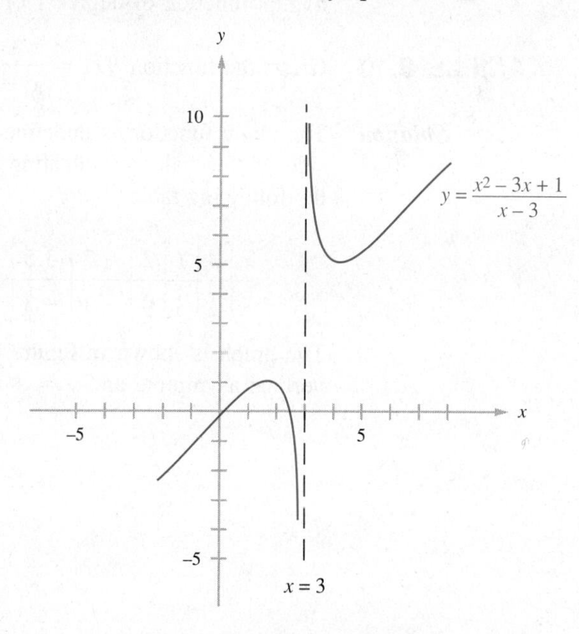

FIGURE 8.12

Practice Problem 12 Graph the rational function $h(x) = \dfrac{1}{(x - 3)^2}$ and identify any vertical and horizontal asymptotes.

APPLICATIONS OF RATIONAL FUNCTIONS

Political analysts are often interested in the size of the "coattails" of the winner of a U.S. presidential election. The following empirical rule called the *cube law* has been formulated by political scientists to estimate the proportion of seats in the House of Representatives won by Democratic candidates:

$$p(x) = \frac{x^3}{x^3 + (1 - x)^3} \quad \text{for} \quad 0 \le x \le 1.$$

Here x denotes the proportion of the popular vote received by the Democratic presidential candidate and $p(x)$ the proportion of seats in the House of Representatives won by Democratic candidates. Thus, for instance, since

$$p(0.55) = \frac{(0.55)^3}{(0.55)^3 + (0.45)^3} = \frac{0.166375}{.2575} \approx 0.646,$$

the Democratic party can expect to win about 64.6% of the seats in the House of Representatives if the Democratic presidential candidate receives 55% of the popular vote in the presidential election.

If we use the binomial theorem to expand algebraically the term $(1 - x)^3$ in the denominator of the cube law, we obtain

$$x^3 + (1 - x)^3 = x^3 + (1 - 3x + 3x^2 - x^3) = 3x^2 - 3x + 1.$$

Therefore

$$p(x) = \frac{x^3}{3x^2 - 3x + 1},$$

and so p is a rational function.

Practice Problem 13 Estimate the proportion of the seats in the House of Representatives that will be won by Democratic candidates if the Democratic presidential candidate receives 60% of the popular vote in the presidential election.

Rational functions also occur in mathematical models for the cost of removing environmental pollutants. In many situations removing each additional percentage of a pollutant becomes increasingly expensive, and in such cases a rational function may serve as a good approximation to the cost $C(x)$ of removing x% of the pollutant. Suppose, for instance, that the annual cost (in millions of dollars) for removing mercury from the waste water discharge of a certain factory has been estimated to be as in the table below.

Percentage of mercury removed	*Cost (in millions)*
50	\$0.6
60	\$0.9
75	\$1.8
80	\$2.4
90	\$5.4
95	\$11.4
98	\$29.4

In this case it is possible to show that the cost of removing $x\%$ of the mercury can be approximated by

$$C(x) = \frac{0.6x}{100 - x} \quad \text{for} \quad 0 \leq x < 100.$$

The graph of C, shown in Figure 8.13, is often called a *cost-benefit curve*.

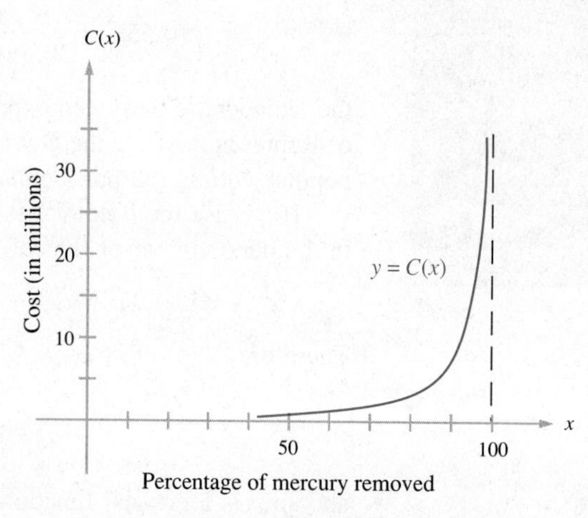

FIGURE 8.13

We will also use rational functions to obtain a model for inventory costs in Section 11.7.

EXERCISES 8.2

In Exercises 1–10 determine if the given function is a polynomial function. If so, give its degree; if not, tell why.

1. $f(x) = 6x^5 - 4x^3 + x$

2. $F(x) = \dfrac{3}{x}$

3. $g(x) = 2\sqrt{x} + 4$

4. $G(x) = 2x^2 + 6x^7 - 3x^4$

5. $h(x) = 8$

6. $H(x) = \dfrac{5x^3}{x^2 + 3x - 4}$

7. $F(x) = (x^2 - 9)^3$

8. $f(x) = \sqrt{3}x^2 - 3x + 6$

9. $G(x) = 7x^3 - 5x^{3/2} + 4x$

10. $g(x) = (3x^4 - x^3 + 2x^2 - 1)^{1/2}$

In Exercises 11–20 determine the domain of the given rational function.

11. $f(x) = \dfrac{x - 3}{x + 7}$

12. $F(x) = \dfrac{2x + 5}{4 - x}$

13. $g(x) = \dfrac{3x^5 - x^3}{9 - 5x}$

14. $G(x) = \dfrac{3 - x^2}{4x - 7}$

15. $h(x) = \dfrac{7x}{x^2 - 16}$

16. $H(x) = \dfrac{3x^3}{x^2 + 6}$

17. $F(x) = \dfrac{3x^2 - 4}{x^2 - 2x - 15}$

18. $f(x) = \dfrac{x^2 - x}{x^2 + x - 12}$

19. $G(x) = \dfrac{x^2 - x - 6}{x^4 + 1}$

20. $g(x) = \dfrac{2x - 5}{x^2 + 9}$

In Exercises 21–38 graph the given function, identifying any vertical and horizontal asymptotes.

21. $f(x) = x^2 + 2$

22. $F(x) = 8 - 2x^2$

23. $g(x) = 7 - 2x - x^2$

24. $G(x) = x^2 - 2x - 5$

25. $h(x) = x^2 - 5x + 3$

26. $H(x) = -x^2 + 2x + 4$

27. $F(x) = 5 - x^3$

28. $f(x) = 4x - x^3$

29. $G(x) = x^3 + 2x^2$

30. $g(x) = x^3 + 3x^2 + 3x + 1$

31. $F(x) = \dfrac{2}{x - 1}$

32. $f(x) = \dfrac{3}{2 - x}$

33. $G(x) = \dfrac{4}{2 - x}$

34. $g(x) = \dfrac{2}{x + 3}$

35. $H(x) = \dfrac{3x}{x + 1}$

36. $h(x) = \dfrac{x - 5}{x}$

37. $f(x) = \dfrac{2}{9 - x^2}$

38. $F(x) = \dfrac{1}{x^2 + 4x}$

39. A glass blower has found that the number of glass sculptures that can be sold per week at a price of p dollars is given by $x = 24 - p$. If the total cost of producing x sculptures is $C = 120 + x$ dollars, determine the price range that will earn a profit for the sculptor.

40. The yearly demand for a certain style of gloves is $x = 20{,}000 - 2000p$, where p denotes the price per pair (in dollars). If the total cost of producing x pairs of gloves is $C = 30{,}000 + 1.50x$ dollars, determine the price range that will earn a profit for the glove manufacturer.

41. The number of units of a certain model of air conditioner that can be sold at a price of p dollars is given by $x = 5000 - 10p$. If $C = 450{,}000 + 50x$ dollars is the total cost of producing x units, determine the price range that will earn a profit for the manufacturer.

42. The number of units of a certain model of stereo receiver that can be sold at a price of p dollars is given by $x = 18{,}000 - 50p$. If $C = 1{,}260{,}000 + 30x$ dollars is the total cost of producing x units, determine the price range that will earn a profit for the manufacturer.

43. If the cost of removing x% of the particulate matter from coal burned by a public utility is approximately

$$C(x) = \frac{48}{101 - x}$$

million dollars, determine the cost of removing the following percentages of the particulate matter.

(a) 26% (b) 61% (c) 76% (d) 91%

44. If the cost of removing x% of the mercury from a polluted river is approximately

$$C(x) = \frac{40x}{120 - x}$$

million dollars, determine the cost of removing the following percentages of mercury.

(a) 60% (b) 70% (c) 80% (d) 90%

45. Suppose that the cost of removing x% of the chemical pollutants from the waste water discharge of a factory is approximately

$$C(x) = \frac{30x}{110 - x}$$

million dollars.

(a) Determine the percentage increase in cost to remove 60% of the pollutants compared to the cost of removing 50% of the pollutants.

(b) Determine the percentage increase in cost to remove 100% of the pollutants compared to the cost of removing 90% of the pollutants.

46. Suppose that the cost of removing x% of the sulfur oxides from the air of a metropolitan region is approximately

$$C(x) = \frac{600}{105 - x}$$

million dollars.

(a) Determine the percentage increase in cost to remove 30% of the sulfur oxides compared to the cost of removing 25% of the sulfur oxides.

(b) Determine the percentage increase in cost to remove 100% of the sulfur oxides compared to the cost of removing 95% of the sulfur oxides.

47. Graph the function $h(x) = \dfrac{x - 1}{x^2 - x}$. What are the vertical asymptotes of this graph?

48. Graph the function $h(x) = \dfrac{(x - 4)^2}{x^2 - 4x}$. What are the vertical asymptotes of this graph?

Answers to Practice Problems

8. The only polynomial functions are f and F, which have degrees 1 and 7, respectively. The function g is not a polynomial because it has a negative exponent, h is not a polynomial because it has a term containing $\sqrt{x}$, and G is not a polynomial because its coefficients are *divided* by nonnegative powers of x.

9.

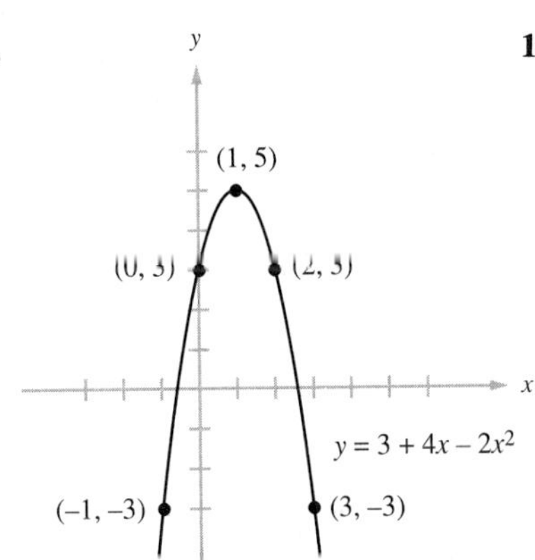

10.

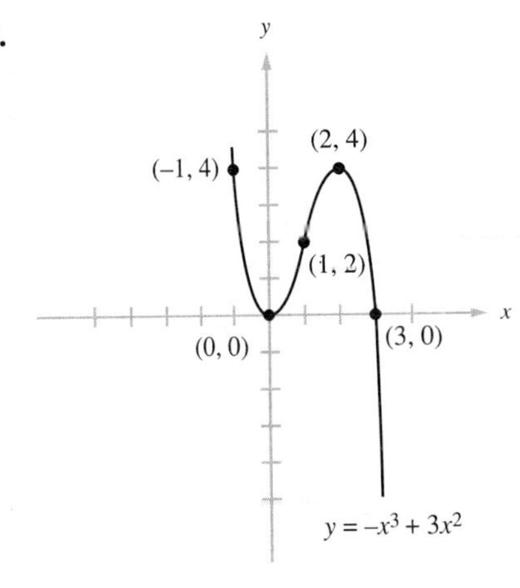

11. (a) $(-\infty, 3) \cup (3, \infty)$ (b) $(-\infty, 0) \cup (0, 1) \cup (1, \infty)$

12. The graph is shown below. The line $x = 3$ is a vertical asymptote, and the line $y = 0$ (the x-axis) is a horizontal asymptote.

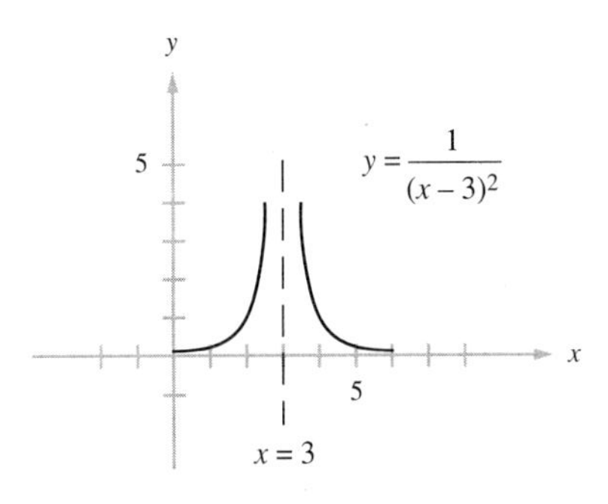

13. 77.1%

8.3 Exponential Functions

We saw in Section 4.2 that if \$1 is deposited in an account paying 6% interest compounded annually, then after n years the value of the account (in dollars) will be

$$A = (1.06)^n$$

In this section we will study functions of the form

$$f(x) = b^x,$$

where $b > 0$ and $b \neq 1$. This function is called an **exponential function,** and the number b is called the **base** of the function.

In algebra courses, expressions with integer and rational number exponents are defined. (See the appendix for a review of these concepts.) Although we will not do so, it is possible to define expressions with irrational exponents x in such a way that if r is a rational number sufficiently close to x, then $b^r \approx b^x$. Thus we will assume that exponential functions are defined for all real numbers. Moreover, we can graph an exponential function by restricting our attention to rational number exponents.

Consider, for example, the exponential function with base 2,

$$g(x) = 2^x.$$

By selecting convenient values of x, we obtain the following table of values.

x	-3	-2	-1	$-\frac{1}{2}$	0	$\frac{1}{2}$	1	2	3
y	$\frac{1}{8}$	$\frac{1}{4}$	$\frac{1}{2}$	$1/\sqrt{2}$	1	$\sqrt{2}$	2	4	8

Plotting the ordered pairs (x, y) and connecting them with a smooth curve produces the graph in Figure 8.14.

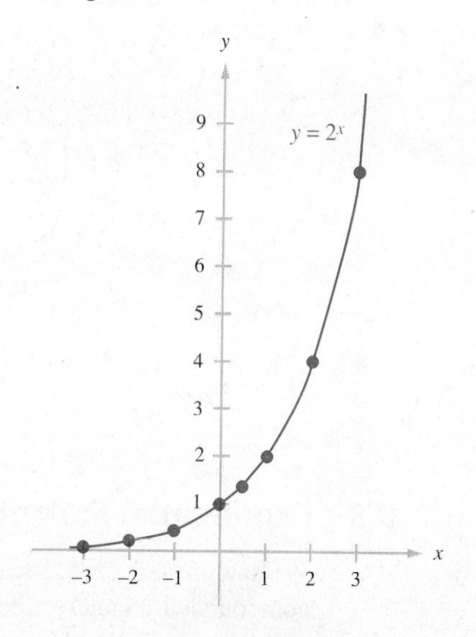

FIGURE 8.14

The shape of the graph in Figure 8.14 is typical of all exponential functions with a base greater than 1. For example, the graph of the exponential function

$$h(x) = 3^x$$

is shown in Figure 8.15(a). It has the same shape as the graph in Figure 8.14 but rises more quickly. Notice that the line $y = 0$ (the x-axis) is a horizontal asymptote for every exponential function. Furthermore, if $b > 1$, then the values of b^x become arbitrarily large as x increases. Figure 8.15(b) shows the graphs of both $g(x) = 2^x$ and $h(x) = 3^x$.

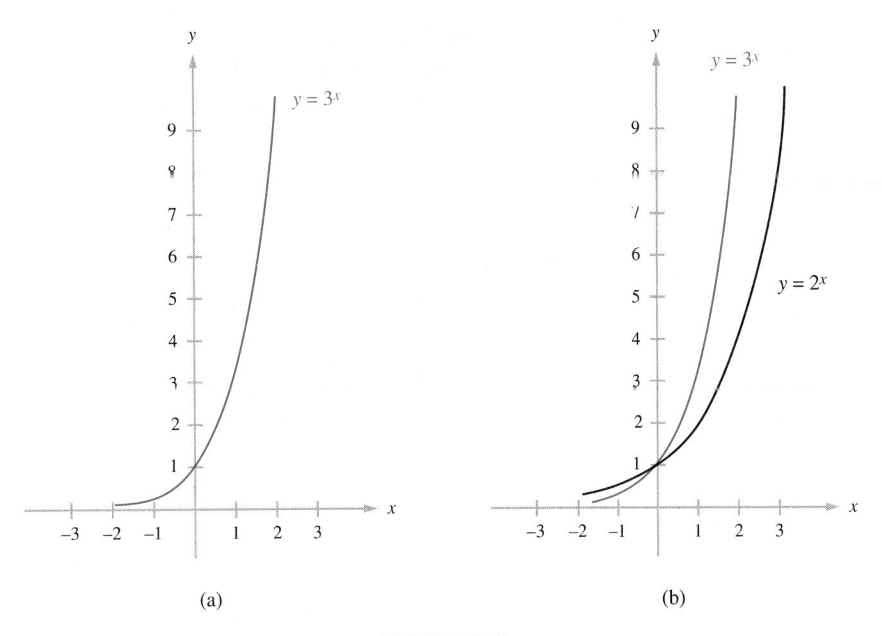

FIGURE 8.15

When $b < 1$, values of the exponential function

$$F(x) = b^x$$

decrease as x increases. The graph of such an exponential function is the mirror image of one having a base greater than 1.

EXAMPLE 8.12 Graph the exponential function $G(x) = (\frac{1}{2})^x$.

Solution Since $2^{-1} = ½$, we can rewrite the function as

$$G(x) = (\tfrac{1}{2})^x = (2^{-1})^x = 2^{-x}.$$

Thus the graph of G is related to the graph of $g(x) = 2^x$ in Figure 8.14. More specifically, a table of values for G can be obtained from the table of values for g above by multiplying all the x values by -1.

x	3	2	1	$\frac{1}{2}$	0	$-\frac{1}{2}$	-1	-2	-3
y	$\frac{1}{8}$	$\frac{1}{4}$	$\frac{1}{2}$	$1/\sqrt{2}$	1	$\sqrt{2}$	2	4	8

The resulting graph is shown in Figure 8.16(a). Figure 8.16(b) shows the graphs of both $g(x) = 2^x$ and $G(x) = (\frac{1}{2})^x$. ■

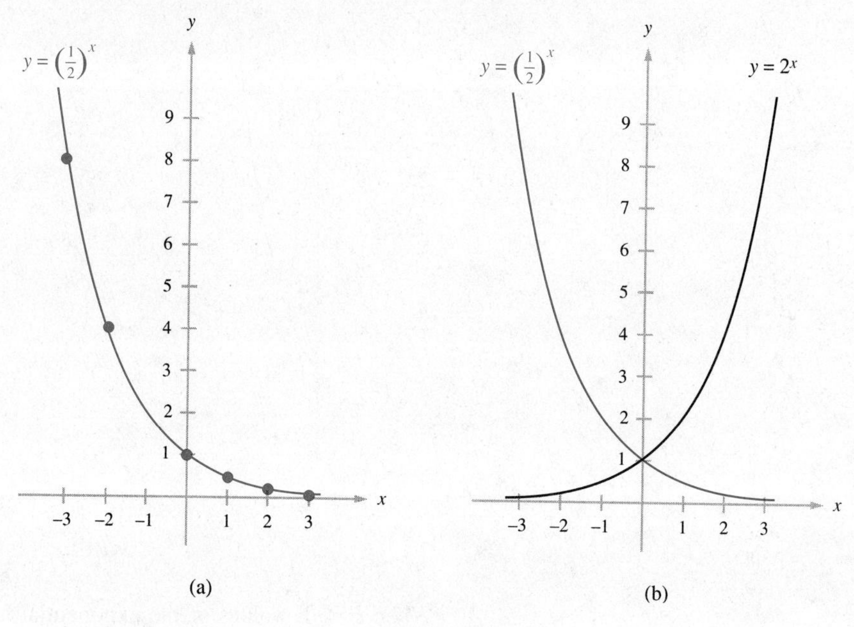

FIGURE 8.16

Practice Problem 14 Graph the function $H(x) = (\frac{1}{3})^x$.

THE EXPONENTIAL FUNCTION WITH BASE *e*

Exponential functions arise naturally in connection with phenomena that involve growth and decay (negative growth). We will discuss some of these applications in Sections 9.3 and 13.3. In most such applications the base of the exponential function is an irrational number denoted e that equals approximately 2.71828. (This is the same number that occurs in the definition of the normal distribution. It will be defined formally in Section 9.3.) Values of the exponential function

$$f(x) = e^x$$

can be computed with a scientific calculator.

EXAMPLE 8.13 Graph the exponential function $f(x) = e^x$.

Solution Using a calculator, we obtain the following table of values for f.

x	-3	-2	-1	$-\frac{1}{2}$	0	$\frac{1}{2}$	1	1.5	2
y	0.05	0.14	0.37	0.61	1	1.65	2.72	4.48	7.39

Plotting the ordered pairs (x, y) produces the graph in Figure 8.17. ■

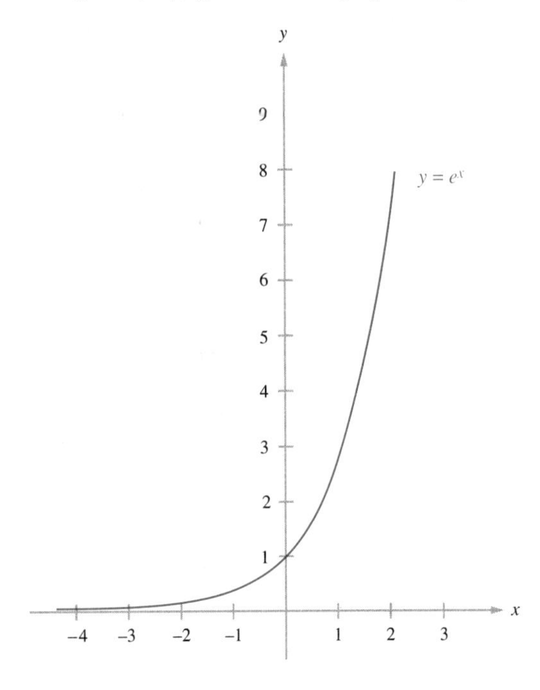

FIGURE 8.17

Practice Problem 15 Graph the exponential function $f(x) = e^{-x}$.

The following examples involving growth and decay all use a function of the form

$$Q(t) = q_0 e^{rt},$$

where $Q(t)$ denotes the quantity of something at time t and q_0 is the quantity at time 0. In this situation r represents the rate of growth or decay. (Positive values of r represent growth, and negative values represent decay.)

EXAMPLE 8.14 In 1970 the population of the earth was approximately 3.7 billion people and was growing at an annual rate of 2%. Write a function that approximates the earth's population t years after 1970 if the 2% annual growth rate continues, and estimate the earth's population in the year 2000.

Solution The earth's population t years after 1970 can be approximated by an exponential function of the form

$$P(t) = p_0 e^{rt},$$

where r is the growth rate and p_0 is the population in 1970. Hence

$$P(t) = 3.7e^{0.02t},$$

where $P(t)$ is measured in billions.

In the year 2000 (when $t = 30$), the earth's population will be about

$$P(30) = 3.7e^{0.02(30)} = 3.7e^{0.6} \approx 3.7(1.82212) \approx 6.7$$

billion people. ■

EXAMPLE 8.15 In reference [1] Casstevens and Denham studied the turnover and retention of members of the Canadian House of Commons during the period 1867–1968. Their work showed that of M_0 members in the House of Commons at a particular time, the number $M(t)$ who remain in office t months later can be approximated quite well by a function of the form

$$M(t) = M_0 e^{rt},$$

where the loss rate r satisfies $-0.019 \leq r \leq -0.011$.

(a) Write a function describing the retention of the 265 members in the 1867 House of Commons if the loss rate was -0.015.

(b) How many of these 265 members could be expected to remain in office 38 months later?

Solution (a) The desired function is

$$M(t) = 265e^{-0.015t}.$$

(b) After 38 months the number remaining in office is approximately

$$M(38) = 265e^{-0.015(38)} = 265e^{-0.57} \approx 265(0.56553) \approx 150. \quad ■$$

Practice Problem 16 Radioactive carbon-14 disintegrates at the rate of 0.012% per year. Suppose that initially there are 100 mg of carbon-14 present. (a) Write a formula that gives the amount present after t years. (b) How much of the carbon-14 will remain undisintegrated after 2000 years?

EXERCISES 8.3

In Exercises 1–12 use a scientific calculator to approximate the given number correct to two decimal places.

1. $e^{3.4}$ **2.** $e^{1.85}$ **3.** $e^{-2.30}$ **4.** $e^{-3.10}$

5. $e^{0.06}$ **6.** $e^{0.32}$ **7.** $e^{-1.28}$ **8.** $e^{-1.64}$

9. $e^{5.1}$ **10.** $e^{7.60}$ **11.** $e^{-0.17}$ **12.** $e^{-0.89}$

Graph each of the functions in Exercises 13–24.

13. $f(x) = 4^x$

14. $F(x) = (0.2)^x$

15. $g(x) = (\frac{1}{4})^x$

16. $G(x) = (1.5)^x$

17. $h(x) = (0.8)^x$

18. $H(x) = 10^x$

19. $F(x) = 3e^x$

20. $f(x) = 2e^{-x}$

21. $G(x) = 5e^{0.2x}$

22. $g(x) = 0.4e^{0.6x}$

23. $H(x) = 10e^{-0.8x}$

24. $h(x) = 20e^{-0.1x}$

25. In 1986 the population of Springfield, Illinois, was 190,000 and was increasing at an annual rate of 0.2%.
(a) Write a function that approximates Springfield's population t years after 1986 if this growth rate continues.
(b) Estimate Springfield's population in the year 2000.

26. In 1986 the population of Decatur, Illinois, was 127,000 and was decreasing at an annual rate of 0.5%.
(a) Write a function that approximates Decatur's population t years after 1986 if this growth rate continues.
(b) Estimate Decatur's population in the year 2010.

27. In reference [4] Vidale and Wolfe state that the sales response to an advertising campaign can be described by a function of the form

$$S(t) = s_0 e^{-\lambda t}.$$

Here s_0 is the sales volume when the advertising is stopped and λ is a positive constant that depends on such factors as the number of competing products and the amount of advertising by competitors. Graph this function if $s_0 = 1000$ and $\lambda = 0.2$.

28. In certain learning situations the rate of learning is proportional to the amount remaining to be learned. (See reference [2].) In such cases the amount learned at time t can be described by the function

$$f(t) = N(1 - e^{-rt}),$$

where N is the amount to be learned and r is the rate at which learning occurs. Graph this function if $N = 100$ and $r = 0.5$.

29. In the definition of an exponential function it was assumed that the base is a number other than 1. Graph $f(x) = b^x$ if $b = 1$. What type of function is f in this case?

Answers to Practice Problems **14.**

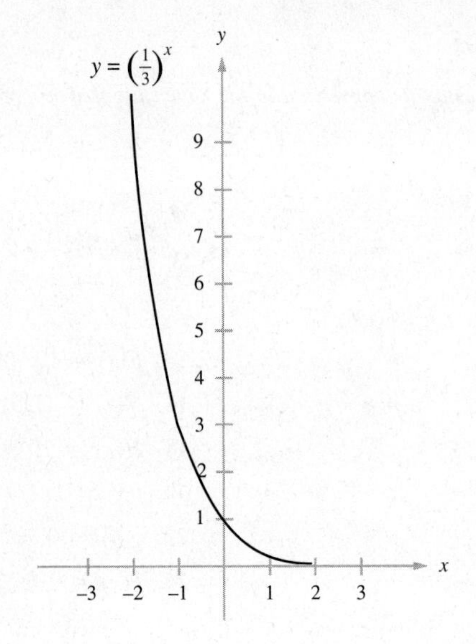

15.

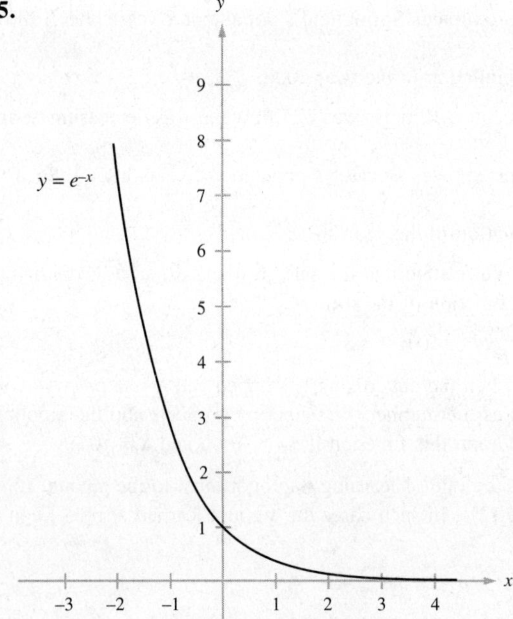

16. (a) $Q(t) = 100e^{-0.00012t}$ (b) about 78.7 mg

8.4 Logarithmic Functions

In Section 8.3 we discussed the exponential function

$$y = b^x,$$

where $b > 0$ and $b \neq 1$. If we reverse the roles of x and y in this equation, we obtain

$$x = b^y.$$

The number y such that $x = b^y$ is called the **logarithm of *x* to the base *b*** and is denoted $\log_b x$. Thus

$$y = \log_b x \qquad \text{means that} \qquad x = b^y.$$

Since $b^y > 0$ for all real numbers y, it follows that $\log_b x$ is defined only if $x > 0$.

As a consequence of this definition, every statement involving logarithms can be rewritten as a statement involving exponents and vice versa. It is often necessary to convert between these two forms. To do so, we need only remember that $\log_b x$ equals the exponent to which the base b must be raised to obtain x.

EXAMPLE 8.16 Rewrite each of the given equations in exponential form.

(a) $\log_3 81 = 4$ (b) $\log_2 \dfrac{1}{8} = -3$ (c) $\log_{100} 1000 = \dfrac{3}{2}$

Solution (a) The given equation states that the exponent to which we must raise the base 3 to get 81 is 4; hence the desired exponential form is

$$3^4 = 81.$$

(b) In this case the given equation means that the exponent to which we must raise the base 2 to get 1/8 is -3; so the desired exponential form is

$$2^{-3} = \frac{1}{8}.$$

(c) As above, we see that the equivalent exponential form is

$$100^{3/2} = 1000. \blacksquare$$

Practice Problem 17 Rewrite the logarithmic equation

$$\log_4 1024 = 5$$

in an equivalent exponential form.

EXAMPLE 8.17 Rewrite each of the given equations in logarithmic form.

(a) $5^3 = 125$ (b) $8^{5/3} = 32$ (c) $6^{-2} = \dfrac{1}{36}$

Solution (a) The first equation states that to write 125 as a power of 5, we need an exponent of 3. Thus the equivalent logarithmic form of this equation is

$$\log_5 125 = 3.$$

(b) In this equation we see that to write 32 as a power of 8, we need an exponent of 5/3. Hence the equivalent logarithmic statement is

$$\log_8 32 = \frac{5}{3}.$$

(c) As above, the equivalent logarithmic equation is

$$\log_6 \frac{1}{36} = -2. \quad ■$$

Practice Problem 18 Rewrite the exponential equation

$$1000^{1/3} = 10$$

in an equivalent logarithmic form.

Because of the relationship between logarithms and exponents, it is natural to expect that there will be properties of logarithms similar to the properties of exponents found in the appendix. We will occasionally need to use these properties in performing calculations with logarithms.

Properties of Logarithms

Let b, r, and s be positive real numbers with $b \neq 1$.

1. $\log_b rs = \log_b r + \log_b s$
2. $\log_b \frac{r}{s} = \log_b r - \log_b s$
3. $\log_b (r^p) = p \log_b r$ for all real numbers p
4. $\log_b 1 = 0$
5. $\log_b (b^x) = x$ for all real numbers x
6. $b^{\log_b x} = x$ for all $x > 0$

EXAMPLE 8.18 Use the properties of logarithms to simplify each of the following expressions.

(a) $\log_3 5 + \log_3 2$ (b) $\log_6 21 - \log_6 3$ (c) $\log_{10} x^8$ (d) $5^{\log_5 11}$

Solution (a) By property 1

$$\log_3 5 + \log_3 2 = \log_3 (5 \cdot 2) = \log_3 10.$$

(b) By property 2

$$\log_6 21 - \log_6 3 = \log_6 \frac{21}{3} = \log_6 7.$$

(c) By property 3

$$\log_{10} x^8 = 8 \log_{10} x.$$

(d) By property 6

$$5^{\log_5 11} = 11. \blacksquare$$

Practice Problem 19 Use the properties of logarithms to simplify each of the following expressions.

(a) $\log_7 8 + \log_7 6$ (b) $\log_e 52 - \log_e 4$ (c) $\log_5 x^{-3}$ (d) $10^{\log_{10} 2x}$

In later chapters we will also need to be able to graph logarithmic functions. As usual, to graph a logarithmic function

$$f(x) = \log_b x,$$

we will plot some ordered pairs (x, y) such that $y = \log_b x$. But this means that we must choose values of x for which the logarithm can be easily found. Property 5 above gives us a simple way to pick such values: *Choose values of x that are powers of the base*. For if we take $x = b^n$, then

$$\log_b (b^n) = n$$

because the exponent to which we must raise b to get b^n is obviously n.

EXAMPLE 8.19 Graph the function $g(x) = \log_2 x$.

Solution As just mentioned, we can evaluate $g(x)$ easily if x is chosen to be a power of the base, which is 2 in this case. Hence we will evaluate g at the values $\frac{1}{8}$, $\frac{1}{4}$, $\frac{1}{2}$, 1, 2, 4, and 8. The following table contains the results.

x	$\frac{1}{8}$	$\frac{1}{4}$	$\frac{1}{2}$	1	2	4	8
y	-3	-2	-1	0	1	2	3

Plotting the ordered pairs (x, y) produces the graph in Figure 8.18. $\blacksquare$

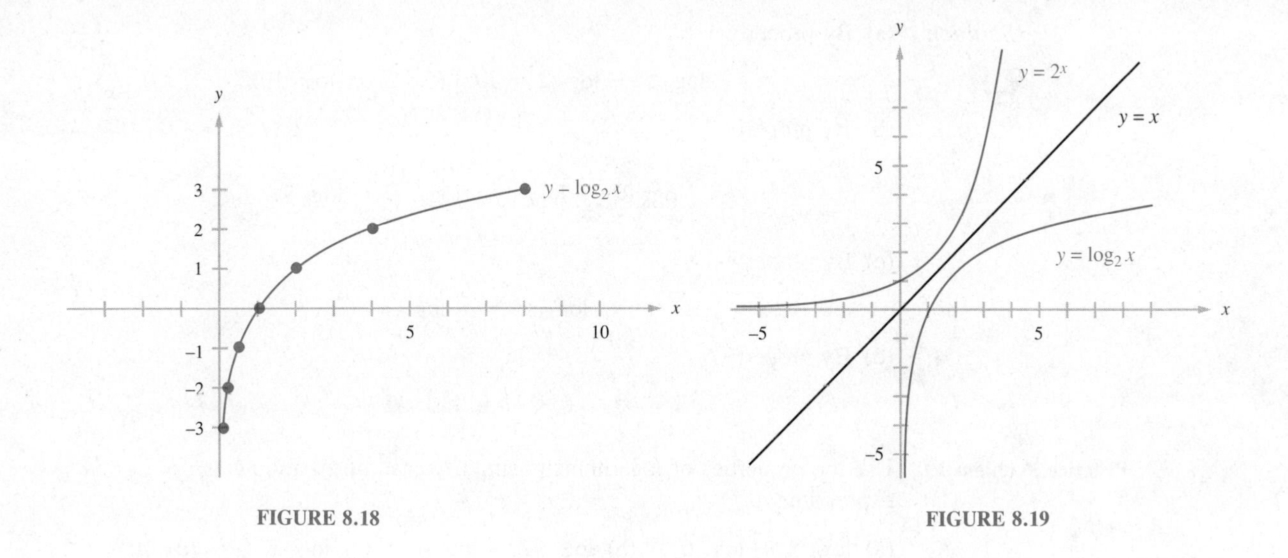

FIGURE 8.18

FIGURE 8.19

The shape of the graph in Figure 8.18 is familiar. As we can see in Figure 8.19, the graph of $g(x) = \log_2 x$ is the mirror image of the graph of $G(x) = 2^x$ across the line $y = x$. This means that if the ordered pair (x, y) lies on the graph of g, then the ordered pair (y, x) lies on the graph of G. The reason for this occurrence is easy to understand. Recall that the equation $y = \log_2 x$ can be rewritten as the exponential equation $2^y = x$. By interchanging the roles of x and y in the equation $2^y = x$, we obtain the equation that defines G; hence the graph of g can also be obtained from the graph of G by interchanging the roles of x and y.

The graph in Figure 8.18 is typical of all logarithmic functions with a base greater than 1. Note also that the domain of every logarithmic function is the set of positive real numbers. Hence the logarithm of a negative number or zero is undefined. Since we will not need to consider logarithmic functions having bases between 0 and 1, we will leave the graphing of such functions to the exercises.

Practice Problem 20 Graph the function $h(x) = \log_3 x$.

THE NATURAL LOGARITHM FUNCTION

As was true for exponential functions, in calculus the most convenient base for a logarithmic function is e. The logarithmic function with base e is called the **natural logarithm function** and is usually denoted $\ln x$ rather than $\log_e x$. Thus $\ln x$ is the exponent to which we must raise e to get x.

Values of the natural logarithm function can be found with a scientific calculator. In this way we obtain the following table of values for the natural logarithm function $y = \ln x$.

x	0.5	1	1.5	2	2.5	3	3.5	4	4.5	5
y	−0.69	0	0.41	0.69	0.92	1.10	1.25	1.39	1.50	1.61

Plotting the ordered pairs (x, y) gives us the graph of the natural logarithm function shown in Figure 8.20.

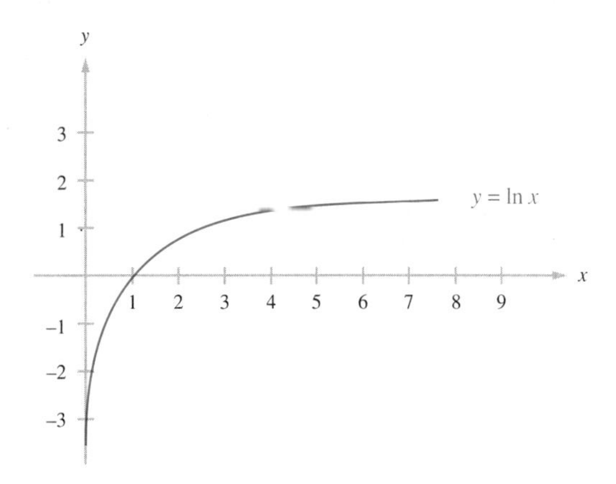

FIGURE 8.20

When applying the techniques of calculus to exponential or logarithmic functions, it will prove to be much simpler to work with functions having e as their base. For this reason we will now describe a method for converting exponential and logarithmic expressions from an arbitrary base b to base e.

Consider, for example, the function

$$F(x) = b^{h(x)},$$

where b satisfies $b > 0$ and $b \neq 1$. By logarithm property 6 on page 478 we have

$$e^{\ln b} = b.$$

Hence

$$F(x) = b^{h(x)} = (e^{\ln b})^{h(x)} = e^{(\ln b)h(x)},$$

which is of the form $e^{g(x)}$, where $g(x) = (\ln b)h(x)$.

EXAMPLE 8.20 Rewrite the function $F(x) = 2^{3x+1}$ with e as its base.

Solution Note that F is of the form $b^{h(x)}$, where $b = 2$ and $h(x) = 3x + 1$. Since

$$b^{h(x)} = e^{(\ln b)h(x)},$$

it follows that

$$F(x) = e^{(\ln 2)(3x+1)}. \quad ■$$

Practice Problem 21 Rewrite the function $F(x) = 10^{4-5x}$ with e as its base.

Just as we are able to write $b^{h(x)}$ as a power of e, we can write $\log_b h(x)$ in terms of $\ln x$. (Of course, we must have $h(x) > 0$ in order that $\log_b h(x)$ be defined.) Let

$$y = \log_b h(x).$$

Then by the definition of a logarithm

$$b^y = h(x).$$

Hence

$$\ln b^y = \ln h(x),$$

and so by the third logarithm property on page 478 we have

$$y \ln b = \ln h(x).$$

Thus

$$\log_b h(x) = y = \frac{1}{\ln b} \ln h(x).$$

EXAMPLE 8.21 Rewrite the function $G(x) = \log_{10}(x^2 + 1)$ with e as its base.

Solution The given function is of the form $\log_b h(x)$, where $b = 10$ and $h(x) = x^2 + 1$. Since $\log_b h(x) = \frac{1}{\ln b} \ln h(x)$, it follows that

$$G(x) = \frac{1}{\ln 10} \ln (x^2 + 1). \quad ■$$

Practice Problem 22 Rewrite the function $G(x) = \log_2(7x - 3)$ with e as its base.

APPLICATIONS OF LOGARITHMS

Logarithms are needed to solve equations in which the unknown occurs as an exponent. The following examples are of this type.

EXAMPLE 8.22 If \$8000 is deposited in an account paying 6% interest compounded quarterly, how long will it take for the account to be worth \$10,000?

Solution Recall from Section 4.2 that the value of the account after n quarters will be

$$A = \$8000(1.015)^n.$$

Since we want to know when the value of the account will be \$10,000, we must solve for n in the equation

$$\$10{,}000 = \$8000(1.015)^n.$$

To do so, we divide both sides of the equation by $8000 to isolate the term involving n:

$$\frac{\$10{,}000}{\$8000} = (1.015)^n$$
$$1.25 = (1.015)^n.$$

Because the unknown occurs as an exponent, logarithms are needed to solve this equation. By taking the natural logarithm of both sides of this equation, we obtain

$$\ln 1.25 = \ln (1.015)^n.$$

Using the third property of logarithms on page 478, we can rewrite this equation as

$$\ln 1.25 = n \ln 1.015.$$

Hence we can solve for n as follows:

$$n = \frac{\ln 1.25}{\ln 1.015} \approx 15.$$

Therefore the value of the account will reach $10,000 in 15 *quarters,* that is, in three years and nine months. ■

EXAMPLE 8.23 In 1986 the population of Springfield, Illinois, was 190,000 and was increasing at an annual rate of 0.2%. If this growth rate continues, when will Springfield's population reach 200,000?

Solution As in Example 8.14, the population of Springfield t years after 1986 can be described by the function

$$P(t) = 190{,}000e^{0.002t}.$$

We want to know the value of t for which $P(t) = 200{,}000$. Hence we must solve for t in the equation

$$200{,}000 = 190{,}000e^{0.002t}.$$

Since the unknown t occurs in an exponent, we will need logarithms to solve this equation.

We begin by dividing both sides by 190,000 to get

$$\frac{200{,}000}{190{,}000} = e^{0.002t}.$$

Taking the natural logarithm of both sides of this equation, we obtain

$$\ln \frac{200{,}000}{190{,}000} = \ln e^{0.002t}.$$

But $\ln e^x = x$, and so

$$\ln \frac{200{,}000}{190{,}000} = 0.002t.$$

Thus

$$t = \frac{\ln \dfrac{200{,}000}{190{,}000}}{0.002} \approx 25.65.$$

Therefore Springfield's population will reach 200,000 about 25.65 years after 1986, that is, during the year 2011. ■

Practice Problem 23 How long will it take for the value of the account in Example 8.22 to reach $14,500?

Archaeologists have used knowledge of radioactive decay to obtain procedures for estimating the age of remnants from ancient civilizations. For dating objects less than 50,000 years old, the most common method is based on the radioactive isotope carbon-14. Carbon-14 is produced in the earth's atmosphere by a chemical reaction that occurs at the same rate as that at which the element decays. Consequently the ratio* of the amount of carbon-14 to the amount of nonradioactive carbon-12 has remained constant over a period of 50,000 years. When alive, most forms of plant and animal life contain the same ratio of carbon-14 to carbon-12 as in the atmosphere. After death, however, the amount of carbon-12 remains constant while the amount of carbon-14 decreases due to radioactive decay. By measuring the ratio of carbon-14 to carbon-12 in a dead organism, it is possible to estimate how long ago it died.

EXAMPLE 8.24 During 1951 and 1952 researchers from the Texas Memorial Museum discovered charred bison bones and dartlike objects that had been used to hunt the animal. (See reference [3] for details.) In the bison bones the ratio of radioactive carbon-14 to carbon-12 was about 30% of the ratio in a living organism. How long ago did the bison hunters, who are known as Folsom man, live?

Solution As in the answer to Practice Problem 16 of Section 8.3, we know that if q_0 represents the amount of carbon-14 present at some original time, then the amount of carbon-14 present t years later is approximately

$$Q(t) = q_0 e^{-0.00012t}.$$

Since the ratio of carbon-14 to carbon-12 in the bison bones was about 30% of the ratio in a living organism, we see that the length of time t since the death of the bison satisfies

$$0.30q_0 = q_0 e^{-0.00012t}.$$

---

*This ratio is about 1 part of carbon-14 to 10^{12} parts of carbon-12.

These artifacts contained about 30% of the ratio of carbon-14 to carbon-12 in a living organism.

To solve this equation, we first divide both sides by q_0 to obtain

$$0.30 = e^{-0.00012t}.$$

Hence

$$\ln 0.30 = \ln e^{-0.00012t},$$

and so

$$\ln 0.30 = -0.00012t.$$

It follows that

$$t = \frac{\ln 0.30}{-0.00012} \approx 10{,}000.$$

Thus Folsom man lived about 10,000 years ago. ■

Practice Problem 24 How long will it take for half of a sample of carbon-14 to decay? (This time is called the *half-life* of carbon-14.)

EXERCISES 8.4

In Exercises 1–8 write the given equation in exponential form.

1. $\log_{11} 121 = 2$

2. $\log_9 \frac{1}{9} = -1$

3. $\log_4 \frac{1}{64} = -3$

4. $\log_{216} 6 = \frac{1}{3}$

5. $\log_{10} 0.0001 = -4$

6. $\log_{49} \frac{1}{7} = -\frac{1}{2}$

7. $\log_{0.25} 32 = -2.5$

8. $\log_{0.4} 0.16 = 2$

In Exercises 9–16 write the given equation in logarithmic form.

9. $13^2 = 169$

10. $12^{-1} = \frac{1}{12}$

11. $3^{-4} = \frac{1}{81}$

12. $(1.5)^2 = 2.25$

13. $1000^{1/3} = 10$

14. $81^{-1/2} = \frac{1}{9}$

15. $(0.4)^{-3} = \frac{125}{8}$

16. $32^{0.6} = 8$

In Exercises 17–24 use the definition of a logarithm to evaluate the given expression.

17. $\log_2 8$

18. $\log_4 64$

19. $\log_3 \frac{1}{9}$

20. $\log_5 \frac{1}{5}$

21. $\log_6 \frac{1}{216}$

22. $\log_2 \frac{1}{32}$

23. $\log_4 8$

24. $\log_8 32$

In Exercises 25–32 use the logarithm properties to combine each expression into a single logarithm.

25. $\log_8 33 - \log_8 3$

26. $\log_5 7 + \log_5 2$

27. $\log_{10} 5 + 2 \log_{10} 4$

28. $2 \log_7 6 - \log_7 4$

29. $\log_3 24 + \log_3 10 - \log_3 15$

30. $\log_8 6 - \log_8 15 - \log_8 10$

31. $\ln 6 + 2 \ln 4 - \ln 8$

32. $\log_9 10 + 2 \log_9 3 - \log_9 6$

In Exercises 33–40 solve the given equation for t.

33. $e^{-0.03t} = 0.40$

34. $e^{0.04t} = 3.50$

35. $200e^{0.28t} = 1000$

36. $400e^{-0.08t} = 100$

37. $(1.08)^t = 1.50$

38. $3^t = 100$

39. $25(10^t) = 5000$

40. $4000(1.06)^t = 5000$

In Exercises 41–48 rewrite the given function with e as its base.

41. $f(x) = 3^{2-5x}$

42. $F(x) = \log_7 (3x^2 - x)$

43. $g(x) = \log_9 (7x - 6)$

44. $G(x) = 4^{2x+1}$

45. $h(x) = \log_{10} (3x^2 + 4)$

46. $H(x) = \log_{16} (\sqrt{x} + 4)$

47. $F(x) = 2^{3x-x^2}$

48. $f(x) = 7^{2/x}$

49. Remnants found in the tomb of Chaldean king Nebuchadnezzar contain a ratio of carbon-14 to carbon-12 that is about 73.7% of the ratio in a living organism. Determine the age of these remnants as in Example 8.24.

50. In Example 8.14 we saw that t years after 1970 the earth's population would be approximately $3.7e^{0.02t}$ billion if the annual growth rate remains at 2%. Determine when the earth's population will reach 7 billion under these circumstances.

51. If \$4000 is deposited in an account paying 8% interest compounded quarterly, how long must the money remain in the account until it grows to \$10,000?

52. If money is deposited in an account paying 6% interest compounded monthly, how long must the money remain in the account until it grows to twice its original value?

53. The radiocarbon dating process described in this section is reliable for determining the age of objects less than 50,000 years old. A similar technique called *potassium-argon dating* can be used to estimate the age of objects up to 3 billion years old. (This process estimates the age of minerals or rocks in a deposit from the ratio of the amount of potassium-40 to the amount of argon-40 present.) If q_0 denotes the original amount of potassium-40 in an object, the amount t years later is given by $Q(t) = q_0e^{-5.2116(10^{-7})t}$. Determine the half-life of potassium-40.

54. Chemists measure the relative acidity of solutions by the quantity

$$\text{pH} = -\log_{10} [\text{H}^+],$$

where $[\text{H}^+]$ denotes the concentration of hydrogen ions in the solution (measured in moles per liter). Most scientific calculators have a key marked "log" for evaluating logarithms to base 10. Use this key to determine pH for each of the following.

(a) distilled water, for which $[H^+] = 10^{-7}$ moles per liter

(b) human blood, for which $[H^+] \approx 3.0 \cdot 10^{-8}$ moles per liter

(c) cow's milk, for which $[H^+] \approx 4.0 \cdot 10^{-7}$ moles per liter

(d) an orange, for which $[H^+] \approx 2.0 \cdot 10^{-4}$ moles per liter

55. The magnitude of an earthquake is measured on a scale called the *Richter scale*. On this scale the magnitude of an earthquake of intensity I is

$$R = \log_{10} \frac{I}{I_0},$$

where I_0 is a standard intensity used for comparison purposes.

(a) The 1906 earthquake in San Francisco had an intensity of $10^{8.2} I_0$. What did it measure on the Richter scale?

(b) How much more intense is an earthquake that measures 8 on the Richter scale compared to one that measures 4?

56. The population of Nevada grew from 291,000 in 1960 to 480,000 in 1970. Assuming that Nevada's population can be described by an exponential function of the form $P(t) = p_0 e^{rt}$, determine

(a) the annual growth rate r

(b) the year in which Nevada's population will reach 2,000,000 if the annual growth rate in part (a) continues.

Answers to Practice Problems

17. $4^5 = 1024$ **18.** $\log_{1000} 10 = \frac{1}{3}$

19. (a) $\log_7 48$ (b) $\log_e 13$ (c) $-3 \log_5 x$ (d) $2x$

20.

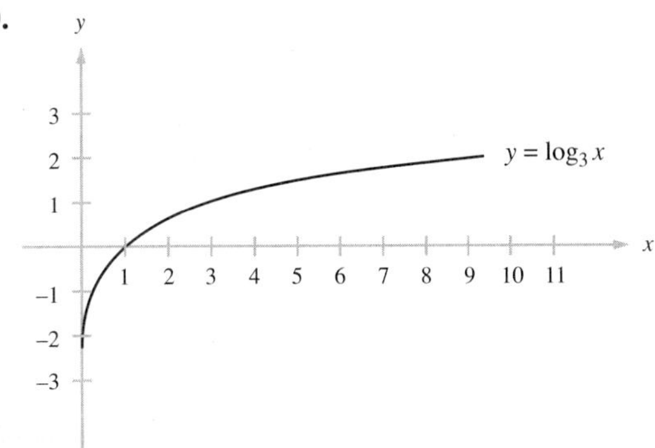

21. $F(x) = e^{(\ln 10)(4-5x)}$

22. $G(x) = \frac{1}{\ln 2} \ln (7x - 3)$

23. approximately 10 years

24. approximately 5776 years

8.5 The Algebra of Functions

The Fun Times catering service will provide food for x children at a birthday party at a price of $f(x)$ dollars, and it will provide balloons and noisemakers at a cost of $g(x)$ dollars. Obviously then the cost of purchasing food, balloons, and noisemakers for x children from Fun Times is

$$f(x) + g(x)$$

dollars.

The simple example above illustrates a situation in which we need to perform arithmetic operations on two functions. These situations often arise in financial matters. For instance, we have already seen that if $R(x)$ is the revenue derived from the sale of x units of some product and $C(x)$ is the cost of producing x units, then

$$R(x) - C(x)$$

is the profit derived from the production and sale of x units.

Although we have encountered such situations previously in this book, we have not actually explained what we mean by the sum, difference, product, and quotient of two functions. The definitions below formally explain these operations.

Operations on Functions

Let f and g be functions.

(a) The **sum** of f and g, denoted $f + g$, is the function defined by

$$(f + g)(x) = f(x) + g(x).$$

(b) The **difference** of f and g, denoted $f - g$, is the function defined by

$$(f - g)(x) = f(x) - g(x).$$

(c) The **product** of f and g, denoted fg, is the function defined by

$$fg(x) = f(x) \cdot g(x).$$

(d) If $g(x) \neq 0$, the **quotient** of f and g, denoted $\frac{f}{g}$ or f/g, is the function defined by

$$\frac{f}{g}(x) = \frac{f(x)}{g(x)}.$$

The definition of $f + g$ states that the value of the function $f + g$ at x is obtained by adding together the value of f at x and the value of g at x. Similar statements apply to the difference, product, and quotient of functions. Thus all the definitions above are completely natural and should be familiar from previous courses in algebra. The example below illustrates these definitions.

EXAMPLE 8.25 If $f(x) = 3x - 5$ and $g(x) = x$, find $f + g$, $f - g$, fg, and $\frac{f}{g}$.

Solution The sum $f + g$ of f and g is the function

$$(f + g)(x) = f(x) + g(x) = (3x - 5) + x = 4x - 5.$$

The difference $f - g$ of f and g is the function

$$(f - g)(x) = f(x) - g(x) = (3x - 5) - x = 2x - 5.$$

The product fg of f and g is the function

$$(fg)(x) = f(x) \cdot g(x) = (3x - 5) \cdot x = 3x^2 - 5x.$$

The quotient f/g of f and g is the function

$$\frac{f}{g}(x) = \frac{f(x)}{g(x)} = \frac{3x - 5}{x}. \quad \blacksquare$$

Practice Problem 25 If $f(x) = 12x^4 - 8x^2$ and $g(x) = -2x^2$, find $f + g$, $f - g$, fg, and f/g.

In order to add, subtract, or multiply $f(x)$ and $g(x)$, both f and g must be defined at x. Hence the domain of $f + g$, $f - g$, and fg is the set of all values that belong to *both* the domain of f and the domain of g. In other words, the domain of $f + g$, $f - g$, and fg is the intersection of the domains of f and g.

For the quotient to be defined, there is one other requirement: We can divide $f(x)$ by $g(x)$ only when both f and g are defined at x and $g(x) \neq 0$. Hence the domain of f/g is the set of all x that satisfy $g(x) \neq 0$ and belong to both the domain of f and the domain of g. This statement is illustrated by the domain of a rational function, which consists of all real numbers except those for which the denominator is zero.

When f is a constant function, say $f(x) = k$, the product of f and g takes the following simpler form:

$$(fg)(x) = f(x) \cdot g(x) = kg(x).$$

In this case we will often write kg rather than fg. We call the function kg the **constant multiple** of k and g. Thus if $g(x) = e^x + x^2$, the function $3g$ is defined by

$$3g(x) = 3(e^x + x^2) = 3e^x + 3x^2.$$

Practice Problem 26 Find $4g$ if $g(x) = x - \ln x$.

THE COMPOSITION OF FUNCTIONS

There is another important way in which certain functions can be combined. Often two functions are applied one after the other. For example, we saw in Section 8.1 that the function

$$F(x) = \tfrac{9}{5}x + 32$$

gives the Fahrenheit temperature $F(x)$ corresponding to a Celsius temperature x. Thus to convert a Celsius temperature x to its Fahrenheit equivalent, we perform the following steps.

1. Multiply x by $\tfrac{9}{5}$.
2. Add 32 to the result of step 1.

We can accomplish each of these steps by applying the appropriate function. Step 1 can be accomplished by applying the function

$$M(x) = \tfrac{9}{5}x,$$

and step 2 can be performed by applying the function

$$A(x) = x + 32$$

to $M(x)$. Thus first performing M and then performing A accomplishes both steps, and so yields the Fahrenheit temperature F. Hence

$$F(x) = A(M(x)).$$

If g and f are functions, the function h defined by

$$h(x) = f(g(x))$$

is called the **composition** of f and g. Thus in the preceding paragraph F is the composition of the functions A and M. Note that the composition of f and g is formed by replacing each occurrence of x in the definition of f by $g(x)$. If h is the composition of f and g, then the domain of h consists of all x in the domain of g for which $g(x)$ lies in the domain of f.

EXAMPLE 8.26 If $f(x) = \sqrt{3x^2 + 1}$ and $g(x) = x - 2$, evaluate each of the following.

(a) $f(g(x))$ (b) $g(f(x))$

Solution (a) To evaluate $f(g(x))$, we replace each occurrence of x in the definition of f by $g(x) = x - 2$. Thus

$$\begin{aligned} f(g(x)) &= \sqrt{3(x - 2)^2 + 1} \\ &= \sqrt{3(x^2 - 4x + 4) + 1} \\ &= \sqrt{3x^2 - 12x + 13}. \end{aligned}$$

(b) On the other hand, to evaluate $g(f(x))$, we replace each occurrence of x in the definition of g by $f(x) = \sqrt{3x^2 + 1}$. Hence

$$g(f(x)) = \sqrt{3x^2 + 1} - 2. \quad ■$$

In Example 8.26 note that $f(g(x)) \neq g(f(x))$. In general, these compositions are almost never equal, and so care must be taken to apply the functions in the proper order.

EXAMPLE 8.27 If $f(x) = 3x^2 - 4x + 2$ and $g(x) = -\dfrac{1}{x}$, evaluate each of the following.

(a) $f(g(x))$ (b) $g(f(x))$

Solution (a) Substituting $g(x) = -\dfrac{1}{x}$ for *each* occurrence of x in the definition of $f(x)$, we find that

$$\begin{aligned} f(g(x)) &= 3\left(-\frac{1}{x}\right)^2 - 4\left(-\frac{1}{x}\right) + 2 \\ &= \frac{3}{x^2} + \frac{4}{x} + 2. \end{aligned}$$

(b) Substituting $f(x) = 3x^2 - 4x + 2$ for each occurrence of x in the definition of $g(x)$, we obtain

$$g(f(x)) = -\frac{1}{3x^2 - 4x + 2}. \quad ■$$

Practice Problem 27 If $f(x) = e^x$ and $g(x) = 2x + 3$, evaluate each of the following.

(a) $f(g(x))$ (b) $g(f(x))$

In Section 9.5 we will need to express a function h as a composition of two functions f and g. The example that follows illustrates this type of problem.

EXAMPLE 8.28 For each of the given functions h, express h as a composition of functions f and g.

(a) $h(x) = \ln (3x^2 + 10)$
(b) $h(x) = 7\sqrt{x^2 + 1} - 8$
(c) $h(x) = 4(e^{0.5x} + 6)^9 - 2$
(d) $h(x) = (3x + 2)^2 + 5(3x + 2) - 7$

Solution (a) Note that if $g(x) = 3x^2 + 10$, then

$$h(x) = \ln g(x).$$

Hence if we choose

$$f(x) = \ln x \qquad \text{and} \qquad g(x) = 3x^2 + 10,$$

then $h(x) = f(g(x))$.

The functions f and g obtained in the preceding paragraph are not unique. For instance, we could also have chosen

$$f(x) = \ln(x + 10) \qquad \text{and} \qquad g(x) = 3x^2.$$

In general there will be many ways to choose functions F and G so that a given function H is expressed in the form $H(x) = F(G(x))$. For our purposes, however, it is preferable to choose F to be a simple function, such as a polynomial, logarithmic, or exponential function.

(b) Taking $g(x) = \sqrt{x^2 + 1}$, we see that

$$h(x) = 7 \cdot g(x) - 8.$$

Hence if

$$f(x) = 7x - 8 \qquad \text{and} \qquad g(x) = \sqrt{x^2 + 1},$$

then $h(x) = f(g(x))$. As in part (a) there are other choices for f and g such as

$$f(x) = 7\sqrt{x} - 8 \qquad \text{and} \qquad g(x) = x^2 + 1.$$

(c) If we let $g(x) = e^{0.5x} + 6$, then

$$h(x) = 4[g(x)]^9 - 2.$$

Therefore if we take

$$f(x) = 4x^9 - 2 \qquad \text{and} \qquad g(x) = e^{0.5x} + 6,$$

then $h(x) = f(g(x))$. Again the answer is not unique.

(d) Let $g(x) = 3x + 2$. Then

$$h(x) = [g(x)]^2 + 5g(x) - 7.$$

Therefore $h(x) = f(g(x))$, where

$$f(x) = x^2 + 5x - 7 \qquad \text{and} \qquad g(x) = 3x + 2. \blacksquare$$

Practice Problem 28 Write $h(x) = 4\sqrt[3]{5x - 6} + 7$ as a composition of functions f and g.

EXERCISES 8.5

In Exercises 1–8 compute the indicated function if $f(x) = 4x^2 - 12x + 8$ *and* $g(x) = 2x$.

1. $f + g$

2. $g - f$

3. $\dfrac{f}{g}$

4. fg

5. $3f - 6g$

6. $2f + 12g$

7. $\left(\dfrac{1}{4}f\right)\left(\dfrac{1}{2}g\right)$

8. $\dfrac{5g}{f}$

In Exercises 9–22 evaluate the given expression if $f(x) = 3x^2 + 2x$ *and* $g(x) = 4x + 6$.

9. $(f + g)(-1)$

10. $(2f)(-4)$

11. $(g - f)(2)$

12. $(f - g)(-3)$

13. $(3g)(10)$

14. $(f + g)(5)$

15. $\dfrac{f}{g}(6)$

16. $gf(-2)$

17. $fg(1)$

18. $\dfrac{g}{f}(1)$

19. $g(f(-2))$

20. $f(g(-2))$

21. $f(g(-1))$

22. $g(f(0))$

In Exercises 23–30 compute $f(g(x))$ *for the given functions f and g.*

23. $f(x) = x^3 - 2x, \quad g(x) = e^x$

24. $f(x) = \log_8 x, \quad g(x) = x^2 + 7$

25. $f(x) = 3x - 2, \quad g(x) = \dfrac{10}{x^2}$

26. $f(x) = \sqrt{x^2 + 1}, \quad g(x) = 2^x$

27. $f(x) = \ln x, \quad g(x) = \dfrac{4x}{\sqrt{x + 3}}$

28. $f(x) = 2x - 3, \quad g(x) = \dfrac{x}{x^2 - x}$

29. $f(x) = \dfrac{x}{x + 1}, \quad g(x) = 5x - 6$

30. $f(x) = \dfrac{3 - x}{x^2}, \quad g(x) = 2x - 1$

In Exercises 31–38 express the given function h as a composition of functions f and g.

31. $h(x) = \dfrac{2}{(3x^2 - 2x + 1)^3}$

32. $h(x) = e^{1-3x}$

33. $h(x) = \sqrt{4 - x^2}$

34. $h(x) = 3(x - 2)^4 + 5(x - 2)^2 - 6(x - 2)$

35. $h(x) = 3 \ln (x^2 + 4x + 4)$

36. $h(x) = \sqrt[3]{5 - x^2}$

37. $h(x) = 2(x^2 - x)e^{x^2 - x}$

38. $h(x) = (2x + 1) \ln (2x + 1)$

39. The weekly revenue and cost functions for a manufacturer of men's socks are

$$R(x) = 10x - 0.0005x^2 \quad \text{and} \quad C(x) = 1.50x + 30{,}000,$$

where x denotes the number of pairs of socks produced. What is the weekly profit function?

40. The monthly revenue and cost functions for a manufacturer of television sets are

$$R(x) = 360x - 0.02x^2 \quad \text{and} \quad C(x) = 30x + 1{,}250{,}000,$$

where x denotes the number of sets produced. What is the monthly profit function?

41. A publisher sells a mathematics textbook at a price of \$40 per copy. The cost of producing x copies of this book is $C(x) = 15x + 400{,}000$ dollars. What is the publisher's profit from the production and sale of x copies of this book?

42. A manufacturer sells baking pans at a price of \$6 apiece. Her cost of producing x pans is $C(x) = 3x + 36{,}000$ dollars. What is the manufacturer's profit from the production and sale of x pans?

43. When the selling price of her product is p dollars, a manufacturer is willing to supply $S(p) = 150p$ units per week. The selling price for this product depends on the weekly demand d according to the formula $p(d) = 8 + 0.002d$. Compute $S(p(d))$ and interpret the result.

44. When the selling price of his product is p dollars, a manufacturer is willing to supply $S(p) = 240p$ units per week. The selling price for this product depends on the weekly demand d according to the formula $p(d) = (2400 + d)/200$. Compute $S(p(d))$ and interpret the result.

45. The number of units of a certain model of dishwasher that can be sold at a price of p dollars is given by $D(p) = 10{,}000 - 25p$. The cost of producing x units of this dishwasher is given by $C(x) = 800{,}000 + 120x$. Compute $C(D(p))$ and interpret the result.

46. The population (in hundred thousands of people) of a metropolitan area t years after 1990 is expected to be approximately $P(t) = 5e^{0.01t}\ t$. Urban planners claim that if the population of a metropolitan area is x hundred thousand people, then the average number of cars (in hundred thousands) traveling downtown on a weekday will be $C(x) = \sqrt{\frac{x}{2}}$. Compute $C(P(t))$ and interpret the result.

Answers to Practice Problems

25. $(f + g)(x) = 12x^4 - 10x^2$, $(f - g)(x) = 12x^4 - 6x^2$, $(fg)(x) = -24x^6 + 16x^4$, and $\frac{f}{g}(x) = -6x^2 + 4$ if $x \neq 0$

26. $4g(x) = 4x - 4 \ln x$

27. (a) $f(g(x)) = e^{2x+3}$ (b) $g(f(x)) = 2e^x + 3$

28. One possible answer is to choose $f(x) = 4x + 7$ and $g(x) = \sqrt[3]{5x - 6}$.

CHAPTER 8 REVIEW

IMPORTANT TERMS

- dependent variable *(8.1)*
- domain of a function
- function
- graph of a function
- independent variable
- degree of a polynomial *(8.2)*
- horizontal asymptote
- parabola
- polynomial function
- quadratic function
- rational function
- vertex of a parabola
- vertical asymptote
- zero polynomial
- base of an exponential function *(8.3)*
- exponential function
- logarithm of x to the base b *(8.4)*
- natural logarithm function
- composition of functions *(8.5)*
- constant multiple of a function
- difference of functions
- product of functions
- quotient of functions
- sum of functions

REVIEW EXERCISES

1. If $f(x) = 2x^2 - 3$, evaluate the following.

(a) $f(-4)$ (b) $f(2)$ (c) $f(5)$ (d) $f(x + h)$

2. If $f(x) = \begin{cases} 2x + 1 & \text{if } x < 1 \\ 4 - x^2 & \text{if } x \geq 1, \end{cases}$ evaluate each of the following.

(a) $f(-3)$ (b) $f(0)$ (c) $f(1)$ (d) $f(5)$

3. Determine which of the following curves can be graphs of a function of x.

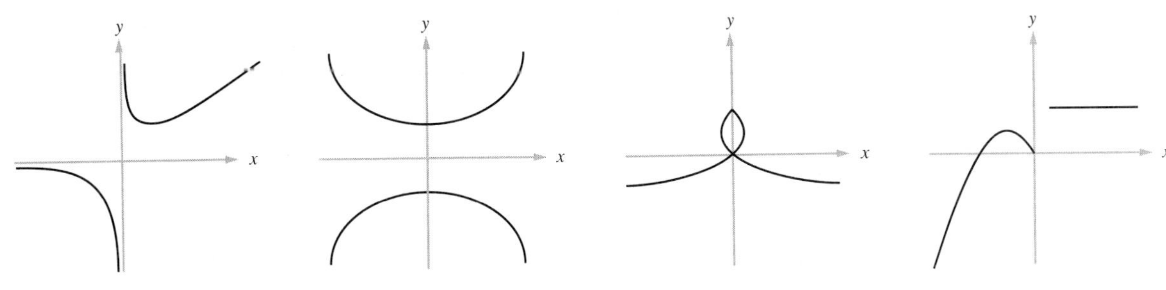

4. Determine the domains of each of the following functions.

(a) $f(x) = \dfrac{1}{2x - 1}$ (b) $F(x) = \sqrt{8 - x}$

(c) $g(x) = 3x^7 - 6x^4 + 2x^3 - 5$ (d) $G(x) = \dfrac{\sqrt{x + 2}}{(x - 3)(x - 5)}$

5. Use interval notation to describe the set of numbers satisfying the given inequality. Then graph each set on a number line.

(a) $x < 23$ (b) $-6 \leq x$

(c) $-9 < x < -3$ (d) $3 < x \leq 17$

(e) $x \leq 12$ (f) $-7 < x$

(g) $-5 \leq x \leq 8$ (h) $-2 \leq x < 11$

6. Identify which of the following functions are polynomial functions, which are rational functions, and which are neither. Give the degree of those that are polynomial functions.

(a) $f(x) = \dfrac{2x - 5}{x^2 + 3}$ (b) $F(x) = 4 - 3x + 6x^2 - 9x^3 + 2x^4$

(c) $g(x) = x^{-2}$ (d) $G(x) = \dfrac{\sqrt{x}}{(x - 6)^3}$

(e) $h(x) = (x + 4)(x - 3)^2$ (f) $H(x) = 4x - 2 + \dfrac{3}{x - 1}$

7. Write the following equations in logarithmic form.

(a) $6^4 = 1296$ (b) $2^{-6} = \dfrac{1}{64}$

8. Write the following equations in exponential form.

(a) $\log_2 1024 = 10$ (b) $\log_8 \frac{1}{4} = -\frac{2}{3}$

9. Write each expression as a single logarithm.

(a) $\log_{16} 24 - \log_{16} 3$ (b) $\log_2 5 + \log_2 4$

10. Rewrite the following functions with e as their base.

(a) $f(x) = 8^{3x+2}$ (b) $g(x) = \log_{10} (2x + 1)$

11. Solve for t in each of the following equations.

(a) $10e^{0.02t} = 18.80$ (b) $2(1.07)^t = 3$

12. Atmospheric pressure lessens at higher altitudes according to the formula

$$h = (30T + 8000) \ln \frac{76}{P},$$

where h denotes the altitude in meters above sea level, T denotes the air temperature in degrees Celsius, and P denotes the atmospheric pressure in centimeters of mercury. If a plane is cruising at an altitude of 12,675 meters above sea level and the air temperature is 15° C, what is the atmospheric pressure on the plane?

13. In 1970 the population of the United States was approximately 200 million, and the annual growth rate was about 1.7%.

(a) Write a function that approximates the population of the United States t years after 1970 if the growth rate remains at 1.7%.

(b) Determine when the population of the United States will reach 300 million if the annual growth rate remains at 1.7%.

14. Radioactive strontium-90 decays in such a way that if there is an amount q_0 originally, then the amount t years later is

$$Q(t) = q_0 e^{-0.0244t}.$$

Determine the half-life of strontium-90 (the time for half of the amount to decay).

In Exercises 15–20 graph the given function.

15. $f(x) = 18 - 2x^2$ **16.** $g(x) = x^2 + 4x + 4$ **17.** $h(x) = \frac{1}{x + 3}$

18. $F(x) = 2^x$ **19.** $G(x) = 3^{-x}$ **20.** $H(x) = \ln x$

REFERENCES

1. Casstevens, Thomas W. and William A. Denham III, "Turnover and Tenure in the Canadian House of Commons, 1867–1968," *Canadian Journal of Political Science,* vol. 3 (December 1970), pp. 655–661.

2. Ettlinger, H. J., "A Curve of Growth to Represent the Learning Process," *Journal of Experimental Psychology,* vol. 9 (October 1926), pp. 409–414.

3. Sellards, E. H., "Age of Folsom Man," *Science,* vol. 115 (January 25, 1952), p. 98.

4. Vidale, M. L. and H. B. Wolfe, "An Operations Research Study of Sales Response to Advertising," *Operations Research,* vol. 5 (June 1957), pp. 370–381.

9

THE DERIVATIVE

This chapter begins the study of calculus, the mathematics of change. In Chapters 9 and 10 we will consider the rate at which a variable is changing. Many practical problems involve this type of measurement. For example, economists are interested in knowing the rate at which a nation's economy is growing, psychologists study the rate at which learning occurs in controlled experiments, and manufacturers must know the rate at which their production costs change with an increase in the number of units produced.

Sections 9.1–9.3 introduce the mathematical concepts needed to study rates of change. Then we present formulas that enable us to determine the rate of change of many common functions. In susbsequent chapters we will encounter a wide variety of applications which require computing the rate of change of a function.

9.1 Instantaneous Rates of Change

In this section we will consider the concept of an instantaneous rate of change. The most familiar example of an instantaneous rate of change arises in regard to speed, and we begin our discussion in this context.

Hours Elapsed	*Miles Traveled*
½	13
1	28
1½	45
2	64
2½	85
3	108

TABLE 9.1

Let us consider a driver who has completed a trip of 108 miles along a two-lane rural road. Suppose that the driver checked the car's odometer every 30 minutes and obtained the readings in Table 9.1.

From these readings we can deduce several facts about the driver's speed. First, the driver's average speed for the trip was

$$\frac{\text{distance traveled}}{\text{time traveling}} = \frac{108 \text{ miles}}{3 \text{ hours}}$$
$$= 36 \text{ miles per hour.}$$

Second, the driver's speed was not constant throughout the trip. Notice that during the first half-hour the car traveled only 13 miles, whereas it traveled $28 - 13 = 15$ miles during the second half-hour. In fact, similar computations show that the driver's average speed increased during each successive 30-minute period. Thus the car's average speed increased from 26 mph during the first half-hour to 46 mph during the last half-hour.

Each of the statements above concerns the driver's *average* speed during some particular interval of time. In many situations it is more important to know the car's *instantaneous* speed (that is, its speedometer reading) at some particular moment. For instance, a police officer who is checking speeds along the road is interested in knowing the car's speed at only one instant, the moment when it passes the radar gun.

To determine the car's instantaneous speed, we must determine the rate of change of the distance traveled with respect to time at a particular moment. This requires more information than is contained in Table 9.1; specifically, we must know the distance traveled at every moment of the trip instead of every 30 minutes. For example, suppose we know that after x hours of the trip the car has traveled a distance of

$$s(x) = 4x^2 + 24x$$

miles. (Note that this formula is consistent with the information in Table 9.1.)

How can we determine the car's instantaneous speed after two hours have elapsed? Since the car's speed cannot change significantly during a very short period, the car's instantaneous speed after two hours can be approximated by computing its average speed over some short period beginning at time $x = 2$. Table 9.2 contains several approximations of this type.

Time Interval	*Average Speed (Miles per Hour)*
$x = 2$ to $x = 2.1$	$\dfrac{s(2.1) - s(2)}{2.1 - 2} = \dfrac{68.04 - 64}{2.1 - 2} = \dfrac{4.04}{0.1} = 40.4$
$x = 2$ to $x = 2.01$	$\dfrac{s(2.01) - s(2)}{2.01 - 2} = \dfrac{64.4004 - 64}{2.01 - 2} = \dfrac{0.4004}{0.01} = 40.04$
$x = 2$ to $x = 2.001$	$\dfrac{s(2.001) - s(2)}{2.001 - 2} = \dfrac{64.040004 - 64}{2.001 - 2} = \dfrac{0.040004}{0.001} = 40.004$
$x = 2$ to $x = 2.0001$	$\dfrac{s(2.0001) - s(2)}{2.0001 - 2} = \dfrac{64.0040004 - 64}{2.0001 - 2} = \dfrac{0.0040004}{0.0001} = 40.0004$
$x = 2$ to $x = 2.00001$	$\dfrac{s(2.00001) - s(2)}{2.00001 - 2} = \dfrac{64.0004000004 - 64}{2.00001 - 2} = \dfrac{0.0004000004}{0.00001} = 40.00004$

TABLE 9.2

The computations in Table 9.2 strongly suggest that at the instant when $x = 2$, the car's instantaneous speed was 40 mph.

Let us generalize the calculations in Table 9.2 to determine the car's average speed over the time interval from x to $x + h$. (In Table 9.2 we had $x =$

2 and $h = 0.1$, $h = 0.01$, $h = 0.001$, $h = 0.0001$, and $h = 0.00001$ in successive lines.) Remember that $s(x) = 4x^2 + 24x$, and so

$$s(x + h) = 4(x + h)^2 + 24(x + h).$$

Then

$$\begin{aligned}
\text{average speed} &= \frac{\text{distance traveled}}{\text{time traveling}} \\
&= \frac{s(x+h) - s(x)}{(x+h) - x} \\
&= \frac{[4(x+h)^2 + 24(x+h)] - (4x^2 + 24x)}{h} \\
&= \frac{[4(x^2 + 2xh + h^2) + 24(x+h)] - (4x^2 + 24x)}{h} \\
&= \frac{4x^2 + 8xh + 4h^2 + 24x + 24h - 4x^2 - 24x}{h} \\
&= \frac{8xh + 4h^2 + 24h}{h} \\
&= 8x + 4h + 24.
\end{aligned}$$

To obtain a good approximation of the instantaneous speed at time x, we let h be a small positive number. Thus we are computing the car's average speed over a very short time interval, from x to $x + h$. If h is very small, then the expression $8x + 4h + 24$ is close to $8x + 24$. Hence a reasonable estimate for the instantaneous speed at time x is the value $8x + 24$. Notice that for $x = 2$ the expression $8x + 24$ gives the instantaneous speed to be 40 mph, in agreement with the calculations in Table 9.2.

In accordance with the preceding example, let us define the **instantaneous rate of change** of a function f at a value x in its domain to be the number obtained as follows.

Instantaneous Rate of Change of a Function f at x

Step 1. Calculate the average rate of change for f over the interval from x to $x + h$, which is

$$\frac{f(x+h) - f(x)}{(x+h) - x} = \frac{f(x+h) - f(x)}{h}.$$

Step 2. After simplifying the expression in Step 1, let h assume values progressively closer to zero.

Step 3. Then the values of

$$\frac{f(x+h) - f(x)}{h}$$

approach the instantaneous rate of change of f at x.

The expression in Step 1

$$\frac{f(x + h) - f(x)}{(x + h) - x} = \frac{f(x + h) - f(x)}{h}$$

is called a **difference quotient.** Thus the instantaneous rate of change of f at x is the value that the difference quotients approach (if such a value exists) as h becomes progressively closer to zero. So our previous calculations show that the instantaneous rate of change of $s(x) = 4x^2 + 24x$ at x is $8x + 24$. Because $s(x)$ represents the distance traveled at time x, the instantaneous rate of change of s at x represents the car's instantaneous speed at time x.

EXAMPLE 9.1 Compute the instantaneous rate of change of f at x if

$$f(x) = 3x^2.$$

Solution We must calculate and simplify the difference quotient for f. The results are shown below.

$$\begin{aligned}\frac{f(x + h) - f(x)}{h} &= \frac{3(x + h)^2 - 3x^2}{h} \\ &= \frac{3(x^2 + 2xh + h^2) - 3x^2}{h} \\ &= \frac{3x^2 + 6xh + 3h^2 - 3x^2}{h} \\ &= \frac{6xh + 3h^2}{h} \\ &= \frac{h(6x + 3h)}{h} \\ &= 6x + 3h\end{aligned}$$

As h assumes values close to zero, the expression $6x + 3h$ becomes very close to $6x$. Hence the instantaneous rate of change of f at x is $6x$. ■

Practice Problem 1 Compute the instantaneous rate of change of f at x if

$$f(x) = 2 - 5x.$$

EXAMPLE 9.2 Compute the instantaneous rate of change of f at $x = 4$ if

$$f(x) = \frac{8}{x}.$$

Solution We begin by determining the instantaneous rate of change of f at x. Thus we compute and simplify the difference quotient:

$$\frac{f(x+h)-f(x)}{h} = \frac{\dfrac{8}{x+h} - \dfrac{8}{x}}{h}$$

$$= \frac{\dfrac{8x - 8(x+h)}{x(x+h)}}{h} = \frac{8x - 8x - 8h}{hx(x+h)}$$

$$= \frac{-8h}{hx(x+h)} = \frac{-8}{x(x+h)}.$$

As h approaches zero, the values of $x + h$ approach x, and so the values of $x(x + h)$ approach x^2. Hence the instantaneous rate of change of f at x is $-8/x^2$.

To determine the instantaneous rate of change of f at $x = 4$, we need only evaluate $-8/x^2$ when $x = 4$. The resulting value is

$$\frac{-8}{4^2} = \frac{-8}{16} = -\frac{1}{2}. \quad ■$$

Practice Problem 2 Compute the instantaneous rate of change of f at $x = 2$ if

$$f(x) = 4x^2 - 5.$$

EXAMPLE 9.3 Sylvia operates a shuttle service between the Astor Hotel and the local airport. This year's van rental, insurance, license costs, and franchise fees total \$4800, and each round trip averages \$2 for gasoline and maintenance costs. Thus Sylvia's costs (in dollars) are $C(x) = 2x + 4800$, where x denotes the number of round-trip runs between the Astor Hotel and the airport. (See Example 1.19 for details.) At what rate are Sylvia's costs changing after 500 runs?

Solution We must determine the instantaneous rate of change of C with respect to the number of runs when $x = 500$. As above this requires computing the difference quotient.

$$\frac{C(x+h) - C(x)}{h} = \frac{[2(x+h) + 4800] - (2x + 4800)}{h}$$

$$= \frac{(2x + 2h + 4800) - (2x + 4800)}{h}$$

$$= \frac{2h}{h} = 2.$$

In this case the simplified difference quotient does not contain h. Hence as h approaches 0, the values of the difference quotient remain 2. Therefore the instantaneous rate of change of C at x is 2. Since this is true for each x, it is certainly true for $x = 500$. Thus Sylvia's costs are changing by \$2 per run after 500 runs. This statement means that the cost of an additional run (the 501st run) will be \$2. This amount is simply Sylvia's variable costs for gasoline and maintenance. ■

THE TANGENT LINE TO A GRAPH

The instantaneous rate of change of a function f at x has a significant geometric interpretation. In Figure 9.1 we see that the difference quotient

$$\frac{f(x + h) - f(x)}{(x + h) - x}$$

represents the slope of the line through the points P and Q having coordinates

$$(x, f(x)) \qquad \text{and} \qquad (x + h, f(x + h)).$$

Any such line joining two points on a graph is called a **secant line.** As h approaches 0, the point Q moves closer to P along the graph of $y = f(x)$. The limiting position of the secant line as Q approaches P is called the **tangent line** to the graph at P. (See Figure 9.2.) *Thus the slope of the tangent line to the graph of $y = f(x)$ at P is the value that the difference quotients approach as h approaches zero.* Just as the slope of a line indicates the direction of the line, the slope of the tangent line to a curve indicates the direction of the curve at that point.

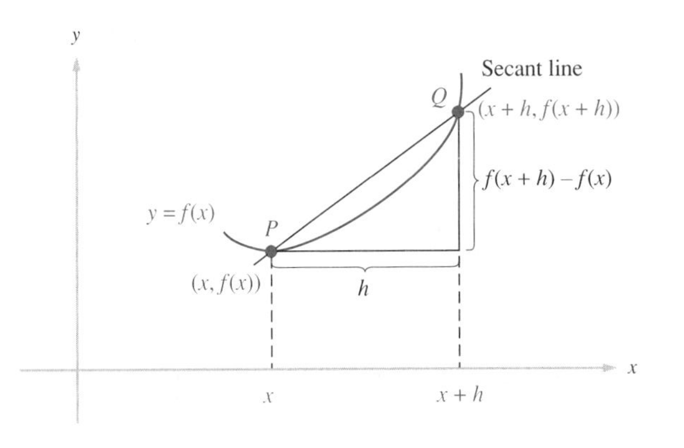

FIGURE 9.1

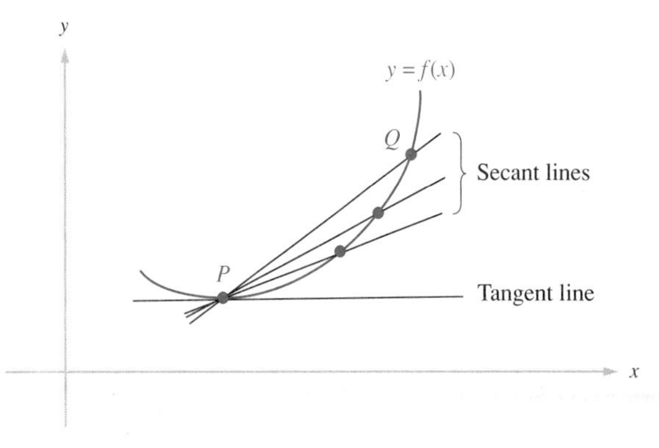

FIGURE 9.2

EXAMPLE 9.4 Determine the equation of the tangent line to the graph of

$$f(x) = \frac{-5}{x + 3}$$

at the point $(-2, -5)$.

Solution To compute the equation of the tangent line, we must know its slope. Thus we compute the difference quotient

$$\begin{aligned}
\frac{f(x + h) - f(x)}{(x + h) - x} &= \frac{f(x + h) - f(x)}{h} \\
&= \frac{\dfrac{-5}{x + h + 3} + \dfrac{5}{x + 3}}{h} \\
&= \frac{\dfrac{-5(x + 3) + 5(x + h + 3)}{(x + 3)(x + h + 3)}}{h} \\
&= \frac{\dfrac{-5x - 15 + 5x + 5h + 15}{(x + 3)(x + h + 3)}}{h} \\
&= \frac{5h}{h(x + 3)(x + 3 + h)} \\
&= \frac{5}{(x + 3)(x + 3 + h)}.
\end{aligned}$$

As h approaches 0, the difference quotient values approach

$$\frac{5}{(x + 3)^2}.$$

At the point $(-2, -5)$, where $x = -2$, we have

$$\frac{5}{(x + 3)^2} = \frac{5}{(-2 + 3)^2} = \frac{5}{1} = 5.$$

This value is the slope of the tangent line at $(-2, -5)$. Using the point-slope form of the equation of a line

$$y - y_1 = m(x - x_1),$$

we find the equation of the tangent line as shown below.

$$\begin{aligned}
y - (-5) &= 5[x - (-2)] \\
y + 5 &= 5(x + 2) \\
y + 5 &= 5x + 10 \\
y &= 5x + 5
\end{aligned}$$

The tangent line is shown in Figure 9.3. ■

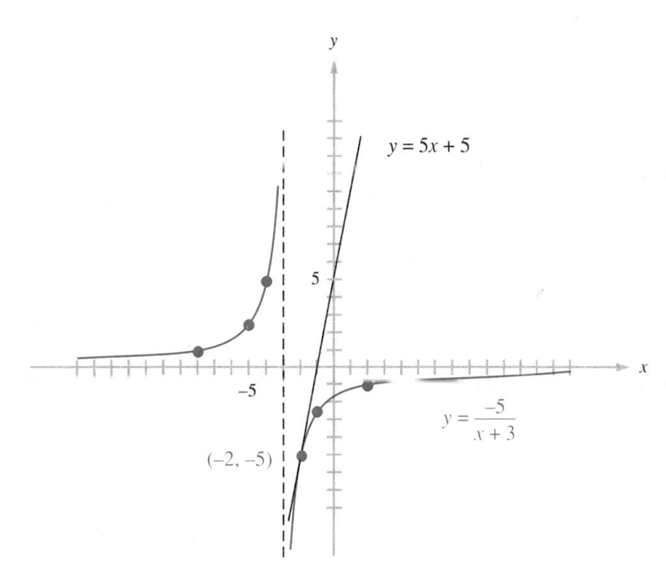

FIGURE 9.3

Practice Problem 3 Find the equation of the tangent line to the graph of $f(x) = \frac{1}{x}$ at the point $(2, \frac{1}{2})$.

EXERCISES 9.1

In Exercises 1–12 compute the instantaneous rate of change of the given function at x.

1. $f(x) = 5$
2. $F(x) = 3x - 4$
3. $g(x) = 12 - 2x$
4. $G(x) = -3$
5. $h(x) = x^2 - 5x$
6. $H(x) = 12 - x^2$
7. $F(x) = 2x^2 + x - 1$
8. $f(x) = 3x^2 - 2x + 5$
9. $G(x) = x^3 - 2$
10. $g(x) = 8 - x^3$
11. $H(x) = \dfrac{1}{x - 2}$
12. $h(x) = \dfrac{2}{x + 3}$

In Exercises 13–20 compute the instantaneous rate of change of f at x = a.

13. $f(x) = -3x, \quad a = 1$
14. $f(x) = 8x, \quad a = 5$
15. $f(x) = 5x + 8, \quad a = -2$
16. $f(x) = 18 - 6x, \quad a = -1$
17. $f(x) = x^2, \quad a = -5$
18. $f(x) = 9 - x^2, \quad a = 2$
19. $f(x) = \dfrac{1}{x - 1}, \quad a = 2$
20. $f(x) = \dfrac{2}{4 - x}, \quad a = 3$

In Exercises 21–30 compute the equation of the tangent line to the graph of $y = f(x)$ at the point (a, b).

21. $f(x) = 4x, \quad (a, b) = (-2, -8)$

22. $f(x) = -2x, \quad (a, b) = (-3, 6)$

23. $f(x) = 12 - 3x, \quad (a, b) = (2, 6)$

24. $f(x) = 5x - 4, \quad (a, b) = (1, 1)$

25. $f(x) - 6 - 2x^2, \quad (a, b) = (2, -2)$

26. $f(x) = 3x^2 + x, \quad (a, b) = (-1, 2)$

27. $f(x) = \dfrac{2}{3 - x}, \quad (a, b) = (1, 1)$

28. $f(x) = \dfrac{1}{x + 5}, \quad (a, b) = (-6, -1)$

29. $f(x) = (x + 1)^3, \quad (a, b) = (1, 8)$

30. $f(x) = (2 - x)^3, \quad (a, b) = (1, 1)$

31. A manufacturer's revenue (in dollars) obtained from selling x units of his product is given by $R(x) = 50x$. How fast is the revenue changing when 100 items are sold?

32. A manufacturer's cost (in dollars) to produce x items is given by

$$C(x) = 100x + 80{,}000.$$

How fast is the cost changing when 500 items are produced?

33. A manufacturer's cost (in dollars) to produce x items is given by

$$C(x) = 2x^2 + 15x + 6000.$$

How fast is the cost changing when 100 items are produced?

If x books are ordered per shipment, then the inventory costs (in dollars) are $C(x) = x + 100{,}000/x$.

34. A manufacturer expects that the revenue (in dollars) obtained from selling x units of his product will be given by $R(x) = 80x - x^2$. How fast is the revenue changing when 30 items are sold?

35. An outbreak of influenza is spreading through a rural school in such a way that the number of sick students after t days is $S(t) = 25t - t^2$. How fast is the influenza spreading after 5 days?

36. A college bookstore expects that during the coming year its inventory costs (in dollars) for a certain textbook will be given by $C(x) = x + \dfrac{100{,}000}{x}$, where x denotes the number of books ordered in a single shipment. How is the inventory cost changing when 500 books are ordered per shipment?

Answers to Practice Problems

1. The instantaneous rate of change of f at x is -5.
2. 16
3. $y = -\frac{1}{4}x + 1$

9.2 Limits

In Section 9.1, when computing the instantaneous rate of change of a function f at x, we calculated the difference quotient

$$\frac{f(x + h) - f(x)}{(x + h) - x} = \frac{f(x + h) - f(x)}{h}$$

and determined the value that it approached as h assumed values progressively closer to zero. This value is called the *limit of the difference quotient as h approaches zero*. The concept of a limit is one of the fundamental ideas of calculus. Indeed, the use of the limit is the most significant way in which calculus differs from algebra. In this section we will discuss this concept and use it to reinterpret the ideas in Section 9.1.

Let us begin by considering the behavior of the function

$$f(x) = \frac{x^3 - 2x^2}{x - 2}$$

near $x = -2$ and $x = 2$. Observe first that $f(x)$ is not defined for $x = 2$ because the denominator is zero there. However, if $x \neq 2$, then we can divide the numerator and denominator of $f(x)$ by $x - 2$ to obtain

$$f(x) = \frac{x^3 - 2x^2}{x - 2} = \frac{x^2(x - 2)}{x - 2} = x^2.$$

Hence if $x \neq 2$, $f(x)$ is identical to x^2; so the graph of $y = f(x)$ is as shown in Figure 9.4. The open circle denotes a point that is not on the graph because $f(x)$ is not defined for $x = 2$.

For values of x close to -2, $f(x)$ is close to 4. In fact, it is clear from Figure 9.4 that $f(x)$ can be made as close to 4 as desired by restricting x to be sufficiently close (but not equal) to -2. We describe this situation by saying that the limit of $f(x)$ is 4 as x approaches -2.

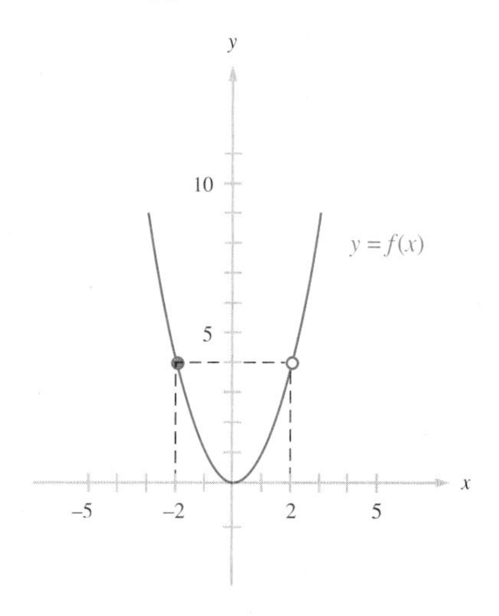

FIGURE 9.4

Although $f(x)$ is not defined for $x = 2$, the behavior of $f(x)$ is exactly the same for x close to 2 as it is for x close to -2. That is, $f(x)$ can be made as close to 4 as desired by restricting x to be sufficiently close (but not equal) to 2. Hence we say that the limit of $f(x)$ is 4 as x approaches 2.

More generally, suppose that f is a function that is defined for all x in some interval containing a except possibly at $x = a$ itself. A real number L is called the **limit of $f(x)$ as x approaches a** if $f(x)$ can be made as close to L as desired for all x sufficiently close (but not equal) to a. To denote that L is the limit of $f(x)$ as x approaches a, we write

$$\lim_{x \to a} f(x) = L.$$

Using this notation with the function f in Figure 9.4, we have

$$\lim_{x \to 2} \frac{x^3 - 2x^2}{x - 2} = 4 \quad \text{and} \quad \lim_{x \to -2} \frac{x^3 - 2x^2}{x - 2} = 4.$$

If $f(x)$ cannot be made arbitrarily close to one specific real number by restricting x to be close to a, then we say that the limit of $f(x)$ does not exist as x approaches a. *Note that whether*

$$\lim_{x \to a} f(x)$$

exists or not is independent of whether $f(a)$ is defined or what its value is. As an example of a function for which the limit does not exist at $x = 2$, consider

$$g(x) = \frac{1}{(x - 2)^2}.$$

The graph of g is shown in Figure 9.5. Notice that $g(x)$ is not defined for $x = 2$. But unlike the function shown in Figure 9.4, $g(x)$ gets increasingly larger as x approaches $a = 2$. Hence there is no real number L to which $g(x)$ gets close as x approaches 2, and so the limit of $g(x)$ does not exist as x approaches 2.

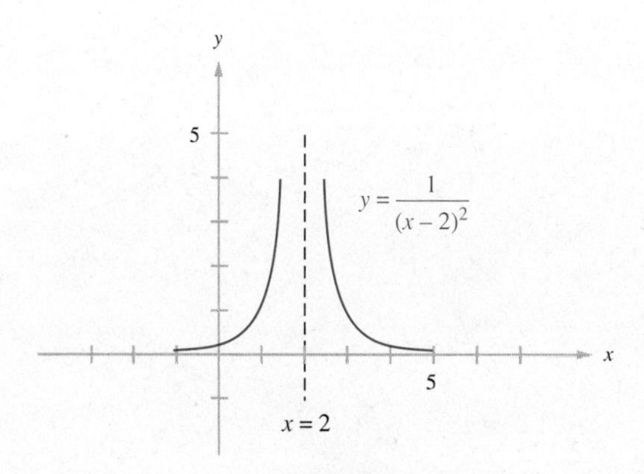

FIGURE 9.5

EXAMPLE 9.5 Determine whether each of the functions below has a limit as x approaches 2.

$$\text{(a) } h(x) = \begin{cases} 2x - 1 & \text{if } x < 2 \\ x + 1 & \text{if } x > 2 \end{cases} \quad \text{and} \quad \text{(b) } H(x) = \begin{cases} x - 1 & \text{if } x < 2 \\ x + 1 & \text{if } x > 2 \end{cases}$$

Solution The graphs of the functions are shown in Figure 9.6.

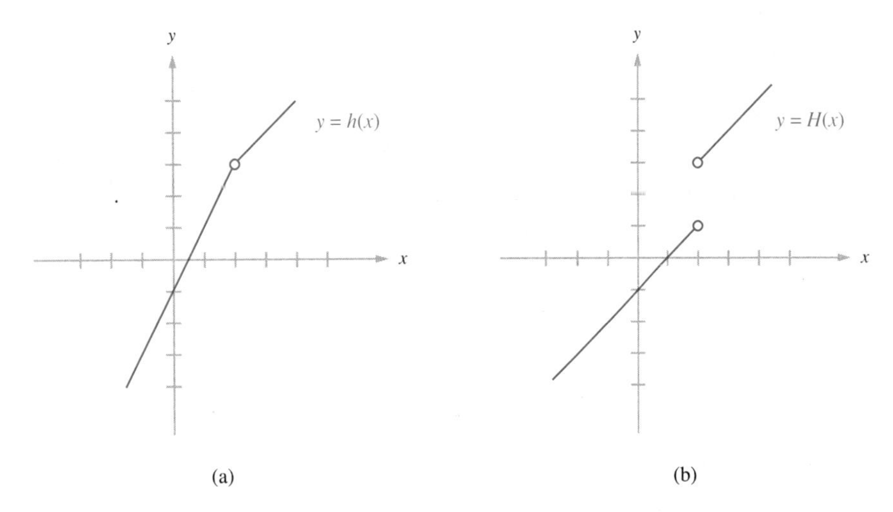

FIGURE 9.6

From Figure 9.6(a) it is clear that $h(x)$ can be made as close to 3 as desired by restricting x to be sufficiently close (but not equal) to 2. Hence the limit of $h(x)$ is 3 as x approaches 2; that is,

$$\lim_{x \to 2} h(x) = 3.$$

On the other hand, Figure 9.6(b) shows that when x is less than 2 but close to 2, $H(x)$ is close to 1. Moreover, when x is greater than 2 but close to 2, $H(x)$ is close to 3. Hence there is no *single* real number to which $H(x)$ can be made close for all x sufficiently close (but not equal) to 2. Thus the limit of $H(x)$ does not exist as x approaches 2. ■

Practice Problem 4 Use Figure 9.7 on page 510 to determine whether each of the functions below has a limit as x approaches 1.

$$\text{(a) } g(x) = \frac{x^2 - 1}{x^2 - x} \qquad \text{(b) } G(x) = \frac{1}{x - 1}$$

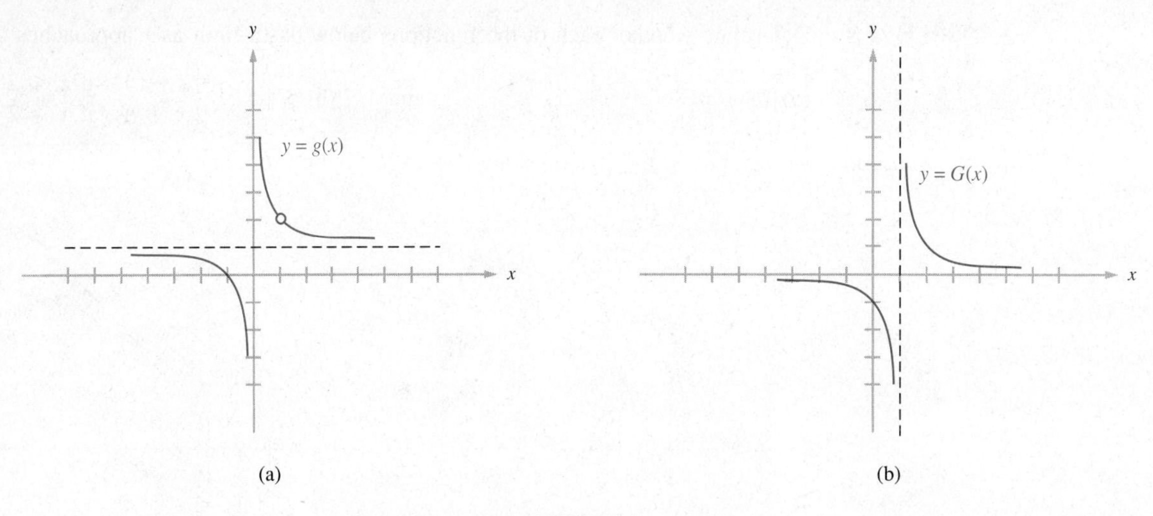

FIGURE 9.7

EVALUATING LIMITS

Fortunately there are a number of results that enable us to compute the limits of many common functions without first drawing their graphs. These results are contained in the following theorem, which we state without proof. Examples 9.6–9.8 illustrate these results.

Theorem 9.1 (limit operations)

Let a and c be constants, and let $\lim_{x \to a} f(x)$ and $\lim_{x \to a} g(x)$ both exist. Then

(a) $\lim_{x \to a} x = a$ *(limit of x rule)*

(b) $\lim_{x \to a} c = c$ *(limit of a constant rule)*

Thus the limit of a constant function equals that constant, no matter what value x is approaching.

(c) For any constant k, $\lim_{x \to a} kf(x) = k\left[\lim_{x \to a} f(x)\right]$. *(constant multiple rule)*

Thus the limit of a constant times a function equals the constant times the limit of the function.

(d) $\lim_{x \to a} [f(x) + g(x)] = \lim_{x \to a} f(x) + \lim_{x \to a} g(x)$ *(sum rule)*

Thus the limit of a sum of functions is the sum of their limits.

(e) $\lim_{x \to a} [f(x) - g(x)] = \lim_{x \to a} f(x) - \lim_{x \to a} g(x)$ *(difference rule)*

Thus the limit of a difference of functions is the difference of their limits.

(f) $\lim_{x \to a} [f(x) \cdot g(x)] = \left[\lim_{x \to a} f(x)\right]\left[\lim_{x \to a} g(x)\right]$ *(product rule)*

Thus the limit of a product of functions is the product of their limits.

(g) If $\lim_{x \to a} g(x) \neq 0$, then

$$\lim_{x \to a} \frac{f(x)}{g(x)} = \frac{\lim_{x \to a} f(x)}{\lim_{x \to a} g(x)}. \qquad \textit{(quotient rule)}$$

Thus the limit of a quotient of functions is the quotient of their limits provided that the limit of the denominator is not zero.

(h) If both $\lim_{x \to a} [f(x)]^r$ and $\left[\lim_{x \to a} f(x)\right]^r$ are defined*, then

$$\lim_{x \to a} [f(x)]^r = \left[\lim_{x \to a} f(x)\right]^r. \qquad \textit{(power rule)}$$

Thus the limit of a power of a function is the power of the limit of the function provided that all the limits and powers are defined.

The examples below demonstrate the use of Theorem 9.1

EXAMPLE 9.6 Evaluate the following limits.

(a) $\lim_{x \to -2} x$ (b) $\lim_{x \to 7} 3$ (c) $\lim_{x \to 1} (3x - 4)$

Solution (a) By the limit of x rule

$$\lim_{x \to -2} x = -2.$$

---

*Recall that even roots of negative numbers are not defined. So, for instance,

$$\left[\lim_{x \to 3} f(x)\right]^r$$

is not defined if $f(x) = 1 - 2x$ and $r = \frac{1}{2}$.

(b) By the limit of a constant rule

$$\lim_{x\to 7} 3 = 3.$$

(c) By the indicated parts of Theorem 9.1, we have

$$\begin{aligned}\lim_{x\to 1}(3x - 4) &= \lim_{x\to 1} 3x - \lim_{x\to 1} 4 && \textbf{(difference rule)}\\ &= 3\lim_{x\to 1} x - \lim_{x\to 1} 4 && \textbf{(constant multiple rule)}\\ &= 3(1) - 4 = -1. && \textbf{(limit of } x \textbf{ rule and limit of a constant rule)}\end{aligned}$$

See Figure 9.8. ■

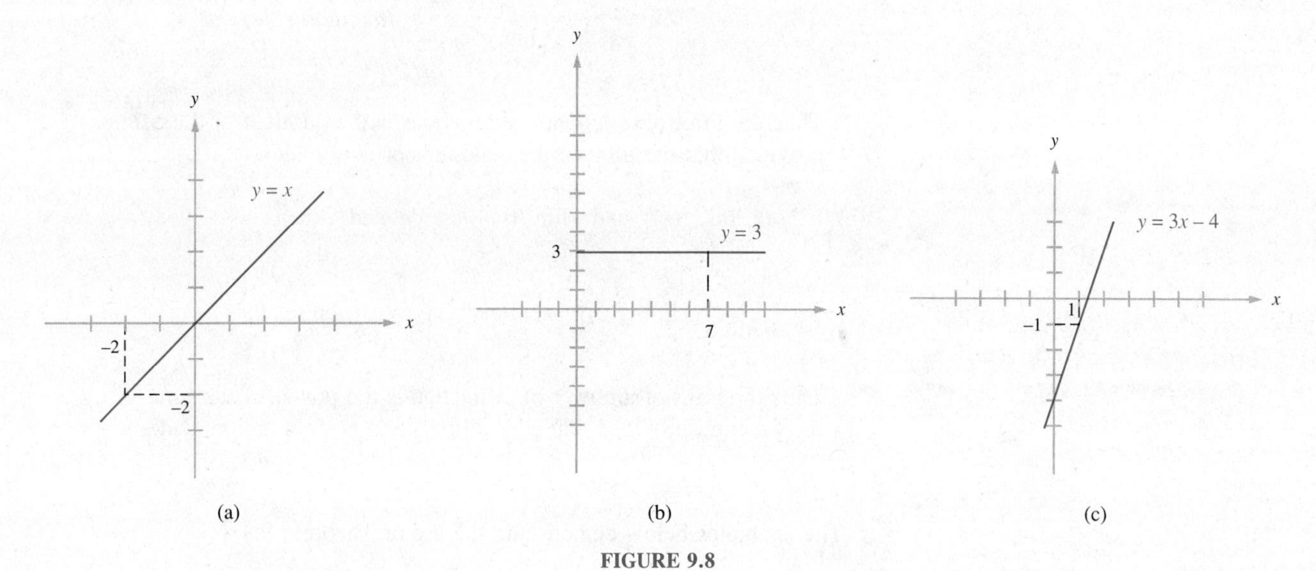

FIGURE 9.8

EXAMPLE 9.7 Evaluate

$$\lim_{x\to -1}(5x^2 - 3x + 6).$$

Solution By using the indicated parts of Theorem 9.1, we see that

$$\begin{aligned}\lim_{x\to -1}(5x^2 - 3x + 6) &= \lim_{x\to -1}(5x^2) - \lim_{x\to -1}(3x) + \lim_{x\to -1} 6 && \textbf{(sum and difference rules)}\\ &= 5\lim_{x\to -1} x^2 - 3\lim_{x\to -1} x + \lim_{x\to -1} 6 && \textbf{(constant multiple rule)}\\ &= 5\lim_{x\to -1} x^2 - 3\lim_{x\to -1} x + 6 && \textbf{(limit of a constant rule)}\end{aligned}$$

$$= 5\left[\lim_{x\to -1} x\right]^2 - 3\lim_{x\to -1} x + 6 \qquad \textbf{(power rule)}$$

$$= 5(-1)^2 - 3(-1) + 6 \qquad \textbf{(limit of } x \textbf{ rule)}$$

$$= 14 \quad ■$$

EXAMPLE 9.8 Evaluate

$$\lim_{x\to 1} \frac{(x+4)(3x-1)}{2x^3}.$$

Solution By using the indicated parts of Theorem 9.1, we see that

$$\lim_{x\to 1} \frac{(x+4)(3x-1)}{2x^3} = \frac{\lim_{x\to 1}(x+4)(3x-1)}{\lim_{x\to 1} 2x^3} \qquad \textbf{(quotient rule)}$$

$$= \frac{\lim_{x\to 1}(x+4)\cdot \lim_{x\to 1}(3x-1)}{\lim_{x\to 1} 2x^3} \qquad \textbf{(product rule)}$$

$$= \frac{\left(\lim_{x\to 1} x + \lim_{x\to 1} 4\right)\left(\lim_{x\to 1} 3x - \lim_{x\to 1} 1\right)}{\lim_{x\to 1} 2x^3} \qquad \textbf{(sum and difference rules)}$$

$$= \frac{\left(\lim_{x\to 1} x + \lim_{x\to 1} 4\right)\left(3\lim_{x\to 1} x - \lim_{x\to 1} 1\right)}{2\lim_{x\to 1} (x^3)} \qquad \textbf{(constant multiple rule)}$$

$$= \frac{\left(\lim_{x\to 1} x + \lim_{x\to 1} 4\right)\left(3\lim_{x\to 1} x - \lim_{x\to 1} 1\right)}{2\left(\lim_{x\to 1} x\right)^3} \qquad \textbf{(power rule)}$$

$$= \frac{(1+4)(3\cdot 1 - 1)}{2(1)^3} = \frac{5(2)}{2} = 5. \qquad \textbf{(limit of } x \textbf{ rule and limit of a constant rule)} \quad ■$$

Notice that in Examples 9.6 and 9.7 we evaluated the limit of polynomial functions. By proceeding as in Example 9.7, it is possible to establish the following useful result.

■ ■ ■ ■ ■ ■

For any polynomial function p and any real number a,

$$\lim_{x\to a} p(x) = p(a).$$

Thus, for instance,

$$\lim_{x \to -1} (5x^2 - 3x + 6) = 5(-1)^2 - 3(-1) + 6 = 14,$$

as we saw in Example 9.7. The following result now follows from the quotient rule of Theorem 9.1.

For any rational function $r = p/q$, where p and q are polynomial functions, and any real number a such that $q(a) \neq 0$,

$$\lim_{x \to a} r(x) = r(a).$$

Therefore in Example 9.8

$$\lim_{x \to 1} \frac{(x + 4)(3x - 1)}{2x^3} = \frac{(1 + 4)[3(1) - 1]}{2(1)^3} = \frac{5(2)}{2} = 5.$$

LIMITS AT INFINITY*

Often it is of interest to know the behavior of a function for large positive or large negative values of the independent variable, that is, as we move far to the right or far to the left along the horizontal axis. For the function

$$f(x) = 2 + \frac{1}{x - 3},$$

we see from the graph in Figure 9.9 that $f(x)$ approaches 2 for large values of x.

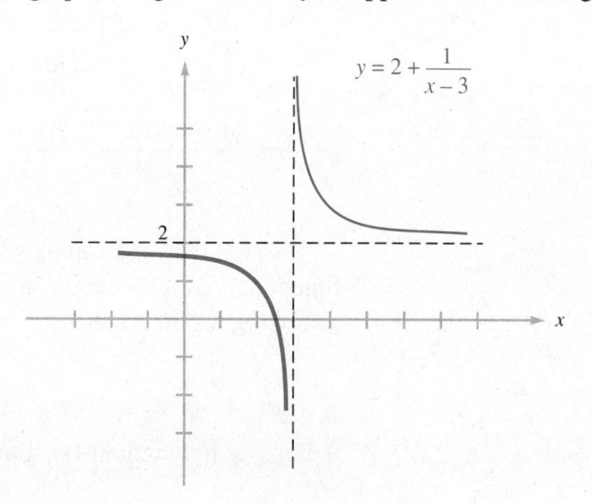

FIGURE 9.9

*Knowledge of limits at infinity is needed only for Sections 13.1 and 14.3.

If $f(x)$ can be made as close as desired to a real number L for all sufficiently large values of x, then we say that the **limit of $f(x)$ as x increases without bound** is L. This situation is also described by saying that *the limit of $f(x)$ as x approaches infinity* is L and is denoted by writing

$$\lim_{x\to\infty} f(x) = L.$$

Similarly, if $f(x)$ can be made as close as desired to a real number L for all sufficiently negative values of x, then we say that the **limit of $f(x)$ as x decreases without bound** is L and write

$$\lim_{x\to-\infty} f(x) = L.$$

Geometrically,

$$\lim_{x\to\infty} f(x) = L \quad \text{or} \quad \lim_{x\to-\infty} f(x) = L$$

means that the graph of $y = f(x)$ has the line $y = L$ as a horizontal asymptote. (See Figure 9.10.)

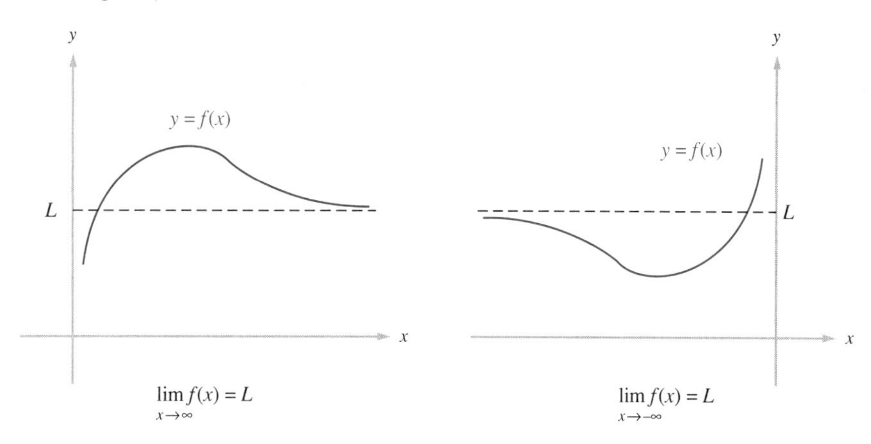

FIGURE 9.10

Since the values of the function in Figure 9.9 approach 2 as we move far to the right along the horizontal axis, we have

$$\lim_{x\to\infty}\left(2 + \frac{1}{x-3}\right) = 2.$$

For this function it is also the case that the values approach 2 as we move far to the left along the horizontal axis; so we also have

$$\lim_{x\to-\infty}\left(2 + \frac{1}{x-3}\right) = 2.$$

Practice Problem 5 By referring to Figure 9.11, determine each of the limits.

(a) $\lim_{x\to\infty} e^x$ (b) $\lim_{x\to-\infty} e^x$

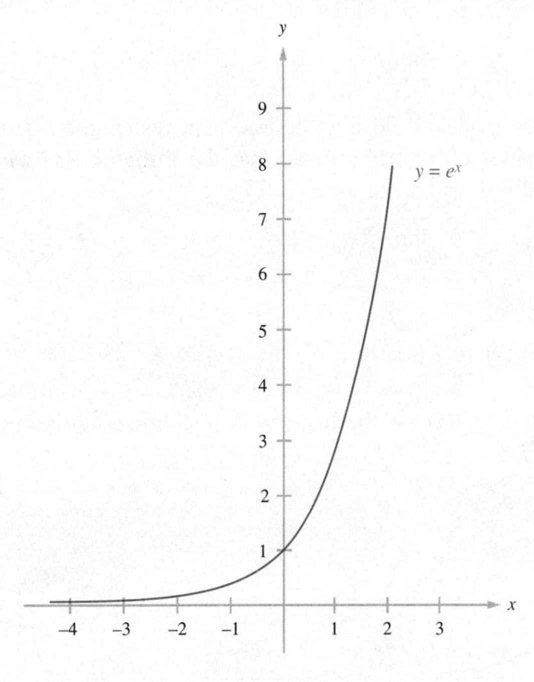

FIGURE 9.11

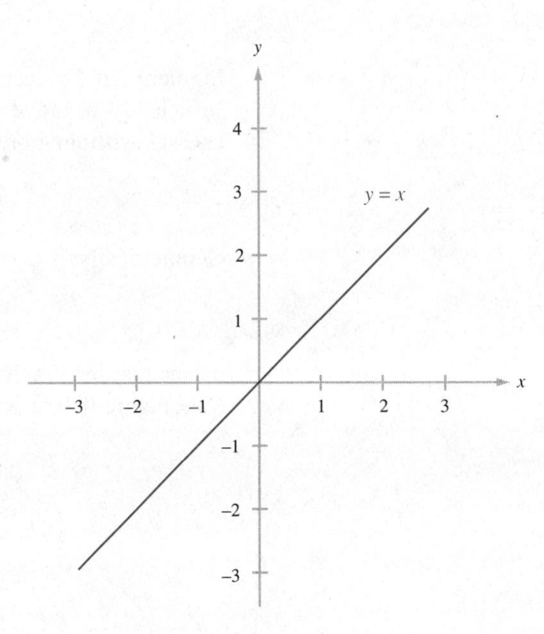

FIGURE 9.12

It is easily seen from the graph of $f(x) = x$ in Figure 9.12 that neither

$$\lim_{x\to\infty} x \quad \text{nor} \quad \lim_{x\to-\infty} x$$

exists. That is, the graph of $f(x) = x$ has no horizontal asymptote. Figure 9.13 illustrates, however, that

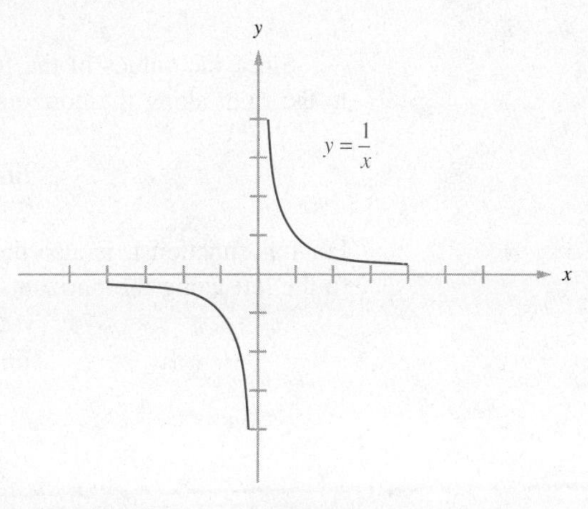

FIGURE 9.13

$$\lim_{x\to\infty} \frac{1}{x} = 0 \quad \text{and} \quad \lim_{x\to-\infty} \frac{1}{x} = 0.$$

Fortunately, parts (b) through (h) of Theorem 9.1 remain true if

$$\lim_{x\to a}$$

is replaced by either

$$\lim_{x\to\infty} \quad \text{or} \quad \lim_{x\to-\infty}.$$

Theorem 9.2

Let c be a constant, and let $\lim_{x\to\infty} f(x)$ and $\lim_{x\to\infty} g(x)$ both exist.

Then

(a) $\lim_{x\to\infty} \frac{1}{x} = 0$ *(limit of $\frac{1}{x}$ rule)*

(b) $\lim_{x\to\infty} c = c$ *(limit of a constant rule)*

(c) For any constant k, $\lim_{x\to\infty} kf(x) = k\left[\lim_{x\to\infty} f(x)\right]$. *(constant multiple rule)*

(d) $\lim_{x\to\infty} [f(x) + g(x)] = \lim_{x\to\infty} f(x) + \lim_{x\to\infty} g(x)$ *(sum rule)*

(e) $\lim_{x\to\infty} [f(x) - g(x)] = \lim_{x\to\infty} f(x) - \lim_{x\to\infty} g(x)$ *(difference rule)*

(f) $\lim_{x\to\infty} [f(x) \cdot g(x)] = \left[\lim_{x\to\infty} f(x)\right]\left[\lim_{x\to\infty} g(x)\right]$ *(product rule)*

(g) If $\lim_{x\to\infty} g(x) \neq 0$, then

$$\lim_{x\to\infty} \frac{f(x)}{g(x)} = \frac{\lim_{x\to\infty} f(x)}{\lim_{x\to\infty} g(x)}. \quad \textit{(quotient rule)}$$

(h) If both $\lim_{x\to\infty} [f(x)]^r$ and $\left[\lim_{x\to\infty} f(x)\right]^r$ are defined, then

$$\lim_{x\to\infty} [f(x)]^r = \left[\lim_{x\to\infty} f(x)\right]^r. \quad \textit{(power rule)}$$

Moreover, each part remains true if

$$\lim_{x\to\infty} \quad \text{is replaced by} \quad \lim_{x\to-\infty}.$$

Before Theorem 9.2 can be used, algebraic manipulations are usually required. The example below demonstrates the technique required.

EXAMPLE 9.9 Evaluate the following limits.

(a) $\lim_{x\to\infty} \dfrac{x - 1}{2x^2 + 3}$ (b) $\lim_{x\to-\infty} \dfrac{x^3 + 5x}{x - 4}$ (c) $\lim_{x\to\infty} \dfrac{4x^2 + 3x - 1}{2x^2 - x + 6}$

Solution Note that Theorem 9.2 cannot be used to evaluate any of these three limits because neither the limit of the numerator nor the limit of the denominator exists. Consequently, in order to evaluate the limit of these rational functions as x increases or decreases without bound, we must rewrite each function by dividing the numerator and denominator by x^k, where k is the largest power of x appearing in the *denominator* of the given function.

Thus in (a) we must divide each term by x^2 and then apply Theorem 9.2. The results are as follows.

$$\lim_{x\to\infty} \frac{x - 1}{2x^2 + 3} = \lim_{x\to\infty} \frac{\left(\frac{1}{x} - \frac{1}{x^2}\right)}{\left(2 + \frac{3}{x^2}\right)}$$

$$= \frac{\lim_{x\to\infty}\left(\frac{1}{x} - \frac{1}{x^2}\right)}{\lim_{x\to\infty}\left(2 + \frac{3}{x^2}\right)} \qquad \text{(quotient rule)}$$

$$= \frac{\lim_{x\to\infty}\frac{1}{x} - \lim_{x\to\infty}\frac{1}{x^2}}{\lim_{x\to\infty} 2 + \lim_{x\to\infty}\frac{3}{x^2}} \qquad \text{(sum and difference rules)}$$

$$= \frac{\lim_{x\to\infty}\frac{1}{x} - \lim_{x\to\infty}\frac{1}{x^2}}{\lim_{x\to\infty} 2 + 3\lim_{x\to\infty}\frac{1}{x^2}} \qquad \text{(constant multiple rule)}$$

$$= \frac{\lim_{x\to\infty}\frac{1}{x} - \left(\lim_{x\to\infty}\frac{1}{x}\right)^2}{\lim_{x\to\infty} 2 + 3\left(\lim_{x\to\infty}\frac{1}{x}\right)^2} \qquad \text{(power rule)}$$

$$= \frac{0 - (0)^2}{2 + 3(0)^2} = 0 \qquad \text{(limit of } 1/x \text{ rule and limit of a constant rule)}$$

For (b) we divide each term by x to obtain

$$\lim_{x \to -\infty} \frac{x^3 + 5x}{x - 4} = \lim_{x \to -\infty} \frac{x^2 + 5}{\left(1 - \frac{4}{x}\right)}.$$

Applying the difference, constant multiple, and limit of $1/x$ rules enables us to evaluate the denominator of the preceding expression as follows.

$$\begin{aligned} \lim_{x \to -\infty} \left(1 - \frac{4}{x}\right) &= \lim_{x \to -\infty} 1 - \lim_{x \to -\infty} \frac{4}{x} \\ &= 1 - 4 \lim_{x \to -\infty} \frac{1}{x} \\ &= 1 - 4(0) = 1 \end{aligned}$$

Now the numerator $x^2 + 5$ becomes larger and larger as x decreases without bound. Since the denominator approaches 1, it follows that the fraction

$$\frac{x^2 + 5}{\left(1 - \frac{4}{x}\right)}$$

grows larger and larger as x decreases without bound, and hence the limit in (b) does not exist.

In (c) we divide each term in the fraction by x^2 and apply the quotient rule to obtain

$$\lim_{x \to \infty} \frac{4x^2 + 3x - 1}{2x^2 - x + 6} = \lim_{x \to \infty} \frac{4 + \frac{3}{x} - \frac{1}{x^2}}{2 - \frac{1}{x} + \frac{6}{x^2}} = \frac{\lim_{x \to \infty} \left(4 + \frac{3}{x} - \frac{1}{x^2}\right)}{\lim_{x \to \infty} \left(2 - \frac{1}{x} + \frac{6}{x^2}\right)}.$$

Applying Theorem 9.2 to the numerator gives

$$\begin{aligned} \lim_{x \to \infty} \left(4 + \frac{3}{x} - \frac{1}{x^2}\right) &= \lim_{x \to \infty} 4 + \lim_{x \to \infty} \frac{3}{x} - \lim_{x \to \infty} \frac{1}{x^2} \\ &= 4 + 3 \lim_{x \to \infty} \frac{1}{x} - \left(\lim_{x \to \infty} \frac{1}{x}\right)^2 \\ &= 4 + 3(0) - (0)^2 = 4. \end{aligned}$$

A similar calculation shows that

$$\lim_{x \to \infty} \left(2 - \frac{1}{x} + \frac{6}{x^2}\right) = 2.$$

Thus

$$\lim_{x\to\infty}\frac{4x^2+3x-1}{2x^2-x+6}=\frac{\lim\limits_{x\to\infty}\left(4+\dfrac{3}{x}-\dfrac{1}{x^2}\right)}{\lim\limits_{x\to\infty}\left(2-\dfrac{1}{x}+\dfrac{6}{x^2}\right)}=\frac{4}{2}=2. \quad \blacksquare$$

Practice Problem 6 Evaluate the following limits.

(a) $\lim\limits_{x\to\infty}\dfrac{x^3+4x}{2x^3-7}$ (b) $\lim\limits_{x\to-\infty}\dfrac{x^2-x+6}{x^5-4x^3+2}$ (c) $\lim\limits_{x\to\infty}\dfrac{4x^3-7x+3}{8x^2+9x-5}$

INSTANTANEOUS RATES OF CHANGE AS LIMITS

As we noted above, the instantaneous rate of change of a function f at x is defined as the limit of the difference quotient for f:

$$\lim_{h\to 0}\frac{f(x+h)-f(x)}{h}.$$

This limit is of such special significance that it has its own name and notation: It is called the **derivative of f at x** and is denoted $f'(x)$.

Derivative of f at x

$$f'(x)=\lim_{h\to 0}\frac{f(x+h)-f(x)}{h}$$

provided that this limit exists.

Thus with a given function f we can associate a function f' called the **derivative of f.** This function has as its value at x the number $f'(x)$, the derivative of f at x. The domain of f' is the set of all x for which

$$f'(x)=\lim_{h\to 0}\frac{f(x+h)-f(x)}{h}$$

exists. (This limit may exist for many or perhaps all real values of x.) The discussion in Section 9.1 shows that $f'(x)$ *represents both the instantaneous rate of change of f at x and the slope of the tangent line to the graph of f at the point $(x, f(x))$.*

EXAMPLE 9.10 Determine the equation of the tangent line to the graph of $y = x^3$ at the point (2, 8).

Solution As in Example 9.4, we must determine the slope of the tangent line to the graph of $f(x) = x^3$ at (2, 8). This requires computing the derivative of f at 2. Recall that

$$(x + h)^3 = x^3 + 3x^2h + 3xh^2 + h^3$$

by the binomial theorem. Hence the derivative of f at x is

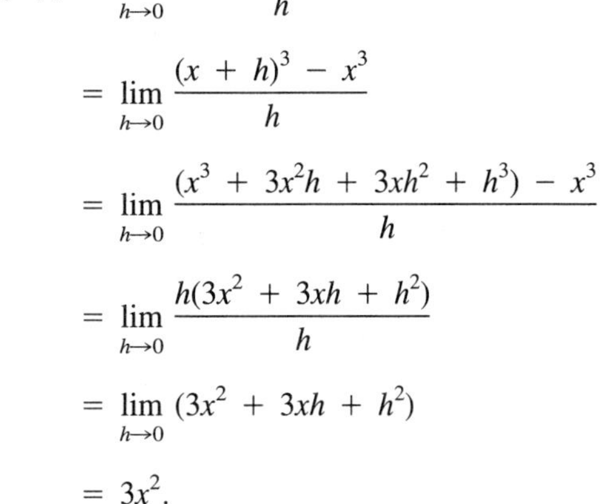

$$\begin{aligned} f'(x) &= \lim_{h\to 0} \frac{f(x + h) - f(x)}{h} \\ &= \lim_{h\to 0} \frac{(x + h)^3 - x^3}{h} \\ &= \lim_{h\to 0} \frac{(x^3 + 3x^2h + 3xh^2 + h^3) - x^3}{h} \\ &= \lim_{h\to 0} \frac{h(3x^2 + 3xh + h^2)}{h} \\ &= \lim_{h\to 0} (3x^2 + 3xh + h^2) \\ &= 3x^2. \end{aligned}$$

Thus the derivative of f at 2 is

$$f'(2) = 3(2)^2 = 12.$$

This value is the slope of the tangent line to $y = f(x)$ at the point (2, 8), where $x = 2$. Hence the desired tangent line passes through (2, 8) and has slope 12; its equation is

$$y - 8 = 12(x - 2)$$

or

$$y = 12x - 16.$$

See Figure 9.14. ■

FIGURE 9.14

In subsequent sections we will present formulas that enable us to compute the derivatives of many common functions.

EXERCISES 9.2

In Exercises 1–8 determine whether or not the function that is graphed has a limit at $x = 2$. If the limit exists, estimate its value from the graph.

1.

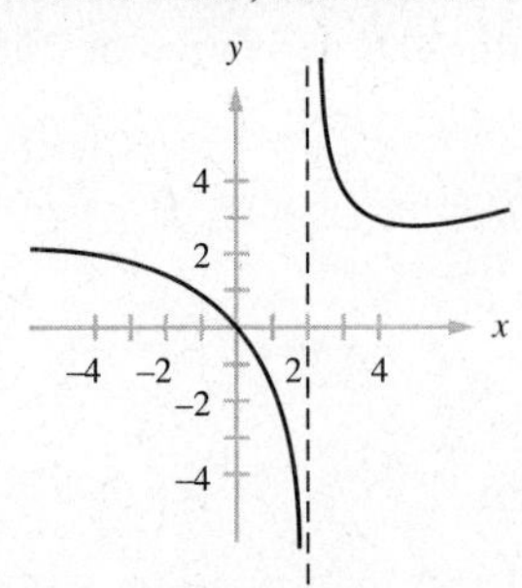

2.

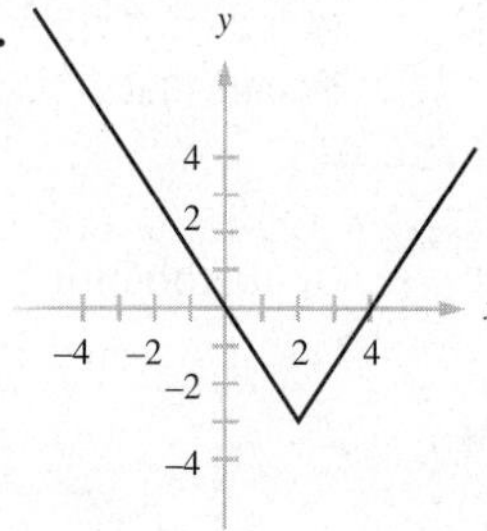

3.

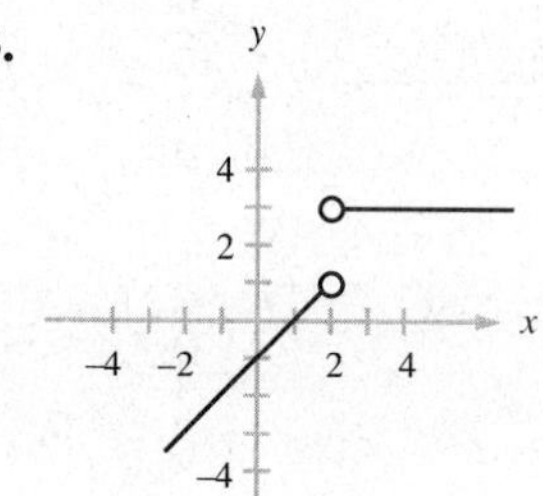

4.

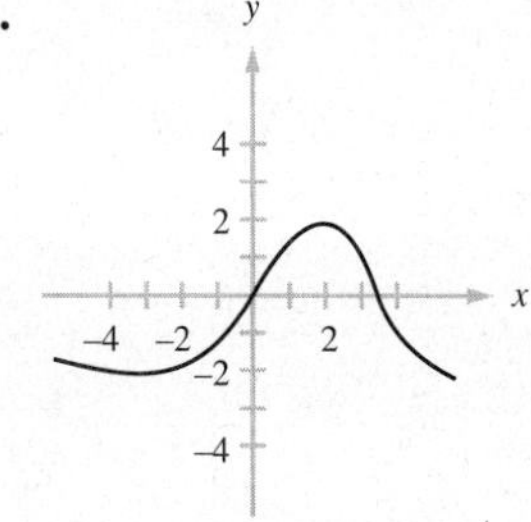

5.

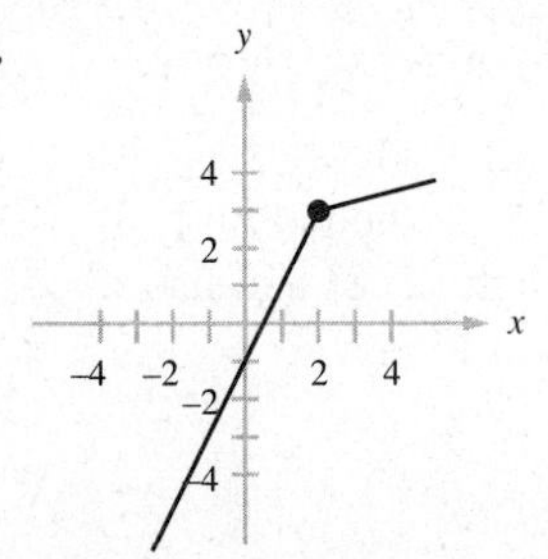

6.

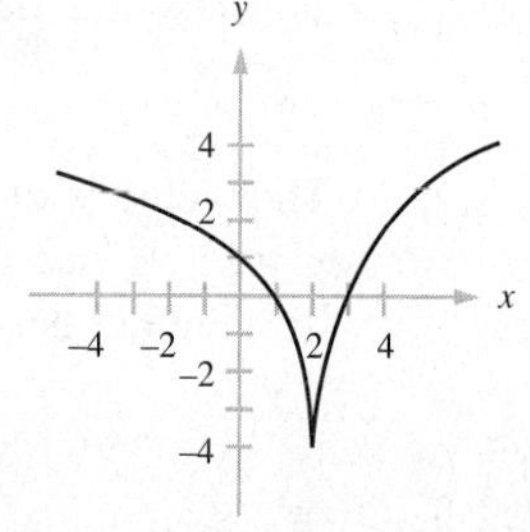

7.

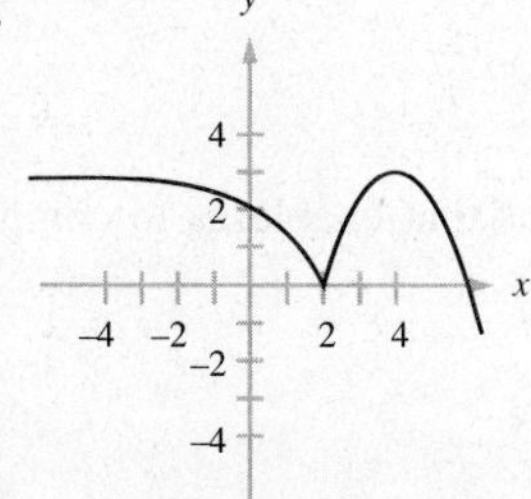

8.

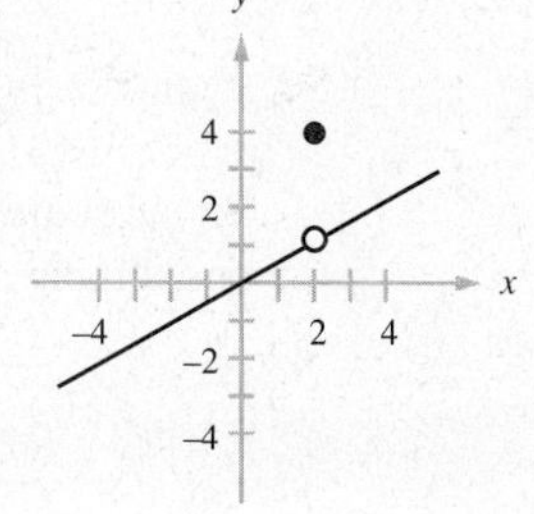

In Exercises 9–32 determine whether or not the given limit exists. Use Theorems 9.1 and 9.2 to evaluate those that exist.

9. $\lim_{x \to -3} (2x + 5)$

10. $\lim_{x \to -8} (4 - 3x)$

11. $\lim_{x \to -2} 6$

12. $\lim_{x \to 2} (x^2 - 3x)$

13. $\lim_{x \to 7} (2x^2 - 10x)$

14. $\lim_{x \to 9} -2$

15. $\lim_{x \to -4} (2x + 3)(x^2 - 1)$

16. $\lim_{x \to 2} (3x - 2)(8 - x^2)$

17. $\lim_{x \to 0} \dfrac{x - 2}{x^2 + 1}$

18. $\lim_{x \to -2} \dfrac{2x + 1}{x^2 - 4}$

19. $\lim_{x \to 3} \sqrt{x^3 - 2}$

20. $\lim_{x \to 0} (4x - 1)^5$

21. $\lim_{x \to 3} \dfrac{x + 4}{x^2 - 9}$

22. $\lim_{x \to 5} \dfrac{x^2 - 1}{2x^2 + 3}$

23. $\lim_{x \to 4} \dfrac{x - 4}{x^2 - x - 12}$

24. $\lim_{x \to 3} \dfrac{x^2 - 4x + 3}{9 - x^2}$

25. $\lim_{x \to -1} \dfrac{x^2 + x}{x^2 - 1}$

26. $\lim_{x \to -2} \dfrac{x^2 - x - 6}{x^2 + x - 2}$

27. $\lim_{x \to \infty} \dfrac{-x^3}{3x^2 + 1}$

28. $\lim_{x \to -\infty} \dfrac{2x}{x^2 - 3x + 1}$

29. $\lim_{x \to -\infty} \dfrac{4x^5 - 3x + 2}{9x^5 - 7x^2}$

30. $\lim_{x \to \infty} \dfrac{x^2 - 1}{3x - 7}$

31. $\lim_{x \to \infty} \dfrac{x^3 - 7x + 4}{9x^4 + 6x^2 - 2}$

32. $\lim_{x \to -\infty} \dfrac{5x^3 + 2x^2}{3x^3 - 4x + 1}$

In Exercises 33–36 compute the derivative of the given function f at x.

33. $f(x) = 3x + 4$

34. $f(x) = 5 - 9x$

35. $f(x) = 4 + 5x - x^2$

36. $f(x) = 2x^2 + 3x$

37. Find the equation of the tangent line to the graph of $y = 2 - 3x$ at $(1, -1)$.

38. Find the equation of the tangent line to the graph of $y = 5x + 3$ at $(-1, -2)$.

39. Find the equation of the tangent line to the graph of $y = \dfrac{2}{3 - x}$ at $(5, -1)$.

40. Find the equation of the tangent line to the graph of $y = \dfrac{4}{x + 1}$ at $(-5, -1)$.

Answers to Practice Problems

4. (a) $\lim_{x \to 1} g(x) = 2$ (b) $\lim_{x \to 1} G(x)$ does not exist.

5. (a) $\lim_{x \to \infty} e^x$ does not exist (b) $\lim_{x \to -\infty} e^x = 0$.

6. (a) ½ (b) 0 (c) does not exist

9.3 Differentiable and Continuous Functions

The derivative of a function f at x is defined as the limit of a difference quotient. At any x for which this limit exists, f is said to be **differentiable.** It may happen, however, that the limit of the difference quotient fails to exist at one or more points. At such places the slope of the tangent line (which is measured by the derivative) is not defined. This situation can occur for several different reasons.

Because the slope of a vertical line is not defined, the derivative of a function is not defined at any point on its graph where there is a vertical tangent line. For example, the function

$$f(x) = 2 + \sqrt[3]{x - 1}$$

has a vertical tangent line at $x = 1$. (See Figure 9.15.) Thus this function is not differentiable at $x = 1$.

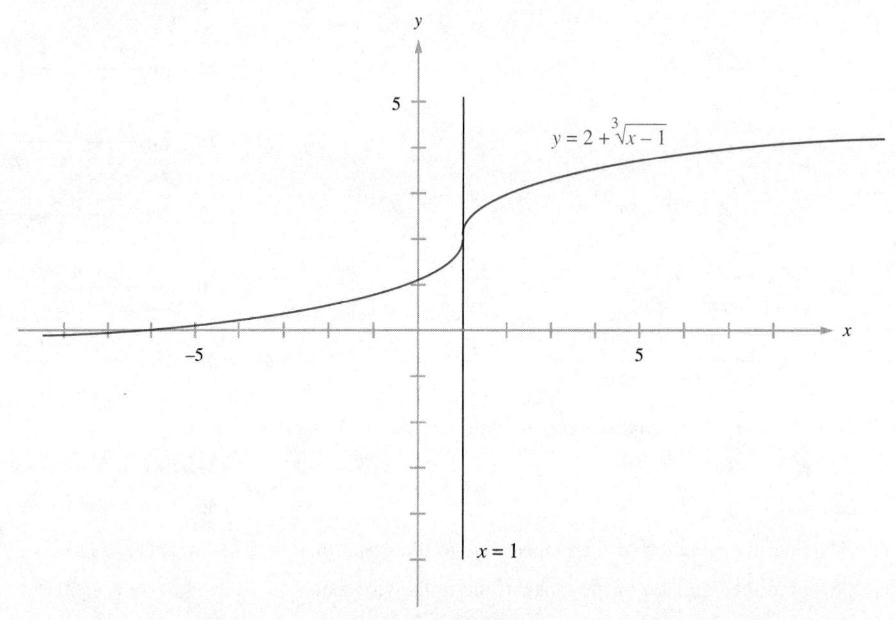

A vertical tangent at $x = 1$

FIGURE 9.15

The derivative of a function can also fail to exist because its graph has no tangent whatsoever at a particular point. This situation arises whenever the graph of the function has an abrupt change of direction. The function

$$g(x) = |x - 2| + 1,$$

for instance, has an abrupt change of direction at $x = 2$. (See Figure 9.16.) Hence g is not differentiable at $x = 2$.

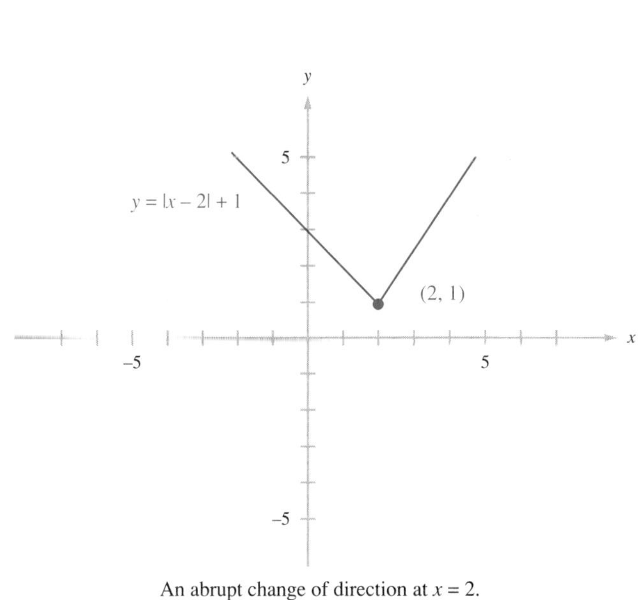

An abrupt change of direction at $x = 2$.

FIGURE 9.16

FIGURE 9.17

Places where two different pieces of a function are joined often are points at which the derivative fails to exist. The following example illustrates this situation.

EXAMPLE 9.11 A student who earns \$6 per hour on a summer job is paid time-and-a-half (\$9 per hour) for work beyond the first eight hours of any day. Graph the function W, where $W(x)$ denotes the student's earnings if she works x hours on a given day.

Solution If the student works x hours, where $x \leq 8$, then her earnings for the day will be $6x$ dollars. If $x > 8$, however, the student will be paid \$6 per hour for the first eight hours of work (\$48 in all) plus \$9 per hour for her work in excess of 8 hours (an additional $\$9(x - 8)$). Thus in this case her daily earnings are

$$48 + 9(x - 8) = 9x - 24$$

dollars. Hence

$$W(x) = \begin{cases} 6x & \text{if } 0 \leq x \leq 8 \\ 9x - 24 & \text{if } 8 < x. \end{cases}$$

The graph of this function is shown in Figure 9.17. Notice that the graph of W is composed of two pieces joined at the point (8, 48) where overtime pay starts. At this point W is not differentiable because the graph has an abrupt change of direction there. ■

Practice Problem 7 For the graph of the function f in Figure 9.18, determine the values of x at which f is not differentiable.

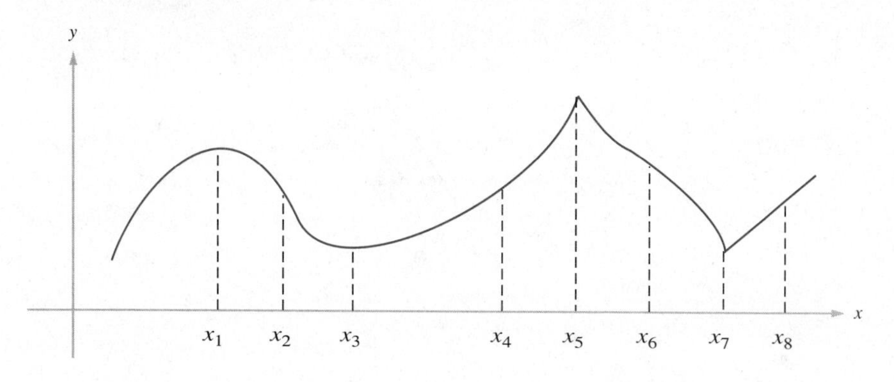

FIGURE 9.18

CONTINUOUS FUNCTIONS

In Section 9.2 we observed that for any polynomial function p and any real number a,

$$\lim_{x \to a} p(x) = p(a).$$

This condition means that $p(x)$ can be made as close to $p(a)$ as desired for x close (but not equal) to a. It can be shown that this condition implies that the graph of p has no jumps or breaks at a.

Functions having this property are of special importance. A function f is called **continuous** at $x = a$ whenever

$$\lim_{x \to a} f(x) = f(a).$$

If f is not continuous at $x = a$, then a is called a **point of discontinuity** of f. In Figure 9.19, f is continuous at $x = a$; on the other hand, $x = b$ is a point of discontinuity of f.

Notice that the definition of continuity involves three separate conditions.

1. The function must be defined at $x = a$; that is, $f(a)$ is defined.
2. The function must have a limit as x approaches a; that is, $\lim_{x \to a} f(x)$ must exist.
3. The values in conditions 1 and 2 must be equal.

A function can fail to be continuous at a point because any of these three conditions is not met. For example, the function

$$f_1(x) = \frac{x^2 - 4}{x - 2}$$

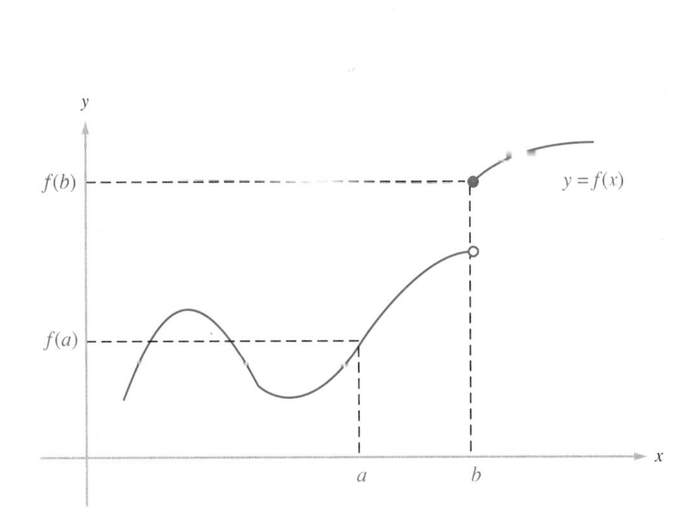

A point of discontinuity at $x = b$

FIGURE 9.19

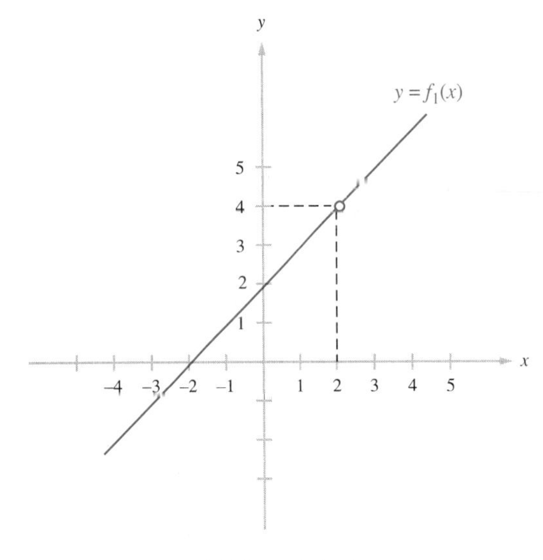

$f_1(2)$ is not defined

FIGURE 9.20

is continuous at each value for which it is defined. (See Figure 9.20.) However, $f(2)$ is not defined, and so the function is not continuous at $x = 2$. Observe that

$$\lim_{x \to 2} f_1(x) = 4.$$

Hence if we define $f_1(x)$ to be 4 at $x = 2$, then the resulting function

$$F(x) = \begin{cases} \dfrac{x^2 - 4}{x - 2} & \text{if } x \neq 2 \\ 4 & \text{if } x = 2 \end{cases}$$

is continuous at every real number x. Notice that defining $F(2) = 4$ fills the hole in the graph shown in Figure 9.20.

By contrast, the function

$$f_2(x) = \begin{cases} \dfrac{1}{x - 2} & \text{if } x \neq 2 \\ 0 & \text{if } x = 2, \end{cases}$$

which is shown in Figure 9.21, is defined for all real numbers. But

$$\lim_{x \to 2} f_2(x)$$

does not exist. Hence f_2 is not continuous at $x = 2$.

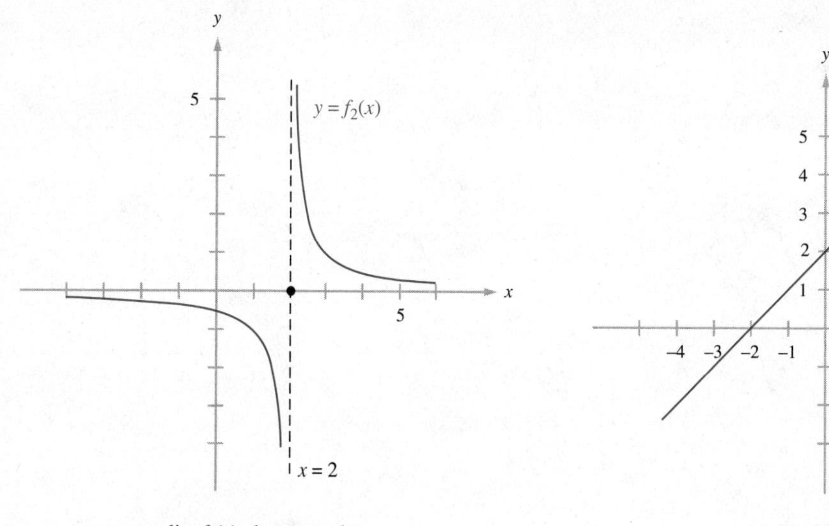

$\lim_{x\to 2} f_2(x)$ does not exist

FIGURE 9.21

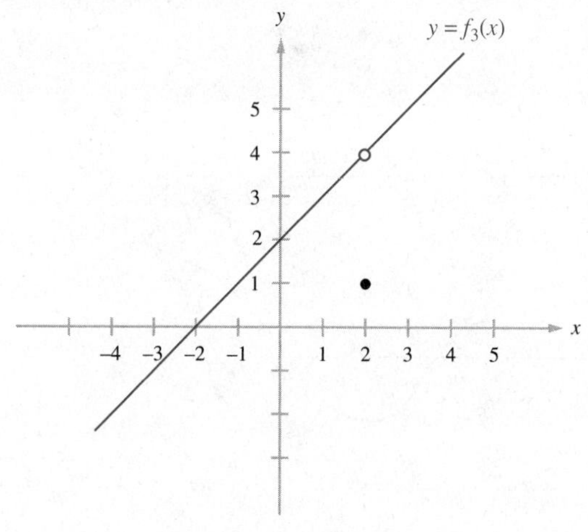

$\lim_{x\to 2} f_3(x) \neq f(2)$

FIGURE 9.22

Now consider the function

$$f_3(x) = \begin{cases} \dfrac{x^2 - 4}{x - 2} & \text{if } x \neq 2 \\ 1 & \text{if } x = 2, \end{cases}$$

which is defined for all real numbers. We see in Figure 9.22 that

$$\lim_{x\to a} f_3(x)$$

exists for all real numbers a. But

$$\lim_{x\to 2} f_3(x) = 4,$$

whereas $f_3(2) = 1$. Therefore f_3 is not continuous at $x = 2$, that is, $x = 2$ is a point of discontinuity for f_3. However f_3 is continuous whenever $x \neq 2$.

Although the elementary functions considered in Chapter 8 are all continuous at each point in their domains, functions that have points of discontinuity often arise in applications.

EXAMPLE 9.12 For first-class mail weighing up to 5 ounces, the U.S. Postal Service charges 25¢ for the first ounce and 20¢ for each additional ounce or fraction thereof. Let $C(x)$ denote the cost in cents to send a letter weighing x ounces via first-class mail. At which x, $0 < x < 5$, does the function C have points of discontinuity?

Solution The cost function is given by

$$C(x) = \begin{cases} 25 & \text{if } 0 < x \leq 1 \\ 45 & \text{if } 1 < x \leq 2 \\ 65 & \text{if } 2 < x \leq 3 \\ 85 & \text{if } 3 < x \leq 4 \\ 105 & \text{if } 4 < x \leq 5. \end{cases}$$

The graph of this function is shown in Figure 9.23. Notice that it has points of discontinuity at the positive integers 1, 2, 3, and 4. ■

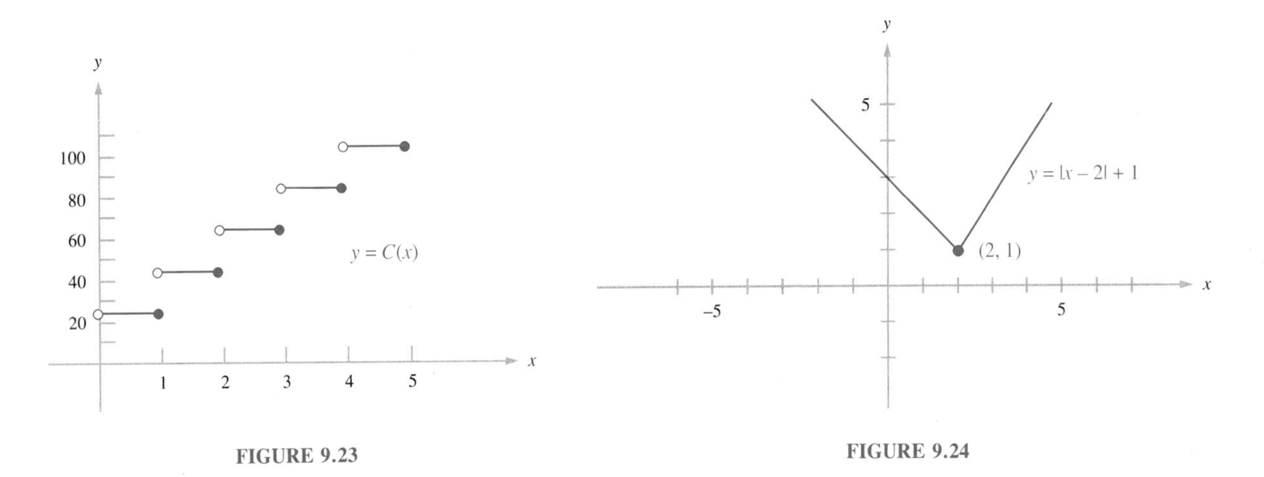

FIGURE 9.23

FIGURE 9.24

In Section 9.2 we saw that if p and q are polynomial functions, the rational function $r = p/q$ satisfies

$$\lim_{x \to a} r(x) = r(a).$$

Hence r is continuous at every real number a for which $q(a) \neq 0$. Thus *the only breaks or jumps in the graph of a rational function occur at points where its denominator is zero.*

Now we will consider the connection between differentiable and continuous functions. Figure 9.24 shows the graph of the function

$$g(x) = |x - 2| + 1,$$

which we discussed earlier in this section. Recall that g is not differentiable at $x = 2$ because its graph has an abrupt change of direction there. However it is easily seen in Figure 9.24 that g is continuous for all values of x. Thus *not every function that is continuous at a point is differentiable there.* The next theorem tells us, however, that differentiable functions are always continuous.

Theorem 9.3

If f is differentiable at $x = a$, then f is continuous at $x = a$.

Geometrically, this result means that if the graph of $y = f(x)$ has a tangent line at $x = a$, then there cannot be a break in the graph at that point. It follows from Theorem 9.3 that *a function cannot be differentiable at a point of discontinuity*. Thus a third reason that a function f may fail to be differentiable at $x = a$ is that a may be a point of discontinuity of f.

Practice Problem 8 For the graph of the function f in Figure 9.25, determine the values of x at which f has a point of discontinuity.

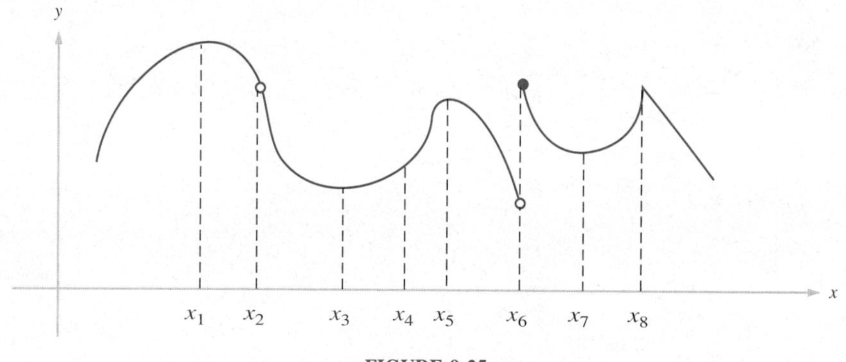

FIGURE 9.25

A SPECIAL LIMIT

Figure 9.26 shows a portion of the graph of the function

$$f(x) = (1 + x)^{1/x}.$$

Although this function is not defined for $x = 0$ (because division by zero is prohibited), it is clear from the graph that

$$\lim_{x \to 0} (1 + x)^{1/x}$$

exists. The value of this limit is one of the most important numbers in mathematics; it is denoted by the letter e. Thus

$$e = \lim_{x \to 0} (1 + x)^{1/x}.$$

In Section 8.3 we encountered the exponential function

$$f(x) = e^x$$

and learned that

$$e \approx 2.71828.$$

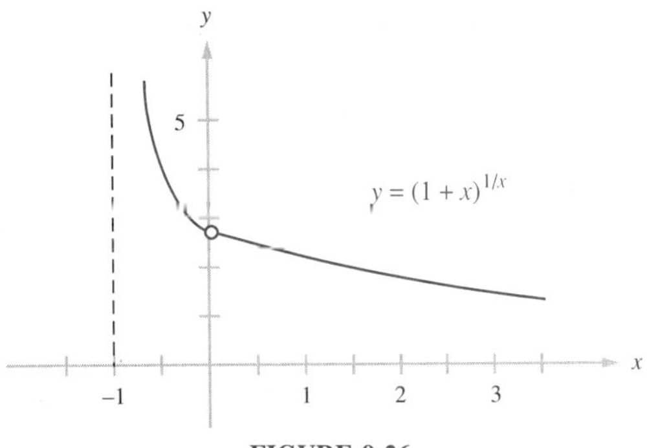

FIGURE 9.26

The limit $\lim_{x \to 0} (1 + x)^{1/x}$ occurs quite often in calculus, and so the number e arises in a wide variety of applications. One of these applications is to the mathematics of finance.

Recall from Exercise 33 in Section 4.2 that if a principal P is invested at an annual interest rate r compounded m times per year, then its value after t years is

$$A = P\left(1 + \frac{r}{m}\right)^{mt}.$$

In Chapter 4 we usually considered interest compounded annually ($m = 1$), quarterly ($m = 4$), or monthly ($m = 12$). But interest may also be compounded *continuously,* that is, at every instant. To compute continuously compounded interest, we must let m increase without bound in the formula above.

$$\begin{aligned} A &= \lim_{m \to \infty} P\left(1 + \frac{r}{m}\right)^{mt} \\ &= P \lim_{m \to \infty} \left[\left(1 + \frac{r}{m}\right)^{m/r}\right]^{rt} \qquad \textbf{(constant multiple rule)} \end{aligned}$$

Now let $x = \dfrac{r}{m}$, and note that x approaches 0 as m increases without bound.

Then

$$\begin{aligned} A &= P \lim_{x \to 0} [(1 + x)^{1/x}]^{rt} \\ &= P\left[\lim_{x \to 0} (1 + x)^{1/x}\right]^{rt}. \qquad \textbf{(power rule)} \end{aligned}$$

But the expression in brackets is the definition of e. Therefore

$$A = Pe^{rt},$$

and thus we have the following formula.

Continuously Compounded Interest

$$A = Pe^{rt}$$

where A denotes the amount after t years,
P denotes the principal,
r denotes the annual interest rate, and
t denotes the number of years for which the money is invested.

EXAMPLE 9.13 Compute the amount when \$1000 is deposited at an annual interest rate of 9% compounded continuously for 5 years.

Solution Since interest is to be compounded continuously, we use the formula above with $P = \$1000$, $r = .09$, and $t = 5$.

$$\begin{aligned} A &= Pe^{rt} \\ &= \$1000e^{.09(5)} \\ &= \$1000e^{.45} \\ &\approx \$1568.31. \end{aligned}$$

Compare this value to those calculated in Example 4.9 when interest is compounded annually, quarterly, and monthly. ■

Practice Problem 9 Compute the amount when \$2500 is deposited at an annual interest rate of 8% compounded continuously for 4 years.

EXERCISES 9.3

For the functions graphed in Exercises 1–10, determine (a) all points of discontinuity and (b) all values of x at which the function is not differentiable.

1.

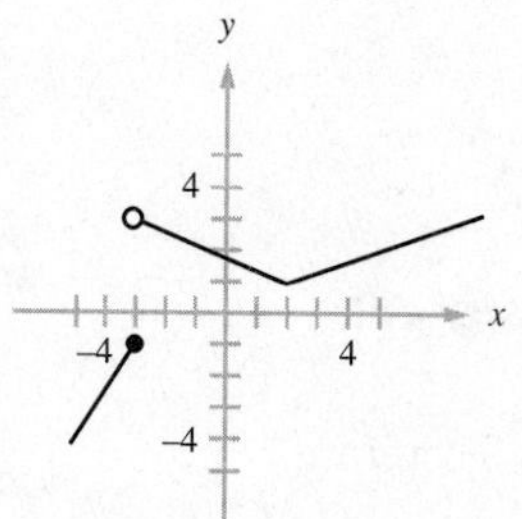

2.

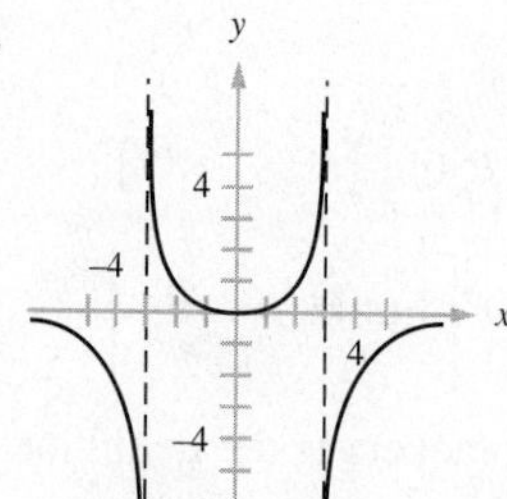

3.

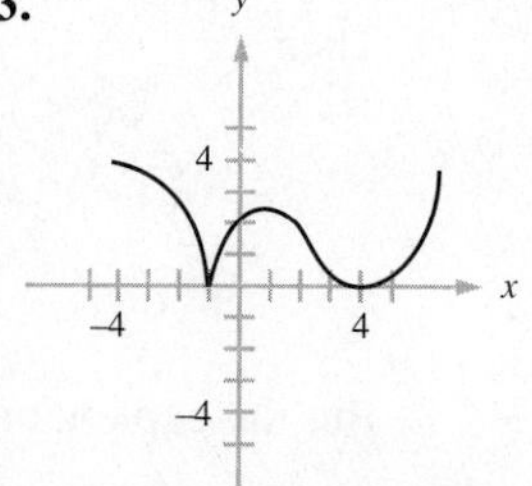

4.

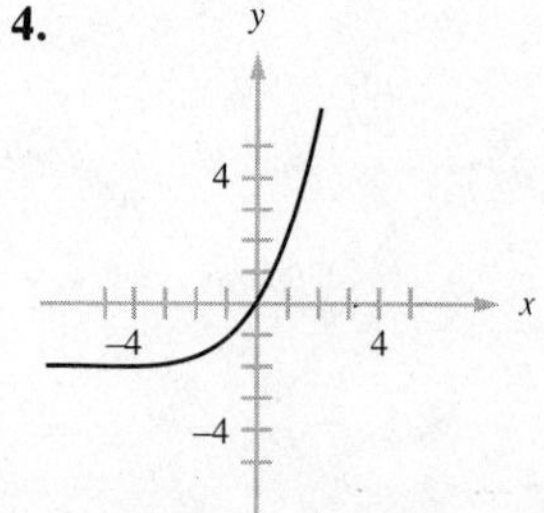

5.

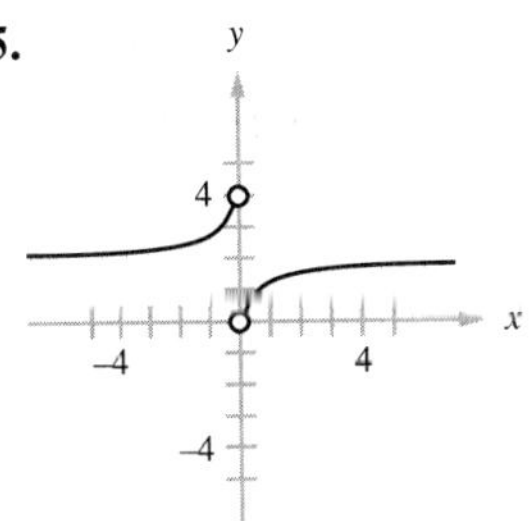

6.

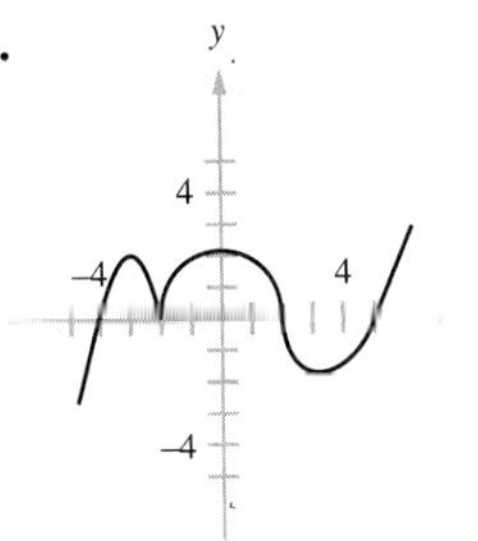

7.

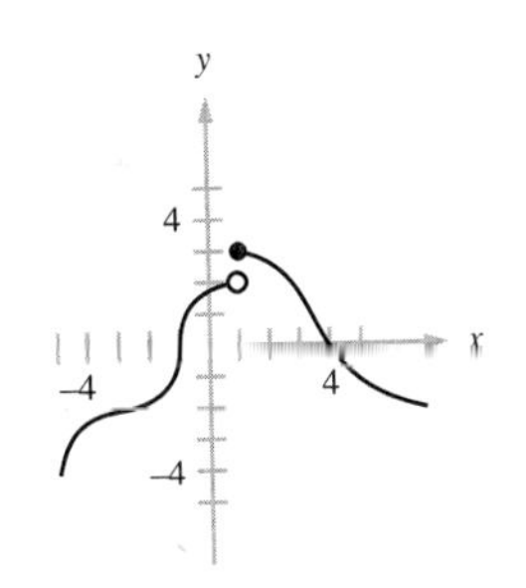

8.

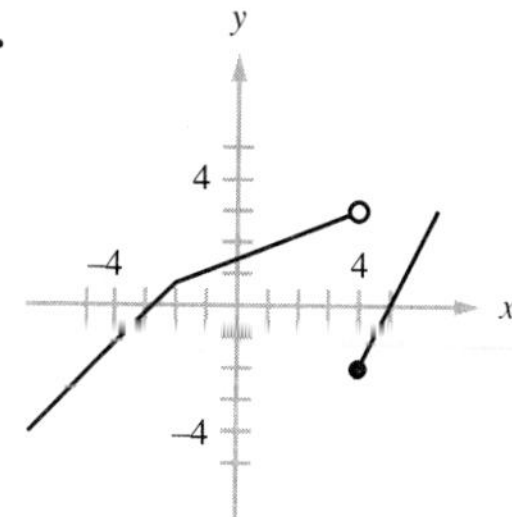

9.

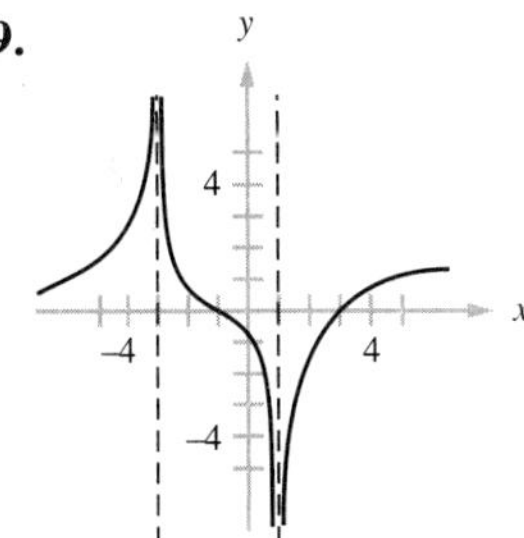

10.

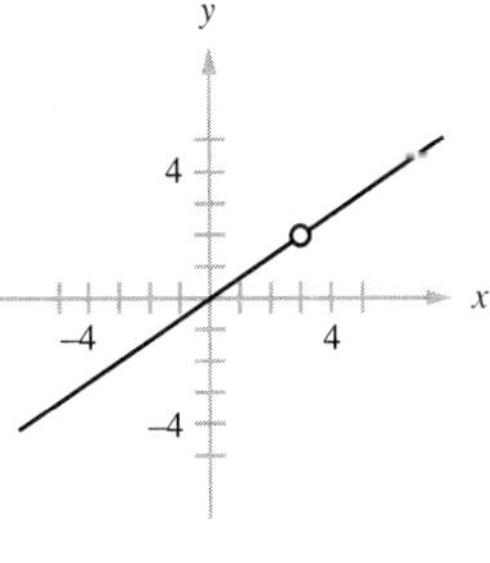

In Exercises 11–20, determine if the given functions are continuous at $x = a$.

11. $a = 4, \quad f(x) = 3x^2 - 5x + 2$

12. $a = 3, \quad g(x) = \begin{cases} \dfrac{x}{3 - x} & \text{if } x \neq 3 \\ -4 & \text{if } x = 3 \end{cases}$

13. $a = 1, \quad h(x) = \begin{cases} 2x + 5 & \text{if } x \leq 1 \\ 3x + 4 & \text{if } x > 1 \end{cases}$

14. $a = -1, \quad F(x) = \begin{cases} \dfrac{x^2 - 1}{x + 1} & \text{if } x \neq -1 \\ -2 & \text{if } x = -1 \end{cases}$

15. $a = -1, \quad G(x) = \begin{cases} \dfrac{1}{x + 1} & \text{if } x \neq -1 \\ 5 & \text{if } x = -1 \end{cases}$

16. $a = -3, \quad H(x) = 2x^3 - 3x^2 + 4x$

17. $a = 1, \quad F(x) = \begin{cases} 3 & \text{if } x < 1 \\ x^2 + 3 & \text{if } x \geq 1 \end{cases}$

18. $a = -2, \quad f(x) = \begin{cases} 4 - x^2 & \text{if } x \leq -2 \\ x^2 & \text{if } x > -2 \end{cases}$

19. $a = 3, \quad g(x) = \begin{cases} \dfrac{x^2 + 3x - 18}{x^2 - 9} & \text{if } x \neq 3 \\ 3/2 & \text{if } x = 3 \end{cases}$

20. $a = 1, \quad G(x) = \begin{cases} 5 - x^3 & \text{if } x < 1 \\ 6 - 2x^2 & \text{if } x \geq 1 \end{cases}$

In Exercises 21–28 determine the amount of a principal P invested in an account paying continuously compounded interest for t years at an annual rate of r.

21. $P = \$1000, \quad t = 1\frac{1}{2}, \quad r = 5\%$

22. $P = \$2000, \quad t = 2\frac{1}{2}, \quad r = 6\%$

23. $P = \$300, \quad t = 2, \quad r = 5\frac{1}{2}\%$

24. $P = \$800, \quad t = 4, \quad r = 5\frac{1}{4}\%$

25. $P = \$500, \quad t = \frac{3}{4}, \quad r = 6\%$

26. $P = \$4000, \quad t = 3, \quad r = 5\%$

27. $P = \$4000, \quad t = 1\frac{1}{4}, \quad r = 8\%$

28. $P = \$1000, \quad t = 2, \quad r = 6\frac{1}{2}\%$

In Exercises 29–32 evaluate the given limits.

29. $\lim_{x \to 0} (1 + x)^{-1/x}$

30. $\lim_{x \to 0} \sqrt{(1 + x)^{1/x}}$

31. $\lim_{x \to \infty} \left(\frac{x + 1}{x}\right)^x$

32. $\lim_{x \to \infty} \left(\frac{x}{x + 1}\right)^{2x}$

Answers to Practice Problems

7. The function f fails to be differentiable at x_2 (where there is a vertical tangent line) and at x_5 and x_7 (where there are abrupt changes of direction).

8. There are points of discontinuity at x_2 (where f is not defined) and at x_6 (where the limit of f does not exist).

9. \$3442.82

9.4 Some Basic Rules of Differentiation

In Section 9.2 we defined the derivative of a function f at x to be

$$\lim_{h \to 0} \frac{f(x + h) - f(x)}{h},$$

provided that this limit exists. To **differentiate** a function means to calculate its derivative, and the process of differentiating a function is called **differentiation.** Note that the derivative of f is a function, which we denote by f'. For instance, in Example 9.10 we determined that if $f(x) = x^3$, then its derivative is the function f' defined for all x by

$$f'(x) = 3x^2.$$

EXAMPLE 9.14 Calculate f' if $f(x) = x^2$.

Solution Using the definition of the derivative, we see that

$$\begin{aligned} f'(x) &= \lim_{h \to 0} \frac{f(x + h) - f(x)}{h} \\ &= \lim_{h \to 0} \frac{(x + h)^2 - x^2}{h} \\ &= \lim_{h \to 0} \frac{(x^2 + 2xh + h^2) - x^2}{h} \\ &= \lim_{h \to 0} \frac{2xh + h^2}{h} \\ &= \lim_{h \to 0} (2x + h) \\ &= 2x. \end{aligned}$$

Hence f' is defined by $f'(x) = 2x$ for all x. ■

Practice Problem 10 Calculate g' if $g(x) = x$.

As we have seen, $f'(x)$ represents the instantaneous rate of change of f at x and the slope of the tangent line to the graph of f at the point (x, y). In the remainder of this book we will see many applications that require knowing the derivative of a function. Fortunately it is not necessary to apply the definition of the derivative each time that we must differentiate a function. In the remainder of this chapter we will develop formulas that enable us to differentiate many of the functions that arise in applications.

THE DERIVATIVES OF CONSTANT FUNCTIONS AND POWER FUNCTIONS

Perhaps the simplest type of function is the constant function defined for all x by

$$f(x) = k.$$

The derivative of any constant function can be computed using the definition of the derivative as follows.

$$\begin{aligned} f'(x) &= \lim_{h \to 0} \frac{f(x + h) - f(x)}{h} \\ &= \lim_{h \to 0} \frac{k - k}{h} \\ &= \lim_{h \to 0} \frac{0}{h} \\ &= \lim_{h \to 0} 0 = 0. \end{aligned}$$

Thus we have obtained the following rule.

Derivative of a Constant Function

If $f(x) = k$, where k is any real number, then $f'(x) = 0$ for all x. That is, the derivative of any constant function is the constant function with value 0.

Recall that the graph of a constant function is a horizontal line. Since $f'(x)$ is the slope of the tangent line to the graph of f at $(x, f(x))$, the rule for differentiating a constant function tells us that the slope of the tangent line to the graph of a constant function is 0. But if the slope is 0, then the tangent line must be a horizontal line, and hence the tangent line to the graph of a constant function coincides with the graph of the function, as shown in Figure 9.27.

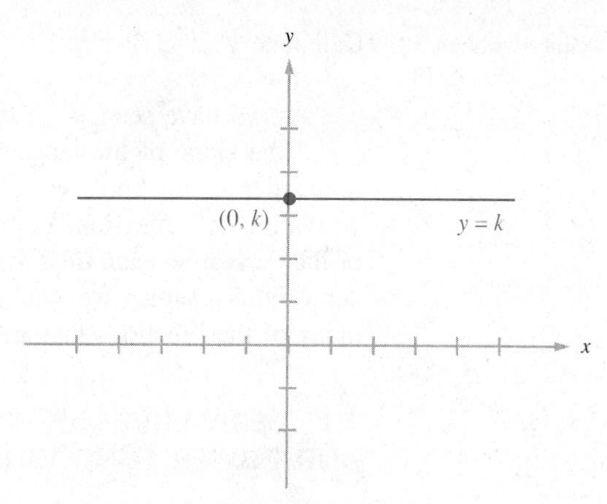

The tangent line to the graph of $y = k$ is $y = k$.

FIGURE 9.27

Practice Problem 11 For each x compute $f'(x)$, $g'(x)$, and $h'(x)$ if

$$f(x) = \frac{3}{8}, \qquad g(x) = e, \qquad \text{and} \qquad h(x) = 5^3.$$

Another common type of function is the power function

$$f(x) = x^r,$$

where the exponent r is any real number. Earlier in this section we discussed the derivative of f when $r = 1, 2,$ or 3. Recall that:

$$\text{If } f(x) = x, \text{ then } f'(x) = 1.$$
$$\text{If } f(x) = x^2, \text{ then } f'(x) = 2x.$$
$$\text{If } f(x) = x^3, \text{ then } f'(x) = 3x^2.$$

The next differentiation formula enables us to compute the derivative of any power function. It can be proved by the binomial theorem when r is a positive integer, but it is in fact true for all real numbers r.

Derivative of a Power Function

If $f(x) = x^r$, where r is any real number, then $f'(x) = rx^{r-1}$. Thus the derivative of x^r is found by decreasing the exponent of x by 1 and multiplying the result by the original exponent.

Observe that for $r = 1, 2,$ or 3 this formula gives the same result as those obtained by using the definition of the derivative.

EXAMPLE 9.15 Compute the derivatives of the following functions:

$$f(x) = x^5, \qquad g(x) = \frac{1}{x^2}, \qquad \text{and} \qquad h(x) = \sqrt{x}.$$

Solution Since $f(x) = x^5$ is a power function, we can apply the rule for the derivative of a power function to obtain

$$f'(x) = 5x^{5-1} = 5x^4.$$

In order to apply the rule for the derivative of a power function to g, we must write the function in the form x^r. To do this, we must use the exponent property

$$x^{-m} = \frac{1}{x^m}.$$

Thus we see that

$$g(x) = \frac{1}{x^2} = x^{-2}.$$

It now follows from the rule for the derivative of a power function that

$$g'(x) = -2x^{-2-1} = -2x^{-3} = -\frac{2}{x^3}.$$

To differentiate $h(x) = \sqrt{x}$, we must write the square root as a fractional exponent:

$$h(x) = \sqrt{x} = x^{1/2}.$$

The rule for the derivative of a power function now gives

$$h'(x) = \frac{1}{2}x^{(1/2)-1} = \frac{1}{2}x^{-1/2} = \frac{1}{2x^{1/2}} = \frac{1}{2\sqrt{x}}. \quad ■$$

Practice Problem 12 Compute $f'(x)$, $g'(x)$, and $h'(x)$, where

$$f(x) = x^7, \qquad g(x) = \sqrt[3]{x}, \qquad \text{and} \qquad h(x) = \frac{1}{x^4}.$$

EXAMPLE 9.16 Determine the equation of the tangent line to the graph of $f(x) = x^4$ at the point $(-1, 1)$.

Solution The slope of the tangent line is measured by the derivative of f. By the rule for the derivative of a power function,

$$f'(x) = 4x^3.$$

At the point $(-1, 1)$, where $x = -1$, we have

$$f'(-1) = 4(-1)^3 = 4(-1) = -4.$$

Thus the tangent line at $(-1, 1)$ has slope -4. Using the point-slope form of the equation of a line as in Example 9.4, we see that the equation of this line is

$$y - 1 = -4[x - (-1)],$$

that is,

$$y = -4x - 3. \quad ■$$

Practice Problem 13 Determine the equation of the tangent line to the graph of $f(x) = \dfrac{1}{x}$ at the point $(2, \frac{1}{2})$.

DERIVATIVES OF SUMS, DIFFERENCES, AND CONSTANT MULTIPLES

Many functions that arise in applications are sums or differences of other functions. For example,

$$f(x) = x^3 + 1$$

is the sum of the power function $g(x) = x^3$ and the constant function $h(x) = 1$. The next pair of formulas enable us to compute the derivative of a sum or difference of two functions in terms of their derivatives.

Derivative of the Sum or Difference of Functions

(a) If $f = g + h$, then $f' = g' + h'$. Thus the derivative of the sum of two functions is the sum of their derivatives.

(b) If $f = g - h$, then $f' = g' - h'$. Thus the derivative of the difference of two functions is the difference of their derivatives.

For the function $f(x) = x^3 + 1$ above,

$$\begin{aligned} f'(x) &= (x^3)' + (1)' \\ &= 3x^2 + 0 \\ &= 3x^2 \end{aligned}$$

by the rules for differentiating a power function and a constant function.

The rules above for computing the sum or difference of functions generalize to an arbitrary number of functions.

EXAMPLE 9.17 Compute the derivatives of the following functions.

(a) $f(x) = x^8 - x + 7$ (b) $g(x) = \frac{1}{x^5} + \sqrt{x^3}$.

Solution (a) Using the rules for the derivative of the sum and difference of functions, we see that

$$\begin{aligned} f'(x) &= (x^8)' - (x^1)' + (7)' \\ &= 8x^7 - 1x^0 + 0 \\ &= 8x^7 - 1. \end{aligned}$$

(b) We can express g algebraically as the sum of two power functions:

$$g'(x) = \frac{1}{x^5} + \sqrt{x^3} = x^{-5} + x^{3/2}.$$

Hence

$$\begin{aligned} g'(x) &= (x^{-5})' + (x^{3/2})' \\ &= -5x^{-6} + \frac{3}{2}x^{1/2}. \quad \blacksquare \end{aligned}$$

Practice Problem 14 Compute the derivatives of the following functions.

(a) $f(x) = x^{12} - x^9 + x$ (b) $g(x) = \sqrt[4]{x^3} + 6$.

Many times we encounter a function that is a constant times another function. For example,

$$f(x) = 3x^4$$

is the constant 3 multiplied by the function $g(x) = x^4$. The next rule enables us to differentiate such functions.

Derivative of a Constant Multiple of a Function

If $f = kg$, where k is any real number, then $f' = kg'$. Thus the derivative of a constant multiple of a function is the constant times the derivative of the function.

As a consequence of this formula, we see that at the point (x, y) the slope of the tangent line to the graph of kf is k times the slope of the tangent line to the graph of f. See Figure 9.28 on page 540 for the case $k = 2$.

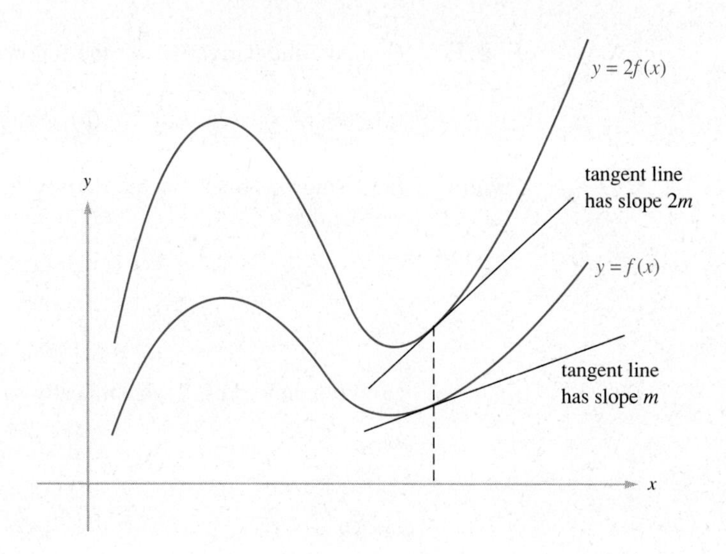

FIGURE 9.28

EXAMPLE 9.18 Compute the derivatives of the following functions.

(a) $f(x) = 3x^{10} + 4x^9 - 7$ (b) $g(x) = 6\sqrt[3]{x^2} - 5x$.

Solution (a) Applying the rules for the sum and difference of functions and then the rule for a constant multiple of a function, we see that

$$\begin{aligned} f'(x) &= (3x^{10})' + (4x^9)' - (7)' \\ &= 3(x^{10})' + 4(x^9)' - (7)' \\ &= 3(10x^9) + 4(9x^8) - 0 \\ &= 30x^9 + 36x^8. \end{aligned}$$

(b) Likewise,

$$\begin{aligned} g'(x) &= \left(6\sqrt[3]{x^2}\right)' - (5x)' \\ &= 6\left(\sqrt[3]{x^2}\right)' - 5(x^1)' \\ &= 6(x^{2/3})' - 5(1x^0) \\ &= 6\left(\frac{2}{3}x^{-1/3}\right) - 5 \\ &= 4x^{-1/3} - 5 \\ &= \frac{4}{\sqrt[3]{x}} - 5. \quad \blacksquare \end{aligned}$$

Practice Problem 15 Compute the derivatives of the following functions.

(a) $f(x) = \frac{1}{2}x^8 - 5x^6 + 9$ (b) $g(x) = -\frac{2}{x^4} + \sqrt{6}$.

Using the rules presented in this section, we can differentiate any polynomial function as illustrated by the computation in Example 9.18 (a).

EXAMPLE 9.19 Compute the instantaneous rate of change of f at $x = 1$ if

$$f(x) = x^3 - 4x^2 + 5x - 6.$$

Solution The instantaneous rate of change of f at x is given by $f'(x)$. Now

$$\begin{aligned} f'(x) &= (x^3)' - (4x^2)' + (5x)' - 6' \\ &= (3x^2) - (8x) + 5 - 0 \\ &= 3x^2 - 8x + 5. \end{aligned}$$

Thus the instantaneous rate of change of f at $x = 1$ is

$$f'(1) = 3(1)^2 - 8(1) + 5 = 0. \quad \blacksquare$$

Practice Problem 16 Compute the instantaneous rate of change of f at $x = -1$ if

$$f(x) = 2x^4 - 5x^3 + x^2 - 4x - 3.$$

To conclude this section, we will derive the formulas for the derivative of a sum of functions and the derivative of a constant multiple of a function. The derivation of the formula for the derivative of the difference of two functions is similar to that of the sum and is left as an exercise.

Let g and h be functions which have derivatives at x, and let $f = g + h$. Then

$$\begin{aligned} f'(x) &= \lim_{r\to 0} \frac{f(x + r) - f(x)}{r} \\ &= \lim_{r\to 0} \frac{[g(x + r) + h(x + r)] - [g(x) + h(x)]}{r} \\ &= \lim_{r\to 0} \frac{[g(x + r) - g(x)] + [h(x + r) - h(x)]}{r}. \end{aligned}$$

Thus by the sum rule for limits we have

$$\begin{aligned} f'(x) &= \lim_{r\to 0} \frac{[g(x + r) - g(x)]}{r} + \lim_{r\to 0} \frac{[h(x + r) - h(x)]}{r} \\ &= g'(x) + h'(x). \end{aligned}$$

This argument shows that the derivative of a sum of functions is the sum of their derivatives.

Now suppose that g has a derivative at x and that $f = kg$ for some constant k. Then

$$\begin{aligned} f'(x) &= \lim_{r \to 0} \frac{f(x + r) - f(x)}{r} \\ &= \lim_{r \to 0} \frac{kg(x + r) - kg(x)}{r} \\ &= \lim_{r \to 0} \frac{k[g(x + r) - g(x)]}{r}. \end{aligned}$$

Thus by the constant multiple rule for limits

$$\begin{aligned} f'(x) &= k \lim_{r \to 0} \frac{[g(x + r) - g(x)]}{r} \\ &= kg'(x). \end{aligned}$$

This argument shows that the derivative of a constant times a function is the constant times the derivative of the function.

EXERCISES 9.4

In Exercises 1–12 compute the derivative of the given function.

1. $f(x) = x^8$

2. $F(x) = 7$

3. $g(x) = 3^5$

4. $G(x) = x^{-4}$

5. $h(x) = x^{-3} + 3x$

6. $H(x) = 4x^7 - 2x^2$

7. $F(x) = 3x^2 - 5x + 2$

8. $f(x) = 6x^3 - 3x^2 + 4x$

9. $G(x) = x^7 - \frac{3}{x^4}$

10. $g(x) = 6\sqrt[3]{x} - 9$

11. $H(x) = 4\sqrt{x} + x$

12. $h(x) = \frac{7}{x^2} + 5x^4$

In Exercises 13–20 compute the instantaneous rate of change of f at x.

13. $f(x) = \sqrt[3]{x}$

14. $f(x) = \frac{1}{\sqrt{x}}$

15. $f(x) = x^2 - \frac{6}{x}$

16. $f(x) = \frac{3}{x^4} + 7x^5$

17. $f(x) = 4 + \frac{2}{x^3}$

18. $f(x) = \frac{5}{x^2} - \frac{x^2}{5}$

19. $f(x) = x(x + 5)$

20. $f(x) = x^2(x - 7)$

In Exercises 21–28 write the equation of the tangent line to the graph of $y = f(x)$ at the point (a, b).

21. $f(x) = x^3 - 3x, \quad (a, b) = (2, 2)$

22. $f(x) = 4x^2 + 3x, \quad (a, b) = (1, 7)$

23. $f(x) = 2\sqrt{x}, \quad (a, b) = (1, 2)$

24. $f(x) = \sqrt[3]{x}, \quad (a, b) = (8, 2)$

25. $f(x) = \dfrac{4}{x}, \quad (a, b) = (-1, -4)$

26. $f(x) = \dfrac{1}{x^2}, \quad (a, b) = \left(\dfrac{1}{2}, 4\right)$

27. $f(x) = 2x + \dfrac{4}{x^2}, \quad (a, b) = (1, 6)$

28. $f(x) = \dfrac{6}{\sqrt{x}} - 5, \quad (a, b) = (4, -2)$

29. If a manufacturer's total cost (in dollars) of producing x items is given by

$$C(x) = 0.000001x^3 - 0.0015x^2 + 240x + 112{,}500,$$

how fast is the total cost changing when 5000 items are being produced?

30. A manufacturer expects that the revenue (in dollars) obtained from selling x units of his product will be given by $R(x) = 80x - x^2$. How fast is the revenue changing when 50 items are sold?

31. A retail store's inventory costs (in dollars) for a certain item are given by the function

$$C(x) = x + \frac{16{,}000{,}000}{x},$$

where x is the number of items received whenever an order is placed. How fast are the inventory costs changing when each order contains 2000 items?

32. A freely falling object falls approximately $16t^2$ feet in t seconds. How fast is the distance fallen changing after 3 seconds?

33. Derive the rule which states that the derivative of a difference of functions is the difference of their derivatives.

34. Use the binomial theorem to derive the rule for the derivative of a power function $f(x) = x^r$ when r is a positive integer.

Answers to Practice Problems

10. $g'(x) = 1$ for all x

11. $f'(x) = 0$, $g'(x) = 0$, and $h'(x) = 0$

12. $f'(x) = 7x^6$, $g'(x) = \dfrac{1}{3}x^{-2/3} = \dfrac{1}{3\sqrt[3]{x^2}}$ and $h'(x) = -4x^{-5} = -\dfrac{4}{x^5}$

13. $y = -\dfrac{1}{4}x + 1$

14. (a) $f'(x) = 12x^{11} - 9x^8 + 1$ (b) $g'(x) = \dfrac{3}{4}x^{-1/4}$

15. (a) $f'(x) = 4x^7 - 30x^5$ (b) $g'(x) = 8x^{-5} = \dfrac{8}{x^5}$

16. $f'(-1) = -29$

9.5 The Chain Rule

A common type of function has the form

$$f(x) = [h(x)]^r$$

for some real number r. That is, f is a power of another function h. For example, the function

$$f(x) = (x^2 - 5)^{30}$$

has this form with $h(x) = x^2 - 5$ and $r = 30$. Recall from Section 8.5 that f is the composition of two functions g and h, where

$$h(x) = x^2 - 5 \quad \text{and} \quad g(x) = x^{30},$$

and that we denote this relationship by writing

$$f(x) = g(h(x)).$$

EXAMPLE 9.20 Express the function

$$f(x) = \ln (5x + 1)$$

as a composition of functions.

Solution Take

$$g(x) = \ln x \quad \text{and} \quad h(x) = 5x + 1.$$

Then $f(x) = g(h(x))$. ■

Practice Problem 17 Find functions g and h that express

$$f(x) = e^{3-2x}$$

in the form $f(x) = g(h(x))$.

For the function

$$f(x) = (x^2 - 5)^{30}$$

described above, the formulas for differentiation presented in Section 9.4 cannot be used to compute f' unless we first expand $(x^2 - 5)^{30}$ algebraically. Even with the binomial theorem to help us, this computation will be very difficult. Notice, however, that we can differentiate both of the functions

$$h(x) = x^2 - 5 \quad \text{and} \quad g(x) = x^{30}$$

from which f is composed. Our next differentiation formula expresses the derivative of the composition of two functions g and h in terms of the derivatives of g and h, and so it enables us to differentiate $f(x) = (x^2 - 5)^{30}$ directly. This formula is called the **chain rule.**

The Chain Rule

If $f(x) = g(h(x))$ is the composition of two differentiable functions g and h, then f is differentiable. To find the derivative of f, first compute $g'(x)$, and then replace each x by $h(x)$ in the result. Finally, multiply the preceding expression by $h'(x)$. Symbolically,

$$f'(x) = g'(h(x)) \cdot h'(x).$$

For the functions

$$h(x) = x^2 - 5 \qquad \text{and} \qquad g(x) = x^{30},$$

we have

$$h'(x) = 2x \qquad \text{and} \qquad g'(x) = 30x^{29}.$$

Thus by the chain rule we see that

$$\begin{aligned} f'(x) &= g'(h(x)) \cdot h'(x) \\ &= 30(x^2 - 5)^{29} \cdot (2x) \\ &= 60x(x^2 - 5)^{29}. \end{aligned}$$

EXAMPLE 9.21 Differentiate $f(x) = (3x^2 - 4x + 1)^{12}$.

Solution Note that $f(x) = g(h(x))$, where

$$h(x) = 3x^2 - 4x + 1 \qquad \text{and} \qquad g(x) = x^{12}.$$

To compute $f'(x)$, we first compute $h'(x)$ and $g'(x)$:

$$h'(x) = 6x - 4 \qquad \text{and} \qquad g'(x) = 12x^{11}.$$

In $g'(x)$ we replace each occurrence of x by $h(x)$ to obtain

$$g'(h(x)) = 12(3x^2 - 4x + 1)^{11}$$

Finally, we multiply the expression above by $h'(x)$ to obtain $f'(x)$:

$$\begin{aligned} f'(x) &= g'(h(x)) \cdot h'(x) \\ &= 12(3x^2 - 4x + 1)^{11}(6x - 4) \\ &= (72x - 48)(3x^2 - 4x + 1)^{11}. \quad \blacksquare \end{aligned}$$

Practice Problem 18 Compute the derivative of $h(x) = (2x^5 - 9x^3 + 4x - 5)^{10}$.

EXAMPLE 9.22 Compute the derivative of $f(x) = \sqrt{9x^2 + 4x}$.

Solution Notice that $f(x) = g(h(x))$, where

$$h(x) = 9x^2 + 4x \qquad \text{and} \qquad g(x) = \sqrt{x} = x^{1/2}.$$

Now

$$h'(x) = 18x + 4 \qquad \text{and} \qquad g'(x) = \frac{1}{2}x^{-1/2} = \frac{1}{2\sqrt{x}}.$$

Hence by the chain rule

$$\begin{aligned} f'(x) &= g'(h(x)) \cdot h'(x) \\ &= \frac{1}{2\sqrt{9x^2 + 4x}}(18x + 4) \\ &= \frac{9x + 2}{\sqrt{9x^2 + 4x}}. \end{aligned}$$ ■

Practice Problem 19 Differentiate $f(x) = (4x^3 - 7x^2 + 6x - 2)^{3/2}$.

EXAMPLE 9.23 If $g'(x) = \frac{1}{x}$, compute $[g(x^2 - 3x)]'$.

Solution If we let

$$h(x) = x^2 - 3x \qquad \text{and} \qquad f(x) = g(h(x)),$$

then the expression to be computed is

$$[g(x^2 - 3x)]' = [g(h(x))]' = f'(x).$$

Hence the chain rule can be used to obtain the answer. Since

$$\begin{aligned} f'(x) &= g'(h(x)) \cdot h'(x) \\ &= \frac{1}{h(x)} h'(x) \\ &= \frac{1}{x^2 - 3x}(2x - 3), \end{aligned}$$

we have

$$[g(x^2 - 3x)]' = f'(x) = \frac{2x - 3}{x^2 - 3x}.$$ ■

Practice Problem 20 If $g'(x) = e^{-x}$, compute $[g(3x^2 + 5x)]'$.

OTHER NOTATIONS FOR THE DERIVATIVE

There are several common notations for the derivative of a function. Although in this book we will usually denote the derivative of a function f by f', we will sometimes use the symbol

$$\frac{d}{dx} f(x)$$

instead. We read this symbol as "the derivative of $f(x)$ with respect to x." For example, the rules for differentiating constant functions and power functions can be written

$$\frac{d}{dx}(k) = 0 \quad \text{and} \quad \frac{d}{dx}(x^r) = rx^{r-1}$$

using this notation.

If $y = f(x)$, then the derivative of f with respect to x is also frequently written

$$y' \quad \text{or} \quad \frac{dy}{dx} \quad \text{or} \quad dy/dx.$$

Although the second and third notations above look like fractions, they are not—both are merely other ways to denote the derivative of y with respect to x. Hence if $y = x^8$, then

$$y' = \frac{dy}{dx} = dy/dx = 8x^7.$$

The advantage of the second and third notations is that they make explicit the independent variable. For instance, if $z = u^{1/2}$, then z is a function of the independent variable u, and

$$\frac{dz}{du} = \frac{1}{2}u^{-1/2}.$$

Naming the independent variable in this manner is helpful when dealing with composite functions. Suppose that $y = g(h(x))$. If we let

$$u = h(x),$$

then $y = g(u)$; so we can regard y as a function of u as well as a function of x. The symbol y' does not indicate whether we are considering the independent variable to be x or u, whereas the symbols dy/dx and dy/du explicitly name the independent variable that we are considering.

Notice that dy/dx and dy/du will not generally be equal. In fact, the relationship between dy/dx and dy/du is given by the chain rule. For the chain rule states that

$$[g(h(x))]' = g'(h(x)) \cdot h'(x),$$

and so yields the following equivalent formulation.

The Chain Rule

If $y = g(u)$ and $u = h(x)$, where g and h are differentiable functions, then

$$\frac{dy}{dx} = \frac{dy}{du}\frac{du}{dx}.$$

EXAMPLE 9.24 Compute the derivative of $y = (3x^4 - 8x^3 + 7x - 5)^{15}$.

Solution Let

$$u = 3x^4 - 8x^3 + 7x - 5.$$

Then $y = u^{15}$, and so the preceding version of the chain rule gives

$$\begin{aligned}\frac{dy}{dx} &= \frac{dy}{du}\frac{du}{dx}\\ &= (15u^{14})(12x^3 - 24x^2 + 7)\\ &= 15(3x^4 - 8x^3 + 7x - 5)^{14}(12x^3 - 24x^2 + 7)\\ &= (180x^3 - 360x^2 + 105)(3x^4 - 8x^3 + 7x - 5)^{14}.\end{aligned}$$ ■

Practice Problem 21 Compute $\frac{dy}{dx}$ if $y = (4x^2 - 6x - 5)^{-3}$.

In the next section we will use the chain rule to obtain several other important differentiation formulas.

EXERCISES 9.5

In Exercises 1–10 find functions g and h such that $f(x) = g(h(x))$.

1. $f(x) = (2x - 1)^5$
2. $f(x) = e^{3x+1}$
3. $f(x) = \ln(4x^2 - 1)$
4. $f(x) = \sqrt{x^3 - x}$
5. $f(x) = e^{\sqrt{x+1}}$
6. $f(x) = \ln(5x + 2)$
7. $f(x) = \dfrac{4}{\sqrt{3x + 5}}$
8. $f(x) = \dfrac{2}{(3x^2 - 2x + 1)^3}$
9. $f(x) = 2^{x^3 - x}$
10. $f(x) = \ln \dfrac{x - 1}{x + 1}$

In Exercises 11–20 differentiate the given function.

11. $f(x) = (3x - 5)^6$
12. $F(x) = (7 - 4x)^8$
13. $g(x) = (x^2 - 3x + 4)^7$
14. $G(x) = (x^4 - 5x^3 + 2)^9$

15. $h(x) = (2x^5 - 4x^2 + 1)^{-3/2}$

16. $H(x) = \sqrt{x^8 + x}$

17. $F(x) = \left(\dfrac{1}{6 - 2x}\right)^5$

18. $f(x) = (5x^6 - 4x^3 + 3)^{-2}$

19. $G(x) = \sqrt[3]{2x^2 - 7x}$

20. $g(x) = \dfrac{3}{\sqrt{2x + 7}}$

21. If $g'(x) = e^{2x}$, what is $\dfrac{d}{dx}g(x^3)$?

22. If $g'(x) = xe^x$, what is $\dfrac{d}{dx}g(\sqrt{x})$?

23. If $g'(x) = \dfrac{1}{\sqrt{1 - x^2}}$, what is $\dfrac{d}{dx}g(\sqrt{x})$?

24. If $g'(x) = x \ln x$, what is $\dfrac{d}{dx}g(x^2)$?

25. If $g'(x) = x \ln (x^2 + 1)$, what is $\dfrac{d}{dx}g(x^4)$?

26. If $g'(x) = \dfrac{\sqrt{x^2 + 9}}{x}$, what is $\dfrac{d}{dx}g(x^3)$?

27. If $g'(x) = \dfrac{\sqrt{x^2 - 4}}{x^2}$, what is $\dfrac{d}{dx}g(x^2)$?

28. If $g'(x) = \dfrac{x}{2 + 3x}$, what is $\dfrac{d}{dx}g(x^4)$?

29. If $g'(8) = 5$, find $\dfrac{d}{dx}g(x^3)$ at $x = 2$.

30. If $g'(2) = 3$, find $\dfrac{d}{dx}g(\sqrt{x})$ at $x = 4$.

31. If $g'(1) = -6$, find $\dfrac{d}{dx}g(x^2)$ at $x = -1$.

32. If $g'(2) = 7$, find $\dfrac{d}{dx}g\left(\dfrac{1}{x}\right)$ at $x = \dfrac{1}{2}$.

33. If $g(9) = 4$ and $g'(9) = -2$, find $\dfrac{d}{dx}[g(x)]^2$ at $x = 9$.

34. If $g(3) = 5$ and $g'(3) = 7$, find $\dfrac{d}{dx}[g(x)]^3$ at $x = 3$.

35. If $g(-2) = 16$ and $g'(-2) = -3$, find $\dfrac{d}{dx}\sqrt{g(x)}$ at $x = -2$.

36. If $g(0) = 2$ and $g'(0) = -20$, find $\dfrac{d}{dx}\dfrac{1}{g(x)}$ at $x = 0$.

37. Let h be a differentiable function. Derive a formula for differentiating the function f, where

$$f(x) = [h(x)]^r$$

and r is any real number. This formula is often called the **generalized power rule.**

Answers to Practice Problems

17. $g(x) = e^x$ and $h(x) = 3 - 2x$

18. $h'(x) = (100x^4 - 270x^2 + 40)(2x^5 - 9x^3 + 4x - 5)^9$

19. $(18x^2 - 21x + 9)\sqrt{4x^3 - 7x^2 + 6x - 2}$

20. $(6x + 5)e^{-(3x^2 + 5x)}$

21. $\dfrac{-24x + 18}{(4x^2 - 6x - 5)^4}$

9.6 Derivatives of Logarithmic and Exponential Functions

Logarithmic and exponential functions were introduced in Chapter 8. These functions arise frequently in applications involving the calculus, and so it is essential that we know how to differentiate them. This section is concerned with obtaining formulas that enable us to calculate derivatives of these functions.

THE DERIVATIVE OF ln x

Recall that by definition

$$e = \lim_{t \to 0} (1 + t)^{1/t}$$

and that the natural logarithm function satisfies

$$\ln u - \ln v = \ln \frac{u}{v} \qquad \text{and} \qquad r \ln u = \ln u^r.$$

Using these facts, the definition of the derivative, and the continuity of the natural logarithm function, we can compute the derivative of the natural logarithm function for a fixed positive value of x as follows.

$$\begin{aligned}(\ln x)' &= \lim_{h \to 0} \frac{\ln (x + h) - \ln x}{h} \\ &= \lim_{h \to 0} \frac{1}{h} \ln \left(\frac{x + h}{x}\right) \\ &= \lim_{h \to 0} \frac{1}{h} \ln \left(1 + \frac{h}{x}\right) \\ &= \lim_{h \to 0} \ln \left(1 + \frac{h}{x}\right)^{1/h}\end{aligned}$$

Now let $t = h/x$, so that $h = tx$, and observe that t approaches 0 as h approaches 0. Thus we see* that

$$\begin{aligned}(\ln x)' &= \lim_{t \to 0} \ln (1 + t)^{1/tx} \\ &= \lim_{t \to 0} \ln (1 + t)^{(1/t)(1/x)} \\ &= \lim_{t \to 0} \frac{1}{x} \ln (1 + t)^{(1/t)}\end{aligned}$$

*That $\lim_{t \to 0} \ln (1 + t)^{1/t} = \ln \left[\lim_{t \to 0} (1 + t)^{1/t}\right]$ follows from the continuity of the natural logarithm function.

$$= \frac{1}{x} \lim_{t \to 0} \ln (1 + t)^{(1/t)}$$

$$= \frac{1}{x} \ln \left[\lim_{t \to 0} (1 + t)^{(1/t)}\right]$$

$$= \frac{1}{x} \ln e$$

$$= \frac{1}{x}.$$

Hence we have the following formula.

Derivative of ln *x*

For $x > 0$,

$$(\ln x)' = \frac{1}{x}.$$

Since the derivative of the natural logarithm function $\ln x$ is $1/x$, the slope of the tangent line to $y = \ln x$ at $x = b$ is the value of the function $y = 1/x$ at b. See Figure 9.29.

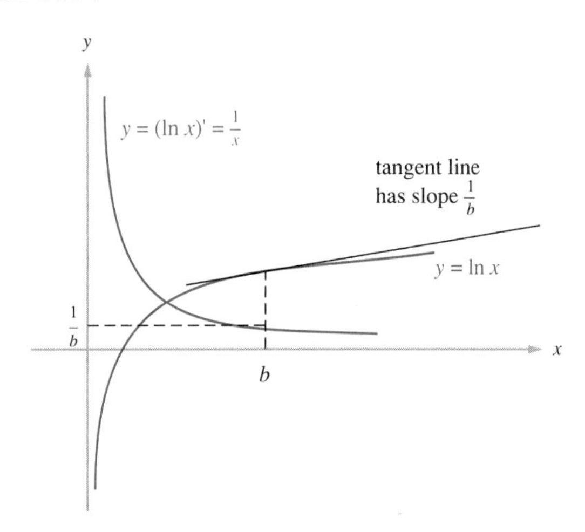

FIGURE 9.29

EXAMPLE 9.25 For $x > 0$ differentiate the following functions.

(a) $f(x) = 4x^3 - \ln x$ (b) $g(x) = (\ln x)^5$.

Solution (a) Since the derivative of the difference of two functions is the difference of their derivatives, we have

$$f'(x) = (4x^3 - \ln x)' = (4x^3)' - (\ln x)' = 12x^2 - \frac{1}{x}.$$

(b) To compute the derivative of $g(x)$, we must use the chain rule.

$$g'(x) = 5(\ln x)^4(\ln x)' = 5(\ln x)^4\left(\frac{1}{x}\right) = 5\,\frac{(\ln x)^4}{x}.$$ ■

Practice Problem 22 For $x > 1$ differentiate the following functions.

(a) $f(x) = 9x^6 + \ln x$ (b) $g(x) = 4x^{-2} + \sqrt{\ln x}$.

EXAMPLE 9.26 Compute the instantaneous rate of change of $f(x) = \ln x$ at $x = 4$.

Solution The instantaneous rate of change of f at $x = 4$ is $f'(4)$, the value of the derivative of f at $x = 4$. Since

$$f'(x) = \frac{1}{x},$$

we have $f'(4) = 1/4$. Thus the instantaneous rate of change of f at $x = 4$ is $1/4$. ■

Practice Problem 23 Determine the equation of the tangent line to the graph of $f(x) = \ln x$ at the point $(1, 0)$.

Often we must compute the derivative of functions of the form $\ln h(x)$, where $h(x)$ is differentiable. Since $\ln h(x)$ is the composition of $g(x) = \ln x$ and $h(x)$, and since $g'(x) = 1/x$, this derivative can be found by using the chain rule as shown below.

$$\begin{aligned}[\ln h(x)]' &= [g(h(x))]' \\ &= g'(h(x)) \cdot h'(x) \\ &= \frac{1}{h(x)} \cdot h'(x) \\ &= \frac{h'(x)}{h(x)}\end{aligned}$$

Thus we have obtained the following result.

Derivative of ln $h(x)$

If h is a differentiable function and x is a value for which $h(x) > 0$, then

$$[\ln h(x)]' = \frac{h'(x)}{h(x)}.$$

Thus the derivative of ln $h(x)$ equals the derivative of $h(x)$ divided by $h(x)$.

EXAMPLE 9.27 Differentiate

$$f(x) = \ln (3x^2 + x).$$

Solution The function f is of the form

$$\ln h(x)$$

with

$$h(x) = 3x^2 + x.$$

Because h is a polynomial function, it is differentiable. Thus the preceding differentiation formula can be used. The result is shown below.

$$\begin{aligned} f'(x) &= [\ln (3x^2 + x)]' \\ &= \frac{(3x^2 + x)'}{3x^2 + x} \\ &= \frac{6x + 1}{3x^2 + x}. \end{aligned}$$ ■

EXAMPLE 9.28 Atmospheric pressure lessens at higher altitudes. The altimeters of most aircraft measure atmospheric pressure and compute the altitude of the aircraft according to the **barometric equation**

$$h = (30T + 8000) \ln \frac{76}{P}.$$

Here h denotes the altitude in meters above sea level, T denotes the air temperature in degrees Celsius, and P denotes the atmospheric pressure in centimeters of mercury. Determine the rate of change of h with respect to P at a constant temperature of 20° C.

Solution If $T = 20$, the barometric equation takes the form

$$h = 8600 \ln 76P^{-1}.$$

Hence the rate of change of h with respect to P is

$$\frac{dh}{dP} = 8600\left(\frac{1}{76P^{-1}}\right) \cdot [76(-1)P^{-2}] = -8600P^{-1} = -\frac{8600}{P}. \blacksquare$$

Practice Problem 24 Compute the derivative of $f(x) = \ln(7x^6 + x^{-3})$.

THE DERIVATIVE OF e^x

Let $u = e^x$. Then by definition of the natural logarithm function,

$$\ln u = x.$$

Hence

$$\frac{d}{dx}(\ln u) = \frac{d}{dx}(x) = 1.$$

Since u is a function of x, we can compute

$$\frac{d}{dx}(\ln u)$$

by the chain rule:

$$\frac{d}{dx}(\ln u) = \frac{d}{du}(\ln u) \cdot \frac{du}{dx} = \frac{1}{u} \cdot \frac{du}{dx}.$$

Combining the two preceding equations yields

$$1 = \frac{1}{u}\frac{du}{dx}.$$

Solving this equation for $\frac{du}{dx}$ and substituting $u = e^x$ yields the following result.

Derivative of e^x

The derivative of e^x is e^x, symbolically

$$(e^x)' = e^x.$$

Thus the exponential function

$$f(x) = e^x$$

is its own derivative! See Figure 9.30.

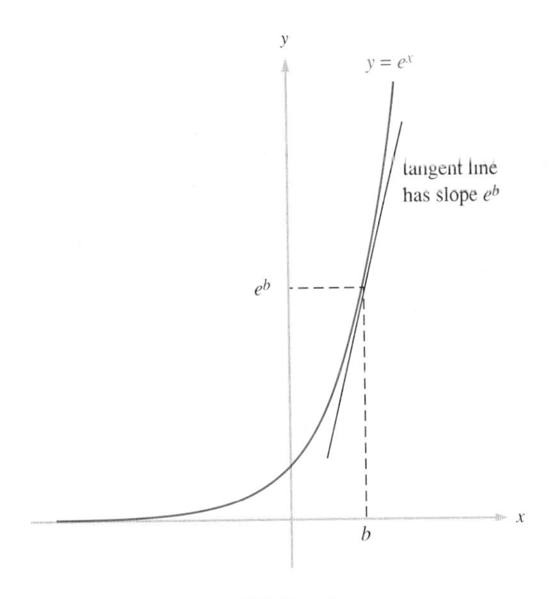

FIGURE 9.30

EXAMPLE 9.29 Compute the equation of the tangent line to the graph of $f(x) = e^x$ at the point (0, 1).

Solution As in Example 9.16 we must first find the slope of the tangent line, which is the value of $f'(x)$ at $x = 0$. Now

$$f'(x) = e^x,$$

and so

$$f'(0) = e^0 = 1.$$

Thus the slope of the tangent line to the graph of $y = f(x)$ at the point (0, 1) is 1. Consequently the equation of the tangent line to the graph of f at (0, 1) is

$$y - 1 = 1(x - 0),$$

that is,

$$y = x + 1. \quad ■$$

Practice Problem 25 Compute the instantaneous rate of change of $f(x) = e^x$ when $x = 1$.

Knowing the derivative of e^x enables us to compute the derivative of any function of the form $e^{h(x)}$ when $h(x)$ is a differentiable function of x. For since

$$e^{h(x)} = g(h(x)),$$

where $g(x) = e^x$, the chain rule produces the following formula.

Derivative of $e^{h(x)}$

If h is a differentiable function, then

$$(e^{h(x)})' = h'(x)e^{h(x)}.$$

Thus the derivative of $e^{h(x)}$ equals the derivative of the exponent times $e^{h(x)}$.

EXAMPLE 9.30 Compute the derivatives of the following functions.

(a) $f(x) = e^{3x+2}$ (b) $F(x) = e^{4x^2-3x}$.

Solution (a) The function $f(x)$ has the form $e^{h(x)}$, where

$$h(x) = 3x + 2.$$

Since $h(x)$ is a differentiable function, we can differentiate f using the preceding formula. The details are as shown below.

$$\begin{aligned} f'(x) &= h'(x)e^{h(x)} \\ &= (3x + 2)'e^{3x+2} \\ &= 3e^{3x+2}. \end{aligned}$$

(b) Likewise, by taking $h(x) = 4x^2 - 3x$, we see that

$$\begin{aligned} F'(x) &= h'(x)e^{h(x)} \\ &= (4x^2 - 3x)'e^{4x^2-3x} \\ &= (8x - 3)e^{4x^2-3x}. \end{aligned}$$ ■

Practice Problem 26 Compute the derivatives of the following functions.

(a) $f(x) = 5e^{-9x}$ (b) $F(x) = e^{7x^3+2x^2-8}$

EXAMPLE 9.31 A principal P is invested in an account paying interest at an annual rate of r compounded continuously. How fast is the value of the account changing after t years?

Solution Recall from Section 9.3 that the value of this account after t years is given by

$$A = Pe^{rt}.$$

The rate at which this value is changing is measured by the derivative dA/dt. Now the preceding formula gives

$$\frac{dA}{dt} = (Pe^{rt})' = P(e^{rt})' = P[(rt)'e^{rt}] = Pre^{rt} = rPe^{rt} = rA.$$

Therefore the rate at which the account is changing after t years is r times its value at time t. This is the case because, at any moment, the rate at which the account is changing per year is the yearly interest at that moment, and the yearly interest is the interest rate r times the amount A in the account at the moment in question. ■

Practice Problem 27 Potassium-42 is a radioactive element that is frequently used as a tracer. If the amount of mass (in mg) of potassium-42 that is present after t hours is given by the equation

$$A = 2e^{-0.05545t},$$

how fast is the mass changing after 10 hours?

DERIVATIVES OF OTHER EXPONENTIAL AND LOGARITHMIC FUNCTIONS

In this section our discussion has been limited to exponential and logarithmic functions with e as their base, that is, to functions of the form

$$f(x) = e^{g(x)} \qquad \text{and} \qquad f(x) = \ln g(x).$$

Recall that in Sections 8.3 and 8.4 we saw that an exponential or logarithmic function involving another base can be rewritten in terms of base e. Specifically

$$b^{h(x)} = (e^{\ln b})^{h(x)} = e^{(\ln b)h(x)},$$

which is of the form $e^{g(x)}$ with $g(x) = (\ln b)h(x)$; and

$$\log_b h(x) = \left(\frac{1}{\ln b}\right) \ln h(x).$$

Therefore being able to differentiate $e^{g(x)}$ and $\ln g(x)$ actually enables us to differentiate exponential or logarithmic functions to any base whatsoever. We will leave the differentiation of $b^{h(x)}$ and $\log_b h(x)$ as an exercise.

Clearly the derivatives of $b^{h(x)}$ and $\log_b h(x)$ are more complicated than those of $e^{h(x)}$ and $\ln x$. As a result, exponential and logarithmic functions with base e are preferred for use in applications requiring calculus. Henceforth in this book we will consider only exponential and logarithmic functions with base e.

EXERCISES 9.6

In Exercises 1–20 differentiate the given function.

1. $f(x) = x^3 + \ln x$

2. $F(x) = 3x^5 - e^x$

3. $g(x) = e^x - 2x^6$

4. $G(x) = 2\sqrt{x} - \ln x$

5. $h(x) = 7e^{-x}$

6. $H(x) = \ln (2x + 5)$

7. $F(x) = \ln (x^2 + 1)$

8. $f(x) = e^{3x+1}$

9. $G(x) = e^{4x}$

10. $g(x) = \ln(x^6 - 3x^3 + 4)$

11. $H(x) = \ln(x^4 - 2x^2 + 8)$

12. $h(x) = e^{2x^2 - 5x}$

13. $f(x) = \sqrt{3x + e^{-2x}}$

14. $F(x) = \ln(16 - x^2)$

15. $g(x) = [\ln(2x^2 + 5x)]^3$

16. $G(x) = \sqrt{9x^2 - e^{-x}}$

17. $h(x) = \sqrt[3]{\ln(x^2 + 4)}$

18. $H(x) = (4x - e^{2x})^3$

19. $F(x) = (e^x + e^{-x})^5$

20. $f(x) = [\ln(x^3 - 7x^2)]^4$

21. Find the instantaneous rate of change of $f(x) = e^x - e^{-x}$ at $x = 0$.

22. Find the instantaneous rate of change of $f(x) = \ln(x^2 - 2x + 3)$ at $x = 4$.

23. Find the instantaneous rate of change of $f(x) = \ln(2x - x^3)$ at $x = 1$.

24. Find the instantaneous rate of change of $f(x) = e^{x^2 - x}$ at $x = 0$.

25. Find the equation of the tangent line to the graph of $y = \ln x$ at $(1, 0)$.

26. Find the equation of the tangent line to the graph of $y = e^{-x}$ at $(0, 1)$.

27. Find the equation of the tangent line to the graph of $y = x + 2e^x$ at $(0, 2)$.

28. Find the equation of the tangent line to the graph of $y = x - \ln x$ at $(1, 1)$.

29. If a principal of \$1000 is invested at an interest rate of 8% compounded continuously, how fast will the value of the account be changing after one year?

30. Humans cannot distinguish small changes in the intensity of a stimulus. (It is impossible, for instance, to distinguish weights of 60 kg and 61 kg by simply lifting the two.) The Weber-Fechner psychophysical law states that the relationship between the actual magnitude r of a stimulus and its *perceived* magnitude $s(r)$ is given by

$$s(r) = k \ln \frac{r}{r_0},$$

where k and r_0 are constants. How fast will the perceived magnitude be changing when r equals 60 kg? (Give your answer in terms of k.)

31. The amount of mass (in kg) of a radioactive substance that will be present after t years is given by $A = 600e^{-0.01t}$. How fast will this mass be changing after 3 years?

32. The amount of mass (in kg) of a radioactive substance that will be present after t years is given by $A = 200e^{-0.005t}$. How fast will this mass be changing after 8 years?

33. The population of Bloomington-Normal t years after 1986 can be approximated by the function $P = 123{,}000e^{0.005t}$. Approximately how fast will Bloomington-Normal's population be changing in 1996?

34. The population of Peoria t years after 1980 can be approximated by the function $P = 366{,}000e^{-0.011t}$. Approximately how fast will Peoria's population be changing in 1992?

35. (a) Show that if h is a differentiable function, then the derivative of $f(x) = b^{h(x)}$ is $f'(x) = (\ln b)h'(x)b^{h(x)}$.

(b) Show that if h is a differentiable function, then the derivative of $F(x) = \log_b h(x)$ is $F'(x) = \dfrac{1}{\ln b}\left(\dfrac{h'(x)}{h(x)}\right)$.

36. Use Exercise 35 to compute the derivatives of the following functions.

(a) $f(x) = 10^{3x^2 - 5x}$ (b) $F(x) = \log_2(x^6 + 4x^2 + 9)$

(c) $g(x) = 8^{3 - 2x^4}$ (d) $G(x) = \log_{10}(x^4 + 3\sqrt{x})$.

Answers to Practice Problems

22. (a) $f'(x) = 54x^5 + \frac{1}{x}$ (b) $g'(x) = -8x^{-3} + \frac{1}{2x\sqrt{\ln x}}$

23. $y = x - 1$

24. $f'(x) = \frac{42x^5 - 3x^{-4}}{7x^6 + x^{-3}}$

25. $f'(1) = e$

26. (a) $f'(x) = -45e^{-9x}$ (b) $F'(x) = (21x^2 + 4x)e^{7x^3+2x^2-8}$

27. The mass is decreasing by approximately 0.0637 mg/hour.

CHAPTER 9 REVIEW

IMPORTANT TERMS

- **difference quotient** *(9.1)*
- **instantaneous rate of change**
- **secant line**
- **tangent line**
- **derivative of a function** *(9.2)*
- **limit of a function**
- **continuous function** *(9.3)*
- **differentiable function**
- **point of discontinuity**
- **differentiation** *(9.4)*

IMPORTANT FORMULAS

Derivative of a Function

$$f'(x) = \lim_{h \to 0} \frac{f(x + h) - f(x)}{h}$$

provided that this limit exists.

Derivative of a Constant Function

If $f(x) = k$, where k is any real number, then $f'(x) = 0$.

Derivative of a Power Function

If $f(x) = x^r$, where r is any real number, then $f'(x) = rx^{r-1}$.

Derivative of the Sum of Functions

If $f = g + h$, then $f' = g' + h'$.

Derivative of the Difference of Functions

If $f = g - h$, then $f' = g' - h'$.

Derivative of a Constant Multiple of a Function

If $f = kg$, where k is any real number, then $f' = kg'$.

The Chain Rule

If $f(x) = g(h(x))$ is the composition of two differentiable functions g and h, then f is differentiable and $f'(x) = g'(h(x)) \cdot h'(x)$.

Derivative of ln $h(x)$

If h is a differentiable function, then $[\ln h(x)]' = \dfrac{h'(x)}{h(x)}$. In particular, $(\ln x)' = \dfrac{1}{x}$.

Derivative of $e^{h(x)}$

If h is a differentiable function, then $(e^{h(x)})' = h'(x)e^{h(x)}$. In particular, $(e^x)' = e^x$.

REVIEW EXERCISES

1. Determine if the function f pictured below has a limit at the following points. If the limit exists, estimate its value from the graph.

(a) -8 (b) -4 (c) 5 (d) 7

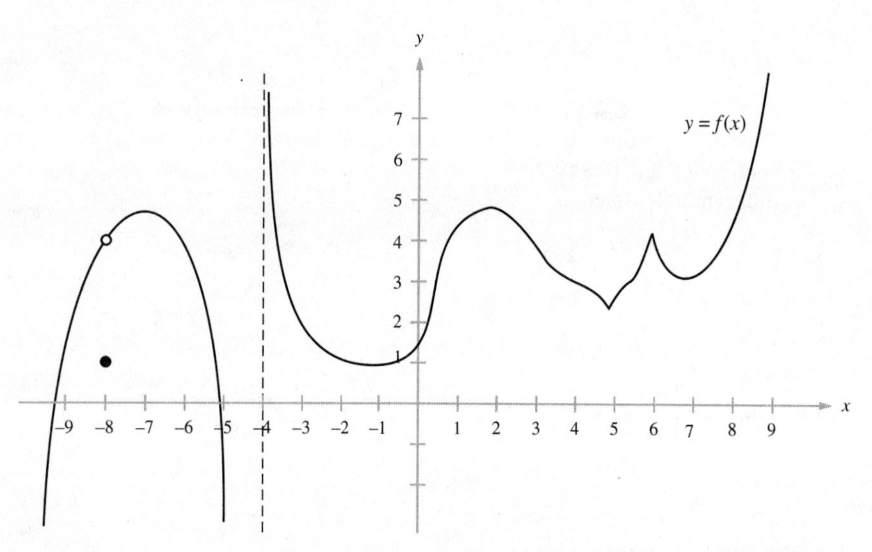

2. Determine if the function f pictured in Exercise 1 is differentiable at the following points.

(a) -8 (b) -1 (c) 5 (d) 7

3. Determine if the function f pictured in Exercise 1 is continuous at the following points.

(a) -8 (b) -4 (c) 4 (d) 6

4. Determine whether or not the given limits exist. Evaluate those that do.

(a) $\displaystyle\lim_{x\to 5} \frac{x^2 + 4x - 3}{2x - 3}$ (b) $\displaystyle\lim_{x\to -1} (2x^3 - 5x + 1)^{3/2}$

(c) $\displaystyle\lim_{x\to 2} (4x - 3)(x^2 - 1)$ (d) $\displaystyle\lim_{x\to 0} (1 + x)^{-1/2x}$

(e) $\displaystyle\lim_{x\to -\infty} \frac{4x^5 - 3x^4 + 8}{-2x^5 + 7x}$ (f) $\displaystyle\lim_{x\to \infty} \frac{7x^3}{5x^2 - 1}$

5. Use the definition of the derivative to compute f' if $f(x) = \dfrac{2}{x} - 1$.

6. Differentiate the given functions using the differentiation rules presented in this chapter.

(a) $f(x) = 7x^6 - 4x^5 + 3x^2 - 9$ (b) $g(x) = \dfrac{1}{x^3} + \ln x$

(c) $h(x) = 8x^7 - 6x^2 + 5 + 9e^x$ (d) $F(x) = \ln(9x^5 - 4x^3)$

(e) $G(x) = \sqrt[3]{3x^4 - 6x^2}$ (f) $H(x) = 4e^{5x^2 - 3x}$

7. A computer store's monthly revenue (in dollars) from the sale of x microcomputers of a certain model is given by $R(x) = 600x - 0.05x^2$. How fast is the revenue from the sale of these microcomputers changing when 100 computers of this model have been sold?

8. If the population of a town t years after 1990 can be approximated by the function $P = 100{,}000e^{-0.04t}$, how fast will the population be changing in 1995?

9. Find the equation of the tangent line to the graph of $y = e^x - x$ at $(0, 1)$.

10. Find the equation of the tangent line to the graph of $y = \sqrt{3x + 10}$ at $(2, 4)$.

10

TECHNIQUES OF DIFFERENTIATION

The derivative of a function was introduced in Chapter 9. We saw there that the derivative measures the instantaneous rate of change of a function at a point and the slope of the tangent line to the graph of the function at a point. We also developed formulas that enable us to differentiate many common functions. In this chapter we will learn how to differentiate other types of functions and will consider further applications of the derivative.

10.1 Derivatives of Products and Quotients

We saw in Section 9.4 that the derivative of the sum of two differentiable functions is the sum of their derivatives and the derivative of the difference of two differentiable functions is the difference of their derivatives. Unfortunately, the rules for differentiating products and quotients of functions are not as natural. In particular, the derivative of the product of two differentiable functions is *not* the product of their derivatives. Instead, the derivative of a product satisfies the following rule, which will be derived later in this section.

Derivative of a Product of Functions

If $f = gh$, then $f' = gh' + hg'$. That is, the derivative of a product of functions equals the first function multiplied by the derivative of the second plus the second function multiplied by the derivative of the first.

EXAMPLE 10.1 Differentiate the following functions.

(a) $F(x) = (3x^2)(5x^6)$ (b) $G(x) = 5x^2e^{-x}$ (c) $H(x) = x^2 \ln x$

Solution (a) We can write F as a product of two functions g and h by taking

$$g(x) = 3x^2 \quad \text{and} \quad h(x) = 5x^6.$$

Then the product rule gives

$$\begin{aligned} F' &= gh' + hg' \\ &= (3x^2)(5x^6)' + (5x^6)(3x^2)' \\ &= (3x^2)(30x^5) + (5x^6)(6x) \\ &= 90x^7 + 30x^7 \\ &= 120x^7. \end{aligned}$$

We can confirm this answer by multiplying out F before differentiating. Since

$$F(x) = (3x^2)(5x^6) = 15x^8,$$

the rule for differentiating a power function gives

$$F'(x) = 15(8x^7) = 120x^7,$$

as we found by using the product rule.

(b) To differentiate

$$G(x) = 5x^2e^{-x},$$

we will regard G as the product of functions g and h, where

$$g(x) = 5x^2 \quad \text{and} \quad h(x) = e^{-x}.$$

The product rule now gives

$$\begin{aligned} G' &= gh' + hg' \\ &= 5x^2(e^{-x})' + e^{-x}(5x^2)' \\ &= 5x^2(-e^{-x}) + e^{-x}(10x) \\ &= (10x - 5x^2)e^{-x}. \end{aligned}$$

(c) To differentiate

$$H(x) = x^2 \ln x,$$

we write $H = gh$, where

$$g(x) = x^2 \quad \text{and} \quad h(x) = \ln x.$$

Then by the product rule

$$\begin{aligned} H' &= gh' + hg' \\ &= x^2(\ln x)' + (\ln x)(x^2)' \\ &= x^2\left(\frac{1}{x}\right) + (\ln x)(2x) \\ &= x + 2x \ln x. \quad \blacksquare \end{aligned}$$

Practice Problem 1 Differentiate the following functions.
(a) $F(x) = (2x^4 - 5x)(6x^3 + 3x^2)$ (b) $G(x) = 4xe^x$
(c) $H(x) = x \ln (4x + 1)$

As Example 10.1 shows, the product rule is often used in combination with other rules from Chapter 9. The next example combines the product rule with the chain rule.

EXAMPLE 10.2 Determine the equation of the tangent line to the graph of the function $f(x) = (2 - x^2)\sqrt{2x + 2}$ at the point (1, 2).

Solution As before, we begin by finding the slope of the tangent line. Now

$$\begin{aligned} f'(x) &= (2 - x^2)(\sqrt{2x + 2})' + \sqrt{2x + 2}(2 - x^2)' \\ &= (2 - x^2)[(2x + 2)^{1/2}]' + \sqrt{2x + 2}(2 - x^2)' \\ &= (2 - x^2)\left[\frac{1}{2}(2x + 2)^{-1/2}(2)\right] + \sqrt{2x + 2}(-2x) \\ &= (2 - x^2)(2x + 2)^{-1/2} - 2x\sqrt{2x + 2}. \end{aligned}$$

Hence the slope of the tangent line at the point (1, 2) is

$$\begin{aligned} f'(1) &= [2 - (1)^2][2(1) + 2]^{-1/2} - 2(1)\sqrt{2(1) + 2} \\ &= (1)\frac{1}{\sqrt{4}} - 2(1)\sqrt{4} \\ &= (1)\left(\frac{1}{2}\right) - 2(1)(2) \\ &= -\frac{7}{2}. \end{aligned}$$

Using the point-slope form of the equation of a line, we see that the equation of the tangent line is

$$y - 2 = -\frac{7}{2}(x - 1)$$

or

$$y = -\frac{7}{2}x + \frac{11}{2}. \quad ■$$

Practice Problem 2 Determine the equation of the tangent line to the graph of the function $g(x) = e^x\sqrt{x^2 + 1}$ at (0, 1).

THE QUOTIENT RULE

Just as the derivative of the product of two differentiable functions is not the product of their derivatives, the derivative of the quotient of two differentiable functions is *not* the quotient of their derivatives. The following rule for differentiating a quotient of differentiable functions will be derived later in this section.

Derivative of a Quotient of Functions

If $f = \frac{g}{h}$, then

$$f' = \frac{hg' - gh'}{h^2}.$$

That is, the derivative of a quotient equals the denominator multiplied by the derivative of the numerator minus the numerator multiplied by the derivative of the denominator, all divided by the square of the denominator.

Note carefully the order of the terms in the numerator of this formula. Putting the minus sign on the wrong term is a common mistake when using the quotient rule.

EXAMPLE 10.3 Differentiate the following functions.

(a) $F(x) = \dfrac{x + 3}{2x - 1}$ (b) $G(x) = \dfrac{e^x}{x^2 + 1}$ (c) $H(x) = \dfrac{\ln x}{x}$

Solution (a) The function F is the quotient of the functions

$$g(x) = x + 3 \quad \text{and} \quad h(x) = 2x - 1,$$

and hence we can differentiate F by the quotient rule.

$$\begin{aligned} F' &= \frac{hg' - gh'}{h^2} \\ &= \frac{(2x - 1)(x + 3)' - (x + 3)(2x - 1)'}{(2x - 1)^2} \\ &= \frac{(2x - 1)(1) - (x + 3)(2)}{(2x - 1)^2} \\ &= \frac{-7}{(2x - 1)^2} \end{aligned}$$

(b) Write $G = \dfrac{g}{h}$, where

$$g(x) = e^x \quad \text{and} \quad h(x) = x^2 + 1.$$

Then by the quotient rule

$$\begin{aligned} G' &= \frac{hg' - gh'}{h^2} \\ &= \frac{(x^2 + 1)(e^x)' - e^x(x^2 + 1)'}{(x^2 + 1)^2} \\ &= \frac{(x^2 + 1)e^x - e^x(2x)}{(x^2 + 1)^2} \\ &= \frac{(x^2 - 2x + 1)e^x}{(x^2 + 1)^2}. \end{aligned}$$

(c) To apply the quotient rule, we take

$$g(x) = \ln x \quad \text{and} \quad h(x) = x.$$

Then $H = \dfrac{g}{h}$, and so

$$\begin{aligned} H' &= \frac{hg' - gh'}{h^2} \\ &= \frac{x(\ln x)' - (\ln x)(x)'}{x^2} \\ &= \frac{x\left(\frac{1}{x}\right) - (\ln x)(1)}{x^2} \\ &= \frac{1 - \ln x}{x^2}. \end{aligned}$$ ■

Practice Problem 3 Compute the derivatives of the following functions.

(a) $F(x) = \dfrac{3x - 1}{2 - 5x}$ (b) $G(x) = \dfrac{e^{-x}}{x}$ (c) $H(x) = \dfrac{\ln x}{2x + 1}$

EXAMPLE 10.4

When pricing a product for sale, a manufacturer must apportion the total cost of producing the product equally among all the units produced. In this context we are interested in the product's cost per unit produced, which is called its **average cost.** To determine the average cost, we need only divide the total cost of production by the number of units produced. Hence the average cost function is given by

$$\text{Average cost} = \frac{C(x)}{x},$$

where $C(x)$ is the total cost of producing x units.

Suppose a manufacturer's accountants have estimated that the cost (in dollars) of producing x microwave ovens will be

$$C(x) = 0.0002x^2 + \ln(10x + 1) + 3000.$$

How fast is the average cost changing when the production level is 100 ovens?

The cost (in dollars) of producing x microwave ovens will be $C(x) = 0.0002x^2 + \ln(10x + 1) + 3000$.

Solution The average cost of a microwave oven at a production level of x units is

$$A(x) = \frac{C(x)}{x} = \frac{0.0002x^2 + \ln(10x + 1) + 3000}{x}$$

$$= 0.0002x + \frac{\ln(10x + 1)}{x} + 3000x^{-1}.$$

The rate at which the average cost is changing is measured by the derivative of the average cost, $A'(x)$. Using the quotient rule enables us to differentiate $A(x)$; the result is

$$A'(x) = 0.0002 + \frac{x[\ln(10x + 1)]' - [\ln(10x + 1)](x)'}{x^2} - 3000x^{-2}$$

$$= 0.0002 + \frac{x\left[\dfrac{10}{10x + 1}\right] - [\ln(10x + 1)](1)}{x^2} - 3000x^{-2}.$$

Hence

$$A'(100) = 0.0002 + \frac{\dfrac{1000}{1001} - \ln 1001}{100^2} - \frac{3000}{100^2}$$

$$\approx -0.300391.$$

Therefore the average cost is *decreasing* by about \$0.30 per unit when the production level is 100 ovens. ■

Practice Problem 4 The cost (in dollars) of manufacturing x light bulbs is

$$C(x) = 0.01x^2 + \ln(1000x + 1) + 20{,}000.$$

How fast is the average cost changing when the production level is 1000 bulbs?

DERIVATION OF THE PRODUCT AND QUOTIENT RULES

Let $f = gh$, where g and h are differentiable at x. Then

$$f'(x) = \lim_{r \to 0} \frac{f(x + r) - f(x)}{r}$$

$$= \lim_{r \to 0} \frac{g(x + r)h(x + r) - g(x)h(x)}{r}$$

We now add zero to the numerator of the preceding expression in the form

$$-g(x + r)h(x) + g(x + r)h(x).$$

Continuing from above, we have

$$
\begin{aligned}
f'(x) &= \lim_{r\to 0} \frac{g(x+r)h(x+r) - g(x+r)h(x) + g(x+r)h(x) - g(x)h(x)}{r} \\
&= \lim_{r\to 0} \left[g(x+r)\frac{h(x+r)-h(x)}{r} + h(x)\frac{g(x+r)-g(x)}{r} \right].
\end{aligned}
$$

Applying the sum and product rules for limits (Theorem 9.1) to the preceding expression gives

$$
\begin{aligned}
f'(x) &= \lim_{r\to 0} g(x+r)\frac{h(x+r)-h(x)}{r} + \lim_{r\to 0} h(x)\frac{g(x+r)-g(x)}{r} \\
&= \lim_{r\to 0} g(x+r) \lim_{r\to 0} \frac{h(x+r)-h(x)}{r} + \lim_{r\to 0} h(x) \lim_{r\to 0} \frac{g(x+r)-g(x)}{r}.
\end{aligned}
$$

Now g is differentiable at x, and so by Theorem 9.3

$$\lim_{r\to 0} g(x+r) = g(x).$$

Moreover

$$\lim_{r\to 0} \frac{h(x+r)-h(x)}{r} = h'(x) \qquad \text{and} \qquad \lim_{r\to 0} \frac{g(x+r)-g(x)}{r} = g'(x)$$

by definition of the derivative. Therefore

$$
\begin{aligned}
f'(x) &= \lim_{r\to 0} g(x+r) \lim_{r\to 0} \frac{h(x+r)-h(x)}{r} + \lim_{r\to 0} h(x) \lim_{r\to 0} \frac{g(x+r)-g(x)}{r} \\
&= g(x)h'(x) + h(x)g'(x),
\end{aligned}
$$

which verifies the product rule.

To verify the quotient rule, note that h^{-1} can be differentiated using the chain rule and the rule for differentiating a power function; the result is

$$(h^{-1})' = (-1)h^{-2} \cdot h'.$$

Thus we can derive the quotient rule from the product rule by writing

$$\frac{g}{h} = gh^{-1}.$$

The details are as follows.

$$
\begin{aligned}
\left(\frac{g}{h}\right)' &= (gh^{-1})' \\
&= g(h^{-1})' + h^{-1}g' \\
&= g[(-1)h^{-2} \cdot h'] + h^{-1}g' \\
&= h^{-2}(-gh' + hg') \\
&= \frac{hg' - gh'}{h^2}
\end{aligned}
$$

EXERCISES 10.1

In Exercises 1–20 differentiate the given function.

1. $f(x) = 2x(x^3 - 1)$
2. $F(x) = (x^3 - 4)(x^2 + 2x)$
3. $g(x) = \dfrac{2x - 5}{3 - x}$
4. $G(x) = \dfrac{x + 1}{2x - 1}$
5. $h(x) = x \ln x$
6. $H(x) = (3x + 1)e^x$
7. $F(x) = \sqrt{x}e^x$
8. $f(x) = \dfrac{\ln x}{x^2}$
9. $G(x) = \dfrac{\ln x}{3x - 1}$
10. $g(x) = \dfrac{x^4 - 4x}{e^x + x}$
11. $H(x) = \dfrac{e^x}{x^2}$
12. $h(x) = 2x^3 \ln x$
13. $f(x) = (x^5 + 3x)(2x^3 - 4)^6$
14. $F(x) = (3 - 5x)^7(2x^2 - 5)$
15. $g(x) = \dfrac{(2x + 1)^4}{3x - 2}$
16. $G(x) = \dfrac{5x - 4}{(3x^2 + 1)^4}$
17. $h(x) = \dfrac{\ln (x^2 + 1)}{3x^4 - 6}$
18. $H(x) = (3x^5 - 7x^3 + 4x - 2)e^{1-x}$
19. $F(x) = \sqrt{2x - 1}e^{-x}$
20. $f(x) = \dfrac{\ln (9 - x^2)}{(x^3 + 1)^5}$
21. Write the equation of the tangent line to the graph of $y = x^2e^{-x}$ at (0, 0).
22. Write the equation of the tangent line to the graph of $y = \dfrac{\ln x}{x + 1}$ at (1, 0).
23. Write the equation of the tangent line to the graph of $y = \dfrac{2x}{x^2 + 1}$ at (0, 0).
24. Write the equation of the tangent line to the graph of $y = (2x - 1)^5 \ln (3x + 1)$ at (0, 0).
25. Write the equation of the tangent line to the graph of $y = x\sqrt{x + 5}$ at $(-1, -2)$.
26. Write the equation of the tangent line to the graph of $y = \dfrac{e^x}{(1 - 2x)^4}$ at (0, 1).
27. A manufacturer's inventory cost (in dollars) for a particular item is
$$C(x) = x + 500{,}000\left(\frac{10{,}000 + 50x}{x}\right)$$
when x items are produced per lot. How fast is the inventory cost changing when 1000 items are produced per lot?
28. The cost (in millions of dollars) to remove $x\%$ of the mercury polluting a particular river is given by
$$C(x) = \frac{0.6x}{100 - x}.$$
How fast is the cost of removing the mercury changing when 90% of the mercury has been removed?

29. If the total cost (in dollars) of producing x items is given by

$$C(x) = 0.0002x^2 + \ln(100x + 200) + 800{,}000,$$

how fast is the average cost changing when 500 items are being produced?

30. After t days of the spring semester, the number of students in a dormitory who have been infected by a certain strain of influenza is approximately

$$f(t) = \frac{300}{1 + 4e^{-0.1t}}.$$

How fast is the number of infected students increasing after 10 days?

31. Let f, g, and h be differentiable functions. Use the product rule to differentiate $F = fgh$. *Hint:* Write F as $(fg)h$.

32. Verify the quotient rule by using the definition of the derivative. *Hint:*

$$\frac{g(x + r)h(x) - g(x)h(x + r)}{r \cdot h(x + r)h(x)}$$
$$= \frac{h(x)}{h(x + r)h(x)} \frac{g(x + r) - g(x)}{r} - \frac{g(x)}{h(x + r)h(x)} \frac{h(x + r) - h(x)}{r}.$$

Answers to Practice Problems

1. (a) $F'(x) = 84x^6 + 36x^5 - 120x^3 - 45x^2$
(b) $G'(x) = (4x + 4)e^x$
(c) $H'(x) = \dfrac{4x}{4x + 1} + \ln(4x + 1)$

2. $y = x + 1$

3. (a) $F'(x) = \dfrac{1}{(2 - 5x)^2}$
(b) $G'(x) = \dfrac{-(x + 1)e^{-x}}{x^2}$
(c) $H'(x) = \dfrac{2x + 1 - 2x\ln x}{x(2x + 1)^2}$

4. The average cost is decreasing by slightly more than \$0.01 per unit at a production level of 1000 bulbs.

10.2 Higher-Order Derivatives

In Section 9.1 we saw that if the number of miles traveled after x hours of a trip is given by

$$s(x) = 4x^2 + 24x,$$

then the instantaneous velocity at time x is given by the derivative

$$s'(x) = 8x + 24.$$

In this case the function s' is a polynomial function and hence is itself differentiable. The function obtained by differentiating s' is denoted s''. It gives the instantaneous rate of change of s', that is, the instantaneous rate of change of velocity, which is the instantaneous acceleration.

More generally, the function obtained by differentiating f' is called the **second derivative** of f and is denoted f''. One context in which the second derivative arises was described above: If $s(x)$ measures the distance traveled at time x, then $s''(x)$ measures the instantaneous acceleration at time x. In Section 11.3 we will see that $f''(a)$ tells us about the shape of the graph of $y = f(x)$ at $x = a$.

We can continue taking derivatives of a function f as long as the resulting functions are differentiable. The successive derivatives of f are denoted

$$f',\ f'',\ f''',\ f^{(4)},\ f^{(5)},\ \ldots,\ f^{(n)}.$$

These are called the first, second, third, fourth, fifth, . . . , and nth derivatives of f, respectively. Note the distinction between

$$f^{(n)} \quad \text{and} \quad f^n;$$

the former denotes the nth *derivative* of f, whereas the latter denotes the nth *power* of f.

Table 10.1 shows the appropriate symbolism for the successive derivatives of $y = f(x)$ using the other notations for the derivative that we have seen.

First derivative	$f'(x)$	y'	$\frac{dy}{dx}$
Second derivative	$f''(x)$	y''	$\frac{d^2y}{dx^2}$
Third derivative	$f'''(x)$	y'''	$\frac{d^3y}{dx^3}$
Fourth derivative	$f^{(4)}(x)$	$y^{(4)}$	$\frac{d^4y}{dx^4}$
.	.	.	.
.	.	.	.
.	.	.	.
nth derivative	$f^{(n)}(x)$	$y^{(n)}$	$\frac{d^ny}{dx^n}$

TABLE 10.1

Note carefully the placement of the numerals in the last column of Table 10.1.

EXAMPLE 10.5 Find the first, second, and third derivatives of f at x if

$$f(x) = 5x^4 - 7x^3 + 9x^2 - 8x - 6.$$

Solution Using the differentiation rules in Chapter 9, we see that the derivative of f at x is

$$f'(x) = 20x^3 - 21x^2 + 18x - 8.$$

To obtain the second derivative of f at x, we differentiate f':

$$f''(x) = (20x^3 - 21x^2 + 18x - 8)' = 60x^2 - 42x + 18.$$

If we now differentiate f'', we obtain the third derivative of f at x:

$$f'''(x) = (60x^2 - 42x + 18)' = 120x - 42. \quad ■$$

Practice Problem 5 Find the first, second, and third derivatives of f at x if

$$f(x) = 2x^6 - 5x^3 + 9x - 7.$$

EXAMPLE 10.6 Determine y', y'', and y''' if $y = \ln x$.

Solution By the rule for differentiating the natural logarithm function, we obtain

$$y' = \frac{1}{x}.$$

Since we can write the first derivative as

$$y' = \frac{1}{x} = x^{-1},$$

the rule for differentiating a power function gives the second derivative of y:

$$y'' = -1(x^{-2}) = -x^{-2}.$$

Likewise the third derivative of y is

$$y''' = -(-2x^{-3}) = 2x^{-3}. \quad ■$$

Practice Problem 6 Determine y', y'', and y''' if

$$y = e^{-2x}.$$

EXAMPLE 10.7 If $y = x^2e^x$, find the following.

(a) $\dfrac{dy}{dx}$ (b) $\dfrac{d^2y}{dx^2}$ (c) $\dfrac{d^3y}{dx^3}$

Solution (a) By the product rule

$$\begin{aligned}\frac{dy}{dx} &= x^2(e^x)' + e^x(x^2)' \\ &= x^2e^x + e^x(2x) \\ &= (x^2 + 2x)e^x.\end{aligned}$$

(b) Similarly,

$$\begin{aligned}\frac{d^2y}{dx^2} &= (x^2 + 2x)(e^x)' + e^x(x^2 + 2x)' \\ &= (x^2 + 2x)e^x + e^x(2x + 2) \\ &= (x^2 + 4x + 2)e^x.\end{aligned}$$

(c) Finally,

$$\begin{aligned}\frac{d^3y}{dx^3} &= (x^2 + 4x + 2)(e^x)' + e^x(x^2 + 4x + 2)' \\ &= (x^2 + 4x + 2)e^x + e^x(2x + 4) \\ &= (x^2 + 6x + 6)e^x. \quad \blacksquare\end{aligned}$$

Practice Problem 7 If $y = \dfrac{\ln x}{x}$, find the following.

(a) $\dfrac{dy}{dx}$ (b) $\dfrac{d^2y}{dx^2}$ (c) $\dfrac{d^3y}{dx^3}$

In this book we will be concerned primarily with the first and second derivatives of a function. We close this section with an application involving acceleration.

EXAMPLE 10.8 If the effects of air resistance are ignored, an object that is falling freely to the earth will drop approximately

$$s(t) = 16t^2$$

feet in t seconds. Determine the acceleration of the object after t seconds.

Solution The acceleration after t seconds is given by $s''(t)$. Now

$$s'(t) = 32t,$$

and so

$$s''(t) = 32.$$

Thus the acceleration of the object at time t is 32 feet per second per second. (This is the acceleration due to gravity.) Hence after t seconds the speed of the object is increasing by 32 feet per second per second. ■

Practice Problem 8 An object that is falling freely to the surface of the moon will drop approximately

$$s(t) = 2.7t^2$$

feet in t seconds. Determine the acceleration of the object after t seconds.

EXERCISES 10.2

In Exercises 1–6 find (a) y', (b) y'', and (c) y'''.

1. $y = 6x^7 - 4x^5 + 3x^4 - 2$

2. $y = 9x^5 - 5x^3 + 2x^2 - 7x$

3. $y = (3x - 2)^5$

4. $y = \sqrt[3]{2x + 1}$

5. $y = e^{1-2x}$

6. $y = \ln (5x + 6)$

In Exercises 7–12 find (a) $\frac{dy}{dx}$, (b) $\frac{d^2y}{dx^2}$, and (c) $\frac{d^3y}{dx^3}$.

7. $y = \ln (2x^2 + 1)$

8. $y = e^{1-x}$

9. $y = \sqrt{2x + 4}$

10. $y = (x^2 - x)^6$

11. $y = x^3 \ln x$

12. $y = 5x^4e^x$

In Exercises 13–18 evaluate (a) $f'(c)$ and (b) $f''(c)$.

13. $f(x) = \frac{2x + 5}{3x + 4}$, $c = -1$

14. $f(x) = \frac{4 - x}{2x + 3}$, $c = -2$

15. $f(x) = (x^2 - 4x)(5 - x^3)$, $c = 2$

16. $f(x) = (2x^3 - 3x)(x^2 - 5x + 2)$, $c = 1$

17. $f(x) = \ln (4x^2 + 1)$, $c = 0$

18. $f(x) = e^{4x-x^2}$, $c = 0$

In Exercises 19–24 evaluate (a) $\frac{df}{dx}$ and (b) $\frac{d^2f}{dx^2}$ at $x = c$.

19. $f(x) = \sqrt{2x^2 + 1}$, $c = 2$

20. $f(x) = \sqrt{13 - x^2}$, $c = -2$

21. $f(x) = \ln \frac{2x + 1}{3x + 1}$, $c = 0$

22. $f(x) = \ln \sqrt{9x + 1}$, $c = 0$

23. $f(x) = e^{2x^2+3x}$, $c = 0$

24. $f(x) = e^{(2x-5)^2}$, $c = 3$

25. If $f(x) = x^3 - 3x^2 - 9x$, find the numbers x where
(a) $f'(x) = 0$ and (b) $f''(x) = 0$.

26. If $f(x) = x^4 - 24x^2$, find the numbers x where
(a) $f'(x) = 0$ and (b) $f''(x) = 0$.

27. A manufacturer's total cost (in dollars) of producing x items is

$$C(x) = 0.000001x^3 - 0.0015x^2 + 240x + 112{,}500.$$

The derivative of C is called the *marginal cost.*

(a) How fast is the marginal cost changing when $x = 0$?

(b) How fast is the marginal cost changing when $x = 2000$?

(c) For what value of x does the rate of change of the marginal cost equal zero?

28. A projectile was thrown into the air in such a way that after t seconds its height in feet was $h(t) = 256 + 96t - 16t^2$. Determine

(a) its velocity after t seconds (b) its acceleration after t seconds.

29. In order to check the amount of drowsiness caused by a new drug, a laboratory measured the length of time that a rat slept after receiving various dosages of the drug. It found that the number of minutes of drowsiness after a dosage of x mg was approximately $f(x) = 12x^2 - x^3$.

(a) Determine the rate of change of the length of drowsiness after administering a dosage of x mg.

(b) Determine the dosage at which the rate of change of the length of drowsiness began to decrease. *Hint:* Determine where $f''(x) = 0$.

30. A toxic gas was released when a train derailed. After t minutes, the number of persons exposed to the gas was approximately $H(t) = 3000 + 180t^2 - t^3$.

(a) Determine the rate of change of the number of exposed persons after t minutes.

(b) Determine when the rate of change of the number of exposed persons began to decrease. *Hint:* Determine when $H''(t) = 0$.

Answers to Practice Problems

5. $f'(x) = 12x^5 - 15x^2 + 9$, $f''(x) = 60x^4 - 30x$, and $f'''(x) = 240x^3 - 30$

6. $y' = -2e^{-2x}$, $y'' = 4e^{-2x}$, and $y''' = -8e^{-2x}$

7. (a) $\dfrac{dy}{dx} = \dfrac{1 - \ln x}{x^2}$ (b) $\dfrac{d^2y}{dx^2} = \dfrac{2 \ln x - 3}{x^3}$ (c) $\dfrac{d^3y}{dx^3} = \dfrac{11 - 6 \ln x}{x^4}$

8. The acceleration of the object after t seconds is approximately 5.4 feet per second per second.

10.3 Implicit Differentiation

In our previous work with functions, the functions were expressed in terms of the independent variable alone, for instance, in the form

$$y = f(x).$$

In this case we say that y is an **explicit function** of x. Sometimes, however, the independent and dependent variables are related by an equation in which it is difficult, or even impossible, to solve for one variable in terms of the other. For example, in the equation

$$y^5 + y - x^2 = -2,$$

it is not possible to solve for y in terms of x. Nevertheless, the given equation defines y as a function of x; in this case we say that y is an **implicit function** of x.

When y is defined implicitly as a function of x, the graph of the equation relating x and y may have a tangent line at a point on the graph. In such a case the derivative y' exists even though our previous techniques of differentiation are inadequate for computing it. In this section we will describe a process called **implicit differentiation** that may enable us to compute y' when y is defined implicitly as a function of x. Throughout this discussion we will assume that x and y are related by an equation that defines y as some function of x (a function that we may or may not be able to determine explicitly). We will also assume that y' exists.

The key to implicit differentiation is the chain rule. Since we are differentiating with respect to x, expressions involving x alone can be differentiated as in Chapter 9. But y is a function of x; so *we must use the chain rule to differentiate expressions involving* y.

EXAMPLE 10.9 Assuming that y is a function of x, differentiate the following with respect to x.

(a) $7x^4$ (b) $2y$ (c) y^8 (d) $\ln y$

Solution (a) The expression to be differentiated involves only x. Hence we can use the differentiation rules in Section 9.4 to obtain

$$\frac{d}{dx}(7x^4) = 28x^3.$$

(b) By definition the derivative of y *with respect to* x is y'. Thus

$$(2y)' = 2y'.$$

(c) To differentiate y^8 with respect to x, let $u = y^8$. Then by the chain rule

$$\begin{aligned}\frac{du}{dx} &= \frac{du}{dy}\frac{dy}{dx} \\ &= (8y^7)y'.\end{aligned}$$

(d) Proceeding as in part (c), we let $u = \ln y$. Then

$$\begin{aligned}\frac{du}{dx} &= \frac{du}{dy}\frac{dy}{dx} \\ &= \frac{1}{y}\frac{dy}{dx} \\ &= \frac{y'}{y}.\end{aligned}$$ ■

Practice Problem 9 Assuming that y is a function of x, differentiate the following with respect to x.

(a) $6x^5 - 3x^{-2}$ (b) $2x - y$ (c) y^{-5} (d) e^{6y}

Let us now consider a simple example involving implicit differentiation. The equation

$$xy = 2$$

defines y implicitly as a function of x. In this case we can write y as an explicit function of x by solving for y:

$$y = \frac{2}{x} = 2x^{-1}.$$

Hence we can compute y' by the techniques of Chapter 9:

$$y' = (2x^{-1})' = -2x^{-2}.$$

But we can also compute y' directly from the equation $xy = 2$ by using implicit differentiation. To do so, we differentiate both sides of the equation with respect to x:

$$\frac{d}{dx}(xy) = \frac{d}{dx}(2)$$

The derivative on the left side of the equation can be computed by the product rule:

$$\begin{aligned}\frac{d}{dx}(xy) &= x\frac{dy}{dx} + y\frac{dx}{dx}\\ &= xy' + y.\end{aligned}$$

Also $\frac{d}{dx}(2) = 0$ by the rule for differentiating constant functions. Substituting these results into the equation

$$\frac{d}{dx}(xy) = \frac{d}{dx}(2)$$

yields

$$xy' + y = 0.$$

This equation can be easily solved for y' as follows.

$$\begin{aligned}xy' &= -y\\ y' &= -\frac{y}{x}\end{aligned}$$

Recall that when we solved for y in terms of x and computed y', the answer was

$$y' = -2x^{-2}.$$

Although the two preceding expressions for y' appear different, they are equivalent. For since $y = 2x^{-1}$,

$$-\frac{y}{x} = -\frac{2x^{-1}}{x} = -(2x^{-1})x^{-1} = -2x^{-2}.$$

Thus the slope of the tangent line to $xy = 2$ at (x, y) is

$$y' = -\frac{y}{x} = -2x^{-2}.$$

Figure 10.1 shows the tangent line $y = -2x + 4$ at $(1, 2)$, which has slope -2.

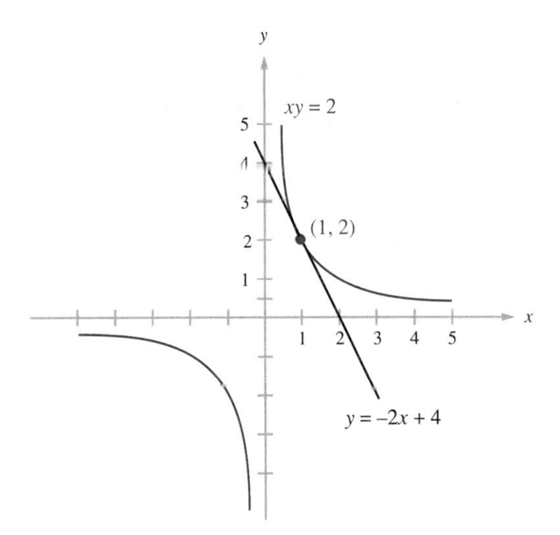

FIGURE 10.1

It is important to realize that *when implicit differentiation is used, the only values of x and y that are meaningful in the formula for y′ are values that satisfy the original equation.* In the example above, for instance, we can substitute $x = 2$ and $y = 4$ into the equation

$$y' = -\frac{y}{x}$$

obtained by implicit differentiation. But the resulting value

$$y' = -\frac{4}{2} = -2$$

is meaningless because $x = 2$ and $y = 4$ do not satisfy the equation $xy = 2$, and hence we cannot talk about the slope of the curve at the point (2, 4).

The steps involved in implicit differentiation are summarized below.

Implicit Differentiation

Given an equation involving x and y that defines y implicitly as a function of x, we can find y' (when it exists) in the following way.

1. Differentiate both sides of the equation *with respect to x.*
2. Collect all the terms involving y' on one side of the equation, and move all terms without y' to the other side.
3. Solve for y'.

In the preceding example, implicit differentiation was not really necessary because it was possible to write y explicitly as a function of x. When y cannot be easily expressed as an explicit function of x, however, the simplest way to compute y' is via implicit differentiation. The remaining examples in this section involve equations for which implicit differentiation is the most convenient method of obtaining y'.

EXAMPLE 10.10 Use implicit differentiation to determine y' if y is defined implicitly by the equation $y^2 - 5x = xy$.

Solution We begin by differentiating both sides of the equation with respect to x; this yields

$$\begin{aligned} (y^2)' - (5x)' &= (xy)' \\ 2yy' - 5 &= xy' + y. \quad \textbf{(chain rule and product rule)} \end{aligned}$$

By adding $5 - xy'$ to both sides of the equation, we can collect all the terms involving y' on the left side of the equation and all the other terms on the right:

$$2yy' - xy' = y + 5.$$

Factoring y' from the left side now produces

$$y'(2y - x) = y + 5.$$

Thus we see that

$$y' = \frac{y + 5}{2y - x}. \quad \blacksquare$$

Practice Problem 10 Use implicit differentiation to determine y' if y is defined implicitly by the equation

$$4y^3 - y - 5x^2 = 3x.$$

EXAMPLE 10.11 Use implicit differentiation to determine y' if y is defined implicitly by the equation $\ln(x^2 + y^2) = y^2$.

Solution If we differentiate both sides of the given equation with respect to x, we obtain

$$\begin{aligned} \frac{1}{x^2 + y^2}(x^2 + y^2)' &= (y^2)' \\ \frac{1}{x^2 + y^2}(2x + 2yy') &= 2yy'. \end{aligned}$$

Multiplying both sides of the equation above by $x^2 + y^2$ produces

$$(2x + 2yy') = 2yy'(x^2 + y^2).$$

To collect all the terms involving y' on the right side of the equation, we must subtract $2yy'$ from both sides.

$$2x = 2yy'(x^2 + y^2) - 2yy'.$$

Factoring y' from each term on the right side now yields

$$2x = [2y(x^2 + y^2) - 2y]y',$$

and so

$$\frac{x}{y(x^2 + y^2) - y} = y'. \quad ■$$

Practice Problem 11 Use implicit differentiation to determine y' if y is defined implicitly by the equation

$$\ln (2x + y) = x^3.$$

EXAMPLE 10.12 Determine the equation of the tangent line to the curve defined by the equation

$$y^5 + y - x^2 = -2$$

at the point (2, 1).

Solution Differentiating both sides of the equation with respect to x gives

$$5y^4y' + y' - 2x = 0.$$

Therefore

$$y'(5y^4 + 1) = 2x,$$

and so

$$y' = \frac{2x}{5y^4 + 1}.$$

Hence at the point (2, 1) the value of y' is

$$\frac{2(2)}{5(1)^4 + 1} = \frac{4}{6} = \frac{2}{3}.$$

Since this is the slope of the tangent line to the given curve at (2, 1), the equation of the tangent line is

$$y - 1 = \frac{2}{3}(x - 2),$$

that is,

$$y = \frac{2}{3}x - \frac{1}{3}.$$

See Figure 10.2. ■

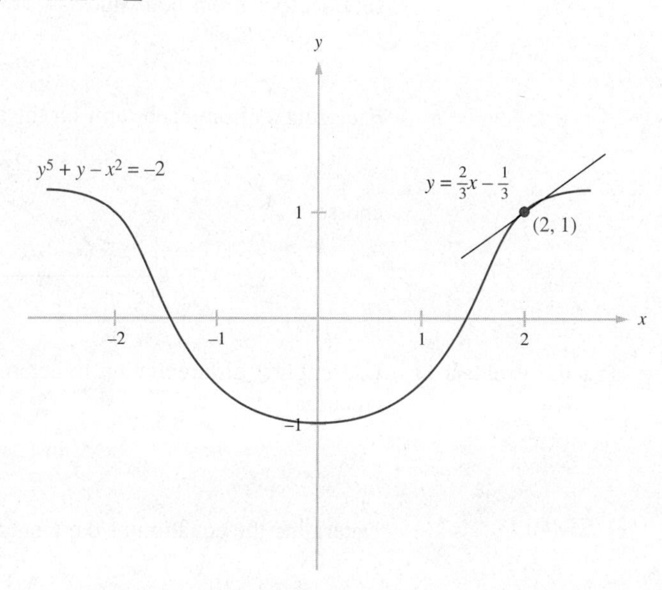

FIGURE 10.2

Practice Problem 12 Find the equation of the tangent line to the curve defined by the equation

$$4x^2 + 3y^2 = 64$$

at the point (2, 4).

EXERCISES 10.3

In Exercises 1–12 differentiate the given expression with respect to x, assuming that y is a function of x.

1. $3\sqrt{x}$

2. e^{-4x}

3. $y^3 - 7y^2 - 5$

4. $y^2 - 3y + 4$

5. e^{3y}

6. $\ln(3y + 2)$

7. x^2y^4

8. x^3e^y

9. $x \ln(x^2 + y)$

10. $\frac{x^3}{y^2}$

11. $\frac{y}{x^2 - 3x}$

12. $\frac{x^3}{x + y^2}$

In Exercises 13–24 determine y′ by implicit differentiation.

13. $y^2 - 3x = 12$

14. $y^2 = x^3 - x$

15. $2x^2 + y^2 = 32$

16. $x^2 - y^2 = 9$

17. $x^2y = 6$

18. $xy^2 = 4$

19. $y^3 = 4x^3 + 4xy$

20. $4x^2 - \sqrt{y} - 6xy$

21. $\ln(x + y) = xy^2$

22. $\ln(x^3 - y^2) = x^2y$

23. $\dfrac{e^y}{x^2} = 3y$

24. $\dfrac{e^{-x}}{y + 1} = 5xy^3$

In Exercises 25–32 write the equation of the tangent line to the graph of the given equation at the point (a, b).

25. $x^2 + y^2 = 100, \quad (a, b) = (8, 6)$

26. $4x^2 + 9y^2 = 72, \quad (a, b) = (3, 2)$

27. $y^2 = x^3 - x^2, \quad (a, b) = (2, 2)$

28. $y^2 = x^3 - 2x, \quad (a, b) = (2, -2)$

29. $x^2y^2 = 1, \quad (a, b) = (-1, 1)$

30. $x^2y^3 = 8, \quad (a, b) = (-8, \frac{1}{2})$

31. $(2x - y)^2 = 3y^2 + 4, \quad (a, b) = (3, 2)$

32. $(xy + y)^2 = 16x^2, \quad (a, b) = (1, 2)$

Answers to Practice Problems

9. (a) $30x^4 + 6x^{-3}$ (b) $2 - y'$ (c) $-5y^{-6}y'$ (d) $6e^{6y}y'$

10. $y' = \dfrac{10x + 3}{12y^2 - 1}$

11. $y' = 3x^2(2x + y) - 2$

12. $y = -\dfrac{2}{3}x + \dfrac{16}{3}$

10.4 Related Rates

Many variable quantities change with time. For example, the amount of a drug in the bloodstream is a function of the time that has elapsed since the drug was taken, and the sales of a particular product often vary with the time of year.

In many practical problems in which two variables are related by an equation, both of the variables change with time. Hence they can be regarded as functions of a third variable, time, denoted t. In this case the derivatives of the variables with respect to t measure the rate of change of the variables with respect to time. Since the variables are related by an equation, implicit differentiation can be used to obtain a relationship between their derivatives. If the value of one of these derivatives is known at a particular moment, it may be possible to use this relationship to find the value of the other. We call such derivatives **related rates.**

EXAMPLE 10.13 If $y = 3x^2 - 4x$, find $\frac{dy}{dt}$ when $x = 2$ and $\frac{dx}{dt} = \frac{1}{2}$.

Solution By differentiating both sides of the given equation with respect to t, we obtain an equation relating $\frac{dy}{dt}$ and $\frac{dx}{dt}$:

$$\frac{dy}{dt} = 6x\frac{dx}{dt} - 4\frac{dx}{dt}$$

$$\frac{dy}{dt} = (6x - 4)\frac{dx}{dt}.$$

Hence when $x = 2$ and $\frac{dx}{dt} = \frac{1}{2}$, we have

$$\frac{dy}{dt} = [6(2) - 4]\left(\frac{1}{2}\right) = 8\left(\frac{1}{2}\right) = 4. \blacksquare$$

Practice Problem 13 If $y = 8x - x^3$, find $\frac{dy}{dt}$ when $x = -1$ and $\frac{dx}{dt} = 2$.

Problems involving related rates require evaluating a rate of change at a specific moment. Since the rate of change of f is given by f', the rate of change of f at $x = a$ is the value $f'(a)$. When the derivative of f is denoted by $\frac{df}{dt}$, the symbol

$$\left.\frac{df}{dt}\right|_{x=a}$$

is used to represent the value of the derivative at $x = a$. This notation is used several times in the examples of this section.

EXAMPLE 10.14 A company manufactures electric toothbrushes which it sells for $20. If the number of toothbrushes made per week, x, satisfies $0 \le x \le 5000$, then its total costs are approximately

$$C = 0.001x^2 + 10x + 25{,}000.$$

Since demand for the toothbrushes is strong, the company intends to increase its production from the current level of 1000 toothbrushes per week. If production is increased by 200 toothbrushes per week for the next two months, at what rate will the company's profit be increasing when production reaches the level of 2000 toothbrushes per week?

Solution We must determine

$$\left.\frac{dP}{dt}\right|_{x=2000},$$

the rate of change in profit at the production level of 2000 toothbrushes per week. Recall that a company's profit is the difference between its revenues and costs. Since the company is producing x toothbrushes per week and selling them for \$20 apiece, its weekly revenue is $20x$ dollars. Hence its weekly profit is given by

$$\begin{aligned} P &= 20x - C \\ &= 20x - (0.001x^2 + 10x + 25{,}000) \\ &= 10x - 0.001x^2 - 25{,}000. \end{aligned}$$

Differentiating implicitly with respect to t gives

$$\begin{aligned} \frac{dP}{dt} &= 10\frac{dx}{dt} - 0.002x\frac{dx}{dt} \\ &= (10 - 0.002x)\frac{dx}{dt}. \end{aligned}$$

Since the company is increasing its weekly production by 200 toothbrushes per week, we have

$$\frac{dx}{dt} = 200.$$

Thus

$$\left.\frac{dP}{dt}\right|_{x=2000} = [10 - 0.002(2000)](200) = 1200.$$

Hence the company's profit will increase at the rate of \$1200 per week when the weekly level of production is 2000 toothbrushes. ■

EXAMPLE 10.15 The creator of a new piece of computer software receives a royalty of

$$1.25 + 0.0001x$$

dollars for each copy sold, where x denotes the number of sales per month. Sales increase at the constant rate of 1000 per month. How much will the creator's monthly royalty check be increasing when 5000 copies of the software are sold per month?

Solution We must find the amount of increase in the monthly royalty check, dR/dt, when the monthly sales x reach 5000. Now the amount R of the monthly royalty check is the product of the royalty received per copy and the number of copies sold; so

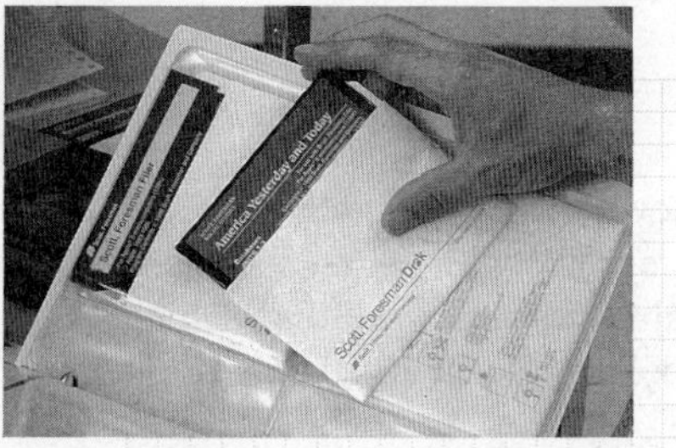

The royalty is 1.25 + 0.0001*x* dollars for each copy sold, where *x* is the number of copies sold in a given month.

$$\begin{aligned} R &= (1.25 + 0.0001x)x \\ &= 1.25x + 0.0001x^2. \end{aligned}$$

Therefore

$$\begin{aligned} \frac{dR}{dt} &= 1.25\frac{dx}{dt} + 0.0002x\frac{dx}{dt} \\ &= (1.25 + 0.0002x)\frac{dx}{dt}. \end{aligned}$$

Since the monthly sales are increasing by 1000, $\dfrac{dx}{dt} = 1000$. Thus when the monthly sales reach 5000 copies, we have

$$\left.\frac{dR}{dt}\right|_{x=5000} = [1.25 + 0.0002(5000)](1000) = 2250.$$

Hence the creator's royalty check is increasing by \$2250 per month when the monthly sales of the software reach 5000 copies. ■

EXAMPLE 10.16 The rate of increase in a town's population is expected to grow by 1% per year for the next ten years because of the construction of a new automobile plant. If the town's population is given by

$$P = 100{,}000e^{50r^2 - .02},$$

where r is the yearly rate of population increase, how large will the yearly population increase be when the yearly rate of increase reaches 5%?

Solution We must determine the yearly population increase $\dfrac{dP}{dt}$ when $r = 0.05$. Now

$$\frac{dP}{dt} = 100{,}000e^{50r^2 - .02}\left(100r\frac{dr}{dt}\right).$$

Therefore since $dr/dt = 0.01$, we have

$$\begin{aligned} \left.\frac{dP}{dt}\right|_{r=0.05} &= 100{,}000e^{0.105}(100)(.05)(.01) \\ &\approx 5554. \end{aligned}$$

Hence the town's population will be increasing at the rate of about 5554 people per year when the rate of increase reaches 5%. ■

EXAMPLE 10.17 After colliding with another ship, an oil tanker began leaking oil that is spreading over the water as a circular slick 1/12 foot thick. If the oil is leaking from the ship at the rate of 100 cubic feet per minute, how fast will the radius of the slick be increasing when the radius is 240 feet?

Solution We must determine dr/dt when $r = 240$. Since oil is leaking from the tanker at the rate of 100 cubic feet per minute, the volume of the oil slick is increasing at this rate. Note that the oil slick is cylindrical in shape with a height of 1/12 foot. Since the volume of a cylinder with radius r and height h is $V = \pi r^2 h$, the formula relating the volume V and the radius r of the oil slick is

$$V = \frac{1}{12}\pi r^2.$$

Thus

$$\frac{dV}{dt} = \frac{1}{6}\pi r \frac{dr}{dt},$$

and so

$$\frac{6}{\pi r}\frac{dV}{dt} = \frac{dr}{dt}.$$

Hence

$$\left.\frac{dr}{dt}\right|_{r=240} = \frac{6}{\pi(240)}(100) = \frac{2.5}{\pi} \approx 0.8.$$

Therefore the radius of the slick will be increasing at approximately 0.8 feet per minute when it reaches 240 feet. ■

Practice Problem 14 Industrial pollution has contaminated the ground around a toxic waste disposal site. The contaminated ground forms a circle centered at the waste disposal site. If the radius of this circle is increasing at the rate of 10 feet per month, determine the rate at which the contaminated area is increasing when the contaminated ground has a 600-foot radius.

EXERCISES 10.4

1. If $y = 6x - 2$, find $\frac{dy}{dt}$ when $\frac{dx}{dt} = \frac{1}{2}$.

2. If $y = 8x^2 - 4x + 5$, find $\frac{dy}{dt}$ when $\frac{dx}{dt} = 2$ and $x = 5$.

3. If $y = \frac{3600}{x}$, find $\frac{dy}{dt}$ when $\frac{dx}{dt} = -2$ and $x = 30$.

4. If $y = 100e^{-0.5x}$, find $\frac{dy}{dt}$ when $\frac{dx}{dt} = 0.1$ and $x = 2$.

5. If $y = \ln(x^2 + 1)$, find $\frac{dy}{dt}$ when $\frac{dx}{dt} = \frac{1}{2}$ and $x = 3$.

6. If $4x^2 + 9y^2 = 72$, find $\frac{dy}{dt}$ when $\frac{dx}{dt} = 2$, $x = 3$, and $y = -2$.

7. If $x^2y = 12$, find $\frac{dy}{dt}$ when $\frac{dx}{dt} = -1$, $x = 2$, and $y = 3$.

8. If $2x^2 - y^2 = 1$, find $\frac{dy}{dt}$ when $\frac{dx}{dt} = 0.35$, $x = 5$, and $y = 7$.

9. A manufacturer's cost (in dollars) of producing x sewing machines per month is given by

$$C(x) = 0.00002x^3 - 0.048x^2 + 300x + 150{,}000.$$

If production is being reduced by 100 sewing machines per month, how fast is the cost changing when the level of production reaches 4000 sewing machines per month?

10. A manufacturer's cost (in dollars) of producing x calculators per month is given by

$$C(x) = 0.00001x^3 - 0.012x^2 + 20x + 10{,}000.$$

If production is being increased by 100 calculators per month, how fast is the manufacturer's cost changing when the level of production reaches 6000 calculators per month?

11. The rate of increase in a town's population is expected to grow by $\frac{1}{2}\%$ per year for the next ten years. If the town's population is given by

$$P = 200{,}000e^{100r^2 - .01},$$

where r is the yearly rate of population increase, how large will the yearly population increase be when the yearly rate of increase reaches 3%?

12. A town's population is decreasing at the rate of $\frac{1}{2}\%$ per year. If this trend continues, the population will be

$$P = 500{,}000e^{-100r^2 + .04}$$

when the yearly rate of decrease is r. How will the town's population be changing when the yearly rate of decrease reaches 4%?

13. The monthly demand for a certain lighting fixture is given by

$$D = 1200 - 150p,$$

where p denotes its selling price. If the price is increasing by \$0.20 per month, how fast is the demand changing?

14. The demand for a calculator is given by

$$D = \frac{32{,}000}{\sqrt{p}},$$

where p denotes its selling price in dollars. If the price is increasing by \$0.10 per year, how fast will the demand be changing when the price reaches \$16?

15. A manufacturer's revenue and cost functions are

$$R(x) = 5x \quad \text{and} \quad C(x) = 3x + 1600,$$

where x denotes the number of items produced. At what rate will the manufacturer's profit be changing if production increases by 10 units per week?

16. A manufacturer's revenue and cost functions are

$$R(x) = 12x \quad \text{and} \quad C(x) = 8x + 4400,$$

where x denotes the number of items produced. At what rate will the manufacturer's profit be changing if production increases by 500 units per month?

17. An American coffee distributor can sell approximately $10{,}000 - 1250p$ pounds of coffee each week at a price of p dollars per pound. Because of a severe frost in South America, the cost of imported coffee beans is increasing at the rate of \$0.10 per pound each week. If the distributor passes the entire cost increase along to his customers, at what rate will the distributor's revenue be changing when the price of coffee reaches \$5.00 per pound?

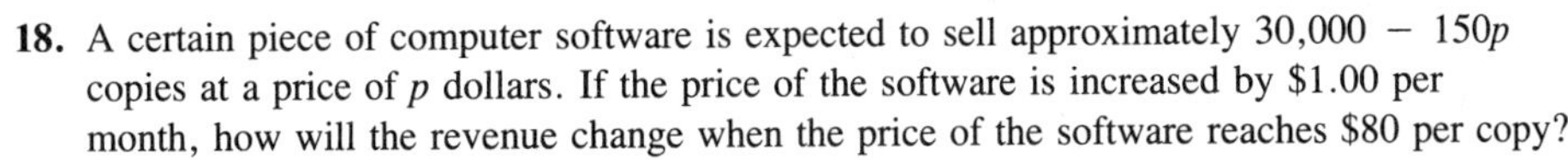

18. A certain piece of computer software is expected to sell approximately $30{,}000 - 150p$ copies at a price of p dollars. If the price of the software is increased by \$1.00 per month, how will the revenue change when the price of the software reaches \$80 per copy?

19. A manufacturer's production output P for a certain product is given by

$$P = 64x^{0.75}y^{0.25},$$

where x and y denote the number of units of labor and capital, respectively, required as production inputs. Suppose that the number of units of labor and capital required as production inputs are increasing by ¼ and ⅓ units per year, respectively. How fast will the production output be changing when 81 units of labor and 16 units of capital are required as inputs?

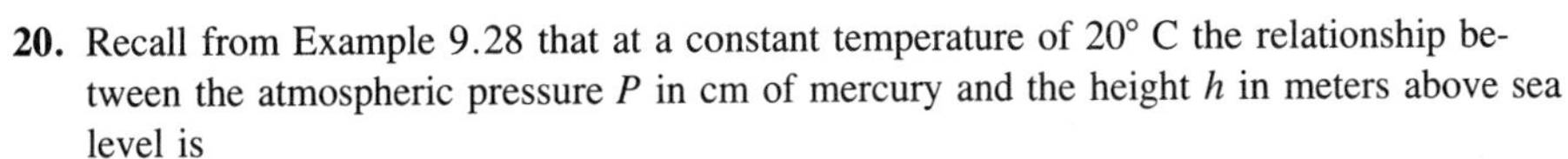

20. Recall from Example 9.28 that at a constant temperature of 20° C the relationship between the atmospheric pressure P in cm of mercury and the height h in meters above sea level is

$$h = 8600 \ln \frac{76}{P}.$$

If the height of an airplane is changing by 200 meters per minute, how fast is the barometric pressure on the plane changing when the pressure is 34.4 cm of mercury?

21. At noon a ship sailed east from a Caribbean port at 32 knots per hour. One hour later another ship left this port sailing north at 30 knots per hour. How fast will the distance between the ships be increasing at 5 P.M.?

22. A child's kite was flying 90 feet above the ground when it got caught in a wind gust that blew the kite horizontally at a rate of 15 feet per second. How fast will the child be letting out kitestring when there are 150 feet of string already out? (Assume that the kitestring is straight.)

23. A metal cube expands uniformly when heated. If the length of a side is increasing at the rate of 0.2 inches per hour, how fast is the volume of the cube changing when the length of a side is 10 inches?

24. A metal disk with a 9-inch radius expands when heated. If its radius increases at the rate of 0.01 inch per second, how fast is the area of its top face increasing when the radius is 9.1 inches?

25. A man 6 feet tall is walking away from a lamppost at a constant speed of 5 feet per second. If the light at the top of the lamppost is 18 feet above the sidewalk, how fast is the length of the man's shadow increasing?

26. Two spotlights, each 60 feet high, are located on poles 100 feet apart. One of the lights is working, and the other is not. A workman repairing the second light lowers his toolkit from the top of the second pole at the rate of 5 feet per second. How fast is the shadow of the toolkit moving when it is 30 feet above the ground?

27. Water is being pumped into the hemispherical tank shown below at the rate of 25π cubic feet per minute. How fast is the water level rising when the water is 5 feet deep at the center of the tank? (The volume of the shaded part of the figure is $V = 10\pi h^2 - \frac{1}{3}\pi h^3$.)

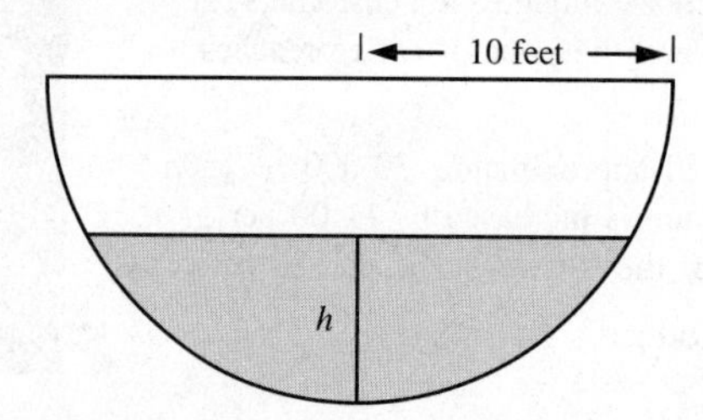

28. A swimming pool has a rectangular surface 30 meters long and 20 meters wide. Its bottom is a slanting plane that is 5 meters deep at one end and 1 meter deep at the other. (See the figure below for a side view of the pool.) If water is pumped into the pool at the rate of 225 cubic meters per hour, how fast will the water level be rising after 1 hour and 20 minutes?

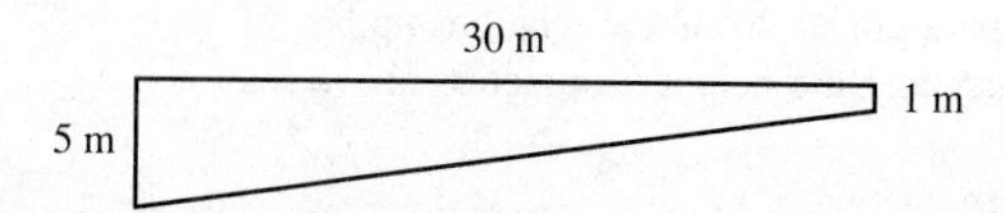

29. A farmer wants to raise a bale of hay into the loft of her barn. To do so, she has run a 59-foot rope through a pulley attached to the side of the barn at a height of 25 feet above the ground. One end of the rope is attached to the hay and the other end to the hitch of a tractor 1 foot above the ground. If the farmer drives the tractor away from the barn at a speed of 10 feet per second, how fast will the hay be rising when the tractor is 32 feet from the barn?

Answers to Practice Problems **13.** $\dfrac{dy}{dt} = 10$

14. $12{,}000\pi$ square feet per month

CHAPTER 10 REVIEW

IMPORTANT TERMS

- **second derivative** *(10.2)*
- **explicit function** *(10.3)*
- **implicit differentiation**
- **implicit function**
- **related rates** *(10.4)*

IMPORTANT FORMULAS

Derivative of the Product of Functions

If $f = gh$, then $f' = gh' + hg'$.

Derivative of the Quotient of Functions

If $f = \frac{g}{h}$, then $f' = \frac{hg' - gh'}{h^2}$.

REVIEW EXERCISES

In Exercises 1–4 differentiate the given functions.

1. $f(x) = (6 - x)^{3/2}(5x^4 - 3x^2 + 7)$

2. $F(x) = \frac{2x^2 - 1}{4x - 5}$

3. $g(x) = \ln \frac{2x + 1}{3x - 2}$

4. $G(x) = (3x^2 - 4x)e^{-2x}$

5. The cost of removing $x\%$ of the chemical contaminants from waste water discharge is approximately

$$C(x) = \frac{30x}{110 - x}$$

million dollars. How fast is the cost changing when 90% of the contaminants have been removed?

6. Find the equation of the tangent line to the graph of $y = (x + 2)e^x$ at the point (0, 2).

7. Find the equation of the tangent line to the graph of $y = \frac{\ln x}{2x - 1}$ at the point (1, 0).

8. Evaluate $\frac{d^2y}{dx^2}$ if $y = x \ln (3x + 1)$.

9. Evaluate $\frac{d^3y}{dx^3}$ if $y = x^3e^{-x}$.

10. Compute y' if $4y^2 - 2x^2 = 16$.

11. Write the equation of the tangent line to the graph of $10x^2 = x^2y + y^3$ at the point $(\frac{1}{3}, 1)$.

12. A California farmer can sell approximately $6000 - 4000p$ heads of lettuce per week at a price of p dollars apiece. In order to combat an outbreak of whiteflies, the farmer is being forced to raise the price of a head of lettuce by \$0.05 per week. At what rate will the farmer's revenue be changing when the price of lettuce is \$1.00 per head?

13. A company manufactures portable radios which it sells for \$30 apiece. Its monthly costs are given by

$$C = 0.0002x^2 + 4x + 50{,}000,$$

where x denotes the number of radios made per month. If production is increased by 100 per month, at what rate will the company's profit be changing when production reaches 10,000 per month?

11

APPLICATIONS OF THE DERIVATIVE

In this chapter additional applications of differentiation will be considered. One of the most important of these applications is to the graphing of functions, and it is this application with which we begin the chapter. From these considerations will come methods for determining maxima and minima of functions that have important practical applications.

11.1 Some Features of Graphs

In this section we will discuss certain important features of the graph of a function. In Sections 11.2 and 11.3 we will learn how to discover their presence by use of differentiation.

We are interested in features of a graph that reveal the behavior of a function. For example, under certain circumstances the rate at which a new product sells can be described by a function having the graph shown in Figure 11.1. From this graph we see that the rate of sales for the new product is small at first, grows to a maximum at time t, then decreases rapidly for a time, and finally diminishes to zero.

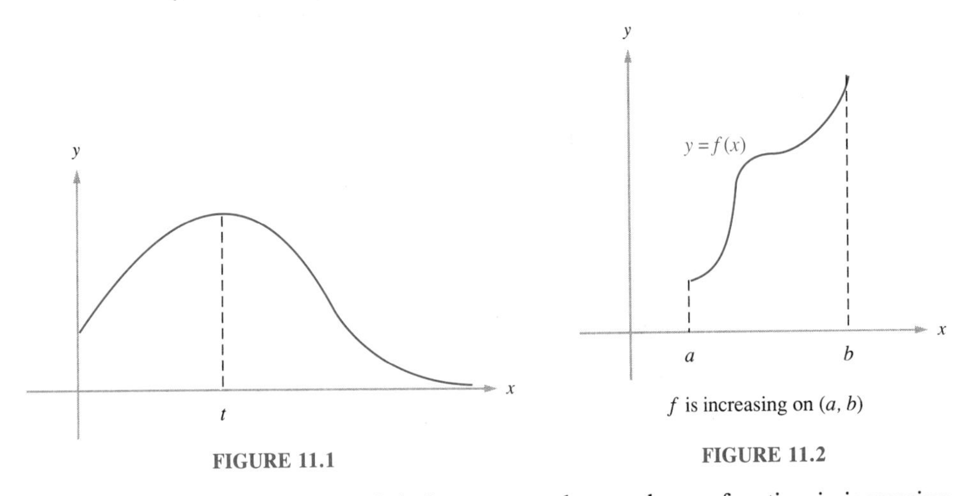

FIGURE 11.1

FIGURE 11.2

In many situations it is important to know where a function is increasing and where it is decreasing. For instance, if the manufacturer of a new product knew when its rate of sales would be increasing and when it would be decreasing, then the number of units produced could be adjusted in accordance with the expected sales. This would prevent an undersupply of the product that would result in lost sales or an oversupply that might require increased storage costs.

A function f is said to be **increasing** on an interval if $f(s) > f(r)$ whenever r and s are numbers in the interval such that $s > r$. Thus for f to be increasing on an interval means that the functional values of f get larger as we move from left to right through the interval. See Figure 11.2.

Likewise f is said to be **decreasing** on an interval if $f(s) < f(r)$ whenever r and s are numbers in the interval such that $s > r$. Thus for f to be decreasing on an interval means that the functional values of f get smaller as we move from left to right through the interval. See Figure 11.3. Hence in Figure 11.1, f is increasing on the interval $(0, t)$ and decreasing on the interval (t, ∞).

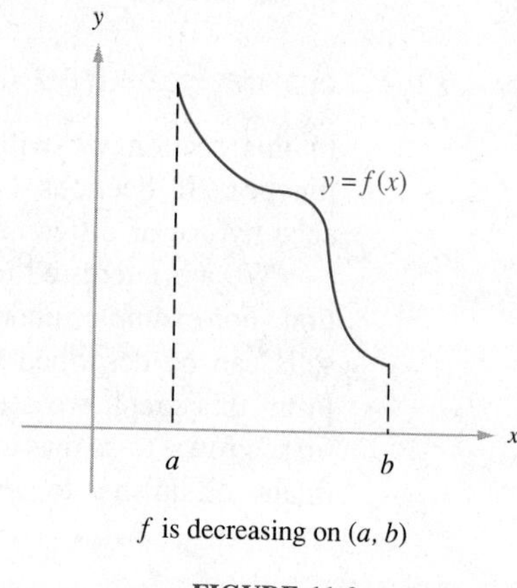

f is decreasing on (a, b)

FIGURE 11.3

EXAMPLE 11.1 For the function f having the graph in Figure 11.4, determine the intervals where f is increasing and the intervals where f is decreasing.

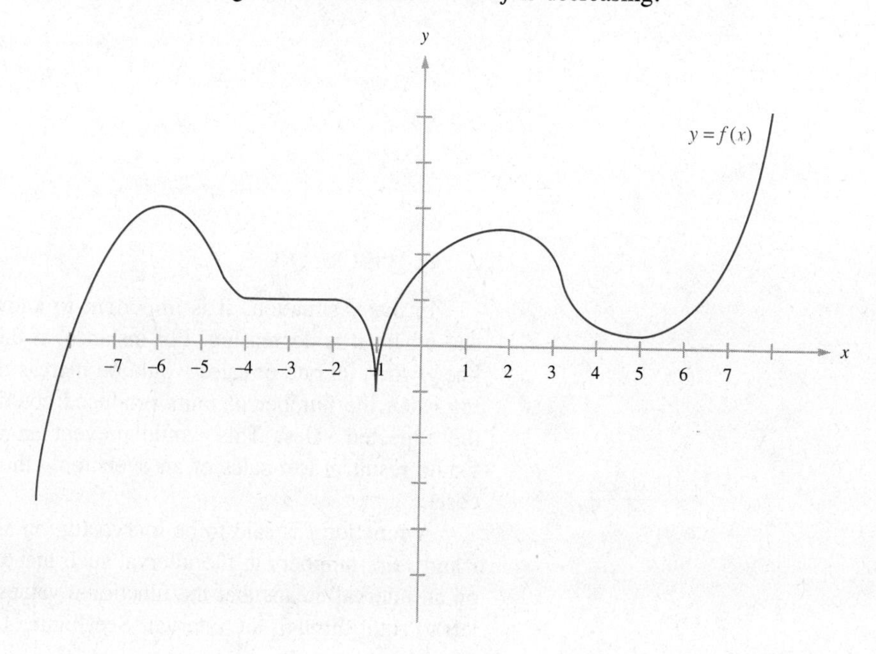

FIGURE 11.4

Solution As we move from left to right, the functional values $f(x)$:

get larger until we reach $x = -6$;
get smaller between $x = -6$ and $x = -4$;
remain constant (neither increase nor decrease) between $x = -4$ and $x = -2$;
get smaller between $x = -2$ and $x = -1$;
get larger between $x = -1$ and $x = 2$;
get smaller between $x = 2$ and $x = 5$; and
get larger from $x = 5$ on.

Hence f is increasing on the intervals $(-\infty, -6)$, $(-1, 2)$, and $(5, \infty)$; f is decreasing on the intervals $(-6, -4)$, $(-2, -1)$, and $(2, 5)$; and f is constant on the interval $(-4, -2)$. ■

Practice Problem 1 For the function g having the graph in Figure 11.5, determine the intervals where g is increasing and the intervals where g is decreasing.

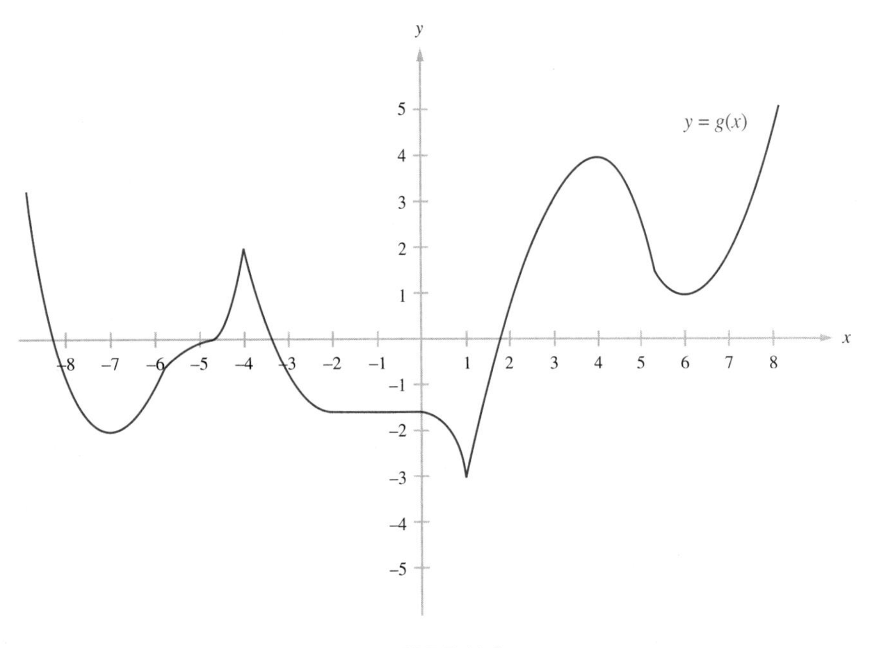

FIGURE 11.5

EXTREMA

Consider the function g pictured in Figure 11.6. In an open interval around $x = -6$, the largest value of $g(x)$ is 6. In this case we say that g has a *local maximum* of 6 at $x = -6$. Likewise g has a local maximum of 7 at $x = 2$

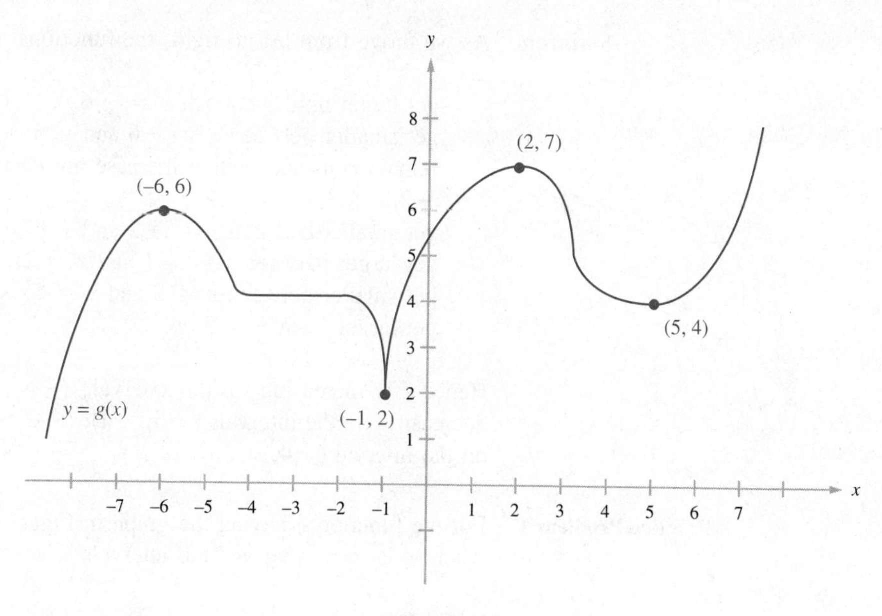

FIGURE 11.6

because 7 is the largest value of g on some open interval around $x = 2$. Notice that *a local maximum of g is a value of y,* whereas we say that the local maximum *occurs* at a value of x.

In general $f(a)$ is a **local** (or **relative**) **maximum** of f if

$$f(x) \leq f(a)$$

for all x in some open interval containing a. In this case we say that the local maximum **occurs** at $x = a$. Hence for a local maximum of f to occur at $x = a$ means that $f(a)$ is the largest value of $f(x)$ on some open interval containing a.

Similarly, in Figure 11.6 we say that g has a local minimum of 2 at $x = -1$ because the smallest value of $g(x)$ is 2 on some open interval around $x = -1$. The number 4 is another local minimum of g; it occurs at $x = 5$.

More generally, $f(a)$ is called a **local** (or **relative**) **minimum** of f if

$$f(x) \geq f(a)$$

for all x in some open interval containing a. Again in this case we say that the local minimum **occurs** at $x = a$. Hence for a local minimum of f to occur at $x = a$ means that $f(a)$ is the smallest value of $f(x)$ on some open interval containing a. A number that is either a local maximum or a local minimum is called a **local** (or **relative**) **extremum.** (The plurals of maximum, minimum, and extremum are *maxima, minima,* and *extrema.*)

Practice Problem 2 For the function g shown in Figure 11.5, determine the local extrema and the numbers at which they occur.

CONCAVITY

Although the graphs of the functions f and g in Figures 11.7(a) and (b) are both increasing on the interval (a, b), they are fundamentally different. Notice that at each point in the interval, the graph of f lies entirely above the tangent to the graph at that point. When this occurs, we say that f is **concave up** on the interval. (See Figure 11.8(a).) On the other hand, at each point in the interval (a, b) the graph of g in Figure 11.7(b) lies entirely below the tangent to the graph at that point. In this case, we say that g is **concave down** on the interval. (See Figure 11.8(b).)

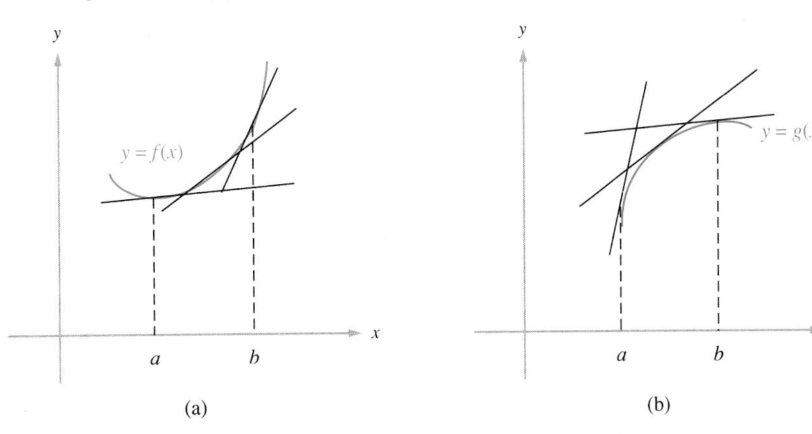

FIGURE 11.7

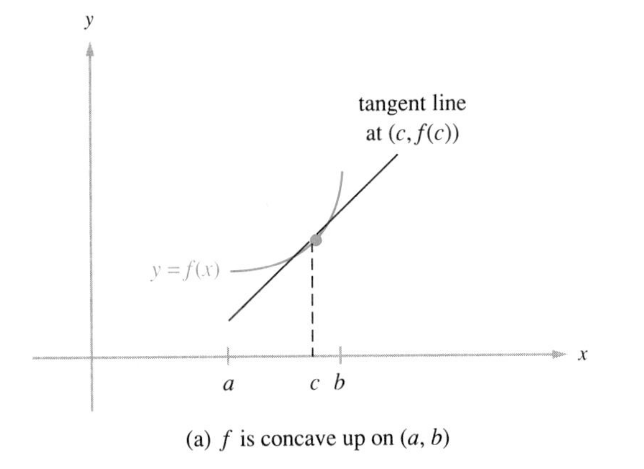

(a) f is concave up on (a, b)

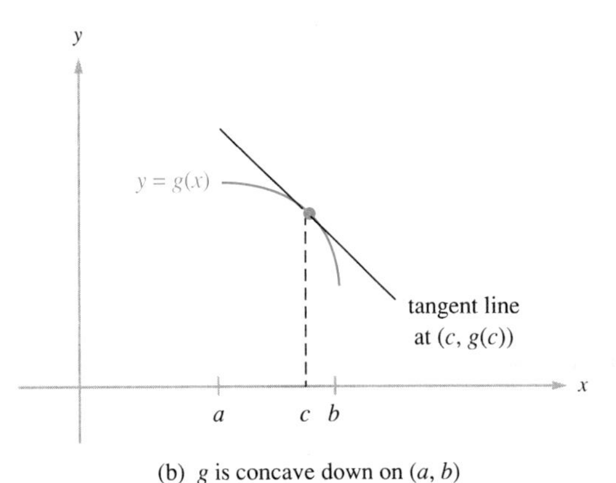

(b) g is concave down on (a, b)

FIGURE 11.8

EXAMPLE 11.2 For the function shown in Figure 11.9, determine the intervals where the function is concave up and the intervals where it is concave down.

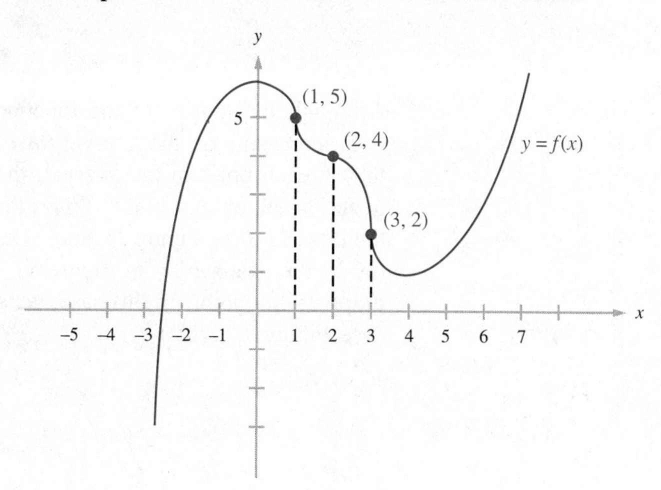

FIGURE 11.9

Solution The graph of f lies above all its tangent lines on the intervals (1, 2) and $(3, \infty)$ and below all its tangent lines on the intervals $(-\infty, 1)$ and (2, 3). Therefore f is concave up on the intervals (1, 2) and $(3, \infty)$, and f is concave down on the intervals $(-\infty, 1)$ and (2, 3). ■

Practice Problem 3 For the function g shown in Figure 11.10, determine the intervals where g is concave up and the intervals where it is concave down.

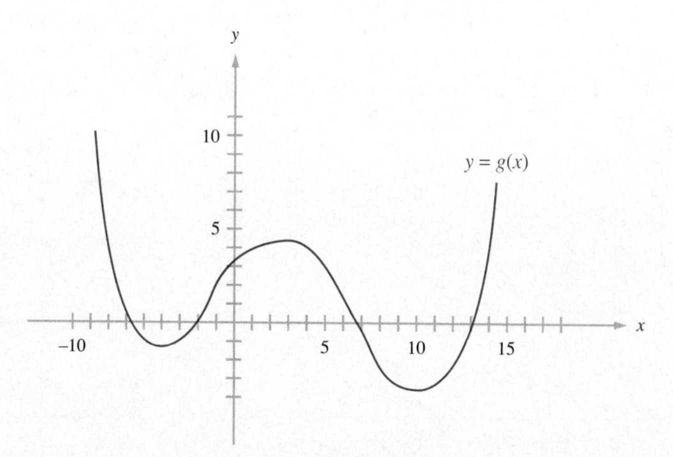

FIGURE 11.10

In Figure 11.9 f is concave down on $(-\infty, 1)$ and concave up on (1, 2). Thus the concavity of f changes at $x = 1$. A point on the graph of a function at which the concavity changes is called a **point of inflection.** (See Figure

11.11.) If the tangent line exists at a point of inflection, the graph of f must cross it. In Figure 11.9 the three labeled points, namely (1, 5), (2, 4), and (3, 2), are points of inflection.

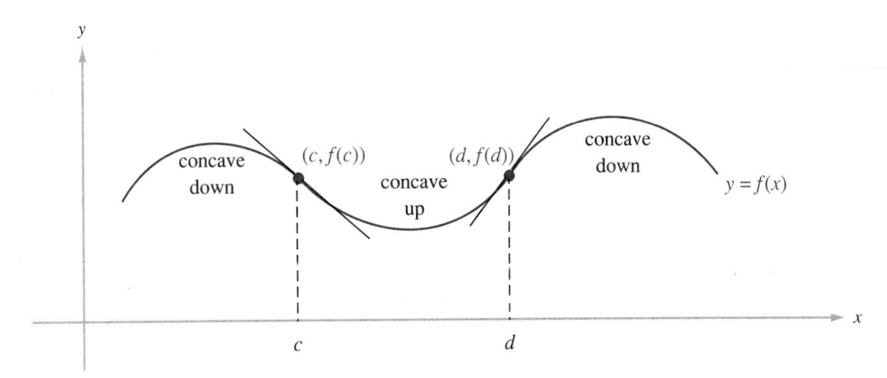

$(c, f(c))$ and $(d, f(d))$ are points of inflection

FIGURE 11.11

Practice Problem 4 Find all the points of inflection for the function g shown in Figure 11.10.

SOLVING INEQUALITIES

In Section 11.2 we will see that determining where f is increasing depends on being able to solve inequalities of the form

$$h(x) > 0 \qquad \text{and} \qquad h(x) < 0.$$

We will now consider a method for solving such inequalities that can be used whenever it can be determined at which numbers x the function h is not continuous or $h(x) = 0$.

Let g be a continuous function. Recall that this implies that the graph of g has no jumps or breaks. Thus if $g(a) > 0$ and $g(b) < 0$, there must be at least one number c between a and b for which $g(c) = 0$. (See Figure 11.12.) That is, if the graph between the points $(a, g(a))$ and $(b, g(b))$ can be traced without

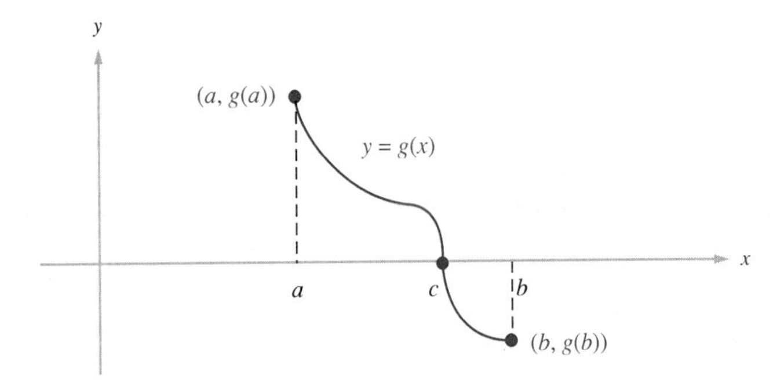

FIGURE 11.12

lifting a pencil from the paper, the curve must cross the x-axis between a and b. A similar conclusion is true if $g(a) < 0$ and $g(b) > 0$. It follows that *if there is an interval containing no number c such that $g(c) = 0$, then g must be positive or negative throughout the interval.*

Let us now consider the general problem of solving the inequalities

$$h(x) > 0 \qquad \text{and} \qquad h(x) < 0$$

for any function h. The discussion above shows that the intervals throughout which h is positive or negative must be separated by numbers x such that

1. h is not continuous at x or
2. $h(x) = 0$.

Consider, for example, the function

$$h(x) = 3x^2 + 6x - 24.$$

Since h is a polynomial function, there are no numbers at which h is not continuous. There are, however, two numbers for which $h(x)$ equals 0, namely, $x = -4$ and $x = 2$. (These numbers can be found by solving the equation

$$3x^2 + 6x - 24 = 0$$

using factoring or the quadratic formula.) Locate these numbers on a number line as in Figure 11.13.

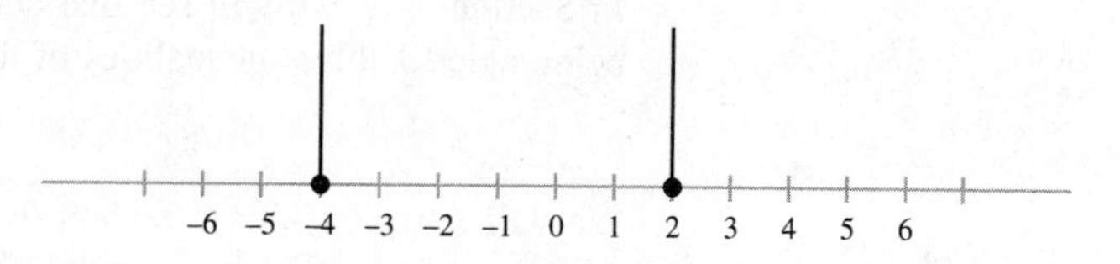

FIGURE 11.13

These two numbers divide the number line into three intervals: $(-\infty, -4)$, $(-4, 2)$, and $(2, \infty)$. Since h is continuous and nonzero on these intervals, it cannot change sign on any of them. To determine the sign of h throughout one of these intervals, we need only evaluate h at any number in it. Choose, for instance, the number -6 in the interval $(-\infty, -4)$. Since

$$h(-6) = 3(-6)^2 + 6(-6) - 24 = 48 > 0,$$

we conclude that h is positive throughout the interval $(-\infty, -4)$. In the interval $(-4, 2)$, choose 0 as the test number. Since

$$h(0) = 3(0)^2 + 6(0) - 24 = -24 < 0,$$

h is negative throughout the interval $(-4, 2)$. In the interval $(2, \infty)$, choose 3 as the test number. Because

$$h(3) = 3(3)^2 + 6(3) - 24 = 21 > 0,$$

h must be positive throughout the interval $(2, \infty)$. We can record all this information on our number line as in Figure 11.14.

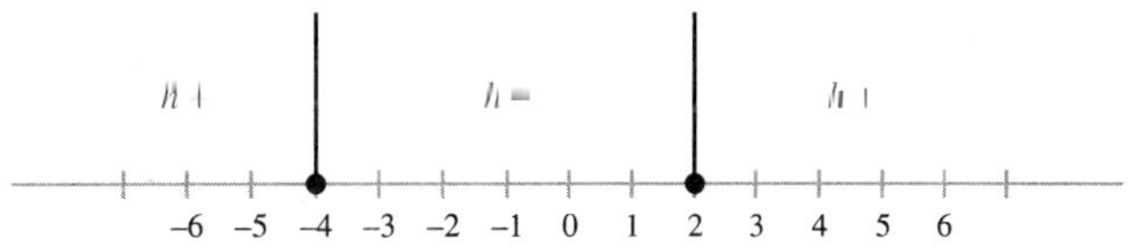

$h +$ | $h -$ | $h +$

−6 −5 −4 −3 −2 −1 0 1 2 3 4 5 6

FIGURE 11.14

The process for determining the intervals where $h(x) > 0$ or $h(x) < 0$ is summarized below.

Procedure for Solving $h(x) > 0$ or $h(x) < 0$

1. Find all the numbers x for which h is not continuous or $h(x) = 0$.
2. Plot the numbers found in step 1 on a number line, thereby dividing the line into intervals.
3. Evaluate h at a test number t in each of the intervals formed in step 2.
4. If $h(t) > 0$, mark the interval containing t with $+$; otherwise mark the interval with $-$.
5. Then $h(x) > 0$ throughout the intervals marked $+$, and $h(x) < 0$ throughout the intervals marked $-$.

EXAMPLE 11.3 Determine the solution of the inequality $g(x) > 0$ if

$$g(x) = \frac{x^2 + 4x - 5}{(x + 2)^2}.$$

Solution Recall that:

1. a rational function equals zero when its numerator equals zero, and
2. a rational function is not continuous when its denominator is zero.

Hence g is not continuous at -2, where the denominator is 0. Also if $g(x) = 0$, the numerator must be 0; that is

$$x^2 + 4x - 5 = 0.$$

Thus, by the quadratic formula, the numbers for which $g(x) = 0$ are $x = -5$ and $x = 1$.

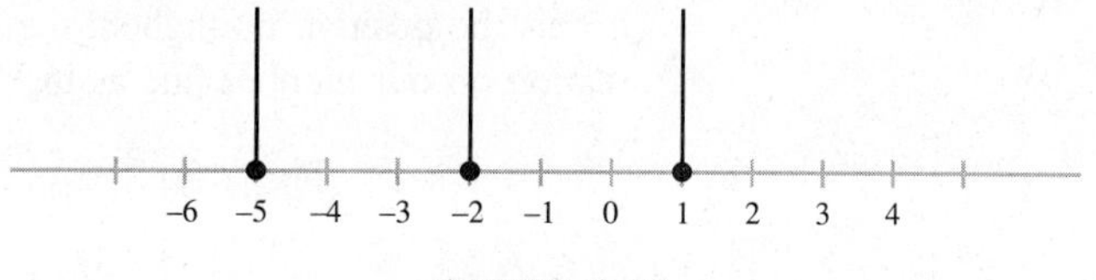

FIGURE 11.15

Hence we plot -2, -5, and 1 on a number line as in Figure 11.15. Evaluating a test number in each interval, we obtain

$$g(-6) = \frac{(-6)^2 + 4(-6) - 5}{[(-6) + 2]^2} = \frac{7}{16} > 0,$$

$$g(-3) = \frac{(-3)^2 + 4(-3) - 5}{[(-3) + 2]^2} = \frac{-8}{1} < 0,$$

$$g(0) = \frac{(0)^2 + 4(0) - 5}{(0 + 2)^2} = \frac{-5}{4} < 0, \qquad \text{and}$$

$$g(2) = \frac{(2)^2 + 4(2) - 5}{(2 + 2)^2} = \frac{7}{16} > 0.$$

Recording these results on our number line produces Figure 11.16. Hence g is positive on the intervals $(-\infty, -5)$ and $(1, \infty)$ and negative on the intervals $(-5, -2)$ and $(-2, 1)$. So the solution of the inequality $g(x) > 0$ is the union of the two intervals $(-\infty, -5)$ and $(1, \infty)$. ■

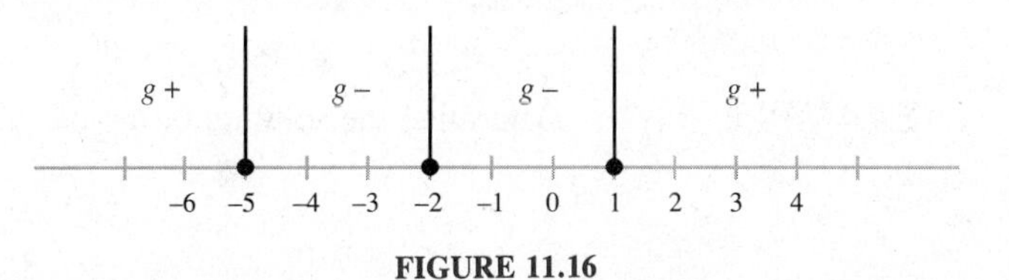

FIGURE 11.16

Practice Problem 5 Determine the solution of the inequality $f(x) < 0$ if

$$f(x) = -x^2 + 8x - 15.$$

EXERCISES 11.1

In Exercises 1–16 the graph of a function f is given. Identify the intervals on which f is increasing, decreasing, concave up, and concave down. Then determine any: (a) local extrema for f and the values of x at which they occur, (b) points of inflection.

1.

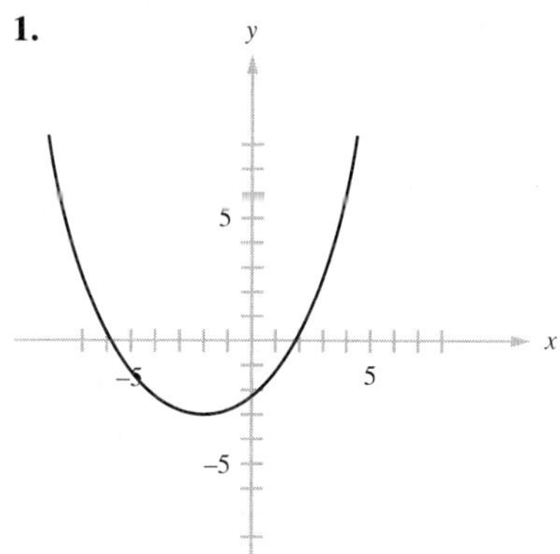

2.

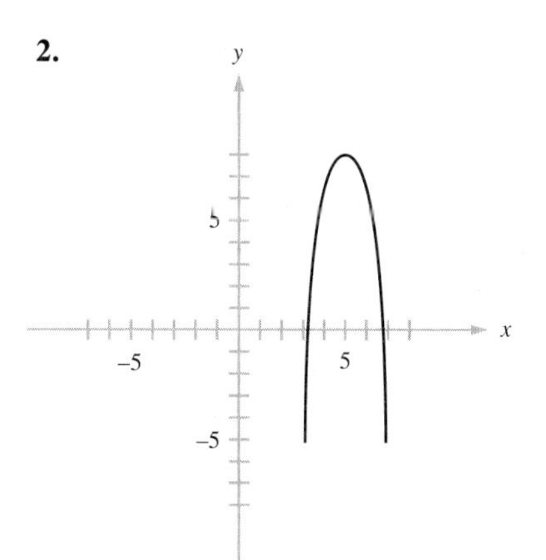

3.

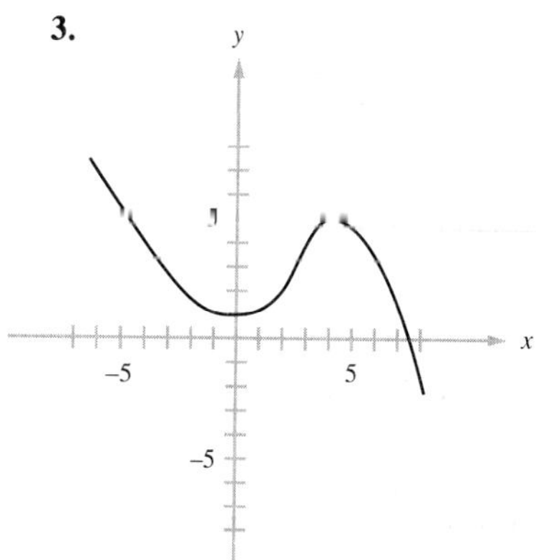

4.

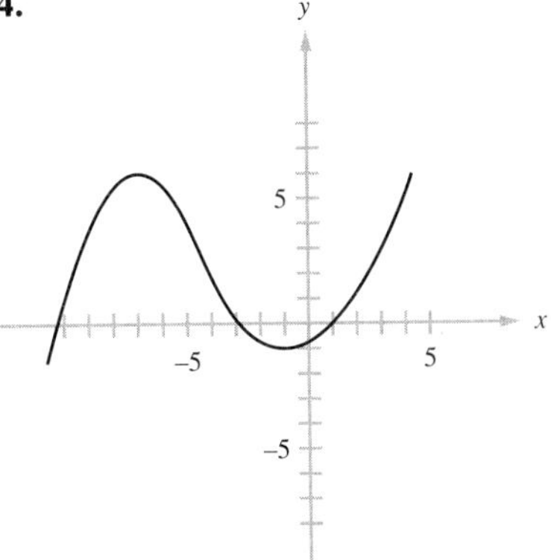

5.

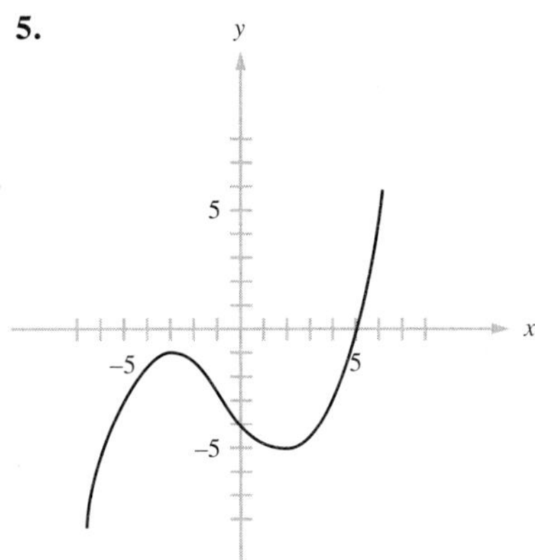

6.

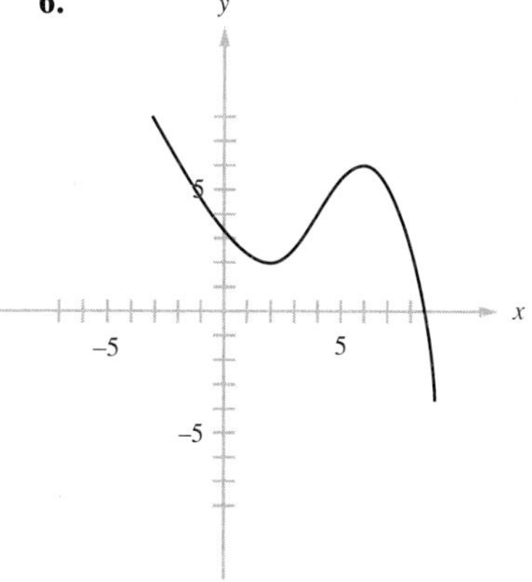

7.

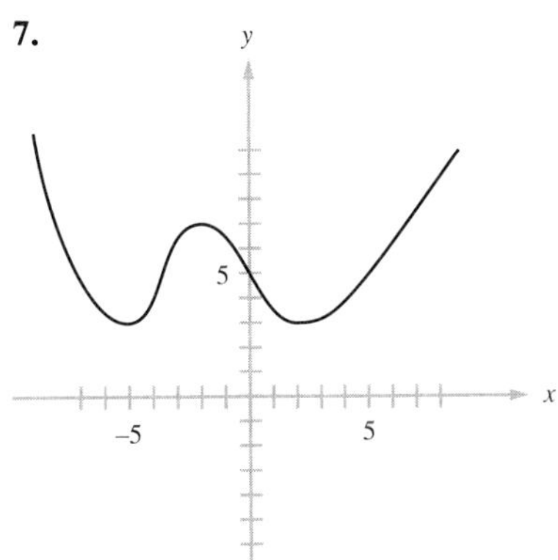

8.

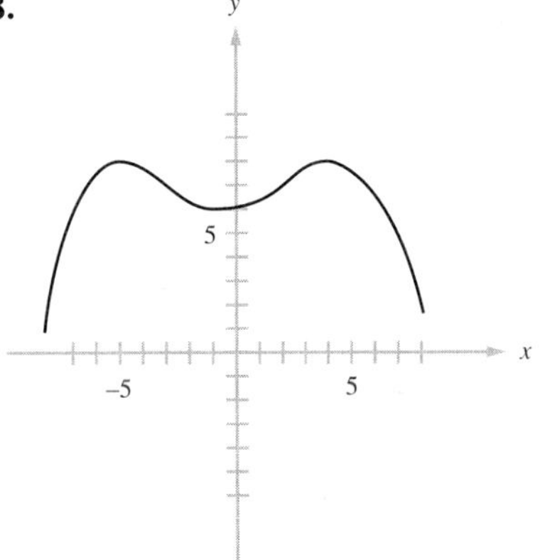

9.

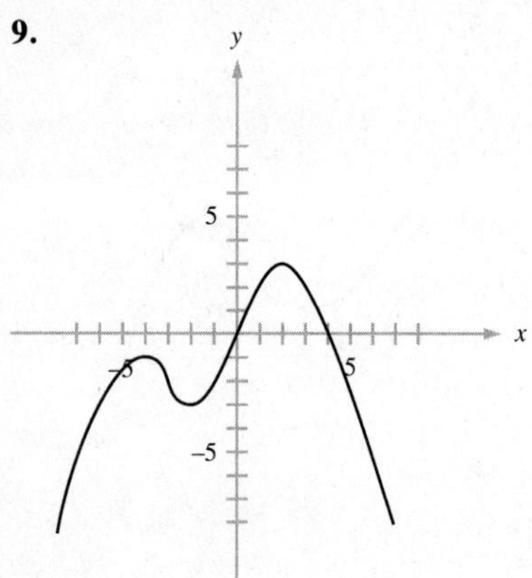

10.

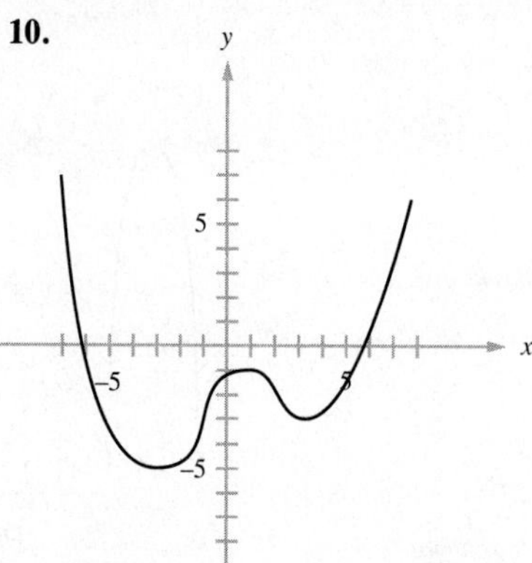

11.

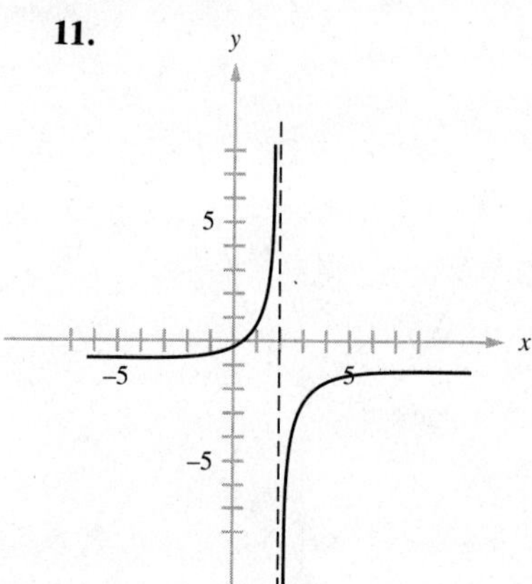

12.

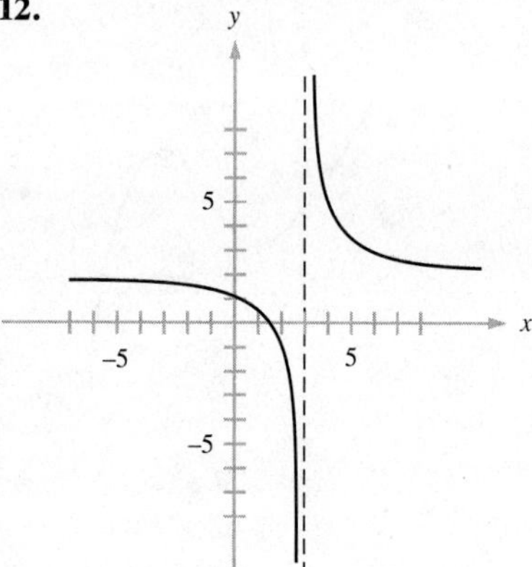

13.

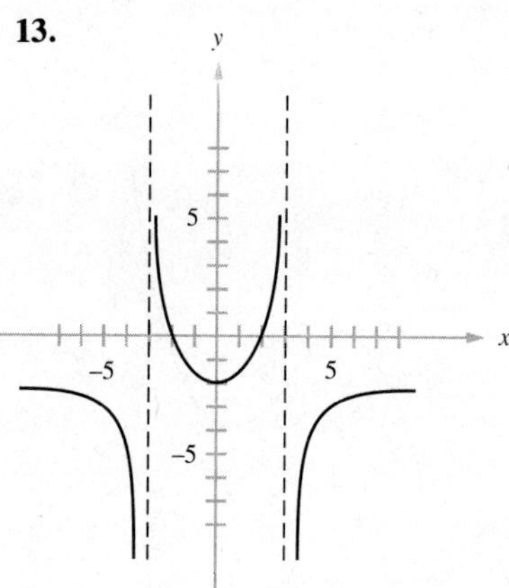

14.

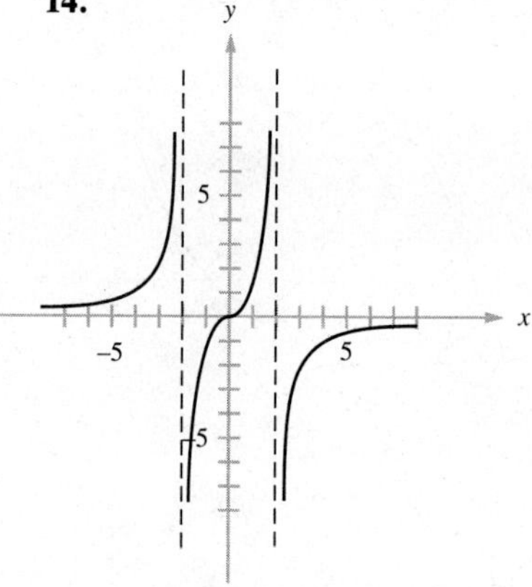

15.

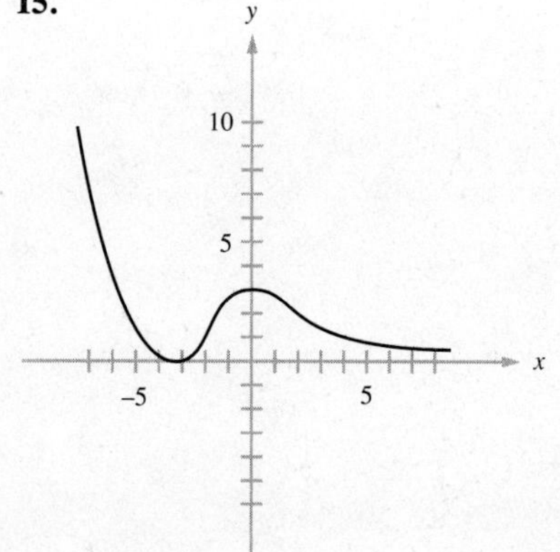

16.

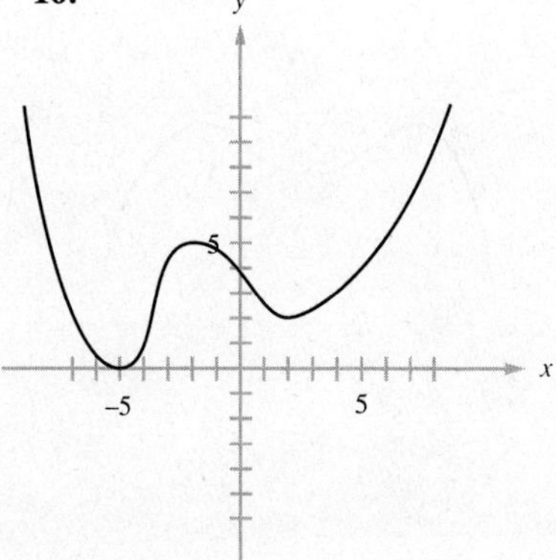

In Exercises 17–32 solve the given inequality using the technique described in this section.

17. $(x - 5)(x + 6) > 0$

18. $(x + 3)(x - 7) < 0$

19. $\dfrac{x + 2}{x - 1} < 0$

20. $\dfrac{x - 4}{x + 5} > 0$

21. $\dfrac{x - 6}{(x + 3)^2} > 0$

22. $\dfrac{x + 2}{(x - 8)(x - 3)} < 0$

23. $\dfrac{(x + 1)(x + 7)}{x - 4} < 0$

24. $\dfrac{(x + 4)(x - 8)}{(x - 1)^2} > 0$

25. $\dfrac{x^2 + x - 12}{x + 1} > 0$

26. $\dfrac{x^2 - 9}{x} < 0$

27. $\dfrac{x + 7}{x^2 - 3x - 10} < 0$

28. $\dfrac{x - 5}{x^2 + 2x - 3} > 0$

29. $\dfrac{x^2 + 3x + 2}{x^2 - 4x + 4} > 0$

30. $\dfrac{x^2 + 2x + 1}{x^2 - x - 20} < 0$

31. $\dfrac{x^2 + 7x + 10}{x^2 - 5x + 4} < 0$

32. $\dfrac{x^2 - 3x - 18}{x^2 + 4x - 5} > 0$

Answers to Practice Problems

1. On the intervals $(-\infty, -7)$, $(-4, -2)$, $(0, 1)$ and $(4, 6)$, g is decreasing; on the intervals $(-7, -4)$, $(1, 4)$, and $(6, \infty)$, g is increasing; and on the interval $(-2, 0)$, g is constant.
2. There are local maxima of g at $x = -4$ and $x = 4$ and local mimima of g at $x = -7$, $x = 1$, and $x = 6$. Thus $g(-4) = 2$ and $g(4) = 4$ are local maxima, and $g(-7) = -2$, $g(1) = -3$, and $g(6) = 1$ are local minima.
3. On $(-\infty, -1)$ and $(6, \infty)$, g is concave up; and on $(-1, 6)$, g is concave down.
4. There are two points of inflection, $(-2, 2)$ and $(6, 2)$.
5. For x in $(-\infty, 3)$ or $(5, \infty)$, $f(x) < 0$.

11.2 Using the First Derivative in Graphing

Our primary aim in this section is to use the derivative of a function to determine where that function is increasing and decreasing without first knowing its graph.

Because the tangent line to a graph is the limiting position of secant lines, the tangent line reflects the direction of the graph near a point. More specifically, if $f'(c) > 0$, then the tangent line at the point $(c, f(c))$ has positive slope as in Figure 11.17. Thus f must be increasing on some interval around c. Likewise if the tangent line at the point $(c, f(c))$ has negative slope as in Figure 11.18, then f must be decreasing on some interval around c.

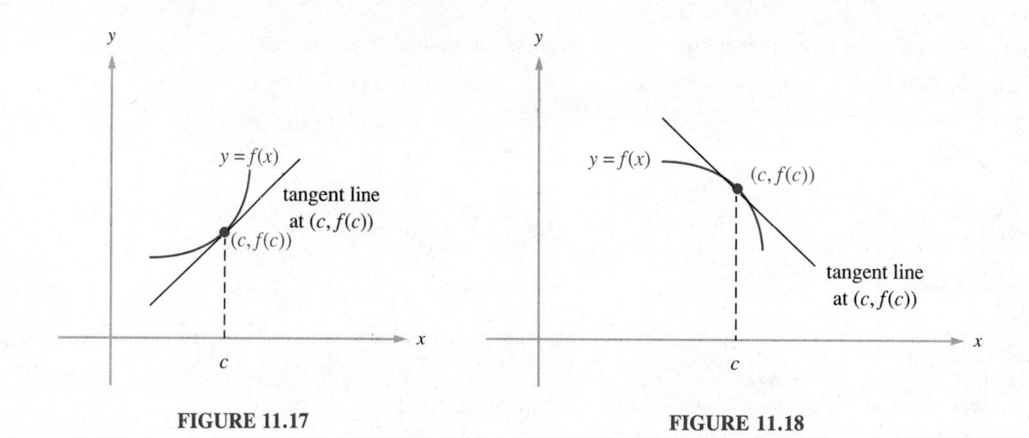

FIGURE 11.17 FIGURE 11.18

Let f be differentiable on an interval.

(a) If f' is positive throughout the interval, then f is increasing on the interval.

(b) If f' is negative throughout the interval, then f is decreasing on the interval.

EXAMPLE 11.4 Determine the intervals where f is increasing and the intervals where f is decreasing if $f(x) = x^2 - x - 6$.

Solution To determine where f is increasing and where it is decreasing, we examine the *derivative* of f. Now

$$f'(x) = 2x - 1.$$

Proceeding as in Section 11.1, we obtain Figure 11.19.

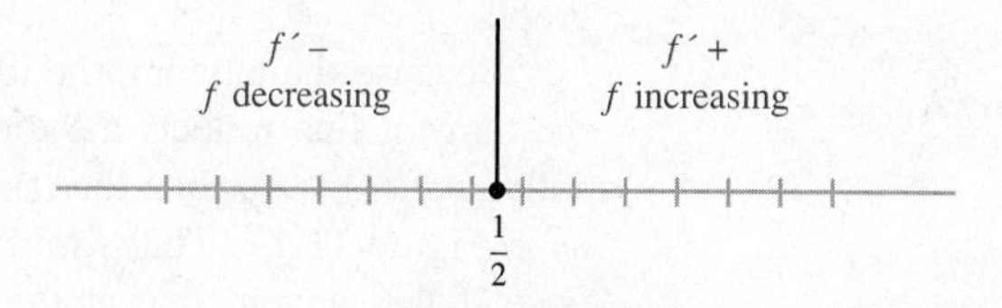

FIGURE 11.19

Since f' is positive for $x > \frac{1}{2}$, that is, on the interval $(\frac{1}{2}, \infty)$, f is increasing there. Similarly f' is negative if $x < \frac{1}{2}$, and so f is decreasing for $x < \frac{1}{2}$, that is, on the interval $(-\infty, \frac{1}{2})$. The graphs of f and f' are shown in Figure 11.20. ■

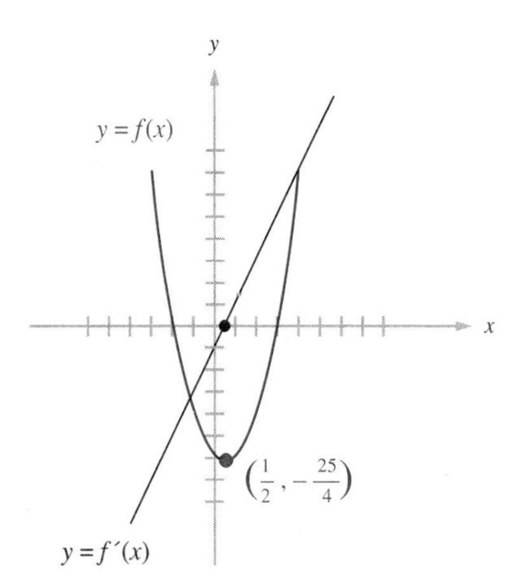

FIGURE 11.20

EXAMPLE 11.5 Determine the intervals where f is increasing and the intervals where f is decreasing if $f(x) = e^{-x^2/2}$.

Solution As above, we will examine the derivative of f, which is

$$f'(x) = -xe^{-x^2/2}.$$

Recall from Section 8.4 that an exponential function is defined for all real numbers and its values are always positive. Therefore

$$e^{-x^2/2} > 0$$

for all x, and so $f'(x)$ is positive or negative according to whether $-x$ is positive or negative. Applying the method of Section 11.1 to $h(x) = -x$ yields Figure 11.21.

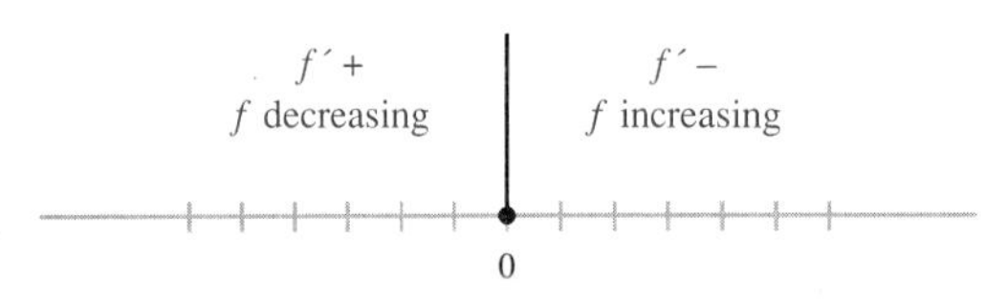

FIGURE 11.21

Thus

$$f'(x) > 0 \text{ if } x < 0 \qquad \text{and} \qquad f'(x) < 0 \text{ if } x > 0.$$

We conclude that f is increasing if $x < 0$ and f is decreasing if $x > 0$. ■

The function $f(x) = e^{-x^2/2}$ discussed in Example 11.5 is quite important. Recall from Section 7.5 that the equation of the normal curve with mean 0 and standard deviation 1 is

$$y = \frac{1}{\sqrt{2\pi}} e^{-x^2/2}$$

Thus the function f in Example 11.5 is a constant multiple of a normal distribution, and so its graph has the shape of a normal curve. See Figure 11.22. Notice again that f is increasing where $f'(x) > 0$, that is, where $x < 0$, and that f is decreasing where $f'(x) < 0$, that is, where $x > 0$.

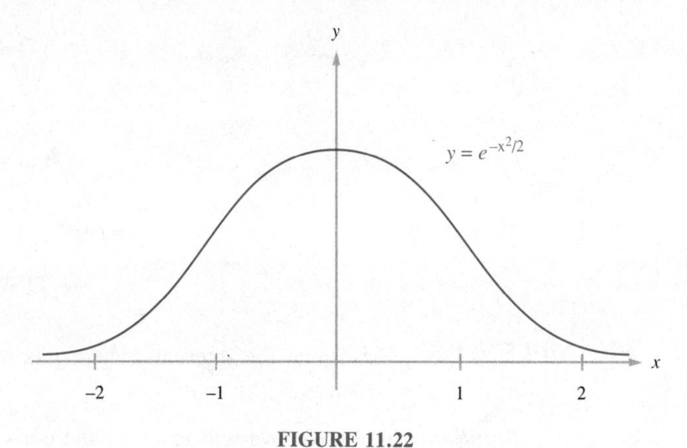

FIGURE 11.22

Practice Problem 6 Determine the intervals where f is increasing and the intervals where f is decreasing if $f(x) = 3 + 4x - x^2$.

EXAMPLE 11.6

Determine the intervals where f is increasing and the intervals where f is decreasing if

$$f(x) = \frac{x^2 + 5}{x + 2}.$$

Solution Since the intervals throughout which f is increasing or decreasing are determined by the intervals where its derivative is positive or negative, we must compute f'. By the quotient rule we find that

$$f'(x) = \frac{x^2 + 4x - 5}{(x + 2)^2}.$$

This is the function g considered in Example 11.3 in the previous section. We saw there that

$$f'(x) > 0 \text{ on the intervals } (-\infty, -5) \text{ and } (1, \infty)$$

and

$$f'(x) < 0 \text{ on the intervals } (-5, -2) \text{ and } (-2, 1).$$

Hence f is increasing on the intervals $(-\infty, -5)$ and $(1, \infty)$ and decreasing on the intervals $(-5, -2)$ and $(-2, 1)$.

The graph of f is shown in Figure 11.23. Observe that $f(x)$ is not defined for $x = -2$. (The line $x = -2$ is a vertical asymptote of the graph of f.) Therefore, even though f is decreasing on $(-5, -2)$ and $(-2, 1)$, it is not decreasing throughout $(-5, 1)$. ■

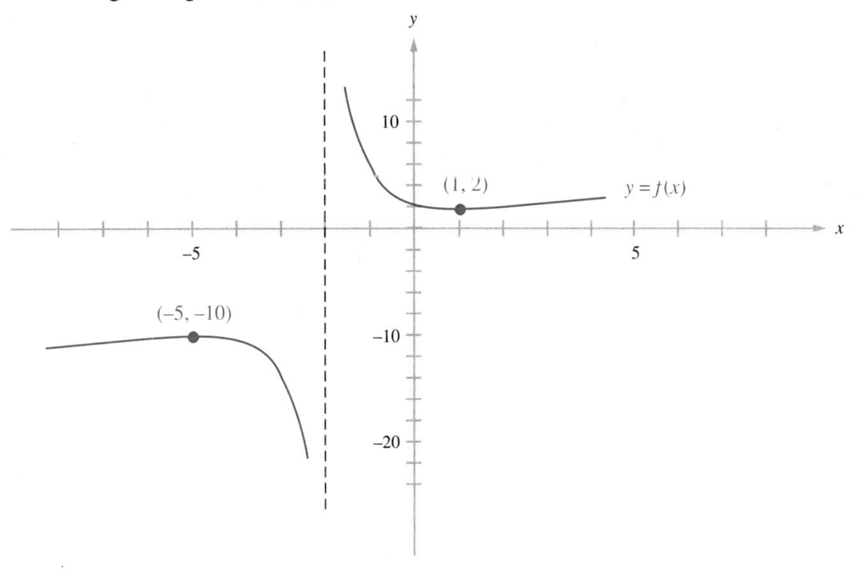

FIGURE 11.23

Practice Problem 7 Determine the intervals where f is increasing and the intervals where f is decreasing if $f(x) = \frac{1}{3}x^3 - x^2 - 15x$.

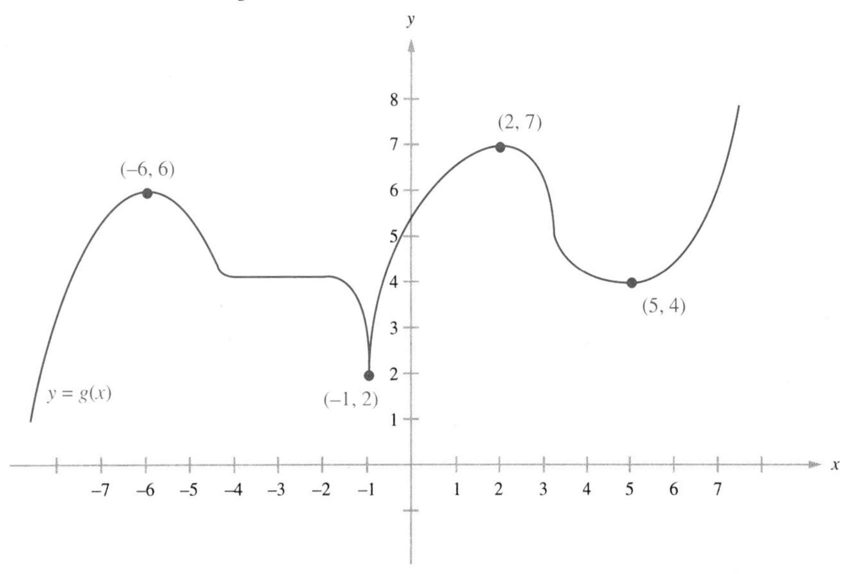

FIGURE 11.24

What can be said about the derivative of a function at a local extremum? At $x = -1$ in Figure 11.24, the graph of g has an abrupt change of direction. Therefore $g'(-1)$ does not exist. But at the other three local extrema of g, namely at the points $(-6, 6)$, $(2, 7)$, and $(5, 4)$, the tangent line to the graph of g is horizontal (has slope 0). Hence $g'(x) = 0$ for $x = -6$, $x = 2$, and $x = 5$. Indeed it is easy to see that this occurrence is true in general: If f has an extremum at $x = c$ and the graph of $y = f(x)$ has a tangent line at $x = c$, then $f'(c) = 0$. Thus we have the following important result, which is illustrated in Figure 11.25.

If a function f has a local extremum at $x = c$, then either

$$f'(c) = 0 \quad \text{or} \quad f'(c) \text{ is undefined.}$$

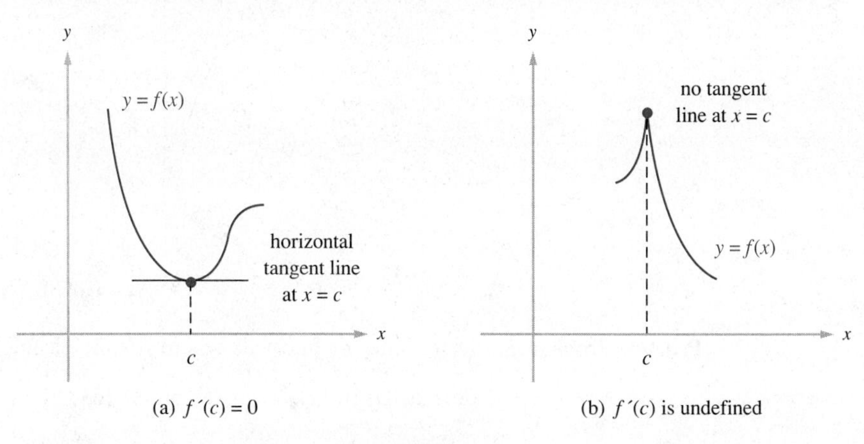

(a) $f'(c) = 0$ (b) $f'(c)$ is undefined

FIGURE 11.25

A number c for which f is defined is called a **critical number** of f if either $f'(c) = 0$ or $f'(c)$ is undefined. (These are the numbers we plot to determine the intervals where f is increasing or decreasing.) Thus the boxed statement above says that every local extremum of f must occur at a critical number. Notice, however, that the statement does *not* say that there is a local extremum at every critical number of f. The critical numbers are merely numbers at which extrema *may* occur.

EXAMPLE 11.7 Determine the critical numbers of $f(x) = x^3$.

Solution The critical numbers are the values for which $f'(x) = 0$ or $f'(x)$ does not exist. Since

$$f'(x) = 3x^2$$

is a polynomial function, $f'(x)$ is defined for all x. Thus the only critical numbers are those for which $f'(x) = 0$. Solving for x produces

$$3x^2 = 0$$
$$x^2 = 0$$
$$x = 0.$$

Thus 0 is the only critical number of f.

The graph of f is shown in Figure 11.26. Notice that f does not have a local extremum at $x = 0$ even though $f'(0) = 0$. In fact, f has no local extrema. ■

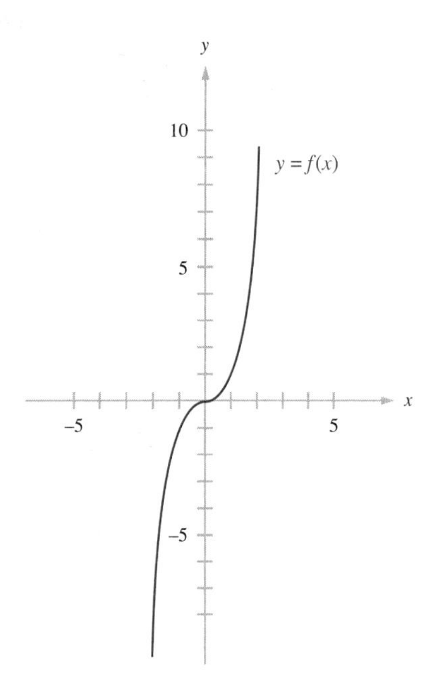

FIGURE 11.26

Practice Problem 8 Determine the critical numbers of the following functions.

(a) $f(x) = x^4 - 12x^3$ (b) $g(x) = \dfrac{x^2 + x + 3}{x - 2}$

THE FIRST-DERIVATIVE TEST

As we noted above, a particular critical number may not yield an extremum. Hence we need a method for determining whether a local maximum, a local minimum, or neither occurs at a critical number. One method is based on the following test.

The First-Derivative Test for Local Extrema

Let c be a critical number of f in the interval (a, b), and let f be differentiable on (a, b) except possibly at $x = c$.

(a) If $f'(x) > 0$ on (a, c) and $f'(x) < 0$ on (c, b), then f has a local maximum at $x = c$.

(b) If $f'(x) < 0$ on (a, c) and $f'(x) > 0$ on (c, b), then f has a local minimum at $x = c$.

It is easy to see why part (a) of the first-derivative test is true. If $f'(x) > 0$ on (a, c) and $f'(x) < 0$ on (c, b), then f is increasing to the left of c and decreasing to the right of c. Hence $f(c)$ must be the largest value of $f(x)$ in the interval (a, b), and so f has a local maximum at $x = c$. See Figure 11.25(b). A similar argument justifies part (b) of the test.

EXAMPLE 11.8 Find all the local extrema of $f(x) = x^3 - 3x^2$.

Solution Since

$$f'(x) = 3x^2 - 6x = 3x(x - 2),$$

the critical numbers of f are 0 and 2. Plotting these numbers on a number line produces Figure 11.27(a).

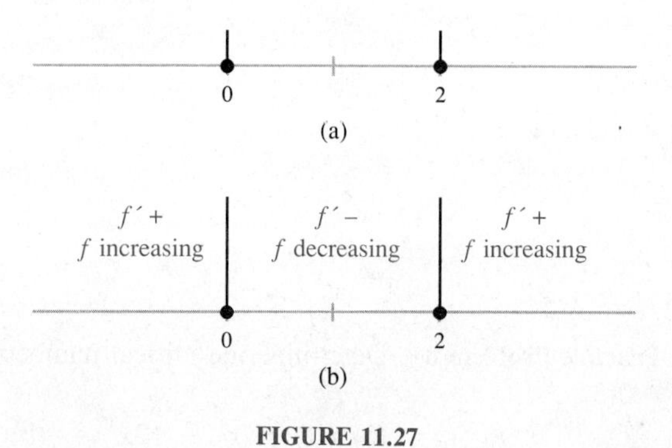

FIGURE 11.27

Now we evaluate f' at a test number in each of the intervals in Figure 11.27(a).

$$f'(-1) = 3(-1)^2 - 6(-1) = 9 > 0$$
$$f'(1) = 3(1)^2 - 6(1) = -3 < 0$$
$$f'(3) = 3(3)^2 - 6(3) = 9 > 0$$

Marking each interval with the sign of f' gives Figure 11.27(b). Hence f is increasing on $(-\infty, 0)$ and $(2, \infty)$ and decreasing on $(0, 2)$. Thus the first-derivative test implies that f has a local maximum at $x = 0$ and a local minimum at $x = 2$. The graph of f is shown in Figure 11.28. ■

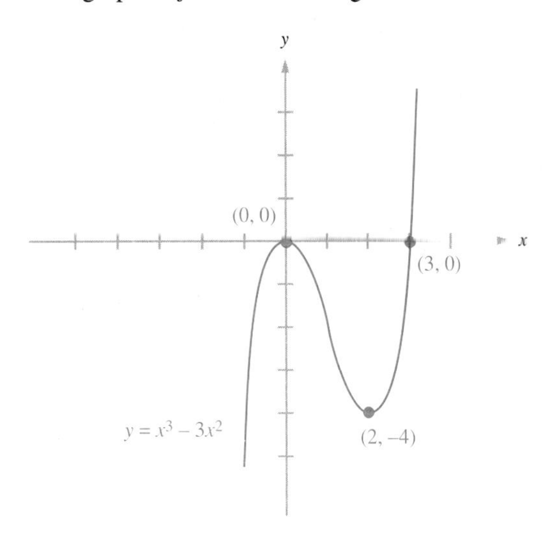

FIGURE 11.28

EXAMPLE 11.9 Find all the local extrema of $g(x) = \ln (x^2 + 1)$.

Solution The derivative of g is

$$g'(x) = \frac{2x}{x^2 + 1},$$

which is defined for all values of x. In order that $g'(x) = 0$, we must have the numerator of g' equal to 0, that is,

$$2x = 0.$$

Thus $x = 0$ is the only critical number of g.

Note that for all real numbers x we have

$$x^2 + 1 > 0.$$

Hence $g'(x)$ is positive or negative according to whether the numerator $2x$ is positive or negative. Applying the method of Section 11.1 to $h(x) = 2x$ gives Figure 11.29.

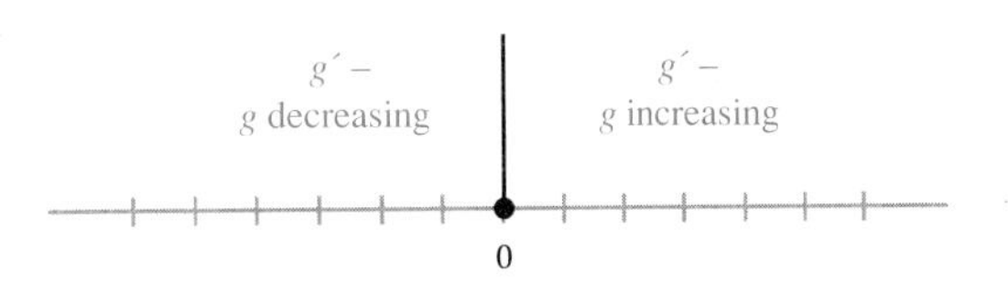

FIGURE 11.29

It follows that

$$g'(x) > 0 \text{ if } x > 0 \quad \text{and} \quad g'(x) < 0 \text{ if } x < 0.$$

Hence g is increasing for $x > 0$ and g is decreasing for $x < 0$, and so g has a local minimum of 0 at $x = 0$. Note these features of the graph of g shown in Figure 11.30. ■

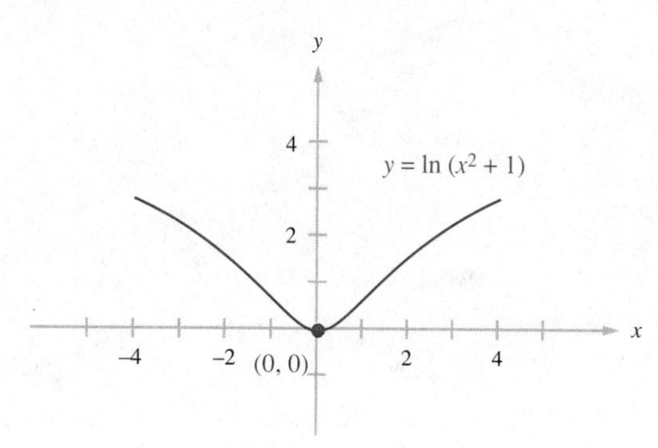

FIGURE 11.30

Practice Problem 9 Find all the local extrema of $f(x) = 4x^3 - 3x^4$ and graph f.

EXERCISES 11.2

In Exercises 1–8 determine the critical numbers of the given function.

1. $f(x) = x^3 - 3x^2 - 45x$

2. $F(x) = -x^3 + 3x^2 + 72x$

3. $g(x) = x^4 + 4x^3 + 4x^2$

4. $G(x) = 32x^2 - 8x^3 - x^4$

5. $F(x) = \dfrac{2x - 5}{3x - 8}$

6. $f(x) = \dfrac{x + 2}{x^2 - 3x + 6}$

7. $G(x) = \dfrac{x^2 + 7x + 8}{x - 1}$

8. $g(x) = \dfrac{x^3}{x^2 - 12}$

In Exercises 9–32 determine the intervals on which the given function is increasing and decreasing. Then determine all the local extrema of the function.

9. $f(x) = -x^2 + 5x - 7$

10. $F(x) = 3x^2 + 18x - 5$

11. $g(x) = 2x^2 - 36x + 24$

12. $G(x) = -x^2 - 7x + 3$

13. $h(x) = x^3 + 3x^2 - 144x + 8$

14. $H(x) = x^3 + 9x^2 - 21x + 10$

15. $F(x) = x^3 + 9x^2 + 15x - 25$

16. $f(x) = x^3 - 18x^2 + 60x - 30$

17. $G(x) = 3x^5 - 20x^3$

18. $g(x) = 6x^4 - 8x^3 + 1$

19. $H(x) = \dfrac{x + 3}{x - 6}$

20. $h(x) = \dfrac{x - 7}{x + 5}$

21. $f(x) = -\dfrac{2}{\sqrt{x}} + \dfrac{2}{x}$ for $x > 0$

22. $F(x) = \dfrac{4}{\sqrt{x}} - \dfrac{6}{x}$ for $x > 0$

23. $g(x) = xe^{-2x}$

24. $G(x) = x^{-1/2} \ln x$ for $x > 0$

25. $h(x) = \dfrac{\ln x}{x}$ for $x > 0$

26. $H(x) = \dfrac{e^x}{x}$

27. $F(x) = \dfrac{x^2 + 5x + 3}{x - 1}$

28. $f(x) = \dfrac{x^2 - x + 2}{x + 1}$

29. $G(x) = \dfrac{x - 1}{x^2 + 2x - 3}$

30. $g(x) = \dfrac{x - 1}{x^2 - 4x + 4}$

31. $H(x) = (x^2 - 2x + 1)e^{-x}$

32. $h(x) = (x^2 - 8x + 14)e^{2x}$

Graph the functions in Exercises 33–38. Your graph should show the intervals on which the function is increasing and decreasing and all local extrema.

33. $F(x) = x^2 - 2x - 15$

34. $f(x) = 8 - x^2$

35. $G(x) = 12x - x^3$

36. $g(x) = x^3 - 6x^2$

37. $H(x) = x^3 - 3x^2 - 24x + 12$

38. $h(x) = 2x^2 - x^4$

39. If a manufacturer of men's socks receives revenue of

$$R(p) = 20{,}000p - 2000p^2$$

when a pair of socks is priced at p dollars, on what intervals is the revenue increasing?

40. If a lawn care service receives revenue of

$$R(p) = 5000p - 10p^2$$

when its service is priced at p dollars per year, on what intervals is the revenue decreasing?

The revenue is $R(p) = 5000p - 10p^2$ if the price of the service is p dollars.

41. If the marginal cost of producing x units of a product is

$$M(x) = 0.00003x^2 - 0.024x + 20,$$

on what intervals is the marginal cost decreasing?

42. If the marginal cost of producing x units of a product is

$$M(x) = 0.00006x^2 - 0.096x + 300,$$

on what intervals is the marginal cost decreasing?

Answers to Practice Problems

6. On the interval $(-\infty, 2)$, f is increasing; and on the interval $(2, \infty)$, f is decreasing.
7. On the intervals $(-\infty, -3)$ and $(5, \infty)$, f is increasing; on the interval $(-3, 5)$, f is decreasing.
8. (a) 0 and 9 (b) -1 and 5
9. The only local extremum of f is a local maximum of 1 at $x = 1$.

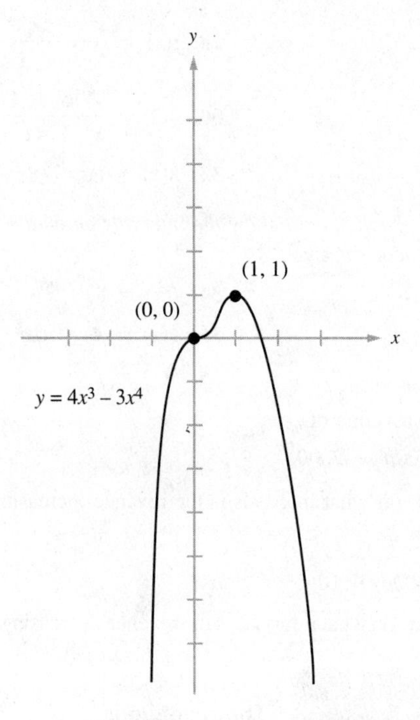

11.3 Using the Second Derivative in Graphing

In the preceding section we saw that f' gives information about the intervals on which f is increasing and the intervals on which f is decreasing. If f' is differentiable, then f'', the second derivative of f, tells us where f' is increasing and decreasing. This enables us to determine the concavity of f. In Figure 11.31(a), for example, f'' is positive on the interval (a, b). This means that f' is increasing throughout the interval; that is, the slope of the tangent line to the graph of f is increasing throughout (a, b). Hence f must be concave up on this interval. Likewise if g'' is negative throughout (a, b) as in Figure 11.31(b), then g' must be decreasing there, and so g will be concave down on this interval. This discussion can be summarized as follows.

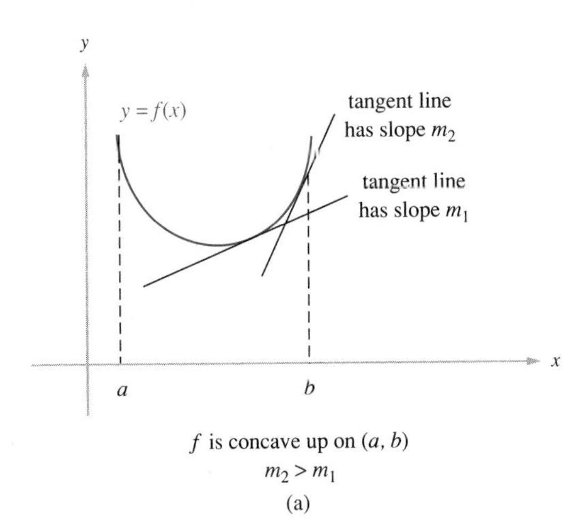

f is concave up on (a, b)
$m_2 > m_1$
(a)

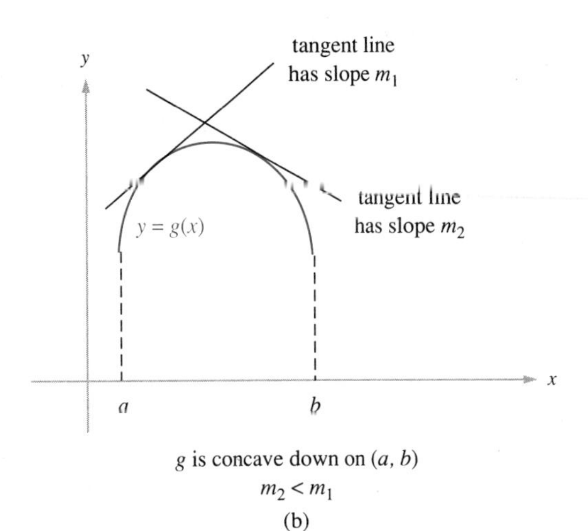

g is concave down on (a, b)
$m_2 < m_1$
(b)

FIGURE 11.31

Let f'' exist throughout the interval (a, b).

(a) If f'' is positive throughout (a, b), then f is concave up on the interval.

(b) If f'' is negative throughout (a, b), then f is concave down on the interval.

Recall that a point on the graph of f at which the concavity changes is called a point of inflection. If f has a point of inflection at $x = c$ and f'' exists on an interval containing c, then its values must change from positive to negative or from negative to positive at c. Hence $f''(c) = 0$.

If $(c, f(c))$ is a point of inflection of a function f, then either

$$f''(c) = 0 \qquad \text{or} \qquad f''(c) \text{ is undefined.}$$

Note that the statement above does *not* say that if $f''(c)$ fails to exist or $f''(c) = 0$, then f has a point of inflection at $x = c$. The points where the second derivative fails to exist or equals 0 are only *possible* points of inflection; to actually be a point of inflection, the concavity must change at the point in question.

EXAMPLE 11.10 For $f(x) = 2 - x^3$, find the intervals where f is concave up and concave down. Then determine any points of inflection.

Solution The concavity of f is determined by f'', the second derivative of f. Now

$$f'(x) = -3x^2,$$

and so

$$f''(x) = -6x.$$

Since we must determine where f'' is positive and where it is negative, we apply the method of Section 11.1. Figure 11.32 shows the result.

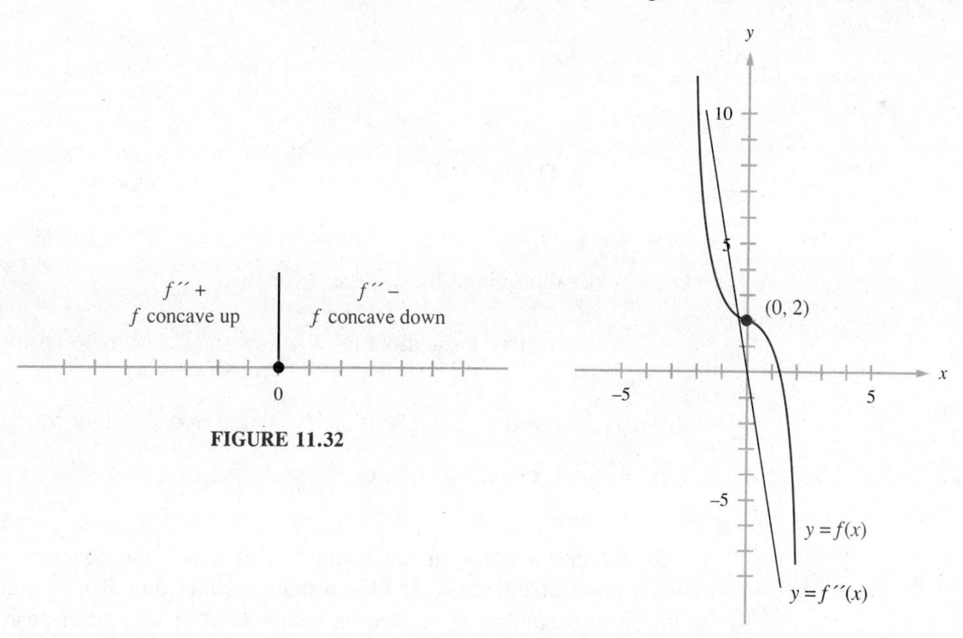

FIGURE 11.32

FIGURE 11.33

Thus f is concave up when

$$x < 0,$$

and f is concave down when

$$x > 0.$$

That is, f is concave up on $(-\infty, 0)$ and concave down on $(0, \infty)$. The graphs of f and f'' are both shown in Figure 11.33. Notice that f is concave up where f'' is positive, and f is concave down where f'' is negative.

Since the concavity of f changes from up to down at $x = 0$, the point $(0, 2)$ is a point of inflection for f. Note that at $x = 0$, the second derivative of f is zero. ■

EXAMPLE 11.11 Determine the concavity of f if

$$f(x) = \frac{1}{\sqrt{2\pi}} e^{-x^2/2}.$$

Solution To determine the concavity of f, we must compute f''. Now

$$f'(x) = -\frac{1}{\sqrt{2\pi}} xe^{-x^2/2},$$

and so the product rule is required to differentiate f'. The result is

$$\begin{aligned} f''(x) &= -\frac{1}{\sqrt{2\pi}} xe^{-x^2/2}(-x) + e^{-x^2/2}\left(-\frac{1}{\sqrt{2\pi}}\right) \\ &= \frac{1}{\sqrt{2\pi}}(x^2 - 1)e^{-x^2/2}. \end{aligned}$$

As in Example 11.5, f'' exists for all real numbers, and the sign of f'' depends only on the sign of $x^2 - 1$ because

$$e^{-x^2/2} > 0$$

for all x.

Thus we must determine the values of x for which

$$x^2 - 1 > 0 \qquad \text{and} \qquad x^2 - 1 < 0.$$

This can be done by the method described in Section 11.1. First we find the solutions of

$$x^2 - 1 = 0,$$

which are $x = -1$ and $x = 1$. Plotting these numbers on a number line produces Figure 11.34(a). Now we evaluate $x^2 - 1$ at a test number in each interval:

$$\begin{aligned} (-2)^2 - 1 &= 4 - 1 = 3 > 0, \\ (0)^2 - 1 &= 0 - 1 = -1 < 0, \\ (2)^2 - 1 &= 4 - 1 = 3 > 0. \end{aligned}$$

The results are shown in Figure 11.34(b).

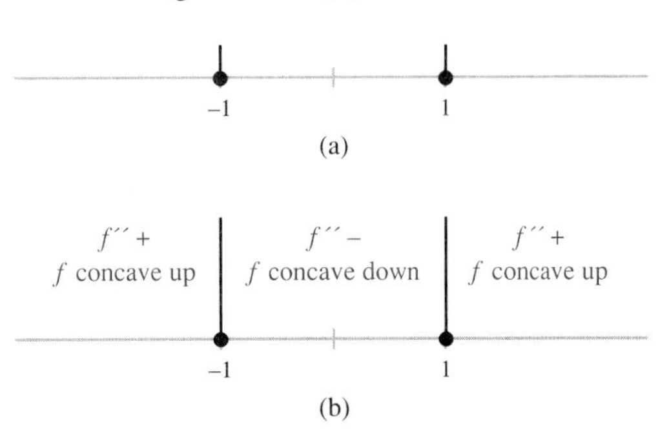

FIGURE 11.34

Thus $x^2 - 1 > 0$ on $(-\infty, -1)$ and $(1, \infty)$, and $x^2 - 1 < 0$ on $(-1, 1)$. Hence $f''(x) > 0$ on $(-\infty, -1)$ and $(1, \infty)$, and $f''(x) < 0$ on $(-1, 1)$. We therefore conclude that f is concave up on $(-\infty, -1)$ and $(1, \infty)$, and concave down on $(-1, 1)$.

We noted in the previous section that the graph of f is a normal curve with mean 0 and standard deviation 1. (See Figure 11.35.) Since the concavity changes from up to down at $x = -1$ and from down to up at $x = 1$, the standard normal curve has points of inflection at $x = \pm 1$, that is, one standard deviation above and below the mean. ■

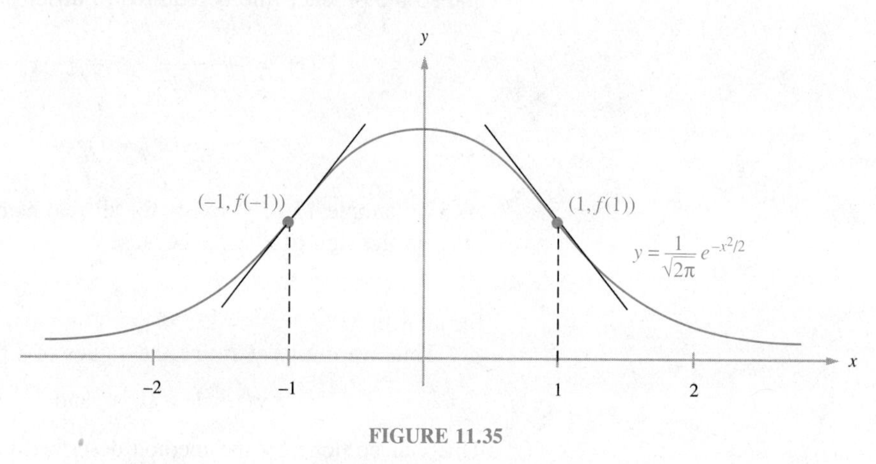

FIGURE 11.35

Practice Problem 10 For $f(x) = \dfrac{2}{x + 1}$, find the intervals where f is concave up and concave down. Then determine any points of inflection.

THE SECOND-DERIVATIVE TEST

In Section 11.2 we presented the first-derivative test, a method for locating local extrema using the first derivative. There is another method for locating local extrema that uses the second derivative. Consider, for instance, the function

$$f(x) = 3x - x^3.$$

Since

$$f'(x) = 3 - 3x^2,$$

the critical numbers of f are -1 and 1. Recall that this tells us that the only possible numbers at which a local extremum can occur is at $x = -1$ or $x = 1$.

Since

$$f''(x) = -6x,$$

the concavity of f is the same as in Example 11.10. From this example we see that f is concave up if $x < 0$, that is, on the interval $(-\infty, 0)$. Similarly f is concave down if $x > 0$, that is, on the interval $(0, \infty)$.

Note that f is concave up on an interval $(-\infty, 0)$ containing a critical number $(x = -1)$. Since the graph is concave up near $x = -1$, it is easy to see that f must have a local minimum there. (See Figure 11.36.) Likewise since f is concave down on an interval containing the critical number $x = 1$, f must have a local maximum there. (See Figure 11.36 again.)

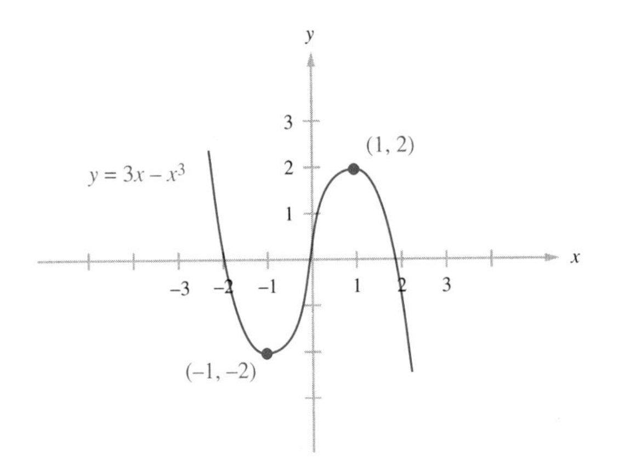

FIGURE 11.36

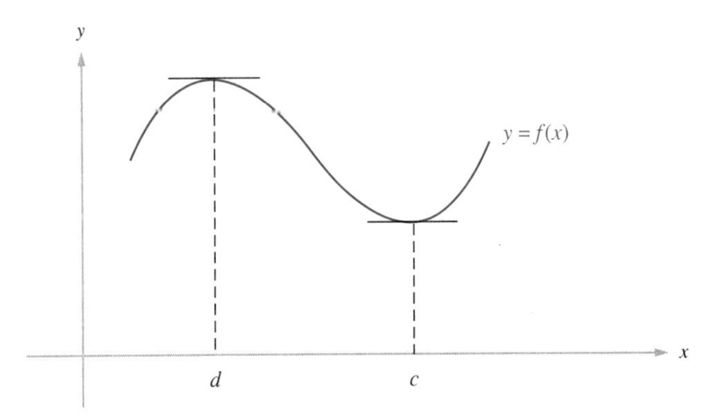

f is concave down around the critical number c and concave up around the critical number d.

FIGURE 11.37

The preceding discussion can be generalized. As Figure 11.37 illustrates, if $f'(c) = 0$ and f is concave down on an interval containing c, then f has a local maximum at $x = c$. Moreover, if $f'(d) = 0$ and f is concave up on an interval containing d, then f has a local minimum at $x = d$. These facts are summarized below.

The Second-Derivative Test for Local Extrema

Let $f'(c) = 0$ and let f'' exist on an interval containing $x = c$.

(a) If $f''(c) > 0$, then f has a local minimum at $x = c$.

(b) If $f''(c) < 0$, then f has a local maximum at $x = c$.

Note that the second-derivative test provides no information if $f''(c) = 0$. In this case there may be a local minimum, a local maximum, or no local extremum at $x = c$.

EXAMPLE 11.12 Find all the local extrema of $f(x) = x^3 - 3x^2$.

Solution Since

$$f'(x) = 3x^2 - 6x = 3x(x - 2),$$

the critical numbers of f are 0 and 2. Now

$$f''(x) = 6x - 6,$$

and so

$$f''(0) = 6(0) - 6 = -6 < 0$$
$$f''(2) = 6(2) - 6 = 6 > 0.$$

Therefore the second-derivative test tells us that f has a local maximum at $x = 0$ and a local minimum at $x = 2$. The graph of f is shown in Figure 11.38. ■

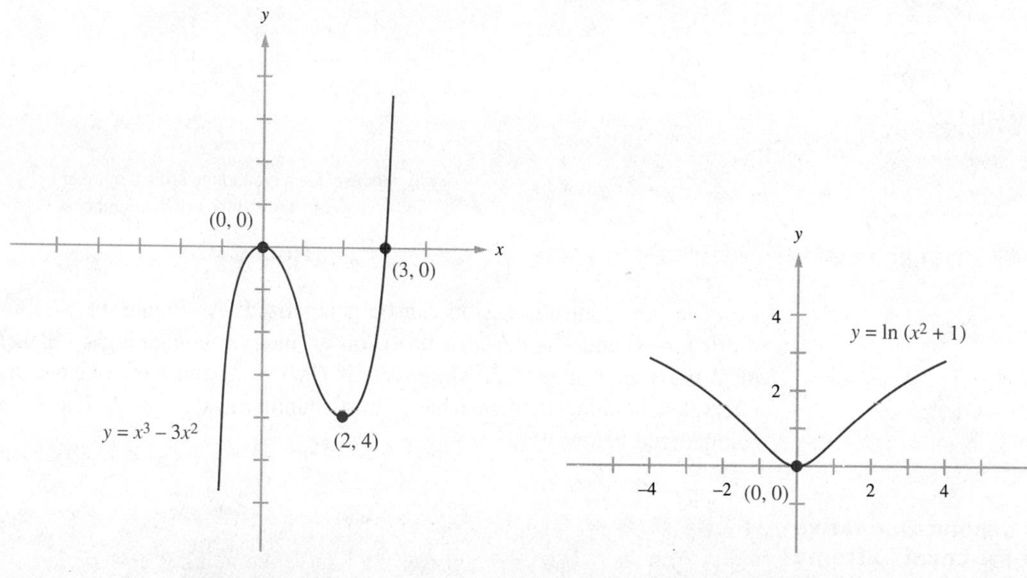

FIGURE 11.38 FIGURE 11.39

EXAMPLE 11.13 Find all the local extrema of $g(x) = \ln(x^2 + 1)$.

Solution The derivative of g is

$$g'(x) = \frac{2x}{x^2 + 1},$$

which is defined for all values of x. In order that $g'(x) = 0$, we must have the numerator of g' equal to 0, that is,

$$2x = 0.$$

Thus $x = 0$ is the only critical number of g.

Using the quotient rule, we find that

$$g''(x) = \frac{(x^2 + 1)(2) - 2x(2x)}{(x^2 + 1)^2} = \frac{2 - 2x^2}{(x^2 + 1)^2}.$$

Because

$$g''(0) = \frac{2 - 2(0)^2}{[(0)^2 + 1]^2} = 2 > 0,$$

the second-derivative test allows us to conclude that g has a local minimum at $x = 0$. See Figure 11.39. ■

Examples 11.12 and 11.13 involve the same functions considered in Examples 11.8 and 11.9. Comparing the corresponding examples, we see that in these examples the second-derivative test is easier to use than the first-derivative test. This will normally be the case. However, the second-derivative test cannot be used to test for a local extremum at $x = c$ if $f''(c) = 0$ or if $f''(c)$ does not exist. In these cases we must rely on the first-derivative test to determine whether or not f has a local extremum at $x = c$, and if so, what type.

Practice Problem 11 Determine where the local extrema of $f(x) = 2x^3 - 3x^2 - 36x$ occur.

EXERCISES 11.3

In Exercises 1–16 find the intervals on which the given function is concave up and concave down. Then determine any points of inflection.

1. $f(x) = 2x^2 - 4x - 1$

2. $g(x) = 3x^2 + 12x + 2$

3. $h(x) = -x^2 - 8x + 2$

4. $F(x) = -2x^2 + 12x + 3$

5. $G(x) = 2x^3 - 3x^2 - 120x + 16$

6. $H(x) = x^3 - 12x^2 + 36x - 15$

7. $F(x) = -x^3 - 6x^2 + 36x + 10$

8. $G(x) = -x^3 - 15x^2 - 48x + 12$

9. $H(x) = x^4 - 4x^3 - 5$

10. $f(x) = 18x^2 - x^4$

11. $g(x) = \ln x \quad \text{for } x > 0$

12. $h(x) = e^{-x}$

13. $f(x) = \dfrac{1}{x - 1} \quad \text{for } x \neq 1$

14. $g(x) = \dfrac{x}{x + 4} \quad \text{for } x \neq -4$

15. $h(x) = x^2e^{-x}$

16. $F(x) = x + \ln (x^2 + 4)$

In Exercises 17–32 use the second-derivative test to find where the local extrema of the given function occur.

17. $f(x) = 3x^2 - 12x + 5$

18. $F(x) = -x^2 - 8x + 6$

19. $g(x) = -x^3 + 3x^2 + 24x - 20$

20. $G(x) = x^3 - 2x^2 - 15x + 30$

21. $h(x) = x^4 + 4x^3 - 8x^2 + 8$

22. $H(x) = x^4 - 4x^3 - 20x^2 + 72$

23. $F(x) = x^5 - 15x^3$

24. $f(x) = x^5 - 20x^2$

25. $G(x) = (x - 2)^2(x + 8)^2$

26. $g(x) = (x + 3)^2(x - 5)^2$

27. $H(x) = \frac{1}{4}x^4 + 3x^3 + 10x^2 + 12x - 1$

28. $h(x) = \frac{1}{4}x^4 - x^3 - 5x^2 + 24x + 9$

29. $f(x) = (x - 3)e^{-x}$

30. $F(x) = (x - 4)e^{4x - x^2/2}$

31. $g(x) = x + 2 \ln (x^2 + 3)$

32. $G(x) = x + \ln (x^2 + 4x + 5)$

In Exercises 33–40 graph the given function. Your graph should show local extrema and points of inflection.

33. $F(x) = x^3 + 3x^2 + 12x - 8$

34. $f(x) = 9x^2 - x^3$

35. $G(x) = 6x^2 - x^4$

36. $g(x) = 3x^4 - 4x^3 - 12x^2$

37. $f(x) = x^4 - 4x^3$

38. $F(x) = x^4 - 8x$

39. $g(x) = 3x^5 - 10x^3$

40. $G(x) = 20x^4 - 4x^5$

41. Show that $f(x) = x^4$ has no points of inflection even though $f''(0) = 0$.

42. Show that the general quadratic function $f(x) = ax^2 + bx + c$, where $a \neq 0$, has the same concavity for all values of x. Hence f has no points of inflection.

43. (a) Show that the general cubic function $f(x) = ax^3 + bx^2 + cx + d$, where $a \neq 0$, has exactly one point of inflection, namely $(p, f(p))$ for $p = -\frac{b}{3a}$.

(b) Show that for any real number r,

$$f(p + r) - f(p) = f(p) - f(p - r).$$

Hence the graph of every cubic function is symmetric with respect to its unique point of inflection. (See the figure below.)

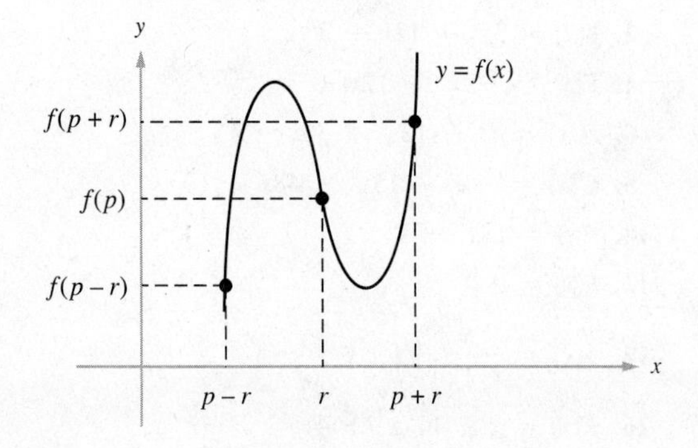

Answers to Practice Problems

10. On $(-\infty, -1)$ the function f is concave down, and on $(-1, \infty)$ it is concave up. Because f is not defined at $x = -1$, there are no points of inflection.

11. There is a local maximum of f at $x = -2$ and a local minimum at $x = 3$.

11.4 Curve Sketching

The most useful tool for understanding the behavior of a function is its graph. For this reason we have discussed the graphing of functions throughout this book. In order to obtain a good representation of a function without plotting a large number of points, we must determine as many special features of the function as possible. The following steps provide a basis for obtaining an accurate sketch of the graph of a function f.

1. Compute f' and f'', the first and second derivatives of f.
2. Locate the extrema of f, the intervals on which f is increasing, and the intervals on which it is decreasing.
3. Determine where f is concave up, where it is concave down, and where there are points of inflection.
4. Find any vertical asymptotes.
5. Determine the intercepts of the graph if it is easy to do so. (Recall that if $f(0)$ is defined, then the y-intercept is $(0, f(0))$. The x-intercepts are found by solving the equation $f(x) = 0$. If this equation is difficult to solve, omit finding the x-intercepts.)

Table 11.1 on page 626 summarizes the information that can be obtained from the first and second derivatives of a function.

EXAMPLE 11.14 Graph the function $f(x) = x^3 - 3x^2 - 9x + 27$.

Solution We begin by computing the first and second derivatives:

$$f'(x) = 3x^2 - 6x - 9 \qquad \text{and} \qquad f''(x) = 6x - 6.$$

Because f' is a polynomial function, it is defined for all values of x. The quadratic formula shows that the solutions of $f'(x) = 0$ are $x = -1$ and $x = 3$. Hence -1 and 3 are the only critical numbers of f. Furthermore, $f''(x)$ is defined for all values of x, and the only solution of $f''(x) = 0$ is $x = 1$. Determining where f' and f'' are positive and negative as in Section 11.1, we obtain Figure 11.40.

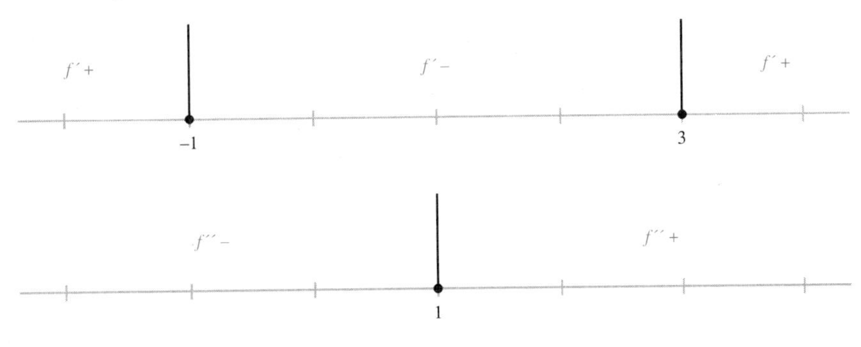

FIGURE 11.40

Values of the first and second derivatives on (a, b)	*Behavior of f on (a, b)*	*Graph of f on (a, b)*
$f'(c) > 0$ $f''(c) > 0$	f is increasing f is concave up	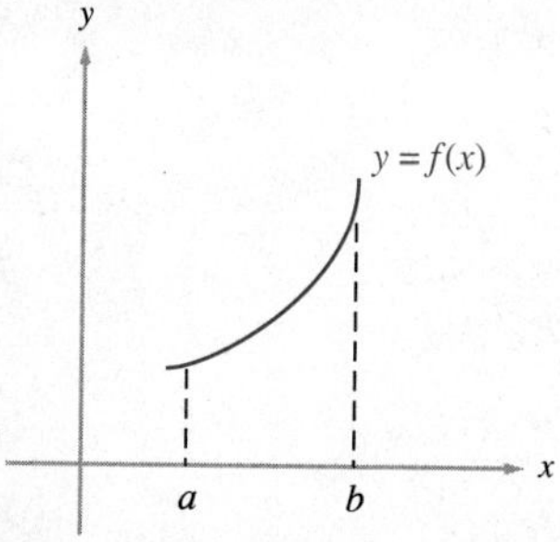 y, $y = f(x)$, a, b, x
$f'(c) > 0$ $f''(c) < 0$	f is increasing f is concave down	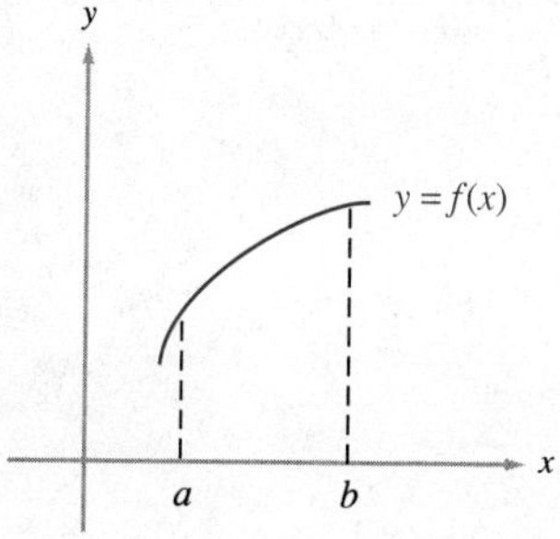 y, $y = f(x)$, a, b, x
$f'(c) < 0$ $f''(c) > 0$	f is decreasing f is concave up	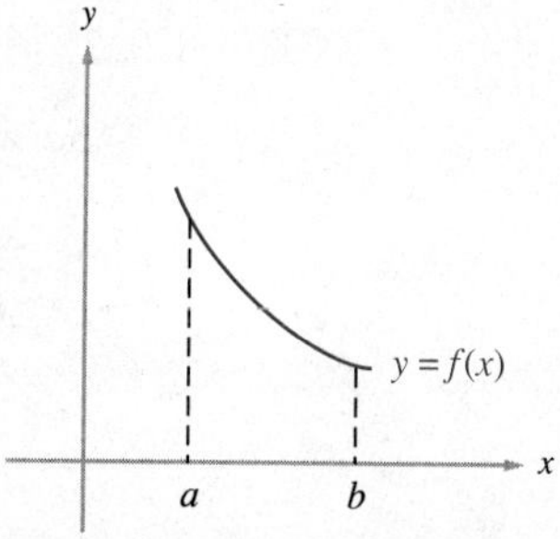 y, $y = f(x)$, a, b, x
$f'(c) < 0$ $f''(c) < 0$	f is decreasing f is concave down	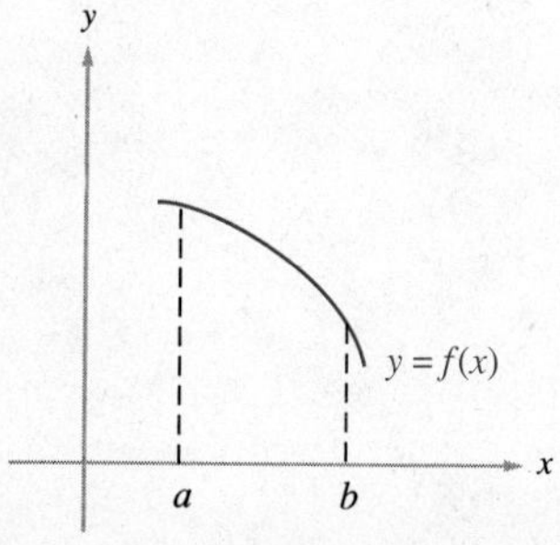 y, $y = f(x)$, a, b, x

TABLE 11.1

We conclude therefore that f is increasing on $(-\infty, -1)$ and $(3, \infty)$, f is decreasing on $(-1, 3)$, f is concave up on $(1, \infty)$, and f is concave down on $(-\infty, 1)$. It is convenient to record this information on a number line as in Figure 11.41. From this figure it is easy to see that f has a local maximum at 1, a point of inflection at 1, and a local minimum at 3.

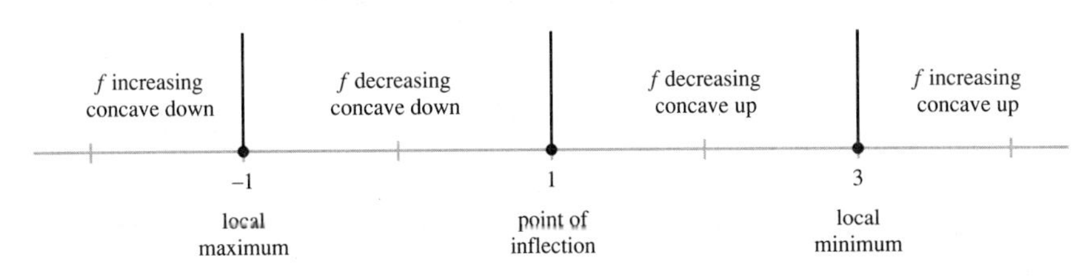

FIGURE 11.41

Finally note that since $f(0) = 27$, the y-intercept of the graph is $(0, 27)$. Because the equation $f(x) = 0$ is difficult to solve in this case, we omit finding the x-intercepts.

The information that we have obtained enables us to graph f accurately by plotting only a few important points on the graph. In addition to the y-intercept, we will plot the points corresponding to the two extrema

$$(-1, f(-1)) = (-1, 32) \qquad \text{and} \qquad (3, f(3)) = (3, 0)$$

and the point of inflection

$$(1, f(1)) = (1, 16).$$

Plotting these four points and using the information in Figure 11.41 leads to the graph in Figure 11.42. ■

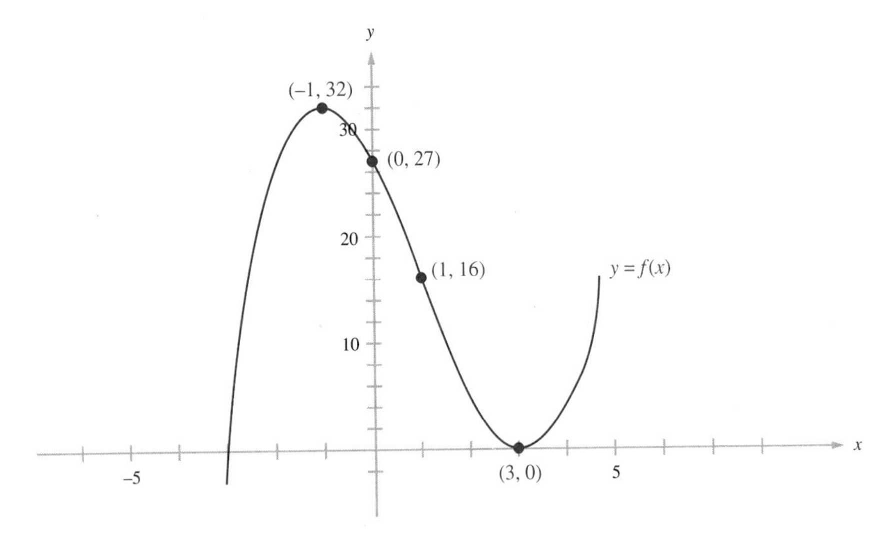

FIGURE 11.42

EXAMPLE 11.15 Graph the function $g(x) = x + \dfrac{1}{x - 3}$.

Solution Writing g in the form

$$g(x) = x + (x - 3)^{-1},$$

we see that the first derivative of g is

$$\begin{aligned} g'(x) &= 1 - (x - 3)^{-2} \\ &= 1 - \frac{1}{(x - 3)^2} \\ &= \frac{(x - 3)^2 - 1}{(x - 3)^2} \\ &= \frac{x^2 - 6x + 8}{(x - 3)^2} \\ &= \frac{(x - 2)(x - 4)}{(x - 3)^2}. \end{aligned}$$

Note that $g'(x)$ is undefined when $x = 3$ and equals zero when $x = 2$ or $x = 4$.

By differentiating $g'(x) = 1 - (x - 3)^{-2}$, we obtain

$$g''(x) = 2(x - 3)^{-3} = \frac{2}{(x - 3)^3}.$$

In this case $g''(x)$ is also undefined if $x = 3$. But the numerator of $g''(x)$ is never zero, and so $g''(x) = 0$ has no solutions.

Determining where g' and g'' are positive and negative produces Figure 11.43.

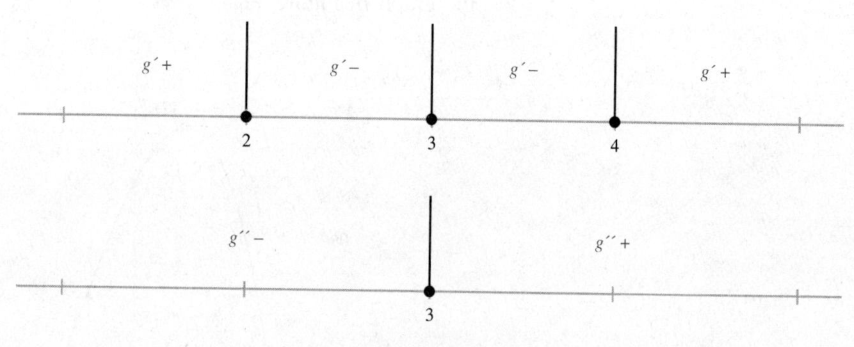

FIGURE 11.43

From Figure 11.43 we can determine where g is increasing and decreasing and where it is concave up and down. This information is recorded in Figure 11.44. Notice that because $g(x)$ is not defined for $x = 3$, g does *not* have a point of inflection there. In fact, the vertical line $x = 3$ is an asymptote of the graph of g.

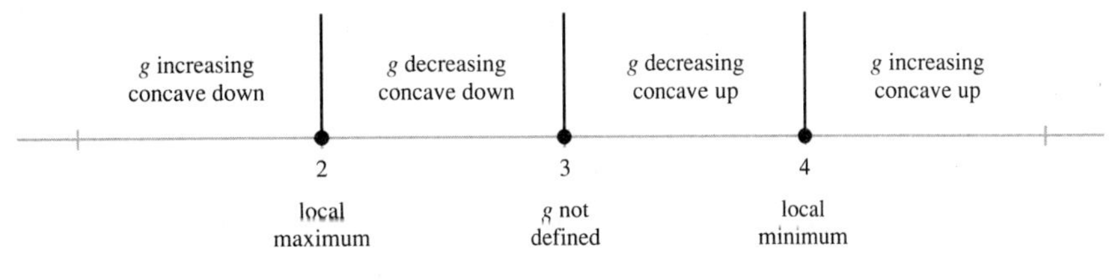

FIGURE 11.44

Since $g(0) = -\frac{1}{3}$, the y-intercept of the graph is $(0, -\frac{1}{3})$. Again the equation $f(x) = 0$ appears difficult to solve, so we omit finding the x-intercepts. Combining all the information we have obtained produces Figure 11.45. ■

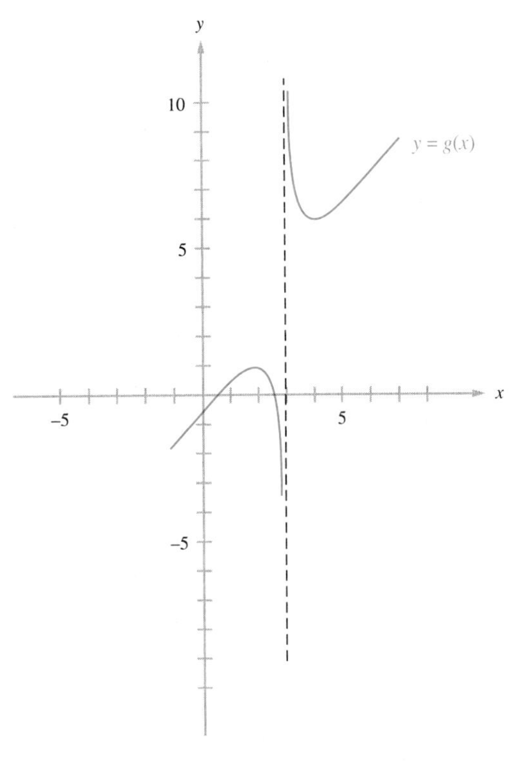

FIGURE 11.45

EXAMPLE 11.16 Graph the function $F(x) = x^2e^{-x}$.

Solution The first derivative of F is

$$
\begin{aligned}
F'(x) &= x^2e^{-x}(-1) + e^{-x}(2x) \\
&= (-x^2 + 2x)e^{-x} \\
&= -x(x - 2)e^{-x},
\end{aligned}
$$

and the second derivative is

$$
\begin{aligned}
F''(x) &= (-x^2 + 2x)e^{-x}(-1) + e^{-x}(-2x + 2) \\
&= (x^2 - 4x + 2)e^{-x}.
\end{aligned}
$$

Note that $F'(x)$ is defined for all x and $F'(x) = 0$ when $x = 0$ or $x = 2$. Furthermore, $F''(x)$ is defined for all x and the quadratic formula shows that $F''(x) = 0$ when $x = 2 - \sqrt{2}$ or $x = 2 + \sqrt{2}$. Proceeding as before, we find where F' and F'' are positive and negative. The results are shown in Figure 11.46.

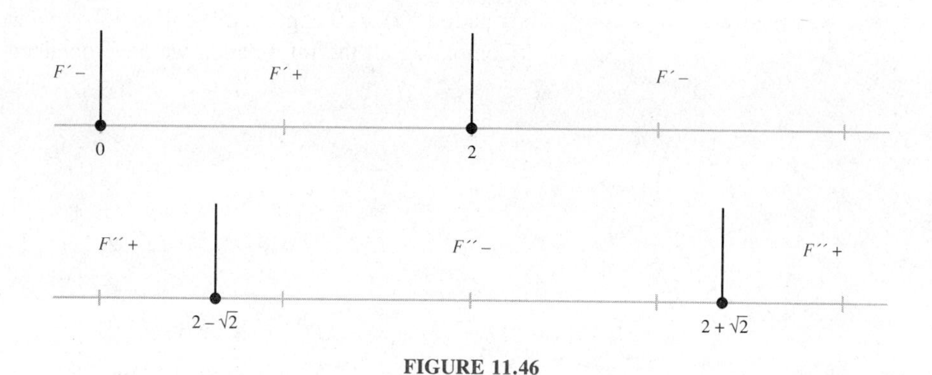

FIGURE 11.46

Figure 11.47 describes the behavior of F.

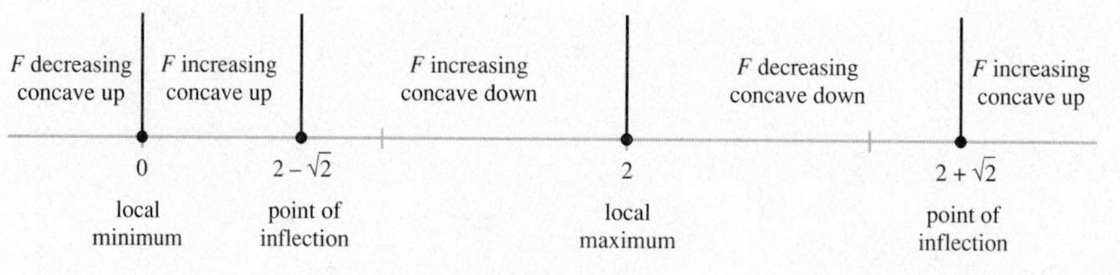

FIGURE 11.47

It follows that the graph of F is as shown in Figure 11.48. ■

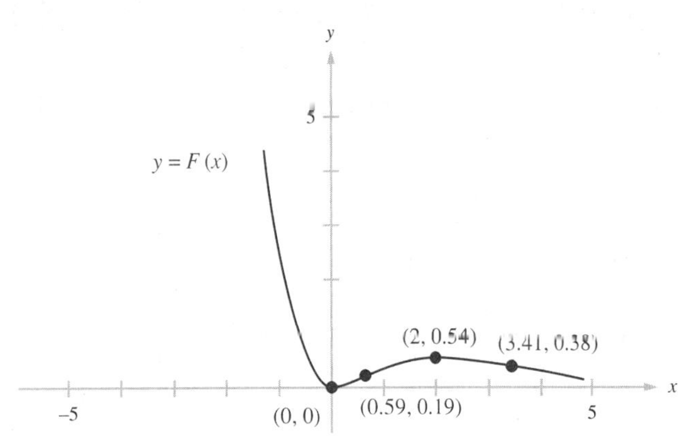

FIGURE 11.48

EXAMPLE 11.17 Graph the function $G(x) = (x^2 - 1)^{2/3}$.

Solution The first derivative of G is

$$\begin{aligned} G'(x) &= \frac{2}{3}(x^2 - 1)^{-1/3}(2x) \\ &= \frac{4x}{3}(x^2 - 1)^{-1/3} \\ &= \frac{4x}{3\sqrt[3]{x^2 - 1}}. \end{aligned}$$

To compute the second derivative, we apply the product rule to

$$G'(x) = \frac{4x}{3}(x^2 - 1)^{-1/3};$$

the result is

$$\begin{aligned} G''(x) &= \frac{4x}{3}\left(-\frac{1}{3}\right)(x^2 - 1)^{-4/3}(2x) + \frac{4}{3}(x^2 - 1)^{-1/3} \\ &= (x^2 - 1)^{-4/3}\left[-\frac{8x^2}{9} + \frac{4}{3}(x^2 - 1)\right] \\ &= (x^2 - 1)^{-4/3}\left(\frac{4}{9}x^2 - \frac{4}{3}\right) \\ &= (x^2 - 1)^{-4/3}\left(\frac{4}{9}\right)(x^2 - 3) \\ &= \frac{4(x^2 - 3)}{9\sqrt[3]{(x^2 - 1)^4}}. \end{aligned}$$

Clearly the numerator of G' equals 0 only if $x = 0$. But $G'(x)$ is undefined if $x = -1$ or $x = 1$ since its denominator is 0 at these numbers. Hence the critical numbers of G are -1, 0, and 1. Likewise G'' equals 0 only if $x = -\sqrt{3}$ or $x = \sqrt{3}$, and G'' is undefined if $x = -1$ or $x = 1$. Figure 11.49 shows where G' and G'' are positive and negative.

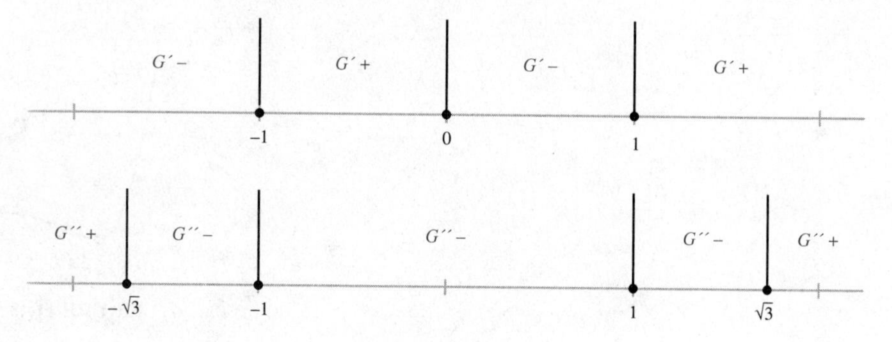

FIGURE 11.49

Thus the behavior of G is shown in Figure 11.50.

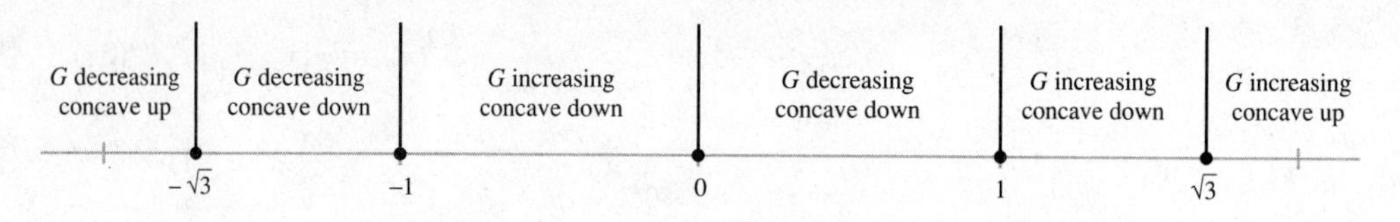

FIGURE 11.50

Finally, note that the intercepts of G are $(-1, 0)$, $(1, 0)$, and $(0, 1)$. Combining all this information gives the graph shown in Figure 11.51. ■

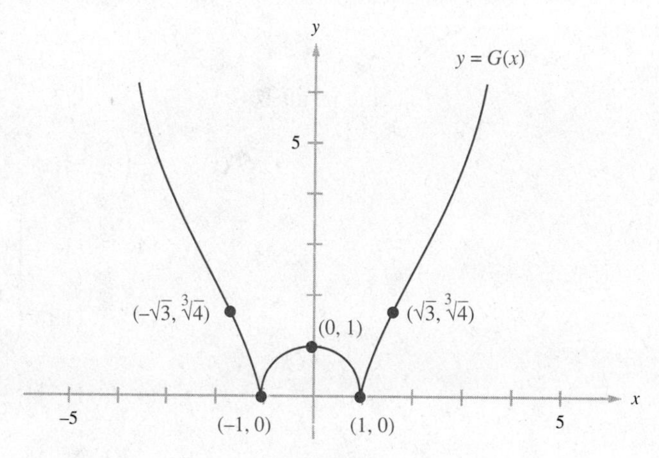

FIGURE 11.51

Practice Problem 12 Graph $f(x) = 1 + 3x - 4x^3$.

EXERCISES 11.4

Graph the functions in Exercises 1–32 using the technique described in this section.

1. $f(x) = 2x^2 - 3x - 20$

2. $F(x) = 2x^2 - x - 21$

3. $g(x) = -x^2 + 6x - 5$

4. $G(x) = -x^2 + 2x + 24$

5. $h(x) = x^3 - 3x^2 - 24x + 30$

6. $H(x) = -x^3 - 3x^2 + 9x + 7$

7. $F(x) = 12x - x^3$

8. $f(x) = 3x^3 - 3x^2 - 8x + 8$

9. $G(x) = x^4 - 8x^2$

10. $g(x) = x^3 - 3x^2$

11. $H(x) = 6x^4 - 8x^3 + 1$

12. $h(x) = 3x^4 - 4x^3$

13. $f(x) = 2x^4 - 4x^2 + 1$

14. $F(x) = x^4 - 4x^3 + 4x^2 - 3$

15. $g(x) = 3x^5 - 5x^3$

16. $G(x) = 6x^4 - 8x^3 + 1$

17. $h(x) = x^{2/3}$

18. $H(x) = 1 - x^{1/3}$

19. $F(x) = \dfrac{x - 2}{x + 1}$

20. $f(x) = \dfrac{x + 1}{x - 2}$

21. $G(x) = \dfrac{x^2}{1 - x}$

22. $g(x) = \dfrac{x^2}{x - 3}$

23. $H(x) = \dfrac{x^2}{x^2 + 4}$

24. $h(x) = \dfrac{-x^3}{3x^2 + 1}$

25. $f(x) = \sqrt{x} \ln x$ for $x > 0$

26. $F(x) = \dfrac{\ln x}{x}$ for $x > 0$

27. $g(x) = xe^{-x}$

28. $G(x) = \sqrt{x}e^{-x/8}$

29. $h(x) = \sqrt[3]{x}(x - 4)$

30. $H(x) = \dfrac{4x}{4 - x^2}$

31. $F(x) = (x + 1)^{2/3}(x - 2)^2$

32. $f(x) = x^{2/3}(x - 5)$

Answer to Practice Problem **12.** The behavior of f is described in Figure (a), and its graph is shown in Figure (b) on page 634.

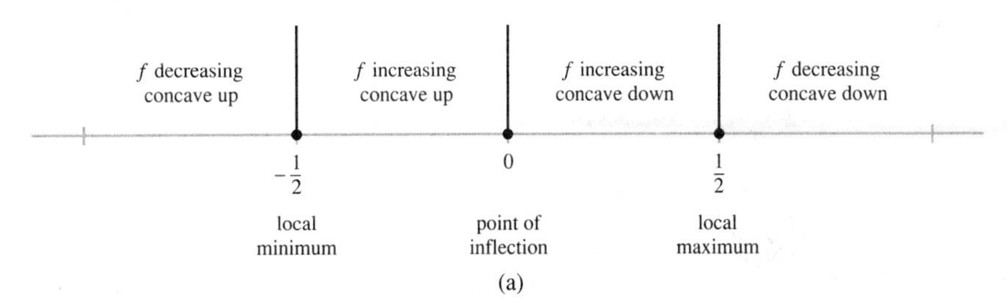

(a)

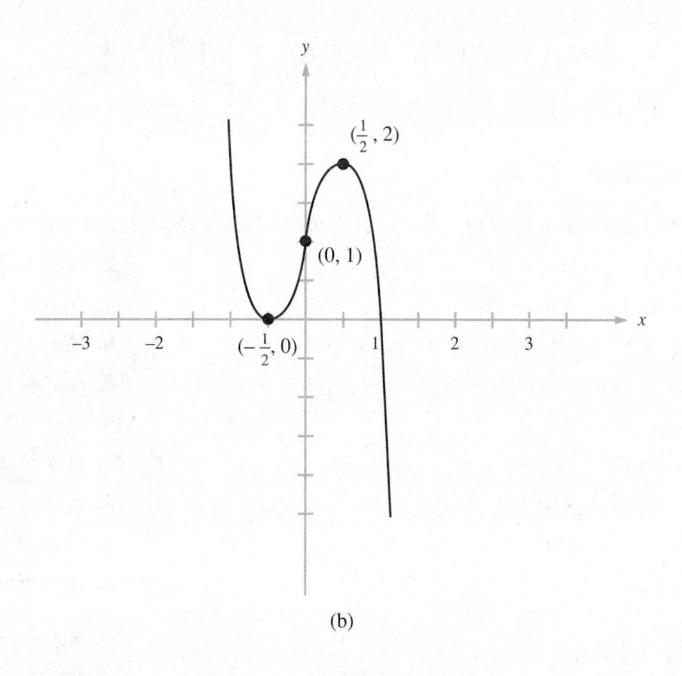

(b)

11.5 Optimization

Practical mathematical problems often require maximizing or minimizing some function. For instance, business executives are concerned with maximizing profits, and government economists are concerned with minimizing unemployment. Here is a simple example of a problem requiring maximization.

> An apple grower intends to increase the number of trees in his orchard. At present there are 1000 trees, but there is room for as many as 4500. Unfortunately, adding additional trees will decrease the yield of the existing trees because the new trees will compete with the present ones for sunlight and water. The average yearly profit from each of the present trees is \$40, but the grower expects that the yearly profit per tree will decrease by 1¢ for each additional tree added to the orchard. How many new trees should the grower add to the orchard in order to make the yearly profit from the orchard as large as possible?

In this example we are interested in maximizing the profit from the orchard, which is the product of the number of trees and the average yield per tree. Suppose that x new trees are added to the orchard. Then the number of trees in the orchard will be $1000 + x$ (the original 1000 trees plus x new ones). Adding x new trees reduces the yearly profit per tree to $40 - 0.01x$ dollars.

Hence the yearly profit (in dollars) from the orchard is

$$\begin{aligned} P(x) &= (1000 + x)(40 - 0.01x) \\ &= 40{,}000 + 30x - 0.01x^2. \end{aligned}$$

The apple grower wishes to maximize this function. Notice that unlike the objective functions in linear programming problems discussed in Chapter 3, this function is not a linear function. In this section we will use techniques from Sections 11.1 and 11.2 to determine the maximum and minimum values of nonlinear functions.

ABSOLUTE MAXIMA AND MINIMA

In the problem stated above there is a limit of 4500 trees in the orchard. Thus the grower can add at most 3500 trees to the 1000 originally present. Hence the grower needs to find the maximum value of the function

$$P(x) = 40{,}000 + 30x - 0.01x^2$$

over the interval $0 \leq x \leq 3500$. In Section 11.2 we learned how to find local maxima and local minima, but the grower's problem is different: He must determine the largest value of $P(x)$ on the interval $[0, 3500]$.

We call the largest value assumed by a function f on an interval the **absolute maximum** of f on the interval in question. Thus for $f(c)$ to be the absolute maximum of f on the interval $[a, b]$ means that

$$f(c) \geq f(x) \qquad \text{for all } x \text{ such that } a \leq x \leq b.$$

Similarly the **absolute minimum** of f on an interval is the smallest value of $f(x)$ when x is in the interval in question. Symbolically, $f(c)$ is the absolute minimum of f on $[a, b]$ if

$$f(c) \leq f(x) \qquad \text{for all } x \text{ such that } a \leq x \leq b.$$

Unfortunately, even well-behaved functions need not have an absolute maximum or an absolute minimum on an interval. Consider, for instance, the simple function

$$g(x) = x$$

on the interval $(1, 2)$. (See Figure 11.52.) The values of $g(x)$ can be made as close to 2 as desired by taking x sufficiently close to 2. But $g(x) \neq 2$ for any x in $(1, 2)$, and so g has no absolute maximum on this interval. Likewise g has no absolute minimum on the interval $(1, 2)$ because the values of $g(x)$ can be made as close to 1 as desired, but $g(x) \neq 1$ for any x in the interval.

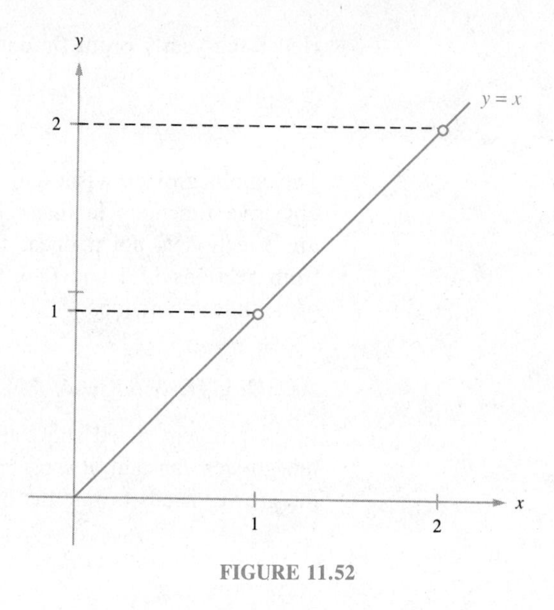

FIGURE 11.52

The following result, however, guarantees the existence of both an absolute maximum and an absolute minimum for continuous functions on intervals of the form $[a, b]$.

> A continuous function on the interval $[a, b]$ has both an absolute maximum and an absolute minimum. Moreover, the absolute maximum occurs at a, at b, or at some number where the function has a local maximum; and the absolute minimum occurs at a, at b, or at some number where the function has a local minimum.

Thus the absolute maximum of a continuous function on $[a, b]$ must occur at either an endpoint of the interval or at a critical number inside the interval. The same can be said about the absolute minimum of a continuous function on this interval. In Figure 11.53, for instance, f has local minima at $x = r$ and $x = t$ and a local maximum at $x = s$. Hence the absolute maximum of f on $[a, b]$ must occur at one of the numbers a, b, or s; and the absolute minimum of f on $[a, b]$ must occur at one of the numbers a, b, r, or t. (In this case the absolute maximum of f occurs at $x = a$ and the absolute minimum at $x = t$.)

The following steps can be used to determine the absolute maximum and absolute minimum of a continuous function on an interval $[a, b]$.

1. Determine the critical numbers of f that lie inside the interval.
2. Evaluate f at each of the critical numbers in step 1 and at the endpoints of the given interval (a and b).
3. The largest of the functional values in step 2 is the absolute maximum of f on the interval, and the smallest of the functional values in step 2 is the absolute minimum on the interval.

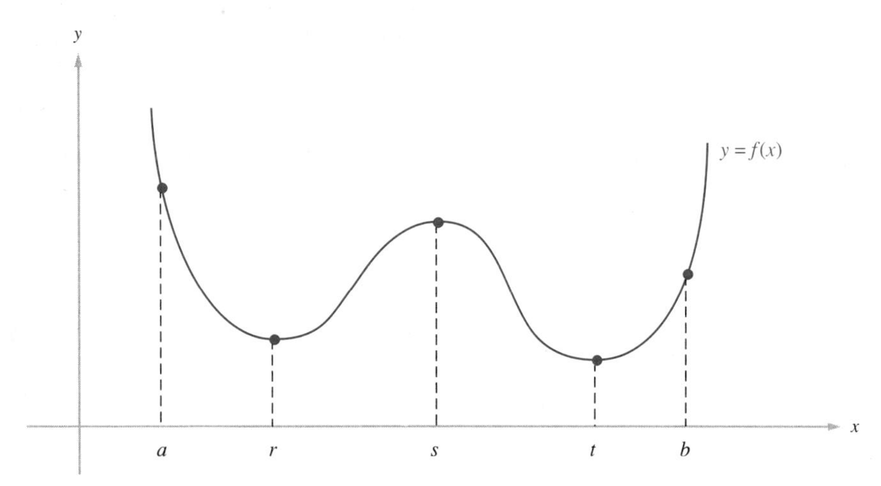

On $[a, b]$ the absolute maximum of f occurs at a and the absolute minimum occurs at t.

FIGURE 11.53

EXAMPLE 11.18 Determine the absolute maximum and the absolute minimum of the function $f(x) = -x^4 + 4x^3 + 20x^2 - 12$ on the interval $[-3, 1]$.

Solution We must determine the values of f at the endpoints of the given interval and at any critical numbers inside the interval. To find the critical numbers, we compute

$$\begin{aligned} f'(x) &= -4x^3 + 12x^2 + 40x \\ &= -4x(x^2 - 3x - 10) \\ &= -4x(x + 2)(x - 5). \end{aligned}$$

Hence the critical numbers of f are -2, 0, and 5. Note that 5 does not lie in the interval $[-3, 1]$, and so we need to consider only the values of f at the critical numbers -2 and 0 and the endpoints of the interval -3 and 1. These values are shown below.

x	$f(x)$	
-2	20	⟵ **absolute maximum**
0	-12	
-3	-21	⟵ **absolute minimum**
1	11	

To find the absolute maximum of f on $[-3, 1]$, we choose the largest of the functional values listed above, which is $f(-2) = 20$. Thus the absolute maximum of f on $[-3, 1]$ is 20 at $x = -2$. Likewise to find the absolute minimum of f on $[-3, 1]$, we choose the smallest of the functional values listed above, which is $f(-3) = -21$. Hence the absolute minimum of f on $[-3, 1]$ is -21 at $x = -3$.

The graph of f on the interval $[-3, 1]$ is shown in Figure 11.54. Observe that in this case the absolute maximum of f occurs at $x = -2$, a number inside the interval where a local maximum occurs, and the absolute minimum of f occurs at $x = -3$, an endpoint of the interval. ■

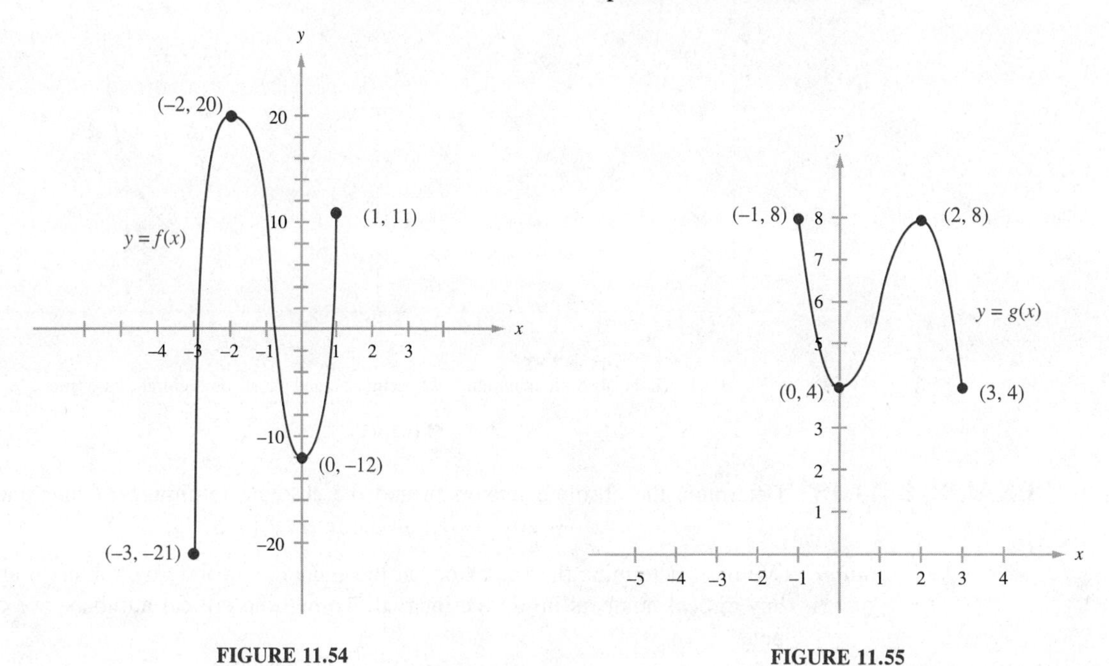

FIGURE 11.54

FIGURE 11.55

Practice Problem 13 Determine the absolute maximum and the absolute minimum of the function f in Example 11.18 on the interval $[-1, 7]$.

EXAMPLE 11.19 Determine the absolute maximum and the absolute minimum of the function $g(x) = 4 + 3x^2 - x^3$ on $[-1, 3]$.

Solution As in Example 11.18 we begin by computing the critical numbers of g that lie inside the given interval. Since

$$g'(x) = 6x - 3x^2 = 3x(2 - x),$$

these critical numbers are 0 and 2. Hence we evaluate g at 0, 2, -1, and 3.

x	$g(x)$	
-1	8	absolute maximum
2	8	
0	4	absolute minimum
3	4	

The largest of these numbers is 8, and so the absolute maximum of g on $[-1, 3]$ is 8. Notice, however, that this absolute maximum occurs at two different numbers, $x = -1$ and $x = 2$. Likewise the smallest of these numbers is 4, and so the absolute minimum of g on $[-1, 3]$ is 4. It also occurs at two different numbers, $x = 0$ and $x = 3$. See Figure 11.55.

This example shows that although a function can have only one absolute maximum and one absolute minimum on an interval, these functional values can occur at more than one value of x. ■

Practice Problem 14 Determine the absolute maximum and the absolute minimum of the function $g(x) = xe^{-x}$ on the interval $[0, 3]$.

EXAMPLES INVOLVING OPTIMIZATION

Let us return to the apple grower's problem stated at the beginning of this section. Recall that the grower needs to find the maximum value of the function

$$P(x) = 40{,}000 + 30x - 0.01x^2$$

over the interval $0 \leq x \leq 3500$. Thus we see that this problem involves finding the absolute maximum of P on the interval $[0, 3500]$. To solve this problem, we proceed as above by finding the critical numbers of P. Since

$$P'(x) = 30 - 0.02x,$$

P has only one critical number, $x = 1500$, where $P'(x) = 0$. Hence we must evaluate P at 1500, 0, and 3500.

x	$P(x)$	
1500	62,500	⟵ absolute maximum
0	40,000	
3500	22,500	

Thus the maximum yearly profit from the apple orchard is $62,500, and it is obtained when $x = 1500$, that is, when 1500 new trees are added to the orchard.

When solving a practical problem involving optimization, the solution typically involves several stages.

1. Formulate the problem in mathematical terms by introducing variables representing the changing quantities and those to be determined.
2. Write a formula for the quantity to be maximized or minimized; a figure is often useful in this process.
3. Reduce the formula in step 2 to one involving a single variable. This may require the use of equations relating two or more independent variables.
4. Determine the interval on which the single variable in step 3 is defined.
5. Use the techniques of calculus to solve the mathematical formulation.
6. Check the mathematical solution in the context of the original problem.

Sometimes the formulation of an optimization problem requires the use of geometric formulas. The most common of these are listed in Figure 11.56.

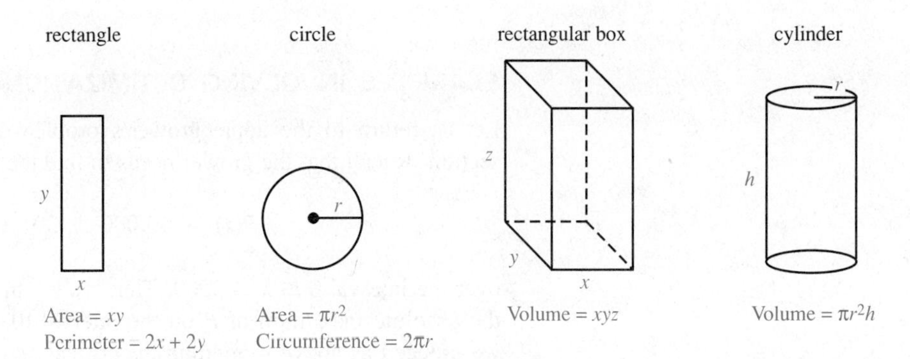

FIGURE 11.56

EXAMPLE 11.20 U.S. postal regulations require that any package mailed by parcel post must have length plus girth (that is, distance around) not exceeding 84 inches. Determine the dimensions of the rectangular box with a square end that satisfies this restriction and has the largest possible volume.

Solution The variables in this problem are the size x of the square end of the box, the length z of the box, and the volume V of the box. We will measure x and z in inches. The function to be maximized is the volume of the box, which is

$$V = x \cdot x \cdot z = x^2 z$$

cubic inches. (See Figure 11.57.)

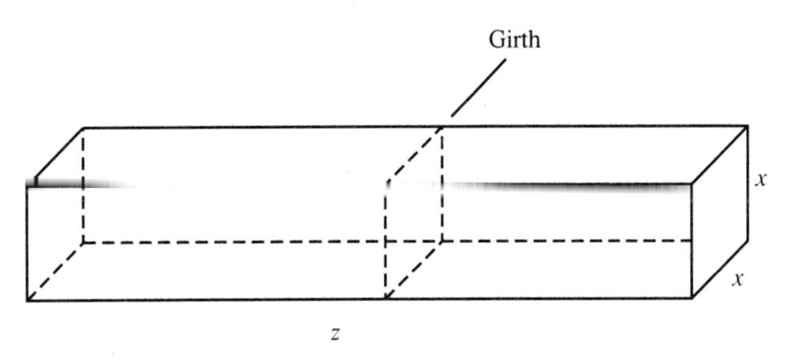

FIGURE 11.57

Since the equation for V involves both x and z, we need to eliminate one of these independent variables. To do so, we must use the condition that the length plus girth not exceed 84 inches. As Figure 11.57 shows, the girth of the box is just the perimeter of its square end, which is $4x$. Hence the condition that the length plus girth not exceed 84 inches translates into the inequality

$$z + 4x \leq 84.$$

In order to obtain the maximum possible volume, we will clearly need to use the largest possible girth of 84 inches; so the preceding inequality gives rise to the equation

$$z + 4x = 84.$$

Thus

$$z = 84 - 4x.$$

Substituting this expression into the formula for V gives

$$V = x^2z = x^2(84 - 4x).$$

Simplifying, we obtain

$$V(x) = 84x^2 - 4x^3,$$

an equation that expresses V in terms of the single variable x.

Now both x and z are lengths of sides of the box, and so they must be nonnegative numbers. Thus

$$0 \leq x \quad \text{and} \quad 0 \leq z.$$

Since $z = 84 - 4x$, we can simplify the latter inequality as follows.

$$\begin{aligned} 0 &\leq 84 - 4x \\ 4x &\leq 84 \\ x &\leq 21 \end{aligned}$$

Therefore x must satisfy $0 \leq x \leq 21$.

In mathematical terminology, we need to determine the absolute maximum of V on the interval $[0, 21]$. Thus we compute

$$V'(x) = 168x - 12x^2 = 12x(14 - x),$$

from which we see that the critical numbers of V are 0 and 14. Hence we must evaluate V at 0, 14, and 21.

x	$V(x)$	
0	0	
14	5488	← **absolute maximum**
21	0	

It follows that the maximum possible volume for the box is 5488 cubic inches, and this is obtained when $x = 14$ (and so $z = 84 - 4x = 28$). Thus the dimensions of the box of greatest volume that can be sent via parcel post are 14 by 14 by 28 inches. ■

EXAMPLE 11.21 A roadside billboard is to be designed with 80 square feet of advertising space plus margins of 2½ feet on the left and right and 2 feet on the top and bottom. State law requires that roadside billboards be between 10 and 20 feet long. What dimensions should the billboard have in order to minimize its total area?

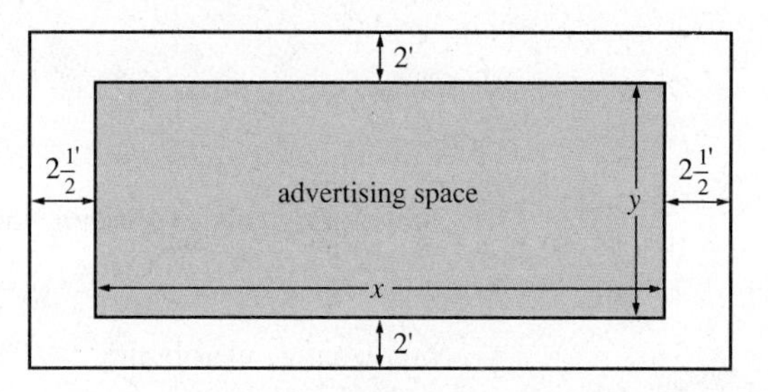

FIGURE 11.58

Solution Let x and y denote the width and height of the advertising space in feet, respectively. Then the billboard is $x + 5$ feet wide and $y + 4$ feet high as we see in Figure 11.58. We wish to minimize the area A of the billboard, which is given by

$$A = (x + 5)(y + 4).$$

Since

$$xy = 80,$$

we can express A as a function of x alone:

$$A = (x + 5)\left(\frac{80}{x} + 4\right)$$
$$= 100 + 4x + \frac{400}{x}.$$

Moreover we must have

$$10 \leq x + 5 \leq 20$$

because roadside billboards are required to be between 10 and 20 feet long. Thus we must determine the absolute minimum of A on the interval $5 \leq x \leq 15$.

Now

$$A' = 4 - \frac{400}{x^2}.$$

Although A' is undefined when $x = 0$, this number is not in the interval $[5, 15]$. Hence the only critical numbers of A in this interval are those for which $A' = 0$.

$$4 - \frac{400}{x^2} = 0$$
$$4 = \frac{400}{x^2}$$
$$4x^2 = 400$$
$$x^2 = 100$$
$$x = \pm 10$$

Therefore A has only one critical number in the interval $[5, 15]$, namely $x = 10$. Hence we must evaluate A at 10, 5, and 15.

x	A	
10	180	⟵ **absolute minimum**
5	200	
15	$186\frac{2}{3}$	

Thus by taking $x = 10$ and $y = 80/x = 8$, we obtain the billboard of minimum area, 180 square feet. The dimensions of this billboard are $x + 5$ by $y + 4$, that is, 15 feet wide by 12 feet high. ■

EXAMPLE 11.22 Morris and Seneca are located 2 miles and 3 miles from the Illinois River, which flows past them along a straight course. (See Figure 11.59.) They have agreed to jointly construct a water treatment facility at the edge of the river for use by both towns. Where should this facility be built so that the total length of the pipelines joining it to the two towns will be a minimum?

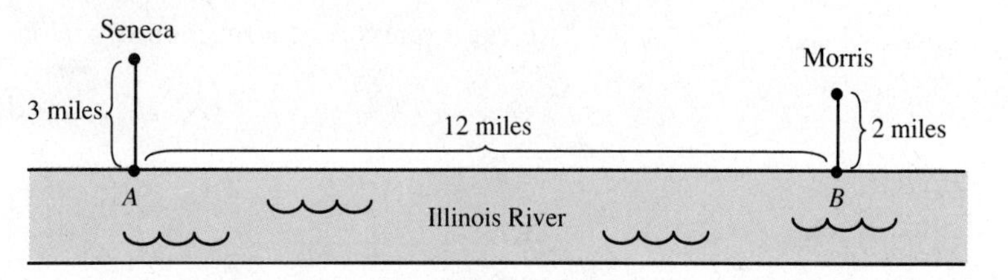

FIGURE 11.59

Solution We must minimize the total length of the pipelines from each town to an arbitrary point along the river's edge. Clearly the minimum length will be obtained by constructing the facility at some point along the river's edge between points A and B in Figure 11.59. Suppose that the facility is constructed at the point F along the river's edge that is x miles from A and y miles from B. (See Figure 11.60.)

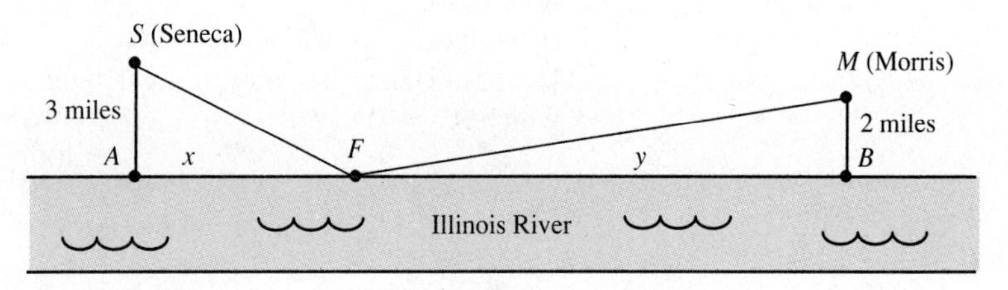

FIGURE 11.60

Applying the Pythagorean theorem to triangles SAF and MFB in Figure 11.60, we see that the length of the pipeline from Seneca to F is $\sqrt{x^2 + 9}$ and the length of the pipeline from Morris to F is $\sqrt{y^2 + 4}$. Hence we must minimize the function

$$L = \sqrt{x^2 + 9} + \sqrt{y^2 + 4}.$$

Since points A and B are 12 miles apart, it follows that

$$x + y = 12,$$

or

$$y = 12 - x.$$

Therefore we can express L as a function of x alone:

$$L = \sqrt{x^2 + 9} + \sqrt{(12 - x)^2 + 4}.$$

Since F is located somewhere between A and B, x must lie in the interval $[0, 12]$. Thus we must find the absolute minimum of L on $[0, 12]$.

We can differentiate L using the chain rule as follows:

$$L' = \frac{1}{2}(x^2 + 9)^{-1/2}(2x) + \frac{1}{2}[(12 - x)^2 + 4]^{-1/2}[2(12 - x)(-1)]$$

$$= \frac{x}{\sqrt{x^2 + 9}} - \frac{12 - x}{\sqrt{(12 - x)^2 + 4}}.$$

Notice that L' is defined throughout the interval [0, 12], and so the only critical numbers of L are those where L' equals 0. We can find these numbers as follows.

$$0 = \frac{x}{\sqrt{x^2 + 9}} - \frac{12 - x}{\sqrt{(12 - x)^2 + 4}}$$

$$\frac{12 - x}{\sqrt{(12 - x)^2 + 4}} = \frac{x}{\sqrt{x^2 + 9}}$$

$$\sqrt{x^2 + 9}(12 - x) = x\sqrt{(12 - x)^2 + 4}$$

Now square both sides of the preceding equation to eliminate the radical signs.

$$(x^2 + 9)(12 - x)^2 = x^2[(12 - x)^2 + 4]$$

$$x^2(12 - x)^2 + 9(12 - x)^2 = x^2(12 - x)^2 + 4x^2$$

$$9(12 - x)^2 = 4x^2$$

$$9x^2 - 216x + 1296 = 4x^2$$

$$5x^2 - 216x + 1296 = 0$$

The quadratic formula gives 7.2 and 36 as solutions of the equation above. Thus we see that the only critical number of L in the interval [0, 12] is $x =$ 7.2, and so we must evaluate L at 7.2, 0, and 12.

x	L	
7.2	13	⟵ **absolute minimum**
0	15.17	
12	14.37	

Therefore the minimum pipe length is 13 miles when the treatment facility is built along the river's edge at the point that is $x = 7.2$ miles from A and $y =$ 4.8 miles from B. ■

Practice Problem 15 An airline has arranged a charter flight that requires at least 60 passengers and can accommodate as many as 120. The price per ticket will be $400 if there are 80 or fewer passengers, but this price will be reduced by $4 for each passenger in excess of 80. (a) Write a formula for $R(x)$, the airline's revenue if there are x passengers on the flight. (b) What is the airline's maximum revenue from this flight?

EXERCISES 11.5

In Exercises 1–12 determine the absolute maximum and absolute minimum of each function on the given interval.

1. $f(x) = x^2 + x - 20$ on $[0, 10]$
2. $F(x) = 2x^2 - x - 3$ on $[0, 5]$
3. $g(x) = -2x^2 + x + 15$ on $[0, 6]$
4. $G(x) = 2x^2 + x - 21$ on $[0, 8]$
5. $h(x) = 2x^5 - 4x^4 + 2x^3$ on $[-1, 2]$
6. $H(x) = 2x^4 - 16x^3 + 32x^2$ on $[1, 6]$
7. $F(x) = 10x^3(x - 5)^2$ on $[-1, 4]$
8. $f(x) = 10x^3(x - 5)^2$ on $[1, 6]$
9. $G(x) = \dfrac{6}{x^2} - \dfrac{6}{x}$ on $[1, 3]$
10. $g(x) = x + \dfrac{9}{x}$ on $[1, 5]$
11. $h(x) = \dfrac{x}{x^2 + 1}$ on $[0, 3]$
12. $H(x) = \dfrac{2\sqrt{x}}{x + 1}$ on $[0, 9]$

13. The alumni association has arranged a bus trip to see the college basketball team play its archrival. The trip can accommodate as many as 40 persons but will be cancelled unless there are at least 20. Cost of the trip plus tickets to the game will be \$12.00 per person if 24 or fewer go on the trip, but the price will be reduced by \$0.25 for each passenger in excess of 24. What is the maximum revenue that the alumni association can earn from this trip?

14. A travel agency can accommodate as many as 120 people on a Hawaiian tour that it will offer if there are at least 60 advance reservations. The tour will cost \$500 each if there are 75 persons or less, but for each person in excess of 75 the fare will be reduced by \$4. What is the maximum revenue that can be earned from this tour?

15. Packages sent by the United Parcel Service must have length plus girth not exceeding 108 inches. Find the dimensions of the rectangular box with a square end that has the largest possible volume of any that can be sent via United Parcel Service.

16. Recall from Example 11.20 that packages sent by parcel post must have length plus girth not exceeding 84 inches. Determine the dimensions of the cylindrical package of greatest volume that can be sent via parcel post.

17. According to reference [2], the body's reaction to a drug is represented by the function

$$R(d) = d^2\left(\frac{c}{2} - \frac{d}{3}\right),$$

where c is the maximum amount of the drug that can be given, d is the dosage administered, and $R(d)$ is a measure of the strength of the body's reaction. (For instance, $R(d)$ can be a change in body temperature or a change in blood pressure.) The body's sensitivity to the drug is measured by the derivative of R with respect to d. Find the dosage for which there is maximum sensitivity to the drug.

18. During a cough the trachea (windpipe) contracts so as to increase the velocity of air emitted. According to reference [3], the velocity v of air flowing through the trachea (in centimeters per second) is approximately

$$v = c(r - r_0)r^2,$$

where r_0 is the normal radius of the trachea, r is its radius during a cough, and c is a negative constant that depends on its length. Find the value of r that maximizes v. (X-ray photography shows that the trachea does contract by this much during a cough.)

19. The cost (in dollars) of constructing an office building containing x floors is given by

$$C(x) = 60{,}000x^2 + 140{,}000x + 3{,}840{,}000.$$

Upon completion, the monthly profit from the building will be $20,000 per floor. The rate at which the construction costs are recovered depends on the ratio R of the monthly profit to $C(x)$. If local zoning laws restrict the height of buildings to at most 15 floors, what value of x maximizes R?

20. Work Exercise 19 if the monthly profit will be $15,000 per floor and

$$C(x) = 40{,}000x^2 + 110{,}000x + 5{,}760{,}000.$$

21. The number of clock radios that a manufacturer can sell each month at a price of p dollars is $x = 2000 - 25p$ and the cost in dollars of manufacturing x radios is $C(x) = 20x + 10{,}000$. How should the radios be priced so as to maximize the manufacturer's profit?

22. The number of pairs of a certain style of men's shoes that a manufacturer can sell each month at a price of p dollars is $x = 1000 - 20p$ and the cost of manufacturing x pairs is $C(x) = 10x + 5000$ dollars. How should the shoes be priced so as to maximize the manufacturer's profit?

23. A cable television service charges $10.00 per month and has 60,000 subscribers. Market surveys show that there are 200,000 potential subscribers for the service, and that for each $0.10 reduction in the monthly charge the number of subscribers will increase by 1000. Under these conditions what monthly charge will yield the greatest revenue?

24. An apartment complex contains 130 units. Records show that there are an average of 20 unoccupied units when the rent is set at $240 per month and that for each $2 reduction in the monthly rent, one additional unit can be filled. Under these conditions what monthly rent will yield the greatest revenue?

25. In order to meet a demand for more electricity, a power company intends to build a generating station at the edge of a river to service two cities. These cities, Algonquin and Boulder, are located 2 miles and 4 miles from the river, which flows past them along a straight easterly course. Where along the 9-mile riverfront between the cities should the generating station be built so that the sum of its distances to the two cities is a minimum?

26. Work Exercise 25 if Algonquin and Boulder are located 4 miles and 3 miles from the river and there is 14 miles of riverfront between them.

27. A printer must produce 100,000 copies of a newsletter. The press used by the printer costs $3.00 per hour to operate and is capable of printing as many as 30 copies simultaneously. Each copy that is printed simultaneously requires a separate master copy costing $0.75. If 1000 copies per hour can be made from each master copy, how many copies of the newsletter should be printed simultaneously in order to minimize the printing costs?

28. The roof of a porch is supported by a cantilever beam of length L. One end of the beam is built into a wall and the other end is supported by a pillar of bricks. The deflection of the beam x feet from the built-in end is given by

$$y = c(8x^4 - 16Lx^3 + 8L^2x^2),$$

where c is a positive constant that depends on the material used in the beam, its shape, and its weight. How far from the built-in end does the maximum deflection of the beam occur?

29. A box is to be constructed with a square base and open top having a volume of 256 square inches. Neglecting the thickness of the box and any waste in construction, determine the dimensions of the box that require the least material.

30. A box is to be made from a piece of cardboard that is 12 inches long and 12 inches wide by cutting four equal squares from the corners and folding up the sides. How long should the sides of the square cuts be in order to maximize the volume of the box?

31. A department store intends to open a garden shop alongside its store. It would like the garden shop to be rectangular in shape and extend the entire 80-foot width of the store (or more). What are the dimensions of the largest area the store can enclose with 100 feet of fencing? (No fencing is required along the wall between the garden shop and the store.)

32. A billboard containing 72 square feet of area is to be designed. The advertiser wants margins of two feet on the left and right sides of the advertisement and margins of one foot on the top and bottom. What dimensions should the billboard have in order to maximize the space for the advertisement?

33. A cylindrical can is to have a volume of v. Find the dimensions of the can (the height and radius of the base) that requires the least material.

34. A Norman window (a window in the shape of a rectangle topped by a semicircle) is to have a perimeter of 12 feet. Find the width and height of such a window that maximizes the amount of light that passes through.

35. An athletic field is to be constructed in the shape of a rectangle with a semicircle at each end. (See the figure below.) If the perimeter of the field is to be a 440-yard track, find the dimensions of the rectangle that yield a field of maximum area.

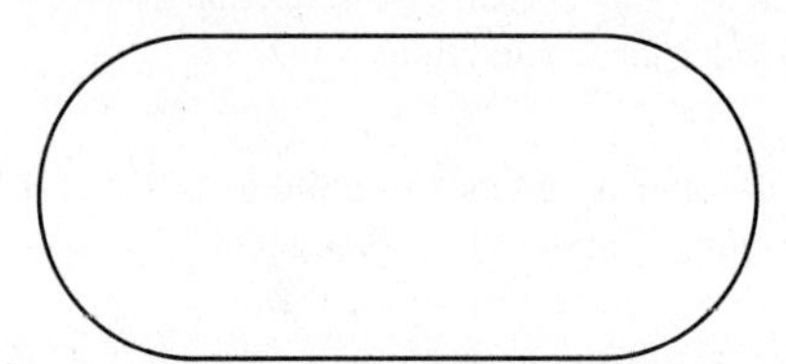

36. A steel mill and a chemical refinery are located 12 miles apart. The steel mill emits 8 times as much particulate matter into the air as the refinery. Assuming that the concentration of particulate matter in the air is proportional to the square of the reciprocal of the distance from the source, at what point on the line between the steel mill and the refinery will the concentration of particulate matter from these two sources be least?

Answers to Practice Problems

13. The absolute maximum of f on $[-1, 7]$ is 363 at $x = 5$, and the absolute minimum of f on this interval is -61 at $x = 7$.

14. The absolute maximum of g on $[0, 3]$ is $\frac{1}{e} \approx 0.368$ at $x = 1$, and the absolute minimum of g on this interval is 0 at $x = 0$.

15. (a) The revenue function is

$$R(x) = \begin{cases} 400x & \text{if } 60 \le x \le 80 \\ (720 - 4x)x & \text{if } 80 < x \le 120. \end{cases}$$

(b) The maximum revenue is $32,400 when there are 90 passengers.

MATHEMATICS IN ACTION

11.6 The Marginal Concept

The concepts introduced in Chapters 9–11 find frequent application in the branch of economics called *the theory of the firm*. This subject is concerned with making production decisions that enable a firm or industry to operate efficiently. Throughout this section we will use the following notation.

$C(x)$ denotes the cost of producing x items.

$R(x)$ denotes the revenue derived from the sale of x items.

$P(x)$ denotes the profit derived from the production and sale of x items.

These functions are related by the condition that profit equals revenue minus cost; hence

$$P(x) = R(x) - C(x).$$

Throughout this section we will measure $C(x)$, $R(x)$, and $P(x)$ in dollars, although any other monetary unit could be used instead, and we will assume that all three of these functions are differentiable.

Economists use the word *marginal* to refer to a rate of change. Thus the **marginal cost** is the rate of change in cost per unit produced, the **marginal revenue** is the rate of change in revenue per unit sold, and the **marginal profit** is the rate of change in profit per unit produced and sold. Because rates of change can be computed by finding derivatives, these marginal functions are simply the derivatives of the functions defined above. That is,

$C'(x)$ is the marginal cost,

$R'(x)$ is the marginal revenue, and

$P'(x)$ is the marginal profit.

EXAMPLE 11.23 A company manufactures tennis balls that it packages in cans of three and sells at a price of \$174 per gross (144 cans). Its cost in dollars of producing x gross per week is given by

$$C(x) = 0.000002x^3 - 0.012x^2 + 144x + 30{,}000.$$

Find the following functions.

(a) marginal cost (b) marginal revenue (c) marginal profit

Solution (a) The marginal cost function is the derivative of the cost function. Hence the marginal cost function is

$$C'(x) = 0.000006x^2 - 0.024x + 144.$$

(b) Likewise the marginal revenue function is the derivative of the revenue function. Since the selling price of the balls is \$174 per gross, the revenue derived from the sale of x gross is

$$R(x) = 174x.$$

The marginal revenue function is therefore

$$R'(x) = 174.$$

(c) The profit function for the tennis balls is

$$\begin{aligned} P(x) &= R(x) - C(x) \\ &= 174x - (0.000002x^3 - 0.012x^2 + 144x + 30{,}000) \\ &= -0.000002x^3 + 0.012x^2 + 30x - 30{,}000. \end{aligned}$$

Thus the marginal profit function is

$$P'(x) = -0.000006x^2 + 0.024x + 30. \quad ■$$

Practice Problem 16 The monthly revenue that a manufacturer of stereophonic speakers derives from the sale of x pairs of speakers is

$$R(x) = 500x - 0.1x^2$$

dollars. If the cost of producing x pairs of speakers per month is

$$C(x) = 50x + 450{,}000$$

dollars, find the following functions.

(a) marginal cost (b) marginal revenue (c) marginal profit

MARGINAL ANALYSIS OF REVENUE FUNCTIONS

In Example 11.23 we assumed that the selling price of the tennis balls is constant. A constant selling price is typical of a competitive market in which there are several manufacturers producing similar products. In this situation each manufacturer is likely to produce only a small portion of the total supply of the product in question, and so changes in any manufacturer's production level have only a minor effect on the available supply of the product. Therefore each manufacturer can assume that the market price of his product is essentially unaffected by modest changes in its level of production. That is, the selling price of the product can be regarded as a constant. In this case the revenue function has the form

$$R(x) = px,$$

where p is the selling price of each unit of the product. Thus

$$R'(x) = p,$$

and so we see that the marginal revenue function is a constant function. This equation tells us that the rate of change of revenue is equal to the selling price of each unit of the product.

On the other hand, a manufacturer whose product has no major competitors produces a large portion of the supply of the product. In this case changes in the level of production can significantly affect the supply of the product, and a change in supply will cause the price that can be charged for the product to move to a point where the demand matches the supply. Thus in this situation the selling price p of the product is a function of x, the quantity produced. (See Figure 11.61.) Moreover, since the price decreases as the production (and supply) increase, the rate of change of p with respect to x is negative; that is, $p'(x) < 0$. Because the revenue derived from the product equals the number sold multiplied by the selling price per item, we have

$$R(x) = xp(x).$$

Therefore since $p'(x) < 0$, the product rule gives

$$R'(x) = xp'(x) + p(x) < p(x).$$

Consequently the product's marginal revenue is always less than its selling price when the manufacturer produces most of the supply of the product.

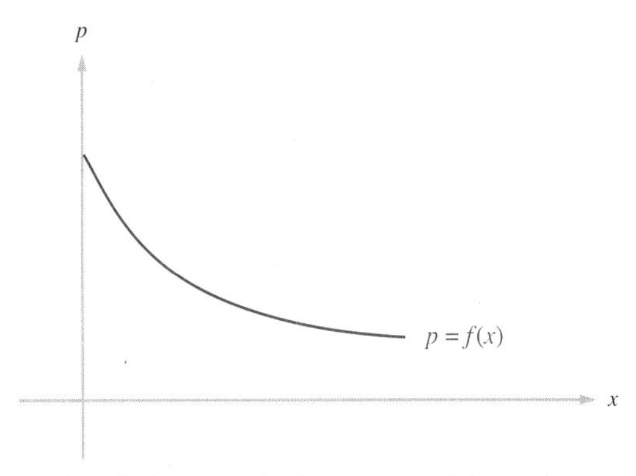

A typical cost function in a noncompetitive market

FIGURE 11.61

MARGINAL ANALYSIS OF COST FUNCTIONS

Economists generally assume that a cost function has the following four properties.

1. There is a positive fixed cost.
2. The cost function is an increasing function of x, the number of units produced.
3. At low levels of production the marginal cost decreases as production increases.
4. At high levels of production the marginal cost increases as production increases.

The reasons for assuming these properties are easy to understand. The first property indicates that in any manufacturing operation there are overhead costs that are independent of the number of items produced. The second property reflects the fact that every product requires some raw materials and labor; and so the more items that are produced, the higher the total cost of production. The third property is a consequence of the fact that at low levels of production, raw materials are likely to be expensive and labor used inefficiently. But as production increases, raw materials can be obtained more cheaply (by purchasing in larger quantities) and labor used more efficiently (by such means as assembly line production). On the other hand, property 4 results from the fact that, at high levels of production, increased expenses are incurred for such things as overtime pay, the hiring of inexperienced workers, and the use of inefficient machinery.

Mathematically, these four properties mean that:

1. $C(0) > 0$.
2. If $x_1 < x_2$, then $C(x_1) < C(x_2)$.
3. At low levels of production the cost function is concave down.
4. At high levels of production the cost function is concave up.

Thus the graph of a typical cost function is shown in Figure 11.62.

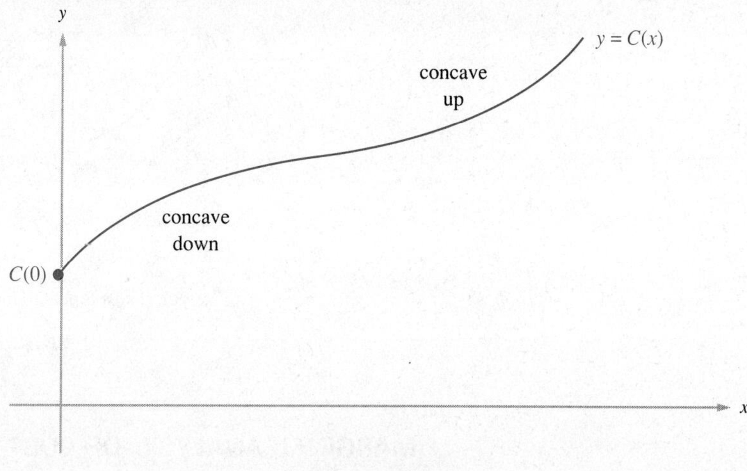

A typical cost function

FIGURE 11.62

Although it is sometimes possible to approximate a cost function by a linear or quadratic function, the simplest functions that exhibit all four of the properties above are polynomial functions of degree three, such as

$$C(x) = 0.000002x^3 - 0.012x^2 + 144x + 30{,}000$$

in Example 11.23.

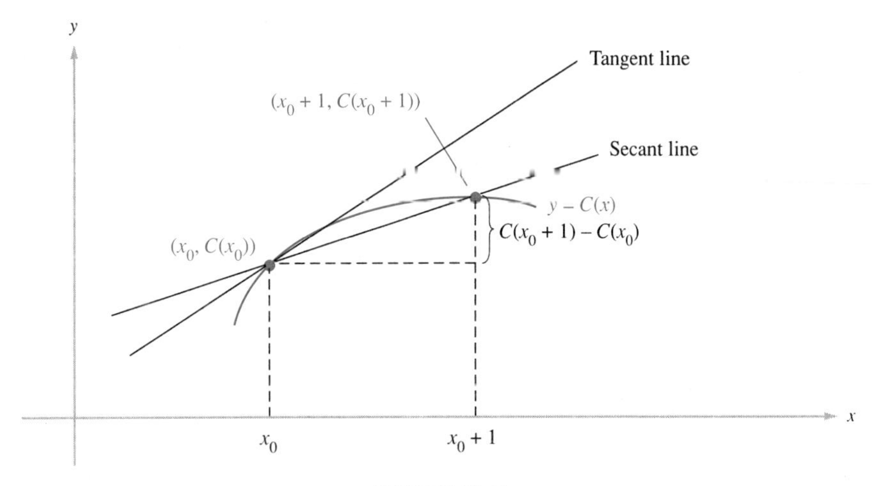

FIGURE 11.63

The marginal cost at a given level of production has a useful interpretation. Recall that the tangent line to the graph of a function is the limiting position of secant lines. Thus the slope of a secant line to the graph can be approximated by the slope of the tangent line. (See Figure 11.63.) But the slope of the secant line in Figure 11.63 is

$$\frac{C(x_0 + 1) - C(x_0)}{(x_0 + 1) - x_0} = C(x_0 + 1) - C(x_0).$$

Since $C(x_0 + 1)$ is the cost of producing $x_0 + 1$ items and $C(x_0)$ is the cost of producing x_0 items, the slope of the secant line

$$C(x_0 + 1) - C(x_0)$$

measures the cost of producing one additional item when the level of production is x_0 units. Hence we have the following useful result.

The value of the marginal cost function at x_0 approximates the cost of manufacturing one additional item when x_0 units are being produced.

This type of information is useful to a manufacturer who is considering an increase in production above some current level, for in this context the difference between the increased revenue from the sale of additional units and the increased production costs will determine whether such an increase in production is desirable.

EXAMPLE 11.24 Approximate the additional cost to produce an extra gross of tennis balls if the manufacturer in Example 11.23 currently produces the following.

(a) 1000 gross of tennis balls per week
(b) 2000 gross of tennis balls per week

Solution (a) The cost of producing an additional gross of balls when current production is 1000 gross per week is approximated by $C'(1000)$, the value of the marginal cost function when $x = 1000$. From Example 11.23 we have

$$C'(x) = 0.000006x^2 - 0.024x + 144.$$

Hence

$$\begin{aligned} C'(1000) &= 0.000006(1000)^2 - 0.024(1000) + 144 \\ &= 6 - 24 + 144 = 126. \end{aligned}$$

Thus making an extra gross per week will cost approximately \$126 at this level of production. Note that the actual cost of producing an extra gross of balls is

$$\begin{aligned} C(1001) - C(1000) &= \$164{,}125.99 - \$164{,}000 \\ &= \$125.99. \end{aligned}$$

(b) Similarly, making an extra gross of tennis balls will cost approximately

$$\begin{aligned} C'(2000) &= 0.000006(2000)^2 - 0.024(2000) + 144 \\ &= 24 - 48 + 144 = 120 \end{aligned}$$

dollars if the manufacturer currently produces 2000 gross per week. In this case the actual cost of producing an extra gross of balls is exactly \$120. ■

Practice Problem 17 Approximate the additional cost to produce an extra gross of tennis balls if the manufacturer in Example 11.24 currently produces the following.

(a) 3000 gross of tennis balls per week
(b) 4000 gross of tennis balls per week

MAXIMIZING PROFITS

Businesses are interested in maximizing their profits. In order for a company to be profitable, the level of production for each of its products must be carefully chosen. Recall that we are assuming the profit function P to be differentiable. Therefore a maximum value of P can occur only at a critical number where $P'(x) = 0$. But since $P(x) = R(x) - C(x)$, we must have

$$0 = P'(x) = R'(x) - C'(x).$$

Hence,

$$C'(x) = R'(x).$$

Thus we have discovered the following general economic principle.*

Profit is maximized when the marginal cost equals the marginal revenue.

EXAMPLE 11.25 Determine the level of production that yields the maximum weekly profit for the manufacturer in Example 11.23. What is this maximum profit?

Solution In Example 11.23 we found that the marginal cost and marginal revenue functions for the manufacturer are

$$C'(x) = 0.000006x^2 - 0.024x + 144 \quad \text{and} \quad R'(x) = 174.$$

The level of production at which a maximum profit occurs is the value of x for which the marginal cost equals the marginal revenue. Hence we equate $C'(x)$ and $R'(x)$ and solve for x.

$$\begin{aligned} C'(x) &= R'(x) \\ 0.000006x^2 - 0.024x + 144 &= 174 \\ 0.000006x^2 - 0.024x - 30 &= 0 \\ 6x^2 - 24{,}000x - 30{,}000{,}000 &= 0 \\ x^2 - 4000x - 5{,}000{,}000 &= 0 \\ (x + 1000)(x - 5000) &= 0 \\ x = -1000 \quad \text{or} \quad x &= 5000 \end{aligned}$$

Since the level of production x must be nonnegative, we see that the maximum profit occurs when $x = 5000$, that is, when the level of production is 5000 gross of tennis balls per week.

Recall from Example 11.23 that

$$P(x) = -0.000002x^3 + 0.012x^2 + 30x - 30{,}000.$$

Thus when the level of production is $x = 5000$, the profit is

$$\begin{aligned} P(5000) &= -0.000002(5000)^3 + 0.012(5000)^2 + 30(5000) - 30{,}000 \\ &= -250{,}000 + 300{,}000 + 150{,}000 - 30{,}000 \\ &= 170{,}000. \end{aligned}$$

*The fact that profit is maximized rather than minimized follows from the nature of revenue and cost functions.

Therefore the manufacturer can earn a maximum weekly profit of \$170,000 by producing 5000 gross of tennis balls per week. Note that in this example

$$P''(x) = -0.000012x + 0.024.$$

Thus

$$P''(5000) = -0.060 + 0.024 = -0.036 < 0,$$

and so the second-derivative test shows that \$170,000 is indeed a maximum value of $P(x)$. ■

Practice Problem 18 Recall the manufacturer of stereophonic speakers (described in Practice Problem 16) for whom the monthly revenue and cost functions are

$$R(x) = 500x - 0.1x^2 \quad \text{and} \quad C(x) = 50x + 450{,}000.$$

Determine the maximum monthly profit and the level of production that yields this profit.

INVESTMENT IN A NEW PRODUCT

A manufacturer who is considering the introduction of a new product into a competitive market must determine if the present market price p will allow operating at a profit. More specifically, the manufacturer must determine if the new product can be produced at a cost of less than p per unit; if not, then the cost per unit will exceed the selling price, and a loss will result.

In this context we are interested in the product's cost per unit produced, which is called its **average cost.** To determine the average cost, we need only divide the cost by the number of units produced. Hence the average cost function A is defined by

$$A(x) = \frac{C(x)}{x}.$$

In order to determine if the proposed new product can be profitably produced, we can compare the minimum value of the average cost function to p, the market price of competing products. As long as the minimum value of the average cost is less than p, it will be possible to produce and sell the new product at a profit. Since we are assuming that C is differentiable, A is defined and differentiable when $x > 0$. Hence we need only determine where $A'(x) = 0$ to find the critical numbers for A. Now

$$A'(x) = \frac{xC'(x) - C(x)}{x^2}.$$

In order that $A'(x) = 0$, the numerator must be 0, that is,

$$xC'(x) - C(x) = 0.$$

It follows that

$$C'(x) = \frac{C(x)}{x} = A(x).$$

We have derived another general economic principle.*

Average cost is minimized when the marginal cost equals the average cost.

EXAMPLE 11.26 A manufacturer of home appliances is considering the production of a new microwave oven. The manufacturer's accountants have estimated that the cost of producing x ovens will be

$$C(x) = 0.000001x^3 - 0.0015x^2 + 240x + 112{,}500.$$

If comparable ovens made by competitors are selling at \$300, can the manufacturer earn a profit by making this oven?

Solution In order to earn a profit from the new oven, the manufacturer must be able to produce the oven at less than \$300, the market price of similar products. Thus we are interested in knowing the minimum value of the average cost function, which occurs where the marginal cost equals the average cost. Consequently we equate the marginal cost and the average cost and solve for x. Now the marginal cost function is

$$C'(x) = 0.000003x^2 - 0.003x + 240,$$

and the average cost function is

$$\frac{C(x)}{x} = 0.000001x^2 - 0.0015x + 240 + \frac{112{,}500}{x}$$

Hence the minimum value of the average cost function occurs where

$$C'(x) = \frac{C(x)}{x}$$

$$0.000003x^2 - 0.003x + 240 = 0.000001x^2 - 0.0015x + 240 + \frac{112{,}500}{x}$$

$$0.000002x^2 - 0.0015x - \frac{112{,}500}{x} = 0$$

$$0.000002x^3 - 0.0015x^2 - 112{,}500 = 0.$$

*The fact that the average cost is minimized rather than maximized follows from the nature of a cost function.

There are various computer techniques for solving such equations. It can be shown that the only positive solution of the preceding equation is approximately $x = 4098.57$. At this level of production the average cost per oven is

$$A(x) = \frac{C(x)}{x} = \frac{C(4098.57)}{4098.57} \approx 278.10.$$

Thus the minimum value of the average cost function is about \$278.10, and it occurs when approximately 4099 ovens are made. Note that in this instance

$$A''(x) = 0.000002 + \frac{225{,}000}{x^3},$$

so that the second-derivative test assures that $A(x)$ does indeed have a minimum value at this level of production.

Since the manufacturer can produce ovens at a cost of \$278.10 per unit and sell them at a cost of \$300, a small profit is possible. Note, however, that the production level must be carefully chosen in order to obtain a profit. At a production level near the value at which the minimum value of the average cost occurs, say at a level of 4100 units, the cost per unit is less than the selling price. But at other levels this relationship may no longer hold; for example, at a production level of 8000 units the average cost per unit is \$306.06, which exceeds the selling price. ■

Practice Problem 19 A manufacturer of paint is considering a new line of quality exterior paint. The cost of producing x thousands of gallons of this paint is expected to be given by

$$C(x) = 0.1x^3 - 5x^2 + 80x + 2.$$

(a) Write an equation for which the solution is the level of production at which the average cost function is minimized.

(b) Show that a solution to the equation in part (a) is approximately $x = 25.016$.

(c) If similar paints are selling for \$17 per gallon, can the manufacturer expect to make a profit on this new line of paint?

EXERCISES 11.6

In Exercises 1–8 compute the marginal cost, marginal revenue, and marginal profit functions.

1. $C(x) = 0.00001x^3 - 0.012x^2 + 20x + 10{,}000$ and $R(x) = 25x$
2. $C(x) = 0.00002x^3 - 0.048x^2 + 300x + 150{,}000$ and $R(x) = 320x$
3. $C(x) = 0.0003x^2 + 12x + 2400\sqrt{x + 1} + 60{,}000$ and $R(x) = 132x$
4. $C(x) = 0.00025x^2 + 8x + 3456\sqrt{x + 1} + 32{,}000$ and $R(x) = 152x$
5. $C(x) = 0.001x^2 + 10x + 2000 \ln (x + 1) + 5000$ and $R(x) = 15x$
6. $C(x) = 0.002x^2 + 20x + 5760 \ln (x + 1) + 12{,}000$ and $R(x) = 32x$
7. $C(x) = 0.0003x^2 + 15x + 5400 \ln (x + 1) + 14{,}000$ and $R(x) = 20x$
8. $C(x) = 0.0002x^2 + 25x + 10{,}000 \ln (x + 1) + 15{,}000$ and $R(x) = 30x$
9. An artist sells reproductions of famous oil paintings. The daily revenue and cost from the production of x copies are approximately $R(x) = 24x - x^2$ and $C(x) = 40 + 6x$, respectively. How many copies should the artist make in order to maximize the daily profit? What is this maximum profit?
10. Rework Exercise 9 if $R(x) = 22x - x^2$.
11. The weekly revenue and cost functions for a manufacturer of men's wallets are approximately $R(x) = 10x - 0.0005x^2$ and $C(x) = 1.50x + 30{,}000$, where x denotes the number of wallets made. How many wallets should be made each week in order to maximize profit? What is this maximum profit?
12. Rework Exercise 11 if $R(x) = 9x - 0.0005x^2$ and $C(x) = 1.50x + 20{,}000$.
13. The monthly revenue and cost functions for a manufacturer of stereophonic speakers are approximately $R(x) = 500x - 0.1x^2$ and $C(x) = 120x + 300{,}000$, where x denotes the number of pairs of speakers made. What level of production maximizes the monthly profit? What is this maximum profit?
14. Rework Exercise 13 if $R(x) = 490x - 0.1x^2$.
15. The monthly revenue and cost functions for a manufacturer of portable television sets are approximately $R(x) = 360x - 0.02x^2$ and $C(x) = 30x + 1{,}260{,}000$, where x denotes the number of sets produced. What level of production maximizes the monthly profit? What is this maximum profit?
16. Rework Exercise 15 if $R(x) = 350x - 0.02x^2$.
17. A manufacturer of electronic equipment is considering the production of a scientific calculator. If x calculators are produced per month, the cost function is expected to be approximately

$$C(x) = 0.00001x^3 - 0.012x^2 + 20x + 10{,}000.$$

   (a) Write an equation for which the solution is the level of production at which the average cost function is minimized.
   (b) Show that a solution to the equation in part (a) is approximately 1052.
   (c) Should the manufacturer produce this calculator if similar competing calculators sell for $25?

18. A manufacturer of portable typewriters is considering the production of a new electric typewriter. If x of these typewriters are produced per year, the cost function is expected to be approximately

$$C(x) = 0.000001x^3 - 0.015x^2 + 180x + 112{,}500.$$

(a) Write an equation for which the solution is the level of production at which the average cost function is minimized.

(b) Show that a solution to the equation in part (a) is approximately 8313.5.

(c) Should the manufacturer produce this typewriter if similar competing models sell for $180?

19. A manufacturer of home appliances is considering the production of a new food processor. If x of these food processors are produced per month, the cost function is expected to be approximately

$$C(x) = 0.000002x^3 - 0.012x^2 + 120x + 30{,}000.$$

(a) Write an equation for which the solution is the level of production at which the average cost function is minimized.

(b) Show that a solution to the equation in part (a) is approximately 3584.

(c) Should the manufacturer produce this food processor if similar competing food processors sell for $150?

20. A manufacturer of home appliances is considering the production of a new sewing machine. If x sewing machines are produced per month, the cost function is expected to be approximately

$$C(x) = 0.00002x^3 - 0.048x^2 + 300x + 150{,}000.$$

(a) Write an equation for which the solution is the level of production at which the average cost function is minimized.

(b) Show that a solution to the equation in part (a) is approximately 2072.8.

(c) Should the manufacturer produce this sewing machine if similar models sell for $350?

Answers to Practice Problems

16. (a) $C'(x) = 50$ (b) $R'(x) = 500 - 0.2x$ (c) $P'(x) = 450 - 0.2x$

17. (a) $126 (b) $144

18. The maximum profit of $56,250 occurs when 2250 pairs of speakers are produced.

19. (a) $A(x) = C'(x)$, which simplifies to $0.2x^3 - 5x^2 - 2 = 0$
(b) $0.2(25.016)^3 - 5(25.016)^2 - 2 \approx 0$
(c) No, the cheapest price at which this paint can be produced is about $17.58 per gallon.

11.7 Further Applications

This section contains three applications that use ideas considered earlier in this chapter. These applications are independent of one another and so can be read in any order.

AN INVENTORY MODEL

During the next year, a college bookstore expects to sell 1000 pen and pencil sets decorated with the school seal. Its supplier is able to fill an order immediately, but the cost of placing each order is \$25. Moreover, the bookstore's average storage cost is \$0.80 per set per year. Assuming that the pen and pencil sets sell at a uniform rate, how many sets should the bookstore order in each shipment?

The problem stated above is an example of an **inventory control** problem. The bookstore has two types of costs to consider—the cost of ordering the pen and pencil sets and the cost of storing them when they arrive. Notice that the larger the number of sets in each shipment, the fewer times the bookstore needs to order from its supplier but the more sets it must store when the order arrives. Hence the two costs are interrelated: A larger shipment size holds down the cost of ordering but raises the cost of storage, whereas a smaller shipment size holds down the storage costs while raising the cost of ordering. By using the techniques discussed in Section 11.5, we can determine the shipment size that minimizes the bookstore's total costs for both ordering and storing.

In order to apply the techniques of calculus, we need to determine the bookstore's total costs for a shipment containing x pen and pencil sets. Consider first the ordering costs. Since each shipment contains x sets, the bookstore will have to place $1000/x$ orders in order to receive a total of 1000 sets. Because the cost of placing each order is \$25, the bookstore's ordering costs during the year will be

$$25\left(\frac{1000}{x}\right) = \frac{25{,}000}{x}$$

dollars.

To determine the bookstore's yearly storage costs, we need to consider the number of sets in storage at any time. Since the sets sell at a uniform rate and the supplier fills orders immediately, Figure 11.64 shows how the inventory of pen and pencil sets varies with time. Note that the number of sets being stored

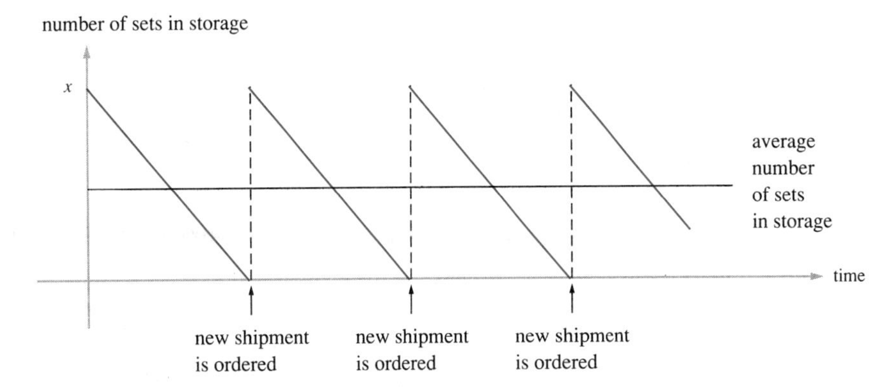

FIGURE 11.64

varies uniformly from x (when a new shipment arrives) to 0 (when a new order is placed). Thus at any given time during the year, the average number of pen and pencil sets being stored is $x/2$. Since it costs \$0.80 per year to store each set, the bookstore's storage costs for the year will be

$$0.80\left(\frac{x}{2}\right) = 0.40x$$

dollars.

From the analysis above we see that the bookstore's total cost (in dollars) for both ordering and storage is

$$C = \frac{25{,}000}{x} + 0.40x.$$

Because the size of a shipment must satisfy $1 \le x \le 1000$, we must determine the absolute minimum of C on the interval [1, 1000]. Now

$$C' = -25{,}000x^{-2} + 0.40.$$

Hence C' is defined throughout the interval [1, 1000]. Setting $C'(x) = 0$ and solving for x yields $x = 250$; so C has a critical number at $x = 250$. Therefore we need to examine the values of C at 1, 250, and 1000.

x	$C(x)$	
1	\$25,000.40	
250	\$200.00	⟵ **absolute minimum**
1000	\$425.00	

Thus we see that the bookstore's minimum costs for ordering and storage are \$200 when $x = 250$ pen and pencil sets are ordered in each shipment. (It follows that the bookstore must place $1000/x = 4$ orders during the year.) The optimal shipment size of 250 sets is called the *economic order quantity*.

Practice Problem 20 A roofing contractor expects to sell 200,000 of a certain brand of shingle during the summer season. There is an \$8 cost to place each order for shingles, and the cost of storing a shingle is \$0.05 per year. How many shingles should be ordered in each shipment so that the contractor minimizes his total costs for ordering and storing?

THE POINT OF DIMINISHING RETURNS

A company has up to \$60,000 to spend in an advertising campaign for one of its products. Experience with similar campaigns shows that if x thousand dollars is spent on advertising, the increase in product sales will be approximately

$$f(x) = 2.4x^2 - 0.025x^3$$

thousand dollars. The company is interested in deciding how much to spend on its advertising campaign. As a step in making this decision, it has graphed the function f. See Figure 11.65.

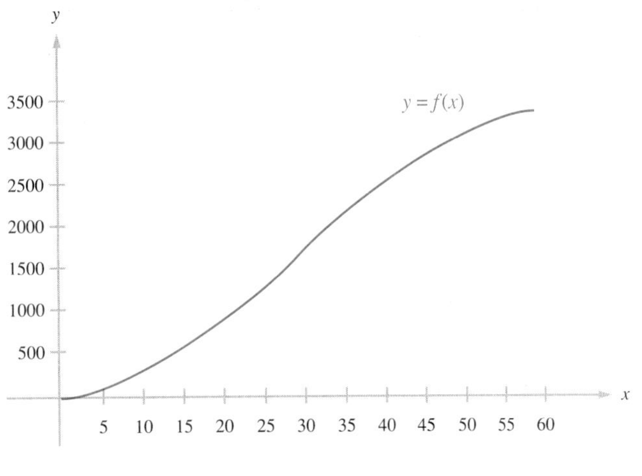

FIGURE 11.65

Since f is increasing on the interval [0, 60], the more that is spent on advertising, the more revenue that the company will receive. Thus we may be tempted to conclude that the company should spend all of the available \$60,000 in its advertising campaign. But a closer analysis shows that there is more to consider than the total revenue received: The company should advertise only as long as the *rate of change* of new sales with respect to the amount spent on advertising (rather than the *amount* of new sales) is increasing. For instance, spending \$1000 on advertising that results in only \$800 worth of increased sales is no bargain. The point at which the rate of additional revenue begins to decrease is called the *point of diminishing returns*.

Thus the company must determine the number at which the *first derivative* of f with respect to x (which measures the rate of increase of f) changes from increasing to decreasing. As in our discussion of concavity in Section 11.3, this will be a number where the *second derivative* of f changes from positive to negative. In other words, the number we seek corresponds to a point of inflection of f. As before, we can find the points of inflection by determining the numbers for which the second derivative is undefined or equals zero. Since

$$f'(x) = 4.8x - 0.075x^2,$$

we have

$$f''(x) = 4.8 - 0.150x.$$

Because f'' is a polynomial function, it is defined for all values of x. Clearly $f''(x) = 0$ precisely when

$$x = \frac{4.8}{0.150} = 32.$$

Since f'' changes from positive to negative at $x = 32$, 32 is the point of diminishing returns. Hence the company should spend 32 thousand dollars on its advertising campaign in order to increase sales of its product by

$$f(32) = 2.4(32)^2 - 0.025(32)^3 = 1638.40$$

thousand dollars (i.e., by $1,638,400).

In Figure 11.66 we see that f is concave up on the interval [0, 32], and f' is increasing on this interval. Hence if $0 \leq x \leq 32$, each additional dollar of advertising results in more revenue from new sales than the previous dollar. On the other hand, f is concave down on the interval [32, 60], and f' is decreasing there. Hence for $32 \leq x \leq 60$ each additional dollar of advertising results in less revenue from new sales than the previous dollar. For this reason any expenditure on advertising beyond the point of diminishing returns is not considered to be a good use of capital.

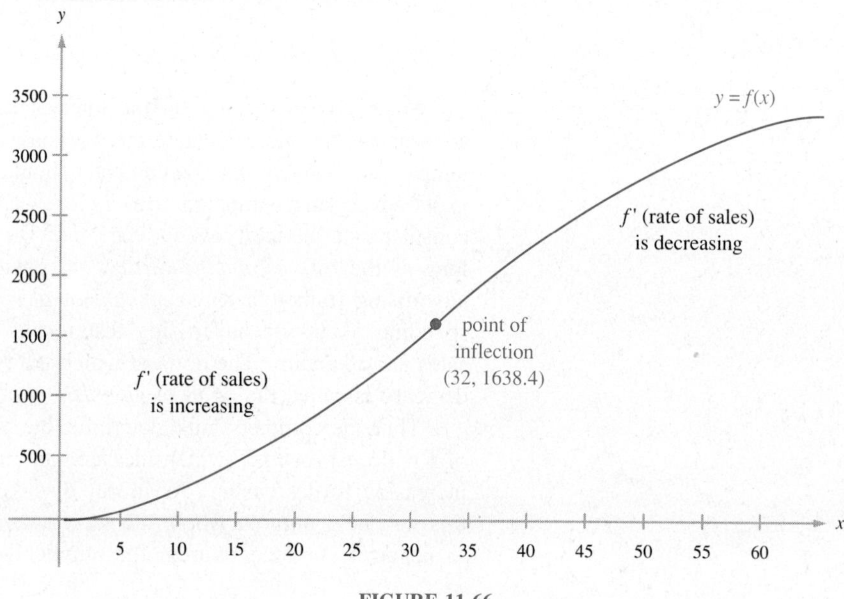

FIGURE 11.66

Practice Problem 21

The state fair has $30,000 available for advertising its grandstand shows. Past experience shows that if x thousand dollars are spent on advertising, then an additional

$$g(x) = \quad 0.01x^3 + 0.6x^2$$

thousand dollars in revenue will be received. Determine how much should be spent on advertising by computing the point of diminishing returns.

REPLACEMENT OF DURABLE GOODS

Durable goods such as automobiles and large appliances eventually must be replaced. Because these items are expensive, it is not economical to replace them frequently. On the other hand, the cost of repairing an item increases as it gets older. Thus consumers often must decide whether to replace an old item or continue to repair it.

In reference [1] Brems discusses a mathematical model for determining the cost of durable goods at any moment during their useful lives. His model assumes (as is typical of most consumers) that little preventive maintenance is performed on the item, and so almost all of the costs incurred after purchasing the item are repair costs. Brems also assumes that the number of repairs n required by an item during t years is given by a function of the form

$$n = \frac{t^{\alpha}}{a},$$

where a is a positive constant that measures the durability of the item and α is a constant greater than 1.

Suppose, for example, that the number of repairs required by a certain model of dishwasher during t years of ownership is given by

$$n = \frac{t^2}{8}.$$

(Thus $\alpha = 2$ and $a = 8$ in the notation above.) If the cost of replacing this dishwasher is $500 and the average cost per repair is $50, when should the dishwasher be replaced by a new one?

Since the cost of a new dishwasher is $500, its cost per year is

$$\frac{500}{t}$$

dollars if it is replaced after t years. Likewise the average number of repairs per year is

$$\frac{n}{t} = \frac{t}{8},$$

and so the average yearly repair cost is

$$50\left(\frac{t}{8}\right)$$

dollars. Hence the total cost per year for the dishwasher, including both its purchase cost and the cost of repairs, is

$$C = \frac{500}{t} + \frac{50t}{8}.$$

To determine the optimal time to replace the dishwasher, we must find the positive value of t that minimizes C. Now

$$C' = -500t^{-2} + \frac{50}{8}.$$

Since C' is defined for all positive values of t, the minimal value of C must occur at a critical number for which $C'(t) = 0$. We can determine this value as follows.

$$\begin{aligned} 0 &= -500t^{-2} + \frac{50}{8} \\ 500t^{-2} &= \frac{50}{8} \\ 4000 &= 50t^2 \\ 80 &= t^2 \\ t &= \sqrt{80} \approx 8.94 \end{aligned}$$

Thus the dishwasher should be replaced about 9 years after it is purchased.

Practice Problem 22 The number of repairs required by a certain model refrigerator in t years is $n = t^{1.25}/1.6$, and the average cost per repair is \$100. If the cost of replacing the refrigerator is \$500, when should the refrigerator be replaced?

EXERCISES 11.7

In Exercises 1–6 determine the shipment size that minimizes the total ordering and storage costs under the given conditions.

1. The demand is 24,000, the yearly storage cost per unit is \$3.20, and the cost per order is \$24.

2. The demand is 3200, the yearly storage cost per unit is \$1.80, and the cost per order is \$45.

3. The demand is 20,000, the yearly storage cost per unit is \$2.50, and the cost per order is \$40.

4. The demand is 60,000, the yearly storage cost per unit is \$2.00, and the cost per order is \$24.

5. The demand is 1500, the yearly storage cost per unit is \$2.00, and the cost per order is \$60.

6. The demand is 4860, the yearly storage cost per unit is \$7.50, and the cost per order is \$25.

7. A company has up to \$50,000 to spend in an advertising campaign. If it spends x thousand dollars on advertising, its increase in product sales will be approximately $f(x) = 0.9x^2 - 0.01x^3$ thousand dollars. Determine the point of diminishing returns for this advertising campaign.

8. A company has up to \$100,000 to spend in an advertising campaign. If it spends x thousand dollars on advertising, its increase in product sales will be approximately $f(x) = 0.36x^2 - 0.003x^3$ thousand dollars. Determine the point of diminishing returns for this advertising campaign.

9. A company has up to \$72,000 to spend in an advertising campaign. If it spends x thousand dollars on advertising, its increase in product sales will be approximately $f(x) = 0.06x^2 - 0.0004x^3$ thousand dollars. Determine the point of diminishing returns for this advertising campaign.

10. A company has up to \$80,000 to spend in an advertising campaign. If it spends x thousand dollars on advertising, its increase in product sales will be approximately $f(x) = 0.45x^2 - 0.0025x^3$ thousand dollars. Determine the point of diminishing returns for this advertising campaign.

11. A company has up to \$60,000 to spend in an advertising campaign. If it spends x thousand dollars on advertising, its increase in product sales will be approximately $f(x) = 0.36x^2 - 0.005x^3$ thousand dollars. Determine the point of diminishing returns for this advertising campaign.

12. A company has up to \$48,000 to spend in an advertising campaign. If it spends x thousand dollars on advertising, its increase in product sales will be approximately $f(x) = 0.33x^2 - 0.0025x^3$ thousand dollars. Determine the point of diminishing returns for this advertising campaign.

13. The purchase price for a certain microwave oven is \$135, its average repair cost is \$25, and the number of repairs required in t years is $t^{1.5}/2.5$. When should the oven be replaced in order to minimize the total purchase and repair costs?

14. The purchase price for a certain furnace is \$6400, its average repair cost is \$50, and the number of repairs required in t years is $t^{2.5}/12$. When should the furnace be replaced in order to minimize the total purchase and repair costs?

15. The purchase price for a certain microcomputer is \$800, its average repair cost is \$80, and the number of repairs required in t years is $t^2/3.6$. When should the microcomputer be replaced in order to minimize the total purchase and repair costs?

16. The purchase price for a certain sump pump is \$200, its average repair cost is \$60, and the number of repairs required in t years is $t^{4/3}/1.6$. When should the sump pump be replaced in order to minimize the total purchase and repair costs?

17. The purchase price for a certain automobile is \$12,000, its average repair cost is \$300, and the number of repairs required in t years is $t^3/50$. When should the automobile be replaced in order to minimize the total purchase and repair costs?

18. The purchase price of a new roof is \$4000, its average repair cost is \$60, and the number of repairs required in t years is $t^{1.8}/3$. When should the roof be replaced in order to minimize the total purchase and repair costs?

Let x denote the number of units of some product that can be sold at a price p. Note that we are regarding x as a function of p. Economists call the expression

$$\eta = \frac{-p}{x}\frac{dx}{dp}$$

the **price elasticity of demand.** *It measures the responsiveness of consumer demand for the product to changes in the price of the product.*

19. Compute the price elasticity of demand if $x = 100 - 2p$.

20. Compute the price elasticity of demand if $x = 10{,}000 - p^2$.

21. Compute the price elasticity of demand if $x = c/p$, where c is a positive constant.

22. Compute the price elasticity of demand if $x = 500 - 50 \ln 20p$.

23. Show that if $\eta < 1$, then a rise in price results in increased revenue. In this case the demand function is called *inelastic.*

24. Show that if $\eta > 1$, then a rise in price results in decreased revenue. In this case the demand function is called *elastic.*

25. If $\eta = 1$, what can be said about the revenue function?

26. Let $x = b - mp$, where b and m are constants and $m > 0$.
(a) Show that if $p < b/2m$, then the demand function is inelastic.
(b) Show that if $p > b/2m$, then the demand function is elastic.

27. If $R = px$ is the revenue function, show that

$$\frac{dR}{dx} = \frac{R}{x}\left(1 - \frac{1}{\eta}\right).$$

This equation states that the marginal revenue equals $\left(1 - \frac{1}{\eta}\right)$ times the average revenue per unit sold. *Hint:* Obtain two expressions for dR/dp, one by using the chain rule and the other by applying the rule for the derivative of a product of functions to $R = px$.

Answers to Practice Problems **20.** 8000 **21.** \$20,000 **22.** after 16 years

CHAPTER 11 REVIEW

IMPORTANT TERMS

- **concave down** *(11.1)*
- **concave up**
- **decreasing function**
- **increasing function**
- **local extremum**
- **local maximum**
- **local minimum**
- **point of inflection**
- **critical number** *(11.2)*
- **first-derivative test**
- **second-derivative test** *(11.3)*
- **absolute maximum** *(11.5)*
- **absolute minimum**
- **average cost** *(11.6)*
- **marginal cost**
- **marginal profit**
- **marginal revenue**

REVIEW EXERCISES

In Exercises 1–4 determine the intervals on which the given function is increasing, decreasing, concave up, and concave down. Also find any points of inflection.

1. $f(x) = x^3 - 3x^2 - 24x - 12$

2. $g(x) = -x^3 - 7x^2 + 5x + 4$

3. $F(x) = 2x^2 - x^4$

4. $G(x) = x^4 + 4x^3$

In Exercises 5–8 determine all the extrema of the given function.

5. $f(x) = -x^3 + 6x^2 + 12x - 10$

6. $F(x) = x^5 - 15x^3$

7. $g(x) = \dfrac{\sqrt{x}}{x + 1}$ for $x \geq 0$

8. $G(x) = xe^{-x}$

In Exercises 9–12 graph the given function.

9. $f(x) = x^3 + 3x^2 - 9x - 7$

10. $g(x) = \dfrac{1}{x^2 - 1}$

11. $F(x) = \dfrac{\ln x}{x^2}$ for $x > 0$

12. $G(x) = \sqrt{x}e^{-\sqrt{x}}$ for $x \geq 0$

13. Packages sent via United Parcel Service must have length plus girth not exceeding 108 inches. Determine the dimensions of the cylindrical package of greatest volume that can be sent via United Parcel Service.

14. Cable must be laid from a generating station on one bank of a river to a new factory on the other bank 10 miles downstream. Cost of laying the cable is \$13,000 per mile under the river and \$5000 per mile on land. If the river is 4.8 miles wide, determine the minimum cost of laying the cable.

15. A watchmaker's revenue and cost (in dollars) are given by

$$R(x) = 15.60x - 0.001x^2 \quad \text{and} \quad C(x) = 0.0001x^3 - 0.004x^2 + 12x + 100,$$

respectively, where x denotes the number of watches produced per day.

(a) Find the marginal cost, marginal revenue, and marginal profit functions.

(b) Determine the daily production level that maximizes the company's profit from the sale of these watches.

16. A company that manufactures lighting fixtures is interested in introducing a new fixture. The company's yearly cost (in dollars) to produce x of these new fixtures is expected to be

$$C = 0.000002x^3 - 0.012x^2 + 78x + 20{,}000.$$

(a) Approximate the cost of producing an additional fixture when the yearly production level is 3000 fixtures.

(b) Write the average cost function for this new fixture.

(c) Write an equation for which the solution is the level of production that minimizes the average cost of each fixture produced.

(d) Show that 3426 is a solution of this equation.

(e) Can the company afford to produce the new fixtures if similar competing fixtures sell for \$80?

17. The automotive department of a retail store expects to sell 60,000 of its most popular line of automobile tires next year. The cost of ordering the tires is \$80 per order, and the yearly storage cost per tire is \$3.75. Determine the shipment size that minimizes the store's total ordering and storage costs under these conditions.

18. A company has up to \$60,000 to spend in an advertising campaign. If it spends x thousand dollars on advertising, its increase in product sales will be approximately $f(x) = 0.12x^2 - 0.002x^3$ thousand dollars. Determine the point of diminishing returns for this advertising campaign.

19. The purchase price for a certain washing machine is \$400, its average repair cost is \$50, and the number of repairs required in t years is $\frac{t^2}{24.5}$. When should the washing machine be replaced in order to minimize the total purchase and repair costs?

REFERENCES

1. Brems, Hans, *Quantitative Economic Theory: A Synthetic Approach*. New York: Wiley, 1968, Chapter 4.
2. Thrall, R. M., J. A. Mortimer, K. R. Rebman, and R. F. Baum, eds., *Some Mathematical Models in Biology,* rev. ed. Washington: U.S. Department of Commerce, 1967, p. 221.
3. Tuchinsky, Philip, *The Human Cough*. Newton, MA: Education Development Center, 1978.

12

INTEGRATION

In previous chapters we studied the techniques and usefulness of differentiation. Now we will look at the reverse process, called integration. We will learn how to find a function which has a given derivative, and investigate the geometric interpretation and real-world applications of this procedure.

12.1 Antidifferentiation

Recall from Section 11.6 that economists often use the word *marginal* to denote a rate of change. In particular, if $C(x)$ is the cost of producing x items, then the **marginal cost** is the derivative $C'(x)$.

A company manufacturing potato mashers has found that the marginal cost of producing x of the utensils is given by $C'(x) = 1/\sqrt{x}$, and that the fixed cost is $10,000. The company would like to determine the cost $C(x)$ of producing x mashers. This is a problem of finding a function when its derivative is given.

In general if f is a given function and F is a function such that $F'(x) = f(x)$ for all x in some set, then we say that F is an **antiderivative** of f on that set. For example $f(x) = 2x$ has $F(x) = x^2$ as an antiderivative on the real numbers, since $F'(x) = (x^2)' = 2x$ for all real numbers x. The function x^2 is not the only antiderivative of $2x$, however. Another is $x^2 + 5$, since

$$(x^2 + 5)' = 2x + 0 = 2x.$$

Clearly any function of the form $x^2 + k$, where k is a constant, is an antiderivative of $2x$. It turns out that every antiderivative of $2x$ is of the form $x^2 + k$. This is a consequence of the following theorem.

Theorem 12.1

If F_1 and F_2 are both antiderivatives of the function f on some interval, then

$$F_2(x) = F_1(x) + k$$

for some constant k.

The fact that the set of antiderivatives of the function $2x$ is the set of functions of the form $x^2 + k$ is symbolized by the notation

$$\int 2x\, dx = x^2 + k. \tag{12.1}$$

Here the symbol $\int$ is called an **integral sign** and the expression on the left is read "the indefinite integral of $2x$, dx." Thus by the **indefinite integral** of a function we mean the set of antiderivatives of the function. The constant k in

(12.1) is called the **constant of integration.** The "dx" on the left side indicates that x is the variable with respect to which we wish to reverse differentiation; this variable is called the **variable of integration.** The function, in this case $2x$, for which we are finding all antiderivatives is called the **integrand.** In Section 12.2 we will define dx as a mathematical variable that multiplies the integrand.

The process of finding antiderivatives is usually called **integration,** and is in general much more difficult than differentiation. For example, although the product rule for differentiation allows us to compute the derivative of fg if we know f' and g', there is no corresponding rule for computing

$$\int f(x)g(x)\,dx,$$

even if we know antiderivatives for f and g.

Nonetheless antiderivatives of certain standard forms are easily computed. For example, if $f(x)$ is a power of x, we have the following result.

Indefinite Integral of a Power

$$\int x^n\,dx = \frac{x^{n+1}}{n+1} + k \qquad \text{if } n \neq -1. \tag{12.2}$$

This result, like other antidifferentiation formulas, is easily checked by differentiating the right side:

$$\left(\frac{x^{n+1}}{n+1} + k\right)' = (n+1)\frac{x^{n+1-1}}{n+1} + 0 = x^n.$$

Notice that when $n = 0$, formula (12.1) gives us

$$\int 1 \cdot dx = \int x^0\,dx = \frac{x^{0+1}}{0+1} + k = x + k,$$

as we would expect. Note that (12.2) does not apply when $n = -1$; this case will be treated later in this section.

EXAMPLE 12.1 Find (a) $\int x^3\,dx$ (b) $\int x^{-2}\,dx$ (c) $\int 5x^3\,dx$.

Solution (a) Applying 12.2 with $n = 3$ we have

$$\int x^3\,dx = \frac{x^{3+1}}{3+1} + k = \frac{x^4}{4} + k.$$

(b) Applying 12.2 with $n = -2$ gives

$$\int x^{-2}\,dx = \frac{x^{-2+1}}{-2+1} + k = \frac{x^{-1}}{-1} + k = -x^{-1} + k.$$

(c) In part (a) we saw that $x^4/4$ has the derivative x^3. Thus a function with derivative $5x^3$ is $5x^4/4$, since

$$\left(\frac{5x^4}{4}\right)' = 5\left(\frac{x^4}{4}\right)' = 5x^3.$$

Therefore

$$\int 5x^3\,dx = \frac{5x^4}{4} + k. \quad \blacksquare$$

Part (c) of the last example illustrates the fact that an antiderivative of a constant times a function is that constant times an antiderivative of the function. This follows from the corresponding rule for derivatives. Likewise the differentiation rules $(F \pm G)' = F' \pm G'$ imply the second formula below.

Linearity Rules for Indefinite Integrals

$$\int cf(x)\,dx = c\int f(x)\,dx \quad \text{if } c \text{ is any constant} \qquad (12.3)$$

$$\int [f(x) \pm g(x)]\,dx = \int f(x)\,dx \pm \int g(x)\,dx \qquad (12.4)$$

Practice Problem 1 Find $\int (3x^5 + 6x^{-3})\,dx$.

Notice that while (12.3) allows us to pull a *constant* "through the integral sign," the corresponding statement is not true for the variable of integration. For example, $\int x^2\,dx \neq x \int x\,dx$, for

$$\int x^2\,dx = \frac{x^3}{3} + k,$$

while

$$x\int x\,dx = x\left(\frac{x^2}{2} + k\right) = \frac{x^3}{2} + kx.$$

Let us use (12.2) to solve the problem at the beginning of this section, which involved finding a cost function C with marginal cost $C'(x) = 1/\sqrt{x}$ and fixed cost \$10,000. We have

$$\begin{aligned} C(x) &= \int \frac{1}{\sqrt{x}}\,dx = \int x^{-1/2}\,dx \\ &= \frac{x^{-1/2+1}}{-\frac{1}{2}+1} + k \\ &= \frac{x^{1/2}}{\frac{1}{2}} + k \\ &= 2x^{1/2} + k. \end{aligned}$$

Now we use our knowledge of the fixed cost, that is, the cost when $x = 0$, to determine k. Since the fixed cost is \$10,000, we have

$$10{,}000 = C(0) = 2(0)^{1/2} + k = k.$$

Thus $C(x) = 2x^{1/2} + 10{,}000$.

In general, for different values of k the antiderivatives $F(x) + k$ of a function have similar graphs. For example the graphs of the three antiderivatives x^2, $x^2 + 1$, and $x^2 + 3$ of $f(x) = 2x$ are pictured in Figure 12.1.

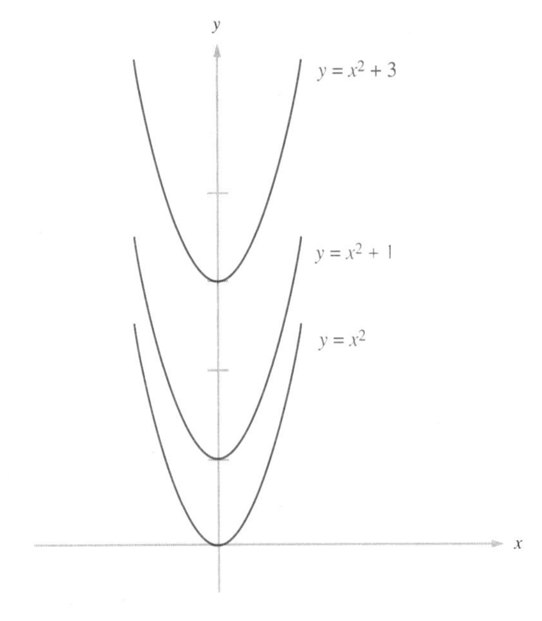

FIGURE 12.1

If a particular point on the antiderivative curve is specified, then k can be determined. Thus in the potato masher example we know by integrating $C'(x) = 1/\sqrt{x}$ that $C(x) = 2x^{1/2} + k$ for some constant k. The fact that $C(x) = 10{,}000$ when $x = 0$ tells us that the point (0, 10,000) is on the graph of $C(x)$, which implies that the particular function we want is $C(x) = 2x^{1/2} + 10{,}000$.

EXAMPLE 12.2 Determine a function F such that $F'(x) = 2x^4 - 5$ and $F(1) = 2$.

Solution We have

$$F(x) = \int (2x^4 - 5)\,dx = \int 2x^4\,dx + \int (-5)\,dx \qquad \textbf{(using 12.4)}$$

$$= 2\int x^4\,dx + (-5)\int 1\,dx \qquad \textbf{(using 12.3)}$$

$$= \frac{2x^5}{5} + (-5)x + k. \qquad \textbf{(using 12.2)}$$

Also

$$2 = F(1) = \frac{2(1)^5}{5} + (-5)(1) + k = \frac{2}{5} - 5 + k,$$

and so

$$k = 2 - \frac{2}{5} + 5 = \frac{33}{5}.$$

Thus we see that

$$F(x) = \frac{2x^5}{5} - 5x + \frac{33}{5}. \quad ■$$

Practice Problem 2 Determine a function F such that $F'(x) = \frac{3}{x^2} + 6$ and $F(2) = 1$.

OTHER INTEGRATION FORMULAS

The formulas we derived for the derivatives of the logarithmic and exponential functions can be turned around to yield antidifferentiation rules. For example the differentiation formula $(\ln x)' = 1/x$ leads to the following:

$$\int x^{-1}\, dx = \ln x + k.$$

A defect of this formula is that the right side only makes sense when $\ln x$ is defined, that is, for $x > 0$. The function x^{-1} is defined for all x except $x = 0$, on the other hand. It turns out that we can find a more comprehensive antiderivative by considering the function $\ln |x|$. Notice that

$$(\ln |x|)' = \begin{cases} (\ln x)' = x^{-1} & \text{if } x > 0 \\ [\ln(-x)]' = (-x)^{-1}(-x)' = -x^{-1}(-1) = x^{-1} & \text{if } x < 0. \end{cases}$$

(See Figure 12.2.) Since in either case $(\ln |x|)' = x^{-1}$, we have the following result, which covers the exponent -1, excluded in (12.2).

Indefinite Integral of x^{-1}

$$\int x^{-1}\, dx = \ln |x| + k \tag{12.5}$$

The formula for the antiderivatives of an exponential function is easily derived from the fact that $(e^{cx})' = ce^{cx}$.

Indefinite Integral of an Exponential Function

$$\int e^{cx}\, dx = \frac{e^{cx}}{c} + k, \quad \text{if } c \text{ is any nonzero constant.} \tag{12.6}$$

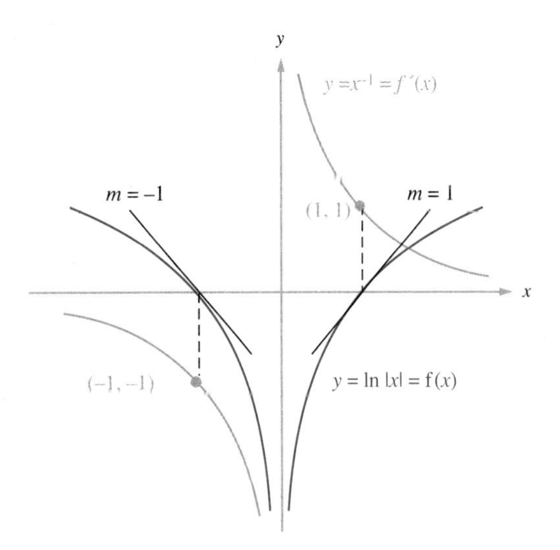

FIGURE 12.2

EXAMPLE 12.3 Find $\int \left(8e^{-2x} + \frac{4}{x} + \frac{6}{x^3}\right) dx.$

Solution We have

$$\begin{aligned}\int \left(8e^{-2x} + \frac{4}{x} + \frac{6}{x^3}\right) dx &= 8\int e^{-2x}\,dx + 4\int x^{-1}\,dx + 6\int x^{-3}\,dx\\ &= \frac{8e^{-2x}}{-2} + 4\ln|x| + \frac{6x^{-2}}{-2} + k\\ &= -4e^{-2x} + 4\ln|x| - 3x^{-2} + k.\end{aligned}$$

Check

$$\begin{aligned}(-4e^{-2x} + 4\ln|x| - 3x^{-2} + k)' &= (-4)(-2)e^{-2x} + 4\left(\frac{1}{x}\right) - 3(-2)x^{-3}\\ &= 8e^{-2x} + \frac{4}{x} + \frac{6}{x^3} \quad ■\end{aligned}$$

Note that *the computation of any indefinite integral can always be checked by differentiation of the answer*. Antidifferentiation can be difficult, and one may not always be able to find an antiderivative for a given function. However, there is no excuse for coming up with a *wrong* antiderivative, since differentiating will always reveal the mistake.

Practice Problem 3 (a) Find $\int \left(3x + 5 + \frac{3}{x}\right) dx.$

(b) Find an antiderivative F of $6e^{3x} + 2$ such that $F(2) = 1$.

EXAMPLE 12.4 It has been determined experimentally that t seconds after being dropped an object will fall with a velocity $v(t) = -16t$ feet per second, where $v(t)$ is negative because we take the positive direction as upward. An object is dropped from a building 500 feet high. (a) How high will it be after falling for 2 seconds? (b) How long will it take to hit the ground?

Solution Let $s(t)$ be the distance the object is from the ground t seconds after being dropped. Then $v(t) = s'(t)$, and s is an antiderivative of v. Thus

$$\begin{aligned} s(t) &= \int (-16t)\,dt \\ &= \frac{-16t^2}{2} + k \\ &= -8t^2 + k. \end{aligned}$$

Since the building is 500 feet high, $s(t) = 500$ when $t = 0$. Thus

$$500 = s(0) = -8(0)^2 + k = k.$$

We see that $s(t) = -8t^2 + 500$. In particular the height of the object after 2 seconds is

$$s(2) = -8(2)^2 + 500 = -32 + 500 = 468 \text{ feet}.$$

This is the answer to question (a).

To answer question (b) we must determine the value of t when $s(t) = 0$. Thus

$$\begin{aligned} -8t^2 + 500 &= 0, \\ 8t^2 &= 500, \\ t^2 &= \frac{500}{8} = 62.5, \\ t &= \sqrt{62.5} \approx 7.91. \end{aligned}$$

The object will hit the ground after approximately 7.91 seconds. ■

Practice Problem 4 Suppose that the marginal cost of making x hair clippers is $C'(x) = 30 - \sqrt[3]{x}$ and the fixed cost is \$20,000. (a) Find a formula for $C(x)$. (b) What is the cost of making 1000 hair clippers?

EXERCISES 12.1

For each of the pairs of functions f and g in Exercises 1–8 answer the following questions. (a) Is f an antiderivative of g? (b) Is g an antiderivative of f?

1. $f(x) = 2x^5, \quad g(x) = 10x^4$

2. $f(x) = 3x^8, \quad g(x) = \frac{1}{3}x^9$

3. $f(x) = \sqrt{x}, \quad g(x) = \frac{1}{\sqrt{x}}$

4. $f(x) = 2x + 3, \quad g(x) = x^2 + 3x + 5$

5. $f(x) = x^4 - x^3, \quad g(x) = 4x^3 - 3x^2 + 2$

6. $f(x) = e^{-x}, \quad g(x) = -e^{-x}$

7. $f(x) = \ln|2x|, \quad g(x) = \dfrac{1}{x}$

8. $f(x) = e^{2x} + 6x, \quad g(x) = 2e^{2x} + 3x^2$

In Exercises 9–30 find the given indefinite integral.

9. $\int x^7\,dx$

10. $\int x^{11}\,dx$

11. $\int x^{-5}\,dx$

12. $\int x^{-3}\,dx$

13. $\int x^{3/5}\,dx$

14. $\int x^{-5/3}\,dx$

15. $\int \sqrt[3]{x}\,dx$

16. $\int \sqrt[4]{x}\,dx$

17. $\int (x + 3)\,dx$

18. $\int (4 - x^2)\,dx$

19. $\int (3x^3 + \sqrt{x} + 5x)\,dx$

20. $\int \left(3x^2 - \dfrac{1}{x}\right)dx$

21. $\int \left(\dfrac{4}{\sqrt{x}} - \dfrac{5}{x}\right)dx$

22. $\int \left(\dfrac{3}{x^2} + \dfrac{2}{\sqrt[3]{x}}\right)dx$

23. $\int (e^{3x} + 2)\,dx$

24. $\int (e^x - e^{-x})\,dx$

25. $\int (x^3 + 3 - x^{-1})\,dx$

26. $\int (3x^{-1} + 2e^{-1})\,dx$

27. $\int x^{-3}(x + x^2 - 2)\,dx$

28. $\int \left(1 + \dfrac{1}{x}\right)(2x + 3)\,dx$

29. $\int (2x + 1)^2\,dx$

30. $\int (3 + e^x)^2\,dx$

31. Find an antiderivative F of $3x + 5$ such that $F(1) = 2$.

32. Find an antiderivative F of $x^2 - \sqrt{x}$ such that $F(4) = 3$.

33. Find an antiderivative F of $\dfrac{2 + \sqrt{x}}{x}$ such that $F(1) = 3$.

34. Find an antiderivative F of $4e^{3x}$ such that $F(1) = 5$.

35. Find an antiderivative F of $\dfrac{1 + 3x^2}{x}$ such that $F(-2) = 7$.

36. Find an antiderivative F of $\dfrac{1}{e^{2x}}$ such that $F(0) = 2$.

37. Suppose that the marginal cost of making x television sets is $C'(x) = 120x + .0001x^2$, and the fixed cost is \$200,000. Find a formula for $C(x)$. What is the cost of making 1000 sets?

38. Suppose that the marginal cost of making x throw rugs is $C'(x) = 10x - \sqrt{x}$, and the fixed cost is \$5,000. Find a formula for $C(x)$. What is the cost of making 100 rugs?

39. Suppose that t seconds after it has been dropped, an iron anchor falling in water has a velocity of $-4t$ feet per second, where the positive direction is taken to be up. How deep will it be after 3 seconds? How long will it take to reach the bottom, if the water is 100 feet deep?

40. After t seconds a cannonball shot upward from the bottom of a well 48 feet deep has a velocity $-32t + 128$ feet per second, where the positive direction is taken to be up. When will it reach the surface? How high will the ball get above the surface? When will it hit the bottom of the well again?

41. A driver accelerates his car so that after x seconds the speedometer shows a speed of $5x$ miles per hour. Determine $s(x)$, the distance in miles he goes in the first x seconds. How far does he go in the first 10 seconds? How long will it take him to go one mile?

42. The depth h of the liquid in a conical tank is 4 feet at noon, and the level of the liquid is rising at the rate $h'(x) = \dfrac{1}{x^{2/3}}$ at x minutes after noon. Determine $h(x)$. How high is the liquid at 1:04 P.M.?

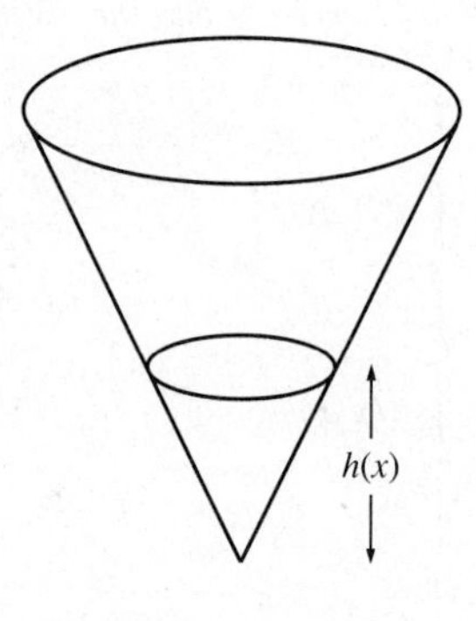

Answers to Practice Problems

1. $\dfrac{x^6}{2} - 3x^{-2} + k$

2. $F(x) = \dfrac{-3}{x} + 6x - \dfrac{19}{2}$

3. (a) $\dfrac{3x^2}{2} + 5x + 3 \ln |x| + k$
   (b) $F(x) = 2e^{3x} + 2x - 3 - 2e^6$

4. (a) $C(x) = 30x - \dfrac{3}{4}x^{4/3} + 20{,}000$ (b) \$42,500

12.2 Integration by Substitution

One method of integration depends on reversing the action of the chain rule for differentiation. Recall that the chain rule says that

$$[g(h(x))]' = g'(h(x))h'(x).$$

For example

$$[(x^3 + 2x)^5]' = 5(x^3 + 2x)^4(3x^2 + 2).$$

Corresponding to the last formula is the integration formula

$$\int 5(x^3 + 2x)^4(3x^2 + 2)\, dx = (x^3 + 2x)^5 + k. \tag{12.7}$$

Of course if we were simply presented with the left side of (12.7), we might not recognize that the function inside the integral sign was the derivative of $(x^3 + 2x)^5$. In this section we will develop a mechanical way of undoing chain-rule differentiations, called the method of **substitution.**

Suppose we let $u = x^3 + 2x$. Then $u' = 3x^2 + 2$ and (12.7) takes the form

$$\int 5u^4u'\, dx = u^5 + k. \tag{12.8}$$

This is quite reminiscent of the integration formula

$$\int 5u^4\,du = u^5 + k,$$

which follows from the rule for the antiderivative of a power. In fact the second formula would result if we could replace $u'dx$ by du in (12.8). Since another notation for u' is du/dx, replacing

$$u'\,dx = \frac{du}{dx}\,dx$$

by du amounts to "cancelling" the dx's in this expression. Of course this is unjustified because du/dx is not really a fraction, but just another way of writing u', the derivative of u with respect to x.

We will legitimatize the above replacement by *defining* the symbols dx and du, called **differentials.** Suppose u is some differentiable function of x, say $u = h(x)$. We let dx be a new variable, and define du by

$$du = u'\,dx = h'(x)\,dx,$$

where the prime denotes differentiation with respect to x. If g is another differentiable function, the chain rule says that

$$[g(h(x))]' = g'(h(x))h'(x),$$

which implies the integration formula

$$\int g'(h(x))h'(x)\,dx = g(h(x)) + k.$$

Using $h(x) = u$ and $h'(x)\,dx = du$ yields the simpler and more obvious formula

$$\int g'(u)\,du = g(u) + k,$$

which follows from the definition of the indefinite integral.

Practice Problem 5 Find du if (a) $u = e^{3x}$ (b) $u = \dfrac{1}{\sqrt{x}}$ (c) $u = x^3 + x^2$.

We will show how the substitution method can be applied to the integral

$$\int 5(x^3 + 2x)^4(3x^2 + 2)\,dx. \tag{12.9}$$

Set $u = x^3 + 2x$. (We will explain the reason for this choice of the function u later.) Then $u' = 3x^2 + 2$, and so $du = u'\,dx = (3x^2 + 2)\,dx$. Substituting these quantities yields

$$\int 5(x^3 + 2x)^4(3x^2 + 2)\,dx = \int 5u^4\,du = u^5 + k.$$

Since the original integral was in terms of the variable x, we must now *undo the substitution* by using $u = x^3 + 2x$. Thus we have

$$\int 5(x^3 + 2x)^4(3x^2 + 2)\,dx = (x^3 + 2x)^5 + k.$$

How did we know to choose $u = x^3 + 2x$ in evaluating (12.9)? In general this method works on integrals of the form

$$\int g(u)u'\,dx = \int g(u)\,du,$$

where g is some function we can integrate. Thus we try to choose u so that the expression to be integrated is some integrable function of u times u', the derivative of u. Actually it is sufficient to have some constant multiple of u' present, as the next example illustrates.

EXAMPLE 12.5 Evaluate $\int 3xe^{x^2}\,dx$.

Solution Because $(x^2)' = 2x$, the derivative of $u = x^2$ is not quite present. Since we are only off by a constant multiple, however, this substitution still will work. Notice that $du = u'\,dx = 2x\,dx$. Then

$$\begin{aligned}\int 3xe^{x^2}\,dx &= \int 3e^u x\,dx \\ &= \frac{3}{2}\int e^u 2x\,dx \\ &= \frac{3}{2}\int e^u\,du \\ &= \frac{3}{2}e^u + k \\ &= \frac{3}{2}e^{x^2} + k.\end{aligned}$$

Notice that in the second line of the calculation we simultaneously multiplied and divided the integral by the constant 2 in order to get the desired form $2x\,dx = du$.

Check $\left(\frac{3}{2}e^{x^2} + k\right)' = \frac{3}{2}e^{x^2}(2x) + 0 = 3xe^{x^2}$. ■

Practice Problem 6 Use the substitution $u = 5 - 3x^2$ to change $\int 2x\sqrt{5 - 3x^2}\,dx$ to an indefinite integral involving only u.

EXAMPLE 12.6 Evaluate $\int \frac{x^2}{x^3 + 1}\,dx$.

Solution Notice that if we let $u = x^3 + 1$, then the differential $du = 3x^2\,dx$ is present except for the constant 3.

$$\int \frac{x^2}{x^3 + 1}\,dx = \int u^{-1}x^2\,dx$$
$$= \frac{1}{3}\int u^{-1}3x^2\,dx$$
$$= \frac{1}{3}\int u^{-1}\,du$$
$$= \frac{1}{3}\ln|u| + k$$
$$= \frac{1}{3}\ln|x^3 + 1| + k \quad ■$$

Practice Problem 7 (a) How would you choose u to evaluate $\int x^3(x^4 + 1)^{1/2}\,dx$?
(b) Evaluate the integral.

EXAMPLE 12.7 Evaluate $\int \frac{x^3}{x^3 + 1}\,dx$.

Discussion If we let $u = x^3 + 1$, as in the previous example, no constant multiple of the differential $du = 3x^2\,dx$ is present. Notice what happens if we try this substitution.

$$\int \frac{x^3}{x^3 + 1}\,dx = \int u^{-1}x^3\,dx = \frac{1}{3}\int u^{-1}\,3x^3\,dx$$

We need $du = 3x^2\,dx$, but have $3x^3\,dx$. Factoring out one of the x's does not help.

$$\int \frac{x^3}{x^3 + 1}\,dx = \frac{1}{3}\int u^{-1}\,3x^3\,dx$$
$$= \frac{1}{3}\int u^{-1}\,x\,3x^2\,dx$$
$$= \frac{1}{3}\int u^{-1}x\,du$$

Since we cannot write the integral entirely in terms of u, we cannot proceed. Note that, as we saw in Section 12.1, it would not be correct to factor the x out of the integral sign. It is not possible for us to evaluate the integral of this example with the techniques we now possess. ■

EXAMPLE 12.8 Evaluate $\int \frac{x^5}{x^3 + 1}\, dx$.

Solution The substitution $u = x^3 + 1$, $du = 3x^2\, dx$ appears not to work with this integral either, as the following computation indicates.

$$\begin{aligned}\int \frac{x^5}{x^3 + 1}\, dx &= \int u^{-1}x^5\, dx \\ &= \frac{1}{3}\int u^{-1}x^3 3x^2\, dx \\ &= \frac{1}{3}\int u^{-1}x^3\, du\end{aligned}$$

Evidently there is a factor of x^3 we cannot eliminate. There is a way out, however, if we notice that $x^3 = u - 1$. Thus the last expression equals

$$\begin{aligned}\frac{1}{3}\int u^{-1}(u - 1)\, du &= \frac{1}{3}\int (1 - u^{-1})\, du \\ &= \frac{1}{3}(u - \ln |u|) + k \\ &= \frac{1}{3}(x^3 + 1 - \ln |x^3 + 1|) + k.\end{aligned}$$

Check

$$\left[\frac{1}{3}(x^3 + 1 - \ln |x^3 + 1|) + k\right]' = \frac{1}{3}\left(3x^2 + 0 - \frac{3x^2 + 0}{x^3 + 1}\right) + 0$$

$$= \frac{1}{3}\,\frac{3x^2(x^3 + 1) - 3x^2}{x^3 + 1} = \frac{1}{3}\cdot\frac{3x^5}{x^3 + 1} = \frac{x^5}{x^3 + 1} \quad ■$$

EXAMPLE 12.9 Evaluate $\int x(x^3 + 1)^2\, dx$.

Solution The substitution $u = x^3 + 1$ does not work here because a constant multiple of the differential $du = 3x^2\, dx$ is not present. In any case a more direct method is available: we simply multiply out the integrand.

$$\begin{aligned}\int x(x^3 + 1)^2\, dx &= \int x(x^6 + 2x^3 + 1)\, dx \\ &= \int (x^7 + 2x^4 + x)\, dx \\ &= \int x^7\, dx + 2\int x^4\, dx + \int x\, dx \\ &= \frac{x^8}{8} + \frac{2x^5}{5} + \frac{x^2}{2} + k \quad ■\end{aligned}$$

PITFALLS OF INTEGRATION

Integration is difficult, and when attempting it carefulness and humility are very useful. We really know how to integrate functions of only two types:

$$\int x^n\,dx = \begin{cases} \dfrac{x^{n+1}}{n+1} + k & \text{if } n \neq -1 \\ \ln|x| + k & \text{if } n = -1, \end{cases}$$

$$\int e^{cx}\,dx = \frac{e^{cx}}{c} + k \qquad \text{if } c \neq 0.$$

To evaluate any integral that is not a sum of constant multiples of one or the other of these forms, we must use either substitution or some algebraic transformation to reduce it to these formulas.

At the risk of giving an impressionable reader dangerous ideas, we list several common fallacies about integration.

The Fallacy of Substitution Without u′ Present Even though

$$\int x^3 dx = \frac{x^4}{4} + k,$$

it is *not* true that

$$\int (x^2 + 1)^3\,dx = \frac{(x^2 + 1)^4}{4} + k. \qquad \textbf{(not true)}$$

Notice that the last equation does not check, since

$$\begin{aligned}\left[\frac{(x^2 + 1)^4}{4}\right]' &= 4(x^2 + 1)^3\,\frac{(2x)}{4} \\ &= 2x(x^2 + 1)^3 \\ &\neq (x^2 + 1)^3.\end{aligned}$$

The Fallacy of Taking the Variable of Integration Outside the Integral Sign To integrate $\int xe^{2x}\,dx$ we *cannot* factor an x outside the integral sign as follows:

$$\begin{aligned}\int xe^{2x}\,dx &= x\int e^{2x}\,dx \qquad \textbf{(not true)} \\ &= \frac{xe^{2x}}{2} + k.\end{aligned}$$

Note that the answer does not check, since by the product rule for differentiation

$$\left(\frac{xe^{2x}}{2}\right)' = \frac{e^{2x}}{2} + x\left(\frac{2e^{2x}}{2}\right) = \frac{e^{2x}}{2} + xe^{2x} \neq xe^{2x}.$$

The Fallacy of Making Up One's Own Integration Rules In attempting to evaluate $\int xe^{2x}\,dx$, we *cannot* proceed as follows:

$$\begin{aligned}\int xe^{2x}\,dx &= \left(\int x\,dx\right)\left(\int e^{2x}\,dx\right) \qquad \textbf{(not true)}\\ &= \left(\frac{x^2}{2}\right)\left(\frac{e^{2x}}{2}\right) + k\\ &= \frac{x^2e^{2x}}{4} + k.\end{aligned}$$

This calculation is based on an invented rule, namely that the integral of a product is the product of the integrals. Rest assured that if the authors of this book knew of such a useful rule, they would pass it on to their readers. Again the end result does not check, since

$$\left(\frac{x^2e^{2x}}{4}\right)' = (2x)\left(\frac{e^{2x}}{4}\right) + (x^2)\left(\frac{2e^{2x}}{4}\right) = (x + x^2)\frac{e^{2x}}{2} \neq xe^{2x}.$$

EXERCISES 12.2

In Exercises 1–8 make the indicated substitution to get an indefinite integral involving only the variable u.

1. $\int x(3x^2 + 1)\,dx, \quad u = 3x^2 + 1$

2. $\int \frac{1}{x + 2}\,dx, \quad u = x + 2$

3. $\int e^{4x+3}\,dx, \quad u = 4x + 3$

4. $\int \frac{x^2}{x^3 - 8}\,dx, \quad u = x^3 - 8$

5. $\int (3 - x)^5\,dx, \quad u = 3 - x$

6. $\int (4x^3 + 3)(x^4 + 3x)\,dx, \quad u = x^4 + 3x$

7. $\int xe^{-x^2}\,dx, \quad u = -x^2$

8. $\int \frac{\ln x}{x}\,dx, \quad u = \ln x$

Evaluate the indefinite integrals in Exercises 9–21.

9. $\int e^{x+1}\,dx$

10. $\int e^{2x+3}\,dx$

11. $\int (x + 2)^{-1}\,dx$

12. $\int \frac{3x^2 + 1}{(x^3 + x - 1)^5}\,dx$

13. $\int \frac{1}{\sqrt{3x - 2}}\,dx$

14. $\int \frac{x}{x^2 + 1}\,dx$

15. $\int x^{-1}(\ln x)^3\,dx$

16. $\int xe^{x^2+1}\,dx$

17. $\int e^{1-x}\,dx$

18. $\int \frac{e^{1/x}}{x^2}\,dx$

19. $\int \frac{\ln (x^2)}{x}\,dx$

20. $\int \frac{1}{x(\ln x)^2}\,dx$

21. $\int (x + 1)\sqrt{3x^2 + 6x}\,dx$

22. $\int x^3(x^4 + 2x)^2\,dx$

23. $\int \sqrt{2 - 3x}\,dx$

24. $\int \frac{e^x - e^{-x}}{e^x + e^{-x}}\,dx$

25. $\int \frac{x}{e^{x^2}}\,dx$

26. $\int \frac{dx}{x^2 + 2x + 1}$

27. $\int \frac{(\sqrt{x} - 1)^3}{\sqrt{x}}\,dx$

28. $\int \frac{3x - 1}{3 + 4x - 6x^2}\,dx$

29. $\int \frac{e^{\sqrt{x}}}{\sqrt{x}}\,dx$

30. Determine all the differentiable functions for which the slope at the point (x, y) is $x\sqrt{9 - x^2}$. Which of these functions has a graph passing through the point $(0, 2)$?

31. Determine all the differentiable functions for which the slope at the point (x, y) is $x/(x^2 + 1)$. Which of these functions has a graph passing through the point $(0, 2)$?

32. Find an antiderivative F for e^{3x+5} such that $F(0) = 2$.

33. Find an antiderivative F for $x(x^2 + 2)^{1/3}$ such that $F(6) = 0$.

34. Suppose that a company's marginal cost in dollars is approximated by the function

$$C'(x) = 0.0006x + 12 + \frac{1200}{\sqrt{2x + 1}},$$

where x denotes the number of units produced. Approximate the company's total cost $C(x)$ of producing x units if its fixed cost is \$60,000.

35. Suppose that a company's marginal cost in dollars is approximated by the function

$$C'(x) = 0.0008x + 50 + \frac{10{,}000}{2x + 1},$$

where x denotes the number of units produced. Approximate the company's total cost of producing x units if its fixed cost is \$15,000.

36. The speed in yards per second of a certain sprinter t seconds after the beginning of a race is given by

$$s'(t) = 9\sqrt{t + 1} - 9.$$

How far does the sprinter run in the first 8 seconds?

Answers to Practice Problems

5. (a) $3e^{3x}\,dx$ (b) $-\frac{1}{2}x^{-3/2}\,dx$ (c) $(3x^2 + 2x)\,dx$

6. $\int -\frac{1}{3}\sqrt{u}\,du$

7. (a) $u = x^4 + 1$ (b) $\frac{(x^4 + 1)^{3/2}}{6} + k$

12.3 The Definite Integral

In this section we will consider a subject that will appear to be unrelated to Sections 12.1 and 12.2—the definition of area. There is a strong connection between area and the indefinite integral, however, as will be explained in Section 12.4.

WHAT IS AREA?

Area is a concept that is often used, but seldom defined. Intuitively the *area* of a set of points in the plane is a number that measures how big it is. (See Figure 12.3.)

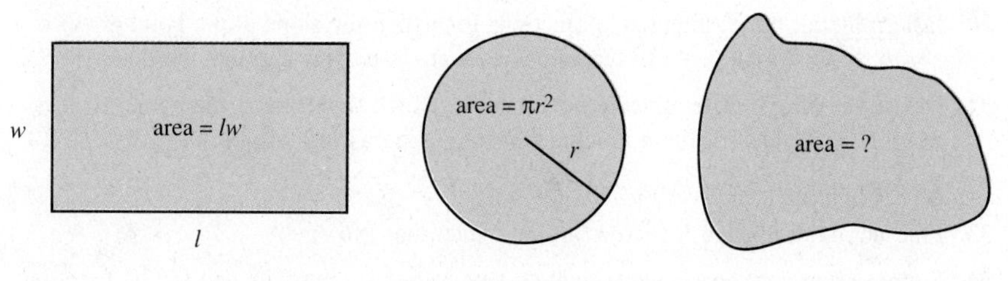

FIGURE 12.3

Whatever area means, most people would probably agree that it should have certain basic properties, for example:

(a) The area of a set is a nonnegative real number.
(b) The area of a rectangle with length l and width w is lw.
(c) If S and T are disjoint sets having areas s and t, then the area of $S \cup T$ is $s + t$. (See Figure 12.4.)
(d) Congruent sets have the same area. (See Figure 12.5.)

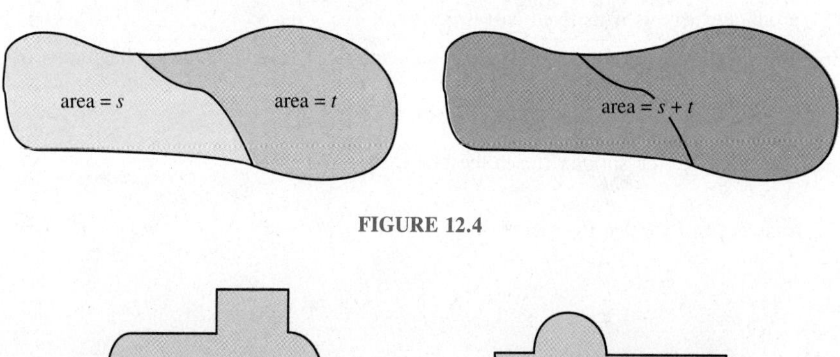

FIGURE 12.4

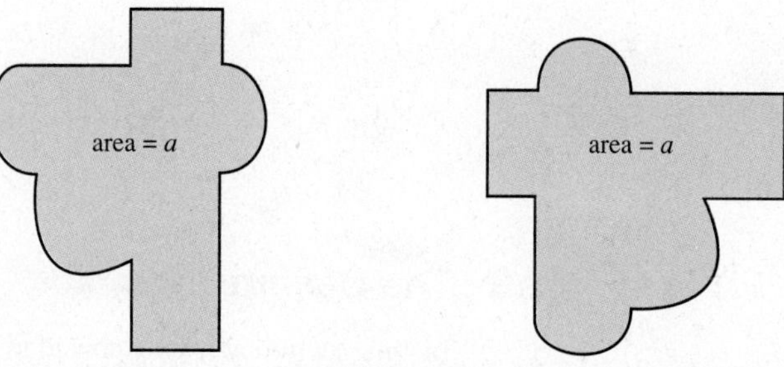

FIGURE 12.5

We will only define the area of sets of a certain type, namely those consisting of all points between two vertical lines, and between the graph of a nonnegative function f and the x-axis, as shown in Figure 12.6.

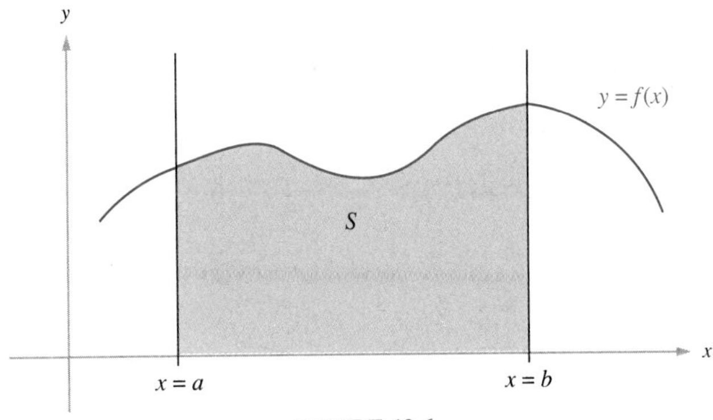

FIGURE 12.6

Thus we will attempt to define the area of a set of the form

$$S = \{(x, y) | \ a \leq x \leq b \text{ and } 0 \leq y \leq f(x)\},$$

where f is a function that is defined and nonnegative on $[a, b]$. A crude first approximation to the number we seek would be the area of the rectangle shown in Figure 12.7, which has width $b - a$ and height $f(a)$, the value of f at the left-hand endpoint of the interval $[a, b]$.

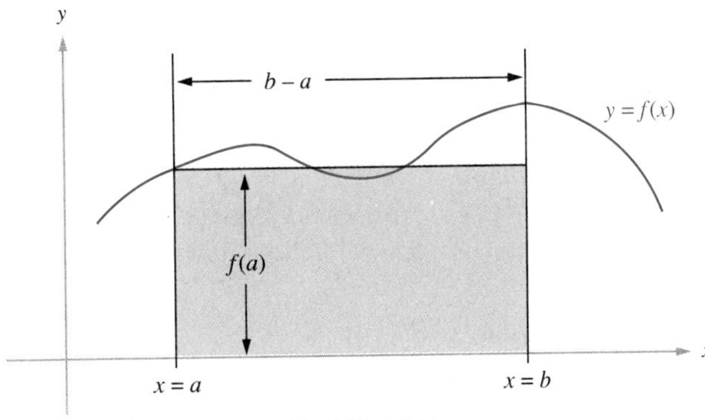

FIGURE 12.7

A seemingly better estimate results from using more rectangles. For example, if we divide $[a, b]$ into 4 equal subintervals and build a rectangle on each with height equal to the value of f at the left endpoint of the subinterval, then we get Figure 12.8.

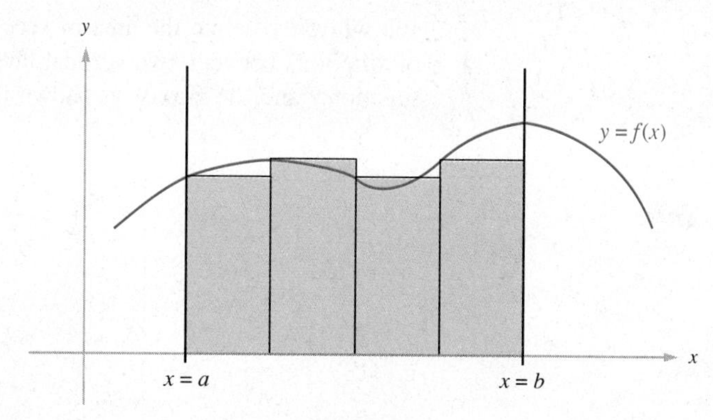

FIGURE 12.8

The width of each of these narrower rectangles is the same, namely $(b - a)/4$. Let us denote this width by Δx. (Think of Δx as a single symbol rather than as a product of Δ and x.) Of course the rectangles have different heights, depending on the values of f at the left-hand endpoints of the corresponding intervals. Let us denote the endpoints of the subintervals of $[a, b]$ by

$$a = x_0 < x_1 < x_2 < x_3 < x_4 = b,$$

so that $x_1 = a + \Delta x$, $x_2 = a + 2\Delta x$, and $x_3 = a + 3\Delta x$. Then the height of the first rectangle is $f(x_0)$, the height of the second rectangle is $f(x_1)$, etc. We see that the sum of the areas of the four approximating rectangles is

$$S_4 = f(x_0)\Delta x + f(x_1)\Delta x + f(x_2)\Delta x + f(x_3)\Delta x.$$

EXAMPLE 12.10 Use four rectangles of the same width to approximate the area under the curve $y = 0.2x^2$ and above the interval $[1, 3]$.

Solution We have $a = 1$, $b = 3$, and $\Delta x = (3 - 1)/4 = .5$. Thus the endpoints of the subintervals are $x_0 = 1$, $x_1 = 1 + 0.5 = 1.5$, $x_2 = 2$, $x_3 = 2.5$, and $x_4 = 3$. The following table gives the values of f at each left-hand endpoint.

x_i	$f(x_i) = 0.2x_i^2$
1	$0.2(1)^2 = 0.20$
1.5	$0.2(1.5)^2 = 0.45$
2	$0.2(2)^2 = 0.80$
2.5	$0.2(2.5)^2 = 1.25$

Thus we have

$$\begin{aligned} S_4 &= f(x_0)\Delta x + f(x_1)\Delta x + f(x_2)\Delta x + f(x_3)\Delta x \\ &= .20(.5) + .45(.5) + .80(.5) + 1.25(.5) = 1.35. \end{aligned}$$

The rectangles are shown in Figure 12.9. ■

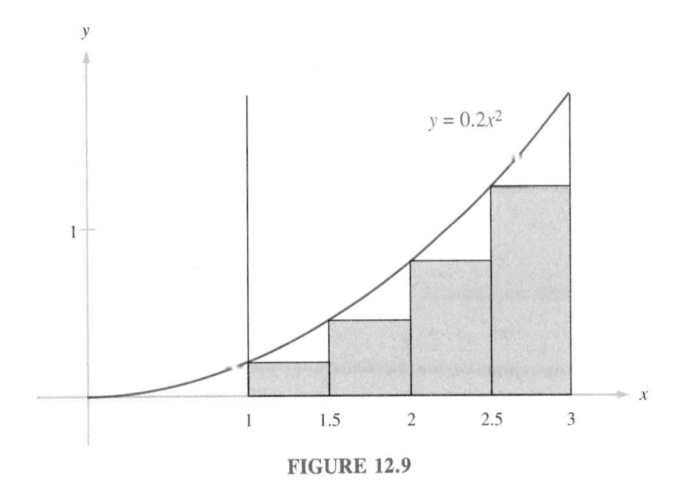

FIGURE 12.9

In general, if we divide $[a, b]$ into n congruent subintervals, then each will have width $\Delta x = (b - a)/n$, and the endpoints of these intervals will be

$$\begin{aligned} x_0 &= a, \\ x_1 &= a + \Delta x, \\ x_2 &= a + 2\Delta x, \\ &\vdots \\ x_{n-1} &= a + (n - 1)\Delta x, \\ x_n &= a + n\Delta x = b. \end{aligned}$$

The heights of the corresponding rectangles will be $f(x_0), f(x_1), \ldots, f(x_{n-1})$, and so a reasonable estimate for what we think of as the area under the curve will be

$$S_n = f(x_0)\Delta x + f(x_1)\Delta x + \cdots + f(x_{n-1})\Delta x.$$

A sum of this form is called a **Riemann sum.** Note that the sum S_n can be computed even if f is not nonnegative throughout the interval $[a, b]$.

EXAMPLE 12.11 Compute S_{10} for the function $f(x) = 0.2x^2$, using the interval $[1, 3]$.

Solution We have $\Delta x = (3 - 1)/10 = .2$, and the endpoints of the 10 subintervals are

$$a = 1, 1.2, 1.4, 1.6, 1.8, 2, 2.2, 2.4, 2.6, 2.8, 3 = b.$$

Thus

$$\begin{aligned} S_{10} &= f(x_0)\Delta x + f(x_1)\Delta x + f(x_2)\Delta x + \ldots + f(x_9)\Delta x \\ &= .2(1^2).2 + .2(1.2^2).2 + .2(1.4^2).2 + \ldots + .2(2.8^2).2 \\ &= 1.576. \end{aligned}$$

The corresponding rectangles are shown in Figure 12.10. ■

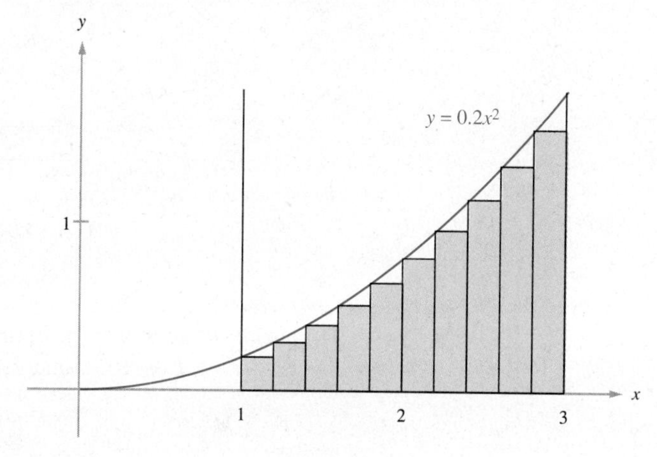

FIGURE 12.10

THE DEFINITE INTEGRAL

We have seen that by dividing the interval $[a, b]$ into n subintervals $[x_0, x_1]$, $[x_1, x_2]$, . . ., $[x_{n-1}, x_n]$, each of width $\Delta x = (b - a)/n$, and erecting a rectangle with height $f(x_{i-1})$ over the ith subinterval, we get a number

$$S_n = f(x_0)\Delta x + f(x_1)\Delta x + \ldots + f(x_{n-1})\Delta x \tag{12.10}$$

that approximates what we think of as the area under the curve $y = f(x)$ and above the interval $[a, b]$. The larger n is, the closer the union of the n rectangles can be expected to approximate the set of points in which we are interested. Thus we define the **area** under the curve $y = f(x)$ above the interval $[a, b]$ to be

$$\lim_{n\to\infty} S_n, \tag{12.11}$$

assuming that this limit exists. Fortunately for almost all functions we will consider, this limit does exist. For example, it is possible to prove that if the function f is continuous on the interval $[a, b]$, then the limit of (12.11) exists.

When this limit of the Riemann sums, $\lim_{n\to\infty} S_n$, exists (and whether f is nonnegative for x in $[a, b]$ or not), we denote it by

$$\int_a^b f(x)\, dx, \tag{12.12}$$

and call it the **definite integral** of f over the interval $[a, b]$. Of course this notation is very similar to that for the indefinite integral

$$\int f(x)\,dx.$$

Only the endpoints a and b of the interval in (12.12) distinguish the symbolism for the definite integral. These are called the **lower** and **upper limits of integration,** respectively. There is a reason for the similarity between the two symbols that will be explained in the next section. Nonetheless the definite and indefinite integrals represent very different mathematical entities.

The definite integral $\int_a^b f(x)\,dx$ stands for a *number,* while the indefinite integral $\int f(x)\,dx$ stands for a *set of functions.*

For example $\int_0^3 x\,dx$ is the limit of the sums of areas of rectangles approximating the region under the line $y = x$ and above the interval $[0, 3]$ (see Figure 12.11).

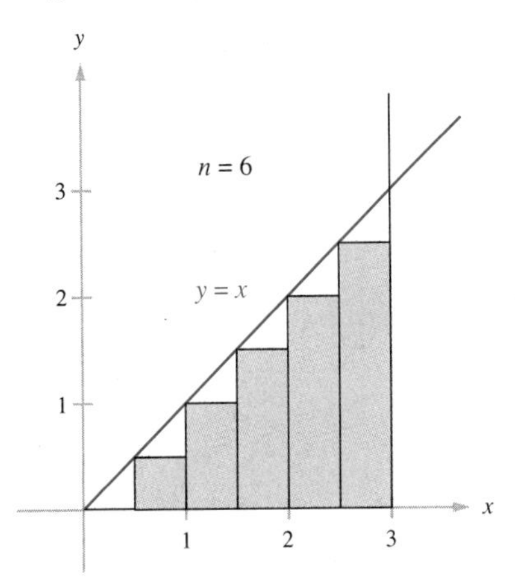

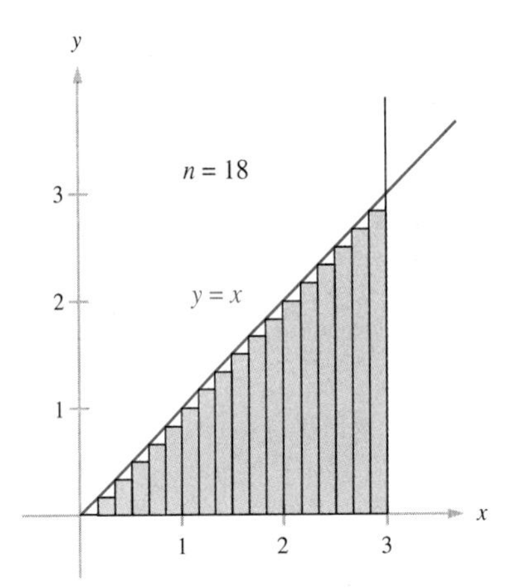

FIGURE 12.11

It can be shown that this limit is 9/2. This is what we would expect, since the triangular region in question is half of a 3-by-3 square. Thus

$$\int_0^3 x\,dx = \frac{9}{2}.$$

On the other hand

$$\int x\,dx = \frac{x^2}{2} + k.$$

This class of functions is illustrated in Figure 12.1.

Notice that the sum (12.10), and so the limit (12.11), may make sense even when the function f is not positive for all x in $[a, b]$. Thus the definition of the definite integral applies to arbitrary functions defined on $[a, b]$ for which the limit (12.11) exists, even though the interpretation as area breaks down. In the next section we will show the connection between area and definite integrals of nonpositive functions.

It can be shown that the definition of area given above is consistent with all the area formulas with which you are familiar. For example the graph of the equation $x^2 + y^2 = r^2$ is the circle with center at the origin and radius r. The graph of the function $y = \sqrt{r^2 - x^2}$ is the top half of this circle. (See Figure 12.12, where the rectangles associated with the Riemann sum S_6 are shown.)

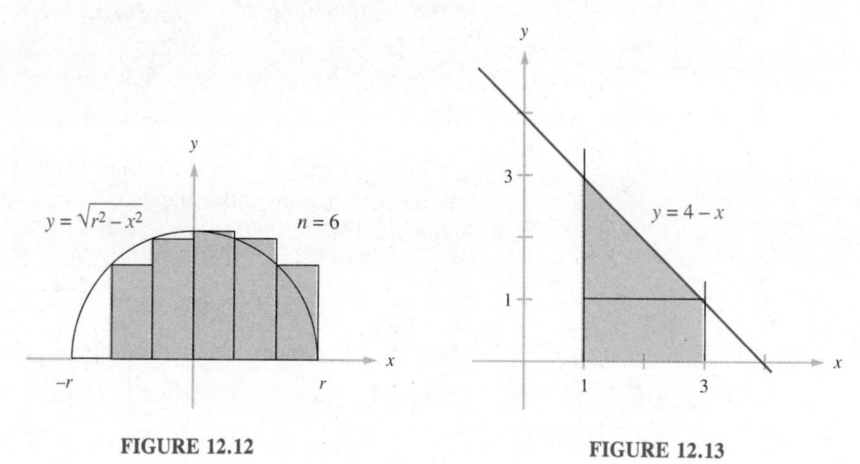

FIGURE 12.12

FIGURE 12.13

It can be shown that in this case the limit of the Riemann sums $\lim_{n \to \infty} S_n$ is $\pi r^2/2$; that is

$$\int_{-r}^{r} \sqrt{r^2 - x^2}\, dx = \frac{\pi r^2}{2}.$$

This is what one obtains by applying the usual formula for the area of half a circle.

EXAMPLE 12.12 Evaluate $\int_1^3 (4 - x)\, dx$.

Solution The graph of $y = 4 - x$ is shown in Figure 12.13, with the region under the graph and above the interval $[1, 3]$ shaded. Instead of using Riemann sums, we will evaluate the integral by using standard formulas to find the area of this region.

Notice that the enclosed region can be divided into a 1-by-2 rectangle, with area 2, and a triangle with base 2 and altitude 2, with area $\frac{1}{2}(2)(2) = 2$. Thus we have

$$\int_1^3 (4 - x)\, dx = 2 + 2 = 4. \blacksquare$$

In the next section we will give a method of evaluating definite integrals that is much easier than finding the limit of Riemann sums. Then we will be able to find the area of regions by evaluating definite integrals, rather than proceeding in the opposite direction as in the last example.

Practice Problem 8 Evaluate $\int_{-1}^{2} (2x + 3)\, dx$ by sketching the corresponding region and finding its area.

EXERCISES 12.3

In Exercises 1–4 an interval $[a, b]$ *is given along with a value of* n. *Compute the left-hand endpoints if* $[a, b]$ *is divided into* n *equal subintervals.*

1. $[a, b] = [1, 5], \quad n = 8$

2. $[a, b] = [-3, 0], \quad n = 5$

3. $[a, b] = [-2, 1], \quad n = 4$

4. $[a, b] = [-4, -1], \quad n = 6$

In Exercises 5–10 compute the Riemann sum S_n *for the given function* f, *interval* $[a, b]$, *and value of* n.

5. $f(x) = 3x - 2, \quad [a, b] = [0, 2], \quad n = 4$

6. $f(x) = 2x^2, \quad [a, b] = [1, 4], \quad n = 3$

7. $f(x) = x(x - 1), \quad [a, b] = [-1, 1], \quad n = 6$

8. $f(x) = x^3, \quad [a, b] = [-2, 6], \quad n = 8$

9. $f(x) = 1/x, \quad [a, b] = [1, 1.5], \quad n = 5$

10. $f(x) = e^x, \quad [a, b] = [1, 9], \quad n = 4$

In Exercises 11–24 compute the given definite integral by sketching the corresponding region and using standard area formulas.

11. $\int_1^4 5\, dx$

12. $\int_3^7 8\, dx$

13. $\int_0^3 2x\, dx$

14. $\int_0^5 3x\, dx$

15. $\int_1^5 (3x - 3)\, dx$

16. $\int_3^6 (12 - 2x)\, dx$

17. $\int_2^5 (3x + 1)\, dx$

18. $\int_3^4 (4x - 10)\, dx$

19. $\int_{-3}^3 (7 + 2x)\, dx$

20. $\int_{-3}^2 (10 - 2x)\, dx$

21. $\int_{-1}^1 \sqrt{1 - x^2}\, dx$

22. $\int_{-2}^2 \sqrt{4 - x^2}\, dx$

23. $\int_0^3 \sqrt{9 - x^2}\, dx$

24. $\int_{-4}^0 \sqrt{16 - x^2}\, dx$

25. Suppose the interval $[1, 3]$ is divided into n equal subintervals $[x_{i-1}, x_i]$, where $i = 1, 2, \ldots, n$. Find a formula for the left-hand endpoint x_{i-1} as a function of i and n. What is Δx?

26. Suppose the interval $[-1, 2]$ is divided into n equal subintervals $[x_{i-1}, x_i]$, where $i = 1, 2, \ldots, n$. Find a formula for the left-hand endpoint x_{i-1} as a function of i and n. What is Δx?

27. Consider the function $f(x) = x + 1$ on the interval $[1, 3]$. Using Exercise 25, show that $f(x_{i-1}) = 2 + 2(i - 1)/n$.

28. Consider the function $f(x) = 3 - x$ on the interval $[-1, 2]$. Using Exercise 26, show that $f(x_{i-1}) = 4 - 3(i - 1)/n$.

Exercises 29–30 require the fact that the sum of the terms of the arithmetic progression

$$x_0,\ x_0 + b,\ x_0 + 2b,\ \ldots,\ x_0 + (n - 1)b$$

is $n[x_0 + (n - 1)b/2]$. *(See Section 4.1.)*

29. Consider the function $f(x) = x + 1$ on the interval $[1, 3]$. Using Exercises 25 and 27, show that $S_n = 4 + 2(n - 1)/n$. Use this formula to evaluate $\int_1^3 (x + 1)\, dx$.

30. Consider the function $f(x) = 3 - x$ on the interval $[-1, 2]$. Using Exercises 26 and 28, show that $S_n = 12 - 9(n - 1)/2n$. Use this formula to evaluate $\int_{-1}^2 (3 - x)\, dx$.

Answer to Practice Problem **8.** $\int_{-1}^2 (2x + 3)\, dx = 12$

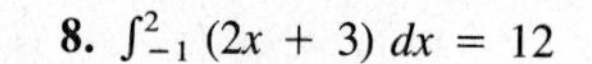

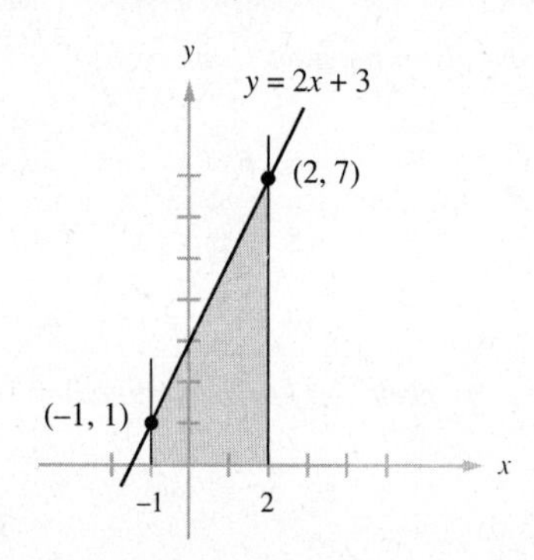

12.4 The Fundamental Theorem of Calculus

We investigated in the previous section the relation between the definite integral and the definition of area; other applications of the definite integral will be considered in Section 12.5. What we are lacking is a reasonable method of computing definite integrals. The following theorem gives such a method, and also reveals the connection between definite and indefinite integrals that is behind their similar notations.

The Fundamental Theorem of Calculus

Suppose F is an antiderivative for the function f on the interval $[a, b]$. Then

$$\int_a^b f(x)\, dx = F(b) - F(a).$$

We illustrate the theorem with the definite integral

$$\int_1^3 (4 - x)\, dx$$

of Example 12.12 of the previous section. This is of the form $\int_a^b f(x)\, dx$, where $a = 1$, $b = 3$, and $f(x) = 4 - x$. One antiderivative of f is $F(x) = 4x - \frac{x^2}{2}$. Thus by the fundamental theorem

$$\begin{aligned}\int_1^3 (4 - x)\, dx &= F(3) - F(1)\\ &= \left(4 \cdot 3 - \frac{3^2}{2}\right) - \left(4 \cdot 1 - \frac{1^2}{2}\right)\\ &= 12 - \frac{9}{2} - 4 + \frac{1}{2} = 4.\end{aligned}$$

This is the same answer we found in Example 12.12 by using standard area formulas.

EXAMPLE 12.13 Use the fundamental theorem to evaluate $\int_{-1}^2 x^2\, dx$.

Solution An antiderivative of x^2 is $F(x) = \frac{x^3}{3} + 5$. Thus by the fundamental theorem we have

$$\begin{aligned}\int_{-1}^2 x^2\, dx &= F(2) - F(-1)\\ &= \left(\frac{2^3}{3} + 5\right) - \left[\frac{(-1)^3}{3} + 5\right]\\ &= \frac{8}{3} + 5 + \frac{1}{3} - 5 = 3.\end{aligned}$$

This is the area of the region shaded in Figure 12.14. ■

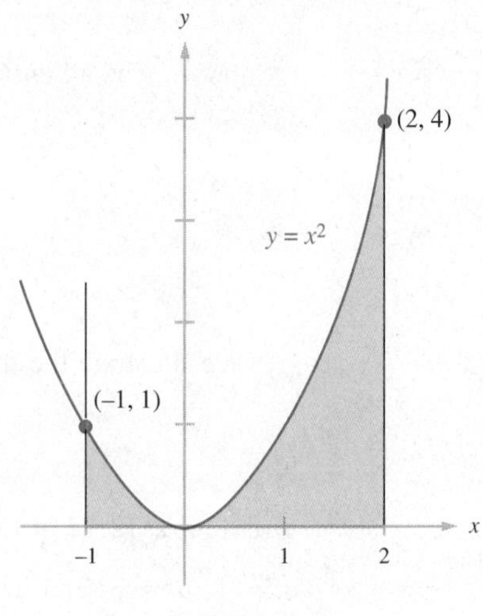

FIGURE 12.14

Notice that in the last example including $+5$ in $F(x)$ had no effect, since the 5's cancelled out when we computed $F(2) - F(-1)$. This points up the fact that *any* antiderivative can be used when computing integrals with the fundamental theorem.

Practice Problem 9 Use the fundamental theorem to compute $\int_{-1}^{2} (2x + 3)\,dx$.

In order to be able to write the computation of definite integrals easily we introduce the notation $F(x)\big|_a^b$ to stand for $F(b) - F(a)$. For example, the calculation of Example 12.13 could be written as follows.

$$\int_{-1}^{2} x^2\,dx = \frac{x^3}{3} + 5\,\Bigg|_{-1}^{2}$$

$$= \left(\frac{2^3}{3} + 5\right) - \left[\frac{(-1)^3}{3} + 5\right] = 3$$

EXAMPLE 12.14

Compute $\displaystyle\int_1^4 e^{3x}\,dx$.

Solution

$$\int_1^4 e^{3x}\,dx = \frac{e^{3x}}{3}\Bigg|_1^4$$

$$= \frac{e^{3(4)}}{3} - \frac{e^{3(1)}}{3}$$

$$= \frac{e^{12} - e^3}{3} \quad ■$$

EXAMPLE 12.15 Find the area under the curve $y = 1/x$, above the x-axis, and between the lines $x = 1$ and $x = 4$.

Solution We want the area of the region shown in Figure 12.15. It is

$$\int_1^4 x^{-1}\,dx = \ln|x| \Big|_1^4$$
$$= \ln 4 - \ln 1$$
$$= \ln 4 \approx 1.39. \quad ■$$

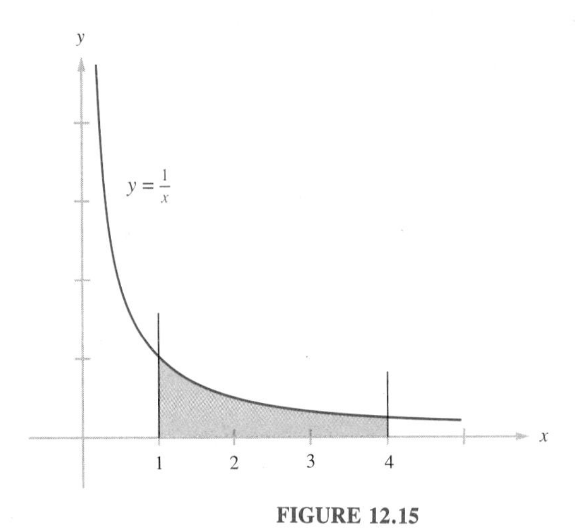

FIGURE 12.15

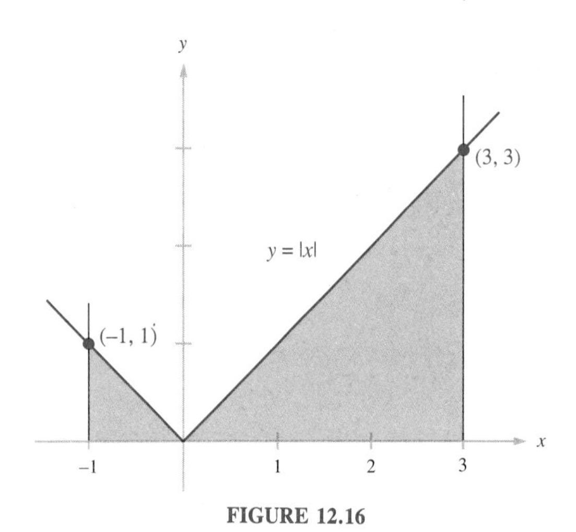

FIGURE 12.16

EXAMPLE 12.16 Evaluate $\int_{-1}^{3} |x|\,dx$.

Solution Instead of looking for an antiderivative for $f(x) = |x|$, we exploit the relation between the definite integral and area. The given integral gives the area of the region pictured in Figure 12.16. By adding the areas of the two triangles we compute

$$\int_{-1}^{3} |x|\,dx = \frac{1}{2}(1)(1) + \frac{1}{2}(3)(3) = \frac{1}{2} + \frac{9}{2} = 5. \quad ■$$

EXAMPLE 12.16 Revisited We could have used the fundamental theorem to solve the last example if we had noticed that $|x| = -x$ for $-1 \leq x \leq 0$, while $|x| = x$ for $0 \leq x \leq 3$.

Thus

$$\int_{-1}^{3} |x|\,dx = \int_{-1}^{0} |x|\,dx + \int_{0}^{3} |x|\,dx$$

$$= \int_{-1}^{0} (-x)\,dx + \int_{0}^{3} x\,dx$$

$$= \frac{-x^2}{2}\Bigg|_{-1}^{0} + \frac{x^2}{2}\Bigg|_{0}^{3}$$

$$= \left[\frac{-0^2}{2} - \frac{-(-1)^2}{2}\right] + \left[\frac{3^2}{2} - \frac{0^2}{2}\right]$$

$$= \left[0 + \frac{1}{2}\right] + \left[\frac{9}{2} - 0\right] = \frac{10}{2} = 5. \blacksquare$$

The last computation illustrates the first of the following list of properties of definite integrals. All can be proved from the definition of the definite integral as the limit of Riemann sums.

Properties of Definite Integrals

(a) If $a < b < c$, then $\int_a^c f(x)\,dx = \int_a^b f(x)\,dx + \int_b^c f(x)\,dx$.
(See Figure 12.17.)

(b) $\int_a^b [f(x) \pm g(x)]\,dx = \int_a^b f(x)\,dx \pm \int_a^b g(x)\,dx$.

(c) If c is any constant, then $\int_a^b cf(x)\,dx = c\int_a^b f(x)\,dx$.

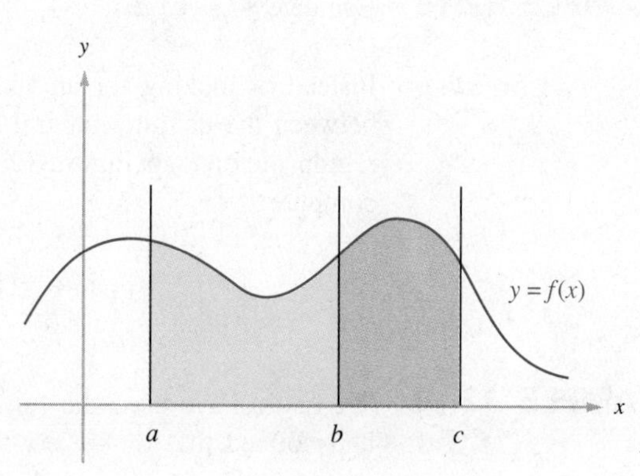

FIGURE 12.17

EXAMPLE 12.17 Evaluate $\int_{-2}^{2} x(x^2 - 4)\,dx$.

Solution We have

$$\begin{aligned}\int_{-2}^{2} x(x^2 - 4)\,dx &= \int_{-2}^{2} (x^3 - 4x)\,dx \\ &= \left.\frac{x^4}{4} - \frac{4x^2}{2}\right|_{-2}^{2} \\ &= \left.\frac{x^4}{4} - 2x^2\right|_{-2}^{2} \\ &= \left[\frac{2^4}{4} - 2(2)^2\right] - \left[\frac{(-2)^4}{4} - 2(-2)^2\right] \\ &= (4 - 8) - (4 - 8) = 0. \quad \blacksquare\end{aligned}$$

The reader may be surprised that the definite integral of a function that is zero at only three points (namely, -2, 0, and 2; see Figure 12.18) is zero. Recall that the interpretation of $\int_a^b f(x)\,dx$ as area is valid only when $f(x)$ is *nonnegative* throughout $[a, b]$. When f is negative, the terms $f(x_i)\Delta x$ in a Riemann sum will be negative, and will represent -1 times the area of the corresponding rectangle. (See Figure 12.19.)

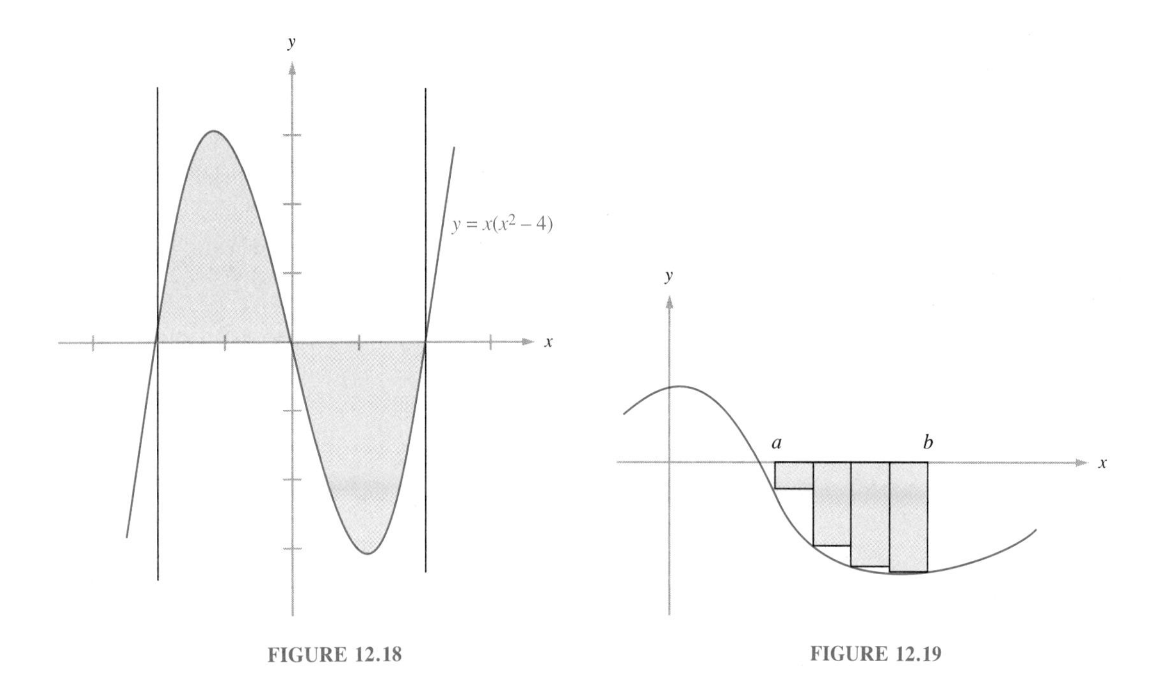

FIGURE 12.18

FIGURE 12.19

In the case of Example 12.17 we can compute

$$\int_0^2 (x^3 - 4x)\,dx = \frac{x^4}{4} - 2x^2 \Big|_0^2$$
$$= \left(\frac{2^4}{4} - 2 \cdot 2^2\right) - \left(\frac{0^4}{4} - 2 \cdot 0^2\right)$$
$$= 4 - 8 - 0 + 0 = -4.$$

This means the area between the graph of $y = x^3 - 4x$ and the interval $[0, 2]$ on the x-axis is 4; the integral turns out to be -4 because the region is below the x-axis. A similar computation shows that $\int_{-2}^{0} (x^3 - 4x)\,dx = 4$. Since the area below the x-axis equals the area above, the integral from -2 to 2 is 0. Hence we see that for a function f that is not nonnegative on $[a, b]$, evaluating $\int_a^b f(x)\,dx$ may not give the area bounded by the graph of f, the x-axis, and the lines $x = a$ and $x = b$.

Practice Problem 10 Find the area between the graph of $y = x^2 - 10$, the x-axis, and the lines $x = 1$ and $x = 3$.

SUBSTITUTION WITH DEFINITE INTEGRALS

Although the method of substitution can be used with definite integrals, some modification is necessary.

EXAMPLE 12.18 Evaluate $\int_2^4 x(x^2 + 1)^{3/2}\,dx$.

Solution The substitution $u = x^2 + 1$ comes to mind, since then $du = 2x\,dx$, and $x\,dx$ is present in the integral. There is a problem with what to do about the limits of integration, however. Of course we could simply make the substitution to find an antiderivative of $f(x) = x(x^2 + 1)^{3/2}$, and then use the fundamental theorem to evaluate the original integral. We will do this as a review of the substitution method.

$$\int x(x^2 + 1)^{3/2}\,dx = \frac{1}{2}\int (x^2 + 1)^{3/2} 2x\,dx = \frac{1}{2}\int u^{3/2}\,du$$
$$= \frac{\frac{1}{2}u^{5/2}}{\frac{5}{2}} + k = \frac{1}{5}u^{5/2} + k$$
$$= \frac{1}{5}(x^2 + 1)^{5/2} + k$$

Then using the antiderivative $(\frac{1}{5})\,(x^2 + 1)^{5/2}$ for $x(x^2 + 1)^{3/2}$ and the fundamental theorem yields

$$\int_2^4 x(x^2 + 1)^{3/2}\,dx = \frac{1}{5}(x^2 + 1)^{5/2}\Big|_2^4$$
$$= \frac{1}{5}[(4^2 + 1)^{5/2} - (2^2 + 1)^{5/2}]$$
$$= \frac{1}{5}(17^{5/2} - 5^{5/2}).$$

There is a less cumbersome method, however, that involves changing the limits of integration. In the original integral x varies from 2 to 4, and our substitution was $u = x^2 + 1$. Thus when $x = 2$ we have $u = 2^2 + 1 = 5$, and when $x = 4$ we have $u = 4^2 + 1 = 17$. Thus we switch to the limits 5 and 17 when changing the variable of integration to u, as follows:

$$\int_2^4 x(x^2 + 1)^{3/2}\,dx = \frac{1}{2}\int_2^4 (x^2 + 1)^{3/2}2x\,dx$$
$$= \frac{1}{2}\int_5^{17} u^{3/2}\,du$$
$$= \frac{1}{2}\left(\frac{u^{5/2}}{5/2}\right)\Big|_5^{17}$$
$$= \frac{1}{5}u^{5/2}\Big|_5^{17}$$
$$= \frac{1}{5}(17^{5/2} - 5^{5/2}). \quad ■$$

Care must be taken when using the substitution method on definite integrals to match the limits of integration properly. In Example 12.18 the original *lower* limit of integration was 2, and upon making the substitution $u = x^2 + 1$ the new *lower* limit became $2^2 + 1 = 5$. Keeping the limits of integration straight is not always as easy as it might seem, as the next example demonstrates.

EXAMPLE 12.19 Evaluate $\int_2^3 e^{7-2x}\,dx$.

Solution If we let $u = 7 - 2x$, then $du = -2\,dx$. The old lower and upper limits of integration are $x = 2$ and $x = 3$, respectively, so the corresponding new lower and upper limits are $u = 7 - 2 \cdot 2 = 3$ and $u = 7 - 2 \cdot 3 = 1$. Thus the substitution proceeds as follows:

$$\int_2^3 e^{7-2x}\,dx = -\frac{1}{2}\int_2^3 e^{7-2x}(-2x\,dx) = -\frac{1}{2}\int_3^1 e^u\,du.$$

Of course the limits of integration on the last integral are backwards; the lower limit is greater than the upper limit. Yet 3 does correspond to the original lower limit 2, and 1 corresponds to the original upper limit 3. What should we do about this problem?

The answer is: *ignore it!* Use the fundamental theorem just as if the new definite integral made sense:

$$\begin{aligned}\int_2^3 e^{7-2x}\,dx &= -\frac{1}{2}\int_3^1 e^u\,du \\ &= -\frac{1}{2}e^u\Big|_3^1 \\ &= -\frac{1}{2}e^1 - \left(-\frac{1}{2}e^3\right) \\ &= \frac{1}{2}(e^3 - e) \approx 8.68.\end{aligned}$$

Notice that if we had mistakenly evaluated the integral

$$-\frac{1}{2}\int_1^3 e^u\,du,$$

we would have gotten the incorrect negative answer $-\frac{1}{2}(e^3 - e) \approx -8.68$. This cannot be correct because the function we were integrating, e^{7-2x}, is positive for all values of x, and so the answer can be interpreted as an area. ■

In order that the computation of the last example make sense we will *define* $\int_a^b f(x)\,dx$ for $a > b$ as follows:

If $a > b$, then we define $\int_a^b f(x)\,dx$ to be $-\int_b^a f(x)\,dx$.

Notice that if F is an antiderivative for f on $[a, b]$, then by the fundamental theorem

$$-\int_b^a f(x)\,dx = -[F(a) - F(b)] = F(b) - F(a),$$

and so we get the correct answer by applying the fundamental theorem in the usual way to the "backwards" integral.

Practice Problem 11 (a) How would you choose u to evaluate $\int_3^5 (6 - x)^{-1}\,dx$?
(b) Evaluate this integral.

THE PLAUSIBILITY OF THE FUNDAMENTAL THEOREM

Although we will not prove the fundamental theorem of calculus, we offer an argument based on intuitive ideas of area that makes it seem reasonable. Let us consider a function f that is continuous and nonnegative at all points in the interval $[a, b]$. We define a new function A on $[a, b]$ by the equation

$$A(t) = \int_a^t f(x)\,dx \qquad \text{if } a \leq t \leq b.$$

Note that $A(t)$ is just our definition of the area of the region under the curve $y = f(x)$ and above the interval $[a, t]$. (See Figure 12.20.)

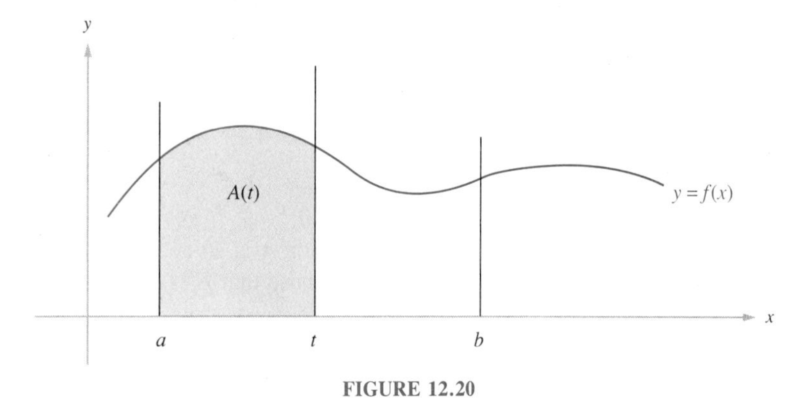

FIGURE 12.20

Now we will attempt to compute the *derivative* of the function A at t. This is defined to be the limit as h goes to 0 of the difference quotient

$$\frac{A(t+h) - A(t)}{h} = \frac{\int_a^{t+h} f(x)\,dx - \int_a^t f(x)\,dx}{h}$$

$$= \frac{\int_t^{t+h} f(x)\,dx}{h},$$

where we have used part (a) of our Properties of Definite Integrals to get the second equality. (See Figure 12.21, in which we have assumed $h > 0$.)

The numerator of this difference quotient represents the area of the shaded region in Figure 12.21. For small values of h this is close to the area of the rectangle with height $f(t)$ and width h shown in the figure, namely $f(t)h$. Thus it is reasonable that as $h \to 0$ the difference quotient approaches $f(t)h/h = f(t)$.

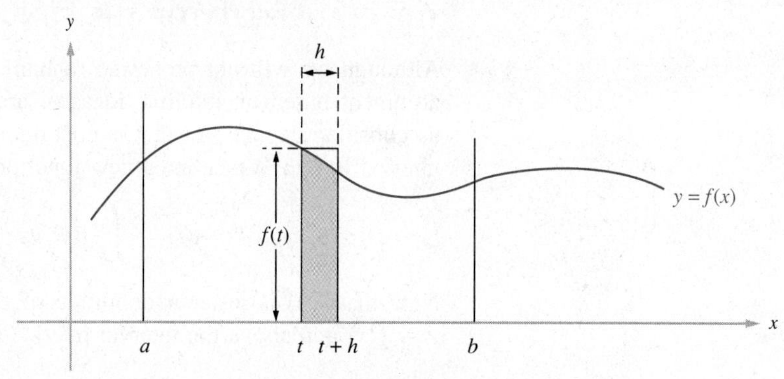

FIGURE 12.21

In other words

$$\frac{dA}{dt} = f(t),$$

and the function A is an antiderivative of f.

Now suppose that F is *any* antiderivative of f. By Theorem 12.1

$$F = A + k$$

for some constant k. Thus in particular

$$\begin{aligned} F(b) - F(a) &= [A(b) + k] - [A(a) + k] \\ &= A(b) - A(a) \\ &= \int_a^b f(x)\,dx - \int_a^a f(x)\,dx \\ &= \int_a^b f(x)\,dx, \end{aligned}$$

which gives the fundamental theorem of calculus.

EXERCISES 12.4

Evaluate the definite integrals in Exercises 1–21.

1. $\int_1^3 dx$ **2.** $\int_4^6 dx$ **3.** $\int_1^5 3\,dx$

4. $\int_{-2}^2 x\,dx$ **5.** $\int_{-4}^6 x^3\,dx$ **6.** $\int_0^4 x^{3/2}\,dx$

7. $\int_1^2 (4 - x^2)\,dx$ **8.** $\int_2^6 e^{2x}\,dx$ **9.** $\int_{-3}^{-1} x^{-1}\,dx$

10. $\int_1^3 \left(x + 1 + \frac{1}{x}\right) dx$ **11.** $\int_1^3 x\sqrt{x^2 + 1}\,dx$ **12.** $\int_{-2}^0 xe^{x^2}\,dx$

13. $\int_{1}^{-2} (x + 3)\, dx$

14. $\int_{6}^{3} (x^2 + e^x)\, dx$

15. $\int_{4}^{4} (x + 1)^{2/3}\, dx$

16. $\int_{-3}^{2} |x|\, dx$

17. $\int_{0}^{6} e^{2-3x}\, dx$

18. $\int_{1}^{2} \sqrt{3 - x}\, dx$

19. $\int_{2}^{5} x^{-1} \ln x\, dx$

20. $\int_{0}^{4} \sqrt{16 - x^2}\, dx$

21. $\int_{0}^{3} (2x - 1)^4\, dx$

In Exercises 22–32 make a rough sketch of the region described and then calculate its area.

22. The region bounded by the curve $y = x^2$ and the lines $x = 1$, $x = 4$, and $y = 0$.

23. The region bounded by the curve $y = e^x$ and the lines $x = -1$, $x = 1$, and $y = 0$.

24. The region bounded by the curve $y = x^2(x^3 + 1)$ and the lines $x = 1$, $x = 2$, and $y = 0$.

25. The region bounded by the curve $y = \dfrac{x}{1 + x^2}$ and the lines $x = 1$, $x = 3$, and $y = 0$.

26. The region bounded by the curve $y = x - x^2$ and the lines $x = 2$, $x = 4$, and $y = 0$.

27. The region bounded by the curve $y = x^2 - e^x$ and the lines $x = 2$, $x = 3$, and $y = 0$.

28. The region bounded by the curve $y = x^3 - x$ and the x-axis.

29. The region bounded by the curve $y = 4 - x^2$ and the x-axis.

30. The region bounded by the curve $y = x(x - 1)(x + 2)$ and the x-axis.

31. The region bounded by the curve $y = x^2 - 1$ and the lines $x = -1$, $x = 2$, and $y = 0$.

32. The region bounded by the curve $y = e^x - 1$ and the lines $x = -1$, $x = 1$, and $y = 0$.

Answers to Practice Problems

9. 12 (Compare with the last practice problem of Section 12.3.)

10. $11\frac{1}{3}$

11. (a) $u = 6 - x$ (b) $\ln 3$

12.5 Applications of the Definite Integral

In this section we will consider several applications of the definite integral. We will start with an example that illustrates the idea behind the fundamental theorem of calculus.

INTEGRATING MARGINAL COST

Let us consider a manufacturer of metal castings with a marginal cost given by $C'(x) = 120/\sqrt{x}$. Recall that $C'(x)$ approximates the cost of producing one more casting when x castings have already been produced. Suppose 10,000 castings have already been produced, and it is decided to manufacture 4400 more. What will the additional cost be?

We will give two approaches to this problem. The fact that they produce the same answer suggests the validity of the fundamental theorem of calculus.

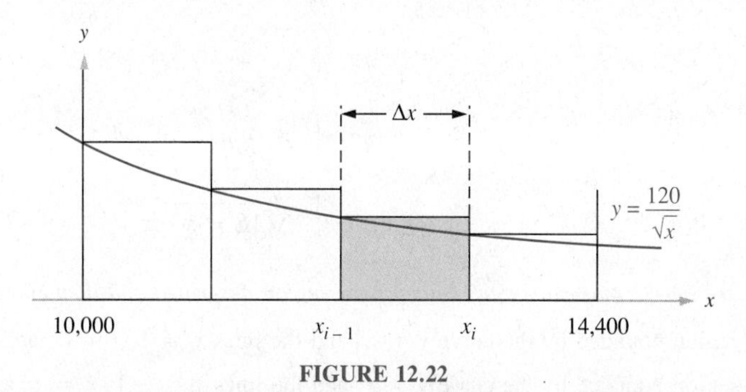

FIGURE 12.22

One approach is to divide the interval [10,000, 14,400] into n equal subintervals $[x_{i-1}, x_i]$, $i = 1, 2, \ldots, n$. This is pictured in Figure 12.22 with $n = 4$. Each subinterval will have length $\Delta x = (14{,}400 - 10{,}000)/n$. Since the cost of one more casting is $C'(x_{i-1})$ when the production is x_{i-1} castings, the cost of Δx more castings will be approximately $C'(x_{i-1})\Delta x$. (This is the area of the shaded rectangle.) Thus the total cost we seek will be approximated by the Riemann sum

$$S_n = C'(x_0)\Delta x + C'(x_1)\Delta x + \cdots + C'(x_{n-1})\Delta x.$$

Because the approximation will be better the larger n is, we can take the answer to our problem to be

$$\lim_{n\to\infty} S_n = \int_{10000}^{14400} C'(x)\,dx, \tag{12.13}$$

since the definite integral on the right is by definition the limit of the Riemann sums.

The second approach is to note that the marginal cost $C'(x)$ is the derivative of some cost function $C(x)$. In fact if F is any antiderivative of $C'(x) = 120/\sqrt{x}$, then we must have $C(x) = F(x) + k$ for some constant k. We seek the additional cost of increasing production from 10,000 to 14,400 castings, that is

$$C(14{,}400) - C(10{,}000).$$

But this is

$$\begin{aligned}[F(14{,}400) + k] - [(F(10{,}000) + k)] &= F(14{,}400) - F(10{,}000)\\ &= F(x)\Big|_{10000}^{14400}.\end{aligned}$$

Of course the fundamental theorem of calculus says that we can evaluate the definite integral given in (12.13) by computing $F(x)\,\big|_{10000}^{14400}$, where F is any antiderivative of $C'(x)$. The fact that we have arrived at these two answers independently adds to the plausibility of the fundamental theorem.

We finish our problem by completing the computation.

$$\begin{aligned}\int_{10000}^{14400} C'(x)\,dx &= \int_{10000}^{14400} 120x^{-1/2}\,dx \\ &= \frac{120x^{1/2}}{\frac{1}{2}}\Bigg|_{10000}^{14400} \\ &= 240x^{1/2}\Bigg|_{10000}^{14400} \\ &= 240(14400^{1/2} - 10000^{1/2}) \\ &= 240(120 - 100) = 4800\end{aligned}$$

Thus the additional cost of increasing production from 10,000 to 14,400 castings is $4800.

EXAMPLE 12.20 For $x \geq 20$ the marginal cost in dollars of producing x ash trays is found to be $100x/(x^2 + 1)$. Find the total cost of increasing production from 40 to 50 ash trays.

Solution The answer is

$$\int_{40}^{50} \frac{100x}{x^2 + 1}\,dx.$$

We make the substitution $u = x^2 + 1$, so that $du = 2x\,dx$. Notice that $u = 40^2 + 1 = 1601$ when $x = 40$, and $u = 50^2 + 1 = 2501$ when $x = 50$. Thus we have

$$\begin{aligned}\int_{40}^{50} \frac{100x}{x^2 + 1}\,dx &= 50\int_{40}^{50} (x^2 + 1)^{-1}2x\,dx \\ &= 50\int_{1601}^{2501} u^{-1}\,du \\ &= 50 \ln u \Bigg|_{1601}^{2501} \\ &= 50(\ln 2501 - \ln 1601) \approx 22.30.\end{aligned}$$

Thus the extra 10 ash trays can be produced for approximately $22.30. ■

Practice Problem 12 Suppose the marginal cost in dollars of producing x baseball gloves is $C'(x) = 0.00003x^2 - 0.024x + 20$. Find the cost of increasing production from 1000 to 1200 gloves.

COMPUTING AREAS

Although the most direct application of the definite integral is in computing the area of a region under a curve and above an interval, the areas of more complicated regions can also be determined.

EXAMPLE 12.21 Find the area of the region bounded by the line $y = x$ and the curve $y = x^2$.

Solution The region is shown in Figure 12.23(a). Notice that its area can be found by subtracting the area of the region shown in (c) (between $y = x^2$ and the x-axis) from the region of (b) (between $y = x$ and the x-axis).

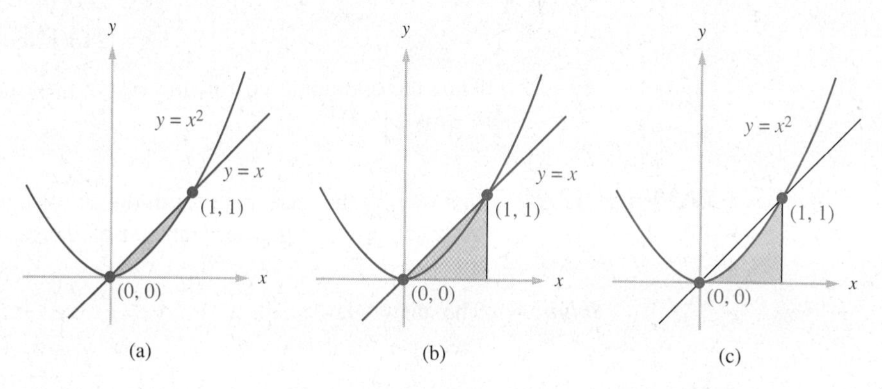

FIGURE 12.23

Thus the area we desire is given by

$$\int_0^1 x\,dx - \int_0^1 x^2\,dx = \int_0^1 (x - x^2)\,dx$$

$$= \frac{x^2}{2} - \frac{x^3}{3}\bigg|_0^1$$

$$= \frac{1^2}{2} - \frac{1^3}{3} - \left(\frac{0^2}{2} - \frac{0^3}{3}\right)$$

$$= \frac{1}{2} - \frac{1}{3} = \frac{1}{6}. \quad ■$$

This example illustrates a general principle:

The Area Between Two Curves

If $f(x) \geq g(x)$ for x in $[a, b]$, then the area between the curves $y = f(x)$ and $y = g(x)$ and between the lines $x = a$ and $x = b$ is given by

$$\int_a^b [f(x) - g(x)]\,dx.$$

This is pictured in Figure 12.24. Note that it is not necessary for the functions f and g to be nonnegative for x in $[a, b]$ for the above formula to be valid.

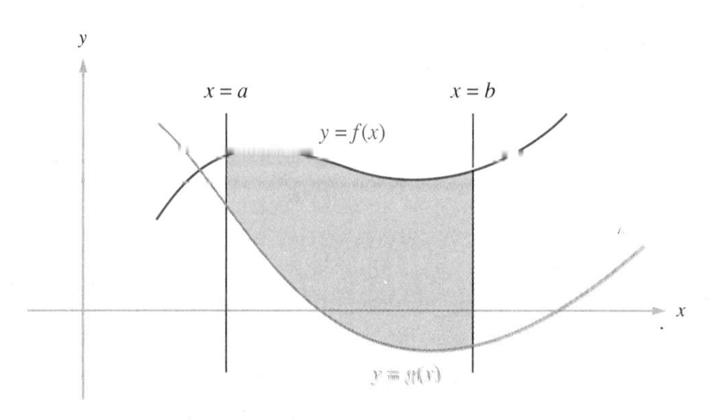

FIGURE 12.24

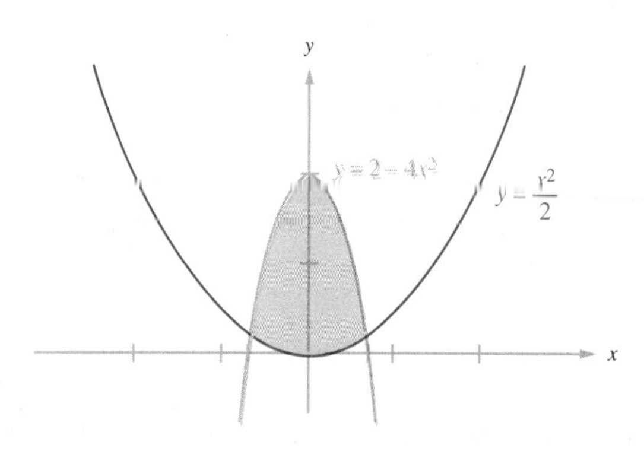

FIGURE 12.25

EXAMPLE 12.22 Find the area bounded by the curves $y = x^2/2$ and $y = 2 - 4x^2$.

Solution Figure 12.25 contains a sketch of the region for which we want the area. In order to determine the proper limits of integration we need to know the x-coordinates of the points where the two curves intersect. These can be found by solving the equations $y = x^2/2$ and $y = 2 - 4x^2$ simultaneously.

$$\begin{aligned} \frac{x^2}{2} &= 2 - 4x^2 \\ x^2 &= 4 - 8x^2 \\ 9x^2 &= 4 \\ x^2 &= \frac{4}{9} \\ x &= \pm\frac{2}{3} \end{aligned}$$

Since from the graph the curve $y = 2 - 4x^2$ is above the curve $y = x^2/2$, the area we want is given by

$$\begin{aligned} \int_{-\frac{2}{3}}^{\frac{2}{3}} \left(2 - 4x^2 - \frac{x^2}{2}\right) dx &= \int_{-\frac{2}{3}}^{\frac{2}{3}} \left(2 - \frac{9x^2}{2}\right) dx \\ &= 2x - \frac{9x^3}{3 \cdot 2}\Bigg|_{-\frac{2}{3}}^{\frac{2}{3}} = 2x - \frac{3x^3}{2}\Bigg|_{-\frac{2}{3}}^{\frac{2}{3}} \\ &= 2(\tfrac{2}{3}) - \frac{3(\frac{2}{3})^3}{2} - \left[2(-\tfrac{2}{3}) - \frac{3(-\frac{2}{3})^3}{2}\right] \\ &= \frac{4}{3} - \frac{4}{9} + \frac{4}{3} - \frac{4}{9} = \frac{16}{9}. \quad \blacksquare \end{aligned}$$

Practice Problem 13 Find the area of the region bounded by the curve $y = x^{1/2}$ and the line $y = x/2$.

CONSUMER'S AND PRODUCER'S SURPLUS

In Chapter 1 we considered how the price of an item determines the supply and the demand for that item. In general, as the price increases, the supply also increases, while the demand becomes less. Now we will turn these relationships around and let the number of items produced be the *independent* variable, denoted by x. We will take as the dependent variable the price of one item—both the price that consumers are willing to pay, and the price for which producers are willing to supply the item.

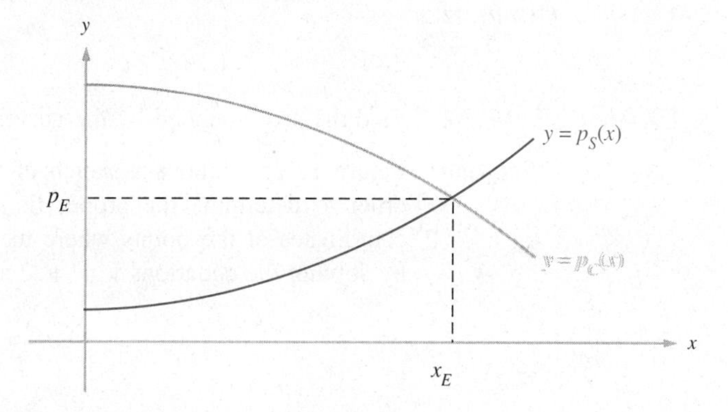

FIGURE 12.26

In general consumers are willing to pay more for an item when it is scarce and less when it is abundant. Thus if p_C represents the price consumers are willing to pay for an item, we expect p_C to be a decreasing function of x, the number of items produced. (See Figure 12.26.) Likewise suppliers will require a higher price per unit to increase their production. (They may have to build new plants, use costlier raw materials, or pay overtime.) Thus if p_S represents the price at which suppliers are willing to supply x items, then p_S is an increasing function of x.

The two curves $y = p_C(x)$ and $y = p_S(x)$ cross at the **equilibrium point** (x_E, p_E), where these two prices (the price consumers are willing to pay and the price suppliers are willing to produce for) are the same. The y-coordinate of the equilibrium point, p_E, is the price determined by the market. As long as the production x satisfies $0 < x < x_E$, the market price is below what consumers are willing to pay, but above what suppliers are willing to supply for, and everyone is happy. If $x > x_E$, however, the market breaks down because the price suppliers require is higher than consumers are willing to pay.

EXAMPLE 12.23 Suppose that for a certain type of custom-made fishing rod we have $p_C(x) = 1300 - x^2$ and $p_S(x) = 100 + 10x$. Find the equilibrium production and price.

Solution The curves $y = 1300 - x^2$ and $y = 100 + 10x$ cross when

$$\begin{aligned} 1300 - x^2 &= 100 + 10x, \\ 0 &= x^2 + 10x - 1200 \\ &= (x + 40)(x - 30). \end{aligned}$$

The solutions to this equation are $x = -40$ and $x = 30$. Since it makes no sense for x to be negative, we see that $x_E = 30$ is the equilibrium production. We can find the equilibrium price by substituting x_E into either function. For example

$$p_E = p_S(30) = 100 + 10(30) = \$400. \quad ■$$

Practice Problem 14 Suppose that when x items are produced consumers are willing to pay $15 - 0.1x$ dollars apiece, and producers are willing to supply the item for $0.05x$ dollars each. Find the equilibrium supply and equilibrium price.

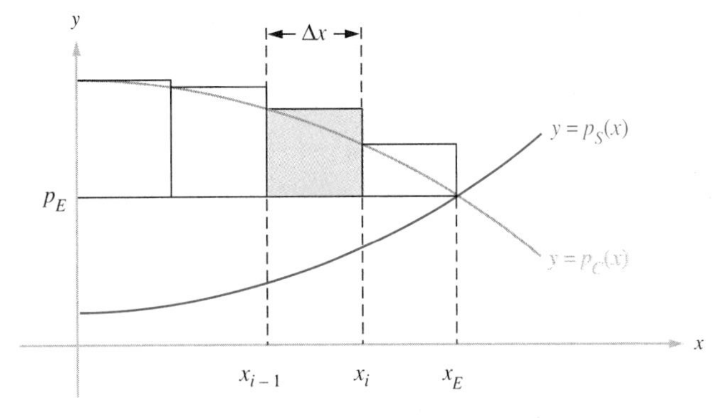

FIGURE 12.27

Economists have calculated the total benefit to consumers when they are able to buy a commodity at a price less than they would be willing to pay. Let the interval $[0, x_E]$ be divided into n equal subintervals $[x_{i-1}, x_i]$, each of width Δx. (See Figure 12.27.) Consumers buying Δx items at the price p_E instead of the price $p_C(x_{i-1})$ that they would be willing to pay save a total of approximately $[p_C(x_{i-1}) - p_E]\Delta x$. This is the area of the shaded rectangle in Figure 12.27. The total savings of consumers as production rises to x_E is approximated by the Riemann sum

$$S_n = [p_C(x_0) - p_E]\Delta x + [p_C(x_1) - p_E]\Delta x + \cdots + [p_C(x_{n-1}) - p_E]\Delta x.$$

Of course as n grows large S_n approaches the definite integral

$$\int_0^{x_E} [p_C(x) - p_E]\, dx.$$

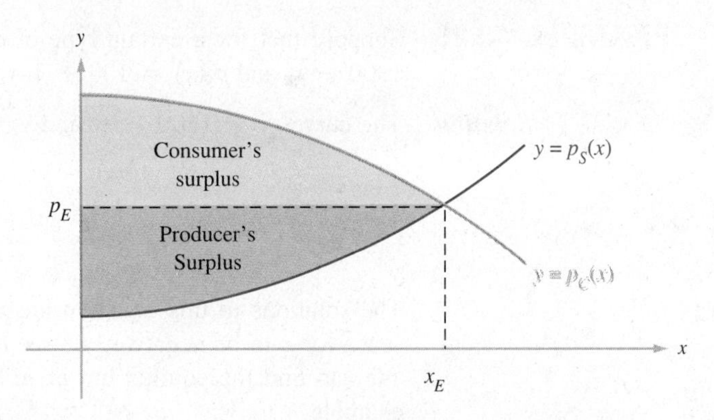

FIGURE 12.28

This quantity is called the **consumer's surplus,** and is the area of the region indicated in Figure 12.28. The **producer's surplus** is calculated in a similar way by integrating from 0 to x_E the excess of the market price p_E over the price $p_S(x)$ at which producers are willing to supply the item.

Consumer's and Producer's Surplus

$$\text{Consumer's surplus} = \int_0^{x_E} [p_C(x) - p_E]\, dx$$

$$\text{Producer's surplus} = \int_0^{x_E} [p_E - p_S(x)]\, dx$$

EXAMPLE 12.23 Revisited Compute the consumer's and producer's surplus for the manufacture of fishing rods in Example 12.23.

Solution Recall that we computed $x_E = 30$ and $p_E = 400$ in Example 12.23. Thus the consumer's surplus is

$$\begin{aligned}
\int_0^{x_E} [p_C(x) - p_E]\, dx &= \int_0^{30} (1300 - x^2 - 400)\, dx \\
&= \int_0^{30} (900 - x^2)\, dx \\
&= 900x - \frac{x^3}{3}\Bigg|_0^{30} \\
&= 900(30) - \frac{30^3}{3} - \left[900(0) - \frac{0^3}{3}\right] \\
&= \$18{,}000.
\end{aligned}$$

Likewise the producer's surplus is

$$\begin{aligned}
\int_0^{x_E} [p_E - p_S(x)]\,dx &= \int_0^{30} [400 - (100 + 10x)]\,dx \\
&= \int_0^{30} (300 - 10x)\,dx \\
&= 300x - \frac{10x^2}{2}\Bigg|_0^{30} \\
&= 300x - 5x^2\Bigg|_0^{30} \\
&= 300(30) - 5 \cdot 30^2 - [300(0) - 5 \cdot 0^2] \\
&= \$4500.
\end{aligned}$$

The interpretation of these figures is that consumers benefit a total of $18,000 from the fact that the market price of $400 per rod is less than what they are willing to pay, and producers benefit a total of $4500 because $400 is more than the price at which they are willing to supply the rods. ■

Practice Problem 15 For p_C and p_S as in Practice Problem 14 compute (a) the consumer's surplus and (b) the producer's surplus.

EXERCISES 12.5

1. Suppose that the marginal cost associated with producing x office chairs of a certain type is approximately $80 + 60x^{-1/2}$ dollars when $0 < x < 600$. What is the total cost of increasing production from 100 to 400 chairs?

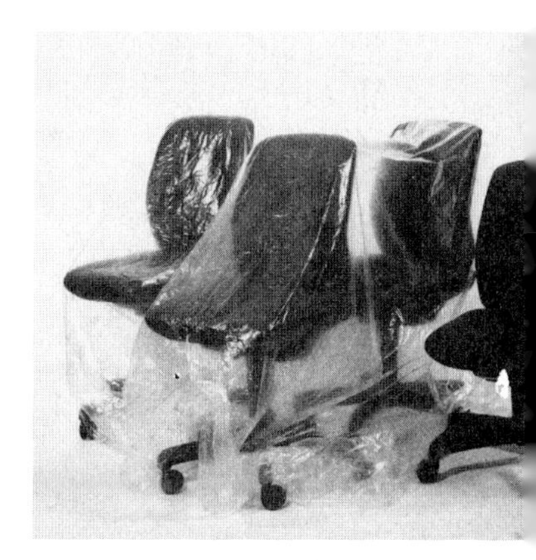

2. Suppose that the marginal cost associated with producing x metal tennis rackets is $0.0003x^2 - 0.024x + 20$ dollars. What is the total cost of increasing production from 1000 to 1200 rackets?

3. If the marginal cost (in dollars) of producing unit number x of a certain model portable radio is $0.00004x^2 - 0.03x + 12$, determine the total cost of producing units 101 through 200.

4. If the marginal cost (in dollars) of producing unit number x of a certain model shower head is $2 + \dfrac{1}{\sqrt{x + 1}}$, determine the total cost of producing units 10,001 through 16,000.

5. Suppose that the marginal revenue function for a certain model calculator is $60 - 0.02x$ dollars per unit. Determine the revenue generated by producing calculators number 21 through 30.

6. Suppose that the rate of sales of a certain product after t weeks of an advertising campaign is $150(1 - 0.8e^{-0.2t})$ units per week. Determine the total number of units sold during the first 10 weeks of the advertising campaign.

7. Suppose that the rate of sales of a certain product after t weeks of an advertising campaign is $100(1 - 0.6e^{-0.1t})$ units per week. Determine the total number of units sold during the eleventh through twentieth weeks of the advertising campaign.

8. A company manufacturing a new product must hire inexperienced workers for its production. Past experience with similar types of work shows that after the workers have produced x units of the product, the number of hours of direct labor required to produce an additional unit is approximated by $2 + e^{-0.02x}$. Find the number of hours of direct labor required to manufacture units 51 through 100.

In Exercises 9–24 find the area of the region bounded by the given curves.

9. $y = x^2$, $x = 3$, $x = 5$, and the x-axis

10. $y = \dfrac{1}{x^2}$, $x = 2$, $x = 4$, and the x-axis

11. $y = 3 - x$, $x = 5$, $x = 8$, and the x-axis

12. $y = \sqrt{x} - x$, $x = 4$, $x = 9$, and the x-axis

13. $y = x^3$, $x = 8$, and the x-axis

14. $y = x^3 - 1$, $x = 3$, and the x-axis

15. $y = x^2$ and $y = 4$

16. $y = x^4$ and $y = 16$

17. $y = 2x^2$ and $y = x^3$

18. $y = \sqrt{x}$ and $y = x$

19. $y = x^2$ and $y = x + 2$

20. $y = 2x^2$ and $y = 1 - x$

21. $y = x^2 + 2$ and $y = 3x$

22. $y = 2x^2 - x + 1$ and $y = x^2 + 2x - 1$

23. $y = \dfrac{1}{x + 2}$ and $x + y = 2$

24. $y = e^{3x}$, $x + y = 1$, and $x = 2$

25. Suppose that when x snowshoes are produced, consumers are willing to pay $25 - 0.002x$ dollars apiece for them, and producers are willing to supply them for $19 + 0.001x$ dollars each. Find the equilibrium supply and equilibrium price.

26. Suppose that when x Halloween make-up kits are made, producers are willing to supply them for $6 + 0.0002x$ dollars each, and consumers are willing to buy them for $9 - 0.0003x$ dollars each. Find the equilibrium demand and equilibrium price.

27. Suppose that when x golf carts are made, consumers are willing to pay $1500 - 0.01x^2$ dollars each for them, and producers are willing to supply them for $300 + x$ dollars each. Find the equilibrium supply and equilibrium price.

28. Suppose that when x triple cassette decks are made, producers are willing to supply them for $300 + x$ dollars each, and consumers are willing to buy them for $1050 - 0.005x^2$ dollars each. Find the equilibrium demand and equilibrium price.

29. Find the producer's surplus and consumer's surplus in Exercise 25.

30. Find the producer's surplus and consumer's surplus in Exercise 26.

31. Find the producer's surplus and consumer's surplus in Exercise 27.

32. Find the producer's surplus and consumer's surplus in Exercise 28.

33. Determine the consumer's and the producer's surplus if the demand and supply functions are $p_C(x) = 15 - 0.1x$ and $p_S(x) = 0.05x$.

34. Determine the consumer's and the producer's surplus if the demand and supply functions are $p_C(x) = 200 - \frac{x^2}{45}$ and $p_S(x) = 2x$.

35. Determine the consumer's and the producer's surplus if the demand and supply functions are $p_C(x) = 8000 - 0.03x^2$ and $p_S(x) = 8x$.

36. Determine the consumer's and the producer's surplus if the demand and supply functions are $p_C(x) = 20{,}000 - 0.006x^2$ and $p_S(x) = 14x$.

Answers to Practice Problems

12. \$6000

13. $\int_0^4 \left(x^{1/2} - \frac{x}{2}\right) dx = \frac{4}{3}$

14. $x_E = 100$ and $p_E = \$5$.

15. (a) \$500 (b) \$250

CHAPTER 12 REVIEW

IMPORTANT TERMS

- **antiderivative** *(12.1)*
- **constant of integration**
- **indefinite integral**
- **integral sign**
- **integrand**
- **marginal cost**
- **variable of integration**
- **differential** *(12.2)*
- **substitution**
- **area** *(12.3)*
- **definite integral**
- **lower and upper limits of integration**
- **Riemann sum**
- **consumer's and producer's surplus** *(12.5)*
- **equilibrium point**

IMPORTANT FORMULAS

Theorem 12.1

If F_1 and F_2 are both antideratives of the function f on some interval, then

$$F_2(x) = F_1(x) + k$$

for some constant k.

Indefinite Integral of a Power

$$\int x^n dx = \frac{x^{n+1}}{n+1} + k \qquad \text{if } n \neq -1$$

Linearity Rules for Indefinite Integrals

$$\int cf(x)\,dx = c\int f(x)\,dx \qquad \text{if } c \text{ is any constant.}$$

$$\int [f(x) \pm g(x)]\,dx = \int f(x)\,dx \pm \int g(x)\,dx.$$

Indefinite Integral of x^{-1}

$$\int x^{-1}\,dx = \ln|x| + k$$

Indefinite Integral of an Exponential Function

$$\int e^{cx}\,dx = \frac{e^{cx}}{c} + k, \qquad \text{if } c \text{ is any nonzero constant}$$

Definition of *du*

If $u = g(x)$, then $du = g'(x)\,dx$.

Riemann Sum

$$S_n = f(x_0)\Delta x + f(x_1)\Delta x + \ldots + f(x_{n-1})\Delta x$$

Definite Integral

$$\int_a^b f(x)\,dx = \lim_{n\to\infty} S_n$$

The Fundamental Theorem of Calculus

Suppose F is an antiderivative for the function f on the interval $[a, b]$. Then

$$\int_a^b f(x)\,dx = F(b) - F(a).$$

The Area Between Two Curves

If $f(x) \geq g(x)$ for x in $[a, b]$, then the area between the curves $y = f(x)$ and $y = g(x)$ and between the lines $x = a$ and $x = b$ is given by

$$\int_a^b [f(x) - g(x)]\,dx.$$

Consumer's and Producer's Surplus

$$\text{Consumer's surplus} = \int_0^{x_E} [p_C(x) - p_E]\,dx$$

$$\text{Producer's surplus} = \int_0^{x_E} [p_E - p_S(x)]\,dx$$

REVIEW EXERCISES

In Exercises 1–6 evaluate the given integral.

1. $\int (3x^2 - 2x^{-1} + 5x^{-1/2})\, dx$

2. $\int e^{7x}\, dx$

3. $\int x^2(2x^3 + 5)^{1/2}\, dx$

4. $\int (2x + 1)(x^2 + x)^{-1}\, dx$

5. $\int_1^8 \sqrt[3]{x}\, dx$

6. $\int_0^3 x^2 e^{x^3}\, dx$

7. Suppose the marginal cost of producing x steam irons is $10 + \dfrac{1}{(2x + 1)}$, with a fixed cost of \$100,000. What is the cost of producing 1000 irons?

8. Find an antiderivative F of e^{3-x} such that $F(1) = 5$.

9. After t seconds the velocity of a sliding rock is $24t$ feet per second. How far will it slide in the first 10 seconds?

10. Find the Riemann sum S_n for the function $f(x) = x^2$ over the interval $[1, 3]$ with $n = 4$.

11. In Exercise 10 what is $\lim_{n \to \infty} S_n$?

12. Find the area of the region bounded by the curve $y = x^3 - x$ and the lines $x = 3/2$ and $y = 0$.

13. Find the area of the region between the curves $y = 2x^2 + x$ and $y = x^2 - x$.

14. If the marginal cost for making x pairs of gloves is $4 - x^{-2}$ dollars, find the total cost of increasing production from 400 to 500 pairs.

15. Suppose that when x riding mowers are made, producers are willing to supply them for $2000 + 3x$ dollars apiece, and consumers are willing to buy them for $3400 - 0.02x^2$ dollars each. Find the equilibrium supply and equilibrium price. What is the consumer's surplus? The producer's surplus?

13

FURTHER TOPICS IN INTEGRATION

In this chapter we consider a variety of more advanced topics relating to definite and indefinite integrals.

13.1 Differential Equations

Suppose we know that the number of compact disc players in use in the United States is growing continuously at the rate of 20% per year. Various natural questions present themselves, such as:

(A) How much will the number of players in use grow in 10 years?
(B) How long will it take for the number of players in use to double?

Before these can be answered we need to express mathematically the statement that the growth rate is 20%. Let x be the time in years after the present, and let y be the number of players in use after x years. (So y will depend on x.) Although the *rate* of growth is given to be a constant 20%, the actual increase in players in use during any time period will depend on the number in use at the start of that period. In fact, since y' represents the rate of change of the number of players in use, y, with respect to x, the condition that y grows continuously at a rate of 20% amounts to the equation

$$y' = 0.20y.$$

This is an example of a **differential equation,** that is, an equation involving one or more derivatives. Here are some other examples of differential equations:

(i) $y' = x$
(ii) $xy' = 2y$
(iii) $y' = y + e^x$
(iv) $3y'' + 2y' + y = 0.$

Equations (i), (ii), and (iii) are examples of **first-order** equations because they involve only a first derivative; and we will only consider such equations in this section. By a **solution** to a first-order differential equation we mean a function $y = f(x)$ such that if $f(x)$ and $f'(x)$ are substituted into the equation for y and y', an identity in x results. For example $y = xe^x$ is a solution to equation (iii). For then $y' = e^x + xe^x$, and substituting these functions into

(iii) $$y' = y + e^x$$

for y and y' yields

$$e^x + xe^x = xe^x + e^x,$$

which holds for all values of x.

Practice Problem 1 (a) To which of (i), (ii), and (iii) is $y = x^2$ a solution?
(b) To which of (i), (ii), and (iii) is $y = e^x(x + 1)$ a solution?
(c) To which of (i), (ii), and (iii) is $y = x^2/2$ a solution?

Solving differential equation (i),

$$y' = x,$$

amounts to finding an antiderivative for the function $g(x) = x$. By Theorem 12.1 every solution is of the form

$$y = \frac{x^2}{2} + k, \tag{13.1}$$

for some constant k. This is the **general solution** to the equation, since as k takes on all constant values we get all solutions of the differential equation. If k is specified, then we have a **particular solution,** for example $y = \frac{x^2}{2} + 5$.

Any differential equation that can be put in the form

$$y' = g(x) \tag{13.2}$$

has the general solution $y = \int g(x)\,dx$. The differential equations we will consider, even when not of the form (13.2), will ordinarily have infinitely many solutions; and their general solutions will involve an arbitrary constant such as the k in (13.1).

A GEOMETRIC INTERPRETATION

Let us consider again the differential equation $y' = x$, this time from a geometric standpoint. If $y = f(x)$ is a solution to this differential equation, then at each point on its graph the slope y' of a tangent line equals the x-coordinate of the point. This is indicated in Figure 13.1.

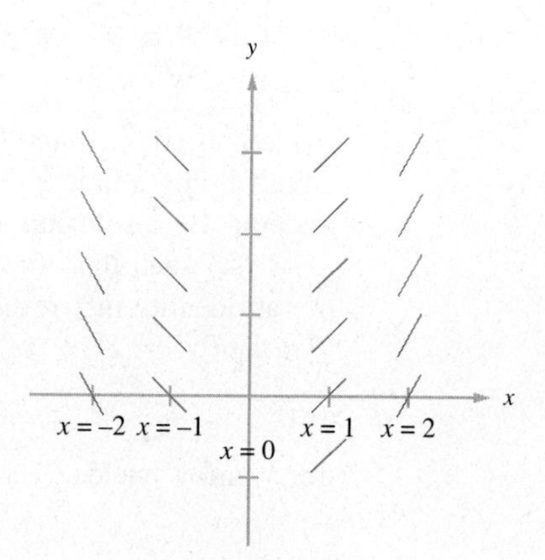

FIGURE 13.1

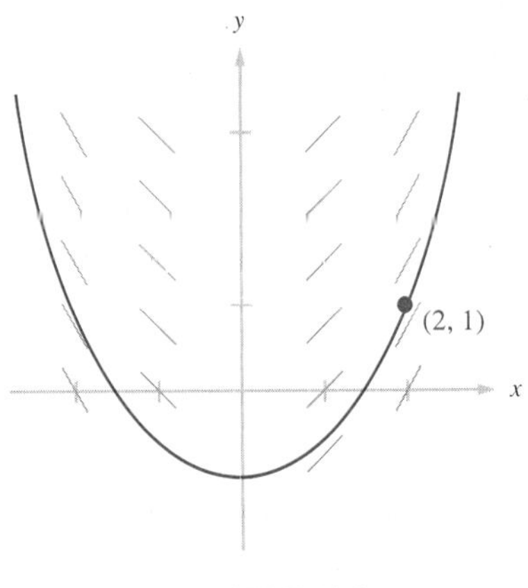

FIGURE 13.2

If we know one point on the graph of the function, then the differential equation determines the rest of the curve. Suppose we know that $y = 1$ when $x = 2$, for example. Then the curve shown in Figure 13.2 is determined. In this case since we know the general solution to $y' = x$ is

$$y = \frac{x^2}{2} + k,$$

we can find the particular solution algebraically. Substituting $x = 2$ and $y = 1$ gives

$$1 = \frac{2^2}{2} + k,$$
$$1 = 2 + k,$$
$$k = -1.$$

Thus the particular solution to $y' = x$ going through the point $(x, y) = (2, 1)$ is $y = \frac{x^2}{2} - 1$.

The condition that $y = 1$ when $x = 2$ in this example is called an **initial condition.** Having an initial condition enables us to determine a particular solution to a differential equation from the general solution.

SEPARABLE EQUATIONS

Let us consider another way of solving the differential equation

$$y' = x.$$

This equation can also be written

$$\frac{dy}{dx} = x. \qquad (13.3)$$

In Section 12.2 the differentials du and dx were defined, with $du = u'\,dx$. Using y in place of u we have $dy = y'\,dx$, and so the expression dy/dx can be interpreted either as y', or as the quotient of dy divided by dx. Thus the differential equation (13.3) is equivalent to

$$dy = x\,dx.$$

We now integrate both sides of this equation as follows:

$$\int dy = \int x\,dx$$
$$y + k_1 = \frac{x^2}{2} + k_2$$
$$y = \frac{x^2}{2} + k_2 - k_1 = \frac{x^2}{2} + k.$$

Here we have replaced the difference $k_2 - k_1$ by the single constant k. Of course this is the same general solution we found before.

This same technique can be used whenever a differential equation can be put in the form

$$f(y)\,dy = g(x)\,dx,$$

and such an equation is said to be **separable.** An example is the equation (ii) $xy' = 2y$ of our previous list. We rewrite the equation as

$$x\frac{dy}{dx} = 2y,$$

$$x\,dy = 2y\,dx,$$

$$y^{-1}\,dy = 2x^{-1}\,dx.$$

Integrating both sides gives

$$\int y^{-1}\,dy = \int 2x^{-1}\,dx,$$

$$\ln|y| = 2\ln|x| + k.$$

(Notice that only one constant need be introduced, as in the previous example.)

In order to solve for y we raise e to the power of both sides of the previous equation.

$$e^{\ln|y|} = e^{2\ln|x|+k} = (e^{\ln|x|})^2e^k$$

Since $e^{\ln z} = z$ by logarithm property 6 in Section 8.4, we can simplify this equation further.

$$|y| = |x|^2e^k = x^2e^k$$

$$y = \pm x^2e^k$$

Because $\pm e^k$ is just some new constant, we write our general solution as

$$y = cx^2.$$

Check If $y = cx^2$, then $y' = 2cx$. Substituting these functions in

(ii) $$xy' = 2y$$

yields

$$x(2cx) = 2(cx^2),$$

which is true for all values of x.

Practice Problem 2 Write each of the following equations that is separable with the variables separated.

(a) $y'y = x + y$

(b) $y'y = x - y'e^y$

(c) $y' \ln x = x + e^yx$

(d) $(1 + x + y)y' = 2$

(e) $x^2y' = 1 + x + y + xy$

EXAMPLE 13.1 At the beginning of this section we let y be the number of compact disc players in use after x years, which satisfies the differential equation $y' = 0.20y$. Find the general solution to this equation.

Solution We write the equation as

$$\frac{dy}{dx} = .2y,$$

or

$$y^{-1}\,dy = .2\,dx.$$

Notice that the left side is a function of y times dy, and the right side is a (constant) function of x times dx. Thus the variables have been separated and we proceed to integrate both sides.

$$\int y^{-1}\,dy = \int .2\,dx$$
$$\ln|y| = .2x + k$$

We can solve for y by raising e to the power of both sides.

$$e^{\ln|y|} = e^{.2x+k} = e^{.2x}e^{k}$$
$$|y| = e^{.2x}e^{k}$$
$$y = \pm e^{.2x}e^{k}$$

If we replace $\pm e^{k}$ by the constant c, we get the general solution

$$y = ce^{.2x}.$$

Check If $y = ce^{.2x}$, then $y' = c(.2)e^{.2x}$. Substituting these functions in the original equation

$$y' = .2y$$

yields

$$c(.2)e^{.2x} = .2(ce^{.2x}),$$

which is true for all values of x. ■

EXAMPLE 13.2 Find the general solution to the differential equation $yy' = x^2$. Then find the particular solution such that $y = 3$ when $x = 0$.

Solution We rewrite the equation and separate the variables as follows.

$$y\frac{dy}{dx} = x^2$$
$$y\,dy = x^2\,dx$$

Now we get the general solution by integrating both sides.

$$\int y\,dy = \int x^2\,dx$$

$$\frac{y^2}{2} = \frac{x^3}{3} + k$$

$$y^2 = \frac{2}{3}x^3 + 2k = \frac{2}{3}x^3 + c$$

$$y = \pm\left(\frac{2}{3}x^3 + c\right)^{1/2}$$

Notice that the domain of the solution functions must be restricted to those values of x such that $\frac{2}{3}x^3 + c > 0$.

Check If $y = \pm\left(\frac{2}{3}x^3 + c\right)^{1/2}$, then

$$y' = \pm\frac{1}{2}\left(\frac{2}{3}x^3 + c\right)^{-1/2}\left(\frac{2}{3}\cdot 3x^2\right) = \pm x^2\left(\frac{2}{3}x^3 + c\right)^{-1/2}$$

Substituting these functions in the original differential equation

$$yy' = x^2$$

yields

$$\left[\pm\left(\frac{2}{3}x^3 + c\right)^{1/2}\right]\left[\pm x^2\left(\frac{2}{3}x^3 + c\right)^{-1/2}\right] = x^2.$$

Since for any particular solution we will have either both plus signs or both minus signs, this amounts to the identity

$$x^2 = x^2.$$

Now we find the particular solution satisfying the initial condition $y = 3$ when $x = 0$. Substituting these values in

$$y = \pm\left(\frac{2x^3}{3} + c\right)^{1/2}$$

gives

$$3 = \pm\left(\frac{2\cdot 0^3}{3} + c\right)^{1/2}$$

$$3 = \pm c^{1/2}.$$

Since by definition $c^{1/2} \geq 0$, the plus sign must hold. Squaring yields

$$9 = c.$$

Thus the particular solution we seek is $y = (\frac{2}{3}x^3 + 9)^{1/2}$. ■

Practice Problem 3 (a) Find the general solution to the separable differential equation $y'x^2 = 4\sqrt{y}$. (b) Find the particular solution satisfying $y = 4$ when $x = 2$.

EXERCISES 13.1

In Exercises 1–4 a differential equation is given, along with three functions. Tell which, if any, of the functions are solutions to the differential equation.

1. $y^3y' = 8x$ (a) $y = \sqrt{x}$ (b) $y = 2\sqrt{x}$ (c) $y = 2\sqrt{x} + 3$

2. $2y' + y = x$ (a) $y = e^{-2x}$ (b) $y = x - 2$ (c) $y = e^{-2x} + x - 2$

3. $y' - y = \frac{1}{x} - \ln x$ (a) $y = \ln 3x$ (b) $y = e^x + \ln x$ (c) $y = \ln ex - 2e^x - 1$

4. $xy' - 2y = x^2$ (a) $y = x^2$ (b) $y = x^2e^x$ (c) $y = x^2 \ln x$

In Exercises 5–8 make a graph similar to Figure 13.1 indicating tangents to curves satisfying the given differential equation.

5. $y' = y$

6. $y' = \frac{y}{x}$

7. $y' = -\frac{x}{y}$

8. $y' = -\frac{1}{x^2}$

In Exercises 9–14 write each differential equation with the variables separated, if possible.

9. $\frac{xy'}{y} = 3 - xy'$

10. $y'e^x = y - y'$

11. $y \ln x + y' = xy$

12. $y^2 = y \ln x + y'$

13. $xy' + ye^{2x} = x^2$

14. $xy' - ye^{3x} = y \ln x$

In Exercises 15–28 find the general solution of the given differential equation.

15. $y' = e^{3x}$

16. $y' = x^2 + 3$

17. $y' = xe^{x^2}$

18. $y' = \sqrt{3x + 2}$

19. $y' = x^{-1} \ln x$

20. $y' = x(x^2 + 1)^{1/2}$

21. $y' = -5y$

22. $y' = 1 - 3y$

23. $y' = -\frac{y}{x}$

24. $y' = \frac{x}{y}$

25. $y' = \frac{e^x}{y}$

26. $y' = \frac{y^2}{x^3}$

27. $x^2y' = xy - y'$

28. $y' - \sqrt{x} = y\sqrt{x}$

In Exercises 29–36 find the particular solution satisfying the given initial condition.

29. $y' = \sqrt{2 - x}$; $y = 4$ when $x = 1$

30. $y' = (3x + 2)^{-1}$; $y = 3$ when $x = 2$

31. $y' = 2y$; $y = 4$ when $x = 0$

32. $y' = \frac{x^2}{y}$; $y = -1$ when $x = 1$

33. $y' = 2x - 3xy$; $y = 2$ when $x = 0$

34. $y' = (y + 1)e^{-x}$; $y = 0$ when $x = 0$

35. $y'(x^3 + 1) = x^2y$; $y = -2$ when $x = 0$

36. $y^2y' = xe^{x^2}$; $y = 1$ when $x = 1$

In Exercises 37–40 find the particular solution satisfying the initial condition and graph it. Compare with the graph of the earlier exercise.

37. Equation of Exercise 5; $y = 1$ when $x = 0$

38. Equation of Exercise 6; $y = 2$ when $x = 1$

39. Equation of Exercise 7; $y = 2$ when $x = 0$

40. Equation of Exercise 8; $y = 1$ when $x = 1$

Answers to Practice Problems

1. (a) (ii) (b) (iii) (c) (i) and (ii)

2. (b) $(y + e^y)\,dy = x\,dx$ (c) $(1 + e^y)^{-1}\,dy = \left(\frac{x}{\ln x}\right)dx$

(e) $(1 + y)^{-1}\,dy = \left(\frac{1}{x^2} + \frac{1}{x}\right)dx$

3. (a) $y = (-2x^{-1} + k)^2$, where $-2x^{-1} + k \geq 0$ (b) $y = (-2x^{-1} + 3)^2$

13.2 Applications of Differential Equations

The ability to solve differential equations is useful in many practical situations. Three types of differential equations arising in many applications will be studied in this section.

EXPONENTIAL GROWTH AND DECAY

We start by answering the questions at the beginning of Section 13.1 about the number of compact disc players in use. Remember that this number, denoted y, satisfies the differential equation $y' = .2y$, which has the general solution $y = ce^{.2x}$.

(A) How much will the number of players in use grow in 10 years?

We compute the ratio of $ce^{.2(10)}$ (the number in 10 years) to $ce^{.2(0)}$ (the present number). This is

$$\frac{ce^2}{c} = e^2 \approx 7.39.$$

Thus in 10 years the number of players will be more than 7 times the present number.

(B) How long will it take for the number of players in use to double?

Since the present number of players is $ce^{.2(0)} = c$, we solve the equation $y = 2c$. This gives

$$\begin{aligned} ce^{.2x} &= 2c, \\ e^{.2x} &= 2, \\ .2x &= \ln 2, \\ x &= \frac{\ln 2}{.2} \approx 3.47. \end{aligned}$$

Thus the number of players will approximately double every $3\frac{1}{2}$ years.

The function $y = ce^{.2x}$ of this example has the form

$$y = ce^{kx}.$$

Such functions are said to describe **exponential growth** if $k > 0$ or **exponential decay** if $k < 0$. They arise in many contexts.

EXAMPLE 13.3 When an element is radioactive, it spontaneously changes to another element at a rate proportional to the amount of the first element present. This is called **radioactive decay.** The rate at which the element decays is usually indicated by giving the **half-life** of the element, which is the time it takes for half of any given amount of the element to change. For example, the radioactive element carbon-14 has a half-life of 5776 years.

Suppose we have 100 pounds of carbon-14. If y is the number of pounds left x years after the present, then y satisfies a differential equation of the form

$$y' = cy,$$

where c is some constant. We separate the variables and solve this equation in the usual way.

$$\begin{aligned} \frac{dy}{dx} &= cy \\ y^{-1}\,dy &= c\,dx \\ \int y^{-1}\,dy &= \int c\,dx \\ \ln|y| &= cx + k \\ |y| &= e^{cx+k} = e^{cx}e^{k} \\ y &= \pm e^{k}e^{cx} = Ke^{cx} \end{aligned}$$

Here we have replaced $\pm e^{k}$ by K. We can determine K from the initial condition that $y = 100$ when $x = 0$.

$$100 = Ke^{0} = K$$

Thus the solution has the form

$$y = 100e^{cx},$$

where c is a constant that depends of the rate of decay of carbon-14.

We can determine c from the information that the half-life of carbon-14 is 5776 years. Because in 5776 years only half of our 100 pounds of carbon-14 will be left, we have

$$50 = 100e^{c(5776)},$$

$$e^{5776c} = \frac{50}{100} = .5,$$

$$5776c = \ln .5,$$

$$c = \frac{\ln .5}{5776}.$$

We see that the amount of carbon-14 left after x years will be

$$y = 100e^{(\ln .5)x/5776}.$$

For example after 40 years the amount left will be

$$100e^{(\ln .5)(40)/5776} \approx 99.52 \text{ pounds.}$$ ■

Practice Problem 4 The half-life of lead-210 is 22 years. (a) If we start with 500 kg of lead-210, give a formula for the amount y left after x years. (b) How much will be left after 100 years?

EXAMPLE 13.4 The number of bacteria in a culture is increasing at a rate proportional to the number present. There are 50,000 bacteria at 1 P.M., and 60,000 at 2 P.M. How many will there be at 5 P.M.? At what time will there be 200,000 bacteria?

Solution If y is the number of bacteria x hours after 1 P.M., we have the differential equation

$$y' = cy,$$

where c is a constant. Just as in Example 13.3 the general solution is

$$y = Ke^{cx},$$

and the initial condition $y = 50{,}000$ when $x = 0$ allows us to find the particular solution

$$y = 50{,}000e^{cx}.$$

Now $y = 60{,}000$ when $x = 1$, which lets us solve for c.

$$60{,}000 = 50{,}000e^c$$

$$e^c = \frac{60{,}000}{50{,}000} = 1.2$$

$$c = \ln 1.2$$

Thus we have the equation

$$y = 50{,}000e^{(\ln 1.2)x}.$$

At 5 P.M. we have $x = 4$, and so

$$y = 50{,}000e^{4 \ln 1.2} \approx 103{,}680.$$

In order to find out when there will be 200,000 bacteria, we solve the equation

$$200{,}000 = 50{,}000e^{(\ln 1.2)x}.$$

We have

$$e^{(\ln 1.2)x} = \frac{200{,}000}{50{,}000} = 4,$$

$$(\ln 1.2)x = \ln 4,$$

$$x = \frac{\ln 4}{\ln 1.2} \approx 7.6.$$

Thus there will be 200,000 at 7.6 hours after 1 P.M., or about 8:36 P.M. ■

GROWTH TO A LIMIT

Clearly the number of compact disc players in use will not increase by 20% per year indefinitely. Eventually everyone will have as many players as he or she wants or can afford, and sales will level off. Thus the model we have developed will only be accurate so long as the players are a fairly new appliance.

A different model is more appropriate for consumer items nearing the saturation level of acceptance. Consider microwave ovens, for example. We will suppose that a reasonable upper limit for the number of such ovens in use in the United States is the number of households, 80 million. Let y be the number of millions of microwave ovens in use x years after 1980, and suppose the yearly rate of increase of y is 10% of the number of households without microwave ovens, that is, 10% of the amount that y is below 80. This leads to the differential equation

$$y' = r(L - y),$$

where $r = 0.10$ and $L = 80$. Such a differential equation models **growth to a limit.** Suppose we have the initial condition that $y = 20$ when $x = 0$; that is, there are 20 million microwave ovens in use in 1980. We solve our differential equation as follows.

$$\frac{dy}{dx} = r(L - y)$$
$$(L - y)^{-1}\,dy = r\,dx$$
$$\int (L - y)^{-1}\,dy = \int r\,dx$$

The indefinite integral on the left may be evaluated by use of the substitution $u = L - y$. Then $du = -dy$, and

$$\begin{aligned}\int (L - y)^{-1}\,dy &= -\int (L - y)^{-1}(-dy)\\ &= -\int u^{-1}\,du\\ &= -\ln|u| + k\\ &= -\ln|L - y| + k.\end{aligned}$$

Thus we have

$$\begin{aligned}-\ln|L - y| &= rx + c,\\ \ln|L - y| &= -rx - c,\\ |L - y| &= e^{-rx-c} = e^{-rx}e^{-c},\\ L - y &= \pm e^{-c}e^{-rx} = Ke^{-rx},\\ y &= L - Ke^{-rx},\end{aligned}$$

where we have renamed the constant $\pm e^{-c}$ as K. Thus our differential equation has the general solution

$$y = 80 - Ke^{-.1x}.$$

We can determine K by using the initial condition $y = 20$ when $x = 0$.

$$20 = 80 - Ke^{0} = 80 - K$$
$$K = 60$$

We see that the particular solution we seek is

$$y = 80 - 60e^{-.1x}.$$

This curve is graphed in Figure 13.3. Notice how y approaches the limiting value 80 as x gets large. For example in the year 2000, when $x = 20$, we can expect the number of microwave ovens in use to be

$$80 - 60e^{-.1(20)} \approx 72,$$

or about 72 million.

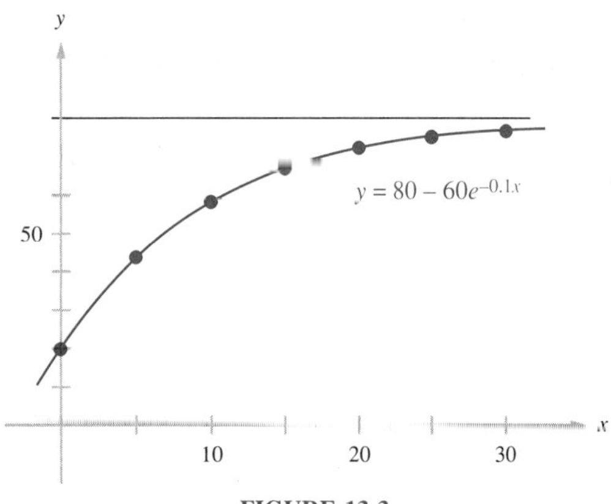

FIGURE 13.3

EXAMPLE 13.5 When an object is in contact with another object having a different, constant temperature, its temperature changes at a rate proportional to the temperature difference between the two objects. Consider a metal plate with temperature 60° F that is placed in contact with a hot water tank having a constant temperature of 200° F. After 1 minute the temperature of the plate has risen to 70°. When will the plate reach 180°?

Solution Let y be the temperature of the plate x minutes after it has been placed against the tank. Then we have the differential equation

$$y' = r(L - y),$$

where $L = 200°$, and r is a constant to be determined. This differential equation has exactly the same form as that describing the growth to a limit of compact disc players, and so has the same general solution, namely

$$y = L - Ke^{-rx} = 200 - Ke^{-rx}.$$

We can determine K by use of the initial condition that $y = 60$ when $x = 0$.

$$60 = 200 - Ke^{0} = 200 - K$$
$$K = 200 - 60 = 140.$$

Thus we have the particular solution

$$y = 200 - 140e^{-rx}.$$

Now we determine r from the information that the temperature of the plate is 70° when $x = 1$.

$$\begin{aligned} 70 &= 200 - 140e^{-r(1)} \\ 140e^{-r} &= 200 - 70 = 130 \\ e^{-r} &= \frac{130}{140} = \frac{13}{14} \\ -r &= \ln \frac{13}{14} \end{aligned}$$

Thus our equation is

$$y = 200 - 140e^{(\ln 13/14)x}.$$

To find when the temperature of the plate will reach 180°, we set $y = 180$ and solve for x.

$$\begin{aligned} 180 &= 200 - 140e^{(\ln 13/14)x} \\ 140e^{(\ln 13/14)x} &= 200 - 180 = 20 \\ e^{(\ln 13/14)x} &= \frac{20}{140} = \frac{1}{7} \\ \left(\ln \frac{13}{14}\right)x &= \ln \frac{1}{7} \\ x &= \frac{\ln \frac{1}{7}}{\ln \frac{13}{14}} \approx 26.3 \end{aligned}$$

Thus it will take approximately 26.3 minutes for the plate to reach 180°. ■

Practice Problem 5 The number of bacteria in an enclosure is increasing at an hourly rate of 20% of the difference between 30,000 (a natural limit for the population in the available space) and the number of bacteria. Let y be the number of bacteria x hours after the present time, when there are 5000 bacteria. (a) Write a differential equation describing the situation. (b) Give its particular solution. (c) How many bacteria will there be after 10 hours?

DIFFUSION

Another application of differential equations is to *diffusion,* the spread of some characteristic throughout a population. (It might be a disease, a political viewpoint, or the possession of some particular knowledge.) Suppose among a population P some characteristic is shared by y individuals at time x. The characteristic is passed from the individuals having it to those who do not. We might expect the increase in y to be proportional to both the number y itself, and the

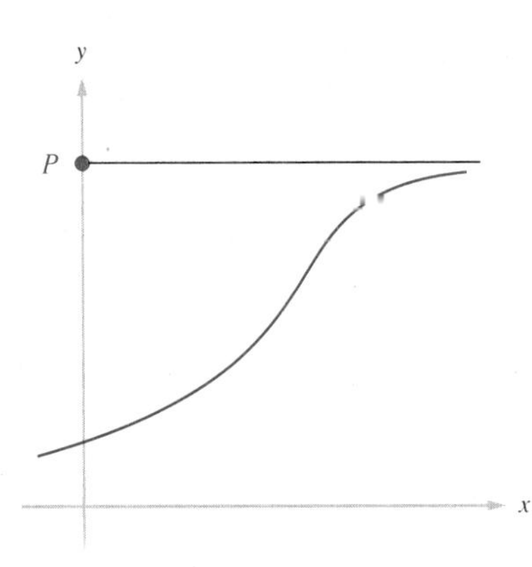

FIGURE 13.4

number $P - y$ of individuals who still do not have the characteristic. Thus we have a differential equation of the form

$$y' = cy(P - y),$$

where c is some constant. A solution can be found by separating the variables as

$$c\,dx = \frac{dy}{y(P - y)} = \frac{1}{P}\left(\frac{1}{y} + \frac{1}{P - y}\right) dy.$$

It can be shown that this equation has the solution

$$y = \frac{P}{1 + ke^{-Pcx}},$$

where k is an arbitrary constant. The graph of this function is known as the **logistic curve,** and its general shape is shown in Figure 13.4.

EXAMPLE 13.6 Suppose that the number y of millions of Americans who have been exposed to a certain type of influenza is growing at a monthly rate given by

$$y' = .002y(P - y),$$

where P is 180 million, and that y is two million at the start of December 1991. Write y as a function of the number of months x after this time. What will y be on December 1, 1992?

Solution We have the equation leading to the logistic curve with $c = .002$ and $P = 180$ million. Thus the solution has the form

$$y = \frac{180}{1 + ke^{-180(.002)x}} = \frac{180}{1 + ke^{-.36x}}$$

for some constant k. Since $y = 2$ when $x = 0$, we have

$$2 = \frac{180}{1 + k},$$

$$1 + k = \frac{180}{2} = 90,$$

$$k = 89.$$

We see that the equation for y is

$$y = \frac{180}{1 + 89e^{-.36x}}.$$

Since x is measured in months, on December 1, 1992 we have $x = 12$, and

$$y = \frac{180}{1 + 89e^{-.36(12)}} \approx 82.4 \text{ million.}$$ ■

Practice Problem 6 On September 15 the number y of millions of teenage boys who had Mohawk haircuts was .5, and growing at a rate

$$y' = .01y(P - y)$$

millions per week, where P is 15 million. (a) Write y as a function of x, the number of weeks after September 15. (b) What will y be after 10 weeks?

EXERCISES 13.2

1. The number y of persons having a certain communicable disease is growing continuously at a yearly rate of 15% of the number having the disease. Suppose 100,000 people have the disease in 1985. Write y as a function of x, the number of years after 1985. How many will have the disease in 1995? When will one million people have the disease?
2. Suppose that the number y of computer-controlled automobile braking systems in use is growing at a continuous rate of 12% per year, and that 50,000 such systems are in use in 1988. Write y as a function of x, the number of years after 1988. How many will be in use in 1998? When will 200,000 be in use?
3. The half-life of thorium-234 is 24.5 days. Let y be the number of grams of thorium-234 left after x days, with $y = 200$ when $x = 0$. Write y as a function of x. What is y when $x = 365$? When is $y = 20$?
4. The half-life of bismuth-214 is 19.7 minutes. Let y be the number of grams of bismuth-214 left from an original amount of one gram after x minutes. Write y as a function of x. What is y after one hour? When is $y = 0.01$ gram?
5. Only 80% of a quantity of polonium-218 will be left after one minute. What is the half-life of this substance?
6. After one year 97% of a quantity of lead-210 will be left. What is the half-life of this substance?
7. The population of an underdeveloped nation is growing continuously at a rate of 4% per year. If the present population is 6 million, what will the population be in 10 years? How long will it take for the population to double?
8. The population of a colony of mice is increasing continuously at 11% per year. If the present population of the colony is 250, what will it be in 3 years? How long will it take to reach 1000?
9. A forest can support 10,000 foxes; there are 500 now. Suppose the number of foxes after x years is y, and that their rate of growth is 5% of the difference between 10,000 and y. Express y as a function of x. What will y be after 5 years? How long will it take for y to reach 5000?
10. In a certain city the number y of cassette players in taxis is growing at a yearly rate of 20% of the difference between the number of taxis, which is fixed at 5000, and y. At present 1000 taxis have the players. Write y as a function of x, the number of years after the present. What will y be in 10 years? How long will it take until half of the taxis have players?

11. A certain country has 10 million coffee drinkers, and this number never changes. Of these, 15% now drink decaffeinated coffee. The number of those who drink decaffeinated coffee is growing at a yearly rate of 25% of the number of drinkers of regular coffee. How many will drink regular coffee after 5 years? When will half of the coffee drinkers drink decaffeinated?

12. A homeowner's lawn can support 5000 dandelions, and has 1000 now. The number of dandelions is increasing at a yearly rate of 30% of the difference between 5000 and the number of dandelions. How many dandelions will there be after 6 years? When will there be 4000 dandelions?

13. An insulated box contains food at 40° F. It is put outside where the temperature is 85°. If the temperature of the food rises at an hourly rate of 20% of the difference between 85° and the food temperature, when will it reach 70°?

14. A test tube containing a liquid at 65° F is put in a water bath at a constant 40°. The temperature of the liquid falls at an hourly rate of 30% of the difference between it and 40°. What will the temperature of the liquid be after 5 hours?

15. The number of bacteria in a dish is increasing at a rate proportional to the number present. There are 20,000 bacteria now, and there will be 45,000 in one hour. How many bacteria will there be after 24 hours?

16. The number of people who have heard a new joke is increasing at a rate proportional to the number who have already heard it. On May 3 one million people had heard the joke, and on May 9 the number was 1.5 million. On what day will 100 million people have heard the joke?

17. The number of fish in a pond grows at a rate proportional to the difference between 500 and the number of fish. At the beginning of 1985 there were 100 fish, and one year later there were 150. When will there be 400 fish?

18. A container of water is placed in a room which is kept at a constant temperature of 20° C. At noon the water is 5°, and at 1 P.M. it is 9°. When will it reach 15°?

19. Suppose the number of horses y infected with an eye disease is growing at a yearly rate given by $y' = 0.00005y(10{,}000 - y)$, and that in 1990 the disease affects 500 horses. Write y as a function of the number of years x after 1990. How many horses will have the disease in 1995?

20. Suppose the number of Americans y owning sunshades for their automobiles is growing at a yearly rate of $y' = 0.10y(P - y)$, where P is 150 million, and that $y = 20$ million in 1985. Write y as a function of x, the number of years after 1985. What will y be in 2000?

21. The number of people y who have tried a new brand of soft drink was 3 million in 1988, and growing at a yearly rate

$$y' = .03y(P - y),$$

where P is 30 million. Write y as a function of x, the number of years after 1988. After how many years will y be 20 million?

22. The number of elm trees y that have been infected by a blight is 10 million in 1975, and growing at a yearly rate

$$y' = .05y(P - y),$$

where P is 100 million. Write y as a function of x, the number of years after 1975. When will y be 90 million?

23. Show that the constant k in the equation of the logistic curve is the ratio of the number of individuals not having the characteristic to the number who do when $x = 0$.

24. Show that the equation

$$c\,dx = \frac{dy}{y(P - y)} = \frac{1}{P}\left(\frac{1}{y} + \frac{1}{P - y}\right) dy,$$

leads to the solution

$$y = \frac{P}{1 + ke^{-Pcx}},$$

where k is a constant.

Answers to Practice Problems

4. (a) $y = 500e^{(\ln .5)x/22}$ (b) approximately 21.4 kg

5. (a) $y' = .2(30{,}000 - y)$ (b) $y = 30{,}000 - 25{,}000e^{-.2x}$ (c) 26,617

6. (a) $y = \dfrac{15}{1 + 29e^{-.15x}}$ (b) about 2 million

13.3 Improper Integrals

We will investigate definite integrals for which the range of integration is unbounded. Evaluating such integrals enables us to measure the effect of processes that continue into the indefinite future, as well as to calculate probabilities connected with continuous distribution functions.

EXAMPLE 13.7 Suppose an agreement is negotiated to rent a piece of property for 5 years at a rate of \$3000 per year. Obviously a total of \$15,000 will eventually be paid in rent, although the payments will be spread over the 5 years. The \$15,000 can be interpreted as the area shaded in Figure 13.5, which is the definite integral of the function $f(x) = 3000$ over the interval $[0, 5]$, that is, $\int_0^5 3000\,dx$.

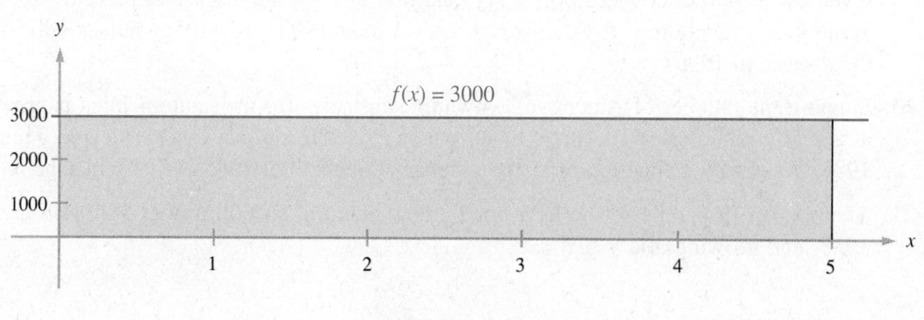

FIGURE 13.5

Of course \$15,000 paid over the course of the next 5 years is not the same as \$15,000 paid at the present time, since if the money were paid at the start it could be earning interest for the person owning the property. Suppose that money can be invested at an annual rate of 12% interest compounded continuously. Then, an investment of P_0 dollars now will be worth $P = P_0e^{0.12x}$ after x years. By solving for P_0 we see that P dollars paid after x years has a value of only $P_0 = P/e^{0.12x} = Pe^{-0.12x}$ now. (This is the continuous version of the concept of "present value" introduced in Section 4.2.) Thus a truer measure of the present value of the rental agreement is

$$\int_0^5 3000e^{-0.12x}\,dx.$$

This is the area of the region pictured in Figure 13.6.

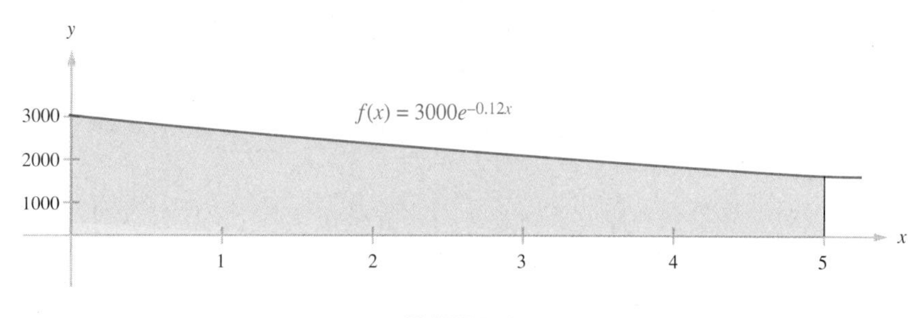

FIGURE 13.6

Now let us suppose that the rental agreement is not just for 5 years, but forever! We will try to make some reasonable estimate of the present value of the agreement. One might expect this value to be infinite, but it turns out that such is not the case. If the agreement were for 10 years instead of 5, its present value would be $\int_0^{10} 3000e^{-0.12x}\,dx$; and, in general an agreement for b years would be worth $\int_0^b 3000e^{-0.12x}\,dx$. Thus a reasonable estimate for the value of an agreement for the indefinite future is

$$\lim_{b\to\infty}\int_0^b 3000e^{-0.12x}\,dx.$$

(See Figure 13.7.) ■

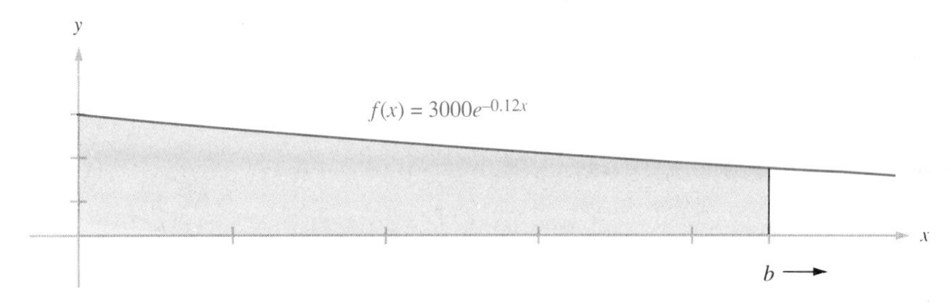

FIGURE 13.7

The answer to Example 13.7,

$$\lim_{b\to\infty} \int_0^b 3000e^{-0.12x}\,dx,$$

is an instance of an **improper integral,** and we denote it by $\int_0^\infty 3000e^{-0.12x}\,dx$. The general definition is as follows.

Improper Integrals

$$\int_a^\infty f(x)\,dx = \lim_{b\to\infty} \int_a^b f(x)\,dx, \qquad \text{provided this limit exists;}$$

$$\int_{-\infty}^b f(x)\,dx = \lim_{a\to-\infty} \int_a^b f(x)\,dx, \qquad \text{provided this limit exists; and}$$

$$\int_{-\infty}^\infty f(x)\,dx = \int_{-\infty}^0 f(x)\,dx + \int_0^\infty f(x)\,dx.$$

When the required limit or limits exist, we say that the improper integral **converges,** otherwise it **diverges.** Note that the improper integral from $-\infty$ to ∞ is defined only when both the integrals $\int_{-\infty}^0 f(x)\,dx$ and $\int_0^\infty f(x)\,dx$ converge. Often convergent improper integrals can be evaluated using the fundamental theorem of calculus, which may also be used to identify divergence.

EXAMPLE 13.8 Evaluate $\int_2^\infty x^{-2}\,dx$.

Solution We have

$$\int_2^\infty x^{-2}\,dx = \lim_{b\to\infty} \int_2^b x^{-2}\,dx = \lim_{b\to\infty} \left.\frac{x^{-1}}{-1}\right|_2^b$$

$$= \lim_{b\to\infty} \left(\frac{b^{-1}}{-1} - \frac{2^{-1}}{-1}\right) = \lim_{b\to\infty} \left(\frac{1}{2} - \frac{1}{b}\right) = \frac{1}{2}.$$

This is illustrated in Figure 13.8. ■

EXAMPLE 13.9 Evaluate $\int_2^\infty x^{-1}\,dx$.

Solution We have

$$\int_2^\infty x^{-1}\,dx = \lim_{b\to\infty} \int_2^b x^{-1}\,dx = \lim_{b\to\infty} \ln x \,\Big|_2^b$$

$$= \lim_{b\to\infty} (\ln b - \ln 2).$$

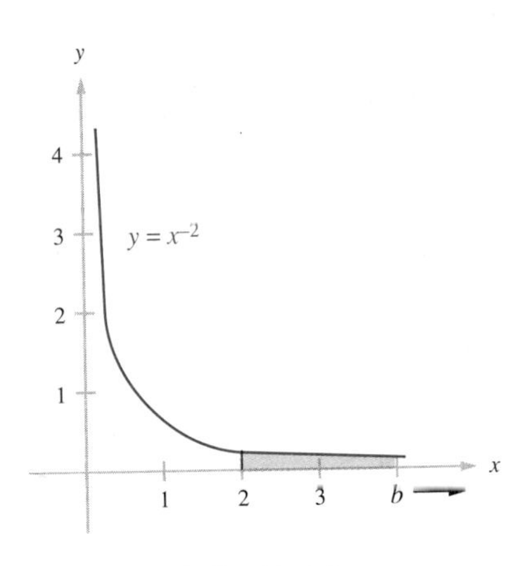

FIGURE 13.8

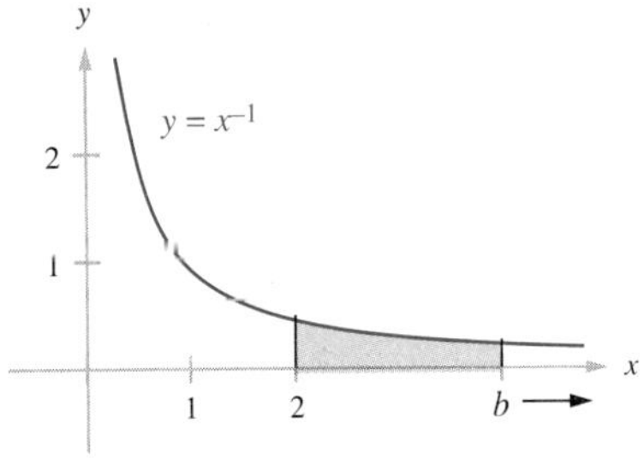

FIGURE 13.9

But as b gets arbitrarily large so does $\ln b$. Thus the limit does not exist, and the integral of this example diverges. Figure 13.9 shows the region corresponding to this integral. Notice that it does not look much different from the region of Figure 13.8, for which the integral converged. ■

Practice Problem 7 Evaluate if possible: (a) $\int_2^{\infty} x^{-3}\, dx$ (b) $\int_2^{\infty} x^{-1/2}\, dx$.

EXAMPLE 13.10

Evaluate $\int_{-\infty}^{1} e^x\, dx$.

Solution We have

$$\int_{-\infty}^{1} e^x\, dx = \lim_{a\to-\infty} \int_a^1 e^x\, dx = \lim_{a\to-\infty} e^x \Big|_a^1$$

$$= \lim_{a\to-\infty} (e^1 - e^a) = e - 0 = e,$$

where we have used the fact that $\lim\limits_{a\to-\infty} e^a = 0$. ■

EXAMPLE 13.11

Evaluate $\int_{-\infty}^{\infty} xe^{-x^2}\, dx$.

Solution First we note that the substitution $u = -x^2$, $du = -2x\, dx$, yields

$$\int xe^{-x^2}\, dx = -\frac{1}{2}\int e^{-x^2}(-2x)\, dx = -\frac{1}{2}\int e^u\, du$$

$$= -\frac{1}{2}e^u + k = -\frac{1}{2}e^{-x^2} + k.$$

Thus we have

$$\begin{aligned}
\int_{-\infty}^{\infty} xe^{-x^2}\,dx &= \int_{-\infty}^{0} xe^{-x^2}\,dx + \int_{0}^{\infty} xe^{-x^2}\,dx \\
&= \lim_{a\to-\infty}\int_{a}^{0} xe^{-x^2}\,dx + \lim_{b\to\infty}\int_{0}^{b} xe^{-x^2}\,dx \\
&= \lim_{a\to-\infty} -\frac{1}{2}e^{-x^2}\Big|_{a}^{0} + \lim_{b\to\infty} -\frac{1}{2}e^{-x^2}\Big|_{0}^{b} \\
&= \lim_{a\to-\infty} -\frac{1}{2}(e^{0} - e^{-a^2}) + \lim_{b\to\infty} -\frac{1}{2}(e^{-b^2} - e^{0}) \\
&= -\frac{1}{2}(1 - 0) - \frac{1}{2}(0 - 1) = 0. \quad ■
\end{aligned}$$

EXAMPLE 13.7 Revisited Recall that in Example 13.7 we considered the problem of estimating the present value of rent payments of \$3000 per year indefinitely, when money can be invested at 12% interest. Our answer was $\int_0^\infty 3000e^{-0.12x}\,dx$. We evaluate this integral as follows.

$$\begin{aligned}
\int_{0}^{\infty} 3000e^{-0.12x}\,dx &= \lim_{b\to\infty}\int_{0}^{b} 3000e^{-0.12x}\,dx \\
&= \lim_{b\to\infty} \frac{3000e^{-0.12x}}{-0.12}\Big|_{0}^{b} = \lim_{b\to\infty} \frac{-3000}{0.12}(e^{-0.12b} - e^{0}) \\
&= -25000(0 - 1) = 25000
\end{aligned}$$

Here we have used the fact that the limit of $e^{-0.12b}$ is 0 as b increases without bound. Thus the present value of the agreement is \$25,000. ■

The amount \$25,000 in Example 13.7 is called the **capitalized value** of the property. In general, if property has a fixed annual rent of R and the annual interest rate is i, then its capitalized value is given by

$$\int_{0}^{\infty} Re^{-ix}\,dx.$$

Practice Problem 8 Find the capitalized value of property earning a fixed rent of \$40,000 per year, if the interest rate is 10% per year.

EXAMPLE 13.12 After x years a polluted river empties a certain chemical into the ocean at the rate of $5000/(1 + .02x)^2$ tons per year. What is the total amount of the chemical that will enter the ocean?

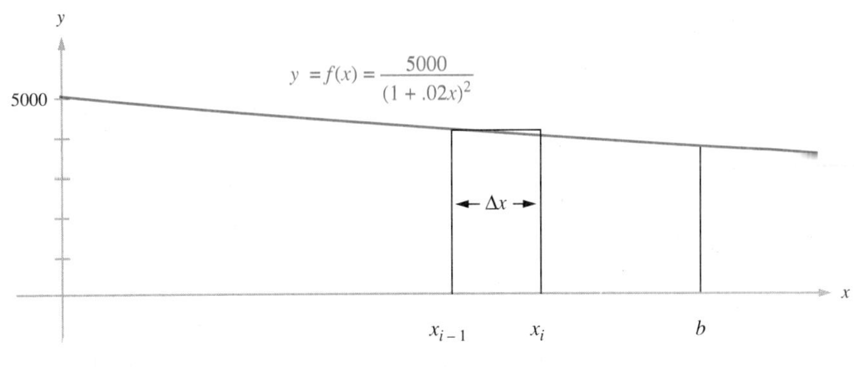

FIGURE 13.10

Solution Let $f(x) = 5000/(1 + .02x)^2$. The graph of $y = f(x)$ is shown in Figure 13.10. If the interval $[0, b]$, where b is some positive real number, is divided into subintervals $[x_{i-1}, x_i]$ of width Δx, then the amount of the chemical entering the ocean over the time interval $[x_{i-1}, x_i]$ will be approximated by $f(x_{i-1})\Delta x$. This is the area of the rectangle shown in the figure.

The sum of such rectangles is a Riemann sum which approaches $\int_0^b f(x)\,dx$ as Δx approaches 0. Thus we see that the amount of the chemical entering the ocean up to time b is $\int_0^b f(x)\,dx$, and the total amount is

$$\int_0^\infty f(x)\,dx = \int_0^\infty 5000(1 + .02x)^{-2}\,dx.$$

Note that if $u = 1 + .02x$, then $du = .02\,dx$, and so

$$\int 5000(1 + .02x)^{-2}dx = \frac{5000}{.02}\int (1 + .02x)^{-2}(.02\,dx)$$

$$= 250{,}000\int u^{-2}\,du = \frac{250{,}000u^{-1}}{-1} + k$$

$$= -250{,}000u^{-1} + k$$

$$= -250{,}000(1 + .02x)^{-1} + k.$$

Thus

$$\int_0^\infty 5000(1 + .02x)^{-2}\,dx = \lim_{b\to\infty}\int_0^b 5000(1 + .02x)^{-2}\,dx$$

$$= \lim_{b\to\infty} -250{,}000(1 + .02x)^{-1}\Big|_0^b$$

$$= \lim_{b\to\infty} -250{,}000\left(\frac{1}{1 + .02b} - 1\right)$$

$$= -250{,}000(0 - 1) = 250{,}000.$$

A total of 250,000 tons of the chemical will enter the ocean. ■

EXERCISES 13.3

In Exercises 1–24 evaluate the given improper integral, if possible.

1. $\int_3^{\infty} x^{-5}\, dx$

2. $\int_4^{\infty} x^{-3/2}\, dx$

3. $\int_1^{\infty} e^{-3x}\, dx$

4. $\int_5^{\infty} x^{-2/3}\, dx$

5. $\int_0^{\infty} \frac{1}{2x+1}\, dx$

6. $\int_{-\infty}^{-1} \frac{1}{\sqrt{(4-x)^3}}\, dx$

7. $\int_{-\infty}^{\infty} \frac{x}{x^2+1}\, dx$

8. $\int_1^{\infty} \frac{1}{\sqrt[3]{x}}\, dx$

9. $\int_{-\infty}^{-1} x^{-4}\, dx$

10. $\int_{-\infty}^{\infty} e^{-x}\, dx$

11. $\int_{16}^{\infty} x^{-5/4}\, dx$

12. $\int_{-\infty}^{1} e^{2x}\, dx$

13. $\int_2^{\infty} \frac{1}{x(\ln x)^2}\, dx$

14. $\int_{-\infty}^{\infty} \frac{x}{(x^2+4)^3}\, dx$

15. $\int_1^{\infty} \frac{dx}{(x+3)^2}$

16. $\int_4^{\infty} \sqrt{x}\, dx$

17. $\int_{-\infty}^{-2} \frac{x^2\, dx}{(x^3+1)^2}$

18. $\int_1^{\infty} xe^{-x^2}\, dx$

19. $\int_{-\infty}^{0} e^{-x}\, dx$

20. $\int_{-\infty}^{-1} \frac{3+5x}{x^4}\, dx$

21. $\int_e^{\infty} \frac{\ln x}{x}\, dx$

22. $\int_e^{\infty} \frac{1}{x(\ln x)^3}\, dx$

23. $\int_1^{\infty} \frac{\sqrt{x}+\sqrt[3]{x}}{x^2}\, dx$

24. $\int_{-\infty}^{0} (1-x)^{-5/2}\, dx$

25. Find the capitalized value of property earning a fixed rent of \$1200 per year, if the interest rate is 8% per year.

26. Find the capitalized value of property earning a fixed rent of \$6000 per year, if the interest rate is 9% per year.

27. A chemical plant empties a pollutant into a lake at the rate of $300/(2 + 3x)^2$ kg per year after x years. What is the total amount of the pollutant that will enter the lake?

28. Oil seeps into an underground stream from a storage facility at the rate of $x/(x^2 + 1)^2$ tons per month after x months. What is the total amount of oil that will enter the stream?

29. A contest winner has a choice of taking \$12,000 per year for the next 20 years or \$10,000 per year indefinitely. If the interest rate is 9%, which should she choose?

30. Is it better to buy a piece of land for \$100,000, or pay \$12,000 per year rent on a perpetual lease? Suppose the interest rate is 8% per year.

31. Find, if possible, the area between the curves $y = 1/x^2$ and $y = 1/x^3$ to the right of the line $x = 1$.

32. Find, if possible, the area between the curves $y = e^{-x}$ and $y = e^{-2x}$ to the right of the y-axis.

33. Find, if possible, the area between the curves $y = 1/x$ and $y = 1/(x + 1)$ to the right of the line $x = 2$.

34. Find, if possible, the area between the curves $y = x^{-1/2}$ and $y = (x + 1)^{-1/2}$ to the right of the line $x = 3$.

35. Give an example of a function f such that the limit of $\int_{-b}^{b} f(x)\,dx$ as b increases without bound exists, yet $\int_{-\infty}^{\infty} f(x)\,dx$ diverges.

Answers to Practice Problems **7.** (a) $\frac{1}{8}$ (b) The integral diverges **8.** \$400,000

13.4 Integration by Parts (Optional)

In Section 12.2 we observed that the chain rule for differentiation leads to the method of integration by substitution. We shall now consider an important integration technique based on the rule for differentiating the product of two functions.

Recall that if u and v are differentiable functions of x, then

$$(uv)' = uv' + vu'.$$

Taking the indefinite integral of both sides of this equation with respect to x yields

$$uv = \int uv'\,dx + \int vu'\,dx.$$

Recalling that $du = u'\,dx$ and $dv = v'\,dx$ by the definition of the differential, we can write this more simply as

$$uv = \int u\,dv + \int v\,du.$$

We solve for $\int u\,dv$ to derive the following formula for **integration by parts.**

Integration by Parts

$$\int u\,dv = uv - \int v\,du \tag{13.4}$$

This formula sometimes allows us to evaluate integrals by converting an integral we cannot handle into one that we can. To illustrate it, let us consider the indefinite integral

$$\int xe^x\,dx.$$

Choosing $u = x$ and $dv = v'\,dx = e^x\,dx$, we obtain $du = dx$ and $v = e^x$. (For v we may use any antiderivative of $v' = e^x$.) Thus by (13.4)

$$\begin{aligned}\int xe^x\,dx &= \int u\,dv = uv - \int v\,du \\ &= xe^x - \int e^x\,dx \\ &= xe^x - e^x + k = (x-1)e^x + k.\end{aligned}$$

In this case the evaluation of the difficult indefinite integral $\int xe^x\,dx$ was accomplished by using integration by parts and evaluation of the simpler integral $\int e^x\,dx$.

Use of the integration by parts formula requires that the factors u and dv be identified. Although selection of u and $dv = v'\,dx$ is often a matter of trial and error, it is helpful to keep the following points in mind.

1. The function u should generally be chosen so that its derivative is simple.
2. An antiderivative v of v' must be calculable.
3. The indefinite integral $\int v\,du$ must be computable.

EXAMPLE 13.13 We will evaluate

$$\int \ln|x|\,dx,$$

using integration by parts. In this case there is only one choice for u and dv: we must let $u = \ln|x|$ and $dv = dx$. Then $du = x^{-1}\,dx$ and $v = x$, so that

$$\int \ln|x|\,dx = \int u\,dv = uv - \int v\,du = x\ln|x| - \int dx = x\ln|x| - x + k.$$

Check $$\begin{aligned}(x\ln|x| - x + k)' &= \ln|x| + x\cdot x^{-1} - 1 + 0 \\ &= \ln|x| + 1 - 1 = \ln|x| \quad ■\end{aligned}$$

EXAMPLE 13.14 Evaluate $\int x\sqrt{x+2}\,dx$ using integration by parts.

Solution In this case there are several possible choices for u and dv.

(a) $u = x\sqrt{x+2}$ and $dv = dx$,
(b) $u = x$ and $dv = \sqrt{x+2}\,dx$,
(c) $u = \sqrt{x+2}$ and $dv = x\,dx$.

Notice, however, that in (a) and (c) the derivative of u is not simple, and therefore the indefinite integral $\int v\,du$ may be difficult to compute. In (b), on the other hand, there is no such problem: If $u = x$ and $dv = \sqrt{x+2}\,dx = (x+2)^{1/2}\,dx$, then $du = dx$ and $v = \frac{2}{3}(x+2)^{3/2}$. Hence

$$\begin{aligned}
\int x\sqrt{x+2}\,dx &= x\left[\frac{2}{3}(x+2)^{3/2}\right] - \int \frac{2}{3}(x+2)^{3/2}\,dx \\
&= \frac{2}{3}x(x+2)^{3/2} - \frac{\frac{2}{3}(x+2)^{5/2}}{\frac{5}{2}} + k \\
&= \frac{2}{3}(x+2)^{3/2}\left[x - \frac{2}{5}(x+2)\right] + k \\
&= \frac{2}{3}(x+2)^{3/2}\left[\frac{3x}{5} - \frac{4}{5}\right] + k \\
&= \frac{2}{15}(3x-4)(x+2)^{3/2} + k. \quad \blacksquare
\end{aligned}$$

EXAMPLE 13.15 Evaluate $\int (x+2)e^{-x}\,dx$.

Solution We will use integration by parts with $u = x + 2$ and $dv = e^{-x}\,dx$. Then $du = dx$ and $v = -e^{-x}$. Hence

$$\begin{aligned}
\int (x+2)e^{-x}\,dx &= -(x+2)e^{-x} - \int (-e^{-x})\,dx \\
&= -(x+2)e^{-x} + \int e^{-x}\,dx \\
&= -(x+2)e^{-x} + \frac{e^{-x}}{-1} + k \\
&= -(x+2)e^{-x} - e^{-x} + k \\
&= -e^{-x}(x+2+1) + k \\
&= -(x+3)e^{-x} + k.
\end{aligned}$$

Check

$$\begin{aligned}
[-(x+3)e^{-x} + k]' &= -[(x+3)(-1)e^{-x} + e^{-x}] + 0 \\
&= -e^{-x}(1 - x - 3) \\
&= -e^{-x}(-x - 2) \\
&= (x+2)e^{-x} \quad \blacksquare
\end{aligned}$$

Practice Problem 9 (a) How would you choose u and dv in order to integrate $\int x \ln |x|\,dx$ by parts? (b) Evaluate the integral.

Sometimes it is necessary to use integration by parts more than once to evaluate a particular integral. This situation often arises when the integrand contains a polynomial factor.

EXAMPLE 13.16 Use integration by parts to compute $\int x^2e^x\,dx$.

Solution Selecting $u = x^2$ and $dv = e^x\,dx$, we have $du = 2x\,dx$ and $v = e^x$. Thus

$$\int x^2e^x\,dx = x^2e^x - \int 2xe^x\,dx = x^2e^x - 2\int xe^x\,dx. \tag{13.5}$$

Although we cannot immediately write an antiderivative of xe^x, the indefinite integral

$$\int xe^x\,dx$$

can itself be computed using integration by parts. In fact this was done at the beginning of this section, where we found

$$\int xe^x\,dx = (x - 1)e^x + k.$$

We substitute this expression into (13.5) to get

$$\begin{aligned}\int x^2e^x\,dx &= x^2e^x - 2[(x - 1)e^x + k]\\ &= e^x(x^2 - 2x + 2) - 2k\\ &= (x^2 - 2x + 2)e^x + c,\end{aligned}$$

where $c = -2k$. ■

Because integration by parts can be difficult, and mistakes are common, we remind the reader that *the computation of an indefinite integral can always be checked by differentiating the answer*.

DEFINITE INTEGRALS

The method of integration by parts can also be applied to definite integrals.

Integration by Parts for a Definite Integral

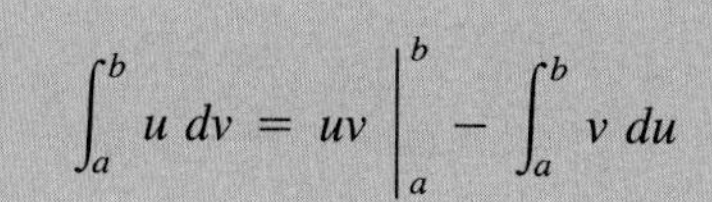

$$\int_a^b u\,dv = uv\Big|_a^b - \int_a^b v\,du$$

EXAMPLE 13.17 Evaluate $\int_1^8 x^{2/3}\ln|x|\,dx$.

Solution The choice $u = \ln|x|$ is usually a good one because then du is a simple power of x. If we let $u = \ln|x|$ and $dv = x^{2/3}\,dx$, we have $du = x^{-1}\,dx$ and $v = \frac{3}{5}x^{5/3}$. Then

$$\int_1^8 x^{2/3} \ln|x|\, dx = uv\Big|_1^8 - \int_1^8 v\, du$$

$$= (\ln|x|)\frac{3}{5}x^{5/3}\Big|_1^8 - \int_1^8 \frac{3}{5}x^{5/3}x^{-1}\, dx$$

$$= \frac{3}{5}[(\ln 8)8^{5/3} - (\ln 1)1^{5/3}] - \frac{3}{5}\int_1^8 x^{2/3}\, dx$$

$$= \frac{3}{5}[(\ln 8)(32) - 0] - \frac{3}{5}\frac{x^{5/3}}{\frac{5}{3}}\Big|_1^8$$

$$= \frac{96}{5}(\ln 8) - \frac{9}{25}[8^{5/3} - 1^{5/3}]$$

$$= \frac{96}{5}(\ln 8) - \frac{9}{25}[32 - 1]$$

$$= \frac{96}{5}\ln 8 - \frac{279}{25} \approx 28.77. \quad ■$$

Practice Problem 10 Evaluate $\int_0^2 xe^{2x}\, dx$.

EXERCISES 13.4

In Exercises 1–4: (a) State what u and dv should be taken to be so that the given integral is of the form ∫ u dv, where v and ∫ v du can be evaluated. (b) What is v du?

1. $\int xe^{-x}\, dx$

2. $\int x^2 \ln|x|\, dx$

3. $\int x(x + 1)^{2/3}\, dx$

4. $\int x^3\sqrt{x^2 + 1}\, dx$

Evaluate the following integrals.

5. $\int x(3x - 1)^7\, dx$

6. $\int x\sqrt{2x - 3}\, dx$

7. $\int \frac{x}{\sqrt{5x + 4}}\, dx$

8. $\int (x + 2)e^x\, dx$

9. $\int \ln(4x)\, dx$

10. $\int x^3\sqrt{x^2 - 1}\, dx$

11. $\int \sqrt{x}\ln x\, dx \quad x > 0$

12. $\int x^3 \ln|x|\, dx$

13. $\int x^3e^{x^2}\, dx$

14. $\int e^{\sqrt{x}}\, dx \quad x > 0$

15. $\int x^3\sqrt{9 - x^2}\, dx$

16. $\int \frac{x + 1}{e^x}\, dx$

17. $\int \frac{x}{5x + 2}\, dx \quad x > 0$

18. $\int x^2e^{-x}\, dx$

19. $\int_2^4 xe^{2x}\,dx$

20. $\int_1^3 x^2 \ln |x|\,dx$

21. $\int_0^2 x^2\sqrt{2x+1}\,dx$

22. $\int_0^1 \ln (3x+1)\,dx$

23. $\int_1^2 x^2(4x-1)^5\,dx$

24. $\int_1^4 x^{-2} \ln x\,dx$

25. $\int \frac{x^2}{(x+2)^2}\,dx \quad x > -2$

26. $\int (\ln |x|)^2\,dx$

27. $\int (\ln |x|)^3\,dx$

28. $\int x(\ln |x|)^2\,dx$

Answers to Practice Problems **9.** (a) $u = \ln |x|$ and $dv = x\,dx$ (b) $\frac{1}{2}x^2 \ln |x| - \frac{1}{4}x^2 + k$ **10.** $\frac{3}{4}e^4 + \frac{1}{4}$

13.5 Integral Tables (Optional)

The rules for differentiating the sum, difference, product, and quotient of functions, along with the chain rule, allow us to differentiate almost any combination of powers and the exponential and logarithmic functions. Unfortunately the same is not true for integration. The indefinite integrals of many seemingly simple expressions cannot be computed by the techniques at our disposal. Others yield only to substitutions involving functions not treated in this text. Still other functions have no antiderivative that can be expressed in terms of familiar functions. An example of such a function is e^{x^2}. It is not that $\int e^{x^2} dx$ does not exist; there do exist functions F such that $F'(x) = e^{x^2}$. But F cannot be written as a combination of the functions for which we have names.

Helpful in the evaluation of more complicated indefinite integrals is an **integral table,** such as appears inside the back cover of this book. Most of the integral formulas there involve constants, so that each formula covers a whole class of integrals.

EXAMPLE 13.18 Evaluate $\int 2x(3-4x)^{-3}\,dx$ using the integral table.

Solution The factor $3 - 4x$ of the integrand is of the form $a + bu$, where $a = 3$, $b = -4$, and $u = x$, so that $du = dx$. Scanning the portion of the table headed "integrands containing $a + bu$", we find that formula 11 applies, except for the constant factor 2. Thus we have

$$\int 2x(3-4x)^{-3}\,dx = 2\int \frac{x\,dx}{(3-4x)^3}$$

$$= 2\left[\frac{1}{(-4)^2}\left(-\frac{1}{3-4x} + \frac{3}{2(3-4x)^2}\right) + k\right]$$

$$= \frac{1}{8}\left(\frac{-1}{3-4x} + \frac{3}{2(3-4x)^2}\right) + K,$$

where $K = 2k$. Of course this can be checked by differentiating the answer. ■

Practice Problem 11 The integral $\int x^{-2}(x - 2)^{-1}dx$ can be evaluated by means of formula 13. (a) What should a, b, and u be? (b) What is the answer?

In some formulas in the table $\pm$ ("plus or minus") or $\mp$ ("minus or plus") signs appear. When applying these formulas either all the top signs or else all the bottom signs should be used, depending on the form of the original integrand. The next example illustrates this.

EXAMPLE 13.19 Evaluate $\int 5x^2(x^2 + 9)^{-1/2}\,dx$.

Solution We look at the formulas under the heading "Integrands containing $u^2 \pm a^2$", and find that, except for the constant 5, the given integrand matches formula 32, with $a = 3$, $u = x$, and the (top) plus sign taken. Thus we have

$$\begin{aligned}\int 5x^2(x^2 + 9)^{-1/2}\,dx &= 5\int \frac{x^2\,dx}{\sqrt{x^2 + 9}} \\ &= 5\left[\frac{x}{2}\sqrt{x^2 + 9} - \frac{3^2}{2}\ln\left|x + \sqrt{x^2 + 9}\,\right| + k\right] \\ &= \frac{5}{2}\left(x\sqrt{x^2 + 9} - 9\ln\left|x + \sqrt{x^2 + 9}\right|\right) + K,\end{aligned}$$

where $K = 5k$. Notice that the minus sign is taken between the main terms in the answer because it is the top sign in $\mp$. The reader should check this answer by differentiation. ■

EXAMPLE 13.20 Evaluate $\int (x^2 - 25)^{-1}\,dx$.

Solution Scanning the list of formulas with "Integrands containing $u^2 \pm a^2$" does not reveal a match. Note, however, that

$$\int (x^2 - 25)^{-1}\,dx = -\int (25 - x^2)^{-1}\,dx,$$

to which formula 23 applies, with $a = 5$ and $u = x$. Thus

$$\begin{aligned}\int (x^2 - 25)^{-1}dx &= -\int \frac{dx}{25 - x^2} \\ &= -\left[\frac{1}{2(5)}\ln\left|\frac{5 + x}{5 - x}\right| + k\right] \\ &= -\frac{1}{10}\ln\left|\frac{5 + x}{5 - x}\right| + K,\end{aligned}$$

where $K = -k$. ■

EXAMPLE 13.21 Evaluate $\int \sqrt{9 - x^2}\, dx$.

Analysis Scanning the formulas headed "Integrands containing $a^2 - u^2$" does not reveal a match. Formula 28 could be used, however, if we had $x^2 - 9$ instead of $9 - x^2$. Unfortunately

$$\int \sqrt{9 - x^2}\, dx \neq -\int \sqrt{x^2 - 9}\, dx,$$

and there is no algebra we can do to make formula 28 apply. We can not evaluate this integral. (There is a formula, but it involves an inverse trigonometric function.) ■

Practice Problem 12 Which formula in our table, if any, could be applied to each of the following?

(a) $\int x^{-2}(5 + x^2)^{1/2}\, dx$ (b) $\int x^2(x + 9)^{-2}\, dx$ (c) $\int x^{-1}(x^2 - 9)^{1/2}\, dx$

EXAMPLE 13.22 Evaluate $\int x(9 + x^2)^{1/2}\, dx$.

Solution The reader scanning the formulas labelled "Integrands containing $u^2 \pm a^2$" may be surprised to find no match for such a simple integrand. The reason is that no formula is needed; the substitution $u = 9 + x^2$ is sufficient to compute the integral. Then $du = 2x\, dx$, and so

$$\int x(9 + x^2)^{1/2}\, dx = \frac{1}{2}\int (9 + x^2)^{1/2}(2x\, dx) = \frac{1}{2}\int u^{1/2}\, du$$

$$= \frac{\frac{1}{2}u^{3/2}}{\frac{3}{2}} + k = \frac{1}{3}u^{3/2} + k = \frac{1}{3}(9 + x^2)^{3/2} + k. \quad ■$$

In general, integrals that can be evaluated by an obvious substitution, such as the one in Example 13.22, are not listed in the table.

In all the examples so far where we have used the table, we have taken $u = x$. Then $du = dx$, which makes applying the formula quite direct. Sometimes we must take u to be a more complicated function of x to use a formula from the table, as in the next example. In such cases du must be adjusted accordingly.

EXAMPLE 13.23 Evaluate $\int (4x^2 - 1)^{-1/2}\, dx$.

Solution The integrand is close to that of formula 30 if we set $u = 2x$. Then $du = 2\, dx$, and so

$$\begin{aligned}\int (4x^2 - 1)^{-1/2}\,dx &= \frac{1}{2}\int (4x^2 - 1)^{-1/2}(2\,dx)\\ &= \frac{1}{2}\int (u^2 - 1)^{-1/2}\,du\\ &= \frac{1}{2}\int \frac{du}{\sqrt{u^2 - 1}}\\ &= \frac{1}{2}\left[\ln|u + \sqrt{u^2 - 1}| + k\right]\\ &= \frac{1}{2}\ln|2x + \sqrt{4x^2 - 1}| + K,\end{aligned}$$

where we have taken $a = 1$ in formula 30 and $K = \frac{1}{2}k$. Notice that we used the bottom signs of the $\pm$ symbol.

An alternative attack is to note that

$$\int (4x^2 - 1)^{-1/2}\,dx = \frac{1}{2}\int \left(x^2 - \frac{1}{4}\right)^{-1/2} dx,$$

then use formula 30 with $u = x$ and $a = \frac{1}{2}$. This produces the answer

$$\tfrac{1}{2}\ln|x + \sqrt{x^2 - \tfrac{1}{4}}| + K.$$

Although this appears different from our previous result, we have

$$\begin{aligned}\ln\left|x + \sqrt{x^2 - \frac{1}{4}}\right| &= \ln\left(\frac{1}{2}\left|2x + \sqrt{4x^2 - 1}\right|\right)\\ &= \ln\frac{1}{2} + \ln|2x + \sqrt{4x^2 - 1}|,\end{aligned}$$

and so the two answers differ only in the constant. ■

Practice Problem 13 Evaluate $\int (7 - 9x^2)^{-3/2}\,dx$ by letting $u = 3x$.

A few of the formulas in the table are **reduction formulas,** that is, formulas that give the desired integral in terms of a simpler integral. Such a formula may have to be applied several times. Before using such a formula one should check that the eventual result will be an integral that can be evaluated.

EXAMPLE 13.24 Evaluate $\int x^2(3 + 5x)^{-1/2}\,dx$.

Solution Notice that the integrand matches that of formula 19 with $n = 2$, and that the right side of this formula gives the same integral except with $n = 1$, which can be evaluated using formula 18. We proceed as follows, with $a = 3$, $b = 5$, and $u = x$, so $du = dx$.

$$\int x^2(3 + 5x)^{-1/2}\,dx = \int \frac{x^2\,dx}{\sqrt{3 + 5x}}$$

$$= \frac{2x^2\sqrt{3 + 5x}}{5(2 \cdot 2 + 1)} - \frac{2 \cdot 3 \cdot 2}{5(2 \cdot 2 + 1)} \int \frac{x\,dx}{\sqrt{3 + 5x}}$$

$$= \frac{2x^2\sqrt{3 + 5x}}{25} - \frac{12}{25}\left[\frac{2(5x - 2 \cdot 3)\sqrt{3 + 5x}}{3 \cdot 5^2} + k\right]$$

$$= \frac{2}{625}(25x^2 - 20x + 24)\sqrt{3 + 5x} + K \quad \blacksquare$$

EXERCISES 13.5

In Exercises 1–6 tell which formula of the integral table, if any, could be used to evaluate the given integral.

1. $\displaystyle\int \frac{\sqrt{x^2 + 4}}{x}\,dx$

2. $\displaystyle\int (x^2 - 9)^{-3/2}\,dx$

3. $\displaystyle\int \frac{x\,dx}{\sqrt{2x - 1}}$

4. $\displaystyle\int x^{-1}(1 + 2x)^{-2}\,dx$

5. $\displaystyle\int 3x(4 + x)^{-3}\,dx$

6. $\displaystyle\int \frac{\sqrt{4x^2 + 1}}{x^2}\,dx$

In Exercises 7–26 evaluate the given integral. Tell which formula from the table or which other method is used.

7. $\displaystyle\int \frac{dx}{x\sqrt{1 + x^2}}$

8. $\displaystyle\int \sqrt{x^2 + 16}\,dx$

9. $\displaystyle\int \frac{\sqrt{9 - x^2}}{2x}\,dx$

10. $\displaystyle\int x\sqrt{3x + 2}\,dx$

11. $\displaystyle\int (x \ln x)^{-1}\,dx$

12. $\displaystyle\int x^{-1}\sqrt{5 + x^2}\,dx$

13. $\displaystyle\int \frac{dx}{8 - x^2}$

14. $\displaystyle\int \frac{x^{-1}\,dx}{\sqrt{1 + 2x}}$

15. $\displaystyle\int \frac{\sqrt{3 - t}}{t}\,dt$

16. $\displaystyle\int \frac{dz}{z(2z - 1)}$

17. $\displaystyle\int_1^4 \frac{x\,dx}{3x + 1}$

18. $\displaystyle\int_0^2 \frac{x^2\,dx}{\sqrt{x^2 + 4}}$

19. $\int_3^4 \frac{dx}{x^2\sqrt{25 - x^2}}$

20. $\int_1^2 x^{-2}(4x + 1)^{-1}\, dx$

21. $\int \frac{x\, dx}{\sqrt{2 + 3x^2}}$

22. $\int \frac{x^{-2}\, dx}{(4x^2 + 1)^{1/2}}$

23. $\int \left(\frac{x}{2 - x}\right)^2 dx$

24. $\int \frac{dx}{(4 - x)^2}$

25. $\int \frac{x^2\, dx}{\sqrt{x + 1}}$

26. $\int x^2\sqrt{3 - x}\, dx$

In Exercises 27–30 verify the given formula from the table of integrals by differentiating the right side.

27. Formula 11

28. Formula 13

29. Formula 34

30. Formula 18

Answers to Practice Problems

11. (a) $a = -2, b = 1, u = x$ (b) $\frac{1}{2x} + \frac{1}{4}\ln\left|\frac{x - 2}{x}\right| + k$

12. (a) 37 (b) 10 (c) none

13. Using formula 27 with $u = 3x$ and $a = \sqrt{7}$ yields the answer $\frac{x(7 - 9x^2)^{-1/2}}{7} + K$.

CHAPTER 13 REVIEW

IMPORTANT TERMS

- **differential equation** *(13.1)*
- **first-order equation**
- **general solution**
- **initial condition**
- **particular solution**
- **separable equation**
- **solution**
- **exponential growth and decay** *(13.2)*
- **half-life**
- **radioactive decay**
- **capitalized value** *(13.3)*
- **convergent improper integral**
- **divergent improper integral**
- **improper integral**
- **integration by parts** *(13.4)*
- **integral table** *(13.5)*
- **reduction formula**

IMPORTANT FORMULAS

Improper Integrals

$$\int_a^\infty f(x)\, dx = \lim_{b\to\infty} \int_a^b f(x)\, dx, \quad \text{provided this limit exists;}$$

$$\int_{-\infty}^b f(x)\, dx = \lim_{a\to-\infty} \int_a^b f(x)\, dx, \quad \text{provided this limit exists; and}$$

$$\int_{-\infty}^\infty f(x)\, dx = \int_{-\infty}^0 f(x)\, dx + \int_0^\infty f(x)\, dx.$$

Capitalized Value

The capitalized value of a property with fixed annual rent R and annual interest rate i is $\int_0^\infty Re^{-ix}\,dx$.

Integration by Parts

$$\int u\,dv = uv - \int v\,du$$

Integration by Parts for a Definite Integral

$$\int_a^b u\,dv = uv\Big|_a^b - \int_a^b v\,du$$

REVIEW EXERCISES

In Exercises 1–4 find the particular solution to the differential equation satisfying the initial condition.

1. $2xy' = \ln x;\quad y = 4$ when $x = 1$

2. $y' = 3y + 1;\quad y = 3$ when $x = 0$

3. $x^2yy' = 1;\quad y = 2$ when $x = 1$

4. $y' + e^x = e^xy;\quad y = 5$ when $x = 1$

5. Suppose that the number y of people who subscribe to a certain magazine is growing continuously at a yearly rate of 6% of the number who subscribe. In 1988 there were 600,000 subscribers. Write y as a function of x, the number of years after 1988. How many subscribe in 1990? When will the number of subscribers reach one million?

6. The half-life of protactinium is 1.14 minutes. How much of 1000 pounds of protactinium will be left after five minutes?

7. The number y of viewers of a certain television program is now 3 million, and is growing at a yearly rate of 20% of the difference between 80 million and y. How many viewers will there be after 4 years?

8. The number y of people owning shower radios is growing at a rate proportional to the difference between 20 million and y. The value of y was 2 million in 1985 and 2.1 million in 1988. How many owners will there be in 1995?

9. In 1955 five million people had hung fuzzy dice from the rear-view mirrors of their cars. If y is the number of millions of people doing this x years after 1955, and if

$$\frac{dy}{dx} = .03y(P - y),$$

where P is 40 million, write y as a function of x. What was y in 1957?

Evaluate the integrals in Exercises 10–13.

10. $\displaystyle\int_2^\infty x^{-3}\,dx$

11. $\displaystyle\int_1^\infty \frac{1}{\sqrt{x}}\,dx$

12. $\displaystyle\int_{-\infty}^0 \frac{1}{(x-1)^2}\,dx$

13. $\displaystyle\int_{-\infty}^\infty \frac{x}{(x^2+1)^2}\,dx$

14. Find the capitalized value of a property earning a fixed rent of \$100,000 per year if the annual interest rate is 6%.

15. A city empties pollutants into a river at a rate of $\frac{1000}{(x+1)^2}$ tons per year after x years. What is the total amount entering the river?

Use integration by parts to evaluate the integrals in Exercises 16–19.

16. $\int x(x+2)^{3/5}\,dx$

17. $\int x^4 \ln|x|\,dx$

18. $\int xe^{3x}\,dx$

19. $\int_1^3 \frac{x}{(2x+1)^2}\,dx$

Evaluate the integrals in Exercises 20–23, using the integral tables if necessary.

20. $\int x\sqrt{x-1}\,dx$

21. $\int \frac{x}{\sqrt{1-x^2}}\,dx$

22. $\int \frac{dx}{x^2\sqrt{4x^2+1}}$

23. $\int (3x+2x^2)^{-1}\,dx$

14

APPLICATIONS OF THE DEFINITE INTEGRAL TO PROBABILITY

In Section 7.6 we studied probability distributions for continuous random variables x, that is, random variables that can take on any value in some interval. Recall that such a distribution is a curve having the property that the probability that x falls between a and b is the area under the curve over the interval $[a, b]$. Such a curve is the graph of some probability density function f. Since the area under the graph of $y = f(x)$ above the interval $[a, b]$ is given by the definite integral $\int_a^b f(x)\, dx$, the integration techniques we have learned may be applied to continuous probability distributions.

14.1 Probability Density Functions

In Section 7.6 we studied normal probability distributions (see Figure 14.1).

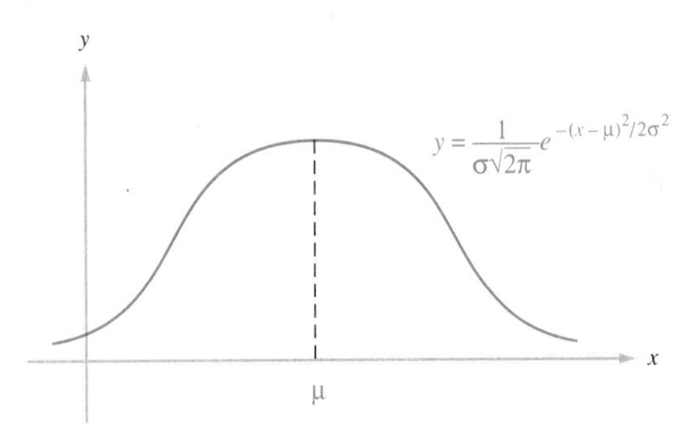

The normal probability distribution with mean μ and standard deviation σ

FIGURE 14.1

Theoretically, we could compute $P(a \leq x \leq b)$ for the normal distribution with mean μ and standard deviation σ by evaluating the integral

$$\int_a^b \frac{1}{\sigma\sqrt{2\pi}} e - \frac{1}{2}^{-(x - \mu)^2 2\sigma^2} dx.$$

Unfortunately, the integrand has no antiderivative that is expressible in terms of familiar functions. Thus to compute such a probability we are still forced to use a table as in Section 7.6.

Many random variables x have density functions f such that the probability that $a \leq x \leq b$ can be computed by means of an integral, however. Recall that for f to be a density function for some continuous random variable x, it must satisfy the following three conditions:

(a) $f(x) \geq 0$ for all x.
(b) The total area under the curve $y = f(x)$ is 1.
(c) $P(a \leq x \leq b)$ equals the area under $y = f(x)$ above the interval $[a, b]$.

If the corresponding definite integrals exist, the last two conditions can be stated as follows:

(b) $\int_{-\infty}^{\infty} f(x)\,dx = 1$

(c) $P(a \le x \le b) = \int_a^b f(x)\,dx.$

An example of a function satisfying conditions (a) and (b) is given by

$$f(x) = \begin{cases} 2x & \text{if } 0 \le x \le 1 \\ 0 & \text{otherwise.} \end{cases}$$

This function is pictured in Figure 14.2. Clearly $f(x) \ge 0$ for all values of x, so condition (a) is satisfied. To check condition (b) we note that $f(x)$ is zero except for $0 \le x \le 1$. Thus the area under the curve is given by

$$\int_0^1 f(x)\,dx = \int_0^1 2x\,dx = x^2\Big|_0^1 = 1^2 - 0^2 = 1.$$

To compute $P(1/4 \le x \le 1/2)$ we evaluate the corresponding integral as follows.

$$\int_{1/4}^{1/2} 2x\,dx = x^2\Big|_{1/4}^{1/2} = \left(\frac{1}{2}\right)^2 - \left(\frac{1}{4}\right)^2 = \frac{1}{4} - \frac{1}{16} = \frac{3}{16}$$

If we want to find the probability that $2/3 \le x \le 4/3$, we first note that $P(2/3 \le x \le 4/3) = P(2/3 \le x \le 1) + P(1 \le x \le 4/3)$. Thus

$$\begin{aligned} P\left(\frac{2}{3} \le x \le \frac{4}{3}\right) &= \int_{2/3}^{1} f(x)\,dx + \int_{1}^{4/3} f(x)\,dx \\ &= \int_{2/3}^{1} 2x\,dx + \int_{1}^{4/3} 0\,dx \\ &= x^2\Big|_{2/3}^{1} + 0 = 1 - \frac{4}{9} = \frac{5}{9}. \end{aligned}$$

In general, when a density function is nonzero only on some part of the interval $[a, b]$, we compute $P(a \le x \le b)$ by integrating over the appropriate subinterval of that interval.

Practice Problem 1 Find $P(-1 \le x \le 0.3)$ for the probability distribution pictured in Figure 14.2.

Verifying condition (c) for a density function may involve deeper considerations. Whether the area under the graph of a function f over the interval $[a, b]$ is the probability that x lies between a and b for some random variable x depends on the real-world interpretation of x. The appropriate density function for a particular random variable can sometimes be determined by theoretical considerations. In other cases certain standard functions have been found empirically to model particular situations accurately. The simplest type of density function is illustrated in the next example.

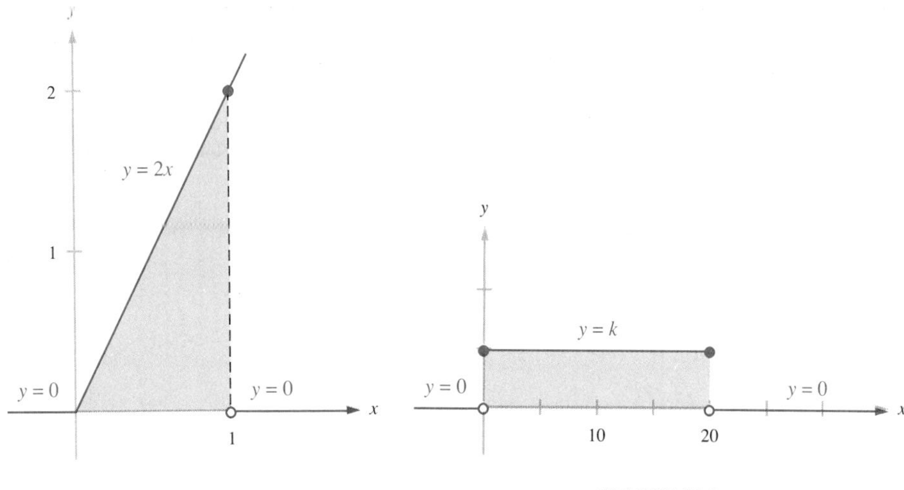

FIGURE 14.2

FIGURE 14.3

EXAMPLE 14.1 Mrs. Snavely is going to town to shop. A bus stops at the corner nearest her house every 20 minutes, but she doesn't know the exact schedule and so arrives at the corner at a random time. Let x be the number of minutes she must wait until a bus arrives. What density function corresponds to the random variable x? What is the probability that she will have to wait no more than 5 minutes?

Solution All we know is that a bus will arrive at some time during the next 20 minutes. This is just as likely to happen during the first minute after Mrs. Snavely arrives as during the twentieth minute. Thus the appropriate distribution is one for which the areas over intervals of equal length are the same. Such a distribution is pictured in Figure 14.3.

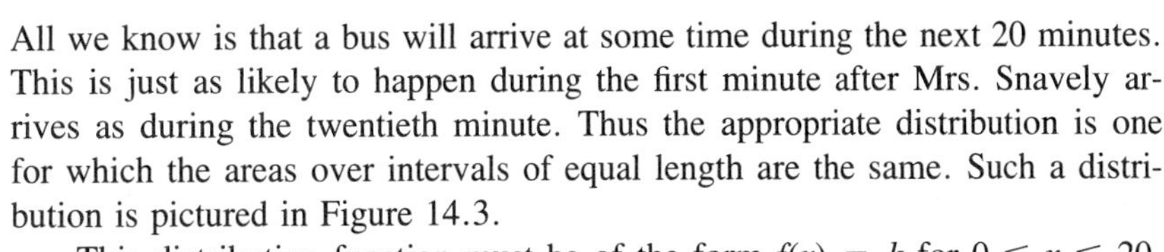

Buses pass every 20 minutes.

This distribution function must be of the form $f(x) = k$ for $0 \le x \le 20$, where k is some constant. It only remains to determine k. Since the area of a rectangle with width 20 and height k is $20k$, we must have $20k = 1$ by condition (b) for density functions. Thus $k = 1/20 = .05$, and so

$$f(x) = \begin{cases} .05 & \text{for } 0 \le x \le 20 \\ 0 & \text{otherwise.} \end{cases}$$

The probability that Mrs. Snavely will have to wait no more than 5 minutes is

$$P(x \le 5) = \int_0^5 .05\, dx = .25.$$

This may also be computed by noting that the area of a rectangle with width 5 and height .05 is $5(.05) = .25$. ■

Example 14.1 illustrates a **uniform probability distribution.** In general such a distribution has a density function of the form

$$f(x) = \begin{cases} \dfrac{1}{b - a} & \text{if } a \leq x \leq b \\ 0 & \text{otherwise.} \end{cases}$$

Such a distribution satisfies conditions (a) and (b) since a rectangle with width $b - a$ and height $1/(b - a)$ has area 1. (See Figure 14.4.)

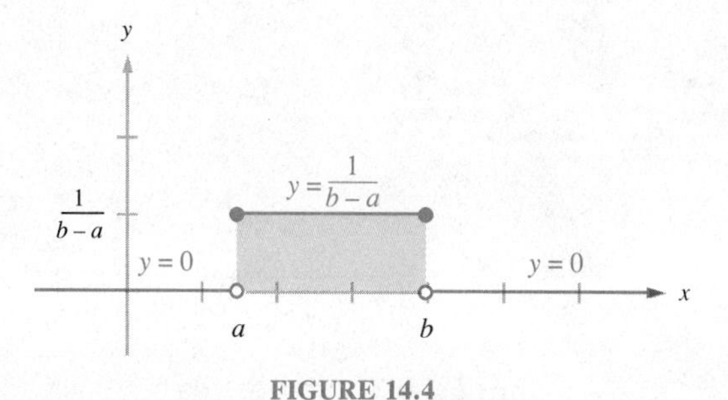

FIGURE 14.4

EXAMPLE 14.2 Give the density function for the uniform probability distribution on the interval [2, 5], and find $P(x \leq 4)$ for this distribution.

Solution Since $a = 2$ and $b = 5$ the density function is given by

$$f(x) = \begin{cases} \dfrac{1}{5 - 2} = \dfrac{1}{3} & \text{if } 2 \leq x \leq 5 \\ 0 & \text{otherwise.} \end{cases}$$

Thus

$$P(x \leq 4) = \int_2^4 \frac{1}{3}\,dx = \frac{1}{3}x \Big|_2^4 = \frac{1}{3}(4 - 2) = \frac{2}{3}. \quad ■$$

EXAMPLE 14.3 Show that the function defined by

$$f(x) = \begin{cases} \dfrac{3}{4}(1 - x^2) & \text{if } -1 \leq x \leq 1 \\ 0 & \text{otherwise} \end{cases}$$

satisfies conditions (a) and (b) for a probability density function.

Solution Since $1 - x^2 \geq 0$ for $-1 \leq x \leq 1$, we see that $f(x) \geq 0$ for all x. It remains to show that the area under the corresponding curve (shown in Figure 14.5) is 1. But we have

$$\int_{-\infty}^{\infty} f(x)\,dx = \int_{-1}^{1} \frac{3}{4}(1 - x^2)\,dx = \frac{3}{4}\left(x - \frac{x^3}{3}\right)\Bigg|_{-1}^{1}$$
$$= \frac{3}{4}\left[\left(1 - \frac{1}{3}\right) - \left(-1 + \frac{1}{3}\right)\right] = 1. \quad ■$$

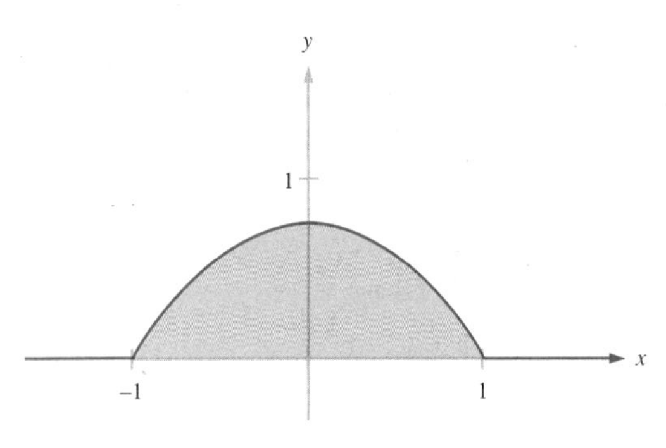

FIGURE 14.5

EXAMPLE 14.4 Find a value for the constant k so that the function

$$f(x) = \begin{cases} kx^3 & \text{if } 0 \leq x \leq 2 \\ 0 & \text{otherwise} \end{cases}$$

satisfies properties (a) and (b) of a probability density function.

Solution Clearly if $k \geq 0$ property (a) will be satisfied. We compute that

$$\int_{-\infty}^{\infty} f(x)\,dx = \int_{0}^{2} kx^3\,dx = \frac{kx^4}{4}\Bigg|_{0}^{2} = \frac{k(2^4)}{4} = 4k.$$

Since this integral must equal 1 for f to satisfy condition (b), we conclude that $k = 1/4$. ■

Practice Problem 2 (a) Find a value for the constant k so that the function defined by

$$f(x) = \begin{cases} ke^x & \text{if } 0 \leq x \leq 3 \\ 0 & \text{otherwise} \end{cases}$$

is a density function for the random variable x.

(b) Find $P(x \geq 2)$ if x has the density function in part (a).

EXAMPLE 14.5 An all-news radio station gives the weather report on the hour and at 20 minutes after the hour, 24 hours per day. Let x represent the time in minutes that a listener tuning in at a random time must wait before hearing a weather report. What is the probability that he or she must wait more than 15 minutes?

Solution Suppose the listener tunes in at t minutes after the hour. Figure 14.6 indicates how the time until the next weather report depends on t.

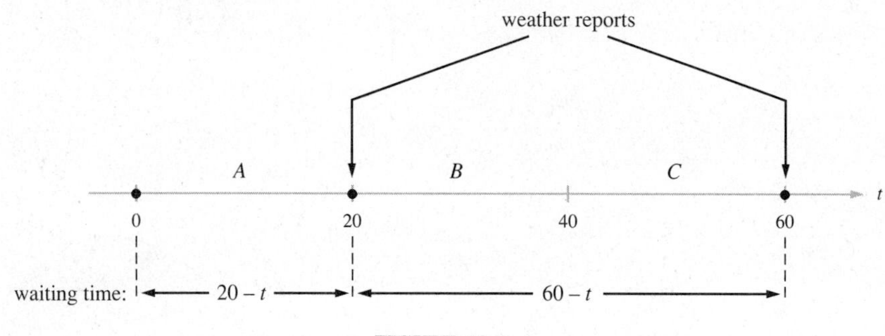

FIGURE 14.6

For example, if $t = 17$, then a listener will only have to wait $3 = 20 - 17$ minutes for a weather report. If $t = 26$, on the other hand, it will be $34 = 60 - 26$ minutes until the weather report on the hour. Note that if the listener tunes in during either of the intervals labeled A or C, the time until the weather report will be between 0 and 20 minutes. However, tuning in during the B interval means a wait of from 20 to 40 minutes. Since tuning in during intervals A or C is twice as likely as tuning in during interval B, the distribution of the waiting time x will be as in Figure 14.7, where the height of the first rectangle must be twice that of the second.

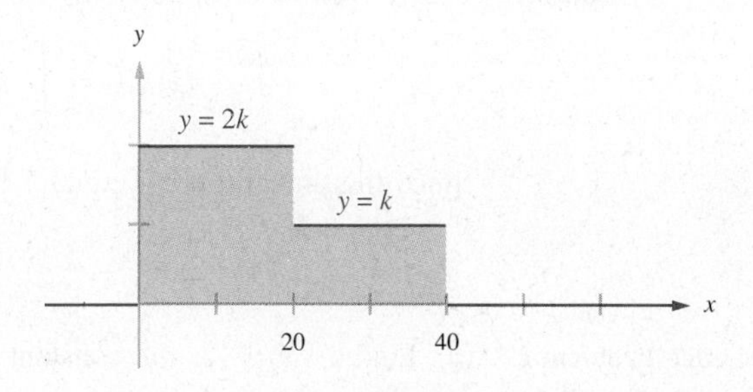

FIGURE 14.7

Since the area under the distribution must be 1 by condition (b), we have

$$20(2k) + 20(k) = 60k = 1,$$

and so $k = 1/60$. Thus the density function for x is given by

$$f(x) = \begin{cases} \dfrac{1}{30} & \text{if } 0 < x \leq 20 \\ \dfrac{1}{60} & \text{if } 20 < x \leq 40 \\ 0 & \text{otherwise.} \end{cases}$$

The probability of a wait of more than 15 minutes is

$$\begin{aligned} \int_{15}^{40} f(x)\,dx &= \int_{15}^{20} \frac{1}{30}\,dx + \int_{20}^{40} \frac{1}{60}\,dx \\ &= \frac{1}{30}x \Big|_{15}^{20} + \frac{1}{60}x \Big|_{20}^{40} \\ &= \frac{1}{30}(20 - 15) + \frac{1}{60}(40 - 20) \\ &= \frac{1}{2}. \end{aligned}$$

In Exercises 1–8 tell which of the conditions (a) $f(x) \geq 0$ *for all* x*, and (b)* $\int_{-\infty}^{\infty} f(x)\,dx = 1$ *the given function satisfies.*

1. $f(x) = \begin{cases} 1/2 & \text{if } -1 \leq x \leq 1 \\ 0 & \text{otherwise} \end{cases}$

2. $f(x) = \begin{cases} x/2 & \text{if } -1 \leq x \leq 1 \\ 0 & \text{otherwise} \end{cases}$

3. $f(x) = \begin{cases} x^2 & \text{if } 0 \leq x \leq \sqrt{3} \\ 0 & \text{otherwise} \end{cases}$

4. $f(x) = \begin{cases} 24x^2 - 12 & \text{if } .5 \leq x \leq 1 \\ 0 & \text{otherwise} \end{cases}$

5. $f(x) = \begin{cases} \dfrac{2 \ln x}{5 \ln 2 - 3} & \text{if } .5 \leq x \leq 2 \\ 0 & \text{otherwise} \end{cases}$

6. $f(x) = \begin{cases} \dfrac{1}{x \ln 2} & \text{if } .5 \leq x \leq 1 \\ 0 & \text{otherwise} \end{cases}$

7. $f(x) = \begin{cases} 1/x^2 & \text{if } x \geq 1 \\ 0 & \text{otherwise} \end{cases}$

8. $f(x) = \begin{cases} 1/x^3 & \text{if } x \geq .5 \\ 0 & \text{otherwise} \end{cases}$

In Exercises 9–16 find the probabilities listed for the given density function.

9. $P(x \geq 2)$ and $P(x < 1)$ for $f(x) = \begin{cases} 1/3 & \text{if } -1 \leq x \leq 2 \\ 0 & \text{otherwise} \end{cases}$

10. $P(1 \leq x \leq 4)$ and $P(x \leq .5)$ for $f(x) = \begin{cases} x/2 & \text{if } 0 \leq x \leq 2 \\ 0 & \text{otherwise} \end{cases}$

11. $P(x > 0)$ and $P(x \leq .5)$ for $f(x) = \begin{cases} 3x^2/2 & \text{if } -1 \leq x \leq 1 \\ 0 & \text{otherwise} \end{cases}$

12. $P(x \geq 1)$ and $P(.5 \leq x \leq 2.5)$ for $f(x) = \begin{cases} .25 & \text{for } 0 \leq x \leq 2 \\ .5 & \text{for } 2 < x \leq 3 \\ 0 & \text{otherwise} \end{cases}$

13. $P(x \geq -.5)$ and $P(|x| > .5)$ for $f(x) = \begin{cases} 1/2 & \text{for } -1 \leq x < 0 \\ x & \text{for } 0 \leq x \leq 1 \\ 0 & \text{otherwise} \end{cases}$

14. $P(x \leq 3)$ and $P(x > 5)$ for $f(x) = \begin{cases} e^{-x} & \text{if } x \geq 0 \\ 0 & \text{otherwise} \end{cases}$

15. $P(x \leq 2)$ and $P(x \geq 3)$ for $f(x) = \begin{cases} 3x^{-4} & \text{if } x \geq 1 \\ 0 & \text{otherwise} \end{cases}$

16. $P(-1 \leq x \leq 3)$ and $P(x \geq 2)$ for $f(x) = |x|e^{-x^2}$

In Exercises 17–20 find the requested probability for a uniform distribution on the given interval.

17. $P(x \geq 3)$ on $[0, 4]$

18. $P(x < 1.5)$ on $[-1, 4]$

19. $P(1 \leq x \leq 3)$ on $[0, 20]$

20. $P(5 < x \leq 15)$ on $[1, 10]$

In Exercises 21–26 find k so that the given function is a probability density function.

21. $f(x) = \begin{cases} k(x+1) & \text{if } 2 \leq x \leq 6 \\ 0 & \text{otherwise} \end{cases}$

22. $f(x) = \begin{cases} kx^2 & \text{if } -1 \leq x \leq 3 \\ 0 & \text{otherwise} \end{cases}$

23. $f(x) = \begin{cases} k\sqrt{x} & \text{for } 0 \leq x \leq 3 \\ 0 & \text{otherwise} \end{cases}$

24. $f(x) = \begin{cases} kx^{-3} & \text{for } x \geq 2 \\ 0 & \text{otherwise} \end{cases}$

25. $f(x) = \begin{cases} x^3 & \text{for } 0 \leq x \leq k \\ 0 & \text{otherwise} \end{cases}$

26. $f(x) = \begin{cases} x^{-3} & \text{for } x \geq k \\ 0 & \text{otherwise} \end{cases}$

27. At an amusement park, shuttles run from the parking lot to the main gate every 10 minutes. A family arrives at the parking lot at a random time. Let x be the time in minutes they must wait for the next shuttle. What is the probability density function for x? Find $P(x \geq 6)$.

28. The office blackboards at a certain university are cleaned every 80 days. Let x be the time in days until Professor Snodgrass's blackboard is next cleaned. What is the probability density function for x? Find the probability that the Professor's blackboard will be cleaned in the next week.

29. Fulsome Prison has a revolving searchlight that scans the prison yard. It makes one revolution each 5 minutes. A convict has tunneled from his cell to a corner of the yard. Let x be the time in minutes until the light shines on him. What is the probability density function for x? If friends plan to pick him up with a helicopter in 3 minutes, what is the probability the light will shine on him before this?

30. Sam's apartment is near the tracks, and every 20 minutes a train comes by and rattles everything. He has just made a house of cards 5 levels high, a new personal record. It will take Sam 7 minutes to get his camera and take a picture of this feat. What is the probability of this?

31. A certain radio station gives the livestock prices on the hour and also at fifteen minutes after the hour. Farmer Blue has just come in from the fields and turns on the radio. Let x be the time in minutes until the livestock prices are given. What is the probability density function for x? Find $P(x > 20)$.

32. Forest rangers visit a fire tower in a wilderness area each Monday, and the boy scouts have a picnic there every Thursday. A lost hiker finds the tower and decides to wait there for help. Let x be the time in days he has to wait. What is the probability density function for x? Find $P(x \leq 2)$.

Answers to Practice Problems

1. $\int_{-1}^{0.3} f(x)\, dx = \int_{0}^{0.3} 2x\, dx = 0.09$

2. (a) $k = \dfrac{1}{e^3 - 1}$ (b) $\dfrac{e^3 - e^2}{e^3 - 1}$

14.2 The Exponential and Poisson Distributions

In this section we will study two special probability distributions that have been found useful for modeling many practical situations.

THE EXPONENTIAL DISTRIBUTION

We define an **exponential distribution** to be a distribution with a density function of the form

$$f(x) = \begin{cases} \lambda e^{-\lambda x} & \text{if } x \geq 0 \\ 0 & \text{otherwise.} \end{cases}$$

Here λ is a positive constant. The general form of an exponential distribution is shown in Figure 14.8.

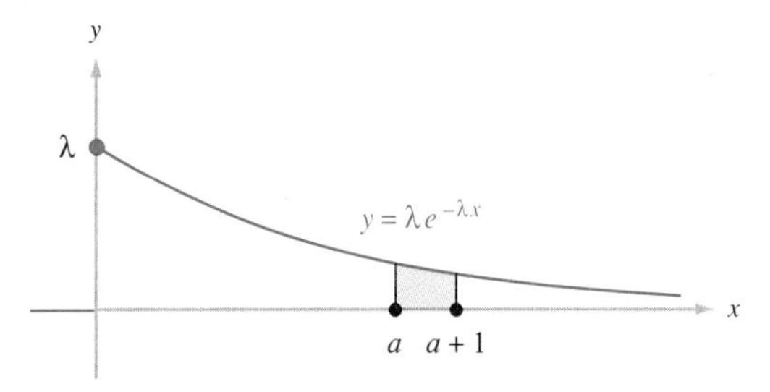

An exponential distribution

FIGURE 14.8

Notice that an exponential density function is positive for $x \geq 0$, but that the probability that x falls in an interval of a given length, say $[a, a + 1]$, gets smaller as a gets larger. This distribution is used to model such random variables as the length of time before a light bulb fails, the length of time a dentist takes with a patient, the length of time before some customer enters a shop, or the length of time before a cosmic ray trips a Geiger counter. We will delay until the end of this section an explanation of why an exponential density function is appropriate for such random variables.

To check that the area under an exponential distribution is 1, we compute

$$\int_0^\infty \lambda e^{-\lambda x}\, dx.$$

Note that

$$\int_0^b \lambda e^{-\lambda x}\, dx = \lambda \frac{1}{-\lambda} e^{-\lambda x} \Big|_0^b = (-1)(e^{-\lambda b} - 1) = 1 - e^{-\lambda b}.$$

Thus

$$\int_0^\infty \lambda e^{-\lambda x}\, dx = \lim_{b\to\infty} (1 - e^{-\lambda b}) = 1,$$

since

$$\lim_{b\to\infty} e^{-\lambda b} = 0.$$

Thus property (b) for probability density functions is satisfied by any exponential distribution.

Knowing that the area under an exponential distribution is 1 and using the formula

$$\int_0^b \lambda e^{-\lambda x}\, dx = 1 - e^{-\lambda b} \tag{14.1}$$

derived above allow us to compute probabilities involving exponential distributions, as the following example demonstrates.

EXAMPLE 14.6 Suppose the random variable x has an exponential distribution with $\lambda = 3$. Find $P(x \le 2)$, $P(1 \le x \le 4)$, and $P(x \ge 5)$.

Solution Using (14.1) yields

$$P(x \le 2) = \int_0^2 3e^{-3x}\, dx = 1 - e^{-3\cdot 2} \approx .998.$$

Likewise

$$\begin{aligned} P(1 \le x \le 4) &= \int_1^4 3e^{-3x}\, dx = \int_0^4 3e^{-3x}\, dx - \int_0^1 3e^{-3x}\, dx \\ &= (1 - e^{-3\cdot 4}) - (1 - e^{-3\cdot 1}) \approx .0498. \end{aligned}$$

Although $P(x \geq 5)$ equals the improper integral $\int_5^\infty 3e^{-3x}\,dx$, we can also use the fact that

$$P(x \geq 5) = 1 - P(x < 5) = 1 - \int_0^5 3e^{-3x}\,dx$$

to compute

$$P(x \geq 5) = 1 - (1 - e^{-3 \cdot 5}) \approx 3.06 \times 10^{-7}. \quad \blacksquare$$

Practice Problem 3 Suppose that the number of hours x before an oil drilling bit wears out has an exponential distribution with $\lambda = 1/2$. What is the probability that a drill bit lasts more than 2 hours?

EXAMPLE 14.7 It is found that half of all inquiries to an Internal Revenue Service hotline are answered in 5 minutes or less. Assuming that the time x in minutes it takes to answer an inquiry has an exponential distribution, what is the probability that an answer takes no more than 12 minutes?

Solution Our first problem is to determine what λ is. We know that

$$\frac{1}{2} = P(x \leq 5) = \int_0^5 \lambda e^{-\lambda x}\,dx.$$

But by (14.1)

$$\int_0^5 \lambda e^{-\lambda x}\,dx = 1 - e^{-5\lambda}.$$

We set this quantity equal to 1/2 and solve for λ.

$$1 - e^{-5\lambda} = \frac{1}{2}$$

$$\frac{1}{2} = e^{-5\lambda}$$

$$\ln\frac{1}{2} = \ln e^{-5\lambda} = -5\lambda$$

$$\lambda = \frac{\ln\frac{1}{2}}{-5} = \frac{\ln 2^{-1}}{-5} = \frac{-\ln 2}{-5} = \frac{\ln 2}{5}$$

Now we compute $P(x \leq 12)$. By (14.1) we have

$$\int_0^{12} \lambda e^{-\lambda x}\,dx = 1 - e^{-12\lambda}.$$

Thus

$$P(x \leq 12) = 1 - e^{-12(\ln 2)/5} \approx .811.$$

Approximately 81% of all inquiries will be answered within 12 minutes. $\blacksquare$

INFINITE DISCRETE RANDOM VARIABLES

In Chapter 7 we studied mostly random variables that took on a finite number of values. For example, suppose two coins are flipped and x represents the number of heads. Then x must have one of the values 0, 1, or 2; its distribution is shown in Figure 14.9.

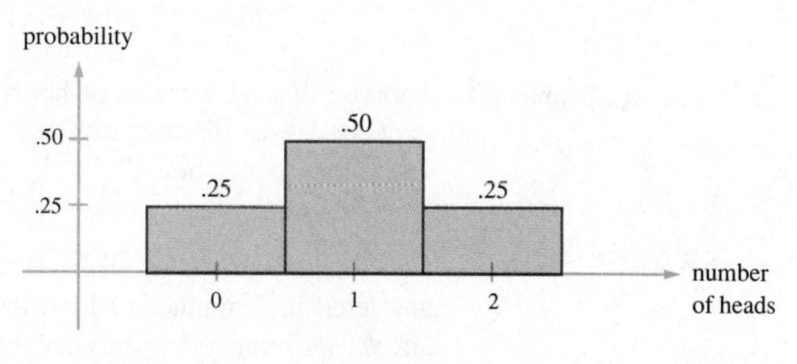

FIGURE 14.9

Normal and exponential distributions, on the other hand, correspond to *continuous* random variables, that is, variables that can take on any value in an interval. A third type of random variable has infinitely many discrete values. The next example illustrates such a random variable.

EXAMPLE 14.8 A coin is flipped until a head comes up. Let x be the number of flips this takes. What is the probability distribution for x?

Solution Since half the time the first flip will be a head, we have

$$P(x = 1) = \frac{1}{2}.$$

Likewise, since $x = 2$ exactly when the first flip is a tail and the second is a head, and since these independent events each have probability 1/2, we see that

$$P(x = 2) = \frac{1}{2} \cdot \frac{1}{2} = \frac{1}{2^2}.$$

A similar argument shows that $P(x = 3) = \frac{1}{2^3}$, and, in general,

$$P(x = k) = \frac{1}{2^k}.$$

Note that x can take on any of the infinitely many values 1, 2, 3, ... The probability distribution for x is illustrated in Figure 14.10.

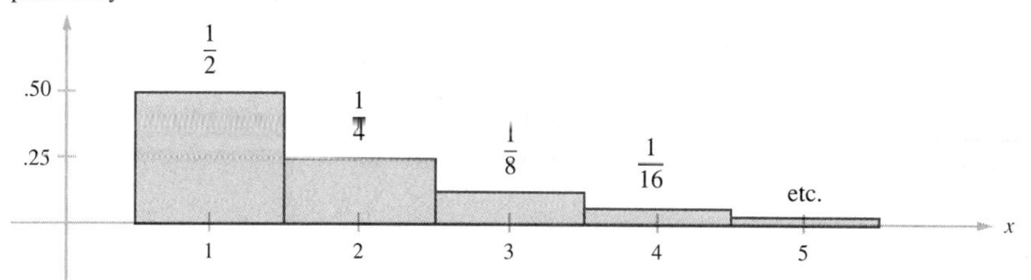

FIGURE 14.10

Here the total area under the distribution is 1 in the sense that it can be shown that

$$\frac{1}{2} + \frac{1}{4} + \frac{1}{8} + \ldots + \frac{1}{2^k}$$

approaches 1 as k gets larger and larger. ■

EXAMPLE 14.9 A coin is flipped until *two* heads come up. Let x be the number of flips this takes. Find the probability distribution for x.

Solution In order that the second head occurs on flip k, we must have a sequence of $k - 1$ flips, exactly one of which is a head, followed by a head. For example if $k = 4$ the possible sequences are

HTT-H

THT-H

TTH-H,

and in general the possible sequences are

HTT . . . T-H

THT . . . T-H

TTH . . . T-H

⋮

TTT . . . H-H,

where there are $k - 1$ flips before the last head. Since the probability of a head or tail on any particular flip is 1/2, the probability of any one of these sequences is $(1/2)^k$. But there are $k - 1$ positions the first head could take among the first $k - 1$ flips, so

$$P(x = k) = \frac{k - 1}{2^k}.$$

The distribution is pictured in Figure 14.11. It can be proved that

$$\frac{0}{2} + \frac{1}{4} + \frac{2}{8} + \ldots + \frac{k - 1}{2^k}$$

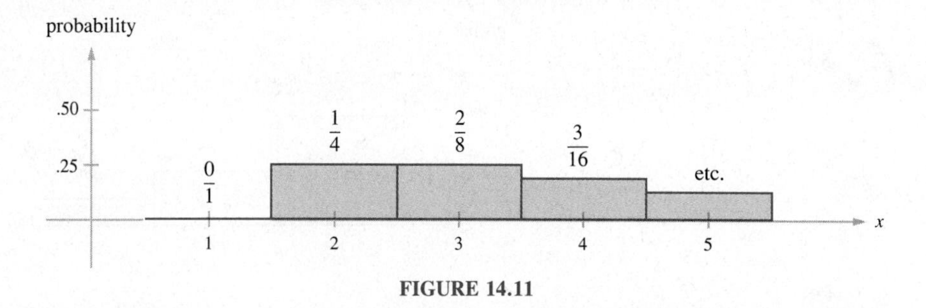

FIGURE 14.11

approaches 1 as k gets larger and larger, so that the area under the distribution is 1. ■

Practice Problem 4 A die is rolled until a 3 comes up. Let x be the number of rolls this takes.
(a) Find $P(x = k)$ in terms of k.
(b) What is the probability that at least 4 rolls are needed?

THE POISSON DISTRIBUTION

One type of infinite discrete distribution that seems to describe many situations is called a **Poisson* distribution.** This distribution depends on a positive constant μ and is nonzero for $x = 0, 1, 2, \ldots$. We have

$$P(x = k) = \frac{e^{-\mu}\mu^k}{k!} \qquad \text{for } k = 0, 1, \ldots.$$

For example, when $\mu = 2$ we have

$$P(x = k) = \frac{e^{-2}2^k}{k!} \qquad \text{for } k = 0, 1, \ldots.$$

The following table shows the probabilities of the first few values of k.

k	$P(x = k)$
0	$\frac{e^{-2}2^0}{0!} = e^{-2} \approx .135$ (recall that $0! = 1$)
1	$\frac{e^{-2}2^1}{1!} = 2e^{-2} \approx .271$
2	$\frac{e^{-2}2^2}{2!} = 2e^{-2} \approx .271$
3	$\frac{e^{-2}2^3}{3!} = \frac{4e^{-2}}{3} \approx .180$
4	$\frac{e^{-2}2^4}{4!} = \frac{2e^{-2}}{3} \approx .090$

*The French mathematician Simeon Denis Poisson (1781–1840) was known for his work in probability and the applications of mathematics to electricity and magnetism.

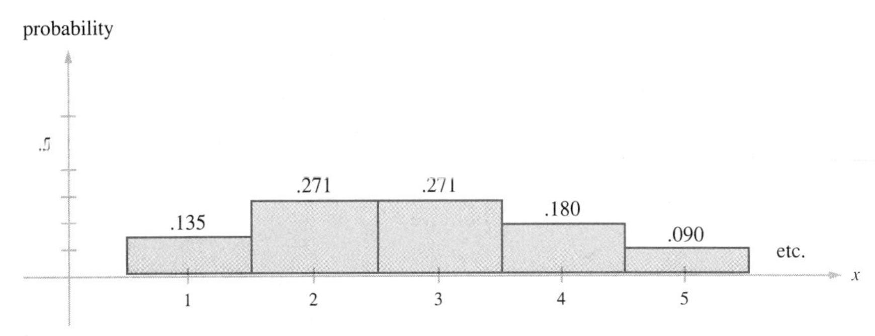

The Poisson distribution for $\mu = 2$

FIGURE 14.12

This particular Poisson distribution is pictured in Figure 14.12.

It can be shown that

$$\lim_{k \to \infty} \left(\frac{e^{-\mu}\mu^0}{0!} + \frac{e^{-\mu}\mu^1}{1!} + \ldots + \frac{e^{-\mu}\mu^k}{k!} \right) = 1;$$

so that the total area under any Poisson distribution is 1. Poisson distributions are used to model such random variables as the number of customers who enter a hamburger restaurant in a one-hour period, the number of insects electrocuted by a bug killer in one minute, and the number of telephone calls coming into a switchboard in a business day. The Poisson distribution is often used to model customer arrivals, and business software is available that constructs simulations based on such a distribution. At the end of this section we will indicate why a Poisson distribution is appropriate for such random variables.

EXAMPLE 14.10 The number x of books checked out from a certain small library in one hour has a Poisson distribution with $\mu = 10$. What is the probability that at most 4 books are checked out between 1:00 and 2:00 P.M.? What is the probability that exactly 9 books are checked out during that hour?

Solution To answer the first question we compute

$$\begin{aligned} P(x \leq 4) &= P(x = 0) + P(x = 1) + P(x = 2) + P(x = 3) + P(x = 4) \\ &= \frac{e^{-10}10^0}{0!} + \frac{e^{-10}10^1}{1!} + \frac{e^{-10}10^2}{2!} + \frac{e^{-10}10^3}{3!} + \frac{e^{-10}10^4}{4!} \\ &= e^{-10}\left(1 + 10 + 50 + \frac{500}{3} + \frac{1250}{3}\right) \approx .029. \end{aligned}$$

Answering the second question is easier, since the probability that exactly 9 books are checked out is

$$P(x = 9) = \frac{e^{-10}10^9}{9!} \approx .125. \quad ■$$

Practice Problem 5 Suppose the daily number x of automobile accidents in a certain city has a Poisson distribution with $\mu = 3$. What is the probability that there are no more than 2 accidents on a given day?

EXAMPLE 14.11 Farmer Brown sets a large cage-type mouse trap every night. He finds that half the time it catches no mice at all during the night. Assuming a Poisson distribution, what is the probability he finds at least 3 mice in his trap on a given morning?

Solution First we must determine the constant μ of the distribution. We know that

$$\frac{1}{2} = P(x = 0) = \frac{e^{-\mu}\mu^0}{0!} = e^{-\mu}.$$

Thus

$$\begin{aligned} \ln\frac{1}{2} &= \ln e^{-\mu}, \\ \ln 2^{-1} &= -\mu, \\ -\ln 2 &= -\mu, \\ \ln 2 &= \mu. \end{aligned}$$

It is easier to finish the problem by computing the probability that Farmer Brown finds *fewer* than 3 mice in his trap. This is

$$\begin{aligned} P(x < 3) &= P(x = 0) + P(x = 1) + P(x = 2) \\ &= \frac{e^{-\ln 2}(\ln 2)^0}{0!} + \frac{e^{-\ln 2}(\ln 2)^1}{1!} + \frac{e^{-\ln 2}(\ln 2)^2}{2!} \\ &= e^{-\ln 2}\left[1 + \ln 2 + \frac{(\ln 2)^2}{2}\right] \approx .967. \end{aligned}$$

Then

$$P(x \geq 3) = 1 - P(x < 3) \approx .033. \quad ■$$

In deciding whether to apply the exponential or Poisson distribution to a particular situation, it should be remembered that the exponential distribution corresponds to a *continuous* random variable that may assume any positive real value. The Poisson distribution, on the other hand, corresponds to a *discrete* random variable taking on only nonnegative integer values. Consider, for example, flies landing on a piece of cake. The length of time until a fly lands on the cake may be any positive real number, and so an exponential distribution is more appropriate to model this time than the Poisson distribution. The number of flies landing on the cake during a certain time period can only be a nonnegative integer, however, and so a Poisson distribution is more appropriate for this random variable.

AN EXPLANATION OF THE EXPONENTIAL AND POISSON DISTRIBUTIONS

Now we indicate *why* the exponential and Poisson distributions are appropriate to model situations such as our example of flies landing on a piece of cake. It seems reasonable to assume that the probability that some fly lands on the cake during the interval $[a, b]$ is proportional to the length, $b - a$, of this interval. From this assumption we have

$$P(\text{a fly lands during the interval } [a, b]) \approx \lambda(b - a), \qquad (14.2)$$

for some positive real number λ.

Notice that this assumption can only be valid for small intervals $[a, b]$, since if b is sufficiently larger than a, then $\lambda(b - a)$ will exceed 1. Thus in our analysis we will let the length of the intervals we consider approach 0.

Let us try to compute the probability that some fly lands on the cake in the interval $[0, x]$. It turns out to be easier to compute the probability that *no* fly lands during this time. We divide this interval into n subintervals

$$\left[0, \frac{x}{n}\right], \left[\frac{x}{n}, \frac{2x}{n}\right], \left[\frac{2x}{n}, \frac{3x}{n}\right], \ldots, \left[\frac{(n-1)x}{n}, \frac{nx}{n}\right],$$

each having length x/n. Then by (14.2) the probability that no fly lands on the cake during some particular one of these intervals is approximately

$$1 - \lambda\left(\frac{x}{n}\right).$$

Since these events are independent, the probability that no fly lands on the cake in any of the n subintervals is

$$P(\text{no fly lands during } [0, x]) \approx \left(1 - \frac{\lambda x}{n}\right)^n.$$

We will compute the limit of this quantity as n increases without bound, thus making our subintervals smaller and smaller.

Let $h = -\lambda x/n$, so that $n = -\lambda x/h$. Notice that as n increases without bound, h approaches 0. Thus

$$\lim_{n\to\infty} \left(1 - \frac{\lambda x}{n}\right)^n = \lim_{h\to 0} (1 + h)^{-\lambda x/h}$$

$$= \left[\lim_{h\to 0} (1 + h)^{1/h}\right]^{-\lambda x},$$

where we have used the power rule for limits. (See Theorem 9.1.) But according to the definition given in Section 9.3,

$$\lim_{h\to 0} (1 + h)^{1/h} = e.$$

Thus we see that the probability that no fly lands on the cake during the interval $[0, x]$ is

$$\lim_{n\to\infty}\left(1 - \frac{\lambda x}{n}\right)^n = e^{-\lambda x}. \tag{14.3}$$

Then

$$P(\text{a fly lands during } [0, x]) = 1 - e^{-\lambda x}.$$

Now let t be the random variable giving the time until a fly lands on the cake. We will show that this variable has an exponential distribution. We would like a distribution function f for t. Such a function should have the property that

$$\int_a^b f(t)\, dt = P(a < t < b)$$

whenever $a < b$. In particular we must have that

$$\int_0^x f(t)\, dt = P(0 < t < x) = 1 - e^{-\lambda x}.$$

To determine the function f, we differentiate both sides of this equation with respect to x. According to the argument at the end of Section 12.4 we have

$$\frac{d}{dx}\int_0^x f(t)\, dt = f(x).$$

Thus we have

$$f(x) = \frac{d}{dx}(1 - e^{-\lambda x}) = \lambda e^{-\lambda x},$$

which is the exponential density function.

Now let us shift gears and make t the *number* of times a fly lands on the cake during the interval $[0, x]$. As before, we divide this interval into n subintervals.

$$\left[0, \frac{x}{n}\right], \left[\frac{x}{n}, \frac{2x}{n}\right], \left[\frac{2x}{n}, \frac{3x}{n}\right], \ldots, \left[\frac{(n-1)x}{n}, \frac{nx}{n}\right],$$

each having length x/n. If n is large, then each subinterval is small, and the probability that a fly lands during a particular subinterval is approximately $\lambda x/n$. Likewise the probability that no fly lands during a particular subinterval is $1 - \lambda x/n$. The probability that a fly lands during exactly k of the subintervals is thus

$$P(t = k) \approx C(n, k)\left(\frac{\lambda x}{n}\right)^k\left(1 - \frac{\lambda x}{n}\right)^{n-k},$$

where we have used Bernoulli's Formula (Theorem 6.7 of Section 6.6). To get the exact probability we must take the limit of the right-hand quantity as n increases without bound.

Notice that

$$C(n, k)\left(\frac{\lambda x}{n}\right)^k\left(1 - \frac{\lambda x}{n}\right)^{n-k}$$

$$= \frac{n(n-1)(n-2)\cdots(n-k+1)}{k!}\frac{(\lambda x)^k}{n^k}\frac{\left(1 - \frac{\lambda x}{n}\right)^n}{\left(1 - \frac{\lambda x}{n}\right)^k}$$

$$= \frac{(\lambda x)^k}{k!\left(1 - \frac{\lambda x}{n}\right)^k} \cdot \frac{n(n-1)\cdots(n-k+1)}{n^k} \cdot \left(1 - \frac{\lambda x}{n}\right)^n.$$

We will find the limit of this quantity as n increases without bound by evaluating the limit of each factor separately. Note that

$$\lim_{n\to\infty} \frac{(\lambda x)^k}{k!\left(1 - \frac{\lambda x}{n}\right)^k} = \frac{(\lambda x)^k}{k!(1-0)^k} = \frac{(\lambda x)^k}{k!}.$$

Likewise

$$\lim_{n\to\infty} \frac{n(n-1)\cdots(n-k+1)}{n^k} = \lim_{n\to\infty} \frac{n(n-1)\cdots(n-k+1)}{n\cdot n\cdot\cdots\cdot n}$$

$$= \lim_{n\to\infty} (1)\left(1 - \frac{1}{n}\right)\cdots\left(1 - \frac{k-1}{n}\right) = (1)(1)\cdots(1) = 1.$$

Finally, we showed in (14.3) above that

$$\lim_{n\to\infty} \left(1 - \frac{\lambda x}{n}\right)^n = e^{-\lambda x}.$$

Now by the product rule for limits in Section 9.2 we have

$$P(t = k) = \frac{(\lambda x)^k}{k!} \cdot 1 \cdot e^{-\lambda x}.$$

Setting $\mu = \lambda x$ gives

$$P(t = k) = \frac{\mu^k e^{-\mu}}{k!},$$

which corresponds to a Poisson distribution.

EXERCISES 14.2

In Exercises 1–4 find the indicated probability for an exponential distribution with the given value of λ.

1. $P(x \leq 5)$; $\lambda = 3$

2. $P(x \leq 2)$; $\lambda = \frac{1}{2}$

3. $P(x > 2)$; $\lambda = 0.2$

4. $P(x \geq 1)$; $\lambda = 3$

In Exercises 5–8 a probability is given that corresponds to an exponential distribution. Find λ.

5. $P(x \leq 2) = 0.3$

6. $P(x < 3) = 0.5$

7. $P(x > 1) = 0.9$

8. $P(x \geq 0.5) = 0.1$

Use the formulas of Examples 14.8 and 14.9 to work Exercises 9–12.

9. A coin is flipped until a head comes up. What is the probability that no more than 3 flips are necessary?

10. A coin is flipped until a head comes up. What is the probability that more than 2 flips are necessary?

11. A coin is flipped until two heads come up. What is the probability that exactly 5 flips are needed?

12. A coin is flipped until two heads come up. What is the probability that more than 3 flips are necessary?

In Exercises 13–16 find the indicated probability for a Poisson distribution with the given value of μ.

13. $P(x < 2)$; $\mu = 3$

14. $P(x \leq 2)$; $\mu = 2$

15. $P(x \geq 3)$; $\mu = 0.5$

16. $P(x > 2)$; $\mu = 0.3$

In Exercises 17–20 a probability is given that corresponds to a Poisson distribution. Find μ.

17. $P(x = 0) = 0.3$

18. $P(x = 0) = 0.5$

19. $P(x > 0) = 0.9$

20. $P(x \geq 1) = 0.2$

In Exercises 21–28 use the appropriate exponential or Poisson distribution to model the situation given.

21. A disk jockey finds that when she offers free concert tickets on her program, the first call comes in within 30 seconds half the time. What is the probability of getting no call in the first minute?

22. When a kernel of popping corn is dropped on a hot skillet, it pops within 10 seconds with probability 0.3. What is the probability that it does not pop within a minute?

23. When Mr. Tang fills his bird feeder, the first bird arrives within one minute 3/4 of the time. What is the probability that no bird arrives for two minutes?

24. When the local K-Mart manager announces a "blue-light special," the first customer arrives within 20 seconds half the time. What is the probability that the first customer arrives within 5 seconds?

25. When Sandy opens her doughnut shop in the morning, there are one or more people waiting at the door 80% of the time. What is the probability that at least 3 people are waiting?

26. It takes 10 minutes to clean the deep fryer at the Arkansas Fried Chicken carry-out. Half the time no one places an order during this time. What is the probability that 3 orders are placed while the deep fryer is being cleaned?

27. Winton checks his crab traps daily, finding no crabs of legal size in 10% of them. What is the probability that he finds more than 2 legal crabs in a trap?

28. A police officer at a speed trap observes no speeders during a given hour 5% of the time. What is the probability of observing 3 speeders during an hour?

29. A die is rolled until a 5 or 6 comes up. Suppose this takes x rolls. Find $P(x = k)$ in terms of k. What is the probability that at least 5 rolls are needed?

30. A deck of cards is cut until a heart shows. Suppose this takes x cuts. Find $P(x = k)$ in terms of k. What is the probability that at least 4 cuts are needed?

31. A die is rolled until two 3's come up. Suppose this takes x rolls. Find $P(x = k)$ in terms of k. What is the probability that no more than 5 rolls are needed?

32. A deck of cards is cut until two clubs show. Suppose this takes x cuts. Find $P(x = k)$ in terms of k. What is the probability that no more than 4 cuts are needed?

33. A coin is flipped until a head and a tail come up consecutively in either order. Let x be the number of flips this takes. Find $P(x = k)$ in terms of k. What is the probability that more than 3 flips are needed?

Answers to Practice Problems

3. $e^{-1} \approx .368$

4. (a) $\left(\frac{5}{6}\right)^{k-1}\left(\frac{1}{6}\right)$ (b) $\frac{125}{216} \approx .579$

5. $8.5e^{-3} \approx .423$

14.3 Expected Value and Variance of a Continuous Random Variable

Figure 14.13 illustrates probability distributions of discrete and continuous random variables.

For the discrete distribution in Figure 14.13(a) the probability that the random variable x takes on the value x_i is p_i for $i = 1, 2, \ldots, n$, and so

$$p_1 + p_2 + \ldots + p_n = \sum p = 1.$$

The corresponding equation for the continuous distribution in Figure 14.13(b) is

$$\int_{-\infty}^{\infty} f(x)\, dx = 1,$$

where f is the probability density function for x. Notice that we replace p (the height of a bar in the discrete distribution) by $f(x)$ (the height of the curve in the continuous distribution) and integrate instead of sum. These equations amount to saying that the shaded area under each distribution is 1.

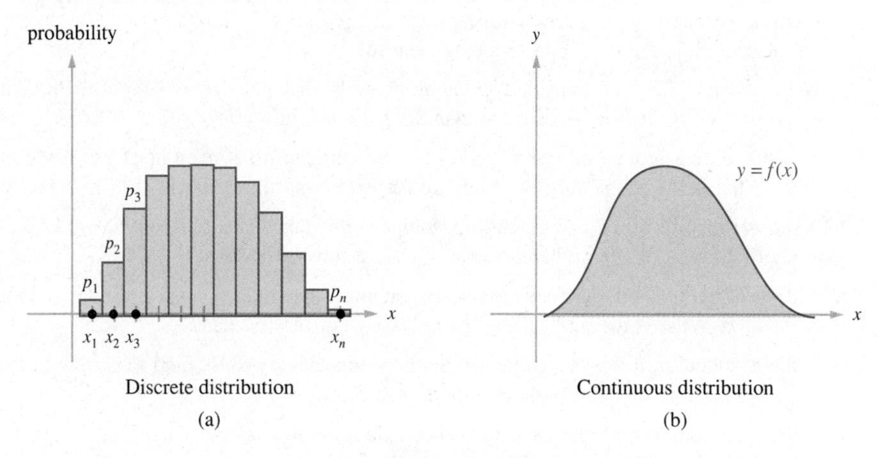

FIGURE 14.13

In Chapter 7 we defined the expected value (or mean) and variance of a discrete random variable. We will now define these quantities for continuous random variables. Recall that the expected value of the discrete random variable pictured in Figure 14.13(a) is

$$\mu = x_1p_1 + x_2p_2 + \ldots + x_np_n = \sum xp.$$

We define the **expected value** of the continuous distribution with density function f by

$$\mu = \int_{-\infty}^{\infty} xf(x)\, dx.$$

When we compare this with the definition of the mean of a discrete random variable, we see that again we have replaced p with $f(x)$ and integrated instead of summed. Of course, if f is zero except on some interval $[a, b]$, the mean may be computed by integrating from a to b.

EXAMPLE 14.12 Find the expected value of the random variable x with density function

$$f(x) = \begin{cases} 2x & \text{if } 0 \le x \le 1 \\ 0 & \text{otherwise,} \end{cases}$$

which was considered in Section 14.1.

Solution

$$\mu = \int_0^1 x(2x)\, dx = \left.\frac{2x^3}{3}\right|_0^1 = \frac{2}{3}(1 - 0) = \frac{2}{3} \quad ■$$

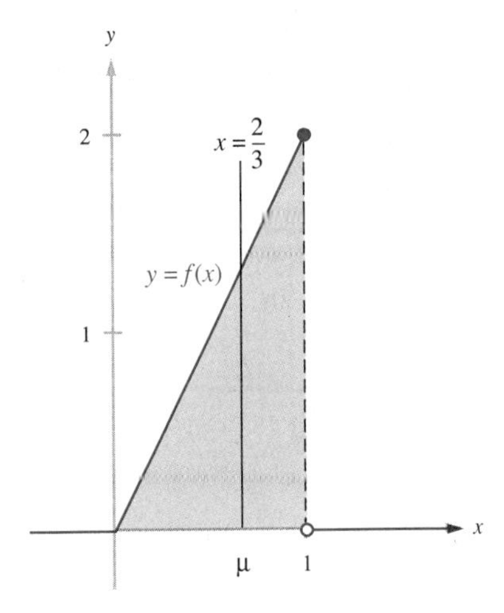

FIGURE 14.14

Remark If μ is the expected value of a random variable, then the region under the distribution of x would balance on the line $x = \mu$ if it were cut out of some uniform material. (See Figure 14.14 for the distribution of the previous example.) This does not necessarily mean that the area under the distribution is the same on both sides of this line. The reader should check that in Figure 14.14 the area to the left of the line $x = 2/3$ is 4/9 and the area to the right of this line is 5/9.

EXAMPLE 14.13 Find the expected value of the random variable with density function

$$f(x) = \begin{cases} \dfrac{1}{30} & \text{if } 0 \le x \le 20 \\ \dfrac{1}{60} & \text{if } 20 < x \le 40 \\ 0 & \text{otherwise,} \end{cases}$$

Solution The expected value is

$$\begin{aligned} \int_0^{40} xf(x)\,dx &= \int_0^{20} x\left(\frac{1}{30}\right) dx + \int_{20}^{40} x\left(\frac{1}{60}\right) dx \\ &= \frac{1}{30}\left(\frac{x^2}{2}\right)\bigg|_0^{20} + \frac{1}{60}\left(\frac{x^2}{2}\right)\bigg|_{20}^{40} \\ &= \frac{1}{30}(200 - 0) + \frac{1}{60}(800 - 200) = \frac{50}{3} = 16^2/_3. \end{aligned}$$

In Example 14.5 the random variable x represented the time a person tuning in a certain radio station had to wait to hear a weather report. Our computation shows that in some sense the average waiting time will be 16⅔ minutes. This holds in spite of our previous computation showing that there is an equal probability of waiting more or less than 15 minutes. ■

Practice Problem 6 Find the expected value of the random variable x with density function

$$f(x) = \begin{cases} \dfrac{3x^2}{26} & \text{if } 1 \le x \le 3 \\ 0 & \text{otherwise.} \end{cases}$$

Recall that a uniform probability distribution has a density function of the form

$$f(x) = \begin{cases} \dfrac{1}{b-a} & \text{if } a \le x \le b \\ 0 & \text{otherwise.} \end{cases}$$

Thus the expected value of the corresponding random variable is

$$\int_a^b \frac{x}{b-a}\,dx = \frac{x^2}{2(b-a)}\bigg|_a^b = \frac{b^2-a^2}{2(b-a)}$$

$$= \frac{(b+a)(b-a)}{2(b-a)} = \frac{b+a}{2}.$$

We see that the expected value μ is halfway between a and b, as might be expected. (See Figure 14.15.)

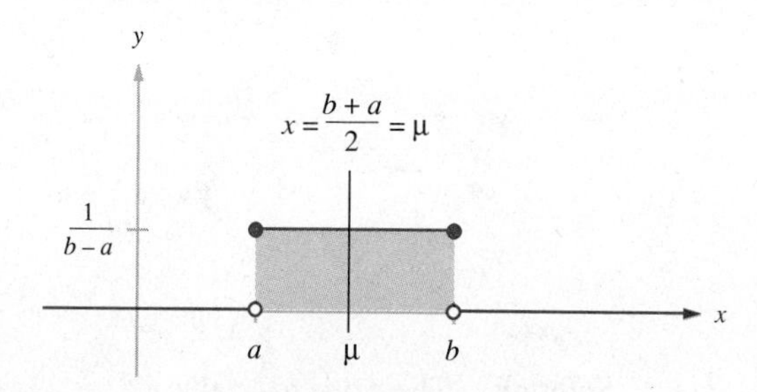

FIGURE 14.15

THE VARIANCE AND STANDARD DEVIATION OF A CONTINUOUS RANDOM VARIABLE

Recall that the variance of the discrete random variable x with distribution pictured in Figure 14.13(a) is given by

$$\sigma^2 = (x_1 - \mu)^2 p_1 + (x_2 - \mu)^2 p_2 + \ldots + (x_n - \mu)^2 p_n = \Sigma\,(x - \mu)^2 p,$$

where μ is the expected value of x. In an analogous way we define the **variance** of the continuous random variable x with density function f by

$$\sigma^2 = \int_{-\infty}^{\infty} (x - \mu)^2 f(x)\, dx,$$

where μ is the expected value of the distribution. The **standard deviation** σ of x is defined, as usual, to be the positive square root of the variance.

EXAMPLE 14.14 Compute the variance and standard deviation of the random variable x with density function

$$f(x) = \begin{cases} 2x & \text{if } 0 \le x \le 1 \\ 0 & \text{otherwise,} \end{cases}$$

which occurs in Example 14.12.

Solution In Example 14.12 we found that $\mu = 2/3$ for this random variable. Thus

$$\begin{aligned} \sigma^2 &= \int_0^1 \left(x - \frac{2}{3}\right)^2 2x\, dx = \int_0^1 \left(x^2 - \frac{4x}{3} + \frac{4}{9}\right) 2x\, dx \\ &= \int_0^1 \left(2x^3 - \frac{8x^2}{3} + \frac{8x}{9}\right) dx = \frac{2x^4}{4} - \frac{8x^3}{9} + \frac{8x^2}{18}\Bigg|_0^1 \\ &= \frac{2}{4} - \frac{8}{9} + \frac{8}{18} = \frac{1}{18}. \end{aligned}$$

Then

$$\sigma = \sqrt{\frac{1}{18}} \approx .236. \quad \blacksquare$$

As the previous example illustrates, calculating the variance for even a simple density function can be fairly messy. There is an alternate formula for the variance that often is easier to apply. Notice that

$$\begin{aligned} \sigma^2 &= \int_{-\infty}^{\infty} (x - \mu)^2 f(x)\, dx = \int_{-\infty}^{\infty} (x^2 - 2\mu x + \mu^2) f(x)\, dx \\ &= \int_{-\infty}^{\infty} x^2 f(x)\, dx - 2\mu \int_{-\infty}^{\infty} x f(x)\, dx + \mu^2 \int_{-\infty}^{\infty} f(x)\, dx. \end{aligned}$$

At this point we note that the second integral, $\int_{-\infty}^{\infty} x f(x)\, dx$, is by definition the expected value μ, and the third integral, $\int_{-\infty}^{\infty} f(x)\, dx$, has the value 1 by property (b) of a probability density function. Thus we have

$$\sigma^2 = \int_{-\infty}^{\infty} x^2 f(x)\, dx - 2\mu^2 + \mu^2 = \int_{-\infty}^{\infty} x^2 f(x)\, dx - \mu^2.$$

Alternate Formula for the Variance

$$\sigma^2 = \int_{-\infty}^{\infty} x^2 f(x)\, dx - \mu^2$$

Applying the alternate formula to the density function of Example 14.14 leads to the following computation.

$$\sigma^2 = \int_0^1 x^2(2x)\, dx - \left(\frac{2}{3}\right)^2 = \int_0^1 2x^3\, dx - \frac{4}{9}$$
$$= \frac{2x^4}{4}\bigg|_0^1 - \frac{4}{9} = \frac{2}{4} - \frac{4}{9} = \frac{1}{18}$$

Notice that this calculation of σ^2 is considerably simpler than the one in Example 14.14.

Practice Problem 7 Use the alternate formula to compute the variance and standard deviation of the random variable x with density function

$$f(x) = \begin{cases} \dfrac{3x^2}{26} & \text{if } 1 \le x \le 3 \\ 0 & \text{otherwise.} \end{cases}$$

(The expected value $\mu = 30/13$ was computed in Practice Problem 6.)

Let us compute the variance and standard deviation for a uniformly distributed random variable with density function

$$f(x) = \begin{cases} \dfrac{1}{b - a} & \text{if } a \le x \le b \\ 0 & \text{otherwise.} \end{cases}$$

Recall that we have already found that the expected value of such a random variable is $\mu = (a + b)/2$. Thus

$$\sigma^2 = \int_a^b x^2(b - a)^{-1} dx - \mu^2$$
$$= \frac{(b - a)^{-1}x^3}{3}\bigg|_a^b - \left(\frac{a + b}{2}\right)^2$$
$$= \frac{b^3 - a^3}{3(b - a)} - \frac{(a + b)^2}{4}.$$

The reader should use the identity $b^3 - a^3 = (b - a)(b^2 + ab + a^2)$ to check that if $a < b$, then this simplifies to $(b - a)^2/12$.

Expected Value and Variance of a Uniform Distribution

If the random variable x has the uniform density function

$$f(x) = \begin{cases} \dfrac{1}{b - a} & \text{if } a \le x \le b \\ 0 & \text{otherwise,} \end{cases}$$

then the expected value of x is $\mu = (a + b)/2$ and the variance of x is $\sigma^2 = (b - a)^2/12$.

Practice Problem 8 Find the expected value, variance, and standard deviation of the uniformly distributed random variable that is nonzero on [1, 5].

THE EXPECTED VALUE AND VARIANCE OF THE NORMAL AND EXPONENTIAL DISTRIBUTIONS

In Section 7.6 we considered the normal probability distribution, which has the density function

$$f(x) = \frac{1}{\sigma\sqrt{2\pi}} e^{-(x - \mu)^2/2\sigma^2}.$$

The constants μ and σ were called the "mean" and "standard deviation" of x, although these concepts were not defined for continuous random variables in Chapter 7. When the definitions of the present section are applied to the density function f above, it turns out that the expected value and standard deviation are indeed the constants μ and σ.

We can also compute the expected value and standard deviation of the exponential distribution, with density function

$$f(x) = \begin{cases} \lambda e^{-\lambda x} & \text{if } x \ge 0 \\ 0 & \text{otherwise.} \end{cases}$$

By definition we have

$$\mu = \int_0^\infty x\lambda e^{-\lambda x}\,dx \qquad \text{and} \qquad \sigma^2 = \int_0^\infty x^2\lambda e^{-\lambda x}\,dx - \mu^2.$$

Each of these improper integrals can be evaluated using either integration by parts or integral tables, but we leave the details to the exercises. The results are summarized below.

The Expected Value and Standard Deviation of an Exponential Distribution

If the random variable x has the density function

$$f(x) = \begin{cases} \lambda e^{-\lambda x} & \text{if } x \ge 0 \\ 0 & \text{otherwise,} \end{cases}$$

then the expected value and standard deviation of x are both $1/\lambda$.

EXAMPLE 14.15 A survey of a chain of pizza restaurants finds that the time x in minutes a party spends ordering and eating dinner averages 50 minutes. Assuming that x is exponentially distributed, what is the probability that a party spends more than one hour and 15 minutes for dinner?

Solution Since we know that the expected value of x is $\mu = 1/\lambda = 50$, we conclude that $\lambda = 1/50 = .02$. Thus the density function for x is

$$f(x) = \begin{cases} .02e^{-.02x} & \text{if } x \geq 0 \\ 0 & \text{otherwise.} \end{cases}$$

To avoid an improper integral, we will compute the probability that a party takes *less* than 75 minutes for dinner. This is

$$P(x < 75) = \int_0^{75} .02e^{-.02x}\, dx = \frac{.02e^{-.02x}}{(-.02)}\Bigg|_0^{75}$$

$$= -[e^{-.02(75)} - 1] = 1 - e^{-1.5}.$$

Hence

$$P(x \geq 75) = 1 - P(x < 75) = e^{-1.5} \approx .223.$$

The pizza chain can expect that about 22% of the parties eating dinner will take more than one hour and 15 minutes. ■

Practice Problem 9 The average waiting time x at Joe's Barber Shop is 9 minutes. If x has an exponential distribution, what percentage of customers wait less than 5 minutes?

THE EXPECTED VALUE AND VARIANCE OF THE POISSON DISTRIBUTION

We need new definitions to define the expected value and variance of a random variable that is discrete but infinite, such as one having the Poisson distribution. In general, if the random variable x takes on the values $x_1, x_2, \ldots$ with probabilities $p_1, p_2, \ldots$, respectively, then we define the expected value μ and variance σ^2 of x by

$$\mu = \lim_{n\to\infty} (x_1p_1 + x_2p_2 + \cdots + x_np_n), \qquad \text{and}$$

$$\sigma^2 = \lim_{n\to\infty} [(x_1 - \mu)^2p_1 + (x_2 - \mu)^2p_2 + \cdots + (x_n - \mu)^2p_n].$$

Unfortunately we have not developed the techniques necessary to compute such limits. In the case of the Poisson distribution, which is defined by

$$P(x = k) = \frac{e^{-\mu}\mu^k}{k!}, \qquad k = 0, 1, 2, \ldots,$$

where μ is a positive constant, the above definitions may be used to show that the expected value of x is the constant μ and the variance is $\sigma^2 = \mu^2$. Since μ is assumed to be positive, this means that $\sigma = \mu$.

Expected Value and Standard Deviation of the Poisson Distribution

If the random variable x has the Poisson distribution with

$$P(x = k) = \frac{e^{-\mu}\mu^k}{k!} \qquad \text{for } k = 0, 1, 2 \ldots,$$

then the expected value and standard deviation of x are both μ.

EXAMPLE 14.16 In a small hospital 730 babies were born last year. Assuming that this is typical, and that the number x of babies born there on any one day has a Poisson distribution, find the percentage of days that no baby is born in the hospital.

Solution Since 730 babies are born per year, there are an average of 730/365 = 2 babies born per day. Thus $\mu = 2$. Then

$$P(x = 0) = \frac{e^{-2}2^0}{0!} = e^{-2} \approx .1353.$$

We see that no baby is born in the hospital on about 14% of the days. ■

EXAMPLE 14.17 A botanist notices that her Venus-flytrap plant catches no flies on one day out of three. Assuming that x, the number of flies it catches in a particular day, has a Poisson distribution, how many flies can she expect the plant to catch in a 30-day period?

Solution Since x has a Poisson distribution, we have

$$P(x = 0) = \frac{e^{-\mu}\mu^0}{0!} = \frac{1}{3}.$$

This means that

$$e^{-\mu} = \frac{1}{3}$$

$$-\mu = \ln e^{-\mu} = \ln\left(\frac{1}{3}\right) = \ln 3^{-1} = -\ln 3$$

$$\mu = \ln 3 \approx 1.10.$$

Thus the plant will catch an average of 1.1 flies per day. In 30 days the botanist can expect it to catch 30 ln 3 ≈ 33 flies. ■

Practice Problem 10 On the average 45 students at a small college will visit the infirmary in any 30-day period. What is the probability that more than 2 students will visit the infirmary on a given day?

EXERCISES 14.3

In Exercises 1–8 find the expected value of the random variable with the given density function.

1. $f(x) = \begin{cases} 0.25 & \text{if } 1 \le x \le 5 \\ 0 & \text{otherwise} \end{cases}$

2. $f(x) = \begin{cases} 1/6 & \text{if } -3 \le x \le 3 \\ 0 & \text{otherwise} \end{cases}$

3. $f(x) = \begin{cases} \dfrac{1}{x \ln 3} & \text{if } 1 \le x \le 3 \\ 0 & \text{otherwise} \end{cases}$

4. $f(x) = \begin{cases} x^2/3 & \text{if } -1 \le x \le 2 \\ 0 & \text{otherwise} \end{cases}$

5. $f(x) = \begin{cases} 4x^3 & \text{if } 0 \le x \le 1 \\ 0 & \text{otherwise} \end{cases}$

6. $f(x) = \begin{cases} 4/3x^2 & \text{if } 1 \le x \le 4 \\ 0 & \text{otherwise} \end{cases}$

7. $f(x) = \begin{cases} 3/x^4 & \text{if } x \ge 1 \\ 0 & \text{otherwise} \end{cases}$

8. $f(x) = \begin{cases} 1/4x^5 & \text{for } x \ge 0.5 \\ 0 & \text{otherwise} \end{cases}$

In Exercises 9–16 find the variance corresponding to the density function of the given exercise.

9. Exercise 1

10. Exercise 2

11. Exercise 3

12. Exercise 4

13. Exercise 5

14. Exercise 6

15. Exercise 7

16. Exercise 8

In Exercises 17–28 give the expected value, variance, and standard deviation for the random variable with the given distribution.

17. Uniform on $[-2, 4]$

18. Uniform on $[0, 10]$

19. Uniform on $[-5, 0]$

20. Uniform on $[1.3, 1.5]$

21. Exponential with $\lambda = 3$

22. Exponential with $\lambda = 8$

23. Exponential with $\lambda = 0.1$

24. Exponential with $\lambda = \dfrac{1}{3}$

25. Poisson with $\mu = 5$

26. Poisson with $\mu = \dfrac{3}{2}$

27. Poisson with $\mu = \dfrac{1}{5}$

28. Poisson with $\mu = 0.25$

In Exercises 29–38 assume a uniform, exponential, or Poisson distribution, as appropriate.

29. A bus stops in front of Mrs. Rivera's house every 24 minutes. If she goes to the bus stop at a random time, how long on the average will she wait for a bus?

30. Every 9 minutes a telethon gives the telephone number to call to pledge a donation. If a person tunes in at random, how long on the average will it be until the telephone number is given?

31. The average time a barber takes to give a haircut is 20 minutes. What is the probability that a haircut takes more than 1/2 hour?

32. An artist requires an average of 6 hours of sitting to complete a portrait. What is the probability that a portrait is done with less than 5 hours of sitting?

33. A hotdog stand averages 15 customers per hour. What is the probability that no more than one customer comes in during a 20-minute period?

34. On the average a camper gets 3 mosquito bites per hour. What is the probability that an hour goes by with no bite?

35. A certain surgical procedure takes more than 30 minutes 25% of the time. What is the probability that it is done in less than 20 minutes?

36. A fast-food restaurant serves lunch in less than 5 minutes to 4 out of 5 customers. What is the probability that it takes more than 8 minutes to serve a lunch?

37. On 2 out of 3 days no flies get into the Murphy house. What is the probability that 2 flies get in on a given day?

38. All of the VCRs rented out by Joe's Rental are returned the next day in good condition on 90% of the days. What is the probability that there is a problem with exactly 2 VCRs on a given day?

39. A radio station broadcasts advertisements for an appliance store on the hour and at 45 minutes after the hour. What is the average time a person tuning into the station must wait to hear such an advertisement?

40. Use either integration by parts or the integral table to show that the expected value of the exponential distribution with constant λ is $1/\lambda$.

41. Use either integration by parts (twice) or the integral table to show that the variance of the exponential distribution with constant λ is $1/\lambda^2$.

Answers to Practice Problems

6. $\frac{30}{13}$

7. $\sigma^2 = \frac{219}{845} \approx .259$ and $\sigma \approx .509$

8. $\mu = 3$, $\sigma^2 = \frac{4}{3}$, and $\sigma = \sqrt{\frac{4}{3}} \approx 1.15$

9. about 43%

10. $1 - 3.625e^{-1.5} \approx .191$

MATHEMATICS IN ACTION

14.4 A Single-Server Queueing System

In this section we will consider a basic problem of queueing theory. A *queue* is simply a line that forms, such as when people wait for service at the post office or cars wait to pass through a toll booth. Humans need not be involved. For example, refrigerators being made on an automated assembly line may have to wait until a machine to paint them is available, and programs input into a university computer have to wait until the central processing unit is free.

We will consider only *single-server queueing systems,* that is, situations in which all *customers* (post-office patrons, cars at the toll booth, refrigerators, programs) must be serviced by a single *server* (postal clerk, toll booth, painting machine, computer).

Although in our assembly-line example the refrigerators may arrive at the painting machine at regular intervals, in most situations the customers will appear at random times. Likewise the service time will vary in most applications. For example, the time a postal customer takes with a clerk will depend on whether the former is just buying a stamp or sending an insured package to Sri Lanka. The Poisson and exponential distributions turn out to be useful in modeling the arrival of customers and the time it takes to service them.

Let us consider the situation at Yetta's Yogurtorium, a small shop selling carry-out frozen yogurt in various flavors. A single clerk waits on customers, so that only one customer can be served at a time. An average of 12 customers enter the shop per hour, and it takes the clerk an average of 3 minutes (1/20 of an hour) to wait on each. Of course some transactions take longer than others. Let us suppose that the time x in hours that it takes to wait on a customer has an exponential distribution with density function

$$f(x) = \begin{cases} \lambda e^{-\lambda x} & \text{if } x \geq 0 \\ 0 & \text{otherwise.} \end{cases}$$

In the previous section we found that the expected value of this distribution is $1/\lambda$. Since this represents the average time to wait on a customer, we have $1/\lambda = 1/20$, and so $\lambda = 20$. We can interpret this as meaning that 20 customers can be served per hour, on the average.

Recall that on the average 12 customers enter Yetta's Yogurtorium per hour. We will use the Poisson distribution to model the number x of customers who enter the shop in any one-hour period. Thus we have the distribution

$$P(x = k) = \frac{e^{-\mu}\mu^k}{k!}, \qquad k = 0, 1, 2, \ldots .$$

In Section 14.3 we found that μ is the expected value of this distribution. It represents the average number of customers entering the shop in one hour. Since the shop averages 12 customers per hour, we have $\mu = 12$. (In the analysis that follows it is important that we use the same time units, in this case hours, for both λ and μ.)

To recap, customers enter the yogurt shop at an average rate of $\mu = 12$ per hour, and can be served at an average rate of $\lambda = 20$ per hour. Note that $12 < 20$. This is as it should be if the system is not to break down. If the rate at which customers entered the shop exceeded the rate at which they were served, longer and longer lines would form; and customers would avoid the shop because of the long wait. *We will only consider situations for which* $\mu < \lambda$.

Even when $\mu < \lambda$, as in our example, queues will sometimes form. Remember that the 12 customers arriving per hour and the 20 customers who can be served per hour are just averages. Sometimes several people will enter the shop at about the same time, and some will have to wait for service. At other times the clerk will have no one to serve. And of course an occasional customer will take an especially long time.

We define the **utilization factor** of a single-server queueing system to be the ratio μ/λ, which we denote by ρ. Since we are assuming that $\mu < \lambda$, we have $\rho < 1$. In the case of Yetta's Yogurtorium we have $\rho = 12/20 = .6$.

Practice Problem 11 Suppose Yetta's Yogurtorium increases in popularity to the extent that 15 customers enter the shop per hour. What is the new value of ρ?

By using methods beyond this course it is possible to answer various natural questions about a single-server queueing system. One such question is how many customers we can expect to be in the system at a given time, either being served or waiting in line. This is called the **expected number of customers in the system,** and is denoted by N. It can be proved that

$$N = \frac{\rho}{1 - \rho}.$$

For example, the average number of customers in Yetta's Yogurtorium at any given time is

$$N = \frac{.6}{1 - .6} = 1.5.$$

A related quantity is the average number of people in line, called the **expected number of customers in the queue.** We denote this quantity by N_q. It can be computed using the following formula.

$$N_q = \frac{\rho^2}{1 - \rho} = \rho N$$

In our example we have

$$N_q = .6(1.5) = .9.$$

One might think that N_q would be just 1 less than N, since everyone except the one person being served is standing in line, but this analysis does not take into account the fact that sometimes no customers are in the shop. At such times the number in the system and the number in the queue are equal.

Note that in this discussion a ‘‘customer’’ is a person or thing that is served in some way and may not be a customer in the usual sense of the word.

Practice Problem 12 Consider Yetta's Yogurtorium when customers are arriving at 15 per hour. What is the average number of customers in the shop and in line? (The server can still service 20 customers per hour.)

Another quantity of interest is the average time a customer spends in the yogurt shop. In general this is called the **expected time in the system,** and we denote it by T. It can be proved that

$$T = \frac{1}{\lambda(1 - \rho)}.$$

In our example we had $\lambda = 20$ and $\rho = .6$, so we compute

$$T = \frac{1}{20(1 - .6)} = \frac{1}{20(.4)} = \frac{1}{8}.$$

Thus the average customer spends 1/8 of an hour, or 60/8 = 7½ minutes in the Yogurtorium.

We can also compute the **expected time in the queue** for a single-server queueing system. This is the average time spent just waiting in line, and is important for consumer satisfaction because people are more aware of the time they spend doing nothing than their time being served. We denote this quantity by T_q, and compute it with the formula

$$T_q = \frac{\rho}{\lambda(1 - \rho)} = \rho T.$$

In Yetta's Yogurtorium we have

$$T_q = .6\left(\frac{1}{8}\right) = .075.$$

Thus a customer spends an average of .075 hours or 60(.075) = 4½ minutes waiting in line. Notice that this is just 3 minutes less than the 7½ minutes which is the average time in the shop. This reflects the fact that being waited on takes an average of 3 minutes per customer. It can be proved that in general

$$T_q = T - \frac{1}{\lambda},$$

where we recall that $1/\lambda$ is the average service time.

Practice Problem 13 Find the expected time in the system and expected time in the queue when Yetta's Yogurtorium gets 15 customers per hour. Give your answers in minutes.

We summarize our formulas for a single-server queueing system in which the number of customers arriving per unit of time has a Poisson distribution with constant μ and the time to serve a customer has an exponential distribution with constant λ.

Utilization factor:

$$\rho = \frac{\mu}{\lambda}$$

Expected number of customers in the system:

$$N = \frac{\rho}{1 - \rho}$$

Expected number of customers in the queue:

$$N_q = \frac{\rho^2}{1 - \rho} = \rho N$$

Expected time in the system:

$$T = \frac{1}{\lambda(1 - \rho)}$$

Expected time in the queue:

$$T_q = \frac{\rho}{\lambda(1 - \rho)} = \rho T$$

Notice that μ is the *average number of customers arriving per unit of time,* while $1/\lambda$ is the *average time spent serving a customer*. This means that λ itself is the *average number of customers who can be served per unit of time*.

EXAMPLE 14.18 The owner of a small printing press can handle only one printing job at a time, and the average job takes 2½ hours to set up and run. (Note that in this example a "customer" is a printing job.) He averages 3 jobs during each 12-hour period his shop is open. Find the utilization factor, the expected number of jobs in the system and in the queue, and the expected time in the system and in the queue. How much will the expected time in the system increase if an average of 4 jobs come in during each 12-hour period?

Solution We will measure time in hours. Since the average job takes 2½ hours, we have $1/\lambda = 2\frac{1}{2} = 5/2$, and so $\lambda = 2/5$. (This means that the printer can do 2/5 of an average job in one hour.) The jobs come in at an average rate of 3 per 12 hours, or $3/12 = 1/4$ per hour. Thus $\mu = 1/4$. Therefore the utilization factor is

$$\rho = \frac{\mu}{\lambda} = \frac{\frac{1}{4}}{\frac{2}{5}} = \frac{5}{8}.$$

Next we compute the expected number of jobs in the system and in the queue. The first of these is

$$N = \frac{\rho}{1 - \rho} = \frac{\frac{5}{8}}{1 - \frac{5}{8}} = \frac{5}{3} \approx 1.67.$$

The second is

$$N_q = \rho N = \frac{5}{8} \cdot \frac{5}{3} = \frac{25}{24} \approx 1.04.$$

Thus there are an average of 1.67 jobs in the shop at any time, and an average of 1.04 jobs waiting to be worked on.

Finally the expected time in the system is

$$T = \frac{1}{\lambda(1 - \rho)} = \frac{1}{\frac{2}{5}\left(1 - \frac{5}{8}\right)} = \frac{20}{3} = 6^{2}/_{3}.$$

and the expected time in the queue is

$$T_q = \rho T = \frac{5}{8} \cdot \frac{20}{3} = \frac{25}{6} = 4^{1}/_{6}.$$

In hours and minutes these times amount to 6:40 and 4:10, respectively. (Note that 6:40 − 4:10 = 2:30, the average service time.) Thus when a job comes into the shop, it takes an average of 4 hours and 10 minutes before it is worked on, for an average total time in the shop of 6 hours and 40 minutes.

Now we consider what happens if an average of 4 jobs come in per 12-hour period. Then μ, the average number of jobs per hour, changes to $4/12 = 1/3$. We still have $\lambda = 2/5$, but now

$$\rho = \frac{\mu}{\lambda} = \frac{\frac{1}{3}}{\frac{2}{5}} = \frac{5}{6}.$$

Then the average time in the system is

$$T = \frac{1}{\lambda(1 - \rho)} = \frac{1}{\frac{2}{5}\left(1 - \frac{5}{6}\right)} = 15.$$

We see that the average time in the system changes from 6 hours and 40 minutes to 15 hours, an increase of 8 hours and 20 minutes. ■

EXAMPLE 14.19 A professional income-tax preparer averages 9 clients per 8-hour day and spends an average of 40 minutes with each. She knows from past experience that clients are unhappy if they spend more than 2 hours in the system, and so would like the average time in the system to be no more than this. Will this be the case? If not, how much time could she spend with an average client to make it so? She does not want to serve any fewer customers.

Solution We will measure time in hours. Since 9 customers arrive per 8 hours, the average number of customers per hour is $9/8 = \mu$. Likewise the average time spent per customer is 40 minutes, or 2/3 of an hour, and this is $1/\lambda$. Thus we have $\lambda = 3/2$. We compute

$$\rho = \frac{\mu}{\lambda} = \frac{\frac{9}{8}}{\frac{3}{2}} = \frac{3}{4}.$$

Then

$$T = \frac{1}{\lambda(1 - \rho)} = \frac{1}{\left(\frac{3}{2}\right)\left(\frac{1}{4}\right)} = \frac{8}{3} = 2^2/_3 \text{ hours.}$$

As this exceeds 2 hours, the tax-preparer needs to work faster.

Since we would like $T = 2$, we will try to find λ satisfying the equation

$$2 = \frac{1}{\lambda(1 - \rho)}.$$

We must be careful, however, because although μ is fixed at 9/8 (assuming no fewer clients are to be served), the value of ρ depends on λ. If we substitute $\rho = \mu/\lambda = 9/8\lambda$ in our equation we get

$$2 = \frac{1}{\lambda\left(1 - \frac{9}{8\lambda}\right)} = \frac{1}{\lambda - \frac{9}{8}},$$

$$2\left(\lambda - \frac{9}{8}\right) = 1,$$

$$\lambda - \frac{9}{8} = \frac{1}{2},$$

$$\lambda = \frac{1}{2} + \frac{9}{8} = \frac{13}{8}.$$

We see that the time to service a client must be exponentially distributed with $\lambda = 13/8$. Since the average service time is $1/\lambda$, this means the tax-preparer must average $1/(13/8) = 8/13$ hours per client, or $60(8/13) \approx 36.9$ minutes each. She must shorten her average time with each client by about 3 minutes in order to make the average customer time in the system no more than two hours. Recall that the previous value of T was $2\frac{2}{3}$, or 2 hours and 40 minutes. Thus in this example shortening the service time by about 3 minutes shortens the total customer time in the system by 40 minutes. ■

Practice Problem 14 The average transaction at the single drive-up window of a small bank takes 5 minutes, and an average of 8 customers arrive per hour. Find the utilization factor, the expected number of customers in the system and in the queue, and the expected time in the system and in the queue. How much will the expected time in the system decrease if only 6 customers arrive per hour?

EXERCISES 14.4

In Exercises 1–6 a single-server queueing system is described. Determine what λ and μ are, using the indicated units.

1. A coffee shop averages 20 customers per hour, and it takes an average of 2 minutes to serve each customer. Measure time in hours.

2. An average job at an automobile body shop takes 3 hours, and the shop averages 2 jobs per 8-hour day. Measure time in hours.

3. A jeweler takes an average of 15 minutes with each customer, and an average of 3 customers enter his shop per hour. Measure time in minutes.

4. A kiln will bake a tray of ceramic figures in an average of 90 minutes, and trays arrive for baking at an average of one every two hours. Measure time in hours.

5. A woman takes an average of 3 minutes to kill a fly, and flies enter her apartment at an average rate of 10 per hour. Measure time in minutes.

6. The electronic control center of a telephone exchange takes an average of 1 second to switch an incoming call to the correct line, and calls come in at the rate of 45 per minute. Measure time in seconds.

In Exercises 7–10 values of λ and μ are given. Find ρ, N, N_q, T, and T_q.

7. $\lambda = 0.9, \quad \mu = 0.8$

8. $\lambda = 8, \quad \mu = 5$

9. $\lambda = \frac{1}{4}, \quad \mu = \frac{1}{6}$

10. $\lambda = \frac{7}{5}, \quad \mu = \frac{5}{6}$

In Exercises 11–20 find the expected number of customers in the system and in the queue, and the expected time in the system and in the queue.

11. A person who cuts paper silhouettes in a mall takes an average of 10 minutes to do a silhouette, and averages 3 customers per hour.

12. A girl playing an arcade game takes an average of 2 seconds to shoot down an invader from Mars. These arrive at an average rate of 20 per minute.

13. It takes the staff of an amusement park an average of 1/2 hour to find the parents of a lost child. The park averages 15 lost children per 12-hour day.

14. A judge takes an average of 3 days to try a case. Her monthly workload averages 5 cases. She holds court 20 days per month.

15. The person at the information kiosk at a department store averages 5 minutes with each person wanting information. There are an average of 3 people at the kiosk throughout the day, either being served or waiting their turn.

16. The single automatic developing and printing machine at a one-hour photo store takes an average of 12 minutes to process a roll of film. There are an average of 5 rolls of film either in the machine or waiting to be processed at any time.

17. An average of 3 scientists per month want to use a certain radio telescope. At any given time an average of 4 scientists are either using the telescope or waiting to use it.

18. A doctor sees an average of 9 patients per 8-hour day, and there are an average of 4 patients in his office, either being treated or waiting.

19. A car-wash crew takes an average of 12 minutes to wash a car. The average time a car takes in the system is 20 minutes.

20. Students find it takes an average of 5 minutes for the university's computer to run one of their programs. Of this, an average of 2 seconds is actual computer time.

Answers to Practice Problems

11. .75

12. 3 and 2.25

13. 12 minutes and 9 minutes

14. We have $\rho = \frac{2}{3}$, $N = 2$, $N_q = \frac{4}{3}$, $T = 15$ minutes, and $T_q = 10$ minutes. If only 6 customers arrive per hour then T decreases by 5 minutes to 10 minutes.

CHAPTER 14 REVIEW

IMPORTANT TERMS

- **uniform probability distribution** *(14.1)*
- **exponential distribution** *(14.2)*
- **Poisson distribution**
- **expected value of a continuous random variable** *(14.3)*
- **expected value of an infinite discrete random variable**
- **variance of a continuous random variable**
- **variance of an infinite discrete random variable**
- **expected number of customers in the queue** *(14.4)*
- **expected number of customers in the system**
- **expected time in the queue**
- **expected time in the system**
- **single-server queueing system**
- **utilization factor**

IMPORTANT FORMULAS

Conditions for a Probability Density Function f

(a) $f(x) \geq 0$ for all x

(b) $\int_{-\infty}^{\infty} f(x)\, dx = 1$

(c) $P(a \leq x \leq b) = \int_a^b f(x)\, dx$

Uniform Distribution on [*a*, *b*]

$$f(x) = \begin{cases} \dfrac{1}{b - a} & \text{if } a \leq x \leq b \\ 0 & \text{otherwise} \end{cases}$$

Exponential Distribution

$$f(x) = \begin{cases} \lambda e^{-\lambda x} & \text{if } x \geq 0 \\ 0 & \text{otherwise} \end{cases}$$

Poisson Distribution

$$P(x = k) = \frac{e^{-\mu}\mu^k}{k!} \qquad \text{for } k = 0, 1, 2, \ldots$$

Expected Value μ and Variance σ² of a Continuous Random Variable *x*

$$\mu = \int_{-\infty}^{\infty} x\, f(x)\, dx$$

$$\sigma^2 = \int_{-\infty}^{\infty} (x - \mu)^2 f(x)\, dx$$

Alternate Formula for the Variance

$$\sigma^2 = \int_{-\infty}^{\infty} x^2 f(x)\, dx - \mu^2$$

Expected Value μ and Variance σ² of an Infinite Discrete Random Variable *x*

$$\mu = \lim_{n\to\infty} (x_1p_1 + x_2p_2 + \ldots + x_np_n)$$

$$\sigma^2 = \lim_{n\to\infty} [(x_1 - \mu)^2p_1 + (x_2 - \mu)^2p_2 + \ldots + (x_n - \mu)^2p_n]$$

Expected Value μ and Variance σ² of a Uniform Distribution on [*a*, *b*]

$$\mu = \frac{a + b}{2}$$

$$\sigma^2 = \frac{(b - a)^2}{12}$$

Expected Value μ and Standard Deviation σ of an Exponential Distribution with Constant λ

$$\mu = \sigma = \frac{1}{\lambda}$$

Expected Value and Standard Deviation of the Poisson Distribution with Constant μ

expected value = standard deviation = μ

Single-server Queueing System Formulas

Utilization factor:

$$\rho = \frac{\mu}{\lambda}$$

Expected number of customers in the system:

$$N = \frac{\rho}{1 - \rho}$$

Expected number of customers in the queue:

$$N_q = \frac{\rho^2}{1 - \rho} = \rho N$$

Expected time in the system:

$$T = \frac{1}{\lambda(1 - \rho)}$$

Expected time in the queue:

$$T_q = \frac{\rho}{\lambda(1 - \rho)} = \rho T$$

REVIEW EXERCISES

1. Find k such that the function

$$f(x) = \begin{cases} kx & \text{if } 1 \leq x \leq 4 \\ 0 & \text{otherwise} \end{cases}$$

satisfies conditions (a) $f(x) \geq 0$ for all x and (b) $\int_{-\infty}^{\infty} f(x)\,dx = 1$ for a probability density function.

2. Consider the function

$$f(x) = \begin{cases} 2(2 + x - x^2)/9 & \text{if } -1 \leq x \leq 2 \\ 0 & \text{otherwise.} \end{cases}$$

Check that f satisfies conditions (a) and (b) for a probability density function, and find $P(1 \leq x \leq 3)$.

3. Find $P(x \geq -1)$ if x has the uniform probability distribution on the interval $[-5, 3]$.

4. Old Faithful Geyser, in Yellowstone National Park, erupts each 65 minutes. A family with restless children who cannot wait for anything for longer than 15 minutes arrives at the geyser at a random time. What is the probability that they will see it erupt?

5. Find $P(x < 3)$ if x has an exponential distribution with $\lambda = 4$.

6. Suppose x has an exponential distribution and $P(x > 2) = 0.3$. Find λ.

7. Find $P(x < 3)$ if x has a Poisson distribution with $\mu = 4$.

8. Suppose that x has a Poisson distribution and $P(x = 0) = 0.8$. Find μ.

9. When Herman puts suet at his bird feeder in the winter, the first chickadee arrives within 2 minutes half the time. Assuming an exponential distribution, what is the probability that no chickadee comes in the first 5 minutes?

10. When the fleet of school buses returns to the depot each afternoon, none need repairs with probability 0.75. Assuming a Poisson distribution, what is the probability that more than one needs repairs?

11. A basketball player makes 30% of her free throws. In practice she shoots free throws until she makes one. Suppose this takes x shots. Find $P(x = k)$ in terms of k. What is the probability that more than 2 shots are needed?

In Exercises 12 and 13 find the expected value, variance, and standard deviation for the random variable with the given density function.

12. $f(x) = \begin{cases} \dfrac{3x^{1/2}}{2} & \text{if } 0 \leq x \leq 1 \\ 0 & \text{otherwise} \end{cases}$

13. $f(x) = \begin{cases} 2.5x^{-3.5} & \text{if } x \geq 1 \\ 0 & \text{otherwise} \end{cases}$

In Exercises 14–16 find the expected value, variance, and standard deviation for the random variable with the given distribution.

14. Uniform on $[1, 5]$

15. Exponential with $\lambda = 6$

16. Poisson with $\mu = 3$

17. The telephone in the mathematics office rings an average of 15 times an hour. What is the probability that it does not ring in a given 5-minute period? Assume a Poisson distribution.

18. A mushroom hunter takes an average of 40 minutes before he finds a morel in the woods. What is the probability that it takes him more than an hour to find a morel? Assume an exponential distribution.

In Exercises 19 and 20 find the expected number of customers in the system and in the queue, and the expected time in the system and in the queue, assuming the single-server queueing system model.

19. Three small planes come into an airport service facility per week, and it takes an average of 1/4 week to service each.

20. Only one child is allowed on the schoolyard slide at a time. An average of 4 children per minute want to use the slide, and each takes an average of 10 seconds on it.

15

MULTIVARIABLE CALCULUS

For the most part, the functions that have been considered in this book have involved only one independent variable. In particular, our discussion of calculus in Chapters 9–14 is limited to such functions.

In this chapter we will generalize differentiation and integration to apply to functions having more than one independent variable. By doing so, we can reconsider some of the topics in Chapters 11 and 12 for such functions.

15.1 Functions of Several Variables

Recall from Example 3.1 in Section 3.1 that Albert's Smoke Shop blends Virginia and Latakia tobacco to make two blends of tobacco, Blend A (Albert's Mixture) and Blend B (Balkan Intrigue). The revenue (in dollars) derived by producing and selling x cans of Blend A and y cans of Blend B is

$$R = 16x + 12y.$$

In this example the revenue R is a function of two independent variables x and y. To denote this functional relationship we write

$$R(x, y) = 16x + 12y.$$

Likewise in Example 3.13 in Section 3.4 we considered a furniture finishing factory that produces desks, buffets, and cocktail tables. Its profit from the sale of x desks, y buffets, and z cocktail tables is

$$P = 10x + 14y + 7z.$$

Here the profit P is a function of three independent variables x, y, and z. We denote this functional relationship by writing

$$P(x, y, z) = 10x + 14y + 7z.$$

More generally, by a **function f of n variables $x_1, x_2, \ldots, x_n$** we mean a rule that assigns to each ordered n-tuple $(x_1, x_2, \ldots, x_n)$ a number denoted

$$f(x_1, x_2, \ldots, x_n).$$

To evaluate a function of n variables at the n-tuple $(a_1, a_2, \ldots, a_n)$, we replace each occurrence of x_1 by a_1, each occurrence of x_2 by a_2, and so forth. Thus for the functions

$$R(x, y) = 16x + 12y \qquad \text{and} \qquad P(x, y, z) = 10x + 14y + 7z$$

mentioned above, we have

$$R(2, 3) = 16(2) + 12(3) = 32 + 36 = 68$$

and

$$P(8, 5, 10) = 10(8) + 14(5) + 7(10) = 80 + 70 + 70 = 220.$$

EXAMPLE 15.1 The IQ (intelligence quotient) of a child is computed by dividing the child's mental age m by the child's chronological age c and multiplying by 100. (a) Write an equation expressing the functional relationship of the IQ to the mental and chronological ages.
(b) Determine the IQ of an 8 year old child who has a mental age of 10.

Solution (a) The functional relationship is given by

$$\text{IQ}(m, c) = \frac{100m}{c}.$$

(b) The child's IQ is

$$\text{IQ}(10, 8) = \frac{100(10)}{8} = 125. \quad \blacksquare$$

In Example 15.1 note that the IQ of a 10-year old with a mental age of 8 is

$$\text{IQ}(8, 10) = 80.$$

Hence $\text{IQ}(10, 8) \neq \text{IQ}(8, 10)$. Thus in evaluating a function of more than one variable, we must be careful to substitute values for the proper variables.

EXAMPLE 15.2 Suppose that a principal P is deposited in an account paying interest at the rate of i per period for n interest periods. (a) Write an equation expressing the amount in the account as a function of the three variables P, i, and n. (b) Determine the amount of an account paying 8% annual interest compounded quarterly if $1000 is deposited for 6 quarters.

Solution (a) According to formula (4.4) in Section 4.2, the amount A of the account is given by

$$A(P, i, n) = P(1 + i)^n.$$

(b) Thus the value of $1000 deposited in an account paying 8% annual interest compounded quarterly for 6 quarters is

$$A(\$1000, 0.02, 6) = \$1000(1.02)^6 \approx \$1126.16. \quad \blacksquare$$

Practice Problem 1 (a) Write a formula expressing the volume V of a rectangular box as a function of its length x, its width y, and its height z. (b) Compute the volume of a box that is 4 feet long, 3 feet wide, and 2 feet high.

By the **graph** of a function $f(x, y)$ of two independent variables we mean the set of all ordered triples (x_0, y_0, z_0) that satisfy the equation

$$z = f(x, y).$$

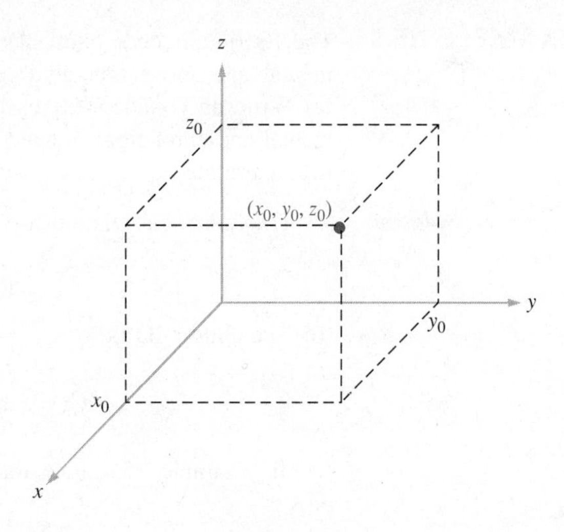

FIGURE 15.1

To draw the graph of such a function, we require a three-dimensional coordinate system having three mutually perpendicular axes labeled x, y, and z. In such a system each point is specified by three coordinates as shown in Figure 15.1. The graph of a function of two variables will be a surface in three-dimensional space. (See Figure 15.2.) For instance, the graph of the function

$$f(x, y) = 3$$

consists of all ordered triples (x_0, y_0, z_0) that satisfy $z_0 = 3$; these triples form a plane 3 units above the x, y plane (the plane containing the x and y axes). The graph of $f(x, y) = 3$ is shown in Figure 15.3.

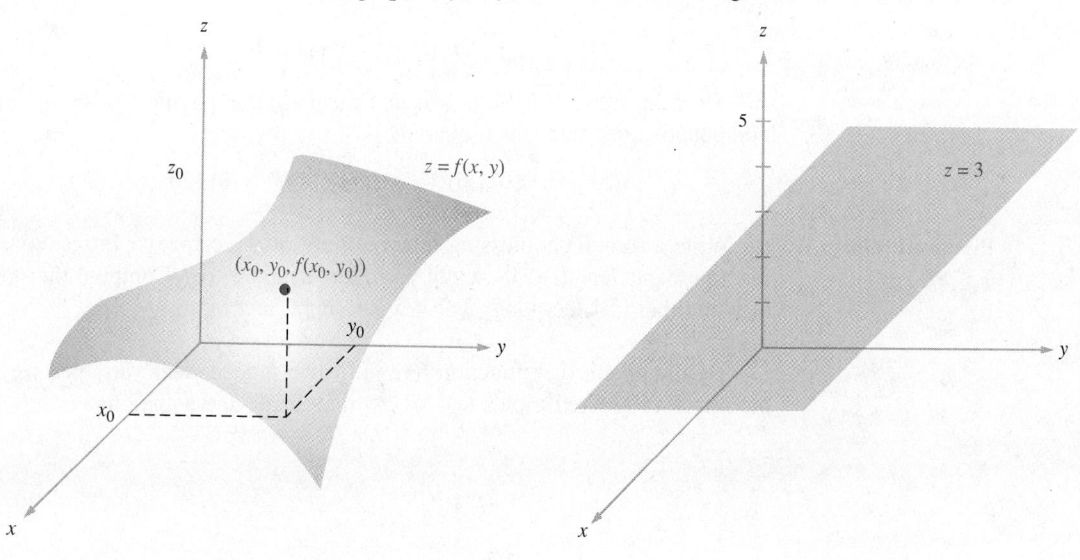

FIGURE 15.2

FIGURE 15.3

Unfortunately, graphing a function of two variables is often quite difficult because of the need to represent three-dimensional space in a two-dimensional picture. Moreover, it is impossible to graph functions of three or more variables in a three-dimensional coordinate system. For these reasons we will not emphasize the graphing of functions of more than one variable. (We will, however, show the graphs of such functions when a figure can help to clarify a concept under consideration.)

PRODUCTION FUNCTIONS

Throughout this chapter we will consider a particular function of several variables called a *production function,* which occurs frequently in economics. In the remainder of this section we will discuss the significance of such a function and introduce an important example.

A manufacturer's total production of a particular item depends on such inputs as labor, capital, machinery, and land. A **production function** is a function that expresses the relationship between the production inputs and the quantity of the item produced. For simplicity we will consider only two inputs, labor and capital. Thus we consider as capital not only the money needed to finance production, but also the equipment or land necessary for production. To some extent a manufacturer is able to control the proportions of labor and capital that are used by the production process. For example, by automating production a manufacturer can reduce the need for labor and increase the need for capital (in the form of machinery). On the other hand, a manufacturer who refuses to automate production will require more labor and less capital.

A **Cobb-Douglas production function** has the form

$$f(x, y) = ax^{\lambda}y^{1-\lambda},$$

where x and y denote the number of units of labor and capital, respectively, and a and λ are constants. This type of function is named after the American economists C. W. Cobb and Paul H. Douglas, who conjectured that the total industrial output of the American economy could be described by such a function. Surprisingly enough, it has been shown that this simple function agrees closely with historical data when $\lambda = 0.75$.

EXAMPLE 15.3

A California farmer finds that the number of heads of lettuce he produces can be described by the Cobb-Douglas production function

$$f(x, y) = 1000x^{0.25}y^{0.75},$$

where x denotes the number of laborers hired to harvest the lettuce and y denotes the number of units of capital used. (Here the capital includes the farmer's expenses for land, farm equipment, and seed.) Determine the number of heads produced when 8 laborers and 648 units of capital are used.

Solution The number of heads produced is

$$\begin{aligned} f(8, 648) &= 1000(8)^{0.25}(648)^{0.75} \\ &= 216{,}000. \quad \blacksquare \end{aligned}$$

Practice Problem 2 For the Cobb-Douglas production function in Example 15.3, determine the number of heads produced when 16 laborers and 1296 units of capital are used.

In Practice Problem 2 the units of labor and capital are twice those in Example 15.3. It turns out that the number of heads produced in the practice problem is also twice the number produced in Example 15.3. One important characteristic of a Cobb-Douglas production function is that if all the production inputs are multiplied by a factor k, then the output is also multiplied by k. Symbolically this property can be written

$$f(kx, ky) = kf(x, y).$$

Economists describe this property by saying that f has *constant returns to scale*. The following calculation verifies that a Cobb-Douglas production function satisfies this condition.

$$\begin{aligned} f(kx, ky) &= a(kx)^{\lambda}(ky)^{1-\lambda} \\ &= a(k^{\lambda}x^{\lambda})(k^{1-\lambda}y^{1-\lambda}) \\ &= a(k^{\lambda}k^{1-\lambda})(x^{\lambda}y^{1-\lambda}) \\ &= ak(x^{\lambda}y^{1-\lambda}) \\ &= k(ax^{\lambda}y^{1-\lambda}) \\ &= kf(x, y) \end{aligned}$$

In later sections we will examine some other properties of Cobb-Douglas production functions.

EXERCISES 15.1

In Exercises 1–6 evaluate the given function f at the indicated point (a, b).

1. $f(x, y) = x\sqrt{y}, \quad (a, b) = (-2, 9)$

2. $f(x, y) = 2ye^{x}, \quad (a, b) = (0, 5)$

3. $f(x, y) = 4x^2 - 9y^2, \quad (a, b) = (-5, 2)$

4. $f(x, y) = 8x^2 + y^2, \quad (a, b) = (-2, 6)$

5. $f(x, y) = (x^2 - 2y^2)e^{2x-y}, \quad (a, b) = (3, 6)$

6. $f(x, y) = 3x^2 - 4xy + 2y^2, \quad (a, b) = (-3, 1)$

In Exercises 7–12 evaluate the given function g at the indicated point (a, b, c).

7. $g(x, y, z) = 2xy + 2xz + yz, \quad (a, b, c) = (10, 20, 30)$

8. $g(x, y, z) = x^2 + y^2 + z^2, \quad (a, b, c) = (-1, 3, -2)$

9. $g(x, y, z) = z^2 - 4x^2 - 9y^2, \quad (a, b, c) = (10, 4, 2)$

10. $g(x, y, z) = x^2 \ln \left|\frac{y}{z}\right|, \quad (a, b, c) = (7, -2, 2)$

11. $g(x, y, z) = y \ln x - 3z^2, \quad (a, b, c) = (1, -5, 2)$

12. $g(x, y, z) = y^2\sqrt{x^2 + z^2}, \quad (a, b, c) = (3, 6, 4)$

13. Determine the total production resulting from 81 units of labor and 256 units of capital if the production function is $f(x, y) = 50x^{0.75}y^{0.25}$.

14. The surface area of a cylinder with a height of h and a base of radius r is given by $S(h, r) = 2\pi rh$. Determine the surface area of a cylinder with height 6 inches and base of radius 10 inches.

15. Use Example 15.1 to determine the IQ of a child aged 11.5 who has a mental age of 13.8.

16. Determine the volume of a cylinder with height 8 inches and base of radius 6 inches.

17. The empirical formula

$$S(w, h) = 0.007w^{0.425}h^{0.725}$$

estimates the surface area S in square meters of the body of a person whose weight is w kilograms and whose height is h centimeters. Estimate the surface area of a man who is 175 centimeters tall and weighs 68 kilograms.

18. The resistance R to blood flowing through a blood vessel of length l and radius r is described by a function of the form

$$R(l, r) = \frac{kl}{r^4}$$

where k is a constant that depends on the viscosity of the blood. (This result is called Poiseuille's law.) Evaluate (in terms of k) the resistance to blood flowing through an arteriole of length 2 centimeters and radius 0.01 centimeters.

19. Use Example 15.2 to compute the value of a principal of $600 deposited for 2 years in an account that compounds interest quarterly at an annual rate of 5%.

20. Use Example 15.2 to compute the value of a principal of $8000 deposited for 2½ years in an account that compounds interest monthly at an annual rate of 6%.

21. Use formula (4.3) in Section 4.1 to determine the amount due after 1½ years if $1000 is borrowed at 12% simple interest per year.

22. Use formula (4.3) in Section 4.1 to determine the amount due after 4 years if $600 is borrowed at 15% simple interest per year.

23. In reference [4], Weinstein and Srinivasan studied the salaries of 136 graduates of M.B.A. programs. They determined that in 1969 the salary (in dollars) of persons such as accountants and analysts whose jobs required no formal managerial responsibility could be predicted by the formula

$$S(x, y) = 1120x + 873y + 10{,}990,$$

where x and y denote the respective number of years of work experience before and after receiving the M.B.A degree. Predict the 1969 salary of an accountant with an M.B.A. degree who has 4 years of prior work experience and 5 years of work experience after receiving this degree.

24. The distance from a point with coordinates (x, y, z) to the origin $(0, 0, 0)$ is

$$d(x, y, z) = \sqrt{x^2 + y^2 + z^2}.$$

Determine the distance from $(2, -1, -2)$ to the origin.

25. According to reference [1], a sound emitted from a source in the atmosphere can be heard for

$$D(s, h, v) = 2\sqrt{\frac{sh}{v}}$$

kilometers, where s denotes the temperature at the surface of the earth (in degrees Kelvin), h denotes the altitude of the source (in kilometers), and v denotes the vertical temperature gradient (in degrees Kelvin per kilometer). How far can a sonic boom caused by a jet be heard if the jet is traveling at an altitude of 12 kilometers, the surface temperature is 288° K, and the vertical temperature gradient is 6° K per kilometer?

Answers to Practice Problems

1. (a) $V(x, y, z) = xyz$ (b) $V(4, 3, 2) = 24$ cubic feet

2. 432,000

15.2 Partial Derivatives

In Chapter 9 we defined the derivative of a function of one variable and saw that it measured the instantaneous rate of change of the function with respect to the independent variable. An analogous concept for functions of several variables will be introduced in this section.

Let us begin by considering a function $f(x, y)$ of two variables. It is often useful to know the instantaneous rate at which f is changing with respect to one of the independent variables when the other is held constant. For instance, suppose that f is a production function and the variables x and y denote units of labor and capital, respectively. The instantaneous rate at which production changes per unit change in labor when capital is constant is a useful economic measure. This measure approximates the change in production that occurs if an additional unit of labor is added and capital remains fixed.

If y is held constant, then $f(x, y)$ can be regarded as a function of x alone. In this case the derivative of f with respect to x is called the **partial derivative of f with respect to x** and is denoted $f_x(x, y)$. Likewise the **partial derivative of f with respect to y,** denoted $f_y(x, y)$, is defined to be the derivative of f with respect to y when x is held constant. Symbolically we have

$$f_x(x, y) = \lim_{h \to 0} \frac{f(x + h, y) - f(x, y)}{h}$$

and

$$f_y(x, y) = \lim_{h \to 0} \frac{f(x, y + h) - f(x, y)}{h}$$

provided that these limits exist. The process of computing a partial derivative is called **partial differentiation.**

Note the similarity between the definitions of the partial derivatives and the definition of the derivative of a function $f(x)$ of one variable:

$$f'(x) = \lim_{h \to 0} \frac{f(x + h) - f(x)}{h}.$$

Because of this similarity we can compute partial derivatives by a simple modification of the methods discussed in Chapters 9 and 10.

(a) To compute $f_x(x, y)$, differentiate f as a function of x while regarding y as a constant.

(b) To compute $f_y(x, y)$, differentiate f as a function of y while regarding x as a constant.

EXAMPLE 15.4 Compute f_x and f_y if $f(x, y) = 7x^3 + 4x^2y - 5y^2$.

Solution To compute f_x, we regard f as a function of x alone and treat y as a constant. Using the familiar differentiation rules from Chapter 9, we obtain

$$\begin{aligned} f_x(x, y) &= (7x^3 + 4x^2y - 5y^2)_x \\ &= (7x^3)_x + (4x^2y)_x - (5y^2)_x \\ &= 7(x^3)_x + 4y(x^2)_x - 5y^2(1)_x \\ &= 7(3x^2) + 4y(2x) - 5y^2(0) \\ &= 21x^2 + 8xy. \end{aligned}$$

To compute f_y, we proceed similarly, regarding f as a function of y alone and treating x as a constant. The result is

$$\begin{aligned} f_y(x, y) &= (7x^3 + 4x^2y - 5y^2)_y \\ &= (7x^3)_y + (4x^2y)_y - (5y^2)_y \\ &= 7x^3(1)_y + 4x^2(y)_y - 5(y^2)_y \\ &= 7x^3(0) + 4x^2(1) - 5(2y) \\ &= 4x^2 - 10y. \quad \blacksquare \end{aligned}$$

Practice Problem 3 Compute both partial derivatives of $f(x, y) = 32x - y^2 + 2x^2y$.

EXAMPLE 15.5 Compute both partial derivatives of $f(x, y) = 3ye^{2x}$.

Solution By regarding y as a constant, we see that

$$\begin{aligned} f_x(x, y) &= (3ye^{2x})_x \\ &= 3y(e^{2x})_x \\ &= 3y(2e^{2x}) \\ &= 6ye^{2x}. \end{aligned}$$

Similarly, by regarding x as a constant, we see that

$$\begin{aligned} f_y(x, y) &= (3ye^{2x})_y \\ &= (3e^{2x})(y)_y \\ &= 3e^{2x}(1) \\ &= 3e^{2x}. \quad \blacksquare \end{aligned}$$

Practice Problem 4 Compute both partial derivatives of the Cobb-Douglas production function

$$f(x, y) = ax^{\lambda}y^{1-\lambda}.$$

Although the basic idea behind partial differentiation is simple, it is nevertheless easy to make mistakes when computing partial derivatives. Carefully study the examples below, making sure that each step is understood.

EXAMPLE 15.6 Compute both partial derivatives of $f(x, y) = (x^2 - 3xy + y^3)^8$.

Solution Using the chain rule, we see that

$$\begin{aligned} f_x(x, y) &= [(x^2 - 3xy + y^3)^8]_x \\ &= 8(x^2 - 3xy + y^3)^7(x^2 - 3xy + y^3)_x \\ &= 8(x^2 - 3xy + y^3)^7(2x - 3y) \\ &= (16x - 24y)(x^2 - 3xy + y^3)^7. \end{aligned}$$

Likewise

$$\begin{aligned} f_y(x, y) &= [(x^2 - 3xy + y^3)^8]_y \\ &= 8(x^2 - 3xy + y^3)^7(x^2 - 3xy + y^3)_y \\ &= 8(x^2 - 3xy + y^3)^7(-3x + 3y^2) \\ &= (-24x + 24y^2)(x^2 - 3xy + y^3)^7. \quad \blacksquare \end{aligned}$$

EXAMPLE 15.7 Compute both partial derivatives of $f(x, y) = \dfrac{x^2y}{6x - y^2}$.

Solution Using the quotient rule, we see that

$$\begin{aligned} f_x(x, y) &= \left[\frac{x^2y}{6x - y^2}\right]_x \\ &= \frac{(6x - y^2)(x^2y)_x - x^2y(6x - y^2)_x}{(6x - y^2)^2} \\ &= \frac{(6x - y^2)(2xy) - x^2y(6)}{(6x - y^2)^2} \\ &= \frac{12x^2y - 2xy^3 - 6x^2y}{(6x - y^2)^2} \\ &= \frac{6x^2y - 2xy^3}{(6x - y^2)^2}. \end{aligned}$$

Similarly

$$\begin{aligned} f_y(x, y) &= \left[\frac{x^2y}{6x - y^2}\right]_y \\ &= \frac{(6x - y^2)(x^2y)_y - x^2y(6x - y^2)_y}{(6x - y^2)^2} \\ &= \frac{(6x - y^2)(x^2) - x^2y(-2y)}{(6x - y^2)^2} \\ &= \frac{6x^3 - x^2y^2 + 2x^2y^2}{(6x - y^2)^2} \\ &= \frac{6x^3 + x^2y^2}{(6x - y^2)^2}. \end{aligned}$$ ■

Practice Problem 5 Compute both partial derivatives of $f(x, y) = xe^{xy}$.

EXAMPLE 15.8 Compute $f_x(1, 2)$ and $f_y(3, -1)$ if $f(x, y) = \ln(x + y^2)$.

Solution To compute $f_x(1, 2)$ and $f_y(3, -1)$, we must begin by determining $f_x(x, y)$ and $f_y(x, y)$. As above

$$f_x(x, y) = \frac{1}{x + y^2}(x + y^2)_x = \frac{1}{x + y^2}$$

and

$$f_y(x, y) = \frac{1}{x + y^2}(x + y^2)_y = \frac{2y}{x + y^2}.$$

Hence

$$f_x(1, 2) = \frac{1}{1 + (2)^2} = \frac{1}{5}$$

and

$$f_y(3, -1) = \frac{2(-1)}{3 + (-1)^2} = \frac{-2}{4} = -\frac{1}{2}. \quad ■$$

Practice Problem 6 Compute $f_x(-2, 1)$ and $f_y(3, -4)$ if $f(x, y) = 10x - 3y^2 + 4x^2y$.

Recall that if g is a function of one variable, then $g'(x_0)$ is the slope of the tangent line to the graph of $y = g(x)$ at $x = x_0$. Similar geometric interpretations can be given to $f_x(x_0, y_0)$ and $f_y(x_0, y_0)$. As we saw in Section 15.1, the graph of $z = f(x, y)$ is a surface in three-dimensional space. If y is held constant at y_0, the graph of $z = f(x, y_0)$ is a curve lying on this surface. In fact, it is the curve formed by the intersection of the surface $z = f(x, y)$ and the plane $y = y_0$. (See Figure 15.4(a).) Then $f_x(x_0, y_0)$ is the slope of the tangent to this curve at (x_0, y_0). Likewise $f_y(x_0, y_0)$ is the slope of the tangent line to the curve formed by the intersection of the surface $z = f(x, y)$ and the plane $x = x_0$. (See Figure 15.4(b).) Because of these interpretations, $f_x(x_0, y_0)$ and $f_y(x_0, y_0)$ are often called the slopes to the surface $z = f(x, y)$ at (x_0, y_0) in the x and y directions.

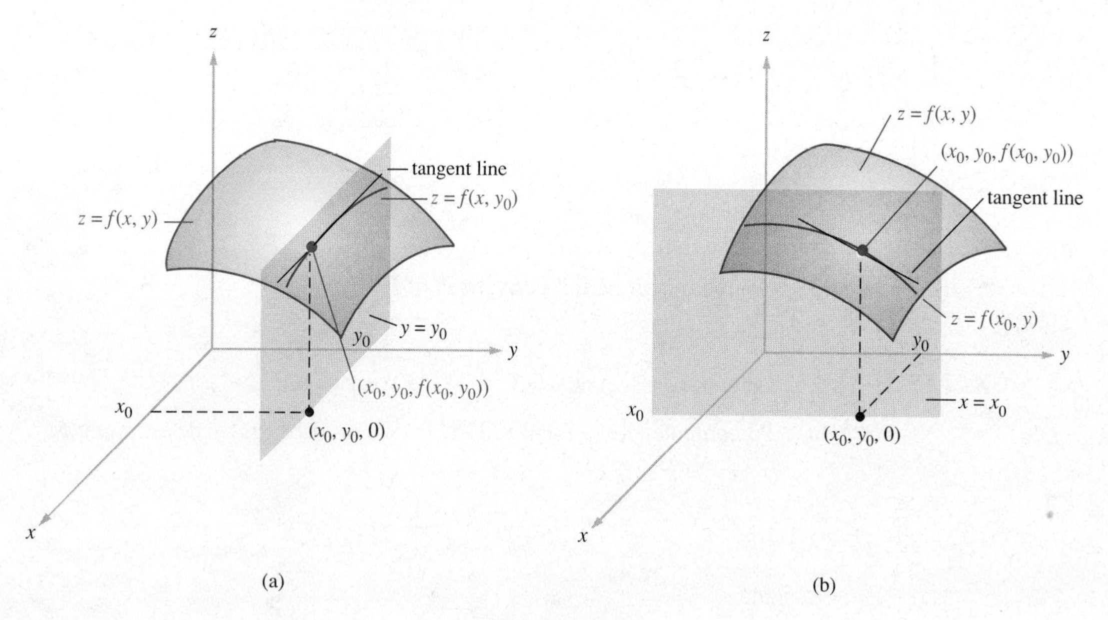

FIGURE 15.4

Partial derivatives can also be interpreted as instantaneous rates of change. The instantaneous rate of change of f at (x_0, y_0) in the x direction when y is held constant is $f_x(x_0, y_0)$, and the instantaneous rate of change of f at (x_0, y_0) in the y direction when x is held constant is $f_y(x_0, y_0)$. As for functions of one variable, partial derivatives approximate the change in f per unit change in one of the variables when the other is held constant. For instance, $f_x(3, 5)$ approximates the change in f when (x, y) changes from (3, 5) to (4, 5), and $f_y(6, 2)$ approximates the change in f when (x, y) changes from (6, 2) to (6, 3).

EXAMPLE 15.9 The *ideal gas law* states that the pressure P exerted by a gas is related to its absolute temperature T and its volume V by the equation $P(T, V) = kT/V$, where k is a constant. If P is measured in pounds per square inch and $k = \frac{1}{2}$, find the instantaneous rate of change of pressure with respect to volume when the volume is 10 cubic inches and the temperature remains constant at 20° K.

Solution The instantaneous rate of change of pressure with respect to volume when the volume is 10 cubic inches and the temperature remains constant at 20° K is $P_V(20, 10)$. Now

$$P(T, V) = \frac{1}{2}\frac{T}{V} = \frac{1}{2}TV^{-1},$$

and so

$$P_V(T, V) = -\frac{1}{2}TV^{-2}.$$

Hence

$$P_V(20, 10) = -\frac{1}{2}(20)(10)^{-2} = -\frac{10}{100} = -0.1$$

pounds per square inch per cubic inch. ■

Practice Problem 7 The volume V of a cylinder depends on r, the radius of its base, and h, its height, according to the formula $V(r, h) = \pi r^2 h$. Determine the instantaneous rate of change of the volume with respect to the radius when the radius is 5 inches and the height remains constant at 8 inches.

For a Cobb-Douglas production function

$$f(x, y) = ax^{\lambda}y^{1-\lambda},$$

the instantaneous rates of change in the x and y directions are

$$f_x(x, y) = a\lambda x^{\lambda-1}y^{1-\lambda} \quad \text{and} \quad f_y(x, y) = a(1 - \lambda)x^{\lambda}y^{-\lambda},$$

respectively. The former may be interpreted as the instantaneous rate of change of production with respect to labor when capital is held constant, and the latter may be interpreted as the instantaneous rate of change of production with respect to capital when labor is held constant. Economists refer to these rates as the *marginal productivity of labor* and the *marginal productivity of capital*, respectively.

EXAMPLE 15.10 Show that for a Cobb-Douglas production function the total output equals the quantity of labor times the marginal productivity of labor plus the quantity of capital times the marginal productivity of capital.

Solution The result to be shown can be written symbolically as

$$f(x, y) = xf_x(x, y) + yf_y(x, y).$$

To verify this equation, we need only use the expressions above for f_x and f_y. Then

$$\begin{aligned} xf_x(x, y) + yf_y(x, y) &= x[a\lambda x^{\lambda-1}y^{1-\lambda}] + y[a(1 - \lambda)x^{\lambda}y^{-\lambda}] \\ &= a\lambda x^{\lambda}y^{1-\lambda} + a(1 - \lambda)x^{\lambda}y^{1-\lambda} \\ &= ax^{\lambda}y^{1-\lambda}[\lambda + (1 - \lambda)] \\ &= ax^{\lambda}y^{1-\lambda}(1) \\ &= f(x, y). \quad \blacksquare \end{aligned}$$

In this section we have so far considered only functions of two variables. The concept of a partial derivative, however, can be easily extended to functions of more than two variables. If

$$g(x_1, x_2, \ldots, x_n)$$

is a function of n variables, then the **partial derivative of g with respect to x_k**, denoted g_{x_k}, is obtained by differentiating g with respect to x_k while regarding all the other $n - 1$ variables as constants. As for functions of two variables, we can interpret g_{x_k} as the instantaneous rate of change of g with respect to x_k when the other variables are held constant.

EXAMPLE 15.11 Recall that the amount A of a principal P compounded continuously for t years at an annual interest rate r is $A = Pe^{rt}$. Compute all the partial derivatives of A.

Solution The amount A is a function of the three variables P, r, and t. To compute the partial derivative of A with respect to P, we treat both r and t as constants. Thus

$$A_P = e^{rt}\frac{d}{dP}(P) = e^{rt}.$$

Likewise, to compute the partial derivative of A with respect to r, we treat both P and t as constants. Therefore

$$A_r = P\frac{d}{dr}(e^{rt}) = P(te^{rt}) = Pte^{rt}.$$

Similarly

$$A_t = P\frac{d}{dt}(e^{rt}) = P(re^{rt}) = Pre^{rt}. \quad ■$$

Practice Problem 8 The enthalpy H of a thermodynamic system is defined by the formula $H = U + PV$, where U is the internal energy of the system, P is its pressure, and V is its volume. Compute all the partial derivatives of H.

HIGHER-ORDER PARTIAL DERIVATIVES

Just as it is possible to compute higher-order derivatives of a function of one variable, it is possible to compute higher-order partial derivatives of a function of several variables. In Section 15.3 we will see an important use of the second-order partial derivatives of a function of two variables. Since in this book we have no need for partial derivatives beyond the second order, we will confine our discussion to this case. Moreover, we will restrict ourselves to functions of two variables. The generalization to higher-order partial derivatives for functions of more than two variables is obvious, however.

If f is a function of two variables x and y, then so are its first-order partial derivatives f_x and f_y. Consequently it is possible to take partial derivatives of these functions with respect to x and y. The partial derivative of f_x with respect to x is denoted f_{xx}, and the partial derivative of f_x with respect to y is denoted f_{xy}. Likewise the partial derivative of f_y with respect to x is denoted f_{yx}, and the partial derivative of f_y with respect to y is denoted f_{yy}. Notice that

$$f_{xy} \quad \text{means} \quad (f_x)_y$$

and

$$f_{yx} \quad \text{means} \quad (f_y)_x.$$

The four partial derivatives f_{xx}, f_{xy}, f_{yx}, and f_{yy} are called **second-order** partial derivatives.

EXAMPLE 15.12 Compute the four second-order partial derivatives of $f(x, y) = 3x^2e^{-y}$.

Solution We must begin by computing the first-order partial derivatives f_x and f_y. As above, we find that

$$f_x(x, y) = 6xe^{-y} \quad \text{and} \quad f_y(x, y) = -3x^2e^{-y}.$$

Since $f_{xx} = (f_x)_x$ and $f_{xy} = (f_x)_y$, we see that

$$f_{xx}(x, y) = 6e^{-y} \quad \text{and} \quad f_{xy}(x, y) = -6xe^{-y}.$$

Likewise $f_{yx} = (f_y)_x$ and $f_{yy} = (f_y)_y$, and so

$$f_{yx}(x, y) = -6xe^{-y} \quad \text{and} \quad f_{yy}(x, y) = 3x^2e^{-y}. \quad ■$$

Notice that in Example 15.12 the "mixed partials" f_{xy} and f_{yx} are equal. This occurrence is not a coincidence: For most of the functions encountered in applications it is true that $f_{xy} = f_{yx}$.

Practice Problem 9 Compute the second-order partial derivatives of $f(x, y) = 10x - 3y^2 + 4x^2y$.

Now let us use partial derivatives to analyze the demand for competing products. Suppose that a popular college textbook, which is available in both a paperback and a more expensive hardcover edition, is expected to sell 12,600 copies during the coming year. Let x and y denote the selling prices (in dollars) of the paperback and hardcover editions, respectively. At these prices the number of paperback copies that can be sold is

$$900 - 100x + 200y,$$

and the number of hardcover copies that can be sold is

$$1200 + 300x - 100y.$$

In this situation we know the demand for the paperback and hardcover editions as functions of their selling prices. Since the yearly demand for the textbook is fixed, increased sales of either edition will result in decreased sales of the other.

Consider the demand function for the paperback edition

$$D(x, y) = 900 - 100x + 200y.$$

The first-order partial derivatives of D are

$$D_x(x, y) = -100 \quad \text{and} \quad D_y(x, y) = 200.$$

We have seen that $D_x(x, y)$ approximates the change in demand for the paperback edition when its price is increased by one dollar and the price of the hardcover edition is held constant. Thus increasing the price of the paperback edition and holding constant the price of the hardcover edition will *decrease* the demand for the paperback edition by about approximately 100 copies. This decrease results from the fact that more people will prefer the hardcover edition if the paperback price is raised. On the other hand, $D_y(x, y)$ approximates the change in demand for the paperback edition when the price of the hardcover edition is increased by one dollar and the price of the paperback edition is held constant. Hence increasing the price of the hardcover edition by one dollar and leaving the paperback price unchanged will *increase* demand for the paperback by 200 copies.

A similar analysis holds for the demand function

$$1200 + 300x - 100y$$

for the hardcover edition. Increasing the price of the hardcover edition by one dollar and leaving the paperback price unchanged will *decrease* demand for the hardcover edition by 100 copies, whereas increasing the price of the paperback edition by one dollar and holding constant the price of the hardcover edition will *increase* the demand for the hardcover edition by 300 copies.

When price changes for two products affect the demand for the products in the manner described above, the products are called **competing products.** Butter and margarine are another example of competing products.

In Chapter 9 we saw that there are several common notations for the derivative of a function of one variable. The same is true about the partial derivatives of a function of several variables. Alternate notations for the first-order and second-order partial derivatives of $z = f(x, y)$ are listed below.

First-order partial derivatives

with respect to x	f_x	$\frac{\partial f}{\partial x}$	$\frac{\partial z}{\partial x}$
with respect to y	f_y	$\frac{\partial f}{\partial y}$	$\frac{\partial z}{\partial y}$

Second-order partial derivatives

with respect to x and then x	f_{xx}	$\frac{\partial^2 f}{\partial x^2}$	$\frac{\partial^2 z}{\partial x^2}$
with respect to x and then y	f_{xy}	$\frac{\partial^2 f}{\partial y \partial x}$	$\frac{\partial^2 z}{\partial y \partial x}$
with respect to y and then x	f_{yx}	$\frac{\partial^2 f}{\partial x \partial y}$	$\frac{\partial^2 z}{\partial x \partial y}$
with respect to y and then y	f_{yy}	$\frac{\partial^2 f}{\partial y^2}$	$\frac{\partial^2 z}{\partial y^2}$

EXERCISES 15.2

Compute all the first-order partial derivatives of the functions in Exercises 1–18.

1. $f(x, y) = 2x^2 + 3y^2 + 9$
2. $g(u, v) = 12uv^2 + 4$
3. $h(r, s) = \frac{r}{s}$
4. $F(x, y) = 4x^3(y^2 - 6y)$
5. $G(u, v) = 3v^2e^{-u}$
6. $H(r, s) = s^5 \ln (1 + r^2)$
7. $g(x, y) = \frac{x^2y}{3x + 5y}$
8. $f(u, v) = (3v + 4u^3)e^{-uv}$
9. $F(r, s) = (r^2 + 3rs - 8s^2)^5$
10. $h(x, y) = \sqrt{9x^2 - 3xy + 4y^2}$
11. $H(u, v) = (6u - 7v^3)(u^2 + 4v^2)$
12. $G(r, s) = \frac{5r^2s^4}{2s + 3r}$
13. $f(x, y, z) = 2xyz^2 + \ln (x^2 + 3y^2)$
14. $g(u, v, w) = 4w^3 - 3uv^2 + u^4w^2$
15. $h(r, s, t) = (r + 2s)e^{2r+3st}$
16. $F(x, y, z) = (9 - z^2)\sqrt{x^2 + y^2}$
17. $G(u, v, w) = \frac{v^2}{3u - 2w}$
18. $H(r, s, t) = \sqrt{rs^3t^2e^{-st}}$

Compute all the second-order partial derivatives of the functions in Exercises 19–24.

19. $f(r, s) = 5r^2s - 3r^3s^6 + 8$

20. $F(u, v) = 10uv^3 + 6v^2 - 8u^2 + 16$

21. $h(u, v) = 2u^3e^{v-u}$

22. $H(x, y) = 4x^2y^3 - \ln \dfrac{y}{x}$

23. $G(x, y) = \ln x^2y$

24. $g(r, s) = (r^2 - 9s)e^{-s}$

25. Suppose that for some commodity the total production resulting from the use of x units of labor and y units of capital is given by the Cobb-Douglas production function

$$f(x, y) = 50x^{0.5}y^{0.5}.$$

(a) Calculate the marginal productivities of labor and capital when 25 units of labor and 100 units of capital are used.

(b) Use your answer to part (a) to approximate the effect on production of keeping labor at 25 units and increasing capital to 101 units.

(c) Use your answer to part (a) to approximate the effect on production of keeping capital at 100 units and decreasing labor to 24 units.

26. Suppose that for some commodity the total production resulting from the use of x units of labor and y units of capital is given by the Cobb-Douglas production function

$$f(x, y) = 64x^{0.75}y^{0.25}.$$

(a) Calculate the marginal productivities of labor and capital when 81 units of labor and 16 units of capital are used.

(b) Use your answer to part (a) to approximate the effect on production of keeping capital at 16 units and increasing labor to 82 units.

(c) Use your answer to part (a) to approximate the effect on production of keeping labor at 81 units and decreasing capital to 15 units.

27. Let x and y denote the average selling prices (in thousands of dollars) of houses and condominiums in a certain city. Suppose that the yearly demand for houses in this city is given by $D(x, y) = 2000 - 0.04x^2 + 0.02y^2$. Compute $D_x(150, 100)$ and $D_y(150, 100)$ and interpret these values.

28. Let x and y denote the average selling prices (in thousands of dollars) of houses and condominiums in a certain city. Suppose that the yearly demand for condominiums in this city is given by $D(x, y) = 800 + 0.03x^2 - 0.05y^2$. Compute $D_x(200, 110)$ and $D_y(200, 110)$, and interpret these values.

The yearly demand for condominiums is $D(x, y) = 800 + 0.03x^2 - 0.05y^2$ where x and y denote the average selling prices of houses and condominiums (in thousands of dollars), respectively.

29. A company manufactures two models of telephones. Each unit of the basic model sells for \$20, and each unit of the deluxe model sells for \$30. The cost of producing x thousand basic phones and y thousand deluxe phones is $C(x, y) = 10x + 16y + 0.2xy$ thousand dollars.

(a) Determine $P(x, y)$, the company's profit when x thousand basic phones and y thousand deluxe phones are produced and sold.

(b) Compute $P_x(18, 12)$ and $P_y(18, 12)$, and interpret these values.

30. A health food store sells x pounds of regular coffee per day at a price of \$2 per pound and y pounds of decaffeinated coffee at a price of \$2.40 per pound. Its daily cost of preparing x pounds of regular coffee and y pounds of decaffeinated coffee is $C(x, y) = 1.50x + 1.80y + 0.01xy$.

(a) Determine $P(x, y)$, the store's profit when x pounds of regular coffee and y pounds of decaffeinated coffee are prepared and sold.

(b) Compute $P_x(100, 200)$ and $P_y(100, 200)$, and interpret these values.

31. Suppose that two manufacturers are the only suppliers of a particular product. Let $f(x, y)$ denote the number of units of this product that the first manufacturer can sell when his product is priced at price x and his competitor's product is priced at price y. If the two manufacturers' current selling prices are p and q, respectively, explain why $f_x(p, q) < 0$ and $f_y(p, q) > 0$.

Answers to Practice Problems

3. $f_x(x, y) = 32 + 4xy$ and $f_y(x, y) = -2y + 2x^2$

4. $f_x(x, y) = a\lambda x^{\lambda-1}y^{1-\lambda}$ and $f_y(x, y) = a(1 - \lambda)x^{\lambda}y^{-\lambda}$

5. $f_x(x, y) = e^{xy} + xye^{xy}$ and $f_y(x, y) = x^2e^{xy}$

6. $f_x(-2, 1) = -6$ and $f_y(3, -4) = 60$

7. $V_r(5, 8) = 80\pi$ cubic inches per inch

8. $H_U = 1$, $H_P = V$, and $H_V = P$

9. $f_{xx}(x, y) = 8y$, $f_{xy}(x, y) = 8x$, $f_{yx}(x, y) = 8x$, and $f_{yy}(x, y) = -6$

15.3 Local Extrema

In Section 11.3 we discussed the second-derivative test for determining local extrema of a function of one variable. In this section a similar result will be presented for determining local extrema of a function of two variables.

Let $f(x, y)$ be a function that is defined in a region containing the point (a, b). The functional value $f(a, b)$ is called a **local minimum** if

$$f(a, b) \leq f(x, y)$$

for all points (x, y) in some circular disk surrounding (a, b). Likewise $f(a, b)$ is called a **local maximum** if

$$f(a, b) \geq f(x, y)$$

for all points (x, y) in some circular disk surrounding (a, b). Figure 15.5 illustrates these definitions. As before, a **local extremum** of f is a value that is

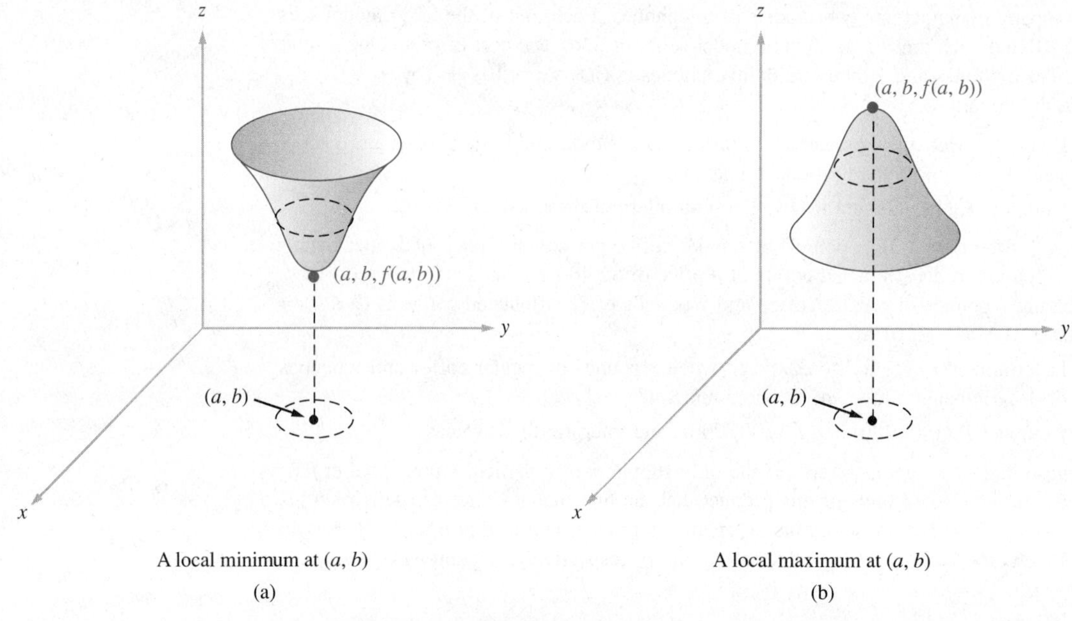

A local minimum at (a, b)
(a)

A local maximum at (a, b)
(b)

FIGURE 15.5

either a local minimum or a local maximum. Again a local extremum is a value of the function f, whereas we say that a local extremum *occurs* at a point.

In Section 11.2 we saw that if a function f of one variable has a local extremum at $x = a$, then either $f'(a)$ is undefined or $f'(a) = 0$. An analogous result is true for functions of two variables.

Theorem 15.1

If f has a local extremum at (a, b) and f_x and f_y are both defined at (a, b), then

$$f_x(a, b) = 0 \quad \text{and} \quad f_y(a, b) = 0.$$

A point (a, b) such that

$$f_x(a, b) = 0 \quad \text{and} \quad f_y(a, b) = 0$$

is called a **critical point** of f. Just as in the one-variable case, the fact that (a, b) is a critical point does *not* guarantee that $f(a, b)$ is a local extremum. In fact, just as the function $f(x) = x^3$ has a zero derivative at $x = 0$ yet $f(0)$ is not a local extremum, so the function $f(x, y) = x^3$ has zero partial derivatives at $(0, 0)$ yet $f(0, 0)$ is not a local extremum. (See Figure 15.6.) Thus the critical points of f are merely points at which local extrema *may* occur.

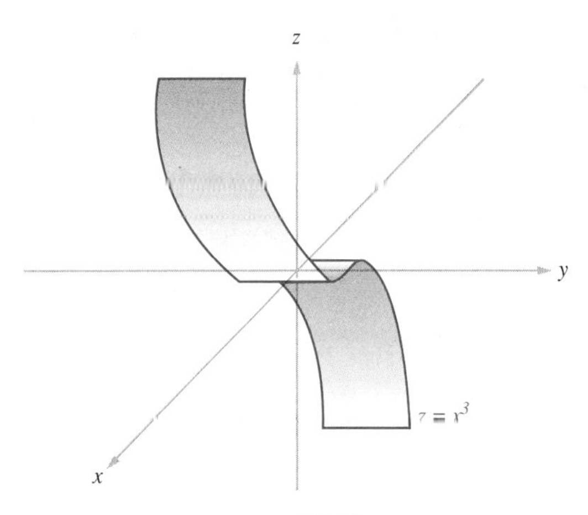

FIGURE 15.6

EXAMPLE 15.13 Determine all the critical points of

$$f(x, y) = -x^2 + 6x - 2y^2 - 8y + 1.$$

Solution The first-order partial derivatives of f are

$$f_x(x, y) = -2x + 6 \quad \text{and} \quad f_y(x, y) = -4y - 8.$$

Since critical points of f occur where both f_x and f_y equal zero, we must have

$$\begin{aligned} -2x + 6 &= 0 \quad &\text{and} \quad -4y - 8 &= 0 \\ -2x &= -6 & -4y &= 8 \\ x &= 3 & y &= -2. \end{aligned}$$

Therefore $(3, -2)$ is the only critical point of f, and so if f has a local extremum, it must occur at $(3, -2)$. ■

Practice Problem 10 Determine all the critical points of the function

$$f(x, y) = x^2 - 8x + y^2 + 6y + 30.$$

EXAMPLE 15.14 Determine all the critical points of $f(x, y) = x^2 - xy + y^2$.

Solution The first-order partial derivatives of f are

$$f_x(x, y) = 2x - y \quad \text{and} \quad f_y(x, y) = -x + 2y.$$

To determine where both f_x and f_y equal zero, we must solve the system of linear equations

$$\begin{aligned} 2x - y &= 0 \\ -x + 2y &= 0. \end{aligned}$$

Using the techniques of Chapter 2, we see that the only solution of this system of equations is $(x, y) = (0, 0)$. Hence $(0, 0)$ is the only critical point of f, and so $f(0, 0) = 0$ is the only possible local extremum of f. ■

Practice Problem 11 Determine all the critical points of the function

$$g(x, y) = \frac{1}{2}x^2 + xy - y^2 + 4y - 5x.$$

Economists have observed that the proportion of the total national income earned by labor has remained fairly constant for many decades. The next example shows that this is to be expected if the total production is described by a Cobb-Douglas production function.

EXAMPLE 15.15 Suppose that, in producing some commodity, the total production is described by the Cobb-Douglas production function

$$Q = ax^{\lambda}y^{1-\lambda}$$

and labor is used optimally to maximize profit. Show that the ratio of the amount paid for labor to the total revenue derived from the commodity is the constant λ in the Cobb-Douglas production function.

Discussion Note that by assuming that labor is used optimally, we assure that everything produced can be sold. Recall from Section 11.6 that in a competitive market the price p at which a commodity sells is essentially independent of the quantity produced. Hence in a competitive market the revenue function for a commodity has the form

$$R = pQ,$$

where p is constant.

Let c_1 and c_2 denote the respective costs of purchasing one unit of labor and one unit of capital. Then the total cost of purchasing x units of labor and y units of capital is

$$C = c_1x + c_2y.$$

Hence the total profit for the commodity under consideration is

$$P = R - C = pQ - (c_1x + c_2y).$$

Now

$$P_x = pQ_x - c_1 \quad \text{and} \quad P_y = pQ_y - c_2$$

exist for all positive x and y because Q_x and Q_y do. Thus by Theorem 15.1 the assumption that profit is maximized implies that $P_x = 0$ and $P_y = 0$. The former equation can be written as

$$pQ_x - c_1 = 0.$$

Recall from Section 15.2 that the marginal productivity of labor is

$$Q_x = a\lambda x^{\lambda-1}y^{1-\lambda} = \frac{\lambda Q}{x}.$$

Substituting this expression for Q_x in the preceding equation gives

$$p\left(\frac{\lambda Q}{x}\right) - c_1 = 0.$$

Therefore

$$\lambda\left(\frac{pQ}{x}\right) = c_1,$$

and so

$$\lambda = \frac{c_1 x}{pQ}.$$

Since c_1 is the cost per unit of labor and x is the number of units of labor used, c_1x is the total amount paid for labor. Moreover pQ is the total revenue obtained from the commodity under consideration. Hence the preceding equation means that the ratio of the amount paid for labor to the total income from the commodity is the constant λ in the Cobb-Douglas production function. ■

In Examples 15.13 and 15.14 we were able to determine points at which local extrema of a given function might occur. In these examples, however, we did not determine whether a critical point actually gave rise to a local extremum and did not distinguish a local maximum from a local minimum. Fortunately there is a result analogous to the second-derivative test for functions of one variable that can often distinguish local minima from local maxima.

Second-Partial-Derivative Test for Functions of Two Variables

Let (a, b) be a critical point of f, and let $f(x, y)$ have continuous* second-order partial derivatives in some circular disk surrounding (a, b). Define

$$D = f_{xx}f_{yy} - f_{xy}^2.$$

(a) If $D(a, b) > 0$ and $f_{xx}(a, b) > 0$, then f has a local minimum at (a, b).

(b) If $D(a, b) > 0$ and $f_{xx}(a, b) < 0$, then f has a local maximum at (a, b).

(c) If $D(a, b) < 0$, then f does not have a local extremum at (a, b).

(d) If $D(a, b) = 0$, then the test gives no information.

---

*To say that a function g of two variables is continuous at (a, b) means that $g(x, y)$ can be made arbitrarily close to $g(a, b)$ whenever (x, y) is sufficiently close to (a, b).

Note that when $D(a, b) > 0$ in the second-partial-derivative test, $f_{xx}(a, b)$ distinguishes local minima from local maxima in the same way that $f''(a)$ distinguishes local minima from local maxima in the one-variable case: If $f_{xx}(a, b)$ is positive, there is a local minimum; and if it is negative, there is a local maximum. When $D(a, b) < 0$ [part (c) of the second-partial-derivative test], the behavior of f near (a, b) is like that shown in Figure 15.7. That is, the curve formed by the intersection of the surface $z = f(x, y)$ and one plane has a local maximum at (a, b), whereas the curve formed by the intersection of the surface $z = f(x, y)$ and another plane has a local minimum at (a, b). This type of behavior is described by saying that f has a **saddle point** at (a, b). Note that if f has a saddle point at (a, b), then a local extremum of f does *not* occur at (a, b). Finally observe that the second-partial-derivative test gives no information if $D(a, b) = 0$; in this case a local extremum of f may or may not occur at (a, b).

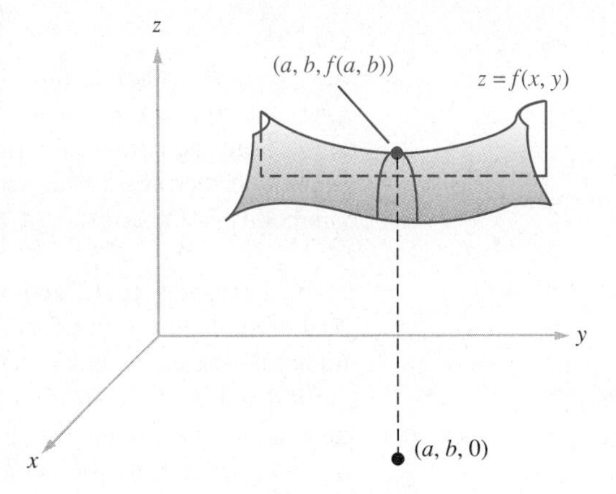

FIGURE 15.7

EXAMPLE 15.13 Revisited Determine all the local extrema of

$$f(x, y) = -x^2 + 6x - 2y^2 - 8y + 1.$$

Solution We must first determine the critical points of f. We saw in Example 15.13 that $(3, -2)$ is the only critical point of f. To determine what type of extremum occurs, we can use the second-partial-derivative test. Since

$$f_x(x, y) = -2x + 6 \quad \text{and} \quad f_y(x, y) = -4y - 8,$$

we see that

$$f_{xx}(x, y) = -2, \quad f_{xy}(x, y) = 0, \quad \text{and} \quad f_{yy}(x, y) = -4.$$

Hence

$$D(x, y) = f_{xx}(x, y)f_{yy}(x, y) - [f_{xy}(x, y)]^2$$
$$= (-2)(-4) - (0)^2$$
$$= 8.$$

Thus

$$D(3, -2) = 8 > 0 \qquad \text{and} \qquad f_{xx}(3, -2) = -2 < 0,$$

so that the conditions of part (b) of the second-partial-derivative test are met. It follows that a local maximum of f occurs at $(3, -2)$. The local maximum is

$$f(3, -2) = -(3)^2 + 6(3) - 2(-2)^2 - 8(-2) + 1 = 18. \quad ■$$

EXAMPLE 15.14 Revisited Determine all the local extrema of

$$f(x, y) = x^2 - xy + y^2.$$

Solution In Example 15.14 the only critical point of f was found to be $(0, 0)$. To see if a local extremum of f occurs at this point, we can use the second-partial-derivative test. Now

$$f_x(x, y) = 2x - y \qquad \text{and} \qquad f_y(x, y) = -x + 2y,$$

and so

$$f_{xx}(x, y) = 2, \qquad f_{xy}(x, y) = -1, \qquad \text{and} \qquad f_{yy}(x, y) = 2.$$

Therefore

$$D(x, y) = f_{xx}(x, y)f_{yy}(x, y) - [f_{xy}(x, y)]^2$$
$$= 2(2) - (-1)^2$$
$$= 3.$$

Since

$$D(0, 0) = 3 > 0 \qquad \text{and} \qquad f_{xx}(0, 0) = 2 > 0,$$

It follows from part (a) of the second-partial-derivative test that $f(0, 0) = 0$ is a local minimum of f. ■

Practice Problem 12 Let $f(x, y) = -9x + 8y - xy + x^2 + 2y^2$.

(a) Compute f_x and f_y, and show that f has a critical point at $(4, -1)$.
(b) Compute the function D in the second-partial-derivative test.
(c) Use the second-partial-derivative test to determine which type of local extremum occurs at $(4, -1)$.

In the two preceding examples we knew the critical points of the given function from an earlier example. Normally this will not be the case. Examples 15.16 and 15.17 illustrate the use of the second-partial-derivative test with a function f for which the critical points are not known. Note that in such cases we must:

1. Determine the critical points of f.
2. Compute the function $D = f_{xx}f_{yy} - f_{xy}^2$.
3. Evaluate D at each critical point of f, and apply the appropriate part of the second-partial-derivative test.

EXAMPLE 15.16 Determine all the local extrema of $f(x, y) = x^2 - 4xy + y^2 + \frac{1}{3}y^3$.

Solution We begin by determining the critical points of f. Now

$$f_x(x, y) = 2x - 4y \qquad \text{and} \qquad f_y(x, y) = -4x + 2y + y^2.$$

Setting $f_x(x, y) = 0$ and $f_y(x, y) = 0$, we obtain the system of equations

$$\begin{aligned} 2x - 4y \qquad &= 0 \\ -4x + 2y + y^2 &= 0. \end{aligned}$$

Solving the first equation for x yields $x = 2y$, and substituting this expression into the second equation produces

$$0 = -4(2y) + 2y + y^2 = y^2 - 6y = y(y - 6).$$

Hence $y = 0$ or $y = 6$. Because $x = 2y$, the critical points of f are $(0, 0)$ and $(12, 6)$.

Next we compute the function D in the second-partial-derivative test. Because

$$f_x(x, y) = 2x - 4y \qquad \text{and} \qquad f_y(x, y) = -4x + 2y + y^2,$$

we see that

$$f_{xx}(x, y) = 2, \qquad f_{xy}(x, y) = -4, \qquad \text{and} \qquad f_{yy}(x, y) = 2 + 2y.$$

Hence

$$\begin{aligned} D(x, y) &= f_{xx}(x, y)f_{yy}(x, y) - [f_{xy}(x, y)]^2 \\ &= 2(2 + 2y) - (-4)^2 \\ &= 4y - 12. \end{aligned}$$

Finally we evaluate D at each critical point of f. Since

$$D(0, 0) = 4(0) - 12 = -12 < 0$$

and

$$D(12, 6) = 4(6) - 12 = 12 > 0,$$

f has a saddle point at $(0, 0)$ and either a local minimum or a local maximum at $(12, 6)$. Since $f_{xx}(12, 6) = 2 > 0$, part (a) of the second-partial-derivative test guarantees that $f(12, 6) = -36$ is a local minimum of f. ■

EXAMPLE 15.17 Determine all the local extrema of

$$f(x, y) = \frac{1}{2}x^2y - \frac{3}{2}x^2 - 2xy - y^2 + 6x.$$

Solution As before, we begin by determining the critical points of f. Now

$$f_x(x, y) = xy - 3x - 2y + 6 \quad \text{and} \quad f_y(x, y) = \frac{1}{2}x^2 - 2x - 2y.$$

Since

$$xy - 3x - 2y + 6 = x(y - 3) - 2(y - 3) = (x - 2)(y - 3),$$

$f_x(x, y) = 0$ precisely when $x = 2$ or $y = 3$. If $f_y(x, y) = 0$ and $x = 2$, then

$$\begin{aligned} 0 &= \frac{1}{2}(2)^2 - 2(2) - 2y \\ 2y &= \frac{1}{2}(4) - 4 \\ 2y &= -2 \\ y &= -1. \end{aligned}$$

Thus $(2, -1)$ is one point at which both f_x and f_y are zero. On the other hand, if $y = 3$ and $f_y(x, y) = 0$, then

$$0 = \frac{1}{2}x^2 - 2x - 2(3) = \frac{1}{2}x^2 - 2x - 6.$$

Applying the quadratic formula to this equation gives $x = 6$ or $x = -2$. Hence $(6, 3)$ and $(-2, 3)$ are two other points at which both f_x and f_y are zero. Therefore the critical points of f are $(2, -1)$, $(6, 3)$, and $(-2, 3)$.

Computing the second-order partial derivatives of f, we obtain

$$f_{xx}(x, y) = y - 3, \quad f_{xy}(x, y) = x - 2, \quad \text{and} \quad f_{yy}(x, y) = -2.$$

Hence the function D in the second-partial-derivative test is

$$\begin{aligned} D(x, y) &= f_{xx}(x, y)f_{yy}(x, y) - [f_{xy}(x, y)]^2 \\ &= (y - 3)(-2) - (x - 2)^2 \\ &= -2y + 6 - (x - 2)^2. \end{aligned}$$

Since

$$D(2, -1) = 8 > 0 \quad \text{and} \quad f_{xx}(2, -1) = -4 < 0,$$

$f(2, -1) = 7$ is a local maximum of f. On the other hand

$$D(6, 3) = -16 < 0 \quad \text{and} \quad D(-2, 3) = -16 < 0,$$

so that local extrema do not occur at either $(6, 3)$ or $(-2, 3)$. Hence the only local extremum of f is the local maximum which occurs at $(2, -1)$. ■

Practice Problem 13 Determine all the local extrema of $f(x, y) = x^2 + 8xy + 8y^4$.

EXERCISES 15.3

In Exercises 1–8 determine all the critical points of the given function.

1. $f(x, y) = 3x^2 + 4y^2 + 24x - 24y$

2. $F(x, y) = x^2 - 14x + 3y^2 + 30y$

3. $g(x, y) = x^3 - 3x + 2y^4 - y^2$

4. $G(x, y) = x^2 - 10x + y^3 - 27y$

5. $h(x, y) = x^2 - 5x - xy + 2y^2 - y$

6. $H(x, y) = 3xy + x - x^2 - \frac{1}{2}y^2 - 12y$

7. $F(x, y) = 2x^3 + 24x^2 - 18xy + 3y^2$

8. $f(x, y) = x^2 + 2xy - 3y^2 + \frac{1}{3}y^3$

In Exercises 9–28 determine all the local extrema of the given function.

9. $f(x, y) = -x^2 + 6x + 2y^2 + 4y - 5$

10. $F(x, y) = x^2 + 12x + 3y^2 - 6y + 4$

11. $g(x, y) = 40x - 5x^2 - 6y - y^2 + 12$

12. $G(x, y) = 3x^2 - 12x - y^2 + 8y + 10$

13. $h(x, y) = 4x^2 + 16x + 3y^2 + 18y - 9$

14. $H(x, y) = 36y - 3y^2 - 20x - 2x^2 - 24$

15. $F(x, y) = x^4 + 4x^2 + y^2 - 6y$

16. $f(x, y) = x^4 - 2x^2 - y^2 + 12$

17. $G(x, y) = x^3 + 3x^2 - y^2 + 5$

18. $g(x, y) = x^4 + y^4 + 4y - 32x$

19. $H(x, y) = x^4 - y^4 - 8x^2 + 2y^2 + 3$

20. $h(x, y) = -x^3 + y^3 - 3xy$

21. $f(x, y) = 6xy - x^3 - y^3$

22. $F(x, y) = x^2y^2 + x^2 - 4y^2 + 68x$

23. $g(x, y) = \frac{9}{x} + \frac{3}{y} + xy$

24. $G(x, y) = (x + 1) \ln y$

25. $h(x, y) = (x^2 + 6x + 12)e^{y^2 - 2y}$

26. $H(x, y) = e^{(x+5)(y-4)}$

27. $F(x, y) = (y^2 - 1) \ln (x + 3)$

28. $f(x, y) = \sqrt{x^2 + y^2 + 1}$

29. A manufacturer produces two styles of men's shirts. When the less expensive shirts are sold at a price of x dollars and the more expensive shirts are sold at a price of y dollars, the number of shirts of each type that can be sold per week is $1200 - 350x + 200y$ and $1500 + 200x - 175y$ respectively. At what price should each style of shirt be sold in order to maximize the manufacturer's revenue?

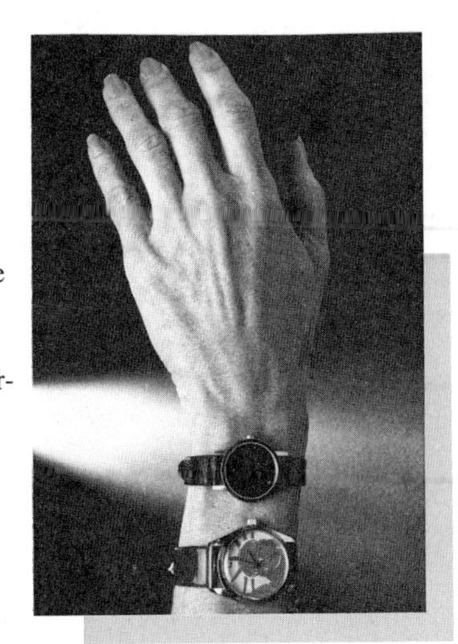

30. A manufacturer produces two types of women's watches. When the less expensive watches are sold at a price of x dollars and the more expensive watches are sold at a price of y dollars, the number of watches of each type that can be sold per month is $1250 - 100x + 55y$ and $1800 + 55x - 80y$ respectively. At what price should each type of watch be sold in order to maximize the manufacturer's revenue?

31. Under certain circumstances the best inventory policy for a manufacturer is to produce periodically a specific number of units of a product and not produce more units until the current supply is exhausted. In the period between the sale of the last unit from one lot and the production of the next lot, customer orders are placed on back-order status and filled immediately when a new lot is made. In this situation the manufacturer must determine the number x of units to produce in each lot and the number y of units to store in inventory after filling back orders. (Thus $x - y$ units are used to fill back-orders.) If a manufacturer's inventory cost in dollars is given by

$$C(x, y) = \frac{2{,}400{,}000}{x} + \frac{20y^2}{x} + 15x - 30y,$$

determine the values of x and y that minimize the inventory cost.

32. Rework Exercise 31 if $C(x, y) = \dfrac{1{,}008{,}000}{x} + \dfrac{30y^2}{x} + 21x - 42y$.

Answers to Practice Problems

10. The only critical point of f is $(4, -3)$.

11. The only critical point of g is $(2, 3)$.

12. (a) $f_x(x, y) = -9 - y + 2x$, $f_y(x, y) = 8 - x + 4y$; $f_x(4, -1) = 0$, $f_y(4, -1) = 0$

(b) $D(x, y) = 7$

(c) Since $D(4, -1) > 0$ and $f_{xx}(4, -1) = 2 > 0$, $f(4, -1) = -22$ is a local minimum.

13. There are local minima of -8 at $(-4, 1)$ and $(4, -1)$.

MATHEMATICS IN ACTION

15.4 The Method of Least Squares

One measure of the toxicity of a pesticide is how much must be given in order to kill half of a population of white rats. For carbaryl, this dosage is 500 mg per kg of body weight. (Note that the higher the dosage required to kill half the white rats, the less toxic is the pesticide.) For carbaryl this measure of toxicity is denoted LD_{50} 500; this symbol means that the lethal dosage of carbaryl for 50% of the white rats in a test population is 500 mg per kg of body weight.

In order to determine this number for a new pesticide, a manufacturer must administer various dosages to populations of 20 white rats. Suppose that the results of such tests are as given in Table 15.1. How can this data be used to predict the dosage of the pesticide that will kill 10 of the rats?

Dosage of pesticide (mg per kg of body weight)	*Number of rats killed*
40	3
60	5
80	6
100	9
120	11

TABLE 15.1

This example requires that we discover a relationship between the dosage of the pesticide and the number of white rats killed. The branch of statistics concerned with making predictions about a variable based on its relationship to a known variable is called *regression analysis*. In this section we will discuss an important special case of regression analysis in which two variables are connected by a relationship that is essentially linear.

In order to investigate the relationship between the dosages and the number of rats killed in Table 15.1, let us denote each of the corresponding measurements as an ordered pair (x_k, y_k), where x_k denotes the kth dosage (in mg per kg of body weight) and y_k denotes the corresponding number of rats killed. The five ordered pairs that result can then be plotted to obtain a picture of the data that is called a **scatter diagram.** The scatter diagram for this data is shown in Figure 15.8.

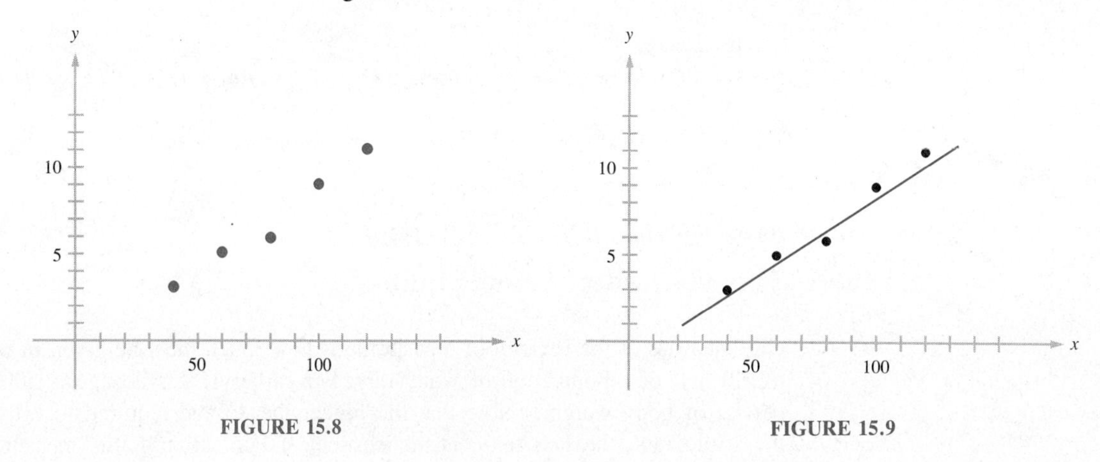

FIGURE 15.8

FIGURE 15.9

The five points shown in Figure 15.8 are clustered closely around the line shown in Figure 15.9. The latter diagram suggests that the relationship between the dosages and the numbers of rats killed is almost linear. Note that the relationship is not exactly linear, for if it were, then the points would lie on the same line. Figure 15.9 suggests, however, that the relationship between the dosages and the numbers of rats killed can be approximated quite well by assuming that the number killed y_k is a linear function of the dosage x_k.

When a scatter diagram reveals that the relationship between two variables is almost linear, there will be many lines that approximate the data reasonably well. One line that has proved to be a good approximation to data that is almost linear is called the **least squares line.** It is the line for which the sum of the squares of the vertical distances from the data points to the line is as small as possible.

To understand better the meaning of the condition that defines the least squares line, we will consider a simpler example. Suppose that we have the four data points

$$(1, 13), \quad (3, 7), \quad (4, 5), \quad \text{and} \quad (7, 3).$$

These points are plotted in Figure 15.10. The least squares line is the line that minimizes the sum of the squares of the vertical distances from the data points to the line, which is

$$E = d_1^2 + d_2^2 + d_3^2 + d_4^2.$$

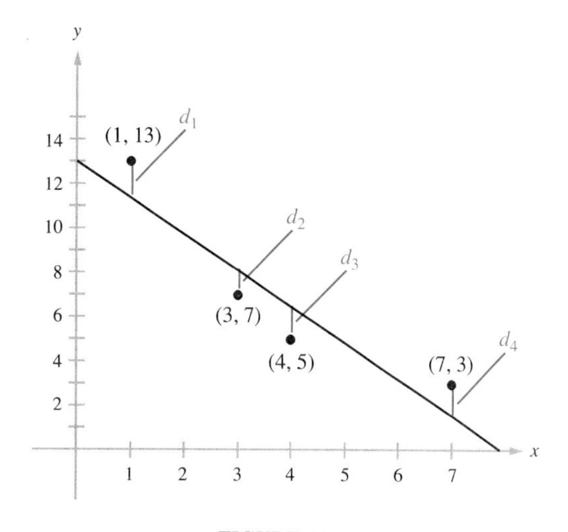

FIGURE 15.10

Suppose that $y = mx + b$ is the equation of the least squares line. Then the point on the line corresponding to $x = 1$ has a y-coordinate of

$$m(1) + b = m + b.$$

Hence the vertical distance from (1, 13) to the line is the absolute value of the difference of the y-coordinates, which is,

$$d_1 = |13 - (m + b)| = |13 - m - b|.$$

Likewise

$$d_2 = |7 - (3m + b)| = |7 - 3m - b|,$$
$$d_3 = |5 - (4m + b)| = |5 - 4m - b|,$$

and

$$d_4 = |3 - (7m + b)| = |3 - 7m - b|.$$

Since $|d_k|^2 = d_k^2$, we have

$$\begin{aligned} E &= d_1^2 + d_2^2 + d_3^2 + d_4^2 \\ &= (13 - m - b)^2 + (7 - 3m - b)^2 + (5 - 4m - b)^2 + (3 - 7m - b)^2. \end{aligned}$$

Note that E is a function of the two variables m and b. By Theorem 15.1 the minimum value of E must occur at a point where

$$E_m(m, b) = 0 \quad \text{and} \quad E_b(m, b) = 0.$$

Taking partial derivatives, we find that

$$\begin{aligned} E_m(m, b) &= 2(13 - m - b)(-1) + 2(7 - 3m - b)(-3) + 2(5 - 4m - b)(-4) + 2(3 - 7m - b)(-7) \\ &= (-26 + 2m + 2b) + (-42 + 18m + 6b) + (-40 + 32m + 8b) + (-42 + 98m + 14b) \\ &= -150 + 150m + 30b \end{aligned}$$

and

$$\begin{aligned} E_b(m, b) &= 2(13 - m - b)(-1) + 2(7 - 3m - b)(-1) + 2(5 - 4m - b)(-1) + 2(3 - 7m - b)(-1) \\ &= (-26 + 2m + 2b) + (-14 + 6m + 2b) + (-10 + 8m + 2b) + (-6 + 14m + 2b) \\ &= -56 + 30m + 8b. \end{aligned}$$

Equating $E_m(m, b)$ and $E_b(m, b)$ to zero produces the system of equations

$$\begin{aligned} 150m + 30b &= 150 \\ 30m + 8b &= 56. \end{aligned}$$

Using the techniques presented in Chapter 2, we find that the solution of this system is $m = -1.6$ and $b = 13$. Thus the equation of the least squares line $y = mx + b$ is

$$y = -1.6x + 13.$$

The graph of this equation is the line shown in Figure 15.10.

THE GENERAL CASE

Now we will determine the least squares line for an arbitrary collection of data points

$$(x_1, y_1), (x_2, y_2), \ldots, (x_n, y_n).$$

If

$$y = mx + b$$

is the equation of the least squares line, then the sum of the squares of the vertical distances from the data points to the line is

$$\begin{aligned} E &= d_1^2 + d_2^2 + \cdots + d_n^2 \\ &= [y_1 - (mx_1 + b)]^2 + [y_2 - (mx_2 + b)]^2 + \cdots + [y_n - (mx_n + b)]^2. \end{aligned}$$

The calculations performed in the preceding example are similar to those in the general case. The partial derivatives of E with respect to m and b are*

$$\begin{aligned} E_m(m, b) &= 2[y_1 - (mx_1 + b)](-x_1) + 2[y_2 - (mx_2 + b)](-x_2) + \cdots + 2[y_n - (mx_n + b)](-x_n) \\ &= -2(x_1y_1 + x_2y_2 + \cdots + x_ny_n) + 2m(x_1^2 + x_2^2 + \cdots + x_n^2) + 2b(x_1 + x_2 + \cdots + x_n) \\ &= -2 \sum xy + 2m \sum x^2 + 2b \sum x \end{aligned}$$

and

$$\begin{aligned} E_b(m, b) &= 2[y_1 - (mx_1 + b)](-1) + 2[y_2 - (mx_2 + b)](-1) + \cdots + 2[y_n - (mx_n + b)](-1) \\ &= -2(y_1 + y_2 + \cdots + y_n) + 2m(x_1 + x_2 + \cdots + x_n) + 2b(1 + 1 + \cdots + 1) \\ &= -2(y_1 + y_2 + \cdots + y_n) + 2m(x_1 + x_2 + \cdots + x_n) + 2bn \\ &= -2 \sum y + 2m \sum x + 2bn. \end{aligned}$$

Equating $E_m(m, b)$ and $E_b(m, b)$ to zero gives the system of equations

$$\begin{aligned} \left(\sum x^2\right)m + \left(\sum x\right)b &= \sum xy \\ \left(\sum x\right)m + \qquad nb &= \sum y. \end{aligned}$$

If we multiply the first equation by n and the second by $\Sigma\, x$ and subtract, the resulting equation is

$$n\left(\sum x^2\right)m - \left(\sum x\right)^2 m = n \sum xy - \left(\sum x\right)\left(\sum y\right).$$

Solving this equation for m, we have

$$m = \frac{n\, \Sigma\, xy - (\Sigma\, x)(\Sigma\, y)}{n\, \Sigma\, x^2 - (\Sigma\, x)^2}.$$

*Recall that $\Sigma\, x$ denotes the sum of the numbers x_i.

Now solving the equation

$$\left(\sum x\right)m + nb = \sum y$$

for b, we obtain

$$b = \frac{\Sigma\, y - m\, \Sigma\, x}{n}.$$

It can be shown that these values of m and b do, in fact, minimize E. Thus we have obtained the following result.

Equation of the Least Squares Line

For data points $(x_1, y_1), (x_2, y_2), \ldots, (x_n, y_n)$, the equation of the least squares line is $y = mx + b$, where

$$m = \frac{n\, \Sigma\, xy - (\Sigma\, x)(\Sigma\, y)}{n\, \Sigma\, x^2 - (\Sigma\, x)^2} \quad \text{and} \quad b = \frac{\Sigma\, y - m\, \Sigma\, x}{n}.$$

Having determined the equation of the least squares line in general, we will not need to use partial differentiation again. For example, we can compute directly the equation of the least squares line using the data points from our earlier example:

$$(1, 13), \quad (3, 7), \quad (4, 5) \quad \text{and} \quad (7, 3).$$

The equation of the least squares line is $y = mx + b$, where m and b are given by the formulas above. In order to evaluate m and b we must determine the following quantities:

$$\sum x, \quad \sum y, \quad \sum xy, \quad \text{and} \quad \sum x^2.$$

If we record the data values in a table as shown below, the column totals are precisely the quantities we need.

	x-coordinate of data points	*y-coordinate of data points*	*Product of coordinates*	*Square of x-coordinate*
	x_k	y_k	$x_k y_k$	x_k^2
	1	13	13	1
	3	7	21	9
	4	5	20	16
	7	3	21	49
Totals	$15 = \Sigma\, x$	$28 = \Sigma\, y$	$75 = \Sigma\, xy$	$75 = \Sigma\, x^2$

Note that since there are 4 data points, $n = 4$. Thus

$$\begin{aligned} m &= \frac{n \, \Sigma \, xy - (\Sigma \, x)(\Sigma \, y)}{n \, \Sigma \, x^2 - (\Sigma \, x)^2} \\ &= \frac{4(75) \qquad 15(20)}{4(75) - (15)^2} \\ &= \frac{-120}{75} = -1.6, \end{aligned}$$

and

$$\begin{aligned} b &= \frac{\Sigma \, y - m \, \Sigma \, x}{n} \\ &= \frac{28 - (-1.6)(15)}{4} \\ &= \frac{52}{4} = 13. \end{aligned}$$

Hence the equation of the least squares line is

$$y = -1.6x + 13,$$

as we found earlier.

EXAMPLE 15.18 Find the equation of the least squares line for the data (2, 10), (4, 20), (6, 20), (8, 30), and (10, 40).

Solution Proceeding as above, we record the data values in the table below.

	x_k	y_k	$x_k y_k$	x_k^2
	2	10	20	4
	4	20	80	16
	6	20	120	36
	8	30	240	64
	10	40	400	100
Totals	$30 = \Sigma \, x$	$120 = \Sigma \, y$	$860 = \Sigma \, xy$	$220 = \Sigma \, x^2$

Note that in this case $n = 5$ because there are 5 data points. Thus

$$\begin{aligned} m &= \frac{n \, \Sigma \, xy - (\Sigma \, x)(\Sigma \, y)}{n \, \Sigma \, x^2 - (\Sigma \, x)^2} \\ &= \frac{5(860) - 30(120)}{5(220) - (30)^2} \\ &= \frac{700}{200} = 3.5, \end{aligned}$$

and

$$\begin{aligned} b &= \frac{\Sigma y - m \Sigma x}{n} \\ &= \frac{120 - 3.5(30)}{5} \\ &= \frac{15}{5} = 3. \end{aligned}$$

Hence the equation of the least squares line is $y = 3.5x + 3$. ■

Practice Problem 14 (a) Construct a scatter diagram for the data

$$(2, 10), (3, 7), (5, 3), \text{ and } (6, 2).$$

(b) Compute Σx, Σy, Σxy, and Σx^2.
(c) Find the equation of the least squares line for this data.

GOODNESS OF FIT

Figure 15.11 shows the scatter diagram from the answer of Practice Problem 14 along with the corresponding least squares line. In this case the least squares line appears to approximate the data points quite well, and so statisticians say that the least squares line fits the data well.

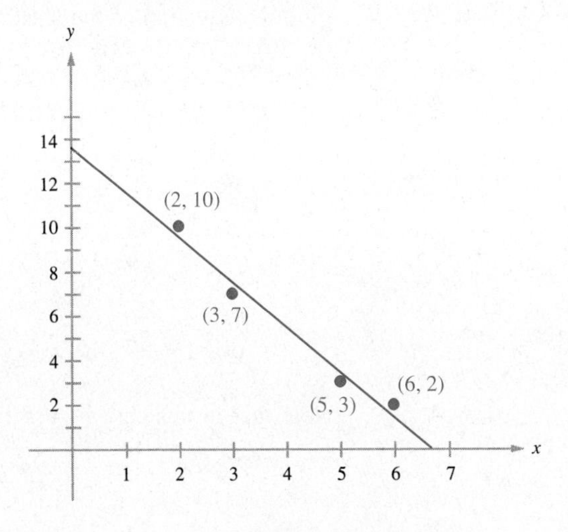

FIGURE 15.11

It is useful to be able to measure how well a least squares line fits the corresponding data. One such measure is the **coefficient of correlation,** usually denoted r, that is defined as follows.

Coefficient of Correlation

For data points $(x_1, y_1), (x_2, y_2), \ldots, (x_n, y_n)$, the coefficient of correlation is given by

$$r = \frac{n \sum xy - (\sum x)(\sum y)}{\sqrt{n \sum x^2 - (\sum x)^2}\sqrt{n \sum y^2 - (\sum y)^2}}.$$

It can be shown that r is a number between -1 and 1 inclusive. Values of r that are close to ± 1 indicate that the least squares line provides an excellent approximation to the data points, whereas values of r close to 0 indicate that the least squares line fits the data points poorly. In the latter case the relationship between the variables x_k and y_k cannot be described well by a linear function.

Note that the numerator in the formula for r is the same as the numerator in the formula for m given earlier. Because the denominator of r will always be positive, this means that m and r will both be positive or both be negative. Hence the slope of the least squares line will be positive or negative according to whether r is positive or negative, and this corresponds to whether y increases or decreases as x increases.

EXAMPLE 15.19 Compute the coefficient of correlation for the data in Figure 15.10.

Solution On page 834 we constructed the following table to help us compute the values of m and b in the equation of the least squares line. If we add another column to this table containing the squares of the values y_k, we will have all the information necessary to compute the coefficient of correlation.

	x_k	y_k	$x_k y_k$	x_k^2	y_k^2
	1	13	13	1	169
	3	7	21	9	49
	4	5	20	16	25
	7	3	21	49	9
Totals	$15 = \sum x$	$28 = \sum y$	$75 = \sum xy$	$75 = \sum x^2$	$252 = \sum y^2$

Hence

$$\begin{aligned} r &= \frac{n \sum xy - (\sum x)(\sum y)}{\sqrt{n \sum x^2 - (\sum x)^2}\sqrt{n \sum y^2 - (\sum y)^2}} \\ &= \frac{4(75) - 15(28)}{\sqrt{4(75) - (15)^2}\sqrt{4(252) - (28)^2}} \\ &= \frac{-120}{\sqrt{75}\sqrt{224}} \approx -.926. \end{aligned}$$

Practice Problem 15 Compute the coefficient of correlation for the data (2, 10), (3, 7), (5, 3), and (6, 2) in Practice Problem 14.

Since the coefficient of correlation for the data in Practice Problem 14 is very close to -1, we see that the least squares line fits its data quite well. This confirms the earlier observation we made by looking at Figure 15.11.

APPLICATIONS

Let us return to the data in Table 15.1 showing the dosages of pesticide and numbers of rats killed. By proceeding as in Example 15.18, we can compute the equation of the least squares line.

	x_k	y_k	$x_k y_k$	x_k^2	y_k^2
	40	3	120	1,600	9
	60	5	300	3,600	25
	80	6	480	6,400	36
	100	9	900	10,000	81
	120	11	1,320	14,400	121
Totals	400 $= \Sigma x$	34 $= \Sigma y$	3,120 $= \Sigma xy$	36,000 $= \Sigma x^2$	272 $= \Sigma y^2$

Thus the slope of the least squares line is

$$\begin{aligned} m &= \frac{n \Sigma xy - (\Sigma x)(\Sigma y)}{n \Sigma x^2 - (\Sigma x)^2} \\ &= \frac{5(3{,}120) - 400(34)}{5(36{,}000) - (400)^2} \\ &= \frac{2{,}000}{20{,}000} = 0.1, \end{aligned}$$

and the y-intercept is

$$\begin{aligned} b &= \frac{\Sigma y - m \Sigma x}{n} \\ &= \frac{34 - 0.1(400)}{5} \\ &= \frac{-6}{5} = -1.2. \end{aligned}$$

Hence the equation of the least squares line is

$$y = 0.1x - 1.2.$$

The coefficient of correlation for this data is

$$\begin{aligned} r &= \frac{n \Sigma xy - (\Sigma x)(\Sigma y)}{\sqrt{n \Sigma x^2 - \Sigma x^2}\sqrt{n \Sigma y^2 - (\Sigma y)^2}} \\ &= \frac{5(3{,}120) - 400(34)}{\sqrt{5(36{,}000) - (400)^2}\sqrt{5(272) - (34)^2}} \\ &= \frac{2{,}000}{\sqrt{20{,}000}\sqrt{204}} \approx .990. \end{aligned}$$

Since the coefficient of correlation is very close to 1, the least squares line approximates the data quite well. We can use the least squares line to predict the dosage of pesticide that will kill 10 of the rats. For if $y = 10$, the dosage x must satisfy

$$10 = 0.1x - 1.2.$$

Solving for x gives

$$\begin{aligned} 11.2 &= 0.1x \\ 112 &= x. \end{aligned}$$

Therefore a dosage of 112 mg per kg of body weight can be expected to kill 10 of the rats, and so the toxicity of this pesticide is LD_{50} 112. ■

EXAMPLE 15.20 Table 15.2* gives the number of automobiles in use in the United States and domestic fuel consumption at five-year intervals. Use this table to predict the amount of motor fuel consumption when there were 90 million automobiles in use in the United States.

Year	*Automobiles in use (millions)*	*Motor fuel consumed (billions of gallons)*
1960	56.9	57.9
1965	68.9	71.1
1970	80.4	92.3
1975	95.2	109.0
1980	104.6	115.0

TABLE 15.2

*Source: *Statistical Abstract of the United States*, 1984. Washington, DC: U.S. Department of Commerce, Bureau of the Census.

Solution We must compute the least squares line for the data in Table 15.2. Proceeding as before, we arrange the data in a table as shown below.

	x_k	y_k	$x_k y_k$	x_k^2	y_k^2
	56.9	57.9	3,294.51	3,237.61	3,352.41
	68.9	71.1	4,898.79	4,747.21	5,055.21
	80.4	92.3	7,420.92	6,464.16	8,519.23
	95.2	109.0	10,376.80	9,063.04	11,881.00
	104.6	115.0	12,029.00	10,941.16	13,225.00
Totals	406.0 $= \Sigma x$	445.3 $= \Sigma y$	38,020.02 $= \Sigma xy$	34,453.18 $= \Sigma x^2$	42,032.91 $= \Sigma y^2$

The slope of the least squares line is

$$\begin{aligned} m &= \frac{n \Sigma xy - (\Sigma x)(\Sigma y)}{n \Sigma x^2 - (\Sigma x)^2} \\ &= \frac{5(38{,}020.02) - 406(445.3)}{5(34{,}453.18) - (406)^2} \\ &= \frac{9308.3}{7429.9} \approx 1.2528, \end{aligned}$$

and the y-intercept is

$$\begin{aligned} b &= \frac{\Sigma y - m \Sigma x}{n} \\ &= \frac{445.3 - 1.2528(406)}{5} \\ &= \frac{-63.343}{5} \approx -12.7. \end{aligned}$$

Hence the equation of the least squares line is approximately

$$y = 1.25x - 12.7.$$

The coefficient of correlation for this data is

$$\begin{aligned} r &= \frac{n \Sigma xy - (\Sigma x)(\Sigma y)}{\sqrt{n \Sigma x^2 - (\Sigma x)^2}\sqrt{n \Sigma y^2 - (\Sigma y)^2}} \\ &= \frac{5(38{,}020.02) - 406(445.3)}{\sqrt{5(34{,}453.18) - (406)^2}\sqrt{5(42{,}032.91) - (445.3)^2}} \\ &= \frac{9308.3}{\sqrt{7429.9}\sqrt{11872.46}} \approx .991. \end{aligned}$$

Since the coefficient of correlation is very close to 1, the least squares line fits the data quite well. Again we can use the least squares line to estimate motor fuel consumption. For example, when there were 90 million cars in use (sometime between 1970 and 1975), the yearly motor fuel consumption was approximately

$$1.25(90) - 12.7 = 99.8 \text{ billion gallons.} \quad \blacksquare$$

EXERCISES 15.4

In Exercises 1–10 plot a scatter diagram for the given data points. Then compute the coefficient of correlation and the equation of the least squares line.

1. (1, 10), (3, 18), (4, 20), (6, 20)

2. (1, 4), (3, 7), (5, 9), (7, 10)

3. (1, 20), (2, 17), (4, 11), (7, 4)

4. (1, 3), (3, 4), (4, 5), (7, 10)

5. (3, 8), (8, 16), (12, 24), (15, 25)

6. (−3, 10), (−2, 7), (0, 3), (1, 2)

7. (1, 30), (3, 26), (7, 13), (8, 11), (10, 3)

8. (1, 9), (2, 7), (3, 6), (4, 4), (5, 2)

9. (1, 5), (2, 6), (3, 8), (4, 10), (5, 11)

10. (1, 2), (2, 4), (3, 7), (4, 8), (5, 10)

11. On a hot August day, a construction worker heard the following temperature readings on his radio.

Time (P.M.)	1	2	4	5
Temperature (° F)	82	87	95	98

(a) Determine the least squares line for these data.

(b) Estimate the temperature at 3 P.M.

12. During a 5-day period when the air was stagnant, Linda heard the following pollen counts on the 10:30 A.M. local news.

Day	1	2	3	5
Pollen count	420	516	610	742

(a) Determine the least squares line for these data.

(b) Estimate the pollen count on day 4.

13. Hanford, Washington, located on the Columbia River, is the site of a storage facility for radioactive wastes. In reference [2], Fadeley investigated the relationship between deaths from cancer in the city of Portland and the Oregon counties bordering the Columbia River and the degree to which the population in these counties is exposed from wastes that may be carried from the Hanford facility. The data that he obtained are shown on the next page.

Location	*Index of exposure*	*Cancer mortality (per 100,000)*
Umatilla County	2.49	147.1
Morrow County	2.57	130.1
Gilliam County	3.41	129.9
Sherman County	1.25	113.5
Wasco County	1.62	137.5
Hood River County	3.83	162.3
City of Portland	11.64	207.5
Columbia County	6.41	177.9
Clatsop County	8.34	210.3

Show that there is a strong linear relationship between the index of exposure and the cancer mortality rate, and then compute the equation of the least squares line for these data.

14. Prevailing currents in the Pacific Ocean normally divert the waters of the Columbia River southward along the coast of Oregon. Thus the population of the Oregon counties bordering the Pacific is also exposed to radioactive wastes from the Hanford storage facility described in Exercise 13. Using the data below, show that there is a strong linear relationship between the index of exposure and the cancer mortality rate, and then compute the equation of the least squares line for these data.

County	*Index of exposure*	*Cancer mortality (per 100,000)*
Clatsop	8.34	210.3
Tillamook	5.51	163.8
Lincoln	6.21	170.4
Coos	1.85	117.6
Curry	2.81	73.6

15. During the years 1970–76, disposable personal income in the United States was as shown in the table* below.

Year	*1970*	*1972*	*1973*	*1974*	*1975*	*1976*
Disposable personal income (billions of $)	685.9	801.3	901.7	982.9	1080.9	1181.7

(a) Letting $x = 0$ denote the year 1970, $x = 2$ denote 1972, and so forth, determine the least squares line that expresses disposable personal income as a function of the year.

(b) Compute the coefficient of correlation for these data.

(c) Predict the U.S. disposable personal income in 1971.

16. Before beginning construction of a dam on the Kootenai River at Newgate, British Columbia, it was necessary to collect data on the rate of stream flow there. Unfortunately, stream records for Newgate were not kept until 1931, but records existed since 1925 at Libby, Montana, further downstream. These data appear in the table on the next page.

**Source: Statistical Abstract of the United States, 1977.* Washington, DC: U.S. Department of Commerce, Bureau of the Census.

(a) Compute the least squares line for the stream-flow rates at Libby and Newgate during the years 1931–43.

(b) Predict the stream-flow rates at Newgate for the years 1925–30. (These predicted rates were actually used in planning construction of the dam at Newgate.)

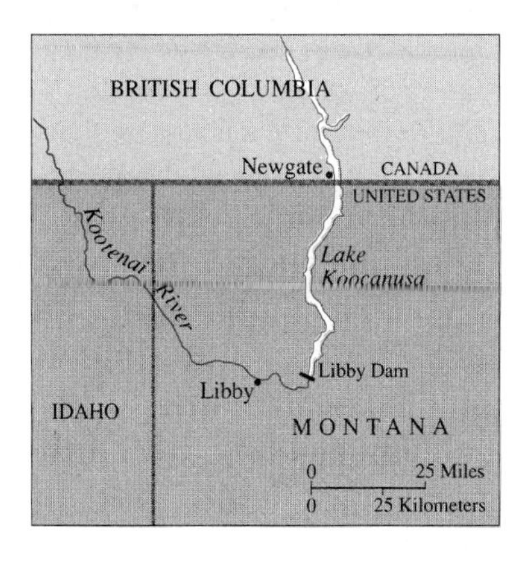

*Water Flow at Two Points on the Kootenai River in January**

Year	*Flow at Libby (hundreds of cubic feet per second)*	*Flow at Newgate (hundreds of cubic feet per second)*
1925	42.0	
1926	24.0	
1927	38.0	
1928	49.4	
1929	24.6	
1930	24.2	
1931	27.1	19.7
1932	20.9	18.0
1933	33.4	26.1
1934	77.6	44.9
1935	37.0	26.1
1936	21.6	19.9
1937	17.6	15.7
1938	35.1	27.6
1939	32.6	24.9
1940	26.0	23.4
1941	27.6	23.1
1942	38.7	31.3
1943	27.8	23.8

Answers to Practice Problems

14. (a) The scatter diagram is shown below. (b) $\Sigma x = 16$, $\Sigma y = 22$, $\Sigma xy = 68$, and $\Sigma x^2 = 74$ (c) The equation of the least squares line is $y = -2x + 13.5$.

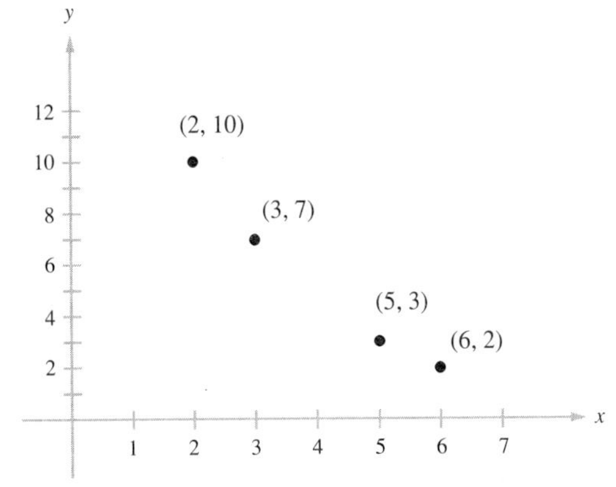

15. $r \approx -.988$

*Source: *Extending Stream-Flow Records,* U.S. Department of the Interior, Geological Survey, Water Resources Branch, September, 1947, pp. 7–8.

15.5 Lagrange Multipliers and Constrained Optimization

In previous sections we encountered problems in which it was necessary to maximize or minimize a function subject to one or more constraints. We have already learned how to solve an important class of problems of this type, namely, linear programming problems, in which the objective function is a linear function and the constraints are linear inequalities.

Frequently, however, we must maximize or minimize a nonlinear function. In Example 11.20 in Section 11.5, for instance, we determined the dimensions of the largest rectangular box with a square end that can be sent by parcel post. Since U.S. Postal regulations require that the length plus the girth of any package not exceed 84 inches (see Figure 15.12), this amounts to maximizing the nonlinear function $V = x^2y$ subject to the constraint $y + 4x = 84$.

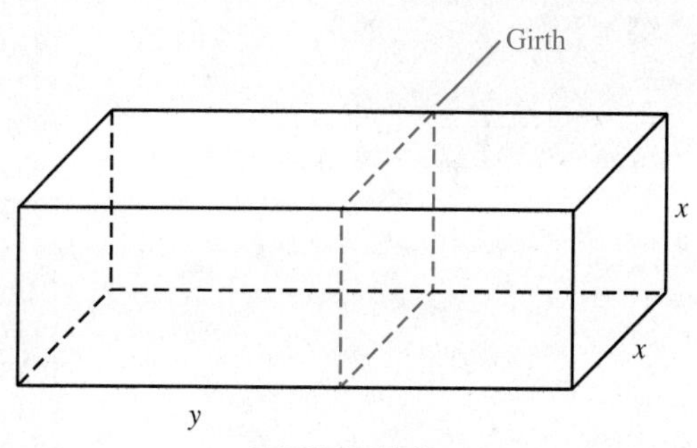

FIGURE 15.12

In working Example 11.20 we solved the constraint equation for y and substituted into the expression for V. Unfortunately, it is often difficult to solve the constraint equation in this manner. In this section we will discuss a useful method for solving problems that require maximizing or minimizing a function subject to one or more constraint equations. This procedure is due to the French mathematician Joseph Louis Lagrange (1736–1813) and is called the **method of Lagrange multipliers.** It is based on the following fact.

Theorem 15.2

Suppose that $f(x, y)$ has a local extremum at (a, b) when subject to the constraint $g(x, y) = 0$. Then there is a constant λ such that

$$f_x(a, b) = \lambda g_x(a, b) \quad \text{and} \quad f_y(a, b) = \lambda g_y(a, b).$$

The method of Lagrange multipliers works as follows. The equations in Theorem 15.2,

$$f_x(a, b) = \lambda g_x(a, b) \quad \text{and} \quad f_y(a, b) = \lambda g_y(a, b),$$

can each be solved for λ, and equating the resulting expressions gives an equation relating x and y. This can be used with the constraint $g(x, y) = 0$ to solve for x and y. The resulting points (x, y) are candidates for local extrema of f. The constant λ in Theorem 15.2 is called a *Lagrange multiplier*.

In applying the method of Lagrange multipliers, we will use the following procedure.

1. Rewrite the constraint equation in the form $g(x, y) = 0$.
2. Compute both partial derivatives of f and g.
3. Set $f_x(x, y) = \lambda g_x(x, y)$ and $f_y(x, y) = \lambda g_y(x, y)$ and solve each equation for λ.
4. Set equal the two values of λ from step 3. Then solve the resulting equation together with the constraint equation for x and y.

EXAMPLE 15.21 Use the method of Lagrange multipliers to solve the problem considered in Example 11.20 on page 640. That is, maximize $f(x, y) = x^2y$ subject to the constraint $y + 4x = 84$.

Solution Writing the constraint equation as

$$y + 4x - 84 = 0$$

yields the constraint function

$$g(x, y) = y + 4x - 84.$$

Now

$$f_x(x, y) = 2xy, \qquad f_y(x, y) = x^2,$$
$$g_x(x, y) = 4, \quad \text{and} \quad g_y(x, y) = 1.$$

Setting

$$f_x(x, y) = \lambda g_x(x, y) \quad \text{and} \quad f_y(x, y) = \lambda g_y(x, y),$$

we obtain the following system of equations.

$$2xy = 4\lambda$$
$$x^2 = \lambda$$

Solving the first equation for λ produces

$$\frac{1}{2}xy = \lambda.$$

Equating this value of λ with that from the second equation in the system above, we obtain

$$\frac{1}{2}xy = x^2$$
$$xy = 2x^2$$
$$0 = 2x^2 - xy$$
$$0 = x(2x - y).$$

Since in the original context of the problem x and y are the lengths of the sides of a box, we must have $x \neq 0$. Hence the preceding equation implies that $y = 2x$. If we substitute this expression for y into the constraint equation, we can solve for x.

$$\begin{aligned} g(x, y) &= 0 \\ y + 4x - 84 &= 0 \\ 2x + 4x - 84 &= 0 \\ 6x &= 84 \\ x &= 14 \end{aligned}$$

Therefore the only possible local extremum of f subject to the constraint $g(x, y) = 0$ occurs at $x = 14$ and $y = 2x = 28$. Since intuitively the box must have a maximum possible volume, we conclude that it occurs when $x = 14$ and $y = 28$. Thus the maximum value of f subject to the given constraint is $f(14, 28) = 5488$. This answer agrees with the one obtained in Example 11.20. ■

EXAMPLE 15.22 Use the method of Lagrange multipliers to find the point on the line $x + 2y = 10$ that is closest to the origin.

Solution The distance from a point (x, y) on the line $x + 2y = 10$ to the origin is $\sqrt{x^2 + y^2}$. It is geometrically clear (see Figure 15.13) that there must be a point on this line that is closest to the origin. On the other hand, there is no point on the line that is farthest from the origin; so f has no local maximum subject to the constraint $x + 2y = 10$.

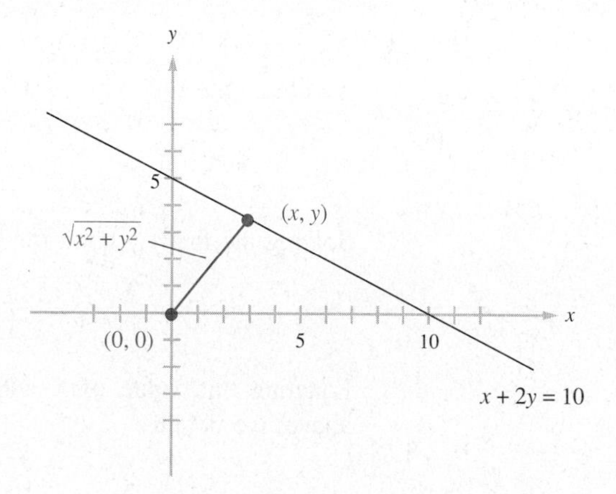

FIGURE 15.13

To find the point closest to the origin, we must minimize the function

$$f(x, y) = \sqrt{x^2 + y^2} = (x^2 + y^2)^{1/2}$$

subject to the constraint

$$x + 2y = 10.$$

As in Example 15.21 the constraint function is

$$g(x, y) = x + 2y - 10.$$

Now

$$f_x(x, y) = \frac{1}{2}(x^2 + y^2)^{-1/2}(2x), \qquad f_y(x, y) = \frac{1}{2}(x^2 + y^2)^{-1/2}(2y),$$

$$g_x(x, y) = 1, \qquad \text{and} \qquad g_y(x, y) = 2.$$

Hence Theorem 15.2 yields the equations

$$x(x^2 + y^2)^{-1/2} = \lambda \qquad \text{and} \qquad y(x^2 + y^2)^{-1/2} = 2\lambda.$$

Solving the second equation for λ and equating it to the value from the first equation, we obtain

$$x(x^2 + y^2)^{-1/2} = \frac{1}{2}y(x^2 + y^2)^{-1/2}.$$

If (x, y) is a point on the line $x + 2y = 10$, then $(x, y) \neq (0, 0)$; so

$$x^2 + y^2 \neq 0.$$

Thus we can divide the preceding equation by $(x^2 + y^2)^{-1/2}$ to produce

$$x = \frac{1}{2}y.$$

Substituting this expression for x in the constraint enables us to solve for y as follows.

$$\begin{aligned} x + 2y - 10 &= 0 \\ \frac{1}{2}y + 2y - 10 &= 0 \\ \frac{5}{2}y &= 10 \\ y &= 4 \end{aligned}$$

Therefore

$$x = \frac{1}{2}y = \frac{1}{2}(4) = 2.$$

Hence the point $(x, y) = (2, 4)$ is the only possible local extremum of f subject to the constraint $g(x, y) = 0$, and so $(2, 4)$ must be the local minimum of f subject to this constraint. Thus $(2, 4)$ is the point on the line $x + 2y = 10$ that is closest to the origin; its distance from the origin is $f(2, 4) = \sqrt{20}$. ■

The method of Lagrange multipliers is generally used in situations like those in Examples 15.21 and 15.22, where the context of the problem makes it clear that the desired type of extremum exists.

EXAMPLE 15.23 The function $f(x, y) = 6xy - x^3 - y^3$ has a local maximum subject to the constraint $x + y = 8$. What is this value, and where does it occur?

Solution As before, the constraint function for the method of Lagrange multipliers is

$$g(x, y) = x + y - 8.$$

Now

$$f_x(x, y) = 6y - 3x^2, \qquad f_y(x, y) = 6x - 3y^2,$$
$$g_x(x, y) = 1, \qquad \text{and} \qquad g_y(x, y) = 1.$$

Equating $f_x(x, y)$ with $\lambda g_x(x, y)$ and $f_y(x, y)$ with $\lambda g_y(x, y)$ produces

$$6y - 3x^2 = \lambda \qquad \text{and} \qquad 6x - 3y^2 = \lambda.$$

Thus

$$\begin{aligned} 6y - 3x^2 &= 6x - 3y^2 \\ 0 &= 3x^2 - 3y^2 + 6x - 6y \\ 0 &= 3(x - y)(x + y) + 3(x - y)(2) \\ 0 &= 3(x - y)(x + y + 2). \end{aligned}$$

It follows that

$$x - y = 0 \qquad \text{or} \qquad x + y + 2 = 0,$$

that is,

$$x = y \qquad \text{or} \qquad x = -y - 2.$$

Let us first consider the possibility that $x = -y - 2$. Substituting this expression into the constraint $g(x, y) = 0$ yields

$$\begin{aligned} x + y - 8 &= 0 \\ (-y - 2) + y - 8 &= 0 \\ -10 &= 0. \end{aligned}$$

Thus the equation $x = -y - 2$ does not lead to any local extrema of f.

Next we will consider the case that $x = y$. Substituting this expression into $g(x, y) = 0$ yields

$$\begin{aligned} x + y - 8 &= 0 \\ y + y - 8 &= 0 \\ 2y &= 8 \\ y &= 4. \end{aligned}$$

Therefore

$$x = y = 4.$$

Thus the local maximum of f subject to the constraint $x + y = 8$ must occur at $(4, 4)$. This value is $f(4, 4) = -32$. ■

EXAMPLE 15.24 A popular college textbook that is expected to sell 12,600 copies during the coming year is available in both a paperback and a more expensive hardcover edition. Let x and y denote the selling prices (in dollars) of the paperback and hardcover editions, respectively. At these prices the number of paperback copies that can be sold is

$$900 - 100x + 200y,$$

and the number of hardcover copies that can be sold is

$$1200 + 300x - 100y.$$

Find the selling prices for the two editions that maximize the publisher's revenue from the sales of this textbook. What is this maximum revenue?

Solution The publisher's revenue from the paperback edition equals the product of the selling price and the number of copies sold, which is

$$x(900 - 100x + 200y).$$

Likewise the publisher's revenue from the hardcover edition is

$$y(1200 + 300x - 100y).$$

Hence the total revenue from this textbook is given by

$$\begin{aligned} R(x, y) &= x(900 - 100x + 200y) + y(1200 + 300x - 100y) \\ &= 900x - 100x^2 + 200xy + 1200y + 300xy - 100y^2 \\ &= 900x - 100x^2 + 500xy + 1200y - 100y^2. \end{aligned}$$

We wish to maximize R subject to the constraint that the total demand for both editions is 12,600. This constraint leads to the equation

$$(900 - 100x + 200y) + (1200 + 300x - 100y) = 12{,}600,$$

which can be rewritten as

$$200x + 100y = 10{,}500$$

or

$$2x + y = 105.$$

Hence we will take

$$g(x, y) = 2x + y - 105$$

as the constraint function for the method of Lagrange multipliers.

The partial derivatives of R and g are

$$R_x(x, y) = 900 - 200x + 500y, \qquad R_y(x, y) = 500x + 1200 - 200y,$$
$$g_x(x, y) = 2, \text{ and} \qquad g_y(x, y) = 1.$$

Solving the equations

$$R_x(x, y) = \lambda g_x(x, y) \quad \text{and} \quad R_y(x, y) = \lambda g_y(x, y)$$

for λ produces

$$450 - 100x + 250y = \lambda \quad \text{and} \quad 500x + 1200 - 200y = \lambda.$$

Thus

$$\begin{aligned} 450 - 100x + 250y &= 500x + 1200 - 200y \\ -600x + 450y &= 750 \\ -4x + \quad 3y &= 5. \end{aligned}$$

Combining the preceding equation with the constraint equation produces the following system of linear equations.

$$\begin{aligned} -4x + 3y &= \quad 5 \\ 2x + \quad y &= 105 \end{aligned}$$

Using the techniques discussed in Chapter 2, we find that the solution of this system is $x = 31$ and $y = 43$. Hence the publisher's maximum revenue is obtained by selling the paperback edition at \$31 and the hardcover edition at \$43. The maximum revenue possible is $R(31, 43) = 465{,}000$ dollars. ■

Practice Problem 16 The function $f(x, y) = xy$ has a local maximum subject to the constraint $2x + y = 12$. What is it, and where does it occur?

EXAMPLE 15.25 Let $f(x, y)$ be the production function for a commodity, where x and y denote the number of units of labor and capital, respectively, and let c_1 and c_2 denote the respective costs per unit of labor and capital. We will use the method of Lagrange multipliers to determine the use of labor and capital that maximizes f subject to a fixed budget for labor and capital. (This problem is sometimes called the *fixed budget problem*.)

Discussion Since there is a fixed budget for labor and capital, the total expenditure for labor and capital must be a constant k. Now the total expenditure for labor is c_1x, and the total expenditure for capital is c_2y. Hence we have the constraint

$$c_1x + c_2y = k.$$

Thus the constraint function for the method of Lagrange multipliers is

$$g(x, y) = c_1x + c_2y - k.$$

Taking partial derivatives, we obtain

$$g_x(x, y) = c_1 \quad \text{and} \quad g_y(x, y) = c_2.$$

Therefore by Theorem 15.2 we see that if f is maximized at (x, y), then

$$f_x(x, y) = c_1\lambda \quad \text{and} \quad f_y(x, y) = c_2\lambda.$$

Hence

$$\frac{f_x(x, y)}{c_1} = \lambda \quad \text{and} \quad \frac{f_y(x, y)}{c_2} = \lambda,$$

and so

$$\frac{f_x(x, y)}{c_1} = \frac{f_y(x, y)}{c_2}.$$

The equation above has a meaningful economic interpretation. Recall that $f_x(x, y)$ and $f_y(x, y)$ are the marginal productivity of labor and capital, respectively. Since c_1 represents the cost per unit for labor (e.g., the prevailing hourly wage), the left side of the equation represents the marginal productivity of one dollar's worth of labor. Similarly, the right side of the equation represents the marginal productivity of one dollar's worth of capital. Thus the preceding equation states that the marginal productivity of one dollar's worth of labor and capital must be equal if production is to be maximized. This conclusion makes sense intuitively: If the marginal productivity of one dollar's worth of labor were not equal to the marginal productivity of one dollar's worth of capital, then production could be increased at no additional cost by using more of the input having the larger marginal productivity and less of the other. ■

EXTENSIONS OF THE METHOD OF LAGRANGE MULTIPLIERS

So far in this section we have considered only functions of two variables. Actually the method of Lagrange multipliers can be applied to a function of any number of variables, and in this more general situation the analog of Theorem 15.2 remains true. The following example demonstrates the use of the method of Lagrange multipliers with a function of three variables.

EXAMPLE 15.26 A glass aquarium is to be constructed that can hold 32 cubic feet of water. If the aquarium is to be in the shape of a rectangular box that is open on the top, what dimensions will minimize the amount of glass needed, and what is this amount?

Solution Let the aquarium be x feet long, y feet wide, and z feet deep as in Figure 15.14. Then the bottom of the aquarium requires xy square feet of glass, the two sides require $2yz$ square feet of glass, and the front and back together require $2xz$ square feet of glass. Thus the total amount of glass needed to make the aquarium is

$$f(x, y, z) = xy + 2yz + 2xz.$$

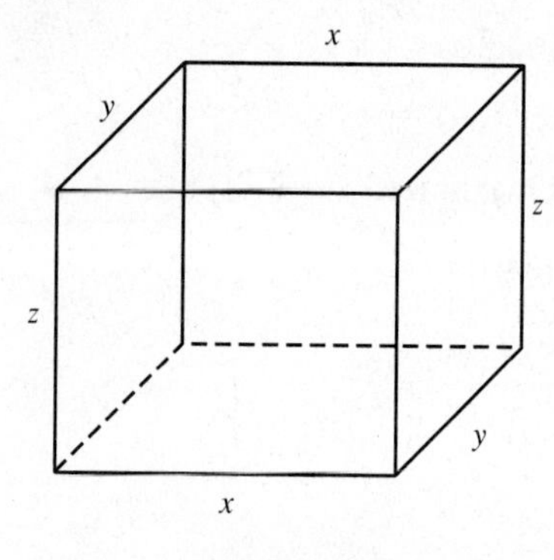

FIGURE 15.14

We wish to minimize f subject to the constraint that the volume of the aquarium be 32 cubic feet, that is, subject to the condition $xyz = 32$. Thus the constraint function for the method of Lagrange multipliers is

$$g(x, y, z) = xyz - 32.$$

Proceeding as described on page 845 for the two-variable case, we first compute

$$f_x(x, y, z) = y + 2z, \quad f_y(x, y, z) = x + 2z, \quad f_z(x, y, z) = 2y + 2x,$$
$$g_x(x, y, z) = yz, \quad g_y(x, y, z) = xz, \quad \text{and} \quad g_z(x, y, z) = xy.$$

Then we equate $f_x(x, y, z)$ with $\lambda g_x(x, y, z)$, $f_y(x, y, z)$ with $\lambda g_y(x, y, z)$, and $f_z(x, y, z)$ with $\lambda g_z(x, y, z)$, to produce the following equations:

$$\lambda = \frac{y + 2z}{yz}, \quad \lambda = \frac{x + 2z}{xz}, \quad \text{and} \quad \lambda = \frac{2x + 2y}{xy}.$$

Setting equal the first two expressions for λ and simplifying gives

$$\begin{aligned} \frac{y + 2z}{yz} &= \frac{x + 2z}{xz} \\ xz(y + 2z) &= yz(x + 2z) \\ xyz + 2xz^2 &= xyz + 2yz^2 \\ 2xz^2 &= 2yz^2. \end{aligned}$$

Since $z \neq 0$, we can divide both sides of this last equation by $2z^2$ to obtain $x = y$. Likewise, setting equal the second and third expressions for λ and simplifying yields

$$\begin{aligned} \frac{x + 2z}{xz} &= \frac{2x + 2y}{xy} \\ xy(x + 2z) &= xz(2x + 2y) \\ x^2y + 2xyz &= 2x^2z + 2xyz \\ x^2y &= 2x^2z \\ y &= 2z. \end{aligned}$$

Thus $x = y = 2z$.

Now we substitute for x and y in the constraint equation.

$$\begin{aligned} g(x, y, z) &= 0 \\ xyz - 32 &= 0 \\ xyz &= 32 \\ (2z)(2z)z &= 32 \\ 4z^3 &= 32 \\ z^3 &= 8 \\ z &= 2 \end{aligned}$$

Thus $x = y = 2z = 2(2) = 4$. Hence the amount of glass needed for the aquarium is minimized when the aquarium is 4 feet long, 4 feet wide, and 2 feet deep. The minimal amount of glass needed is $f(4, 4, 2) = 48$ square feet. ■

Practice Problem 17 The function $f(x, y, z) = x^2yz$ has a local maximum subject to the constraint $x + 2y + z = 8$. Use the method of Lagrange multipliers to determine this value and where it occurs.

Although we will not do so in this book, the method of Lagrange multipliers can also be applied to problems involving more than one constraint. In this case there must be a Lagrange multiplier λ_k for each constraint function g_k, but the method of Lagrange multipliers is basically the same as when only one constraint is involved.

EXERCISES 15.5

In Exercises 1–12 use the method of Lagrange multipliers to determine the location of all possible local extrema of f subject to the constraint $g(x, y) = 0$.

1. $f(x, y) = x^2 + xy + y^2, \quad g(x, y) = x - 2y - 14$

2. $f(x, y) = 6xy - 4x^2 - 4x - 3y^2 + 6y, \quad g(x, y) = -x + 2y + 4$

3. $f(x, y) = -x^2 + 6x - y^2 - 4y, \quad g(x, y) = x - y + 5$

4. $f(x, y) = x^2 + 4y^2 - xy + 3x - 6y, \quad g(x, y) = x + y - 5$

5. $f(x, y) = 2x^2 - 4x + y^2 + 6y, \quad g(x, y) = -2x + y - 7$

6. $f(x, y) = x^2 + 4y^2, \quad g(x, y) = x - 4y - 15$

7. $f(x, y) = x\sqrt{y}, \quad g(x, y) = x^2 + 2y^2 - 96$

8. $f(x, y) = x + y, \quad g(x, y) = x^2 + y^2 - 8$

9. $f(x, y) = 2x + y - 1, \quad g(x, y) = x^2 + 2y^2 - 18$

10. $f(x, y) = 3x - y + 6, \quad g(x, y) = x^2 + y^2 - 10$

11. $f(x, y) = x^2 + y^2, \quad g(x, y) = 2x - y + 15$

12. $f(x, y) = 3x^2 - 2xy + y^2, \quad g(x, y) = 2x + y - 11$

13. The function $f(x, y) = x^3 + 2y^3$ assumes local maximum and local minimum values subject to the constraint $2x + y = 15$. What are these values, and where do they occur?

14. The function $f(x, y) = x^3 - 6y^2$ assumes local maximum and minimum values subject to the constraint $4y - x = 14$. What are these values, and where do they occur?

15. The function $f(x, y) = x^2 + 4xy + y^2$ assumes a local minimum value subject to the constraint $x - y = 6$. What is this value, and where does it occur?

16. The function $f(x, y) = xy$ assumes a local maximum value subject to the constraint $2x + 3y = 12$. What is this value, and where does it occur?

17. The function $f(x, y) = x^2 + y^2$ assumes a local minimum value subject to the constraint $xy = 9$. What is this value, and where does it occur?

18. For $x > 0$ and $y > 0$ the function $f(x, y) = x^3 + 2y^{3/2}$ assumes a local minimum value subject to the constraint $xy = 1$. What is this value, and where does it occur?

19. The function $f(x, y) = 4x^2 - 4xy + y^2$ assumes a local maximum value subject to the constraint $x^2 + y^2 = 80$. What is this value, and where does it occur?

20. The function $f(x, y) = xy$ assumes local maximum values subject to the constraint $x^2 + 9y^2 = 18$. What are these values, and where do they occur?

21. The function $f(x, y, z) = 4x - 2y + z$ assumes a local minimum value subject to the constraint $2x^2 + y^2 = z$. What is this value, and where does it occur?

22. The function $f(x, y, z) = -6x + 12y - z$ assumes a local maximum value subject to the constraint $x^2 + 3y^2 = z$. What is this value, and where does it occur?

23. The function $f(x, y, z) = -x^2 - y^2 - z^2$ assumes a local maximum value subject to the constraint $x + 3y - 2z = 14$. What is this value, and where does it occur?

24. The function $f(x, y, z) = 2x^2 - 18y + z^2$ assumes a local minimum value subject to the constraint $2x^2 + y^2 = z^2$. What is this value, and where does it occur?

25. For $x > 0$, $y > 0$, and $z > 0$ the function $f(x, y, z) = x^2y^2z^2$ assumes a local maximum value subject to the constraint $x^2 + y^2 + z^2 = 12$. What is this value, and where does it occur?

26. The function $f(x, y, z) = 16x + y^2 - 2z$ assumes maximum values of 43 subject to the constraint $4x^2 + y^2 + z^2 = 26$. Where do these values occur?

27. The number of apartments that a builder can construct per year is given by the Cobb-Douglas production function $f(x, y) = 1.05x^{2/3}y^{1/3}$, where x and y denote the numbers of units of labor and capital, respectively. If each unit of labor costs \$80, each unit of capital costs \$60, and there is a total of \$36,000 available for labor and capital, determine the numbers of units of labor and capital that maximize production. Show that this answer agrees with the conclusion reached in Example 15.25.

28. Use the method of Lagrange multipliers to determine the points on the circle with equation $x^2 + y^2 = 45$ that are closest to and farthest from the point (2, 4). *Hint:* Find the local minimum and local maximum values of the *square* of the distance from (2, 4) to a point on the circle.

29. A box sent via United Parcel Service must have length plus girth not exceeding 108 inches. Determine the dimensions of the rectangular box with the greatest volume that can be sent via United Parcel Service.

30. A rectangular box with an open top is to have a surface area of 300 square inches. Determine the dimensions of the box with a maximum volume.

31. The prices at which a carpenter can sell x small coffee tables and y large coffee tables per year are $125 - x + 0.5y$ and $65 + 0.5x - 0.4y$ dollars, respectively. If each small table requires wood costing \$100, each large table requires wood costing \$200, and the carpenter's budget for wood is \$16,500 per year, how many tables of each size should the carpenter make in order to maximize his yearly profit?

32. The prices at which a leather artisan can sell x small handbags and y large handbags per month are $56 - x + 0.5y$ and $59 + 0.5x - y$ dollars, respectively. If each small handbag requires leather costing $20, each large handbag requires leather costing $30, and the artisan's budget for leather is $600 per month, how many handbags of each size should the artisan make in order to maximize her monthly profit?

Answers to Practice Problems

16. The maximum value of f subject to the constraint $2x + y = 12$ is 18 when $x = 3$ and $y = 6$.

17. The local maximum of 32 occurs at $(4, 1, 2)$.

15.6 Iterated Integrals

In previous sections of this chapter we have generalized the concept of differentiation to apply to functions of several variables. This generalization amounts to holding all but one of the variables constant and differentiating with respect to the remaining variable. In this section we will generalize the process of integration to functions of several variables in a similar manner. Although many of the ideas to be presented apply to functions of n variables, we will limit our discussion to functions of two variables.

The terminology in Section 12.1 is also used for functions of several variables. Thus

$$\int (8x^3y^2 - 6xy^3 + 3y + e^{2x})\, dx$$

is called an *indefinite integral,* and dx signifies that the *variable of integration* is x. Hence we hold all the variables constant except x, and find an antiderivative of the integrand just as we did in Chapter 12. Using the rules for indefinite integrals presented in Section 12.1, we see that

$$\begin{aligned} \int (8x^3y^2 - 6xy^3 + 3y + e^{2x})\, dx &= 8y^2 \int x^3\, dx - 6y^3 \int x\, dx + 3y \int dx + \int e^{2x}\, dx \\ &= 8y^2\left(\frac{x^4}{4}\right) - 6y^3\left(\frac{x^2}{2}\right) + 3yx + \frac{1}{2}e^{2x} + C(y) \\ &= 2x^4y^2 - 3x^2y^3 + 3xy + \frac{1}{2}e^{2x} + C(y). \end{aligned}$$

Notice that because y is being held constant, we can factor terms involving y outside the integral sign. However, *we are still prohibited from taking the variable of integration outside the integral sign.* Also note that instead of the usual constant of integration, in this calculation the answer contains an arbitrary function of y. Since y is being held constant, the function $C(y)$ is analogous to the constant of integration in the case of a function of one variable. In other words, since $\frac{\partial}{\partial x} C(y) = 0$, the presence of the term $C(y)$ does not change the partial derivative with respect to x of

$$2x^4y^2 - 3x^2y^3 + 3xy + \frac{1}{2}e^{2x} + C(y).$$

As with functions of one variable, it is possible to check the answer by differentiating, although in this case we must take the *partial* derivative of the answer with respect to the variable of integration x.

Check

$$\frac{\partial}{\partial x}\left[2x^4y^2 - 3x^2y^3 + 3xy + \frac{1}{2}e^{2x} + C(y)\right]$$
$$= 8x^3y^2 - 6xy^3 + 3y + e^{2x}$$

EXAMPLE 15.27 Evaluate $\int (15x^3y^4 - 8x^2y + 4e^{-x})\, dy$.

Solution In this case the differential dy signifies that the variable of integration is y, and so we treat x as a constant. Hence

$$\begin{aligned}\int (15x^3y^4 - 8x^2y + 4e^{-x})\, dy &= 15x^3 \int y^4\, dy - 8x^2 \int y\, dy + 4e^{-x} \int dy \\ &= 15x^3\left(\frac{y^5}{5}\right) - 8x^2\left(\frac{y^2}{2}\right) + (4e^{-x})y + C(x) \\ &= 3x^3y^5 - 4x^2y^2 + 4ye^{-x} + C(x).\end{aligned}$$

Since x is being held constant, the indefinite integral involves an arbitrary function of x instead of a constant of integration.

Check $\frac{\partial}{\partial y}[3x^3y^5 - 4x^2y^2 + 4ye^{-x} + C(x)] = 15x^3y^4 - 8x^2y + 4e^{-x}$ ■

Practice Problem 18 Evaluate $\int (24x^5y^8 - 9x^2y^5 + 6e^{-2x})\, dx$.

Integration by substitution, as described in Section 12.2, can also be used with functions of several variables. The next example demonstrates this technique.

EXAMPLE 15.28 Evaluate $\int 15x^2y(x^3 + y^2)^4\, dx$.

Solution Here the variable of integration is x. If we let $u = x^3 + y^2$, then $du = 3x^2\, dx$. Since we can factor $5y$ outside of the integral sign, this substitution will enable us to write the integrand in the form $u\, du$. The steps are as follows.

$$\int 15x^2y(x^3 + y^2)^4\,dx = 5y\int 3x^2(x^3 + y^2)^4\,dx$$
$$= 5y\int u^4\,du = 5y\left(\frac{u^5}{5}\right) + C(y)$$
$$= y(x^3 + y^2)^5 + C(y)$$

Check

$$\frac{\partial}{\partial x}[y(x^3 + y^2)^5 + C(y)] = 5y(x^3 + y^2)^4(3x^2) = 15x^2y(x^3 + y^2)^4 \quad ■$$

Practice Problem 19 Evaluate $\int 4x^2y(x^4 + y^2)^6\,dy$.

We can also use the fundamental theorem of calculus to evaluate definite integrals as in Section 12.4. The following pair of examples demonstrate the method. Note that the limits of integration are values of the variable of integration, which is x in Example 15.29 and y in Example 15.30.

EXAMPLE 15.29 Evaluate $\int_0^1 f(x, y)\,dx$ if $f(x, y) = 10x^4y^2 - 3x^2y + 2x$.

Solution First we compute an antiderivative of the integrand:

$$\int f(x, y)\,dx = \int (10x^4y^2 - 3x^2y + 2x)\,dx$$
$$= 2x^5y^2 - x^3y + x^2 + C(y).$$

Let us denote this function by $F(x, y)$. Now we evaluate

$$F(x, y)\Big|_0^1 = F(1, y) - F(0, y)$$

in order to obtain the answer.

$$\int_0^1 (10x^4y^2 - 3x^2y + 2x)\,dx = [2x^5y^2 - x^3y + x^2 + C(y)]\Big|_0^1$$
$$= [2(1)^5y^2 - (1)^3y + (1)^2 + C(y)] - [2(0)^5y^2 - (0)^3y + (0)^2 + C(y)]$$
$$= 2y^2 - y + 1$$

Observe that $C(y)$ drops out. Hence as in Section 12.4 we can use *any* antiderivative of the integrand to evaluate a definite integral. ■

EXAMPLE 15.30 Evaluate $\int_{-1}^2 f(x, y)\,dy$ if $f(x, y) = 8x^2y - 6xy^2$.

Solution The indefinite integral of f is

$$\int f(x, y)\,dy = \int (8x^2y - 6xy^2)\,dy = 4x^2y^2 - 2xy^3 + C(x).$$

Since we can use any antiderivative of f, we will take $C(x) = 0$. Hence

$$\begin{aligned}\int_{-1}^{2} (8x^2y - 6xy^2)\,dy &= (4x^2y^2 - 2xy^3)\Big|_{-1}^{2} \\ &= [4x^2(2)^2 - 2x(2)^3] - [4x^2(-1)^2 - 2x(-1)^3] \\ &= (16x^2 - 16x) - (4x^2 + 2x) \\ &= 12x^2 - 18x. \quad ■\end{aligned}$$

In Examples 15.29 and 15.30 notice that $\int_a^b f(x, y)\,dx$ is a function of y whereas $\int_a^b f(x, y)\,dy$ is a function of x. These functions of one variable can therefore be integrated with respect to the single variable present.

EXAMPLE 15.31 Evaluate $\int_{-2}^{4} \left[\int_0^1 f(x, y)\,dx\right] dy$ if $f(x, y) = 4xy^3 - 3x^2y$.

Solution

$$\begin{aligned}\int_0^1 f(x, y)\,dx &= \int_0^1 (4xy^3 - 3x^2y)\,dx \\ &= (2x^2y^3 - x^3y)\Big|_0^1 \\ &= [2(1)^2y^3 - (1)^3y] - [2(0)^2y^3 - (0)^3y] \\ &= 2y^3 - y\end{aligned}$$

Thus

$$\begin{aligned}\int_{-2}^{4} \left[\int_0^1 f(x, y)\,dx\right] dy &= \int_{-2}^{4} (2y^3 - y)\,dy \\ &= \left(\frac{1}{2}y^4 - \frac{1}{2}y^2\right)\Big|_{-2}^{4} \\ &= \left[\frac{1}{2}(4)^4 - \frac{1}{2}(4)^2\right] - \left[\frac{1}{2}(-2)^4 - \frac{1}{2}(-2)^2\right] \\ &= (128 - 8) - (8 - 2) \\ &= 120 - 6 = 114. \quad ■\end{aligned}$$

EXAMPLE 15.32 Evaluate $\int_0^1 \left[\int_{-2}^{4} f(x, y)\,dy\right] dx$ if $f(x, y) = 4xy^3 - 3x^2y$.

Solution

$$\begin{aligned}\int_{-2}^{4} f(x, y)\,dy &= \int_{-2}^{4} (4xy^3 - 3x^2y)\,dy \\ &= \left[xy^4 - \frac{3}{2}x^2y^2\right]\Bigg|_{-2}^{4} \\ &= \left[x(4)^4 - \frac{3}{2}x^2(4)^2\right] - \left[x(-2)^4 - \frac{3}{2}x^2(-2)^2\right] \\ &= (256x - 24x^2) - (16x - 6x^2) \\ &= 240x - 18x^2\end{aligned}$$

Thus

$$\begin{aligned}\int_0^1 \left[\int_{-2}^{4} f(x, y)\,dy\right] dx &= \int_0^1 (240x - 18x^2)\,dx \\ &= (120x^2 - 6x^3)\Bigg|_0^1 \\ &= (120 - 6) = 114. \quad \blacksquare\end{aligned}$$

We see in Examples 15.31 and 15.32 that

$$\int_0^1 \left[\int_{-2}^{4} f(x, y)\,dy\right] dx = \int_{-2}^{4} \left[\int_0^1 f(x, y)\,dx\right] dy.$$

This equality is not a coincidence. For most of the functions encountered in applications it is true that

$$\int_a^b \left[\int_c^d f(x, y)\,dy\right] dx = \int_c^d \left[\int_a^b f(x, y)\,dx\right] dy.$$

An expression of either of these forms is called an **iterated integral** because it requires that we integrate twice, first with respect to one variable and then with respect to the other. We will follow the customary practice of omitting the brackets in an iterated integral; thus the iterated integrals above will be written as

$$\int_a^b \int_c^d f(x, y)\,dy\,dx \qquad \text{and} \qquad \int_c^d \int_a^b f(x, y)\,dx\,dy.$$

Practice Problem 20 Evaluate the iterated integral $\int_{-1}^{1} \int_0^2 (3x^2y + 6xy^2)\,dx\,dy$.

VOLUME

We saw in Section 12.3 that if $f(x) \geq 0$ for $a \leq x \leq b$, then the definite integral

$$\int_a^b f(x)\,dx$$

represents the area under the curve $y = f(x)$, above the x-axis, and between the vertical lines $x = a$ and $x = b$. Therefore it is not surprising that an iterated integral has a similar geometric interpretation.

In either of the iterated integrals

$$\int_a^b \int_c^d f(x, y)\, dy\, dx = \int_c^d \int_a^b f(x, y)\, dx\, dy,$$

the variables x and y satisfy the inequalities

$$a \leq x \leq b \qquad \text{and} \qquad c \leq x \leq d.$$

The set of points (x, y) in the coordinate plane satisfying these conditions is the rectangle R in Figure 15.15.

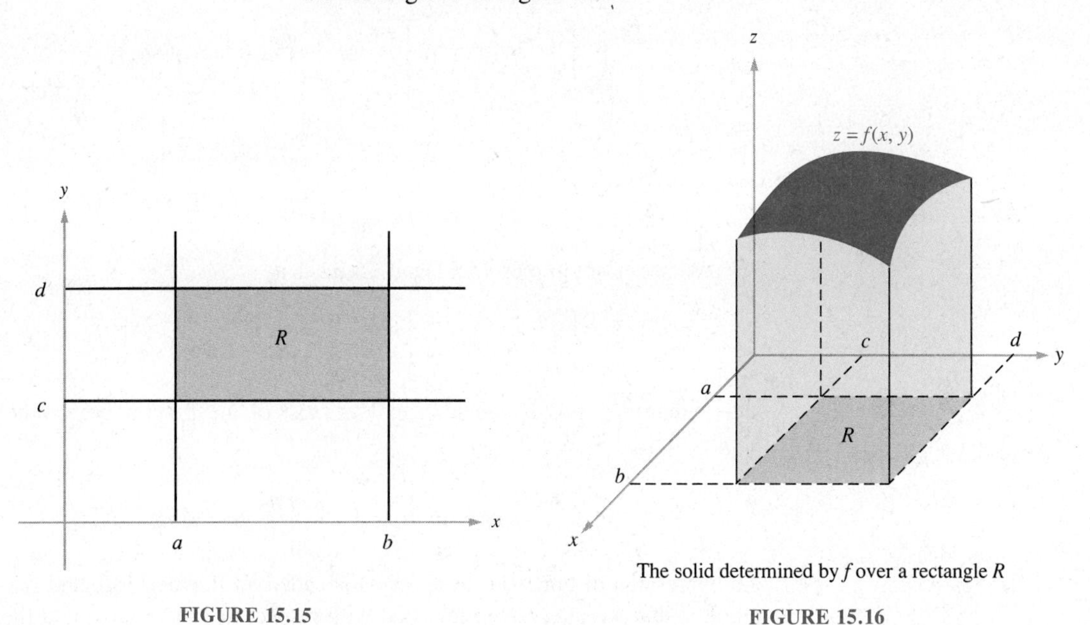

The solid determined by f over a rectangle R

FIGURE 15.15

FIGURE 15.16

If $f(x, y)$ is a function of two variables such that

$$f(x, y) \geq 0 \quad \text{for} \quad (x, y) \in R,$$

then on rectangle R the graph of f lies entirely above the plane $z = 0$. Hence the graph of f, which is the surface $z = f(x, y)$, forms the top of a solid figure having the plane $z = 0$ as its bottom, the planes $x = a$ and $x = b$ as its front and back, and the planes $y = c$ and $y = d$ as its sides. We will call this solid figure the **solid determined by f over R.** See Figure 15.16.

The theorem below states that the volume of the solid determined by a nonnegative function of two variables over a rectangle can be computed as an iterated integral.

Theorem 15.3

Let f be a function of two variables that is continuous and nonnegative on the rectangle

$$\{(x, y) | a \le x \le b \quad \text{and} \quad c \le y \le d\}.$$

Then the volume of the solid determined by f is given by either of the iterated integrals

$$\int_a^b \int_c^d f(x, y)\, dy\, dx = \int_c^d \int_a^b f(x, y)\, dx\, dy.$$

EXAMPLE 15.33 Determine the volume of the solid determined by $f(x, y) = x + 2y$ over the rectangle $\{(x, y) | 1 \le x \le 4 \quad \text{and} \quad 2 \le y \le 3\}$.

Solution The volume of the solid determined by f is given by either of the iterated integrals

$$\int_1^4 \int_2^3 f(x, y)\, dy\, dx = \int_2^3 \int_1^4 f(x, y)\, dx\, dy.$$

To evaluate the former, we begin by computing the inner integral

$$\begin{aligned}
\int_2^3 f(x, y)\, dy &= \int_2^3 (x + 2y)\, dy \\
&= (xy + y^2)\Big|_2^3 \\
&= (3x + 9) - (2x + 4) \\
&= x + 5.
\end{aligned}$$

Thus

$$\begin{aligned}
\int_1^4 \int_2^3 f(x, y)\, dy\, dx &= \int_1^4 (x + 5)\, dx \\
&= \left(\frac{1}{2}x^2 + 5x\right)\Big|_1^4 \\
&= (8 + 20) - \left(\frac{1}{2} + 5\right) \\
&= 22\frac{1}{2}.
\end{aligned}$$

Therefore the volume of the solid determined by f is 22.5. ■

Practice Problem 21 Determine the volume of the solid determined by $f(x, y) = 6x^2y$ over the rectangle $\{(x, y)|0 \leq x \leq 3 \text{ and } 2 \leq y \leq 5\}$.

When it is not true that $f(x, y) \geq 0$ for all (x, y) in rectangle

$$\{(x, y)|a \leq x \leq b \quad \text{and} \quad c \leq y \leq d\},$$

the value of the iterated integrals

$$\int_a^b \int_c^d f(x, y)\, dy\, dx = \int_c^d \int_a^b f(x, y)\, dx\, dy$$

can no longer be interpreted as the volume of a solid. But the graph of f that is above the plane $z = 0$ determines a solid figure F_+ as before. In addition, the graph of f that is below the plane $z = 0$ determines another solid figure F_-; this solid is bounded on the bottom by the graph of f and on the top by the plane $z = 0$. In this case the value of the iterated integrals

$$\int_a^b \int_c^d f(x, y)\, dy\, dx = \int_c^d \int_a^b f(x, y)\, dx\, dy,$$

equals the volume of F_+ minus the volume of F_-. Note the similarity between this interpretation and that of

$$\int_a^b f(x)\, dx$$

in Section 12.4 when f fails to be nonnegative on the interval $[a, b]$.

EXERCISES 15.6

Evaluate the indefinite integrals in Exercises 1–8.

1. $\int 12x^3y^2\, dx$

2. $\int 12x^3y^2\, dy$

3. $\int (3x - 5y + 2xy)\, dy$

4. $\int (x^3y^2 - 4x + e^{2y})\, dx$

5. $\int (x^4y - x^2y^3)\, dx$

6. $\int (xy^3 + ye^{-x})\, dy$

7. $\int 2xe^{xy}\, dy$

8. $\int x(x^2 + y)^7\, dx$

Evaluate the iterated integrals in Exercises 9–20.

9. $\int_{-1}^1 \int_{-1}^2 3x^2y\, dx\, dy$

10. $\int_0^4 \int_1^3 6x^2y\, dy\, dx$

11. $\int_0^1 \int_1^2 (1 - x^2)\, dy\, dx$

12. $\int_0^1 \int_1^4 9\sqrt{xy}\, dx\, dy$

13. $\int_0^2 \int_{-1}^2 (2x + y^2)\, dx\, dy$

14. $\int_1^e \int_1^3 \frac{2y}{x}\, dy\, dx$

15. $\int_0^2 \int_0^1 xe^{xy}\, dy\, dx$

16. $\int_{-5}^5 \int_0^2 \frac{5xy}{(1 + x^2)^2}\, dx\, dy$

17. $\int_0^1 \int_{-1}^4 (3x - 6y)\, dy\, dx$

18. $\int_1^2 \int_0^3 (4x - 2y)\, dx\, dy$

19. $\int_{-3}^4 \int_1^2 \frac{1}{y}\, dx\, dy$

20. $\int_{-2}^3 \int_1^4 4xy\, dy\, dx$

In Exercises 21–32 compute the volume of the solid determined by f over R.

21. $f(x, y) = 5x + y, \quad R = \{(x, y) | 1 \le x \le 3 \text{ and } 0 \le y \le 2\}$

22. $f(x, y) = 2x + 3y, \quad R = \{(x, y) | -1 \le x \le 1 \text{ and } 2 \le y \le 5\}$

23. $f(x, y) = 3x^2, \quad R = \{(x, y) | 1 \le x \le 2 \text{ and } 2 \le y \le 4\}$

24. $f(x, y) = 4y^3, \quad R = \{(x, y) | 1 \le x \le 3 \text{ and } 0 \le y \le 1\}$

25. $f(x, y) = x^2y, \quad R = \{(x, y) | 1 \le x \le 3 \text{ and } 0 \le y \le 4\}$

26. $f(x, y) = xy^3, \quad R = \{(x, y) | 0 \le x \le 2 \text{ and } -1 \le y \le 3\}$

27. $f(x, y) = 2xy, \quad R = \{(x, y) | -1 \le x \le 1 \text{ and } 2 \le y \le 5\}$

28. $f(x, y) = \frac{1}{y}, \quad R = \{(x, y) | -2 \le x \le -1 \text{ and } 1 \le y \le 3\}$

29. $f(x, y) = \sqrt{xy}, \quad R = \{(x, y) | 4 \le x \le 9 \text{ and } 1 \le y \le 4\}$

30. $f(x, y) = x^2 + y^2, \quad R = \{(x, y) | 1 \le x \le 2 \text{ and } 0 \le y \le 1\}$

31. $f(x, y) = 2ye^x, \quad R = \{(x, y) | 0 \le x \le 1 \text{ and } -1 \le y \le 2\}$

32. $f(x, y) = e^{-y}, \quad R = \{(x, y) | 2 \le x \le 3 \text{ and } -1 \le y \le 0\}$

Answers to Practice Problems

18. $4x^6y^8 - 3x^3y^5 - 3e^{-2x} + C(y)$

19. $\frac{2}{7}x^2(x^4 + y^2)^7 + C(x)$

20. 8

21. 567

CHAPTER 15 REVIEW

IMPORTANT TERMS

- **function of *n* variables** *(15.1)*
- **graph of a function**
- **production function**
- **partial derivative** *(15.2)*
- **local extremum** *(15.3)*
- **local maximum**
- **local minimum**
- **saddle point**
- **coefficient of correlation** *(15.4)*
- **least squares line**
- **scatter diagram**
- **Lagrange multiplier** *(15.5)*
- **iterated integral** *(15.6)*
- **solid determined by a function**

REVIEW EXERCISES

1. Evaluate $f(-2, 1)$ if $f(x, y) = 3x^2y - 4xy^2$.

2. Evaluate $g(1, -1, 2)$ if $g(x, y, z) = 2x^2y^2 - 3x^2z^2 + y^2z^2 - 4xyz$.

3. Compute both first-order partial derivatives of $f(x, y) = (x^2 + 3x) \ln y$.

4. Compute both first-order partial derivatives of $F(x, y) = x^2e^{xy}$.

5. Compute the four second-order partial derivatives of $g(x, y) = 3xy^4 - 2x^2y$.

6. Compute the four second-order partial derivatives of $G(x, y) = 6x^5y^3 - 4x^3y^8$.

7. Determine all the critical points of $f(x, y) = 2x^2 + xy - y^2 + 4x + 10y$.

8. Determine all the critical points of $g(x, y) = 9 + \sqrt{2x^2 + 3y^2 - 12y + 13}$.

In Exercises 9–12 determine all the local extrema of the given function.

9. $f(x, y) = -2x^2 + 16x - y^2 - 10y - 12$

10. $g(x, y) = x^2 + 14x + 3y^2 - 24y + 5$

11. $F(x, y) = x^4 - xy + y^4$

12. $G(x, y) = xy + \dfrac{4}{y} - \dfrac{16}{x}$

In Exercises 13–16 determine the location of all possible local extrema of f subject to the constraint $g(x, y) = 0$.

13. $f(x, y) = 2x^2 + 2xy - 12x + 4y^2$, $g(x, y) = x + 2y - 12$

14. $f(x, y) = 6x - x^2 + 20y - 2y^2$, $g(x, y) = x - y + 2$

15. $f(x, y) = 3x - y$, $g(x, y) = 3x^2 + y^2 - 16$

16. $f(x, y) = x^2 + y^2 - 2$, $g(x, y) = x^2 + y^2 + 2x - 4y$

17. (a) Plot a scatter diagram for the data (2, 3), (6, 7), (8, 10), and (10, 12).

(b) Compute the coefficient of correlation for the data in part (a).

(c) Find the equation of the least squares line for the data in part (a).

18. In reference [3], Sharpe and Johnsgard investigate the relationship between the physical and behavioral characteristics of 11 offspring resulting from the mating of pintail and mallard ducks. These characteristics were measured by a plumage index and a behavioral index, respectively. Their data are shown below.

Plumage index	7	7	6	4	9	8	13	14	14	14	15
Behavioral index	3	4	5	7	9	10	10	11	11	15	15

Plot a scatter diagram, compute the coefficient of correlation, and find the equation of the least squares line for these data.

19. Evaluate $\int (6x^2y - 12x^5y^3)\,dy$.

20. Evaluate $\int (6x^2y - 12x^5y^3)\,dx$.

21. Evaluate $\int_{-1}^{1}\int_{1}^{2} (6x^2y - 2xy)\,dy\,dx$.

22. Evaluate $\int_{1}^{e}\int_{0}^{9} \frac{\sqrt{x}}{y}\,dx\,dy$.

23. Find the volume of the solid determined by $f(x, y) = \frac{12x}{y^2}$ over the rectangle

$$\{(x, y) | 1 \leq x \leq 3 \quad \text{and} \quad 4 \leq y \leq 16\}.$$

24. Find the volume of the solid determined by $f(x, y) = xe^y$ over the rectangle

$$\{(x, y) | 0 \leq x \leq 2 \quad \text{and} \quad 0 \leq y \leq 1\}.$$

25. A manufacturer produces two models of clock radios. When the less expensive model is sold at a price of x dollars and the more expensive model is sold at a price of y dollars, the number of radios of each model that can be sold per month are $3000 - 80x + 30y$ and $2750 + 30x - 50y$, respectively. At what prices should the two models of radios be sold in order to maximize the manufacturer's revenue?

26. A manufacturer's inventory cost (in dollars) is given by

$$C(x, y) = \frac{1{,}000{,}000}{x} + \frac{25y^2}{x} + 20x - 40y,$$

where x denotes the number of units to be produced per lot and y denotes the number of units to be stored in inventory. Determine the values of x and y that minimize the inventory cost.

27. If the material for the bottom of a rectangular box costs \$3 per square foot and the material for the top and sides costs \$1 per square foot, find the dimensions of the box with greatest volume that can be constructed for \$12.

28. Determine the points on the curve with equation $x^2 - y^2 = 1$ that are closest to the point (0, 1). *Hint:* Find the local minimum value of the *square* of the distance from (0, 1) to a point on the curve.

REFERENCES

1. Balachandran, N. K., W. L. Donn, and D. H. Rind, "Concorde Sonic Booms as an Atmospheric Probe," *Science,* vol. 197 (July 1, 1977), p. 97.
2. Fadeley, Robert Cunningham, "Oregon Malignancy Pattern Physiographically Related to Hanford Washington Radioisotope Storage," *Journal of Environmental Health,* vol. 27 (May-June 1965), pp. 883–897.
3. Sharpe, Roger S. and Paul Johnsgard, "Inheritance of Behavioral Characters in F_2 Mallard × Pintail (*Anas platyrhynchos* L. × *Anas acuta* L.) Hybrids," *Behaviour,* vol. 27 (1966), pp. 259–272.
4. Weinstein, Alan G. and V. Srinivasan, "Predicting Managerial Success of Master of Business Administration (MBA) Graduates," *Journal of Applied Psychology,* vol. 59, no. 2, pp. 207–212.

APPENDIX

ALGEBRA REVIEW

Throughout this book we use basic algebraic techniques from intermediate algebra. In this appendix we review some of these techniques. No attempt will be made to present a complete development of each topic, but this material should serve to recall your previous algebra training and to refresh basic skills. This appendix may be studied as a complete unit or used as a reference when needed.

A.1 The Real Number System

The only numbers we will be concerned with in this book are **real numbers,** which can be thought of as corresponding to the points on a line. Some real numbers are labeled in Figure A.1.

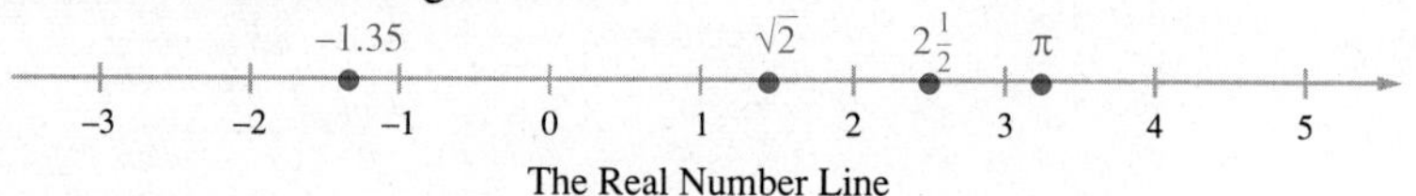

The Real Number Line

FIGURE A.1

FUNDAMENTAL OPERATIONS

Real numbers may be combined using four basic **binary operations,** that is, a pair of real numbers may be added, subtracted, multiplied, or divided, producing another real number. These operations obey various algebraic rules, the most basic of which are the following.

For all real numbers a, b, and c:

(A1) $a + b = b + a$ *(commutative law of addition)*

(A2) $ab = ba$ *(commutative law of multiplication)*

(A3) $a + (b + c) = (a + b) + c$ *(associative law of addition)*

(A4) $a(bc) = (ab)c$ *(associative law of multiplication)*

(A5) $a(b + c) = (ab) + (ac)$ *(distributive law)*

(A6) $a + 0 = a$ *(additive identity)*

(A7) $1a = a$ *(multiplicative identity)*

(A8) There exists a real number denoted $-a$ such that

$a + (-a) = 0$. *(additive inverse)*

(A9) If $a \neq 0$, there exists a real number denoted $\frac{1}{a}$ such that

$a\left(\frac{1}{a}\right) = 1$. *(multiplicative inverse)*

The rules just listed involve only addition and multiplication. We can define subtraction and division from these operations as follows:

$b - a = b + (-a)$, where $-a$ is the element given in (A8).

If $a \neq 0$, then $\dfrac{b}{a} = b\left(\dfrac{1}{a}\right)$, where $\dfrac{1}{a}$ is the element given in (A9).

Other rules which are useful in computation follow from those just listed. Examples are

$$-(a - b) = -a + b,$$

$$0a = 0, \qquad \text{and}$$

$$\frac{\frac{a}{b}}{\frac{c}{d}} = \frac{ad}{bc} \qquad \text{if none of } b,\ c, \text{ and } d \text{ is } 0.$$

The last rule amounts to the familiar fact that to divide fractions we invert and multiply. We assume the reader can perform computations involving the four basic operations correctly.

ORDER OF OPERATIONS

Rules (A3) and (A4) mean that we can omit parentheses when indicating a sequence of additions or multiplications. For example, it does not matter whether we interpret $3 + 5 + 2$ as $3 + (5 + 2)$ or $(3 + 5) + 2$, since (A3) tells us that both computations give the same answer. The operations of subtraction and division are *not* associative, however, as the following examples illustrate.

$$3 - (5 - 2) = 0, \qquad \text{while} \qquad (3 - 5) - 2 = -4$$

$$\frac{3}{\left(\frac{5}{2}\right)} = \frac{6}{5}, \qquad \text{while} \qquad \frac{\left(\frac{3}{5}\right)}{2} = \frac{3}{10}$$

For subtraction the convention is that operations are performed from left to right, so that we would interpret

$$3 - 5 - 2$$

as $(3 - 5) - 2 = -4$. In fact this convention applies to any sequence of additions and subtractions. For example

$$\begin{aligned} 4 - 7 + 10 - 3 - 1 &= [((4 - 7) + 10) - 3] - 1 \\ &= [(-3 + 10) - 3] - 1 \\ &= (7 - 3) - 1 = 4 - 1 = 3. \end{aligned}$$

There is no such convention for division. Thus an expression such as

$$3/5/2$$

has no standard interpretation, and should be avoided.

When an expression involves both addition or subtraction and multiplication or division, it is a convention that *unless the contrary is indicated by parentheses, multiplication or division is performed before addition or subtraction*. For example, $3 + 5/2$ means $3 + (5/2) = 5.5$, and not $(3 + 5)/2 = 4$. With this convention the parentheses could have been omitted in the right side of (A5), making it

$$a(b + c) = ab + ac.$$

The conventions we have been discussing are universal and useful for efficient communication. Thus $(x + 3)(x + 4)$ *cannot* be written as

$$x + 3 \cdot x + 4$$

because the conventional interpretation of the latter expression is

$$x + (3 \cdot x) + 4 = 4x + 4.$$

Many mistakes are caused by insufficient use of parentheses. For example, if we know that $y = 2x$ and $z = x - 3$, then

$$y - z = 2x - (x - 3) = 2x - x + 3 = x + 3.$$

Carelessly writing

$$y - z = 2x - x - 3 = x - 3$$

leads to the wrong result.

Practice Problem 1 Evaluate $1 - (2 \cdot 3 - 4)/5 - 6/(7 - 8)$.

INTEGERS AND RATIONAL NUMBERS

Certain subsets of the real numbers have been named. Most fundamental are the **natural numbers** 1, 2, 3, ... These are the numbers used to count nonempty sets. Since the sum or product of two natural numbers is another natural number, we say that the natural numbers are **closed** under addition and multiplication. If we include zero and the additive inverses of the natural numbers, we get the **integers:**

$$\ldots, -3, -2, -1, 0, 1, 2, 3, 4, \ldots.$$

The integers are closed under addition, multiplication, and subtraction, but not division. For example, 3 and 5 are integers, but 3/5 is not. To get a set also closed under division (except by 0) we must consider the **rational numbers,** that is, those numbers a/b that can be written as the quotient of two integers a and b with $b \neq 0$. Examples of rational numbers are

$$\frac{2}{3}, \quad 1\frac{1}{4} = \frac{5}{4}, \quad -4 = \frac{(-4)}{1}, \quad \text{and} \quad 3.79 = \frac{379}{100}.$$

Every integer is also a rational number, since if n is an integer, then $n = n/1$.

It can be proved that every rational number has a decimal expansion that is either **terminating** (such as $5/4 = 1.25$), or else **repeating** (such as $1/3 = .3333\ldots$ or $7981/4950 = 1.61232323\ldots$, where the 3's and 23's go on forever). There are real numbers that cannot be written as the quotient of integers, however, and these are called **irrational numbers.** An irrational number has a decimal expansion that does not terminate or repeat. An example is the number

$$.12345678910111213\ldots,$$

where the natural numbers are listed in order. Others are $\sqrt{2}$, π (the ratio of the circumference of a circle to its diameter), and the number e discussed in Chapters 8 and 9.

The relationships among the various sets we have been discussing is illustrated in Figure A.2.

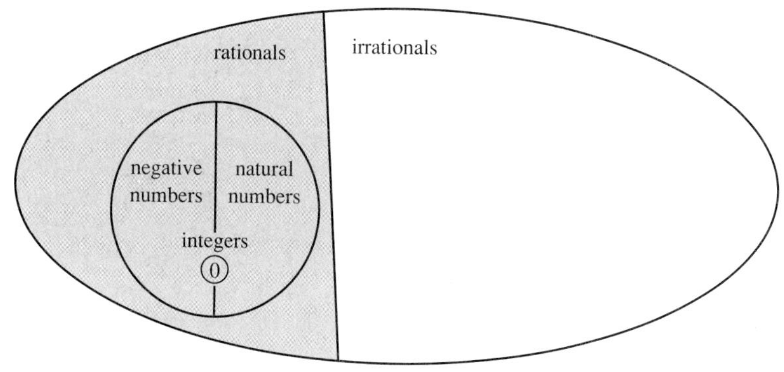

FIGURE A.2

EXERCISES A.1

In Exercises 1–10 evaluate the given expression as a single number. If the answer is a fraction, write it in lowest terms.

1. $5 + 3 - 4 - 1$

2. $-3 + 7 - (-2) - 5$

3. $7 - \frac{6}{2} + 3$

4. $3 \cdot 5 - 4 \cdot 2$

5. $\frac{3}{5} + \frac{7}{5}$

6. $\frac{5}{7} + 3 - \frac{1}{7}$

7. $3 + 1 \cdot 4 + 1$

8. $(3 + 1) \cdot 4 + 1$

9. $\frac{3}{7} + \frac{2}{3}$

10. $\frac{5}{6} - \frac{7}{4} + 1$

In Exercises 11–20 compute the two expressions given. If the answers are equal, tell which of rules (A1) through (A9) guarantees their equality.

11. $5(2 + 7)$ and $5 \cdot 2 + 5 \cdot 7$

12. $1(3 - 6)$ and $3 - 6$

13. $5 - (3 - 1)$ and $(5 - 3) - 1$

14. $4\left(\frac{1}{4}\right)$ and 1

15. $4(3 \cdot 2)$ and $(4 \cdot 3)2$

16. $3 + 5 \cdot 2$ and $(3 + 5)(3 + 2)$

17. $\frac{2}{3} + \frac{1}{2}$ and $\frac{2 + 1}{3 + 2}$

18. $\frac{3}{4} + \frac{1}{5}$ and $\frac{1}{5} + \frac{3}{4}$

19. $\frac{16}{\left(\frac{8}{2}\right)}$ and $\frac{\left(\frac{16}{8}\right)}{2}$

20. $(5 \cdot 2) + 4$ and $5 \cdot (2 + 4)$

In Exercises 21–30 state whether or not each equation is true for all values of the variables.

21. $a(b - c) = ab - ac$

22. $-\left(\frac{a}{b}\right) = \frac{-a}{-b}$, for $b \neq 0$

23. $\frac{a}{b} + \frac{c}{d} = \frac{a + c}{b + d}$, for $b \neq 0, d \neq 0, b + d \neq 0$

24. $\frac{\frac{a}{b}}{\frac{c}{d}} = \frac{\frac{a}{c}}{\frac{b}{d}}$, for $b \neq 0, c \neq 0, d \neq 0$

25. $\frac{a}{b} - \frac{c}{d} = \frac{ad - bc}{bc}$, for $b \neq 0, c \neq 0$

26. $(a + b)(c - d) = ac - ad + bc - bd$

27. $(a + b)(a - b) = a \cdot a - b \cdot b$

28. $\frac{a + c}{b + c} = \frac{a}{b}$, for $b + c \neq 0, b \neq 0$

29. $\frac{1}{a} + \frac{1}{b} = \frac{1}{ab}$, for $a \neq 0, b \neq 0$

30. $\frac{1}{a} + a = \frac{a \cdot a + 1}{a}$, for $a \neq 0$

In Exercises 31–40 tell which of the following sets the given number is in: N = natural numbers, Z = integers, Q = rational numbers, R = real numbers.

31. $\frac{91}{7}$

32. $5\frac{1}{3}$

33. $\frac{-1}{0.5}$

34. -4.66

35. $\sqrt{3}$

36. $\sqrt{6.25}$

37. $3.121212\ldots$

38. $\sqrt{-4}$

39. $-\sqrt{9}$

40. $0.101001000100001\ldots$ (one more 0 each time)

Answer to Practice Problem **1.** 6.6

A.2 Linear Equations

Letters, called **variables,** are often used to stand for real numbers. In the expression

$$x(5 + 7x) - 3,$$

for example, x is a variable, while the numbers 5, 7, and 3 are called **constants.** An equation involving a variable will be true or false depending on the number for which the variable stands. For example, the equation

$$(x - 1)x + 3x - 7 = 1$$

is true for $x = 2$, since

$$(2 - 1)2 + 3 \cdot 2 - 7 = 2 + 6 - 7 = 1,$$

but false for $x = 3$, since

$$(3 - 1)3 + 3 \cdot 3 - 7 = 6 + 9 - 7 = 8 \neq 1.$$

The **solution set** of an equation involving a variable is the set of all numbers that, when assigned to the variable, make the equation true. For example the solution set of the equation

$$3x = 6$$

can easily be seen to be $\{2\}$, while the solution set of the equation

$$(x - 4)(x - 5) = 0$$

is given by $\{4, 5\}$. The latter fact follows from the property of real numbers that

$$ab = 0 \quad \text{if and only if} \quad a = 0 \text{ or } b = 0.$$

Thus either $x - 4 = 0$, and so $x = 4$; or else $x - 5 = 0$, and so $x = 5$.

Two equations are said to be **equivalent** if they have the same solution set. One method of finding the solution set of an equation is to transform it into equivalent equations until arriving at an equation whose solution set is obvious. Fundamental to this method are the following facts.

An equivalent equation results if:

an expression is added to both sides of an equation, or both sides of an equation are multiplied by a nonzero constant.

In this section we will only consider **linear equations** in a single variable, say x. Such equations involve only sums and differences of constants and constant multiples of x. The following example illustrates how the solution set of a linear equation is found by replacing it with equivalent equations.

EXAMPLE A.1 Find the solution set of the equation $5x - 3 = 2x + 7$.

Solution To solve the given equation, we must isolate the variable x on one side of an equation equivalent to the one given. The operation producing the equivalent equation will be listed in parentheses after each step.

$$\begin{aligned} 5x - 3 &= 2x + 7 \\ 5x - 3 + 3 &= 2x + 7 + 3 && \text{(adding 3 to both sides)} \\ 5x &= 2x + 10 && \text{(simplifying)} \\ 5x + (-2x) &= 2x + 10 + (-2x) && \text{(adding } -2x \text{ to both sides)} \\ 3x &= 10 && \text{(simplifying)} \\ \left(\frac{1}{3}\right)3x &= \left(\frac{1}{3}\right)10 && \left(\text{multiplying both sides by } \frac{1}{3}\right) \\ x &= \frac{10}{3} && \text{(simplifying)} \end{aligned}$$

Clearly the solution set for the last equation is $\{10/3\}$. ■

EXAMPLE A.2 A man is considering setting up a poster shop in a shopping mall. It will cost him \$8,000 per year to lease the selling space. The posters sell for \$6 each, and cost him \$2 each. He estimates yearly costs for labor and other expenses to be \$30,000. How many posters must he sell per year to break even?

Solution Suppose the man sells x posters. Then his yearly expenses will be

$$8{,}000 + 30{,}000 + 2x,$$

while his revenue will be $6x$. He will break even if these are equal. Thus we solve the equation below for x.

$$\begin{aligned} 8{,}000 + 30{,}000 + 2x &= 6x \\ 38{,}000 + 2x &= 6x && \text{(simplifying)} \\ 38{,}000 + 2x + (-2x) &= 6x + (-2x) && \text{(adding } -2x \text{ to both sides)} \\ 38{,}000 &= 4x && \text{(simplifying)} \\ \left(\frac{1}{4}\right)38{,}000 &= \left(\frac{1}{4}\right)4x && \left(\text{multiplying both sides by } \frac{1}{4}\right) \\ 9{,}500 &= x && \text{(simplifying)} \end{aligned}$$

We see that the man must sell 9,500 posters per year to break even. ■

Practice Problem 2 Find the solution set of the linear equation $3x - 4 = 5x + 1$.

EXERCISES A.2

In Exercises 1–20 solve the given linear equation.

1. $5x - 2 = 28$

2. $7x + 3 = 17$

3. $3A + 5 = 12$

4. $6k - 7 = 13$

5. $5 - 4x = 13$

6. $13 - 8x = 43$

7. $5 + 7u = 3u$

8. $4v - 7 = -17v$

9. $5x - 2 = 3 + 3x$

10. $4 - 3x = 2 + 12x$

11. $3(x - 1) = 2x$

12. $x = 2x + 5(3 - x)$

13. $3x + 4 = x + \frac{1}{2}$

14. $\frac{2}{3} - x = 5x - 4$

15. $\frac{x}{2} + 3 = x - 5$

16. $\frac{2x}{3} - 5 = 2x + 1$

17. $\frac{t}{3} + \frac{t}{7} = 4$

18. $\frac{y}{4} + 3 = \frac{2y + 1}{5}$

19. $\dfrac{4 - \frac{x}{3}}{2} = 7x$

20. $\frac{x}{4} + 7 = \frac{3x}{14} - \frac{1}{3}$

21. A newspaper seller sells papers for 50¢ each. The papers cost him 30¢ each and daily expenses are \$15. How many papers must he sell to break even?

22. Suzy likes green jelly beans and Sally likes red. Suzy traded Sally 25 red jelly beans for 19 green jelly beans and 15¢. If this is a fair trade, how much is a jelly bean worth?

23. After winning the lottery, Sam could have bought several \$25,000 Corvettes and had \$250,000 left over, but instead he bought the same number of \$40,000 Porsches and had only \$100,000 left over. How many cars did he buy?

24. Betty has an opportunity to invest in a new company with a sum of money she has just inherited. She could buy a 1/4 interest and have \$100,000 left over, or else buy a 1/5 interest and have \$130,000 left over. What is the value of the company?

Answer to Practice Problem **2.** $\left\{-\frac{5}{2}\right\}$

A.3 Linear Inequalities

Consider the positions of two real numbers a and b on the real line, where, as is conventional, the positive real numbers are on the right. (See Figure A.3.)

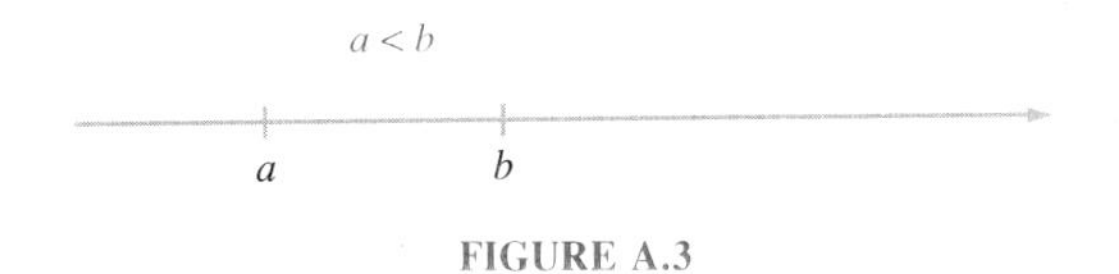

FIGURE A.3

If b is to the right of a, we write

$$a < b.$$

An equivalent notation is $b > a$. This gives an order relation between certain pairs of real numbers. This relation satisfies the following rules.

For all real numbers a, b, and c:

(O1) Exactly one of the following holds:

$$a < b, \qquad a = b, \qquad a > b.$$

(O2) If $a < b$ and $b < c$, then $a < c$.

We use the notation $a \leq b$ to mean that $a < b$ *or* $a = b$. Another way to indicate the same thing is $b \geq a$. Statements involving $<$, $>$, $\leq$, or $\geq$ are called **inequalities.**

Practice Problem 3 Which of the following inequalities are true?

(a) $3 < 5$ (b) $2 < -6$ (c) $4 > 7$

(d) $9 > 9$ (e) $4 \leq 3$ (f) $-7 \geq -3$

(g) $2 \leq 2$ (h) $-3 \leq 5$ (i) $\frac{3}{8} < \frac{1}{3}$

Of course whether a statement such as

$$x \geq 3$$

is true or not depends on what x is. The set of real numbers x satisfying $x \geq 3$ is indicated in part (a) of Figure A.4, while those x satisfying $x < 2$ are indicated in part (b). Notice that a heavy dot is used to indicate that $x = 3$ is included in the first set, and an open dot indicates that $x = 2$ is not included in the second.

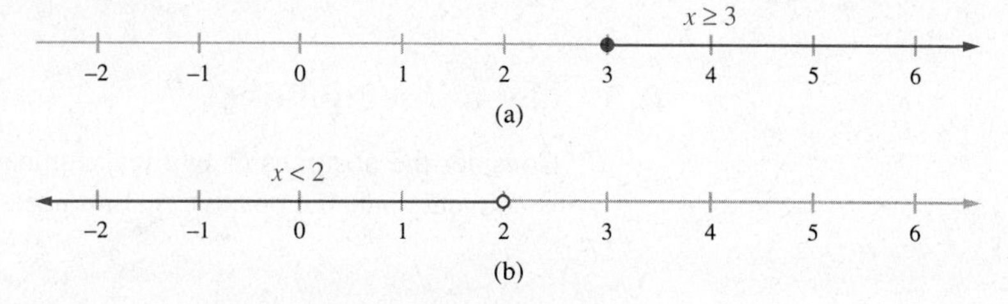

FIGURE A.4

A statement of the form

$$3 < x \leq 5$$

is a short way of indicating that

$$3 < x \quad \textit{and} \quad x \leq 5.$$

The set of numbers x satisfying this statement is indicated in Figure A.5.

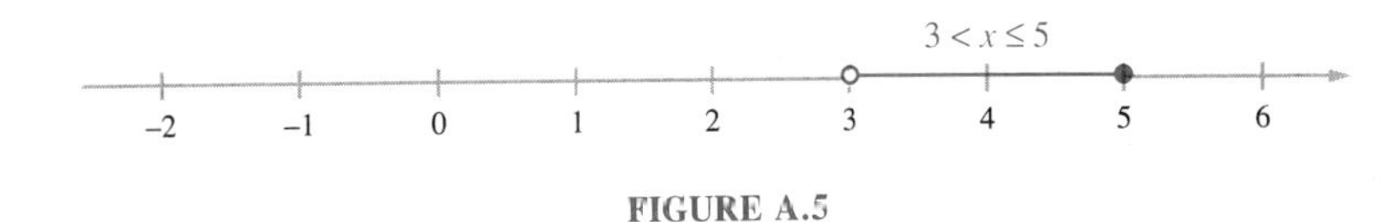

FIGURE A.5

The set of numbers x satisfying the statement

$$3x + 2 > -5x - 10$$

is not so obvious. The solution set of an inequality of this sort can be found in a way similar to how we solved equations. That is, we replace the inequality with equivalent inequalities until we get one whose solution set is obvious. The following rules give the allowable operations.

An equivalent inequality is obtained if:
the same expression is added to both sides of an inequality, or both sides of an inequality are multiplied by the same *positive* constant.

An equivalent inequality is also obtained if:
both sides of an inequality are multiplied by the same *negative* constant, and the sense of the inequality is reversed.

By "reversing the sense of an inequality" we mean replacing $<$ with $>$, $\leq$ with $\geq$, $>$ with $<$, or $\geq$ with $\leq$. For example

$$3x + 5 < 2x - 3 \quad \text{is equivalent to} \quad (3x + 5) + 3 < (2x - 3) + 3,$$

and

$$2x - 4 \geq \frac{1}{2} \quad \text{is equivalent to} \quad 2(2x - 4) \geq 2\left(\frac{1}{2}\right),$$

while

$$5 - 3x > 6 \quad \text{is equivalent to} \quad -2(5 - 3x) < -2(6),$$

and

$$-5x + 10 \geq 12 \quad \text{is equivalent to} \quad \left(-\frac{1}{5}\right)(-5x + 10) \leq \left(-\frac{1}{5}\right)(12).$$

Practice Problem 4 Perform the indicated operation or operations on each given inequality and then simplify both sides.

(a) Add -5 to $2x + 5 > 7x - 1$.

(b) Multiply $3x - 6 \leq 18$ by $\frac{1}{3}$.

(c) Multiply $-.25x \geq x + .5$ by -4.

(d) Multiply $-5x + 35 < 10$ by $-\frac{1}{5}$, and then add 7.

(e) Add 2 to $3x - 2 \leq 7$, and then multiply by $\frac{1}{3}$.

EXAMPLE A.3

Find the set of real numbers satisfying the inequality

$$3x + 2 \geq -5x - 10.$$

Solution As with equations we try to isolate the variable x on one side of the inequality. We will indicate in parentheses the operations performed.

$$\begin{aligned} 3x + 2 + (-2) &\geq -5x - 10 + (-2) && \textbf{(adding } -2 \textbf{ to both sides)} \\ 3x &\geq -5x - 12 && \textbf{(simplifying)} \\ 3x + 5x &\geq -5x - 12 + 5x && \textbf{(adding } 5x \textbf{ to both sides)} \\ 8x &\geq -12 && \textbf{(simplifying)} \\ \left(\frac{1}{8}\right)8x &\geq \left(\frac{1}{8}\right)(-12) && \left(\textbf{multiplying both sides by } \frac{1}{8}\right) \\ x &\geq -\frac{3}{2} && \textbf{(simplifying)} \end{aligned}$$

We see that x satisfies the original inequality if and only if $x \geq -\frac{3}{2}$. ■

EXAMPLE A.4

An inventor has two companies who want to buy her design for a new carburetor. Company A offers \$100,000 plus a royalty of 50¢ per carburetor sold, while Company B offers \$75,000 plus a royalty of 60¢ per carburetor sold. How many carburetors would have to be sold for the offer from Company B to be better for her?

Solution Suppose x carburetors are sold. Then Company A would pay the inventor $100{,}000 + .50x$ dollars, while Company B would pay her $75{,}000 + .60x$ dollars. We must decide the set of values of x such that

$$100{,}000 + .50x < 75{,}000 + .60x.$$

We solve this inequality as follows.

$$100{,}000 + .50x + (-100{,}000) < 75{,}000 + .60x + (-100{,}000) \qquad \textbf{(adding } -100{,}000\textbf{)}$$

$$.50x < .60x - 25{,}000 \qquad \textbf{(simplifying)}$$

$$.50x + (-.60x) < .60x - 25{,}000 + (-.60x) \qquad \textbf{(adding } -.60x\textbf{)}$$

$$.10x < -25{,}000 \qquad \textbf{(simplifying)}$$

$$(-10)(-.10x) > (-10)(-25{,}000) \qquad \textbf{(multiplying by } -10\textbf{)}$$

$$x > 250{,}000 \qquad \textbf{(simplifying)}$$

We see that more than 250,000 carburetors must be sold in order for the offer from Company B to be better for the inventor. ■

EXERCISES A.3

In Exercises 1–10 tell whether the given inequality is true or false.

1. $3.46 > 4.21$

2. $0.3 > \frac{1}{3}$

3. $-7.5 > -5.7$

4. $-\frac{1}{3} < -\frac{1}{4}$

5. $\frac{3}{4} \geq .75$

6. $3 \leq 7$

7. $-5.5 \leq 3.3$

8. $-\frac{3}{5} \leq -0.6$

9. $\frac{2}{5} < \frac{1}{3}$

10. $\frac{3}{7} \geq \frac{4}{9}$

In Exercises 11–16 graph on a number line the set of points x satisfying the inequality or inequalities given.

11. $3 \leq x < 5$

12. $-1 < x \leq 4$

13. $4 > x > -2$

14. $-1 \geq x \geq -3$

15. $x > 2$

16. $x \leq \frac{3}{2}$

In Exercises 17–26 perform the indicated operation to get an equivalent inequality, and then simplify both sides.

17. $3x + 4 \geq 7x - 5$; add -4

18. $5x - 2 > 2x + 1$; add 2

19. $3x \leq x + 7$; add $-x$

20. $-2x \geq 5 - 3x$; add $3x$

21. $\frac{x}{2} > 5$; multiply by 2

22. $3x \leq 9$; multiply by $\frac{1}{3}$

23. $-4x < 18$; multiply by $-\frac{1}{4}$

24. $-\frac{x}{5} \geq 7$; multiply by -5

25. $-7x < 21$; multiply by $\frac{1}{7}$

26. $\frac{x}{4} \leq \frac{3}{8}$; multiply by -4

In Exercises 27–38 solve the given inequality.

27. $3x + 1 > 16$

28. $4 + 5x \le 19$

29. $\frac{x}{2} + 5 < 3$

30. $\frac{3x}{5} - 4 \ge -3$

31. $5x - 1 > 2x + 7$

32. $8x - 6 \le 3x + 4$

33. $3(x - 2) \ge x - 9$

34. $7x + 2 < 3(x - 4)$

35. $2x + 2 > 5x + 1$

36. $x + 1 \ge 6x + \frac{1}{3}$

37. $-5x \le 12$

38. $-\frac{x}{7} > \frac{3}{14}$

39. A vendor sells x balloons for 50¢ each that cost her 15¢ each. Her expenses are $25. How big must x be in order that she make a profit?

40. A lottery winner can have $1,000,000 cash, or else $500,000 cash and $20,000 per year for life. He takes the $1,000,000 and comes out ahead. What does this say about x, the number of additional years he lives? (Ignore the fact that money received sooner is worth more.)

41. A novelist is offered two contracts, one paying $20,000 plus $1.40 per book sold, the other paying $30,000 plus 90¢ per book sold. If he is not hurt by choosing the second contract, what does this say about the size of x, the number of books sold?

42. It costs an appliance company $11 in materials and labor to make a toaster, plus $400,000 per year in fixed costs. It expects to sell 50,000 toasters next year and wants a profit of at least $300,000. What does this say about the size of x, the selling price of a toaster?

Answers to Practice Problems

3. Statements (a), (g), and (h) are true; the others are false.

4. (a) $2x > 7x - 6$
(b) $x - 2 \le 6$
(c) $x \le -4x - 2$
(d) $x > 5$
(e) $x \le 3$

A.4 Absolute Value

We define the **absolute value** of a real number x, denoted by $|x|$, as follows:

$$|x| = \begin{cases} x & \text{if } x \ge 0 \\ -x & \text{if } x < 0. \end{cases}$$

For example $|5| = 5$ (since $5 \ge 0$), but $|-3| = -(-3) = 3$.

EXAMPLE A.5 Make a table of the values of $|x - 3|$ for $x = 0, 1, 2, 3, 4, 5$.

Solution We have

$$\begin{aligned}|0 - 3| &= |-3| = 3\\ |1 - 3| &= |-2| = 2\\ |2 - 3| &= |-1| = 1\\ |3 - 3| &= |0| = 0\\ |4 - 3| &= |1| = 1\\ |5 - 3| &= |2| = 2.\end{aligned}$$

Thus our table is as follows.

x	0	1	2	3	4	5
$\|x - 3\|$	3	2	1	0	1	2

■

Practice Problem 5 Make a table of the values of $|x| - |x + 2|$ for $x = -4, -3, -2, -1, 0, 1, 2$.

It can be shown that absolute value has the following properties.

For all real numbers a and b:

(AV1) $|a| \geq 0$; and $|a| = 0$ if and only if $a = 0$.

(AV2) $|ab| = |a||b|$

(AV3) If $b \neq 0$, then $\left|\frac{a}{b}\right| = \frac{|a|}{|b|}$.

(AV4) $|a + b| \leq |a| + |b|$ *(triangle inequality)*

(AV5) If $a > 0$, then

$|x| = a$ if and only if $x = a$ or $x = -a$,

$|x| < a$ if and only if $-a < x < a$, and

$|x| \leq a$ if and only if $-a \leq x \leq a$.

EXAMPLE A.6 Solve $|3 - x| = 4$.

Solution By the first part of (AV5) we have $|3 - x| = 4$ if and only if $3 - x = 4$ or $3 - x = -4$. Solving these equations for x gives $x = -1$ or $x = 7$. ■

EXERCISES A.4

Evaluate the quantity given in Exercises 1–6.

1. $|3.6|$
2. $|-7.3|$
3. $-|-3|$
4. $|3||-5|$
5. $|-3| - |-5|$
6. $|-3 - (-5)|$

In Exercises 7–10 give the values of the given expression for $x = -3, -2, -1, 0, 1, 2,$ and 3.

7. $|x| + |x + 2|$
8. $|3 - x| - |x|$
9. $|2 - |x||$
10. $|1 - x| - |(2|x| - 5)|$

In Exercises 11–14 solve the given equation.

11. $|x| = 3$
12. $|x| = -5$
13. $|x - 2| = 7$
14. $|4 - 2x| = 3$

Answer to Practice Problem **5.**

x	-4	-3	-2	-1	0	1	2
$\|x\| - \|x + 2\|$	2	2	2	0	-2	-2	-2

A.5 Integer Exponents

It is conventional to write

$$a \cdot a = a^2,$$
$$a \cdot a \cdot a = a^3,$$

and, in general,

$$\underbrace{a \cdot a \cdot a \cdot \ldots \cdot a}_{n \text{ times}} = a^n,$$

where n is any positive integer. In the expression a^n, called the **nth power of a**, a is the **base**, and n is the **exponent.** The following rules hold.

For all nonzero numbers a and b and all positive integers m and n:

(E1) $a^m a^n = a^{m+n}$ (E2) $(a^m)^n = a^{mn}$

(E3) $(ab)^n = a^n b^n$ (E4) $\left(\dfrac{a}{b}\right)^n = \dfrac{a^n}{b^n}$.

If $a \neq 0$, the expression a^n can also be defined for 0 and negative integer exponents as follows:

$$a^0 = 1$$

$$a^{-n} = \frac{1}{a^n} \quad \text{if } n \text{ is a positive integer.}$$

Note in particular that a^{-1} is another name for $1/a$. With these definitions it can be proved that rules (E1) through (E4) hold for all integers m and n, positive, negative, and zero, as does the following rule.

For any nonzero real number a and all integers m and n:

(E5) $\dfrac{a^m}{a^n} = a^{m-n}$.

EXAMPLE A.7 Simplify the following expressions, writing them with only positive exponents.

(a) $\dfrac{(x^{-3}y^2)^{-1}}{(x^3y^0)^{-2}}$ (b) $\dfrac{y^4x}{y^{-3}} \cdot \dfrac{x(xy)^{-2}}{yx^3}$

Solution (a) We have

$$\begin{aligned}
\frac{(x^{-3}y^2)^{-1}}{(x^3y^0)^{-2}} &= \frac{(x^{-3})^{-1}(y^2)^{-1}}{(x^3)^{-2}(y^0)^{-2}} && \text{(using (E3) twice)} \\
&= \frac{x^{(-3)(-1)}y^{2(-1)}}{x^{3(-2)}y^{0(-2)}} && \text{(using (E2) four times)} \\
&= \frac{x^3y^{-2}}{x^{-6}y^0} && \text{(simplifying)} \\
&= \frac{x^3y^{-2}}{x^{-6}(1)} && \text{(using the definition of } y^0\text{)} \\
&= x^{3-(-6)}y^{-2} && \text{(using (E5))} \\
&= x^9y^{-2} && \text{(simplifying)} \\
&= x^9 \cdot \frac{1}{y^2} && \text{(definition of negative exponents)} \\
&= \frac{x^9}{y^2}. && \text{(simplifying)}
\end{aligned}$$

(b) We have

$$\frac{y^4x}{y^{-3}} \cdot \frac{x(xy)^{-2}}{yx^3} = y^{4-(-3)}x \cdot \frac{x(x^{-2}y^{-2})}{yx^3} \qquad \text{((E5) and (E3))}$$

$$= \frac{y^7x^1x^1x^{-2}y^{-2}}{yx^3} \qquad \text{(simplifying)}$$

$$= \frac{y^{7+(-2)}x^{1+1+(-2)}}{yx^3} \qquad \text{((E1) several times)}$$

$$= \frac{y^5x^0}{yx^3} \qquad \text{(simplifying)}$$

$$= \frac{y^{5-1}(1)}{x^3} \qquad \text{((E5) and the definition of } x^0\text{)}$$

$$= \frac{y^4}{x^3}. \qquad \text{(simplifying)}$$ ■

Practice Problem 6 Simplify as much as possible and write with only positive exponents.

(a) $\dfrac{u^2(u/v^2)^{-3}}{(v^2u)^{-1}}$ (b) $\dfrac{[u(v^{-2}u)^{-1}]^3}{\left(\dfrac{u}{v}\right)^{-3}\left(\dfrac{v^0u}{u^{-2}}\right)^2}$

It is a convention that, unless the contrary is indicated with parentheses, exponents are applied before the operations of addition, subtraction, multiplication, division, or negation. Thus

$$-2^2 \text{ means } -(2^2) = -4, \quad \text{and } \textit{not } (-2)^2 = 4.$$

Likewise

$$2x^3 \text{ means } 2(x^3), \quad \text{and } \textit{not } (2x)^3 = 8x^3.$$

EXERCISES A.5

In Exercises 1–12 evaluate the given expression.

1. 3^3

2. 4^2

3. 2^{-2}

4. $\left(\dfrac{1}{3}\right)^{-1}$

5. $\left(\dfrac{3}{2}\right)^{-2}$

6. $\left(\dfrac{3}{2^2}\right)^{-1}$

7. 0.25^{-2}

8. $\dfrac{3^{-1}}{6^{-2}}$

9. $\left(\dfrac{1}{4^{-2}}\right)^{-1}$

10. $\dfrac{2^8}{2^5}$

11. $\dfrac{6^3}{2^2}$

12. $2^5(4^2)^{-1}$

In Exercises 13–22 simplify the expression and write it with only positive exponents. Assume all variables represent nonzero numbers.

13. $\dfrac{u^5v^{-3}}{u^8v^{-7}}$

14. $\dfrac{x^5y^{-3}x^{-2}}{(xy)^2}$

15. $\dfrac{s^6(st^2)^{-3}}{(s^{-1}t^{-2})^3}$

16. $\dfrac{Q^4[(R^2)^0]^5}{RQ^3/(QR)^{-2}}$

17. $\dfrac{[(ab^3)/(ba^{-1})]^2}{ba^3}$

18. $\left(\dfrac{a^0b^1a^2b^3}{b^1a^3b^{-5}}\right)^{-1}$

19. $\dfrac{xy^2z^3}{(zx^2y^3)^0}$

20. $\left(\dfrac{r}{s}\right)^2\left(\dfrac{s}{t}\right)^3\left(\dfrac{t}{r}\right)^{-4}$

21. $\dfrac{X^2(YZ^3)^{-2}}{[X^2(YZ^3)^{-2}]^{-1}}$

22. $\dfrac{([(rs^3)^{-2}t]^{-1}rt^0)^2}{[(t^2)^2s]^2}$

Answer to Practice Problem **6.** (a) v^8 (b) $\dfrac{v^3}{u^3}$

A.6 Roots and Rational Exponents

If n is a positive integer and b is a nonzero real number, we call r an ***n*th root** of b if $r^n = b$. The terms **square root** and **cube root** are used in the special cases $n = 2$ and $n = 3$, respectively. The number of nth roots of b depends on whether b is positive or negative and also whether n is odd or even. For example the equation

$$r^2 = 4$$

has exactly two real solutions: $r = 2$ and $r = -2$. Thus $b = 4$ has exactly 2 square roots. The equation

$$r^3 = 8,$$

on the other hand, has only one real solution, namely $r = 2$. Thus $b = 8$ has exactly 1 cube root.

When b is negative, there may be no nth root at all. For example the equation

$$r^4 = -16$$

has no real solutions, since the 4th power of no real number is negative. The equation

$$r^5 = -32,$$

however, has the solution $r = -2$; and this is the only real solution.

The four examples just given illustrate the general situation, which is summarized in the following chart.

	n even	*n odd*
$b > 0$	**two** real nth roots of b (one positive, one negative)	**one** real nth root of b (positive)
$b < 0$	**no** real nth root of b	**one** real nth root of b (negative)
$b = 0$	**one** real nth root of b (namely, 0)	**one** real nth root of b (namely, 0)

Practice Problem 7 How many

(a) 7th roots does -3 have?
(b) 6th roots does -64 have?
(c) 14th roots does 5 have?
(d) 17th roots does 1/2 have?

If b is positive or n is odd, then b has at least one nth root, and sometimes two. The symbol

$$\sqrt[n]{b}$$

is used to denote an nth root of b, with the *positive* root chosen when there are two. For example

$$\sqrt[2]{4} = 2, \quad \sqrt[3]{8} = 2, \quad \text{and} \quad \sqrt[5]{-32} = -2,$$

while

$$\sqrt[4]{-16}$$

is undefined. When $n = 2$ the small raised number indicating the root is omitted, so that, for example, $\sqrt{9} = 3$. An expression of the form

$$\sqrt[n]{b}$$

is called a **radical.**

Practice Problem 8 Evaluate, if possible, the following.

(a) $\sqrt[3]{-8}$ (b) $\sqrt[6]{64}$ (c) $\sqrt{\frac{1}{4}}$ (d) $\sqrt[5]{32}$

Another notation for

$$\sqrt[n]{b}$$

is $b^{1/n}$. The motivation for this notation is that if we assume rule (E2) holds for such an exponent, then

$$(b^{1/n})^n = b^{(1/n)n} = b^1 = b,$$

which means that $b^{1/n}$ satisfies the definition of an nth root of b.

Practice Problem 9 Evaluate, if possible, the following.

(a) $125^{1/3}$ (b) $\left(-\frac{1}{9}\right)^{1/2}$ (c) $(-1000)^{1/3}$ (d) $\left(\frac{1}{16}\right)^{1/4}$

Generalizing the $b^{1/n}$ notation allows us to define fractional powers.

Definition of $b^{m/n}$

If m and n are integers such that $n > 0$ and the fraction m/n is in lowest terms, then we define

$$b^{m/n} = (b^{1/n})^m = (\sqrt[n]{b})^m$$

whenever $b^{1/n}$ exists.

EXAMPLE A.8 Evaluate (a) $8^{5/3}$ (b) $(-27)^{2/3}$ (c) $(-25)^{3/2}$.

Solution (a) $8^{5/3} = (\sqrt[3]{8})^5 = 2^5 = 32$.

(b) $(-27)^{2/3} = (\sqrt[3]{-27})^2 = (-3)^2 = 9$.

(c) $(-25)^{3/2}$ is undefined because $(-25)^{1/2}$ does not exist. ■

Practice Problem 10 Evaluate, if possible, (a) $(-64)^{5/6}$ (b) $\left(\frac{4}{9}\right)^{3/2}$ (c) $(-32)^{3/5}$.

It can be proved that the equations (E1) through (E5) continue to hold for rational exponents, provided that the base is positive. For example

$$\begin{aligned}
4^{2/3}54^{5/3} &= (2^2)^{2/3}(2 \cdot 3^3)^{5/3} \\
&= 2^{2(2/3)}2^{5/3}(3^3)^{5/3} && \text{(using (E2) and (E3))} \\
&= 2^{4/3}2^{5/3}3^{3(5/3)} && \text{(simplifying and using (E2))} \\
&= 2^{4/3+5/3}3^5 && \text{(using (E1) and simplifying)} \\
&= 2^3 3^5 = 1944. && \text{(simplifying)}
\end{aligned}$$

Practice Problem 11 Simplify $(-16)^{7/3}\left(\frac{1}{4}\right)^{7/6}$.

The rules for exponents can be translated into rules for radicals by replacing $a^{1/k}$ with $\sqrt[k]{a}$. For example, if we take $n = 1/k$ in (E3) we get

$$(ab)^{1/k} = a^{1/k}b^{1/k},$$

or

$$\sqrt[k]{ab} = \sqrt[k]{a}\sqrt[k]{b}.$$

We list some of the rules for radicals obtained this way. These hold so long as both sides are defined.

(R1) $\sqrt[n]{ab} = \sqrt[n]{a}\,\sqrt[n]{b}$

(R2) $\sqrt[n]{\frac{a}{b}} = \frac{\sqrt[n]{a}}{\sqrt[n]{b}}$

(R3) $\sqrt[m]{\sqrt[n]{a}} = \sqrt[mn]{a}$

(R4) $(\sqrt[n]{a})^n = a$

Note that, in spite of (R4), we do not claim that $\sqrt[n]{a^n} = a$. This is false if a is negative and n is even. For example

$$\sqrt[4]{(-2)^4} = \sqrt[4]{16} = 2 \neq -2.$$

The correct rule is as follows.

(R5) $\sqrt[n]{a^n} = \begin{cases} |a| & \text{if } n \text{ is even} \\ a & \text{if } n \text{ is odd} \end{cases}$

In particular (for $n = 2$) we have

$$\sqrt{a^2} = |a|.$$

Often the best way to simplify expressions involving radicals is to convert to exponential notation and use rules (E1) through (E5).

EXAMPLE A.9 Simplify $\dfrac{\sqrt{5}}{\sqrt[6]{5}}$.

Solution We have

$$\begin{aligned}\frac{\sqrt{5}}{\sqrt[6]{5}} &= \frac{5^{1/2}}{5^{1/6}}\\ &= 5^{1/2-1/6} && \text{(using (E5))}\\ &= 5^{1/3} && \text{(simplifying)}\\ &= \sqrt[3]{5}. && \blacksquare\end{aligned}$$

Practice Problem 12 Simplify $\sqrt{18}\sqrt[4]{64}$.

EXERCISES A.6

1. How many 5th roots does 32 have?

2. How many cube roots does -125 have?

3. How many square roots does 49 have?

4. How many 4th roots does -16 have?

5. How many 7th roots does -8 have?

6. How many 4th roots does 25 have?

7. How many 6th roots does -27 have?

8. How many 11th roots does 121 have?

In Exercises 9–40 evaluate the given expression, if possible.

9. $\sqrt{64}$

10. $\sqrt{81}$

11. $\sqrt[3]{64}$

12. $\sqrt[5]{243}$

13. $\sqrt[3]{-125}$

14. $\sqrt{-9}$

15. $\sqrt[4]{-16}$

16. $\sqrt[7]{-128}$

17. $\sqrt{\dfrac{9}{4}}$

18. $\sqrt{0.64}$

19. $\sqrt[3]{-.125}$

20. $\sqrt[5]{0.01024}$

21. $\sqrt[4]{-0.0081}$

22. $\sqrt[3]{\dfrac{-24}{81}}$

23. $49^{1/2}$

24. $-9^{1/2}$

25. $-16^{1/4}$

26. $27^{1/3}$

27. $(-32)^{1/5}$

28. $(-25)^{1/2}$

29. $0.25^{1/2}$

30. $\left(\dfrac{81}{16}\right)^{1/4}$

31. $(-625)^{1/4}$

32. $(-0.343)^{1/3}$

33. $125^{4/3}$

34. $27^{5/3}$

35. $(-27)^{5/3}$

36. $.0625^{3/4}$

37. $(-0.25)^{3/4}$

38. $-\left(\dfrac{1}{8}\right)^{2/3}$

39. $\left(\dfrac{1}{8}\right)^{2/3}$

40. $\left(-\dfrac{1}{8}\right)^{2/3}$

In Exercises 41–67 simplify the given expressions, if possible.

41. $\sqrt{3}\,\sqrt{12}$

42. $\sqrt{5}\,\sqrt{125}$

43. $\dfrac{\sqrt{8}}{\sqrt{2}}$

44. $\dfrac{\sqrt{18}}{\sqrt{8}}$

45. $\sqrt{(1.4)(1.4)}$

46. $\sqrt{(-3)(-3)}$

47. $\sqrt{-3}\,\sqrt{-3}$

48. $\sqrt[3]{4}\sqrt[3]{2}$

49. $\sqrt{2} + \sqrt{8}$

50. $\sqrt{3} + \sqrt{12}$

51. $\sqrt{4 + 9}$

52. $\sqrt{9 + 16}$

53. $\sqrt{4} + \sqrt{9}$

54. $\sqrt{3} + \sqrt{2}$

55. $(\sqrt{3})^4$

56. $\sqrt{5}\,(\sqrt{5} + \sqrt{0.2})$

57. $(\sqrt{7})^{1/3}$

58. $(7^{1/3})^6$

59. $\sqrt{x \cdot x}$

60. $\dfrac{\sqrt{2} + \sqrt{50}}{\sqrt{8}}$

61. $\sqrt[3]{6} \cdot \sqrt[6]{6}$

62. $\dfrac{7}{\sqrt{7} \cdot \sqrt[4]{7}}$

63. $\sqrt{11} \cdot \sqrt[5]{11}$

64. $\dfrac{\sqrt{13}}{\sqrt[10]{13}}$

65. $\dfrac{(\sqrt[3]{-6})^2}{\sqrt{-6}}$

66. $[(-4)^{2/3}]^{3/2}$

67. $[(-4)^{3/2}]^{2/3}$

Answers to Practice Problems

7. (a) 1 (b) 0 (c) 2 (d) 1

8. (a) -2 (b) 2 (c) 1/2 (d) 2

9. (a) 5 (b) undefined (c) -10 (d) 1/2

10. (a) undefined (b) 8/27 (c) -8

11. -128

12. 12

A.7 Quadratic Equations

An equation of the form

$$ax^2 + bx + c = 0,$$

where a, b, and c are constants and $a \neq 0$, is called a **quadratic equation.** An example is

$$x^2 - 9x + 20 = 0.$$

This equation can also be written as

$$(x - 4)(x - 5) = 0,$$

since

$$\begin{aligned}(x - 4)(x - 5) &= (x - 4)x - (x - 4)5 \\ &= x^2 - 4x - (x \cdot 5 - 4 \cdot 5) \\ &= x^2 - 4x - 5x + 20 \\ &= x^2 - 9x + 20.\end{aligned}$$

From the form $(x - 4)(x - 5) = 0$ we see that the solution set of this equation is $\{4, 5\}$, since the product of two real numbers is zero if and only if at least one of the factors is zero.

In a similar way we can check that the equation

$$3x^2 + 12x + 12 = 0$$

can also be written as

$$3(x + 2)(x + 2) = 0.$$

Since this holds if and only if $x + 2 = 0$, which means $x = -2$, the solution set of this quadratic equation is $\{-2\}$.

Finally, the quadratic equation

$$x^2 + 4 = 0$$

is equivalent to

$$x^2 = -4.$$

Since the square of no real number is -4, this equation has no solutions.

These examples show that a quadratic equation may have either 2, 1, or no solutions. It turns out that these are all the possibilities, and all solutions can be found from what is known as the **quadratic formula.**

The Quadratic Formula

The quadratic equation

$$ax^2 + bx + c = 0, \qquad a \neq 0,$$

has 2, 1, or no solutions according as

$$b^2 - 4ac$$

is positive, zero, or negative. In the first two cases all solutions are given by

$$x = \frac{-b \pm \sqrt{b^2 - 4ac}}{2a}.$$

The sign "$\pm$" in the quadratic formula, read "plus or minus", is to be taken first as a "$+$" and then as a "$-$" in order to get two solutions when $b^2 - 4ac > 0$. Notice that if $b^2 - 4ac = 0$, then only one solution results, since $+\sqrt{0} = -\sqrt{0} = 0$.

EXAMPLE A.10 Apply the quadratic formula to solve the equations

(a) $x^2 - 9x + 20 = 0$

(b) $3x^2 + 12x + 12 = 0$

(c) $x^2 + 4 = 0$

used as examples at the beginning of this section.

Solution (a) The equation $x^2 - 9x + 20 = 0$ is of the form $ax^2 + bx + c = 0$, with $a = 1$, $b = -9$, and $c = 20$. Then

$$x = \frac{-b \pm \sqrt{b^2 - 4ac}}{2a} = \frac{-(-9) \pm \sqrt{(-9)^2 - 4(1)(20)}}{2(1)}$$

$$= \frac{9 \pm \sqrt{81 - 80}}{2} = \frac{9 \pm \sqrt{1}}{2} = \frac{9 \pm 1}{2}.$$

Taking the plus sign gives $10/2 = 5$, while the minus sign yields $8/2 = 4$. As before, we see the solution set is $\{4, 5\}$.

(b) In the equation $3x^2 + 12x + 12 = 0$ we have $a = 3$, $b = 12$, and $c = 12$. Thus

$$x = \frac{-b \pm \sqrt{b^2 - 4ac}}{2a} = \frac{-(12) \pm \sqrt{(12)^2 - 4(3)(12)}}{2(3)}$$

$$= \frac{-12 \pm \sqrt{144 - 144}}{6} = \frac{-12 \pm \sqrt{0}}{6} = \frac{-12}{6} = -2.$$

Since $b^2 - 4ac = 0$, the solution set contains the single element -2.

(c) Finally $x^2 + 4 = 0$ has the form $ax^2 + bx + c = 0$ with $a = 1$, $b = 0$, and $c = 4$. If we try the quadratic formula, we get

$$x = \frac{-b \pm \sqrt{b^2 - 4ac}}{2a} = \frac{-0 \pm \sqrt{0^2 - 4(1)(4)}}{2(1)} = \frac{\pm\sqrt{-4}}{2}.$$

Since $\sqrt{-4}$ does not exist, the equation $x^2 + 4 = 0$ has no real solution. ■

Practice Problem 13 Use the quadratic formula to solve the following equations.

(a) $4x^2 - 4x + 1 = 0$
(b) $x^2 + 3x + 5 = 0$
(c) $2x^2 - 5x - 3 = 0$

EXAMPLE A.11 Use the quadratic formula to solve the equation

$$x^2 + 3x = 5.$$

Solution Before we can use the quadratic formula we must add -5 to both sides of this equation so as to put it in the form $ax^2 + bx + c = 0$. This produces the equivalent equation

$$x^2 + 3x - 5 = 0,$$

and so $a = 1$, $b = 3$, and $c = -5$. Then

$$x = \frac{-b \pm \sqrt{b^2 - 4ac}}{2a} = \frac{-3 \pm \sqrt{3^2 - 4(1)(-5)}}{2(1)}$$

$$= \frac{-3 \pm \sqrt{9 + 20}}{2} = \frac{-3 \pm \sqrt{29}}{2}.$$

We see that the solution set is

$$\left\{\frac{-3 + \sqrt{29}}{2}, \frac{-3 - \sqrt{29}}{2}\right\}.$$

The approximate values of the solutions are 1.19 and -4.19. ■

Practice Problem 14 Use the quadratic formula to solve the following equations.

(a) $x^2 + 3x = 3 + 2x$ (b) $x^2 + x = 3 - x^2$

Many simple quadratic equations can be solved by **factoring.** For example we can solve

$$x^2 + 7x + 12 = 0$$

quickly if we notice that the left side equals $(x + 3)(x + 4)$, and so the solution set is $\{-3, -4\}$.

Since

$$(x + r)(x + s) = x^2 + (r + s)x + rs,$$

when confronted with a quadratic equation of the form

$$x^2 + Ax + B = 0,$$

we look for two numbers r and s whose product is B and sum is A.

Given the equation

$$x^2 + 5x - 14 = 0,$$

for example, we note that if $rs = -14$, then we could have the following values for r and s.

r	1	-1	2	-2	7	-7	14	-14
s	-14	14	-7	7	-2	2	-1	1

Taking $r = 7$ and $s = -2$ gives $r + s = 5$, and so the equation can be written as

$$(x + 7)(x - 2) = 0.$$

Thus the solution set is $\{-7, 2\}$.

If our equation were

$$x^2 + 6x - 14 = 0,$$

however, we would find no simple values of r and s so that

$$(x + r)(x + s) = x^2 + 6x - 14.$$

Using the quadratic formula shows that the solution set of this equation is $\{-3 + \sqrt{23}, -3 - \sqrt{23}\}$.

EXERCISES A.7

Find all solutions to the equations in Exercises 1–26.

1. $x^2 + x - 6 = 0$

2. $x^2 + 6x - 7 = 0$

3. $4x^2 + 12x + 9 = 0$

4. $2x^2 - x - 6 = 0$

5. $9x^2 - 9x + 2 = 0$

6. $9x^2 - 12x + 4 = 0$

7. $2x^2 - 3x = 0$

8. $5x^2 - 2x = 0$

9. $x^2 + 3x + 4 = 0$

10. $(x - 2)^2 = 4$

11. $x(x + 1) = 6$

12. $2x^2 - 5x - 4 = 0$

13. $x^2 + 2x - 5 = 0$

14. $x^2 - 5x - 5 = 0$

15. $3x^2 + 3.5x - 1.5 = 0$

16. $-3x^2 + 5x + 2 = 0$

17. $3x^2 + 10x = 3x - 5$

18. $5x^2 + 2 = 7x$

19. $(x - 2)^2 = (x - 3)^2$

20. $(x - 1)x = (x - 1)4$

21. $x^2 + 4 = 3\sqrt{2}x$

22. $x^3 + 125 = 0$

23. $3x^3 - 8x^2 - 3x = 0$

24. $(x - 1)(7x^2 + 11x - 6) = 0$

25. $(x^2 - 5x + 6)^2 = 0$

26. $(x^2 - 5x + 6)^2 = x^2 - 5x + 6$

Answers to Practice Problems **13.** (a) $x = \frac{1}{2}$ (b) no solution (c) $x = -\frac{1}{2}$ or $x = 3$

14. (a) $x = \dfrac{-1 \pm \sqrt{13}}{2}$ (b) $x = 1$ or $x = -\frac{3}{2}$

A.8 Polynomial Equations

If n is a nonnegative integer, then any expression of the form

$$a_n x^n + a_{n-1} x^{n-1} + \ldots + a_1 x + a_0,$$

where $a_n, a_{n-1}, \ldots, a_1$, and a_0 are constants and $a_n \neq 0$, is called a **polynomial in x of degree n.** For example

$$5x^3 - 2x + 7$$

is a polynomial of degree 3, with $a_3 = 5$, $a_2 = 0$, $a_1 = -2$, and $a_0 = 7$. Note that only positive integer powers of x are allowed, so that neither

$$5x^4 + 3x - 4 + 2x^{-1}$$

nor

$$3x^2 - x + x^{1/2} - 6$$

are polynomials. Polynomials of degrees 1, 2, and 3 are called **linear, quadratic,** and **cubic** polynomials, respectively.

If c is a constant, then we **evaluate a polynomial at c** by letting $x = c$ in the polynomial and computing the result. For example, evaluating the polynomial $5x^3 - 2x + 3$ at 2 produces

$$5(2^3) - 2(2) + 3 = 39.$$

We may name a polynomial using a notation such as $P(x)$, which indicates the polynomial involves the variable x. The result of evaluating the polynomial at c is denoted $P(c)$. For example, if

$$P(x) = 5x^3 - 2x + 3,$$

then, as we have seen,

$$P(2) = 39,$$

and

$$P(-1) = 5(-1)^3 - 2(-1) + 3 = 0.$$

It is often necessary to solve equations of the form

$$P(x) = 0,$$

where $P(x)$ is some polynomial. We saw how to do this when $P(x)$ is linear and quadratic in Sections A.2 and A.7. When the degree of $P(x)$ is more than 2 the problem is harder. Often a solution involves factoring $P(x)$, that is, writing it as a product of polynomials of lower degree.

EXAMPLE A.12 Find the solution set of the following equations.

(a) $x^3 - 2x^2 - 3x = 0$ (b) $x^3 + 5x^2 - x - 5 = 0$

Solution (a) Since each term on the left side contains x as a factor, we can rewrite the equation as

$$x(x^2 - 2x - 3) = 0.$$

Thus x is a solution if and only if either $x = 0$ or $x^2 - 2x - 3 = 0$. The quadratic formula gives the solutions to the second equation.

$$x = \frac{-b \pm \sqrt{b^2 - 4ac}}{2a} = \frac{-(-2) \pm \sqrt{(-2)^2 - 4(1)(-3)}}{2(1)}$$

$$= \frac{2 \pm \sqrt{16}}{2} = \frac{2 \pm 4}{2} = 3 \text{ or } -1$$

These solutions could also be found by noticing that $x^2 - 2x - 3 = (x - 3)(x + 1)$. Thus the solution set to the original equation is $\{0, 3, -1\}$.

(b) The polynomial $x^3 + 5x^2 - x - 5$ can be factored by grouping the terms as follows:

$$x^3 + 5x^2 - x - 5 = x^2(x + 5) - 1(x + 5) = (x^2 - 1)(x + 5).$$

Since $x^2 - 1 = 0$ is equivalent to $x^2 = 1$, its solutions are easily seen to be 1 and -1. Thus the solution set is $\{1, -1, -5\}$. ■

Practice Problem 15 Find the solution sets of the following.

(a) $x^4 + 3x^3 - 10x^2 = 0$ (b) $x^3 - 3x^2 + 4x - 12 = 0$

Sometimes a substitution allows us to solve a polynomial equation. Consider, for example, the equation

$$x^4 - 4x^2 + 3 = 0.$$

Since only even powers of x are involved, the substitution $x^2 = t$ yields the simpler equation

$$t^2 - 4t + 3 = 0,$$

which can be solved with the quadratic formula. We have

$$t = \frac{-b \pm \sqrt{b^2 - 4ac}}{2a} = \frac{-(-4) \pm \sqrt{(-4)^2 - 4(1)(3)}}{2(1)}$$

$$= \frac{4 \pm \sqrt{4}}{2} = \frac{4 \pm 2}{2} = 3 \text{ or } 1.$$

These solutions could also be found by noticing that

$$t^2 - 4t + 3 = (t - 3)(t - 1).$$

Since we had $t = x^2$, we have $x^2 = 3$ or $x^2 = 1$. Thus $x = \pm\sqrt{3}$ or $x = \pm 1$. The solution set to the original equation is $\{-1, 1, -\sqrt{3}, \sqrt{3}\}$.

Practice Problem 16 Use the substitution $t = x^3$ to find the solution set of the polynomial equation $8x^6 + 7x^3 - 1 = 0$.

According to the following theorem, knowing one solution to

$$P(x) = 0$$

allows us to factor $P(x)$.

The Factor Theorem

Suppose c is a solution to $P(x) = 0$, where $P(x)$ is a polynomial. Then

$$P(x) = (x - c)Q(x),$$

where $Q(x)$ is a polynomial of degree one less than that of $P(x)$.

For example, suppose we notice that the polynomial equation

$$2x^3 - x - 1 = 0$$

has $x = 1$ as one solution, since $2(1)^3 - (1) - 1 = 0$. Then from the factor theorem the equation can be written in the form

$$(x - 1)Q(x) = 0,$$

where $Q(x)$ is a polynomial of degree 2. If we knew what $Q(x)$ was, we could find any additional solutions to

$$2x^3 - x - 1 = 0$$

by solving $Q(x) = 0$ with the quadratic formula.

Note that since

$$2x^3 - x - 1 = (x - 1)Q(x),$$

we have

$$Q(x) = \frac{2x^3 - x - 1}{x - 1}.$$

The polynomial $Q(x)$ can be found by dividing $2x^3 - x - 1$ by $x - 1$ using a method similar to long division. We illustrate the process below. (Notice that a space is reserved for the missing x^2 term in the dividend.)

$$\begin{array}{r}
2x^2 + 2x + 1 \\
x - 1\overline{)2x^3 \qquad\qquad - x - 1} \\
\underline{2x^3 - 2x^2 \qquad\qquad\quad} \\
2x^2 - x - 1 \\
\underline{2x^2 - 2x \qquad} \\
x - 1 \\
\underline{x - 1} \\
0
\end{array}$$

This calculation shows that

$$2x^3 - x - 1 = (x - 1)(2x^2 + 2x + 1),$$

which can be checked by multiplying out the factors on the right. Thus any solutions to

$$2x^3 - x - 1 = 0$$

besides $x = 1$ must be solutions to

$$Q(x) = 2x^2 + 2x + 1 = 0.$$

Applying the quadratic formula shows that there are no additional solutions, since $b^2 - 4ac = 2^2 - 4(2)(1) = -4 < 0$. Thus the only solution to

$$2x^3 - x - 1 = 0$$

is $x = 1$.

EXAMPLE A.13 Find the solution set to the equation

$$P(x) = 3x^3 + 5x^2 + x - 1 = 0,$$

given that $x = -1$ is a solution.

Solution First we check that indeed

$$P(-1) = 3(-1)^3 + 5(-1)^2 + (-1) - 1 = -3 + 5 - 1 - 1 = 0.$$

Thus we divide $P(x)$ by $x - (-1) = x + 1$.

$$\begin{array}{r}
3x^2 + 2x - 1 \\
x + 1 \overline{)3x^3 + 5x^2 + x - 1} \\
\underline{3x^3 + 3x^2} \qquad\qquad \\
2x^2 + x - 1 \\
\underline{2x^2 + 2x} \quad\;\; \\
-x - 1 \\
\underline{-x - 1} \\
0
\end{array}$$

Then our equation can be written

$$P(x) = (x + 1)(3x^2 + 2x - 1) = 0.$$

It remains to use the quadratic formula to find when the second factor is 0. We have

$$x = \frac{-b \pm \sqrt{b^2 - 4ac}}{2a} = \frac{-2 \pm \sqrt{2^2 - 4(3)(-1)}}{2(3)}$$

$$= \frac{-2 \pm \sqrt{16}}{6} = \frac{-2 \pm 4}{6} = \frac{1}{3} \text{ or } -1.$$

Of course we already have -1 as a solution to the original equation. Thus the solution set is $\{-1, 1/3\}$. ■

Practice Problem 17 Check that $x = 2$ is a solution to

$$6x^3 - 7x^2 - 9x - 2 = 0,$$

and use this fact to find the solution set of the equation.

EXERCISES A.8

In Exercises 1–6 tell whether the given expression is a polynomial or not.

1. $5x^3 + x - 3$

2. $x^{10} + 3\sqrt{x} + 12$

3. $x^3 + x^2 + x + 1 + \frac{1}{x}$

4. x

5. 13

6. $\frac{1}{x^2 + 3}$

In Exercises 7–10 evaluate the polynomial at the given numbers.

7. $x^3 - 2x + 5$ at 0, 1, and 2

8. $2x^2 - x + 3$ at -3, 0, and 3

9. $x^3 + x^2 + x + 1$ at 0, -2, and 4

10. $x^5 - x^4 + 2x^3 - 2x^2 + 3x - 3$ at -2, -1, and 1

11. Suppose $P(x) = 2x^3 - 3x + 5$. What is $P(-2)$? $P(0)$? $P(3)$?

12. Suppose $P(x) = x^4 + 12x^2 + 25$. What is $P(-3)$? $P(0)$? $P(3)$?

In Exercises 13–32 find the solution set of the given equation.

13. $3x^4 - x^3 + 3x^2 - x = 0$

14. $x^3 - 7x^2 + 12x = 0$

15. $x^4 + 2x^3 - 3x^2 = 0$

16. $x^3 + 3x^2 - 2x - 6 = 0$

17. $2x^3 - 3x^2 + 14x - 21 = 0$

18. $x^3 + 5x^2 - 9x - 45 = 0$

19. $x^3 + 2x^2 - 4x - 8 = 0$

20. $x^4 + x^3 + x + 1 = 0$

21. $x^4 + x^3 - x - 1 = 0$

22. $x^4 - x^3 + x - 1 = 0$

23. $x^4 + 5x^2 + 6 = 0$

24. $x^4 - 5x^2 + 6 = 0$

25. $x^4 + 5x^2 + 7 = 0$

26. $x^4 + x^2 - 6 = 0$

27. $x^4 - 8x^2 + 16 = 0$

28. $x^6 - x^3 - 6 = 0$

29. $x^8 - 17x^4 + 16 = 0$

30. $x^3 - \sqrt{2}x^2 + 3x - 3\sqrt{2} = 0$

31. $x^3 + 3x^2 - \sqrt{2}x - 3\sqrt{2} = 0$

32. $x^7 + x^6 - x^5 + 32x^2 + 32x - 32 = 0$

In Exercises 33–40 find the polynomial $Q(x)$ satisfying the given equation.

33. $x^3 - 4x^2 + 5x - 2 = (x - 2)Q(x)$

34. $2x^3 + 3x^2 - x - 2 = (x + 1)Q(x)$

35. $3x^3 - 2x - 20 = (x - 2)Q(x)$

36. $-x^3 + 5x^2 - 18 = (x - 3)Q(x)$

37. $x^4 - 3x^3 + 2x^2 - 7x + 3 = (x - 3)Q(x)$

38. $x^4 + 5x^3 + 6x^2 - 5x - 10 = (x + 2)Q(x)$

39. $2x^3 + 5x^2 - 5x + 1 = \left(x - \frac{1}{2}\right)Q(x)$

40. $2x^3 + x^2 + x + 6 = \left(x + \frac{3}{2}\right)Q(x)$

In Exercises 41–48 verify that the given value of x satisfies the polynomial equation. Then use this fact to find all solutions to the equation.

41. $x^3 + 5x^2 + 3x - 9 = 0, \quad x = 1$

42. $x^3 + 5x^2 + 8x + 6 = 0, \quad x = -3$

43. $x^3 + x^2 - 14x - 24 = 0, \quad x = -2$

44. $x^3 - 37x + 84 = 0, \quad x = 3$

45. $2x^4 - 3x^3 - 3x - 2 = 0, \quad x = 2$

46. $x^5 + 4x^4 - 7x^3 - 28x^2 + 12x + 48 = 0, \quad x = -4$

47. $2x^3 + 5x^2 - x - 1 = 0, \quad x = \frac{1}{2}$

48. $3x^3 + 5x^2 - 7x - 6 = 0, \quad x = -\frac{2}{3}$

Answers to Practice Problems

15. (a) $\{0, 2, -5\}$ (b) $\{3\}$

16. $\left\{\frac{1}{2}, -1\right\}$

17. $\left\{2, -\frac{1}{2}, -\frac{1}{3}\right\}$

TABLES

■ TABLE 1 Compound Interest

$(1 + i)^n$

n \ i	1%	$1\frac{1}{2}$%	2%	3%	4%	5%	6%	8%
1	1.01000	1.01500	1.02000	1.03000	1.04000	1.05000	1.06000	1.08000
2	1.02010	1.03023	1.04040	1.06090	1.08160	1.10250	1.12360	1.16640
3	1.03030	1.04568	1.06121	1.09273	1.12486	1.15763	1.19102	1.25971
4	1.04060	1.06136	1.08243	1.12551	1.16986	1.21551	1.26248	1.36049
5	1.05101	1.07728	1.10408	1.15927	1.21665	1.27628	1.33823	1.46933
6	1.06152	1.09344	1.12616	1.19405	1.26532	1.34010	1.41852	1.58687
7	1.07214	1.10984	1.14869	1.22987	1.31593	1.40710	1.50363	1.71382
8	1.08286	1.12649	1.17166	1.26677	1.36857	1.47746	1.59385	1.85093
9	1.09369	1.14339	1.19509	1.30477	1.42331	1.55133	1.68948	1.99900
10	1.10462	1.16054	1.21899	1.34392	1.48024	1.62889	1.79085	2.15892
11	1.11567	1.17795	1.24337	1.38423	1.53945	1.71034	1.89830	2.33164
12	1.12683	1.19562	1.26824	1.42576	1.60103	1.79586	2.01220	2.51817
13	1.13809	1.21355	1.29361	1.46853	1.66507	1.88565	2.13293	2.71962
14	1.14947	1.23176	1.31948	1.51259	1.73168	1.97993	2.26090	2.93719
15	1.16097	1.25023	1.34587	1.55797	1.80094	2.07893	2.39656	3.17217
16	1.17258	1.26899	1.37279	1.60471	1.87298	2.18287	2.54035	3.42594
17	1.18430	1.28802	1.40024	1.65285	1.94790	2.29202	2.69277	3.70002
18	1.19615	1.30734	1.42825	1.70243	2.02582	2.40662	2.85434	3.99602
19	1.20811	1.32695	1.45681	1.75351	2.10685	2.52695	3.02560	4.31570
20	1.22019	1.34686	1.48595	1.80611	2.19112	2.65330	3.20714	4.66096
21	1.23239	1.36706	1.51567	1.86029	2.27877	2.78596	3.39956	5.03383
22	1.24472	1.38756	1.54598	1.91610	2.36992	2.92526	3.60354	5.43654
23	1.25716	1.40838	1.57690	1.97359	2.46472	3.07152	3.81975	5.87146
24	1.26973	1.42950	1.60844	2.03279	2.56330	3.22510	4.04893	6.34118
25	1.28243	1.45095	1.64061	2.09378	2.66584	3.38635	4.29187	6.84848
26	1.29526	1.47271	1.67342	2.15659	2.77247	3.55567	4.54938	7.39635
27	1.30821	1.49480	1.70689	2.22129	2.88337	3.73346	4.82235	7.98806
28	1.32129	1.51722	1.74102	2.28793	2.99870	3.92013	5.11169	8.62711
29	1.33450	1.53998	1.77584	2.35657	3.11865	4.11614	5.41839	9.31727
30	1.34785	1.56308	1.81136	2.42726	3.24340	4.32194	5.74349	10.06266
31	1.36133	1.58653	1.84759	2.50008	3.37313	4.53804	6.08810	10.86767
32	1.37494	1.61032	1.88454	2.57508	3.50806	4.76494	6.45339	11.73708
33	1.38869	1.63448	1.92223	2.65234	3.64838	5.00319	6.84059	12.67605
34	1.40258	1.65900	1.96068	2.73191	3.79432	5.25335	7.25103	13.69013
35	1.41660	1.68388	1.99989	2.81386	3.94609	5.51602	7.68609	14.78534
36	1.43077	1.70914	2.03989	2.89828	4.10393	5.79182	8.14725	15.96817
37	1.44508	1.73478	2.08069	2.98523	4.26809	6.08141	8.63609	17.24563
38	1.45953	1.76080	2.12230	3.07478	4.43881	6.38548	9.15425	18.62528
39	1.47412	1.78721	2.16474	3.16703	4.61637	6.70475	9.70351	20.11530
40	1.48886	1.81402	2.20804	3.26204	4.80102	7.03999	10.28572	21.72452
41	1.50375	1.84123	2.25220	3.35990	4.99306	7.39199	10.90286	23.46248
42	1.51879	1.86885	2.29724	3.46070	5.19278	7.76159	11.55703	25.33948
43	1.53398	1.89688	2.34319	3.56452	5.40050	8.14967	12.25045	27.36664
44	1.54932	1.92533	2.39005	3.67145	5.61652	8.55715	12.98548	29.55597
45	1.56481	1.95421	2.43785	3.78160	5.84118	8.98501	13.76461	31.92045
46	1.58046	1.98353	2.48661	3.89504	6.07482	9.43426	14.59049	34.47409
47	1.59626	2.01328	2.53634	4.01190	6.31782	9.90597	15.46592	37.23201
48	1.61223	2.04348	2.58707	4.13225	6.57053	10.40127	16.39387	40.21057
49	1.62835	2.07413	2.63881	4.25622	6.83335	10.92133	17.37750	43.42742
50	1.64463	2.10524	2.69159	4.38391	7.10668	11.46740	18.42015	46.90161

TABLE 2 Present Value

$$\frac{1}{(1+i)^n}$$

n \ i	1%	1½%	2%	3%	4%	5%	6%	8%
1	.99010	.98522	.98039	.97087	.96154	.95238	.94340	.92593
2	.98030	.97066	.96117	.94260	.92456	.90703	.89000	.85734
3	.97059	.95632	.94232	.91514	.88900	.86384	.83962	.79383
4	.96098	.94218	.92385	.88849	.85480	.82270	.79209	.73503
5	.95147	.92826	.90573	.86261	.82193	.78353	.74726	.68058
6	.94205	.91454	.88797	.83748	.79031	.74622	.70496	.63017
7	.93272	.90103	.87056	.81309	.75992	.71068	.66506	.58349
8	.92348	.88771	.85349	.78941	.73069	.67684	.62741	.54027
9	.91434	.87459	.83676	.76642	.70259	.64461	.59190	.50025
10	.90529	.86167	.82035	.74409	.67556	.61391	.55839	.46319
11	.89632	.84893	.80426	.72242	.64958	.58468	.52679	.42888
12	.88745	.83639	.78849	.70138	.62460	.55684	.49697	.39711
13	.87866	.82403	.77303	.68095	.60057	.53032	.46884	.36770
14	.86996	.81185	.75788	.66112	.57748	.50507	.44230	.34046
15	.86135	.79985	.74301	.64186	.55526	.48102	.41727	.31524
16	.85282	.78803	.72845	.62317	.53391	.45811	.39365	.29189
17	.84438	.77639	.71416	.60502	.51337	.43630	.37136	.27027
18	.83602	.76491	.70016	.58739	.49363	.41552	.35034	.25025
19	.82774	.75361	.68643	.57029	.47464	.39573	.33051	.23171
20	.81954	.74247	.67297	.55368	.45639	.37689	.31180	.21455
21	.81143	.73150	.65978	.53755	.43883	.35894	.29416	.19866
22	.80340	.72069	.64684	.52189	.42196	.34185	.27751	.18394
23	.79544	.71004	.63416	.50669	.40573	.32557	.26180	.17032
24	.78757	.69954	.62172	.49193	.39012	.31007	.24698	.15770
25	.77977	.68921	.60953	.47761	.37512	.29530	.23300	.14602
26	.77205	.67902	.59758	.46369	.36069	.28124	.21981	.13520
27	.76440	.66899	.58586	.45019	.34682	.26785	.20737	.12519
28	.75684	.65910	.57437	.43708	.33348	.25509	.19563	.11591
29	.74934	.64936	.56311	.42435	.32065	.24295	.18456	.10733
30	.74192	.63976	.55207	.41199	.30832	.23138	.17411	.09938
31	.73458	.63031	.54125	.39999	.29646	.22036	.16425	.09202
32	.72730	.62099	.53063	.38834	.28506	.20987	.15496	.08520
33	.72010	.61182	.52023	.37703	.27409	.19987	.14619	.07889
34	.71297	.60277	.51003	.36604	.26355	.19035	.13791	.07305
35	.70591	.59387	.50003	.35538	.25342	.18129	.13011	.06763
36	.69892	.58509	.49022	.34503	.24367	.17266	.12274	.06262
37	.69200	.57644	.48061	.33498	.23430	.16444	.11579	.05799
38	.68515	.56792	.47119	.32523	.22529	.15661	.10924	.05369
39	.67837	.55953	.46195	.31575	.21662	.14915	.10306	.04971
40	.67165	.55126	.45289	.30656	.20829	.14205	.09722	.04603
41	.66500	.54312	.44401	.29763	.20028	.13528	.09172	.04262
42	.65842	.53509	.43530	.28896	.19257	.12884	.08653	.03946
43	.65190	.52718	.42677	.28054	.18517	.12270	.08163	.03654
44	.64545	.51939	.41840	.27237	.17805	.11686	.07701	.03383
45	.63905	.51171	.41020	.26444	.17120	.11130	.07265	.03133
46	.63273	.50415	.40215	.25674	.16461	.10600	.06854	.02901
47	.62646	.49670	.39427	.24926	.15828	.10095	.06466	.02686
48	.62026	.48936	.38654	.24200	.15219	.09614	.06100	.02487
49	.61412	.48213	.37896	.23495	.14634	.09156	.05755	.02303
50	.60804	.47500	.37153	.22811	.14071	.08720	.05429	.02132

■ TABLE 3 Amount of an Annuity

$$s_{\overline{n}|i} = \frac{(1+i)^n - 1}{i}$$

n \ i	1%	$1\frac{1}{2}$%	2%	3%	4%	5%	6%	8%
1	1.00000	1.00000	1.00000	1.00000	1.00000	1.00000	1.00000	1.00000
2	2.01000	2.01500	2.02000	2.03000	2.04000	2.05000	2.06000	2.08000
3	3.03010	3.04523	3.06040	3.09090	3.12160	3.15250	3.18360	3.24640
4	4.06040	4.09090	4.12161	4.18363	4.24646	4.31013	4.37462	4.50611
5	5.10101	5.15227	5.20404	5.30914	5.41632	5.52563	5.63709	5.86660
6	6.15202	6.22955	6.30812	6.46841	6.63298	6.80191	6.97532	7.33593
7	7.21354	7.32299	7.43428	7.66246	7.89829	8.14201	8.39384	8.92280
8	8.28567	8.43284	8.58297	8.89234	9.21423	9.54911	9.89747	10.63663
9	9.36853	9.55933	9.75463	10.15911	10.58280	11.02656	11.49132	12.48756
10	10.46221	10.70272	10.94972	11.46388	12.00611	12.57789	13.18079	14.48656
11	11.56683	11.86326	12.16872	12.80780	13.48635	14.20679	14.97164	16.64549
12	12.68250	13.04121	13.41209	14.19203	15.02581	15.91713	16.86994	18.97713
13	13.80933	14.23683	14.68033	15.61779	16.62684	17.71298	18.88214	21.49530
14	14.94742	15.45038	15.97394	17.08632	18.29191	19.59863	21.01507	24.21492
15	16.09690	16.68214	17.29342	18.59891	20.02359	21.57856	23.27597	27.15211
16	17.25786	17.93237	18.63929	20.15688	21.82453	23.65749	25.67253	30.32428
17	18.43044	19.20136	20.01207	21.76159	23.69751	25.84037	28.21288	33.75023
18	19.61475	20.48938	21.41231	23.41444	25.64541	28.13238	30.90565	37.45024
19	20.81090	21.79672	22.84056	25.11687	27.67123	30.53900	33.75999	41.44626
20	22.01900	23.12367	24.29737	26.87037	29.77808	33.06595	36.78559	45.76196
21	23.23919	24.47052	25.78332	28.67649	31.96920	35.71925	39.99273	50.42292
22	24.47159	25.83758	27.29898	30.53678	34.24797	38.50521	43.39229	55.45676
23	25.71630	27.22514	28.84496	32.45288	36.61789	41.43048	46.99583	60.89330
24	26.97346	28.63352	30.42186	34.42647	39.08260	44.50200	50.81558	66.76476
25	28.24320	30.06302	32.03030	36.45926	41.64591	47.72710	54.86451	73.10594
26	29.52563	31.51397	33.67091	38.55304	44.31174	51.11345	59.15638	79.95442
27	30.82089	32.98668	35.34432	40.70963	47.08421	54.66913	63.70577	87.35077
28	32.12910	34.48148	37.05121	42.93092	49.96758	58.40258	68.52811	95.33883
29	33.45039	35.99870	38.79223	45.21885	52.96629	62.32271	73.63980	103.96594
30	34.78489	37.53868	40.56808	47.57542	56.08494	66.43885	79.05819	113.28321
31	36.13274	39.10176	42.37944	50.00268	59.32834	70.76079	84.80168	123.34587
32	37.49407	40.68829	44.22703	52.50276	62.70147	75.29883	90.88978	134.21354
33	38.86901	42.29861	46.11157	55.07784	66.20953	80.06377	97.34316	145.95062
34	40.25770	43.93309	48.03380	57.73018	69.85791	85.06696	104.18375	158.62667
35	41.66028	45.59209	49.99448	60.46208	73.65222	90.32031	111.43478	172.31680
36	43.07688	47.27597	51.99437	63.27594	77.59831	95.83632	119.12087	187.10215
37	44.50765	48.98511	54.03425	66.17422	81.70225	101.62814	127.26812	203.07032
38	45.95272	50.71989	56.11494	69.15945	85.97034	107.70955	135.90421	220.31595
39	47.41225	52.48068	58.23724	72.23423	90.40915	114.09502	145.05846	238.94122
40	48.88637	54.26789	60.40198	75.40126	95.02552	120.79977	154.76197	259.05652
41	50.37524	56.08191	62.61002	78.66330	99.82654	127.83976	165.04768	280.78104
42	51.87899	57.92314	64.86222	82.02320	104.81960	135.23175	175.95054	304.24352
43	53.39778	59.79199	67.15947	85.48389	110.01238	142.99334	187.50758	329.58301
44	54.93176	61.68887	69.50266	89.04841	115.41288	151.14301	199.75803	356.94965
45	56.48107	63.61420	71.89271	92.71986	121.02939	159.70016	212.74351	386.50562
46	58.04589	65.56841	74.33056	96.50146	126.87057	168.68516	226.50812	418.42607
47	59.62634	67.55194	76.81718	100.39650	132.94539	178.11942	241.09861	452.90015
48	61.22261	69.56522	79.35352	104.40840	139.26321	188.02539	256.56453	490.13216
49	62.83483	71.60870	81.94059	108.54065	145.83373	198.42666	272.95840	530.34274
50	64.46318	73.68283	84.57940	112.79687	152.66708	209.34800	290.33590	573.77016

■ TABLE 4 Present Value of an Annuity

$$a_{\overline{n}|i} = \frac{1 - (1 + i)^{-n}}{i}$$

n \ i	1%	$1\frac{1}{2}$%	2%	3%	4%	5%	6%	8%
1	.99010	.98522	.98039	.97087	.96154	.95238	.94340	.92593
2	1.97040	1.95588	1.94156	1.91347	1.88609	1.85941	1.83339	1.78326
3	2.94099	2.91220	2.88388	2.82861	2.77509	2.72325	2.67301	2.57710
4	3.90197	3.85438	3.80773	3.71710	3.62990	3.54595	3.46511	3.31213
5	4.85343	4.78264	4.71346	4.57971	4.45182	4.32948	4.21236	3.99271
6	5.79548	5.69719	5.60143	5.41719	5.24214	5.07569	4.91732	4.62288
7	6.72819	6.59821	6.47199	6.23028	6.00205	5.78637	5.58238	5.20637
8	7.65168	7.40503	7.32548	7.01969	6.73274	6.46321	6.20979	5.74664
9	8.56602	8.36052	8.16224	7.78611	7.43533	7.10782	6.80169	6.24689
10	9.47130	9.22218	8.98259	8.53020	8.11090	7.72173	7.36009	6.71008
11	10.36763	10.07112	9.78685	9.25262	8.76048	8.30641	7.88687	7.13896
12	11.25508	10.90751	10.57534	9.95400	9.38507	8.86325	8.38384	7.53608
13	12.13374	11.73153	11.34837	10.63496	9.98565	9.39357	8.85268	7.90378
14	13.00370	12.54338	12.10625	11.29607	10.56312	9.89864	9.29498	8.24424
15	13.86505	13.34323	12.84926	11.93794	11.11839	10.37966	9.71225	8.55948
16	14.71787	14.13126	13.57771	12.56110	11.65230	10.83777	10.10590	8.85137
17	15.56225	14.90765	14.29187	13.16612	12.16567	11.27407	10.47726	9.12164
18	16.39827	15.67256	14.99203	13.75351	12.65930	11.68959	10.82760	9.37189
19	17.22601	16.42617	15.67846	14.32380	13.13394	12.08532	11.15812	9.60360
20	18.04555	17.16864	16.35143	14.87747	13.59033	12.46221	11.46992	9.81815
21	18.85698	17.90014	17.01121	15.41502	14.02916	12.82115	11.76408	10.01680
22	19.66038	18.62082	17.65805	15.93692	14.45112	13.16300	12.04158	10.20074
23	20.45582	19.33086	18.29220	16.44361	14.85684	13.48857	12.30338	10.37106
24	21.24339	20.03041	18.91393	16.93554	15.24696	13.79864	12.55036	10.52876
25	22.02316	20.71961	19.52346	17.41315	15.62208	14.09394	12.78336	10.67478
26	22.79520	21.39863	20.12104	17.87684	15.98277	14.37519	13.00317	10.80998
27	23.55961	22.06762	20.70690	18.32703	16.32959	14.64303	13.21053	10.93516
28	24.31644	22.72672	21.28127	18.76411	16.66306	14.89813	13.40616	11.05108
29	25.06579	23.37608	21.84438	19.18845	16.98371	15.14107	13.59072	11.15841
30	25.80771	24.01584	22.39646	19.60044	17.29203	15.37245	13.76483	11.25778
31	26.54229	24.64615	22.93770	20.00043	17.58849	15.59281	13.92909	11.34980
32	27.26959	25.26714	23.46833	20.38877	17.87355	15.80268	14.08404	11.43500
33	27.98969	25.87895	23.98856	20.76579	18.14765	16.00255	14.23023	11.51389
34	28.70267	26.48173	24.49859	21.13184	18.41120	16.19290	14.36814	11.58693
35	29.40858	27.07559	24.99862	21.48722	18.66461	16.37419	14.49825	11.65457
36	30.10751	27.66068	25.48884	21.83225	18.90828	16.54685	14.62099	11.71719
37	30.79951	28.23713	25.96945	22.16724	19.14258	16.71129	14.73678	11.77518
38	31.48466	28.80505	26.44064	22.49246	19.36786	16.86789	14.84602	11.82887
39	32.16303	29.36458	26.90259	22.80822	19.58448	17.01704	14.94907	11.87858
40	32.83469	29.91585	27.35548	23.11477	19.79277	17.15909	15.04630	11.92461
41	33.49969	30.45896	27.79949	23.41240	19.99305	17.29437	15.13802	11.96723
42	34.15811	30.99405	28.23479	23.70136	20.18563	17.42321	15.22454	12.00670
43	34.81001	31.52123	28.66156	23.98190	20.37079	17.54591	15.30617	12.04324
44	35.45545	32.04062	29.07996	24.25427	20.54884	17.66277	15.38318	12.07707
45	36.09451	32.55234	29.49016	24.51871	20.72004	17.77407	15.45583	12.10840
46	36.72724	33.05649	29.89231	24.77545	20.88465	17.88007	15.52437	12.13741
47	37.35370	33.55319	30.28658	25.02471	21.04294	17.98102	15.58903	12.16427
48	37.97396	34.04255	30.67312	25.26671	21.19513	18.07716	15.65003	12.18914
49	38.58808	34.52468	31.05208	25.50166	21.34147	18.16872	15.70757	12.21216
50	39.19612	34.99969	31.42361	25.72976	21.48218	18.25593	15.76186	12.23348

■ TABLE 5 Powers of *e* and Natural Logarithms

x	e^x	e^{-x}	$\ln x$
.00	1.00000	1.00000	
.01	1.01005	.99004	−4.6052
.02	1.02020	.98019	−3.9120
.03	1.03045	.97044	−3.5066
.04	1.04081	.96078	−3.2189
.05	1.05127	.95122	−2.9957
.06	1.06183	.94176	−2.8134
.07	1.07250	.93239	−2.6593
.08	1.08328	.92311	−2.5257
.09	1.09417	.91393	−2.4079
.10	1.10517	.90483	−2.3026
.11	1.11628	.89583	−2.2073
.12	1.12750	.88692	−2.1203
.13	1.12883	.87810	−2.0402
.14	1.15027	.86936	−1.9661
.15	1.16183	.86071	−1.8971
.16	1.17351	.85214	−1.8326
.17	1.18530	.84366	−1.7720
.18	1.19722	.83527	−1.7148
.19	1.20925	.82696	−1.6607
.2	1.22140	.81873	−1.6094
.3	1.34985	.74081	−1.2040
.4	1.49182	.67032	−.9163
.5	1.64872	.60653	−.6931
.6	1.82211	.54881	−.5108
.7	2.01375	.49658	−.3567
.8	2.22554	.44932	−.2231
.9	2.45960	.40656	−.1054
1.0	2.71828	.36787	.0000
1.1	3.00416	.33287	.0953
1.2	3.32011	.30119	.1823
1,3	3.66929	.27253	.2624
1.4	4.05519	.24659	.3365
1.5	4.48168	.22313	.4055
1.6	4.95302	.20189	.4700
1.7	5.47394	.18268	.5306
1.8	6.04964	.16529	.5878
1.9	6.68589	.14956	.6419
2.0	7.38905	.13533	.6931
2.1	8.16616	.12245	.7419
2.2	9.02500	.11080	.7885
2.3	9.97417	.10025	.8329
2.4	11.0231	.09071	.8755
2.5	12.1824	.08208	.9163
2.6	13.4637	.07427	.9555
2.7	14.8797	.06720	.9933
2.8	16.4446	.06081	1.0296
2.9	18.1741	.05502	1.0647
3.0	20.0855	.04978	1.0986
3.1	22.1979	.04505	1.1314
3.2	24.5325	.04076	1.1632
3.3	27.1126	.03688	1.1939
3.4	29.9641	.03337	1.2238
3.5	33.1154	.03020	1.2528
3.6	36.5982	.02732	1.2809
3.7	40.4473	.02472	1.3083
3.8	44.7012	.02237	1.3350
3.9	49.4024	.02024	1.3610

x	e^x	e^{-x}	$\ln x$
4.0	54.5981	.01832	1.3863
4.1	60.3402	.01657	1.4110
4.2	66.6863	.01500	1.4351
4.3	73.6997	.01357	1.4586
4.4	81.4508	.01228	1.4816
4.5	90.0170	.01111	1.5041
4.6	99.4842	.01005	1.5261
4.7	109.947	.00910	1.5476
4.8	121.510	.00823	1.5686
4.9	134.290	.00745	1.5892
5.0	148.413	.00674	1.6094
5.1	164.022	.00610	1.6292
5.2	181.272	.00552	1.6487
5.3	200.336	.00499	1.6677
5.4	221.406	.00452	1.6864
5.5	244.691	.00445	1.7047
5.6	270.426	.00370	1.7228
5.7	298.867	.00335	1.7405
5.8	330.299	.00303	1.7579
5.9	365.036	.00274	1.7750
6.0	403.428	.00248	1.7918
6.1	445.856	.00224	1.8083
6.2	492.748	.00203	1.8245
6.3	544.570	.00184	1.8405
6.4	601.843	.00166	1.8563
6.5	665.139	.00150	1.8718
6.6	735.093	.00136	1.8871
6.7	812.403	.00123	1.9021
6.8	897.844	.00111	1.9169
6.9	992.271	.00101	1.9315
7.0	1096.63	.00091	1.9459
7.1	1211.96	.00083	1.9601
7.2	1339.43	.00075	1.9741
7.3	1480.29	.00068	1.9879
7.4	1635.98	.00061	2.0015
7.5	1808.03	.00055	2.0149
7.6	1998.19	.00050	2.0281
7.7	2208.34	.00045	2.0412
7.8	2440.59	.00041	2.0541
7.9	2697.27	.00037	2.0669
8.0	2980.94	.00034	2.0794
8.1	3294.45	.00030	2.0919
8.2	3640.94	.00027	2.1041
8.3	4023.86	.00025	2.1163
8.4	4447.05	.00022	2.1282
8.5	4914.75	.00020	2.1401
8.6	5431.65	.00018	2.1518
8.7	6002.90	.00017	2.1633
8.8	6634.23	.00015	2.1748
8.9	7331.96	.00014	2.1861
9.0	8103.08	.00012	2.1972
9.1	8955.29	.00011	2.2083
9.2	9897.13	.00010	2.2192
9.3	10938.0	.00009	2.2300
9.4	12088.4	.00008	2.2407
9.5	13359.7	.000075	2.2513
9.6	14764.8	.000068	2.2618
9.7	16317.6	.000061	2.2721
9.8	18033.8	.000055	2.2824
9.9	19930.4	.000050	2.2925
10.0	22026.5	.000045	2.3026

ANSWERS TO SELECTED EXERCISES

TO THE STUDENT

If you want further help with this course, you may want to obtain a copy of the *Student's Solutions Manual* that accompanies this textbook. This manual provides detailed step-by-step solutions to the odd-numbered exercises in the textbook and can help you study and understand the course material. Your college bookstore either has this manual or can order it for you.

■ Chapter 1 Section 1.1 (page 8)

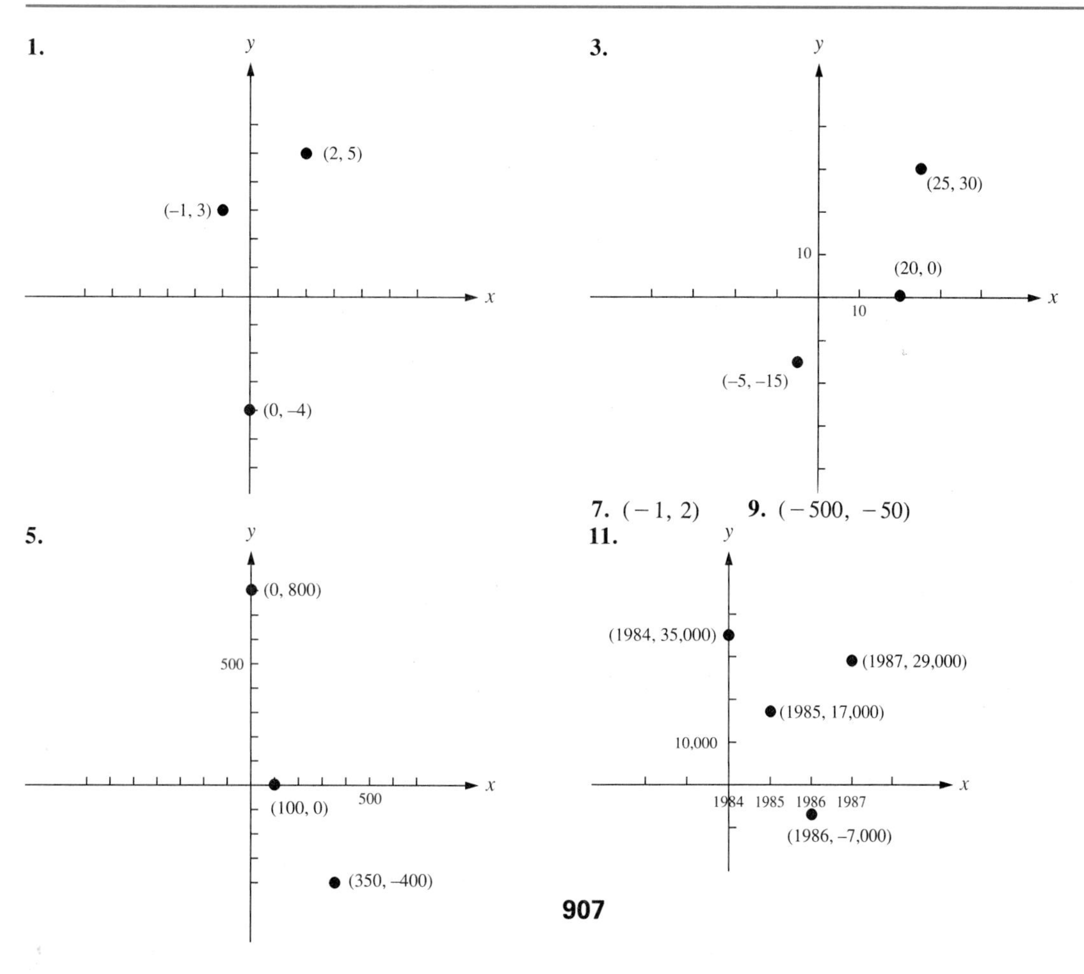

7. $(-1, 2)$ **9.** $(-500, -50)$

13.

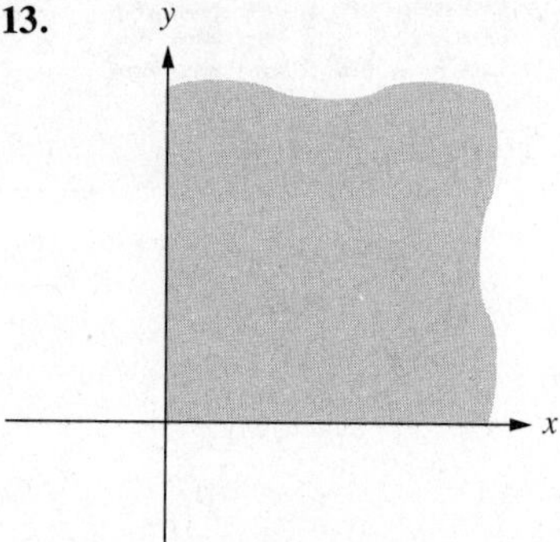

15.

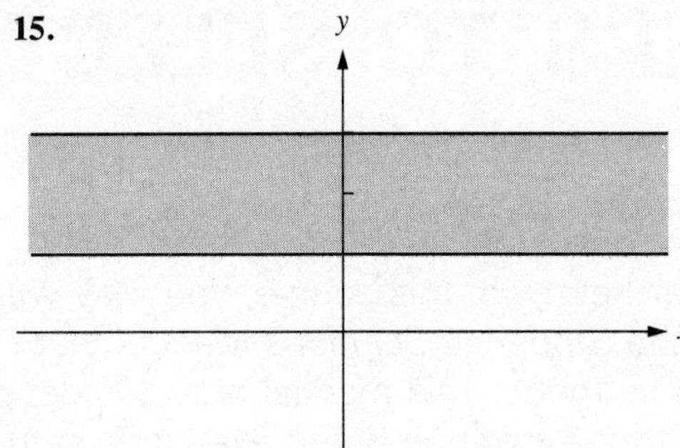

17.

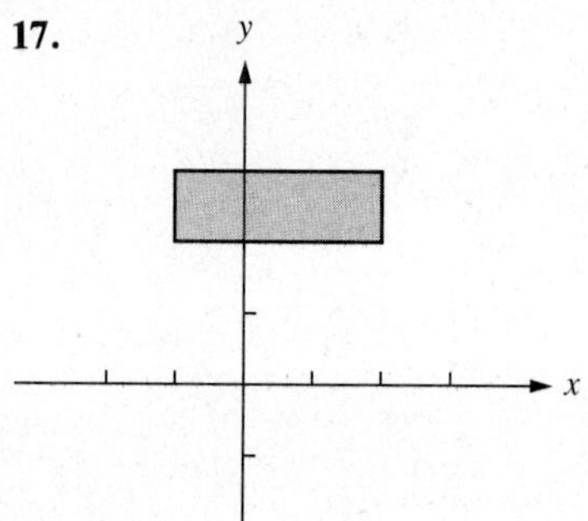

19. $-5x + y = 7$ 21. $x + 4y = -4$

23.

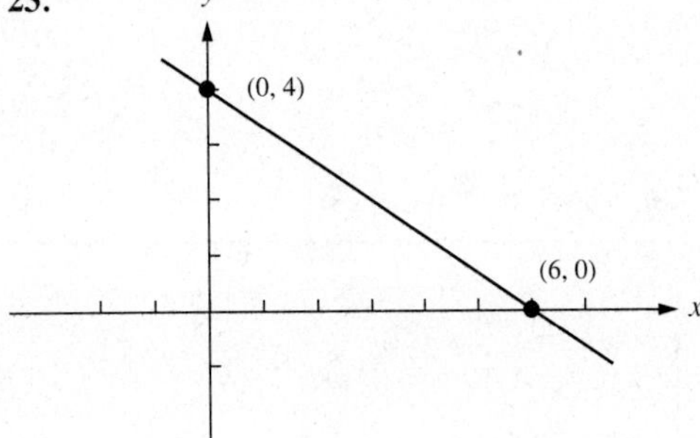

25.

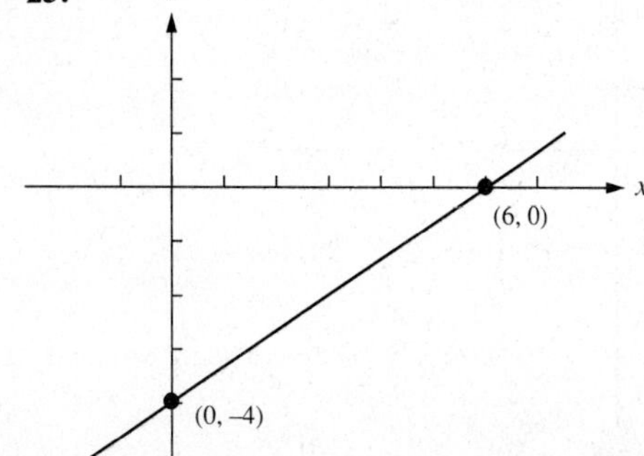

27.

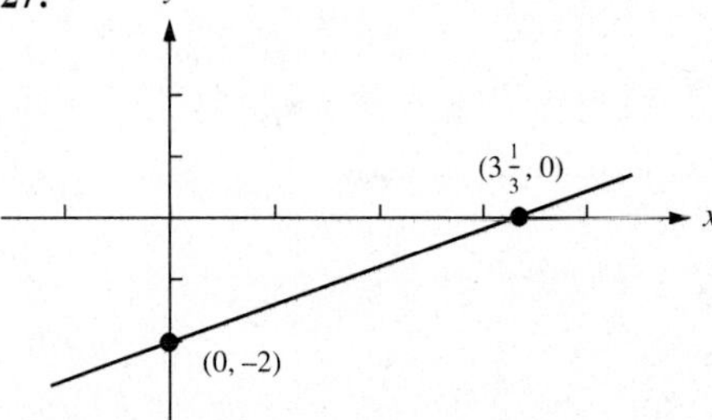

29.

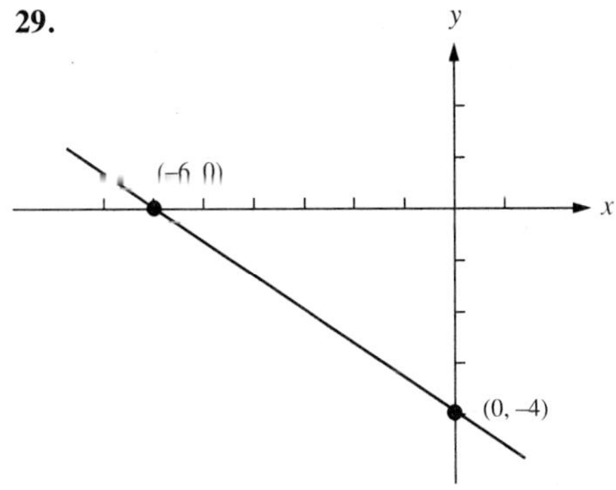

31.

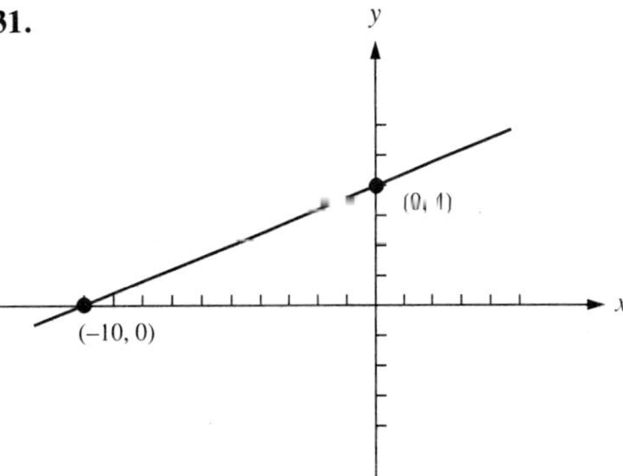

33.

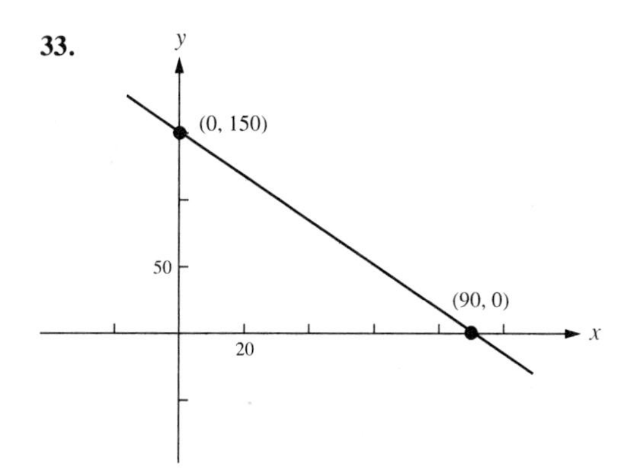

35.

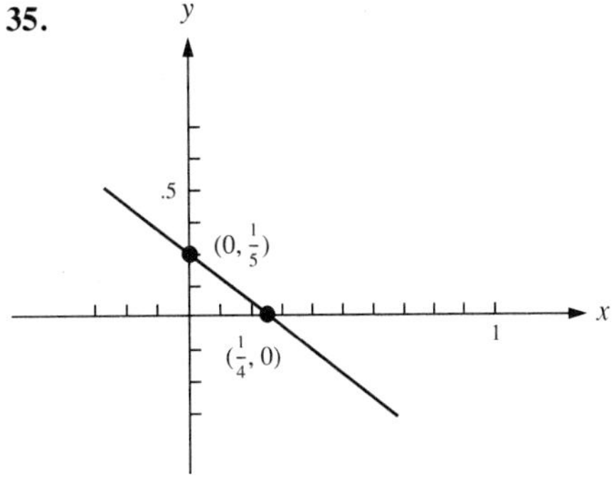

Section 1.2 (page 15)

1. $y = -\frac{3}{4}x + \frac{9}{2}$ **3.** $y = \frac{1}{3}x - \frac{5}{3}$ **5.** $y = 0x - \frac{7}{4}$ **7.** $-\frac{7}{8}, \left(0, \frac{21}{4}\right)$ **9.** $0, \left(0, -\frac{5}{3}\right)$
11. $2, \left(0, -\frac{3}{2}\right)$ **13.** no slope or y-intercept **15.** 2 **17.** $-\frac{3}{4}$ **19.** undefined **21.** $y = 4x - 3$
23. $y = -3x - 1$ **25.** $y = x - 2$ **27.** $x = 7$ **29.** $y = -x + 7$ **31.** $y = -\frac{8}{3}x + \frac{1}{3}$
33. $y = 0x + 2$ **35.** (a) 325,000 (b) \$3,700,000 **37.** 47.5
39. (a) $V = -500t + 10{,}000$ (b) \$8500, \$7500 (c) 12 years

Section 1.3 (page 24)

1. $(2, -2)$ **3.** $(23, 10)$ **5.** $(2, 1)$ **7.** $(6, -3)$ **9.** $(3, 4)$ **11.** $(64, 18)$ **13.** $(5.5, 2.5)$
15. no solutions **17.** $(30, 5)$ **19.** $(8, -2)$ **21.** 20 oz Food A, 5 oz Food B
23. (a) 1440 lbs Blend A, 1260 lbs Blend B (b) no **25.** 80 liters Solution A, 20 liters Solution B
27. 200 rings, 80 bracelets **29.** \$80 for stock A, \$50 for stock B

Section 1.4 (page 33)

1. 11,200 units

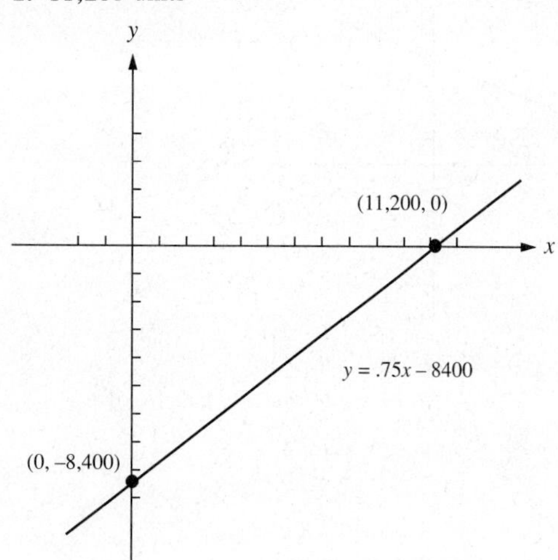

3. 1500 chips

5. (a) 19,200 (b) 16,800 (c) 33,600

7. (a) 5250 (b) 4000 (c) 5000

9. (a) 11,200 units, 12,000 units
(b) Method 2 is better for $x > 16{,}000$.

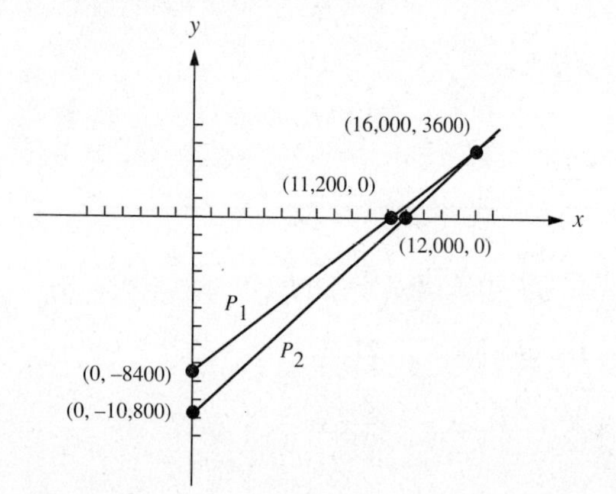

11. (a) 48 cars, 65 cars
(b) When > 100 cars are made.

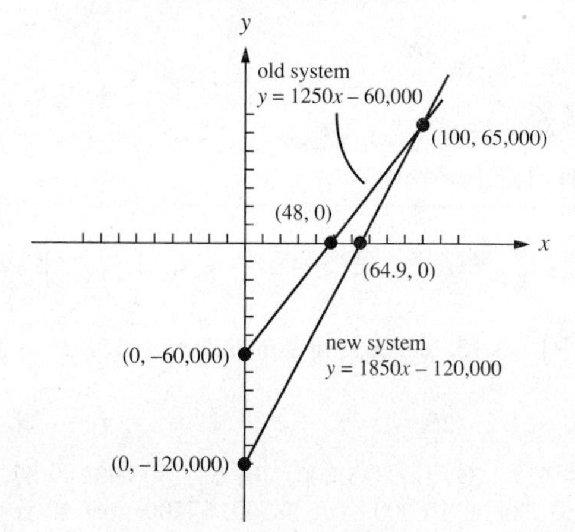

13. Printer 2 is better for > 3000 copies.

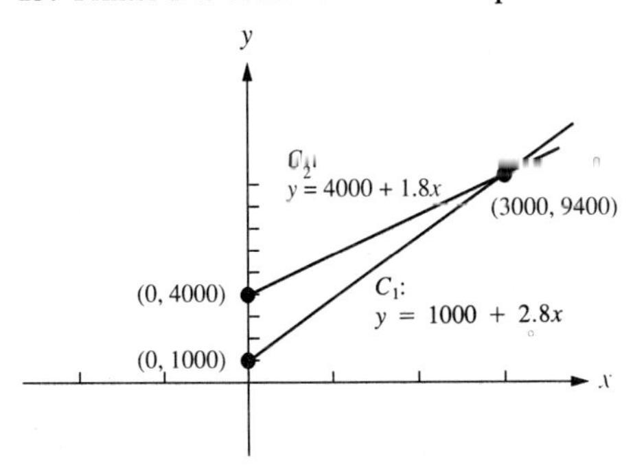

15. Use Red Cab for $x \le 4$, White Cab for $4 \le x \le 10$, and Blue Cab for $x \ge 10$.

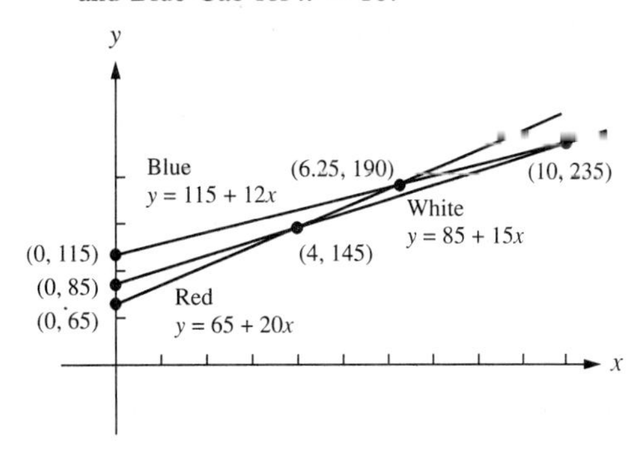

17. First company is better for $x <$ \$4.5 million, second for $x >$ \$4.5 million.

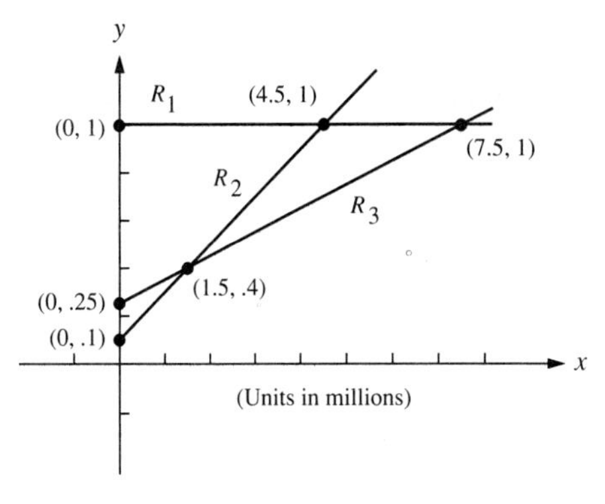

Section 1.5 (page 40)

1. \$6, 1500 **3.** \$7.50, 750 **5.** \$7, 800 **7.** \$13.40, 1288 **9.** \$5.63, 1406, \$791.02 **11.** \$6.67, 667, \$1111.11 **13.** \$11.47, 864, \$2973.60 **15.** \$5.50, 1375, \$1031.25 **17.** \$6.90, 760, \$380 **19.** \$12.90, 1177, \$1059.41 **21.** \$10.40, 880 **23.** \$7, 437 **25.** The factory gets \$7.38 for each box, which sells for \$6.38, and 646 boxes are sold.

Chapter 1 Review Exercises (page 42)

1.
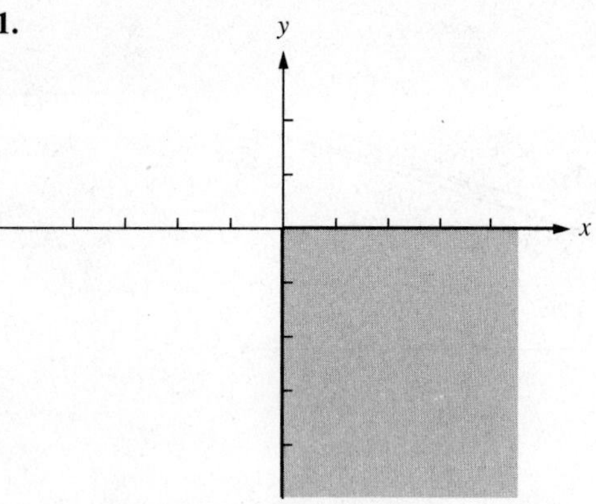

2.
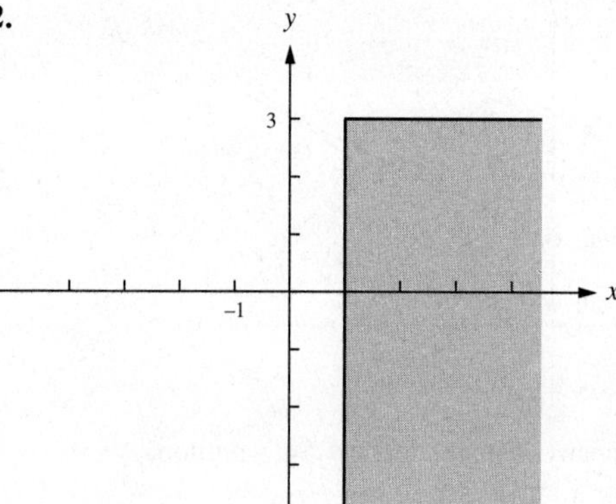

3.
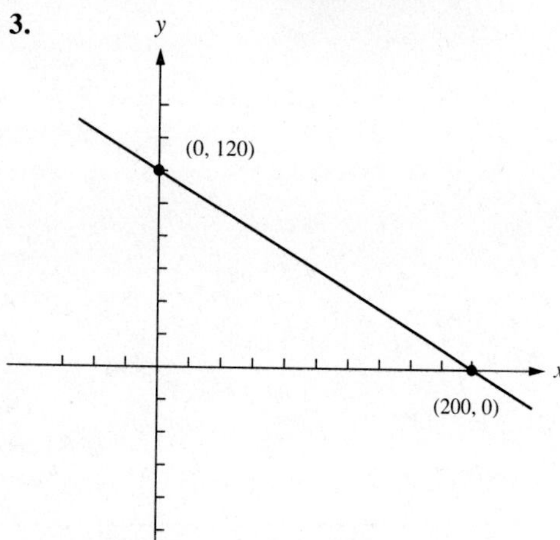

4.
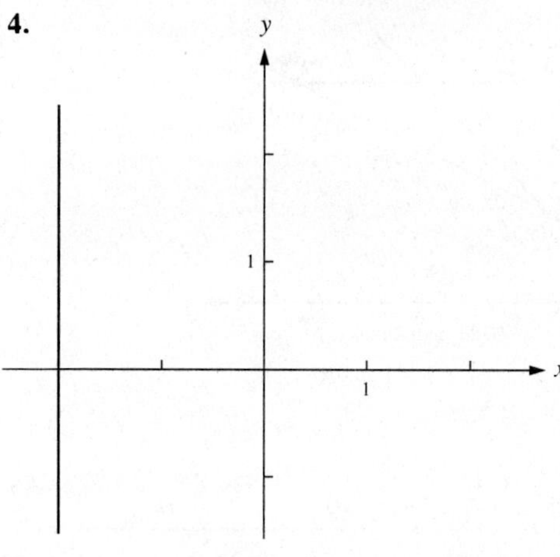

5.
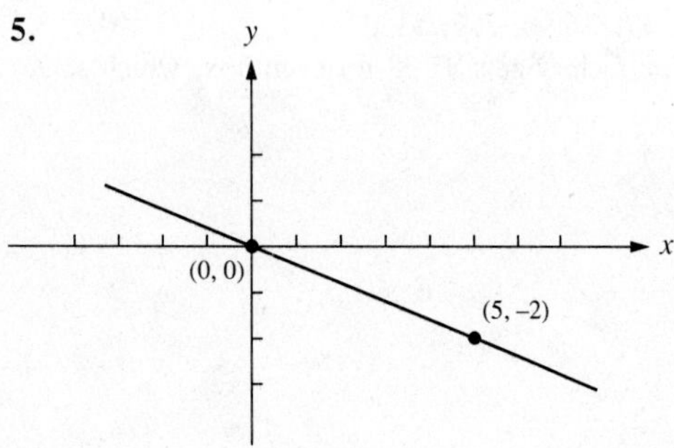

6.
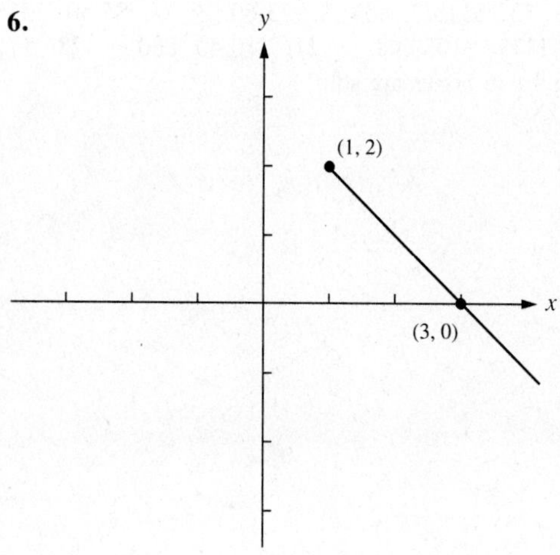

7. $-\frac{3}{2}, (1, 0), \left(\frac{2}{3}, 0\right)$ **8.** $y = -8x + 13$ **9.** 25 **10.** $x = 2.5, y = -3$ **11.** $a = .1, b = .2$
12. 12 Fun-Packs, 1 Economy Box **13.** \$4000 in investment A and \$6000 in investment B
14. (a)

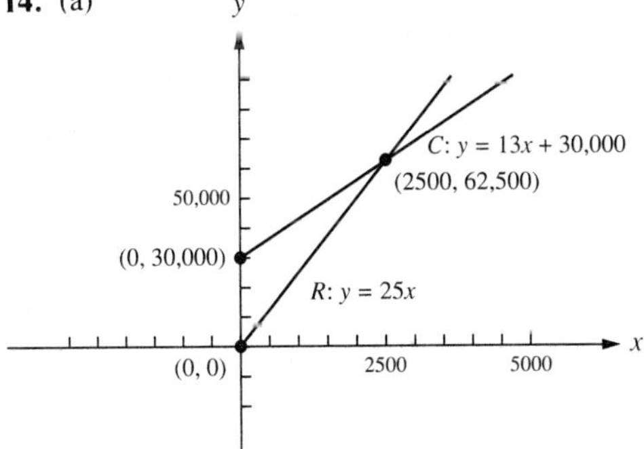

(b) 2500 (c) 3500 (d) 3750 (e) 5000

15. (a)

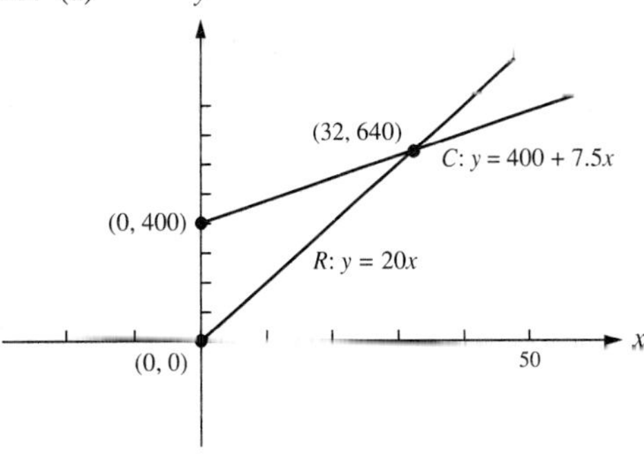

(b) 32 (c) 46 (d) 40 (e) 50

16. \$3.75, 550 **17.** \$3.64, 519 **18.** \$3.60, 508

Chapter 2 Section 2.1 (page 54)

1. $\left[\begin{array}{cc|c} 2 & 3 & 12 \\ 1 & -\frac{1}{4} & \frac{3}{4} \end{array}\right]$ **3.** $\left[\begin{array}{ccc|c} 1 & -2 & \frac{1}{2} & 4 \\ 0 & \frac{1}{2} & -1 & -5 \\ 0 & 4 & -\frac{5}{3} & -2 \end{array}\right]$ **5.** $\left[\begin{array}{cccc|c} 2 & -1 & 1 & 0 & 6 \\ 0 & 1 & \frac{5}{2} & -2 & 4 \\ \frac{3}{4} & -1 & 1 & 2 & 2 \\ -1 & 0 & 1 & 1 & -5 \end{array}\right]$ **7.** $\left[\begin{array}{cc|c} 14 & 0 & 21 \\ -4 & 1 & -3 \end{array}\right]$

9. $\left[\begin{array}{ccc|c} 2 & -4 & 1 & 8 \\ 2 & -\frac{7}{2} & 0 & 3 \\ \frac{10}{3} & -\frac{8}{3} & 0 & \frac{34}{3} \end{array}\right]$ **11.** $\left[\begin{array}{cccc|c} 2 & -1 & 1 & 0 & 6 \\ \frac{3}{8} & 0 & \frac{7}{4} & 0 & 3 \\ \frac{3}{8} & -\frac{1}{2} & \frac{1}{2} & 1 & 1 \\ -\frac{11}{8} & \frac{1}{2} & \frac{1}{2} & 0 & -6 \end{array}\right]$ **13.** $x = 8, y = 5$ **15.** $x = 2, y = 1, z = 4$
17. $x = 0, y = 1, z = 0$ **19.** $x = 1, y = 3, z = 2$ **21.** $x = -1, y = 0, z = 1, w = 2$
23. 9 gal. of the 1st, 5 of the 2nd, 2 of the 3rd **25.** 50 slices pie, 90 shortcakes, 20 baked Alaska
27. Bus 200 whites and 350 blacks from Eastern to Western, and bus 300 whites and 150 blacks from Western to Eastern.
29. $(1/a)R_i$ **31.** $R_i + R_j, (-1)R_j, R_j + R_i, R_i + (-1)R_j$

Section 2.2 (page 65)

1. $(x, y, z) = (3z, 2 - 3z, z)$ **3.** $(x, y, z) = (2 - 2y, y, -1)$ **5.** $(w, x, y, z) = (1 - 3y - 2z, -1 + 5y + 5z, y, z)$ **7.** $(u, v, w) = (1, 2, 3)$ **9.** $(r, s, t) = (8/7 + (11/7)t, -3/7 + (2/7)t, t)$
11. inconsistent **13.** $(x, y, z, u, v) = (-5 + 5z + 2u + 4v, 3 - 2z - u - 3v, z, u, v)$ **15.** $(x, y, z) = \left(-\frac{3}{5} - \frac{1}{5}z, -\frac{4}{5} + \frac{7}{5}z, z\right)$ **17.** (a) $(x, y, z) = (4 + z, -1 - 2z, z)$ (b) $(x, y, z) = \left(\frac{7}{2} - \frac{1}{2}y, y, -\frac{1}{2} - \frac{1}{2}y\right)$
(c) $(x, y, z) = (x, 7 - 2x, -4 + x)$ **19.** $x = -50 + \frac{1}{2}z + w, y = 250 - \frac{3}{2}z - 2w$
21. no. lbs peanuts $= \frac{88}{9} + \frac{5}{9}$(no. lbs almonds), no. lbs cashews $= \frac{272}{9} - \frac{14}{9}$(no. lbs almonds)

Section 2.3 (page 74)

1. Mine 1, 12 days; mine 2, 16 days; mine 3, 3 days **3.** 450 lbs Type A, 100 lbs Type B, 50 lbs Type C **5.** (a) $180{,}000 + 20x + 25y + 40z + 50w = 25(12{,}000 + x + y + z + w)$; $2{,}700{,}000 + 125x + 100y + 75z + 25w = 150(12{,}000 + x + y + z + w)$ (b) $x = 3z + 5w - 24{,}000$, $y = -3z - 5w + 30{,}000$ (c) Those such that $z \geq 0$, $w \geq 0$, and $24{,}000 \leq 3z + 5w \leq 30{,}000$ **7.** \$8000, \$11,000, \$5000, \$10,800, \$10,700 **9.** \$40,000, \$60,000, \$20,000, \$42,000, \$36,000, \$56,000 **11.** 200

Section 2.4 (page 84)

1. $\begin{bmatrix} 9 & 2 & 11 & 12 \\ -1 & 4 & 2 & 4 \end{bmatrix}$ **3.** undefined **5.** $\begin{bmatrix} 3 & 4 & 11 & 4 \\ 5 & -6 & -2 & 2 \end{bmatrix}$ **7.** $\begin{bmatrix} 18 & 9 & 33 & 24 \\ 6 & -3 & 0 & 9 \end{bmatrix}$
9. $\begin{bmatrix} 12 & 1 & 11 & 16 \\ -4 & 9 & 4 & 5 \end{bmatrix}$ **11.** $\begin{bmatrix} 24 & 2 & 22 & 32 \\ -8 & 18 & 8 & 10 \end{bmatrix}$ **13.** $\begin{bmatrix} 0 & 5 & 11 & 0 \\ 8 & -11 & -4 & 1 \end{bmatrix}$
15. $\begin{bmatrix} \frac{15}{2} & \frac{15}{2} & 22 & 10 \\ \frac{17}{2} & -\frac{19}{2} & -3 & \frac{9}{2} \end{bmatrix}$ **17.** $\left[\begin{array}{ccc|c} 1 & 3 & 1 & 5 \\ 1 & 7 & -1 & 0 \end{array}\right]$ **19.** $\left[\begin{array}{ccc|c} 1 & 3 & 0 & 9 \\ 1 & 0 & 7 & 12 \end{array}\right]$ **21.** $\left[\begin{array}{cccc|c} 1 & 0 & 2 & 0 & -3 \\ 0 & 1 & 0 & -1 & 5 \\ 1 & -1 & 4 & 0 & 7 \end{array}\right]$
23. $[3 \ \ 3 \ \ 3 \ \ 3 \ \ 3]$ **25.** $\begin{bmatrix} 5 & 5 \\ 5 & 5 \\ 5 & 5 \end{bmatrix}$ **27.** \$300 **29.** $E - F$ **31.** $3E + 2F$

Section 2.5 (page 91)

1. undefined **3.** $[15]$ **5.** $\begin{bmatrix} 5 \\ 15 \\ 16 \end{bmatrix}$ **7.** $\begin{bmatrix} 8 \\ 35 \\ 15 \end{bmatrix}$ **9.** $\begin{bmatrix} 9 \\ 26 \end{bmatrix}$ **11.** $\begin{bmatrix} 17 & -4 & -11 \\ 8 & 9 & -9 \\ 17 & 3 & -16 \end{bmatrix}$ **13.** $\begin{bmatrix} -7 & 32 \\ 12 & -2 \\ 7 & 18 \end{bmatrix}$
15. undefined **17.** $\begin{bmatrix} 0 & 12 & -12 \\ -10 & -4 & -16 \\ 30 & 3 & 57 \end{bmatrix}$ **19.** $[18 \ \ 1 \ \ 10]$ **21.** $\begin{bmatrix} 12 & -8 & 8 \\ 9 & -6 & 6 \\ 0 & 0 & 0 \end{bmatrix}$ **23.** $\begin{bmatrix} 101 & -58 \\ 59 & -72 \\ 116 & -102 \end{bmatrix}$
25. $AB = \begin{bmatrix} 34.5 \\ 77 \\ 47 \end{bmatrix}$ cups sugar / cups flour / number eggs **27.** $CAB = \begin{bmatrix} 12.43 \\ 12.56 \end{bmatrix}$ Chicago cost / Detroit cost **29.** $ZY = \begin{bmatrix} 190 \\ 285 \end{bmatrix}$ wholesale / retail
31. $WXY = [47.75]$

Section 2.6 (page 100)

1. $I_3 = \begin{bmatrix} 1 & 0 & 0 \\ 0 & 1 & 0 \\ 0 & 0 & 1 \end{bmatrix}$ **3.** $A = \begin{bmatrix} 1 & 2 & 1 \\ 2 & 0 & 2 \\ 0 & 3 & -1 \end{bmatrix}$, $X = \begin{bmatrix} x \\ y \\ z \end{bmatrix}$, $B = \begin{bmatrix} 3 \\ 5 \\ 0 \end{bmatrix}$ **5.** $A = \begin{bmatrix} 1 & 1 & 1 \\ 0 & 1 & 1 \end{bmatrix}$, $X = \begin{bmatrix} r \\ s \\ t \end{bmatrix}$, $B = \begin{bmatrix} 5 \\ 8 \end{bmatrix}$

7. $A = \begin{bmatrix} 1 & 0 & 1 \\ 1 & 1 & -1 \\ 1 & 0 & 1 \end{bmatrix}, X = \begin{bmatrix} x_1 \\ x_2 \\ x_3 \end{bmatrix}, B = \begin{bmatrix} 5 \\ 0 \\ 0 \end{bmatrix}$ **9.** $2x - 2y = 0, 3x + 5y = 2$

11. $u + 2v = 0, 2u + v = 2, 3v = 5$ **13.** $\begin{bmatrix} -5 & 3 \\ 2 & -1 \end{bmatrix}$ **15.** $\begin{bmatrix} 1 & -2 & 0 \\ -1 & 3 & 0 \\ -1 & 0 & 1 \end{bmatrix}$ **17.** $\begin{bmatrix} 1 & \frac{1}{2} & -\frac{3}{2} \\ 0 & \frac{1}{2} & \frac{3}{2} \\ 1 & 0 & 1 \end{bmatrix}$

19. no inverse **21.** $\begin{bmatrix} 1 & -1 & 0 & 0 \\ 0 & 1 & 0 & 0 \\ 1 & 0 & 1 & -1 \\ -1 & 0 & 0 & 1 \end{bmatrix}$ **23.** $x = -8, y = 3, z = 11$ **25.** $x = -16, y = 11, z = -11$

27. $u = 4, v = -2, w = -7$ **29.** $x_1 = 16, x_2 = -3, x_3 = -5$

Section 2.7 (page 109)

1. nonmetals, services **3.** \$300,000, \$1.65 million, \$225,000, \$825,000 **5.** agriculture, machinery **7.** \$5 billion, \$0, \$400 million, \$1 billion, \$1.4 billion

9. $\begin{matrix} & \textit{mfg} & \textit{engy} \\ \textit{mfg} & .3 & .2 \\ \textit{engy} & .4 & .1 \end{matrix}$ **11.** (a) $\begin{bmatrix} 1250 \\ 1800 \end{bmatrix}$ (b) $\begin{bmatrix} 169 \\ 204 \end{bmatrix}$ (c) $\begin{bmatrix} 1081 \\ 1596 \end{bmatrix}$ (d)

$.12x_1$
$.02x_1$
1st sector
$.08x_2$
2nd sector
$.03x_2$

13. (a) \$10,600 goods, \$11,600 services (b) \$34,000 goods, \$24,000 services
15. 40 units housing, 40 units clothing, 35 units food **17.** 440 units electricity, 540 units oil, 360 units coal

Chapter 2 Review Exercises (page 112)

1. $x = -1, y = 2, z = 3$ **2.** $(x, y, z, w) = (1.2 + .4z - .6w, .6 + .2z + .2w, z, w)$ **3.** no solutions

4. 300 lbs 5-15-10, 100 lbs 20-5-5, 200 lbs 10-10-10 **5.** $\begin{bmatrix} -4 & -1 & 1 \\ 0 & -2 & 3 \\ -1 & 5 & -6 \end{bmatrix}$ **6.** $\begin{bmatrix} 0 & -1 & 2 \\ 3 & 0 & 3 \\ -5 & -2 & -1 \end{bmatrix}$

7. $\begin{bmatrix} 0 \\ -1 \\ 3 \end{bmatrix}$ **8.** undefined **9.** $\begin{bmatrix} 1 & 1 & 1 \\ 2 & 1 & 1 \\ -3 & 1 & 1 \end{bmatrix}$ **10.** $\begin{bmatrix} 2 & 1 & 0 \\ 3 & -1 & 5 \end{bmatrix}$ **11.** $\begin{bmatrix} 1 & -1 & -1 \\ 3 & -3 & -2 \\ -2 & 3 & 2 \end{bmatrix}$ **12.** no inverse

13. $A = \begin{bmatrix} 3 & 1 & 1 \\ 1 & -1 & 1 \\ -1 & 4 & -2 \end{bmatrix}, X = \begin{bmatrix} x \\ y \\ z \end{bmatrix}, B = \begin{bmatrix} 2 \\ 0 \\ 3 \end{bmatrix}$

$A = \begin{bmatrix} 2 & 1 & -1 & 1 \\ 1 & -2 & 0 & 1 \\ -1 & 7 & -1 & -2 \end{bmatrix}, X = \begin{bmatrix} x \\ y \\ z \\ w \end{bmatrix}, B = \begin{bmatrix} 3 \\ 0 \\ 3 \end{bmatrix}$

$$A = \begin{bmatrix} 1 & 2 & -3 & 1 \\ 2 & -1 & 1 & -1 \\ 4 & 3 & -5 & 1 \end{bmatrix}, X = \begin{bmatrix} a \\ b \\ c \\ d \end{bmatrix}, B = \begin{bmatrix} 6 \\ -1 \\ 2 \end{bmatrix}$$

14. (a) $(x, y, z) = (-4.5, -2.5, 7.5)$ (b) $(x, y, z) = (-175, -125, 525)$ **15.** (a) $S = \begin{bmatrix} 60 \\ 80 \end{bmatrix}$ (b) 9400 units goods, 5800 units services

Chapter 3 Section 3.1 (page 123)

1. Maximize $x + 1.25y$,
subject to $8x + 4y \leq 1000$
$8x + 12y \leq 1500$
$x \geq 0, y, \geq 0$

3.

	Big Barrel	*Supersack*	*Needed*
White pieces	10	7	100
Dark pieces	10	4	80
Cost	\$9	\$6	
Number bought	x	y	

Minimize $9x + 6y$
subject to $10x + 7y \geq 100$
$10x + 4y \geq 80$
$x \geq 0, y \geq 0$

5. Maximize $110x + 120y$
subject to $90x + 80y \leq 72{,}000$
$10x + 20y \leq 10{,}000$
$x \geq 0, y \geq 0$

7. Maximize $3.5x + 4y$
subject to $.25x + .6y \leq 300$
$.5x + .2y \leq 250$
$.25x + .2y \leq 150$
$x \geq 0, y \geq 0$

9. Let x amethysts and y beryls be finished.
Maximize $75x + 100y$
subject to $x + 2y \leq 40$
$x + 5y \leq 40$
$2x + 2y \leq 40$
$x \geq 0, y \geq 0$

11. Let x rolls from Company A and y rolls from Company B be ordered.
Minimize $100x + 300y$
subject to $2x + 5y \geq 1000$
$35x + 25y \leq 10{,}500$
$x \geq 0, y \geq 0$

13. (3, 5) **15.** (5, 1) **17.** (1, 4, 0)

19.

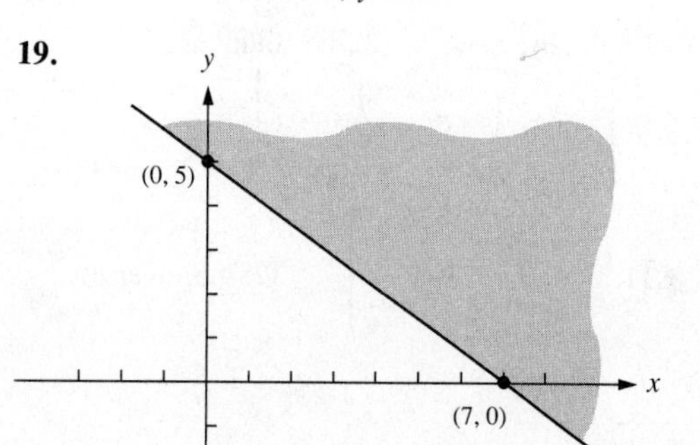

21.

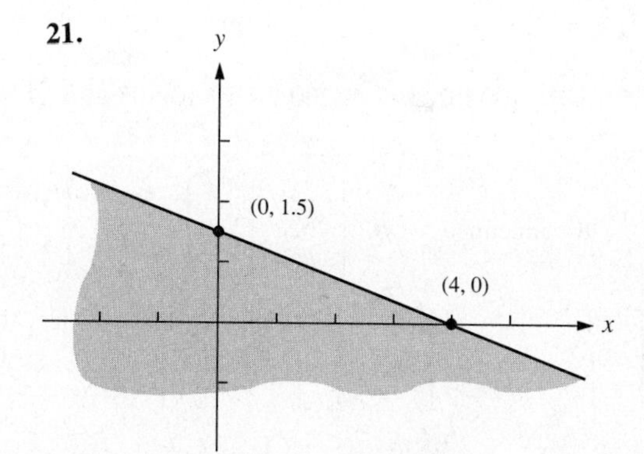

23.

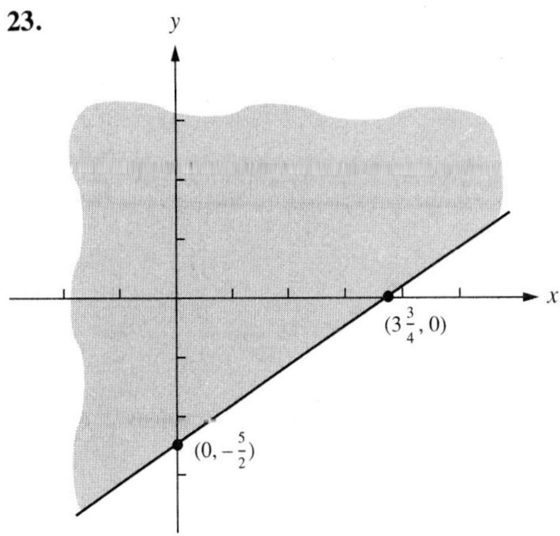

25.

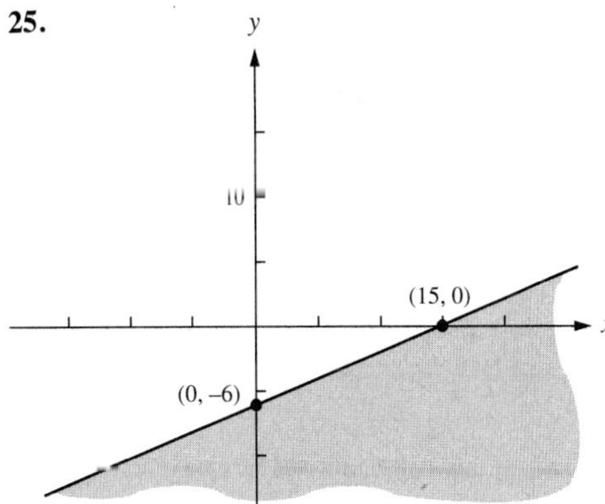

27.

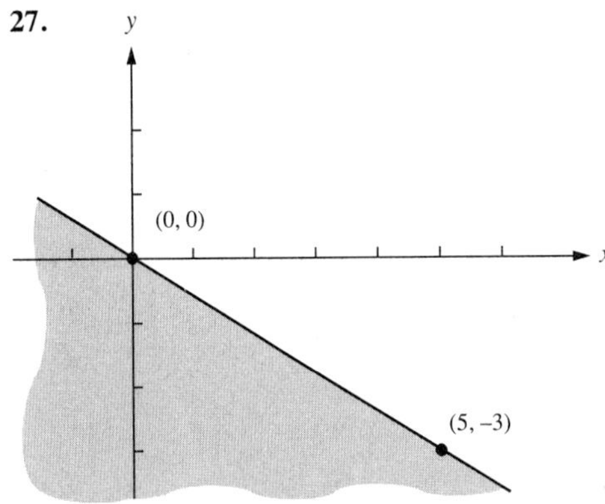

29.

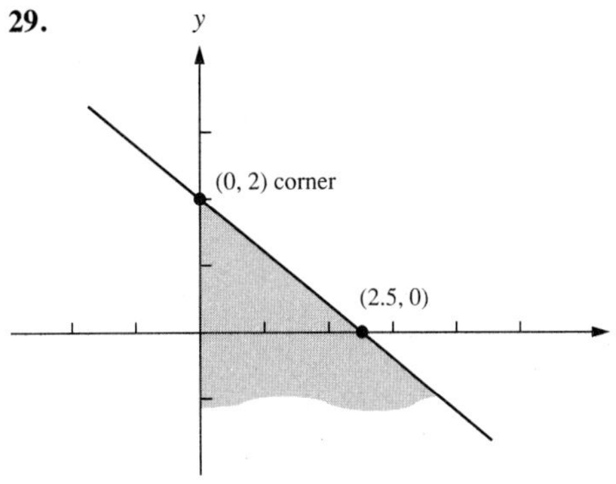

31.

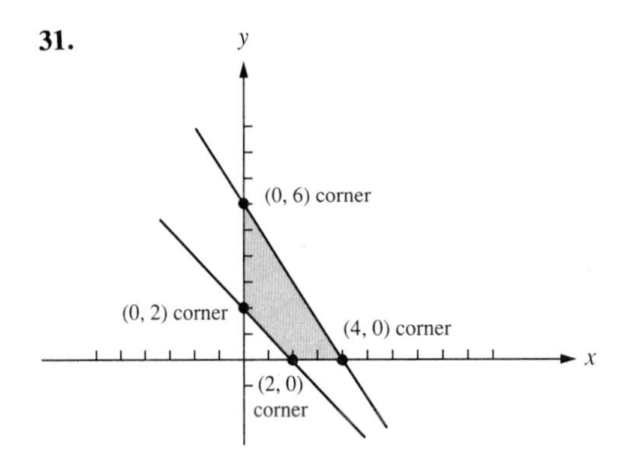

33.

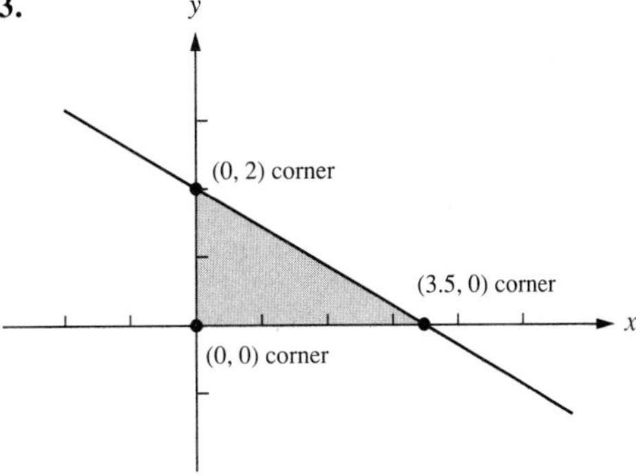

35.

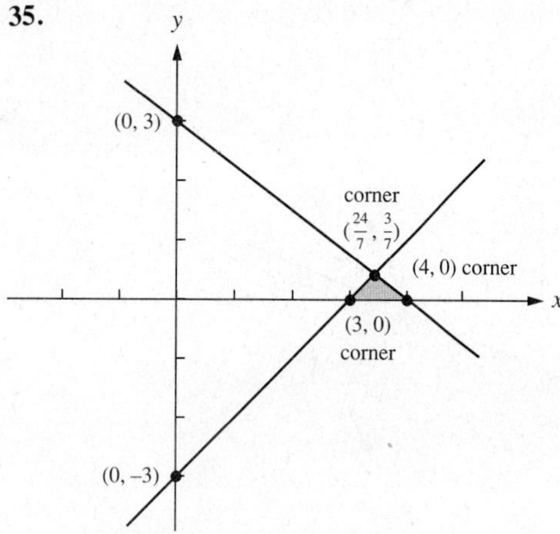

Section 3.2 (page 136)

1. 100 at (20, 20) **3.** $18\frac{8}{19}$ at $(\frac{45}{19}, \frac{20}{19})$ **5.** 90 at (10, 40) **7.** 92,000 at (640, 180) **9.** 2550 at (300, 375) **11.** 1625 at (15, 5) **13.** 55,600 at (220, 112) **15.** 24 animals, using 12 rats and 12 mice **17.** 1800 calories, using 6 oz of Food 1 and 4 oz of Food 2 **19.** \$242,400 with Albany 96 days, Boston 12 days

Section 3.3 (page 147)

1.
$$\left[\begin{array}{cccc|c} 1 & 3 & 1 & 0 & 200 \\ 2 & 5 & 0 & 1 & 500 \\ \hline 6 & 4 & 0 & 0 & R \end{array}\right]$$

3.
$$\left[\begin{array}{cccc|c} 2 & 4 & 1 & 0 & 1200 \\ 1 & 3 & 0 & 1 & 1000 \\ \hline 3 & 8 & 0 & 0 & R \end{array}\right]$$

5.
$$\left[\begin{array}{ccccc|c} 3 & 4 & 1 & 0 & 0 & 300 \\ 1 & 2 & 0 & 1 & 0 & 120 \\ 1 & 0 & 0 & 0 & 1 & 30 \\ \hline 5 & 12 & 0 & 0 & 0 & R \end{array}\right]$$

7.
$$\left[\begin{array}{ccccc|c} 2 & 1 & 1 & 1 & 0 & 30 \\ 0 & 1 & 3 & 0 & 1 & 40 \\ \hline 5 & 4 & 3 & 0 & 0 & R \end{array}\right]$$

9.
$$\begin{array}{c} \\ x \\ t \\ \\ \end{array}\begin{array}{c} \begin{array}{cccc} x & y & s & t \end{array} \\ \left[\begin{array}{cccc|c} 1 & \frac{4}{3} & \frac{1}{3} & 0 & \frac{7}{3} \\ 0 & \frac{2}{3} & -\frac{1}{3} & 1 & \frac{23}{3} \\ \hline 0 & -2 & -1 & 0 & R-7 \end{array}\right] \end{array}$$

$(x, y) = (7/3, 0)$

11.
$$\begin{array}{c} \\ s \\ x \\ \\ \end{array}\begin{array}{c} \begin{array}{cccc} x & y & s & t \end{array} \\ \left[\begin{array}{cccc|c} 0 & -2 & 1 & -3 & -23 \\ 1 & 2 & 0 & 1 & 10 \\ \hline 0 & -4 & 0 & -3 & R-30 \end{array}\right] \end{array}$$

$(x, y) = (10, 0)$

13.
$$\begin{array}{c} \\ y \\ t \\ \\ \end{array}\begin{array}{c} \begin{array}{cccc} x & y & s & t \end{array} \\ \left[\begin{array}{cccc|c} \frac{3}{2} & 1 & \frac{1}{2} & 0 & 4 \\ -4 & 0 & -2 & 1 & -4 \\ \hline 1 & 0 & -1 & 0 & R-8 \end{array}\right] \end{array}$$

$(x, y) = (0, 4)$

15.
$$\begin{array}{c} \\ x \\ y \\ \\ \end{array}\begin{array}{c} \begin{array}{cccc} x & y & s & t \end{array} \\ \left[\begin{array}{cccc|c} 1 & 0 & \frac{1}{2} & -\frac{1}{4} & 1 \\ 0 & 1 & -\frac{1}{4} & \frac{3}{8} & \frac{5}{2} \\ \hline 1 & 0 & -1 & 0 & R-8 \end{array}\right] \end{array}$$

$(x, y) = (1, 5/2)$

17. $R = 1200$ at $(x, y) = (200, 0)$ **19.** $R = 2400$ at $(x, y) = (0, 300)$ **21.** $R = 720$ at $(x, y) = (0, 60)$ **23.** $R = 120$ at $(x, y, z) = (0, 30, 0)$

25. $\left[\begin{array}{cccc|c} 6 & 5 & 1 & 0 & 600 \\ 3 & 4 & 0 & 1 & 240 \\ \hline 4.5 & 3.5 & 0 & 0 & R \end{array}\right]$

Pivoting on the 3 in row 2, column 1 shows $R = 360$ at $(x, y) = (80, 0)$.

Section 3.4 (page 156)

1. row 2, column 1 **3.** row 1, column 1 **5.** row 1, column 1 **7.** row 1, column 2 **9.** no pivot
11. $R = 13, x = 0, y = 7$ **13.** $R = 22, x = 7, y = 0, z = 5$

15.

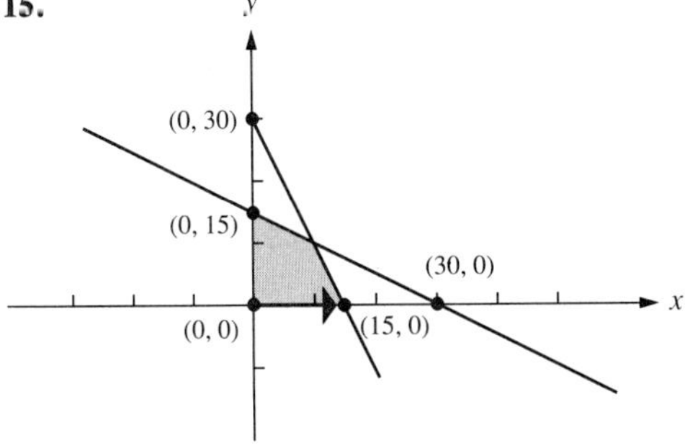

$R = 75$ at $(x, y) = (15, 0)$

17.

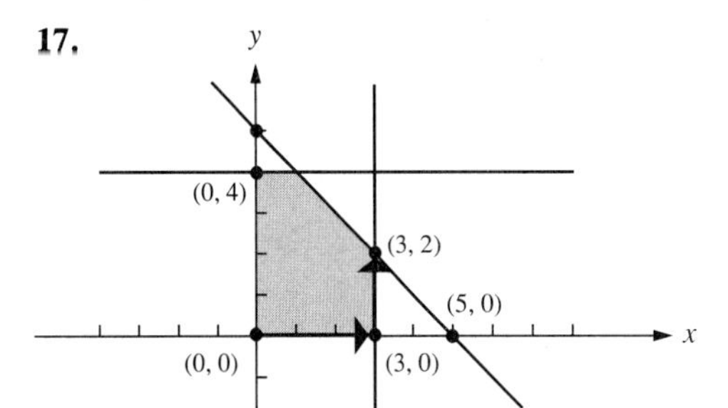

$R = 40$ at $(x, y) = (3, 2)$

19. $R = 40$ at $(x, y) = (20, 10)$ **21.** $R = 120$ at $(x, y, z) = (60, 0, 0)$ **23.** $R = 7$ at $(x, y, z) = (3, 1, 0)$
25. $R = 10$ at $(x, y, z, w) = (2, 4, 0, 0)$ **27.** 500 acres of asparagus, 500 acres of brussels sprouts, no cauliflower
29. $P = 1980$ at $(x, y, z) = (100, 0, 140)$

Section 3.5 (page 167)

1. Maximize $-5X + 20Y + 15Z = R$
subject to
$$\begin{aligned} X + Y + Z &\leq 5 \\ -X + 2Y + Z &\leq 4 \\ X \geq 0, Y \geq 0, Z &\geq 0 \end{aligned}$$

3. Maximize $40X + 40Y + 25Z = R$
subject to
$$\begin{aligned} X + 4Y + Z &\leq 6 \\ 4X + Y + Z &\leq 2 \\ X \geq 0, Y \geq 0, Z &\geq 0 \end{aligned}$$

5. Maximize $5X + 20Z = R$
subject to
$$\begin{aligned} -Y + 5Z &\leq 10 \\ X + Y + 2Z &\leq 20 \\ X \geq 0, Y \geq 0, Z &\geq 0 \end{aligned}$$

7. $C = 65$ at $(x, y) = (5, 10)$ **9.** $C = 70$ at $(x, y) = (5, 20)$ **11.** $C = 120$ at $(x, y) = (2, 5)$ **13.** $C = 110$ at $(x, y, z) = (10, 0, 20)$ **15.** $C = 100$ at $(x, y, z, w) = (0, 40, 60, 0)$ **17.** \$1775 for 5 tons oil, 5 tons coal
19. \$66,000 with 6 Type A and 12 Type B planes **21.** \$3.30 with 70 g of the first food, 30 g of the second food

Section 3.6 (page 183)

1. $\begin{array}{c} \begin{array}{ccccc} x & y & s & t & a \end{array} \\ \left[\begin{array}{ccccc|c} 1 & 1 & 1 & 0 & 0 & 20 \\ 3 & 1 & 0 & -1 & 1 & 36 \\ \hline 3 & 5 & 0 & 0 & 0 & R \\ 0 & 0 & 0 & 0 & -1 & Q \end{array}\right] \end{array}$

3. $\begin{array}{c} \begin{array}{ccccc} x & y & s & t & a \end{array} \\ \left[\begin{array}{ccccc|c} 1 & 2 & 1 & 0 & 0 & 200 \\ 2 & 1 & 0 & -1 & 1 & 250 \\ \hline 3 & 1 & 0 & 0 & 0 & R \\ 0 & 0 & 0 & 0 & -1 & Q \end{array}\right] \end{array}$

5. $\begin{array}{ccccc|c} x & y & s & t & a & \\ 2 & 1 & 1 & 0 & 0 & 75 \\ 1 & 1 & 0 & -1 & 1 & 60 \\ \hline -3 & -6 & 0 & 0 & 0 & R \\ 0 & 0 & 0 & 0 & -1 & Q \end{array}$ **7.** $\begin{array}{ccccc|c} x & y & s & t & a & \\ 1 & 2 & 1 & 0 & 0 & 150 \\ 2 & 3 & 0 & -1 & 1 & 240 \\ \hline -6 & -4 & 0 & 0 & 0 & R \\ 0 & 0 & 0 & 0 & -1 & Q \end{array}$

9. $R = -5$ at $(x, y) = (0, 5)$ **11.** $R = 84$ at $(x, y) = (8, 12)$ **13.** $R = 600$ at $(x, y) = (200, 0)$ **15.** $C = 315$ at $(x, y) = (15, 45)$ **17.** $C = 420$ at $(x, y) = (30, 60)$ **19.** $R = 800$ at $(x, y) = (0, 400)$ **21.** $C = 320$ at $(x, y) = (160, 0)$ **23.** $C = 95$ at $(x, y, z) = (35, 0, 5)$ **25.** Profit is \$216 with 22 of 1st style, none of 2nd style, and 25 of 3rd style. **27.** Cost is \$650 for 50 lbs of mixture A and 75 lbs of mixture B. **29.** Maximum $\frac{11}{3}$ at $(x, y, z) = (\frac{7}{3}, \frac{4}{3}, 0)$ **31.** Minimum 4 at $(x, y, z) = (4, 0, 0)$

Chapter 3 Review Exercises (page 186)

1. Let x bears and y dogs be made.
Maximize $3x + 3.2y$
subject to
$$\begin{aligned} 15x + 18y &\le 300 \\ 6x + 5y &\le 120 \\ 12x + 10y &\le 240 \\ x \ge 0, y &\ge 0 \end{aligned}$$

2. Let x cups of snap beans, y cups of kidney beans, and z cups of baked beans with pork and molasses be used.
Minimize $30x + 230y + 385z$
subject to
$$\begin{aligned} 63x + 74y + 161z &\ge 350 \\ 2x + 15y + 16z &\ge 50 \\ x \ge 0, y \ge 0, z &\ge 0 \end{aligned}$$

3.

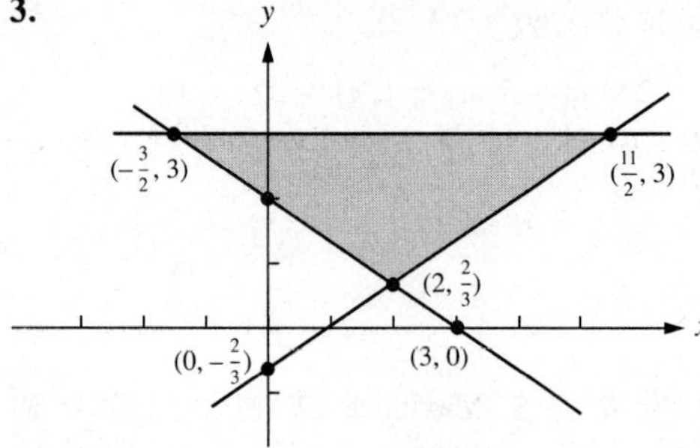

4.

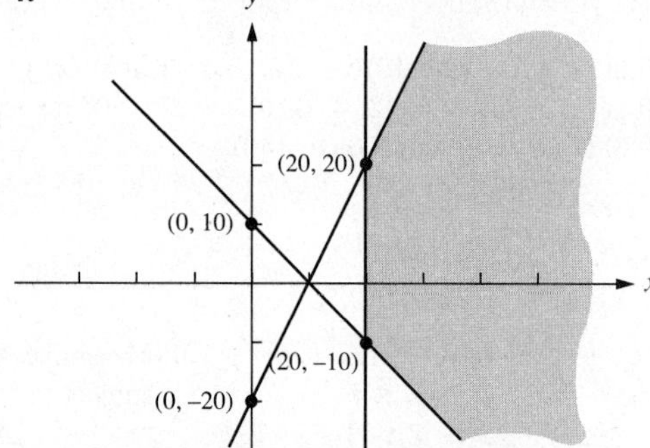

5. $C = 16$ for $(x, y) = (1, 2.4)$ **6.** No solution. Critical region unbounded for decreasing C. **7.** No solution. Critical region is empty. **8.** $R = 32/3$ at $(x, y) = (\frac{44}{15}, \frac{14}{15})$ **9.** $R = 60$ at $(x, y, z) = (0, 12, 0)$ **10.** $R = 12$ at $(x, y, z) = (0, \frac{14}{3}, \frac{4}{3})$ **11.** $C = 13$ at $(x, y, z) = (0, 4, 5)$ **12.** $C = 50$ at $(x, y, z) = (25, 0, 25)$ **13.** $R = 400$ at $(x, y) = (0, 200)$ **14.** $R = 720$ at $(x, y) = (180, 0)$ **15.** 13 tons of titanium with 100 tons from Mine A and 400 tons from Mine C. **16.** \$184 with 4 super economy books and 20 "B" tickets

Chapter 4 Section 4.1 (page 194)

1. 3, 14, 69, 344, 1719, 8594 neither **3.** 0, 1.5, 3, 4.5, 6, 7.5 arithmetic progression **5.** 625, −1125, 2025, −3645, 6561, −11809.8 geometric progression **7.** 5000, 5600, 6260, 6986, 7784.6, 8663.06 neither **9.** $x_n = 6n + 2$; $x_{10} = 62$; $x_{20} = 122$ **11.** $x_n = -2.5n + 35$; $x_{10} = 10$; $x_{20} = -15$ **13.** $x_n = (1.5)^n$;

$x_{10} \approx 57.665$; $x_{20} \approx 3325.257$ **15.** $x_n = 100(0.9)^n$; $x_{10} \approx 34.868$; $x_{20} \approx 12.158$ **17.** \$525 **19.** \$24,000 **21.** \$1440 **23.** \$4200 **25.** \$8228; \$1428 **27.** 20% **29.** 16% **31.** 16%

Section 4.2 (page 200)

1. \$7321.14 **3.** \$323.21 **5.** \$862.30 **7.** \$2827.79 **9.** 6.17% **11.** 7.25% **13.** \$261.42 **15.** (a) \$739.96 (b) \$743.23 (c) \$744.18 (d) \$744.65 **17.** \$1,068,843.68 **19.** \$144,892.92 **21.** \$91.74 **23.** \$3297.19 **25.** 10.97% **27.** \$121.67 **29.** about 3.74% **31.** about 3.42%

Section 4.3 (page 209)

1. $x_n = 7(3^n) - 2$; $x_{10} = 413{,}341$ **3.** $x_n = -5(-0.6)^n + 15$; $x_{10} \approx 4.97$ **5.** $x_n = 6 - 5n$; $x_{10} = -44$ **7.** $x_n = 2600(1.01)^n - 2500$ **9.** \$2682.42 **11.** \$3933.61 **13.** \$16,573.64 **15.** \$8516.06 **17.** \$49,054.30 **19.** \$391.81 **21.** (a) $B_{n+1} = 1.02B_n + 50$ (b) $B_n = \$3000(1.02)^n - \$2500$ (c) \$1957.84 **23.** \$49,723.17 **25.** \$25,168.33 **27.** \$98,225.80 **29.** \$271,545.95 **31.** \$13,131.70 **33.** not replacing the machinery **35.** \$14,214.37 **37.** $x_n = x_0 + nb$ **39.** \$18,033.42

Section 4.4 (page 216)

1. \$278.83 **3.** \$233.02 **5.** \$6832.89 **7.** \$199.29 **9.** \$138.39 **11.** \$1022.74 **13.** \$21,254.37 **15.** (a) \$947.90 (b) \$194,370 **17.** \$9307.30 **19.** \$288,652.40 **21.** \$252.34 **23.** \$44,201.35 **25.** \$1179.20 **27.** \$5964.77

29.

Month	*Old Balance*	*Interest*	*Payment*	*New Balance*
1	\$1000.00	\$15.00	\$175.53	\$839.47
2	\$839.47	\$12.59	\$175.53	\$676.53
3	\$676.53	\$10.15	\$175.53	\$511.15
4	\$511.15	\$7.67	\$175.53	\$343.29
5	\$343.29	\$5.15	\$175.53	\$172.91
6	\$172.91	\$2.59	\$175.50	\$0

31. \$1029.34 **33.** \$2218.10

35. (a) $B_{k+1} = (1 + i)B_k - p$ (b) $B_k = \left(L - \frac{p}{i}\right)(1 + i)^k + \frac{p}{i}$

Chapter 4 Review Exercises (page 220)

1. $x_n = 2n - 1$; $x_8 = 15$ **2.** $x_n = 125(0.8)^n$; $x_8 \approx 20.97$ **3.** $x_n = 3(4^n) - 3$; $x_8 = 196{,}605$ **4.** $x_n = 254(0.5)^n + 2$; $x_8 \approx 2.99$ **5.** $x_n = 3(-2)^n$; $x_8 = 768$ **6.** $x_n = -1500(1.02)^n + 2500$; $x_8 \approx 742.51$ **7.** $x_n = 7(-\frac{2}{3})^n + 1$; $x_8 \approx 1.27$ **8.** $x_n = 0.75n + 0.5$; $x_8 = 6.5$ **9.** (a) \$81,000 (b) \$6000 **10.** 18% **11.** (a) \$2970.25 (b) \$2987.08 (c) \$2991.03 **12.** 16.075% **13.** \$13,112.30 **14.** \$12,093.83 **15.** (a) $B_{n+1} = 1.007B_n - \$700$ (b) $x_n = -\$75{,}000(1.007)^n + \$100{,}000$ (c) \$3589.97 **16.** \$184.46 **17.** \$85,111.81 **18.** \$12,416.81 **19.** \$17,289,588.07 **20.** \$16,299.52 **21.** \$420,162.64 **22.** \$44,399.00 **23.** \$1488.89 **24.** \$2051.75

Chapter 5 Section 5.1 (page 229)

1. $\{3, 4, 5, 6, 7, 8\}$ **3.** $\{5, 6\}$ **5.** $\{1, 2, 5\}$ **7.** $\{1, 2, 5, 10\}$ **9.** $\varnothing$ **11.** $\{u, y\}$ **13.** $\varnothing$
15. $\{u, v, w, y, z\}$ **17.** $\{z\}$ **19.** $\{u, v, w, x, y\}$ **21.** true **23.** true **25.** true **27.** false **29.** false
31. true **33.** (a) $\{a\}, \{b\}, \{c\}, \{d\}$ (b) $\{a, b\}, \{a, c\}, \{a, d\}, \{b, c\}, \{b, d\}, \{c, d\}$
(c) $\{a, b, c\}, \{a, b, d\}, \{a, c, d\}, \{b, c, d\}$

35. (a) $M \cap \overline{F}$

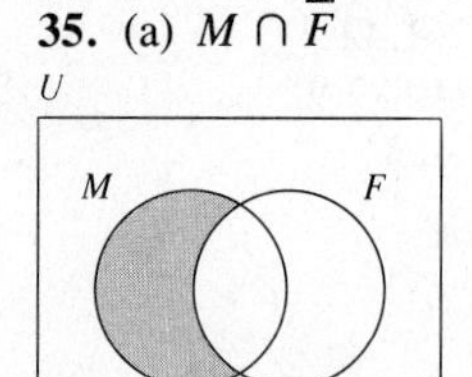

(b) $M \cap F$

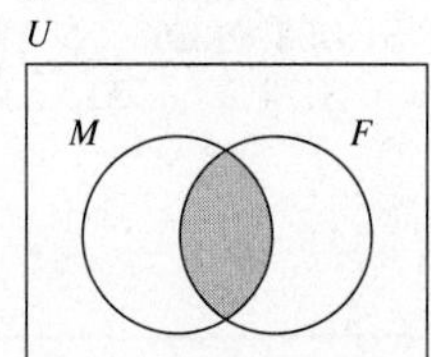

(c) $(M \cap \overline{F}) \cup (\overline{M} \cap F)$

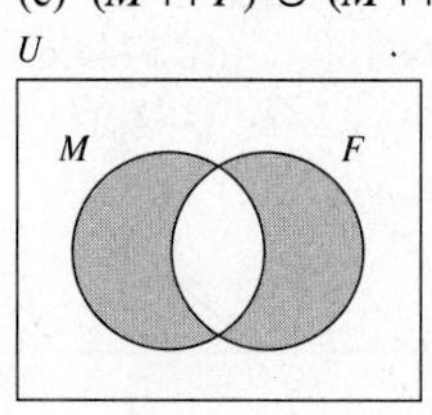

(d) $\overline{M} \cap \overline{F}$

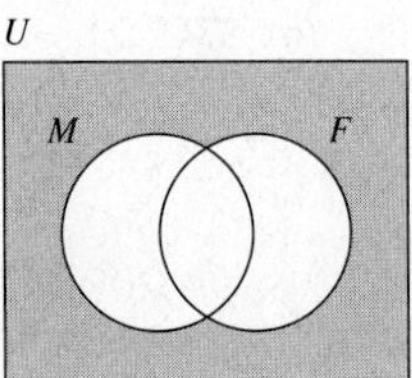

37. (a) $F \cap \overline{S} \cap E$

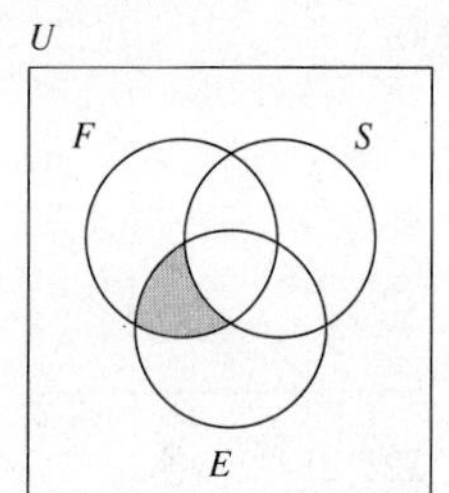

(b) $\overline{F} \cap S \cap \overline{E}$

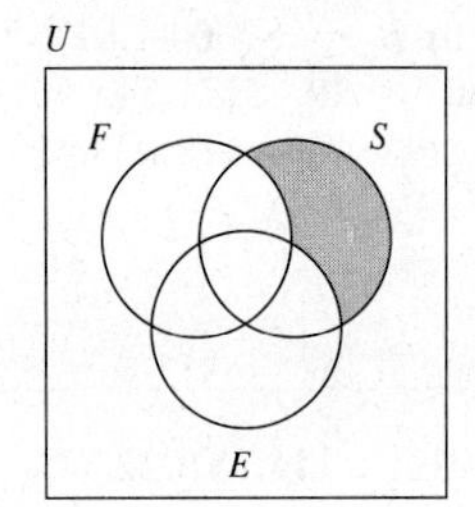

(c) $F \cap S$

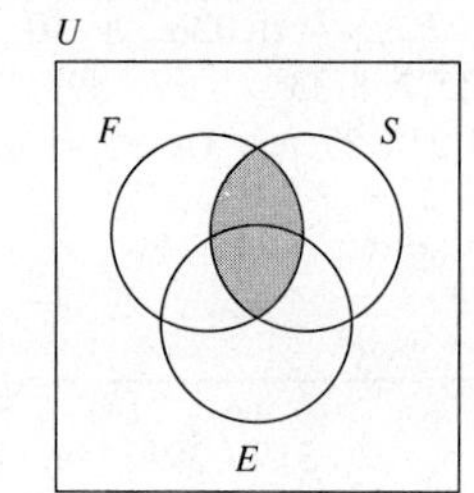

(d) $S \cup E$

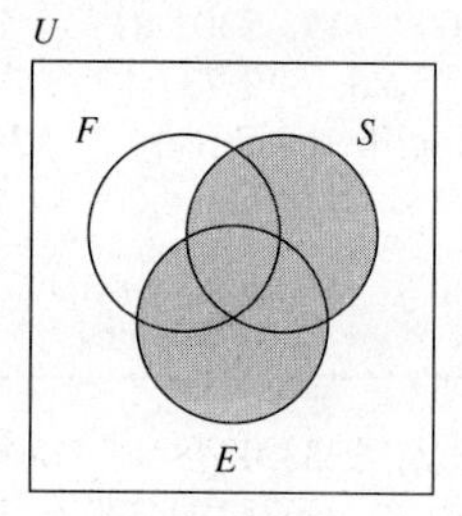

Section 5.2 (page 237)

1.

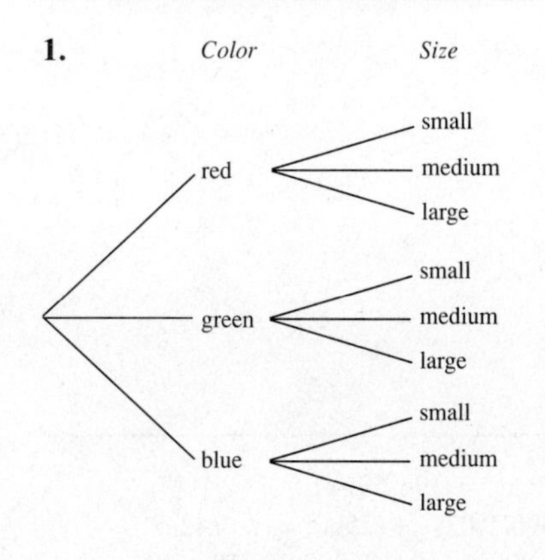

3.

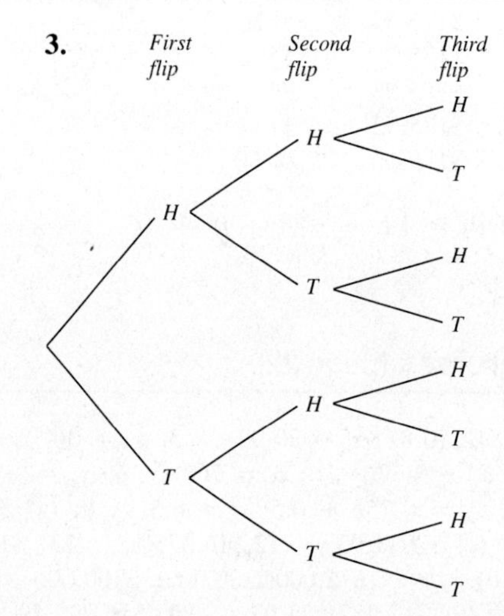

5.

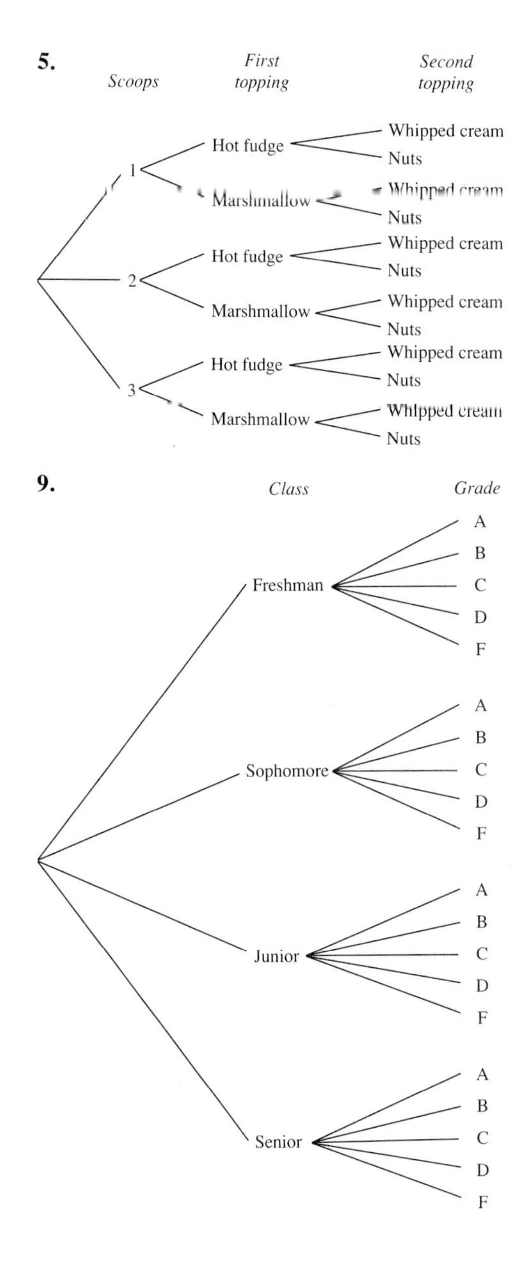

9.

7.

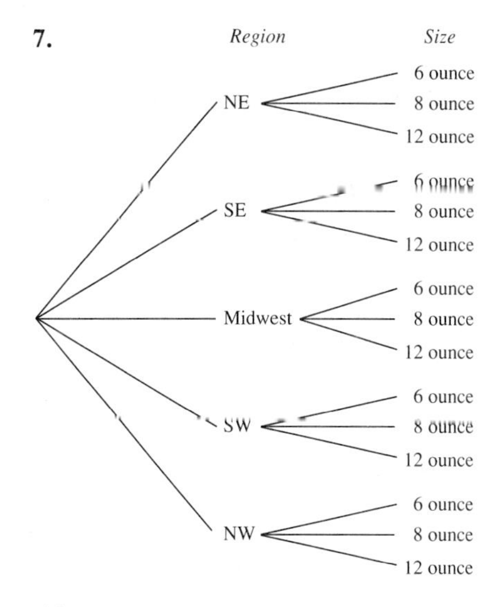

11. yes **13.** 896 **15.** (a) 7 (b) 5 **17.** (a) 167 (b) 1059 **19.** 17 are AB+, 3 are AB−, 102 are A+, 18 are A−, 68 are B+, 12 are B−, 153 are O+, and 27 are O−. **21.** The data indicates that there are 910 new policyholders, but the company claims to have only 900. **23.** $S = \varnothing$ **25** (a) 64 (b) 39 (c) 74 **27.** 430 husbands and 215 wives

■ Section 5.3 (page 249)

1. 70 **3.** 23 **5.** 15 **7.** 21 **9.** 24 **11.** 15 **13.** 1320 **15.** 4096 **17.** 60 **19.** (a) 4096 (b) 244,140,625 **21.** 384 **23.** (a) 6,400,000 (b) 1,024,000,000 **25.** 1267 **27.** (a) 720 (b) 144 (c) 48 (d) 72 (e) 72 **29.** (a) 196 (b) 182 (c) 63 **31.** (a) 192 (b) 48 (c) 240 **33.** (a) 90,000 (b) 3125 (c) 120 (d) 20,000 (e) 37,512

Section 5.4 (page 260)

1. 2520 **3.** 3360 **5.** 24 **7.** 40,320 **9.** 84 **11.** 70 **13.** 1 **15.** 1
17. (a) 20; *ab, ac, ad, ae, ba, bc, bd, be, ca, cb, cd, ce, da, db, dc, de, ea, eb, ec, ed*
(b) 10; $\{a, b\}$, $\{a, c\}$, $\{a, d\}$, $\{a, e\}$, $\{b, c\}$, $\{b, d\}$, $\{b, e\}$, $\{c, d\}$, $\{c, e\}$, $\{d, e\}$ **19.** (a) 1,860,480 (b) 15,504
21. 720 **23.** 15,120 **25.** 56 **27.** (a) 38,760 (b) 10,010 **29.** 27,720 **31.** (a) 126 (b) 40
(c) 60 (d) 45 **33.** $C(2170, 1438) \cdot C(732, 522) \cdot C(210, 210)$ **35.** 11 **37.** (a) 2^n (b) $C(n, k)$
(c) Permutations can only be used for sequences of *distinct* objects.

39. $r \cdot C(n, r) = r \cdot \dfrac{n!}{r!(n-r)!} = \dfrac{n!}{(r-1)!(n-r)!} = n\dfrac{(n-1)!}{(r-1)!(n-r)!} = n \cdot C(n-1, r-1)$

41. (a) 1728 (b) 576 **43.** $(n-1)!$

Section 5.5 (page 266)

1. 1 9 36 84 126 126 84 36 9 1 **3.** $x^6 + 6x^5y + 15x^4y^2 + 20x^3y^3 + 15x^2y^4 + 6xy^5 + y^6$
5. $x^7 - 7x^6y + 21x^5y^2 - 35x^4y^3 + 35x^3y^4 - 21x^2y^5 + 7xy^6 - y^7$ **7.** $a^4 + 16a^3b + 96a^2b^2 + 256ab^3 + 256b^4$
9. $243u^5 - 405u^4v + 270u^3v^2 - 90u^2v^3 + 15uv^4 - v^5$ **11.** $64x^6 - 96x^5 + 60x^4 - 20x^3 + \dfrac{15}{4}x^2 - \dfrac{3}{8}x + \dfrac{1}{64}$
13. $1 - 8x + 28x^2 - 56x^3 + 70x^4 - 56x^5 + 28x^6 - 8x^7 + x^8$ **15.** 125,970 **17.** $-119{,}759{,}850$ **19.** 319,770
21. $-347{,}373{,}600$ **23.** 1,188,096 **25.** $-\dfrac{3003}{32}$ **27.** 92 **29.** Take $x = y = 1$ in the binomial theorem.
31. $C(n, k)$ is the number of k-element subsets of a set with n elements. So the left side of the equation counts all the subsets of a set with n elements; there are 2^n such subsets.

33.
$$C(n, r-1) + C(n, r) = \frac{n!}{(r-1)!(n-r+1)!} + \frac{n!}{r!(n-r)!} = \frac{r(n!)}{r!(n-r+1)!} + \frac{(n-r+1)n!}{r!(n-r+1)!}$$
$$= \frac{(n+1)n!}{r!(n+1-r)!} = \frac{(n+1)!}{r!(n+1-r)!} = C(n+1, r)$$

Chapter 5 Review Exercises (page 269)

1. (a) $\{1, 2, 3, 4, 5, 6, 8\}$ (b) $\{2, 5, 6\}$ (c) $\{1, 3, 4, 7, 9, 10\}$ (d) $\{2, 5, 6, 7, 8, 9, 10\}$ **2.** (a) true
(b) true (c) false (d) true (e) false (f) true (g) true (h) true **3.** 138 **4.** (a) 38 (b) 12
(c) 260

5.

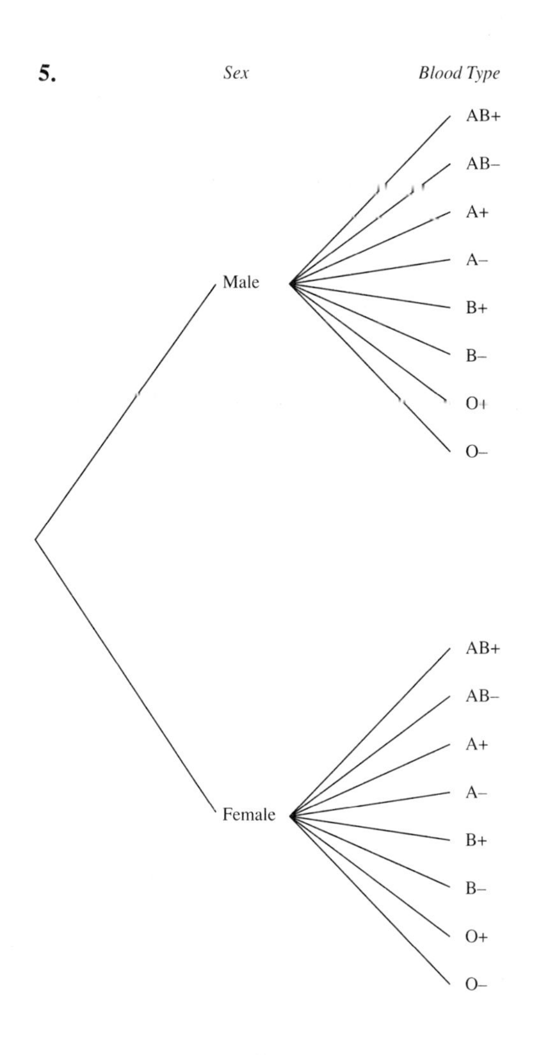

6. (a) 120 (b) 362,880 (c) 1 (d) 9 (e) 1680 (f) 35 (g) 5040 (h) 1 **7.** 6720 **8.** 440 **9.** 720 **10.** 4845 **11.** 1320 **12.** 1260 **13.** (a) 39,916,800 (b) 330 (c) 15 (d) 150 (e) 265

14.

1
1 1
1 2 1
1 3 3 1
1 4 6 4 1
1 5 10 10 5 1
1 6 15 20 15 6 1

15. $729r^6 + 1458r^5s + 1215r^4s^2 + 540r^3s^3 + 135r^2s^4 + 18rs^5 + s^6$
16. $-96{,}096$

■ Chapter 6 Section 6.1 (page 283)

1. $\{H, T\}$ **3.** {Monday, Tuesday, Wednesday, Thursday, Friday, Saturday, Sunday} **5.** {Republican, Democrat} **7.** {0, 1, 2, 3, . . .} **9.** $S = \{0, 1, 2, 3, 4, 5, 6, 7, 8, 9, 10, 11, 12, 13, 14, 15\}$ (a) {0} (b) {0, 1, 2} (c) {8, 9, 10, 11, 12, 13, 14, 15} **11.** $S = \{HH, HT, TH, TT\}$ (a) $\{HH, TH\}$ (b) $\{HH, HT, TH\}$ (c) $\varnothing$ **13.** (a) 1/2 (b) 1/3 (c) 1/6 (d) 0 **15.** (a) 31/365 (b) 46/365 **17.** (a) .18 (b) .46 (c) .70 (d) .72 **19.** (a) .25 (b) .35 (c) .05 (d) .44 **21.** 1/9 **23.** 1/6 **25.** 4/9 **27.** 0 **29.** 5/12 **31.** 1/1000 **33.** (a) 5/16 (b) 3/8 (c) 1/2 **35.** 1/7,059,052 **37.** 1/24 **39.** 1/10 **41.** (a) 1/33 (b) 5/11 **43.** (a) 14/323 (b) 160/323 (c) 135/323

Section 6.2 (page 293)

1. yes **3.** no **5.** yes **7.** no **9.** (a) 3/8 (b) 5/8 (c) 1 (d) 1/2 **11.** (a) 0 (b) .25 (c) .65 (d) .90 **13.** 93/568

15. (a)

Outcome	*Probability*
0 years	.050
1 year	.200
2 years	.325
3 years	.250
4 years	.175

(b) .175 (c) .250 (d) .425

17. (a)

Outcome	*Probability*
$45 \le s < 55$	.15
$55 \le s < 65$	.31
$65 \le s < 75$	.35
$75 \le s < 85$	.17
$85 \le s < 95$	.02

(b) .31 (c) .46 (d) .19

19. (a)

Outcome	*Probability*
0 requests	.11
1 request	.20
2 requests	.32
3 requests	.19
4 requests	.12
5 requests	.05
6 requests	.01

(b) .19 (c) .63 (d) .31 (e) 2

	Probability	*Odds in favor*	*Odds against*
21.	3/4	3 to 1	1 to 3
23.	7/25	7 to 18	18 to 7
25.	4/9	4 to 5	5 to 4

27. 1 to 2 **29.** 1 to 5 **31.** 1 to 2 **33.** 1/3 **35.** .28

Section 6.3 (page 307)

1. (a) $\overline{A}$ (b) $A \cap B$ (c) $A \cap \overline{B}$ (d) $\overline{A} \cap \overline{B}$

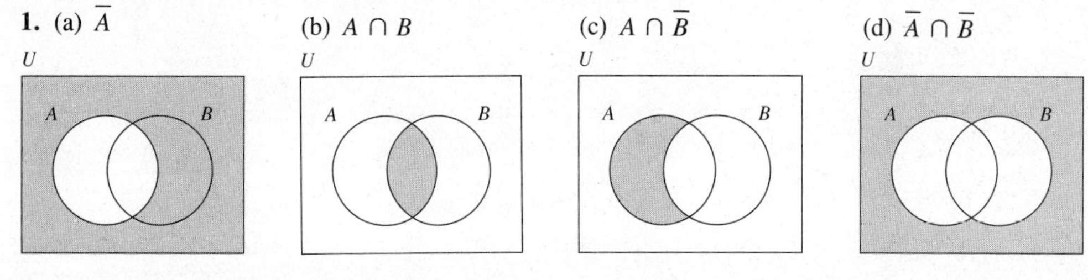

3. (a) A student is not chosen. (b) A female is not chosen. (c) A female student is chosen. (d) A female or a student is chosen.
5. (a) .6 (b) .2 (c) .3 (d) .1 **7.** (a) .47 (b) .28 (c) .77 (d) .23 **9.** (a) .77 (b) .96 (c) .04 (d) 0 **11.** not mutually exclusive **13.** not mutually exclusive **15.** mutually exclusive **17.** mutually exclusive **19.** 7/8 **21.** $\dfrac{7{,}797{,}275}{11{,}881{,}376} \approx .656$ **23.** 137/228 **25.** .78 **27.** (a) .95 (b) .05 **29.** 11/12 **31.** .45 **33.** (a) .46 (b) .90 (c) .92 **35.** 31/42 **37.** (a) 13/14 (b) 1/2

Section 6.4 (page 321)

1. (a) 2/3 (b) 48/53 **3.** (a) 5/12 (b) 7/11 (c) 1/2 **5.** (a) 7/8 (b) 1/2 (c) 3/4 (d) 3/4 **7.** (a) 94/209 (b) 63/209 (c) 45/82 (d) 12/47 **9.** 191/626 **11.** 62/107 **13.** 67/191 **15.** no **17.** no **19.** 8/17 **21.** 733/864 **23.** $P(H|M) = \dfrac{331}{1600} \approx .207$ and $P(H|A) = \dfrac{334}{3600} \approx .093$ **25.** .02

27. .88 **29.** .72 **31.** (a) .15 (b) .25 (c) .50 (d) .40 **33.** 25/609 **35.** (a) $P(E \cap F) = .12$, $P(E \cup F) = .68$ (b) $P(E \cap F) = 0$, $P(E \cup F) = .8$ **37.** (a) .0592 (b) .0008 **39.** .3376
41. (a) .125 (b) .3

Section 6.5 (page 334)

1. .40 **3.** .05 **5.** .70 **7.** .25 **9.** .31 **11.** .38 **13.** .37 **15.** .25 **17.** 41/158 **19.** .24
21. 14/19 **23.** (a) .44 (b) 27/44, 3/11, 5/44
25. (a) .0106 (b) 25/53 under 25, 6/53 between 25 and 34, 10/53 between 35 and 60, and 12/53 over 60
27. 2/3 **29.** 2/3 **31.** $P(\text{pregnant}|\text{test is positive}) = \dfrac{.5p}{.59 - .09p} < \dfrac{.5p}{.50} = p$.

Section 6.6 (page 344)

1. (a) .18522 (b) .117649 (c) .420175 (d) .882351 **3.** (a) 125/216 (b) 75/216 (c) 15/216
(d) 1/216 **5.** (a) $\dfrac{6561}{130{,}321}$ (b) $\dfrac{35{,}721}{130{,}321}$ **7.** $\dfrac{1{,}082{,}565}{4{,}194{,}304}$ **9.** about .91386 **11.** 81/256
13. about .62419 **15.** about .16729 **17.** 7 **19.** The probabilities are: (a) $1 - (5/6)^4 \approx .5178$
(b) $1 - \left(\dfrac{35}{36}\right)^{24} \approx .4914$ **21.** 16/81 **23.** (a) about .07019 (b) 7/128

Section 6.7 (page 357)

1. **3.**

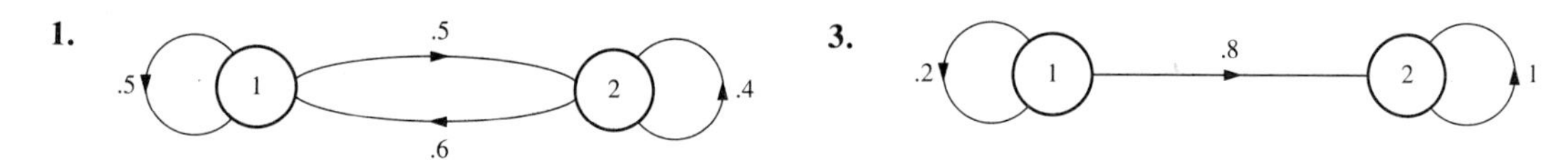

5.

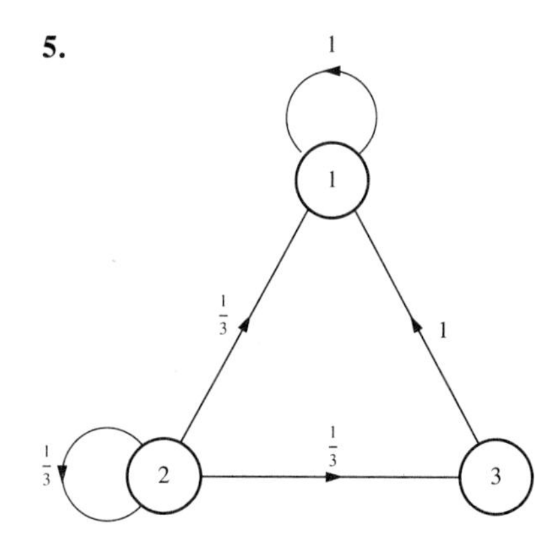

7. $\begin{bmatrix} .8 & .2 \\ .1 & .9 \end{bmatrix}$ **9.** $\begin{bmatrix} 0 & .5 & .5 \\ .5 & 0 & .5 \\ 0 & 0 & 1 \end{bmatrix}$ **11.** (a) [.3 .7]
(b) [.2375 .7625] **13.** .8752 **15.** It will drop to 42.92%.
17. In three years, 50.29% will live in the cities. **19.** .375
21. 26.5% agricultural, 38% blue-collar, and 35.5% white-collar
23. 44.509% large, 21.875% mid-size, and 33.616% small
25. [5/9 4/9] **27.** no **29.** no **31.** [5/28 5/7 3/28]
33. (a) [1 0], [0 1], [1 0], [0 1] (b) [0 1], [1 0], [0 1], [1 0]
(c) [.5 .5] (d) no **35.** 87.5% **37.** 25% **39.** 2/3 **41.** 1/3
43. 27.5% agricultural, 37.5% blue-collar, and 35% white-collar
45. 30% large, 20% mid-size, and 50% small

Chapter 6 Review Exercises (page 363)

1. (a) .12 (b) .33 (c) .67 (d) .91 **2.** 18/25 **3.** 3 to 2 **4.** 1/7 **5.** 2/7 **6.** 1/6 **7.** $\frac{485{,}875}{2{,}035{,}852}$ **8.** $\frac{21{,}879}{92{,}378}$ **9.** (a) .6 (b) .4 (c) .5 **10.** (a) 54% (b) .45 **11.** .58 **12.** 47/95 **13.** .75 **14.** .18 **15.** .44 **16.** 8/27 **17.** (a) 60% (b) 20% **18.** (a) 3/8 (b) 7/8 **19.** (a) about .81707 (b) about .98382 **20.** (a) about .83223 (b) about .49668 **21.** (a) $\begin{bmatrix} .52 & .48 \\ .02 & .98 \end{bmatrix}$, [.05 .95] (b) 4.5% (c) 4% **22.** (a) $\begin{bmatrix} .7 & .2 & .1 \\ 0 & .8 & .2 \\ .2 & .2 & .6 \end{bmatrix}$, [.4 .5 .1] (b) [.225 .500 .275] (c) [.2 .5 .3]

Chapter 7 Section 7.1 (page 375)

1.

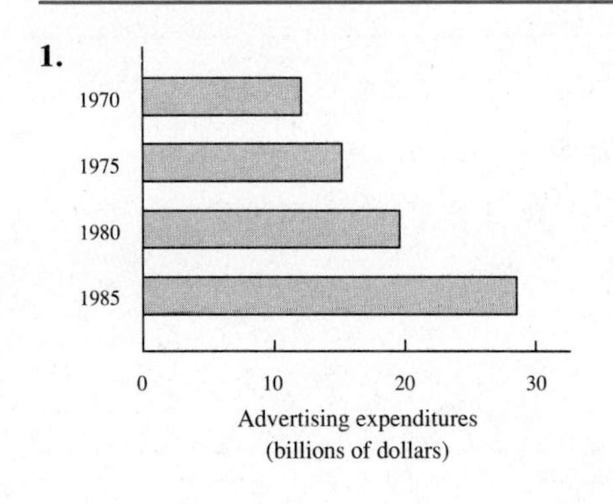

3. Frequency

30
20
10
29
26
11
9
5
1400 1600 1800
Tensile strength (lbs)

5. (a) 6 (b) 7 (c) about 2.19
7. (a) 28 (b) 29 (c) about 8.124
9. (a) 7 (b) 7.5 (c) about 2.45
11. (a) 5 (b) 6 (c) about 2.398
13. (a) 53 (b) 52.5 (c) about 3.38
15. The mean is $389,000, and the median is $262,500. The median is a better measure of a representative salary because the one atypically large value distorts the mean.

Section 7.2 (page 383)

1. $\bar{x} = 0.9$ and $s \approx 0.995$
3. $\bar{x} = 1610$ and $s \approx 105.59$
5. (a) For A, $\bar{x} = 1.5$ and $s \approx 0.84$; for B, $\bar{x} = 2$ and $s \approx 1.45$.
(b) Company B had the better record, but company A was more consistent.

7. One possibility is shown below.

(a)

Interval	*Frequency*
30–35	2
35–40	4
40–45	7
45–50	8
50–55	6
55–60	3
Total	30

(c) $\bar{x} \approx 46$ and $s \approx 6.85$

(b)

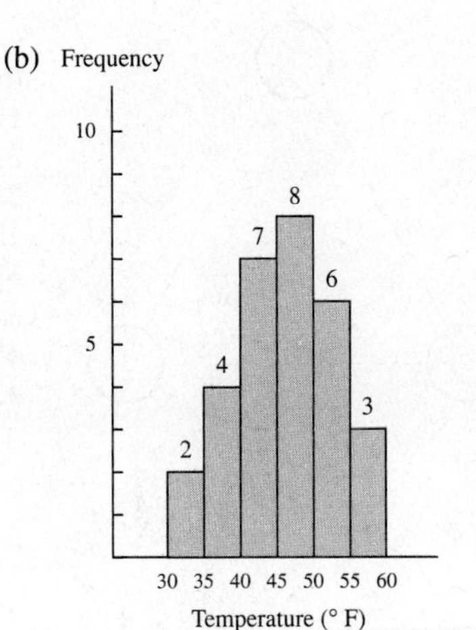

9. One possibility is shown below.

(a)

Interval	*Frequency*
80.5– 90.5	1
90.5–100.5	4
100.5–110.5	4
110.5–120.5	8
120.5–130.5	6
130.5–140.5	2
140.5–150.5	1
Total	26

(c) $\bar{x} \approx 115.50$ and $s \approx 14.23$

(b)

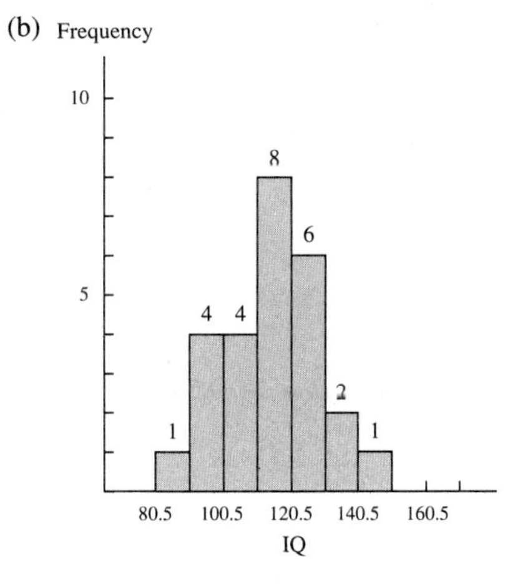

11. One possibility is shown below.

(a)

Interval	*Frequency*
41.5–44.5	2
44.5–47.5	2
47.5–50.5	6
50.5–53.5	6
53.5–56.5	10
56.5–59.5	5
59.5–62.5	4
62.5–65.5	2
65.5–68.5	1
68.5–71.5	1
Total	39

(c) $\bar{x} \approx 54.69$ and $s \approx 6.07$

(b)

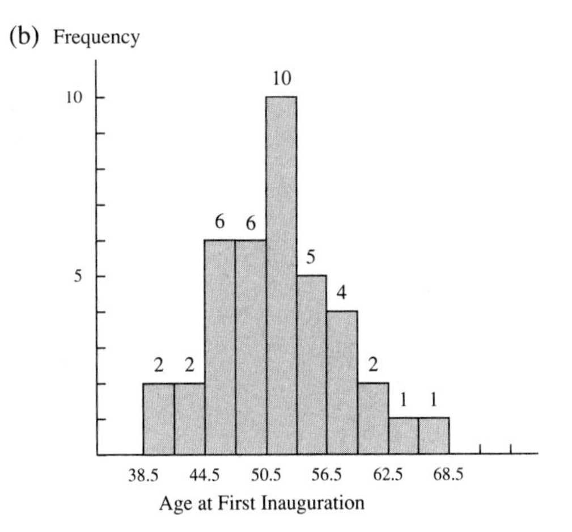

13. One possibility is shown below.

(a)

Interval	*Frequency*
21.5–25.5	3
25.5–29.5	0
29.5–33.5	9
33.5–37.5	8
37.5–41.5	9
41.5–45.5	12
45.5–49.5	6
49.5–53.5	1
53.5–57.5	1
57.5–61.5	1
Total	50

(c) $\bar{x} \approx 39.34$ and $s \approx 7.55$

(b)

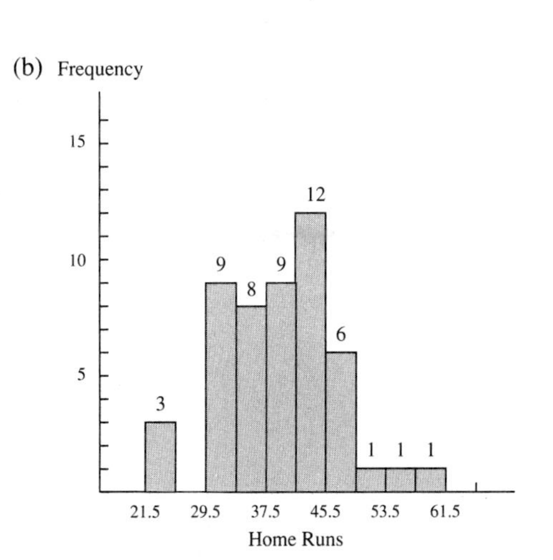

Section 7.3 (page 394)

1. (a)

Violations	*Probability*
0	.425
1	.350
2	.150
3	.050
4	.025

(b) $\bar{x} = 0.9$ and $s \approx 0.995$

3. (a)

Tensile strength (lb.)	*Probability*
1400–1500	.1375
1500–1600	.3625
1600–1700	.3250
1700–1800	.1125
1800–1900	.0625

(b) $\bar{x} = 1610$ and $s \approx 105.59$

5. $\bar{x} = 2.5$ and $s \approx 1.148$ **7.** (a) $0.16 (b) −$0.84 **9.** $6.00 **11.** $2.5 million **13.** $40,000
15. $194 **17.** 1.66

19. (a)

Number of Marbles	*Probability*
1	.40
2	.30
3	.20
4	.10

(b) $\mu = 2$ and $\sigma = 1$ **21.** 0.75 **23.** 12/7

Section 7.4 (page 408)

1. (a) a_3 (b) a_3 (c) a_3 **3.** The dealer should buy 2 papers for an expected profit of 64¢. **5.** The dealer should buy 8 pounds of salmon for an expected profit of $19.60. **7.** (a) Set up the equipment. (b) 70% **9.** The values of nodes *A, B, C, D,* and *E* are 58, 58, 72, 57, and 57, respectively. The expected payoff is 58 if an optimal strategy is followed. **11.** Hire the lobbyist for an expected gain of $650,000. **13.** Look for a space on the street, and pay the meter if successful. (The expected loss is 69¢.) **15.** File the suit, and accept a settlement if it is offered. (The expected gain is $1950.) **17.** 10 **19.** 41¢

Section 7.5 (page 420)

1. (a) .6 (b) .5 (c) .7

3. (a)

Violations	*Probability*
0	.425
1	.350
2	.150
3	.050
4	.025

(b) .925 (c) .550 (d) .575

5. (a)

Probability
.4
.3
.2
.1
.218
.318
.272
.130
.062
1 2 3 4 5
Delivery time (days)

(b) .720 (c) .808

7.

Successes	*Probability*
0	.216
1	.432
2	.288
3	.064

9.

Successes	*Approximate Probability*
0	.2373
1	.3955
2	.2637
3	.0879
4	.0146
5	.0010

11. $\mu = 15$ and $\sigma \approx 3.54$ **13.** $\mu = 900$ and $\sigma \approx 28.62$ **15.** $\mu = 35$ and $\sigma \approx 5.61$ **17.** $\mu = 150$ and $\sigma \approx 6.12$ **19.** (a) $\mu = 40$ and $\sigma \approx 2.83$ (b) $\mu = 30$ and $\sigma \approx 3.46$

Section 7.8 (page 434)

1. .7190 **3.** .8043 **5.** .0336 **7.** .3153 **9.** −0.98 **11.** 1.13 **13.** −1.50 **15.** 1.80 **17.** 0.86 **19.** −2.60 **21.** 0.0401 **23.** 0.3413 **25.** 0.9192 **27.** 0.6730 **29.** 0.1605 **31.** 9.94% **33.** 0.0808 **35.** 98.76% **37.** .8788 **39.** .9608 **41.** .8365 **43.** .9146 **45.** (a) 30.85% (b) 628 **47.** from 67.3 to 70.7 inches

Chapter 7 Review Exercises (page 438)

1. \$6800, \$7000, $s \approx \$1029.56$ **2.** The median of 10.5% better reflects a typical increase than the mean of 22%, which is distorted by one atypically large value.

3. (a)

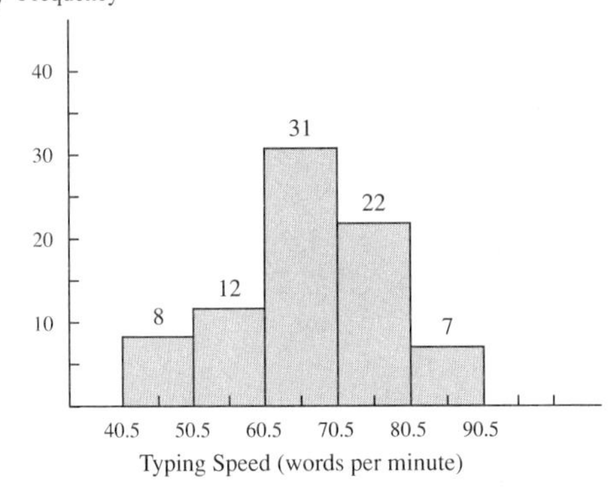

(b) $\bar{x} = 66$ and $s \approx 10.79$

4. (a) One possibility is given below.

Interval	*Frequency*
300.5–400.5	3
400.5–500.5	3
500.5–600.5	8
600.5–700.5	8
700.5–800.5	3
Total	25

(b)

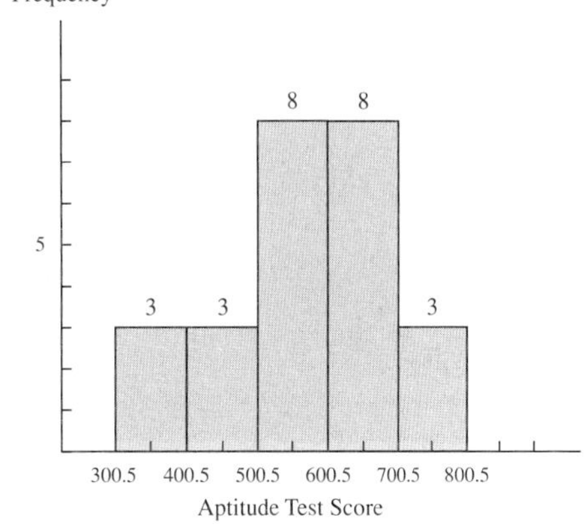

(c) $\bar{x} \approx 570$ and $s \approx 116.62$

5. $\bar{x} = 1.6$ and $s \approx 1.23$ **6.** $\mu = 0$ and $\sigma \approx 2.24$ **7.** $\mu = 5$ and $\sigma \approx 1.83$ **8.** $\mu = 15$ and $\sigma \approx 3.24$ **9.** 25.78% **10.** 11.51% **11.** .7286 **12.** .0102 **13.** 9% **14.** \$500 **15.** Yes, the farmer's expected gain is \$13,000 more than the cost of cloud-seeding. **16.** Print 10,000 copies for an expected profit of \$101,000. **17.** Make two apple pies per day for an expected profit of \$1.60. **18.** Try to answer the jackpot question. If successful, keep the money already won. (The contestant's expected winnings are \$4600 with this strategy.)

Chapter 8 Section 8.1 (page 451)

1. function **3.** not a function **5.** not a function **7.** function **9.** function

11. $[3, \infty)$ **13.** $(-4, 8)$ **15.** $(-\infty, 13]$

17. $[-1, 5)$ **19.** $(2, 15]$

21. $[1, 10]$

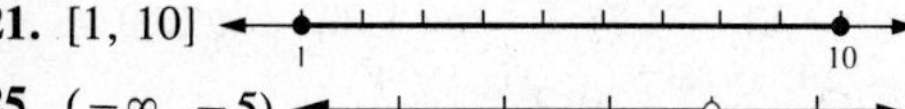

23. $(-6, \infty)$

25. $(-\infty, -5)$ **27.** $\frac{3}{20}$ **29.** $\frac{1}{9}$ **31.** 13 **33.** 19 **35.** -6 **37.** $(-\infty, \infty)$

39. $(-\infty, -3) \cup (-3, 3) \cup (3, \infty)$ **41.** $(-\infty, 2]$ **43.** $(-\infty, 0) \cup (0, \infty)$ **45.** $(-3, 0) \cup (0, 5) \cup (5, \infty)$

47.

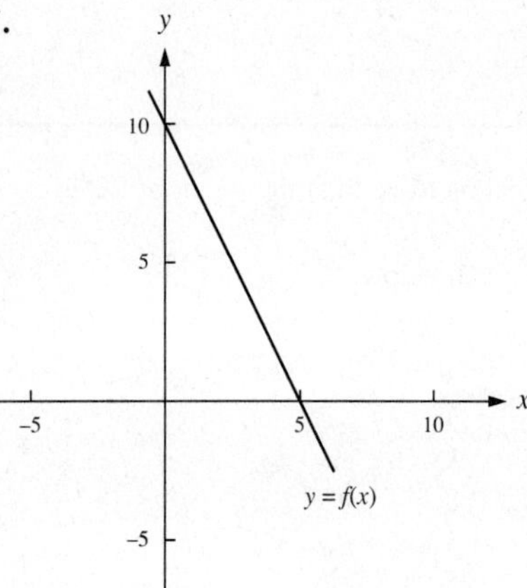

49.

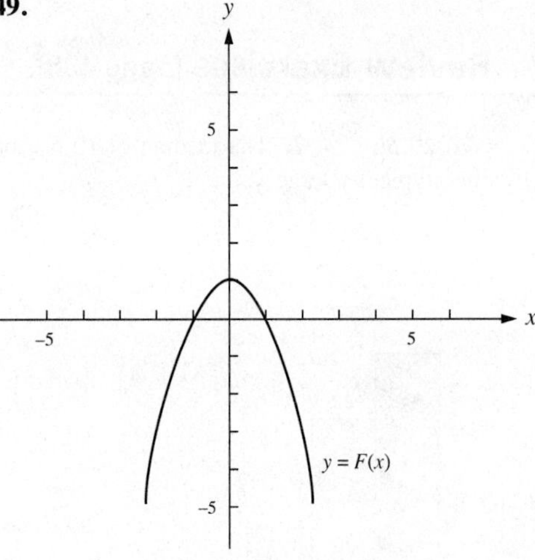

51.

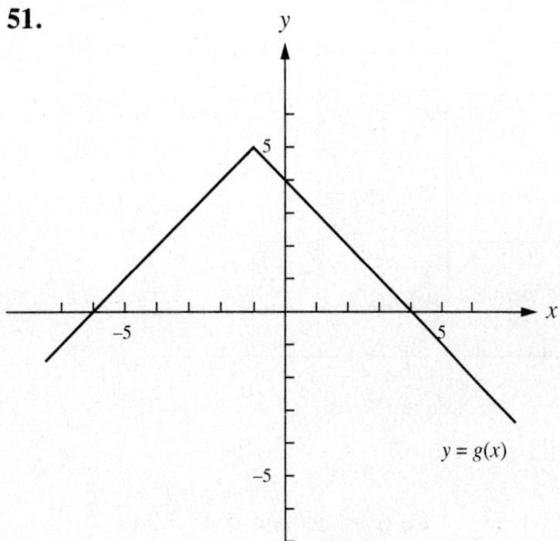

53.

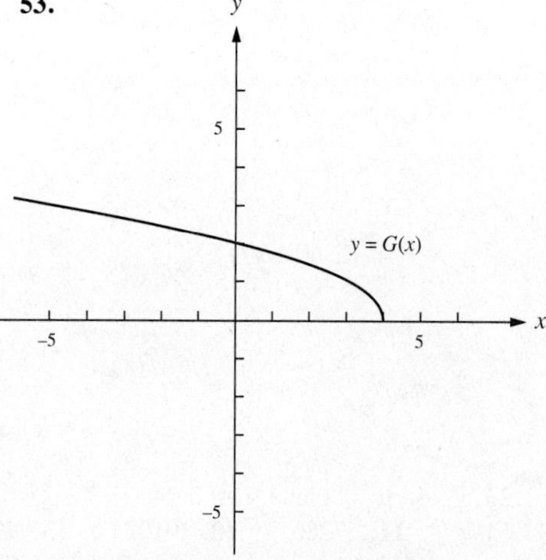

55.

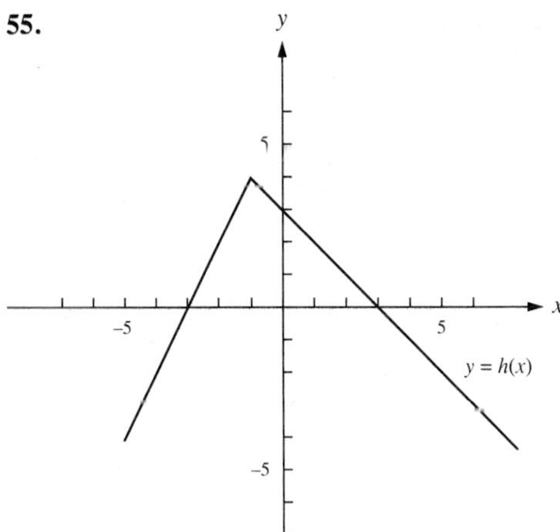

57.

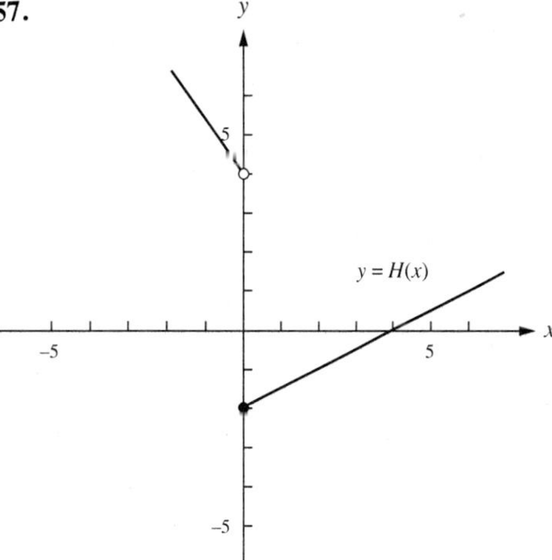

59.

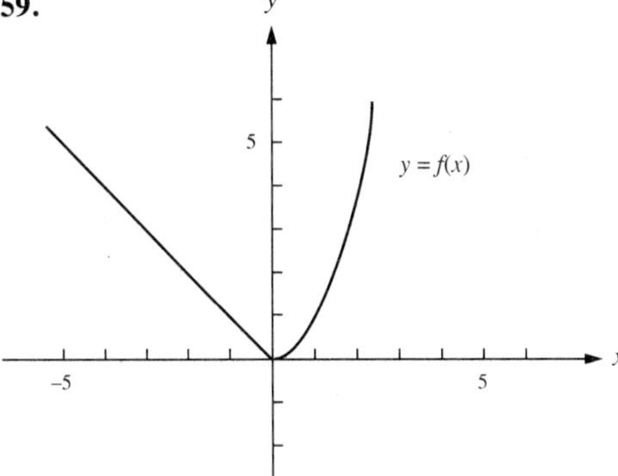

■ Section 8.2 (page 466)

1. polynomial of degree 5 **3.** not a polynomial
5. polynomial of degree 0 **7.** polynomial of degree 6
9. not a polynomial **11.** $(-\infty, -7) \cup (-7, \infty)$
13. $(-\infty, 1.8) \cup (1.8, \infty)$ **15.** $(-\infty, -4) \cup (-4, 4) \cup (4, \infty)$
17. $(-\infty, -3) \cup (-3, 5) \cup (5, \infty)$ **19.** $(-\infty, \infty)$

21. no asymptotes

23. no asymptotes

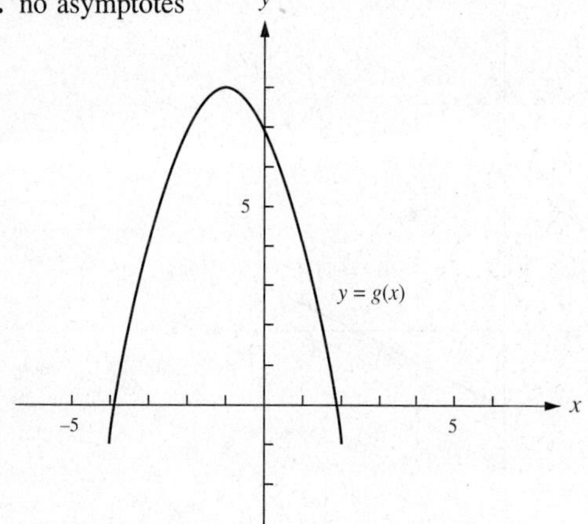

25. no asymptotes

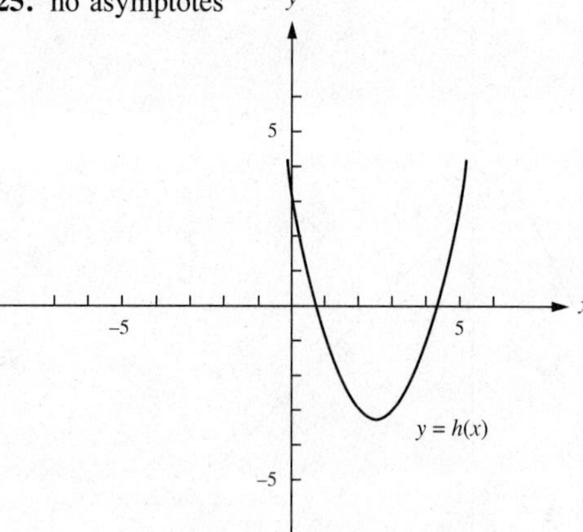

27. no asymptotes

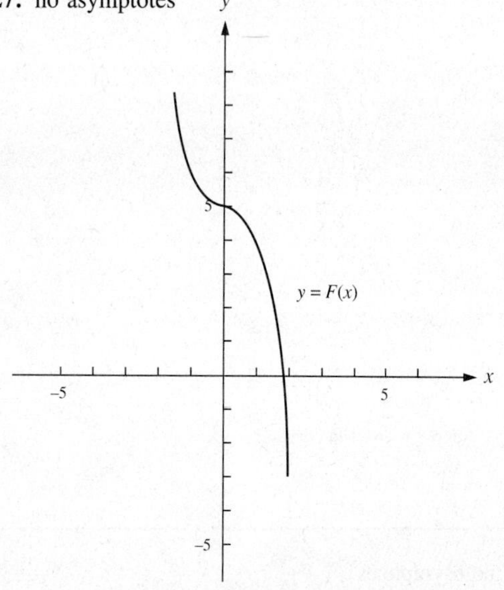

29. no asymptotes

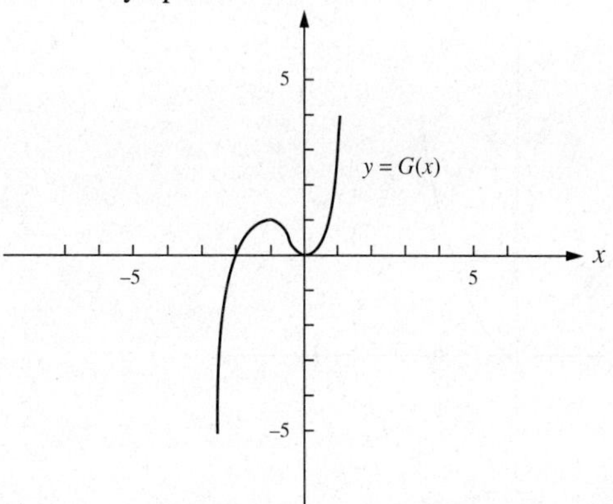

31. vertical asymptote at $x = 1$, horizontal asymptote at $y = 0$

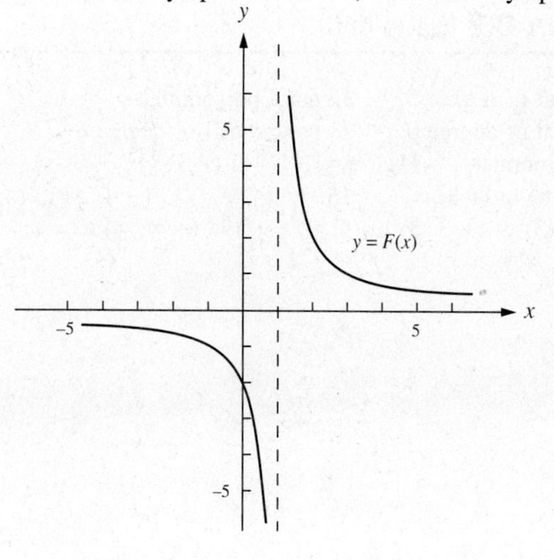

33. vertical asymptote at $x = 2$, horizontal asymptote at $y = 0$

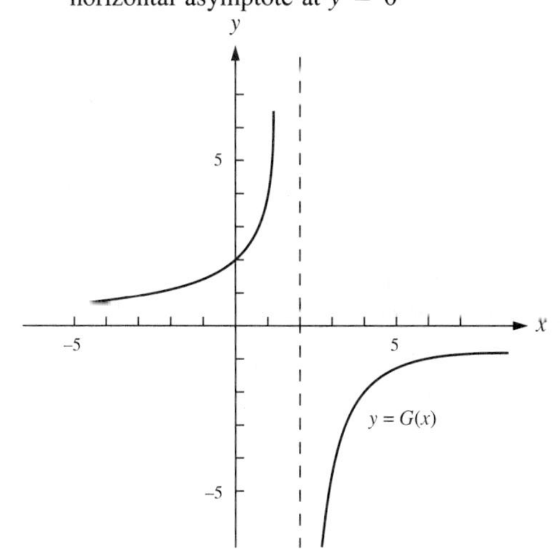

35. vertical asymptote at $x = -1$, horizontal asymptote at $y = 3$

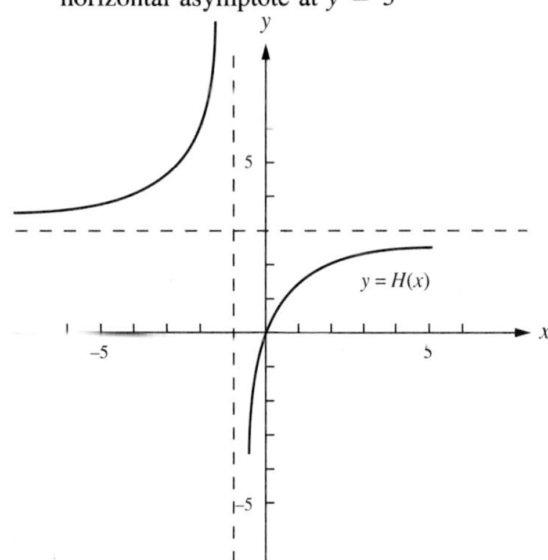

37. vertical asymptotes at $x = -3$ and $x = 3$, horizontal asymptote at $y = 0$

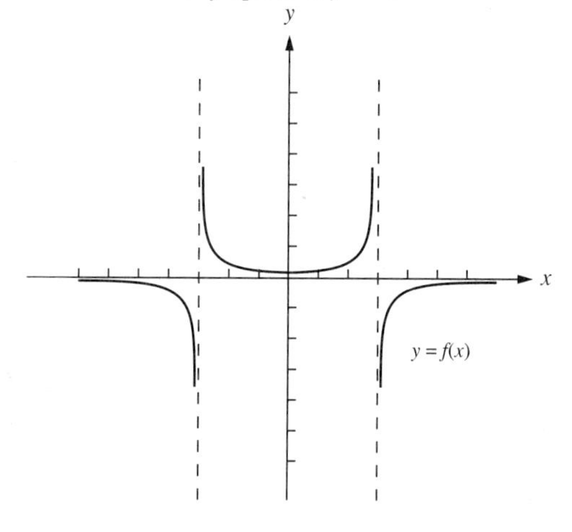

39. $9 < p < 16$ **41.** $200 < p < 350$ **43.** (a) \$0.64 million (b) \$1.2 million (c) \$1.92 million (d) \$4.8 million

45. (a) 44% increase (b) $122\frac{2}{9}$% increase

47. $x = 0$ is the only vertical asymptote

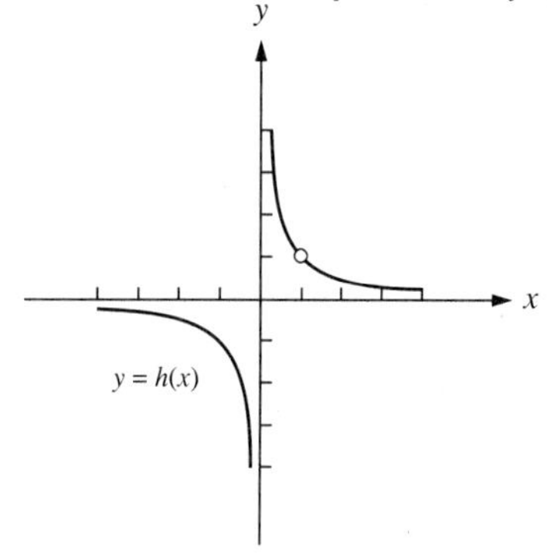

■ Section 8.3 (page 475)

1. 29.96 **3.** 0.10 **5.** 1.06 **7.** 0.28 **9.** 164.02 **11.** 0.84

13.

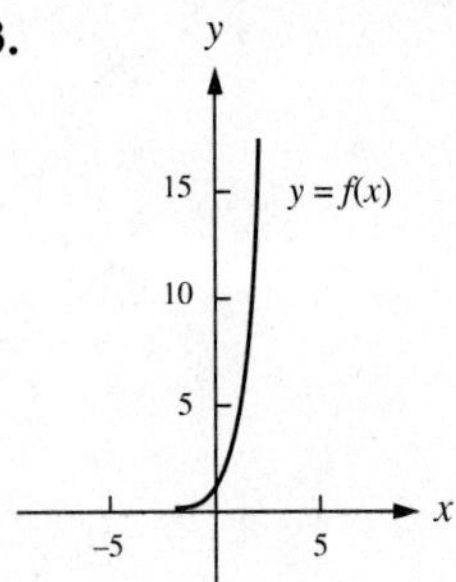

15.

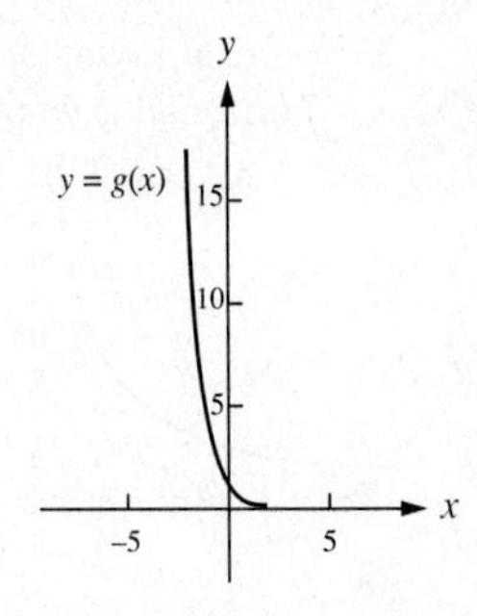

17.

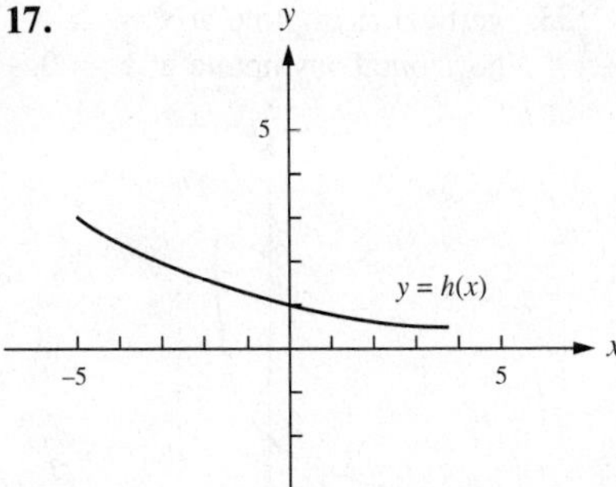

19.

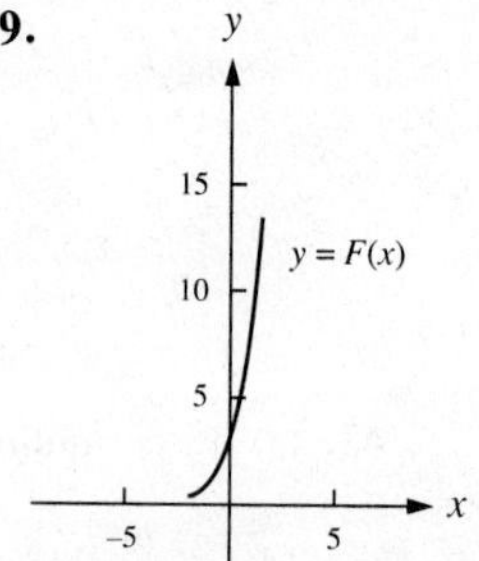

21.

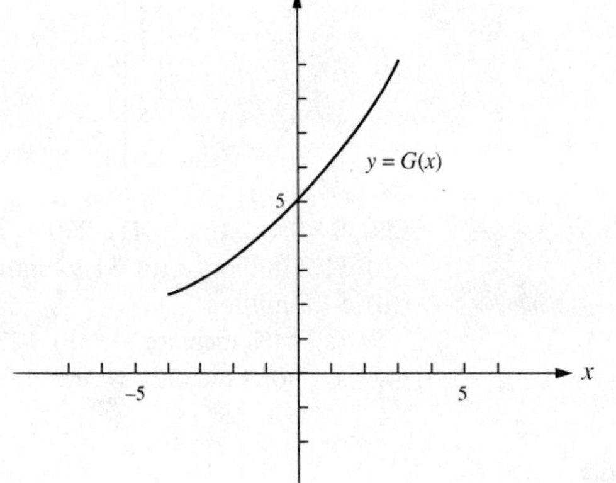

23.

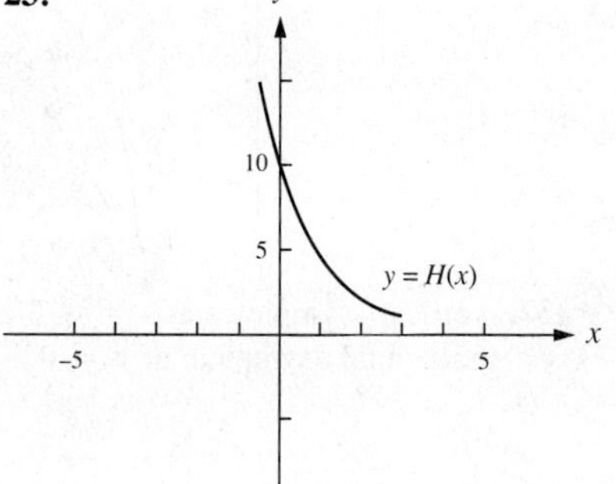

25. (a) $P(t) = 190{,}000e^{0.002t}$
(b) 195,395

27.

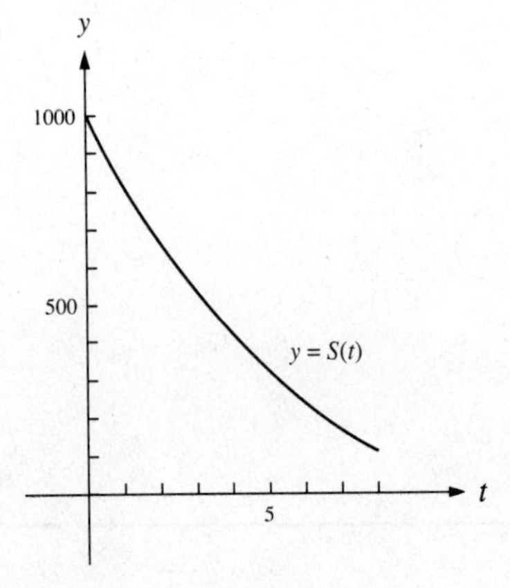

29. f is a constant function.

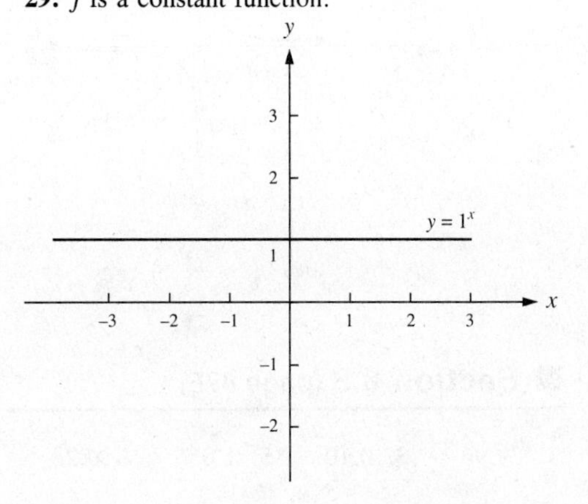

Section 8.4 (page 485)

1. $11^2 = 121$ **3.** $4^{-3} = \dfrac{1}{64}$ **5.** $10^{-4} = 0.0001$ **7.** $(0.25)^{-2.5} = 32$ **9.** $\log_{13} 169 = 2$

11. $\log_3 \dfrac{1}{81} = -4$ **13.** $\log_{1000} 10 = \dfrac{1}{3}$ **15.** $\log_{0.4} \dfrac{125}{8} = -3$ **17.** 3 **19.** -2 **21.** -3 **23.** $\dfrac{3}{2}$

25. $\log_8 11$ **27.** $\log_{10} 80$ **29.** $\log_3 16$ **31.** $\ln 12$ **33.** about 30.543 **35.** about 5.748 **37.** about 5.268 **39.** about 2.301 **41.** $f(x) = e^{(2-5x)(\ln 3)}$ **43.** $g(x) = \dfrac{1}{\ln 9}\ln(7x - 6)$ **45.** $h(x) = \dfrac{1}{\ln 10}\ln(3x^2 + 4)$ **47.** $F(x) = e^{(3x-x^2)(\ln 2)}$ **49.** about 2543 years **51.** about 11.6 years **53.** about 1,330,008 years
55. (a) 8.2 (b) The intensity of an earthquake that measures r on the Richter scale is $10^r I_0$. Hence an earthquake measuring 8 is $10^4 = 10{,}000$ times more intense than an earthquake measuring 4.

Section 8.5 (page 492)

1. $(f + g)(x) = 4x^2 - 10x + 8$ **3.** $\dfrac{f}{g}(x) = 2x - 6 + \dfrac{4}{x}$ **5.** $(3f - 6g)(x) = 12x^2 - 48x + 24$
7. $(\frac{1}{4}f)(\frac{1}{2}g)(x) = x^3 - 3x^2 + 2x$ **9.** 3 **11.** -2 **13.** 138 **15.** $20/3$ **17.** 50 **19.** 38 **21.** 16
23. $f(g(x)) = e^{3x} - 2e^x$
25. $f(g(x)) = \dfrac{30}{x^2} - 2$ **27.** $f(g(x)) = \ln\left(\dfrac{4x}{\sqrt{x+3}}\right)$ **29.** $f(g(x)) = \dfrac{5x-6}{5x-5}$ **31.** $f(x) = \dfrac{2}{x^3}$ and $g(x) = (3x^2 - 2x + 1)$ **33.** $f(x) = \sqrt{x}$ and $g(x) = 4 - x^2$ **35.** $f(x) = 3\ln x$ and $g(x) = x^2 + 4x + 4$
37. $f(x) = 2xe^x$ and $g(x) = x^2 - x$ **39.** $P(x) = R(x) - C(x) = 8.50x - 0.0005x^2 - 30{,}000$ **41.** $P(x) = R(x) - C(x) = 40x - (15x + 400{,}000) = 25x - 400{,}000$ **43.** $S(p(d)) = 1200 + 0.3d$ expresses the number of units supplied per week when the weekly demand is d units. **45.** $C(D(p)) = 2{,}000{,}000 - 3000p$ is the cost of producing all of the dishwashers that can be sold at a price of p dollars.

Chapter 8 Review Exercises (page 495)

1. (a) 29 (b) 5 (c) 47 (d) $2x^2 + 4hx + 2h^2 - 3$ **2.** (a) -5 (b) 1 (c) 3 (d) -21
3. (a) function (b) not a function (c) not a function (d) function **4.** (a) $(-\infty, \frac{1}{2}) \cup (\frac{1}{2}, \infty)$
(b) $(-\infty, 8]$ (c) $(-\infty, \infty)$ (d) $[-2, 3) \cup (3, 5) \cup (5, \infty)$ **5.** (a) $(-\infty, 23)$
(b) $[-6, \infty)$ (c) $(-9, -3)$
(d) $(3, 17]$  (e) $(-\infty, 12]$ 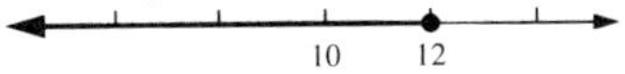
(f) $(-7, \infty)$ 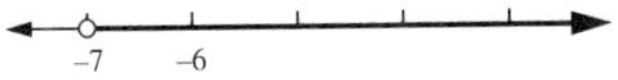 (g) $[-5, 8]$ 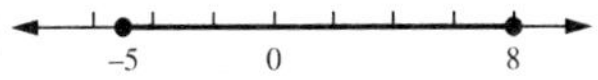
(h) $[-2, 11)$ **6.** (a) rational function (b) polynomial of degree 4
(c) rational function (d) neither (e) polynomial of degree 3 (f) rational function **7.** (a) $\log_6 1296 = 4$
(b) $\log_2 \frac{1}{64} = -6$ **8.** (a) $2^{10} = 1024$ (b) $8^{-2/3} = \frac{1}{4}$ **9.** (a) $\log_{16} 8$ (b) $\log_2 20$ **10.** (a) $f(x) = e^{(3x+2)\ln 8}$
(b) $g(x) = \dfrac{1}{\ln 10}\ln(2x + 1)$ **11.** (a) about 31.564 (b) about 5.993 **12.** about 16.96 cm of mercury
13. (a) $P(t) = 200e^{0.017t}$ million (b) during October, 1993 **14.** about 28.41 years

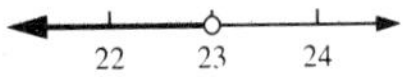

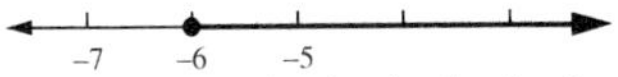

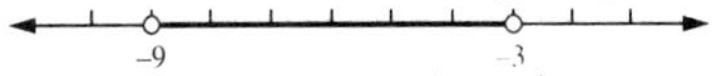

15.

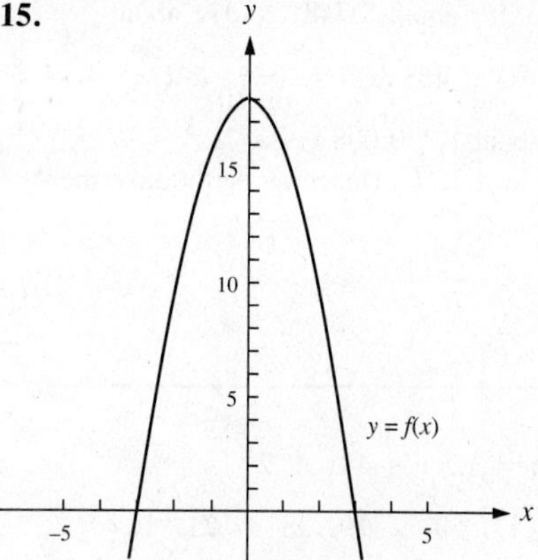

16.

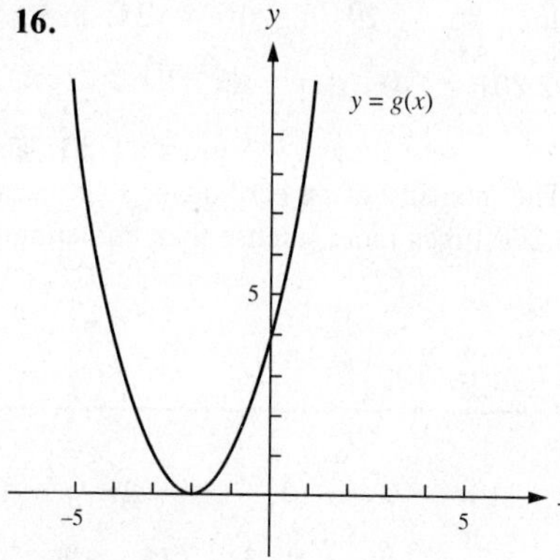

17.

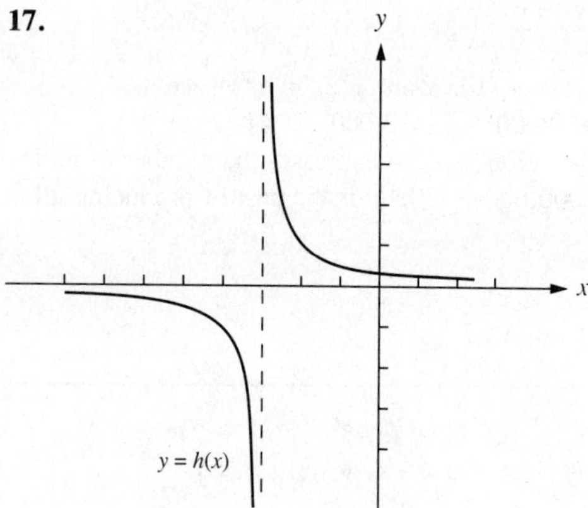

18.

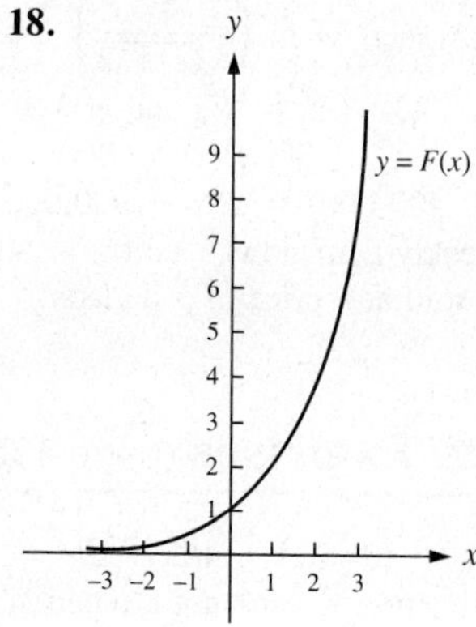

19.

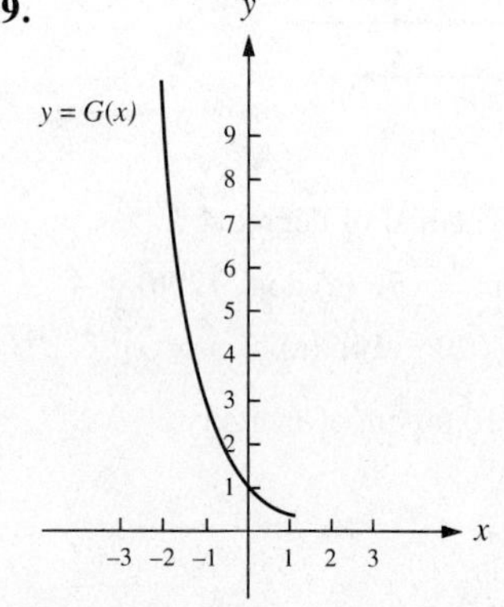

20.

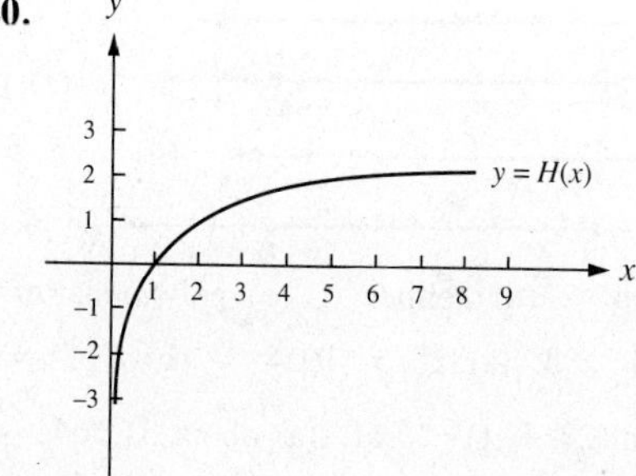

■ Chapter 9 Section 9.1 (page 505)

1. 0 **3.** -2 **5.** $2x - 5$ **7.** $4x + 1$ **9.** $3x^2$ **11.** $\dfrac{-1}{(x-2)^2}$ **13.** -3 **15.** 5 **17.** -10
19. -1 **21.** $y = 4x$ **23.** $y = -3x + 12$ **25.** $y = -8x + 14$ **27.** $y = 0.5x + 0.5$ **29.** $y = 12x - 4$
31. increasing by \$50 per unit sold **33.** increasing by \$415 per unit produced **35.** increasing by 15 students per day

Section 9.2 (page 522)

1. the limit does not exist **3.** the limit does not exist **5.** 3 **7.** 0 **9.** -1 **11.** 6 **13.** 28 **15.** 75 **17.** -2 **19.** 5 **21.** does not exist **23.** $\frac{1}{7}$ **25.** $\frac{1}{2}$ **27.** does not exist **29.** $\frac{4}{9}$ **31.** 0 **33.** $f'(x) - 3$ **35.** $f'(x) = 5 - 2x$ **37.** $y = -3x + 2$ **39.** $y = 0.3x - 3.5$

Section 9.3 (page 532)

1. (a) -3 (b) $-3, 2$ **3.** (a) none (b) -1 **5.** (a) 0 (b) 0 **7.** (a) 1 (b) $-1, 1$ **9.** (a) $-3, 1$ (b) $3, 1$ **11.** yes **13.** yes **15.** no **17.** no **19.** yes **21.** \$1077.88 **23.** \$334.88 **25.** \$523.01 **27.** \$4420.68 **29.** $\frac{1}{e}$ **31.** e

Section 9.4 (page 542)

1. $f'(x) = 8x^7$ **3.** $g'(x) = 0$ **5.** $h'(x) = -3x^{-4} + 3$ **7.** $F'(x) = 6x - 5$ **9.** $G'(x) = 7x^6 + \frac{12}{x^5}$ **11.** $H'(x) = \frac{2}{\sqrt{x}} + 1$ **13.** $f'(x) = \frac{1}{3}x^{-2/3}$ **15.** $f'(x) = 2x + \frac{6}{x^2}$ **17.** $f'(x) = -\frac{6}{x^4}$ **19.** $f'(x) = 2x + 5$ **21.** $y = 9x - 16$ **23.** $y = \frac{1}{4}x + 1$ **25.** $y = -4x - 8$ **27.** $y = -6x + 12$ **29.** increasing by \$300 per item produced **31.** decreasing by \$3 per item received in each order

Section 9.5 (page 548)

1. $g(x) = x^5$ and $h(x) = 2x - 1$ **3.** $g(x) = \ln x$ and $h(x) = 4x^2 - 1$ **5.** $g(x) = e^x$ and $h(x) = \sqrt{x + 1}$ **7.** $g(x) = \frac{4}{\sqrt{x}}$ and $h(x) = 3x + 5$ **9.** $g(x) = 2^x$ and $h(x) = x^3 - x$ **11.** $f'(x) = 18(3x - 5)^5$ **13.** $g'(x) = 7(2x - 3)(x^2 - 3x + 4)^6$ **15.** $h'(x) = -1.5(10x^4 - 8x)(2x^5 - 4x^2 + 1)^{-5/2}$ **17.** $F'(x) = 10(6 - 2x)^{-6}$ **19.** $G'(x) = \frac{1}{3}(4x - 7)(2x^2 - 7x)^{-2/3}$ **21.** $3x^2e^{2x^3}$ **23.** $\frac{1}{2\sqrt{x - x^2}}$ **25.** $4x^7 \ln (x^8 + 1)$ **27.** $\frac{2\sqrt{x^4 - 4}}{x^3}$ **29.** 60 **31.** 12 **33.** -16 **35.** $-\frac{3}{8}$ **37.** $f'(x) = r[h(x)]^{r-1}[h'(x)]$

Section 9.6 (page 557)

1. $f'(x) = 3x^2 + \frac{1}{x}$ **3.** $g'(x) = e^x - 12x^5$ **5.** $h'(x) = -7e^{-x}$ **7.** $F'(x) = \frac{2x}{x^2 + 1}$ **9.** $G'(x) = 4e^{4x}$ **11.** $H'(x) = \frac{4x^3 - 4x}{x^4 - 2x^2 + 8}$ **13.** $f'(x) = \frac{1}{2}(3x + e^{-2x})^{-1/2}(3 - 2e^{-2x})$ **15.** $g'(x) = 3[\ln (2x^2 + 5x)]^2 \frac{4x + 5}{2x^2 + 5x}$ **17.** $h'(x) = \frac{1}{3}[\ln (x^2 + 4)]^{-2/3} \frac{2x}{x^2 + 4}$ **19.** $F'(x) = 5(e^x + e^{-x})^4(e^x - e^{-x})$ **21.** $f'(0) = 2$ **23.** $f'(1) = -1$ **25.** $y = x - 1$ **27.** $y = 3x + 2$ **29.** increasing by \$86.66 per year **31.** decreasing by 5.82 kg per year **33.** increasing by about 647 persons per year **35.** (a) $f'(x) = (\ln b)h'(x)b^{h(x)}$ (b) $F'(x) = \frac{1}{\ln b} \cdot \frac{h'(x)}{h(x)}$

Chapter 9 Review Exercises (page 560)

1. (a) 4 (b) does not exist (c) 2 (d) 3 **2.** (a) no (b) yes (c) no (d) yes **3.** (a) no (b) no (c) yes (d) yes **4.** (a) 6 (b) 8 (c) 15 (d) $e^{-1/2}$ (e) -2 (f) does not exist
5. $f'(x) = -\dfrac{2}{x^2}$ **6.** (a) $f'(x) = 42x^5 - 20x^4 + 6x$ (b) $g'(x) = -\dfrac{12}{x^4} + \dfrac{1}{x}$ (c) $h'(x) = 56x^6 - 12x + 9e^x$
(d) $F'(x) = \dfrac{45x^4 - 12x^2}{9x^5 - 4x^3}$ (e) $G'(x) = (4x^3 - 4x)(3x^4 - 6x^2)^{-2/3}$ (f) $H'(x) = (40x - 12)e^{5x^2-3x}$ **7.** increasing by \$590 per unit sold **8.** decreasing by 3275 people per year **9.** $y = 1$ **10.** $y = \dfrac{3}{8}x + \dfrac{13}{4}$

Chapter 10 Section 10.1 (page 570)

1. $f'(x) = 8x^3 - 2$ **3.** $g'(x) = \dfrac{1}{(3-x)^2}$ **5.** $h'(x) = \ln x + 1$ **7.** $F'(x) = \left(\sqrt{x} + \dfrac{1}{2\sqrt{x}}\right)e^x$
9. $G'(x) = \dfrac{3x - 1 - 3x\ln x}{x(3x-1)^2}$ **11.** $H'(x) = \dfrac{(x-2)e^x}{x^3}$ **13.** $f'(x) = 46x^7 - 20x^4 + 114x^3 - 12)(2x^3 - 4)^5$
15. $g'(x) = \dfrac{(18x - 19)(2x+1)^3}{(3x-2)^2}$ **17.** $h'(x) = \dfrac{6x^5 - 12x - (12x^5 + 12x^3)\ln(x^2+1)}{(x^2+1)(3x^4-6)^2}$
19. $F'(x) = \dfrac{2 - 2x}{\sqrt{2x-1}}e^{-x}$ **21.** $y = 0$ **23.** $y = 2x$ **25.** $y = 1.75x - 0.25$ **27.** decreasing by \$4999 per item **29.** decreasing by \$3.20 per item **31.** $F' = fgh' + fg'h + f'gh$

Section 10.2 (page 575)

1. (a) $y' = 42x^6 - 20x^4 + 12x^3$ (b) $y'' = 252x^5 - 80x^3 + 36x^2$ (c) $y''' = 1260x^4 - 240x^2 + 72x$ **3.** (a) $y' = 15(3x-2)^4$ (b) $y'' = 180(3x-2)^3$ (c) $y''' = 1620(3x-2)^2$ **5.** (a) $y' = -2e^{1-2x}$ (b) $y'' = 4e^{1-2x}$
(c) $y''' = -8e^{1-2x}$ **7.** (a) $y' = \dfrac{4x}{2x^2+1}$ (b) $y'' = \dfrac{4 - 8x^2}{(2x^2+1)^2}$ (c) $y''' = \dfrac{32x^3 - 48x}{(2x^2+1)^3}$ **9.** (a) $y' = (2x+4)^{-1/2}$
(b) $y'' = -(2x+4)^{-3/2}$ (c) $y''' = 3(2x+4)^{-5/2}$ **11.** (a) $y' = x^2(1 + 3\ln x)$ (b) $y'' = x(5 + 6\ln x)$
(c) $y''' = 11 + 6\ln x$ **13.** (a) $f'(-1) = -7$ (b) $f''(-1) = 42$ **15.** (a) $f'(2) = 48$ (b) $f''(2) = 42$
17. (a) $f'(0) = 0$ (b) $f''(0) = 8$ **19.** (a) $\frac{4}{3}$ (b) $\frac{2}{27}$ **21.** (a) -1 (b) 5 **23.** (a) 3 (b) 13
25. (a) -1, 3 (b) 1 **27.** (a) increasing by \$240 per unit (b) increasing by \$246 per unit (c) 500
29. (a) $f'(x) = 24x - 3x^2$ (b) 4 mg

Section 10.3 (page 582)

1. $\frac{3}{2}x^{-1/2}$ **3.** $3y^2y' - 14yy'$ **5.** $3y'e^{3y}$ **7.** $4x^2y^3y' + 2xy^4$ **9.** $\dfrac{x(2x+y')}{x^2+y} + \ln(x^2+y)$
11. $\dfrac{x^2y' - 3xy' - 2xy + 3y}{(x^2-3x)^2}$ **13.** $y' = \dfrac{3}{2y}$ **15.** $y' = -\dfrac{2x}{y}$ **17.** $y' = -\dfrac{2y}{x}$ **19.** $y' = \dfrac{12x^2 + 4y}{3y^2 - 4x}$

21. $y' = \dfrac{1 - xy^2 - y^3}{2x^2y + 2xy^2 - 1}$ **23.** $y' = \dfrac{6xy}{e^y - 3x^2}$ **25.** $y = -\dfrac{4x}{3} + \dfrac{50}{3}$ **27.** $y = 2x - 2$

29. $y = x + 2$ **31.** $y = \dfrac{4}{5}x - \dfrac{2}{5}$

Section 10.4 (page 587)

1. 3 **3.** 8 **5.** 0.3 **7.** 3 **9.** decreasing by $87,600 per month **11.** increasing by 6500 people per year **13.** decreasing by 30 units per month **15.** increasing by $20 per week **17.** decreasing by $250 per week **19.** increasing by 26 per year **21.** 43.6 knots per hour **23.** 60 cubic inches per hour **25.** 2.5 feet per second **27.** 1/3 feet per minute **29.** 8 feet per second

Chapter 10 Review Exercises (page 591)

1. $f'(x) = \sqrt{6 - x}\,(-27.5x^4 + 120x^3 + 10.5x^2 - 36x - 10.5)$ **2.** $F'(x) = \dfrac{8x^2 - 20x + 4}{(4x - 5)^2}$

3. $g'(x) = \dfrac{2}{(2x + 1)(3x - 2)} - \dfrac{3 \ln (2x + 1)}{(3x - 2)^2}$ **4.** $G'(x) = (-6x^2 + 14x - 4)e^{-2x}$ **5.** When 90% of the contaminants have been removed, the cost is increasing by $8.25 million per percent of contaminants removed.

6. $y = 3x + 2$ **7.** $y = x - 1$ **8.** $y'' = \dfrac{9x + 6}{(3x + 1)^2}$ **9.** $y''' = (-x^3 + 9x^2 - 18x + 6)e^{-x}$ **10.** $y' = \dfrac{x}{2y}$

11. $y = \dfrac{27}{14}x + \dfrac{5}{14}$ **12.** decreasing by $100 per week **13.** increasing by $2200 per month

Chapter 11 Section 11.1 (page 602)

1. increasing on $(-2, \infty)$; decreasing on $(-\infty, -2)$; concave up on $(-\infty, \infty)$; $f(-2) = -3$ is a local minimum; no points of inflection **3.** increasing on $(0, 4)$; decreasing on $(-\infty, 0)$ and $(4, \infty)$; concave up on $(-\infty, 2)$; concave down on $(2, \infty)$; $f(0) = 1$ is a local minimum and $f(4) = 5$ is a local maximum; $(2, 2)$ is a point of inflection **5.** increasing on $(-\infty, -3)$ and $(2, \infty)$; decreasing on $(-3, 2)$; concave up on $(-1, \infty)$; concave down on $(-\infty, -1)$; $f(-3) = -1$ is a local maximum; $f(2) = -5$ is a local minimum; $(-1, -2)$ is a point of inflection **7.** increasing on $(-5, -2)$ and $(3, \infty)$; decreasing on $(-\infty, -5)$ and $(-2, 3)$; concave up on $(-\infty, -4)$ and $(0, \infty)$; concave down on $(-4, 0)$; $f(-5) = 3$ and $f(3) = 3$ are local minima, and $f(-2) = 7$ is a local maximum; $(-4, 4)$ and $(0, 5)$ are points of inflection **9.** increasing on $(-\infty, -4)$ and $(-2, 2)$; decreasing on $(-4, -2)$ and $(2, \infty)$; concave up on $(-3, 0)$; concave down on $(-\infty, -3)$ and $(0, \infty)$; $f(-4) = -1$ and $f(2) = 3$ are local maxima, and $f(2) = -3$ is a local minimum; $(-3, -2)$ and $(0, 0)$ are points of inflection **11.** increasing on $(-\infty, 2)$ and $(2, \infty)$; concave up on $(-\infty, 2)$; concave down on $(2, \infty)$; no local extrema; no points of inflection **13.** increasing on $(0, 3)$ and $(3, \infty)$; decreasing on $(-\infty, -3)$ and $(-3, 0)$; concave up on $(-3, 3)$; concave down on $(-\infty, -3)$ and $(3, \infty)$; $f(0) = -2$ is a local minimum; no points of inflection **15.** increasing on $(-3, 1)$; decreasing on $(-\infty, -3)$ and $(1, \infty)$; concave up on $(-\infty, -2)$ and $(2, \infty)$; concave down on $(-2, 2)$; $f(-3) = 0$ is a local minimum and $f(1) = 3$ is a local maximum; $(-2, 1)$ and $(2, 2)$ are points of inflection **17.** $(-\infty, -6) \cup (5, \infty)$ **19.** $(-2, 1)$ **21.** $(6, \infty)$ **23.** $(-\infty, -7) \cup (-1, 4)$ **25.** $(-4, -1) \cup (3, \infty)$ **27.** $(-\infty, -7) \cup (-2, 5)$ **29.** $(-\infty, -2) \cup (-1, 2) \cup (2, \infty)$ **31.** $(-5, -2) \cup (1, 4)$

Section 11.2 (page 614)

1. $-3, 5$ **3.** $-2, -1, 0$ **5.** none **7.** $-3, 5$ **9.** increasing on $(-\infty, 2.5)$; decreasing on $(2.5, \infty)$; $f(2.5) = -0.75$ is a local maximum **11.** increasing on $(9, \infty)$; decreasing on $(-\infty, 9)$; $g(9) = -138$ is a local minimum **13.** increasing on $(-\infty, -8)$ and $(6, \infty)$; decreasing on $(-8, 6)$; $h(-8) = 840$ is a local maximum, and $h(6) = -532$ is a local minimum **15.** increasing on $(-\infty, -5)$ and $(-1, \infty)$; decreasing on $(-5, -1)$; $F(-5) = 0$ is a local maximum, and $F(-1) = -32$ is a local minimum **17.** increasing on $(-\infty, -2)$ and $(2, \infty)$; decreasing on $(-2, 2)$; $G(-2) = 64$ is a local maximum, and $G(2) = -64$ is a local minimum **19.** decreasing on $(-\infty, 6)$ and $(6, \infty)$; no local extrema **21.** increasing on $(4, \infty)$; decreasing on $(0, 4)$; $f(4) = -\frac{1}{2}$ is a local minimum **23.** increasing on $(-\infty, 0.5)$; decreasing on $(0.5, \infty)$; $g(0.5) = 0.5e^{-1} \approx 0.184$ is a local maximum **25.** increasing on $(0, e)$; decreasing on (e, ∞); $h(e) = e^{-1} \approx 0.368$ is a local maximum **27.** increasing on $(-\infty, -2)$ and $(4, \infty)$; decreasing on $(-2, 1)$ and $(1, 4)$; $F(-2) = 1$ is a local maximum, and $F(4) = 13$ is a local minimum **29.** decreasing on $(-\infty, -3)$, $(-3, 1)$, and $(1, \infty)$; no local extrema **31.** increasing on $(1, 3)$; decreasing on $(-\infty, 1)$ and $(3, \infty)$; $H(1) = 0$ is a local minimum, and $H(3) = 4e^{-3} \approx 0.199$ is a local maximum

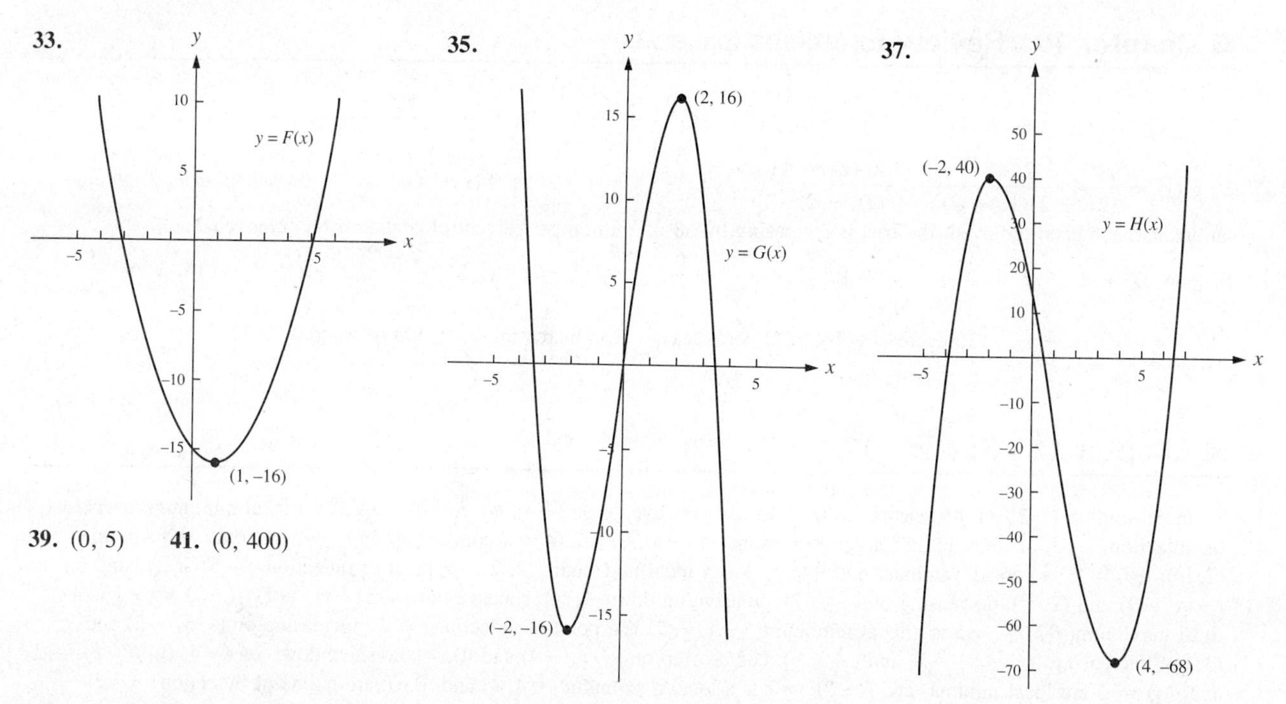

39. $(0, 5)$ **41.** $(0, 400)$

Section 11.3 (page 623)

1. concave up on $(-\infty, \infty)$; no points of inflection **3.** concave down on $(-\infty, \infty)$; no points of inflection **5.** concave up on $(\frac{1}{2}, \infty)$; concave down on $(-\infty, \frac{1}{2})$; $(0.5, -44.5)$ is a point of inflection **7.** concave up on $(-\infty, -2)$; concave down on $(-2, \infty)$; $(-2, -78)$ is a point of inflection **9.** concave up on $(-\infty, 0)$ and $(2, \infty)$; concave down on $(0, 2)$; $(0, -5)$ and $(2, -21)$ are points of inflection **11.** concave down on $(0, \infty)$; no points of inflection **13.** concave up on $(1, \infty)$; concave down on $(-\infty, 1)$; no points of inflection **15.** concave up on $(-\infty, 2 - 2\sqrt{2})$ and $(2 + \sqrt{2}, \infty)$; concave down on $(2 - \sqrt{2}, 2 + \sqrt{2})$; $(341, 0.38)$ and $(0.59, 0.19)$ are points of inflection **17.** $f(2) = -7$ is a local minimum **19.** $g(-2) = -48$ is a local minimum and $g(4) = 60$ is a local maximum **21.** $h(-4) = -120$ is a local mininum, $h(0) = 8$ is a local maximum, and $h(1) = 5$ is a local minimum **23.** $F(-3) = 162$ is a local maximum, and $F(3) = -162$ is a local minimum **25.** $G(2) = 0$ and $G(-8) = 0$ are local minima, and $G(-3) = 625$ is a local maximum **27.** $H(-6) = -37$ and $H(-1) = -5.75$ are local minima, and $H(-2) = -5$ is a local maximum **29.** $f(4) = e^{-4}$ is a local maximum **31.** $g(-3) \approx 1.97$ is a local maximum, and $g(-1) \approx 1.77$ is a local minimum

33.

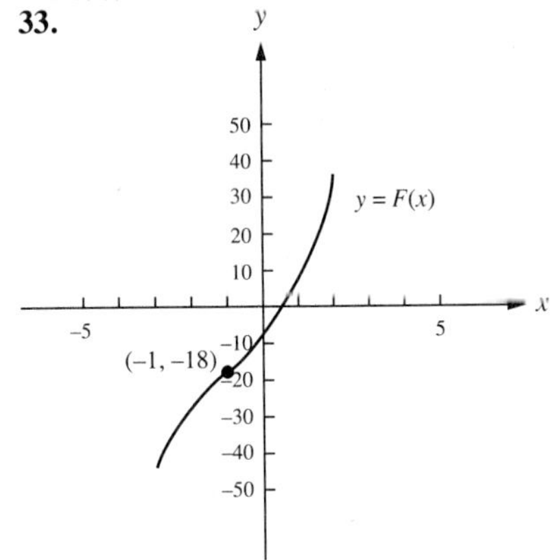

35.

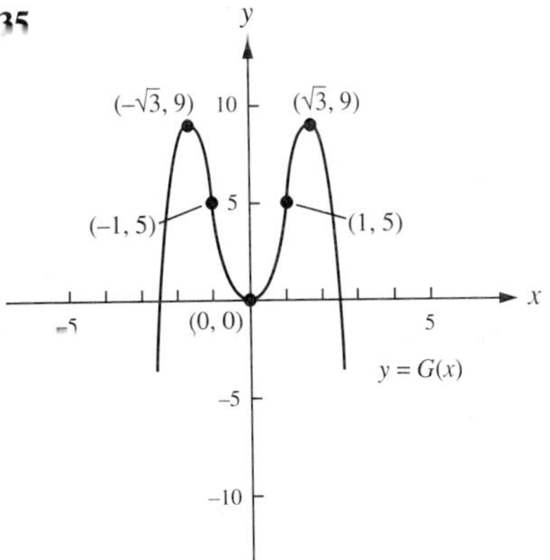

37.

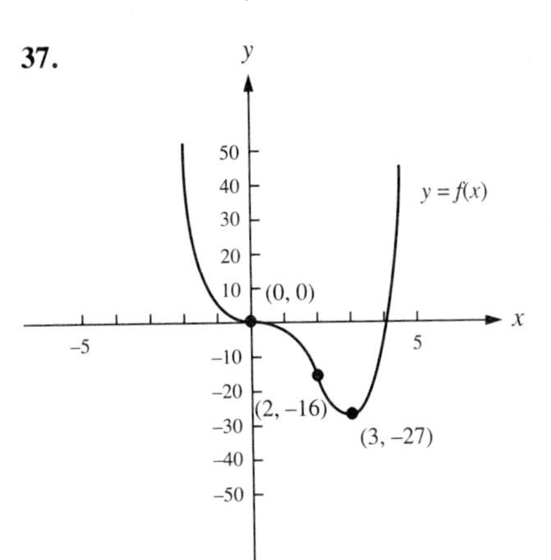

39.

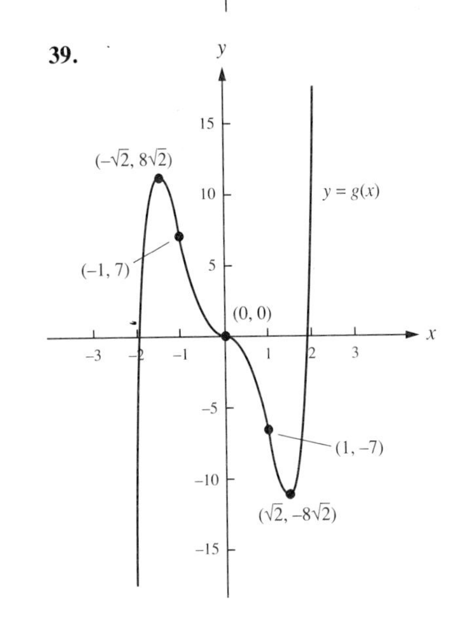

41. Since $f''(x) = 12x^2$ is never negative, f is concave up on $(-\infty, \infty)$.

43. (a) $f'(x) = 3ax^2 + 2bx + c$

$f''(x) = 6ax + 2b$

$f''(x) = 0$ if and only if $x = -\dfrac{b}{3a}$

Section 11.4 (page 633)

1.

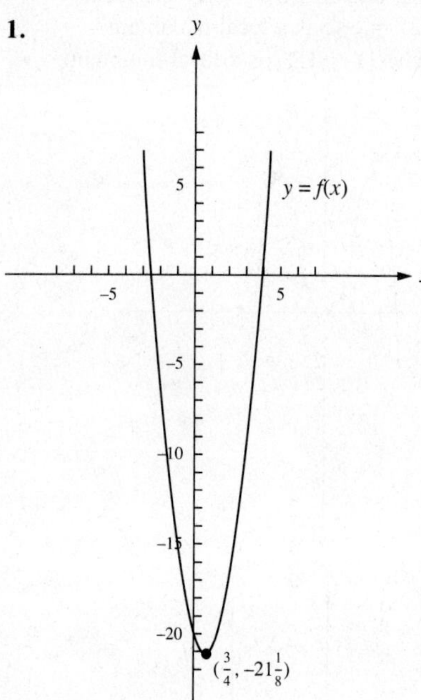

3.

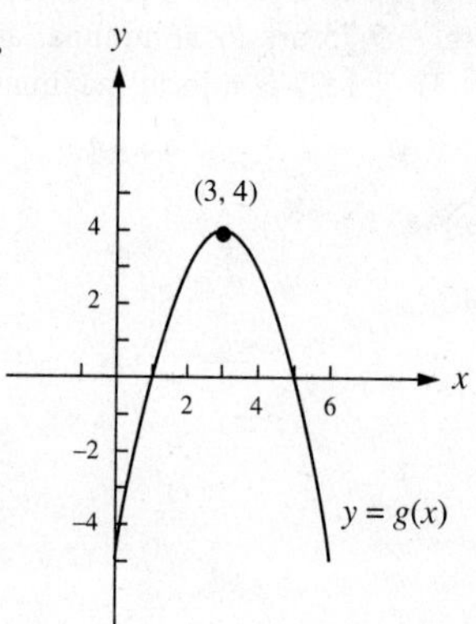

5.

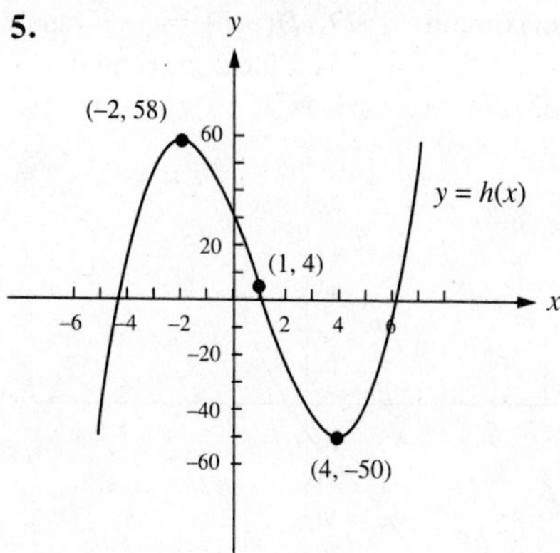

7.

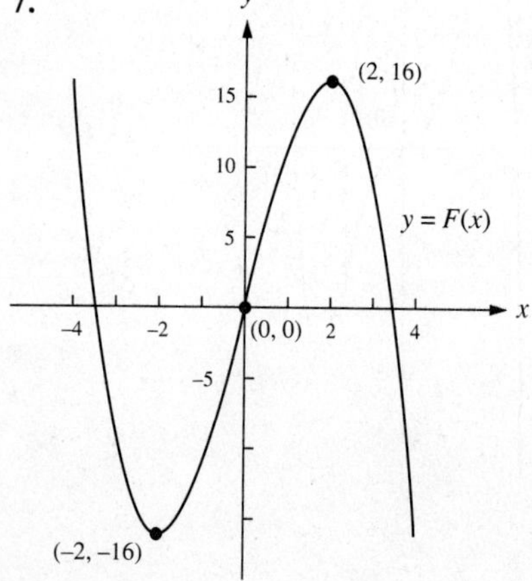

9.

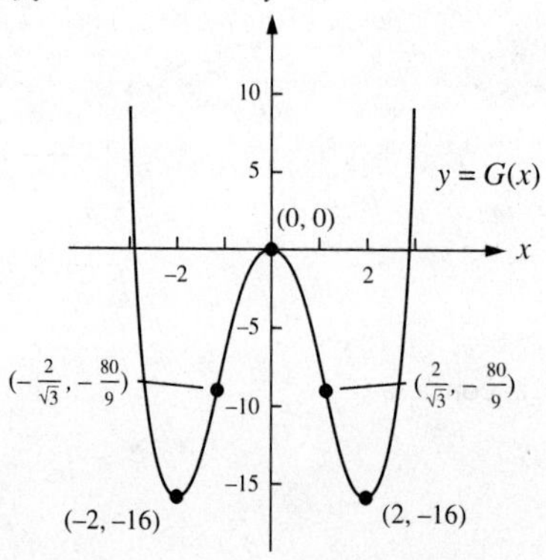

11.

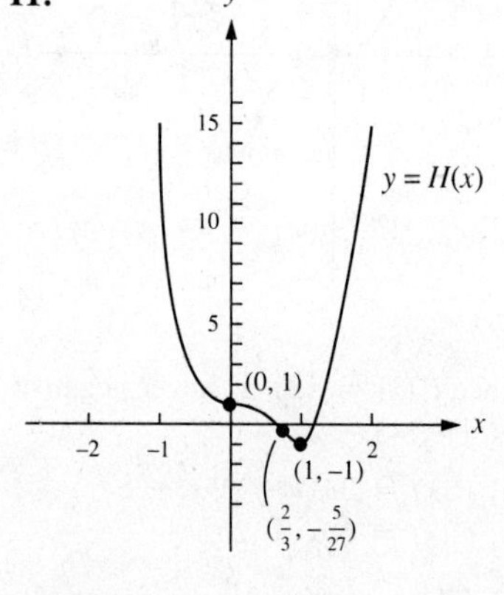

13.

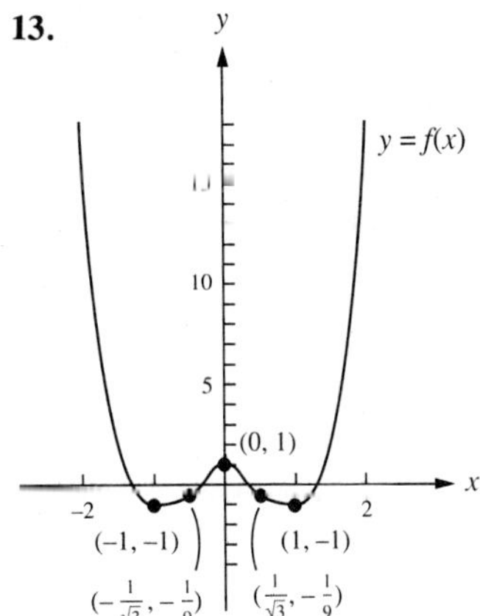

15.

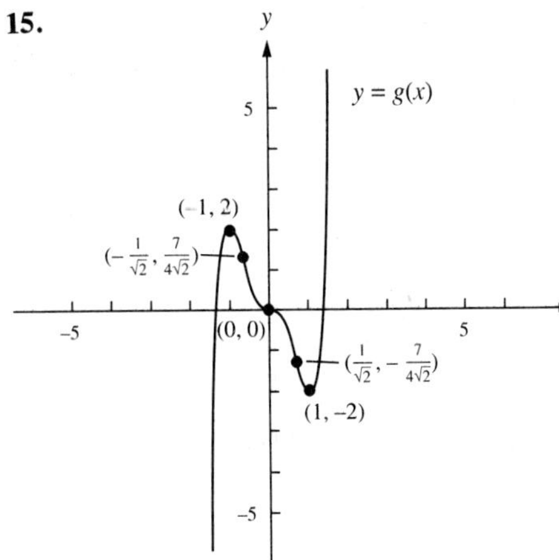

17.

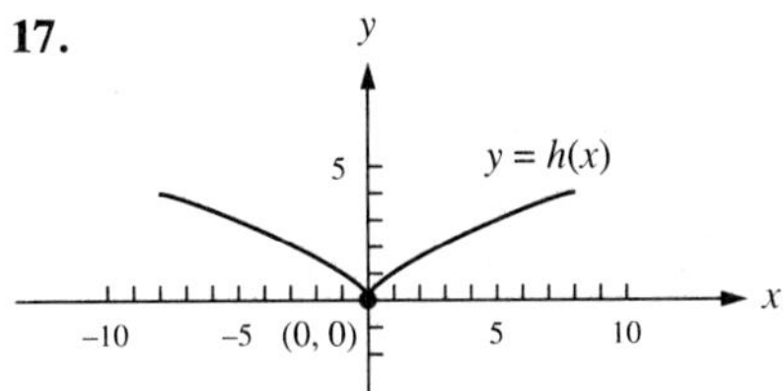

19.

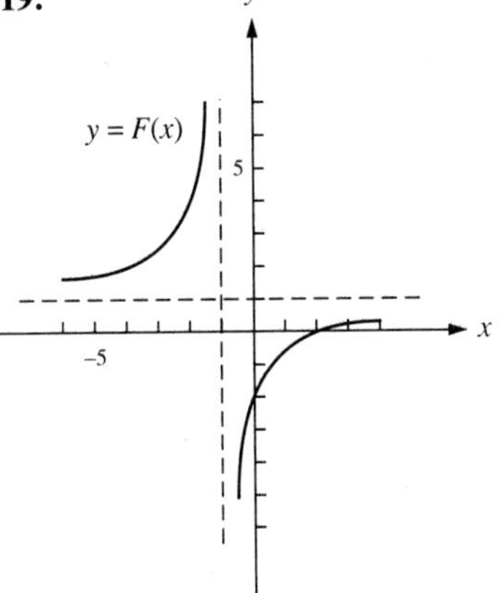

21.

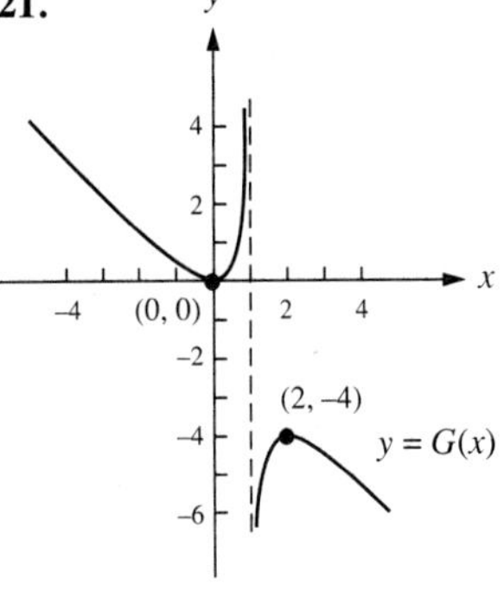

23.

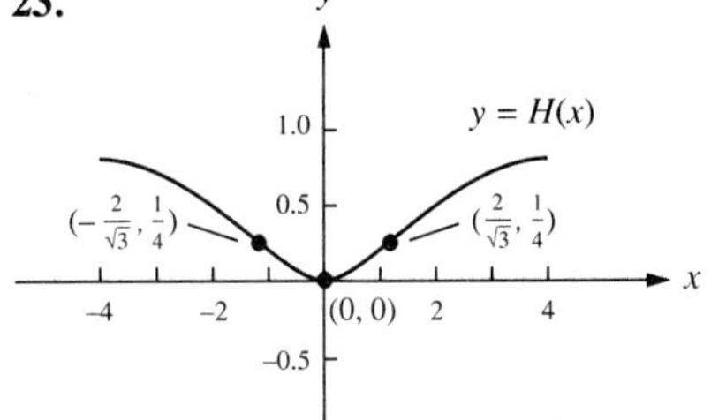

25.

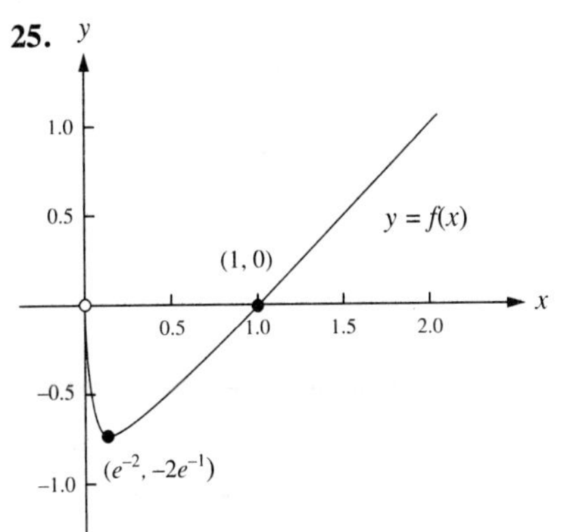

27.

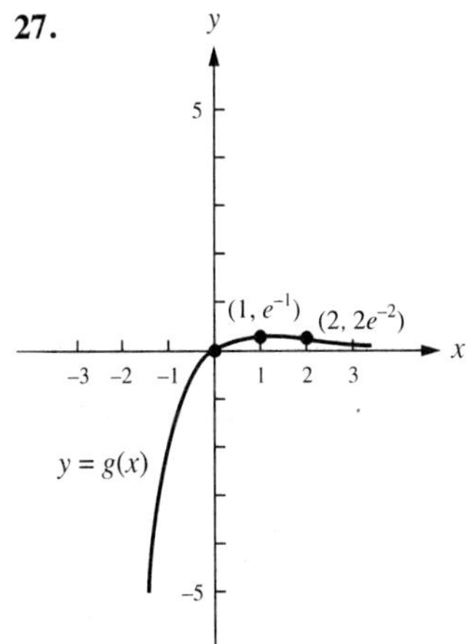

29.

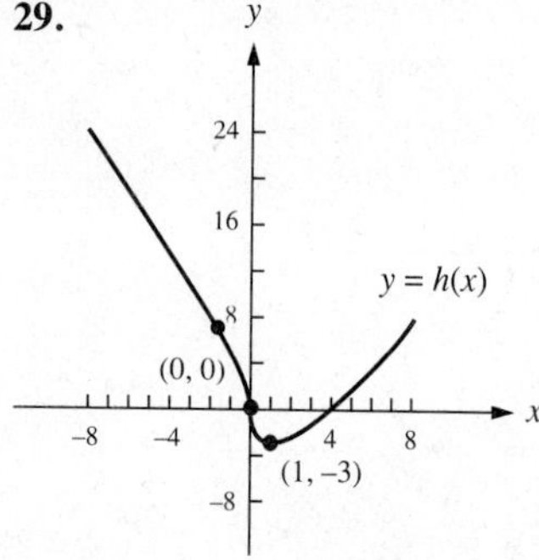

31.

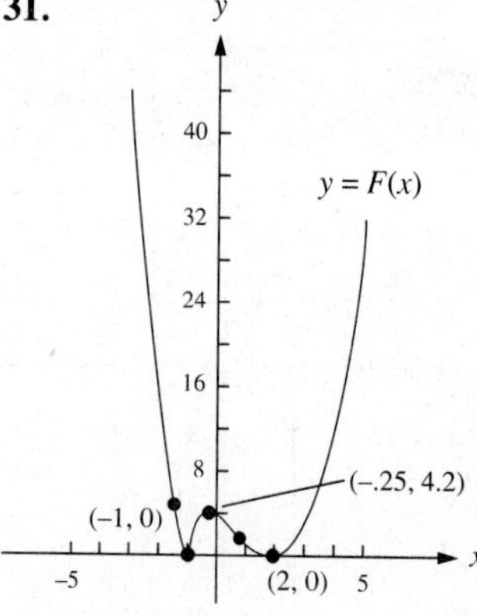

Section 11.5 (page 646)

1. $f(10) = 90$ is the absolute maximum, and $f(0) = 0$ is the absolute minimum **3.** $g(1/4) = 15.125$ is the absolute maximum, and $g(6) = -51$ is the absolute minimum **5.** $h(2) = 16$ is the absolute maximum, and $h(-1) = -8$ is the absolute minimum **7.** $F(3) = 1080$ is the absolute maximum, and $F(-1) = -360$ is the absolute minimum **9.** $G(1) = 0$ is the absolute maximum, and $G(2) = -1.5$ is the absolute minimum **11.** $h(1) = 0.5$ is the absolute maximum, and $h(0) = 0$ is the absolute minimum **13.** \$324 **15.** 18 inches by 18 inches by 36 inches **17.** $c/2$ **19.** 8 **21.** The radios should be priced at \$50 apiece. **23.** \$8.00 **25.** The generating station should be built at point G in the figure below.

27. 20 **29.** 8 inches by 8 inches by 4 inches **31.** 80 feet by 10 feet

33. height $= \sqrt[3]{\dfrac{4v}{\pi}}$ and radius $= \sqrt[3]{\dfrac{v}{2\pi}}$ **35.** 110 yards by $\dfrac{220}{\pi}$ yards

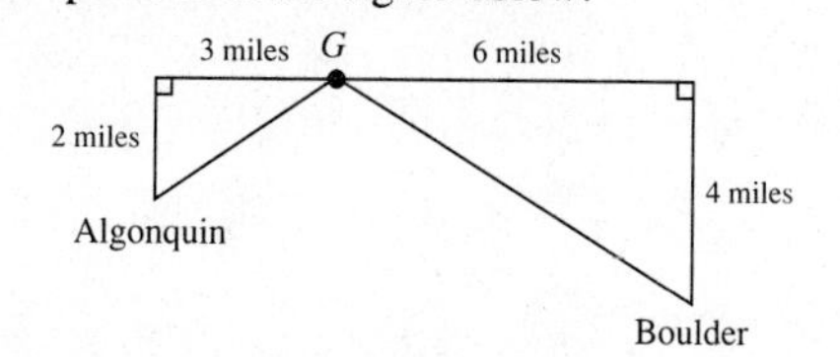

Section 11.6 (page 659)

1. $C'(x) = 0.00003x^2 - 0.024x + 20$, $R'(x) = 25$, and $P'(x) = 5 + 0.024x - 0.00003x^2$ **3.** $C'(x) = 0.0006x + 12 + \dfrac{1200}{\sqrt{x+1}}$, $R'(x) = 132$, and $P'(x) = 120 - 0.0006x - \dfrac{1200}{\sqrt{x+1}}$ **5.** $C'(x) = 0.002x + 10 + \dfrac{2000}{x+1}$, $R'(x) = 15$, and $P'(x) = 5 - 0.002x - \dfrac{2000}{x+1}$

7. $C'(x) = 0.0006x + 15 + \dfrac{5400}{x+1}$, $R'(x) = 20$, and $P'(x) = 5 - 0.0006x - \dfrac{5400}{x+1}$ **9.** The maximum profit of \$41 is obtained by making 9 copies. **11.** The maximum profit of \$6125 is obtained by making 8500 wallets. **13.** The maximum profit of \$56,250 is obtained by making 2250 pairs of speakers. **15.** The maximum profit of \$101,250 is obtained by making 8250 sets.

17. (a) $0.00003x^2 - 0.024x + 20 = 0.00001x^2 - 0.012x + 20 + \dfrac{10{,}000}{x}$ (c) no

19. (a) $0.000006x^2 - 0.024x + 120 = 0.000002x^2 - 0.012x + 120 + \dfrac{30{,}000}{x}$ (c) yes

Section 11.7 (page 666)

1. 600 **3.** 800 **5.** 300 **7.** \$30,000 **9.** \$50,000 **11.** \$24,000 **13.** after 9 years **15.** after 6 years **17.** after 10 years **19.** $\eta = \dfrac{p}{50 - p}$ **21** $\eta = 1$ **23.** $R = px$. Hence $\dfrac{dR}{dp} = p\dfrac{dx}{dp} + x = -\eta x + x = x(1 - \eta) > 0$ if $\eta < 1$. **25.** If $\eta = 1$, then a rise in price leaves revenue constant.

Chapter 11 Review Exercises (page 669)

1. increasing on $(-\infty, -2)$ and $(4, \infty)$; decreasing on $(-2, 4)$; concave up on $(1, \infty)$; concave down on $(-\infty, 1)$; $(1, -38)$ is a point of inflection **2.** increasing on $(-5, 1/3)$; decreasing on $(-\infty, -5)$ and $(1/3, \infty)$; concave up on $(-\infty, -7/3)$; concave down on $(-7/3, \infty)$; $(-7/3, -893/27)$ is a point of inflection **3.** increasing on $(-\infty, -1)$ and $(0, 1)$; decreasing on $(-1, 0)$ and $(1, \infty)$; concave up on $(-\sqrt{3}/3, \sqrt{3}/3)$; concave down on $(-\infty, -\sqrt{3}/3)$ and $(\sqrt{3}/3, \infty)$; $(-\sqrt{3}/3, 5/9)$ and $(\sqrt{3}/3, 5/9)$ are points of inflection **4.** increasing on $(-3, \infty)$; decreasing on $(-\infty, -3)$; concave up on $(-\infty, -2)$ and $(0, \infty)$; concave down on $(-2, 0)$; $(0, 0)$ and $(-2, -16)$ are points of inflection **5.** $f(2 + 2\sqrt{2}) = 30 + 32\sqrt{2}$ is a local maximum, and $f(2 - 2\sqrt{2}) = 30 - 32\sqrt{2}$ is a local minimum **6.** $F(-3) = 162$ is a local maximum, and $F(3) = -162$ is a local minimum **7.** $g(1) = 1/2$ is a local maximum and $g(0) = 0$ is a local minimum
8. $G(1) = e^{-1}$ is a local maximum

9.

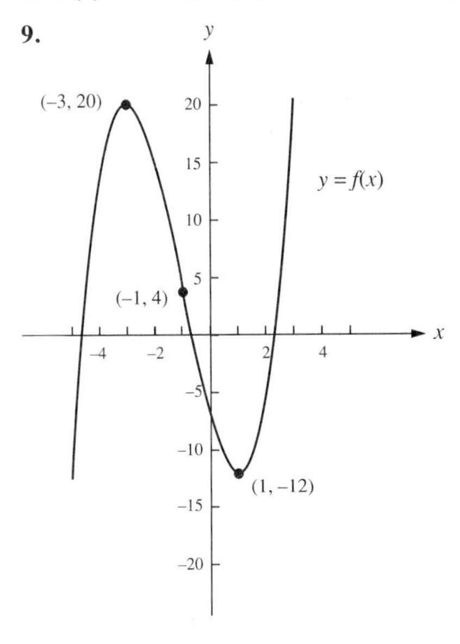

10.

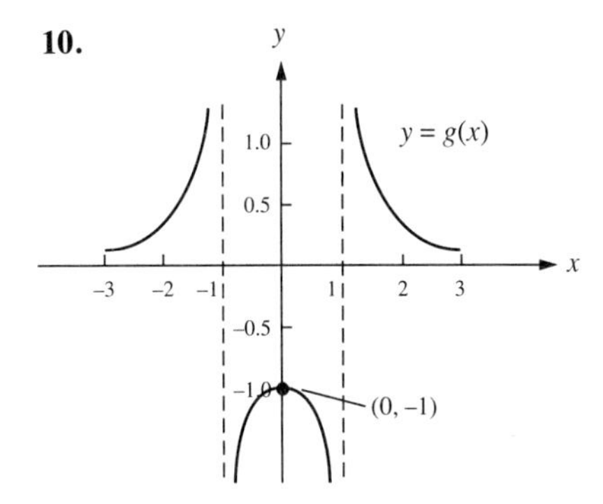

11.

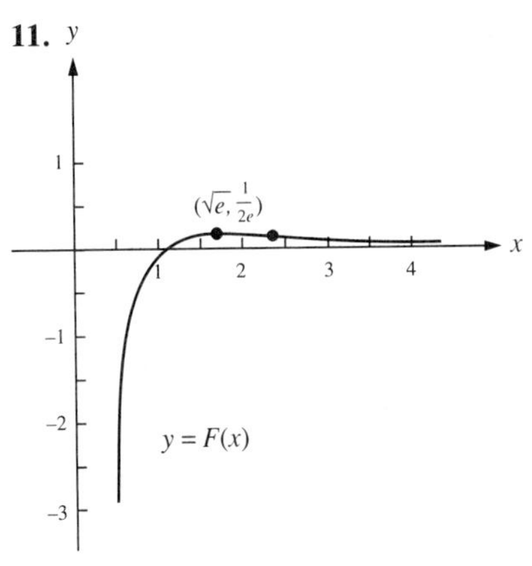

12.

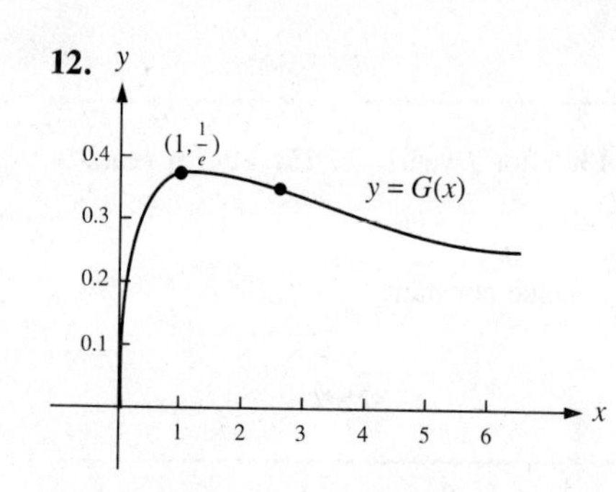

13. radius of $\frac{36}{\pi}$ inches and length of 36 inches **14.** \$107,600

15. (a) $C'(x) = 0.0003x^2 - 0.008x + 12$, $R'(x) = 15.60 - 0.002x$, $P'(x) = 3.60 + 0.006x - 0.0003x^2$ (b) 120 watches

16. (a) \$60 (b) $A(x) = 0.000002x^2 - 0.012x + 78 + \frac{20{,}000}{x}$

(c) $0.000006x^2 - 0.024x + 78 = 0.000002x^2 - 0.012x + 78 + \frac{20{,}000}{x}$

(d) Yes **17.** 1600 **18.** \$20,000

19. after 14 years

Chapter 12 Section 12.1 (page 678)

1. (a) yes (b) no **3.** (a) no (b) no **5.** (a) no (b) no **7.** (a) yes (b) no **9.** $\frac{x^8}{8} + k$

11. $-\frac{x^{-4}}{4} + k$ **13.** $5x^{8/5} + k$ **15.** $\frac{3x^{4/3}}{4} + k$ **17.** $\frac{x^2}{2} + 3x + k$ **19.** $\frac{3x^4}{4} + \frac{2x^{3/2}}{3} + \frac{5x^2}{2} + k$

21. $8x^{1/2} - 5 \ln |x| + k$ **23.** $\frac{e^{3x}}{3} + 2x + k$ **25.** $\frac{x^4}{4} + 3x - \ln |x| + k$ **27.** $-x^{-1} + \ln |x| + x^{-2} + k$

29. $\frac{4}{3}x^3 + 2x^2 + x + k$ **31.** $\frac{3x^2}{2} + 5x - \frac{9}{2}$ **33.** $2 \ln |x| + 2\sqrt{x} + 1$ **35.** $\ln |x| + \frac{3x^2}{2} + 1 - \ln 2$

37. $C(x) = 60x^2 + \frac{.0001x^3}{3} + 200{,}000$, \$60,233, 333.33 **39.** 18 feet, 7.07 seconds **41.** $s(x) = \frac{x^2}{1440}$, $\frac{5}{72}$ miles, 37.95 seconds

Section 12.2 (page 686)

1. $\int \frac{1}{6} u \, du$ **3.** $\int \frac{1}{4} e^u \, du$ **5.** $\int (-u^5) \, du$ **7.** $\int \left(-\frac{1}{2} e^u\right) du$ **9.** $e^{x+1} + k$ **11.** $\ln |x + 2| + k$

13. $\frac{2}{3}\sqrt{3x - 2} + k$ **15.** $\frac{(\ln x)^4}{4} + k$ **17.** $-e^{1-x} + k$ **19.** $(\ln x)^2 + k$ **21.** $\frac{(3x^2 + 6x)^{3/2}}{9} + k$

23. $-\frac{2}{9}(2 - 3x)^{3/2} + k$ **25.** $-1/(2e^{x^2}) + k$ **27.** $\frac{(\sqrt{x} - 1)^4}{2} + k$ **29.** $2e^{\sqrt{x}} + k$

31. $y = \frac{1}{2} \ln (x^2 + 1) + k$, $y = \frac{1}{2} \ln (x^2 + 1) + 2$ **33.** $F(x) = \frac{3}{8}(x^2 + 2)^{4/3} - \frac{3}{8}(38)^{4/3}$

35. $C(x) = .0004x^2 + 50x + 5000 \ln |2x + 1| + 15{,}000$

Section 12.3 (page 695)

1. 1, 1.5, 2, 2.5, 3, 3.5, 4, 4.5 **3.** −2, −1.25, −.5, .25 **5.** .5 **7.** 28/27 **9.** 25381/60060 ≈ .4226

11. 15 **13.** 9 **15.** 24 **17.** 34.5 **19.** 42 **21.** $\frac{\pi}{2}$ **23.** $\frac{9\pi}{4}$ **25.** $x_{i-1} = 1 + \frac{2(i - 1)}{n}$, $\Delta x = \frac{2}{n}$

29. 6

Section 12.4 (page 706)

1. 2 **3.** 12 **5.** 260 **7.** 5/3 **9.** $-\ln 3$ **11.** $\frac{1}{3}(10^{3/2} - 2^{3/2})$ **13.** -7.5 **15.** 0
17. $\frac{1}{3}(e^{-4} - e^{-16})$ **19.** $\frac{1}{2}[(\ln 5)^2 - (\ln 2)^2]$ **21.** 312.6 **23.** $e - e^{-1}$ **25.** $\frac{1}{2}\ln 3$ **27.** $e^3 - e^{[illegible]}$ [illegible]
29. $\frac{32}{3}$ **31.** $\frac{8}{3}$

Section 12.5 (page 715)

1. \$25,200 **3.** \$843.33 **5.** \$595 **7.** $1000 + 600(e^{-2} - e^{-1})$ **9.** $\frac{98}{3}$ **11.** 10.5 **13.** 1024 **15.** $\frac{32}{3}$
17. $\frac{4}{3}$ **19.** 4.5 **21.** $\frac{1}{6}$ **23.** $4\sqrt{3} - 2\ln(\sqrt{3} + 2)$ **25.** $x_E = 2000$, $p_E = \$21$ **27.** $x_E = 300$, $p_E = \$600$
29. \$2000, \$4000 **31.** \$45,000, \$180,000 **33.** \$500, \$250 **35.** \$1,280,000, \$640,000

Chapter 12 Review Exercises (page 718)

1. $x^3 - 2\ln|x| + 10x^{1/2} + k$ **2.** $\frac{e^{7x}}{7} + k$ **3.** $\frac{(2x^3 + 5)^{3/2}}{9} + k$ **4.** $\ln|x^2 + x| + k$ **5.** 45/4
6. $\frac{1}{3}(e^{27} - 1)$ **7.** $\frac{1}{2}\ln 2001 + 110{,}000 \approx \$110{,}003.80$ **8.** $F(x) = -e^{3-x} + 5 + e^2$ **9.** 1200 feet
10. 6.75 **11.** $\frac{26}{3}$ **12.** $\frac{57}{64}$ **13.** $\frac{4}{3}$ **14.** \$399.9995 **15.** $x_E = 200$, $p_E = \$2600$, \$106,666.67, \$60,000

Chapter 13 Section 13.1 (page 727)

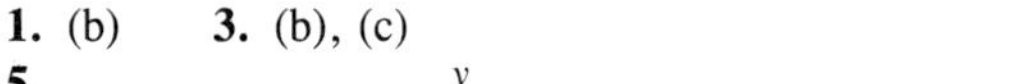

1. (b) **3.** (b), (c)

5. **7.**

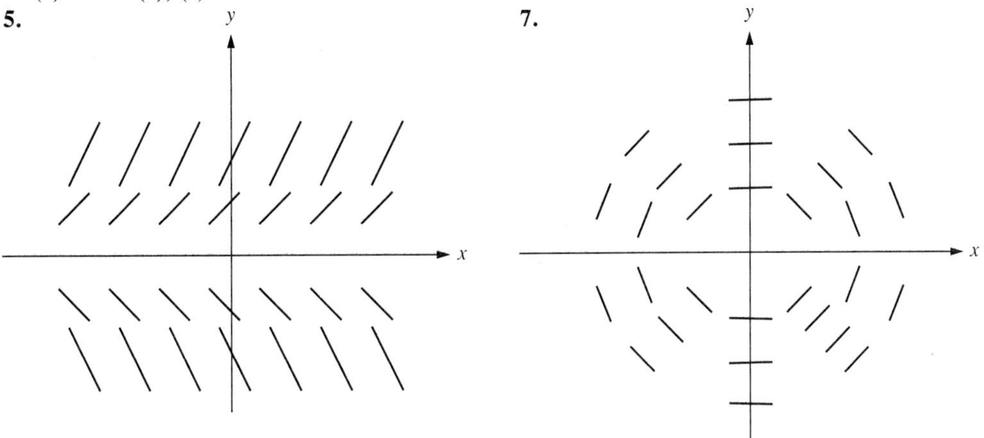

9. $(y^{-1} + 1)\,dy = 3x^{-1}\,dx$ **11.** $y^{-1}\,dy = (x - \ln x)\,dx$ **13.** not separable **15.** $y = \frac{1}{3}e^{3x} + k$
17. $y = \frac{1}{2}e^{x^2} + k$ **19.** $y = \frac{1}{2}(\ln x)^2 + k$ **21.** $y = Ce^{-5x}$ **23.** $y = \frac{C}{x}$ **25.** $y = (2e^x + k)^{1/2}$
27. $y = C(x^2 + 1)^{1/2}$ **29.** $y = -\frac{2}{3}(2 - x)^{3/2} + \frac{14}{3}$ **31.** $y = 4e^{2x}$ **33.** $\frac{2}{3}(1 + 2e^{-3x^2/2})$
35. $y = -2(x^3 + 1)^{1/3}$

37. $y = e^x$

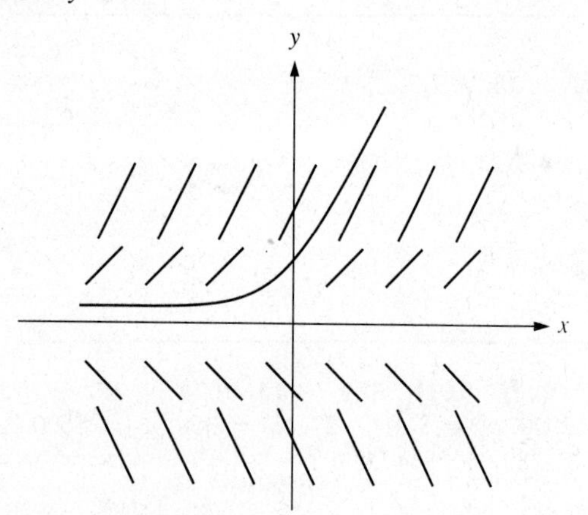

39. $x^2 + y^2 = 4, y \neq 0$

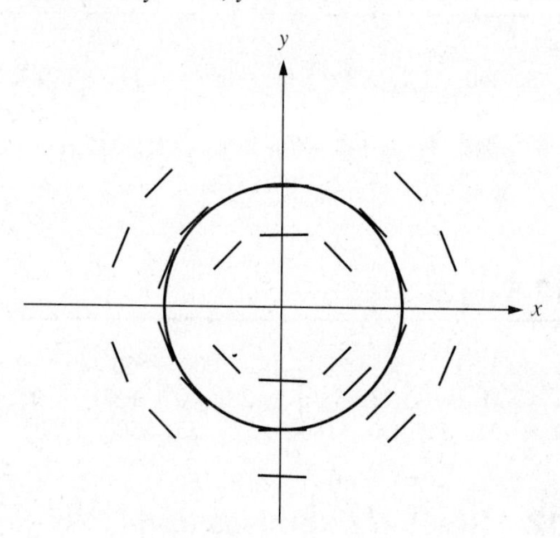

Section 13.2 (page 736)

1. $y = 100{,}000e^{.15x}$, 448,169, 15.4 years **3.** $y = 200e^{(\ln .5/24.5)x}$, .00655 lb, after 81.4 days **5.** 3.1 minutes **7.** 8.95 million, 17.3 years **9.** $y = 10{,}000 - 9{,}500e^{-.05x}$, 2601, 12.8 years **11.** 698,000, after 2.12 years **13.** after 5.49 hours **15.** 4.55 million **17.** about halfway through 1995 **19.** $y = 10{,}000/(1 + 19e^{-.5x})$, 3907 **21.** $y = 30/(1 + 9e^{-.9x})$, 3.21 years

Section 13.3 (page 744)

1. $\frac{1}{324}$ **3.** $\frac{1}{3}e^{-3}$ **5.** diverges **7.** diverges **9.** $\frac{1}{3}$ **11.** 2 **13.** $\frac{1}{\ln 2}$ **15.** $\frac{1}{4}$ **17.** $\frac{1}{21}$ **19.** diverges **21.** diverges **23.** $\frac{7}{2}$ **25.** $15,000 **27.** 50 kg **29.** $12,000 per year for 20 years **31.** $\frac{1}{2}$ **33.** $\ln 3 - \ln 2$ **35.** $f(x) = x$

Section 13.4 (page 749)

1. (a) $u = x, dv = e^{-x}\,dx$ (b) $-e^{-x}\,dx$ **3.** (a) $u = x, dv = (x + 1)^{2/3}\,dx$ (b) $\frac{3}{5}(x + 1)^{5/3}\,dx$ **5.** $\frac{1}{24}x(3x - 1)^8 - \frac{1}{648}(3x - 1)^9 + k$ **7.** $\frac{2}{5}x(5x + 4)^{1/2} - \frac{4}{75}(5x + 4)^{3/2} + k$ **9.** $x \ln (4x) - x + k$ **11.** $\frac{2}{3}x^{3/2} \ln x - \frac{4}{9}x^{3/2} + k$ **13.** $\frac{1}{2}e^{x^2}(x^2 - 1) + k$ **15.** $-\frac{1}{3}x^2(9 - x^2)^{3/2} - \frac{2}{15}(9 - x^2)^{5/2} + k$ **17.** $\frac{1}{5}x - \frac{2}{25}\ln (5x + 2) + k$ **19.** $\frac{1}{4}(7e^8 - 3e^4)$ **21.** $\frac{2}{105}(5^{7/2} - 1)$ **23.** 319574/21 **25.** $-x^2(x + 2)^{-1} - 4 \ln (x + 2) + 2(x + 2) + k$ **27.** $x[(\ln |x|)^3 - 3(\ln |x|)^2 + 6 \ln |x| - 6] + k$

Section 13.5 (page 754)

1. formula 36 **3.** formula 18 **5.** formula 11 **7.** $-\ln\left|\dfrac{1 + \sqrt{1 + x^2}}{x}\right| + k$ by formula 33

9. $\dfrac{1}{2}\left(\sqrt{9 - x^2} - 3\ln\left|\dfrac{3 + \sqrt{9 - x^2}}{x}\right|\right) + k$ by formula 24 **11.** $\ln|\ln x| + k$ by formula 42

13. $\dfrac{1}{2\sqrt{8}}\ln\left|\dfrac{\sqrt{8} + x}{\sqrt{8} - x}\right| + k$ by formula 23 **15.** $2\sqrt{3 - t} + \sqrt{3}\ln\left|\dfrac{\sqrt{3 - t} - \sqrt{3}}{\sqrt{3 - t} + \sqrt{3}}\right| + k$ by formula 22

17. $1 + \dfrac{1}{9}\ln\dfrac{4}{13}$ by formula 8 **19.** 7/300 by formula 26 **21.** $\frac{1}{3}\sqrt{2 + 3x^2}$ by the substitution $u = 2 + 3x^2$

23. $-2 + x + 4(2 - x)^{-1} + 4\ln|2 - x| + k$ by formula 10

25. $\dfrac{2}{5}x^2\sqrt{x + 1} - \dfrac{8}{15}(x - 2)\sqrt{x + 1} + k$ by formulas 19 and 18

Chapter 13 Review Exercises (page 756)

1. $y = \dfrac{1}{4}(\ln x)^2 + 4$ **2.** $y = \dfrac{10}{3}e^{3x} - \dfrac{1}{3}$ **3.** $y = \sqrt{-2x^{-1} + 6}$ **4.** $y = 4e^{-e}e^{e^x} + 1$

5. $y = 600{,}000e^{.06x}$, 676,498, after 8.5 years **6.** 47.8 lbs **7.** 45.4 million **8.** 2.33 million

9. $y = 40/(1 + 7e^{-1.2x})$, 24.5 million **10.** $\frac{1}{8}$ **11.** diverges **12.** 1 **13.** 0 **14.** $1,666,666.67

15. 1000 tons **16.** $\dfrac{5}{8}x(x + 2)^{8/5} - \dfrac{25}{104}(x + 2)^{13/5} + k$ **17.** $\dfrac{1}{5}x^5\ln|x| - \dfrac{1}{25}x^5 + k$ **18.** $\dfrac{e^{3x}}{9}(3x - 1) + k$

19. $-\dfrac{1}{21} + \dfrac{1}{4}(\ln 7 - \ln 3)$ **20.** $\dfrac{1}{15}(6x + 4)(x - 1)^{3/2} + k$ **21.** $-\sqrt{1 - x^2} + k$ **22.** $\dfrac{\sqrt{4x^2 + 1}}{x} + k$

23. $\dfrac{1}{3}\ln\left|\dfrac{x}{3 + 2x}\right| + k$

Chapter 14 Section 14.1 (page 765)

1. (a), (b) **3.** (a) **5.** (b) **7.** (a), (b) **9.** 0, $\frac{2}{3}$ **11.** $\frac{1}{2}$, $\frac{9}{16}$ **13.** $\frac{3}{4}$, $\frac{5}{8}$ **15.** $\frac{7}{8}$, $\frac{1}{27}$ **17.** $\frac{1}{4}$ **19.** $\frac{1}{10}$

21. $\frac{1}{20}$ **23.** $\dfrac{\sqrt{3}}{6}$ **25.** $\sqrt{2}$ **27.** $f(x) = \begin{cases} .1 \text{ if } 0 \le x \le 10 \\ 0 \text{ otherwise} \end{cases}$, $P(x \ge 6) = .4$

29. $f(x) = \begin{cases} .2 \text{ if } 0 \le x \le 5 \\ 0 \text{ otherwise} \end{cases}$, $P(x \le 3) = .6$ **31.** $f(x) = \begin{cases} 1/30 \text{ if } 0 \le x \le 15 \\ 1/60 \text{ if } 15 < x < 45, \\ 0 \text{ otherwise} \end{cases}$ $P(x > 20) = 7/12$

Section 14.2 (page 778)

1. $1 - e^{-15}$ **3.** $e^{-.4}$ **5.** $-\frac{1}{2}\ln .7$ **7.** $-\ln .9$ **9.** $\frac{7}{8}$ **11.** $\frac{1}{8}$ **13.** $4e^{-3}$ **15.** $1 - \dfrac{13}{8}e^{-.5}$

17. $-\ln .3$ **19.** $\ln 10$ **21.** .25 **23.** .0625 **25.** $.8 - .2\ln 5 - .1(\ln 5)^2 \approx .219$

27. $.9 - .1\ln 10 - .05(\ln 10)^2 \approx .405$ **29.** $P(x = k) = 2^{k-1}/3^k$, $P(x \ge 5) = \frac{16}{81}$

31. $P(x = k) = (k - 1)5^{k-2}/6^k$, $P(x \le 5) = \frac{1526}{7776}$ **33.** $P(x = k) = 1/2^{k-1}$ for $k \ge 2$, $P(x > 3) = \frac{1}{4}$

Section 14.3 (page 788)

1. 3 **3.** 2/ln 3 **5.** $\frac{4}{5}$ **7.** $\frac{3}{2}$ **9.** $\frac{4}{3}$ **11.** $4(\ln 3 - 1)/(\ln 3)^2$ **13.** $\frac{2}{75}$ **15.** $\frac{3}{4}$ **17.** 1, 3, $\sqrt{3}$
19. -2.5, 25/12, $5\sqrt{3}/6$ **21.** $\frac{1}{3}, \frac{1}{9}, \frac{1}{3}$ **23.** 10, 100, 10 **25.** 5, 25, 5 **27.** $\frac{1}{5}, \frac{1}{25}, \frac{1}{5}$ **29.** 12 minutes
31. $e^{-1.5} \approx .223$ **33.** $6e^{-5} \approx .04$ **35.** $1 - e^{-2 \ln 4/3} \approx .603$ **37.** $(\ln 1.5)^2/3 \approx .055$ **39.** 18.75 minutes

Section 14.4 (page 796)

1. $\lambda = 30$, $\mu = 20$ **3.** $\lambda = \frac{1}{15}$, $\mu = \frac{1}{20}$ **5.** $\lambda = \frac{1}{3}$, $\mu = \frac{1}{6}$ **7.** $\rho = \frac{8}{9}$, $N = 8$, $N_q = \frac{64}{9}$, $T = 10$, $T_q = \frac{80}{9}$ **9.** $\rho = \frac{2}{3}$, $N = 2$, $N_q = \frac{4}{3}$, $T = 12$, $T_q = 8$ **11.** $N = 1$, $N_q = \frac{1}{2}$, $T = 20$ minutes, $T_q = 10$ minutes
13. $N = \frac{5}{3}$, $N_q = \frac{25}{24}$, $T = 1\frac{1}{3}$ hours, $T_q = 50$ minutes **15.** $N = 3$, $N_q = \frac{9}{4}$, $T = 20$ minutes, $T_q = 15$ minutes
17. $N = 4$, $N_q = \frac{16}{5}$, $T = \frac{4}{3}$ months, $T_q = \frac{16}{15}$ months **19.** $N = \frac{2}{3}$, $N_q = \frac{4}{15}$, $T = 20$ minutes, $T_q = 8$ minutes

Chapter 14 Review Exercises (page 799)

1. $\frac{1}{8}$ **2.** $\frac{7}{27}$ **3.** $\frac{1}{2}$ **4.** $\frac{3}{13}$ **5.** $1 - e^{-12}$ **6.** $-(\ln .3)/2$ **7.** $13e^{-4}$ **8.** $-\ln .8$ **9.** $e^{-2.5 \ln 2}$
10. $(1 + 3 \ln .75)/4 \approx .0342$ **11.** $3 \cdot 7^{k-1}/10^k$, $\frac{49}{100}$ **12.** $\frac{3}{5}, \frac{12}{175}, \sqrt{12/175} \approx .262$ **13.** $\frac{5}{3}, \frac{20}{9}, 2\sqrt{5}/3$
14. 3, $\frac{4}{3}$, $2/\sqrt{3}$ **15.** $\frac{1}{6}, \frac{1}{36}, \frac{1}{6}$ **16.** 3, 9, 3 **17.** $e^{-5/4}$ **18.** $e^{-3/2}$
19. $N = 3$, $N_q = \frac{9}{4}$, $T = 1$ week, $T_q = \frac{3}{4}$ week **20.** $N = 2$, $N_q = \frac{4}{3}$, $T = 2$ minutes, $T_q = 1.5$ minutes

Chapter 15 Section 15.1 (page 806)

1. -6 **3.** 64 **5.** -63 **7.** 1600 **9.** -540 **11.** -12 **13.** 5400 **15.** 120
17. 1.78 square meters **19.** \$662.69 **21.** \$1180 **23.** \$19,835 **25.** 48 km

Section 15.2 (page 817)

1. $f_x(x, y) = 4x$ and $f_y(x, y) = 6y$ **3.** $h_r(r, s) = \dfrac{1}{s}$ and $h_s(r, s) = -\dfrac{r}{s^2}$ **5.** $G_u(u, v) = -3v^2e^{-u}$ and $G_v(u, v) = 6ve^{-u}$ **7.** $g_x(x, y) = \dfrac{3x^2y + 10xy^2}{(3x + 5y)^2}$ and $g_y(x, y) = \dfrac{3x^3}{(3x + 5y)^2}$ **9.** $F_r(r, s) = 5(2r + 3s)(r^2 + 3rs - 8s^2)^4$, $F_s(r, s) = 5(3r - 16s)(r^2 + 3rs - 8s^2)^4$ **11.** $H_u(u, v) = 18u^2 - 14uv^3 + 24v^2$ and $H_v(u, v) = 48uv - 140v^4 - 21u^2v^2$ **13.** $f_x(x, y, z) = 2yz^2 + \dfrac{2x}{x^2 + 3y^2}$, $f_y(x, y, z) = 2xz^2 + \dfrac{6y}{x^2 + 3y^2}$, and $f_z(x, y, z) = 4xyz$
15. $h_r(r, s, t) = (2r + 4s + 1)e^{2r+3st}$, $h_s(r, s, t) = (3rt + 6st + 2)e^{2r+3st}$, and $h_t(r, s, t) = (3rs + 6s^2)e^{2r+3st}$
17. $G_u(u, v, w) = -\dfrac{3v^2}{(3u - 2w)^2}$, $G_v(u, v, w) = \dfrac{2v}{3u - 2w}$, and $G_w(u, v, w) = \dfrac{2v^2}{(3u - 2w)^2}$
19. $f_{rr}(r, s) = 10s - 18rs^6$, $f_{rs}(r, s) = 10r - 54r^2s^5$, $f_{sr}(r, s) = 10r - 54r^2s^5$, and $f_{ss}(r, s) = -90r^3s^4$
21. $h_{uu}(u, v) = (2u^3 - 12u^2 + 12u)e^{v-u}$, $h_{uv}(u, v) = (6u^2 - 2u^3)e^{v-u}$, $h_{vu}(u, v) = (6u^2 - 2u^3)e^{v-u}$, and $h_{vv}(u, v) = 2u^3e^{v-u}$
23. $G_{xx}(x, y) = -\dfrac{2}{x^2}$, $G_{xy}(x, y) = 0$, $G_{yx}(x, y) = 0$,and $G_{yy}(x, y) = -\dfrac{1}{y^2}$ **25.** (a) $f_x(25, 100) = 50$ and $f_y(25, 100) =$

12.5 (b) Production increases by about 12.5 units. (c) Production decreases by about 50 units. **27.** $D_x(150, 100) = -12$, and $D_y(150, 100) = 4$. When the average selling price of a house increases from \$150,000 to \$151,000, the demand for housing decreases by 12. When the average selling price of a condominium increases from \$100,000 to \$101,000, the demand for houses increases by 4. **29.** (a) $P(x, y) = 10x + 14y - 0.2xy$ (b) $P_x(18, 12) = 7.6$ and $P_y(18, 12) = 10.4$. When production and sale of basic phones increases from 18,000 units to 19,000 units and production and sale of deluxe phones is 12,000 units, profit increases by \$7600. When production and sale of basic phones is 18,000 units and production and sale of deluxe phones increases from 12,000 to 13,000 units, profit increases by \$10,400. **31.** If the first manufacturer increases the selling price from p to $p + 1$, then demand for his product will decrease. On the other hand, an increase from q to $q + 1$ in the selling price of the second manufacturer's product will increase demand for the first manufacturer's product.

Section 15.3 (page 828)

1. $(-4, 3)$ **3.** $(-1, -\frac{1}{2})$, $(-1, 0)$, $(-1, \frac{1}{2})$, $(1, -\frac{1}{2})$, $(1, 0)$, $(1, \frac{1}{2})$ **5.** $(3, 1)$ **7.** $(0, 0)$, $(1, 3)$ **9.** none **11.** $g(4, -3) = 101$ is a local maximum **13.** $h(-2, -3) = -52$ is a local minimum **15.** $F(0, 3) = -9$ is a local minimum **17.** $G(-2, 0) = 9$ is a local maximum **19.** $H(0, -1) = 4$ and $H(0, 1) = 4$ are local maxima; $H(-2, 0) = -13$ and $H(2, 0) = -13$ are local minima **21.** $f(2, 2) = 8$ is a local maximum **23.** $g(3, 1) = 9$ is a local minimum **25.** $h(-3, 1) = 3/e$ is a local minimum **27.** none **29.** \$12 and \$18, respectively **31.** $x = 800$, $y = 600$

Section 15.4 (page 841)

1. $r \approx .874$
$y = 2x + 10$

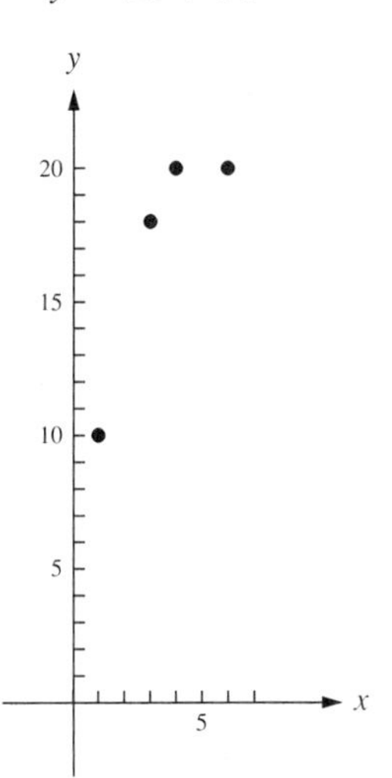

3. $r \approx -.998$
$y = -\frac{8}{3}x + \frac{67}{3}$

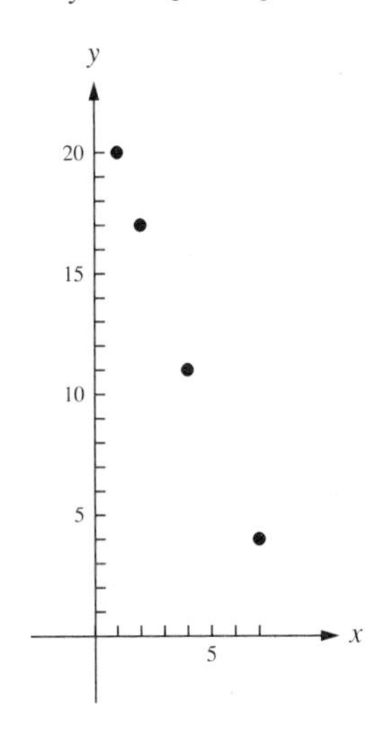

5. $r \approx .983$
$y = 1.5x + 4$

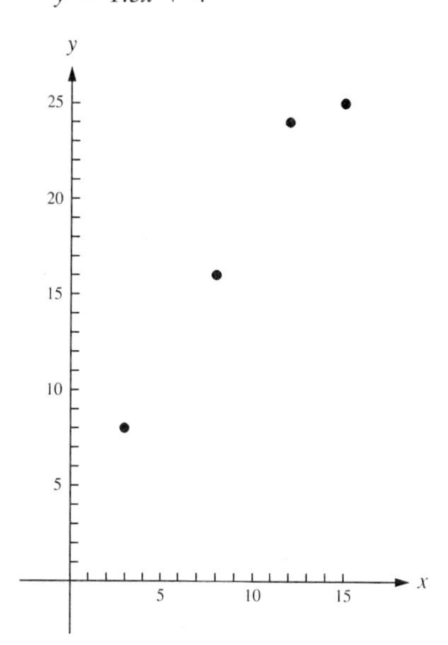

7. $r \approx -.996$
$y = -3x + 34$

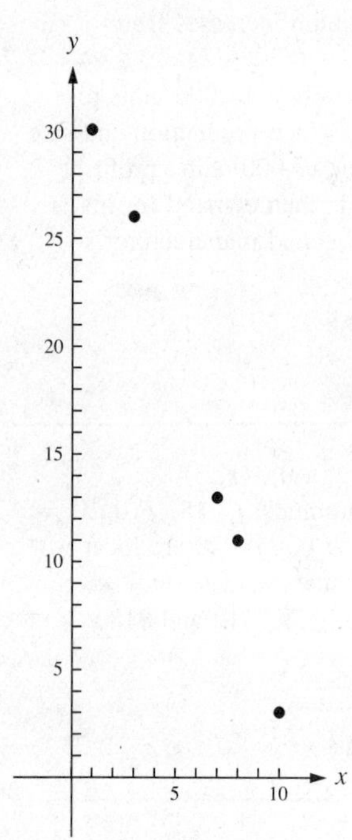

9. $r \approx .992$
$y = 1.6x + 3.2$

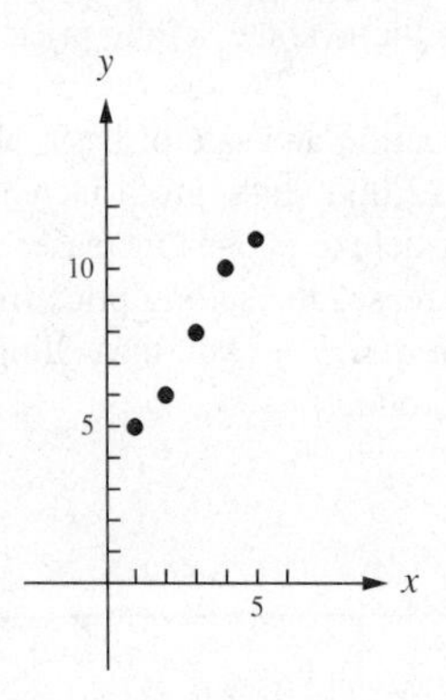

11. (a) $y = 4x + 78.5$ (b) 90.5°
13. $r \approx .926$; $y = 9.23x + 114.72$
15. (a) $y = 83.69x + 660.11$
(b) $r \approx .994$ (c) \$743.8 billion

Section 15.5 (page 853)

1. (4, −5) **3.** (−2, 3) **5.** (−3, 1) **7.** (−8, 4) and (8, 4) **9.** (−4, −1) and (4, 1) **11.** (−6, 3) **13.** $f(10, -5) = 750$ is a maximum, and $f(6, 3) = 270$ is a minimum **15.** $f(3, -3) = -18$ **17.** $f(-3, -3) = f(3, 3) = 18$ **19.** $f(-8, 4) = f(8, -4) = 80$ **21.** $f(-1, 1, 3) = -3$ **23.** $f(1, 3, -2) = -14$ **25.** $f(2, 2, 2) = 64$ **27.** 300 units of labor and 200 units of capital **29.** 18 inches by 18 inches by 36 inches **31.** 65 small tables and 50 large tables

Section 15.6 (page 862)

1. $3x^4y^2$ **3.** $3xy - 2.5y^2 + xy^2$ **5.** $\frac{1}{5}x^5y - \frac{1}{3}x^3y^3$ **7.** $2e^{xy}$ **9.** 0 **11.** ⅔ **13.** 14 **15.** $e^2 - 3$ **17.** −34.5 **19.** ln ⁴⁄₃ **21.** 44 **23.** 14 **25.** 208/3 **27.** 0 **29.** 532/9 **31.** $3e - 3$

Chapter 15 Review Exercises (page 864)

1. 20 **2.** 2 **3.** $f_x(x, y) = (2x + 3) \ln y$ and $f_y(x, y) = \dfrac{x^2 + 3x}{y}$ **4.** $F_x(x, y) = (x^2y + 2x)e^{xy}$ and $F_y(x, y) = x^3e^{xy}$ **5.** $g_{xx}(x, y) = -4y$, $g_{xy}(x, y) = 12y^3 - 4x$, $g_{yx}(x, y) = 12y^3 - 4x$, and $g_{yy}(x, y) = 36xy^2$ **6.** $G_{xx}(x, y) =$

$120x^3y^3 - 24xy^8$, $G_{xy}(x, y) = 90x^4y^2 - 96x^2y^7$, $G_{yx}(x, y) = 90x^4y^2 - 96x^2y^7$, and $G_{yy}(x, y) = 36x^5y - 214x^3y^6$
7. $(-2, 4)$ **8.** $(0, 2)$ **9.** $f(4, -5) = 69$ is a local maximum **10.** $g(-7, 4) = -92$ is a local minimum
11. $F(-1/2, -1/2) = -1/8$ and $F(1/2, 1/2) = -1/8$ are local minima **12.** none **13.** $f(6, 3) = 72$ is a minimum
14. $f(3, 5) = 59$ is a maximum **15.** $f(2, -2) = 8$ is a maximum and $f(-2, 2) = -8$ is a minimum
16. $f(-2, 4) = 18$ is a maximum and $f(0, 0) = -2$ is a minimum
17. $r \approx .997$; $y = \frac{8}{7}x + \frac{4}{7}$ **18.** $r \approx .825$; $y = 0.84x + 0.61$

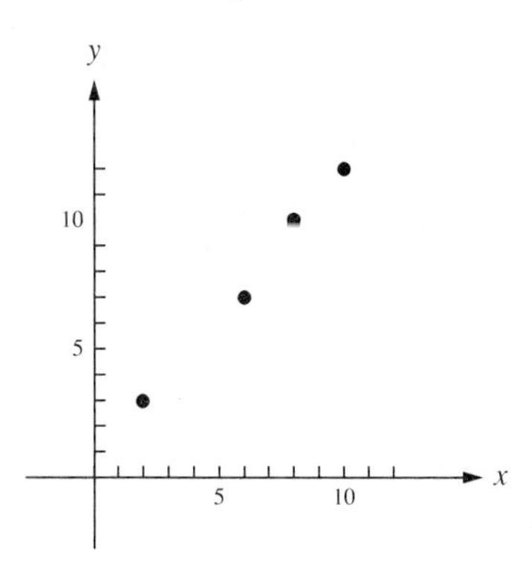

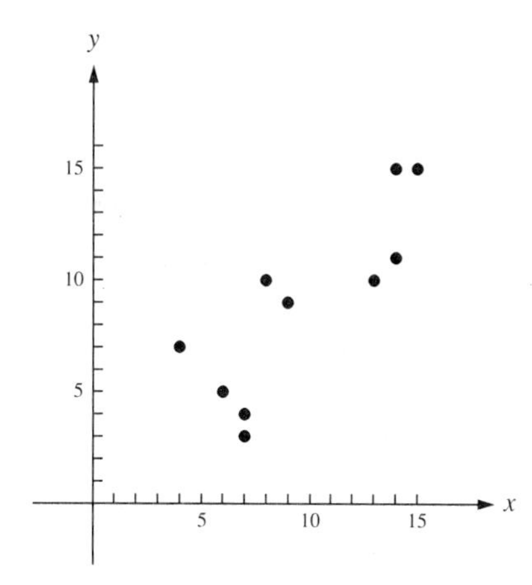

19. $3x^2y^2 - 3x^5y^4$ **20.** $2x^3y - 2x^6y^3$ **21.** 6 **22.** 18 **23.** 9 **24.** $2e - 2$ **25.** \$37.50 and \$50
26. $x = 500$, $y = 400$ **27.** 1 foot by 1 foot by 2 feet

Appendix Section A.1 (page 871)

1. 3 **3.** 7 **5.** 2 **7.** 8 **9.** $\frac{23}{21}$ **11.** 45, (A5) **13.** 3, 1 **15.** 24, (A4) **17.** $\frac{7}{6}$, $\frac{3}{5}$ **19.** 4, 1
21. yes **23.** no **25.** no **27.** yes **29.** no **31.** *N, Z, Q, R* **33.** *Z, Q, R* **35.** *R* **37.** *Q, R*
39. *Z, Q, R*

Section A.2 (page 875)

1. $x = 6$ **3.** $A = \frac{7}{3}$ **5.** $x = -2$ **7.** $u = -\frac{5}{4}$ **9.** $x = 5/2$ **11.** $x = 3$ **13.** $x = -\frac{7}{4}$
15. $x = 16$ **17.** $t = 42/5$ **19.** $x = 12/43$ **21.** 75 **23.** 10

Section A.3 (page 879)

1. false **3.** false **5.** true **7.** true **9.** false **11.** 3 5
13. −2 0 4 **15.** 0 2

17. $3x \geq 7x - 9$ **19.** $2x \leq 7$ **21.** $x > 10$ **23.** $x > -\frac{9}{2}$ **25.** $-x < 3$ **27.** $x > 5$ **29.** $x < -4$
31. $x > \frac{8}{3}$ **33.** $x \geq -\frac{3}{2}$ **35.** $x < \frac{1}{3}$ **37.** $x \geq -\frac{12}{5}$ **39.** $x > 71\frac{3}{7}$ **41.** $x \leq 20,000$

Section A.4 (page 882)

1. 3.6 **3.** -3 **5.** -2 **7.** 4, 2, 2, 2, 4, 6, 8 **9.** 1, 0, 1, 2, 1, 0, 1 **11.** $\{3, -3\}$ **13.** $\{-5, 9\}$

Section A.5 (page 884)

1. 27 **3.** $\frac{1}{4}$ **5.** $\frac{4}{9}$ **7.** 16 **9.** $\frac{1}{16}$ **11.** 54 **13.** v^4/u^3 **15.** s^6 **17.** ab^3 **19.** xy^2z^3
21. X^4/Y^4Z^{12}

Section A.6 (page 889)

1. 1 **3.** 2 **5.** 1 **7.** 0 **9.** 8 **11.** 4 **13.** -5 **15.** undefined **17.** $\frac{3}{2}$ **19.** $-.5$
21. undefined **23.** 7 **25.** -2 **27.** -2 **29.** .5 **31.** undefined **33.** 625 **35.** -243
37. undefined **39.** $\frac{1}{4}$ **41.** 6 **43.** 2 **45.** 1.4 **47.** undefined **49.** $3\sqrt{2}$ **51.** $\sqrt{13}$ **53.** 5
55. 9 **57.** $\sqrt[6]{7}$ **59.** $|x|$ **61.** $\sqrt{6}$ **63.** $11^{7/10}$ **65.** undefined **67.** undefined

Section A.7 (page 894)

1. $x = 2, -3$ **3.** $x = -\frac{3}{2}$ **5.** $x = \frac{1}{3}, \frac{2}{3}$ **7.** $x = 0, \frac{3}{2}$ **9.** no solutions **11.** $x = -3, 2$
13. $x = -1 \pm \sqrt{6}$ **15.** $x = \frac{1}{3}, -\frac{3}{2}$ **17.** no solutions **19.** $x = \frac{5}{2}$ **21.** $x = 2\sqrt{2}, \sqrt{2}$
23. $x = 0, 3, -\frac{1}{3}$ **25.** $x = 2, 3$

Section A.8 (page 899)

1. yes **3.** no **5.** yes **7.** 5, 4, 9 **9.** 1, -5, 85 **11.** -5, 5, 50 **13.** $\{0, \frac{1}{3}\}$ **15.** $\{0, 1, -3\}$
17. $\{\frac{3}{2}\}$ **19.** $\{2, -2\}$ **21.** $\{-1, 1\}$ **23.** no solutions **25.** no solutions **27.** $\{-2, 2\}$
29. $\{1, -1, 2, -2\}$ **31.** $\{-3, \sqrt[4]{2}, -\sqrt[4]{2}\}$ **33.** $x^2 - 2x + 1$ **35.** $3x^2 + 6x + 10$ **37.** $x^3 + 2x - 1$
39. $2x^2 + 6x - 2$ **41.** $x = 1, -3$ **43.** $x = -2, -3, 4$ **45.** $x = 2, -\frac{1}{2}$ **47.** $x = \frac{1}{2}, \frac{1}{2}(-3 + \sqrt{5}),$
$\frac{1}{2}(-3 - \sqrt{5})$

SOLUTIONS TO PRACTICE PROBLEMS

Chapter 1

Problem 1 If $3x - 4(0) = 6$, then

$$3x = 6$$
$$x = 2,$$

so the x-intercept is $(2, 0)$.

If $3(0) - 4y = 6$, then

$$-4y = 6$$
$$y = -\tfrac{6}{4} = -\tfrac{3}{2},$$

so the y-intercept is $(0, -3/2)$.

Problem 2 Since $m = -2$ and $(x_1, y_1) = (5, 3)$, we have

$$y - 3 = -2(x - 5)$$
$$y - 3 = -2x + 10$$
$$y = -2x + 13.$$

Problem 3 Let $(x_1, y_1) = (1, 3)$ and $(x_2, y_2) = (2, -5)$. Then

$$m = \frac{-5 - 3}{2 - 1} = -8.$$

Thus the equation of the line is

$$y - 3 = -8(x - 1)$$
$$y - 3 = -8x + 8$$
$$y = -8x + 11.$$

Problem 4 Multiplying the first equation by -3 gives

$$\begin{aligned} -3x - 9y &= 15 \\ 3x - 2y &= 7. \end{aligned}$$

Adding yields

$$\begin{aligned} -11y &= 22 \\ y &= -2. \end{aligned}$$

Substituting this in the first equation gives

$$\begin{aligned} x + 3(-2) &= -5 \\ x - 6 &= -5 \\ x &= 1. \end{aligned}$$

The point of intersection is $(1, -2)$.

Problem 5 Suppose the man buys x bags of apples and y bags of pears. Then the total weight is

$$6x + 4y = 112,$$

and the total cost is

$$4x + 3y = 79.$$

We multiply the first equation by 2 and the second by -3.

$$\begin{aligned} 12x + 8y &= 224 \\ -12x - 9y &= -237 \end{aligned}$$

Adding gives

$$\begin{aligned} -y &= -13 \\ y &= 13. \end{aligned}$$

We substitute this in the first equation.

$$\begin{aligned} 6x + 4(13) &= 112 \\ 6x + 52 &= 112 \\ 6x &= 60 \\ x &= 10 \end{aligned}$$

He bought 10 bags of apples and 13 bags of pears.

Problem 6 (a) We have

$$\begin{aligned} R &= .50x \\ C &= .10x + 44, \end{aligned}$$

where x is the number of glasses sold. To break even we need $R = C$, or

$$.50x = .10x + 44$$
$$.40x = 44$$
$$x = \frac{44}{.40} = 110 \text{ glasses.}$$

(b) The profit is

$$P = R - C = .50x - (.10x + 44)$$
$$= .40x - 44.$$

(c) Setting $P = \$100$ gives

$$.40x - 44 = 100$$
$$.40x = 144$$
$$x = \frac{144}{.40} = 360 \text{ glasses.}$$

Problem 7 Let x wigs be sold per week. Then

$$R = 60x$$
$$C = 25x + 21{,}000.$$

The company breaks even when $R = C$, or

$$60x = 25x + 21{,}000$$
$$35x = 21{,}000$$
$$x = \frac{21{,}000}{35} = 600 \text{ wigs.}$$

Problem 8 (a) If x_0 is the equilibrium price, then

$$10{,}000x_0 - 7000 = -8000x_0 + 20{,}000$$
$$18{,}000x_0 = 27{,}000$$
$$x_0 = \frac{27{,}000}{18{,}000} = 1.5 = \$1.50.$$

(b) The number of chips supplied at \$1.50 is

$$10{,}000(1.50) - 7000 = 8000.$$

Problem 9 Under a 10% sales tax the price paid by the consumer will be $x + .10x$, and so the demand will be

$$D_1 = 1200 - 150(x + .10x) = 1200 - 165x.$$

Under a 20¢ stamp tax the consumer will pay $x + .20$, and so the demand will be

$$D_2 = 1200 - 150(x + .20) = 1170 - 150x.$$

Chapter 2

Problem 1

$$\left[\begin{array}{rrr|r} 1 & 2 & 5 & 0 \\ 0 & \textcircled{3} & -6 & 9 \\ 0 & -1 & -1 & 2 \end{array}\right]$$

$$\left[\begin{array}{rrr|r} 1 & 2 & 5 & 0 \\ 0 & \textcircled{1} & -2 & 3 \\ 0 & -1 & -1 & 2 \end{array}\right] \quad (\tfrac{1}{3}R_2)$$

$$\left[\begin{array}{rrr|r} 1 & 0 & 9 & -6 \\ 0 & 1 & -2 & 3 \\ 0 & 0 & -3 & 5 \end{array}\right] \quad \begin{array}{l} (R_1 + (-2)R_2) \\ \\ (R_3 + R_2) \end{array}$$

Problem 2

$$\left[\begin{array}{rrr|r} 3 & 0 & 1 & 0 \\ \textcircled{1} & 1 & 1 & 4 \\ 0 & 3 & -1 & 3 \end{array}\right] \qquad \left[\begin{array}{rrr|r} 0 & \textcircled{-3} & -2 & -12 \\ 1 & 1 & 1 & 4 \\ 0 & 3 & -1 & 3 \end{array}\right] \quad (R_1 + (-3)R_2)$$

$$\left[\begin{array}{rrr|r} 0 & \textcircled{1} & \frac{2}{3} & 4 \\ 1 & 1 & 1 & 4 \\ 0 & 3 & -1 & 3 \end{array}\right] \quad (-\tfrac{1}{3}R_1) \qquad \left[\begin{array}{rrr|r} 0 & 1 & \frac{2}{3} & 4 \\ 1 & 0 & \frac{1}{3} & 0 \\ 0 & 0 & \textcircled{-3} & -9 \end{array}\right] \quad \begin{array}{l} \\ (R_2 + (-1)R_1) \\ (R_3 + (-3)R_1) \end{array}$$

$$\left[\begin{array}{rrr|r} 0 & 1 & \frac{2}{3} & 4 \\ 1 & 0 & \frac{1}{3} & 0 \\ 0 & 0 & \textcircled{1} & 3 \end{array}\right] \quad (-\tfrac{1}{3}R_3) \qquad \left[\begin{array}{rrr|r} 0 & 1 & 0 & 2 \\ 1 & 0 & 0 & -1 \\ 0 & 0 & 1 & 3 \end{array}\right] \quad \begin{array}{l} (R_1 + (-\frac{2}{3})R_3) \\ (R_2 + (-\frac{1}{3})R_3) \\ \\ \end{array}$$

$$\left[\begin{array}{rrr|r} 1 & 0 & 0 & -1 \\ 0 & 1 & 0 & 2 \\ 0 & 0 & 1 & 3 \end{array}\right] \quad \begin{array}{l} (R_2) \\ (R_1) \\ \\ \end{array}$$

Thus $x = -1$, $y = 2$, $z = 3$.

Problem 3 Let there be x toasters, y blenders, and z mixers. Then

$$\begin{aligned} x + y + z &= 12 \\ 6x + 5y + 3z &= 57 \\ 20x + 15y + 15z &= 205. \end{aligned}$$

We give the result of each pivot.

$$\left[\begin{array}{rrr|r} \textcircled{1} & 1 & 1 & 12 \\ 6 & 5 & 3 & 57 \\ 20 & 15 & 15 & 205 \end{array}\right] \qquad \left[\begin{array}{rrr|r} 1 & 1 & 1 & 12 \\ 0 & \textcircled{-1} & -3 & -15 \\ 0 & -5 & -5 & -35 \end{array}\right]$$

$$\left[\begin{array}{rrr|r} 1 & 0 & -2 & -3 \\ 0 & 1 & 3 & 15 \\ 0 & 0 & \textcircled{10} & 40 \end{array}\right] \qquad \left[\begin{array}{rrr|r} 1 & 0 & 0 & 5 \\ 0 & 1 & 0 & 3 \\ 0 & 0 & 1 & 4 \end{array}\right]$$

Thus $x = 5$, $y = 3$, and $z = 4$.

Problem 4

$$\left[\begin{array}{ccc|c} \textcircled{1} & 2 & -1 & 3 \\ 2 & 0 & 1 & 2 \\ 4 & 4 & -1 & 6 \end{array}\right] \qquad \left[\begin{array}{ccc|c} 1 & 2 & -1 & 3 \\ 0 & \textcircled{-4} & 3 & -4 \\ 0 & -4 & 3 & -6 \end{array}\right] \quad \begin{array}{l} \\ (R_2 + (-2)R_1) \\ (R_3 + (-4)R_1) \end{array}$$

$$\left[\begin{array}{ccc|c} 1 & 2 & -1 & 3 \\ 0 & \textcircled{1} & -\frac{3}{4} & 1 \\ 0 & -4 & 3 & -6 \end{array}\right] \quad (-\tfrac{1}{4}R_2) \qquad \left[\begin{array}{ccc|c} 1 & 0 & \frac{1}{2} & 1 \\ 0 & 1 & -\frac{3}{4} & 1 \\ 0 & 0 & 0 & -2 \end{array}\right] \quad \begin{array}{l} (R_1 + (-2)R_2) \\ \\ (R_3 + 4R_2) \end{array}$$

Since the last row has the form $0\ 0\ .\ .\ .\ 0 \mid a$, with $a = -2 \neq 0$, there is no solution.

Problem 5

$$\left[\begin{array}{ccc|c} 1 & 2 & 1 & 3 \\ 0 & \textcircled{1} & -1 & -2 \end{array}\right] \qquad \left[\begin{array}{ccc|c} 1 & 0 & 3 & 7 \\ 0 & 1 & -1 & -2 \end{array}\right] \quad (R_1 + (-2)R_2)$$

The last tableau corresponds to

$$\begin{aligned} x \quad + 3z &= 7 \\ y - z &= -2. \end{aligned}$$

Thus $(x, y, z) = (-3z + 7, z - 2, z)$, where z is any real number.

Problem 6

Suppose she buys x standard, y large, and z jumbo rolls.

$$\begin{aligned} x + y + z &= 30 \\ 9x + 12y + 18z &= 357 \end{aligned}$$

$$\left[\begin{array}{ccc|c} \textcircled{1} & 1 & 1 & 30 \\ 9 & 12 & 18 & 357 \end{array}\right] \qquad \left[\begin{array}{ccc|c} 1 & 1 & 1 & 30 \\ 0 & \textcircled{3} & 9 & 87 \end{array}\right] \quad (R_2 + (-9)R_1)$$

$$\left[\begin{array}{ccc|c} 1 & 1 & 1 & 30 \\ 0 & \textcircled{1} & 3 & 29 \end{array}\right] \quad (\tfrac{1}{3}R_2) \qquad \left[\begin{array}{ccc|c} 1 & 0 & -2 & 1 \\ 0 & 1 & 3 & 29 \end{array}\right] \quad (R_1 + (-1)R_2)$$

The corresponding equations are $x - 2z = 1$, $y + 3z = 29$. Thus she must buy $1 + 2z$ standard rolls and $29 - 3z$ large rolls.

Problem 7

(a) Calories provided $= 70x + 80y = 1090$,
grams of protein provided $= 2x + 6y = 46$.

(b)

$$\left[\begin{array}{cc|c} 70 & 80 & 1090 \\ \textcircled{2} & 6 & 46 \end{array}\right] \qquad \left[\begin{array}{cc|c} 70 & 80 & 1090 \\ \textcircled{1} & 3 & 23 \end{array}\right] \quad (\tfrac{1}{2}R_2)$$

$$\left[\begin{array}{cc|c} 0 & \textcircled{-130} & -520 \\ 1 & 3 & 23 \end{array}\right] \quad (R_1 + (-70)R_2) \qquad \left[\begin{array}{cc|c} 0 & \textcircled{1} & 4 \\ 1 & 3 & 23 \end{array}\right] \quad (-\tfrac{1}{130}R_1)$$

$$\left[\begin{array}{cc|c} 0 & 1 & 4 \\ 1 & 0 & 11 \end{array}\right] \quad (R_2 + (-3)R_1)$$

The solution $y = 4$, $x = 11$ tells us that 11 slices of bread and 4 eggs should be provided per week.

Problem 8 (a) From the first two rows of the table

$$x_1 = 400 + 0.2x_1 + 0.1x_2$$
$$x_2 = 210 + 0.2x_1 + 0.2x_2.$$

(b)

$$\begin{aligned} 0.8x_1 - 0.1x_2 &= 400 \\ -0.2x_1 + 0.8x_2 &= 210 \end{aligned} \qquad \left[\begin{array}{cc|c} .8 & -.1 & 400 \\ -.2 & .8 & 210 \end{array}\right]$$

$$\left[\begin{array}{cc|c} 0 & 3.1 & 1240 \\ 1 & -4 & -1050 \end{array}\right] \qquad \left[\begin{array}{cc|c} 0 & 1 & 400 \\ 1 & 0 & 550 \end{array}\right]$$

We see that $x_2 = 400$ and $x_1 = 550$. The equations corresponding to (2.11) are

$$x_3 = 500 + 0.2x_1 + 0.4x_2,$$
$$x_4 = 600 + 0.4x_1 + 0.3x_2.$$

Substituting the above values for x_2 and x_1 gives $x_3 = 770$ and $x_4 = 940$. Thus \$770 should be alloted to Direct Sales and \$940 to Mail Order.

Problem 9 Let u, v, w, x, y, and z denote the number of cars per hour on certain sections of streets, and let A, B, C, D, and E denote the certain intersections, as shown on the diagram.

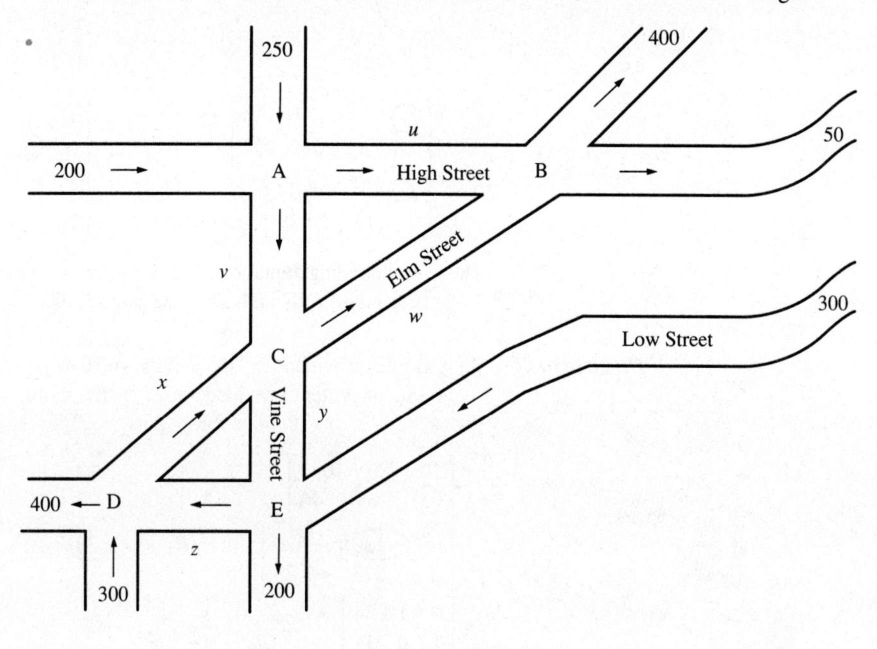

A: $250 + 200 = u + v$
B: $u + w = 400 + 50$
C: $v + x = w + y$
D: $300 + z = 400 + x$
E: $300 + y = 200 + z$

or

$$\begin{aligned} u + v \qquad\qquad\qquad &= 450 \\ u \quad + w \qquad\qquad &= 450 \\ v - w + x - y \quad &= 0 \\ x \qquad - z &= -100 \\ y - z &= -100. \end{aligned}$$

$$\left[\begin{array}{cccccc|c} 1 & 1 & 0 & 0 & 0 & 0 & 450 \\ 1 & 0 & 1 & 0 & 0 & 0 & 450 \\ 0 & 1 & -1 & 1 & -1 & 0 & 0 \\ 0 & 0 & 0 & 1 & 0 & -1 & -100 \\ 0 & 0 & 0 & 0 & 1 & -1 & -100 \end{array}\right]$$

Gauss-Jordan elimination produces

$$\left[\begin{array}{cccccc|c} 1 & 0 & 1 & 0 & 0 & 0 & 450 \\ 0 & 1 & -1 & 0 & 0 & 0 & 0 \\ 0 & 0 & 0 & 1 & 0 & -1 & -100 \\ 0 & 0 & 0 & 0 & 1 & -1 & -100 \\ 0 & 0 & 0 & 0 & 0 & 0 & 0 \end{array}\right]$$

which gives the equations

$$\begin{aligned} u &= 450 - w \\ v &= w \\ x &= -100 + z \\ y &= -100 + z. \end{aligned}$$

Since no variable can be negative, we have $w \le 450$ and $z \ge 100$. In particular, the number of cars per hour between Vine and Elm on Low must be at least 100.

Problem 10

$$\begin{bmatrix} 3 & 0 \\ -1 & 5 \\ -2 & 1 \end{bmatrix} + \begin{bmatrix} 2 & 2 \\ 6 & 4 \\ 5 & -3 \end{bmatrix} = \begin{bmatrix} 3+2 & 0+2 \\ -1+6 & 5+4 \\ -2+5 & 1+(-3) \end{bmatrix} = \begin{bmatrix} 5 & 2 \\ 5 & 9 \\ 3 & -2 \end{bmatrix}$$

Problem 11

$$\begin{aligned} 2A - 3B &= 2\begin{bmatrix} 1 & 3 & -1 \\ 0 & 2 & 1 \end{bmatrix} - 3\begin{bmatrix} -2 & 0 & 1 \\ 3 & -3 & 5 \end{bmatrix} \\ &= \begin{bmatrix} 2 & 6 & -2 \\ 0 & 4 & 2 \end{bmatrix} - \begin{bmatrix} -6 & 0 & 3 \\ 9 & -9 & 15 \end{bmatrix} \\ &= \begin{bmatrix} 8 & 6 & -5 \\ -9 & 13 & -13 \end{bmatrix} \end{aligned}$$

Problem 12

$$\begin{bmatrix} 1 & 3 & 2 \\ 0 & -1 & 1 \end{bmatrix}\begin{bmatrix} -1 \\ 3 \\ 0 \end{bmatrix} = \begin{bmatrix} 1(-1) + 3 \cdot 3 + 2 \cdot 0 \\ 0(-1) + (-1)3 + 1 \cdot 0 \end{bmatrix} = \begin{bmatrix} 8 \\ -3 \end{bmatrix}$$

Problem 13

$$AB = \begin{bmatrix} 1 \cdot 1 + 2 \cdot 2 & 1 \cdot 0 + 2 \cdot 3 & 1(-1) + 2 \cdot 0 \\ 3 \cdot 1 + (-1)2 & 3 \cdot 0 + (-1)3 & 3(-1) + (-1)0 \end{bmatrix} = \begin{bmatrix} 5 & 6 & -1 \\ 1 & -3 & -3 \end{bmatrix}$$

Problem 14

$$\left[\begin{array}{ccc|ccc} 0 & 1 & 3 & 1 & 0 & 0 \\ 0 & \textcircled{1} & 2 & 0 & 1 & 0 \\ 1 & 0 & -1 & 0 & 0 & 1 \end{array}\right] \qquad \left[\begin{array}{ccc|ccc} 0 & 0 & \textcircled{1} & 1 & -1 & 0 \\ 0 & 1 & 2 & 0 & 1 & 0 \\ 1 & 0 & -1 & 0 & 0 & 1 \end{array}\right] \quad (R_1 + (-1)R_2)$$

$$\left[\begin{array}{ccc|ccc} 0 & 0 & 1 & 1 & -1 & 0 \\ 0 & 1 & 0 & -2 & 3 & 0 \\ 1 & 0 & 0 & 1 & -1 & 1 \end{array}\right] \quad \begin{array}{l} \\ (R_2 + (-2)R_1) \\ (R_3 + R_1) \end{array} \qquad \left[\begin{array}{ccc|ccc} 1 & 0 & 0 & 1 & -1 & 1 \\ 0 & 1 & 0 & -2 & 3 & 0 \\ 0 & 0 & 1 & 1 & -1 & 0 \end{array}\right] \quad \begin{array}{l} (R_3) \\ \\ (R_1) \end{array}$$

We see that the inverse is

$$\begin{bmatrix} 1 & -1 & 1 \\ -2 & 3 & 0 \\ 1 & -1 & 0 \end{bmatrix}.$$

Problem 15

$$X = A^{-1}B = \begin{bmatrix} 1 & -1 & 1 \\ -2 & 3 & 0 \\ 1 & -1 & 0 \end{bmatrix}\begin{bmatrix} 3 \\ -1 \\ 5 \end{bmatrix}$$

$$= \begin{bmatrix} 1 \cdot 3 + (-1)(-1) + 1 \cdot 5 \\ (-2)3 + 3(-1) + 0 \cdot 5 \\ 1 \cdot 3 + (-1)(-1) + 0 \cdot 5 \end{bmatrix} = \begin{bmatrix} 9 \\ -9 \\ 4 \end{bmatrix}$$

Problem 16 (a)

$$\left[\begin{array}{ccc|ccc} 2 & 1 & 0 & 1 & 0 & 0 \\ 0 & 1 & -1 & 0 & 1 & 0 \\ \textcircled{1} & 0 & 1 & 0 & 0 & 1 \end{array}\right] \qquad \left[\begin{array}{ccc|ccc} 0 & 1 & -2 & 1 & 0 & -2 \\ 0 & \textcircled{1} & -1 & 0 & 1 & 0 \\ 1 & 0 & 1 & 0 & 0 & 1 \end{array}\right]$$

$$\left[\begin{array}{ccc|ccc} 0 & 0 & \textcircled{-1} & 1 & -1 & -2 \\ 0 & 1 & -1 & 0 & 1 & 0 \\ 1 & 0 & 1 & 0 & 0 & 1 \end{array}\right] \qquad \left[\begin{array}{ccc|ccc} 0 & 0 & \textcircled{1} & -1 & 1 & 2 \\ 0 & 1 & -1 & 0 & 1 & 0 \\ 1 & 0 & 1 & 0 & 0 & 1 \end{array}\right]$$

$$\left[\begin{array}{ccc|ccc} 0 & 0 & 1 & -1 & 1 & 2 \\ 0 & 1 & 0 & -1 & 2 & 2 \\ 1 & 0 & 0 & 1 & -1 & -1 \end{array}\right] \qquad \left[\begin{array}{ccc|ccc} 1 & 0 & 0 & 1 & -1 & -1 \\ 0 & 1 & 0 & -1 & 2 & 2 \\ 0 & 0 & 1 & -1 & 1 & 2 \end{array}\right]$$

The inverse is

$$\begin{bmatrix} 1 & -1 & -1 \\ -1 & 2 & 2 \\ -1 & 1 & 2 \end{bmatrix}.$$

(b)

$$\left[\begin{array}{ccc|ccc} 2 & 1 & 0 & 1 & 0 & 0 \\ \textcircled{1} & 2 & 1 & 0 & 1 & 0 \\ 4 & 5 & 2 & 0 & 0 & 1 \end{array}\right] \quad \left[\begin{array}{ccc|ccc} 0 & \textcircled{-3} & -2 & 1 & -2 & 0 \\ 1 & 2 & 1 & 0 & 1 & 0 \\ 0 & -3 & -2 & 0 & -4 & 1 \end{array}\right]$$

$$\left[\begin{array}{ccc|ccc} 0 & \textcircled{1} & \frac{2}{3} & -\frac{1}{3} & \frac{2}{3} & 0 \\ 1 & 2 & 1 & 0 & 1 & 0 \\ 0 & -3 & -2 & 0 & -4 & 1 \end{array}\right] \quad \left[\begin{array}{ccc|ccc} 0 & 1 & \frac{2}{3} & -\frac{1}{3} & \frac{2}{3} & 0 \\ 1 & 0 & -\frac{1}{3} & \frac{2}{3} & -\frac{1}{3} & 0 \\ 0 & 0 & 0 & -1 & -2 & 1 \end{array}\right]$$

From the row of zeros on the left of the last tableau, we see that the matrix has no inverse.

Problem 17

$$U = AX = \begin{bmatrix} .2 & .1 & .3 \\ .4 & .2 & .1 \\ .3 & .5 & .5 \end{bmatrix} \begin{bmatrix} 1000 \\ 1200 \\ 2000 \end{bmatrix} = \begin{bmatrix} 920 \\ 840 \\ 1900 \end{bmatrix}$$

$$S = X - U = \begin{bmatrix} 1000 \\ 1200 \\ 2000 \end{bmatrix} - \begin{bmatrix} 920 \\ 840 \\ 1900 \end{bmatrix} = \begin{bmatrix} 80 \\ 360 \\ 100 \end{bmatrix}$$

Problem 18

$$X = (I - A)^{-1}D = \begin{bmatrix} 2.80 & 1.60 & 2.0 \\ 1.84 & 2.48 & 1.6 \\ 3.52 & 3.44 & 4.8 \end{bmatrix} \begin{bmatrix} 200 \\ 100 \\ 300 \end{bmatrix} = \begin{bmatrix} 1320 \\ 1096 \\ 2488 \end{bmatrix}$$

Thus 1320 units of agricultural products, 1096 units of manufactured goods, and 2488 units of energy are needed.

Chapter 3

Problem 1 (a) Since $8(50) + 3(30) = 490 > 480$, the point (50, 30) is not feasible. Since $8(30) + 3(50) = 390 \leq 480$, $4(30) + 9(50) = 570 \leq 720$, $30 \geq 0$, and $50 \geq 0$, the point (30, 50) is feasible. Since $8(45) + 3(40) = 480 \leq 480$, $4(45) + 9(40) = 540 \leq 720$, $45 \geq 0$, and $40 \geq 0$, the point (45, 40) is feasible.
(b) For (30, 50) we have $R = 16(30) + 12(50) = 1080$, and for (45, 40) we have $R = 16(45) + 12(40) = 1200$. Thus the point (45, 40) is better.

Problem 2 If

x = the number of TV sets carried, and

y = the number of VCRs carried,

then the information can be summarized in the following table.

	TV	*VCR*	*Available*
Weight	40 lbs	25 lbs	4000 lbs
Volume	3 cu ft	2 cu ft	2000 cu ft
Revenue	\$5	\$4	
Number	x	y	

The linear program is:

$$\begin{aligned} \text{Maximize } 5x + 4y &= R \\ \text{subject to } 40x + 25y &\leq 4000 \\ 3x + \ \ 2y &\leq 2000 \\ x &\geq 0 \\ y &\geq 0. \end{aligned}$$

Problem 3 The line $2x + 3y = 12$ has the intercepts (6, 0) and (0, 4), while the line $x = 2$ is vertical and goes through (2, 0). Note that (0, 0) satisfies $2x + 3y \leq 12$ but not $x \geq 2$, so the arrows indicate the desired region.

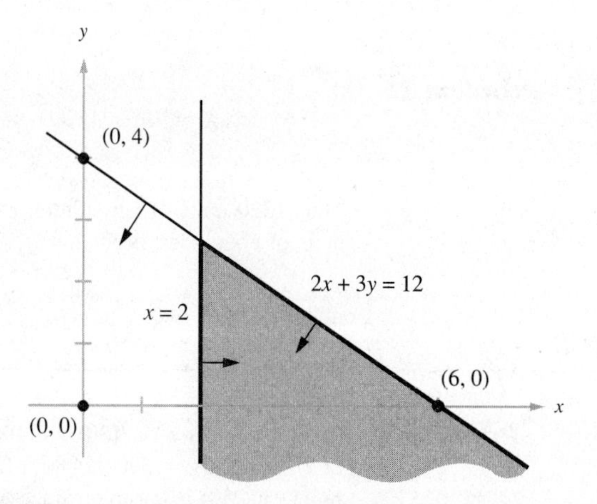

Problem 4 Notice that $3x + 3y = 75$ has intercepts (25, 0) and (0, 25), $2x + 4y = 60$ has intercepts (30, 0) and (0, 15), and $4x + 2y = 60$ has intercepts (15, 0) and (0, 30). The point (0, 0) satisfies none of the inequalities $3x + 3y \geq 75$, $2x + 4y \geq 60$, $4x + 2y \geq 60$.

Problem 5 The four corner points are at the intersections of the line (3) and the y-axis, lines (3) and (1), lines (1) and (2), and line (2) and the x-axis.

Problem 6

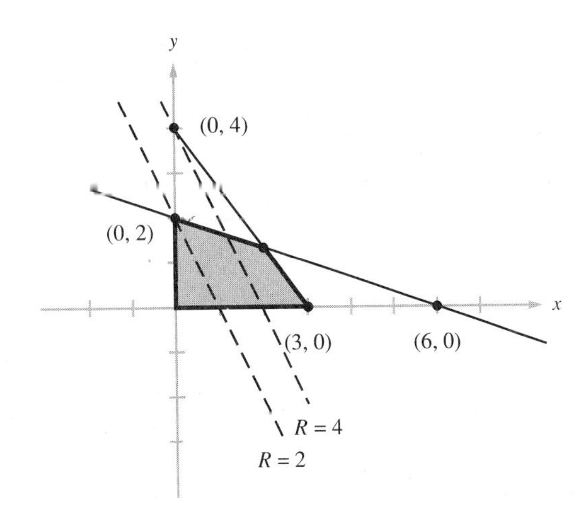

From the sketch the optimal point is at the intersection of the lines $4x + 3y = 12$ and $y = 0$. This is the point (3, 0). Then $R = 2(3) + 0 = 6$.

Problem 7 Let $x + 2y + s = 320$, $5x + 4y + t = 1000$, $s \geq 0$, $t \geq 0$.

$$\begin{array}{c} \\ s \\ t \\ \\ \end{array}\left[\begin{array}{cccc|c} x & y & s & t & \\ 1 & 2 & \boxed{1} & 0 & 320 \\ \textcircled{5} & 4 & 0 & \boxed{1} & 1000 \\ \hline 4 & 3 & 0 & 0 & R \end{array}\right]$$

$$\begin{array}{c} \\ s \\ x \\ \\ \end{array}\left[\begin{array}{cccc|c} x & y & s & t & \\ 0 & 1.2 & \boxed{1} & -.2 & 120 \\ \boxed{1} & .8 & 0 & .2 & 200 \\ \hline 0 & -.2 & 0 & -.8 & R - 800 \end{array}\right]$$

The last row corresponds to

$$R = 800 - .2y - .8t.$$

Thus the maximum is 800 for $y = t = 0$. Then $x = 200$ and $y = 0$.

Problem 8 We choose column 3 because 7 is the largest positive indicator. Dividing the other entries in column 3 into the corresponding rightmost entries gives $34/5 = 6.8$, $16/2 = 8$, and $70/8 = 8.75$. Since 6.8 is the smallest of these, we choose row 1.

Problem 9 The row labels of the tableau give the basic variables x, z, and s. The last row corresponds to

$$R = 22 - 3y - 5t,$$

so $R = 22$ when $y = t = 0$. Then $x = 10$ from row 3, and $z = 4$ from row 1.

Problem 10 The compact tableau is

$$\begin{array}{ccc|c} x & y & z & \\ 1 & 2 & 4 & (\geq)\ 12 \\ 3 & 1 & 2 & (\geq)\ 15 \\ \hline 2 & 5 & 3 & (\text{minimize})\ C \end{array}$$

with dual compact tableau

$$\begin{array}{cc|c} X & Y & \\ 1 & 3 & (\leq)\ 2 \\ 2 & 1 & (\leq)\ 5 \\ 4 & 2 & (\leq)\ 3 \\ \hline 12 & 15 & (\text{maximize})\ R \end{array}.$$

Thus we should maximize $12X + 15Y = R$, subject to

$$X + 3Y \leq 2$$
$$2X + \ \ Y \leq 5$$
$$4X + 2Y \leq 3$$
$$X \geq 0,\ Y \geq 0.$$

Problem 11 In order to satisfy (a) we replace

$$\text{minimize } -3x + y = C$$

with

$$\text{maximize } 3x - y = R.$$

In order to satisfy (d) we replace

$$-x + y \geq -2 \quad \text{and} \quad -2x - 3y \geq -6$$

with

$$x - y \leq 2 \quad \text{and} \quad 2x + 3y \leq 6.$$

Problem 12 The linear program is

$$\begin{array}{lrcl} \text{Maximize} & x + 2y - 3z & = R & \\ \text{subject to} & x \quad\ + \ z + s & = & 30 \\ & 2x + \ y + \ z \quad - t \quad + a & = & 10 \\ & x + 3y + 2z \quad - u \quad + b & = & 15 \\ & x,\ y,\ z,\ s,\ t,\ u,\ a,\ b \geq 0. & & \end{array}$$

We want to maximize $-a - b = Q$, which leads to the tableau

$$\left[\begin{array}{cccccccc|c} 1 & 0 & 1 & 1 & 0 & 0 & 0 & 0 & 30 \\ 2 & 1 & 1 & 0 & -1 & 0 & 1 & 0 & 10 \\ 1 & 3 & 2 & 0 & 0 & -1 & 0 & 1 & 15 \\ \hline 1 & 2 & -3 & 0 & 0 & 0 & 0 & 0 & R \\ 0 & 0 & 0 & 0 & 0 & 0 & -1 & -1 & Q \end{array}\right].$$

Since rows 2 and 3 correspond to artificial variables, we will add them to the bottom row.

Chapter 4

Problem 1 The first term in the sequence is the initial value x_0, which is 4. The next term x_1 is found by taking $n = 0$ in the defining equation $x_{n+1} = -3x_n + 10$; this yields

$$x_1 = -3x_0 + 10 = -3(4) + 10 = -12 + 10 = -2.$$

Similarly, by taking n to be 1, 2, 3, and 4 in the defining equation, we obtain

$$x_2 = -3x_1 + 10 = -3(-2) + 10 = 6 + 10 = 16,$$
$$x_3 = -3x_2 + 10 = -3(16) + 10 = -48 + 10 = -38,$$
$$x_4 = -3x_3 + 10 = -3(-38) + 10 = 114 + 10 = 124, \text{ and}$$
$$x_5 = -3x_4 + 10 = -3(124) + 10 = -372 + 10 = -362.$$

Problem 2 Applying Theorem 4.1 with $b = -3$ and $x_0 = 36$, we obtain

$$x_n = 36 - 3n.$$

Problem 3 Applying Theorem 4.2 with $a = 2$ and $x_0 = 5$, we obtain

$$x_n = 2^n(5).$$

Problem 4 The amount of interest due is

$$I = Prt = \$8000(0.16)(2.5) = \$3200.$$

Problem 5 The amount owed after 18 months is

$$A = P(1 + rt) = \$10{,}000[1 + (0.15)(1.5)] = \$12{,}250.$$

The amount of interest paid is

$$A - P = \$12{,}250 - \$10{,}000 = \$2250.$$

Problem 6 We must solve for r in formula (4.3) given that $A = \$22{,}650$, $P = \$20{,}000$, and $t = \frac{15}{12} = \frac{5}{4}$.

$$\begin{aligned} A &= P(1 + rt) \\ \$22{,}650 &= \$20{,}000[1 + r(\tfrac{5}{4})] \\ 1.1325 &= 1 + r(\tfrac{5}{4}) \\ 0.1325 &= r(\tfrac{5}{4}) \\ r &= 0.1325(\tfrac{4}{5}) = 0.106 \end{aligned}$$

Thus the bond is paying 10.6% simple interest.

Problem 7 Using formula (4.4), we obtain

$$A = P(1 + i)^{20} = \$6000(1.02)^{20} \approx \$8915.68.$$

Problem 8 If we invest \$100 at 6% interest compounded monthly, then after 1 year we will have

$$A = P(1 + i)^{12} = \$100(1.005)^{12} \approx \$106.17.$$

This is an increase of approximately 6.17%, and hence the effective rate of interest is about 6.17%.

Problem 9 (a) Using formula (4.5) with $A = \$8000$, $r = 0.15$, and $t = 6$, we obtain

$$P = \frac{A}{1 + rt} = \frac{\$8000}{1 + (0.15)(6)} = \frac{\$8000}{1.9} \approx \$4210.53.$$

(b) Using formula (4.6) with $A = \$8000$, $i = 0.10/4 = 0.025$, and $n = 6(4) = 24$, we obtain

$$P = \frac{A}{(1 + i)^n} = \frac{\$8000}{(1.025)^{24}} \approx \$4423.00.$$

Problem 10 Taking $a = 3$, $b = -4$, and $x_0 = 8$ in Theorem 4.3 gives

$$c = \frac{b}{1 - a} = \frac{-4}{1 - 3} = 2, \quad \text{and}$$

$$x_n = a^n(x_0 - c) + c = 3^n(8 - 2) + 2 = 6(3^n) + 2.$$

Problem 11 Using formula (4.8) with $p = \$150$, $i = 0.005$, and $n = 60$ gives

$$F = \$150\left[\frac{(1.005)^{60} - 1}{0.005}\right] \approx \$10{,}465.50.$$

Problem 12 Using formula (4.9) with $p = \$50{,}000$, $i = 0.06$, and $n = 20$ gives

$$P = \$50{,}000\left[\frac{1 - (1.06)^{-20}}{0.06}\right] \approx \$573{,}496.06.$$

Problem 13 Using formula (4.10) with $F = \$120{,}000{,}000$, $i = 0.0225$, and $n = 32$ gives

$$P = \frac{0.0225(\$120{,}000{,}000)}{(1.0225)^{32} - 1} \approx \$2{,}600{,}897.92.$$

Problem 14 The amount to be financed is \$9946 − \$1500 = \$8,446. Using formula (4.11) with $A = \$8446$, $i = 0.00825$, and $n = 60$ gives

$$P = \frac{0.00825(\$8446)}{1 - (1.00825)^{-60}} \approx \$179.04.$$

Chapter 5

Problem 1 (a) Since $x^2 - 2x + 1 = (x - 1)^2 \geq 0$ for all real numbers x,

$$\{x|x^2 - 2x + 1 < 0\} = \emptyset.$$

(b) Since $x^2 = 1$ is equivalent to $0 = x^2 - 1 = (x + 1)(x - 1)$,

$$\{x|x^2 = 1\} = \{-1, 1\}.$$

(c) Since $(x - 2)^2 \geq 0$ for all real numbers x,

$$\{x|(x - 2)^2 \geq 0\} = U.$$

Problem 2 Since $A \cup B$ consists of the elements in A or B or both, we have

$$A \cup B = \{1, 2, 3, 4, 6, 8\}.$$

Since $A \cap B$ consists of the elements in both A and B, we have

$$A \cap B = \{2\}.$$

Since $\overline{B}$ consists of the elements in U but not in B, we have

$$\overline{B} = \{1, 5, 7, 8\}.$$

Since $A \cap C$ consists of the elements in both A and C, we have

$$A \cap C = \emptyset.$$

Since $B \cup C$ consists of the elements in B or C or both, we have

$$B \cup C = \{2, 3, 4, 6, 7\}.$$

Using the set $B \cup C$ computed above, we see that

$$\overline{(B \cup C)} = \{1, 5, 8\}.$$

The set $A \cup B \cup C$ consists of the elements in one or more of the sets A, B, or C; so

$$A \cup B \cup C = \{1, 2, 3, 4, 6, 7, 8\}.$$

Problem 3 (a) Region III lies in B but not in A or C; so region III denotes the set

$$\overline{A} \cap B \cap \overline{C}.$$

Region IV lies in both A and C but not in B; so region IV denotes the set

$$A \cap \overline{B} \cap C.$$

(b) Region II consists of athletes in A and B but not in C; region IV consists of athletes in B and C but not in A; and region VI consists of athletes in A and C but not in B. So the Venn diagram showing the athletes who take exactly two of the three vitamins is shown below.

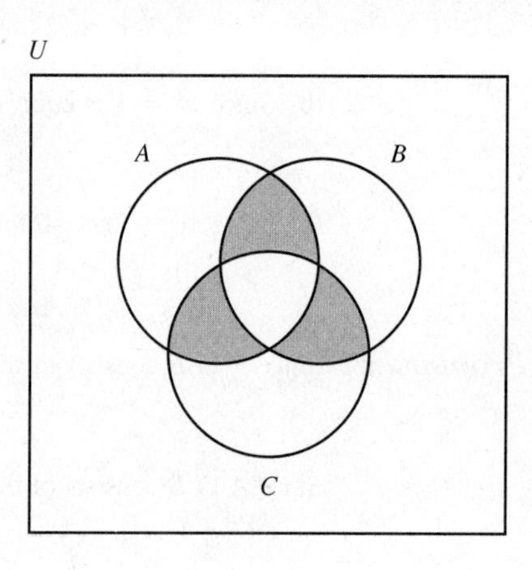

Problem 4 (a) Since 36 executives read *Forbes*,

$$100 - 36 = 64$$

do not read *Forbes*.

(b) By Theorem 5.1(b), there are

$$36 + 28 - 7 = 57$$

executives who read *Forbes* or *Business Week*.

(c) Since 57 executives read *Forbes* or *Business Week*,

$$100 - 57 = 43$$

read neither.

Problem 5 Let B, M, and E denote the sets of students who are taking courses in business, mathematics, and economics, respectively. Proceeding as in Figures 5.14–5.17, we obtain the Venn diagram below.

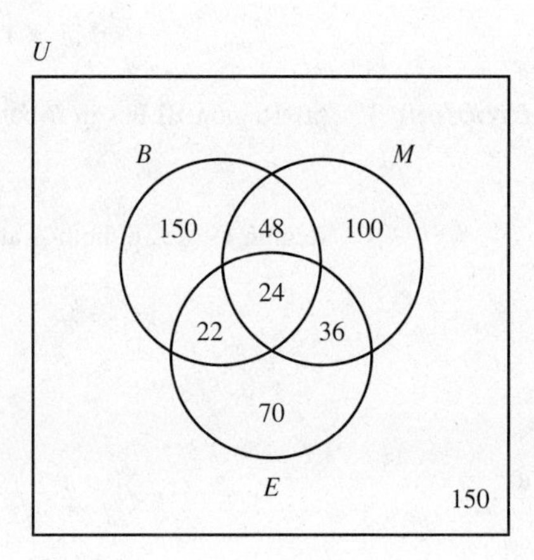

(a) The number of students taking a business course but not taking a course in mathematics or economics is 150.
(b) The number of students taking courses in exactly two of the three disciplines is

$$48 + 36 + 22 = 106.$$

(c) The number of students who were surveyed is the total of all the numbers in the Venn diagram, which is 600.

Problem 6 We can create a tree diagram by successively deciding whether to include each element of S. One possibility is shown in the figure below. The last column contains the eight possible subsets:

$$\varnothing, \{a\}, \{b\}, \{c\}, \{a, b\}, \{a, c\}, \{b, c\}, \{a, b, c\}.$$

Include a?	*Include b?*	*Include c?*	*Subset*
yes	yes	yes	$\{a, b, c\}$
		no	$\{a, b\}$
	no	yes	$\{a, c\}$
		no	$\{a\}$
no	yes	yes	$\{b, c\}$
		no	$\{b\}$
	no	yes	$\{c\}$
		no	$\varnothing$

Problem 7 The multiplication principle shows that the number of possible tests equals the product of the numbers of variations for each question, which is

$$8 \cdot 7 \cdot 10 \cdot 5 \cdot 6 \cdot 8 = 134{,}400.$$

Problem 8 Since each question can be answered in two different ways, there are

$$2^{10} = 1024$$

ways in which all ten questions can be answered.

Problem 9 (a) Any one of the eight people may sit in the first seat, but only one person (the spouse of the person in the first seat) may sit in the second seat. Any one of the remaining six people may sit in the third seat, but only one person may sit in the fourth seat. Any one of the remaining four people may sit in the fifth seat, but only one person may sit in the sixth seat. Finally either of the two remaining people may sit in the seventh seat, leaving only one person to sit in the eighth seat. Thus by the multiplication principle the number of possible seatings is

$$8 \cdot 1 \cdot 6 \cdot 1 \cdot 4 \cdot 1 \cdot 2 \cdot 1 = 384.$$

(b) Choose a man to sit in the first seat, and then have his wife sit in the second seat. Then choose one of the three remaining men to sit in the eighth seat, and have his wife sit in the seventh seat. Any of the four remaining people may sit in the third seat, but only that person's spouse may sit in the fourth seat. Finally either of the two remaining people may sit in the fifth seat, leaving only one person to sit in the sixth seat. Thus the number of possible seatings is

$$4 \cdot 1 \cdot 4 \cdot 1 \cdot 2 \cdot 1 \cdot 1 \cdot 3 = 96.$$

Problem 10 By the addition principle, the number of courses that can be selected is the sum of the number of business courses, the number of physical education courses, and the number of economics courses. This number is

$$4 + 7 + 3 = 14.$$

Problem 11 With each of the three meat entrees, there are four choices of salad dressing and five choices of vegetable. With each of the two seafood dishes there is a choice of four salad dressings. The number of possible choices (which equals the number of possible ways to order dinner) is

$$3 \cdot 4 \cdot 5 + 2 \cdot 4 = 60 + 8 = 68.$$

Problem 12 The number of ways in which the medals can be awarded equals the number of ways to choose three winners (in order) from among the nine contestants; this number is

$$P(9, 3) = 504.$$

Problem 13 The number of clinks equals the number of different pairs of people, which is

$$C(8, 2) = 28.$$

Problem 14 To form a list of five letters containing three different consonants and two different vowels, we must choose 3 different consonants (from among 21) and 2 different vowels (from among 5). Then we must arrange the five chosen letters in order. These operations can be performed in

$$C(21, 3) \cdot C(5, 2) \cdot P(5, 5) = 1{,}596{,}000$$

different ways.

Problem 15 Row 8 of Pascal's triangle begins and ends with 1. The other entries are obtained by adding pairs of adjacent entries in row 7, which is listed in Example 5.27. The resulting numbers are shown below.

1 8 28 56 70 56 28 8 1

Problem 16 Apply the binomial theorem with $n = 5$ and y replaced by $-2y$. The result is shown below.

$$(x - 2y)^5 = x^5 - 10x^4y + 40x^3y^2 - 80x^2y^3 + 80xy^4 - 32y^5$$

Problem 17 The term involving p^5q^{11} in the expansion of $(2p - q)^{16}$ is

$$C(16, 11)(2p)^5(-q)^{11} = 4368(32p^5)(-q^{11}) = -139{,}776p^5q^{11}.$$

Chapter 6

Problem 1 (a) The tree diagram is shown below.

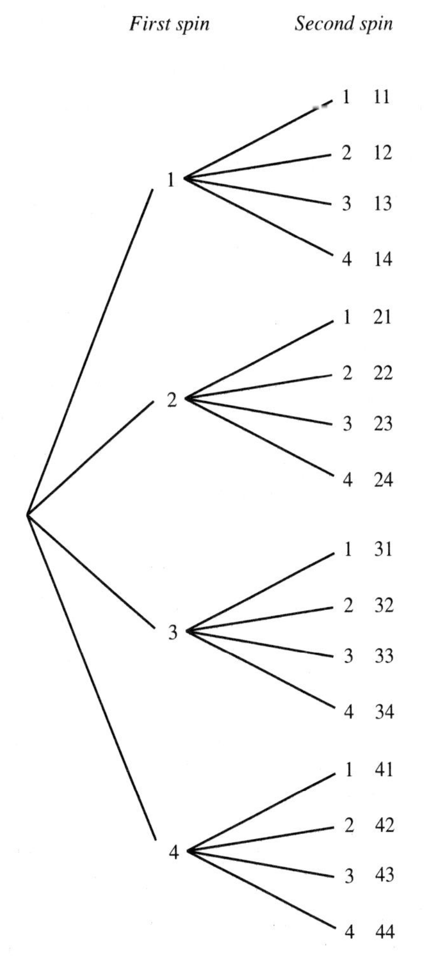

(b) Since the 16 outcomes in the tree diagram are equally likely and the outcome 22 occurs once, the probability of obtaining this outcome is 1/16.

Problem 2 Since the number of calls is a nonnegative integer not exceeding 15, a sample space for this experiment is

$$\{0, 1, 2, \ldots, 15\}.$$

Problem 3 (a) The event of receiving exactly 10 calls is the set $\{10\}$.
(b) The event of receiving at least 12 calls is the set $\{12, 13, 14, 15\}$.
(c) The event of receiving fewer than five calls is the set $\{0, 1, 2, 3, 4\}$.
(d) The event of receiving 0 calls is the set $\{0\}$.
(e) Since it is impossible to receive 16 calls, this set is $\varnothing$.

Problem 4 Of the five doorways leading from the starting room, two lead to room C. Thus the probability that the mouse will next enter room C is 2/5.

Problem 5 The number of possible outcomes when two dice are rolled is $6 \cdot 6 = 36$. Of this number, there are $P(6, 2) = 6 \cdot 5 = 30$ outcomes in which different numbers are rolled. Hence the probability of obtaining two different numbers is

$$\frac{30}{36} = \frac{5}{6}.$$

Problem 6 The number of possible subcommittees is $C(22, 5) = 26{,}334$. Of these, the number of subcommittees containing exactly 3 representatives of management and 2 representatives of labor is $C(12, 3) \cdot C(10, 2) = 220 \cdot 45 = 9900$. Hence the probability that a randomly selected subcommittee contains exactly 3 representatives of management and 2 representatives of labor is

$$\frac{9900}{26{,}334} = \frac{50}{133}.$$

Problem 7 The number of possible sequences in which the interviews can be conducted is $P(6, 6) = 720$. The number of possible sequences in which the four men are interviewed before the two women is $P(4, 4) \cdot P(2, 2) = 24 \cdot 2 = 48$. Therefore the probability that the four men are interviewed before the two women is

$$\frac{48}{720} = \frac{1}{15}.$$

Problem 8 (a) The probability distribution is formed by dividing each transaction's frequency by 4000, the total number of transactions. It is shown below.

Type of Transaction	*Probability*	
Cashed a check	.398	
Made a deposit	.286	
Made a withdrawal	.234	
Paid a bill	.064	
Requested change	.011	
Bought a money order	.007	
	1.000	Total

(b) The probability that a transaction was either a deposit or withdrawal is the sum of the probabilities of these two types of transactions, which is

$$.286 + .234 = .520.$$

Problem 9 (a) If the odds in favor of E are 7 to 4, then the probability of E is

$$\frac{7}{7 + 4} = \frac{7}{11}.$$

(b) If the odds against E are 9 to 5, then the probability of E is

$$\frac{5}{5 + 9} = \frac{5}{14}.$$

Problem 10 (a) If $P(E) = .75 = 3/4$, then the odds in favor of E are 3 to $4 - 3$, that is,

3 to 1.

(b) If $P(E) = 2/9$, then the odds in favor of E are 2 to $9 - 2$. Hence the odds against E are

7 to 2.

Problem 11 In Figure 6.10 let the outcomes be denoted (w, b), where w is the number rolled on the white die and b is the number rolled on the black die. Then the outcomes corresponding to a sum of 6 are

$$(1, 5), (2, 4), (3, 3), (4, 2) \text{ and } (5, 1).$$

Thus of the 36 possible outcomes there are 5 corresponding to a sum of 6; so

$$P(A) = \frac{5}{36}.$$

The outcomes in which at least one die comes up greater than 4 are

$$(1, 5), (1, 6), (2, 5), (2, 6), (3, 5), (3, 6), (4, 5), (4, 6),$$
$$(5, 1), (5, 2), (5, 3), (5, 4), (5, 5), (5, 6),$$
$$(6, 1), (6, 2), (6, 3), (6, 4), (6, 5), \text{ and } (6, 6).$$

Hence

$$P(B) = \frac{20}{36} = \frac{5}{9}.$$

By similar reasoning, we obtain

$$P(A \cup B) = \frac{23}{36}, \qquad P(A \cap B) = \frac{2}{36} = \frac{1}{18},$$
$$P(\overline{A}) = \frac{31}{36}, \qquad \text{and} \qquad P(\overline{B}) = \frac{16}{36} = \frac{4}{9}.$$

Problem 12 (a) Let E be the event "at most 8 calls are received." Then $\overline{E}$ is the event "9 or 10 calls are received." Hence

$$P(\overline{E}) = P(9) + P(10) = .008 + .004 = .012.$$

Therefore

$$P(E) = 1 - P(\overline{E}) = 1 - .012 = .988.$$

(b) If F is the event "no more than 9 calls are received," then $\overline{F}$ is the event "10 calls are received." Hence $P(\overline{F}) = .004$, and so

$$P(F) = 1 - .004 = .996.$$

Problem 13 If a fair coin is flipped five times, the number of possible sequences of heads and tails that can occur is $2^5 = 32$. Of these, each sequence but one contains at least one head. Hence the probability of obtaining at least one head is

$$\frac{31}{32}.$$

Problem 14 Let A and S be the events of being short of aluminum and steel, respectively. Then the event of having an adequate supply of both metals is $\overline{(A \cup S)}$. Now $P(A) = .06$, $P(S) = .05$, and $P(A \cap S) = .01$, and so

$$P(A \cup S) = P(A) + P(S) - P(A \cap S) = .06 + .05 - .01 = .10,$$

by the union rule. Thus by the complement rule

$$P(\overline{A \cup S}) = 1 - P(A \cup S) = 1 - .10 = .90.$$

Problem 15 Let A, R, and K denote the events that a house was built by the Armstrong Company, Rave Brothers Construction, and Kaisner Construction, respectively. Since the events A, R, and K are mutually exclusive, the probability that a house was built by a member of the Ark Corporation is

$$P(A \cup R \cup K) = P(A) + P(R) + P(K) = .35 + .20 + .19 = .74.$$

Problem 16 Of the 250 patients who received a placebo, 133 felt no better. Hence the conditional probability that a patient who received a placebo felt no better is

$$\frac{133}{250} = .532.$$

Problem 17 We see from Figure 6.2 that the reduced sample space consisting of outcomes in which exactly two of the three flips are heads is $\{HHT, HTH, THH\}$. Hence the conditional probability that the first flip is heads given that exactly two of the three flips are heads is 2/3.

Problem 18 Let A and L denote the events of having purchased automobile and life insurance, respectively, from the agent. Then $P(A) = .60$, $P(L) = .40$, and $P(A \cap L) = .15$.

Hence the conditional probability that someone who has purchased life insurance from this agent will also have purchased automobile insurance from her is

$$P(A|L) = \frac{P(A \cap L)}{P(L)} = \frac{.15}{.40} = .375.$$

Problem 19 Let A and B denote the events that a defective engine passes the first and second inspections, respectively. Then

$$P(A) = 1 - .70 = .30 \quad \text{and} \quad P(B|A) = 1 - .80 = .20.$$

Hence the probability that a defective engine passes both inspections is

$$P(A \cap B) = P(A) \cdot P(B|A) = .30(.20) = .06.$$

Problem 20 Let A and B denote the events "the first transistor is defective" and "the second transistor is defective," respectively. If we select two transistors in sequence from a box containing three defective and seven nondefective transistors, the probability that both of the selected transistors are defective is

$$P(A \cap B) = P(A) \cdot P(B|A) = (\tfrac{3}{10})(\tfrac{2}{9}) = \tfrac{6}{90} = \tfrac{1}{15}.$$

Problem 21 Let A, B, and C denote the events "the turntable fails," "the amplifier fails," and "the speakers fail," respectively. Now

$$P(A) = .05, \quad P(B) = .02, \quad \text{and} \quad P(C) = .03.$$

We are interested in determining the probability that none of these components fail. Because the turntables, amplifiers, and speakers are manufactured independently, we will assume that $\overline{A}$, $\overline{B}$, and $\overline{C}$ are independent events. Hence

$$\begin{aligned} P(\text{system does not fail}) &= P(\text{no component fails}) \\ &= P(\overline{A} \cap \overline{B} \cap \overline{C}) \\ &= P(\overline{A}) \cdot P(\overline{B}) \cdot P(\overline{C}) \\ &= (.95)(.98)(.97) \\ &\approx .903. \end{aligned}$$

Problem 22 The stochastic diagram for this situation is shown below.

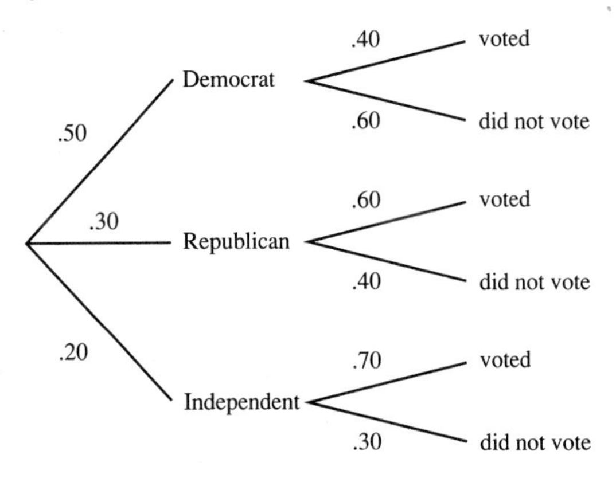

From this diagram we see that

$$P(\text{Democrat}|\text{voted}) = \frac{(.5)(.4)}{(.5)(.4) + (.3)(.6) + (.2)(.7)} = \frac{.20}{.52} = \frac{5}{13},$$

$$P(\text{Republican}|\text{voted}) = \frac{(.3)(.6)}{(.5)(.4) + (.3)(.6) + (.2)(.7)} = \frac{.18}{.52} = \frac{9}{26}, \quad \text{and}$$

$$P(\text{independent}|\text{voted}) = \frac{(.2)(.7)}{(.5)(.4) + (.3)(.6) + (.2)(.7)} = \frac{.14}{.52} = \frac{7}{26}.$$

Problem 23 For unacquainted married couples the events of getting divorced can be assumed to be independent. Hence by Bernoulli's formula the probability of having exactly one divorce among 10 couples during the next year is

$$C(10, 1)(.02)^1(.98)^9 = 10(.02)(.98)^9 \approx .167.$$

Problem 24 Since the calls are made at random, we can assume that the events of making a sale are independent. We will use Bernoulli's formula to compute the probability of *not* making at least one sale, that is, the probability of making no sales. This probability is

$$C(8, 0)(.10)^0(.90)^8 = (.90)^8 \approx .430.$$

Hence the probability of making at least one sale is approximately

$$1 - .430 = .570.$$

Problem 25 In this process there are three states.

State 1: belonging to the upper class

State 2: belonging to the middle class

State 3: belonging to the lower class

The time period between trials is one generation, and the entries in each row of the transition matrix are the probabilities that the children of parents in a particular class will belong to each of the three classes. Hence the transition matrix for this process is

$$\begin{array}{cc} & \text{Next state} \\ & \begin{array}{ccc} 1 & 2 & 3 \end{array} \\ \text{Current state} \begin{array}{c} 1 \\ 2 \\ 3 \end{array} & \begin{bmatrix} .5 & .5 & 0 \\ .1 & .7 & .2 \\ 0 & .5 & .5 \end{bmatrix} \end{array}.$$

Problem 26 Multiplying the initial state vector times the transition matrix, we obtain the state vector after one generation:

$$[.20 \quad .40 \quad .40]\begin{bmatrix} .5 & .5 & 0 \\ .1 & .7 & .2 \\ 0 & .5 & .5 \end{bmatrix} = [.14 \quad .58 \quad .28].$$

Multiplying the state vector computed above times the transition matrix, we obtain the state vector after two generations:

$$[.14 \quad .58 \quad .28]\begin{bmatrix} .5 & .5 & 0 \\ .1 & .7 & .2 \\ 0 & .5 & .5 \end{bmatrix} = [.128 \quad .616 \quad .256].$$

Hence after two generations the distribution of people among the classes will be 12.8% upper class, 61.6% middle class, and 25.6% lower class.

Problem 27 We will proceed as in Example 6.40. Let the equilibrium vector be

$$X = [x_1 \quad x_2 \quad x_3].$$

We must solve the matrix equation $X(I_3 - P) = O$, where P is the transition matrix of the Markov chain, which was obtained in Practice Problem 25.

$$[x_1 \quad x_2 \quad x_3]\begin{bmatrix} .5 & -.5 & 0 \\ -.1 & .3 & -.2 \\ 0 & -.5 & .5 \end{bmatrix} = [0 \quad 0 \quad 0].$$

Multiplying out the left side leads to the system of equations below.

$$\begin{aligned} .5x_1 - .1x_2 \qquad\quad &= 0 \\ -.5x_1 + .3x_2 - .5x_3 &= 0 \\ -.2x_2 + .5x_3 &= 0 \end{aligned}$$

By including the condition that the sum of the unknowns is 1, we obtain the following system.

$$\begin{aligned} .5x_1 - .1x_2 \qquad\quad &= 0 \\ -.5x_1 + .3x_2 - .5x_3 &= 0 \\ -.2x_2 + .5x_3 &= 0 \\ x_1 + x_2 + x_3 &= 1 \end{aligned}$$

We now solve this system by the techniques of Chapter 2, as follows.

$$\left[\begin{array}{ccc|c} .5 & -.1 & 0 & 0 \\ -.5 & .3 & -.5 & 0 \\ 0 & -.2 & .5 & 0 \\ 1 & 1 & 1 & 1 \end{array}\right] \quad \left[\begin{array}{ccc|c} 0 & -.6 & -.5 & -.5 \\ 0 & .8 & 0 & .5 \\ 0 & -.2 & .5 & 0 \\ 1 & 1 & 1 & 1 \end{array}\right] \quad \left[\begin{array}{ccc|c} 0 & -.6 & -.5 & -.5 \\ 0 & .8 & 0 & .5 \\ 0 & 1 & -2.5 & 0 \\ 1 & 1 & 1 & 1 \end{array}\right]$$

$$\begin{bmatrix} 0 & 0 & -2 & | & -.5 \\ 0 & 0 & 2 & | & .5 \\ 0 & 1 & -2.5 & | & 0 \\ 1 & 0 & 3.5 & | & 1 \end{bmatrix} \quad \begin{bmatrix} 0 & 0 & 1 & | & .25 \\ 0 & 0 & 2 & | & .5 \\ 0 & 1 & -2.5 & | & 0 \\ 1 & 0 & 3.5 & | & 1 \end{bmatrix} \quad \begin{bmatrix} 0 & 0 & 1 & | & .25 \\ 0 & 0 & 0 & | & 0 \\ 0 & 1 & 0 & | & .625 \\ 1 & 0 & 0 & | & .125 \end{bmatrix}$$

Hence in the long run 12.5% of the people will be in the upper class, 62.5% in the middle class, and 25% in the lower class.

Chapter 7

Problem 1 The mean is

$$\bar{x} = \frac{\Sigma x}{n} = \frac{26 + 38 + 17 + 32 + 41}{5} = \frac{154}{5} = 30.8.$$

Problem 2 The sum of the 25 given numbers is 19,884. Hence the mean number of tornadoes reported in the United States during the years 1960–84 is

$$\frac{19{,}884}{25} = 795.36.$$

Problem 3 The numbers in Practice Problem 2 listed in increasing order are:

461, 570, 604, 618, 649, 658, 660, 683, 713, 741, 783, 788, 835,
852, 852, 866, 888, 898, 907, 912, 920, 931, 947, 1046, 1102.

The thirteenth number in this list, 835, is the median.

Problem 4 In Example 7.1 we found that the mean of the examination scores is 86. Thus the variance of the four scores is

$$\begin{aligned} s^2 &= \frac{\Sigma(x - \bar{x})^2}{n} \\ &= \frac{(84 - 86)^2 + (94 - 86)^2 + (87 - 86)^2 + (79 - 86)^2}{4} \\ &= \frac{4 + 64 + 1 + 49}{4} = \frac{118}{4} = 29.5. \end{aligned}$$

Hence the standard deviation of the scores is

$$s = \sqrt{29.5} \approx 5.43.$$

Problem 5 (a) There are many acceptable choices of intervals; we will use 8 intervals of width 100: 399.5–499.5, 499.5–599.5, 599.5–699.5, 699.5–799.5, 799.5–899.5, 899.5–999.5, 999.5–1099.5, and 1099.5–1199.5.

(b) The frequency distribution is formed by tallying the number of years in which the number of tornadoes lies within each of the intervals above. The results are contained in the table below.

Number of tornadoes	*Frequency*
399.5–499.5	1
499.5–599.5	1
599.5–699.5	6
699.5–799.5	4
799.5–899.5	6
899.5–999.5	5
999.5–1099.5	1
1099.5–1199.5	1
	25 Total

The histogram using the intervals in part (a) is shown in the figure below.

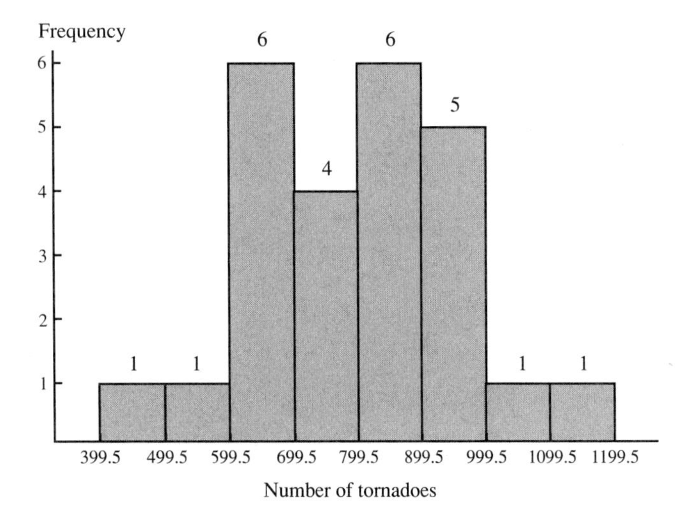

Problem 6 Using the frequency distribution in the answer to Practice Problem 5, we obtain the following table.

Number of tornadoes	*Midpoint* m	*Frequency* f	mf
399.5–499.5	449.5	1	449.5
449.5–599.5	549.5	1	549.5
599.5–699.5	649.5	6	3897.0
699.5–799.5	749.5	4	2998.0
799.5–899.5	849.5	6	5097.0
899.5–999.5	949.5	5	4747.5
999.5–1099.5	1049.5	1	1049.5
1099.5–1199.5	1149.5	1	1149.5
		25	19,937.5 Totals

Thus the mean number of tornadoes reported in the United States during the years 1960–1984 is approximately

$$\frac{19{,}937.5}{25} = 797.5.$$

Problem 7 From Practice Problem 6 we see that the mean of these data is approximately 797.5. To approximate the standard deviation, we construct the following table.

Number of tornadoes	*Midpoint* m	*Frequency* f	$m - \bar{x}$	$(m - \bar{x})^2$	$(m - \bar{x})^2 f$	
399.5–499.5	449.5	1	−348	121,104	121,104	
499.5–599.5	549.5	1	−248	61,504	61,504	
599.5–699.5	649.5	6	−148	21,904	131,424	
699.5–799.5	749.5	4	−48	2,304	9,216	
799.5–899.5	849.5	6	52	2,704	16,224	
899.5–999.5	949.5	5	152	23,104	115,520	
999.5–1099.5	1049.5	1	252	63,504	63,504	
1099.5–1199.5	1149.5	1	352	123,904	123,904	
		25			642,400	Totals

Thus we see that

$$s^2 \approx \frac{642{,}400}{25} = 25{,}696,$$

and so the standard deviation is approximately

$$\sqrt{25{,}696} \approx 160.$$

Problem 8 When a fair die is rolled, each of the outcomes 1, 2, 3, 4, 5, and 6 is equally likely to occur. Thus the probability histogram is shown below.

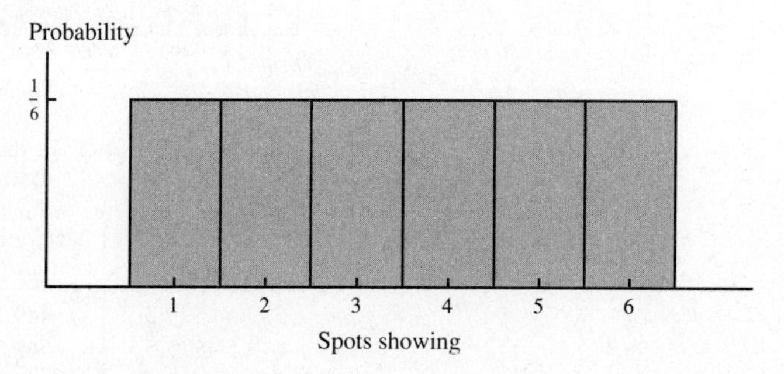

By formula (7.5) the mean is

$$\mu = \Sigma\, xp = 1\left(\frac{1}{6}\right) + 2\left(\frac{1}{6}\right) + 3\left(\frac{1}{6}\right) + 4\left(\frac{1}{6}\right) + 5\left(\frac{1}{6}\right) + 6\left(\frac{1}{6}\right)$$

$$= \frac{21}{6} = 3.5.$$

By formula (7.6) the variance is

$$\sigma^2 = (1 - 3.5)^2\left(\frac{1}{6}\right) + (2 - 3.5)^2\left(\frac{1}{6}\right) + \cdots + (6 - 3.5)^2\left(\frac{1}{6}\right)$$

$$= (6.25 + 2.25 + 0.25 + 0.25 + 2.25 + 6.25)\left(\frac{1}{6}\right)$$

$$= \frac{17.5}{6} = \frac{35}{12};$$

so

$$\sigma = \sqrt{\frac{35}{12}} \approx 1.71$$

Problem 9 If severe winter weather occurs (and the probability of this happening is .4), the contractor's profit will be $30,000. Otherwise (with probability 1 − .4 = .6), the contractor's profit will be $50,000. Hence the contractor's expected profit is

$$.4(\$30{,}000) + .6(\$50{,}000) = \$12{,}000 + \$30{,}000 = \$42{,}000.$$

Problem 10 (a) There are four sensible actions for the farmer—sending 2, 3, 4, or 5 truckloads of eggs to the city. There are also four possible states of nature, according to whether the demand for eggs is 2, 3, 4, or 5 truckloads. Consider the action of sending 4 truckloads of eggs to town. If the demand for eggs is only 2 truckloads, then the farmer will sell 2 truckloads and lose 2 truckloads to spoilage; his profit in this case is

$$2(\$150) + 2(-\$50) = \$300 - \$100 = \$200.$$

If the demand for eggs is 3 truckloads, then the farmer will sell 3 truckloads and lose 1 truckload to spoilage; his profit in this case is

$$3(\$150) + 1(-\$50) = \$450 - \$50 = \$400.$$

Finally, if the demand for eggs is at least 4 truckloads, then the farmer will sell all 4 of the truckloads that he sent, with a resulting profit of

$$4(\$150) + 0(-\$50) = \$600 + \$0 = \$600.$$

These numbers are the entries of the third row in the payoff table below; the other entries of the table are computed similarly.

Action	*State of nature*			
	Demand for eggs			
Number of	*(number of truckloads)*			
truckloads sent	*2*	*3*	*4*	*5*
2	300	300	300	300
3	250	450	450	450
4	200	400	600	600
5	150	350	550	750

(b) To use the maximin criterion, we choose the action corresponding to the row of the payoff table in which the smallest entry is as large as possible. This is row 1, and so the maximin criterion chooses the action of sending 2 truckloads.
(c) To use the maximax criterion, we choose the action corresponding to the row of the payoff table containing the largest entry. This is row 5, and so the maximin criterion chooses the action of sending 5 truckloads.
(d) To use Bayes's criterion, we compute the expected profit for each action. These expected profits are $300, $410, $440, and $410, respectively. Hence Bayes's criterion chooses the action of sending 4 truckloads.

Problem 11

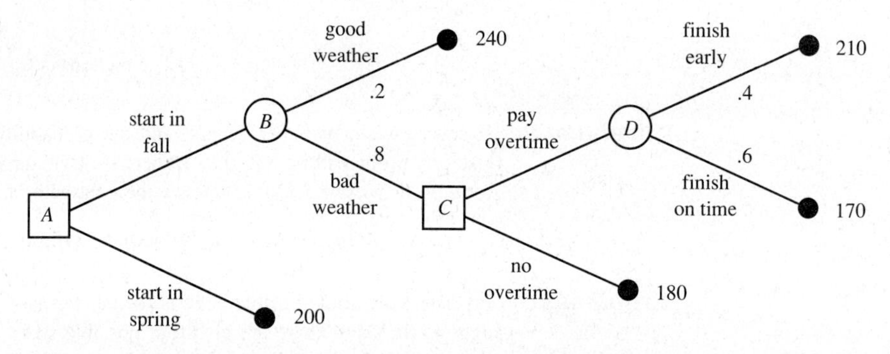

(a) The decision tree is shown in the figure above, where profits are denoted in thousands of dollars. Note that the payoffs are reduced by $20,000 in the event of bad weather, and by an additional $10,000 if overtime is paid.
(b) Working backwards from the end-nodes, we assign to chance-node D the expected value of the nodes to which it branches, which is

$$.4(210) + .6(170) = 186.$$

To choice-node C, we assign the larger of 186 and 180, which is 186. To chance-node B we assign the expected value of the nodes to which it branches, which is

$$.2(240) + .8(186) = 196.80.$$

To choice-node A, we assign the larger of 196.80 and 200, which is 200.
(c) Thus the maximum expected payoff for the contractor is $200,000 when he starts construction in the spring.

Problem 12 (a) The probability of receiving between 3 and 4 service calls, inclusive, is the area under the probability distribution in Figure 7.14 between 2.5 and 4.5. This area is

$$.20 + .10 = .30.$$

(b) The probability of receiving at most 1 service call is the area under the probability distribution in Figure 7.14 between -0.5 and 1.5. This area is

$$.10 + .25 = .35.$$

Problem 13 (a) To determine the distribution of the random variable that counts the number of sixes rolled when a fair die is tossed five times, we apply Bernoulli's formula with "success" meaning that a six is rolled. Then the probability of success is $p = 1/6$, and the probability of failure is $q = 5/6$. Thus the probability of obtaining exactly one six in five rolls is

$$C(5, 1)\left(\frac{1}{6}\right)^1\left(\frac{5}{6}\right)^4 = 5\left(\frac{1}{6}\right)\left(\frac{625}{1296}\right) = \frac{3125}{7776}.$$

Similarly we can compute the other probabilities in the probability density function, which is shown below.

Number of sixes	*Probability*
0	$\frac{3125}{7776}$
1	$\frac{3125}{7776}$
2	$\frac{1250}{7776}$
3	$\frac{250}{7776}$
4	$\frac{25}{7776}$
5	$\frac{1}{7776}$

(b) The probability of obtaining more than one six in five tosses is

$$\begin{aligned} P(2, 3, 4, \text{ or } 5 \text{ sixes}) &= 1 - P(0 \text{ or } 1 \text{ six}) \\ &= 1 - [P(0 \text{ sixes}) + P(1 \text{ six})] \\ &= 1 - \left(\frac{3125}{7776} + \frac{3125}{7776}\right) \\ &= \frac{1526}{7776}. \end{aligned}$$

Problem 14 We must determine the mean and standard deviation of the random variable that counts the number of contracts obtained. Since this random variable is binomially distributed, its mean and standard deviation can be computed by using Theorem 7.2 with $p = .2$, $q = .8$, and $n = 30$. Then

$$\mu = np = 30(.2) = 6$$

and

$$\sigma = \sqrt{npq} = \sqrt{30(.2)(.8)} = \sqrt{4.8} \approx 2.19.$$

Problem 15 (a) The probability is ½ of identifying the psychotic person in each pair by chance alone. Hence the probability of identifying the psychotic person in exactly six of the ten pairs by chance alone is given by Bernoulli's formula as

$$C(10, 6)\left(\frac{1}{2}\right)^6\left(\frac{1}{2}\right)^4 = 210\left(\frac{1}{64}\right)\left(\frac{1}{16}\right) = \frac{210}{1024}.$$

(b) As in (a) above we see that the probability of identifying the psychotic person in exactly seven of the ten pairs by chance alone is

$$C(10, 7)\left(\frac{1}{2}\right)^7\left(\frac{1}{2}\right)^3 = 210\left(\frac{1}{128}\right)\left(\frac{1}{8}\right) = \frac{120}{1024}.$$

(c) By calculations similar to those in (a) and (b), we see that the probabilities of identifying the psychotic person in exactly eight, nine, and ten of the pairs by chance alone are 45/1024, 10/1024, and 1/1024, respectively. Hence the probability of identifying the psychotic person in at least six of the ten pairs by chance alone is

$$\frac{210}{1024} + \frac{120}{1024} + \frac{45}{1024} + \frac{10}{1024} + \frac{1}{1024} = \frac{386}{1024} \approx .377.$$

(d) Since the probability in (c) is so large, the graphologist's claim should not be accepted.

Problem 16 (a) In standard units a score of 48 becomes

$$z = \frac{48 - 54}{8} = -\frac{6}{8} = -.75.$$

Hence, as in Example 7.22(a), the area to the right of 48 is

$$1 - A(-.75) = 1 - .2266 = .7734.$$

(b) In standard units scores of 68 and 72 become $z = 1.75$ and $z = 2.25$, respectively. Hence, as in Example 7.22(b), we see that the area between 68 and 72 is

$$A(2.25) - A(1.75) = 0.9878 - 0.9599 = .0279.$$

Problem 17 The random variable that counts the number of defective bottles is binomially distributed with a mean of

$$\mu = np = 3000(.005) = 15$$

and a standard deviation of

$$\sigma = \sqrt{npq} = \sqrt{3000(.005)(.995)} = \sqrt{14.925} \approx 3.86.$$

Since both np and nq exceed 5, we can approximate this binomial distribution by a normal distribution having the same mean and standard deviation. Now the probability of having 20 or more defective bottles is approximately equal to the area under the normal curve to the right of 19.5. Because 19.5 converts to about 1.17 standard units, we see that

$$P(\text{at least 20 defectives}) \approx 1 - A(1.17) = 1 - .8790 = .1210.$$

■ Chapter 8

Problem 1 Substituting -2 for x in the function g gives

$$g(-2) = (-2)^3 - 2(-2)^2 + (-2) - 4 = -8 - 8 - 2 - 4 = -22.$$

Likewise substituting 3 for x in the function g gives

$$g(3) = (3)^3 - 2(3)^2 + (3) - 4 = 27 - 18 + 3 - 4 = 8.$$

Problem 2 (a) If x dollars are invested, the sales charge $S(x)$ is given by

$$S(x) = \begin{cases} 0.08x & \text{if } \quad 0 \le x < 10{,}000 \\ 0.06x & \text{if } 10{,}000 \le x < 25{,}000 \\ 0.05x & \text{if } 25{,}000 \le x. \end{cases}$$

(b) Evaluating S for $x = 8000$, $x = 12{,}000$, and $x = 30{,}000$, we obtain

$$S(8000) = 0.08(8000) = 640,$$
$$S(12{,}000) = 0.06(12{,}000) = 720, \quad \text{and}$$
$$S(30{,}000) = 0.05(30{,}000) = 1500.$$

Problem 3 (a) The inequality $-3 \le x < 8$ describes the points in the interval $[-3, 8)$. Note that the bracket is used to denote an endpoint that is included in the interval, and the parenthesis is used to denote an endpoint that is not included in the interval. The graph of this inequality is shown below.

(b) The inequality $-5 \le x \le 0$ describes the points in the interval $[-5, 0]$. The graph of this inequality is shown below.

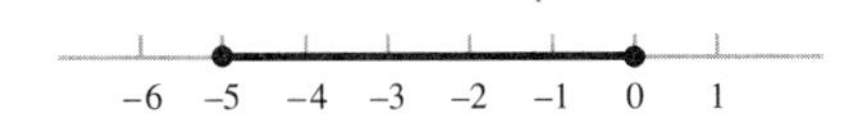

(c) The inequality $-4 \le x$ describes the points in the interval $[-4, \infty)$. The graph of this inequality is shown below.

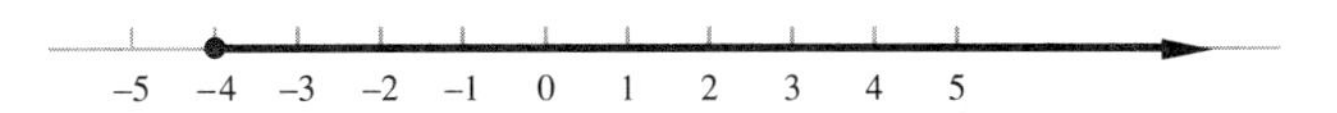

(d) The inequality $x < 6$ describes the points in the interval $(-\infty, 6)$. The graph of this inequality is shown below.

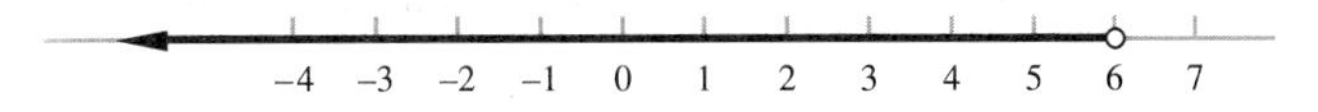

Problem 4 The domain of the function is the set of acceptable values for x. Because $\sqrt{4 - x}$ is defined only when $4 - x \ge 0$, we must have $4 \ge x$. But division by 0 is prohibited, and so we cannot allow $x = 4$. Thus the domain is $(-\infty, 4)$, the set of real numbers such that $4 > x$.

Problem 5 The following table gives points (x, y) that lie on the graph of g.

x	-3	-2	-1	0	1	2	3	4	5
y	10	3	-2	-5	-6	-5	-2	3	10

The graph is shown in the figure below.

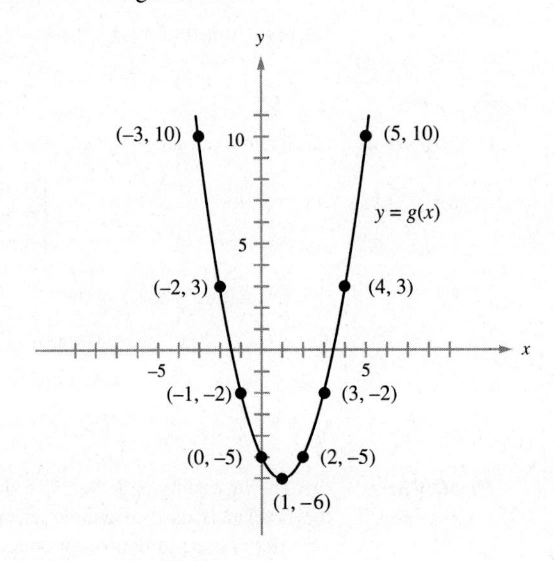

Problem 6 The graph of C is constant on the intervals (0, 1], (1, 2], (2, 3], (3, 4], and (4, 5]. It is shown in the following figure.

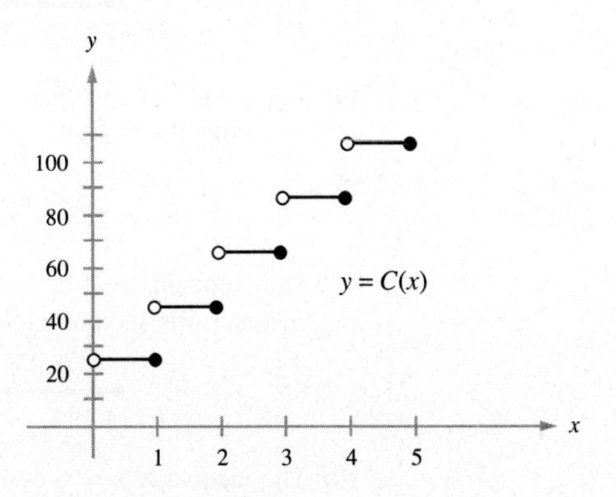

Problem 7 The following table gives points (x, y) that lie on the graph of F.

x	-1	0	1	2	3	4	5
y	4	3	2	1	2	3	4

The graph is shown in the figure below.

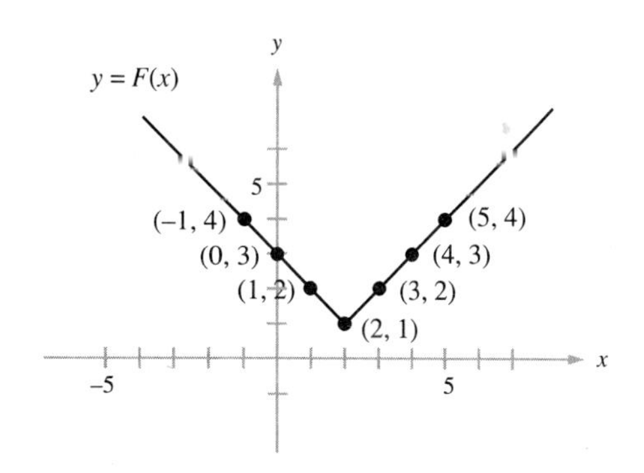

Problem 8 The only polynomial functions are f and F, which have degrees 1 and 7, respectively. The function g is not a polynomial because it has a negative exponent, h is not a polynomial because it has a term containing $\sqrt{x}$, and G is not a polynomial because its coefficients are *divided* by nonnegative powers of x.

Problem 9 Since f is a quadratic function in which the coefficient of the x^2 term is negative, its graph is a downward-opening parabola. The following table gives points (x, y) that lie on the graph of f.

x	-1	0	1	2	3
y	-3	3	5	3	-3

The graph is shown in the figure below.

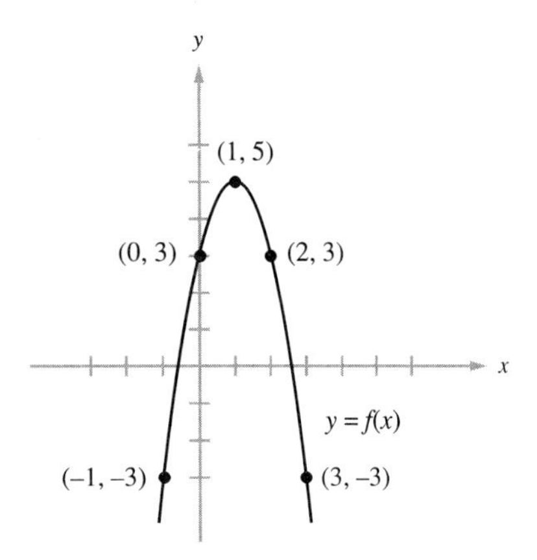

Problem 10 The following table gives points (x, y) that lie on the graph of f.

x	-1	0	1	2	3
y	4	0	2	4	0

The graph is shown in the figure below.

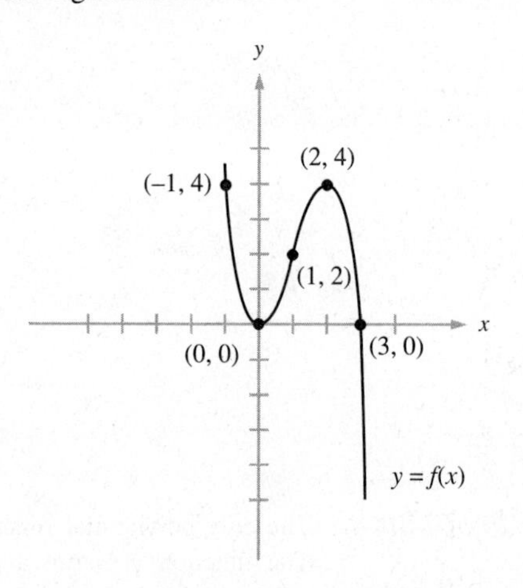

Problem 11 A rational function is defined for all values except those that make its denominator zero. Thus in (a) the function h is defined unless $x = 3$, and so the domain of h is

$$(-\infty, 3) \cup (3, \infty).$$

In (b) the function H is defined unless $x(x - 1) = 0$, that is, unless $x = 0$ or $x = 1$. Hence the domain of H is

$$(-\infty, 0) \cup (0, 1) \cup (1, \infty).$$

Problem 12 The function h is defined for all real numbers except $x = 3$, and the line $x = 3$ is a vertical asymptote. From the graph below we see that $y = 0$ (the x-axis) is a horizontal asymptote. The following table gives points (x, y) that lie on the graph of h.

x	0	1	2	2.5	3.5	4	5	6
y	$\frac{1}{9}$	$\frac{1}{4}$	1	4	4	1	$\frac{1}{4}$	$\frac{1}{9}$

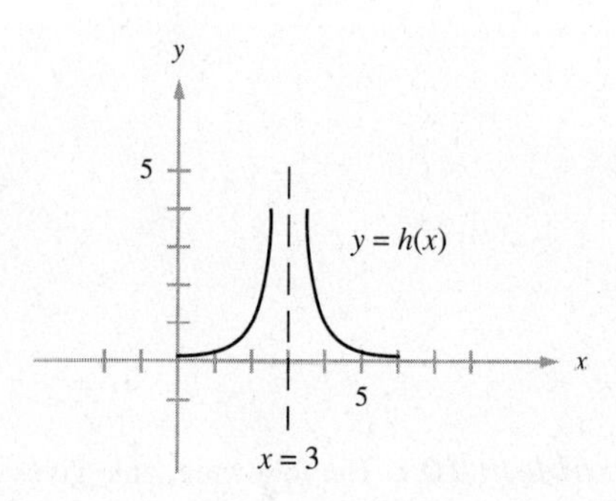

Problem 13 If the Democratic presidential candidate receives 60% of the popular vote, then the cube law gives the proportion of seats in the House of Representatives won by Democratic candidates as

$$p(.6) = \frac{(.6)^3}{(.6)^3 + (.4)^3} = \frac{.216}{.280} = .771.$$

Therefore approximately 77.1% of the seats in the House of Representatives would be won by Democratic candidates.

Problem 14 Because H is an exponential function with base ⅓, the graph of H will have the same shape as the graph in Figure 8.16(a). It will be the mirror image of the graph of $y = 3^x$ reflected about the y-axis. (See Figure 8.15(a).) The following table gives points (x, y) that lie on the graph of H.

x	-2	-1	$-\frac{1}{2}$	0	$\frac{1}{2}$	1	2
y	9	3	$\sqrt{3}$	1	$\frac{1}{\sqrt{3}}$	$\frac{1}{3}$	$\frac{1}{9}$

The graph of H is shown below.

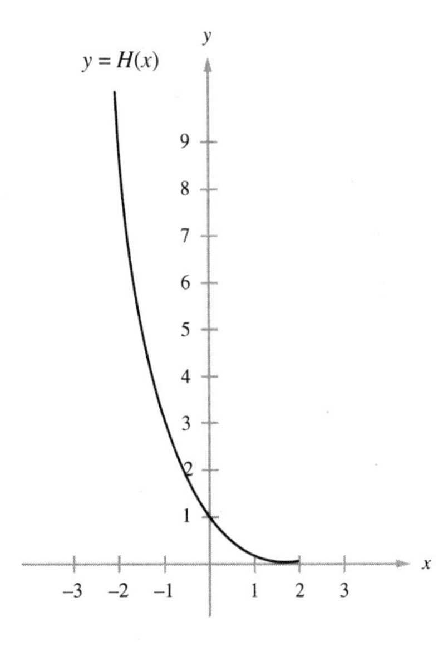

Problem 15 The graph of f is the mirror image of the graph of $y = e^x$ shown in Figure 8.17. The following table gives points (x, y) that lie on the graph of f.

x	-2	-1.5	-1	$-\frac{1}{2}$	0	$\frac{1}{2}$	1	2	3
y	7.39	4.48	2.72	1.65	1	0.61	0.37	0.14	0.05

The graph of f is shown below.

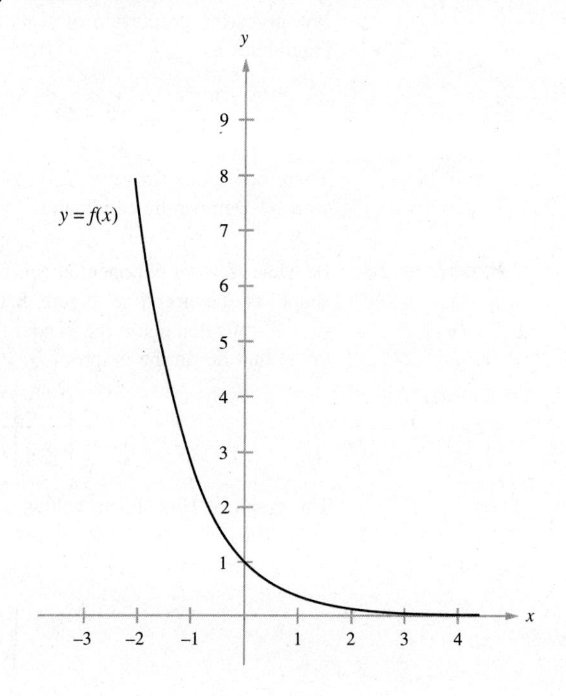

Problem 16 (a) Radioactive decay is described by a function of the form

$$Q(t) = q_0e^{rt},$$

where $r < 0$ because mass is decreasing. In this problem the initial quantity of carbon-14 is 100 mg and the decay rate is $0.012\% = 0.00012$; so the desired function is

$$Q(t) = 100e^{-0.00012t}.$$

(b) The amount of carbon-14 that will remain undisintegrated after 2000 years is

$$Q(2000) = 100e^{-0.00012(2000)} = 100e^{-0.24} \approx 78.7 \text{ mg}.$$

Problem 17 The given equation states that the exponent to which we must raise the base 4 to get 1024 is 5; thus the equivalent exponential form is

$$4^5 = 1024.$$

Problem 18 To write 10 as a power of 1000, we need a base of $\frac{1}{3}$. Hence the logarithmic form of the given equation is

$$\log_{1000} 10 = \tfrac{1}{3}.$$

Problem 19 (a) Using logarithm property 1, we obtain

$$\log_7 8 + \log_7 6 = \log_7 (8 \cdot 6) = \log_7 48.$$

(b) Using logarithm property 2, we obtain

$$\log_e 52 - \log_e 4 = \log_e \left(\frac{52}{4}\right) = \log_e 13.$$

(c) Using logarithm property 3, we obtain

$$\log_5 x^{-3} = -3 \log_5 x.$$

(d) Using logarithm property 6, we obtain

$$10^{\log_{10} 2x} = 2x.$$

Problem 20 The graph of h is the mirror image of the graph of $y = 3^x$ reflected about the line $y = x$. See Figure 8.15(a) for the graph of $y = 3^x$. The following table gives points (x, y) that lie on the graph of h.

x	$\frac{1}{9}$	$\frac{1}{3}$	1	3	9
y	-2	-1	0	1	2

The graph of h is shown below.

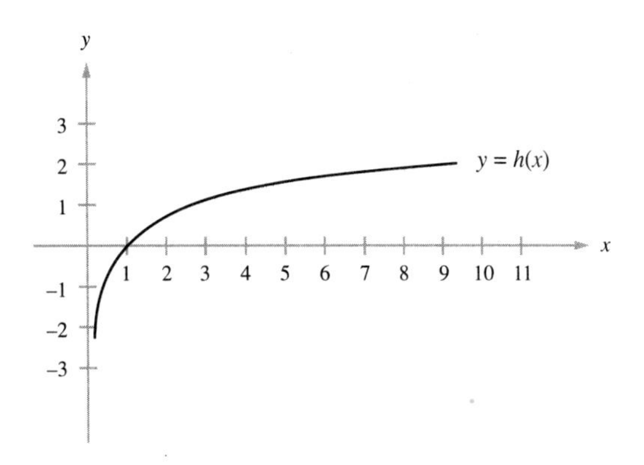

Problem 21 The function F has the form $b^{h(x)}$ with $b = 10$ and $h(x) = 4 - 5x$. Hence

$$F(x) = e^{(\ln 10)(4-5x)}.$$

Problem 22 The function G has the form $\log_b h(x)$ with $b = 2$ and $h(x) = 7x - 3$. Thus

$$G(x) = \frac{1}{\ln 2} \ln (7x - 3).$$

Problem 23 To determine the amount of time required for the account in Example 8.22 to reach \$14,500, we must solve for n in the equation

$$\$14{,}500 = \$8000(1.015)^n.$$

Dividing both sides by \$8000 gives

$$1.8125 = (1.015)^n.$$

Taking the natural logarithm of both sides of this equation and applying the third logarithm property on page 478, we obtain

$$\ln 1.8125 = n \ln 1.015.$$

Hence

$$n = \frac{\ln 1.8125}{\ln 1.015} \approx 39.94.$$

Therefore the account will be worth \$14,500 in approximately 39.94 *quarters,* which is about 10 years.

Problem 24 The quantity of carbon-14 that remains after t years is given by the formula in Example 8.24:

$$Q(t) = q_0 e^{-0.00012t}.$$

We must solve this formula to find the value of t for which $Q(t) = \frac{1}{2}q_0$. Thus

$$\begin{aligned}
\tfrac{1}{2}q_0 &= q_0 e^{-0.00012t} \\
\tfrac{1}{2} &= e^{-0.00012t} \\
\ln \tfrac{1}{2} &= -0.00012t \\
t &= \frac{\ln 0.50}{-0.00012} \approx 5776 \text{ years.}
\end{aligned}$$

Problem 25 Using the definitions of the sum, difference, product, and quotient of functions, we obtain

$$\begin{aligned}
(f + g)(x) &= f(x) + g(x) = (12x^4 - 8x^2) + (-2x^2) = 12x^4 - 10x^2, \\
(f - g)(x) &= f(x) - g(x) = (12x^4 - 8x^2) - (-2x^2) = 12x^4 - 6x^2, \\
(fg)(x) &= f(x) \cdot g(x) = (12x^4 - 8x^2)(-2x^2) = -24x^6 + 16x^4, \text{ and} \\
\frac{f}{g}(x) &= \frac{f(x)}{g(x)} = \frac{12x^4 - 8x^2}{-2x^2} = -6x^2 + 4 \text{ if } x \neq 0.
\end{aligned}$$

Problem 26 $4g(x) = 4(x - \ln x) = 4x - 4 \ln x$

Problem 27 (a) To compute $f(g(x))$, we replace each occurrence of x in $f(x)$ by $g(x) = 2x + 3$. Thus we obtain

$$f(g(x)) = e^{2x+3}.$$

(b) To compute $g(f(x))$, we replace each occurrence of x in $g(x)$ by $f(x) = e^x$. Thus we obtain

$$g(f(x)) = 2e^x + 3.$$

Problem 28 There are many possible ways in which f and g can be defined. One possibility is to choose $f(x) = 4x + 7$ and $g(x) = \sqrt[3]{5x - 6}$.

Chapter 9

Problem 1 Proceeding as in Example 9.1, we compute the difference quotient.

$$\begin{aligned}\frac{f(x + h) - f(x)}{(x + h) - x} &= \frac{f(x + h) - f(x)}{h}\\ &= \frac{[2 - 5(x + h)] - (2 - 5x)}{h}\\ &= \frac{2 - 5x - 5h - 2 + 5x}{h}\\ &= \frac{-5h}{h}\\ &= -5\end{aligned}$$

Since the difference quotient does not contain h, its values are -5 as h approaches zero. Hence the instantaneous rate of change of f at x is -5.

Problem 2 As in Example 9.2, we must first compute the instantaneous rate of change of f at x. The calculation of the difference quotient is shown below.

$$\begin{aligned}\frac{f(x + h) - f(x)}{(x + h) - x} &= \frac{f(x + h) - f(x)}{h}\\ &= \frac{[4(x + h)^2 - 5] - (4x^2 - 5)}{h}\\ &= \frac{[4(x^2 + 2hx + h^2) - 5] - (4x^2 - 5)}{h}\\ &= \frac{4x^2 + 8hx + 4h^2 - 5 - 4x^2 + 5}{h}\\ &= \frac{8hx + 4h^2}{h}\\ &= 8x + 4h\end{aligned}$$

As h approaches zero, the values of the difference quotient approach $8x$. Hence the instantaneous rate of change of f at x is $8x$.

To determine the instantaneous rate of change of f at $x = 2$, we need only evaluate $8x$ when $x = 2$. Thus the instantaneous rate of change of f at $x = 2$ is

$$8(2) = 16.$$

Problem 3 We proceed as in Example 9.4 to compute the difference quotient for f.

$$\begin{aligned}\frac{f(x+h)-f(x)}{(x+h)-x} &= \frac{f(x+h)-f(x)}{h}\\ &= \frac{\dfrac{1}{x+h}-\dfrac{1}{x}}{h}\\ &= \frac{\dfrac{x}{x(x+h)}-\dfrac{x+h}{x(x+h)}}{h}\\ &= \frac{-h}{hx(x+h)}\\ &= \frac{-1}{x(x+h)}\end{aligned}$$

As h approaches 0, the difference quotient values approach $-1/x^2$. Now the slope of the tangent line to $y = f(x)$ at $(2, \frac{1}{2})$ is the instantaneous rate of change of f at $x = 2$. Hence the slope of the desired tangent line is

$$\frac{-1}{x^2} = \frac{-1}{(2)^2} = -\frac{1}{4}.$$

Using the point-slope form of the equation of a line, we find the equation of the tangent line as shown below.

$$\begin{aligned}y - y_1 &= m(x - x_1)\\ y - \tfrac{1}{2} &= -\tfrac{1}{4}(x-2)\\ y - \tfrac{1}{2} &= -\tfrac{1}{4}x + \tfrac{1}{2}\\ y &= -\tfrac{1}{4}x + 1\end{aligned}$$

Problem 4 (a) In Figure 9.7(a) we see that $g(x)$ can be made as close to 2 as desired by restricting x to be sufficiently close (but not equal) to 1. Hence

$$\lim_{x\to 1} g(x) = 2.$$

On the other hand, there is no real number to which $G(x)$ can be made close for all x sufficiently close (but not equal) to 1. Hence

$$\lim_{x\to 1} G(x)$$

does not exist.

Problem 5 The values of e^x get increasingly larger as x increases. Hence

$$\lim_{x\to\infty} e^x$$

does not exist.

On the other hand, the values of e^x become arbitrarily close to 0 as x decreases without bound. Therefore

$$\lim_{x\to-\infty} e^x = 0.$$

Problem 6 (a) We must divide the numerator and denominator by x^3, because x^3 is the largest power of x appearing in the denominator of the given expression. Then we apply Theorem 9.2 as in Example 9.9. In this way we obtain

$$\lim_{x\to\infty} \frac{x^3 + 4x}{2x^3 - 7} = \lim_{x\to\infty} \frac{\left(1 + \frac{4}{x^2}\right)}{\left(2 - \frac{7}{x^3}\right)}$$

$$= \frac{\lim_{x\to\infty}\left(1 + \frac{4}{x^2}\right)}{\lim_{x\to\infty}\left(2 - \frac{7}{x^3}\right)}$$

$$= \frac{\lim_{x\to\infty} 1 + \lim_{x\to\infty} \frac{4}{x^2}}{\lim_{x\to\infty} 2 - \lim_{x\to\infty} \frac{7}{x^3}}$$

$$= \frac{\lim_{x\to\infty} 1 + 4 \lim_{x\to\infty} \frac{1}{x^2}}{\lim_{x\to\infty} 2 - 7 \lim_{x\to\infty} \frac{1}{x^3}}$$

$$= \frac{\lim_{x\to\infty} 1 + 4\left(\lim_{x\to\infty} \frac{1}{x}\right)^2}{\lim_{x\to\infty} 2 - 7\left(\lim_{x\to\infty} \frac{1}{x}\right)^3}$$

$$= \frac{1 + 4(0)^2}{2 - 7(0)^3} = \frac{1}{2}.$$

(b) In this case we must divide the numerator and denominator by x^5. Applying Theorem 9.2 as above, we obtain

$$\lim_{x\to-\infty}\frac{x^2 - x + 6}{x^5 - 4x^3 + 2} = \frac{\left(\lim\limits_{x\to-\infty}\frac{1}{x}\right)^3 - \left(\lim\limits_{x\to-\infty}\frac{1}{x}\right)^4 + 6\left(\lim\limits_{x\to-\infty}\frac{1}{x}\right)^5}{\lim\limits_{x\to-\infty} 1 - 4\left(\lim\limits_{x\to-\infty}\frac{1}{x}\right)^2 + 2\left(\lim\limits_{x\to-\infty}\frac{1}{x}\right)^5}$$

$$= \frac{(0)^3 - (0)^4 + 6(0)^5}{1 - 4(0)^3 + 2(0)^5} = 0.$$

(c) Dividing the numerator and denominator by x^2 produces

$$\frac{4x^3 - 7x + 3}{8x^2 + 9x - 5} = \frac{\left(4x - \frac{7}{x} + \frac{3}{x^2}\right)}{\left(8 + \frac{9}{x} - \frac{5}{x^2}\right)}.$$

As x increases without bound, the numerator increases without bound and the denominator approaches 8. Hence

$$\lim_{x\to\infty}\frac{4x^3 - 7x + 3}{8x^2 + 9x - 5}$$

does not exist.

Problem 7 The function f fails to be differentiable at x_2 (where there is a vertical tangent line) and at x_5 and x_7 (where there are abrupt changes of direction). It is differentiable at all other points.

Problem 8 Because there is a "hole" in the graph at x_2, the function f is not defined at x_2. Since a function must be defined at x_2 in order to be continuous there, f has a point of discontinuity at x_2. At the point x_6 the limit of f does not exist because there is no *single* real number to which $f(x)$ is close for all x close to x_6. Hence f also has a point of discontinuity at x_6. But f is continuous at all other points.

Problem 9 Applying the formula for continuously compounded interest with $P = \$2500$, $r = .08$, and $t = 4$, we obtain

$$A = Pe^{rt} = \$2500e^{0.08(4)} = \$2500e^{0.32} \approx \$3442.82.$$

Problem 10 Since

$$\frac{g(x+h) - g(x)}{(x+h) - x} = \frac{g(x+h) - g(x)}{h} = \frac{(x+h) - x}{h} = \frac{h}{h} = 1,$$

we see that $g'(x) = 1$ for all x.

Problem 11 Since f, g, and h are all constant functions, we have

$$f'(x) = 0, \qquad g'(x) = 0, \qquad \text{and} \qquad h'(x) = 0.$$

Problem 12 Since f is a power function, we see that

$$f'(x) = 7x^6.$$

Writing $g(x) = x^{1/3}$, we see that g is a power function. Hence

$$g'(x) = \frac{1}{3}x^{-2/3} = \frac{1}{3\sqrt[3]{x^2}}.$$

Because $h(x) = x^{-4}$, we have

$$h'(x) = -4x^{-5} = -\frac{4}{x^5}.$$

Problem 13 Since $f(x) = x^{-1}$, f is a power function. Hence

$$f'(x) = -x^{-2} = -\frac{1}{x^2}.$$

Therefore the slope of the tangent line to the graph of f at $(2, \frac{1}{2})$ is

$$f'(2) = -\tfrac{1}{4}.$$

The point-slope form of the equation of a line now gives the desired equation.

$$\begin{aligned} y - y_1 &= m(x - x_1) \\ y - \tfrac{1}{2} &= -\tfrac{1}{4}(x - 2) \\ y - \tfrac{1}{2} &= -\tfrac{1}{4}x + \tfrac{1}{2} \\ y &= -\tfrac{1}{4}x + 1 \end{aligned}$$

Problem 14 As in Example 9.17, we see that

$$\begin{aligned} f'(x) &= (x^{12} - x^9 + x)' \\ &= (x^{12})' - (x^9)' + x' \\ &= 12x^{11} - 9x^8 + 1. \end{aligned}$$

The function g can be written as $g(x) = x^{3/4} + 6$. Therefore

$$\begin{aligned} g'(x) &= (x^{3/4} + 6)' \\ &= (x^{3/4})' + (6)' \\ &= \tfrac{3}{4}x^{-1/4} + 0 \\ &= \tfrac{3}{4}x^{-1/4}. \end{aligned}$$

Problem 15 As in Example 9.18 we have

$$\begin{aligned} f'(x) &= (\tfrac{1}{2}x^8 - 5x^6 + 9)' \\ &= (\tfrac{1}{2}x^8)' - (5x^6)' + (9)' \\ &= \tfrac{1}{2}(x^8)' - 5(x^6)' + 0 \\ &= \tfrac{1}{2}(8x^7) - 5(6x^5) \\ &= 4x^7 - 30x^5. \end{aligned}$$

Likewise since $g(x) = -2x^{-4} + \sqrt{6}$, we have

$$\begin{aligned} g'(x) &= (-2x^{-4} + \sqrt{6})' \\ &= (-2x^{-4})' + (\sqrt{6})' \\ &= -2(x^{-4})' + 0 \\ &= 8x^{-5} \\ &= \frac{8}{x^5}. \end{aligned}$$

Problem 16 The instantaneous rate of change of f at $x = -1$ is $f'(-1)$. Now

$$\begin{aligned} f'(x) &= (2x^4 - 5x^3 + x^2 - 4x - 3)' \\ &= (2x^4)' - (5x^3)' + (x^2)' - (4x)' - (3)' \\ &= 2(x^4)' - 5(x^3)' + (x^2)' - 4(x)' - 0 \\ &= 2(4x^3) - 5(3x^2) + (2x) - 4(1) \\ &= 8x^3 - 15x^2 + 2x - 4. \end{aligned}$$

Hence

$$\begin{aligned} f'(-1) &= 8(-1)^3 - 15(-1)^2 + 2(-1) - 4 \\ &= -8 - 15 - 2 - 4 \\ &= -29. \end{aligned}$$

Problem 17 If we take

$$g(x) = e^x \quad \text{and} \quad h(x) = 3 - 2x,$$

then $f(x) = g(h(x))$.

Problem 18 For the function $f(x) = (2x^5 - 9x^3 + 4x - 5)^{10}$, we have $f(x) = g(h(x))$, where

$$h(x) = 2x^5 - 9x^3 + 4x - 5 \quad \text{and} \quad g(x) = x^{10}.$$

Now

$$h'(x) = (2x^5 - 9x^3 + 4x - 5)' = 10x^4 - 27x^2 + 4,$$

and

$$g'(x) = 10x^9.$$

Thus, by the chain rule,

$$\begin{aligned} f'(x) &= g'(h(x)) \cdot h'(x) \\ &= 10(2x^5 - 9x^3 + 4x - 5)^9(10x^4 - 27x^2 + 4) \\ &= (100x^4 - 270x^2 + 40)(2x^5 - 9x^3 + 4x - 5)^9 \end{aligned}$$

Problem 19 Let

$$h(x) = 4x^3 - 7x^2 + 6x - 2 \qquad \text{and} \qquad g(x) = x^{3/2}.$$

Then $f(x) = g(h(x))$, and so

$$\begin{aligned} f'(x) &= g'(h(x)) \cdot h'(x) \\ &= \tfrac{3}{2}(4x^3 - 7x^2 + 6x - 2)^{1/2} \cdot (12x^2 - 14x + 6) \\ &= (18x^2 - 21x + 9)\sqrt{4x^3 - 7x^2 + 6x - 2}. \end{aligned}$$

Problem 20 Let

$$h(x) = 3x^2 + 5x \qquad \text{and} \qquad f(x) = g(h(x)).$$

Then the expression to be differentiated is $f(x)$. By the chain rule

$$\begin{aligned} f'(x) &= g'(h(x)) \cdot h'(x) \\ &= e^{-(3x^2+5x)} \cdot (6x + 5) \\ &= (6x + 5)e^{-(3x^2+5x)}. \end{aligned}$$

Problem 21 Here dy/dx is the derivative of y with respect to x. Since $y = (4x^2 - 6x - 5)^{-3}$ is the composition of $h(x) = 4x^2 - 6x - 5$ and $g(x) = x^{-3}$, we must apply the chain rule to compute the derivative. The result is

$$\frac{dy}{dx} = -3(4x^2 - 6x - 5)^{-4}(8x - 6) = \frac{-24x + 18}{(4x^2 - 6x - 5)^4}.$$

Problem 22 Because the derivative of a sum of functions is the sum of their derivatives, we have

$$\begin{aligned} f'(x) &= (9x^6 + \ln x)' \\ &= (9x^6)' + (\ln x)' \\ &= 54x^5 + \frac{1}{x} \end{aligned}$$

and

$$\begin{aligned} g'(x) &= (4x^{-2} + \sqrt{\ln x})' \\ &= (4x^{-2})' + [(\ln x)^{1/2}]' \\ &= -8x^{-3} + \tfrac{1}{2}(\ln x)^{-1/2}\left(\frac{1}{x}\right) \\ &= -8x^{-3} + \frac{1}{2x\sqrt{\ln x}}. \end{aligned}$$

Problem 23 The slope of the tangent line to the graph of f at $(1, 0)$ is $f'(1)$. Since

$$f'(x) = (\ln x)' = \frac{1}{x},$$

we see that $f'(1) = 1$. Therefore the desired line has slope 1 and passes through the point $(1, 0)$. Using the point-slope form of the equation of a line, we find that its equation is

$$y - 0 = 1(x - 1),$$

that is,

$$y = x - 1.$$

Problem 24 Note that $f(x)$ has the form $\ln h(x)$, where $h(x) = 7x^6 + x^{-3}$. Hence

$$f'(x) = \frac{h'(x)}{h(x)} = \frac{42x^5 - 3x^{-4}}{7x^6 + x^{-3}}.$$

Problem 25 The instantaneous rate of change of f when $x = 1$ is $f'(1)$. Since

$$f(x) = (e^x)' = e^x,$$

we see that $f'(1) = e$. Thus the instantaneous rate of change of f at $x = 1$ is e.

Problem 26 The function $f(x)$ has the form $5e^{h(x)}$ with $h(x) = -9x$. Thus

$$f'(x) = (5e^{h(x)})' = 5(e^{h(x)})' = 5h'(x)e^{h(x)} = 5(-9)e^{-9x} = -45e^{-9x}.$$

Likewise $F(x)$ has the form $e^{h(x)}$ with $h(x) = 7x^3 + 2x^2 - 8$. Hence

$$F'(x) = (e^{h(x)})' = h'(x)e^{h(x)} = (21x^2 + 4x)e^{7x^3+2x^2-8}.$$

Problem 27 The rate of change of mass with respect to time is given by dA/dt. Now

$$\frac{dA}{dt} = (2e^{-0.05545t})' = 2(e^{-0.05545t})' = 2(-0.05545)e^{-0.05545t}.$$

Hence after 10 hours (when $t = 10$), the rate of change of mass is approximately

$$2(-0.05545)e^{-0.05545(10)} = -0.1109e^{-0.5545} \approx -0.0637.$$

Thus the mass is decreasing by approximately 0.0637 mg/hour.

Chapter 10

Problem 1 (a) By applying the product rule, we obtain

$$\begin{aligned} F'(x) &= [(2x^4 - 5x)(6x^3 + 3x^2)]' \\ &= (2x^4 - 5x)(6x^3 + 3x^2)' + (6x^3 + 3x^2)(2x^4 - 5x)' \\ &= (2x^4 - 5x)(18x^2 + 6x) + (6x^3 + 3x^2)(8x^3 - 5) \\ &= 84x^6 + 36x^5 - 120x^3 - 45x^2. \end{aligned}$$

(b)
$$\begin{aligned} G'(x) &= (4xe^x)' \\ &= 4x(e^x)' + e^x(4x)' \\ &= 4xe^x + e^x(4) \\ &= (4x + 4)e^x \end{aligned}$$

(c)
$$\begin{aligned} H'(x) &= [x \ln (4x + 1)]' \\ &= x[\ln (4x + 1)]' + [\ln (4x + 1)](x)' \\ &= x\left(\frac{4}{4x + 1}\right) + [\ln (4x + 1)](1) \\ &= \frac{4x}{4x + 1} + \ln (4x + 1) \end{aligned}$$

Problem 2 The slope of the tangent line to the graph of g at (0, 1) is $g'(0)$. Now

$$\begin{aligned} g'(x) &= (e^x\sqrt{x^2 + 1})' \\ &= e^x(\sqrt{x^2 + 1})' + \sqrt{x^2 + 1}(e^x)' \\ &= e^x(\tfrac{1}{2})(x^2 + 1)^{-1/2}(2x) + \sqrt{x^2 + 1}(e^x), \end{aligned}$$

and so

$$g'(0) = 1(\tfrac{1}{2})(1)(0) + 1(1) = 1.$$

We can use the point-slope form of the equation of a line to write the equation of the tangent line. The result is

$$y - 1 = 1(x - 0),$$

that is,

$$y = x + 1.$$

Problem 3 (a) By applying the quotient rule, we obtain

$$\begin{aligned} F'(x) &= \left(\frac{3x - 1}{2 - 5x}\right)' \\ &= \frac{(2 - 5x)(3x - 1)' - (3x - 1)(2 - 5x)'}{(2 - 5x)^2} \end{aligned}$$

$$= \frac{(2 - 5x)(3) - (3x - 1)(-5)}{(2 - 5x)^2}$$

$$= \frac{6 - 15x + 15x - 5}{(2 - 5x)^2}$$

$$= \frac{1}{(2 - 5x)^2}$$

(b)

$$G'(x) = \left(\frac{e^{-x}}{x}\right)'$$

$$= \frac{x(e^{-x})' - (e^{-x})(x)'}{x^2}$$

$$= \frac{x(-e^{-x}) - (e^{-x})(1)}{x^2}$$

$$= \frac{-(x + 1)e^{-x}}{x^2}$$

(c)

$$H'(x) = \left(\frac{\ln x}{2x + 1}\right)'$$

$$= \frac{(2x + 1)(\ln x)' - (\ln x)(2x + 1)'}{(2x + 1)^2}$$

$$= \frac{(2x + 1)\left(\dfrac{1}{x}\right) - (\ln x)(2)}{(2x + 1)^2}$$

$$= \frac{(2x + 1) - 2x \ln x}{x(2x + 1)^2}$$

Problem 4 The average cost function is

$$A(x) = \frac{C(x)}{x} = \frac{0.01x^2 + \ln(1000x + 1) + 20{,}000}{x}$$

$$= 0.01x + \frac{\ln(1000x + 1)}{x} + 20{,}000x^{-1}.$$

Hence

$$A'(x) = 0.01 + \frac{x[\ln(1000x + 1)]' - [\ln(1000x + 1](x)'}{x^2} - 20{,}000x^{-2}$$

$$= 0.01 + \frac{x\left(\dfrac{1000}{1000x + 1}\right) - \ln(1000x + 1)}{x^2} - 20{,}000x^{-2}$$

Therefore

$$A'(1000) = 0.01 - 0.000012816 - 0.02 \approx -0.01,$$

and so the average cost is decreasing by slightly more than \$0.01 per unit at a production level of 1000 bulbs.

Problem 5 If $f(x) = 2x^6 - 5x^3 + 9x - 7$, then

$$f'(x) = 12x^5 - 15x^2 + 9.$$

The second derivative f'' is the derivative of f':

$$f''(x) = 60x^4 - 30x.$$

The third derivative f''' is the derivative of f'':

$$f'''(x) = 240x^3 - 30.$$

Problem 6 If $y = e^{-2x}$, then the first, second, and third derivatives of y are

$$y' = -2e^{-2x}, \qquad y'' = 4e^{-2x}, \qquad \text{and} \qquad y''' = -8e^{-2x}.$$

Problem 7 If $y = (\ln x)/x$, then the first, second, and third derivatives of y are found by using the quotient rule to be:

$$\frac{dy}{dx} = \frac{1 - \ln x}{x^2}, \qquad \frac{d^2y}{dx^2} = \frac{2 \ln x - 3}{x^3}, \qquad \text{and}$$

$$\frac{d^3y}{dx^3} = \frac{11 - 6 \ln x}{x^4}.$$

Problem 8 The acceleration of the object after t seconds is given by the second derivative of s with respect to t. Now

$$s'(t) = 5.4t \qquad \text{and} \qquad s''(t) = 5.4.$$

Hence the acceleration of the object after t seconds is approximately 5.4 feet per second per second. (Thus the acceleration due to the moon's gravitational attraction is about one-sixth that of the earth.)

Problem 9 (a) The expression in (a) involves only x, and so its derivative can be found by the rules in Section 9.4:

$$(6x^5 - 3x^{-2})' = 30x^4 + 6x^{-3}.$$

(b) Since the derivative of y with respect to x is y', we have

$$(2x - y)' = (2x)' - (y)' = 2 - y'.$$

(c) To differentiate y^{-5} with respect to x, we let $u = y^{-5}$. Then

$$\frac{du}{dx} = \frac{du}{dy}\frac{dy}{dx}$$
$$= -5y^{-6}\frac{dy}{dx}$$
$$= -5y^{-6}y'.$$

(d) To differentiate e^{6y} with respect to x, we let $u = e^{6y}$. Then

$$\frac{du}{dx} = \frac{du}{dy}\frac{dy}{dx}$$
$$= 6e^{6y}\frac{dy}{dx}$$
$$= 6e^{6y}y'.$$

Problem 10 We begin by differentiating both sides of the given equation with respect to x:

$$(4y^3)' - y' - (5x^2)' = (3x)'$$
$$12y^2y' - y' - 10x = 3$$

Collect the terms involving y' on the left side of the equation and the other terms on the right side:

$$12y^2y' - y' = 10x + 3.$$

Hence

$$y' = \frac{10x + 3}{12y^2 - 1}.$$

Problem 11 Differentiating both sides of the equation with respect to x, we obtain

$$\frac{1}{2x + y}(2 + y') = 3x^2.$$

Multiplying both sides by $2x + y$ now yields

$$2 + y' = 3x^2(2x + y),$$

and so

$$y' = 3x^2(2x + y) - 2.$$

Problem 12 Differentiating both sides of the equation with respect to x, we obtain

$$8x + 6yy' = 0.$$

Therefore

$$y' = -\frac{8x}{6y} = -\frac{4x}{3y}.$$

Thus the value of y' at the point (2, 4) is

$$-\frac{4(2)}{3(4)} = -\frac{8}{12} = -\frac{2}{3},$$

and so the desired tangent line has slope $-2/3$. Its equation can be found using the point-slope form of the equation of a line:

$$y - 4 = -\frac{2}{3}(x - 2),$$

that is,

$$y = -\frac{2}{3}x + \frac{16}{3}.$$

Problem 13 Taking the derivative of both sides of the given equation with respect to t, we find that

$$\frac{dy}{dt} = 8\frac{dx}{dt} - 3x^2\frac{dx}{dt}.$$

Substituting $x = -1$ and $dx/dt = 2$ gives

$$\frac{dy}{dt} = 8(2) - 3(-1)^2(2) = 16 - 6 = 10.$$

Problem 14 The area of a circle of radius r is given by the formula $A = \pi r^2$. Differentiating this formula implicitly with respect to t produces

$$\frac{dA}{dt} = 2\pi r\frac{dr}{dt}.$$

Now $dr/dt = 10$ feet per month. Therefore when the radius is 600 feet, the contaminated area is increasing at the rate of

$$\frac{dA}{dt} = 2\pi(600)(10) = 12{,}000\pi$$

square feet per month.

Chapter 11

Problem 1 As we move from left to right, the functional values $g(x)$:

get smaller until we reach $x = -7$;

get larger until we reach $x = -4$;

get smaller until we reach $x = -2$;

remain constant between $x = -2$ and $x = 0$;

get smaller until we reach $x = 1$;

get larger until we reach $x = 4$;

get smaller until we reach $x = 6$; and

get larger from $x = 6$ on.

Therefore g is decreasing on the intervals $(-\infty, -7)$, $(-4, -2)$, $(0, 1)$ and $(4, 6)$; g is increasing on the intervals $(-7, -4)$, $(1, 4)$, and $(6, \infty)$; and g is constant on the interval $(-2, 0)$.

Problem 2 In Figure 11.5 we see that there are local maxima of g at $x = -4$ and $x = 4$ and local minima of g at $x = -7$, $x = 1$, and $x = 6$. Thus $g(-4) = 2$ and $g(4) = 4$ are local maxima, and $g(-7) = -2$, $g(1) = -3$, and $g(6) = 1$ are local minima.

Problem 3 In Figure 11.5 we see that g is concave up on $(-\infty, -1)$ and $(6, \infty)$, whereas g is concave down on $(-1, 6)$.

Problem 4 In Figure 11.5 the concavity changes from up to down at $x = -1$ and from down to up at $x = 6$. Thus there are two points of inflection, $(-1, 2)$ and $(6, 2)$.

Problem 5 The numbers for which $f(x) = 0$ are 3 and 5. Hence we plot these values on a number line as in the figure below.

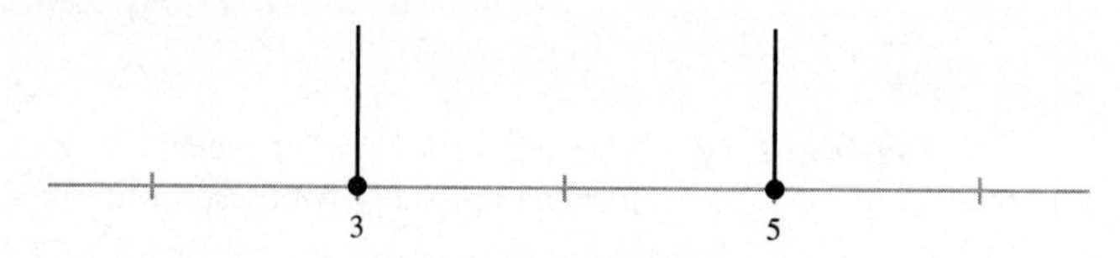

Now we evaluate f at a test number in each of the intervals $(-\infty, 3)$, $(3, 5)$, and $(5, \infty)$. For example,

$$f(2) = -3 < 0,$$
$$f(4) = 1 > 0, \quad \text{and}$$
$$f(6) = -3 < 0.$$

Recording these results on our number line produces the figure below.

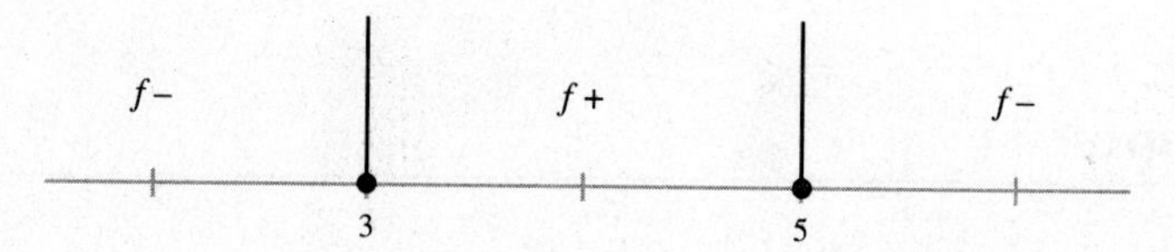

Thus we see that $f(x) < 0$ for x in $(-\infty, 3) \cup (5, \infty)$.

Problem 6 To determine where f is increasing and where it is decreasing, we must find where f' is positive and negative. Now $f'(x) = 4 - 2x$, and so the behavior of f' is as in the figure below.

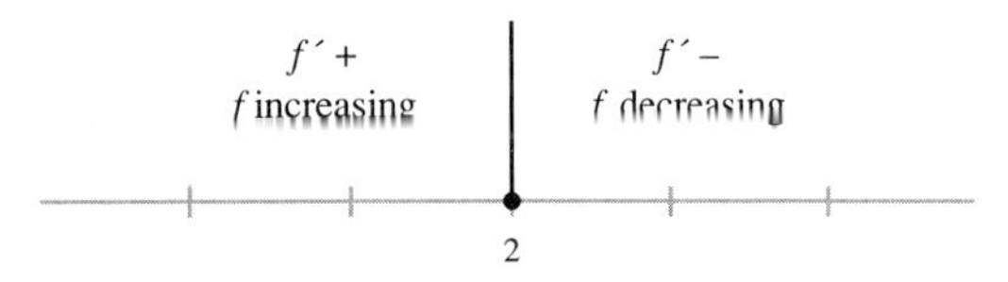

Therefore f is increasing on the interval $(-\infty, 2)$, and f is decreasing on the interval $(2, \infty)$.

Problem 7 We proceed as in Practice Problem 6 to find where f' is positive and where it is negative. Now $f'(x) = x^2 - 2x - 15$, and so the quadratic formula shows that $f'(x) = 0$ for $x = -3$ and $x = 5$. The behavior of f' is shown in the figure below.

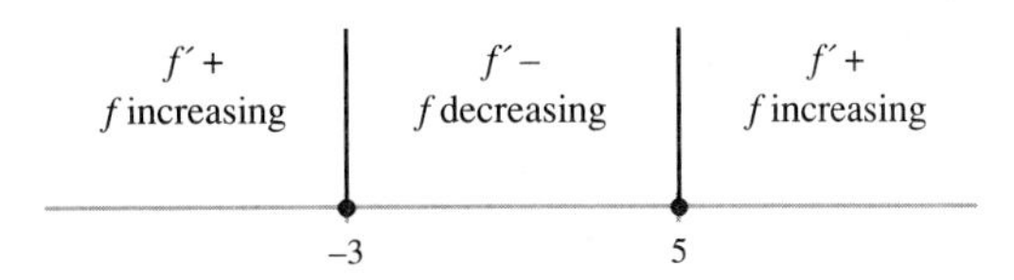

Thus f is increasing on the intervals $(-\infty, -3)$ and $(5, \infty)$, and f is decreasing on the interval $(-3, 5)$.

Problem 8 (a) Since $f(x) = x^4 - 12x^3$, we see that

$$f'(x) = 4x^3 - 36x^2 = 4x^2(x - 9).$$

Since f' is a polynomial function, it is defined for all real numbers, and clearly $f'(x) = 0$ precisely when $x = 0$ or $x = 9$. Thus 0 and 9 are the critical numbers of f.

(b) Applying the quotient rule to g gives

$$\begin{aligned} g'(x) &= \frac{(x - 2)(x^2 + x + 3)' - (x^2 + x + 3)(x - 2)'}{(x - 2)^2} \\ &= \frac{(x - 2)(2x + 1) - (x^2 + x + 3)(1)}{(x - 2)^2} \\ &= \frac{2x^2 - 3x - 2 - x^2 - x - 3}{(x - 2)^2} \\ &= \frac{x^2 - 4x - 5}{(x - 2)^2}. \end{aligned}$$

Now g' is a rational function, and so g' is continuous for all real numbers except $x = 2$, where its denominator is 0. Applying the quadratic formula to the numerator of g', we see that $g'(x) = 0$ precisely when $x = -1$ or $x = 5$. Because g is not defined at $x = 2$, it follows that -1 and 5 are the critical numbers of g.

Problem 9 Differentiating f, we obtain

$$f'(x) = 12x^2 - 12x^3 = 12x^2(1 - x).$$

Thus 0 and 1 are the critical numbers of f. By applying the method of Section 11.1 to f', we obtain the figure below.

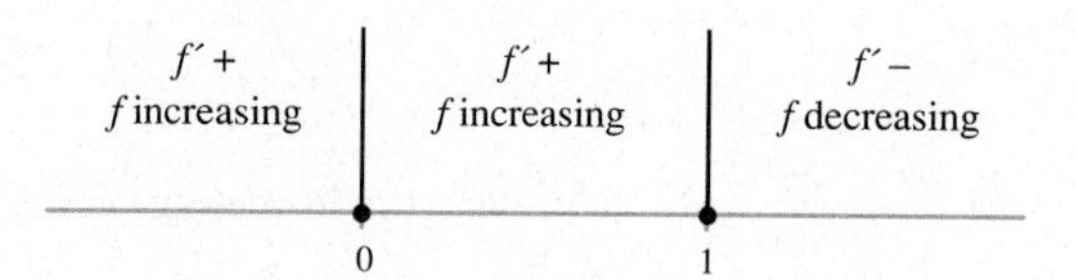

Thus the only local extremum of f is a local maximum of 1 at $x = 1$. The graph of f is shown below.

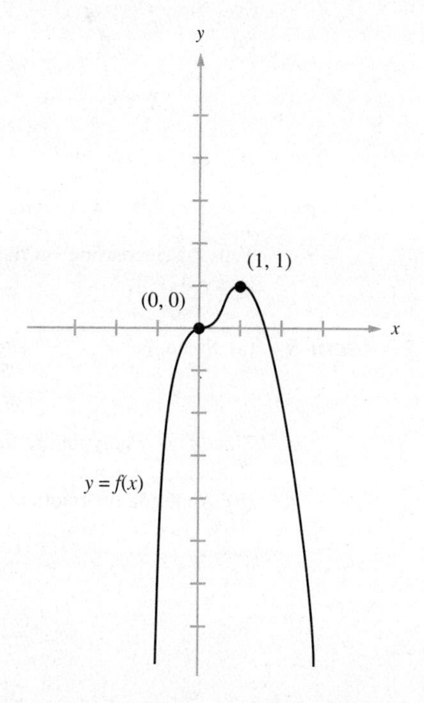

Problem 10 Write f in the form $f(x) = 2(x + 1)^{-1}$. Then

$$f'(x) = -1(2)(x + 1)^{-2} = -2(x + 1)^{-2}$$

and

$$f''(x) = -2(-2)(x + 1)^{-3} = 4(x + 1)^{-3}.$$

Applying the method of Section 11.1, we can determine where f'' is positive and where it is negative. The figure below shows the results.

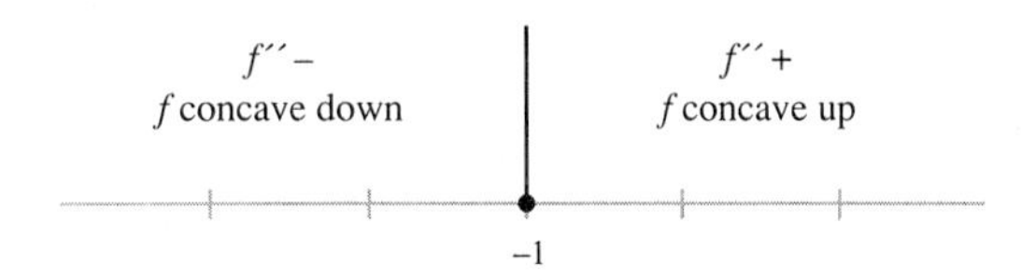

Thus f is concave down on $(-\infty, -1)$, and f is concave up on $(-1, \infty)$. Because f is not defined at $x = -1$, there are no points of inflection.

Problem 11 The first and second derivatives of f are

$$f'(x) = 6x^2 - 6x - 36$$

and

$$f''(x) = 12x - 6.$$

Since $f'(x) = 0$ precisely when $x = -2$ or $x = 3$, these are the only critical numbers of f. Now

$$f''(-2) = -30$$

and

$$f''(3) = 30.$$

Therefore the second-derivative test shows that f has a local maximum at $x = -2$ and a local minimum at $x = 3$.

Problem 12 The first and second derivatives of f are

$$f'(x) = 3 - 12x^2$$

and

$$f''(x) = -24x.$$

Thus the critical numbers of f are $-\frac{1}{2}$ and $\frac{1}{2}$. The behavior of f is described in the figure below.

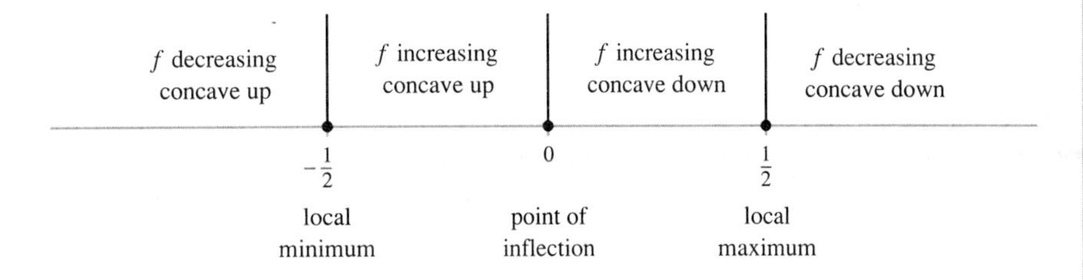

Since f is a polynomial function, it has no vertical asymptotes. Its graph is shown below.

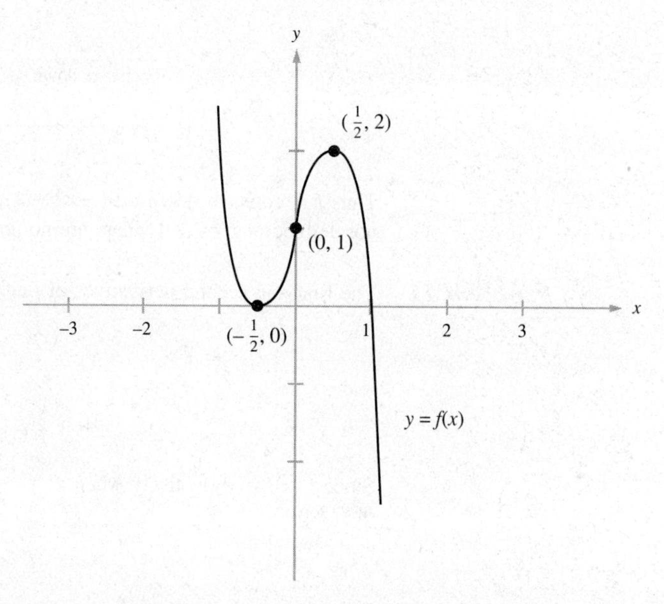

Problem 13 In Example 11.18 we found that the critical numbers of f are -2, 0, and 5. Of these, only 0 and 5 lie in the interval $[-1, 7]$. Hence we need only consider the values of f at the critical numbers 0 and 5 and at the endpoints of the interval -1 and 7. The table below shows these values.

x	$f(x)$	
-1	3	
0	-12	
5	363	⟵ **absolute maximum**
7	-61	⟵ **absolute minimum**

Thus the absolute maximum of f on $[-1, 7]$ is 363 at $x = 5$, and the absolute minimum of f on this interval is -61 at $x = 7$.

Problem 14 The derivative of g is

$$g'(x) = (xe^{-x})' = x(e^{-x})' + e^{-x}(x)' = -x(e^{-x}) + e^{-x} = (1 - x)e^{-x}.$$

Since g' is defined for all real numbers, the only critical number of g is 1, where $g'(x) = 0$. Thus we must examine the values of g at 0, 1, and 3. The table below contains these values.

x	$f(x)$	
0	0	⟵ **absolute minimum**
1	$e^{-1} \approx 0.368$	⟵ **absolute maximum**
3	$3e^{-3} \approx 0.149$	

Thus we see that absolute maximum of g on $[0, 3]$ is $e^{-1} \approx 0.368$ at $x = 1$, and on this interval the absolute minimum of g is 0 at $x = 0$.

Problem 15 If the number of passengers x is between 60 and 80 inclusive, then the ticket price will be \$400 per person. If there are more than 80 passengers, then the ticket price will be reduced by $4(x - 80)$ dollars per person. Thus the ticket price in this case will be

$$400 - 4(x - 80) = 400 - 4x + 320 = 720 - 4x.$$

The revenue function is obtained by multiplying the ticket price by the number of passengers x; it is

$$R(x) = \begin{cases} 400x & \text{if } 60 \le x \le 80 \\ (720 - 4x)x & \text{if } 80 < x \le 120. \end{cases}$$

(b) We must determine the absolute maximum of R on the interval $[60, 120]$. Now

$$R'(x) = \begin{cases} 400 & \text{if } 60 \le x < 80 \\ 720 - 8x & \text{if } 80 < x \le 120. \end{cases}$$

Hence the only critical numbers of R are 90, where $R'(x) = 0$, and 80, where $R'(x)$ is undefined. Examining the values of R at 90, 80, 60, and 120, we find that the maximum revenue is \$32,400 when there are 90 passengers.

x	$R(x)$	
60	24,000	
80	32,000	
90	32,400	← **absolute maximum**
120	28,800	

Problem 16 The marginal cost function is the derivative of the cost function, which is

$$C'(x) = 50.$$

Likewise the marginal revenue function is the derivative of the revenue function, which is

$$R'(x) = 500 - 0.2x.$$

The profit function is the difference of the revenue and cost functions; it is

$$\begin{aligned} P(x) &= R(x) - C(x) \\ &= (500x - 0.1x^2) - (50x + 450{,}000) \\ &= 450x - 0.1x^2 - 450{,}000. \end{aligned}$$

Hence the marginal profit function is

$$P'(x) = 450 - 0.2x.$$

Problem 17 If the present level of production is x gross, the additional cost of producing an extra gross of tennis balls is approximately $C'(x)$. From Example 11.24 we have

$$C'(x) = 0.000006x^2 - 0.024x + 144.$$

Thus

$$C'(3000) = 54 - 72 + 144 = 126$$

and

$$C'(4000) = 96 - 96 + 144 = 144.$$

Therefore the approximate cost of producing another gross of tennis balls is \$126 in case (a) and \$144 in case (b).

Problem 18 The maximum profit occurs where the marginal cost equals the marginal revenue.

$$\begin{aligned} C'(x) &= R'(x) \\ 50 &= 500 - 0.2x \\ -450 &= -0.2x \\ 2250 &= x \end{aligned}$$

Thus the maximum profit occurs when 2250 pairs of speakers are produced. This profit (in dollars) is

$$P(2250) = R(2250) - C(2250) = 618{,}750 - 562{,}500 = 56{,}250.$$

Problem 19 (a) The average cost function is

$$A(x) = \frac{C(x)}{x} = 0.1x^2 - 5x + 80 + \frac{2}{x},$$

and it is minimized when the marginal cost equals the average cost. Hence the desired equation is

$$\begin{aligned} C'(x) &= A(x) \\ 0.3x^2 - 10x + 80 &= 0.1x^2 - 5x + 80 + \frac{2}{x} \\ 0.2x^2 - 5x - \frac{2}{x} &= 0. \end{aligned}$$

Multiplying both sides by x, we can write this equation in the form

$$0.2x^3 - 5x^2 - 2 = 0.$$

(b) Substituting $x = 25.016$ in the equation above, we see that

$$0.2(25.016)^3 - 5(25.016)^2 - 2 \approx 0.$$

Hence $x = 25.016$ is a solution to this equation.

(c) We must determine the smallest cost at which a gallon of paint can be produced, which is the minimum value of the average cost function. From (b) we see that this minimum value occurs when $x = 25.016$. Now

$$\begin{aligned} A(25.016) &= 0.1(25.016)^2 - 5(25.016) + 80 + \frac{2}{25.016} \\ &\approx 17.58. \end{aligned}$$

Thus the cheapest price at which this paint can be produced is about \$17.58 per gallon. Because similar paints are selling for \$17 per gallon, it will not be possible for the manufacturer to make a profit on this line of paint.

Problem 20 Let x denote the number of shingles received in each order. Since the total number of shingles required is 200,000, the contractor must place $200{,}000/x$ orders in all. The total cost (in dollars) of placing these orders is

$$8\left(\frac{200{,}000}{x}\right) = 1{,}600{,}000x^{-1}.$$

The average number of shingles in storage is $x/2$, and so the storage cost per year (in dollars) is

$$0.05\left(\frac{x}{2}\right) = 0.025x.$$

Hence the total cost (in dollars) for ordering and storing the shingles is

$$C(x) = \frac{1{,}600{,}000}{x} + 0.025x.$$

We must determine the absolute minimum of C on the interval $1 \le x \le 200{,}000$. Consequently we must find the critical numbers of C. Now

$$C'(x) = -1{,}600{,}000x^{-2} + 0.025,$$

and so $C'(x)$ is defined for all x in the interval under consideration. Therefore the only critical numbers of C are those for which $C'(x) = 0$. This equation can be solved as follows.

$$\begin{aligned} 0 &= -1{,}600{,}000x^{-2} + 0.025 \\ 1{,}600{,}000x^{-2} &= 0.025 \\ 1{,}600{,}000 &= 0.025x^2 \\ x^2 &= \frac{1{,}600{,}000}{0.025} = 64{,}000{,}000 \end{aligned}$$

Thus the only critical number of C in the interval $1 \le x \le 200{,}000$ is $x = 8000$. From the table below, we see that the absolute minimum of C occurs when $x = 8000$. Thus the contractor should order 8000 shingles per shipment.

x	$C(x)$	
1	1,600,000.025	
8000	400	← **absolute minimum**
200,000	5008	

Problem 21 To find the point of diminishing returns, we must compute the second derivative of g. Now

$$g'(x) = -0.03x^2 + 1.2x \quad \text{and} \quad g''(x) = -0.06x + 1.2.$$

Solving $g''(x) = 0$ gives

$$\begin{aligned} -0.06x + 1.2 &= 0 \\ 1.2 &= 0.06x \\ 20 &= x. \end{aligned}$$

Thus \$20,000 should be spent on advertising for the state fair.

Problem 22 Since the cost of a new refrigerator is \$500, its cost per year is $500/t$ dollars if replaced after t years. Also the average number of repairs per year is

$$\frac{n}{t} = \frac{t^{1.25}}{1.6t} = \frac{t^{0.25}}{1.6},$$

and so the average yearly repair cost is $100t^{0.25}/1.6$. Therefore the total cost of the refrigerator, including both its purchase price and the cost of repairs, is

$$C(t) = \frac{500}{t} + \frac{100t^{0.25}}{1.6}.$$

We must find the positive value of t that minimizes C. Now

$$C'(t) = -500t^{-2} + \frac{100(0.25t^{-0.75})}{1.6}.$$

Since C' is defined for all positive values of t, the only critical numbers of C are those for which $C'(t) = 0$. We can find these values of t as follows.

$$\begin{aligned} 0 &= -500t^{-2} + \frac{100(0.25t^{-0.75})}{1.6} \\ 500t^{-2} &= \frac{100(0.25t^{-0.75})}{1.6} \\ 800t^{-2} &= 25t^{-0.75} \end{aligned}$$

$$\begin{aligned} 32t^{-2} &= t^{-0.75} \\ 32 &= t^{1.25} \\ (32)^{0.8} &= t \\ 16 &= t \end{aligned}$$

Hence 16 is the only critical number of C, and so the refrigerator should be replaced after 16 years.

Chapter 12

Problem 1

$$\begin{aligned} \int (3x^5 + 6x^{-3})\,dx &= \int 3x^5\,dx + \int 6x^{-3}\,dx && \text{(by (12.4))} \\ &= 3\int x^5\,dx + 6\int x^{-3}\,dx && \text{(by (12.3))} \\ &= \frac{3x^6}{6} + \frac{6x^{-2}}{-2} + k && \text{(by (12.2))} \\ &= \frac{x^6}{2} - 3x^{-2} + k \end{aligned}$$

Problem 2 We have

$$\begin{aligned} F(x) &= \int (3x^{-2} + 6)\,dx \\ &= \frac{3x^{-1}}{-1} + 6x + k = -\frac{3}{x} + 6x + k. \end{aligned}$$

Also

$$1 = F(2) = -\frac{3}{2} + 6 \cdot 2 + k = -\frac{3}{2} + 12 + k,$$

so

$$k = 1 + \frac{3}{2} - 12 = -\frac{19}{2}.$$

Thus $F(x) = -3/x + 6x - 19/2$.

Problem 3 (a)

$$\begin{aligned} \int \left(3x + 5 + \frac{3}{x}\right) dx &= 3\int x\,dx + 5\int 1 \cdot dx + 3\int x^{-1}\,dx \\ &= 3 \cdot \frac{x^2}{2} + 5x + 3\ln|x| + k \end{aligned}$$

(b)

$$
\begin{aligned}
F(x) &= \int (6e^{3x} + 2)\,dx \\
&= 6\int e^{3x}\,dx + 2\int 1 \cdot dx \\
&= \frac{6e^{3x}}{3} + 2x + k \\
&= 2e^{3x} + 2x + k
\end{aligned}
$$

Also

$$1 = F(2) = 2e^{3\cdot 2} + 2 \cdot 2 + k = 2e^6 + 4 + k,$$

so

$$k = 1 - 2e^6 - 4 = -3 - 2e^6.$$

Thus $F(x) = 2e^{3x} + 2x - 3 - 2e^6$.

Problem 4 (a) We have

$$
\begin{aligned}
C(x) &= \int (30 - \sqrt[3]{x})\,dx \\
&= 30\int 1 \cdot dx - \int x^{1/3}\,dx \\
&= 30x - \frac{x^{4/3}}{(4/3)} + k \\
&= 30x - \frac{3x^{4/3}}{4} + k.
\end{aligned}
$$

Also

$$20{,}000 = C(0) = 30(0) - \frac{3 \cdot 0^{4/3}}{4} + k = k.$$

Thus

$$C(x) = 30x - \tfrac{3}{4}x^{4/3} + 20{,}000.$$

(b)

$$
\begin{aligned}
C(1000) &= 30 \cdot 1000 - \tfrac{3}{4} \cdot 1000^{4/3} + 20{,}000 \\
&= 30{,}000 - \tfrac{3}{4} \cdot 10{,}000 + 20{,}000 \\
&= 30{,}000 - 7{,}500 + 20{,}000 = \$42{,}500
\end{aligned}
$$

Problem 5 (a) $du = (e^{3x})'\,dx = 3e^{3x}\,dx$

(b) $du = (1/\sqrt{x})'\,dx = (x^{-1/2})'\,dx = -\frac{1}{2}x^{-3/2}\,dx$

(c) $du = (x^3 + x^2)'\,dx = (3x^2 + 2x)\,dx$

Problem 6 We have $du = -6x\,dx$ and so

$$\int 2x\sqrt{5 - 3x^2}\,dx = \int -\tfrac{1}{3}\sqrt{5 - 3x^2}(-6x)\,dx = \int -\tfrac{1}{3}\sqrt{u}\,du.$$

Problem 7 (a) Since $(x^4 + 1)' = 4x^3$, the derivative of $x^4 + 1$ is present in the integrand except for a constant. Thus we take $u = x^4 + 1$.

(b)

$$\int x^3(x^4 + 1)^{1/2}\,dx = \tfrac{1}{4}\int (x^4 + 1)^{1/2}(4x^3)\,dx$$
$$= \tfrac{1}{4}\int u^{1/2}\,du = \frac{\tfrac{1}{4}u^{3/2}}{(3/2)} + k$$
$$= \frac{u^{3/2}}{6} + k = \frac{(x^4 + 1)^{3/2}}{6} + k$$

Problem 8 The region bounded by the vertical lines $x = -1$ and $x = 2$, the x-axis, and the curve $y = 2x + 3$ is as follows.

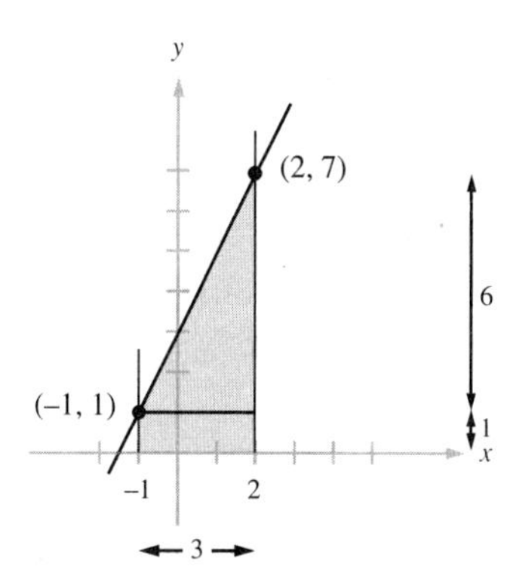

This region can be broken into a rectangle with area $1 \cdot 3 = 3$ and a triangle with area $\frac{1}{2} \cdot 3 \cdot 6 = 9$ as shown. Thus

$$\int_{-1}^{2} (2x + 3)\,dx = 3 + 9 = 12.$$

Problem 9 Since an antiderivative of $2x + 3$ is $F(x) = x^2 + 3x$, we have

$$\int_{-1}^{2} (2x + 3)\,dx = F(2) - F(-1) = 2^2 + 3 \cdot 2 - [(-1)^2 + 3(-1)]$$
$$= 4 + 6 - 1 + 3 = 12.$$

Problem 10 Since $x^2 - 10 < 0$ for $1 \le x \le 3$, the desired area is

$$-\int_1^3 (x^2 - 10)\,dx = -\left(\frac{x^3}{3} - 10x\right)\Bigg|_1^3$$

$$= -\left(\frac{3^3}{3} - 10\cdot 3\right) - \left[-\left(\frac{1^3}{3} - 10\cdot 1\right)\right]$$

$$= -9 + 30 + 1/3 - 10 = 11\,1/3.$$

Problem 11 (a) If $u = 6 - x$, then $du = -dx$, which is present in the integrand except for the minus sign.
(b) If $x = 3$, $u = 6 - 3 = 3$, and if $x = 5$, $u = 6 - 5 = 1$. Thus

$$\int_3^5 (6 - x)^{-1}\,dx = -\int_3^1 u^{-1}\,du = -\ln|u|\,\Bigg|_3^1$$

$$= -\ln 1 - (-\ln 3) = \ln 3.$$

$$\int_{1000}^{1200} (.00003x^2 - .024x + 20)\,dx = .00001x^3 - .012x^2 + 20x\,\Bigg|_{1000}^{1200}$$

$$= .00001(1200)^3 - .012(1200)^2 + 20(1200) - [.00001(1000)^3 - .012(1000)^2 + 20(1000)]$$

$$= \$6000$$

Problem 13

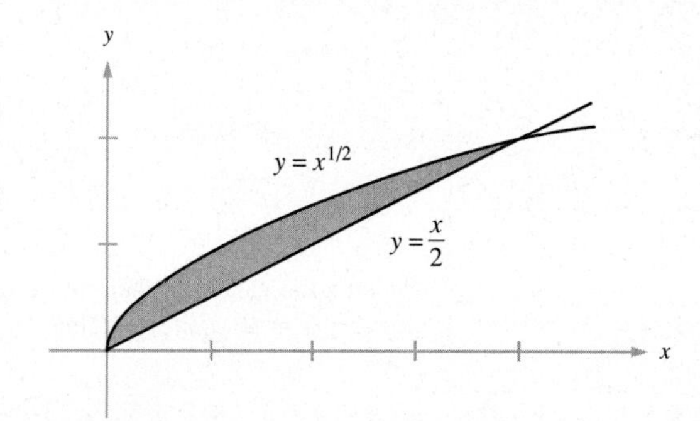

The two curves are graphed above. They intersect when

$$x^{1/2} = \frac{x}{2},$$

$$x = \frac{x^2}{4},$$

$$4x = x^2,$$

$$x^2 - 4x = 0,$$

$$x(x - 4) = 0,$$

$$x = 0 \text{ or } x = 4$$

Thus the area is

$$\int_0^4 \left(x^{1/2} - \frac{x}{2}\right) dx = \frac{4}{3}.$$

Problem 14 The curves $y = 15 - 0.1x$ and $y = 0.05x$ cross when

$$15 - 0.1x = 0.05x,$$
$$15 = .15x,$$
$$x = \frac{15}{.15} = 100.$$

Thus the equilibrium supply is $x_E = 100$. The equilibrium price is then

$$p_E = p_S(100) = .05(100) = \$5.$$

Problem 15 (a) The consumer's surplus is

$$\int_0^{100} (15 - 0.1x - 5)\, dx = 500.$$

(b) The producer's surplus is

$$\int_0^{100} (5 - 0.05x)\, dx = 250.$$

■ Chapter 13

Problem 1 (a) Substituting $y = x^2$ and $y' = 2x$ into (i), (ii), and (iii) gives

$$\text{(i)} \quad 2x = x$$
$$\text{(ii)} \quad x(2x) = 2x^2$$
$$\text{(iii)} \quad 2x = x^2 + e^x.$$

Only (ii) yields an identity in x.
(b) Substituting $y = e^x(x + 1)$ and $y' = e^x(x + 2)$ into (i), (ii), and (iii) gives

$$\text{(i)} \quad e^x(x + 2) = x$$
$$\text{(ii)} \quad xe^x(x + 2) = 2e^x(x + 1)$$
$$\text{(iii)} \quad e^x(x + 2) = e^x(x + 1) + e^x.$$

Only (iii) yields an identity in x.

(c) Substituting $y = x^2/2$ and $y' = x$ into (i), (ii), and (iii) gives

$$\text{(i)} \quad x = x$$

$$\text{(ii)} \quad x \cdot x = 2\left(\frac{x^2}{2}\right)$$

$$\text{(iii)} \quad x = \frac{x^2}{2} + e^x.$$

Thus (i) and (ii) yield identities in x.

Problem 2 (a) Not separable.

(b)

$$\frac{dy}{dx} y = x - \frac{dy}{dx} e^y$$

$$\frac{dy}{dx}(y + e^y) = x$$

$$(y + e^y)\, dy = x\, dx$$

(c)

$$\frac{dy}{dx} \ln x = x + e^y x$$

$$\frac{dy}{dx} \ln x = x(1 + e^y)$$

$$\frac{dy}{1 + e^y} = \frac{x\, dx}{\ln x}$$

(d) Not separable.

(e)

$$x^2 \frac{dy}{dx} = 1 + x + y + xy$$

$$x^2 \frac{dy}{dx} = 1 + y + x(1 + y) = (1 + x)(1 + y)$$

$$\frac{dy}{(1 + y)} = (1 + x)\frac{dx}{x^2} = \left(\frac{1}{x^2} + \frac{1}{x}\right) dx$$

Problem 3 (a)

$$y'x^2 = 4\sqrt{y}$$

$$\frac{dy}{dx} x^2 = 4y^{1/2}$$

$$y^{-1/2}\, dy = 4x^{-2}\, dx$$

$$\int y^{-1/2}\, dy = \int 4x^{-2}\, dx$$

$$\frac{y^{1/2}}{(1/2)} = \frac{4x^{-1}}{(-1)} + k_1$$

$$y^{1/2} = -2x^{-1} + k \qquad (\text{so } -2x^{-1} + k \geq 0)$$

$$y = (-2x^{-1} + k)^2$$

(b) Since $4 = (-2 \cdot 2^{-1} + k)^2 = (k - 1)^2$, and the quantity in the parentheses must be nonnegative, we have

$$2 = k - 1,$$

$$k = 3.$$

Thus $y = (-2x^{-1} + 3)^2$.

Problem 4 (a) The differential equation $y' = cy$ has the solution $y = Ke^{cx}$, and since $y = 500$ when $x = 0$, we have $K = 500$. Since the half-life is 22 years,

$$250 = 500e^{c(22)}.$$

Solving for c gives $c = (\ln .5)/22$. Thus

$$y = 500e^{(\ln .5)x/22}.$$

(b) When $x = 100$, $y = 500e^{(\ln .5)\cdot 100/22} \approx 21.4$ kg.

Problem 5 (a) Since y' is 20% of $30{,}000 - y$, we have

$$y' = .2(30{,}000 - y).$$

(b) Our differential equation has the form $y' = r(L - y)$, where $r = .2$ and $L = 30{,}000$. Thus we have

$$y = L - Ke^{-rx} = 30{,}000 - Ke^{-.2x}.$$

Since $y = 5000$ when $x = 0$, we have

$$5000 = 30{,}000 - K,$$

$$K = 30{,}000 - 5000 = 25{,}000.$$

Thus

$$y = 30{,}000 - 25{,}000e^{-.2x}.$$

(c) When $x = 10$ we have

$$y = 30{,}000 - 25{,}000e^{-.2(10)} \approx 26{,}617.$$

Problem 6 (a) We have $c = .01$ and $P = 15$, so

$$y = \frac{15}{1 + ke^{-15(.01)x}} = \frac{15}{1 + ke^{-.15x}}.$$

Since $y = 1/2$ when $x = 0$, we have

$$\frac{1}{2} = \frac{15}{1 + k},$$

which leads to $k = 29$. Thus $y = 15/(1 + 29e^{-.15x})$.
(b) When $x = 10$,

$$y = \frac{15}{1 + 29e^{-.15(10)}} \approx 2.008.$$

Problem 7 (a)

$$\int_2^{\infty} x^{-3}\,dx = \lim_{b\to\infty} \int_2^b x^{-3}\,dx = \lim_{b\to\infty} \frac{x^{-2}}{-2}\Bigg|_2^b$$

$$= \lim_{b\to\infty} \left(\frac{1}{8} - \frac{1}{2b^2}\right) = \frac{1}{8}$$

(b)

$$\int_2^{\infty} x^{-1/2}\,dx = \lim_{b\to\infty} \int_2^b x^{-1/2}\,dx$$

$$= \lim_{b\to\infty} 2x^{1/2}\Bigg|_2^b = \lim_{b\to\infty} (2b^{1/2} - 2 \cdot 2^{1/2})$$

This limit does not exist.

Problem 8

$$\int_0^{\infty} 40{,}000e^{-.1x}\,dx = \lim_{b\to\infty} \int_0^b 40{,}000e^{-.1x}\,dx = \lim_{b\to\infty} \frac{40{,}000e^{-.1x}}{-.1}\Bigg|_0^b$$

$$= \lim_{b\to\infty} -400{,}000(e^{-.1b} - e^0) = -400{,}000(0 - 1) = \$400{,}000$$

Problem 9 Since $\ln|x|$ has a simple derivative, we set $u = \ln|x|$ and $dv = x\,dx$. Then $du = x^{-1}\,dx$ and $v = x^2/2$. Thus

$$\int x \ln|x|\,dx = (\ln|x|)\frac{x^2}{2} - \int \left(\frac{x^2}{2}\right)x^{-1}\,dx$$

$$= \tfrac{1}{2}x^2 \ln|x| - \tfrac{1}{2}\int x\,dx = \tfrac{1}{2}x^2 \ln|x| - \frac{\frac{1}{2}x^2}{2} + k$$

$$= \tfrac{1}{2}x^2 \ln|x| - \tfrac{1}{4}x^2 + k.$$

Problem 10 Let $u = x$ and $dv = e^{2x}\,dx$. Then $du = dx$ and $v = \frac{1}{2}e^{2x}$. Thus

$$\int_0^2 xe^{2x}\,dx = \tfrac{1}{2}xe^{2x}\Bigg|_0^2 - \int_0^2 \tfrac{1}{2}e^{2x}\,dx$$

$$= \tfrac{1}{2}(2e^4 - 0) - \frac{\frac{1}{2}e^{2x}}{2}\Bigg|_0^2 = e^4 - \tfrac{1}{4}(e^4 - e^0) = \tfrac{3}{4}e^4 + \tfrac{1}{4}.$$

Problem 11 (a)

$$\int x^{-2}(x-2)^{-1}\,dx = \int \frac{dx}{x^2(-2+1\cdot x)},$$

so $a = -2$, $b = 1$, $u = x$.

(b)

$$\int \frac{dx}{x^2(-2+1\cdot x)} = -\frac{1}{-2x} + \frac{1}{(-2)^2}\ln\left|\frac{-2+1\cdot x}{x}\right| + k$$

$$= \frac{1}{2x} + \frac{1}{4}\ln\left|\frac{x-2}{x}\right| + k$$

Problem 12 (a)

$$\int x^{-2}(5+x^2)^{1/2}\,dx = \int \frac{\sqrt{u^2 \pm a^2}}{u^2}\,du$$

with $u = x$, $a = \sqrt{5}$, and the top sign, so formula 37 applies.

(b)

$$\int x^2(x+9)^{-2}\,dx = \int \frac{u^2\,du}{(a+bu)^2}$$

with $u = x$, $a = 9$, and $b = 1$, so formula 10 applies.

Problem 13 If $u = 3x$, then $du = 3\,dx$. Thus by formula 27

$$\int (7-9x^2)^{-3/2}\,dx = \tfrac{1}{3}\int (7-9x^2)^{-3/2}\,3\,dx$$

$$= \tfrac{1}{3}\int \frac{du}{(\sqrt{7}^2 - u^2)^{3/2}} = \tfrac{1}{3}\left(\frac{u}{\sqrt{7}^2\sqrt{\sqrt{7}^2 - u^2}} + k\right)$$

$$= \tfrac{1}{3}\left(\frac{3x}{7\sqrt{7-9x^2}} + k\right) = \frac{x}{7\sqrt{7-9x^2}} + K.$$

Chapter 14

Problem 1

$$\int_{-1}^{0.3} f(x)\,dx = \int_0^{0.3} 2x\,dx = x^2\Big|_0^{0.3} = (0.3)^2 - 0^2 = 0.09$$

Problem 2 (a) We must have

$$1 = \int_0^3 ke^x\,dx = ke^x\Big|_0^3 = k(e^3 - 1),$$

and so $k = 1/(e^3 - 1)$.

(b) $$P(x \geq 2) = \int_2^3 \frac{e^x\,dx}{e^3 - 1} = \frac{e^x}{e^3 - 1}\Big|_2^3 = \frac{e^3 - e^2}{e^3 - 1}.$$

Problem 3

$$\int_2^{\infty} \tfrac{1}{2}e^{-.5x}\,dx = 1 - \int_0^2 \tfrac{1}{2}e^{-.5x}\,dx = 1 - (1 - e^{-.5(2)}) = e^{-1}$$

Problem 4 (a) The probability of rolling a 3 is 1/6, and the probability of rolling some other number is 5/6. Thus $P(x = 1) = 1/6$, $P(x = 2) = (5/6)(1/6) = 5/6^2$, and in general $P(x = k) = 5^{k-1}/6^k = (5/6)^{k-1}(1/6)$.

(b)

$$\begin{aligned} P(x \geq 4) &= 1 - P(x < 4) = 1 - P(x = 1) - P(x = 2) - P(x = 3) \\ &= 1 - \frac{1}{6} - \frac{5}{36} - \frac{25}{216} = \frac{125}{216} \end{aligned}$$

Problem 5

$$\begin{aligned} P(x = 0) + P(x = 1) + P(x = 2) &= \frac{e^{-3}3^0}{0!} + \frac{e^{-3}3^1}{1!} + \frac{e^{-3}3^2}{2!} \\ &= e^{-3}\left(1 + 3 + \frac{9}{2}\right) = 8.5e^{-3} \end{aligned}$$

Problem 6

$$\int_1^3 x\left(\frac{3x^2}{26}\right) dx = \frac{3}{26}\left(\frac{x^4}{4}\right)\Bigg|_1^3 = \frac{3}{104}(81 - 1) = \frac{30}{13}$$

Problem 7 We have

$$\begin{aligned} \sigma^2 &= \int_1^3 x^2\left(\frac{3x^2}{26}\right) dx - \left(\frac{30}{13}\right)^2 = \frac{3}{26}\int_1^3 x^4\,dx - \frac{900}{169} \\ &= \left(\frac{3}{26}\right)\left(\frac{x^5}{5}\right)\Bigg|_1^3 - \frac{900}{169} = \left(\frac{3}{130}\right)(243 - 1) - \frac{900}{169} = \frac{219}{845} \approx .259. \end{aligned}$$

Then $\sigma = \sqrt{219/845} \approx .509$.

Problem 8 Since $a = 1$ and $b = 5$, we have $\mu = (1 + 5)/2 = 3$,

$$\sigma^2 = \frac{(5 - 1)^2}{12} = \frac{4}{3}, \qquad \text{and} \qquad \sigma = \sqrt{\frac{4}{3}} \approx 1.15.$$

Problem 9 Since $\mu = 1/\lambda = 9$, we have $\lambda = 1/9$. Thus

$$P(x < 5) = \int_0^5 \frac{1}{9}e^{-x/9}\,dx = 1 - e^{-5/9} \approx .43.$$

Problem 10 We assume a Poisson distribution for x, the number of students visiting the infirmary in a day. Since 45 students visit the infirmary each 30 days, $\mu = 45/30 = 1.5$. Then

$$\begin{aligned} P(x > 2) &= 1 - P(x = 0) - P(x = 1) - P(x = 2) \\ &= 1 - \frac{e^{-1.5}1.5^0}{0!} - \frac{e^{-1.5}1.5^1}{1!} - \frac{e^{-1.5}1.5^2}{2!} \\ &= 1 - e^{-1.5}(1 + 1.5 + 1.125) = 1 - 3.625e^{-1.5}. \end{aligned}$$

Problem 11

$$\rho = \frac{\mu}{\lambda} = \frac{15}{20} = .75$$

Problem 12 From Practice Problem 11 $\rho = .75$. Then

$$N = \frac{\rho}{1 - \rho} = \frac{.75}{1 - .75} = 3, \quad \text{and} \quad N_q = \rho N = .75(3) = 2.25.$$

Problem 13 From previous practice problems $\rho = .75$ and $\lambda = 20$. Then $T = 1/\lambda(1 - \rho) = 1/20(1 - .75) = .2$ hours, and $T_q = \rho T = .75(.2) = .15$ hours. Multiplying by 60, we get 12 minutes and 9 minutes.

Problem 14 We will measure time in hours. Then $\mu = 8$ (customers per hour) and $1/\lambda = 1/12$ hour ($= 5$ minutes), so $\lambda = 12$. Thus $\rho = \mu/\lambda = 8/12 = 2/3$. Then $N = \rho/(1 - \rho) = (2/3)/(1 - 2/3) = 2$, $N_q = \rho N = (2/3)2 = 4/3$, $T = 1/\lambda(1 - \rho) = 1/12(1 - 2/3) = 1/4$ hour, or 15 minutes, and $T_q = \rho T = \frac{2}{3}(15 \text{ minutes}) = 10$ minutes.

If 6 customers arrive per hour, then we have $\mu = 6$ and $\rho = \mu/\lambda = 6/12 = 1/2$. Then $T = 1/\lambda(1 - \rho) = 1/12(1 - 1/2) = 1/6$ hour, or 10 minutes.

Chapter 15

Problem 1 (a) The volume of a rectangular box equals the product of its length, width, and height. Hence

$$V(x, y, z) = xyz.$$

(b) Taking $x = 4$ feet, $y = 3$ feet, and $z = 2$ feet in the formula above gives

$$V(4, 3, 2) = 24 \text{ cubic feet}.$$

Problem 2 The number of heads of lettuce produced when 16 laborers and 1296 units of capital are used is

$$f(16, 1296) = 1000(16)^{0.25}(1296)^{0.75} = 1000(2)(216) = 432{,}000.$$

Problem 3 To compute f_x, we differentiate f with respect to x, while treating y as a constant. Hence

$$\begin{aligned} f_x(x, y) &= (32x - y^2 + 2x^2y)_x \\ &= (32x)_x - (y^2)_x + (2x^2y)_x \\ &= 32(x)_x - y^2(1)_x + 2y(x^2)_x \\ &= 32(1) - y^2(0) + 2y(2x) \\ &= 32 + 4xy. \end{aligned}$$

Likewise, to compute f_y, we differentiate f with respect to y, while treating x as a constant. The result is

$$\begin{aligned} f_y(x, y) &= (32x - y^2 + 2x^2y)_y \\ &= (32x)_y - (y^2)_y + (2x^2y)_y \\ &= 32x(1)_y - (2y) + 2x^2(y)_y \\ &= 32x(0) - 2y + 2x^2(1) \\ &= -2y + 2x^2. \end{aligned}$$

Problem 4 Taking the partial derivative of f with respect to x yields

$$\begin{aligned} f_x(x, y) &= (ax^{\lambda}y^{1-\lambda})_x \\ &= ay^{1-\lambda}(x^{\lambda})_x \\ &= ay^{1-\lambda}(\lambda x^{\lambda-1}) \\ &= a\lambda x^{\lambda-1}y^{1-\lambda}. \end{aligned}$$

Taking the partial derivative of f with respect to y, we obtain

$$\begin{aligned} f_y(x, y) &= (ax^{\lambda}y^{1-\lambda})_y \\ &= ax^{\lambda}(y^{1-\lambda})_y \\ &= ax^{\lambda}[(1 - \lambda)y^{-\lambda}] \\ &= a(1 - \lambda)x^{\lambda}y^{-\lambda}. \end{aligned}$$

Problem 5

$$\begin{aligned} f_x(x, y) &= (xe^{xy})_x \\ &= x(e^{xy})_x + e^{xy}(x)_x \\ &= x(ye^{xy}) + e^{xy}(1) \\ &= e^{xy} + xye^{xy} \\ f_y(x, y) &= (xe^{xy})_y \\ &= x(e^{xy})_y \\ &= x(xe^{xy}) \\ &= x^2e^{xy} \end{aligned}$$

Problem 6 We begin by computing f_x and f_y. Now

$$f_x(x, y) = 10(x)_x - 3y^2(1)_x + 4y(x^2)_x = 10 + 8xy, \text{ and}$$

$$f_y(x, y) = 10x(1)_y - 3(y^2)_y + 4x^2(y)_y = -6y + 4x^2.$$

Therefore

$$f_x(-2, 1) = 10 + 8(-2)(1) = 10 - 16 = -6,$$

and

$$f_y(3, -4) = -6(-4) + 4(3)^2 = 24 + 36 = 60.$$

Problem 7 The instantaneous rate of change of the volume with respect to the radius when the radius is 5 inches and the height is held constant at 8 inches is $V_r(5, 8)$. Now

$$V_r(r, h) = 2\pi rh,$$

and so

$$V_r(5, 8) = 80\pi \text{ cubic inches per inch.}$$

Problem 8 Because H is a function of three variables, it has three partial derivatives, which can be computed as above. The results are

$$H_U = 1, \qquad H_P = V, \qquad \text{and} \qquad H_V = P.$$

Problem 9 In Practice Problem 6 we computed both of the first-order partial derivatives of f:

$$f_x(x, y) = 10 + 8xy \qquad \text{and} \qquad f_y(x, y) = -6y + 4x^2.$$

To find f_{xx}, we take the partial derivative of f_x with respect to x:

$$f_{xx}(x, y) = (10 + 8xy)_x = 8y.$$

To find f_{xy}, we take the partial derivative of f_x with respect to y:

$$f_{xy}(x, y) = (10 + 8xy)_y = 8x.$$

To find f_{yx}, we take the partial derivative of f_y with respect to x:

$$f_{yx}(x, y) = (-6y + 4x^2)_x = 8x.$$

Note that $f_{xy}(x, y) = f_{yx}(x, y)$. Finally, to compute f_{yy}, we take the partial derivative of f_y with respect to y:

$$f_{yy}(x, y) = (-6y + 4x^2)_y = -6.$$

Problem 10 The partial derivatives of f are

$$f_x(x, y) = 2x - 8 \qquad \text{and} \qquad f_y(x, y) = 2y + 6.$$

Because these partial derivatives exist everywhere, critical points of f occur only where both f_x and f_y equal zero. Equating f_x and f_y to zero, we obtain

$$\begin{aligned} 0 &= 2x - 8 &\quad \text{and} \quad& 0 = 2y + 6 \\ 8 &= 2x &\quad \text{and} \quad& -6 = 2y \\ 4 &= x &\quad \text{and} \quad& -3 = y. \end{aligned}$$

Thus the only critical point of f is $(4, -3)$.

Problem 11 As in Practice Problem 10, we compute the partial derivatives of g:

$$g_x(x, y) = x + y - 5 \qquad \text{and} \qquad g_y(x, y) = x - 2y + 4.$$

Because these partial derivatives exist everywhere, critical points of g occur only where both g_x and g_y equal zero. Equating g_x and g_y to zero, we obtain the following system of linear equations.

$$\begin{aligned} x + y &= 5 \\ x - 2y &= -4. \end{aligned}$$

Solving this system, we find that $x = 2$ and $y = 3$. Hence the only critical point of g is $(2, 3)$.

Problem 12 (a) The first-order partial derivatives of f are

$$f_x(x, y) = -9 - y + 2x \quad \text{and} \quad f_y(x, y) = 8 - x + 4y.$$

Equating these to zero produces the following system of linear equations:

$$\begin{aligned} 2x - y &= 9 \\ x - 4y &= 8. \end{aligned}$$

The unique solution to this system is $x = 4$ and $y = -1$, so that $(4, -1)$ is the only critical point of f.
(b) Using the first-order partial derivatives of f computed in (a), we obtain

$$f_{xx}(x, y) = 2, \quad f_{xy}(x, y) = -1, \quad \text{and} \quad f_{yy}(x, y) = 4.$$

Therefore the function D in the second-partial-derivative test is

$$D = f_{xx}f_{yy} - f_{xy}^2 = (2)(4) - (-1)^2 = 7.$$

(c) Since $D(4, -1) = 7 > 0$ and $f_{xx}(4, -1) = 2 > 0$, the second-partial-derivative test shows that $f(4, -1) = -22$ is a local minimum.

Problem 13 The first-order partial derivatives of f are

$$f_x(x, y) = 2x + 8y \quad \text{and} \quad f_y(x, y) = 8x + 32y^3.$$

Equating these to zero produces the following system of equations:

$$2x + 8y = 0 \quad \text{and} \quad 8x + 32y^3 = 0.$$

From the first of these equations, we see that $2x = -8y$, that is, $x = -4y$. Substituting this expression for x in the second equation above produces

$$0 = 8(-4y) + 32y^3 = -32y + 32y^3 = 32y(y^2 - 1) = 32y(y + 1)(y - 1).$$

Hence $y = 0$, $y = -1$, or $y = 1$. Since $x = -4y$, it follows that the critical points of f are $(0, 0)$, $(4, -1)$, and $(-4, 1)$.

Using the first-order partial derivatives of f computed above, we obtain

$$f_{xx}(x, y) = 2, \quad f_{xy}(x, y) = 8, \quad \text{and} \quad f_{yy}(x, y) = 96y^2.$$

Therefore the function D in the second-partial-derivative test is

$$D = f_{xx}f_{yy} - f_{xy}^2 = 2(96y^2) - (8)^2 = 192y^2 - 64.$$

Because

$$D(0, 0) = -64 < 0,$$

the second-partial-derivative test shows that (0, 0) is not a local extremum of f. However

$$D(4, -1) = 128 > 0 \quad \text{and} \quad f_{xx}(4, -1) = 2 > 0,$$

so that $f(4, -1) = -8$ is a local minimum. Likewise

$$D(-4, 1) = 128 > 0 \quad \text{and} \quad f_{xx}(-4, 1) = 2 > 0,$$

and so $f(-4, 1) = -8$ is also a local minimum of f.

Problem 14 (a) The scatter diagram is shown below.

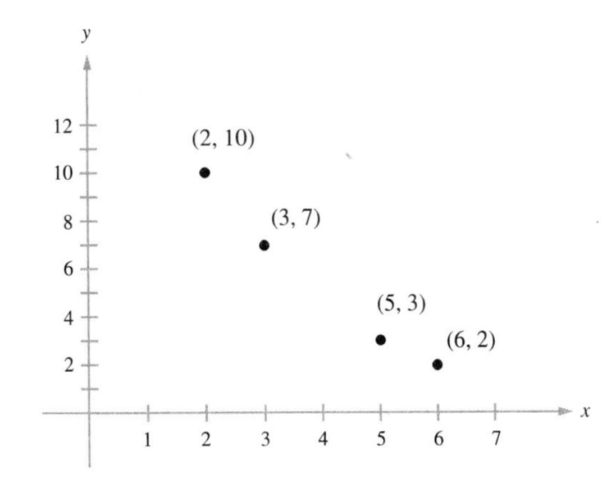

(b) As in Example 15.18, we will construct a table containing the values of x_k, y_k, $x_k y_k$, and x_k^2. This table is shown below.

x_k	y_k	$x_k y_k$	x_k^2	
2	10	20	4	
3	7	21	9	
5	3	15	25	
6	2	12	36	
$16 = \Sigma x$	$22 = \Sigma y$	$68 = \Sigma xy$	$74 = \Sigma x^2$	Totals

Here $\Sigma x = 16$, $\Sigma y = 22$, $\Sigma xy = 68$, and $\Sigma x^2 = 74$.

(c) Note that in this case $n = 4$ because there are 4 data points. Thus

$$\begin{aligned} m &= \frac{n \Sigma xy - (\Sigma x)(\Sigma y)}{n \Sigma x^2 - (\Sigma x)^2} \\ &= \frac{4(68) - 16(22)}{4(74) - (16)^2} \\ &= \frac{-80}{40} = -2, \end{aligned}$$

and

$$b = \frac{\Sigma\, y - m\, \Sigma\, x}{n}$$
$$= \frac{22 - (-2)(16)}{4}$$
$$= \frac{54}{4} = 13.5.$$

Thus the equation of the least squares line is $y = -2x + 13.5$.

Problem 15 We extend the table in Practice Problem 14 to include a column for y_k^2.

x_k	y_k	$x_k y_k$	x_k^2	y_k^2	
2	10	20	4	100	
3	7	21	9	49	
5	3	15	25	9	
6	2	12	36	4	
$16 = \Sigma\, x$	$22 = \Sigma\, y$	$68 = \Sigma\, xy$	$74 = \Sigma\, x^2$	$162 = \Sigma\, y^2$	Totals

Hence

$$r = \frac{n\, \Sigma\, xy - (\Sigma\, x)(\Sigma\, y)}{\sqrt{n\, \Sigma\, x^2 - (\Sigma\, x)^2}\, \sqrt{n\, \Sigma\, y^2 - (\Sigma\, y)^2}}$$
$$= \frac{4(68) - 16(22)}{\sqrt{4(74) - (16)^2}\, \sqrt{4(162) - (22)^2}}$$
$$= \frac{-80}{\sqrt{40}\, \sqrt{164}} \approx -.988.$$

Therefore the coefficient of correlation is about $-.988$.

Problem 16 The constraint function is $g(x, y) = 2x + y - 12$. Now

$$f_x(x, y) = y, \qquad f_y(x, y) = x, \qquad g_x(x, y) = 2, \qquad \text{and} \qquad g_y(x, y) = 1.$$

Setting $f_x(x, y) = \lambda g_x(x, y)$ and $f_y(x, y) = \lambda g_y(x, y)$, we obtain

$$y = 2\lambda \qquad \text{and} \qquad x = \lambda.$$

Hence $\lambda = \frac{1}{2}y$ and $\lambda = x$. Equating these two values of λ gives $y = 2x$. Substituting $2x$ for y in the constraint equation, we see that

$$12 = 2x + y = 2x + 2x = 4x.$$

Therefore $x = 3$, and so $y = 2x = 2(3) = 6$. Thus the maximum value of f subject to the constraint $2x + y = 12$ occurs when $x = 3$ and $y = 6$; the maximum value is $f(3, 6) = 3(6) = 18$.

Problem 17 The constraint function is $g(x, y) = x + 2y + z - 8$. Now

$$f_x(x, y, z) = 2xyz, \qquad f_y(x, y, z) = x^2z, \qquad f_z(x, y, z) = x^2y,$$
$$g_x(x, y, z) = 1, \qquad g_y(x, y, z) = 2, \qquad \text{and} \qquad g_z(x, y, z) = 1.$$

Setting $f_x(x, y, z) = \lambda g_x(x, y, z)$, $f_y(x, y, z) = \lambda g_y(x, y, z)$, and $f_z(x, y, z) = \lambda g_z(x, y, z)$, we obtain

$$2xyz = \lambda, \qquad x^2z = 2\lambda, \qquad \text{and} \qquad x^2y = \lambda.$$

Hence $\lambda = 2xyz$, $\lambda = \frac{1}{2}x^2z$, and $\lambda = x^2y$. Equating the first two values of λ and dividing both sides by xz gives $2y = \frac{1}{2}x$, that is, $y = \frac{1}{4}x$. Likewise, equating the first and third values of λ and dividing both sides by xy gives $2z = x$, that is, $z = \frac{1}{2}x$. Substituting these values into the constraint equation, we see that

$$8 = x + 2y + z = x + 2(\tfrac{1}{4}x) + (\tfrac{1}{2}x) = 2x.$$

Hence $x = 4$, and so $y = \frac{1}{4}x = 1$ and $z = \frac{1}{2}x = 2$. Thus the local maximum of f subject to the constraint $x + 2y + z = 8$ occurs at $(4, 1, 2)$; this maximum value is $f(4, 1, 2) = (4)^2(1)(2) = 32$.

Problem 18 In this indefinite integral we treat y as a constant and integrate with respect to x:

$$\begin{aligned}\int (24x^5y^8 - 9x^2y^5 + 6e^{-2x})\,dx &= 24y^8\int x^5\,dx - 9y^5\int x^2\,dx + 6\int e^{-2x}\,dx\\ &= 24y^8\left(\frac{x^6}{6}\right) - 9y^5\left(\frac{x^3}{3}\right) + 6(-\tfrac{1}{2})e^{-2x} + C(y)\\ &= 4x^6y^8 - 3x^3y^5 - 3e^{-2x} + C(y).\end{aligned}$$

Notice that the answer contains an arbitrary function of y.

Problem 19 In this problem integration by substitution is required. Letting $u = x^4 + y^2$, we obtain $du = 2y\,dy$. Hence

$$\begin{aligned}\int 4x^2y(x^4 + y^2)^6\,dy &= 2x^2\int 2y(x^4 + y^2)^6\,dy = 2x^2\int u^6\,du = 2x^2\left(\frac{u^7}{7}\right) + C(x)\\ &= \frac{2}{7}x^2(x^4 + y^2)^7 + C(x).\end{aligned}$$

Problem 20 Since

$$\int_{-1}^{1}\int_{0}^{2}(3x^2y + 6xy^2)\,dx\,dy = \int_{-1}^{1}\left[\int_{0}^{2}(3x^2y + 6xy^2)\,dx\right]dy,$$

we must evaluate

$$\begin{aligned}\int_0^2 (3x^2y + 6xy^2)\,dx &= 3y\int_0^2 x^2\,dx + 6y^2\int_0^2 x\,dx \\ &= 3y\left(\frac{x^3}{3}\right)\Bigg|_0^2 + 6y^2\left(\frac{x^2}{2}\right)\Bigg|_0^2 \\ &= (8y - 0) + (12y^2 - 0) \\ &= 12y^2 + 8y.\end{aligned}$$

Thus

$$\begin{aligned}\int_{-1}^1\int_0^2 (3x^2y + 6xy^2)\,dx\,dy &= \int_{-1}^1 (12y^2 + 8y)\,dy \\ &= \left[12\left(\frac{y^3}{3}\right) + 8\left(\frac{y^2}{2}\right)\right]\Bigg|_{-1}^1 \\ &= (4y^3 + 4y^2)\Bigg|_{-1}^1 \\ &= 8 - 0 = 8.\end{aligned}$$

Problem 21 The volume of the solid determined by $f(x, y) = 6x^2y$ over the given rectangle is equal to either of the iterated integrals

$$\int_0^3\int_2^5 f(x, y)\,dy\,dx = \int_2^5\int_0^3 f(x, y)\,dx\,dy.$$

We will evaluate the latter. Now

$$\int_0^3 6x^2y\,dx = 6y\left(\frac{x^3}{3}\right)\Bigg|_0^3 = 54y.$$

Hence

$$\begin{aligned}\int_2^5\int_0^3 f(x, y)\,dx\,dy &= \int_2^5 54y\,dy \\ &= 27y^2\Bigg|_2^5 \\ &= 27(25) - 27(4) \\ &= 675 - 108 = 567.\end{aligned}$$

Appendix

Problem 1

$$1 - (2 \cdot 3 - 4)/5 - 6/(7 - 8)$$
$$= 1 - (6 - 4)/5 - 6/(-1)$$
$$= 1 - 2/5 - (-6)$$
$$= 1 - 2/5 + 6 = 3/5 + 6 = 6.6$$

Problem 2

$$3x - 4 = 5x + 1$$
$$3x - 4 + (-1) = 5x + 1 + (-1) \quad \text{or } 3x - 5 = 5x$$
$$3x - 5 + (-3x) = 5x + (-3x) \quad \text{or } -5 = 2x$$
$$\tfrac{1}{2}(-5) = \tfrac{1}{2}(2x) \qquad \text{or} \qquad -\frac{5}{2} = x$$

Problem 4

(a)
$$2x + 5 + (-5) > 7x - 1 + (-5)$$
$$2x > 7x - 6$$

(b)
$$\tfrac{1}{3}(3x - 6) \leq \tfrac{1}{3}(18)$$
$$x - 2 \leq 6$$

(c)
$$-4(-.25x) \leq -4(x + .5)$$
$$x \leq -4x - 2$$

(d)
$$-\frac{1}{5}(-5x + 35) > -\frac{1}{5}(10)$$
$$x - 7 > -2$$
$$x - 7 + 7 > -2 + 7$$
$$x > 5$$

(e)
$$3x - 2 + 2 \leq 7 + 2$$
$$3x \leq 9$$
$$\tfrac{1}{3}(3x) \leq \tfrac{1}{3}(9)$$
$$x \leq 3$$

Problem 5

$$|-4| - |-4 + 2| = 4 - 2 = 2$$
$$|-3| - |-3 + 2| = 3 - 1 = 2$$
$$|-2| - |-2 + 2| = 2 - 0 = 2$$
$$|-1| - |-1 + 2| = 1 - 1 = 0$$
$$|0| - |0 + 2| = 0 - 2 = -2$$
$$|1| - |1 + 2| = 1 - 3 = -2$$
$$|2| - |2 + 2| = 2 - 4 = -2$$

Problem 6 (a)

$$\frac{u^2(u/v^2)^{-3}}{(v^2u)^{-1}} = \frac{u^2(uv^{-2})^{-3}}{(v^2)^{-1}u^{-1}} = \frac{u^2u^{-3}(v^{-2})^{-3}}{v^{2(-1)}u^{-1}}$$

$$= \frac{u^{2-3}v^{(-2)(-3)}}{v^{-2}u^{-1}} = \frac{u^{-1}v^6}{v^{-2}u^{-1}} = v^{6-(-2)} = v^8$$

(b)

$$\frac{[u(v^{-2}u)^{-1}]^3}{(u/v)^{-3}(v^0u/u^{-2})^2} = \frac{[u^1v^{(-2)(-1)}u^{-1}]^3}{(uv^{-1})^{-3}(1 \cdot u^{1-(-2)})^2} = \frac{[u^{1-1}v^2]^3}{u^{-3}v^{(-1)(-3)}(u^3)^2}$$

$$= \frac{[u^0v^2]^3}{u^{-3}v^3u^{3(2)}} = \frac{[1 \cdot v^2]^3}{u^{-3}v^3u^6} = \frac{v^{2(3)}}{u^{-3+6}v^3} = \frac{v^6}{u^3v^3} = \frac{v^{6-3}}{u^3} = \frac{v^3}{u^3}$$

Problem 7 (a) -3 has one 7th root, since $-3 < 0$ and 7 is odd.
(b) -64 has no 6th root, since $-64 < 0$ and 6 is even.
(c) 5 has two 14th roots, since $5 > 0$ and 14 is even
(d) 1/2 has one 17th root, since $1/2 > 0$ and 17 is odd.

Problem 8 (a) $\sqrt[3]{-8} = -2$, since $(-2)^3 = -8$.
(b) $\sqrt[6]{64} = 2$, since $2^6 = 64$ and $2 > 0$.
(c) $\sqrt{1/4} = 1/2$, since $(1/2)^2 = 1/4$ and $1/2 > 0$.
(d) $\sqrt[5]{32} = 2$, since $2^5 = 32$.

Problem 9 (a) $125^{1/3} = \sqrt[3]{125} = 5$, since $5^3 = 125$.
(b) $(-1/9)^{1/2} = \sqrt{-1/9}$ is undefined, since $-1/9 < 0$ and 2 is even.
(c) $(-1000)^{1/3} = \sqrt[3]{-1000} = -10$, since $(-10)^3 = -1000$.
(d) $(1/16)^{1/4} = \sqrt[4]{1/16} = 1/2$, since $(1/2)^4 = 1/16$ and $1/2 > 0$.

Problem 10 (a) $(-64)^{5/6}$ is undefined because $(-64)^{1/6}$ is undefined.
(b) $(4/9)^{3/2} = (\sqrt{4/9})^3 = (2/3)^3 = 8/27$.
(c) $(-32)^{3/5} = (\sqrt[5]{-32})^3 = (-2)^3 = -8$.

Problem 11

$$(-16)^{7/3}(1/4)^{7/6} = (-1)^{7/3}16^{7/3}(2^{-2})^{7/6}$$
$$= (\sqrt[3]{-1})^7(2^4)^{7/3}2^{-2(7/6)} = (-1)^7 2^{28/3}2^{-7/3} = (-1)2^{28/3-7/3}$$
$$= -2^{21/3} = -2^7 = -128$$

Problem 12

$$\sqrt{18}\sqrt[4]{64} = 18^{1/2}64^{1/4} = (2 \cdot 3^2)^{1/2}(2^6)^{1/4}$$
$$= 2^{1/2}3^1 2^{3/2} = 2^{1/2+3/2}3 = 2^2 3 = 12$$

Problem 13

(a) $x = \dfrac{-(-4) \pm \sqrt{(-4)^2 - 4(4)(1)}}{2 \cdot 4} = \dfrac{4 \pm \sqrt{16 - 16}}{8} = \dfrac{4 \pm 0}{8} = \dfrac{1}{2}$

(b) Since $b^2 - 4ac = 3^2 - 4(1)(5) = -11 < 0$, there are no solutions.

(c) $x = \dfrac{-(-5) \pm \sqrt{(-5)^2 - 4(2)(-3)}}{2 \cdot 2} = \dfrac{5 \pm \sqrt{49}}{4} = \dfrac{5 \pm 7}{4} = 3 \text{ or } -\dfrac{1}{2}$

Problem 14 (a) We write the equation as $x^2 + x - 3 = 0$. Then

$$x = \frac{-1 \pm \sqrt{1^2 - 4(1)(-3)}}{2 \cdot 1} = \frac{-1 \pm \sqrt{13}}{2}.$$

(b) We write the equation as $2x^2 + x - 3 = 0$. Then

$$x = \frac{-1 \pm \sqrt{1^2 - 4(2)(-3)}}{2 \cdot 2} = \frac{-1 \pm \sqrt{25}}{4} = \frac{-1 \pm 5}{4} = 1 \text{ or } -\frac{3}{2}$$

Problem 15 (a)

$$\begin{aligned} 0 &= x^4 + 3x^3 - 10x^2 \\ &= x^2(x^2 + 3x - 10) \\ &= x^2(x - 2)(x + 5) \end{aligned}$$

Thus $x = 0, 2$, or -5.

(b)

$$\begin{aligned} 0 &= x^3 - 3x^2 + 4x - 12 \\ &= x^2(x - 3) + 4(x - 3) \\ &= (x^2 + 4)(x - 3) \end{aligned}$$

Since $x^2 + 4 = 0$ is impossible, the only solution is $x = 3$.

Problem 16 If $t = x^3$, then

$$8x^6 + 7x^3 - 1 = 8t^2 + 7t - 1 = (8t - 1)(t + 1).$$

Thus $8t - 1 = 0$ and $t = 1/8$, or $t + 1 = 0$ and $t = -1$. Then $x^3 = 1/8$ or -1, so $x = 1/2$ or -1.

Problem 17 Dividing $x - 2$ into $6x^3 - 7x^2 - 9x - 2$ gives $6x^2 + 5x + 1 = (3x + 1)(2x + 1)$. The latter is 0 when $3x + 1 = 0$ and $x = -1/3$, or $2x + 1 = 0$ and $x = -1/2$.

INDEX

INDEX

PHOTO CREDITS

Page	Credit
16	Library of Congress
25	Amoco Oil Corp.
91, 108	Cameramann Int'l., Ltd.
124	Rod Planck/TSW-Click/Chicago Ltd.
158	Julie Houck/TSW-Click/Chicago Ltd.
195	Jim Pickerell/TSW-Click/Chicago Ltd.
204	Clyde Thomas
213	Cameramann Int'l., Ltd.
217	Kenneth Hayden/TSW-Click/Chicago Ltd.
230	Jean-Claude Lejeune
281	Courtesy Cullinet Software
321	Courtesy Apple Computer, Inc.
327	Cameramann Int'l., Ltd.
331	Brian Seed/TSW-Click/Chicago Ltd.
376	Cameramann Int'l., Ltd.
380	NOAA
402	Charles Gupton/TSW-Click/Chicago Ltd.
435	Focus On Sports
465	Mary Ann Fackelman/The White House
475	Cheryl Woike Kucharzak
484	Texas Archeological Survey/Texas Archeological Research Laboratory
493	Martin Rogers/TSW-Click/Chicago Ltd.
496	USAF photo
498	Cameramann Int'l., Ltd.
525	Courtesy IBM
543	UPI/Bettmann Newsphotos
558	Terry Farmer/TSW-Click/Chicago Ltd.
567	Jim Pickerell/TSW-Click/Chicago Ltd.
634	George Mars Cassidy/TSW-Click/Chicago Ltd.
646	Focus On Sports
662	Glennon Donahue/TSW-Click/Chicago Ltd.
665	George Mars Cassidy/TSW-Click/Chicago Ltd.
679	Robert Frerck/TSW-Click/Chicago Ltd.
778	Cheryl Woike Kucharzak
805	Diane Johnson/TSW-Click/Chicago Ltd.
808	Courtesy Aluminum Company of America
818	Cheryl Woike Kucharzak
864	James P. Rowan/TSW-Click/Chicago Ltd.

All other photos are the property of Scott, Foresman and Company.

A BRIEF TABLE OF INDEFINITE INTEGRALS

General formulas

1. $\int c\, f(u)\, du = c \int f(u)\, du$

2. $\int [f(u) \pm g(u)]\, du = \int f(u)\, du \pm \int g(u)\, du$

3. $\int u\, dv = uv - \int v\, du$

Basic integrals

4. $\int u^n\, du = \dfrac{u^{n+1}}{n+1} + k \qquad (n \neq -1)$

5. $\int \dfrac{du}{u} = \ln |u| + k$

6. $\int e^{cu}\, du = \dfrac{e^{cu}}{c} + k$

7. $\int a^u\, du = \dfrac{a^u}{\ln a} + k \qquad (a > 0)$

Integrands containing $a + bu$

8. $\int \dfrac{u\, du}{a + bu} = \dfrac{1}{b^2}(a + bu - a \ln |a + bu|) + k$

9. $\int \dfrac{u\, du}{(a + bu)^2} = \dfrac{1}{b^2}\left(\dfrac{a}{a + bu} + \ln |a + bu|\right) + k$

10. $\int \dfrac{u^2\, du}{(a + bu)^2} = \dfrac{1}{b^3}\left(a + bu - \dfrac{a^2}{a + bu} - 2a \ln |a + bu|\right) + k$

11. $\int \dfrac{u\, du}{(a + bu)^3} = \dfrac{1}{b^2}\left(-\dfrac{1}{a + bu} + \dfrac{a}{2(a + bu)^2}\right) + k$

12. $\int \dfrac{du}{u(a + bu)} = \dfrac{1}{a} \ln \left|\dfrac{u}{a + bu}\right| + k$

13. $\int \dfrac{du}{u^2(a + bu)} = -\dfrac{1}{au} + \dfrac{b}{a^2} \ln \left|\dfrac{a + bu}{u}\right| + k$

14. $\int \dfrac{du}{u(a + bu)^2} = \dfrac{1}{a(a + bu)} - \dfrac{1}{a^2} \ln \left|\dfrac{a + bu}{u}\right| + k$

15. $\int u^n(a + bu)^m\, du = \dfrac{u^n(a + bu)^{m+1}}{b(m + n + 1)} - \dfrac{an}{b(m + n + 1)} \int u^{n-1}(a + bu)^m\, du$
$(m + n \neq -1)$

16. $\int u\sqrt{a + bu}\, du = \dfrac{(6bu - 4a)(a + bu)^{3/2}}{15b^2} + k$

17. $\int u^n\sqrt{a + bu}\, du = \dfrac{2u^n(a + bu)^{3/2}}{b(2n + 3)} - \dfrac{2an}{b(2n + 3)} \int u^{n-1}\sqrt{a + bu}\, du$

18. $\int \dfrac{u\, du}{\sqrt{a + bu}} = \dfrac{2(bu - 2a)\sqrt{a + bu}}{3b^2} + k$

19. $\int \dfrac{u^n\, du}{\sqrt{a + bu}} = \dfrac{2u^n\sqrt{a + bu}}{b(2n + 1)} - \dfrac{2an}{b(2n + 1)} \int \dfrac{u^{n-1}\, du}{\sqrt{a + bu}}$

20. $\int \dfrac{du}{u\sqrt{a + bu}} = \dfrac{1}{\sqrt{a}} \ln \left|\dfrac{\sqrt{a + bu} - \sqrt{a}}{\sqrt{a + bu} + \sqrt{a}}\right| + k \qquad (a > 0)$

21. $\int \dfrac{du}{u^n\sqrt{a + bu}} = -\dfrac{\sqrt{a + bu}}{a(n - 1)u^{n-1}} - \dfrac{b(2n - 3)}{2a(n - 1)} \int \dfrac{du}{u^{n-1}\sqrt{a + bu}} \qquad (n \neq 1)$

22. $\int \dfrac{\sqrt{a + bu}}{u}\, du = 2\sqrt{a + bu} + \sqrt{a} \ln \left|\dfrac{\sqrt{a + bu} - \sqrt{a}}{\sqrt{a + bu} + \sqrt{a}}\right| + k$